# PROCEEDINGS OF THE 14th WORLD CONGRESS

## International Federation of Automatic Control

*Beijing, P.R. China, 5 - 9 July 1999*

*(in 18 volumes)*

Edited by

Han-Fu CHEN
*Chinese Academy of Sciences*
*Beijing, P.R.China*

Dai-Zhan CHENG
*Chinese Academy of Sciences*
*Beijing, P.R.China*

and

Ji-Feng ZHANG
*Chinese Academy of Sciences*
*Beijing, P.R.China*

## VOLUME F

**NONLINEAR SYSTEM I**

Volume Editors

A. ISIDORI (Italy), T. GLAD (Sweden),
R. BARS (Hungary) and C.V. HOLLOT (USA)

Published for the

INTERNATIONAL FEDERATION OF AUTOMATIC CONTROL

by

PERGAMON
An Imprint of Elsevier Science

UK Elsevier Science Ltd, The Boulevard, Langford Lane, Kidlington, Oxford, OX5 1GB, UK
USA Elsevier Science Inc., 660 White Plains Road, Tarrytown, New York 10591-5153, USA
JAPAN Elsevier Science Japan, Tsunashima Building Annex, 3-20-12 Yushima, Bunkyo-ku, Tokyo 113, Japan



First edition 1999

**Library of Congress Cataloging in Publication Data**

A catalogue record for this book is available from the Library of Congress

**British Library Cataloguing in Publication Data**

A catalogue record for this book is available from the British Library

ISBN 0-08-043217 4



Transferred to Digital Printing 2009

**14th Triennial World Congress of IFAC,**
**the International Federation of Automatic Control**

**Hosted by**
The Chinese Association of Automation

**International Program Committee**
Han-Fu Chen (China), *Chairman*
Luis Basanez (Spain), *Vice-Chair*
Mogens Blanke (Denmark)
Xiren Cao (Hong Kong, China), *Vice-Chair*
Tian-You Chai (China)
Jose B. Cruz, Jr. (USA), *Vice-Chair*
Manfred Deistler (Austria)
Tyrone Duncan (USA)
Lei Guo (China)
Yasushi Hashimoto (Japan)
Alberto Isidori (Italy)
Laszlo Keviczky (Hungary)
Vladimir Kucera (Czech Republic)
Lennart Ljung (Sweden)
Lena Martensson (Sweden)
Micheal K. Masten (USA)
Thomas McAvoy (USA)
Umit Ozguner (USA)
Juan de la Puente (Spain)
Tzyh-Jong Tarn (USA)

**National Organizing Committee**
Yong-Xiang Lu (China), *Chairman*
Han-Fu Chen (China), *Co-Chair*
Ru-Wei Dai (China), *Co-Chair*
Bo-Lin Sun (China), *Secretary General*
Gong-Qing Zhang (China), *Treasurer*
Shu-Hui Deng (China)
Yan-Da Li (China)
Song-De Ma (China)
K. L. Mak (Hong Kong, China)
Yu-Geng Xi (China)
Jing-Heng Yang (China)
Kan Zhang (China)

**National Congress Advisory Board**
Jian-Song (China), *Chairman*
Yu-Chi Ho (USA)
Qi-Heng Hu (China)
Tzyh-Jong Tarn (USA)
Jia-Chi Yang (China)
Guang-Ya Zhu (China)

**The Chinese Association of Automation**
Ru-Wei Dai, *Co-President, (1997-2002)*
Han-Fu Chen, *Co-President, (1997-2002)*

**IFAC President (1996-1999)**
Yong-Zai Lu (China)

# CONTENTS OF VOLUMES

# PREFACE

As the last triennial World Congress of the International Federation of Automatic Control (IFAC) of this century, the 14th IFAC Congress is recognized as a preeminent event for the worldwide community of control systems science and engineering. It is a great pleasure and honor for Chinese control researchers and engineers to host the Congress in Beijing.

Beijing is one of the oldest cities in the world. Its civilization can be traced back to 1000 BC when it was a small town. Since the Jin Dynasty (AD 1115), and during the Yuan, Ming and Qing Dynasties, Beijing has been the capital. It is famous partly for its many historical sites, including the Great Wall, the Forbidden City, the Summer Palace, the Ming Tombs, etc. Beijing is also famous for the development of automatic machines over many centuries, including the armillary sphere for simulating the Universe and the seismograph for detecting earthquakes, making the city one of great world centres for automation.

Beijing is also a young city; every day it shows a new face to the world. The fast pace of change in Beijing is fuelled by the rapid growth of the economy and the development of science and technology, including automation. China has vast potential for the development of automatic control. Beijing cordially welcomes you!

In the light of the outstanding efforts of our predecessors, our goal in organizing this Congress is to continue to increase the quality of the papers and of the Congress activities. We hope this Congress will be remembered as a showcase for the fruitful results on control theory and application achieved since the beginning of this century. In addition, we expect that several of the results and ideas demonstrated in this Congress will have lasting influence in the next century on the directions taken by the field of control systems engineering.

The response from the international control community to the call for papers was strong. The International Program Committee (IPC) received 2169 submitted papers, 53 invited session proposals and eight panel discussion proposals from 68 countries and regions. We were impressed by the high quality of the submissions. The technical program for the Congress covers a broad range of topics, with sessions arranged into nine congress tracks and subdivided into 45 technical areas. Reviews were conducted by all of the IFAC Technical Committees, with the help of hundreds of worldwide experts. Through this rigorous review procedure 1556 papers, one keynote speech, five plenary lectures, 47 invited session proposals and seven panel discussion proposals were accepted. They came from 59 countries and regions. The IPC is confident that the program that has been selected will provide significant contributions to each of the subject areas of the congress tracks.

There are five plenary lectures scheduled, one at the start of each of the five days of the Congress. The plenary speakers include some of the world's most outstanding authorities in various control fields. The topics of the plenary lectures include surveys, prospects and introductions to advanced control theory and applications. In selecting the plenary speakers, we have been guided by a desire to provide perspectives on theoretical topics, important applications, and other issues of common interest. It was a difficult task to select five plenary speakers among so many outstanding candidates who were nominated by our colleagues.

Following the plenary addresses, the congress program is composed of more than 1500 papers, to be presented in 214 lecture-type sessions and 47 poster-type sessions, the latter arranged in three time clusters. The paper sessions are supplemented by seven panel discussions. No distinction is made between poster papers and lecture papers in the Proceedings.

The Proceedings of the Congress are published in two forms, printed and CD-ROM. The Proceedings in its printed form consists of 18 volumes: 17 subject volumes, which contain all lecture and poster papers and a plenary volume. The plenary volume contains the keynote speech, plenary papers, panel discussions, and the table of contents, author index and subject index for the full set of subject volumes.

The CD-ROM, containing all the material in the 18 volumes, is provided to each registered participant as part of the registration material. Both the printed Proceedings and the CD-ROM will be marketed after the Congress, as the official Congress Proceedings, by the IFAC Publisher, Elsevier Science Ltd.

We expect that the reader, whether a participant in the Congress, or a researcher or practising engineer studying the material afterwards, will find the Congress publications useful, informative and helpful in their related creative works.

The Congress organizers have made every effort to create a high quality environment for the Congress, but China, as a developing country, still may seem unsatisfactory in many respects, including the language barrier. You may feel inconvenienced and you may need to be patient in communicating your needs. Imagination, plus a little spirit of adventure, may help you to have some unexpected and unforgettable experiences in this part of the world, which, in the end, makes life colorful.

Finally, we would like to thank all IPC members, all IFAC Technical Committee chairs and all reviewers for their valuable contribution to the Congress. We are also indebted to all our friends from different parts of the world who have given support, help and useful suggestions.

Han-Fu CHEN, Daizhan CHENG and Jifeng ZHANG

# CONTENTS OF VOLUME F

## NONLINEAR SYSTEMS I
## (Technical Areas 2c-III, 2d and 2e-I)

# OPTIMAL OUTPUT TRACKING FOR DISCRETE TIME NONLINEAR SYSTEMS

**Chong-Ho Choi * Gu-Min Jeong * Tae-Jeong Jang ****

* *School of Electrical Eng.,ASRI,ERC-ACI*
*Seoul National University, KOREA*
*chchoi@csl.snu.ac.kr*
** *Dept. of Control and Instrumentation Eng.*
*Kangwon National University, KOREA*
*jangtj@cc.kangwon.ac.kr*

Abstract: A method to solve the optimal control for discrete time nonlinear systems is proposed based on the ILC (iterative learning control). This method updates the input with consideration to the future output as well as the current output. It is shown that the input update law can be derived from an objective function consisted of output error and input energy. The convergence condition of the input update law is given and its convergence is proved. It is tested for some illustrative examples and is found to be numerically robust. 

Keywords: Optimal Control; Nonlinear Systems; Iterative Learning Control

## 1. INTRODUCTION

Optimal control problems for linear systems have been studied extensively since 1960 and many theoretical results has been introduced (Anderson,1989). For nonlinear systems, the optimal problem is solved by using the Hamilton-Jacobi(H-J) equation or the maximum principle. Since it is difficult to solve the H-J equation or to use the maximum principle, many numerical methods have been introduced.

Other control theories for linear systems have been extended for nonlinear systems too. Isidori and Byrnes (Isidori, 1990) showed that the results for linear systems (Francis, 1976) can be generalized to a first order PDE and proposed the Byrnes-Isidori regulator. In addition to these, Devasia et. al. (Devasia, 1996) introduced an inversion-based approach for the exact nonlinear output tracking control.

On the other hand, the iterative learning control(ILC) has been studied to enhance the performance of repeated tasks (Uchiyama, 1978). Also there was a work in the ILC on how to make the magnitude of input within a saturation bound (Choi, 1997).

Recently, Gorinevsky (Gorinevsky, 1997) studied the optimal control problem for uncertain nonlinear systems over a fixed time interval. The algorithm is based on the Levenberg-Marquardt algorithm, which requires the inversion of matrices. As the interval for control becomes longer, the dimension of matrices becomes large. Hence the algorithm becomes numerically sensitive and the computational burden increases greatly.

In this paper, a new scheme to solve the optimal control problem for discrete time nonlinear systems based on the ILC is proposed. It basically utilizes the conventional approach of the ILC, but extending the previous results by taking the output errors and input energy over a finite time interval into consideration. Under certain conditions, the global convergence of the proposed method is shown. This method does not need a matrix inversion which makes it less sensitive numerically. This method is applied to the nonminimum-phase problem in (Devasia, 1996) to test its applicability.

## 2. PRELIMINARIES AND PROBLEM FORMULATIONS

Consider a class of discrete time nonlinear systems described by

$$x(i+1) = f(x(i)) + g(x(i))u(i)$$
$$y(i) = h(x(i)), \qquad (1)$$

where, $u = [u_1, \cdots, u_m]^T \in \Re^m$, $x = [x_1, \cdots, x_n]^T \in \Re^n$, and $y = [y_1, \cdots, y_m]^T \in \Re^m$ are the plant input, the system state, and the plant output of the system, respectively. The functions $f$, $g$ and $h$ are $f : \Re^n \to \Re^n$, $g : \Re^n \to \Re^{n\times m}$, and $h = [h_1, \cdots, h_m]^T : \Re^n \to \Re^m$. Let's introduce some notations for later use. For an n-dimensional vector $v = [v_1, \cdots, v_n]^T$ and an $n \times m$ matrix $W = \{w_{ij}\}$, let us denote 2-norms of these as $||v||$ and $||W||$ respectively. The relative degree $r_1, \cdots, r_m$ for continuous time systems is defined in (Isidori, 1989) and its counterpart in the discrete time systems is the characteristic number $\rho_1, \cdots, \rho_m$ in (Nijmeijer, 1990). We will define the relative degree $\sigma_1, \cdots, \sigma_m$ for discrete time systems as $\sigma_i = \rho_i + 1, i = 1, \cdots, m$.

We use the superscript $k$ to denote the iteration number of updates and the superscript $d$ to denote the vectors corresponding to the desired trajectory. We will solve an optimal control problem for a finite time interval $[0, N-1]$ by the similar approach taken in the ILC. In the ILC, the interest of control is usually concentrated on one point in the interval $[0, N-1]$ at a time. This makes it possible to achieve a perfect tracking for a finite time interval. But this also makes the input excessively large when the control system is nonminimum phase. Hence a compromise between the tracking performance and the input energy is needed. To solve this problem, an input update law is proposed and the convergence of the proposed algorithm is proven. For the proof , the time interval $[i, i+N-1]$ under consideration is moved as $i$ increases and this kind of approach is called the receding horizon control. To prove the convergence, we utilize the proof of ILC algorithm. Therefore we consider the vectors in the time interval $[i, i+N-1], i \in [0, N-1]$ as one big vector. Note that the input to be found is for the time interval $[0, N-1]$. We assume the following properties for the system (1).

(**A1**) The relative degree of the system (1) is $\{\sigma_1, \sigma_2, \cdots, \sigma_m\}$. $\forall (x(i), u(i)) \in \Re^n \times \Re^m, i \in [0, 2N + \sigma_{max} - 2]$. where $\sigma_{max} = \max_i \sigma_i$.

Remark) From the definition of the relative degree, $y_j(i)$, $0 \le i \le \sigma_j - 1$ are influenced only by the initial state and are not influenced by the input $u(i), 0 \le i \le N-1$. To denote a vector consisting of $y_j(i+\sigma_j)$, $j = 1, \cdots, m$, which is influenced by the input $u(i)$ for the first time, we define $y_\sigma(i)$ as $y_\sigma(i) = \left(y_1^T(i+\sigma_1),\ y_2^T(i+\sigma_2),\ \cdots,\ y_m^T(i+\sigma_m)\right)^T$.

(**A2**) $y_\sigma^d(i)$, $i \in [0, N-1]$, which is the desired output for $y_\sigma(i)$ to follow is given. And $y_\sigma^d(i) = c_i$ for $i \in [N, 2N-2]$ for some constant vector $c_i \in \Re^m$. The constant vectors $c_i$ needs not be known beforehand.

Remark) The condition for $y_\sigma^d(i) = c_i$, $N \le i \le 2N-2$ is needed later only for the proof of the convergence.

Remark) The input $u(i)$ for $i \in [N, 2N-2]$ does not influence the output $y_\sigma(i)$, $i \in [0, N-1]$ and we set these inputs to be zero.

$$u^*(i) = u^k(i) = 0, \quad \forall i \in [N, 2N-2] \qquad (2)$$

(**A3**) The functions $f$, $g$ and $h$ are analytic in their domains of definition .

Also let us define $\mathbf{u}(i)$, $\mathbf{y}_\sigma(i)$ as follows.

$$\mathbf{u}(i) = \left(u^T(i),\ \cdots,\ u^T(i+N-1)\right)^T$$
$$\mathbf{y}_\sigma(i) = \left(y_\sigma^T(i), \cdots, y_\sigma^T(i+N-1)\right)^T \qquad (3)$$

With these notations, consider the following objective function $V_0$.

$$V_0 = \frac{1}{2}[\mathbf{y}_\sigma^d(0) - \mathbf{y}_\sigma(0)]^T \mathbf{Q}[\mathbf{y}_\sigma^d(0) - \mathbf{y}_\sigma(0)]$$
$$+\frac{1}{2}[\mathbf{u}(0)]^T \mathbf{R}[\mathbf{u}(0)] \qquad (4)$$
$$\mathbf{Q} = \mathrm{diag}\,[Q(0), \cdots,\ Q(N-1)]\,,$$
$$\mathbf{R} = \mathrm{diag}\,[R(0), \cdots,\ R(N-1)]\,.$$

Here, $Q(i) \in \Re^{m\times m}$ is a positive semi-definite matrix, and $R(i) \in \Re^{m\times m}$ is a positive semi-definite matrix. Let $\mathbf{u}^*(0)$ and $\mathbf{y}_\sigma^*(0)$ be the optimal solution which minimizes $V_0$ for the nonlinear system (1). Also let $\mathbf{x}(i)$ be the state corresponding to $\mathbf{u}(i)$ i.e.,

$$\mathbf{x}(i) = \left(x^T(i),\ \cdots,\ x^T(i+N+\sigma_{max}-1)\right)^T.$$

Then $\mathbf{x}^*(i)$ is the state corresponding to the optimal solution. Now let us describe the problem we are going to solve.

**Problem A**

For the system (1), find the optimal control input $\mathbf{u}^*(0)$ which minimizes $V_0$.

The input $u^*(i), i \in [0, N-1]$ which minimizes the objective function $V_0$ in (4) satisfies

$$\left.\frac{\partial V_0}{\partial \mathbf{u}(0)}\right|_{\mathbf{u}(0)=\mathbf{u}^*(0)} \equiv \left[\frac{\partial V_0}{\partial u(0)}, \cdots, \frac{\partial V_0}{\partial u(N-1)}\right]^T$$
$$= \left[-\mathbf{J}^T\mathbf{Q}\left[\mathbf{y}_\sigma^d(0) - \mathbf{y}_\sigma(0)\right] + \mathbf{R}\mathbf{u}\right]_{\mathbf{u}(0)=\mathbf{u}^*(0)} = 0.$$

It implies

$$[\mathbf{J}^T]_{\mathbf{u}(0)=\mathbf{u}^*(0)}\mathbf{Q}[\mathbf{y}_\sigma^d(0) - \mathbf{y}_\sigma^*(0)] - \mathbf{R}\mathbf{u}^*(0) = 0 \quad (5)$$

where

$$\mathbf{J}^T = \begin{bmatrix} [\frac{\partial y_\sigma(0)}{\partial u(0)}]^T & [\frac{\partial y_\sigma(1)}{\partial u(0)}]^T & \cdots & [\frac{\partial y_\sigma(N-1)}{\partial u(0)}]^T \\ 0 & [\frac{\partial y_\sigma(1)}{\partial u(1)}]^T & \cdots & [\frac{\partial y_\sigma(N-1)}{\partial u(1)}]^T \\ \vdots & \vdots & \ddots & \vdots \\ 0 & 0 & \cdots & [\frac{\partial y_\sigma(N-1)}{\partial u(N-1)}]^T \end{bmatrix}.$$

It is a necessary condition for the optimal control input $\mathbf{u}^*(0)$.

We will introduce some more notations and assumptions for the system. Let us define a matrix function $\mathbf{S}$, which will be used later in the input update law in finding the optimal control input.

$$\mathbf{S}(i) = \begin{bmatrix} S_{11}(i) & S_{12}(i) & \cdots & S_{1N}(i) \\ 0 & S_{22}(i) & \cdots & S_{2N}(i) \\ \vdots & \vdots & \ddots & \vdots \\ 0 & 0 & \cdots & S_{NN}(i) \end{bmatrix}$$

The components of $\mathbf{S}$, $S_{jl}, 1 \le j \le N, 1 \le l \le N$ are matrices of size $m \times m$ and all the components of $S_{jl}$ are analytic. And $S_{jl}(i)$ is a function of $\mathbf{x}(i)$, i.e., $S_{jl}(i) = S_{jl}(\mathbf{x}(i))$. Sometimes $S(\mathbf{x}(i))$ will be used to emphasize that $S(i)$ is a function of $\mathbf{x}(i)$. It will be shown later how $\mathbf{S}(i)$ is selected. The relation between $\mathbf{S}(i-1)$ and $\mathbf{S}(i)$ is made to satisfy

$$S_{jl}(i) = S_{j+1,l+1}(i-1) = S_{j+i,l+i}(0),$$
$$S_{jl}(i) = 0, \forall l, j, N+1 \le j+i, N+1 \le l+i \quad (6)$$

This relation describes that as the receding horizon moves from $i-1$ to $i$, the left upper $(N-1) \times (N-1)$ block matrices of $\mathbf{S}(i)$ is equal to the right lower $(N-1) \times (N-1)$ block matrices of $\mathbf{S}(i-1)$, and all the other blocks are zero matrices. Also let us define a matrix function $\Lambda$ which is a weight function for the input. It is a diagonal matrix of size $mN \times mN$.

$$\Lambda(i) = \mathrm{diag}\,[\Lambda_1(i), \cdots, \Lambda_N(i)]$$

Here $\Lambda_j(i), 1 \le j \le N$, is a matrix of size $m \times m$ and its diagonal elements are between 0 and 1. And $\Lambda_j(i)$ satisfies the following properties.

$$\Lambda_j(i) = \Lambda_{i+j}(0), \; i+j \le N$$
$$\Lambda_j(i) = \epsilon I_{m\times m}, \; i+j \ge N+1$$

where $\epsilon$ is a small positive constant.

This relationship implies that, as the receding horizon of size $N$ moves from $i-1$ to $i$, the left upper $(N-1) \times (N-1)$ block matrices of $\Lambda(i)$ is equal to the right lower $(N-1) \times (N-1)$ block matrices of $\Lambda(i-1)$, and all the other diagonal blocks are $\epsilon I$.

And more assumptions are made for the system (1).

(**A4**) For the desired trajectory $y_\sigma^d$ in (**A2**), there exists a unique $\mathbf{u}^*(0)$ which satisfies

$$\mathbf{S}(\mathbf{x}^*(0))(\mathbf{y}_\sigma^d(0) - \mathbf{y}_\sigma^*(0)) - \Lambda(0)\mathbf{u}^*(0) = 0. \quad (7)$$

Remark) Since $V_0$ is a quadratic objective function, $\mathbf{u}^*(0)$ which satisfies (7) is the optimal solution by this assumption when $\mathbf{S}(0) = \mathbf{J}^T\mathbf{Q}$ and $\mathbf{R} = \Lambda(0)$. Also note that $y^*(i)$ is determined from $\mathbf{y}_\sigma^*(0)$. Therefore (7) can be expressed with $y^*(i)$ instead of $\mathbf{y}_\sigma^*(0)$.

(**A5**) For the system (1), let $\Phi$ denote a mapping from $(x(0), u(i), i \in [0, 2N-2])$ to $(x(i), i \in [0, 2N + \sigma_{max} - 2])$ and $\Psi$ be a mapping from $(x(0), u(i), i \in [0, 2N-2])$ to $(y(i), i \in [0, 2N + \sigma_{max} - 2])$. For each $x(0)$, the input-to-state map $\Phi$ and the input-to-output map $\Psi$ are one-to-one.

(**A6**) Let $U \subset R^m$ be a bounded plant input space. For a given $y_\sigma^d(i), i \in [0, N-1]$ and an initial state $x(0)$, there exists a compact set $X \subset R^n$ such that the plant state $x(i)$, generated by the input trajectories in $\overbrace{U \times U \times \cdots \times U}^{2N-1}$, belongs to $X$,i.e., $x(i) \in X$ for all $i \in [0, 2N + \sigma_{max} - 2]$. So there exists a compact set $\mathbf{X}$ such that $\forall \mathbf{x}(i) \in \mathbf{X}$.

Instead of solving $\mathbf{u}^*(0)$ directly in Problem A, we will solve the following modified problem.

**Problem B** For the system (1), find the control input $u(i), i \in [0, N-1]$ which makes the output error $y^*(i) - y(i), i \in [0, N + \sigma_{max} - 1]$ smaller than an arbitrary small constant $\epsilon_1$ for all time interval $i \in [0, N-1]$ i.e.,

$$\|y^*(i) - y(i)\| \le \epsilon_1, \; \forall i \in [0, N + \sigma_{max} - 1]$$

where $y^*(i)$ satisfies (7).
It will bw shown that the solution for Problem B leads to the solution for Problem A. Let's show that a relation similar to (7) holds for $i = 1, 2, \cdots, N-1$.

**Lemma 1**
Let the system (1) and the desired trajectory

$y^d(i)$ satisfy **(A1)** $\sim$ **(A6)**. If $u^*(j), j \leq i-1$ is determined, the following relation which is similar to (7) holds.

$$\mathbf{S}(\mathbf{x}^*(i))(\mathbf{y}_\sigma^d(i) - \mathbf{y}_\sigma^*(i)) - \Lambda(i)\mathbf{u}^*(i) = 0 \quad 1 \leq i \leq N-1 \qquad (8)$$

Proof) It is omitted for space limitation.

## 3. ITERATIVE LEARNING CONTROL AND OPTIMAL CONTROL

In order to solve Problem B, we will use the similar approach taken in the iterative learning control. In the iterative learning control, it is usually assumed that the system to be controlled is unknown and control $u^{k+1}(i)$ is updated based on $u^k(i)$ and $y^k(i)$. In this paper, we are concerned with controlling known systems. Note that $u^k(i), i \in [0, N-1]$ is an input at time $i$ of $k$-th iteration in the time interval $[0, N-1]$. And $x^k(1), \cdots, x^k(N + \sigma_{max} - 1)$ are the states generated by $x(0)$ and $u^k(0), \cdots, u^k(N-1)$. Also $y_\sigma^k(0), \cdots, y_\sigma^k(N-1)$ are the outputs generated by $x(0)$ and $u^k(0), \cdots, u^k(N-1)$. Using $\mathbf{u}^k(i)$ and $\mathbf{y}_\sigma^k(i)$, $\mathbf{u}^{k+1}(0)$ is updated based on $\mathbf{u}^k(0)$ and $\mathbf{y}_\sigma^k(0)$ as in the ILC. We propose an input update law as

$$\mathbf{u}^{k+1}(0) = (I - \Lambda(0))\mathbf{u}^k(0) + \mathbf{S}(\mathbf{x}^k(0))(\mathbf{y}_\sigma^d(0) - \mathbf{y}_\sigma^k(0)). \qquad (9)$$

With this input update law, it will be shown that the input converges to the solution which gives $\mathbf{y}_\sigma^*(0)$ in (7). From (2) and (6), it can be shown that the input update law (9) holds for an arbitrary $i \in [0, N-1]$. Hence, $\mathbf{u}^{k+1}(i)$ can be written as in the same form as (9).

$$\mathbf{u}^{k+1}(i) = (I - \Lambda(i))\mathbf{u}^k(i) + \mathbf{S}(\mathbf{x}^k(i))(\mathbf{y}_\sigma^d(i) - \mathbf{y}_\sigma^k(i)). \qquad (10)$$

Lemma 2 and Lemma 3 are introduced for later use. By the mean value theorem, the error $\mathbf{y}_\sigma^*(i) - \mathbf{y}_\sigma^k(i)$ can be divided into two parts, the one due to the initial state error $\delta x^k(i)$ and the other one due to input error $\delta\mathbf{u}(i)$.

**Lemma 2**

Let us assume that a nonlinear MIMO system (1) has relative degree $\{\sigma_1, \sigma_2, \cdots, \sigma_m\}$. Then by the mean value theorem the output errors can be expressed as

$$\begin{aligned}
&\mathbf{y}_\sigma^*(i) - \mathbf{y}_\sigma^k(i) \\
&= \begin{pmatrix} A_1^k(i) \\ \vdots \\ A_N^k(i) \end{pmatrix} \delta x^k(i) + \begin{pmatrix} J_{11}^k(i) & \cdots & 0 \\ \vdots & \ddots & \vdots \\ J_{N1}^k(i) & \cdots & J_{NN}^k(i) \end{pmatrix} \delta\mathbf{u}(i) \\
&= \mathbf{A}^k(i)\delta x^k(i) + \mathbf{J}^k(i)\delta\mathbf{u}(i). \qquad (11)
\end{aligned}$$

Here $\delta x^k(i) = x^*(i) - x^k(i)$, $\delta\mathbf{u}^k(i) = \mathbf{u}^*(i) - \mathbf{u}^k(i)$. $A_p^k(i) = [A_{p1}^k(i), \cdots, A_{pN}^k(i)]^T$ and $J_{pq}^k(i) = [J_{pq1}^k(i), \cdots, J_{pqN}^k(i)]^T$, which are $m \times n$ and $m \times m$ respectively, are

$$\begin{aligned}
A_{pr}^k(i) &= h_r'(x(i+\sigma_r+p-1, \gamma^k(i+\sigma_r+p-1))) \\
&\times \prod_{l=\sigma_r+1}^{\sigma_r+p} \{f'(x^*(i+\sigma_r+p-l), x^k(i+\sigma_r+p-l)) \\
&\quad + g'(x^*(i+\sigma_r+p-l), x^k(i+\sigma_r+p-l)) \\
&\quad \times u^*(i+\sigma_r+p-l)\} \\
&\times \prod_{l=2}^{\sigma_r} \{f'(x^*(i+\sigma_r+p-l), x^k(i+\sigma_r+p-l))\}, \\
J_{pqr}^k(i) &= h_r'(x(i+\sigma_r+p-1, \gamma^k(i+\sigma_r+p-1))) \\
&\times \prod_{l=\sigma_r+1}^{\sigma_r+p-q} \{f'(x^*(i+\sigma_r+p-l), x^k(i+\sigma_r+p-l)) \\
&\quad + g'(x^*(i+\sigma_r+p-l), x^k(i+\sigma_r+p-l)) \\
&\quad \times u^*(i+\sigma_r+p-l)\} \\
&\times \prod_{l=2}^{\sigma_r} \{f'(x^*(i+\sigma_r+p-l), x^k(i+\sigma_r+p-l))\} \\
&\times g(x^k(i+q-1)), \qquad r = 1, \cdots, N,
\end{aligned}$$

where

$$\begin{aligned}
&f'(x^d(i), x^k(i)) \\
&\equiv [f_1'(x(i, \alpha_1^k(i))), \cdots, f_n'(x(i, \alpha_n^k(i)))]^T \\
&g'(x^d(i), x^k(i)) \\
&\equiv [g_1'(x(i, \beta_1^k(i))), \cdots, g_n'(x(i, \beta_n^k(i)))]^T. \qquad (12)
\end{aligned}$$

$$\begin{aligned}
f_m'(x(i, \alpha_m^k(i))) &= f_m'(\alpha_m^k(i)x^d(i) + (1 - \alpha_m^k(i))x^k(i)) \\
g_m'(x(i, \beta_m^k(i))) &= g_m'(\beta_m^k(i)x^d(i) + (1 - \beta_m^k(i))x^k(i)) \\
h'(x(i, \gamma^k(i))) &= h'(\gamma^k(i)x^d(i) + (1 - \gamma^k(i))x^k(i)) \\
&0 \leq \alpha_m^k(i), \beta_m^k(i), \gamma^k(i) \leq 1 \forall m, k, i.
\end{aligned}$$

When $\sigma_r = 1$, note that $J_{ppr}^k(i) = h_r'(x(i + \sigma_r + p - 1, \gamma^k(i + \sigma_r + p - 1)) \times g(x^k(i + p - 1))$ in the above equation.

Proof) It is omitted for space limitation.

In Lemma 3, it will be shown that the convergence condition is satisfied for $i \in [0, N-1]$ if it is satisfied for $i = 0$. So the convergence condition needs to be checked only for $i = 0$.

**Lemma 3 (Property of the convergence**

**condition)**
Let the system (1) and the desired trajectory $y^d(i)$ satisfy (**A1**) $\sim$ (**A6**). If the condition $||I - \Lambda(i) - \mathbf{S}(\mathbf{x}^k(i))\mathbf{J}^k(i)|| \le \rho < 1$ holds for $i = 0$, then this condition holds for all $i \in [0, N-1]$.
Proof) It is omitted for space limitation.

We solve Problem B by taking the similar approach in the ILC for $[0, N-1]$. If the convergence condition is satisfied, we can obtain the input $u^*(i)$ which gives the output $y^*(i)$.

**Theorem 1 (Iterative learning control over a finite time interval)**
The system (1) and the desired trajectory $y^d(i)$ satisfy (**A1**) $\sim$ (**A6**) and $x^*(0) = x^k(0)$ and the input is updated as in (9). If $\mathbf{S}(\mathbf{x}^k(0))$ satisfies

$$||I - \Lambda(0) - \mathbf{S}(\mathbf{x}^k(0))\mathbf{J}^k(0)|| \le \rho < 1, \qquad (13)$$

$u^k(i)$ and $y^k(i)$ converge to $u^*(i)$ and $y^*(i)$ respectively for $i \in [0, N + \sigma_{max} - 1]$ as $k$ goes to $\infty$.
Proof) It is omitted for space limitation.

Now consider the ILC in Theorem 1 with respect to an objective function.

**Theorem 2 (Iterative learning control and optimal control)**
Let the system (1) and the desired trajectory $y^d(i)$ satisfy (**A1**) $\sim$ (**A6**). If we set $\mathbf{S}(x^k(0)) = \eta(\mathbf{J}^k)^T\mathbf{Q}$ and $\Lambda(0) = \eta\mathbf{R}$, Then, the input obtained in Theorem 1 minimizes the objective function (4).

Proof) It is omitted for space limitation.

## 4. ILLUSTRATIVE EXAMPLES

To see the applicability and performance of the proposed method, we consider the tracking problem of the nonlinear continuous nonminimum-phase system in (Devasia, 1996). The desired trajectory $y^d$ is

$$y^d(t) = \begin{cases} 0.1 \times (1 - \cos(t)), & \text{if } t \in [0, 2\pi], \\ 0, & \text{otherwise.} \end{cases}$$

Discretizing the system by the Euler's method with the sampling period $\Delta t = 0.04$, it becomes

$$\begin{aligned} x_1(i+1) &= x_1(i) + \Delta t(-x_1(i) + x_2(i)) \\ x_2(i+1) &= x_2(i) + \Delta t(-3x_2(i) + x_1^3(i)) \\ &\quad + \Delta t(2 + \sin^2(x_4(i)))u(i))) \\ x_3(i+1) &= x_3(i) + \Delta t(x_1(i) - 2x_3(i)) \\ x_4(i+1) &= x_4(i) + \Delta t(-x_4(i) + x_3^2(i)) \\ y(i) &= x_1(i) - 3x_3(i). \end{aligned} \qquad (14)$$

So $y^d(i)$ becomes

$$y^d(i) = \begin{cases} 0.1 \times (1 - \cos(i)), & \text{if } i \in [0, 157], \\ 0, & \text{otherwise.} \end{cases}$$

Note that the relative degree of the system is 2. For this system, (**A1**), (**A2**) and (**A3**) are satisfied. But it is difficult to check whether (**A4**), (**A5**) and (**A6**) are satisfied. Since there is no explicit evidence that the system does not satisfy (**A4**), (**A5**) and (**A6**), we solve the problem by the proposed method. If the solution gives a reasonable answer, then we can guess that the system satisfies the assumptions.

We set $\mathbf{Q} = \alpha I$ and $\Lambda = \tau I$. We like to set $\mathbf{S} = \alpha(\mathbf{J}^k)^T$ as in Theorem 2 if possible. But $u^*$ is not known beforehand, and neither is $\mathbf{J}^k$. Therefore we set $S = \alpha(\hat{\mathbf{J}}^k)^T$ with its element given as

$$\hat{J}_{jl}(i) = \left. \frac{\partial y_\sigma(i+j-1)}{\partial u(i+l-1)} \right|_{\mathbf{u}(0)=\mathbf{u}^k(0)}. \qquad (15)$$

Hence the convergence condition becomes $||I - \Lambda - \alpha\hat{\mathbf{J}}^T\mathbf{J}|| < \rho$. Note that $\hat{\mathbf{J}}$, whose elements are given in (15) converges to $\mathbf{J}^k(i)$ as $u^k$ goes to $u^*$. $\hat{\mathbf{J}}$ is a block lower triangular matrix. The nonzero elements of $\hat{\mathbf{J}}$ can be factored by $(\Delta t)^2$.

It is difficult to know whether the convergence condition is satisfied. If $u^k$ is sufficiently close to $u^*$, the convergence condition can be approximated as

$$||I - \Lambda - \alpha\hat{\mathbf{J}}^T\hat{\mathbf{J}}|| < \rho. \qquad (16)$$

Therefore by choosing $\Delta t$ to be sufficiently small, the convergence condition is satisfied when $u^k$ is sufficiently close to $u^*$ and (16) is satisfied.

**Example 1**

The optimal output tracking problem is solved by the proposed method with $\Lambda = 0.001$. Similar to (Devasia, 1996) $y^d(i)$ is set as

$$y^d(i) = \begin{cases} 0, & i \in [-286, -1], \\ 0.1 \times (1 - \cos(i)), & i \in [0, 157], \\ 0, & i \in [158, 314]. \end{cases} \qquad (17)$$

So the input starts at $i = -286$ and the length of horizon $N$ is 600. First, we set $\mathbf{S} = 10\hat{\mathbf{J}}^T$. The corresponding output is shown in Fig. 1. In this case, $y(i)$ tracks $y^*(i)$ near perfectly and it is hard to distinguish one from the other in Fig. 1. The maximum of the error is less than 0.00022. It shows that proposed method works. Fig. 2 shows the input corresponding to $u^*$. In this case the convergence condition is checked and found that $||I - \Lambda - 10\hat{\mathbf{J}}^T\hat{\mathbf{J}}|| \le 0.999$ is always satisfied. Fig. 3 shows that the objective function converges very rapidly.

**Example 2**

To investigate the effect of $\Lambda$ on the control, we solve the same problem as in Example 1 with $\Lambda = 0.1I$, $\Lambda = 0.4I$ and $S = 10\mathbf{J}^T$. Fig. 1 shows the outputs and Fig. 2 shows the inputs. Fig 3 shows the convergence of the objective functions. They converge very rapidly. After 5 iterations, they change very little. Note that the objective function is near zero for $\Lambda = 0$. As expected, the output error becomes bigger and the input becomes smaller for a bigger $\Lambda$. The convergence condition is satisfied as $||I - \Lambda - 10\hat{\mathbf{J}}^{\mathbf{T}}\hat{\mathbf{J}}||$ equals 0.9 and 0.6 for $\Lambda = 0.1I$, $\Lambda = 0.4I$ respectively.

In the above results, the value of the objective function in each case seems to converge to its minimum value. As in the LQR problems, the tracking performance for the desired trajectories changes as the ratio of $\mathbf{S}$ to $\Lambda$ changes. Since perfect modeling is impossible and the calculation of $\mathbf{J}^k$ is very difficult, it is a difficult problem to check whether the convergence condition is satisfied. In the simulation, we checked the convergence condition with (16).

## 5. CONCLUSION

The nonlinear optimal control over a finite time interval has been studied in this paper. A numerical method for finding the optimal solution has been proposed. It is based on the approach taken in the ILC. The control performance over a finite time interval is considered at each time in the proposed method, whereas only the current performance is considered in the ILC at each time. By taking a set of vectors( e.g. inputs over the control horizon) as one big vector, the approach used in the ILC can be applied with minor modifications. The proposed input update law is shown to converge under a certain condition. By considering the output error and the input energy over a control horizon, the proposed method can avoid the situations where the input becomes very big in controlling nonminimum phase systems.

It has also been shown that the proposed method can be derived from the steepest descent method of a quadratic objective function. It shows clearly how the optimal control and the ILC are related. This method has been tested for the example in (Devasia, 1996), and has found the solutions very well in several different cases.

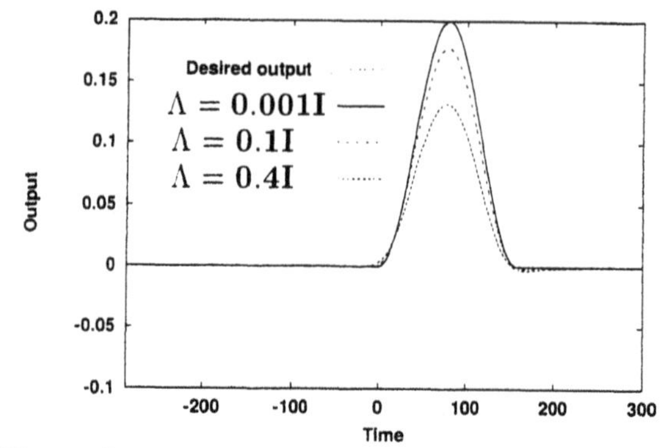


Fig. 1. Output

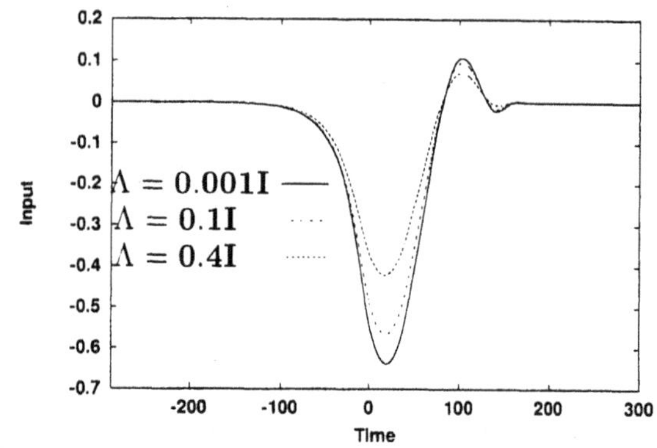


Fig. 2. Input

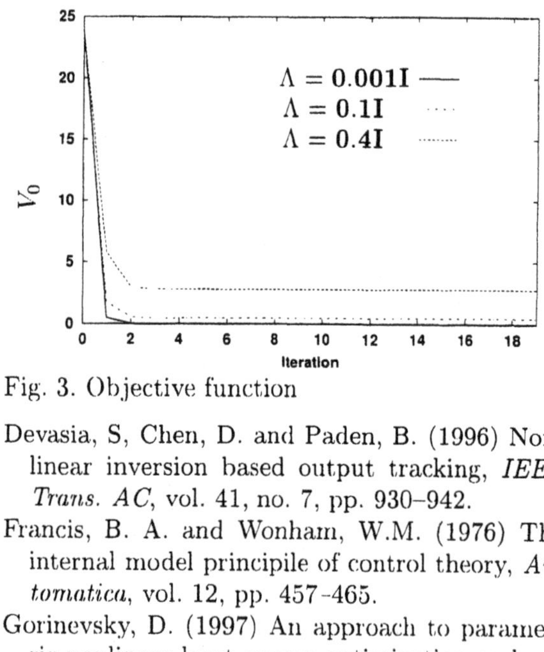


Fig. 3. Objective function

### References

Anderson, B. D. O. and Moore, J. B. (1989) *Optimal Control*. Prentice-Hall.

Choi, C.-H and Jang, T.-J. (1997) Iterative learning control of nonlinear systems with consideration on input magnitude, *Proc. ASCC*, pp. 37-40.

Devasia, S, Chen, D. and Paden, B. (1996) Nonlinear inversion based output tracking, *IEEE Trans. AC*, vol. 41, no. 7, pp. 930–942.

Francis, B. A. and Wonham, W.M. (1976) The internal model principile of control theory, *Automatica*, vol. 12, pp. 457-465.

Gorinevsky, D. (1997) An approach to parametric nonlinear least square optimization and application to task-level learning control , *IEEE Trans. AC*, vol. 42, no. 7, pp. 912 927.

Isidori, A. (1989) *Nonlinear Control Systems: An Introduction*, Springer-Verlag, Berlin.

Isidori, A. and Byrnes, C. I. (1990) Output regulation of nonlinear systems, *IEEE Trans. AC*, vol. 35, no. 3, pp. 131–140.

Nijmeijer, H and A. J. van der shaft (1990) *Nonlinear Dynamical Control Systems*, Springer-Verlag, Berlin.

Uchiyama, M. (1978) Formulation of high-speed motion pattern of a mechanical arm by trial, *Trans. SICE*, vol. 14, no. 6, pp. 706-712.

14th Triennial World Congress, Beijing, P.R. China



F-2c-09-2

# A TITO DESIGN FOR SISO SYSTEMS WITH RATE-AMPLITUDE CONSTRAINTS

**K. Hui**

*Department of Mechanical Engineering, University of Hong Kong*
*Email: khui@hkumea.hku.hk*

Abstract: This paper proposes two frequency domain design methods of saturation compensators for single-input single-output systems subject to rate and amplitude constraints. The SISO nonlinear system is treated in a two-input two-output manner. Characteristic loci are analyzed and modified by phase shifting filters, through which the compensators are parametrized to ensure stability and performance of the nonlinear systems. The first method designs the same rate and amplitude compensators for integral controllers. The second one designs different compensators for controllers without integrators. Simulated examples illustrate the achievable results from these simple practical design methods. *Copyright © 1999 IFAC*

Keywords: nonlinear systems, actuator saturation, compensator designs

## 1. INTRODUCTION

When single-input single-output (SISO) systems are subject to both rate and amplitude constraints on the actuators, existing theories on anti-reset windup or general saturation compensation (Kothare, *et al.*, 1994; Walgama and Sternby, 1993) fail to provide satisfactory solutions to tackle the two nonlinearities. Horowitz (1984) extended 3-Degrees of Freedom (3DoF) controllers for single saturation system to two nonlinearity ones, and specified globally stable equivalent systems to design the 3DoF controllers. This method, however, cannot guarantee that global stability for the two-nonlinearity system is ensured.

Others analyses exist for diagonally dominant systems with multiple saturation nonlinearities in parallel (Park and Choi, 1997; Walgama and Sternby, 1993). These are in stark contrast to rate and amplitude constrained ones, for which the two nonlinearities occur in series and their immediate application are not readily possible.

Among the major works on linear systems subject to rate and amplitude constraints (Horowitz, 1984; Liao and Horowitz, 1986; Kapasouris and Athans, 1990; Hui and Chan, 1997a), it is observed that design methods for the saturation compensators are much less developed than for single saturation ones. This paper proposes a two-input two-output (TITO) configuration to design the rate and amplitude saturation compensators. The graphical nature of the design methods provides a visual means for tackling the saturation problems in practice.

In the following, a TITO compensation structure is formulated in §2. Characteristic loci of the TITO systems are analyzed in §3. Considerations on compensator designs in frequency domains utilizing the characteristic loci analysis are presented in §4. A simulated example is illustrated in §5, showing that for multi-nonlinearity systems, it is insufficient to ensure global stability by considering the linear equivalent systems alone.

## 2. A GENERAL RATE AND AMPLITUDE COMPENSATOR

Let the 2DoF controller for plant $G$ be described by

$$v \;=\; [T/R]\,w \;-\; [S/R]\,y \qquad (1)$$

where $y$ is the system output, $w$ the reference input and $v$ the controller output. $R$, $S$, $T$ are polynomials in Laplace transform variable $s$. $R$ is monic. The arguments are omitted for convenience. The amplitude and rate constraints for the actuator are

$$u(t) \;=\; \begin{cases} u_{\max} \;, & v \ge u_{\max} \\ v \;, & u_{\min} < v < u_{\max} \\ u_{\min} \;, & v \le u_{\min} \end{cases} \qquad (2a)$$

$$\dot u(t) \;=\; \begin{cases} \dot u_{\max} \;, & \dot v \ge \dot u_{\max} \\ \dot v \;, & \dot u_{\min} < \dot v < \dot u_{\max} \\ \dot u_{\min} \;, & \dot v \le \dot u_{\min} \end{cases} \qquad (2b)$$

where $\{u_{\max}, u_{\min}\}$ are amplitude limits and $\{\dot u_{\max}, \dot u_{\min}\}$ rate limits. All the limits are assumed known.

An actuator model is inserted after the controller and before the actuator, as in (Liao and Horowitz, 1986; Kapasouris and Athans, 1990). The constrained system is compensated by two linear blocks within the actuator model, Fig.1 (Hui and Chan, 1998). The rate saturation is compensated by $P_r$ and the amplitude one by $P_a$. $R_2(s)=s$. The uncompensated system is regenerated by $P_r=P_a=0$. For controllers with integral actions, $\Delta v$ in Fig.1 naturally arises as the control increment and modelling of the actuator need not involve differentiators.

For $y=Gu$ and from Fig.1, the rate and amplitude outputs of the actuator model are given by

$$\left.\begin{aligned} v_r &= R_2\,[T/R]\,w - R_2\,[GS/R]\,u + P_r\,\delta_r \\ v &= [1/R_2]\,u_r \;+\; P_a\,\delta_a \end{aligned}\right\} \qquad (3)$$

and $\quad \delta_a(t) \triangleq u(t) - v(t) \;;\; \delta_r(t) \triangleq \dot u(t) - \dot v(t) \qquad (4)$

Using (4) in (3) and let $R_1=R/R_2$, then rearrange,

$$\left.\begin{aligned} v_r &= \frac{P_r}{1+P_r}\,u_r - \frac{GS}{R_1(1+P_r)}\,u + \frac{T}{R_1(1+P_r)}\,w \\ v &= \frac{1}{R_2(1+P_a)}\,u_r + \frac{P_a}{1+P_a}\,u \end{aligned}\right\} \qquad (5)$$

which may be written as

$$\begin{bmatrix} v_r \\ v \end{bmatrix} = -G_P(s)\begin{bmatrix} u_r \\ u \end{bmatrix} + F_T(s)\;w \qquad (6)$$

$$\text{where} \quad G_P(s) = \begin{bmatrix} -\dfrac{P_r}{1+P_r} & \dfrac{GS}{R_1(1+P_r)} \\ -\dfrac{1}{R_2(1+P_a)} & -\dfrac{P_a}{1+P_a} \end{bmatrix} \qquad (7a)$$

$$\text{and} \quad F_T(s) = \begin{bmatrix} T/R_1(1+P_r) \\ 0 \end{bmatrix} \qquad (7b)$$

The system representing (6) is shown in Fig.2, with $K_U=[0\ 1]$ and $K_N$ is the saturation nonlinearity of (2):

$$K_N = diag\{\,sat(v_r),\, sat(v)\,\} \qquad (8)$$

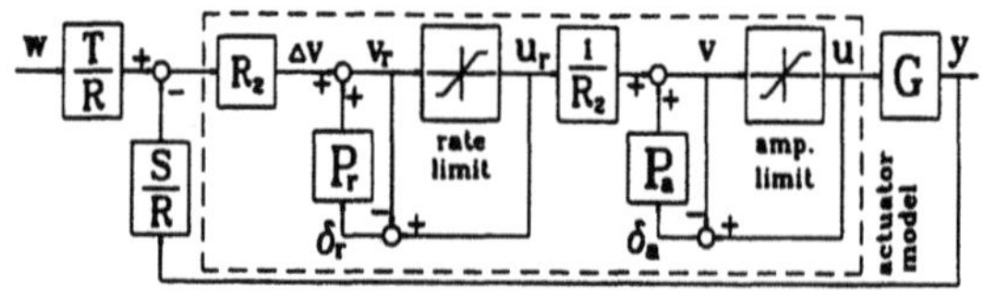


Fig.1 Rate-amplitude saturation compensated system.

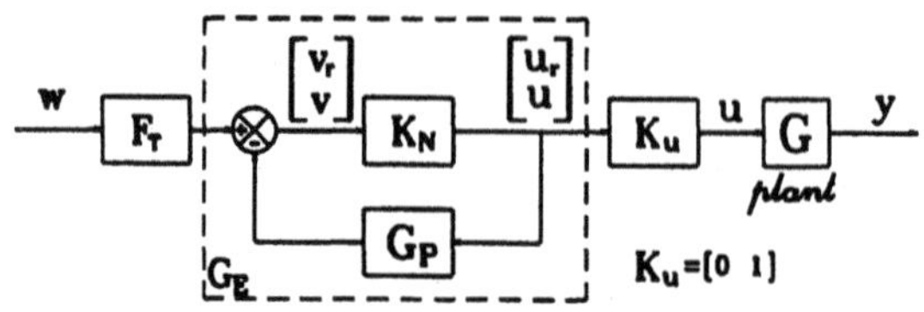


Fig.2 Reconfiguration of Fig.1 into a TITO system

By considering the compensation as in (7), the overall system is studied as a TITO entity, instead of two SISO subsystems. The closed-loop output is

$$y \;=\; Gu \;\triangleq\; y_0 + \Delta y_r + \Delta y_a \qquad (9a)$$

$$\left.\begin{aligned} y_0 &= \frac{GT}{R(1+G_C)}\,w \quad ; \quad G_C = \frac{GS}{R} \\ \Delta y_r &= \frac{G(1+P_r)}{R_2(1+G_C)}\,\delta_r \;,\; \Delta y_a = \frac{G(1+P_a)}{1+G_C}\,\delta_a \end{aligned}\right\} \qquad (9b)$$

$y_0$ is the unconstrained linear system response when $\delta_a=\delta_r\equiv 0$. $\Delta y_a$ is the output variation due to amplitude saturation and $\Delta y_r$ that due to rate saturation. The uncompensated system $[P_a=P_r=0]$ is still affected by $\Delta y_a$ and/or $\Delta y_r$, due to non-zero disturbances $\{\delta_a, \delta_r\}$.

The compensators need to satisfy the following conditions (Hui and Chan, 1998):

**(P1)** $P_a(s)$ and $P_r(s)$ are proper or strictly proper.

**(P2)** poles of $RP_a$ and $R(1+P_r)/R_2$ are asymptotically stable.

**(P3)** zeros of $(1+P_a)$ and $(1+P_r)$ are asymptotically stable.

**(P1)** ensures that the compensators are physically realizable. **(P2)** is necessary for $\{\Delta y_r, \Delta y_a\}$ in (9b) to converge back to zero: i.e., for $y(t)\to y_0(t)\to w(t)$ as $t\to\infty$. **(P3)** is only needed for global stability. It is assumed that performance of the linear system $G_C(s)$ is satisfactory in the absence of saturation.

## 3. CHARACTERISTIC LOCI ANALYSIS

All the investigation of system stability and robustness for the TITO compensated system can in general be carried out using (6)-(7) with existing tools in multivariable frequency domains (Freudenberg and Looze, 1988). This work only considers the analysis of characteristic loci (Maciejowski, 1989), with an emphasis on the intended design of the compensators. The proposed treatment of characteristic loci analysis for rate and amplitude constrained systems has not yet been reported elsewhere in the literature.

By defining the following prefilters:

$$F_r(s) = \frac{1}{1+P_r(s)} \quad \text{and} \quad F_a(s) = \frac{1}{1+P_a(s)} \qquad (10)$$

the compensators are parametrized as

$$P_r = 1/F_r - 1 \quad \text{and} \quad P_a = 1/F_a - 1 \qquad (11)$$

The system matrix $G_P(s)$ in (7) can be expressed as

$$G_P(s) = \begin{bmatrix} F_r - 1 & G_{C1}F_r \\ -F_a/R_2 & F_a - 1 \end{bmatrix}; \quad G_{C1}(s) = \frac{GS}{R_1} \qquad (12)$$

For uncompensated systems, $F_a = F_r = 1$, giving

$$G_P(s) = \begin{bmatrix} 0 & G_{C1} \\ -1/R_2 & 0 \end{bmatrix} \qquad (13)$$

The off-diagonal dominance of the system matrix (13), as a result to preserve the diagonality of (8), renders existing theories on multivariable designs of saturation compensations (Kothare, *et al.*, 1994; Walgama and Sternby, 1993) ineffective.

For general design of compensators, a frequency response analysis of (12) is extremely useful. The characteristic loci, as eigenvalues of $G_P(j\omega)$, are calculated from

$$[(\lambda+1) - F_r][(\lambda+1) - F_a] + G_C F_r F_a = 0 \quad (14)$$

$$\lambda + 1 = \tfrac{1}{2}[(F_a + F_r) \pm \sqrt{(F_a - F_r)^2 - 4G_C F_a F_r}] \quad (15)$$

For the special case $F_a = F_r = F$, (15) simplifies to

$$[1 + \lambda(j\omega)] = F(j\omega)[1 + \lambda_0(j\omega)] \qquad (16)$$

and $\lambda_0(j\omega) \triangleq \pm\sqrt{-G_C(j\omega)}$ is the characteristic loci for the uncompensated system (13). Denote the positive root as $\lambda_{0+}(j\omega)$ and the negative one as $\lambda_{0-}(j\omega)$, and hence $\lambda_+(j\omega)$ as from $\lambda_{0+}(j\omega)$ and $\lambda_-(j\omega)$ with $\lambda_{0-}(j\omega)$ respectively.

(16) gives a simple and clear graphical interpretation: when both rate and amplitude compensators are the same $[P_a = P_r = P]$, then saturation compensation is just the rotation of the characteristic loci of the uncompensated system, with respect to point $-1+j0$, by the vector $F(j\omega)$ $[=1/(1+P)]$. This observation is consistent with compensation of single saturation systems (Hui and Chan, 1997b). From (15), it is noted that the case of $F_a = F_r$ endows the compensator with some flavour of 'optimality' in a least square sense.

This special case of adopting $F_a = F_r$ is equivalent to the synthesis of 3DoF controller proposed by Horowitz (Hui and Chan, 1998). Therefore, more insights can now be gained by the characteristic loci analysis on properties of the 3DoF controllers. It is demonstrated in §5 that, contrary to belief, specifications of the linear equivalent system with global stability, as done by Horowitz (1984), cannot guarantee the same when rate and amplitude saturation occur simultaneously.

Apart from this special case, unfortunately, analysis of (15) does not enlighten the selection of rate and amplitude compensators in general. It is impossible to modify one branch of $\lambda(j\omega)$ without affecting the other. However, interpretation of (16) does inspire the extension that *all* saturation compensators eventually shift the phase angles of some representative points on the characteristic loci. The whole exercises of compensator designs then boil down to the selection of these representative points on $\lambda(j\omega)$, and the amount of phase shifts, as follows.

## 4. DESIGN OF SATURATION COMPENSATORS

Depending on whether the linear controller (1) contains an integrator, two design procedures to select $\{P_a, P_r\}$ are presented.

The symmetry of $\lambda(j\omega)$ in $F_a$ and $F_r$ indicates that both rate and amplitude saturation compensators have the same degree of influence on the system performance, and hence it is of paramount importance to identify and properly match the type of saturation with its compensator. For example, a rate saturated system with $P_r = 0$ will not benefit from any amplitude saturation compensator when $\delta_a \equiv 0$, and vice versa, see (9).

Reflections on (16) and its graphical interpretations suggest that all compensator designs can be selected by the specification of some phase shifting filters. Thus, a simple lead filter such as

$$\left.\begin{aligned} F(s) &= [s + 1/\beta]/[s + 1/\alpha\beta] \\ \alpha &= \frac{1-\sin\phi}{1+\sin\phi}, \quad \beta = \frac{1}{\omega_m\sqrt{\alpha}} \end{aligned}\right\} \qquad (17)$$

so that $F(s)$ introduces a phase shift $\phi$ at frequency $\omega_m$, may be adequate in saturation compensation. Specification of the representative points, and hence $\{\phi, \omega_m\}$, are determined below.

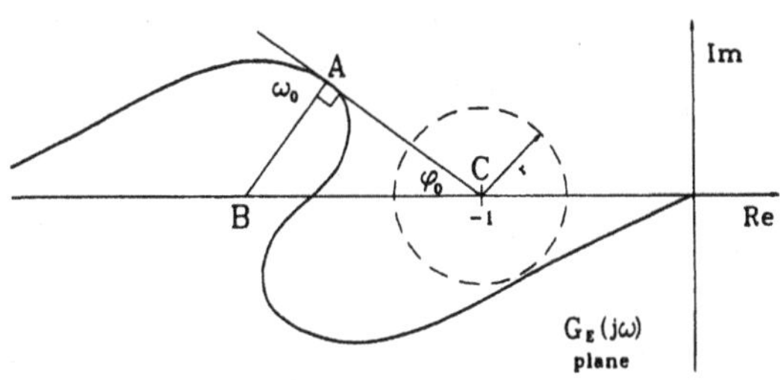


Fig.3 Determination of filter parameters w.r.t. $-1+j0$

Assume that for a given linear plant-controller design $G_C(j\omega)$, the stability behaviour of the uncompensated system is not satisfactory when both rate and amplitude saturation occur. Let the Nyquist curve of $G_C(j\omega)$ be of the shape as shown in Fig.3. This implies that $\lambda_{0-}(j\omega)$ is of the same shape as well. Let point A at frequency $\omega_0$, which subtends the largest angle $\varphi_0$ at point $-1+j0$, be identified. Then filter (17) is uniquely determined by

$$\omega_m = \omega_0 \quad \text{and} \quad \phi = \varphi_0 + \Delta\phi \qquad (18)$$

The only degree of freedom left is the phase margin $\Delta\phi$, which becomes the designer's choice and can be on-line tuned to achieve satisfactory responses. The underlying argument is to shift all the characteristic loci $\lambda(j\omega)$ clear away from the -ve real axis, so that multivariable circle criterion can be applied.

If point A is in the third quadrant, then the design procedure can still be used to increase $\Delta\phi$, so that robustness of the saturated system can be enhanced with a larger phase margin. For those systems without the shape of Fig.3 and point A not determined, other selections of representative points will be needed as in single saturation systems (Hui and Chan, 1997b).

Once $F$ is defined, the compensators are obtained as

$$P(s) = \frac{(1/\alpha - 1)/\beta}{s + 1/\beta} = \frac{2\ \omega_m\ \tan\phi}{s + \omega_m\ (1 - \sin\phi)/\cos\phi} \qquad (19)$$

Since $-90° < \phi < 90°$ for (17), the range of adjustment for $\Delta\phi$ is limited. If large phase shifts are demanded, a cascade of (17) can be applied.

**Design I** (for $P_a = P_r$)

1) Plot $\lambda_0(j\omega)$ of (16);
2) Identify point A of $\omega_0$ and $\varphi_0$ with respect to point $-1+j0$ as in Fig.3;
3) Adjust $\Delta\phi$ in (18) to calculate compensator (19).
4) If the compensated response is satisfactory, exit design loop. Else return to Step (3). §

The above procedure simultaneously designs the same rate and amplitude compensators, by taking advantage of the interpretation of (16). If sufficient phase margin of $\Delta\phi$ can be achieved for the characteristic loci, then by virtue of the multivariable circle criterion, global asymptotic stability for the compensated system can be guaranteed in the presence of both rate and amplitude constraints.

It is easily seen that compensator (19) satisfies conditions (**P1**) and (**P3**). However, condition (**P2**) may not always be true if the 2DoF controller does not contain an integrator. From (9),

$$\Delta y_r = \frac{BR}{AR + BS} \frac{s + 1/\alpha\beta}{s + 1/\beta} \frac{1}{s} \delta_r \qquad (20)$$

for $G = B/A$. The presence of integrator in $\Delta y_r$ shall result in a steady state offset. This is normally not acceptable, even though the saturation system is stabilized. To cater for this situation, filter (17) is reselected as

$$F_r(s) = (s + 1/\beta)^2 / [s(s + 1/\alpha\beta)] \qquad (21)$$

The maximum phase advance of (21), given by

$$\phi = \tan^{-1}[\,k\sqrt{\frac{k}{\alpha}}\,]\ ,\ \omega_m = \frac{1}{\beta\sqrt{\alpha k}}\ ;\ k = \frac{1-2\alpha}{2-\alpha} \qquad (22)$$

is too complicated to solve for the filter parameters $\{\alpha,\beta\}$. A reasonable approximation is to use the same expressions for $\{\alpha,\beta\}$ as in (17), but with a different phase angle specification, as below.

Let $\phi_0$ $[=\varphi_0+\Delta\phi]$ be the desired net phase angle advance at frequency $\omega_0$. The maximum phase lag due to the integrator in (21) is 90°. Hence the total phase shift required is $(\phi_0+90°)$. This is obtained by two first order filters, each advancing $\phi = (\phi_0+90°)/2$ at frequency $\omega_0$. $\{\alpha,\beta\}$ of (21) are thus determined as those given by (17). The compensator follows as

$$P_r(s) = \frac{s[\,1/\alpha - 2\,]/\beta\ -\ 1/\beta^2}{(s + 1/\beta)^2} \qquad (23)$$

The resultant phase shift would not be exactly $\phi_0$ from this approximation. Nevertheless, the difference would not be too great if $\omega_0$ is small.

**Design II** (for $P_a \neq P_r$)

1) Plot $\lambda_0(j\omega)$ of (16);
2) Identify point A of $\omega_0$ and $\varphi_0$ with respect to point $-1+j0$ as in Fig.3;
3) Adjust $\Delta\phi$ in $\phi = (\varphi_0+\Delta\phi)/2+45°$ to calculate $\{\alpha,\beta\}$ per (17) for compensator $P_r$ of (23);
4) Obtain $P_a$ as in Design I or from (19) above;
5) If the compensated response is satisfactory, exit design loop. Else return to Step (3). §

Since $G_C(j\omega)$ and $\lambda_{0-}(j\omega)$ are similar and any offending interception of the characteristic loci with the -ve real axis occurs on the branch of $\lambda_{0-}(j\omega)$, it is feasible to replace $\lambda_0(j\omega)$ in Step (1) of Design II by $G_C(j\omega)$. This enables design of prefilter by directly utilizing the Nyquist curves of the open-loop linear system, instead of plotting the characteristic loci with additional efforts. Designs using $G_C(j\omega)$ and $\lambda_{0-}(j\omega)$ approximate each other reasonably well, provided proper angles of $\varphi_0$ and frequencies of $\omega_0$ are used in the calculations.

From the above analysis, it is obvious that there are infinitely many other choices for the prefilters, apart from (17) and (21). These two are chosen because of their simplicity and practicality. There is only one variable to adjust, namely, the phase margin $\Delta\phi$, as defined in (18). Excellent results, either guaranteeing global asymptotic stability, or ensuring satisfactory transient responses, of the rate and amplitude constrained systems, can be achieved by proper tuning of the phase margin.

## 5. EXAMPLE

This example was studied by Horowitz (1984) and is

borrowed to demonstrate the TITO design methods. The obtained results illustrate the principal ideas presented in the text above.

The plant and the controller are respectively

$$G(s)=\frac{1}{s} \quad ; \quad \frac{S}{R}=\frac{T}{R}=\frac{13(s+5)^3}{(s+0.7)^3(s/250+1)} \qquad (24)$$

Notice the lack of integrator in $R(s)$ of (24).

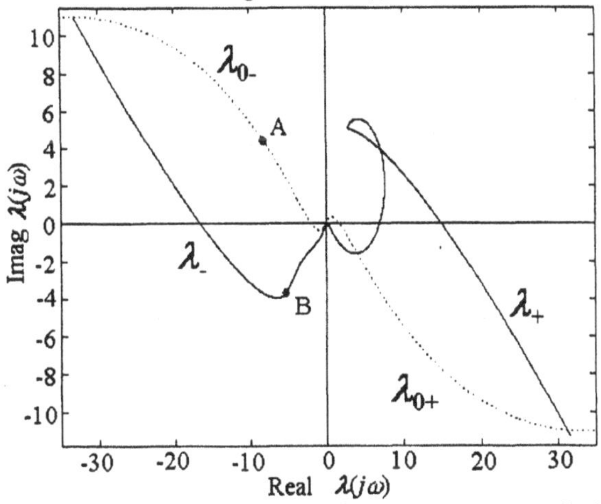


Fig.4 Characteristic loci of the 3DoF controlled system: $\phi$ = 72.4° at $\omega$=2.1286 from A to B.

The characteristic loci, $\lambda_{0+}(j\omega)$ and $\lambda_{0-}(j\omega)$, for the uncompensated rate-amplitude constrained system of (24) are shown in Fig.4. With $\lambda_{0-}(j0.5031)$=-71.54+$j$0 and $\lambda_{0-}(j7.122)$=-1.81+$j$0 (Fig.7), the uncompensated system is not asymptotically stable when sufficient saturation occurs, e.g., at $\{u_{\text{limit}}=\pm1;\ \dot{u}_{\text{limit}}=\pm0.5\}$, the unit step response limit-cycles (Horowitz, 1984; Hui and Chan, 1998). The maximum subtended angle at -1+$j$0, point A in Fig.4, is $\varphi_0=\angle\lambda_{0-}(j2.1286)$=31.56°. The $G_C(j\omega)$-plot for the SISO system is similar to $\lambda_{0-}(j\omega)$ of Fig.4, but point A then makes an angle of 57.73° at $\omega_0$=1.9088.

Using an asymptotically stable linear equivalent system, the 3DoF control synthesis (Horowitz, 1984) resulted in the following high order prefilter, and the corresponding compensator, as

$$F_{3\text{DoF}}=\frac{(s+0.7)^3(s+16)^2(s+250)}{s\left(\begin{array}{c}s^5+252.1s^4+3776.5s^3+49117.8s^2\\+243835.75s+406250\end{array}\right)} \qquad (25)$$

$$P_{3\text{DoF}}=\frac{\left(\begin{array}{c}-32s^5-5073.2s^4-32634.64s^3+\\97202.704s^2+309338.192s-21952\end{array}\right)}{(s+0.7)^3(s+16)^2(s+250)} \qquad (26)$$

The maximum phase shift is $\angle F_{3\text{DoF}}(j\omega)$=72.7° at $\omega_m$=2.36, which is very close to 2.1286 for point A. Fig.4 also shows the characteristic loci of the compensated system using (25)-(26). Point A is moved to B. The range of stability is enlarged for the compensated system; however, it is still locally stable only since $\lambda_{-}(j0.4539)$=-16.55+$j$0. If saturation is sufficiently severe, the compensated system using (25)-(26) shall become unstable. This prediction is confirmed by the simulated system output, $y_{3\text{DoF}}$, shown in Fig.5 on phase diagrams [$e$=$y$-$w$; $\dot{e}$=$de/dt$] for $\{u_{\text{limit}}=\pm0.4668;\ \dot{u}_{\text{limit}}=\pm0.5\}$. This instability was not recognized, nor explainable with the analysis performed, by Horowitz (1984).

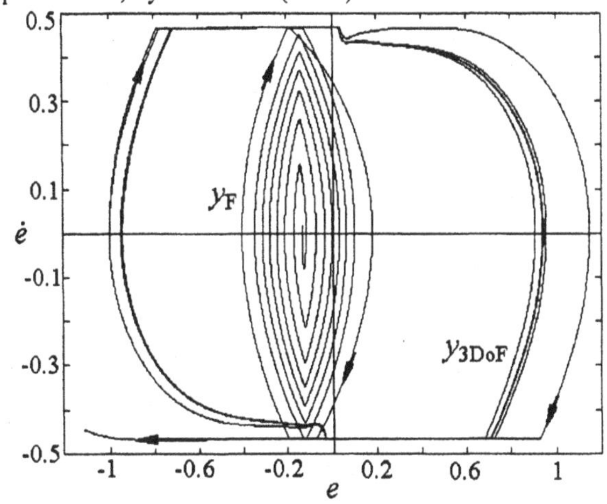


Fig.5 The 3DoF controlled system is unstable for $\{u_{\text{limit}}=\pm0.466;\ \dot{u}_{\text{limit}}=\pm0.5\}$. Response $y_F$ is designed per Fig.6.

If Design I is applied to select $F(s)$, then from Fig.4, choose $\phi$=60° (>31.56°) at frequency $\omega_0$=2.1286. This gives the following prefilter and compensator:

$$F(s)=\frac{s+0.5703}{s+7.9438} \quad \text{and} \quad P(s)=\frac{7.3735}{s+0.5703} \qquad (27)$$

The characteristic loci (CL) of the compensated system using (27) is shown in Fig.6. There is no longer any interception of $\lambda_{-}(j\omega)$ with the -ve real axis and thus the compensated system is asymptotically stable for any combinations of $u_{\text{limit}}$ and $\dot{u}_{\text{limit}}$. However, (27) does not satisfy condition **(P2)** and from (20), the integrator accumulates all the disturbances of $\delta_r(t)$ into a nonzero value of $\Delta y_r(t)$ as $t\to\infty$. This is shown as $y_F$ in Fig.5. The system error settles but does not converge back to the origin.

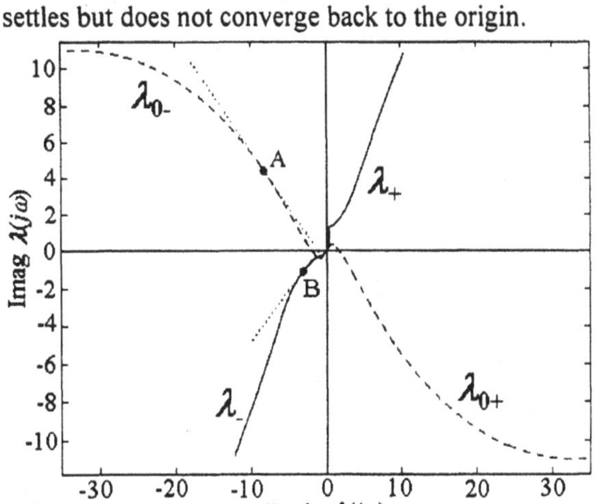


Fig.6 CL of the prefilter compensated system: $\angle F(j2.1286)$=60° from A to B. Stabilized response as $y_F$ in Fig.5.

Design II is applied to select the rate and amplitude compensators by using the Nyquist curve of the open-loop linear system. A $G_C(j\omega)$-plot reveals that $\{\varphi_0=57.7°$ at $\omega_0=1.9\}$. Thus $\phi$ may take 60° and both $F_a(j\omega)$ and $F_r(j\omega)$ are to advance this amount. From (19), $P_a$ is immediately designed as

$$P_a(s) = 6.6122 \;/\; (s+0.5115) \quad (28)$$

$F_r(s)$ of (21) is designed for a phase shift of 150° [=60°+90°=2x75°] and from Design II, it gives

$$F_r(s)=\frac{(s+0.25)^2}{s(s+14.43)}\;;\; P_r(s)=\frac{13.93s-0.0625}{(s+0.25)^2} \quad (29)$$

The characteristic loci $\lambda(j\omega)$ of (15) using compensators {(28), (29)} are shown in Fig.7. The resultant phase shift $\phi$=64° at $\omega_0$=2.1286 is slightly larger than specified. Interception of $\lambda_-(j\omega)$ with the -ve real axis is removed and global stability is guaranteed. The system response, without steady state offset, for $\{u_{\text{limit}}=\pm0.4668;\ \dot{u}_{\text{limit}}=\pm0.5\}$, is shown in Fig.8 and improvement over $y_F$ of Fig.5 is significant.

This example clearly demonstrates the simplicity, practicality, efficiency and robustness of the proposed design methods, compared with the 3DoF control synthesis theory which, in addition, could not even explain the observed instability when severe rate and amplitude saturation occur simultaneously.

## 6. CONCLUSION

This paper presents a general saturation compensation framework for linear systems subject to rate and amplitude constraints on the actuators. The analysis is performed in a two-input two-output setting for the single-input single-output linear system with saturation compensators. This shall enable sufficient conditions for global stability to be studied via the characteristic loci, as well as the design of compensators by interpretations as phase shifting filters. The main contribution is in the establishment of two practical design methods utilizing the characteristic loci of the TITO system. A simulated example illustrates both design procedures and the obtained results of the rate and amplitude saturation compensated system.

## REFERENCES


Freudenberg, J.S. and D.P. Looze (1988). *Frequency Domain Properties of Scalar and Multivariable Feedback Systems*. Springer-Verlag, Berlin.

Horowitz, I. (1984). Feedback systems with rate and amplitude limiting. *Int. J. Control*, **40**, 1215-1229.

Hui, K. and C.W. Chan (1997a). A Stability Analysis of Controllers subject to Amplitude and Rate Constraints. *Proc. IEEE CCA*, 259-264.

Hui, K. and C.W. Chan (1997b). New Design Methods of Actuator Saturation Compensators for PID Controllers, *J. Syst. Cont. Engg.* **211**(I4) 269-280.

Hui, K. and C.W. Chan (1998). Control Synthesis versus Saturation Compensation for Systems with Rate and Amplitude Constraints. *Proc. ACC*, 1473-1477.

Kapasouris, P. and M. Athans (1990). Control Systems with Rate and Magnitude Saturation for Neutrally Stable Open Loop Systems. *Proc. IEEE CDC*, 3404-3409.

Kothare, M.V., P.J. Campo, M. Morari and C.N. Nett (1994). A Unified Framework for the Study of Anti-Windup Designs. *Automatica*, **30**, 1869-1883.

Liao, Y.K. and I.M. Horowitz (1986). Unstable uncertain plants with rate and amplitude saturations. *Int. J. Control*, **44**, 1147-1159.

Maciejowski, J.M. (1989). *Multivariable Feedback Design*. Addison-Wesley, Harlow.

Park, J.K. and C.H. Choi (1997). Dynamical Anti-Reset Windup Method for Discrete-Time Saturating Systems. *Automatica*, **33**, 1055-1072.

Walgama, K.S. and J. Sternby (1993). Conditioing technique for multiinput multioutput processes with input saturation. *IEE Proc.D*, **140** 231-241


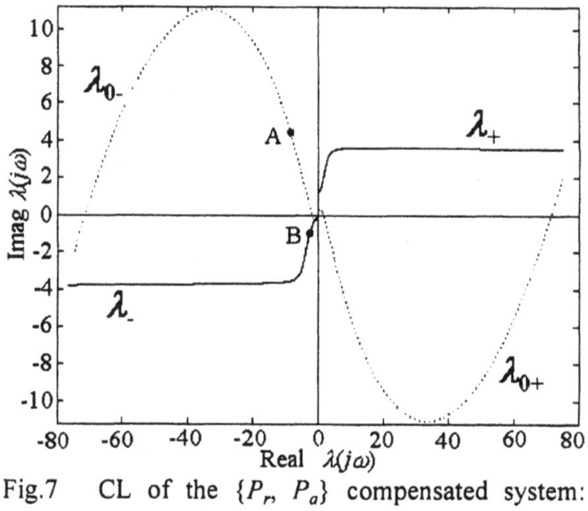


Fig.7 CL of the $\{P_r, P_a\}$ compensated system: $\phi$=64.0° at $\omega$=2.1286 from A to B w.r.t. -1+j0.

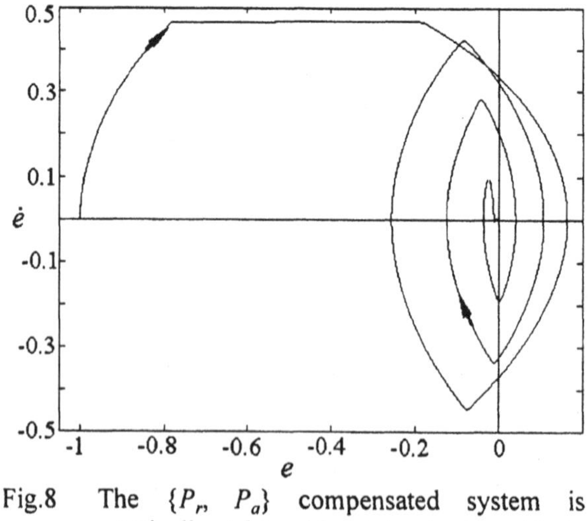


Fig.8 The $\{P_r, P_a\}$ compensated system is asymptotically stable without steady state offset.

14th Triennial World Congress, Beijing, P.R. China

F-2c-09-3

# GLOBAL NUMERICAL APPROACH TO CONSTRAINED NONLINEAR GENERALIZED PREDICTIVE CONTROL

Jianjun Yang, Wei Wang

*Research Center of Automation, Northeastern University, Shenyang, 110006, P.R. China, Email:wangw@mail.neu.edu.cn*

Abstract: A global numerical approach is proposed to solve the nonlinear generalized predictive control (GPC) law. Set inversion via interval analysis (SIVIA) is used to approximate the input sequences with a given length that drive a nonlinear discrete-time state-space system from a given initial state to a given terminal state in a guaranteed way. No requirement other than computability is put on the state equation and objective function. *Copyright ©1999 IFAC*

Keywords: generalized predictive control, constraint, interval analysis, set inversion.

## 1. INTRODUCTION

Generalized predictive control(GPC) has enjoyed great popularity for more than a decade since GPC was originally proposed in (Clarke, *et al.*, 1987). The power of GPC is that it can successfully control 'difficult' objective such as non-minimum phase systems, open-loop unstable systems, systems with unknown or changing delay and so on. It also can deal with the constraint problem. Successful applications have already been reported in some fields (M'saad, *et al.*, 1993; Tham, *et al.*, 1991), but most of success is limited to time-invariant linear systems, few generalized predictive controllers for time-varying linear systems, and fewer still for nonlinear systems. The major difficulty in the design of generalized predictive control law for nonlinear systems arises from the complex of the system structure. A general approach to control nonlinear systems is to linearize the nonlinear model and then use the GPC technique on resulted linear model, but it only in the case of systems which are feedback equivalent to linear systems. Some other approaches have also been proposed for special nonlinear model, such as Hammerstein model (Zhu, *et al.*, 1991), Wiener model (Sandra, *et al.*, 1996) and so on.

The constraint is a general problem in many industry applications. The input constraints are always present and are imposed by physical limitations of the actuators which cannot be exceeded under any circumstance. GPC algorithm considers the constraint in the objective function. Quadratic programming(QP) is used to optimal the objective function for linear systems, but QP cannot deal with the nonlinear systems, Nonlinear programming(NP) should be used. The results obtained by nonlinear programming are often local, because of the nonconvexity of the objective function.

In this paper, a global solution to get a approximate optimal input sequence for constrained nonlinear systems is proposed, it does not require system model to have some specific characteristic, such as analyticity, continuity or differentiability, other than being computable by a finite algorithm. Which permits one to take into account nonlinear such as saturations or thresholds. Although the method is numerical, it provides global and guaranteed results.

## 2. FORMULATION OF THE CONSTRAINED GPC IN TERMS OF SET INVERSION

Consider an $m$th-order discrete-time system described by the state equation

$$x(t+1) = f(x(t), u(t)) \qquad x(0) = x_0 \tag{1}$$

subject to hard constraints at the input

$$\Omega_u = \{u : \underline{u} \leq u \leq \overline{u}\} \tag{2}$$

where $u(t)$ and $x(t)$ are the input and state of the system at time $t$. $f$ is a known nonlinear vector function. $\underline{u}$ and $\overline{u}$ represent lower and upper bounds respectively. Equation (1) and (2) describe the physical systems of interest in many engineering applications.

**A1**: Assume $f$ satisfies $f(0,0) = 0$.

In general, GPC algorithm solves an optimization problem at each sampling time $t$, minimizing the objective functions $J(t)$ as following

$$\begin{aligned} J(t) &= \sum_{i=1}^{N} x(t+i|t)^T Q x(t+i|t) \\ &+ \sum_{i=0}^{N_u-1} u(t+i|t)^T R u(t+i|t) \end{aligned} \tag{3}$$

subject to (2) and

$$\begin{aligned} x(t+i|t) = 0 \quad for \quad i \geq N \\ u(t+i|t) = 0 \quad for \quad i \geq N_u \end{aligned} \tag{4}$$

Where $x(t+i|t)$ the i-step-ahead prediction, at time $t$, of the state and $u(t+i|t)$ denotes the ith postulated control at time $t$. $N$ and $N_u$ are predictive horizon and control horizon, $Q$ and $R$ are positive-defined state weight and input weight. Only the first control variation $u(t|t)$ is applied according to the receding horizon strategy.

A control sequence $U$ of length $N_u$ will be said to be feasible if $U$ satisfies (2) and

$$\begin{aligned} h(U) &= f(f \cdots (f(x(t), u(t|t)) \cdots) \\ &\quad u(t+N_u-2|t))u(t+N_u-1|t))0) \cdots 0) \\ &= O \end{aligned} \tag{5}$$

Where $U = [u(t|t), \cdots, u(t+N_u-1|t)]$. Thus, the set of all feasible input sequences of length $N_u$ is given by

$$U = h^{-1}(O) \tag{6}$$

Where $O$ is a reference vector of length $m$, for the regulation problem suppose $O$ to be a zero vector, $h^{-1}$ is the reciprocal function of $h$ in a set-theoretic sense. $U$ is Solved from equation (6) and an input sequence $u^*$ is selected which minimizes the objective function $J$ from all solutions, so $u^*$ is the GPC law for constrained nonlinear systems. Many methods have been proposed to solve the equation (6), but to the best of our knowledge, almost all of the methods require $h$ to satisfy some conditions such as analyticity, continuity and differentiability (Garcia, and Zangwill, 1979; Mayne, and Michalska, 1990). If the conditions are not satisfied, the above methods will be invalid. For such problem, the set inversion via interval analysis(SIVIA) (Hansen, 1992; Jaulin, and Walter, 1997) will be used to approximate $U$ numerically in this paper. The SIVIA algorithm is introduced in the context of bounded-error estimation (Jaulin, and Walter, 1993) and characterization of stability domains (Walter, and Jaulin, 1994).

## 3. INTERVAL ANALYSIS

Interval analysis is a numerical tool in scientific computation and has been used to solve sets of nonlinear equations or inequalities and optimizing nonconvex criteria in a guaranteed way. Now we use it to solve the problem of set inversion.

Define the box $[B]$ of $R^n$ (or vector interval) as the Cartesian product of $n$ real interval $[b_i]$

$$\begin{aligned} [B] &= [b_1^-, b_1^+] \times \cdots \times [b_n^-, b_n^+] \\ &= [b_1] \times \cdots \times [b_n] \end{aligned} \tag{7}$$

Where $b_i, i = 1, \cdots, n$ is the element of vector $B$, $[b_i^-, b_i^+]$ is a value interval of $b_i$. Denote the set of all boxes of $R^n$ by $IR^n$. The boxes in the context of set inversion make it possible to replace point values of vectors by vector interval, thereby allow a global analysis of infinite sets of point with finite number of operations. Vector $B$ of $R^n$ also can be considered a belonging to $IR^n$ if $B^- = B^+ = B$.

Define the width of box $[B]$

$$w([B]) = max_{i=1,\cdots,n}\{b_i^+ - b_i^-\} \tag{8}$$

Let $g : R^n \to R^p$ be a vector function and $G : \mathcal{R}^n \to \mathcal{R}^p$ be a set-valued function, if $g$ and $G$ satisfy the following two conditions

$$\begin{aligned} g([B]) &\subset G([B]) \\ w([B]) \to 0 &\Rightarrow w(G([B])) \to 0 \end{aligned} \tag{9}$$

$G$ is called an inclusion function of $g$. $G$ is not unique but the minimal inclusion function i.e. the smallest box that contains $g([B])$ is unique. It is easy to obtain an inclusion function $G$ by replacing each of the elementary operators, such as $+, -, *, /, sin, cos, exp, \cdots$, by its minimal inclusion function in the formal expression $g$, which usually far from the most effection. The resulting inclusion function is called natural inclusion function in the literatures. An example is given to explain how to calculate an inclusion function.

**Example 1.**

$$f(x,u) = 0.9x\sin(x) + (2 + \frac{1.5xu}{1+x^2u^2})u \quad (10)$$

$$\begin{aligned} F([X],[U])] &= 0.9[X]\sin([X]) + (2 \\ &+ \frac{1.5[X][U]}{1+[X]^2[U]^2})[U] \quad (11) \end{aligned}$$

If, for instance, $[X] = [0, \pi/3], [U] = [0,1]$, then the box $F([X],[U])$ is computed as follows:

$$\begin{aligned} F([X],[U]) &= 0.9[0,\pi/3]\sin[0,\pi/3] + (2 \\ &+ \frac{1.5[0,\pi/3][0,1]}{1+[0,\pi/3]^2[0,1]^2})[0,1] \\ &= 0.9[0,\sqrt{3}\pi/6] + [2, 2+0.5\pi][0,1] \\ &= [0, 2+(1+\sqrt{3})\pi/2] \quad (12) \end{aligned}$$

## 4. ALGORITHM FOR SET INVERSION

In this paper, set inversion aims to find $U = h^{-1}(O)$, $O$ is a vector, it also can be a box. Although $U$ is exactly described by $N_u$ equations, such equations cannot be conveniently solved on a computer, so we approximate $U$ by subpavings, i.e. sets of nonoverlapping box $U^-$ and $U^+$, which bracket $U$ defined by (6) between an inner and an outer subpaving

$$U^- \subset U \subset U^+ \quad (13)$$

Because $O$ is a singleton, no inner subpaving $U^-$ can be obtained. SIVIA is used to characterize the outer subpaving $U^+$ which contain many feasible boxes $[U] = [u_1] \times \cdots \times [u_{N_u}]$. There is a control sequence

$$u = [\frac{u_1^+ - u_1^-}{2}, \cdots, \frac{u_{N_u}^+ - u_{N_u}^-}{2}]$$

corresponding to each box included in $U^+$. The control sequence $u^*$ which makes the value of objective function smallest is considered as the control law.

SIVIA uses a stack of boxes on which three operation are done, namely stacking, unstacking and testing the stack for emptiness. Stacking is putting the box in the stack, unstacking is removing the box from the stack. The initial box $[U_0]$ is taken equal to a large box in which feasible input sequences are included. If the width of the box is not smaller than some given required accuracy $\epsilon$, then bisect $[U]$ along a symmetry plane perpendicular to one of its largest edges and result two boxes.

SIVIA can be summarized as follows
**Algorithm 1.**
Initialization:

Initial box: $[U_0]$

Inclusion function for h(.): H(.)

Required accuracy: $\epsilon$

Stack=$\emptyset$, [U]=$[U_0]$, $U^+ = \emptyset$
Iteration:

**step1.** If $O \notin H([U])$, then go to step 3.

**step2.** If $w([U]) \leq \epsilon$ then $U^+ = U^+ \cup [U]$, else bisect $[U]$ along a symmetry plane perpendicular to one of its largest edges and stack the two resulted box.

**step3.** If the stack is not empty, then unstack into [U] and go to step 1.

**step4.** Calculate the value of objective function for every control sequence which corresponds to every box included in $U^+$.

**step5.** Select $u^*$ that minimizes the objective function and input $u^*(t|t)$ to the system. Set $t = t+1$ and return to step 1.

## 5. PROPERTIES OF SIVIA

SIVIA is a finite algorithm, the convergence property of $U^+$ towards $U$ when $\epsilon$ tends to zero have been proved in (Jaulin, and Walter, 1993), in the case which satisfies (9). Under continuity condition, the characterization of $[U]$ can then be made as precise as desired, at the cost of increasing the computation. Note that if $U^+$ turns out to be empty, the set-inversion problem is guaranteed to have no solution, i.e. the control strategy is not feasible.

SIVIA terminates in less than $(w([U_0]/\epsilon + 1))^n$ iterations, the main limitation of SIVIA is that the compute time is increased exponentially with $N_u$. In order to decrease the compute burden, some improvement is made to above algorithm.

In algorithm 1, the outer subpaving is computed respectively in every iteration and does not consider the solution of last iteration,thus the compute time of every iteration is almost equal, it will increase the compute burden. The control accuracy of algorithm 1 is also limited by $\epsilon$. Take into account that the objective function is a decreasing function of time in GPC, we consider the optimal objective function value $J_{min}(t-1)$ and $\hat{J}(t+1)$ at time $t$, $\hat{J}(t+1)$ denotes the objective function value of input sequence $\hat{u}(t+1) = [u^*(t+1|t), \cdots, u^*(t+N_u-1|t), 0]$ at time $t+1$.

The modified algorithm can be concluded as follows:

**Algorithm 2.**
Initialization:

Initial box: $[U_0]$

Inclusion function for $h(.)$: H(.)

Inclusion function for $J$: J(.)

Required accuracy: $\epsilon$

Stack=0, $[U] = [U_0], U^+ = 0, J_{min}(0)$
Iteration:

**step1.** If $O \notin H([U])$ then go to step 5.

**step2.** If $J_{min}(t-1) \leq J([U])^-$ then go to 5.

**step3.** If $w(U) \leq \epsilon$ then $U^+ = U^+ \cup [U]$, go to step 5.

**step4.** Bisect $[U]$ along a symmetry plane perpendicular to one of its largest edges and stack the two resulted boxes.

**step5.** If the stack is not empty, then unstack into $[U]$ and go to step 1.

**step6.** Calculate the value of objective function for every control sequence which correspond to every box included in $U^+$, select $u^*$ and calculate $J_{min}(t)$

**step7.** If $\hat{J}(t+1) > J_{min}(t)$ or $J_{min}(t) > \hat{J}(t)$, then $\epsilon = \epsilon/10$, stack the boxes included in $U^+$ and go to step 5.

**step8.** Input $u^*(t|t)$ to the system. Set $t = t+1$ and return to step1.

If the condition of step1 is satisfied, no $u$ in $[U]$ is feasible. If the condition of step2 is satisfied, all $u$ in $[U]$ are not optimal solutions. If the condition of step7 is satisfied, then decreasing the accuracy $\epsilon$ and computing optimal input sequence again guarantee the monotonically decreasing of the objective function. The improved algorithm speeds up the compute and the state converge to its target when $\epsilon$ tends to zero.

A stability conclusion from the modified algorithm is given under the following assumption.

**A2**: There exists a constant $k$ and a radius $r > 0$, such that

$$||f(x,u)||^2 \leq k(||x||^2 + ||u||^2)$$

$$for\ every\ (x,u) \in B(0,r)$$

where $B(0,r)$ denotes an closed ball with center 0 and radius $r$.

**Theorem:** If condition (9) and assumption **A2** are satisfied and the algorithm is feasible(i.e. the space allowed for the control sequence may be a nonempty convex set) then the closed-loop system is globally asymptotically stable.

**Proof:** If condition (9) is satisfied, the convergence of algorithm is given in (Jaulin, and Walter, 1993). Feasibility of the algorithm implies that $U^+$ is not empty. At sampling time $t$, we assume

$$u^*(t) = [u^*(t|t), \cdots, u^*(t+N_u-1|t)] \quad (14)$$

at time $t+1$, let

$$\hat{u}(t+i|t+1) = u^*(t+i|t), \quad i = 1, \cdots N_u \quad (15)$$

Because system satisfies the condition (4) and assumption **A2**, thus

$$\begin{aligned} ||\hat{x}(t+N+1|t)||^2 &= ||f(x(t+N|t),0)||^2 \\ &\leq k||x(t+N|t)||^2 \end{aligned} \quad (16)$$

$x(t+N|t)$ will converge to zero when $\epsilon$ tends to zero, so if $\epsilon$ is selected small enough, $\hat{u}$ also can be a feasible solution but may not be optimal. We have

$$\begin{aligned} & J_{min}(t) - \hat{J}(t+1) = \\ & \quad x^T(t+1|t)Qx(t+1|t) + u^{*T}(t|t)Ru^*(t|t) \\ & \quad - x^T(t+N+1|t)Qx(t+N+1|t) \\ & \quad \geq x^T(t+1|t)Qx(t+1|t) + u^{*T}(t|t)Ru^*(t|t) \\ & \quad - kx^T(t+N|t)Qx(t+N|t) \end{aligned} \quad (17)$$

If $\epsilon$ is small enough, then $kx^T(t+N|t)Qx(t+N|t) < x^T(t+1|t)Qx(t+1|t)$, which yields

$$J_{min}(t) - \hat{J}(t+1) \geq 0 \quad (18)$$

Since $\hat{u}$ is not necessary optimal for $J(t+1)$, it follows that $J_{min}(t) \geq J_{min}(t+1)$ unless the system is stable.

We know from the algorithm that the objective function monotonically decreasing with time $t$ and the state will be stable when $\epsilon$ tends to zero. The result is proved.

## 6. SIMULATION RESULT

To show the validity of the algorithm, an example is given in this paper. The example has been treated with a program written in MATLAB. Consider the discrete-time state-space model

$$\begin{cases} x_1(t+1) = x_1(t)\cos(x_1(t)x_2(t)) + u(t) \\ x_2(t+1) = 3x_1^2(t) - \sin(u(t)x_2(t)) \end{cases} \quad (19)$$

We choose $Q = \begin{bmatrix} 0.1 & 0 \\ 0 & 0.1 \end{bmatrix}$ and $R = 1$. The parameters are $N = 4, N_u = 2, \underline{u} = -1, \overline{u} = 1, \epsilon = 10^{-4}$. The simulation result is given in Fig.1. (a) is output response curve, (b) is input response curve. The solid line represents the response for initial state $x(0) = [1,-1]$, the dashed line represents the response for initial state $x(0) = [-0.5, 1.5]$. It shows that the system will be stable after five steps without constraint violations through the new method.

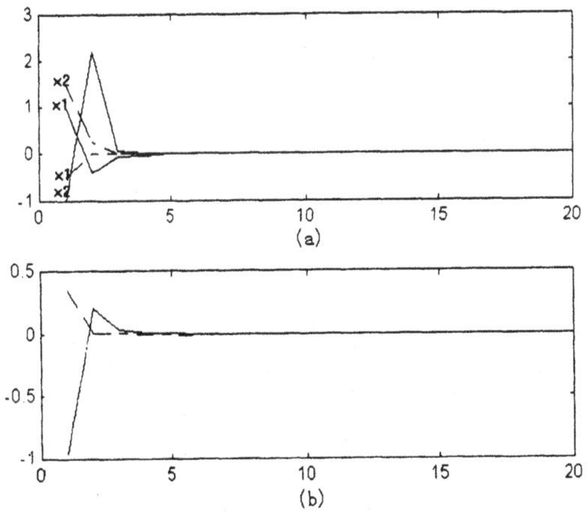


Fig.1. The response curve of system output and input.(a) Output response, (b) Input response.

## 7. CONCLUSION

In this paper, set inversion via interval analysis is proposed to solve the optimization problem of constrained nonlinear generalized predictive control. SIVIA does not require system model and objective function to have some specific characteristic. It provides global and guaranteed results, The method to be presented is not limited to the case where state target is a singleton, when state target is a box, the method is also valid. The method obviously can apply to vector inputs and longer input sequences, but the exponential complexity of the algorithm with respect to length of control sequence limit the dimension of the problems that can be handled.

*Acknowledgements:* This work is supported by grant from K. C. Wong Education Foundation and the National Natural Science Foundation (69674010) of China

## REFERENCE

Clarke, D.W., D. Mohtadi and P. S. Tuffs (1987). Generalized predictive control part 1 and 2. *Automatica*, **23**, 137-162.

Garcia, C.B. and W. I. Zangwill (1979). Finding all solutions to polynomial systems and other systems of equations. *Math. Programming*, **16**, 159-176.

Hansen, E. (1992). *Global optimization using interval analysis*, Marcel Dekker, New York.

Jaulin, L. and E. Walter (1993). Set inversion via interval analysis for nonlinear bounded error estimation. *Automatica*, **29**, 1053-1064.

Jaulin, L. and E. Walter (1997). Global numerical approach to nonlinear discrete -time control, *IEEE Trans. on Autom. Contr.*, **42**, 872-875.

Mayne, D.Q. and H. Michalska (1990). Receding horizon control of nonlinear system. *IEEE Trans. Autom. Contr.*, **35**, 814-824.

M'Saad, M., L. Dugard and S. Hammad (1993). A suitable generalized predictive adaptive controller case study: control of a flexibly arm. *Automatica*, **29**, 585-608.

Sandra, J.N., P. Ahmet and A. R. Jose (1996). Nonlinear model predictive control of pH neutralization using wiener models. *13th IFAC*, San Francisco,USA, **7a-01**, 31-36.

Tham, M.T., F. Vagi, A. J. Morris and R. K. Wood (1991). Multivariable and multirate self-tuning control: a distillon column case Study. *IEE Proc.-D*, **138**, 9-24.

Walter, E. and L. Jaulin (1994). Guaranteed characterization of stability domains via set inversion. *IEEE Trans. Autom. Contr.*, **39**, 886-889.

Zhu, Q.M., K. Warwich and J. L. Donce (1991). Adaptive general predictive controller for nonlinear systems. *IEE Proc.-D*, **138**, 33-40.

# A METHOD FOR DIRECT PID DESIGN USING I/O DATA: THE VIRTUAL MODEL DIRECT DESIGN APPROACH

**Sergio M. Savaresi , Guido O. Guardabassi**

*Politecnico di Milano, Dipartimento di Elettronica e Informazione, Piazza Leonardo da Vinci,32, 20133, Milano, ITALY. E-Mail: {savaresi|guardaba}@elet.polimi.it. Tel. ++39.223993542. Fax. ++39.223993412.*

Abstract. In this paper, an approach for the design of feedback controllers using only I/O data is presented. Such an approach - called "Virtual Model Direct Design" ($VMD^2$) - can be used when the control problem is stated by means of a reference model and I/O measurements of the plant, its main feature being that it reduces the design of the controller to a simple identification problem.

The main goal of this paper is to show how such a design philosophy can be successfully used for the direct (data-based) tuning of PID controllers; in particular, it will be shown how the choice of the PID parameters can be reduced, under quite conceivable assumptions, to the unconstrained minimization of a 3-dimensional quadratic function. 

Key words. Direct digital control; model reference control; identification algorithms; PID control.

## 1. INTRODUCTION

Even though PID controllers are the simplest and - so to say - roughest class of controllers for the regulation of dynamical plants, it is well known that, due to their simplicity and effectiveness, they are the most used regulators in industrial applications. As a consequence of their popularity, several strategies and techniques for the design of their parameters have been searched for and developed; as usual, such methods may be classified into two main classes, namely:

(i) **model-based methods**, i.e. methods which make use of a *mathematical model* of the plant;

(ii) **data-based methods**, i.e. methods which make only use of I/O measurements, corresponding to simple *ad-hoc* experiments made on the plant (or on a plant simulator).

Thanks to the current availability of cheap and powerful μ-processor-based digital controllers (Åström and Wittenmark, 1990), in the last decade the largest effort - in the PID-tuning research field - has been done in the area of data-based methods: as a matter of fact, the combination of a powerful digital processor and a wise data-based technique makes possible, in principle, to achieve the "main-goal" of any control system, i.e. the *completely autonomous tuning* of the controller.

In this paper, we present a new method for the direct design of the parameters of standard PID controllers; we shall call it the "*Virtual Model Direct Design*" ($VMD^2$) approach, since it is based upon an unusual - and, in some sense, "virtual" - modeling scheme; as it will be shown in the next section, the peculiar features of the $VMD^2$ technique can be summarized as follows:

- It is a *direct* method (Narendra and Parthasarathy, 1990), in the sense that it does not require an algebraic model of the plant (or, alternatively, of an *identification step*, which is typical of the *indirect methods*).
- It is a data-based method, i.e. only I/O measurements of the plant are needed.
- It can make use of a large class of excitation-

signals, collected both in open-loop and in closed-loop configurations.

- The closed-loop performances of the control system have to be specified by means of a model of reference.
- It is not computationally demanding, since the parameter design can be reduced to the unconstrained minimization of a quadratic loss function in $\Re^3$.

Thanks to these features, the VMD$^2$ method can be easily implemented within an unsupervised auto-tuning digital controller, and can constitute an appealing alternative to more classical PID-design techniques.

The outline of the paper is as follows: in Sect.2, the VMD$^2$ method is presented and discussed, with reference to the particular case of PID design. Sect.3 is devoted to the presentation and discussion of a numerical example, while Sect.4 ends the paper with some conclusive remarks.

## 2. THE VMD$^2$-BASED PID DESIGN

Consider a Single Input ($u(\cdot)$) Single Output ($y(\cdot)$) (SISO) time-invariant dynamical linear plant, say $\mathcal{G}$, complemented (if necessary) with A/C and C/A converters (see Fig.1), and suppose that the design task is to make the I/O behavior of the control system, say $\mathcal{T}$, as close as possible to that of a "model of reference" $\mathcal{T}^0$, namely

$$\mathcal{T}^0: y(t) = F^0(z^{-1})y^0(t)$$

($F^0(z^{-1})$ being a stable discrete-time rational transfer function), by means of a PID controller $\mathcal{K}(\theta)$, given by:

$$\mathcal{K}(\theta): \; u(t) = R_1(z^{-1},\theta)(y^0(t) - y(t)) - R_2(z^{-1},\theta)y(t), \; (1)$$

where $R_1(z^{-1},\theta) = (a + bz^{-1})/(1 - z^{-1})$ is the Proportional-Integral controller, whereas $R_2(z^{-1},\theta) = c(1 - z^{-1})/(1 + \beta z^{-1})$ is the Derivative part of the controller, which receives $y(\cdot)$ only as input; notice that the zero in (1;0) is (as usual) complemented with a first order low-pass filter $1/(1 + \beta z^{-1})$. The *parameter vector* of the PID is given by $\theta = \{a,b,c\}$, whereas $\beta$ is an a-priori selected parameter (the performance of the control system loosely depend on its value). Moreover, assume that *no mathematical model of the plant is available*, but only I/O measurements, say $\{\tilde{u}(t), \tilde{y}(t)\}_{t=0,1,2,\dots,N}$, are given.

As it is well known, the *direct* design of the controller using I/O data is, in general, a very difficult problem, since the lack of a mathematical model of the plant makes impossible the evaluation of the I/O behavior of the control system $\mathcal{T}=(\mathcal{G}, \mathcal{K}(\theta))$, as a function of $\theta$. Hence, most of the data-based design methods are *indirect* methods, i.e. are characterized by a *preliminary black-box plant-identification step*, which makes easier the solution of the subsequent model-matching problem.

The basic idea of the VMD$^2$ approach, instead, is to take a "*virtual model of the plant*", say $\mathcal{G}$, *constituted by the reference model* $\mathcal{T}^0$ *complemented with a parametric feedback controller* $\mathcal{S}(\theta)$, given by (it should be apparent that the following expression of $\mathcal{S}(\theta)$ can be easily obtained by inverting the $y^0$-$u$ relationship of (1) - see also Fig.1):

$$\mathcal{S}(\theta): \; y^0(t) = R_1^{-1}(z^{-1},\theta)u(t) + R_1^{-1}(z^{-1},\theta)R_2(z^{-1},\theta)y(t), \; (2)$$

and, using (2), the transfer function of the "virtual model" $\mathcal{G}$ can be easily computed as follows:

$$\mathcal{G}: \; y(t) = \frac{F^0(z^{-1})}{R_1(z^{-1},\theta) - F^0(z^{-1})R_1(z^{-1},\theta) - F^0(z^{-1})R_2(z^{-1},\theta)} u(t). \; (3)$$

Hence, the design of the controller parameters $\theta = \{a,b,c\}$ can be reduced to an identification problem (see Fig.1), where the class of parametrized models (3) is forced to fit the I/O data measured on the plant $\mathcal{G}$. To this purpose, it is interesting to notice that such an identification problem (Ljung, 1987) is *parametrized in the controller coefficients*, so that its solution directly provides the PID feedback controller, without passing through a plant-modeling step.

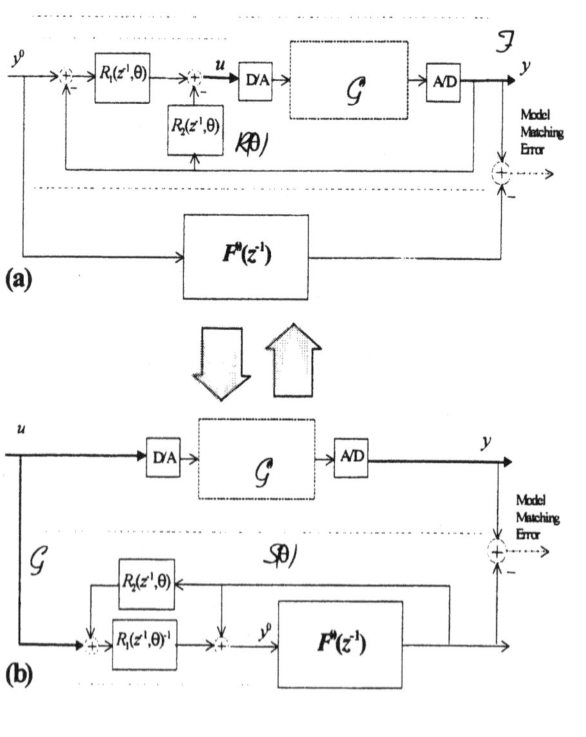


Fig.1. PID design. (a): standard model matching scheme; (b): VMD$^2$ scheme.

Using The VMD[2]-based procedure above outlined, the direct design of the PID from data can be reduced to the following three simple steps:

❶ Compute the signals $\{\tilde{u}_F(t), \tilde{y}_F(t)\}_{t=0,1,2,\ldots,N}$, by filtering $\{\tilde{u}(t), \tilde{y}(t)\}_{t=0,1,2,\ldots,N}$ with the digital filter $F^0(z^{-1})$, namely

$$\begin{cases} \tilde{u}_F(t) = F^0(z^{-1})\tilde{u}(t) \\ \tilde{y}_F(t) = F^0(z^{-1})\tilde{y}(t) \end{cases}. \qquad (4)$$

❷ By expanding the expression (3), and using the definitions (4), after trivial but cumbersome computations it can be shown that the following relationship between the signals $\tilde{u}_F(\cdot)$, $\tilde{y}_F(\cdot)$, and $\tilde{y}(\cdot)$ must hold:

$$a\Psi_a(t) + b\Psi_b(t) + c\Psi_c(t) = \Psi_0(t),\ t = 2,\ldots,N, \quad (5)$$

where:

$$\begin{cases} \Psi_a(t) = \tilde{y}(t) + \beta\tilde{y}(t-1) - \tilde{y}_F(t) - \tilde{y}_F(t-1) \\ \Psi_b(t) = \Psi_a(t-1) \\ \Psi_c(t) = -\tilde{y}_F(t) + 2\tilde{y}_F(t-1) - \tilde{y}_F(t-2) \\ \Psi_0(t) = \tilde{u}_F(t) + (\beta - 1)\tilde{u}_F(t-1) - \beta\tilde{u}_F(t-2) \end{cases}$$

The parameter vector, say $\hat{\theta} = \{\hat{a}, \hat{b}, \hat{c}\}$, which solves the PID-design problem, can thus be computed as the parameter vector which minimizes the following quadratic performance index:

$$J(\theta) = \frac{1}{2}\left(Q\theta^T - K\right)^T\left(Q\theta^T - K\right), \qquad (6)$$

where:

$$Q = \begin{bmatrix} \Psi_a(2) & \Psi_b(2) & \Psi_c(2) \\ \Psi_a(3) & \Psi_b(3) & \Psi_c(3) \\ \Psi_a(4) & \Psi_b(4) & \Psi_c(4) \\ \vdots & \vdots & \vdots \\ \Psi_a(N) & \Psi_b(N) & \Psi_c(N) \end{bmatrix},\ K = \begin{bmatrix} \Psi_0(2) \\ \Psi_0(3) \\ \Psi_0(4) \\ \vdots \\ \Psi_0(N) \end{bmatrix}$$

Hence, the second step of the procedure consists in the computation of the matrices $Q$ and $K$, using the measured data and the signals computed in the previous step.

❸ Minimize the loss function (6). Notice that such an optimization problem is a standard unconstrained minimization of a quadratic function in $\Re^3$; as such, its solution can be explicitly given as:

$$\hat{\theta} = \arg\min_{\theta}\{J(\theta)\} = \left(QQ^T\right)^{-1}QK. \qquad (7)$$

It should be apparent that the procedure ❶-❸ above (which implements the VMD[2] method for the special case of PID tuning) is extremely simple and low-demanding in terms of computational effort. Moreover, notice that the computations involved in the procedure (such as digital filtering and quadratic optimization) are well-studied standard operations, which can be solved using extremely efficient and reliable numerical procedures and software tools. As such, this procedure is well-suited to be implemented within low-cost μ-processor-based digital controllers.

Finally, it is interesting to observe that the method proposed for the PID-tuning can be used, in principle, for the design of both linear and non-linear discrete-time controllers. To this purpose, it is worth reminding that the basic design philosophy of such an approach is strictly related to the "Nicely-Nonlinear" approach (a technique for the enlargement of the "range of validity" of first-order approximations of non-linear plants - see Guardabassi, 1992), and, to a less extent, even to the recently developed "Virtual Input Direct Design" (VID[2]) approach (see Guardabassi and Savaresi, 1997, 1998).

## 3. NUMERICAL RESULTS

In this section, we present a numerical example, in order to show how the parameters of a digital PID can be tuned via the VMD[2] method. To this end - as a "benchmark" for testing such a technique - we have considered the problem of regulating the following (continuous-time) linear plant (see K.J. Åstrom, T. Hägglund: "*Automatic Tuning of PID Controllers*", Instrument Society of America, 1988):

$$\mathcal{G}: \begin{cases} y(t) = G^0(s)u(t), G^0(s) = \\ = \dfrac{1}{(1+s)(1+0.2s)(1+0.5s)(1+0.01s)} \end{cases}$$

(the corresponding Nyquist and Bode plots are depicted in Fig.2), which has been complemented with a Zero Order Hold (ZOH), and a sampling unit at the sampling frequency of 50Hz.

Our goal is to choose the three parameters of a PID digital controller (1), so that the closed-loop I/O behavior of the control system is (approximately) given by the digital counterpart of the linear system

$$F^0(s) = \frac{\omega_n^2}{s^2 + 2\xi\omega_n s + \omega_n^2},\ \omega_n = 5, \xi = 0.6,$$

namely (if $T_c = 0.02$ sec.):

$$\mathcal{F}: \begin{cases} y(t) = F^0(z^{-1})y^0(t), F^0(z^{-1}) = \\ = \dfrac{0.0048z^{-1} + 0.0046z^{-2}}{1 - 1.8775z^{-1} + 0.8869z^{-2}} \end{cases}.$$

Using the above model of reference, we have excited the plant $\mathcal{G}$ in different ways, and we have designed the PID parameters using the corresponding I/O measurements collected over a 20-seconds time interval, via the VMD² method.

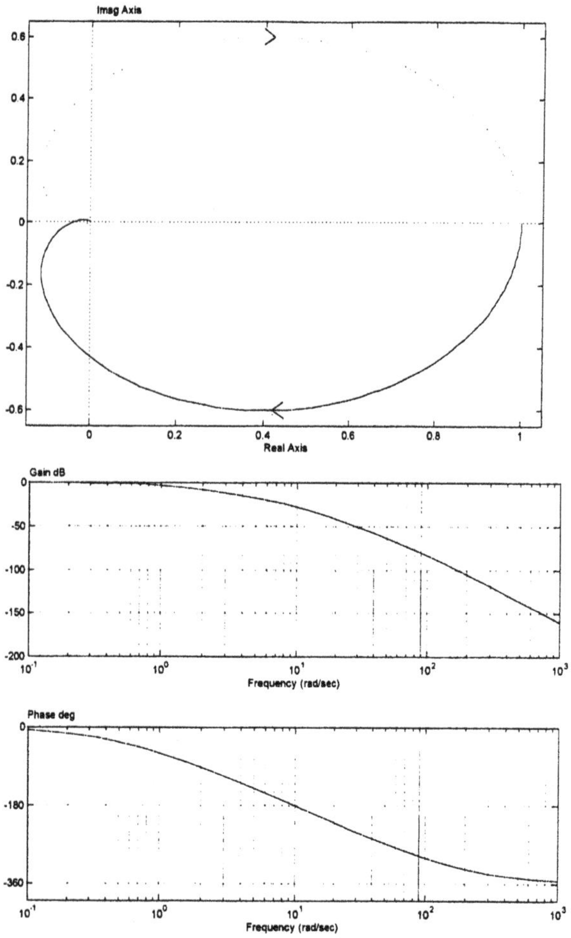


Fig.2. Nyquist and Bode plots of the plant $\mathcal{G}^o$

For the sake of completeness, the performances of the so-obtained control systems (which will be evaluated by means of their step responses) have been compared with those obtainable with the following two controllers:

$$
(\text{Å-H}): \quad \begin{cases} R_1(s) = \dfrac{K_p}{T_i}\dfrac{1+sT_i}{s}, \\ R_2(s) = \dfrac{K_p T_d s}{1+(T_d/N_d)s}, \\ \{K_p, T_i, T_d, N_d\} = \{10, 0.6, 0.1, 10\} \end{cases}
$$

(such a PID controller has been designed using the well-known Åstrom-Hägglund method – see Åstrom, and Hägglund, 1998);

$$
(\text{Z-N}): \quad \begin{cases} R_1(s) = \dfrac{K_p}{T_i}\dfrac{1+sT_i}{s}, \\ R_2(s) = \dfrac{K_p T_d s}{1+(T_d/N_d)s}, \\ \{K_p, T_i, T_d, N_d\} = \{15, 0.31, 0.08, 10\} \end{cases}
$$

(such a PID controller has been designed using the classical Zigler-Nichols rules).

As a first trial, we have fed the plant $\mathcal{G}^o$ with a random sequence of 0.5sec.-long steps; in Fig.3(I), such an input signal $\tilde{u}(t)$, $t$=0,1,2,...,1000, and the corresponding signal $\tilde{y}(t)$ are depicted. Using such an excitation signal, the VMD² method provides the parameter vector $\hat{\theta} = \{\hat{a}, \hat{b}, \hat{c}\} = \{6.29, -6.21, 33.01\}$ (when $\beta = 0.1$ is used).

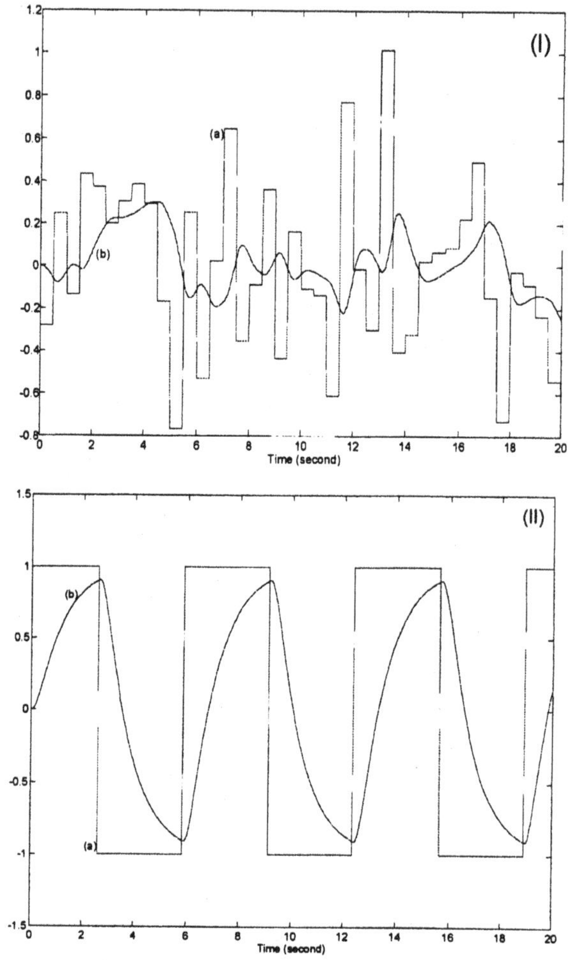


Fig.3. Training signals ((I): open-loop configuration; (II) closed-loop relay-based configuration). (a): $\tilde{u}(t)$; (b): $\tilde{y}(t)$.

The step response of the corresponding control system is depicted in Fig.4(I). Notice the high "degree-of-similarity" with the reference model so-achieved. As for the comparison with the (Å-H) and (Z-N) controllers, observe that the (Å-H) controllers provides good performances (which are slightly different from those of the VMD²-based controller, since we have chosen a reference model slightly more damped), whereas the (Z-N) controller provides an under-damped control system, this fact showing that the design of a smart controller of $\mathcal{G}^o$ is not a trivial task.

As a second experiment, we have tested the sensitivity

of the VMD² design technique, with respect to the choice of the parameter β. To this purpose, using again the training signals of Fig.3(I), the PID has been designed using β=0.4 and β=0.025, instead of β=0.1. Such an experiment has revealed that the method is quite insensitive to β, since the obtained step responses (the plots of which are here omitted) are almost indistinguishable each other.

As a third experiment, we have excited the plant in a closed-loop configuration, using the classical relay-based technique (see Åstrom, and Hägglund, 1998 and Fig.5). In Fig.3(II), the measured and filtered signal, gathered in correspondence of a relay characterized by ρ=0.9 (where ρ is the half-width of the relay hysteresis - see Fig.5), are depicted, whereas in Fig.4(II) the step-response of the control system, designed by means of such signals, is displayed. From Fig.4(II) it should be apparent that, even in this case, the VMD² method provides a control system having a behavior very close to the reference model.

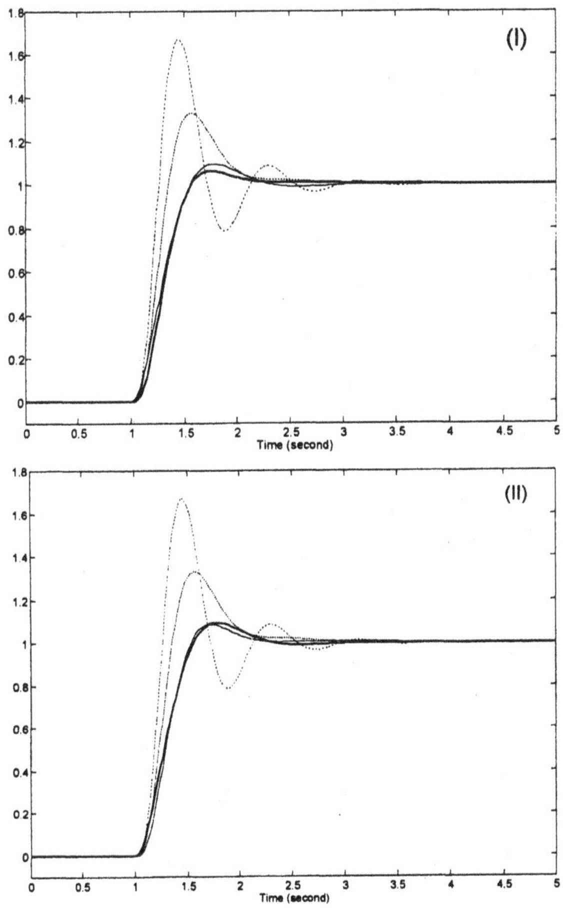


Fig.4. Control systems step responses ((I): control system designed using open-loop training signals; (II) control system designed using the relay-based training signals, ρ=0.9). (thin line): ref. model; (bold line): VMD² controller; (dashed line): (Å-H) controller; (dashed-dotted line): (Z-N) controller.

As a final attempt, we have progressively shortened the size of the hysteresis in the relay function; the so-obtained results are displayed in Fig.6. Notice that, for ρ=0.6 and ρ=0.3, the control system is still very close to the reference model, whereas for ρ=0.1 the control system is over-damped, and for ρ=0 (no hysteresis) it becomes unstable.

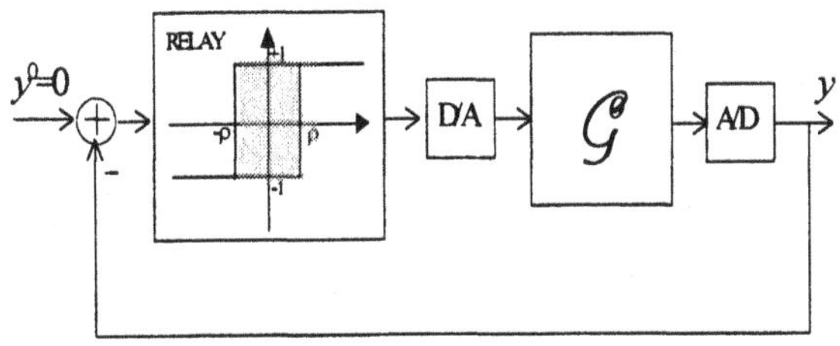


Fig.5. The relay-based excitation scheme.

This fact can be easily interpreted as a "lack of information" in the training signals: as a matter of fact, when ρ=0, the corresponding input is a square wave having the fundamental frequency at about 2Hz, whereas the reference model has a cutting frequency at about 1Hz. As it is well known, the 2Hz square wave cannot provide information on the plant behavior in the frequency range 0÷2Hz, and this makes (obviously) unreliable the design of a control system having its main dynamics within that frequency range.

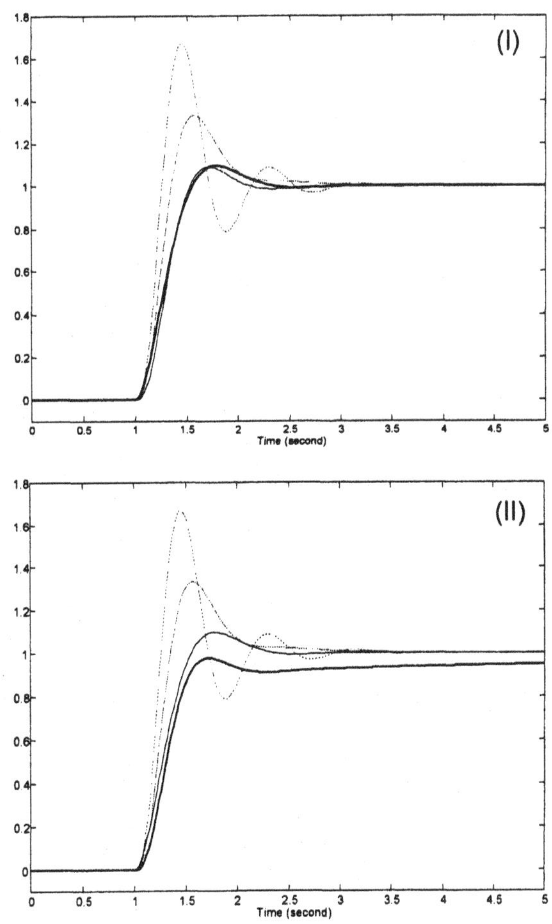

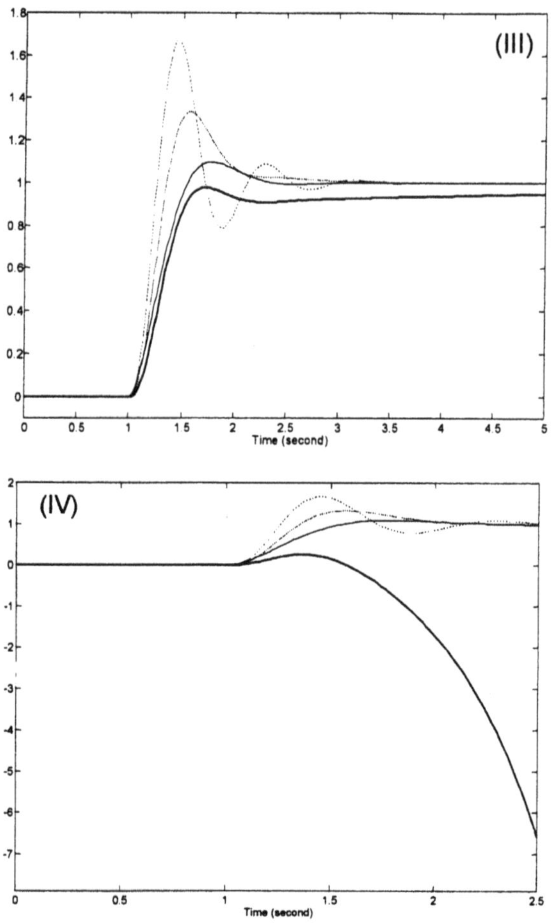


Fig.6. Control systems step responses designed using the relay-based training signals ((I): ρ=0.6; (II): ρ=0.3; (III): ρ=0.1; (IV): ρ=0). (thin line): ref. model; (bold line): VMD² controller; (dashed line): (Å-H) controller; (dashed-dotted line): (Z-N) controller.

## 4. CONCLUSIONS

In this paper, we have presented a new technique for the direct design of PID-based controllers, using only I/O measurements of the plant. Such a method, called the "Virtual Model Direct Design" approach, is an approach which may reveal extremely useful and effective in model matching problems, when the model of reference is a simple algebraic model, and only I/O measurements of the plant are available.

The most interesting features of this technique are its simplicity and flexibility; moreover, it is very low-demanding in terms of computational effort, since it reduces the PID design to the (unconstrained) minimization of a quadratic function in $\Re^3$. As such, it may be an appealing alternative to standard data-based PID-tuning methods.

## ACKNOWLEDGMENTS

Paper supported by the European *HCM* project *"Nonlinear and Adaptive Control"*, the Italian *MURST* project *"Model Identification, System Control, Signal Processing"* and the *CNR Center for System Theory* of Milano.

## REFERENCES

Åstrom K.J., and T. Hägglund (1998). Automatic Tuning of PID Controllers, *Instrument Society of America.*

Åström, K.J., and B. Wittenmark (1990). *Computer Controlled Systems, Theory and Design,* (2nd Edition). Prentice-Hall.

Guardabassi G.O. (1992). Nicely Nonlinear Modelling and Control, In: *Systems, Models and Feedback* (A. Isidori and T.J. Tarn Eds.), pp.265-274. Birkäuser, Boston.

Guardabassi G.O., and Savaresi S.M. (1997). Approximate Feedback Linearization of Discrete-Time Non-Linear Systems Using Virtual Input Direct Design. System & Control Letters, vol.32, pp.63-74.

Guardabassi G.O., and Savaresi S.M. (1998). Approximate I/O feedback Linearization of Discrete-Time Non-Linear Systems Via Virtual Input Direct Design. Automatica, Vol.34, No.6, pp.715-722.

Ljung, L. (1987). System Identification: Theory for the user. Prentice-Hall.

Narendra, K.S., and Parthasarathy (1990). Identification and Control of Dynamical Systems Using Neural Networks. IEEE Trans. on Neural Networks, Vol.1, No.1, pp.4-27.

14th Triennial World Congress, Beijing, P.R. China

F-2c-09-5

# VS CONTROL OF NONLINEAR SYSTEM USING SLIDING SECTOR

**Yaodong Pan * Katsuhisa Furuta ****

* *Department of Computers and Systems Engineering*
*Tokyo Denki University, Hiki-gun, Saitama 350-0394, Japan*
*pan@k.dendai.ac.jp*
** *Department of Mechanical and Environmental Informatics*
*Tokyo Institute of Technology*
*2-12-1 Oh-okayama, Meguro-ku, Tokyo 152, Japan*
*furuta@mei.titech.ac.jp*

Abstract: A variable structure (VS) controller is designed for a class of nonlinear systems with parameter uncertainty by using sliding sector. The control problem is to let the output track a desired trajectory by state feedback. Based on a linear autonomous system, the sliding sector is defined as a subset of the state space $R^n$, where some norm of the state decreases. The VS control law designed in the paper transfers the system state from the outside to the inside of the sliding sector and ensures that some norm of the tracking error keeps decreasing in all of the state space even if there exists parameter uncertainty. Simulation results are given to shown the robustness of the proposed VS controller. 

Keywords: Variable structure control, Sliding sector, Quadratic stability, Nonlinear system, Uncertainty

## 1. INTRODUCTION

The variable structure control (VSC) with sliding mode has been researched for several decades since 1960's (Utkin, 1964) (DeCarlo *et al.*, 1988) (Utkin, 1992) (Hung *et al.*, 1993). To due with the chattering phenomena, which inherently exists in a practical realized VS control system and to implement the VS control algorithm to discrete-time system, VS controllers with the sliding sector were proposed in (Furuta, 1990) and (Furuta and Pan, 1995). The sliding sector (Furuta and Pan, 1995) is defined such that a given norm of the system state is decreasing inside it without any control action if there is no parameter uncertainty and external disturbance.

In this paper, a class of nonlinear systems with parameter uncertainty (Kachroo and Tomizuka, 1996) will be taken into consideration. The control problem is to enable the output to track a desired trajectory by state feedback. The sliding sector designed in (Furuta and Pan, 1995) for linear systems will be used to design a VS controller for nonlinear systems. The defined norm of the tracking error will decrease in all of the state space. The VS control law will ensure that the system state move from the outside to the inside of the sliding sector.

The organization of the paper is as follows: Section 2 describes the considered nonlinear system. Section 3 presents the sliding sector designed in (Furuta and Pan, 1995). Section 4 proposes the VS control law for nonlinear system with the sliding sector. In Section 5, Second order nonlinear systems are taken as simulation model. Various kinds of parameter uncertainty is considered. The simulation results are given.

## 2. PROBLEM FORMULATION

Consider a class of nonlinear system (Kachroo and Tomizuka, 1996):

$$y^{(n)} = f(x,t) + g(x,t)u(t) \tag{1}$$

where, $x = \begin{bmatrix} y(t) & \dot{y}(t) & \cdots & y^{(n-1)}(t) \end{bmatrix}^T$ is the state vector, $u(t)$ and $y(t)$ are the input and output vector, respectively.

$f(x(t),t) : R^n \times R \to R$ and $g(x(t),t) : R^n \times R \to R$ are nonlinear time varying functions. $f(x(t),t)$ and $g(x(t),t)$ are estimated by $\hat{f}(x(t),t)$ and $\hat{g}(x(t),t)$, respectively. The estimation error on $f(x(t),t)$ is assumed to be bounded by some known function $F(x(t),t) : R^n \times R \to R$, i.e.

$$|f(x(t),t) - \hat{f}(x(t),t)| \le F(x(t),t). \tag{2}$$

Without losing of generality, $g(x(t),t)$ is assumed to be a positive function and bounded by two known positive functions $g_{min}(x(t),t) : R^n \times R \to R$ and $g_{max}(x(t),t) : R^n \times R \to R$, that is

$$0 \le g_{min}(x(t),t) \le g(x(t),t) \le g_{max}(x(t),t). \tag{3}$$

The estimated function $\hat{g}(x(t),t)$ is choose as the geometric mean of the above bounds:

$$\hat{g}(x(t),t) = \sqrt{g_{min}(x(t),t)g_{max}(x(t),t)} \tag{4}$$

Thus the following inequalities hold (Kachroo and Tomizuka, 1996).

$$\beta^{-1}(x(t),t) \le \hat{g}(x(t),t)/g(x(t),t) \le \beta(x(t),t) \tag{5}$$
$$\beta^{-1}(x(t),t) \le g(x(t),t)/\hat{g}(x(t),t) \le \beta(x(t),t) \tag{6}$$

where $\beta(x(t),t) : R^n \times R \to R$ is defined as:

$$\beta(x(t),t) = \sqrt{\frac{g_{max}(x(t),t)}{g_{min}(x(t),t)}}$$

To design a VS controller, the nonlinear system (1) is rewritten as the following state equation:

$$\begin{cases} \dot{x}(t) = Ax(t) + B(f(x,t) + g(x,t)u(t)) \\ y(t) = Cx(t) \end{cases} \tag{7}$$

where

$$A = \begin{bmatrix} 0 & 1 & 0 & \cdots & 0 \\ 0 & 0 & 1 & \cdots & 0 \\ \vdots & \vdots & \vdots & \ddots & \vdots \\ 0 & 0 & 0 & \cdots & 1 \\ 0 & 0 & 0 & \cdots & 0 \end{bmatrix} \quad B = \begin{bmatrix} 0 \\ 0 \\ \vdots \\ 0 \\ 1 \end{bmatrix} \tag{8}$$

$$C = \begin{bmatrix} 1 & 0 & \cdots & 0 \end{bmatrix} \tag{9}$$

The control problem is to let the output $y(t)$ track a desired trajectory $y_d(t) \in R^n$ $(t \in [0,+\infty))$ by state feedback with parameter uncertainty on $f(x(t),t)$ and $g(x(t),t)$.

Define the tracking error to be $e(t) = x(t) - x_d(t)$, then the state equation on $e(t)$ is written as follows.

$$\dot{e}(t) = Ae(t) + B(f(x(t),t) - u_d(t) + g(x(t),t)u(t)). \tag{10}$$

where $u_d(t) = y_d^{(n)}(t)$ satisfying

$$\begin{cases} \dot{x}_d(t) = Ax_d(t) + Bu_d(t) \\ y_d(t) = Cx_d(t) \end{cases} \tag{11}$$

Thus the control problem turns to stabilize the nonlinear system (10) to let the tracking error $e(t)$ converge to zero.

## 3. *PR*-SLIDING SECTOR

Define an auxiliary input $v(t)$ as

$$v(t) = f(x(t),t) - u_d(t) + g(x(t),t)u(t), \tag{12}$$

then the state equation (10) on the tracking error can be written as following linear one.

$$\dot{e}(t) = Ae(t) + Bv(t), \tag{13}$$

**Definition 1:** The *P-Norm* $||\cdot||_p$ of the tracking error is defined as

$$||e||_p = (e^TPe)^{\frac{1}{2}}, \quad e \in R^n \tag{14}$$

where $P \in R^{n\times n}$ is a positive definite symmetric matrix.

The square of the $P$-norm is denoted as

$$L = ||e||_p^2 = e^TPe > 0, \quad \forall e \in R^n, e \neq 0.$$

which will be considered as a Lyapunov function candidate.

**Definition 2:** The *PR-Sliding Sector* is a subset of $R^n$ defined as

$$S = \{e|\ e^T(A^TP + PA)e \le -e^TRe, e \in R^n\ \} \tag{15}$$

where $P \in R^{n\times n}$ is a positive definite symmetric matrix, $R \in R^{n\times n}$ is a positive semi-definite symmetric matrix, and $R = C^TC$, $C \in R^{l\times n}$, $l \ge 1$, $(C, A)$ is observable pair.

Inside the $PR$-sliding sector, the $P$-norm $||e||_p$ of the plant (13) with the auxiliary control input $v(t) = 0$ decreases because

$$\dot{L} = e^T(A^TP + PA)e \le -e^TRe \le 0, \quad \forall e \in S.$$

*Theorem 1.* (Furuta and Pan, 1995) For any control system (13) and for any positive definite symmetric matrix $P$ and any positive semi-definite symmetric matrix $R$ described in Definition 2, the $PR$-sliding sector defined by (15) can be rewritten as

$$\mathcal{S} = \{e|\ s^2(e) \leq \delta^2(e)\ \} \tag{16}$$

where

$$s^2(e) = e^T P_1 e \geq 0 \tag{17}$$
$$\delta^2(e) = e^T P_2 e \geq 0, \tag{18}$$

$P_1$ and $P_2$ are $n \times n$ positive semi-definite symmetric matrices.

If the $P$-norm is chosen to be the solution of the following Riccati equation

$$PA + A^T P - PBB^T P = -Q. \tag{19}$$

where $Q \in R^{n\times n}$ is a symmetric positive definite matrix, then it follows for the system (13) with the auxiliary control input $v(t) = 0$ that

$$\begin{aligned}\dot{L}(t) &= e^T(t)(A^T P + PA)e(t)\\ &= e^T(t)PBB^T Pe(t) - e^T(t)Qe(t)\end{aligned} \tag{20}$$

which gives a design algorithm for the $PR$-sliding sector as described in the following theorem.

*Theorem 2.* (Furuta and Pan, 1995) If the solution $P$ of the Riccati equation (19) is used to define the $P$-norm and the matrix $R$ is chosen to be $(1-r)Q$ $(0 < r < 1)$, then the $PR$-sliding sector defined by (15) can be rewritten as

$$\mathcal{S} = \{e|\ |s(e)| \leq |\delta(e)|\ \} \tag{21}$$

where

$$s(e) = Ce(t) \quad , \quad C = B^T P$$
$$\delta(e) = \sqrt{e^T(t)rQe(t)}$$

$Q \in R^{n\times n}$ is a symmetric positive definite matrix.

## 4. DESIGN OF VS CONTROLLER

Based on the $PR$-sliding sector (21), a VS controller for the nonlinear control system (10) should be designed

(1) to move the system state from the outside to the inside of the $PR$-sliding sector when the system state is outside the $PR$-sliding sector;

(2) to let the $P$-norm decease in all of the state space.

To design a chattering free VS controller, an inner sector $\mathcal{S}_i$ of the $PR$-sliding sector is defined as

$$\mathcal{S}_i = \{e|\ |s(e)| \leq \alpha|\delta(e)|\} \tag{22}$$

which is a subset of the $PR$-sliding sector as shown in Figure 1 where $\alpha$ $(0 < \alpha < 1)$ is a constant.

*Theorem 3.* The VS control law

$$u(t) = (1-\sigma(e))u_i(t) + \sigma(e)u_0(t) \tag{23}$$

quadratically stabilizes the nonlinear system (10) without chattering. Where the binary function $\sigma(e)$ is defined as

$$\sigma(x) = \begin{cases} 0 & e(t) \in \mathcal{S}_i \\ \text{unchange} & e(t) \in \mathcal{S} - \mathcal{S}_i \\ 1 & e(t) \bar{\in} \mathcal{S}. \end{cases} \tag{24}$$

the $PR$-sliding sector $\mathcal{S}$ and the inner sector $\mathcal{S}_i$ are defined in (21) and (22), respectively. $u_i(t)$ and $u_o(t)$ are defined as

$$\begin{aligned} u_i(t) &= -\hat{g}^{-1}(x,t)(\hat{f}(x,t) - u_d(t)\\ &\quad + k_1(x,t)sgn(s(e))) \end{aligned} \tag{25}$$
$$\begin{aligned} u_o(t) &= -\hat{g}^{-1}(x,t)((CB)^{-1}CAe(t)\\ &\quad + \hat{f}(x,t) - u_d(t) + ((CB)^{-1}k\beta(x,t)|s(e)|\\ &\quad + k_2(x,t))sgn(s(e))) \end{aligned} \tag{26}$$

$k_1(x,t)$, $k_2(x,t)$, and $k$ satisfy the following inequalities.

$$k_1(x,t) > \beta(x,t)F(x,t) + (\beta(x,t)-1)|\hat{f}(x,t) - u_d(t)|$$
$$k_2(x,t) > \beta(x,t)F(x,t) + (\beta(x,t)-1)$$
$$|\hat{f}(x,t) - u_d(t) + (CB)^{-1}CAe|$$
$$k > \max\{\frac{(CB)^{-1}}{2}, \frac{k_0}{\alpha}\}$$

where the positive constant $k_0$ enables the following inequality hold.

$$2k_0 rQ + (C^T CA + A^T C^T C) > 0.$$

**Proof:** When the control law $u_o(t)$ defined in (26) is used, the system state is outside the inner sector, i.e. $|s(e)| > \alpha|\delta(e)|$. In this case, consider a Lyapunov function

$$V(t) = \frac{1}{2}s^2(e), \tag{27}$$

then,

$$\begin{aligned}\dot{V}(t) &= s(e)\dot{s}(e)\\ &= s(e)(CAe(t) + CB(f(x,t) - u_d(t) + g(x,t)u_o(t)))\end{aligned}$$

$$< -g(x,t)\hat{g}^{-1}(x,t)k\beta(x,t)s^2(e)$$
$$< -ks^2(e)$$

which means that the Lyapunov function $V(t)$ keeps decreasing outside the inner sector, i.e., $|s(e)|$ will decrease until the system state moves into the inner sector. It maybe needs infinite time for $s(e)$ to converge to 0 because $|s(e)|$ decreases in exponential. But for $|s(e)|$ to converge from $|s(e)| > |\delta(e)|$ (or $|s(e)| > \alpha|\delta(e)|$) to $|s(e)| \leq \alpha|\delta(e)|$, it costs only a finite time if the positive constant $k$ is large enough to let the convergence speed of $s(e)$ be higher than the one of $|\delta(x)|$.

While the system state is moving toward the inside of the inner sector with the control law $u_o(t)$, the derivation of the square of the $P$-norm is

$$\begin{aligned}\dot{L}(t) &= e^T(t)(A^TP+PA)e(t) + 2e^T(t)PB(f(x(t),t) \\ &\quad -u_d(t) + g(x(t),t)u_o(t)) \\ &< s^2(e) - \delta^2(e) - 2(CB)^{-1}ks^2(e) \\ &\quad -2(CB)^{-1}CAe(t) - e^T(t)Re(t) \\ &< -2(CB)^{-1}k\alpha\delta^2(e) \\ &\quad -2(CB)^{-1}s(e)CAe(t) - e^T(t)Re(t) \\ &< -e^T(t)Re(t) \\ &< 0 \end{aligned} \tag{28}$$

where $CB = B^TPB > 0$. This inequality means that the $P$-norm decreases with the control law $u_o(t)$ outside the inner sector.

When the control law $u_i(t)$ defined in (25) is used, the system state is inside the $PR$-sliding sector, i.e. $|s(e)| \leq |\delta(e)|$. The derivation of the square of the $P$-norm is

$$\begin{aligned}\dot{L}(t) &= e^T(t)(A^TP+PA)e(t) + 2e^T(t)PB(f(x(t),t) \\ &\quad -u_d(t) + g(x(t),t)u_i(t)) \\ &\leq 2e^T(t)PB(f(x(t),t) - u_d(t) \\ &\quad +g(x(t),t)u_i(t)) - e^T(t)Re(t) \\ &< -e^T(t)Re(t) \\ &< 0 \end{aligned} \tag{29}$$

Thus the $P$-norm will decrease inside the $PR$-sliding sector even if there exists parameter uncertainty on $f(x,t)$ and $g(x,t)$.

Therefore according to equations (28) and (29), the $P$-norm will decrease in all of the state space, i.e.

$$\begin{aligned}\dot{L}(t) &< -e^T(t)Re(t) \\ &< 0 \quad \forall e(t) \in R^n\end{aligned}$$

which means that the designed VS control law (23) quadratically stabilizes the nonlinear system (10). ∎

## 5. EXAMPLES

*EXAMPLE 1:* Consider a second order nonlinear system:

$$\dot{x}(t) = \begin{bmatrix} 0 & 1 \\ 0 & 0 \end{bmatrix} x(t) + \begin{bmatrix} 0 \\ 1 \end{bmatrix} (f(x,t) + g(x,t)u(t)) \tag{30}$$

where the nonlinear functions $f(x,t)$ and $g(x,t)$ are defined as

$$\begin{aligned} f(x,t) &= 2x_1^2(t)sin(\pi t/3) + 3x_2^2(t)cos(\pi t/2) \\ &\quad +5x_1(t)x_2(t) \\ g(x,t) &= 1 + 0.5sin(\pi t/6). \end{aligned}$$

Their bound functions $F(x,t)$, $g_{max}(x,t)$, and $g_{min}(x,t)$ are defined as

$$\begin{aligned} F(x,t) &= 2x_1^2(t) + 3x_2^2(t) + 5|x_1(t)x_2(t)| \\ g_{max}(x,t) &= 1.5 \\ g_{min}(x,t) &= 0.5 \end{aligned}$$

The $PR$-sliding sector is of the form in (21) where parameter matrices $P$, $R$, $C$, $r$ and $Q$ are determined by

$$\begin{aligned} C &= \begin{bmatrix} 10 & 10.9544512 \end{bmatrix} \\ r &= 0.3 \\ R &= 0.7Q \\ Q &= 100 * I(2) \end{aligned}$$

Figure 1 shows the $PR$-sliding sector (21) and the inner sector (22) with the coefficient $\alpha$ being 0.1. The simulation result is shown in Figure 2 where the desired output $y_d(t)$ is assumed to be zero.

*EXAMPLE 2:* Consider a second order linear system:

$$\dot{x}(t) = \begin{bmatrix} 0 & 1 \\ a_1 & a_2 \end{bmatrix} x(t) + \begin{bmatrix} 0 \\ 1 \end{bmatrix} u(t). \tag{31}$$

where $a_1$ and $a_2$ are bounded unknown parameters. In the simulation, their bounds are assumed to be

$$\begin{aligned} |a_1| &< 2.0 \\ |a_2| &< 3.0 \end{aligned}$$

That is

$$\begin{aligned} f(x,t) &= a_1x_1(t) + a_2x_2(t) \\ \hat{f}(x,t) &= 0 \\ F(x,t) &= 2|x_1(t)| + 3|x_2(t)| \end{aligned}$$

Two kind of systems are considered in the simulation with the desired output being zero.

(1) $a_1 = -0.5$ and $a_2 = 1.5$ with poles in 1.0 and 0.5,

(2) $a_1 = -2sin(\pi t)$ and $a_2 = 3cos(\pi t/2)$.

The $PR$-sliding sector is designed as the same as the one in Example 1. The simulation results with the same VS controller and the same control parameters are given in Figure 3 and 4.

All of the simulation results in Example 1 and Example 2 show that after initial transient of a finite time interval, the system state converges to origin inside the $PR$-sliding sector which determines the control performance whenever the plant is stable or unstable, linear or nonlinear, and time invarying or time varying.

*EXAMPLE 3:* To discuss the effect of the proposed VS controller in the tracking control, consider the second order system used in (Kachroo and Tomizuka, 1996) with

$$\begin{aligned} f(x,t) &= -(1+|sint|)\dot{x}^2 cos3x \\ g(x,t) &= 1 \\ \hat{f}(x,t) &= -1.5\dot{x}^2 cos3x \\ F(x,t) &= 0.5\dot{x}^2|cos3x| \\ y_d(t) &= sin(\pi t/2) \end{aligned}$$

In this case, the sampling frequency is taken as 100Hz, which is rather low. The $PR$-sliding sector is designed as the same as the one in Example 1. The simulation result given in Figure 5 shows the good tracking characteristics of the proposed VS controller.

## 6. CONCLUSION

A VS controller was designed for a class of nonlinear systems. The $PR$-sliding sector designed for linear systems in (Furuta and Pan, 1995) was applied to nonlinear systems. The proposed VS controller is robust stable to parameter uncertainty. The simulation results shown that the control performance of the VS control system is determined by the $PR$-sliding sector whenever the plant is nonlinear or linear, stable or unstable.

## 7. REFERENCES


DeCarlo, R.A., S.H. Zak and G.P. Matthews (1988). Variable structure control of nonlinear multivariable systems: A tutorial. *Proc. of the IEEE* **76**, 212 232.

Furuta, K. (1990). Sliding mode control of a discrete system. *System & Control Letters* **14**, 145 152.

Furuta, K. and Y. Pan (1995). A new approach to design a sliding sector for vss controller. In: *Proc. of the American Control Conference*. Seattle. pp. 1304 1308.

Hung, J.Y., W. Gao and J.C. Hung (1993). Variable structure control: A survey. *IEEE Trans. on Industrial Electronics* **40**, 2–22.

Kachroo, P. and M. Tomizuka (1996). Chattering reduction and error convergence in the sliding-mode control of a class of nonlinear systems. *IEEE Transactions on Automatic Control* **AC-41**, 1063–1068.

Utkin, U.I. (1992). *Sliding Modes in Control and Optimization.* Springer-Verlag.

Utkin, V.I. (1964). Variable structure system with sliding mode:a survey. *IEEE Transactions on Automatic Control* **AC-22**, 212 222.


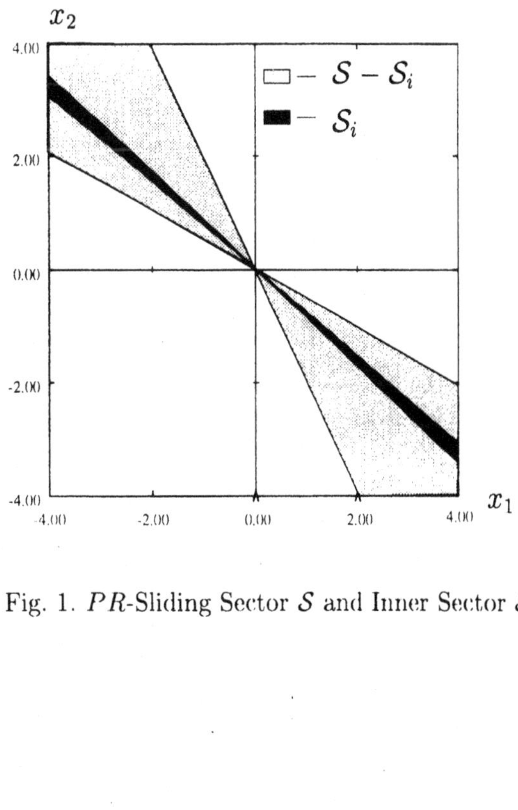


Fig. 1. $PR$-Sliding Sector $\mathcal{S}$ and Inner Sector $\mathcal{S}_i$

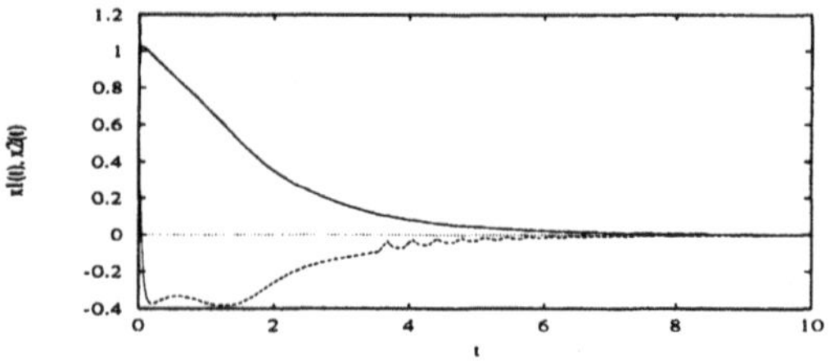

Evolution of State Variables $x_1(t)$ and $x_2(t)$

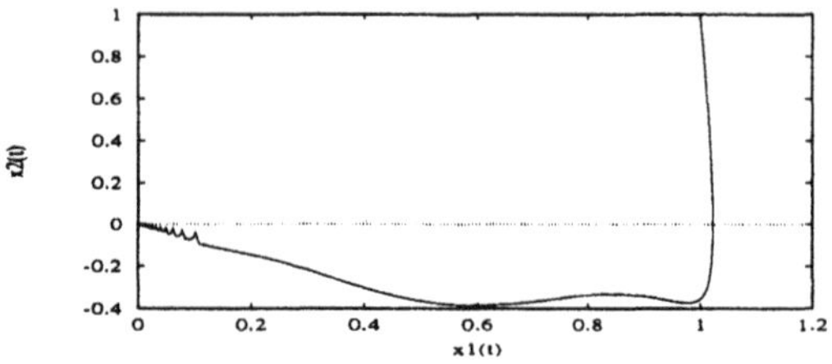

Phase Plane Diagram of $x_2(t)$ versus $x_1(t)$

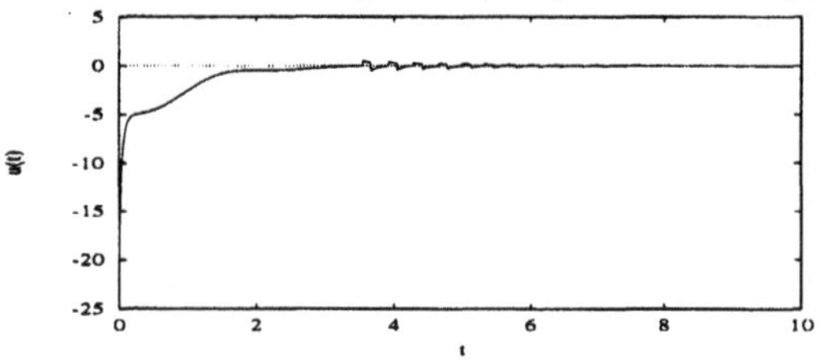

Evolution of Input $u(t)$

Fig. 2. Nonlinear System

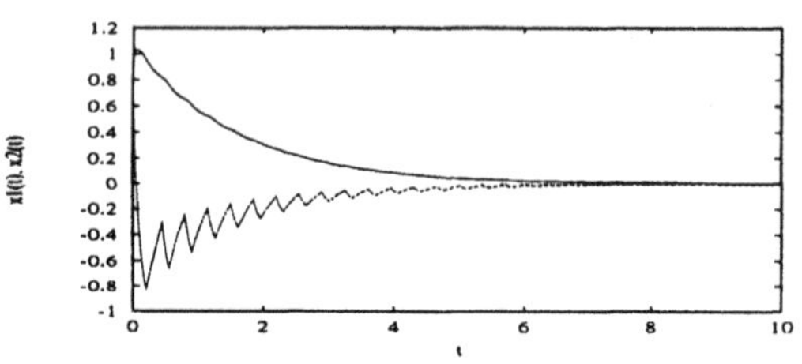

Evolution of State Variables $x_1(t)$ and $x_2(t)$

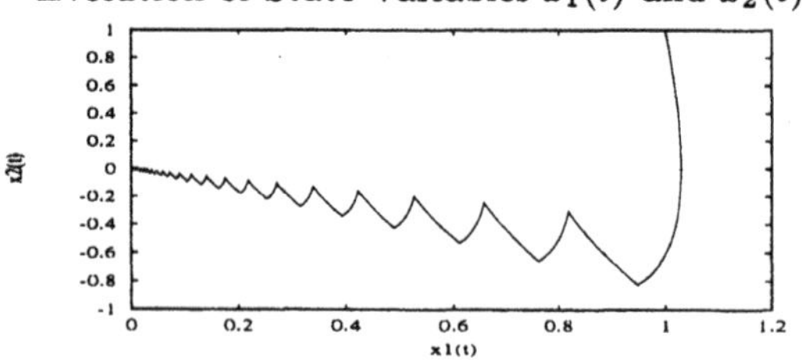

Phase Plane Diagram of $x_2(t)$ versus $x_1(t)$

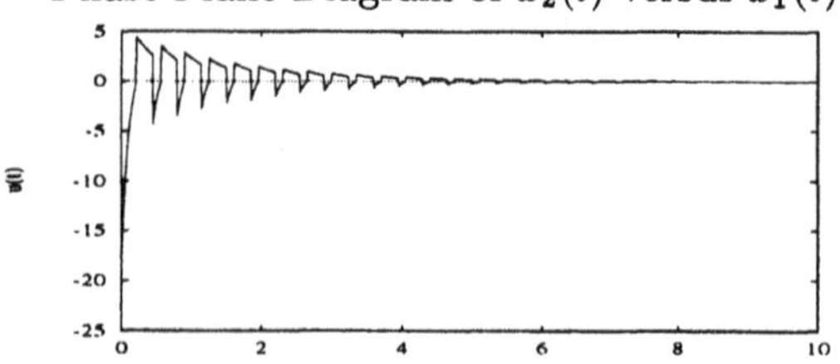

Evolution of Input $u(t)$

Fig. 3. Linear System with poles in 1.0 and 0.5

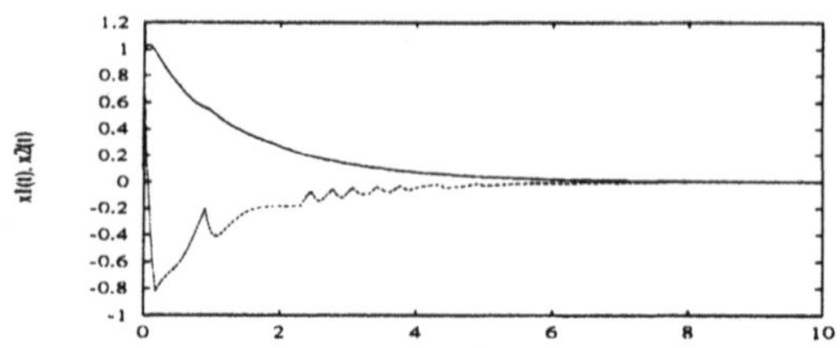

Evolution of State Variables $x_1(t)$ and $x_2(t)$

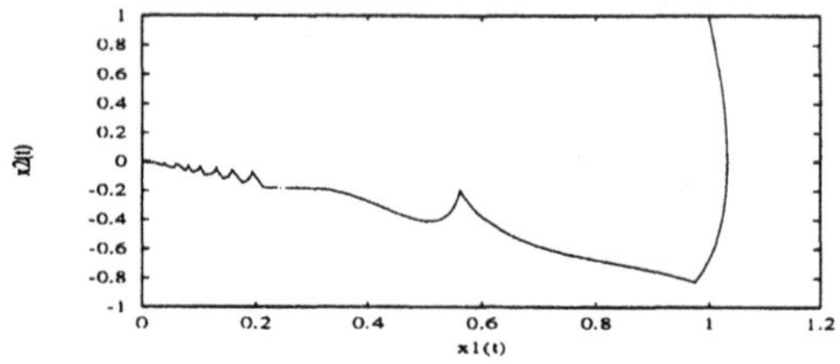

Phase Plane Diagram of $x_2(t)$ versus $x_1(t)$

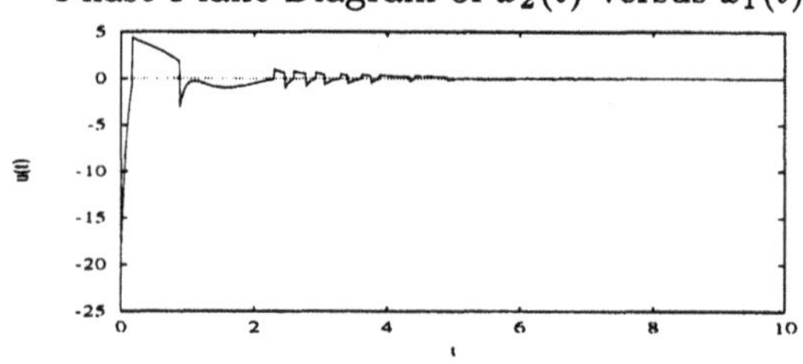

Evolution of Input $u(t)$

Fig. 4. Linear System with $a_1 = -2sin(\pi t)$ and $a_2 = 3cos(\pi t/2)$

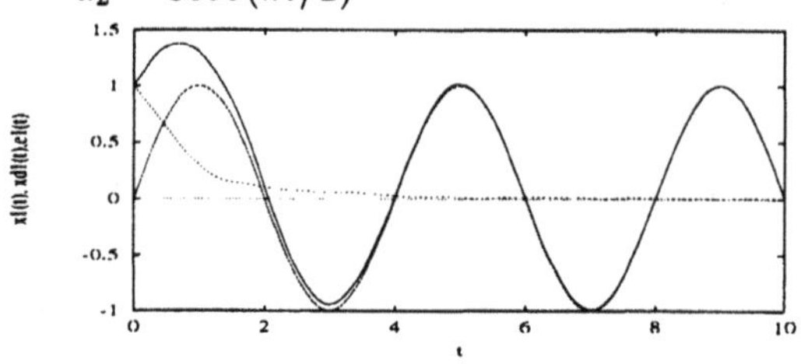

Evolution of $x_1(t)$, $x_{d1}(t)$, and $e_1(t)$

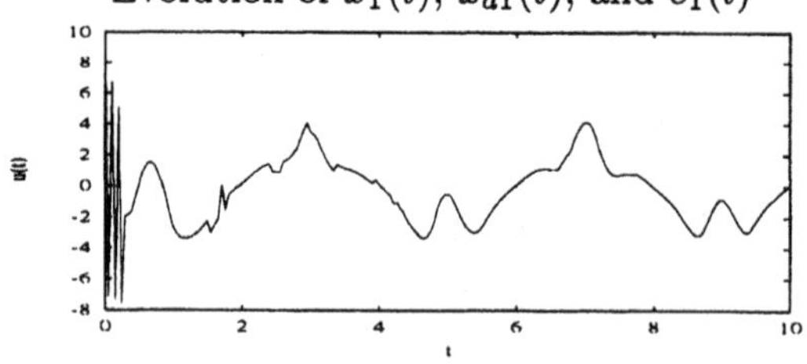

Evolution of Input $u(t)$

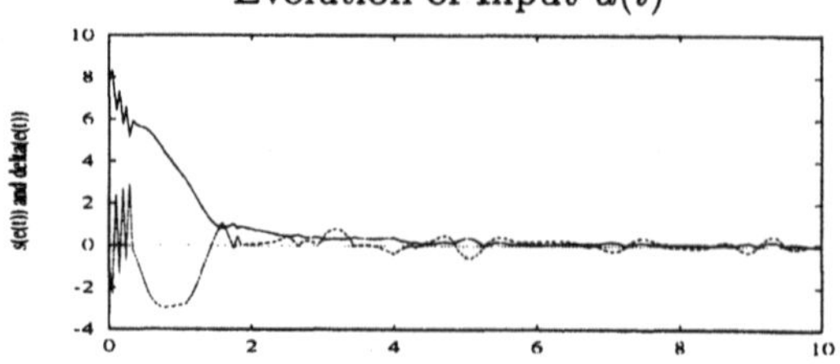

Evolution of $s(e(t))$ and $\delta(e(t))$

Fig. 5. Tracking Control for Nonlinear System with Parameter Uncertainty

14th Triennial World Congress, Beijing, P.R. China

F-2c-09-6

# GLOBAL REGULATION PROBLEM FOR NONLINEAR SYSTEMS [1]

**Shumin Fei** and **Chun-Bo Feng**

*Research Institute of Automation*
*Southeast University, Nanjing 210018 , China*
*E-mail: infcon@seu.edu.cn*

Abstract: In this paper, we consider the global regulation problem of nonlinear systems. A new definition of the global regulation for nonlinear systems is given. The sufficient and necessary conditions for solving the global regulation problem for nonlinear systems are obtained. Finally an example is given. 

Keywords: Global regulation, global asymptotic stability, state feedback, invariant manifold

## 1.INTRODUCTION

In the past decades, many scholars were interested in the tracking and regulation problem for nonlinear systems[1−6]. In the local case, Isidori and Byrnes[6,2] have considered the regulation problem of the following systems:

$$\begin{cases} \dot{x} = f(x) + g(x)u + p(x)\omega , \\ \dot{\omega} = r(\omega) , \\ e = h(x) + q(\omega) , \end{cases} \tag{1}$$

and

$$\begin{cases} \dot{x} = f(x,\omega,u) , \\ \dot{\omega} = r(\omega) , \\ e = h(x,\omega) , \end{cases} \tag{2}$$

respectively. By assuming that the dynamics $\dot{\omega} = r(\omega)$ is the Poisson stable and the $\dot{\bar{x}} = \frac{\partial f}{\partial x}(0)\bar{x} + g(0)u$ or $\dot{\bar{x}} = \frac{\partial f}{\partial x}(0,0,0)\bar{x} + \frac{\partial f}{\partial u}(0,0,0)u$ is exponentially stabilizable they have obtained that the local regulation problem of (1) (or (2)) is solvable via the state feedback if and only if there are a neighborhood $V$ of the origin and mappings $\pi(\omega)$ and $c(\omega)$ defining on $V$, such that on $V$

$$\begin{cases} \frac{\partial \pi}{\partial \omega} r(\omega) = f(\pi(\omega)) + g(\pi(\omega))c(\omega) + p(\pi(\omega))\omega , \\ h(\pi(\omega)) + q(\omega) = 0 , \end{cases} \tag{3}$$

or

$$\begin{cases} \frac{\partial \pi}{\partial \omega} r(\omega) = f(\pi(\omega),\omega,c(\omega)) , \\ h(\pi(\omega),\omega) = 0 \end{cases} \tag{4}$$

In the global case, Dayawansa et al [4] have discussed how to make the systems

$$\begin{cases} \dot{x} = \phi(x.y) , \\ \dot{y}_i = y_{i+1} \quad i = 1,\cdots,m-1 , \\ \dot{y}_m = \alpha(x,y) + \beta(x,y)u \end{cases} \tag{5}$$

to track a reference ( or exogenous ) signal $r(t)$. Assuming that $|r(t)| \leq L$, where $L$ is a constant and $|r^q(t)| \leq \epsilon$ with $1 \leq q \leq p$ and $p \geq m+1$, $\epsilon$ is sufficiently small, the tracking problem of (5) is solvable by feedback with the data $x(t), r(t), \cdots, r^{(p)}(t)$. If the internal flow of the regulator problem of (2) is $C^1$ flow equivalent to the exosystems flow, then in [9] there are the mappings $c(\omega)$ and $S(\omega)$ such that

[1] This work is supportted by National Key Project and National Natural Sciences Foundation of China

$$\begin{cases} \frac{\partial S(\omega)}{\partial \omega} r(\omega) = f(S(\omega), \omega, c(\omega)) \ , \\ h(S(\omega), \omega) \ = 0 \ . \end{cases} \tag{6}$$

for all $\omega$. In this paper, we will prove that (6) is the sufficient and necessary conditions for solving the global regulation problem of (2) under the suitable hypotheses on (2).

The paper is organized as follows: the problem statement and the basic assumption are stated in Section 2. Section 3 reviews briefly the concept and results about invariant manifold without the proof. Section 4 is our main results. Finally, an examples is given.

## 2. PROBLEM STATEMENT

Consider the following systems:

$$\begin{cases} \dot{x} = f(x, \omega, u) \ , \\ \dot{\omega} = r(\omega) \ , \\ e = h(x, \omega) \ , \end{cases} \tag{7}$$

where $x \in R^n$ is the internal state, $\omega \in R^r$ is the external state , $u$ is the control input and $e$ is the controlled or the tracked output error. The functions appeared in (7) are assumed to be of class $C^\infty$ and without loss of the generality, assume that $f(0,0,0) = 0, r(0) = 0, h(0,0) = 0$.

**H.1.** For any invariant submanifold $M \subset R^{n+r}$ of (7) $\lim_{|(x,\omega)| \to \infty} h(x, \omega) = 0, (x, \omega) \in M$, implies $h(x, \omega) = 0$, where the invariant submanifold $M$ of (7) means that $\forall (x_0, \omega_0) \in M$, the solution $(x(t), \omega(t))$ of (7) with $x(0) = x_0, \omega(0) = \omega_0$ satisfies $(x(t), \omega(t)) \in M, \forall t \geq 0$.

**Remark 1:** If $e$ is considered as the output of the full systems (7) (interior and exterior system), then H.1 implies that some observability on the invariant submanifold for nonlinear systems in global sense. In fact, H.1 shows that it is practically impossible that as the state of the system approaches large on the invariant submanifold $M$, but its output conversely approach small. This may be possible only when the state itself on $M$ is unobservable. For instance, the regulation problem for the following linear systems

$$\begin{cases} \dot{x} = Ax + Bu + P\omega \ , \\ \dot{\omega} = S\omega \ , \\ e = h(x.\omega) = Cx + D\omega \ . \end{cases} \tag{8}$$

If take the feedback $u = F_1 x + F_2 \omega$, then the solution of (8) is $\begin{pmatrix} x \\ \omega \end{pmatrix} = e^{\bar{A}t} \begin{pmatrix} x_0 \\ \omega_0 \end{pmatrix}$, where $\bar{A} = \begin{pmatrix} A + BF_1 & P + BF_2 \\ 0 & S \end{pmatrix}$. Suppose that $V$ is invariant subspace (invariant submanifold), i.e. $\forall \begin{pmatrix} x_0 \\ \omega_0 \end{pmatrix} \in V, \begin{pmatrix} x \\ \omega \end{pmatrix} = e^{\bar{A}t} \begin{pmatrix} x_0 \\ \omega_0 \end{pmatrix} \in V, \ t \geq 0$. It is easy to see that if $\lim_{|(x,\omega)| \to \infty} (Cx + D\omega) = 0$, for any $(x, \omega) \in V$ then it must be $Cx + D\omega \equiv 0, \ \forall (x, \omega) \in V$.

**H.2.** $\dot{x} = f(x, 0, u)$ is globally asymptotically stabilizable via state feedback, i.e. there is $k(x)$ such that the zero solution of $\dot{x} = f(x, 0, k(x))$ is globally asymptotically stable (**GAS**).

**H.3.** Let a subset $L \subset C^1$ such that for any $l \in L$ the solution $(x(t), \omega(t))$ of Cauchy problem of the following equations

$$\begin{cases} \dot{x} \quad = f(x, \omega, l(\omega)) \ , \\ \dot{\omega} \quad = r(\omega) \ , \\ x(0) = 0, \omega(0) = \omega_0 \in R^r \ , \end{cases} \tag{9}$$

satisfies the following statement: if we denote $\Phi_\mu(t,s), \Psi_\eta(t,s)$ the fundamental matrix solution of the following partial variation equation of (9)

$$\dot{\mu} = \frac{\partial f}{\partial x}(x(t), \omega(t), l(\omega(t)))\mu$$

and

$$\dot{\eta} = \frac{\partial r}{\partial \omega}(\omega(t))\eta$$

then

$$|\Phi_\mu(t,s)| \leq \gamma_\mu e^{-\beta(t-s)} \quad t \geq s \geq 0 \ , \tag{10}$$

$$|\Psi_\eta(t,s)| \leq \gamma_\eta e^{-\alpha(t-s)} \quad s \geq t \geq 0 \tag{11}$$

where $\alpha, \beta, \gamma_\mu, \gamma_\eta$ are positive constants which are independent upon $l$.

We can now state our problem as follows: To find the state feedback law $u = \alpha(x, \omega), \alpha(0,0) = 0$ and the open set $\Sigma \subset R^{n+r}$ such that

i) The zero solution of $\dot{x} = f(x, 0, \alpha(x, 0))$ in (7) is **GAS** and the solution $(x(t), \omega(t))$ of the Cauchy problem of the equations (12) below satisfies (10) and (11)

$$\begin{cases} \dot{x} &= f(x,\omega,\alpha(x,\omega)) \\ \dot{\omega} &= r(\omega) \\ x(0) &= 0, \omega(0) = \omega_0 \end{cases} \qquad (12)$$

ii) $\forall(x_0,\omega_0) \in \Sigma$ the solution $(x(t),\omega(t))$ of the closed loop (7) with $x(0) = x_0, \omega(0) = \omega_0$ makes $\lim_{t\to\infty} e(t) = \lim_{t\to\infty} h(x(t),\omega(t)) = 0$, under the feedback $u = \alpha(x,\omega)$.

## 3. BRIEF REVIEW OF THE INVARIANT MANIFOLD

In this section we will briefly state the concepts and the results about invariant manifold without proof (see [8] in detail ).

Consider the following autonomous equations

$$\begin{cases} \dot{x} = f(x,y) , \\ \dot{y} = g(x,y) , \end{cases} \qquad (13)$$

where $(x,y) \in R^{n+m}$. We will assume from now on that $f$ and $g$ are of class $C^\infty$ on the whole $(x,y)$-space. Given a bounded subset $W_0$ of the $x$–space, called initial subset, and a mapping $s : W_0 \ni x \mapsto s(x) \in R^n$. In order to state our prpblem we further assume that

(1): if denote $\Phi_\mu, \Psi_\eta$ the principle matrices solution of the linear de's

$$\begin{aligned} \dot{\mu} &= f_x(x(t,x_0),y(t,x_0))\mu , \\ \dot{\eta} &= g_y(x(t,x_0),y(t,x_0))\eta , \end{aligned} \qquad (14)$$

respectively , then $\Phi_\mu, \Psi_\eta$ satisfy

$$|\Phi_\mu(t,s)| \le \gamma_\mu e^{-\beta(t-s)} \quad t \ge s \ge 0 , \qquad (15)$$

$$|\Psi_\eta(t,s)| \le \gamma_\eta e^{-\alpha(t-s)} \quad s \ge t \ge 0 , \qquad (16)$$

where $\alpha, \beta, \gamma_\mu, \gamma_\eta$ are constants and may be dependent on $x_0$.

(2): On $\overline{W} \triangleq \{(x,y) :$ there is $x_0$ such that $x = x(t,x_0), y = y(t,x_0), y(0,x_0) = s(x_0)\}$ there are constants $G$ and $H$ such that

$$|f_y(x,y)| \le G , \quad |g_x(x,y)| \le H , \forall(x,y)$$

(3): the mapping $s$ satisfies

$$|\frac{\partial s}{\partial x}| \le \sigma, \quad x \in cl(W_0) , \sigma \text{ is some constant}$$

where $cl(W_0)$ is the closure of $W_0$.

For each constant $\sigma' > \sigma$ we denote by $\mathcal{N}'(\sigma')$ the subset of the $(x,y)$-space which is defined as follows:

$\forall(x_0,y_0) \in \mathcal{N}'(\sigma')$ , if $x_0 \in W_0$ and if there exists a smooth mapping $x \mapsto \tilde{s}(x) \in R^m$ into the $y$-space such that the following conditions are satisfied:

$$s(x) = \tilde{s}(x), \quad \forall x \in \partial W_0, \quad y_0 = \tilde{s}(x_0) ,$$
$$|\tilde{s}(x) - s(x)| \le |\tilde{s}(x_0) - s(x_0)|, \quad \forall x \in W_0 ,$$
$$|\tilde{s}_x(x) - s_x(x)| \le \sigma' - \sigma, \quad \forall x \in W_0 .$$

It is easy to see that $\forall\sigma' > \sigma$, $\mathcal{N}'(\sigma')$ is not empty and actually an open neighborhood (with respect to the $(x,y)$-space) of the set $\{(x,y) : x \in W_0, y = s(x)\}$. $\forall\epsilon > 0$, denote

$$\begin{aligned} \mathcal{N}(\epsilon,\sigma') &= \mathcal{N}'(\sigma') \cap P_\epsilon \\ &\triangleq \mathcal{N}'(\sigma') \cap \{(x,y) : x \in W_0, |y - s(x)| < \epsilon\} \end{aligned}$$

**Lemma 3.1.** (see [8] Lemma B.3.2 and Thm. B.4.1) Let $\alpha < \beta, \gamma_\mu^2\gamma_\eta \frac{2G}{\beta-\alpha}[\frac{2H}{\beta-\alpha} + \sigma] < 1$, $\beta > \gamma_\mu\gamma_\eta^2 G[\frac{2H}{\beta-\alpha} + \sigma]$, $\beta - \alpha > (\gamma_\mu + \gamma_\eta)\gamma_\mu\gamma_\eta G[\frac{2H}{\beta-\alpha} + \sigma]$ and $\forall\beta' > 0$, $\alpha' = \alpha + \gamma_\mu^2\gamma_\eta \frac{2G}{\beta-\alpha}[\frac{2H}{\beta-\alpha} + \sigma] < \beta' < \beta - \gamma_\mu^2\gamma_\eta \frac{2G}{\beta-\alpha} \cdot [\frac{2H}{\beta-\alpha} + \sigma]$. Let $\epsilon > 0$, $\sigma' > \sigma$ be such that the inequalities above remain valid if $\sigma$ is replaced by $\sigma'$. where the selection for $\epsilon$ requires that when $(x_0,y_0) \in P_\epsilon$, the solution $(x(t),y(t))$ such that the (15) and (16) are also valid if $\sigma$ is replaced by $\sigma'$. Then for any solution $(x(t),y(t))$ of (13) with initial condition $(x(0),y(0)) \in \mathcal{N}(\epsilon,\sigma')$ the following statement is true:

a) $x(t) \in W_t = \{x : (t,x) \in W\}, \forall t \in [0,T]$, where $W = \{(t,x) : t \in [0,T], x = x(t,x_0)$ for some $x_0 \in W_0\}$;

b) if $\tilde{x}(\cdot)$ is the solution of $\dot{\tilde{x}} = f(t,\tilde{x},S(t,\tilde{x})), \tilde{x}(T) = x(T)$. $\tilde{y}(t) = S(t,\tilde{x}(t))$, then $\forall t \in [0,T], \tilde{x}_t \in W_t$

$$|x(t) - \tilde{x}(t)| \le \gamma_\mu\gamma_\eta \frac{G}{\beta' - \alpha'}|y(0) - S(0,x(0))|e^{-\beta' t}$$

$$\begin{aligned} &|y(t) - \tilde{y}(t)| \\ &\le \gamma_\eta\{1 + \gamma_\mu\gamma_\eta^2 \frac{G}{\beta' - \alpha'}[\frac{2H}{\beta-\alpha} + \sigma]\} \\ &\cdot |y(0) - S(x_0)|e^{-\beta' t} \end{aligned}$$

$$\begin{aligned} &|y(t) - S(t,x(t))| \\ &\le \gamma_\eta\{1 + \gamma_\mu\gamma_\eta^2[\frac{2H}{\beta-\alpha} + \sigma]\} \\ &\cdot |y(0) - S(x_0)|e^{-\beta' t} \end{aligned}$$

In addition, if the parameters $\alpha, \beta, \cdots$ are not dependent on $T$ and $\rho > 0$ is selected such that $\alpha < v\rho < \beta, \gamma_\mu \gamma_\eta GH(v\rho - \alpha)(\beta - v\rho)^{-1} < 1, v = 1, 2, \cdots, N, N \geq 2$. Let $\mathcal{X} \subset R^n$ be nonempty subset such that there is $\bar{t} \geq 0$, when $t \geq \bar{t}$, $\mathcal{X} \subset W_t$, then there exists a mapping $S_0 : x \mapsto S_0(x)$, $\forall x \in \mathcal{X}$, and $S_0(x) \in R^m$ is of $C^{N-1}$ which satisfies

c) $\lim_{t \to \infty} S(t, x) = S_0(x)$, if $x \in \mathcal{X}$, with the convergence being uniform with respect to $x$;

d) $S_0(x)$ solves the system of first order peds.

$$\frac{\partial S_0(x)}{\partial x} f(x, S_0(x)) = g(x, S_0(x)) \; . \qquad \square$$

## 4. MAIN RESULTS

At first, we notice that the dynamics of the exosystems $\dot{\omega} = r(\omega)$ in (7) is unforced and is uncontrollable. If we denote its flow by $\Phi_t^r(\omega)$, $\forall \omega \in R^r$ and the collection of all $\omega$-limit set of $\Phi_t^r(\omega)$ by $\Omega$, then any bounded solution $\omega(t)$ of the exosystem follows $\omega(t) \to \Omega \;\; (t \to \infty)$ and $\Omega$ is an invariant submanifold of exosystem. Therefore we first obtain

**Proposition 4.1.** Assume H.2–H.3 hold and $\alpha < \beta$. The global regulation problem of (7) is solvable via state feedback only if there exist mappings $c(\omega) \in L$ and $S(\omega) \in C^1$ such that

$$\begin{cases} \dfrac{\partial S(\omega)}{\partial \omega} r(\omega) = f(S(\omega), \omega, c(\omega)) & \omega \in R^r \, , \\ h(S(\omega), \omega) = 0 & \omega \in \Omega \end{cases} \tag{17}$$

**Proof** : If $u = \alpha(x, \omega) \in C^1$ and the open set $\Sigma \subset R^{n+r}$ such that (7) satisfies i) and ii) (see Section 2). By H.2 and lemma 3.1 , there is a $S(\omega) \in C^1$ so that

$$\frac{\partial S(\omega)}{\partial \omega} r(\omega) = f(S(\omega), \omega, \alpha(S(\omega), \omega)) \;\; \omega \in R^r$$

and when $(x_0, \omega_0) \in \mathcal{N}(\epsilon, \sigma')$, $|x(t) - S(\omega(t))| \to 0 \;\; (t \to \infty)$. Obviously, $c(\omega) = \alpha(S(\omega), \omega) \in L$. i.e. the first form of (17) follows. Now we prove that $\forall \omega \in \Omega, h(S(\omega), \omega) = 0$. Since $\forall (x_0, \omega_0) \in \mathcal{N}(\epsilon, \sigma'), \lim_{t \to \infty} h(x(t), \omega(t)) = 0$, if $\omega(t)$ is bounded, then $\omega(t) \to \Omega \; (t \to \infty)$ and there exist $\overline{\omega} \in \Omega$ and $\{t_k\}, t_k \to \infty (k \to \infty)$, such that $\omega(t_k) \to \overline{\omega} \; (k \to \infty)$. Thus $S(\omega(t_k)) \to S(\overline{\omega}) \; (k \to \infty)$ by the continunity of $S$. Therefore, $\lim_{k \to \infty} x(t_k) = \lim_{k \to \infty} (x(t_k) - S(\omega(t_k)) + \lim_{k \to \infty} S(\omega(t_k))$ $= S(\overline{\omega})$. Consequently, $h(S(\overline{\omega}), \overline{\omega}) = 0$. Conversely, $\forall \omega_0 \in \Omega$, there must exist a bounded trajectory $\omega(t)$ and $\{t_k\}, t_k \to \infty (k \to \infty)$, such that $\omega(t_k) \to \omega_0, (k \to \infty)$. It is implied that $h(S(\omega_0), \omega_0) = 0$. According to the arbitrarity of $\omega_0$, we know that the second form of (17) follows. $\square$

**Corollary 4.1.** Suppose the conditions of proposition 4.1 hold. If $cl(\Omega) = R^r$, $cl(\Omega)$ is the closure of $\Omega$, then the the global regulation problem of (7) is solvable only if (17) holds for all $\omega \in R^r$.

**Proof** : By the proposition 4.1 , there are $S$ and $c \in C^1$ such that the first form of (17) holds for any $\omega \in R^r$ and $\forall \omega \in \Omega, h(S(\omega), \omega) = 0$. Since $cl(\Omega) = R^r$, $\forall \omega \in R^r$ there is $\{\omega_k\} \subset \Omega, \omega_k \to \omega \; (k \to \infty)$, using the contiunuity of $h$ it follows $h(S(\omega), \omega) = 0, \forall \omega \in R^r$. $\square$

**Theroem 4.1.** Suppose H.1–H.4 hold. The global regulation problem of (7) is solvable if and only if there exist $S \in C^1, c \in L, S(0) = 0, c(0) = 0$, such that (17) is valid.

**Proof** : From proposition 4.1 the necessity is valid. Now we prove the sufficiency. Suppose the (17) hold and by H.2 there is $k(x)$ such that the zero solution of $\dot{x} = f(x, 0, k(x))$ is **GAS**. Let $\alpha(x, \omega) = c(\omega) + k(x - S(\omega))$, then $\forall (x_0, \omega_0) \in \mathcal{N}(\epsilon, \sigma')$, from lemma 3.2, the solution $(x(t), \omega(t))$ of (7) with $x(0) = x_0, \omega(0) = \omega_0$ satisfies $\lim_{t \to \infty} |x(t) - S(\omega(t)| = 0$, where we take $\Sigma = \mathcal{N}(\epsilon, \sigma')$. By H.1 it only needs to prove that $\lim_{t \to \infty} e(t) = 0$ for the bounded trajectory $\omega(t)$. If $\omega(t)$ is bounded, by the continuity of $S$ and $\lim_{t \to \infty} |x(t) - S(\omega(t)| = 0$, then $x(t)$ is also bounded. Therefore, the differentiability of $h$ implies that there exist a constant $K$ such that $|h(x(t), \omega(t)) - h(S(\omega(t)), \omega(t))| \leq K|x(t) - S(\omega(t))|$. So,

$$
\begin{aligned}
&|h(x(t),\omega(t))| \\
&\le |h(x(t),\omega(t)) - h(S(\omega(t)),\omega(t))| \\
&\quad +|h(S(\omega(t)),\omega(t))| \\
&\le K|x(t)-S(\omega(t))| + |h(S(\omega(t)),\omega(t))| \, .
\end{aligned} \tag{18}
$$

According to (17) and (18) and $\omega(t) \to \Omega \ (t \to \infty)$ and the continuity of $h$ we gain that $\lim_{t\to\infty} e(t) = \lim_{t\to\infty} h(x(t),\omega(t)) = 0$. □

For the local case, according to Theroem 4.1 we can obtain the more extended result than that of [6]. i.e. if we denote $A = \frac{\partial f}{\partial x}(0,0,0), B = \frac{\partial f}{\partial u}(0,0,0)$, then we have

**Corollary 4.2.** Suppose $(A, B)$ be stabilizable. If the dynamics of $\omega$ is the locally bounded (not asymptotically stable), then the local regulation problem of (7) is solvable if and only if there are a neighborhood $V \in R^r$ of the origin ( $\omega = 0$ ) and the mappings $\pi(\omega), c(\omega) \in C^1, \pi(0) = 0, c(0) = 0$ such that

$$
\begin{cases}
\frac{\partial \pi(\omega)}{\partial \omega} r(\omega) = f(\pi(\omega),\omega,c(\omega)) & \forall \omega \in V \, , \\
h(\pi(\omega),\omega) = 0 & \forall \omega \in \Omega \cap V \, .
\end{cases} \tag{19}
$$

**Example 1**: Consider the special model as follows

$$
\begin{cases}
\dot{x} = f(x,u) \, , \\
z = h(x) \in R \, .
\end{cases} \tag{20}
$$

to track a constant exogenous signal $r(t) = H, H$ is a constant. In accordance with Theroem 4.1, if (20) satisfies the hypotheses of Theroem, then the tracking problem of (20) is solvable if and only if there are $S(H)$ and $c(H)$ such that

$$
\begin{aligned}
f(S(H), c(H)) = 0 \, , \quad (21) \\
h(S(H)) = H \, ,
\end{aligned}
$$

where the exogenous signal can be considered that it is produced by the dynamical equation $\dot{\omega} = 0$.

In paper [4], Dayawansa et al. have discussed in detail that (20) possessing global canonical form (5) can track a constant external signal. However, our result indicate that if $\phi$ in (5) satisfies the condition that $\dot{x} = \phi(x, H, 0)$ has a unique globally exponentially stable equilibrium $S_1(H)$, then (5) satisfies the hypotheses of Theroem 4.1. In this case the condition (21) is $y_1 = H, y_i = 0, (i = 2, \cdots, m), S(H) = (S_1(H), H, 0), c(H) = -\frac{\alpha(S_1(H), H, 0)}{\beta(S_1(H), H, 0)}$. So, we easily obtain that (5) can track a constsnt exogenous signal by the Theorem 4.1.

**Example 2**: Consider the following nonlinear system

$$
\begin{cases}
\dot{x} = A_1 x + B_1 u_1 + B_2 x u_2 + P(x,\omega) \, , \\
\dot{\omega} = A_2 \omega + \Pi_{i=0}^{l}(|\omega|^2 - r_i^2)\omega \, , \\
e = h(x,\omega) \in R \, ,
\end{cases} \tag{22}
$$

where $x = (x_1 \ x_2 \ x_3)^T, \omega = (\omega_1 \ \omega_2)^T, r_0 > r_1 > \cdots > r_l > 0$ are real constants, $l$ is an even number. $A_1 = \begin{pmatrix} -1 & 0 & 2 \\ 0 & 0 & 1 \\ 0 & -1 & 0 \end{pmatrix}, B_1 = \begin{pmatrix} 1 \\ 0 \\ 0 \end{pmatrix}$, $B_2 = \begin{pmatrix} 0 & 0 & 0 \\ 0 & 1 & 0 \\ 0 & 0 & 1 \end{pmatrix}, A_2 = \begin{pmatrix} 0 & -1 \\ 1 & 0 \end{pmatrix}$, $P(x,\omega) = (p_1 \ p_2 \ p_3)^T(x,\omega)$ is continuous function vector, $P(x,0) \equiv 0, \forall x$ and $\max_{(x,\omega)\in R^5} |P| \le r_0^3$. $h$ is cotinuous function, $h(0,0) = 0$, and $\frac{\partial h}{\partial x_1}(x,\omega) \neq 0, \forall (x,\omega) \in R^5$.

Obviously, the system (22) does not satisfy the condition in [6] in local case, because its interior system is not locally exponentially stabilizable, but only locally asymptotically stabilizable, and also the exterior system is not Poisson stable. In matter of fact, $\forall \omega_0, |\omega_0| < r_l$ , $\omega(t) \to \{\omega : |\omega| = r_l\}, (t \to +\infty)$, and $\forall \omega_0 \in R^2$ , $\omega(t) \to \{\omega : |\omega| = r_0\}, (t \to +\infty)$, the origin $(\omega = 0)$ is not stable. Since $\frac{\partial h}{\partial x_1}(x,\omega) \neq 0$, the coordinate transformation $\psi : (x^T, \omega^T)^T \mapsto (e, x_2, x_3, \omega_1, \omega_2)^T$ is a local differomorphism.

From the inversability of $\psi$ there is a function $\alpha$ such that in inverse transformation $\psi^{-1}$ : $(e, x_2, x_3, \omega_1, \omega_2) \mapsto (x, \omega)$, $x_1 = \alpha(e, x_2, x_3, \omega)$. Let $u_2 = -(x_2^2 + x_3^2), u_1 = \alpha - 2x_3 - p_1(x,\omega) - (\frac{\partial h}{\partial x_1})^{-1}[\frac{\partial h}{\partial x_2}(x_3 + x_2 u_2 + p_2(x,\omega)) + \frac{\partial h}{\partial x_3}(-x_2 + x_3 u_2 + p_3(x,\omega)) + \frac{\partial h}{\partial \omega}\dot{\omega} + e]$ , then the closed-loop system is

$$\begin{cases} \dot{e} &= -e\ , \\ \dot{x}_2 &= x_3 - x_2(x_2^2 + x_3^2) + \bar{p}_2(e, x_2, x_3, \omega)\ , \\ \dot{x}_3 &= -x_2 - x_3(x_2^2 + x_3^2) + \bar{p}_3(e, x_2, x_3, \omega)\ , \\ \dot{\omega} &= A_2\omega + \Pi_{i=0}^{l}(|\omega|^2 - r_i^2)\omega\ , \end{cases} \tag{23}$$

where $\bar{p}_i = p_i(\alpha(e, x_2, x_3, \omega), x_2, x_3, \omega), i = 1, 2.$ Therefore, $\forall e_0 \in R$ we have $|e(t)| \leq |e_0|$ and $\lim_{t\to\infty} e(t) = 0.$

Since $\forall \omega_0 \in \Omega_i = \{\omega : |\omega| = r_i\}, i = 0, 1, \cdots, l, \omega(t) \in \Omega_i$ , there must be a mapping $S$ such that $h(S(\omega), \omega) = 0, \forall \omega \in \Omega_i, i = 0, 1, \cdots, l.$ And from the system (22) we have $\Omega = \sum_0^l \Omega_i,$ where $\Omega$ is the $\omega$-limit set of the dynamics $\omega$. Therefore, the mapping $c$ also exists. So by the Thm.4.1 the global regulation problem of the system (22) is solvable.

In fact, if we take $V(x_2, x_3) = \frac{1}{2}(x_2^2 + x_3^2)$, then

$$\begin{aligned} \dot{V}_{(21)} &= -(x_2^2 + x_3^2)^2 + (x_2\ x_3)\begin{pmatrix} \bar{p}_2 \\ \bar{p}_3 \end{pmatrix} \\ &\leq -r^4 + r|P| \leq -r(r^3 - r_0^3)\ , \end{aligned}$$

where $r = \sqrt{(x_2^2 + x_3^2)}$. This implies that $\forall\ (x_2\ x_3)_0^T \in R^3$, we have $|(x_2\ x_3)^T(t)| \leq |(x_2\ x_3)_0^T|$. That is $(x_2\ x_3)^T(t)$ is bounded $\forall t \geq 0$.

When $\omega = 0$, the interior system of the closed-loop system is

$$\begin{cases} \dot{h}(x, 0) = -h(x, 0)\ , \\ \dot{x}_2 \quad = x_3 - x_2(x_2^2 + x_3^2)\ , \\ \dot{x}_3 \quad = -x_2 - x_3(x_2^2 + x_3^2)\ . \end{cases}$$

Thus $h(x, 0) \to 0, x_2 \to 0, x_3 \to 0(t \to \infty)$. Consequently, $x \to 0(t \to \infty)$. That is to say that the global regulation problem of the system (22) is solvable according to the definition i) and ii) in Section 2.

## CONCLUSION

In this paper, the global regulation problem for general nonlinear systems is discussed. The new definition of the global regulation problem for nonlinear systems is given. The conditions of solvability for global regulation problem of nonlinear systems are obtained by using the invariant manifold theory. In local case, our results extended that of [2,6]. Finally, one case (tracking a constant signal) in [4] is considered as particularly example in this paper, according to our result the conditions of the solvability are easily obtained and another example is given.

## REFERENCES

M.D. Di Benedetto, Synthesis of an Internal Model for Nonlinear Output Regulation , Int.J. Contr. 45(1987) 1023–1034

C.I. Byrnes, and A. Isidori, Steady State Response, Seperation Principle and the Output Regulation of Nonlinear Systems , Proc. of 28th CDC, Tampa , Florida Dec. 1989 2247–2251

C.I. Byrnes, and A. Isidori, Stabilization and Output Regulation of Nonlinear Systems in the Large , Proc. of 29th CDC, Honolulu, Hawail Dec. 1990 1255–1261

W.P. Dayawansa, C.F. Martin, and G. Knowles, A Global Nonlinear Tracking Problem , Proc. of 29th CDC, Honolulu, Hawail Dec. 1990 1268–1271

H. Jie, and W.J. Rugu, Stabilization on Zero-Error Manifold and the Nonlinear Servomechanism Problem , Proc. of 29th CDC, Honolulu, Hawail Dec. 1990 1262–1267

A. Isidori, and C.I. Byrnes, Output Regulation of Nonlinear Systems , IEEE Trans. Auto. Contr. 35(1990) 130–140

A. Isidori, Nonlinear Control Systems ,2nd Edition, Springer–Verlag 1989

H.W. Knobloch, A. Isidori, and D. Flockerzi, Topics in Control Theory, Birkhaüser Verlag , 1993

P. Lucibello, A Note on a Necessary Condition for Output Regulation, IEEE Trans. Auto.Contr. 39(1994) 558–559

14th Triennial World Congress, Beijing, P.R. China


# ROBUST OUTPUT TRACKING OF CONSTRAINED NONLINEAR SYSTEMS

Jing Wang†, Yan Zheng‡, Xiaoping Liu†

*†Dept. of Auto. Contr., Box 135, Northeastern University, Shenyang, 110006, China, Email:liuxp@mail.neu.edu.cn ‡Gold Technology College, Northeastern University, Shenyang, 110006, China*

Abstract: The problem of output tracking for constrained nonlinear systems in the presence of uncertainties is studied. The concepts of input/output (I/O) linearization, relative degree, hyperbolically minimum phase and matching condition are reviewed. A sliding mode control (SMC) law is developed which achieves asymptotically tracking despite the uncertainties while maintaining the boundness of all signals inside the loop. *Copyright ©1999 IFAC*

Keywords: I/O linearization, hyperbolically minimum phase, matching condition, sliding surface, output tracking.

## 1. INTRODUCTION

For many physical systems, there are inevitable uncertainties in their constructed models. In general, two classes of uncertainties (Behtash,1990) is dealt with: those arising from unknown or imperfectly parametric in the model vector fields, and those due to the changing environment as well as to the unpredicted disturbances.

So far, to deal with uncertain nonlinear systems, three main approaches have been proposed: 1) adaptive control (Sastry, and Isidori, 1989; Taylor, *et al.*, 1988), 2) Lyapunov-based control (Barmish, *et al.*, 1983; Chen, 1987), and 3) variable structure control (VSC) (Fernandez, and Hedrick, 1987; Slotine, 1984). The first one is applied to systems mainly with parameterized uncertainties, and the other two allow nonparameterized (unstructured) uncertainties. Specifically for VSC, it first defines the sliding surface in the error state space and forms a switching state feedback control also using the same information on the bounds. The high-speed switching forces the error state to slide along the surface until it converges and the tracking is attained (Chen, and Chang, 1992).

A class of nonlinear constrained dynamic systems (CDS) is considered in this paper. Towards the robust output tracking control of this system, an I/O linearization is performed as (Chen, and Shayman, 1992), first. Then a SMC is applied on the linearized system. However, in the presence of modeling uncertainties and disturbances, the linearization may not be possible. But for a certain class of uncertainties and disturbances which satisfy the so-called "matching condition", I/O linearization is guaranteed. It is important to note that there may be a remainder of the state dynamics (i.e. zero dynamics) which does not appear in the linearized structure. It is desired that the zero dynamics is exponentially stable. This point is discussed in Section 2.

A characteristic component of SMC applications is $s(x,t) = 0$, which represents the slide hyperplanes. We then present modifications to the basic tracking control law of unperturbed system, using

SMC schemes, in order to achieve robustness with respect to the class of uncertainties and disturbances satisfying the so-called "matching condition" in Section 3. Finally, a conclusion is reached in Section 4.

## 2. I/O LINEARIZATION OF NONLINEAR CDS WITH UNCERTAINTIES

Consider a class of nonlinear CDS as the following

$$\begin{array}{ll} \dot{x} = f(x) + \Delta f(x,t) + \sum_{i=1}^{m}(g_i(x) + \Delta g_i(x))u_i & \\ 0 = k_i(x) & i = 1,\ldots,l \\ y = h_i(x) & i = 1,\ldots,p \end{array} \quad (1)$$

where $f(x)$ and $g_i(x)$ are smooth vector fields, $k_i(x)$ and $h_i(x)$ are smooth functions. $\Delta f(x,t)$ and $\Delta g_i(x)$, $i = 1,\ldots,m$, represent the disturbances and modeling uncertainties. For convenience denote

$$\begin{array}{rcl} G(x) & = & [g_1(x),\ldots,g_m(x)] \\ \Delta G(x) & = & [\Delta g_1(x),\ldots,\Delta g_m(x)] \\ k(x) & = & [k_1(x),\ldots,k_l(x)]^T \\ h(x) & = & [h_1(x),\ldots,h_p(x)]^T \end{array}$$

Suppose that for all $x \in U$, $g_1(x),\ldots,g_m(x)$ are linear independent vector fields, $\{dk_1(x),\ldots,dk_l(x)\}$ and $\{dh_1(x),\ldots,dh_p(x)\}$ are each linearly independent sets of covector fields, and $m = l + p$.

The unperturbed system is then defined as follows (i.e. with $\Delta f \equiv 0$, $\Delta G \equiv 0$) in (1):

$$\begin{array}{rcl} \dot{x} & = & f(x) + G(x)u \\ 0 & = & k(x) \\ y & = & h(x) \end{array} \quad (2)$$

Assume that the unperturbed system (2) satisfies:

**A1** Both the constraint relative degree vector $\{r_1^c,\ldots,r_l^c\}$ with respect to $(f,G,k)$ and the output relative degree vector $\{r_1^o,\ldots,r_p^o\}$ with respect to $(f,G,h)$ exist.

**A2** $[A^c(x)^T, A^o(x)^T] \in R^{m\times m}$ is nonsingular where $A^c(x)$ is the decoupling matrix for $(f,G,k)$ and $A^o(x)$ is the decoupling matrix for $(f,G,h)$.

Under the assumptions $A1$ and $A2$, system (2) can be I/O linearized (Chen, and Shayman, 1992). It is straightforward to shown that the constraint relative degree $r_i^c$ and the output relative degree $r_i^o$ are unchanged by the addition of uncertainties and perturbations if

**A3**

$$\begin{array}{l} \Delta f(x,t), \Delta g_j(x) \in \\ \quad \mathrm{Ker}[dk_i, dL_f k_i,\ldots,dL_f^{r_i^c-2}k_i], i = 1,\ldots,l \\ \Delta f(x,t), \Delta g_j(x) \in \\ \quad \mathrm{Ker}[dh_i, dL_f h_i,\ldots,dL_f^{r_i^c-2}h_i], i = 1,\ldots,p \end{array}$$

for $j = 1,\ldots,m$. This is so-called "matching condition". It assures that $\Delta f$ and $\Delta G$ terms do not appear in derivatives of $k_i$ and $y_i$ of order less than $r_i^c$ and $r_i^o$.

If the perturbation vector fields satisfy assumption $A3$, there exists a diffeomorphic coordinate change $\Phi(\cdot): R^n \to R^n$:

$$\Phi = \Phi(x) = (\chi(x)^T, \xi(x)^T, \eta(x)^T)^T$$

where

$$\begin{array}{l} \chi(x) = (k_1,\ldots,L_f^{r_1^c-1}k_1,\ldots,k_l,\ldots,L_f^{r_l^c-1}k_l)^T \\ \xi(x) = (h_1,\ldots,L_f^{r_1^o-1}h_1,\ldots,h_p,\ldots,L_f^{r_p^o-1}h_l)^T \\ \eta(x) = (\eta_1,\ldots,\eta_{n-r^c-r^o})^T \end{array}$$

for $r^c = \sum_{i=1}^{l} r_i^c$ and $r^o = \sum_{i=1}^{p} r_i^o$, which transforms the system (1) into the following

$$\begin{array}{rcl} \dot{\chi}_1^i & = & \chi_2^i \\ & \cdots & \\ \dot{\chi}_{r_i^c-1}^i & = & \chi_{r_i^c}^i \\ \dot{\chi}_{r_i^c}^i & = & b_i^c(x) + \Delta b_i^c(x) + \sum_{j=1}^{m}(a_{ij}^c(x) \\ & + & \Delta a_{ij}^c(x))u_j|_{x=\Phi^{-1}(\chi,\xi,\eta)}, \ i = 1,\ldots,l \\ \dot{\xi}_1^i & = & \xi_2^i \\ & \cdots & \\ \dot{\xi}_{r_i^o-1}^i & = & \chi_{r_i^o}^i \\ \dot{\xi}_{r_i^o}^i & = & b_i^o(x) + \Delta b_i^o(x) + \sum_{j=1}^{m}(a_{ij}^o(x) \\ & + & \Delta a_{ij}^o(x))u_j|_{x=\Phi^{-1}(\chi,\xi,\eta)}, \ i = 1,\ldots,p \\ \dot{\eta}_i & = & q(\chi,\xi,\eta), \ 1 \le i \le n - r^c - r^o \\ \chi_1^i & = & 0 \\ y_i & = & \xi_1^i, \ i = 1,\ldots,p \end{array} \quad (3)$$

where for $j = 1,\ldots,m$

$$\begin{array}{rcl} a_{ij}^c(x) & = & L_{g_j}L_f^{r_i^c-1}k_i(x) \\ \Delta a_{ij}^c & = & L_{\Delta g_j}L_f^{r_i^c-1}k_i(x) \\ b_i^c(x) & = & L_f^{r_i^c}k_i(x) \\ \Delta b_i^c & = & L_{\Delta f}L_f^{r_i^c-1}k_i(x), \ i = 1,\ldots,l \\ a_{ij}^o(x) & = & L_{g_j}L_f^{r_i^o-1}h_i(x) \\ \Delta a_{ij}^o & = & L_{\Delta g_j}L_f^{r_i^o-1}h_i(x) \\ b_i^o(x) & = & L_f^{r_i^o}h_i(x) \\ \Delta b_i^o & = & L_{\Delta f}L_f^{r_i^o-1}h_i(x), \ i = 1,\ldots,p \end{array}$$

Equation (3) is written in the following compact form:

$$\begin{array}{rcl} 0 & = & b^c + \Delta b^c + (A^c + \Delta A^c)u \\ \xi^{(r^o)} & = & b^o + \Delta b^o + (A^o + \Delta A^o)u \\ \dot{\eta} & = & q \end{array} \quad (4)$$

where for $j = 1, \ldots, m$

$$\begin{aligned}
\xi^{(r^o)} &= \text{col}[(\xi_1^1)^{(r_1^o)}, \ldots, (\xi_1^p)^{(r_p^o)}] \\
b^c &= \text{col}[b_1^c, \ldots, b_l^c] \\
\Delta b^c &= \text{col}[\Delta b_1^c, \ldots, \Delta b_l^c] \\
b^o &= \text{col}[b_1^o, \ldots, b_p^o] \\
\Delta b^o &= \text{col}[\Delta b_1^o, \ldots, \Delta b_p^o] \\
A^c &= [a_{ij}^c] \\
\Delta A^c &= [\Delta a_{ij}^c],\ i = 1, \ldots, l \\
A^o &= [a_{ij}^o] \\
\Delta A^o &= [\Delta a_{ij}^o],\ i = 1, \ldots, p
\end{aligned}$$

Then the control

$$u = \begin{bmatrix} A^c \\ A^o \end{bmatrix}^{-1} \left( \begin{bmatrix} 0 \\ v \end{bmatrix} - \begin{bmatrix} b^c \\ b^o \end{bmatrix} \right) \tag{5}$$

linearizes the nominal part of equation (4)

$$\begin{aligned}
\begin{bmatrix} 0 \\ \xi^{(r^o)} \end{bmatrix} &= \begin{bmatrix} \Delta b^c \\ \Delta b^o \end{bmatrix} + \begin{bmatrix} \Delta A^c \\ \Delta A^o \end{bmatrix} \begin{bmatrix} A^c \\ A^o \end{bmatrix}^{-1} \\
&\quad \left( \begin{bmatrix} 0 \\ v \end{bmatrix} - \begin{bmatrix} b^c \\ b^o \end{bmatrix} \right) + \begin{bmatrix} 0 \\ v \end{bmatrix}
\end{aligned} \tag{6}$$

where $v$ is the "synthetic input".

Note that the control law (5) make the state vector $\eta$ completely unobservable at the output. With an aim to tracking, it is desired that the dynamics $\dot{\eta} = q(0, \xi, \eta)$ is bounded-input bounded-state (BIBS) stable, with $\xi$ considered as an external input. As discussed in (Byrnes, and Isidori, 1988), this is guaranteed if

**A4** The equilibrium point 0 of the zero dynamics

$$\dot{\eta} = q(0, 0, \eta)$$

is exponentially stable.

Such systems are called hyperbolically minimum phase. The consideration in this paper is for this type of systems only.

## 3. SLIDING MODE CONTROL

To achieve zero tracking error despite the uncertainties and disturbances in system (4), a sliding model control law is proposed using the linearization control law (5) along with a high-speed switching control. The former aims to obtain a set of linear constrain and output equations, whereas the latter tries to compensate for the uncertainties and disturbances and at the same time to provide an asymptotic tracking force. To construct the sliding mode control, two steps will be involved: first, a sliding surface which achieves the desired tracking objective is constructed, second is to derive a suitable switching feedback gain so as to attain the so-called sliding condition.

Now let $y_d = (y_{d1}, \ldots, y_{dp})^T$ be the desired output trajectories where $y_{di}^{(j)}$, $j = 1, \ldots, r_i^o$, $i = 1, \ldots, p$, are bounded signals. Denote $e_i = y_i - y_{di}$, $i = 1, \ldots, p$, the sliding surface is then defined as

$$0 = s_i = e_i^{(r_i^o - 1)} + \sum_{j=1}^{r_i^o - 1} \beta_{ij} e_i^{(r_i^o - j - 1)} \tag{7}$$

where $\beta_{ij}, i = 1, \ldots, p, j = 1, \ldots, r_i^o - 1$, are chosen so that the following polynomials:

$$H_i(s) = s^{(r_i^o - 1)} + \beta_{i1} s^{(r_i^o - 2)} + \ldots + \beta_{i(r_i^o - 1)}$$

for $i = 1, \cdots, p$, are Hurwitz. Now let the input $v$ in (6) be chosen such that

$$\begin{aligned}
v &= -\dot{S} + \begin{bmatrix} y_1^{(r_1^o)} \\ \vdots \\ y_p^{(r_p^o)} \end{bmatrix} - \lambda(x)\text{sgn}(S) \\
&= \theta(x, t) - \lambda(x)\text{sgn}(S)
\end{aligned} \tag{8}$$

where $S = [s_1, \ldots, s_p]^T$ and $\lambda(x)$ is a switching state feedback gain yet to be determined, and $\text{sgn}(S) = [\text{sgn}(s_1), \ldots, \text{sgn}(s_p)]^T$ with

$$\text{sgn}(s_i) = \begin{cases} 1 & s_i > 0 \\ 0 & s_i = 0 \\ -1 & s_i < 0 \end{cases} \quad i = 1, \ldots, p$$

then $\theta(x, t)$ can be shown that

$$\theta(x, t) = \begin{bmatrix} y_{d1}^{(r_1^o)} \\ \vdots \\ y_{dp}^{(r_p^o)} \end{bmatrix} - \begin{bmatrix} \sum_{i=1}^{r_1^o - 1} \beta_{1i} e^{(r_1^o - i)} \\ \vdots \\ \sum_{i=1}^{r_p^o - 1} \beta_{pi} e^{(r_p^o - i)} \end{bmatrix} \tag{9}$$

With the construction of $v$ in (8), (6) becomes

$$\begin{aligned}
\begin{bmatrix} 0 \\ \dot{S} \end{bmatrix} &= \begin{bmatrix} \Delta b^c \\ \Delta b^o \end{bmatrix} - \begin{bmatrix} \Delta A^c \\ \Delta A^o \end{bmatrix} \begin{bmatrix} A^c \\ A^o \end{bmatrix}^{-1} \begin{bmatrix} b^c \\ b^o \end{bmatrix} \\
&\quad + \begin{bmatrix} \Delta A^c \\ \Delta A^o \end{bmatrix} \begin{bmatrix} A^c \\ A^o \end{bmatrix}^{-1} \begin{bmatrix} 0 \\ \theta(x, t) \end{bmatrix} \\
&\quad - \left( I + \begin{bmatrix} \Delta A^c \\ \Delta A^o \end{bmatrix} \begin{bmatrix} A^c \\ A^o \end{bmatrix}^{-1} \right) \begin{bmatrix} 0 \\ \lambda(x)\text{sgn}(S) \end{bmatrix}
\end{aligned} \tag{10}$$

A Lyapunov function is selected as $V(S) = \frac{1}{2} S^T S$, then

$$\dot{V}(S) = S^T\dot{S} = \begin{bmatrix} 0 \\ S \end{bmatrix}^T \begin{bmatrix} 0 \\ \dot{S} \end{bmatrix}$$
$$= S^T \left( \Delta b^o - \Delta A^o \begin{bmatrix} A^c \\ A^o \end{bmatrix}^{-1} \begin{bmatrix} b^c \\ b^o \end{bmatrix} \right)$$
$$+ S^T \left( \Delta A^o \begin{bmatrix} A^c \\ A^o \end{bmatrix}^{-1} \begin{bmatrix} 0 \\ \theta(x,t) \end{bmatrix} - \lambda(x)\mathrm{sgn}(S) \right)$$
$$- \begin{bmatrix} 0 & S^T \end{bmatrix} \begin{bmatrix} \Delta A^c \\ \Delta A^o \end{bmatrix} \begin{bmatrix} A^c \\ A^o \end{bmatrix}^{-1} \begin{bmatrix} 0 \\ \lambda(x)\mathrm{sgn}(S) \end{bmatrix} \quad (11)$$

In order to meet the Negative semidefiniteness condition of the Lyapunov stability criterion, $dV/dt \leq 0$, a suitable choice of the switching gain $\lambda(x)$ has to be chosen. Before this becomes possible, an additional assumption will be require.

**A5** There exist known functions $\Psi_1(\cdot) : R^n \to R_+$, $\Psi_2(\cdot,\cdot) : R^n \times R_+ \to R_+$ and $\Psi_0 < 1$ such that for all $x \in R^n$ and $t \geq 0$

$$\left\| \begin{bmatrix} \Delta A^c \\ \Delta A^o \end{bmatrix} \begin{bmatrix} A^c \\ A^o \end{bmatrix}^{-1} \right\|_{i\infty} \leq \Psi_1(x) < \Psi_0 \quad (12)$$

$$\left\| \Delta b^o - \Delta A^o \begin{bmatrix} A^c \\ A^o \end{bmatrix}^{-1} \begin{bmatrix} b^c \\ b^o \end{bmatrix} + \Delta A^o \begin{bmatrix} A^c \\ A^o \end{bmatrix}^{-1} \begin{bmatrix} 0 \\ \theta(x,t) \end{bmatrix} \right\|_2 \leq \Psi_2(x,t) \quad (13)$$

where $\|\cdot\|_{i\infty}$ means the induced infinite norm in $R^{m\times m}$ space.

With this assumption, (11) is reiterated using vector norms:

$$\dot{V} \leq \|\Psi_2(x,t)\| \, \|S\|_2 - \lambda(x)(1-\Psi_1(x))\|S\|_2 \leq 0 \quad (14)$$

for $S^T\mathrm{sgn}(S) \geq \|S\|_2$, which leads to the acceptable values of $\lambda(x)$ as

$$\lambda(x) > (1-\Psi_1(x))^{-1}(\epsilon + \Psi_2(x,t)) \quad (15)$$

If the condition (15) is true, then the sliding condition

$$S^T\dot{S} < -\epsilon\|S\|_2, \ \epsilon > 0 \quad (16)$$

will be met. This leads to our main theorem of this note.

**Theorem 1** Assumption A1-A5 hold. Given desired output trajectories $y_d(t) \in R^p$ with the aforementioned properties, the sliding mode control law in (5) and (8) with the feedback switching gain as given in (15) will achieve the asymptotic tracking and all signals remain bounded.

**Proof** The proof is a straightforward derivation from the construction and assumptions and is, hence, omitted here.

**Remark 1** Sliding mode control law (8) is discontinuous across the surface $S$, thus leads to the control chattering (Slotine, 1984). To eliminate this, a saturation function is replaced instead of sgn($S$), namely

$$\mathrm{sat}(s_i) = \begin{cases} 1 & s_i \geq \mu \\ \frac{s_i}{\mu} & -\mu \leq s_i \leq \mu \\ -1 & s_i \leq -\mu \end{cases}$$

where $\mu$ is the boundary layer thickness in $s_i$ space. During transient phase of the dynamics, $s_i$ may move in and out of this boundary layer. However, once the sliding condition is reached, $s_i$ remains entrapped in this layer.

## 4. CONCLUSION

A combination of I/O feedback linearization of constrained nonlinear system and a sliding mode control strategy is performed. This is very strong tool for robust output tracking problem of nonlinear CDS with uncertainties satisfying so-called "matching condition". Apparent further enhancements in response characteristics are currently studies.

*Acknowledgements:* This work is supported in part by grants from Fok Ying Tung Education Foundation and Trans-Century Training Program Foundation for the Talents by State Education Commission of China.

## REFERENCE

Barmish, B.B., M. Corless and G. Leitmann (1983). A new class of stabilizing controller for uncertain dynamical system. *SIAM J. Contr. Optimiz.*, 246-255.

Behtash, S. (1990). Robust output tracking of non-linear systems. *Int. J. Contr.*, 1381-1407.

Byrnes, C. and A. Isidori (1988). Local stabilization of minimum phase nonlinear systems. *Syst. Contr. Lett.*, 9-17.

Chen, X.L. and M.A. Shayman (1992). Dynamics and control of constrained nonlinear systems with application to robotics. *Proc. of Amer. Contr. Conf.*, 2962-2966.

Chen, Y.C. and S. Chang (1992). Output tracking design of affine nonlinear plant via variable structure system. *IEEE Trans Automat. Contr.*, 1823-1828.

Chen, Y.H. (1987). Adaptive robust model-following control and application to robot manipulators. *ASME J. Dynam. Syst., Measure., Contr.*, 209-215.

Fernandez, B.R. and J.K. Hedrick (1987). Control of multivariable non-linear systems by the sliding mode method. *Int. J. Contr.*, 1019-1040.

Sastry, S.S. and A. Isidori (1989). Adaptive control of linearizable systems. *IEEE Trans Automat. Contr.*, 1123-1131.

Slotine, J.J. (1984). Sliding controller design for non-linear systems. *Int. J. Contr.*, 421-434.

Taylor, D.G., P.V. Kokotovic, R. Marino and I. Kannellakopoulos (1988). Adaptive regulation of nonlinear systems with unmodelled dynamics. *Proc. of Amer. Contr. Conf.*, 360-365.

14th Triennial World Congress, Beijing, P.R. China

F-2c-22-2

# HANDLING ACTUATOR NON-SMOOTH NONLINEARITIES USING VARIABLE STRUCTURE CONTROL

**Cem Hatipoğlu**[†] **and Ümit Özgüner**[‡]

[†] *AlliedSignal Truck Brake Systems*
*901 Cleveland Street*
*Elyria, OH 44035, USA*
Cem.Hatipoglu@alliedsignal.com

[‡] *The Ohio State University*
*2015 Neil Ave. 205 Dreese Labs,*
*Columbus, OH 43210, USA*
Ozguner.2@osu.edu

Abstract: This paper deals with robust variable structure control design for continuous-time linear systems involving non-smooth actuator nonlinearities. Also examined in this paper are global boundedness, stabilizability to origin, non-zero set point regulation and arbitrary reference tracking topics when discontinuous controllers are employed within the class of systems studied. Advantages of full-state feedback and robust observer design issues to overcome difficulties introduced into control due to the existence of non-smooth nonlinearities are addressed and the benefits of not having to identify the actual nonlinearity to achieve robust performance within the framework of sliding mode controllers are clarified. 

Keywords: Nonlinearity, Nonlinear control, Input nonlinearity, Variable structure control, Tracking control.

## 1. INTRODUCTION

Many physical systems involve an actuation assembly that has some sort of non-smooth characteristics and a plant which is either linear or nonlinear with all the nonlinearities embedded in the plant being smooth. In such a system the non-smooth nonlinearity can be put in a separate block which pre-shapes the designed control input that effectively drives the smooth plant. Handling such systems within the framework of available control theory varies significantly in nature. The major means of compensation for uncertainties at the input channel are the adaptive schemes (Tao, 1996), fuzzy logic controllers and variable structure control schemes (Drakunov *et.al.*, 1997). Being alternatives for one another in many aspects of design and analysis phases, all of those methods demonstrate to yield desired tracking and regulation performance under certain assumptions which differ in each approach.

If the exact pre-inverse exists for a non-smooth nonlinearity at the input channel of a system, a pre-filtering process can compensate for the anomalies caused by the non-smooth nature of the nonlinearity to an extent that can vary from "exact" to "none" depending on the completeness of the information available about the nonlinearity. If the non-smooth nonlinearity can be identified perfectly, a pre-filtering process can be utilized by means of the exact-pre-inverse function so as to annihilate its effects which in return yields an identical response from input to output to that of the system without the nonlinearity. Availability of the exact knowledge of the nonlinearity is relatively unrealistic of an assumption since even though such identification may be possible, it would be costly, unit dependent and possibly time-varying. Nevertheless, the properties of the non-smoothness are likely to change from mildly to drastically (in time) resulting in mismatches between the fixed pre-inverse-nonlinearity model and the nonlinearity itself.

The objective of the classification of non-smooth nonlinearities in a standard form as summarized in (Hatipoğlu, 1997, 1998) was to handle the same control problem utilizing variable structure control laws which would solve the regulation and tracking problems in systems with non-smooth nonlinearities without having to identify the parameters of the nonlinearity -not even the type of it- explicitly. The objective is similar to that of (Tao, 1996), however, the assumptions and the approach remain significantly different.

## 2. NON-SMOOTH ACTUATORS

**Inverse Definitions:** *For a non-smooth nonlinearity given in the form $V = \phi(U)$, the exact pre-inverse is defined as $\phi_p^{-1}(\cdot)$ satisfying $\phi(\phi_p^{-1}(U)) = U$ on $E \subseteq \mathbb{R}$ For a non-smooth nonlinearity given in the form $V = \phi(U)$, the exact post-inverse is defined as $\phi_r^{-1}(\cdot)$ satisfying $\phi_r^{-1}(\phi(U)) = U$ on $E \subseteq \mathbb{R}$ A pseudo-pre-inverse $\phi_{pp}^{-1}(\cdot)$ of a non-smooth nonlinearity $\phi(\cdot)$ on $E \subset \mathbb{R}$ is defined such that for all $U \in E$, there exists $\phi_{pp}^{-1}(U)$ in $E \subset \mathbb{R}$ satisfying $U\phi(\phi_{pp}^{-1}(U)) \geq U^2$.*

Pseudo-pre-inverse can also be referred to as "inverse in the sliding mode sense" because it states that you can find a pre-filtering function that guarantees to achieve a signal of equal or greater magnitude at the output of the nonlinearity. In the case of input nonlinearities where sliding mode type control is used, employment of such a pre-inverse filter would by-pass the effects of the non-smoothness at the input.

**Stability Definitions:** *The motions of a system is called Lagrange stable if starting from any initial conditions $x(t = 0) = x_o$, there exists a $\beta(x_o) \in \mathbb{R}$ such that $||x(t)|| < \beta(x_o)$* Lagrange stability describes the boundedness of all state trajectories in $\mathbb{R}^n$. *The motions of a system is called uniformly ultimately bounded with a bound $\sigma$, if there exists positive constants $\bar{x}, \sigma, t_1 \in \mathbb{R}$ for any initial condition $x(0) = x_o \in E \subset \mathbb{R}^n$ such that for all $t > t_1(x_o)$, $|x| \leq \bar{x}$ and hence $y(t) \leq \sigma$. If, $E = \mathbb{R}^n$ then, the motions of the system is called globally uniformly ultimately bounded g.u.u.b..* Uniform ultimate boundedness imply that after a finite time transient, motions of the system will be contained in a $\bar{x}$ neighborhood, and output will be contained in a $\sigma$ neighborhood.

Consider for now the following SISO linear system

$$\dot{x} = Ax + bu; \qquad y = cx \tag{1}$$

where $x \in \mathbb{R}^n$, $u \in \mathbb{R}$ and $y \in \mathbb{R}$. This system is globally uniformly asymptotically stabilizable to the origin with output feedback $u = -ky = -kcx$ if $\mathcal{K} = \{k \in \mathbb{R} \ : \ (A - bkc) \text{ is Hurwitz}\}$ is nonempty.

**Theorem 1** *The motions of the system (1) which is controllable and observable with a non-smooth actuator nonlinearity of type $u = \phi(v,t)$ can be g.u.u.b. if there exists $k_\ell, k_u \in \mathbb{R}$ with $0 < k_\ell < k_u$ such that $[k_\ell, k_u] \in \mathcal{K}$, $v^* > 0$ and*

$$k_\ell v^2 \leq \phi(v,t)v \leq k_u v^2 \tag{2}$$

*for all $v \in \mathbb{R}$ s.t. $|v| > v^*$ with a linear output feedback law, $v = -ky(t)$ if there exists a single positive definite matrix $P$ that satisfies,*

$$P(A - b\tilde{k}c) + (A - b\tilde{k}c)^T P < -\epsilon I \tag{3}$$

*simultaneously for $\tilde{k} = k_\ell$ and $\tilde{k} = k_u$, $(\exists \epsilon > 0)$.*

Proof is similar in nature to the simultaneous Lyapunov function generation problem, (Khalil, 1992) Consider the Lyapunov function candidate $V(x) = x^T P x$ where $P$ is a symmetric positive definite matrix. Consider $\frac{d}{dt}V(x,t) = x^T(PA + A^T P)x - 2x^T Pb\phi(ky,t)$ For all $v \in \mathbb{R}$ s.t. $|v| > v^*$ for some finite $v^*$, the sector condition (2) is satisfied. Define the time varying gain by,

$$u(t) = \phi(\underbrace{v(t)}_{-ky(t)}, t) = -\kappa(t)y(t) \tag{4}$$

Note that for all $ky > v^*$, $k_\ell \leq \kappa(t) \leq k_u$ which yields $\frac{d}{dt}V(x,t) = x^T[P(A - b\kappa(t)c) + (A - b\kappa(t)c)^T P]x$ which in return reduces the problem into the stability of the time varying system $\dot{x} = (A - b\kappa(t)c)x$ for $\kappa(t)$ taking values in the convex region $k_\ell \leq \kappa(t) \leq k_u$ for all $\{x \in \mathbb{R}^n \ : \ kcx \leq v^*\}$. There exists an $\epsilon > 0$ such that $\frac{d}{dt}V(x,t) < -\epsilon x^T x \ \forall \kappa(t) \in [k_\ell, k_u]$ if and only if $\dot{V}(x,t) < -\epsilon x^T x$ is satisfied for $\kappa(t) = k_\ell$ and $\kappa(t) = k_u$ since $\frac{d}{dt}V(x,t)$ is linear in $\kappa(t)$ and attains a maximum at either of the end-points. Therefore if there exists a positive definite matrix $P$ that satisfies (3) at $k_\ell$ and $k_u$, any the same $P$ will satisfy simultaneously satisfy the same inequality for any gain that lies in the convex region $[k_\ell, k_u]$. Therefore, $\forall v \in \mathbb{R}$ with $|v| > v^*$, motions of (1) can be concluded to exhibit stable dynamics in the Lyapunov sense with $u = \phi(-ky(t))$ type output feedback. Consider the same law for $|ky(t)| \leq v^* < \infty$. The worst case is to have an unstable system for that interval. Then, the unstable modes will be reflected to the output since $(A, c)$ is observable resulting in an instability at the output causing $y$ to grow accordingly. Eventually $\kappa(t)y(t)$ will have to become larger than $v^*$ and the state trajectories can be proven to be stable in the Lyapunov sense once again. For a continuous time system, an infinite frequency switching between stable and unstable structures might occur resulting in the bordering line $y_b \leq \frac{v^*}{k}$ to become an equilibrium point of the nonlinear system. The equilibrium point $y_o$ of the stable system for $|ky| > v^*$ has to satisfy $|ky_o| \leq v^*$ because of (3). Then, the nonlinear system is said to be *g.u.u.b.*.

**Remark:** The inequality (2) is a less restrictive condition than classically defined sector condition. We do not seek asymptotic stabilizability around the origin, therefore, we do not require the non-smooth nonlinearity to be trapped globally by linear functions. This condition only requires $\phi(\cdot,\cdot)$

to be upper and lower bounded by linear functions everywhere except on an interval around the origin of finite Lebesgue measure. The other difference is that, we have very little constraint on the non-smooth nonlinearity. It is not required that $\phi(0,t) = 0$ and also it is not required that $\phi(\cdot,\cdot)$ be memoryless. Actually the sector condition (2) can be replaced with a more generalized form utilizing deadzone type bounding functions. Moreover, the conclusions are valid for any monotonic, unbounding non-smooth nonlinearity with or without dynamics.

**Theorem 2** *Consider the linear SISO system (1) where $(A,b)$ is controllable and $(A,c)$ is observable. Assume that this system has a non-smooth actuator nonlinearity at its input channel. If the motions of this system is g.u.u.b. with linear output feedback $v = -ky$ and stable in the Lyapunov sense for $|v| > v^*$, then the same nonlinear system can be stabilized to the origin by means of an additionally injected discontinuous term, i.e.*

$$v = -ky + w; \qquad w = -W\mathrm{Sign}(S(\cdot)) \qquad (5)$$

*for some large enough $W > 0$ and some manifold $S(\cdot)$ if the underlying plant zero dynamics are stable.*

Note that (5) makes the domain of the inputs to the non-smooth nonlinearity $\mathbb{R} - (-W, W) \subset \mathbb{R}$. Recall that a pseudo-pre-inverse exists for the non-smooth nonlinearity $\phi(\cdot,\cdot)$ meaning that by making $W$ large, one can shrink the image space as well. If $\tilde{W}$ is picked larger than $v^*$ (which is achieved with $w = 0$), then it is guaranteed to have the non-smooth nonlinearity operate exclusively in the region $R = \{v \in \mathbb{R} \ : \ v(t) \geq v^*\}$. If $\tilde{W}$ is picked to ensure this, $W = \tilde{W} + \gamma$ for some $\gamma > 0$ will ensure a discontinuous term at the output of the non-smooth nonlinearity as well due to monotonicity of the signal and existence of a sector condition as in (2). This condition yields upper and lower bounds for the effective magnitude of the discontinuous term at the output of the non-smooth nonlinearity,

$$k_\ell |W| \leq |\phi(W,t)| \leq k_u |W|. \qquad (6)$$

The lower bound is significant. Since a pseudo-pre-inverse exists for unbounding nonlinearities, by making $|W|$ large, $|\phi(W,t)|$ can be made as large as desired. As a result, with the addition of the discontinuous term, output may be steered to zero on the designed manifold $S(\cdot)$ if it is picked such that $\dot{S}(\cdot)$ explicitly depends on the control input, and on the manifold the dynamics stably take the output to zero. The motion of the trajectories are bounded by the linear output feedback term, and hence, bounded discontinuous control is capable of dominating $S\dot{S}$. Since zero dynamics of the underlying plant are assumed stable, regulating the output to zero will yield stable internal dynamics.

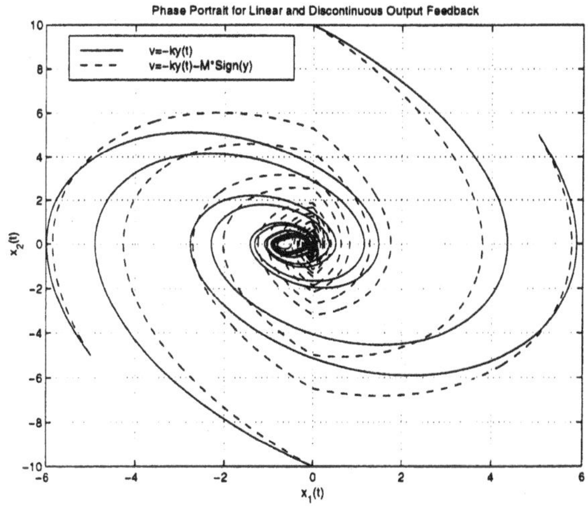


**Figure 1:** The phase portraits for the linear output feedback and discontinuous feedback injection.

**Example:** Consider

$$\begin{aligned} \dot{x}_1 &= x_2; \quad \dot{x}_2 = x_1 - x_2 + u; \qquad (7) \\ u &= \mathrm{backlash}(v, \frac{1}{2}, \frac{3}{2}, 1); \quad y = x_1, \end{aligned}$$

where $x_1$ and $x_2$ are the state variables, $u$ and $v$ are the control inputs as seen by the plant and as designed respectively, $y$ is the output. Note that $A$ is not Hurwitz, however, $(A,b)$ is controllable, $(A,c)$ is observable. Consider a fixed output feedback law $u = -ky$. Then, $(A - bkc)$ is Hurwitz for all $k \in \mathcal{K}$ where $\mathcal{K} = \{k \in \mathbb{R} : k > 1\}$. The linear system with $k = 3$ is asymptotically stable. For this second order system Aizerman's conjecture holds true. Therefore, the linear system with the non-smooth biased backlash type actuator nonlinearity as given in (7) can be concluded to be g.u.u.b. and stable in the Lyapunov sense for large enough $v$ with $k = 3$ for $|v| > v^* \geq \frac{9}{4}$ which is obtained from the intersection of $k_\ell y$ line with the backlash nonlinearity's directional graph. By means of a discontinuous term with $W = \frac{9}{4} + \epsilon$ for any $\epsilon > 0$, origin can be made a globally attractive stable point (See Figure 1).

### *2.1. Decentralized Non-smooth Nonlinearities*

**Corollary 1** *Consider the decentralized MIMO system,*

$$\dot{\boldsymbol{x}} = A\boldsymbol{x} + B\boldsymbol{u}; \quad \boldsymbol{y} = C\boldsymbol{x}; \quad \boldsymbol{u} = \Phi(\boldsymbol{v}, t) \qquad (8)$$

*where $\boldsymbol{x} \in \mathbb{R}^n$, $\boldsymbol{u}, \boldsymbol{y} \in \mathbb{R}^p$ and $\Phi(\boldsymbol{v},t) = \mathrm{diag}(\phi_1(v_1), \phi_2(v_2), \cdots, \phi_p(v_p))$. Assume for the linear part that $(A,B)$ is controllable, $(A,C)$ is observable, and $(A,B,C)$ is decentralized output feedback stabilizable with $\boldsymbol{u} = -\tilde{K}\boldsymbol{y}$ where $\tilde{K}$ is diagonal. Then, the nonlinear system with the decentralized non-smooth actuator nonlinearities*

*$\phi_i(\cdot,\cdot)$ at the input channels can be uniformly ultimately bounded in $\mathbb{R}^n$ with a linear output feedback law $u = -Ky$ for some diagonal $K$ if the condition (9)*

$$k_{\ell_i} v_i^2 \leq \phi_i(v_i, t) v_i \leq k_{u_i} v_i^2 \tag{9}$$

*for all $v_i \in \mathbb{R}$ with $|v_i| \geq v_i^*$ for some $v_i^* > 0$ and for all $i = 1, \cdots, p$ is satisfied and if there exists $P > 0$ satisfying*

$$P(A - Bk^{(i)}C) + (A - Bk^{(i)}C)^T P < -\epsilon I \tag{10}$$

*at the vertices of the convex hypercube*

$$\mathcal{H} = \{k \in \mathbb{R}^p \ : \ k_{\ell_i} \leq k_i \leq k_{u_i}\} \tag{11}$$

*namely $k^{(i)}$ for $i = 1, \cdots, 2^p$, Then, the nonlinear system can be stabilized to the origin with decentralized discontinuous output injection if the underlying plant has stable zero dynamics.*

Simultaneous Lyapunov function generation problem satisfying the negative definiteness property at all vertices can be transformed into a minimax problem, (Khalil, 1996). This method assumes that at least one of the matrices on one of the vertices of the convex hypercube is Hurwitz. Starting from there, a minimization process can be employed for the maximum eigenvalue of the respective matrices, simultaneously. By means of cutting-plane algorithm within nonlinear programming framework and the use of linear matrix inequalities (LMIs), minimax problem can be solved yielding the corresponding $P > 0$.

### *2.2. Finite Domain Results*

All the results reported (as summarized in Theorems 1, 2, and Corollary 1) require a linear control input term to assure global stability of the given system. If the system itself is stable (i.e. $A$ is Hurwitz), then purely discontinuous, bounded control input of sufficient magnitude would stabilize the nonlinear system around the origin. However, unstable plants can not be stabilized globally with bounded inputs, therefore, for such systems if linear term is to be neglected, only "Finite Domain Results" can be achieved. On the other hand, if a pseudo-pre-inverse does not exist for a non-smooth actuator nonlinearity (corresponding to a "bounding" non-smoothness) the sector conditions may not be satisfied globally for all $v > v^*$ for some finite $v^*$. Actually these two problems possess similarities in nature: The first corresponds to a user bounded, the other to a system bounded control input. Nevertheless, the effective control input available to drive the linear system is bounded in either case.

### *2.3. Output Regulation to a Non-zero Set-point*

Consider the SISO linear system (1). Assume that $(A, b)$ is controllable and $(A, c)$ is observable. Additionally assume that $(A, c)$ is reachable. Note that for a nonsingular $A$ matrix, controllability to the origin implies reachability. Let $y_d$ be the desired output set-point for the actual output to follow. Define error as: $\epsilon_y = y - \gamma y_d$ For the linear system a feedback law $u = -k\epsilon_y$ results in the following closed loop system dynamics:

$$\dot{x} = (A - bkc)x - bk\gamma y_d, \qquad y = cx \tag{12}$$

which exhibits the same stability properties around $x_{eq} = (A - bkc)^{-1} bk\gamma y_d$ as does the linear system (1) with $u = -ky$ around $x_{eq} = 0$. Then, the equilibrium point for $y$ becomes $y_{eq} = cx_{eq} = c(A - bkc)^{-1} bk\gamma y_d$. For $y_{eq}$ to be identical to $y_d$, $\gamma$ should be picked as $\gamma = \frac{1}{c(A-bkc)^{-1}bk}$ for some output feedback gain $k$ that stabilizes the unforced system. The observation is that if $\phi(\cdot,\cdot)$ satisfies (2) with the listed assumptions thereafter, the motions of the closed loop nonlinear system can be made *g.u.u.b.* around $y = y_d$ with $u = -k(y - \gamma y_d)$. The proof becomes the same as given for Theorem 1 after making the change of variables: $\tilde{x} = x - (A - bkc)^{-1} bk\gamma y_d$, $\tilde{y} = y - y_d$ and the input defined as $v = -k(\tilde{y})$ which yields,

$$\dot{\tilde{x}} = (A - bkc)\tilde{x}, \qquad \tilde{y} = c\tilde{x} \tag{13}$$

for the linear SISO system. Note that $(x, y)$ to be *g.u.u.b.* around the origin is equivalent to $(\tilde{x}, \tilde{y})$ to be *g.u.u.b.* around $(x_{eq}, y_d)$ with the non-smooth nonlinearity at its input channel. By Theorem 2 the motions of the nonlinear system that can be *g.u.u.b.* around $y = y_d$ by means of a linear output feedback $v = -k(y - \gamma y_d)$ can be globally stabilized to $y = y_d$ by means of a discontinuous term injection based on the error as follows:

$$v = -k(y - \gamma y_d) - W\text{Sign}(\underbrace{y - y_d}_{\epsilon_y}) \tag{14}$$

where $W$ is to be picked exactly the same way as in the controllability to the origin case. Proof is similar to the proof of Theorem 2 with the transformed states. In the case that plant parameters are not known exactly, $\gamma$ can not be determined explicitly. It is always possible to lump the uncertainty into the non-smooth characteristics and simply pick $\gamma = 1$. As long as the non-smoothness satisfies the condition (2) the statements still hold true.

Decentralized MIMO case whose controllability and stability to the origin were studied in Section 2.1 can be shown to exhibit similar properties for non-zero output regulation if reachability assumption is also imposed onto the the underlying linear plant.

### *2.4. Advantages of Full State Feedback*

Should the matching conditions be met, it is shown that a variable structure observer scheme yields in asymptotically convergent error dynamics making full state information available to the controller.

When state feedback is available either explicitly or by means of a robust observer, given a reference signal $y_d$ and all its derivatives, one can define a sliding manifold:

$$S = \epsilon_y^{(p)} + \kappa_{p-1}\epsilon_y^{(p-1)} + \cdots + \kappa_1\dot{\epsilon}_y + \kappa_0\epsilon_y \quad (15)$$

where $\epsilon_y = y - y_d$ is the tracking error, and the polynomial $s^p + \kappa_{p-1}s^{p-1} + \cdots + \kappa_1 s + \kappa_0 = 0$ is asymptotically stable and $p+1$ is the relative degree of the system. Since the relative degree of the system is $p+1$, $\dot{S}$ is dependent on the control input explicitly and a discontinuous control input can enforce $S\dot{S} < 0$ which forces the system trajectories onto the manifold $\sigma = \{x \in \mathbb{R}^n : S(x) = 0\}$ exhibiting asymptotically stable dynamics while steering the error term $\epsilon_y$ to zero.

**Example:** Consider the following linear system with a backlash type non-smooth nonlinearity at its input

$$\begin{aligned} \dot{x}_1 &= x_2;\ \dot{x}_2 = x_3;\ y = x_1, \\ \dot{x}_3 &= -a_1x_1 - a_2x_2 - a_3x_3 + bu \qquad (16) \\ u &= \text{backlash}(v, \delta_1, \delta_2, g) = \phi(v,t). \end{aligned}$$

where $x = [x_1\ x_2\ x_3]^T$ is the state vector, $u$ is the input seen by the linear system, $v$ is the user generated control input, $y$ is the output. Clearly the relative degree of this system is 3. Let the reference signal be given by $y_d(t)$. Assume that full state feedback is available. Then the manifold: $S = \ddot{\epsilon}_y + \kappa_1\dot{\epsilon}_y + \kappa_2\epsilon_y$ with $\epsilon_y = y - y_d$, and a control input $v = -k\epsilon_y - \Omega\text{Sign}(S)$ would guarantee reachability to $S = 0$ in the sliding mode sense, and to $\epsilon_y = 0$ asymptotically if $\Omega$ is picked according to $\Omega >> |k| \cdot ||\epsilon_y||$ and if (since $\phi$ has a pseudo-pre-inverse) it is made sure that $v(t)\phi(v(t)) \geq v(t)\tilde{\Omega}\text{Sign}(S(t))$ and if $\tilde{\Omega} > |a_1| \cdot ||x_1|| + |\kappa_2 - a_2| \cdot ||x_2|| + |\kappa_1 - a_3| \cdot ||x_3|| + |y_d^{(3)} + \kappa_1\ddot{y}_d + \kappa_2\dot{y}_d|$ is satisfied. In the case of non-available full-state feedback, consider the following variable structure observer, (Haskara, 1998):

$$\begin{aligned} \dot{\hat{x}}_1 &= \hat{x}_2 + \ell_1\text{Sign}(y - \hat{y}) \qquad (17) \\ \dot{\hat{x}}_2 &= \hat{x}_3 + \ell_2\text{Sign}([\text{Sign}(y - \hat{y})]_{eq}) \\ \dot{\hat{x}}_3 &= -a_1\hat{x}_1 - a_2\hat{x}_2 - a_3\hat{x}_3 + b\hat{\phi}(v) + \cdots \\ &\cdots + \ell_3\text{Sign}([\text{Sign}([\text{Sign}(y - \hat{y})]_{eq})]_{eq}) \\ \hat{y} &= \hat{x}_1 \end{aligned}$$

where $[\cdot]_{eq}$ denotes the equivalent control of its argument. Let $e$ denote the estimation error. Then, the error dynamics become: $\dot{e}_{x_1} = e_{x_2} - \ell_1\text{Sign}(e_y)$; $\dot{e}_{x_2} = e_{x_3} - \ell_2\text{Sign}([\text{Sign}(e_y)]_{eq})$; $\dot{e}_{x_3} = -a_1e_{x_1} - a_2e_{x_2} - a_3e_{x_3} + be_\phi - \ell_3\text{Sign}([\text{Sign}[\text{Sign}(e_y)]_{eq}]_{eq})$.

If $\ell_1$ is picked large enough, such that $\ell_1 > |e_{x_2}|$ dominating the first equation, $\dot{e}_{x_1}e_{x_1} < 0$ can be enforced and the manifold $e_{x_1} = 0$ can be reached in finite time. Then the equivalent control associated with this equation yields $[\text{Sign}(e_y)]_{eq} = \frac{e_{x_2}}{\ell_1}$. Then, the second equation in the error dynamics above can be forced to exhibit sliding mode like behavior yielding $e_{x_2} = 0$ if $\ell_2$ is picked such that $\ell_2 > |e_{x_3}|$ resulting the other equivalent control of $[\text{Sign}[\text{Sign}(e_y)]_{eq}]_{eq} = \frac{e_{x_3}}{\ell_2}$. Similarly if $\ell_3$ dominates the uncertainty in the equation governing the motions of $e_{x_3}$ stemming from the imprecisely known input nonlinearity, $e_{x_3}$ can also be forced to go to zero in the sliding mode sense. The equivalent controls can be implemented by means of appropriately selected low-pass filters considering the frequency response characteristics of the system in hand. If the gains and the filters are selected appropriately, the errors $e_{x_1}$, $e_{x_2}$ and $e_{x_3}$ go to zero sequentially and all in finite time. Moreover, the equivalent control obtained from the last equation can be used to estimate the non-smooth nonlinearity characteristics since:

$$[\text{Sign}([\text{Sign}[\text{Sign}(e_y)]_{eq}]_{eq})]_{eq} = b(\phi(v) - \hat{\phi}(v)). \quad (18)$$

Assume that the reference signal to be tracked is given by $r(t)$. Then, design a sliding manifold,

$$S = (\hat{x}_3(t) - \ddot{r}(t)) + \kappa_1(\hat{x}_2(t) - \dot{r}(t)) + \kappa_2(y(t) - r(t)) \quad (19)$$

and a discontinuous control input, $u = -ky(t) - M\text{Sign}(S)$, where the linear term serves for global stabilization. Assume that $M$ is picked such that $M > v^*$ and that $S\dot{S} < 0$ is satisfied for the given reference signal. Then, $\dot{S}(t)$ is found as $\dot{S} = -a_1\hat{x}_1 - (a_2 - \kappa_2)\hat{x}_2 - (a_3 - \kappa_1)\hat{x}_3 + b\hat{\phi}(v,t) - (r^{(3)} + \kappa_1\ddot{r} + \kappa_2\dot{r}) + \Psi(e_{x_1}, e_{x_2}, e_{x_3}, e_\phi)$ where $\Psi(e_{x_1}, e_{x_2}, e_{x_3}, e_\phi) \to 0$ in finite time as the estimates of the states converge to their actual counterparts.

In the design, if $\phi(v,t)$ dominates the right hand side in a discontinuous manner, $S\dot{S} < 0$ can be achieved. If the nonlinearity $\phi(v,t)$ has a pseudo-pre inverse, then, picking $|v|$ large enough can make $\phi(v,t)$ magnitude-wise as large as desired by the definition of pseudo-pre inverse. Therefore, tracking error dynamics will eventually satisfy $S = 0$ in Equation (19).

Figure (2) displays the stabilizability to the origin results on a manifold $S = e_{x_3} + 4e_{x_2} + 3e_{x_1}$ starting from an arbitrary initial condition with a backlash type nonlinearity at the input channel. $M = 4$ can not stabilize the system globally. The sole linear output feedback with $k = 3$ though exhibits bounded trajectories, yields oscillatory response when a non-smooth backlash type nonlinearity is present at the input.

Figure (3) displays the simulation results for tracking a sinusoidal reference signal of $r(t) = 4Sin(0.25t)$ in the existence of a backlash nonlinearity at the input. The control law $u = -3(y - r) - M\text{Sign}(S)$ with $S = (\hat{x}_3 - \ddot{r}) + 4(\hat{x}_2 - \dot{r}) + 3(y - r)$ is used where $M = 20$ ensures convergence to the sliding manifold and the $S = 0$ ensures stable asymptotically decaying response in sliding mode

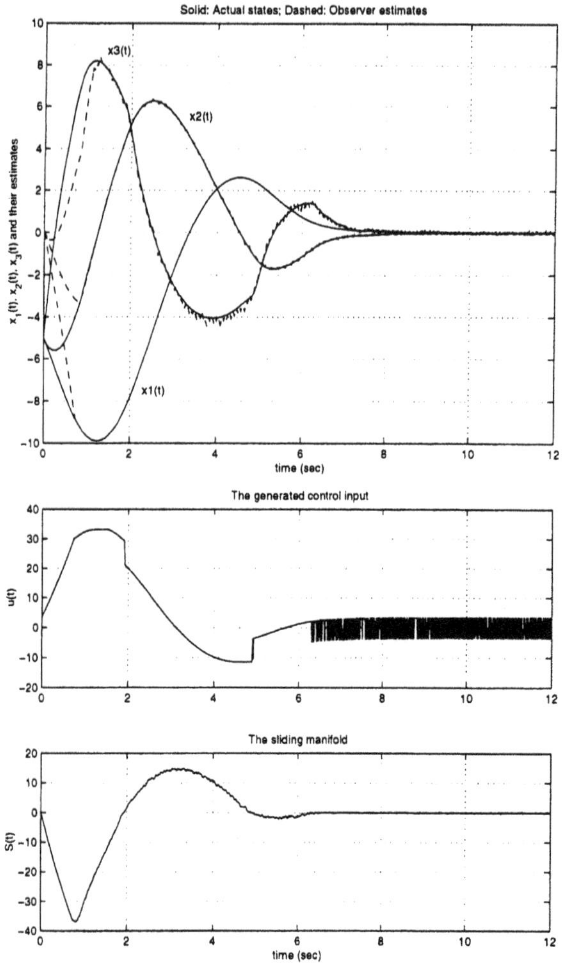


**Figure 2:** The stabilizability to the origin problem: The state trajectories, control input and the sliding manifold.

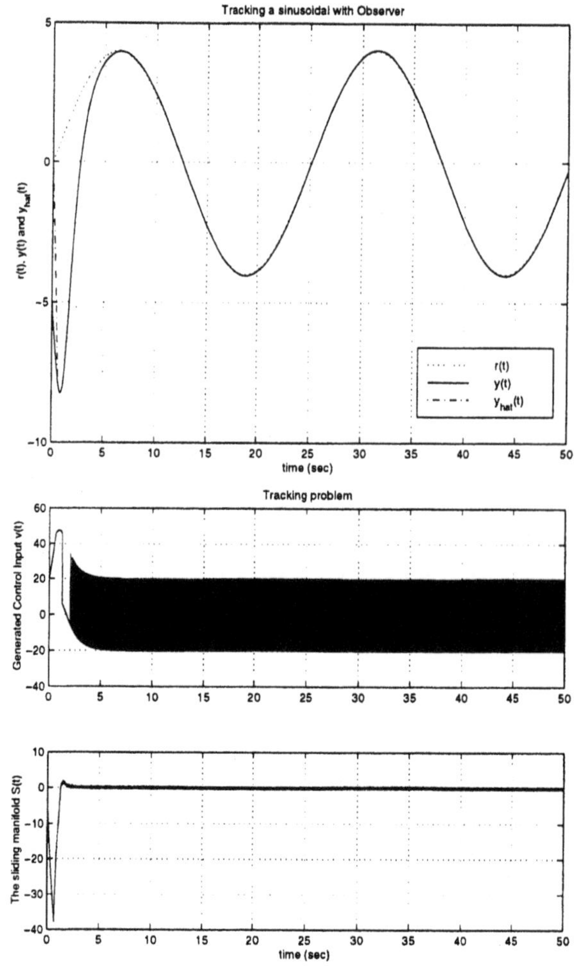


**Figure 3:** The tracking simulations: The output $y(t)$ tracks the reference signal $r(t)$ effectively.

as $y(t) \to r(t)$. Linear term helps bound the trajectories globally.

## 3. CONCLUSIONS

Existence of input nonlinearities often cause undesirable oscillations or stiction at a non-designed manifold with conventional output feedback laws. Input nonlinearities have been shown to be very well suited for variable structure control as the nonlinearity input (and hence its output) can be made discontinuous and rejection of the uncertainty at the input may be attained by means of rich enough control authority. A modified sector condition is required to conclude global attractiveness of the origin along with the existence of a simultaneously satisfied negative definite condition. Robust observer design subject is also investigated and concluded to be feasible by means of equivalent control based VSC observers. Generalized tracking problem is solved by coupled observer/controller design. Some simulation examples are provided to display the performance of the VSC. The conservativeness of the approach is significantly narrowed by means of robust observer coupling and large control authority.

## 4. REFERENCES


S. Drakunov, D. Hanchin, W. C. Su, Ü. Özgüner(1997). "Nonlinear control of a rod-less pneumatic servo-actuator or sliding modes vs coulomb friction", *Automatica*, Vol 33, No 7, pp. 1401-1406.

İ. Haskara, Ü. Özgüner, V.I. Utkin(1998). "On sliding mode observers via equivalent control approach", to appear in *Int. J. of Ctrl.*

C. Hatipoğlu and Ü. Özgüner(1997). *"Controller Design for a Class of Systems with Inherent Right Hand Side Discontinuities"*, Proc. of '97 CDC, San Diego, CA, pp. 4024-4025.

C. Hatipoğlu and Ü. Özgüner(1998). *"Robust Control of Systems Involving Non-smooth Nonlinearities using Multi-Layer Sliding Manifolds"*, Proc. of '98 ACC, Philadelphia, PA.

H.K. Khalil(1992). *Nonlinear Systems*, MacMillan Publishing Company.

G. Tao, P. Kokotovic(1996). *Adaptive Control of Systems with Actuator and Sensor Nonlinearities*, Wiley-Interscience Publication.

F-2c-22-3
14th Triennial World Congress, Beijing, P.R. China

# ON THE OUTPUT REGULATION OF NONLINEAR SYSTEMS BY SLIDING MODE

Claudio Bonivento* Lorenzo Marconi* Zanasi Roberto*

* *University of Bologna, DEIS*
*V. Risorgimento 2, 40136 Bologna, Italy*
*Phone: +39-51-644 30 45, Fax: +39-51-644 30 73*
*E-mail:* cbonivento, lmarconi, rzanasi@deis.unibo.it

Abstract: In this paper the problem of output regulation by error feedback by adopting a Sliding Mode controller is addressed. The reference signal to track, or equivalently the disturbance to reject, is supposed to be generated by a *neutrally stable* autonomous system whose state is supposed to be known. The design procedure of the controller, which holds for nonlinear systems (both minimum and non-minimum phase) having unitary relative degree, as well as simulation results are presented. *Copyright © 1999 IFAC*

Keywords: Sliding Mode Control, Nonlinear Control, Robust Control, Output Feedback.

## 1. INTRODUCTION

This paper deals with the design of a Sliding Mode (SM) controller able to solve the asymptotic output regulation problem for nonlinear systems.

Roughly speaking the problem in question (in the local version) can be stated as follows. Given a nonlinear system and a reference/disturbance signal generated by a known autonomous system (the so-called *exosystem*), design a feedback control law such that, for all sufficiently small values of the overall initial state (namely the initial state of the system/exosystem), the tracking error asymptotically tends to zero with bounded state behaviour. Indeed to solve this problem by adopting a SM strategy, means also to be able to satisfy the above requirements despite the presence of a *matched* disturbance.

In general terms (i.e. without any connection with a SM approach) this problem has been completely addressed in (Isidori and Byrnes, 1990) where the important concept of *internal model* has been extended to a nonlinear framework. In this work, by assuming only knowledge of the tracking error, it is shown that the controller solving the problem is the parallel connection of a *dynamic linear system*, intuitively aiming to force a *steady state* for the overall closed loop system, and a *static nonlinear system* (the internal model) whose goal is to shape such a steady state so that the tracking error is zero.

As far as a SM controller is sought, the works which, by drawing inspiration from (Isidori and Byrnes, 1990), have been proposed in a nonlinear framework, limit the analysis to the so-called *full information* case, namely to the simpler case in which both the state of the system and of the exosystem are accessible for feedback, see, besides others, (Gopalswamy and Hedrick, 1993) and (Marconi *et al.*, 1998). Basically the main idea is to design a sliding surface whose internal dynamics are bounded and admit a *steady state* characterised by tracking error identically zero. In analogy to the solution of the general problem the *static* controller achieving this goal is composed by a linear term of the state of the system (able to enforce a steady state) and a nonlinear term of the state of the exosystem (which aims to shape the steady state to exhibit a tracking error zero). Although the SM approach introduces robustness against matched disturbances, the major criticism which can be done to this result is to require the knowledge of the state of the system (not always available).

This paper aims to solve an interesting (from a practical viewpoint) "intermediate" situation (hereafter called *partial information*) in which only the tracking error and the state of the exosystem (namely only the output of the system and the trajectory to track) are supposed to be known. As commonly done in SM theory, we shall concentrate on nonlinear systems (both minimum and non-minimum phase) characterised by unitary vector relative degree, see (Isidori, 1995). The structure of the controller is composed, in analogy to the results above cited, by a *dynamic linear* unit designed, once the sliding mode has been reached, to enforce a steady state for the overall closed loop system, and by a static nonlinear function of the state of the exosystem (known by assumption) which aim to shape such a steady state to exhibit a zero tracking error.

The paper is organised as follows. In the next Section the statement of the problem of output regulation is proposed together with the basic assumptions done throughout the paper. Section 3 is dedicated to the solution of the "partial information" problem while Section 4 shows simulation results obtained by adopting the proposed approach. Here, in order to reduce the possible influence of chattering due to the sliding control, the DIC technique is implemented, see (Zanasi, 1993). Section 5 concludes the paper with final comments and future activities.

## 2. PROBLEM STATEMENT

In this paper we shall concentrate on the following nonlinear system

$$\mathcal{P}: \begin{cases} \dot{e} = \alpha(e, z, w) + \beta(e, z, w)[u + \delta] \\ \dot{z} = p(e, z, w) + l(e, z, w)[u + \delta] \end{cases} \quad (1)$$

where $e \in \mathbb{R}^m$ denotes the tracking error, $u \in \mathbb{R}^m$ the input, $z \in \mathbb{R}^{n-m}$ the internal state and $w$ the exogenous variable whose dynamics is described by the *exosystem*

$$\dot{w} = s(w) \qquad w \in \mathbb{R}^r$$

assumed *neutrally stable*, namely Lyapunov stable in forward and backward time. The disturbance $\delta$, satisfying the matching condition, takes into account both parametric time-varying uncertainties and external disturbances. It is assumed that $(e, z, w) = (0, 0, 0)$ is an equilibrium point for the overall system and that

$$\beta(0,0,0) = B, \qquad l(0,0,0) = 0$$

where matrix $B$ is non singular. Note that in the formalism (1) can be framed both the "classical" tracking problem, in which a nonlinear system with unitary relative degree, see (Isidori, 1995), is required to track by the output a reference signal $y_d(w)$, as well as the stabilisation problem of uncertain systems in which the uncertain parameter is the exogenous variable $w$.
The local asymptotic output regulation problem refers to the problem of designing a feedback control law such that, for all sufficiently small initial states $(e(0), z(0), w(0))$ and despite the presence of the matched disturbance,

i) the tracking error tends to zero asymptotically, namely $\lim_{t\to\infty} e(t) = 0$;
ii) the overall state behaviour is bounded.

Depending on the amount of information available for feedback, basically three different solutions can be imagined, usually referred as "full information" (in case $(e, z, w)$ are available), "error feedback" (only $e$ available) and "partial information" (both $e$ and $w$ are available). In this paper the latter case is addressed. By referring to the *classical* tracking problem, this means that the goal is the design of a feedback control law aiming to force $e$ to zero provided that the reference trajectory is known. By denoting with

$$\left.\frac{\partial p}{\partial z}\right|_0 = F, \qquad \left.\frac{\partial p}{\partial e}\right|_0 = G, \qquad \left.\frac{\partial \alpha}{\partial z}\right|_0 = C$$

it is assumed that the pairs $(F, G)$ and $(F, C)$ are respectively controllable and observable, namely that system (1) is stabilisable in first approximation by the output $e$ in case the exosystem is disconnected (i.e. with $w = 0$). This assumption, similarly to the analysis carried out in (Isidori and Byrnes, 1990), is necessary to be able to enforce a *steady state* on the closed loop system.

Moreover we shall assume that the following PDE (the so called *regulator equation*)

$$\frac{\partial \lambda}{\partial w} s(w) = p(0, \lambda(w), w) + l(0, \lambda(w), w)c(w)$$

admits a solution for all sufficiently small $w$, namely that "potentially" there exists a steady state $z = \lambda(w)$ for (1), generated by the control input $u = c(w)$, characterised by tracking error identically zero. Note that, since the system in question admits a relative degree (and thus, in particular, it is *left invertible*), the solution $z = \lambda(w)$ is necessarily unique.

As far as the disturbance $\delta$ is concerned, we shall assume that it is bounded by a known bound $\delta$, namely $|\delta(t)| < \Delta \; \forall t$.

## 3. OUTPUT REGULATION BY SLIDING MODE

As above specified, we shall concentrate on the case in which the exogenous variable $w$ is available, namely on the case in which the reference trajectory is perfectly known. Let us refer to the control scheme sketched in Fig. 1: system $\mathcal{R}$

$$\mathcal{R}: \begin{cases} \dot{\tilde{x}} = \tilde{A}\tilde{x} + \tilde{B}\tilde{u} & \tilde{x} = \mathbb{R}^{n-m} \\ u = \tilde{C}\tilde{x} + \tilde{D}\tilde{u} & \tilde{u} = \mathbb{R}^{m} \end{cases}$$

is a linear dynamic system whose state dimension $q = n - m$ is equal to the dimension of the internal dynamic $z$ of (1), while $g(\tilde{x}, w)$ is a smooth nonlinear function of $\tilde{x}$ (which is accessible because it is part of the controller) and $w$ (accessible by assumption).

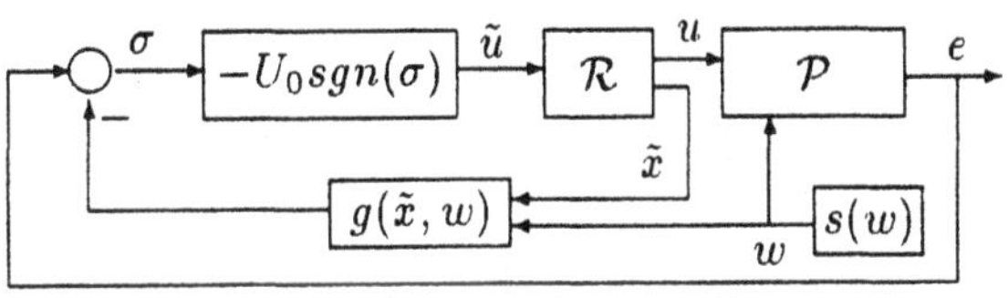


Fig. 1. Control scheme for output regulation.

Roughly speaking, the design methodology hereafter presented is basically based on two steps. In the first step we shall show that with a proper choice of system $\mathcal{R}$ (namely of matrices $\tilde{A}$, $\tilde{B}$, $\tilde{C}$ and $\tilde{D}$) and function $g(\tilde{x}, w)$ (namely of $\partial g/\partial\tilde{x}|_0$) it is possible to stabilise in first approximation the overall closed loop system in the case of exosystem disconnected (i.e. with $w = 0$). This implies that the overall system admits a steady state whose dynamics is a diffeomorphic copy of that of the exosystem and thus bounded by assumption. Then the second step will show how to shape the "remaining structure" of function $g(\tilde{x}, w)$ such that, during the steady state, the tracking error is zero.

By using standard arguments regarding sliding mode theory, see (Utkin, 1992), it is possible to state that, in finite time, the above control scheme is able to force the state on the sliding surface

$$\sigma = e - g(\tilde{x}, w) = 0 \qquad (2)$$

namely to constrain the tracking error $e(t)$ to behave as $g(\tilde{x}(t), w(t))$. The *equivalent control* $\tilde{u}_{eq}$ (representing the equivalent low-frequency control $\tilde{u}$ enforcing $\sigma = 0$) can be obtained from (2) by imposing $\dot{\sigma} = \dot{e} - \dot{g}(\tilde{x}, w) = 0$, namely

$$\dot{\sigma} = \alpha(g, z, w) + \beta(g, z, w)[\tilde{C}\tilde{x} + \tilde{D}\tilde{u}_{eq}] - \frac{\partial g}{\partial \tilde{x}}[\tilde{A}\tilde{x} + \tilde{B}\tilde{u}_{eq}] - \frac{\partial g}{\partial w}s(w) = 0$$

where $g = g(\tilde{x}, w)$, from which

$$\tilde{u}_{eq}(\tilde{x}, z, w) = \left[\frac{\partial g}{\partial \tilde{x}}\tilde{B} - \beta(g, z, w)\tilde{D}\right]^{-1} \cdot \left[\alpha(g, z, w) + \beta(g, z, w)\tilde{C}\tilde{x} - \frac{\partial g}{\partial w}s(w) - \frac{\partial g}{\partial \tilde{x}}\tilde{A}\tilde{x}\right]$$

As above said, the first step is to choose the system $\mathcal{R}$ and the nonlinear function $g(\cdot,\cdot)$ such to asymptotically stabilise in first approximation the overall closed loop system

$$\begin{aligned} \dot{z} &= p(g, z, w) + l(g, z, w)[\tilde{C}\tilde{x} + \tilde{D}\tilde{u}_{eq}] \\ \dot{\tilde{x}} &= \tilde{A}\tilde{x} + \tilde{B}\tilde{u}_{eq} \\ \dot{w} &= s(w) \end{aligned} \qquad (3)$$

where $\tilde{u}_{eq} = \tilde{u}_{eq}(\tilde{x}, z, w)$. To this end, let us define the following matrices

$$\left.\frac{\partial p}{\partial w}\right|_0 = M, \quad \left.\frac{\partial \alpha}{\partial e}\right|_0 = H, \quad \left.\frac{\partial \alpha}{\partial w}\right|_0 = N$$

$$\left.\frac{\partial g}{\partial \tilde{x}}\right|_0 = S, \quad \left.\frac{\partial g}{\partial w}\right|_0 = S_w, \quad \left.\frac{\partial s}{\partial w}\right|_0 = R$$

$$\begin{aligned} \Gamma &= [S\tilde{B} - B\tilde{D}]^{-1} \\ X_x &= HS + B\tilde{C} - S\tilde{A} \\ X_w &= N + HS_w - S_w R \end{aligned}$$

From (3) one obtains the linearised system

$$\begin{bmatrix} \dot{z} \\ \dot{\tilde{x}} \\ \dot{w} \end{bmatrix} = \begin{bmatrix} F & GS & M+GS_w \\ \tilde{B}\Gamma C & \tilde{A}+\tilde{B}\Gamma X_x & \tilde{B}\Gamma X_w \\ 0 & 0 & R \end{bmatrix} \begin{bmatrix} z \\ \tilde{x} \\ w \end{bmatrix}$$

or equivalently, by introducing the new state variables $q = \tilde{x} - z$, the system

$$\begin{bmatrix} \dot{z} \\ \dot{q} \\ \dot{w} \end{bmatrix} = \begin{bmatrix} F+GS & GS & M+GS_w \\ Q_z & Q_q & Q_w \\ 0 & 0 & R \end{bmatrix} \begin{bmatrix} z \\ q \\ w \end{bmatrix} \qquad (4)$$

where

$$\begin{aligned} Q_z &= \tilde{B}\Gamma[C + X_x] + \tilde{A} - F - GS \\ Q_q &= \tilde{A} + \tilde{B}\Gamma X_x - GS \\ Q_w &= \tilde{B}\Gamma X_w - M - GS_w \end{aligned}$$

If matrices $\tilde{A}$ and $\tilde{C}$ are chosen as

$$\begin{aligned} \tilde{A} &= F + GS \\ \tilde{C} &= B^{-1}[S(F + GS) - C - HS] \end{aligned} \qquad (5)$$

then $X_x = -C$, and system (4) becomes

$$\begin{bmatrix} \dot{z} \\ \dot{q} \\ \dot{w} \end{bmatrix} = \begin{bmatrix} F+GS & GS & M+GS_w \\ 0 & F-\tilde{B}\Gamma C & Q_w \\ 0 & 0 & R \end{bmatrix} \begin{bmatrix} z \\ q \\ w \end{bmatrix}$$

Since, by assumption, pairs $(F, G)$ and $(F, C)$ are respectively controllable and observable, the eigenvalues of matrices $F + GS$ and $F - LS$ (with $L = \tilde{B}\Gamma$) can be arbitrarily assigned by suitably choosing matrices $S$ (namely $\partial g/\partial\tilde{x}|_0$) and $L$ (namely $\tilde{B}\Gamma$). Indeed, for a given $L$, it is always possible to determine matrices $\tilde{B}$ and $\tilde{D}$ such that

$$L = \tilde{B}[S\tilde{B} - B\tilde{D}]^{-1} \tag{6}$$

simply by choosing

$$\tilde{B} = LK \qquad \tilde{D} = B^{-1}[SL - I_m]K \tag{7}$$

where $K$ is an arbitrary full rank matrix. Clearly, if one chooses $K = [SL - I_m]^{-1}B$, matrix $\tilde{D}$ becomes the identity matrix, i.e. $\tilde{D} = I_m$. This completes the first part of the controller design. Indeed it has been shown that it is always possible to fix the structure of $\mathcal{R}$ and of the first partial derivative of $g$ with respect to $\tilde{x}$, in such a way to asymptotically stabilise in first approximation the internal dynamics (4) when the exosystem is disconnected (i.e. $w = 0$). Thus, since the exosystem is assumed neutrally stable, the block triangular structure of (4) implies the existence of a steady state dynamics $\tilde{x} = \tilde{\lambda}_g(w)$, $z = \lambda_g(w)$, see (Isidori, 1995), which is solution of the PDEs

$$\begin{aligned}\frac{\partial\lambda_g}{\partial w}s(w) &= p(g(\tilde{\lambda}_g, w), \lambda_g, w) + l(g(\tilde{\lambda}_g, w), \lambda_g, w)\\ &\quad [\tilde{C}\tilde{\lambda}_g + \tilde{D}\tilde{u}_{eq}(\tilde{\lambda}_g, \lambda_g, w)] \\ \frac{\partial\tilde{\lambda}_g}{\partial w}s(w) &= \tilde{A}\tilde{\lambda}_g + \tilde{B}\tilde{u}_{eq}(\tilde{\lambda}_g, \lambda_g, w)\end{aligned} \tag{8}$$

The subscript $g$ in the above steady state function highlights the fact that the solution of (8) necessarily depends on function $g$ whose structure is still partially undetermined. As a matter of fact, remember that up to now only $\partial g/\partial\tilde{x}|_0$ is constrained to assume a well-defined value while the remaining structure of $g(\cdot,\cdot)$ is free. Thus the main effort is now directed to individuate a suitably $g(\cdot,\cdot)$ such that, on one hand, $\partial g/\partial\tilde{x}|_0 = S$ (with $S$ above computed) and, on the other hand, the solutions $\tilde{\lambda}_g(\cdot)$ and $\lambda_g(\cdot)$ comply with a steady state error $e = g(\tilde{\lambda}_g(w), w)$ identically zero. As far as the latter requirement is concerned, the nonlinear function $g(\cdot,\cdot)$ must be chosen so that

$$g(\tilde{\lambda}_g(w), w) = 0$$

for all sufficiently small $w$. To this end, let us choose for $g$ the following structure

$$g(\tilde{x}, w) = S[\tilde{x} - \tilde{\lambda}_g(w)] \tag{9}$$

where $\tilde{\lambda}_g$ is solution of (8) when $g$ is specialised as in (9), namely solution of the PDEs

$$\begin{aligned}\frac{\partial\lambda_g}{\partial w}s(w) &= p(0, \lambda_g, w) + l(0, \lambda_g, w)[\tilde{C}\tilde{\lambda}_g\\ &\quad + \tilde{D}\tilde{u}_{eq}(\tilde{\lambda}_g, \lambda_g, w)] \\ \frac{\partial\tilde{\lambda}_g}{\partial w}s(w) &= \tilde{A}\tilde{\lambda}_g + \tilde{B}\tilde{u}_{eq}(\tilde{\lambda}_g, \lambda_g, w)\end{aligned} \tag{10}$$

where $\lambda_g = \lambda_g(w)$, $\tilde{\lambda}_g = \tilde{\lambda}_g(w)$ and the equivalent control $\tilde{u}_{eq} = \tilde{u}_{eq}(\tilde{\lambda}_g, \lambda_g, w)$ is equal to

$$\begin{aligned}\tilde{u}_{eq} = \left[S\tilde{B} - \beta(0, \lambda_g, w)\tilde{D}\right]^{-1} \Big[\alpha(0, \lambda_g, w) + \\ + \beta(0, \lambda_g, w)\tilde{C}\tilde{\lambda}_g + S\frac{\partial\tilde{\lambda}_g}{\partial w}s(w) - S\tilde{A}\tilde{\lambda}_g\Big]\end{aligned} \tag{11}$$

Note that the choice (9) on one side does not compromise the solvability of (10) (since all the conditions assuring a steady state, and thus the solvability of (8), are still satisfied) and, on the other side, completely fix the structure of the above PDEs. The solution of equations (10) and (11) seems to be quite complicated, but it can be simplified as follows. Combining (11) with the second equation of (10) we obtain the expression

$$\tilde{u}_{eq} = \tilde{D}^{-1}\left[-\beta^{-1}(0, \lambda_g, w)\alpha(0, \lambda_g, w) - \tilde{C}\tilde{\lambda}_g\right] \tag{12}$$

Substituting in (10) we get the following cascade-decoupled system

$$\begin{aligned}\frac{\partial\lambda_g}{\partial w}s(w) &= p(0, \lambda_g, w) - l(0, \lambda_g, w)\beta^{-1}\alpha \\ \frac{\partial\tilde{\lambda}_g}{\partial w}s(w) &= \tilde{A}\tilde{\lambda}_g - \tilde{B}\tilde{D}^{-1}\left[\beta^{-1}\alpha + \tilde{C}\tilde{\lambda}_g\right]\end{aligned}$$

where $\alpha = \alpha(0, \lambda_g, w)$, $\beta = \beta(0, \lambda_g, w)$ and which is equivalent to system

$$\begin{aligned}\frac{\partial\lambda}{\partial w}s(w) &= p(0, \lambda, w) + l(0, \lambda, w)c(w) \\ \frac{\partial\tilde{\lambda}}{\partial w}s(w) &= \tilde{A}\tilde{\lambda} + \tilde{B}\tilde{D}^{-1}\left[c(w) - \tilde{C}\tilde{\lambda}\right]\end{aligned} \tag{13}$$

where $\lambda = \lambda(w) = \lambda_g(w)$, $\tilde{\lambda} = \tilde{\lambda}(w) = \tilde{\lambda}_g(w)$ and

$$c(w) = -\beta^{-1}(0, \lambda(w), w)\alpha(0, \lambda(w), w) \tag{14}$$

The equation (13.a) represents the regulator equation of the original plant and its solution (which exists by assumption), namely $z = \lambda(w)$ and $u = c(w)$, identifies respectively the unique steady state and the unique control input compatible with a steady state error identically zero. Similarly, the equation (13.b) represents the regulator equation for the linear system $\mathcal{R}$, namely identifies the unique steady state $\tilde{x} = \tilde{\lambda}(w)$ compatible with a steady state error identically zero. Note that the uniqueness of the solutions of system (13) derives for the first one from the assumption on the existence of the relative degree (which implies the left invertibility of the system) and, for the second one, from the nonsingularity of the matrix $\tilde{D}$ (always achieved by a proper choice of the matrix $K$ in (7)) which implies the invertibility of $\mathcal{R}$. Thus, since by construction also (10) provide a steady state for the internal dynamics $(\tilde{x}, z)$ characterised by a tracking error identically zero, it turns out that necessarily $\tilde{\lambda}_g(\cdot)$ $\lambda_g(\cdot)$ solution of the (coupled) equations (10) locally coincides respectively with $\lambda(\cdot)$ $\tilde{\lambda}(\cdot)$ solution of the (decoupled) equations (13).

Summarising, all the steps needed to computed the sliding mode controller able to achieve the output regulation in case of reference signal completely known are:

1) Choose matrices $S$, $L$, $\tilde{A}$, $\tilde{B}$, $\tilde{C}$ and $\tilde{D}$ as in (5), (6) and (7) in order to asymptotically stabilise in first approximation the internal dynamics of the system when the exosystem is disconnected;
2) Solve equation (13.a) and in particular find the unique control law $u = c(w)$ able to enforce a steady state error identically zero;
3) Solve equation (13.b) and find the unique steady state solution $\tilde{x} = \tilde{\lambda}(w)$ which is compatible with a steady state error identically zero;
4) Complete the structure of $g(\cdot,\cdot)$ as follows:

$$g(\tilde{x}, w) = S[\tilde{x} - \tilde{\lambda}(w)]. \qquad (15)$$

**Remark.** Note that, in general, the linear system $\mathcal{R}$, which is part of the regulator, *does not* act as observer of the unknown internal states $z$ of the controlled system because of the presence of *higher order terms* and exosystem state $w$, but it simply acts as a local stabiliser for the overall closed loop system.

## 4. SIMULATION RESULTS

Let us consider the following single-input single-output nonlinear system

$$\begin{cases} \dot{x} = -x - 2x^2 + z + u \\ \dot{z} = z + x \end{cases} \qquad (16)$$

having unstable internal dynamics, relative degree one and output $y = x$. Let $y_d = w_1$ denote the sinusoidal reference signal generated by the following linear exosystem $\dot{w} = s(w)$:

$$\begin{bmatrix} \dot{w}_1 \\ \dot{w}_2 \end{bmatrix} = \begin{bmatrix} 0 & \Omega \\ -\Omega & 0 \end{bmatrix} \begin{bmatrix} w_1 \\ w_2 \end{bmatrix} \qquad (17)$$

and let $e = y_d - y = w_1 - x$ denote the tracking error. Accordingly, system (16) can be rewritten as follows

$$\begin{cases} \dot{e} = -e + 2e^2 - 4ew_1 + w_1 \\ \qquad +\Omega w_2 + 2w_1^2 - z - u \\ \dot{z} = z - e + w_1 \end{cases} \qquad (18)$$

This system is characterised by the matrices $B = -1$, $F = 1$, $G = -1$, $C = -1$ and $H = -1$. The internal dynamics of the system is made asymptotically stable in first approximation with eigenvalues $\lambda_{1,2} = -3$ by choosing: $S = 4$, $L = -4$, $K = 0.0588$, $\tilde{A} = -3$, $\tilde{B} = -0.2353$, $\tilde{C} = 7$ and $\tilde{D} = 1$. The solution $z = \lambda(w)$ of equation

$$\frac{\partial \lambda(w)}{\partial w} s(w) = \lambda(w) + w_1$$

is $\lambda(w) = \alpha w_1 + \beta w_2$ where

$$\alpha = \frac{-1}{1+\Omega^2}, \qquad \beta = \frac{-\Omega}{1+\Omega^2}$$

while, from (14), the control law $u = c(w)$ is

$$c(w) = 2w_1^2 + (1-\alpha)w_1 + (\Omega - \beta)w_2$$

The solution $\tilde{x} = \tilde{\lambda}(w)$ of the equation (13.b) is

$$\tilde{\lambda}(w) = a_1 w_1 w_2 + a_2 w_1 + a_3 w_2 + a_4 w_1^2 + a_5 w_2^2$$

where the parameters $a_i$ $(i = 1, 2, \ldots, 5)$ are uniquely defined as functions of parameters $\tilde{A}$, $\tilde{B}$, $\tilde{C}$, $\tilde{D}$ and $\Omega$. When $\Omega = 2$ it results $a_1 = 0.1056$, $a_2 = -0.2592$, $a_3 = -0.0342$, $a_4 = -0.1918$, $a_5 = -0.1561$, and the sliding surface $\sigma = e - g(\tilde{x}, \tilde{\lambda}(w))$ is completely determined according to definition (15). The simulation results reported in the following refer to the case of zero initial conditions for the system and initial conditions $w(0) = [0, 0.01]^T$ for the exosystem. Two different implementations of the switching control law $\tilde{u}(t)$ have been considered and compared in simulation: a) the classical Variable Structure (VS) control law, and b) the modified Discontinuous Integral Control law (DIC) described in (Zanasi, 1993).

The simulation results obtained when the DIC implementation is used with $k = 0.001$, $h = 10$, and $\lambda = 2000$, are shown in Fig. 2, Fig. 3 and Fig. 4.

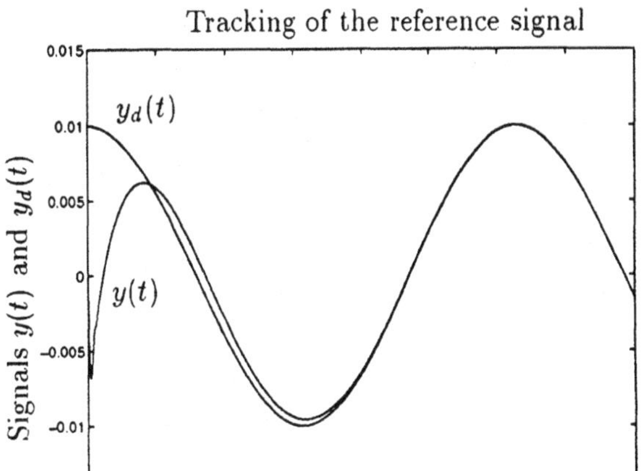


Fig. 2. DIC implementation: output variable $y = x$ tracking the reference signal $y_d = w_1$.

The output variable $y = x$ and the reference signal $y_d = w_1$ are shown in Fig. 2 from which it is evident that the tracking error $e(t)$ tends to zero.

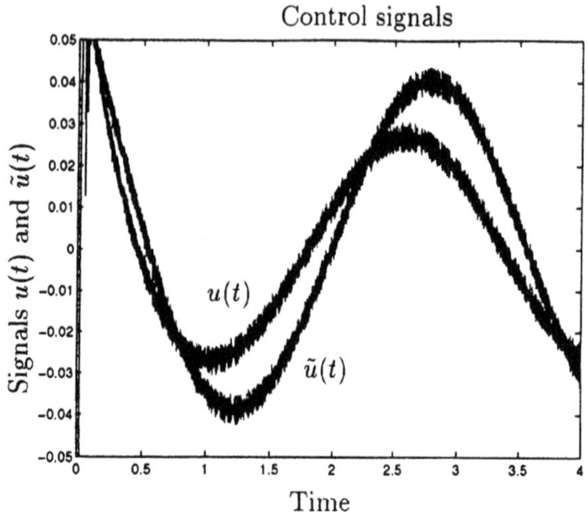


Fig. 3. DIC implementation: control signals $u(t)$ and $\tilde{u}(t)$.

The control signals $u(t)$ and $\tilde{u}(t)$ are reported in Fig. 3: note that the sliding condition $\sigma = 0$ is reached and maintained even if the amplitude $k$ of the switching term is much smaller than the amplitude of the equivalent control $\tilde{u}_{eq}$. This is possible because of the presence in the DIC structure of the integral term which provides a good estimate of $\tilde{u}_{eq}$ so allowing its partial feedforward compensation.

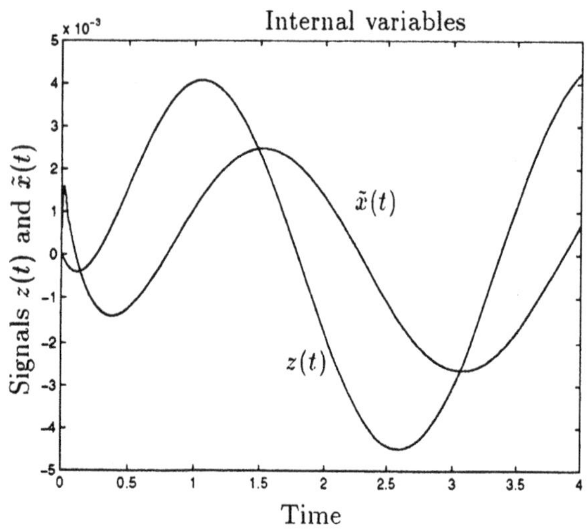


Fig. 4. DIC implementation: internal variable $z(t)$ of the controlled system and state $\tilde{x}$ of the linear system $\mathcal{R}$.

The fact that the linear system $\mathcal{R}$ does not act as observer of the unknown internal states of the controlled system is evident in Fig. 4 where the two variables $z(t)$ and $\tilde{x}(t)$ are reported.

For reasons of space, the simulation results obtained by adopting the classical VS implementation are not reported. The major difference with respect to the above behaviours, is in the amplitude $U_0$ of the switching control action which, for the classical VS implementation, turns out to be approximatively fifty times the DIC one.

## 5. CONCLUSIONS

The paper has addressed the problem of output regulation of nonlinear systems in Sliding Mode terms, both in minimum and non-minimum phase case. It has been shown that the controller solving the problem is the parallel connection of a dynamic linear system, intuitively aiming to force a steady state for the overall closed loop system, and a static nonlinear system whose goal is to shape such a steady state so that the tracking error is zero.

While this paper has referred to systems with relative degree one, the case of higher relative degree represents an interesting point to be investigated in the future by using a similar approach.

## 6. REFERENCES


Bonivento C., R. Zanasi (1996). Advances in Variable Structure Control. In *Colloquium on Automatic Control.* (C. Bonivento, G. Marro and R. Zanasi Eds), Springer, LNCIS, No. 215.

Bonivento C., M. Sandri, R. Zanasi (1998). Discrete Variable Structure Integral Controllers. *Automatica*, Vol. 54, No. 3, pp. 355–361.

Byrnes C.I., F. Delli Priscoli, A. Isidori (1997). *Output Regulation of Uncertain Nonlinear Systems*, Birkhäuser, Boston.

Gopalswamy S., J. K. Hedrick (1993). Tracking Nonlinear Non-Minimum Phase Systems Using Sliding Control. *Int. J. Control.* Vol. 57, No. 5, pp. 1141–1158.

Isidori A., C.I. Byrnes (1990). Output Regulation of Nonlinear System. *IEEE Trans. on Aut. Cont.*. Vol. 35, No. 2, pp. 131–140.

Isidori A. (1995). *NonLinear Control System.* Springer-Verlag, NewYork, 3rd Edition.

Marconi L., C. Melchiorri, R. Zanasi (1998). Sliding Mode Output Regulation of Linear and Nonlinear Systems With Relative Degree One. *IEEE 6th Mediterranean Conference on Control and Automation.* Alghero.

Nersisian A. and R. Zanasi (1993). A Modified Variable Structure Control Algorithm for Stabilisation of Uncertain Dynamical Systems. *Int. J. of Robust and Nonlinear Control*, Vol. 3, pp. 199–209.

Sira-Ramirez H. (1993). On the Dynamical Sliding mode Control of Nonlinear Systems. *Int. J. Control.* Vol. 57, No. 5, pp. 1039–1061.

Utkin V.I. (1992). *Sliding Modes in Control Optimization*, Springer-Verlag, Berlin.

Zanasi R. (1993). Sliding Mode Using Discontinuous Control Algorithms of Integral Type. *Int. Journal of Control*, Special Issue on Sliding Mode Control. Vol. 57, No. 5, pp. 1079–1099.

14th Triennial World Congress, Beijing, P.R. China





# NONLINEAR SLIDING SURFACE DESIGN IN THE PRESENCE OF UNCERTAINTY*

**A. Loukianov*, B. Castillo-Toledo* and S. J. Dodds+**

**CINVESTAV-IPN, Unidad Guadalajara*
*A. P. 31-438, Pza. La Luna, Col. Verde Valle,*
*Guadalajara, Jalisco, México. C. P. 44550.*
*E-mail: toledo@gdl.cinvestav.mx*
+*Department of Electrical and Electronic Engineering,*
*University of East London*
*Longbridge Road, Dagenham, Essex, RM8 2AS U.K.*
*E-mail: dodds@uel.ac.uk*

**Abstract.** This paper combines the block control, sliding mode and high gain robust control techniques to derive a stabilizing control law for nonlinear systems with matched and unmatched uncertainties. *Copyright © 1999 IFAC*

**Keywords.** Nonlinear uncertain systems, decomposition methods, sliding mode control.

## 1. INTRODUCTION

Sliding mode control enables decomposition of the design problem into two independent stages - (i) selection of those discontinuity surfaces with the desired sliding motion, and (ii) enforcing the sliding mode along this manifold using discontinuous control (Utkin, 1981). This paper is dedicated to the first stage - selection of discontinuity surfaces for nonlinear time-varying systems with modelling uncertainties and unknown disturbances.

Consider the following general nonlinear system subject to a disturbance

$$\dot{\mathbf{x}} = \mathbf{f}(\mathbf{x},t)+\mathbf{B}(\mathbf{x},t)\mathbf{u}+\Delta\mathbf{f}(\mathbf{x},t)+\mathbf{D}(\mathbf{x},t)\mathbf{w}(t) \quad (1)$$

where $\mathbf{x} \in X \subset \Re^n, \mathbf{u} \in U \subset \Re^m, \mathbf{w} \in W \subset \Re^s$. The vector field $\mathbf{f}$, and the columns of $\mathbf{B}$, and $\mathbf{D}$ are smooth mappings of class $C^{\infty}_{[t,\infty)}$, and $\mathbf{f}(0,t) = \Delta\mathbf{f}(0,t) = 0$. The unknown mapping $\Delta\mathbf{f}$ represents the model uncertainties and $\mathbf{w}(t)$ is a vector of external disturbances.

Sliding mode should be generated on a smooth manifold $\mathbf{s}(\mathbf{x}) = 0$, $\mathbf{s} \in S \subset \Re^m$ so that the sliding mode dynamics

$$\dot{\mathbf{x}} = [\mathbf{I}_n + \mathbf{B}(\mathbf{GB})^{-1}](\mathbf{f}+\Delta\mathbf{f}+\mathbf{Dw}), \mathbf{G} = \frac{\partial \mathbf{s}}{\partial \mathbf{x}} \quad (2)$$

obtained by using *equivalent control method* (Utkin, 1981), to be stable. In this work, a solution to the problem of sliding dynamics (2) stabilization is presented. The block control method (Loukianov and Utkin, 1981, 1982) is applied to reduce the original system (1) by a nonsingular transformation $\bar{\mathbf{x}} = \mathbf{\Phi}(\mathbf{x},t)$ to the *block controllable form* (BC-form) with uncertainty:

$$\begin{aligned} \dot{\mathbf{x}}_i &= \mathbf{f}_i(\bar{\mathbf{x}}_i,t) + \mathbf{B}_i(\bar{\mathbf{x}}_i,t)\mathbf{x}_{i-1} \\ &\quad +\Delta\mathbf{f}_i(\bar{\mathbf{x}}_i,t) + \mathbf{D}_i(\bar{\mathbf{x}}_i,t)\mathbf{w}(t), \\ i &= 2,...,r, \qquad (3) \\ \dot{\mathbf{x}}_1 &= \mathbf{f}_1(\bar{\mathbf{x}}_1,t) + \mathbf{B}_1(\bar{\mathbf{x}}_1,t)\mathbf{u} \\ &\quad +\Delta\mathbf{f}_1(\bar{\mathbf{x}}_1,t) + \mathbf{D}_1(\bar{\mathbf{x}}_1,t)\mathbf{w}(t), \end{aligned}$$

*Work supported by Mexican Consejo Nacional de Ciencia y Tecnología under grant 26358-A where $\bar{\mathbf{x}} = (\mathbf{x}_1, \mathbf{x}_2, ...., \mathbf{x}_r)^T$, and $\mathbf{x}_i$ is a $n_i \times 1$ vector, and $\bar{\mathbf{x}}_i = (\mathbf{x}_i, ..., \mathbf{x}_r)^T$, $i = 1, ..., r-1$ and $\bar{\mathbf{x}}_r = \mathbf{x}_r$ ; rank $\mathbf{B}_i = n_i\ \forall \mathbf{x},\ t \geq 0$. The integers $(n_1, n_2, ..., n_r)$ are defined as the controllability indices of the system (1). The control law synthesis problem can be solved for each block of BC-form (3) independently, and then, the Lyapunov technique is used to achieve stability of the sliding mode dynamics.

## 2. BC-FORM ALGORITHM

The procedure for obtaining the BC-form (3) from the system (1) consists of a series of steps.

STEP1. Assume, without loss of generality, that rank $\mathbf{B}(\mathbf{x}, t) = n_1 < m\ \forall x$, and $t \geq 0$ and, possibly after reordering, there exists an $(n_1 \times m)$ block $\mathbf{B}_1(\mathbf{x}_1, \mathbf{x}_{12}, t)$ such that

$$\mathbf{B}(\mathbf{x}, t) = \begin{pmatrix} \mathbf{B}_{12}(\mathbf{x}_1, \mathbf{x}_{12}, t) \\ \mathbf{B}_1(\mathbf{x}_1, \mathbf{x}_{12}, t) \end{pmatrix}$$

with rank $\mathbf{B}_1(\mathbf{x}_1, \mathbf{x}_{12}, t) = n_1\ \forall \mathbf{x}$, and $t \geq 0$, where $\mathbf{x}_1$ and $\mathbf{x}_{12}$ are $n_1 \times 1$ and $(n - n_1) \times 1$ vectors, respectively. At this point, the following assumptions will be carried for each step of the procedure.

*A11.* The Pfaffian system

$$\mathbf{dx}_{12} - \mathbf{B}_{12}\mathbf{B}_1^{+}\mathbf{dx}_1 = 0 \tag{4}$$

where $\mathbf{B}_1^{+}$ is the right pseudoinverse of $\mathbf{B}_1$ and $t$ is a parameter, is completely integrable.

*A12.* The unknown mappings $\Delta\mathbf{f}(\mathbf{x}, t)$ and $\mathbf{D}(\mathbf{x}, t)$ can be decomposed in the form

$$\begin{aligned} \Delta\mathbf{f}(\mathbf{x}, t) &= \Delta\mathbf{f}^m(\mathbf{x}_1, \mathbf{x}_{12}, t) + \Delta\mathbf{f}^u(\mathbf{x}_{12}, t) \\ \mathbf{D}(\mathbf{x}, t) &= \mathbf{D}^m(\mathbf{x}_1, \mathbf{x}_{12}, t) + \mathbf{D}^u(\mathbf{x}_{12}, t) \end{aligned}$$

where $\Delta\mathbf{f}^m(\mathbf{x}, t)$ and $\mathbf{D}^m(\mathbf{x}, t)$ satisfy the matching condition, namely

$$\Delta\mathbf{f}^m(x, t), \mathbf{D}^m(\mathbf{x}, t) \in \text{span}\ \mathbf{B}(\mathbf{x}, t).$$

Under assumption A11, it is possible to show that a solution of the equation (4) is given by $\mathbf{x}_{12} = \varphi_1(\mathbf{x}, t, \mathbf{c})$ where $\mathbf{c} = (c_1, c_2, ...., c_{n-n_1})^T$ is a integration constants vector. From this solution, the vector $\mathbf{c}$, can be obtained as $\mathbf{c} = \varphi_1(\mathbf{x}_1, \mathbf{x}_{12}, t)$, and taken as a local change of coordinates given by $\mathbf{x}_2' = \varphi_1(\mathbf{x}_1, \mathbf{x}_{12}, t)$. Now, using assumption A12, and partitioning $\mathbf{f}$, $\Delta\mathbf{f}$ and $\mathbf{D}$ as

$$\begin{aligned} \mathbf{f} &= \begin{pmatrix} \mathbf{f}_{12}(\mathbf{x}, t) \\ \mathbf{f}_1(\mathbf{x}, t) \end{pmatrix}; \Delta\mathbf{f} = \begin{pmatrix} \Delta\mathbf{f}_{12}(\mathbf{x}, t) \\ \Delta\mathbf{f}_1(\mathbf{x}, t) \end{pmatrix} \\ &= \begin{pmatrix} \Delta\mathbf{f}_{12}^m(\mathbf{x}_1, \mathbf{x}_{12}, t) + \Delta\mathbf{f}_{12}^u(\mathbf{x}_{12}, t) \\ \Delta\mathbf{f}_1^m(\mathbf{x}_1, \mathbf{x}_{12}, t) + \Delta\mathbf{f}_1^u(\mathbf{x}_{12}, t) \end{pmatrix}; \\ \mathbf{D} &= \begin{pmatrix} \mathbf{D}_{12}(\mathbf{x}, t) \\ \mathbf{D}_1(\mathbf{x}, t) \end{pmatrix} \\ &= \begin{pmatrix} \mathbf{D}_{12}^m(\mathbf{x}_1, \mathbf{x}_{12}, t) + \mathbf{D}_{12}^u(\mathbf{x}_{12}, t) \\ \mathbf{D}_1^m(\mathbf{x}_1, \mathbf{x}_{12}, t) + \mathbf{D}_1^u(\mathbf{x}_{12}, t) \end{pmatrix}, \end{aligned}$$

it is possible to show that the system (1) under the new coordinates is described by

$$\begin{aligned} \dot{\mathbf{x}}_2' &= \mathbf{f}_2'(\mathbf{x}_1, \mathbf{x}_2', t) \\ &\quad + \Delta\mathbf{f}_2'(\mathbf{x}_2', t) + \mathbf{D}_2'(\mathbf{x}_2', t)\mathbf{w}(t), \qquad (5) \\ \dot{\mathbf{x}}_1 &= \mathbf{f}_1(\mathbf{x}_1, \mathbf{x}_2', t) + \mathbf{B}_1(\mathbf{x}_1, \mathbf{x}_2', t)\mathbf{u} \\ &\quad + \Delta\mathbf{f}_1(\mathbf{x}_1, \mathbf{x}_2', t) + \mathbf{D}_1(\mathbf{x}_1, \mathbf{x}_2', t)\mathbf{w}(t). \end{aligned}$$

The following assumption is fundamental to derive the BC-form:

*A13.* The mapping $\mathbf{f}_2'(\mathbf{x}_1, \mathbf{x}_2', t)$ (5) is affine on its first argument, namely

$$\mathbf{f}_2'(\mathbf{x}_1, \mathbf{x}_2', t) = \mathbf{f}_2''(\mathbf{x}_2', t) + \mathbf{B}_2'(\mathbf{x}_2', t)\mathbf{x}_1.$$

At this point, the following three cases are possible:

(*i*) rank $\mathbf{B}_2'(\mathbf{x}_2', t) = 0, \forall \mathbf{x},\ t \geq 0$, which is equivalent to have an uncontrollable system. For the purposes of this work, assume in the following that the system is locally controllable.

(*ii*) rank $\mathbf{B}_2'(\mathbf{x}_2', t) = n_2\ \ \forall \mathbf{x}_2', t \geq 0$ and $n_2 = n - n_1$. In this case, after defining $\mathbf{x}_2 = \mathbf{x}_2'$, $\mathbf{B}_2 = \mathbf{B}_2'$, $\mathbf{f}_2 = \mathbf{f}_2'$, $\Delta\mathbf{f}_2 = \Delta\mathbf{f}_2'$, $\mathbf{D}_2 = \mathbf{D}_2'$, the BC-form is

$$\begin{aligned} \dot{\mathbf{x}}_2 &= \mathbf{f}_2(\mathbf{x}_2, t) + \mathbf{B}_2(\mathbf{x}_2, t)\mathbf{x}_1 \\ &\quad \Delta\mathbf{f}_2(\mathbf{x}_2, t) + \mathbf{D}_2(\mathbf{x}_2, t)\mathbf{w}(t) \\ \dot{\mathbf{x}}_1 &= \mathbf{f}_1(\mathbf{x}_1, \mathbf{x}_2, t) + \mathbf{B}_1(\mathbf{x}_1, \mathbf{x}_2, t)\mathbf{u} \\ &\quad + \Delta\mathbf{f}_1(\mathbf{x}_1, \mathbf{x}_2, t) + \mathbf{D}_1(\mathbf{x}_1, \mathbf{x}_2, t)\mathbf{w}. \end{aligned}$$

(*iii*) In the case of $n_2 < n - n_1$, subsequent step is necessary, and the system (5) with state $\mathbf{x}_2'$ and input $\mathbf{x}_1$ must be further decomposed and transformed.

STEP K. Consider the system obtained at $(k-1)$ step

$$\begin{aligned} \dot{\mathbf{x}}_k' &= \mathbf{f}_k'(\mathbf{x}_k', t) + \mathbf{B}_k'(\mathbf{x}_k', t)\mathbf{x}_{k-1} \\ &\quad + \Delta\mathbf{f}_k'(\mathbf{x}_k', t) + \mathbf{D}_k'(\mathbf{x}_k', t)\mathbf{w}(t) \qquad (6) \\ \dot{\mathbf{x}}_i &= \mathbf{f}_i(\hat{\mathbf{x}}_i, t) + \mathbf{B}_i(\hat{\mathbf{x}}_i, t)\mathbf{x}_{i-1} \\ &\quad + \Delta\mathbf{f}_i(\hat{\mathbf{x}}_i, t) + \mathbf{D}_i(\hat{\mathbf{x}}_i, t)\mathbf{w}(t) \\ i &= 2, ..., k-1, \qquad (7) \\ \dot{\mathbf{x}}_1 &= \mathbf{f}_1(\hat{\mathbf{x}}_1, t) + \mathbf{B}_1(\hat{\mathbf{x}}_1, t)\mathbf{u} \\ &\quad + \Delta\mathbf{f}_1(\hat{\mathbf{x}}_1, t) + \mathbf{D}_1(\hat{\mathbf{x}}_1, t)\mathbf{w}(t). \qquad (8) \end{aligned}$$

where $\hat{\mathbf{x}}_i = (\mathbf{x}_i, ..., \mathbf{x}_{k-1}, \mathbf{x}_k')^T$, $i = 1, ..., k-1$. If $n_k$ =rank $\mathbf{B}_k(\mathbf{x}_k', t)$ is such that $n_k = n - \sum_{j=1}^{k-1} n_j$, then define $\mathbf{x}_k = \mathbf{x}_k'$, $\mathbf{f}_k = \mathbf{f}_k'$, $\Delta\mathbf{f}_k = \Delta\mathbf{f}_k'$, $\mathbf{D}_k = \mathbf{D}_k'$, and the algorithm terminates giving equations (6)-(8) as the desired BC-form. If $n_k < n - \sum_{j=1}^{k-1} n_j$, the subsystem (6) is partitioned as

$$\begin{aligned} \dot{\mathbf{x}}_{k2} &= \mathbf{f}_{k2}(\mathbf{x}_k, \mathbf{x}_{k2}, t) + \mathbf{B}_{k2}(\mathbf{x}_k, \mathbf{x}_{k2}, t)\mathbf{x}_{k-1} \\ &\quad + \Delta\mathbf{f}_{k2}(\mathbf{x}_k, \mathbf{x}_{k2}, t) + \mathbf{D}_{k2}(\mathbf{x}_k, \mathbf{x}_{k2}, t)\mathbf{w}(t), \\ \dot{\mathbf{x}}_k &= \mathbf{f}_k(\mathbf{x}_k, \mathbf{x}_{k2}, t) + \mathbf{B}_k(\mathbf{x}_k, \mathbf{x}_{k2}, t)\mathbf{x}_{k-1} \\ &\quad + \Delta\mathbf{f}_k(\mathbf{x}_k, \mathbf{x}_{k2}, t) + \mathbf{D}_k(\mathbf{x}_k, \mathbf{x}_{k2}, t)\mathbf{w}(t). \end{aligned}$$

where rank $\mathbf{B}_k = n_k$, $\mathbf{x}'_k = (\mathbf{x}_k, \mathbf{x}_{k2})^T$ and $\mathbf{x}_k$ and $\mathbf{x}_{k2}$ are $n_k \times 1$ and $(n - \sum_{j=1}^{k-1} n_j - n_k) \times 1$ vectors, respectively. For this step, the assumptions A11 and A12 are generalized as follows:

*Ak1.* The Pfaffian system

$$\mathbf{dx}_{k2} - \mathbf{B}_{k2}\mathbf{B}_k^+\mathbf{dx}_k = 0 \tag{9}$$

is completely integrable.

*Ak2.* The unknown mappings $\Delta\mathbf{f}'_k$ and $\mathbf{D}'_k$ can be decomposed in the form

$$\begin{aligned}\Delta\mathbf{f}'_k(\mathbf{x}'_k,t) &= \Delta\mathbf{f}^m_k(\mathbf{x}_k,\mathbf{x}_{k2},t) + \Delta\mathbf{f}^u_k(\mathbf{x}_{k2},t)\\ \mathbf{D}'_k(\mathbf{x}'_k,t) &= \mathbf{D}^m_k(\mathbf{x}_k,\mathbf{x}_{k2},t) + \mathbf{D}^u_k(\mathbf{x}_{k2},t)\end{aligned}$$

where $\Delta\mathbf{f}^m_k$ and $\mathbf{D}^m_k$ satisfy the matching conditions, namely

$$\Delta\mathbf{f}^m_k, \mathbf{D}^m_k \in \text{span } \mathbf{B}'_k(\mathbf{x}'_k,t).$$

Proceeding as in the first step, under the previous assumptions, it is found a local change of coordinates $\mathbf{x}'_{k+1} = \varphi_k(\mathbf{x}_k,\mathbf{x}_{k2},t)$ where $\varphi_k$ is computed from the solution of (9), such that the system is described by

$$\begin{aligned}\dot{\mathbf{x}}'_{k+1} &= \mathbf{f}'_{k+1}(\mathbf{x}_k,\mathbf{x}'_{k+1},t)\\ &\quad +\Delta\mathbf{f}'_{k+1}(\mathbf{x}'_{k+1},t) + \mathbf{D}'_{k+1}(\mathbf{x}'_{k+1},t)\mathbf{w}(t)\\ \dot{\mathbf{x}}_k &= \mathbf{f}_k(\mathbf{x}_k,\mathbf{x}'_{k+1},t) + \mathbf{B}_k(\mathbf{x}_k,\mathbf{x}'_{k+1},t)\mathbf{x}_{k-1}\\ &\quad +\Delta\mathbf{f}_k(\mathbf{x}_k,\mathbf{x}'_{k+1},t) + \mathbf{D}_k(\mathbf{x}_k,\mathbf{x}'_{k+1},t)\mathbf{w}(t)\\ \dot{\mathbf{x}}_i &= \mathbf{f}_i(\hat{\mathbf{x}}_i,t) + \mathbf{B}_i(\hat{\mathbf{x}}_i,t)\mathbf{x}_{i-1}\\ &\quad +\Delta\mathbf{f}_i(\hat{\mathbf{x}}_i,t) + \mathbf{D}_i(\hat{\mathbf{x}}_i,t)\mathbf{w}(t)\ ,\\ &\quad i = 2,...,k-1,\\ \dot{\mathbf{x}}_1 &= \mathbf{f}_1(\hat{\mathbf{x}}_1,t) + \mathbf{B}_1(\hat{\mathbf{x}}_1,t)\mathbf{x}_{r-1}\\ &\quad +\Delta\mathbf{f}_1(\hat{\mathbf{x}}_1,t) + \mathbf{D}_1(\hat{\mathbf{x}}_1,t)\mathbf{w}(t).\end{aligned}$$

In the same way, assumption A13 for step $k$ is stated as:

*Ak3.* The mapping $\mathbf{f}'_{k+1}(\mathbf{x}_k,\mathbf{x}'_{k+1},t)$ is affine on its first argument, namely

$$\mathbf{f}'_{k+1}(\mathbf{x}_k,\mathbf{x}'_{k+1},t) = \mathbf{f}''_{k+1}(\mathbf{x}'_{k+1},t) + \mathbf{B}'_{k+1}(\mathbf{x}'_{k+1},t)\mathbf{x}_k.$$

From the previous algorithm, the following result may be stated:

**Theorem 1** *Assume that the system is locally controllable and at each step of the BC-form algorithm assumptions Ak1, Ak2 and Ak3 hold. Then, there exists an integer $r \le n$ such that the system (1) takes the form (3).*

## 3. DESIRED SLIDING DYNAMICS

It is more conveniently to renumber the state variables of (3) as

$$\begin{aligned}\dot{\mathbf{x}}_1 &= \mathbf{f}_1(\mathbf{x}_1,t) + \mathbf{B}_1(\mathbf{x}_1,t)\mathbf{x}_2\\ &\quad +\Delta\mathbf{f}_1(\mathbf{x}_1,t) + \mathbf{D}_1(\mathbf{x}_1,t)\mathbf{w}(t)\\ \dot{\mathbf{x}}_i &= \mathbf{f}_i(\bar{\mathbf{x}}_i,t) + \mathbf{B}_i(\bar{\mathbf{x}}_i,t)\mathbf{x}_{i+1}\\ &\quad +\Delta\mathbf{f}_i(\bar{\mathbf{x}}_i,t) + \mathbf{D}_i(\bar{\mathbf{x}}_i,t)\mathbf{w}(t),\\ i &= 2,...,r-1,\\ \dot{\mathbf{x}}_r &= \mathbf{f}_r(\bar{\mathbf{x}},t) + \mathbf{B}_r(\bar{\mathbf{x}},t)\mathbf{u}\\ &\quad +\Delta\mathbf{f}_r(\bar{\mathbf{x}},t) + \mathbf{D}_r(\bar{\mathbf{x}},t)\mathbf{w}(t)\end{aligned} \tag{10}$$

At this stage, a nonlinear sliding surface for system (10) can be designed considering the states $\mathbf{x}_{i+1}$, $i = 1,...,r-1$ as a fictitious control vector in the $i^{th}$ block. This procedure is outlined in the following.

On the *first* step, let be the fictitious control $\mathbf{x}_2$ in the $1^{st}$ block (10) chosen as

$$\mathbf{x}_2 = \mathbf{x}_2^c(\mathbf{z}_1,t) + \mathbf{B}_1^+(\mathbf{z}_1,t)(-\mathbf{k}_1\mathbf{z}_1 + \mathbf{E}_{1,1}\mathbf{z}_2), \tag{11}$$

where $\mathbf{z}_1 := \mathbf{x}_1$, $\mathbf{z}_2$ is $n_2 \times 1$ a new variables vector; $k_1 > 0$; $\mathbf{E}_{1,1} = \begin{bmatrix} \mathbf{I}_{n_1} & 0 \end{bmatrix}$, and $\mathbf{x}_2^c(\mathbf{z}_1,t)$ is calculated from the equation $\dot{\mathbf{z}}_1 = 0$ along the trajectories of (10) without uncertainty, namely,

$$\mathbf{x}_2^c = -\mathbf{B}_1^+(\mathbf{z}_1,t)\mathbf{f}_1(\mathbf{z}_1,t) \tag{12}$$

The transformed $1^{st}$ block with new coordinates $\mathbf{z}_1$ and $\mathbf{z}_2$ , and input (11) and (12) has the form

$$\dot{\mathbf{z}}_1 = -\mathbf{k}_1\mathbf{z}_1 + \mathbf{E}_{1,1}\mathbf{z}_2 + \Delta\mathbf{f}_1(\mathbf{z}_1,t) + \mathbf{D}_1(\mathbf{z}_1,t)\mathbf{w}(t) \tag{13}$$

Assume now, without loss of generality, that the elements of matrix $\mathbf{B}_1(\mathbf{z}_1,t)$ can be ordered such that the square matrix

$$\bar{\mathbf{B}}_2(\mathbf{z}_1,t) := \begin{bmatrix} \mathbf{B}_1(\mathbf{z}_1,t) \\ \mathbf{E}_{1,2} \end{bmatrix}$$

with $\mathbf{E}_{1,2} = \begin{bmatrix} 0 & \mathbf{I}_{n_2-n_1} \end{bmatrix}$, has rank $n_2$. Then, $z_2$ can be obtained using (11) as

$$\mathbf{z}_2 = \alpha_1(\mathbf{z}_1,\mathbf{x}_2,t) := \bar{\mathbf{B}}_2(\mathbf{z}_1,t)\mathbf{x}_2 + \begin{bmatrix} \mathbf{f}_1(\mathbf{z}_1,t) + k_1\mathbf{z}_1 \\ 0 \end{bmatrix}. \tag{14}$$

On the *second* step, taking the derivative of (14) along the trajectories of the system (10) yields:

$$\begin{aligned}\dot{\mathbf{z}}_2 &= \bar{\mathbf{f}}_2(\bar{\mathbf{z}}_2,t) + \bar{\mathbf{B}}_2(\bar{\mathbf{z}}_2,t)\mathbf{x}_3\\ &\quad +\Delta\bar{\mathbf{f}}_2(\bar{\mathbf{z}}_2,t) + \bar{\mathbf{D}}_2(\bar{\mathbf{z}}_2,t)\mathbf{w}(t),\end{aligned} \tag{15}$$

where $\bar{\mathbf{z}}_2 = (\mathbf{z}_1,\mathbf{z}_2)^T$, $\bar{\mathbf{f}}_2 = \frac{\partial\alpha_1}{\partial\mathbf{z}_1}(-k_1\mathbf{z}_1 + \mathbf{E}_{1,1}\mathbf{z}_2) + \frac{\partial\alpha_1}{\partial\mathbf{x}_2}\mathbf{f}_2$, $\bar{\mathbf{B}}_2 = \bar{\mathbf{B}}_2\mathbf{B}_2$, $\Delta\bar{\mathbf{f}}_2 = \frac{\partial\alpha_1}{\partial\mathbf{z}_1}\Delta\mathbf{f}_1 + \frac{\partial\alpha_1}{\partial\mathbf{x}_2}\Delta\mathbf{f}_2$, $\bar{\mathbf{D}}_2 = \frac{\partial\alpha_1}{\partial\mathbf{z}_1}\mathbf{D}_1 + \frac{\partial\alpha_1}{\partial\mathbf{x}_2}\mathbf{D}_2$. Note that rank $\bar{\mathbf{B}}_2 = \mathbf{B}_2 = n_2$, $\forall\mathbf{z}_1$, $\mathbf{z}_2$, $t \ge 0$. As in the first step, the fictitious input vector $\mathbf{x}_3$ in (15) is chosen similar to (11) and (12):

$$\begin{aligned}\mathbf{x}_3 &= \mathbf{x}_3^c(\bar{\mathbf{z}}_2,t) + \bar{\mathbf{B}}_2^+(\bar{\mathbf{z}}_2,t)(-k_2\mathbf{z}_2 + \mathbf{E}_{2,1}\mathbf{z}_3),\\ \mathbf{x}_3^c &= -\bar{\mathbf{B}}_2^+(\bar{\mathbf{z}}_2,t)\bar{\mathbf{f}}_2(\bar{\mathbf{z}}_2,t),\end{aligned} \tag{16}$$

where $\mathbf{z}_3$ is $n_3 \times 1$ vector, $k_2 > 0$, and $\mathbf{E}_{2,1} = \begin{bmatrix} \mathbf{I}_{n_2} & 0 \end{bmatrix}$. Thus, equation (15) with (16) takes the form similar to (13), namely

$$\dot{\mathbf{z}}_2 = -k_2\mathbf{z}_2 + \mathbf{E}_{2,1}\mathbf{z}_3 + \Delta\bar{\mathbf{f}}_2(\bar{\mathbf{z}}_2, t) + \bar{\mathbf{D}}_{r-1}(\bar{\mathbf{z}}_2, t)\mathbf{w}(t)$$

This procedure may be performed iteratively obtaining on the $i^{th}$ step, $i = 3, ..., r-1$,

$$\mathbf{x}_{i+1} = \mathbf{x}^c_{i+1}(\bar{\mathbf{z}}_i, t) + \bar{\mathbf{B}}_i^+(\bar{\mathbf{z}}_i, t)(-k_i\mathbf{z}_i + \mathbf{E}_{i,1}\mathbf{z}_{i+1}),$$

with

$$\mathbf{x}^c_{i+1} = -\bar{\mathbf{B}}_i^+(\bar{\mathbf{z}}_i, t)\bar{\mathbf{f}}_i(\bar{\mathbf{z}}_i, t)$$

where $\bar{\mathbf{z}}_i = (\mathbf{z}_1, \mathbf{z}_2, ..., \mathbf{z}_i)$, $\bar{\mathbf{z}}_1 = \mathbf{z}_1$, $\bar{\mathbf{B}}_i = \bar{\mathbf{B}}_i\mathbf{B}_i$, $\mathbf{E}_{i,1} = \begin{bmatrix} \mathbf{I}_{n_i} & 0 \end{bmatrix}$, $\mathbf{E}_{i,1} \in R^{n_i \times n_{i+1}}$. The variables $\mathbf{z}_{i+1}$, $i = 3, ..., r-2$ and discontinuous function $\mathbf{s}$ obtained from this procedure, form a nonlinear transformation of the state variables given by

$$\begin{aligned}
\mathbf{z}_1 &= \mathbf{x}_1, \\
\mathbf{z}_{i+1} &= \bar{\mathbf{B}}_{i+1}(\bar{\mathbf{z}}_i, t)\mathbf{x}_{i+1} + \begin{bmatrix} \bar{\mathbf{f}}_i(\bar{\mathbf{z}}_i, t) + k_i\mathbf{z}_i \\ 0 \end{bmatrix}, \\
i &= 2, ..., r-2, \\
\mathbf{s} &= \bar{\mathbf{B}}_r(\bar{\mathbf{z}}_{r-1}, t)\mathbf{x}_r \\
&\quad + \begin{bmatrix} \bar{\mathbf{f}}_{r-1}(\bar{\mathbf{z}}_{r-1}, t) + k_{r-1}\mathbf{z}_{r-1} \\ 0 \end{bmatrix},
\end{aligned}$$

where $\mathbf{z}_{i+1}$ is a $n_{i+1} \times 1$ vector, $k_i > 0$, $\bar{\mathbf{B}}_{i+1} = \begin{bmatrix} \bar{\mathbf{B}}_i \\ \mathbf{E}_{i,2} \end{bmatrix}$, $\mathbf{E}_{i,2} = \begin{bmatrix} 0 & \mathbf{I}_{n_{i+1}-n_i} \end{bmatrix}$, $\mathbf{E}_{i,2} \in \Re^{(n_{i+1}-n_i) \times n_{i+1}}$. From this, system (10) can be represented in the transformed coordinates as:

$$\begin{aligned}
\dot{\mathbf{z}}_i &= -k_i\mathbf{z}_i + \mathbf{E}_{i,1}\mathbf{z}_{i+1} \\
&\quad + \Delta\bar{\mathbf{f}}_i(\bar{\mathbf{z}}_i, t) + \bar{\mathbf{D}}_i(\bar{\mathbf{z}}_i, t)\mathbf{w}(t), \\
i &= 1, ..., r-2, \\
\dot{\mathbf{z}}_{r-1} &= -k_{r-1}\mathbf{z}_{r-1} + \mathbf{E}_{r-1,1}\mathbf{s} \\
&\quad + \Delta\bar{\mathbf{f}}_{r-1}(\bar{\mathbf{z}}_{r-1}, t) + \bar{\mathbf{D}}_{r-1}(\bar{\mathbf{z}}_{r-1}, t)\mathbf{w}(t), \\
\dot{\mathbf{s}} &= \bar{\mathbf{f}}_r(\bar{\mathbf{z}}_{r-1}, \mathbf{s}, t) + \bar{\mathbf{B}}_r(\bar{\mathbf{z}}_{r-1}, \mathbf{s}, t)\mathbf{u} + \\
&\quad \Delta\bar{\mathbf{f}}_r(\bar{\mathbf{z}}_{r-1}, \mathbf{s}, t) + \bar{\mathbf{D}}_r(\bar{\mathbf{z}}_{r-1}, \mathbf{s}, t)\mathbf{w}(t).
\end{aligned}$$

At this point, it is possible to show that the control input

$$\mathbf{u} = \mathbf{u}^c - k_r\bar{\mathbf{B}}_r^+(\bar{\mathbf{z}}_{r-1}, \mathbf{s}, t)\text{sign}(\mathbf{s}), \quad \mathbf{u}^c = -\bar{\mathbf{B}}_r^+\bar{\mathbf{f}}_r,$$

with $k_r > 0$ under condition $k_r > \left\|\Delta\bar{\mathbf{f}}_r(\bar{\mathbf{z}}_{r-1}, \mathbf{s}, t) + \bar{\mathbf{D}}_r(\bar{\mathbf{z}}_{r-1}, \mathbf{s}, t)\mathbf{w}(t)\right\|$, guarantees a sliding mode on the surface $\mathbf{s} = 0$. Then, the sliding motion is described by the quasi-linear system of $n - n_r$ order described by

$$\begin{aligned}
\dot{\mathbf{z}}_i &= -k_i\mathbf{z}_i + \mathbf{E}_{i,1}\mathbf{z}_{i+1} \\
&\quad + \Delta\bar{\mathbf{f}}_i(\bar{\mathbf{z}}_i, t) + \bar{\mathbf{D}}_i(\bar{\mathbf{z}}_i, t)\mathbf{w}(t), \\
i &= 1, ..., r-2, \qquad (17) \\
\dot{\mathbf{z}}_{r-1} &= -k_{r-1}\mathbf{z}_{r-1} + \Delta\bar{\mathbf{f}}_{r-1}(\bar{\mathbf{z}}_{r-1}, t) \\
&\quad + \bar{\mathbf{D}}_{r-1}(\bar{\mathbf{z}}_{r-1}, t)\mathbf{w}(t),
\end{aligned}$$

Now, the required bound on the uncertainty is given in the following assumption.

*H1.* There exist nonnegative scalar functions $q_{i,j}(t)$ and $d_i(t)$, $i = 1, ..., r-1$, $j = 1, ..., i$, continuous with respect to $t$, satisfying

$$\sup_{t \geq 0} q_{i,j}(t) = \bar{q}_{i,j}, \text{ and } \sup_{t \geq 0} d_i(t) = \bar{d}_i$$

with $\bar{q}_{i,j}$ and $\bar{d}_i$, positive constants, such that

$$\begin{aligned}
&\left\|\Delta\bar{\mathbf{f}}_i(\bar{\mathbf{z}}_i, t) + \bar{\mathbf{D}}_i(\bar{\mathbf{z}}_i, t)w(t)\right\| \\
&\leq \sum_{j=1}^{i} k_j^{(i-j)} q_{i,j}(t) \left\|\mathbf{z}_j\right\| + d_i(t).
\end{aligned}$$

The property of sliding mode motion on $\mathbf{s} = 0$ is expressed in terms of the following result.

**Theorem 2** *Let the assumption H1 holds, then there exist positive $k_1, ..., k_{r-1}$ and $h_1, ..., h_{r-1}$ such that the states of system (17) are uniformly ultimately bounded, i.e.*

$$\lim_{t \to \infty} \sup \left\|\mathbf{z}_i(t)\right\| \leq h_i, \quad i = 1, ..., r-1$$

**Proof.** Choose a Lyapunov candidate function $V$ for(17) as

$$V = \sum_{i=1}^{r-1} V_i, \quad V_i = \frac{1}{2}\mathbf{z}_i^T\mathbf{z}_i, \quad i = 1, ..., r-1,$$

and calculating the derivative $\dot{V}_i$ step by step from $1^{st}$ block to $(r-1)^{th}$ block of (17). On the *first* step, differentiating the Lyapunov candidate function $V_1 = \frac{1}{2}\mathbf{z}_1^T\mathbf{z}_1$, using assumption H1 it comes out that

$$\begin{aligned}
\dot{V}_1 &= -k_1\mathbf{z}_1^T\mathbf{z}_1 \\
&\quad + \mathbf{z}_1^T\left[\mathbf{E}_{1,1}\mathbf{z}_2 + \Delta\bar{\mathbf{f}}_1(\mathbf{z}_1, t) + \bar{\mathbf{D}}_1(\mathbf{z}_1, t)\mathbf{w}(t)\right] \\
&\leq -\left\|\mathbf{z}_1\right\|\left[(k_1 - q_{1,1}(t))\left\|\mathbf{z}_1\right\| - \left\|\mathbf{z}_2\right\| - d_1(t)\right] \\
&\leq -\left\|\mathbf{z}_1\right\|\left[(k_1 - \bar{q}_{1,1})\left\|\mathbf{z}_1\right\| - \left\|\mathbf{z}_2\right\| - \bar{d}_1\right]
\end{aligned}$$

which is negative if

$$(k_1 - \bar{q}_{1,1})\left\|\mathbf{z}_1\right\| - \left\|\mathbf{z}_2\right\| - \bar{d}_1 > 0.$$

Therefore, the state ultimately enter the region in subspace $(\mathbf{z}_1, \mathbf{z}_2)$ defined by

$$\left\|\mathbf{z}_1\right\| \leq \alpha_{1,2}\left\|\mathbf{z}_2\right\| + \beta_{1,2}, \qquad (18)$$

where $\alpha_{1,2} = \frac{1}{k_1 - \bar{q}_{1,1}}$ and $\beta_{1,2} = \alpha_{1,2}\bar{d}_1$ are positive if

$$k_1 > \bar{q}_{1,1} \qquad (19)$$

On the *second* step, following similar lines to those taken for the first block, the derivative $\dot{V}_2$ of the Lyapunov candidate function $V_2 = \frac{1}{2}\mathbf{z}_2^T\mathbf{z}_2$ calculated along the trajectories of the second block, under H1 and (18) is given by

$$\begin{aligned}
\dot{V}_2 &= -k_2\mathbf{z}_2^T\mathbf{z}_2 \\
&\quad +\mathbf{z}_2^T\left[\mathbf{E}_{2,1}\mathbf{z}_3+\Delta\bar{\mathbf{f}}_2(\bar{\mathbf{z}}_2,t)+\bar{\mathbf{D}}_1(\bar{\mathbf{z}}_2,t)\mathbf{w}(t)\right] \\
&\le -\|\mathbf{z}_2\|\left[(k_2-\bar{q}_{2,2})\|\mathbf{z}_2\|\right. \\
&\quad \left. -\|\mathbf{z}_3\|-k_1\bar{q}_{2,1}\|\mathbf{z}_1\|-\bar{d}_2\right] \\
&\le -\|\mathbf{z}_2\|\left[\frac{1}{\alpha_{2,3}}\|\mathbf{z}_2\|-\|\mathbf{z}_3\|-\frac{\beta_{2,3}}{\alpha_{2,3}}\right]
\end{aligned}$$

which is negative if $0 < \frac{1}{\alpha_{2,3}}\|\mathbf{z}_2\| - \|\mathbf{z}_3\| - k_1\bar{q}_{2,1}\beta_{1,2}-\bar{d}_2$, where $\alpha_{2,3} = \frac{1}{k_2-\bar{q}_{2,2}-k_1\bar{q}_{2,1}\alpha_{1,2}}$ and $\beta_{2,3} = \alpha_{2,3}(k_1\bar{q}_{2,1}\beta_{1,2}+\bar{d}_2)$. Hence, the state ultimately enter the region in subspace $(\mathbf{z}_1,\mathbf{z}_2,\mathbf{z}_3)$ defined by

$$\|\mathbf{z}_2\| \le \alpha_{2,3}\|\mathbf{z}_3\| + \beta_{2,3}$$

and consequently

$$\|\mathbf{z}_1\| \le \alpha_{1,3}\|\mathbf{z}_3\| + \beta_{1,3}$$

where $\alpha_{2,3}$, $\beta_{2,3}$, $\alpha_{1,3} = \alpha_{1,2}\alpha_{2,3}$ and $\beta_{1,3} = \alpha_{1,2}\beta_{2,3}+\beta_{1,2}$ are positive if $k_1$ satisfies (19) and

$$k_2 > \bar{q}_{2,2} + k_1\bar{q}_{2,1}\alpha_{1,2} \tag{20}$$

Proceeding in the same fashion, for the step $i$, the convergence region in the subspace $(\mathbf{z}_1,...,\mathbf{z}_i)$ is

$$\begin{aligned}
\|\mathbf{z}_{i-1}\| &\le \alpha_{i-1,i}\|\mathbf{z}_i\| + \beta_{i-1,i} \\
\|\mathbf{z}_{i-2}\| &\le \alpha_{i-2,i}\|\mathbf{z}_i\| + \beta_{i-2,i}, \\
&\vdots \\
\|\mathbf{z}_1\| &\le \alpha_{1,i}\|\mathbf{z}_i\| + \beta_{1,i}.
\end{aligned} \tag{21}$$

On the next step, the derivative of the Lyapunov function $V_i = \frac{1}{2}\mathbf{z}_i^T\mathbf{z}_i$ along the trajectories of the $i^{th}$ block under assumption H1 gives

$$\begin{aligned}
\dot{V}_i &= -k_i\mathbf{z}_i^T\mathbf{z}_i \\
&\quad +\mathbf{z}_i^T\left[\mathbf{E}_{i,1}\mathbf{z}_{i+1}+\Delta\bar{\mathbf{f}}_i(\bar{\mathbf{z}}_i,t)+\bar{\mathbf{D}}_i(\bar{\mathbf{z}}_i,t)\mathbf{w}(t)\right] \\
&\le -\|\mathbf{z}_i\|^2\left(k_i - \sum_{j=1}^{i} k_j^{(i-j)}\bar{q}_{i,j}\alpha_{j,i}\right) \\
&\quad +\|\mathbf{z}_i\|\left(+\|\mathbf{z}_{i+1}\| + \sum_{j=1}^{i} k_j^{(i-j)}\bar{q}_{i,j}\beta_{j,i} - \bar{d}_i\right)
\end{aligned}$$

From this equation it follows

$$\|\mathbf{z}_i\| \le \alpha_{i,i+1}\|\mathbf{z}_{i+1}\| + \beta_{i,i+1} \tag{22}$$

where $\alpha_{i,i+1} = \frac{1}{k_i - \sum_{j=1}^{i} k_j^{(i-j)}\bar{q}_{i,j}\alpha_{j,i}}$ and $\beta_{i,i+1} = \alpha_{i,i+1}\left[\sum_{j=1}^{i} k_j^{(i-j)}\bar{q}_{i,j}\beta_{j,i} - \bar{d}_i\right]$ are positive if

$$k_i > \sum_{j=i}^{r} k_j^{(i-j)}\bar{q}_{i,j}\alpha_{j,i}, \tag{23}$$

Substituting (22) in (21) gives the set of inequalities

$$\|\mathbf{z}_j\| \le \alpha_{j,i+1}\|\mathbf{z}_{i+1}\| + \beta_{j,i+1},\ j = 1,2,...,i,$$

for the subspace $(\mathbf{z}_1,...,\mathbf{z}_i,\mathbf{z}_{i+1})$, where $\alpha_{j,i+1} = \alpha_{j,i}\alpha_{i,i+1}$ and $\beta_{j,i+1} = \alpha_{j,i}\beta_{i,i+1} + \beta_{j,i}$.

On the last step the region of convergence in subspace $(\mathbf{z}_1,\mathbf{z}_2,...,\mathbf{z}_{r-2},\mathbf{z}_{r-1})$ is defined by

$$\|\mathbf{z}_i\| \le \alpha_{i,r-1}\|\mathbf{z}_{r-1}\| + \beta_{i,r-1},\ i = 1,...,r-2, \tag{24}$$

These inequalities are used to evaluate the derivative of the Lyapunov candidate function $V_{r-1} = \frac{1}{2}\mathbf{z}_{r-1}^T\mathbf{z}_{r-1}$ along the trajectories of $(r-1)^{th}$ block of (17):

$$\begin{aligned}
\dot{V}_{r-1} &= -k_{r-1}\mathbf{z}_{r-1}^T\mathbf{z}_{r-1} \\
&\quad +\mathbf{z}_{r-1}^T\left[\Delta\bar{\mathbf{f}}_{r-1}(\bar{\mathbf{z}}_{r-1},t)+\bar{\mathbf{D}}_{r-1}(\bar{\mathbf{z}}_{r-1},t)\mathbf{w}\right] \\
&\le -k_{r-1}\|\mathbf{z}_{r-1}\|^2 \\
&\quad +\|\mathbf{z}_{r-1}\|\sum_{j=1}^{r-1} k_j^{(r-1-j)}q_{r-1,j}(t)\|\mathbf{z}_j\| \\
&\quad +\|\mathbf{z}_{r-1}\|d_{r-1}(t) \\
&\le -\alpha_{r-1}\|\mathbf{z}_{r-1}\|^2 + \beta_{r-1}\|\mathbf{z}_{r-1}\|,
\end{aligned}$$

where $\alpha_{r-1} = k_{r-1} - \sum_{j=1}^{r-1} k_j^{(r-1-j)}\bar{q}_{r-1,j}\alpha_{j,r-1}$ and $\beta_{r-1} = \sum_{j=1}^{r-1} k_j^{(r-1-j)}\bar{q}_{r-1,j}\beta_{j,r-1} + \bar{d}_{r-1}$. If $k_{r-1}$ is chosen such that

$$k_{r-1} > \sum_{j=1}^{r-1} k_j^{(r-1-j)}\bar{q}_{r-1,j}\alpha_{j,r-1}$$

then

$$\dot{V}_{r-1} \le -2\alpha_{r-1}V_{r-1} + \beta_{r-1}\sqrt{2V_{r-1}}$$

with positive $\alpha_{r-1}$ and $\beta_{r-1}$. Using comparison method (Khalil, 1996), take $W_{r-1} = \sqrt{2V_{r-1}}$ so that

$$\begin{aligned}
\dot{W}_{r-1} &= \frac{1}{2\sqrt{2V_{r-1}}}\dot{V}_{r-1} \\
&\le -\frac{\alpha_{r-1}}{\sqrt{2V_{r-1}}}V_{r-1} + \beta_{r-1}\sqrt{2V_{r-1}}
\end{aligned}$$

or

$$\dot{W}_{r-1} \le -\frac{\alpha_{r-1}}{2}W_{r-1} + \frac{\beta_{r-1}}{2}.$$

From this linear differential inequality follows that

$$W_{r-1}(t) \le \left(W_{r-1}(t_0) + \frac{\beta_{r-1}}{\alpha_{r-1}}\right)e^{-\frac{\alpha_{r-1}}{2}(t-t_0)} + \frac{\beta_{r-1}}{\alpha_{r-1}}$$

and, since $\|\mathbf{z}_{r-1}\| = W_{r-1}$, it follows that

$$\|\mathbf{z}_{r-1}(t)\| \le \left[\|\mathbf{z}_{r-1}(t_0)\| - \frac{\beta_{r-1}}{\alpha_{r-1}}\right] e^{-\frac{\alpha_{r-1}}{2}(t-t_0)} + \frac{\beta_{r-1}}{\alpha_{r-1}}, \quad (25)$$

so

$$\limsup_{t\to\infty} \|\mathbf{z}_{r-1}(t)\| \le h_{r-1},\ h_{r-1} = \frac{\beta_{r-1}}{\alpha_{r-1}} \quad (26)$$

Now the evaluation of $\|\mathbf{z}_{r-2}(t)\|$ using the inequality (25) for $\dot{V}_{r-2}$ can be found as

$$\begin{aligned}\dot{V}_{r-2} &\le -\alpha_{r-2}\|\mathbf{z}_{r-2}\|^2 \\ &\quad + (\|\mathbf{z}_{r-1}\| + \beta_{r-2})\|\mathbf{z}_{r-2}\| \\ &\le -2\alpha_{r-2}V_{r-2} \qquad (27) \\ &\quad + (\|\mathbf{z}_{r-1}\| + \beta_{r-2})\sqrt{2V_{r-2}}\end{aligned}$$

where
$\alpha_{r-2} = k_{r-2} - \sum_{j=1}^{r-1} k_j^{(r-1-j)}\bar{q}_{r-2,j}\alpha_{j,r-2}$,
$\beta_{r-2} = \sum_{j=1}^{r-1} k_j^{(r-1-j)}\bar{q}_{r-2,j}\beta_{j,r-2} + \bar{d}_{r-2}$.

Substitution (25) in (27) and applying again comparison method yields

$$\begin{aligned}\|\mathbf{z}_{r-2}(t)\| &\le \gamma_{r-2,r-2}e^{-\frac{\alpha_{r-2}}{2}(t-t_0)} \\ &\quad + \gamma_{r-2,r-1}e^{-\frac{\alpha_{r-1}}{2}(t-t_0)} + \gamma_{r-2}\end{aligned}$$

where $\gamma_{r-2,1} = \|\mathbf{z}_{r-2}(t_0)\| - \frac{\beta_{r-2}+h_{r-1}}{\alpha_{r-2}} - \frac{\|\mathbf{z}_{r-1}(t_0)\|-h_{r-1}}{\alpha_{r-1}-\alpha_{r-1}}$, $\gamma_{r-2,2} = \frac{\|\mathbf{z}_{r-1}(t_0)\|-h_{r-1}}{\alpha_{r-1}-\alpha_{r-1}}$, $\gamma_{r-2} = \frac{\beta_{r-2}+h_{r-1}}{\alpha_{r-2}}$. Remember that $\alpha_{r-2} = \frac{1}{\alpha_{r-2,r-1}}$ and $\beta_{r-2} = \frac{\beta_{r-2,r-1}}{\alpha_{r-2,r-1}}$

$$\limsup_{t\to\infty} \|z_{r-1}(t)\| \le h_{r-2},$$

where $h_{r-2} = \frac{\beta_{r-2}+h_{r-1}}{\alpha_{r-2}} = \alpha_{r-2,r-1}h_{r-1} + \beta_{r-2,r-1}$.

This result for $\|\mathbf{z}_{r-2}(t)\|$, as well as for $\|z_i(t)\|$, $i = 1, ..., r-3$, it is possible to get using inequalities (24) and (26). So, using inequalities (24) and (26) and the fact that the function $V = \sum_{i=1}^{r-1} V_i$ is negative outside the region (24), gives

$$\limsup_{t\to\infty} \|\mathbf{z}_i\| \le h_i,$$

where $h_i = \alpha_{i,r-1}h_{r-1} + \beta_{i,r-1}$, $i = 3, ..., r$.

Hence, if $k_2, , ..., k_{r-1}$ and $k_r$ are chosen so that to satisfy the recurrent inequalities

$$k_i > \sum_{j=i}^{r} k_j^{(i-j)}\bar{q}_{i,j}\alpha_{j,i}, i = 2, ..., r-1, r,$$

then the stated result holds.

**Remark 1.** It can be easy to see, in the case of vanishing uncertainties, that is, $d_i(t) \equiv 0$, system (17) is exponentially stable.

**Remark 2.** The procedure of choosing of $k_i$ is that. First, it is need to choose value of $k_1$ such that to satisfy the inequality (19). Then, with fixed value of $k_1$, it needs to choose the value of $k_2$ to satisfy the inequality (20), and so on.

**Remark 3**. It is interest to note that

$$h_{r-1} = \frac{\sum_{j=1}^{r-1} k_j^{(r-1-j)}\bar{q}_{r-1,j}\beta_{j,r-1} + \bar{d}_{r-1}}{k_{r-1} - \sum_{j=1}^{r-1} k_j^{(r-1-j)}\bar{q}_{r-1,j}\alpha_{j,r-1}}$$

So, with fixed $k_i$, $i = 1, 2, ...r-2$ and increasing the value of $k_{r-1}$, the bound $h_{r-1}$ becomes arbitrarily small. Also, from expression for $h_{r-2}$ :

$$\begin{aligned}h_{r-2} = &\frac{1}{k_{r-2} - \sum_{j=1}^{r-2} k_j^{(r-2-j)}\bar{q}_{r-2,j}\alpha_{j,r-2}} \\ &\times \left[\frac{\sum_{j=1}^{r-1} k_j^{(r-1-j)}\bar{q}_{r-1,j}\beta_{j,r-1} + \bar{d}_{r-1}}{k_{r-1} - \sum_{j=1}^{r-1} k_j^{(r-1-j)}\bar{q}_{r-1,j}\alpha_{j,r-1}}\right] \\ &+ \frac{\sum_{j=1}^{r-1} k_j^{(r-1-j)}\bar{q}_{r-2,j}\beta_{j,r-2} + \bar{d}_{r-2}}{k_{r-2} - \sum_{j=1}^{r-1} k_j^{(r-1-j)}\bar{q}_{r-2,j}\alpha_{j,r-2}}\end{aligned}$$

it can be seen that with fixed $k_i$, $i = 1, 2, ...r-3$ and increasing $k_{r-2}$ and $k_{r-1}$ such that $\frac{k_{r-1}}{k_{r-2}} \to \infty$ it follows that $\lim h_{r-2} = 0$. It can be verified step by step that

$$\lim h_i = 0 \text{ with } \frac{k_i}{k_{i+1}} \to \infty,\ i = 1, 2, ...r-3.$$

It is means, that system (17) is exponentially stable.

## CONCLUSIONS

A sliding mode controller has been developed for nonlinear systems that can be transformed into the nonlinear block controllable form. The controller insures overall stability with sufficiently high state feedback control law. The developed control law is predicted to yield both satisfactory steady-state and transient performance in the presence of dynamic uncertainties and reject bounded disturbances.

## REFERENCES


Utkin, V.I. (1981). *Sliding modes in control and optimization* (in Russian), Nauka. Moscow.

Loukianov A.G. and Utkin, V.I. (1982) Optimal linear systems with degenerate criteria. *Automation and Remote Control*, Vol.43, No.7, (P.1), pp.872-879.

Loukianov A.G. and Utkin V.I. (1981). Methods for reducing dynamic systems to regular form, *Automation and Remote Control*, Vol. 42, No.4, (P.1), pp.413-420.

Hassan K. Khalil 1996. *Nonlinear systems*. Prentice-Hall, Inc. Simon and Schuster, New-Jersey.1107-1123.

# ON THE SLIDING MODE REGULATOR PROBLEM*

**Alexander Loukianov, B. Castillo-Toledo and R. García**

*CINVESTAV-IPN, Unidad Guadalajara*
*A. P. 31-438, Pza. La Luna, Col. Verde Valle,*
*Guadalajara, Jalisco, México. C. P. 44550.*
*E-mail: louk@gdl.cinvestav.mx*

**Abstract.** In this paper we study the problem of tracking asymptotically a reference signal while rejecting the effect of a disturbance signal by means of the sliding mode technique. Analogously to the formulation of the so-called regulator problem, we define the sliding mode regulator problem and derive conditions for the existence of a solution. *Copyright © 1999 IFAC*

**Keywords.** Nonlinear systems, uncertain dynamic systems, decomposition methods, sliding mode control.

## 1. INTRODUCTION

The well-known servomechanism or regulator problem i.e. the problem of tracking asymptotically a reference signal in the presence of a disturbance signal has been extensively studied in the linear case (see for example Davison and Goldenberg, 1975; Sebakhy and Wonhan, 1976). In Francis (1977), a complete solution was presented based on the existence of a solution for a set of matrix equations. For the nonlinear case, it was shown in Isidori and Byrnes (1990) that the solution can be found by solving also a set of nonlinear differential equations, which represents a generalization of the Francis conditions. This set of equations is known as the Francis.-Isidori-Byrnes (FIB) equations. Basically, the regulator solution can be viewed as finding a *steady-state* surface on which the output error map is zero, and which can be made attractive and invariant by feedback. In this work, a discontinuous regulator is proposed using the sliding mode control technique (Utkin, 1981). The underlying idea is to design a sliding surface on which the dynamics of the system is constrained to evolve by means of a discontinuous control law, instead of designing a continuous stabilizing feedback, as in the case of the classical regulator problem. The sliding surface contains the steady-state surface, and the dynamics of the systems tend asymptotically, along the sliding surface, to the steady-state behavior. To make precise the ideas, some basic facts on regulation theory are we briefly recalled. Consider a nonlinear system

$$\begin{aligned} \dot{\mathbf{x}} &= f(\mathbf{x}) + g(\mathbf{x})\mathbf{u} + d(\mathbf{x})\mathbf{w}, & (1)\\ \mathbf{e} &= h(\mathbf{x}) - q(\mathbf{w}), & (2)\\ \dot{\mathbf{w}} &= s(\mathbf{w}), & (3) \end{aligned}$$

where $\mathbf{x} \in \Re^n, \mathbf{u} \in \Re^m$, $\mathbf{e} \in \Re^p$, $\mathbf{w} \in \Re^s$. The vector $f$, the columns of $g$ and $d$ are smooth vector fields, and in addition, it is assumed that $f(0) = h(0) = 0$ . The output tracking error is defined as the difference between the output of system, $h(\mathbf{x})$ and a reference signal, $q(\mathbf{w})$, generated by an external generator described by equation (3), which generates the references and perturbation signals as well..

It has been shown in Isidori and Byrnes (1990) that if the following assumptions hold:

**H1.** The Jacobian matrix

$$\mathbf{S} = \frac{\partial s}{\partial \mathbf{w}}(0)$$

*Work supported by Mexican Consejo Nacional de Ciencia y Tecnología under grant 26358-A at the equilibrium point $\mathbf{w} = 0$ has all eigenvalues on the imaginary axis.

**H2.** The pair $\{\mathbf{A} := \frac{\partial f}{\partial \mathbf{x}}(0), \mathbf{B} := g(0)\}$ is stabilizable

then, providing there exists a pair of mappings $\mathbf{x} = \pi(\mathbf{w})$ and $\mathbf{u} = c(\mathbf{w})$, with $\pi(0) = 0$ and $c(0) = 0$ solving the partial differential equations (FIB equations)

$$\begin{aligned} f(\pi) + g(\pi)c(\mathbf{w}) + d(\pi)\mathbf{w} &= \frac{\partial \pi(\mathbf{w})}{\partial \mathbf{w}} s(\mathbf{w}) \quad (4) \\ h(\pi(\mathbf{w})) - q(\mathbf{w}) &= 0, \quad (5) \end{aligned}$$

a controller of the form

$$\mathbf{u} = c(\mathbf{w}) + K(\mathbf{x} - \pi(\mathbf{w})) \quad (6)$$

where the matrix $K$ stabilizes the linear approximation of system ((1) is such that the equilibrium point $(\mathbf{x}, \mathbf{w}) = (0, 0)$ of the closed loop system (1)-(6) is asymptotically stable, and the output tracking error goes asymptotically to zero, i.e. $\lim_{t \to \infty} \mathbf{e}(t) = 0$.

**Remark 1** *Considering the linear approximation of the system (1) , (2), and (3) at the equilibrium point* $(\mathbf{x}, \mathbf{w}) = (0, 0)$, *namely*

$$\begin{aligned} \dot{\mathbf{x}} &= \mathbf{A}\mathbf{x} + \mathbf{B}\mathbf{u} + \mathbf{D}\mathbf{w}, \quad (7) \\ \mathbf{e} &= \mathbf{C}\mathbf{x} - \mathbf{Q}\mathbf{w}, \quad (8) \\ \dot{\mathbf{w}} &= \mathbf{S}\mathbf{w}, \quad (9) \end{aligned}$$

*where* $\mathbf{D} = d(0)$, *and* $\mathbf{Q} = \frac{\partial q}{\partial \mathbf{w}}(0)$, *the equations (4) and (5) takes the form of the Sylvester matrix equation*

$$\Pi \mathbf{S} = \mathbf{A}\Pi + \mathbf{B}\Gamma + \mathbf{D} \quad (10)$$

$$\mathbf{C}\Pi - \mathbf{Q} = 0 \quad (11)$$

*where* $\Gamma = \frac{\partial c}{\partial \mathbf{w}}(0)$ *and* $\Pi = \frac{\partial \pi}{\partial \mathbf{w}}(0)$.

In the following, the regulator problem with the sliding modifications is introduced, and a solution for the linear and nonlinear case are presented.

## 2. FORMULATION OF THE PROBLEM

Analogously to the State Feedback Regulator Problem, the *Sliding Mode Regulator Problem* **(SMRP)** is defined as the problem of finding a a sliding surface $\sigma(\mathbf{x}, \mathbf{w}) = 0$, $\sigma \in \Re^m$ and a discontinuous controller

$$\mathbf{u} = \mathbf{u}_{eq} - k_0 \mathbf{M}(\mathbf{x}, \mathbf{w}) \text{sign}\sigma, \ k_0 > 0 \quad (12)$$

where the equivalent control $\mathbf{u}_{eq}$ is derived from $\dot{\sigma} = 0$ (Utkin, 1981), and $\mathbf{M}(\mathbf{x}, \mathbf{w})$ is a nonsingular matrix chosen to induce local asymptotic convergence to the surface $\sigma(\mathbf{x}, \mathbf{w}) = 0$, so that the following conditions hold:

**SS** The equilibrium point $\mathbf{x} = 0$ of the closed loop system (1)-(12) is asymptotically stable,

**SR** The output tracking error goes asymptotically to zero, i.e. $t \to \infty \lim \mathbf{e}(t) = 0$.

### *2.1 Linear Case*

Let us consider a switching function defined as

$$\sigma = \Sigma \mathbf{z}, \quad (13)$$

where $\mathbf{z}$ is defined as $\mathbf{z} = \mathbf{x} - \Pi\mathbf{w}$ with $\Sigma$ and $\Pi$ constant matrices of appropriate dimensions. In order to prove that it is possible to induce a sliding regime on this surface, the following result is proven.

**Lemma 1** *Let us define the operator $P$ as* $\mathbf{P} = (I_n - \mathbf{B}(\Sigma\mathbf{B})^{-1}\Sigma)$. *Then, the equality*

$$\mathbf{P}(\mathbf{A}\Pi - \Pi\mathbf{S} + \mathbf{D}) = 0 \quad (14)$$

*holds if and only if there exists a solution* $\Pi, \Gamma$ *of the matrix equation*

$$\mathbf{A}\Pi - \Pi\mathbf{S} + \mathbf{D} = \mathbf{B}\Gamma. \quad (15)$$

**Proof:** It is straighforward to verify that $\mathbf{P}$ is a projection operator along the range space of $\mathbf{B}$ onto the null space of $\Sigma$ (El-Chesawi *et al*, 1983) i.e.

$$\begin{aligned} \mathbf{P}\mathbf{B} &= (I_n - \mathbf{B}(\Sigma\mathbf{B})^{-1}\Sigma)\mathbf{B} = 0 \text{ and} \\ \mathbf{P}\mathbf{z} &= \mathbf{z} \ \forall \mathbf{z} \in \mathcal{N}, \\ \mathcal{N} &= \{\mathbf{z} \in R^n \mid \Sigma\mathbf{z} = 0\} \end{aligned}$$

so, if condition (15) holds, then it follows that $\mathbf{P}(\mathbf{A}\Pi - \Pi\mathbf{S} + \mathbf{D}) = \mathbf{P}\mathbf{B}\Gamma = 0$. Now, if condition (14) is satisfied, then $(\mathbf{A}\Pi - \Pi\mathbf{S} + \mathbf{D})$ must be in the image of $\mathbf{B}$, i.e., $(\mathbf{A}\Pi - \Pi\mathbf{S} + \mathbf{D}) = \mathbf{B}\Gamma$ for some matrix $\Gamma$.

**Remark 2** *Note that condition (14), together with condition (11), guarantees the existence of a solution to the regulator problem for the system (7)-(8). In the following it is shown that (14) guarantees also the existence of a solution to the SMRP.*

**Definition 1** *Let us consider the system*

$$\begin{aligned} \dot{\mathbf{z}} &= \mathbf{P}\mathbf{A}\mathbf{z} \quad (16) \\ \sigma &= \Sigma\mathbf{z}. \end{aligned}$$

*The sliding controlled zero dynamics for this system is defined as*

$$\dot{\mathbf{z}} = \mathbf{P}\mathbf{A}\mathbf{z} \mid_{\Sigma\mathbf{z}=0}, \quad (17)$$

*this is, it represents the dynamics of system (16) restricted to the surface* $\sigma = 0$.

**Remark 3** *From the characterization of a zero dynamic of a system (Isidori and Byrnes, 1990), the matrix* $\mathbf{PA}$ *of system (17) has* $m$ *zero eigenvalues. Also, from the construction of matrix* $\mathbf{P}$*, it is possible to show that the other* $n-m$ *eigenvalues can be placed arbitrarily by means of* $\Sigma$ *(El-Chesawi et al, 1983).*

**Proposition 1** *Suppose assumptions H1 and H2 hold. Let* $\Sigma$ *be chosen such that the sliding controlled zero dynamics is asymptotically stable and suppose there exists a matrix* $\Pi$ *which solves the linear equations*

$$[I_n - \mathbf{B}(\Sigma\mathbf{B})^{-1}\Sigma] \ [\mathbf{A}\Pi - \Pi\mathbf{S} + \mathbf{D}] = 0 \quad (18)$$

$$\mathbf{C}\Pi - \mathbf{Q} = 0, \quad (19)$$

*then the SMRP is solvable.*

**Proof.** Choosing $\mathbf{M} = k(\Sigma\mathbf{B})^{-1}$, $k > 0$, and setting

$$\mathbf{u}_{eq} = -(\Sigma\mathbf{B})^{-1}\Sigma[\mathbf{Az} + (\mathbf{A}\Pi - \Pi\mathbf{S} + \mathbf{D})\mathbf{w}] \quad (20)$$

the control action (12) guarantees a sliding mode motion on the surface $\sigma = 0$ which is described by

$$\dot{\mathbf{z}} = \mathbf{PAz} + \mathbf{P}(\mathbf{A}\Pi - \Pi\mathbf{S} + \mathbf{D})\mathbf{w}, \ \Sigma\mathbf{z} = 0. \quad (21)$$

Now, by using condition (18), it follows that

$$\dot{\mathbf{z}} = \mathbf{PAz}, \Sigma\mathbf{z} = 0$$

which is precisely the sliding controlled zero dynamics. Since matrix $\Sigma$ can be chosen such that det $\Sigma\mathbf{B} \neq 0$, and $n-m$ eigenvalues of $\mathbf{PA}$ are in $\mathbf{C}^-$, then when the sliding mode occurs, $\mathbf{z}(t) \to 0$ and rewriting the output tracking error as

$$\mathbf{e} = \mathbf{Cz} + (\mathbf{C}\Pi - \mathbf{Q})\mathbf{w},$$

if condition (5) is satisfied, then the error goes asymptotically to zero.

**Corollary 1** *If the perturbations matrix* $\mathbf{D}$ *in (7) satisfies the matching condition, i.e. there exist a* $m \times m$ *matrix* $\Omega$ *such that*

$$\mathbf{D} = \mathbf{B}\Omega \quad (22)$$

*then the condition (18) can be trivially satisfied with* $\Pi = 0$.

**Proof.** If $\mathbf{D} = \mathbf{B}\Omega$, then $\mathbf{P}(\mathbf{A}\Pi - \Pi\mathbf{S} + \mathbf{D})$ $=\mathbf{P}(\mathbf{A}\Pi - \Pi\mathbf{S}) + \mathbf{PB}\Omega$, and since $\mathbf{PB} = 0$,it follows inmediately that

$$\mathbf{P}(\mathbf{A}\Pi - \Pi\mathbf{S}) = 0 \quad (23)$$

is given trivially by $\Pi = 0$.

*2.2 Linear systems in regular form*

In order to show the explicit form of the sliding controlled zero dynamics, it is assumed that, possibly after reordering, the system (7) and (8) can be described in the form

$$\begin{aligned} \begin{pmatrix} \dot{\mathbf{x}}_1 \\ \dot{\mathbf{x}}_2 \end{pmatrix} &= \begin{pmatrix} \mathbf{A}_{11} & \mathbf{A}_{12} \\ \mathbf{A}_{21} & \mathbf{A}_{22} \end{pmatrix} \begin{pmatrix} \mathbf{x}_1 \\ \mathbf{x}_2 \end{pmatrix} + \\ & \begin{pmatrix} \mathbf{B}_1 \\ \mathbf{B}_2 \end{pmatrix} \mathbf{u} + \begin{pmatrix} \mathbf{D}_1 \\ \mathbf{D}_2 \end{pmatrix} \mathbf{w} \\ \mathbf{y} &= \begin{pmatrix} \mathbf{C}_1 & \mathbf{C}_2 \end{pmatrix} \mathbf{x} \end{aligned}$$

where dim $(\mathbf{x}_2)=m$ and rank$\mathbf{B}_2 = m$. By using the transformation

$$\begin{pmatrix} \mathbf{x}'_1 \\ \mathbf{x}'_2 \end{pmatrix} = \begin{pmatrix} I_{n-m} & -\mathbf{B}_1\mathbf{B}_2^{-1} \\ 0 & I_m \end{pmatrix} \begin{pmatrix} \mathbf{x}_1 \\ \mathbf{x}_2 \end{pmatrix}$$

a regular form of the system given by

$$\begin{aligned} \begin{pmatrix} \mathbf{x}'_1 \\ \mathbf{x}'_2 \end{pmatrix} &= \begin{pmatrix} \mathbf{A}'_{11} & \mathbf{A}'_{12} \\ \mathbf{A}'_{21} & \mathbf{A}'_{22} \end{pmatrix} \begin{pmatrix} \mathbf{x}'_1 \\ \mathbf{x}'_2 \end{pmatrix} + \\ & \begin{pmatrix} 0 \\ \mathbf{B}_2 \end{pmatrix} \mathbf{u} + \begin{pmatrix} \mathbf{D}'_1 \\ \mathbf{D}_2 \end{pmatrix} \mathbf{w} \end{aligned}$$

$$\mathbf{e} = \mathbf{C}_1\mathbf{x}'_1 + \mathbf{C}'_2\mathbf{x}'_2 - \mathbf{Qw}$$

where
$\mathbf{A}'_{11} = \mathbf{A}_{11} - \mathbf{B}_1\mathbf{B}_2^{-1}\mathbf{A}_{21}$,
$\mathbf{A}'_{12} = (\mathbf{A}_{11} - \mathbf{B}_1\mathbf{B}_2^{-1}\mathbf{A}_{21})\mathbf{B}_1\mathbf{B}_2^{-1} + (\mathbf{A}_{12} - \mathbf{B}_1\mathbf{B}_2^{-1}\mathbf{A}_{22})$,
$\mathbf{A}'_{22} = \mathbf{A}_{21}\mathbf{B}_1\mathbf{B}_2^{-1} + \mathbf{A}_{22}$,
$\mathbf{D}'_1 = \mathbf{D}_1 - \mathbf{B}_1\mathbf{B}_2^{-1}\mathbf{D}_2$,
$\mathbf{C}'_2 = \mathbf{C}_1\mathbf{B}_1\mathbf{B}_2^{-1} + \mathbf{C}_2$.
is derived. Choosing the sliding surface (13) as

$$\begin{aligned} \sigma &= \Sigma_1\mathbf{z}_1 + \mathbf{z}_2 = 0, \quad \text{with} \\ \mathbf{z}_1 &= \mathbf{x}'_1 - \Pi_1\mathbf{w}, \quad \mathbf{z}_2 = \mathbf{x}'_2 - \Pi_2\mathbf{w} \end{aligned}$$

and $\Sigma_1$, $\Pi_1$ and $\Pi_2$, constant matrices, then

$$\begin{aligned} \dot{\mathbf{z}}_1 &= (\mathbf{A}'_{11} + \mathbf{A}_{12}\Sigma_1)\mathbf{z}_1 + \\ & (\mathbf{A}'_{11}\Pi_1 + \mathbf{A}'_{12}\Pi_2 - \Pi_1\mathbf{S} + \mathbf{D}'_1)\mathbf{w} \\ \mathbf{e} &= (\mathbf{C}_1 - \mathbf{C}'_2\Sigma_1)\mathbf{z}_1 + (\mathbf{C}_1\Pi_1 + \mathbf{C}'_2\Pi_2 - \mathbf{Q})\mathbf{w}. \end{aligned}$$

and, if conditions (18) and (19) hold, i.e.

$$\begin{aligned} \mathbf{A}'_{11}\Pi_1 + \mathbf{A}'_{12}\Pi_2 - \Pi_1\mathbf{S} + \mathbf{D}'_1 &= 0 \quad (24) \\ \mathbf{C}_1\Pi_1 + \mathbf{C}'_2\Pi_2 - Q &= 0 \end{aligned}$$

the sliding controlled zero dynamics will be

$$\dot{\mathbf{z}}_1 = (\mathbf{A}'_{11} + \mathbf{A}'_{12}\Sigma_1)\mathbf{z}_1.$$

Now, it is known (Utkin and Young, 1978) that if pair $\{\mathbf{A}, \mathbf{B}\}$ is controllable (stabilizable) then the pair $\{\mathbf{A}'_{11}, \mathbf{A}'_{12}\}$ controllable (stabilizable) as well, therefore there exists a matrix $\Sigma_1$such that the matrix $(\mathbf{A}'_{11}+\mathbf{A}'_{12}\Sigma_1)$ is stable and $\mathbf{z}_1$ tends to zero and in consequence, the output tracking error tends also to zero.

*2.3. Nonlinear Case*

Consider now the nonlinear system (1), (2) and (3). In this case it is assumed that the system is partitioned as

$$\begin{aligned}\mathbf{x}_1 &= f_1(\mathbf{x}_1,\mathbf{x}_2)+g_1(\mathbf{x}_1,\mathbf{x}_2)\mathbf{u}+\\ &\quad d_1(\mathbf{x}_1,\mathbf{x}_2)\mathbf{w} \quad (25)\\ \mathbf{x}_2 &= f_2(\mathbf{x}_1,\mathbf{x}_2)+g_2(\mathbf{x}_1,\mathbf{x}_2)\mathbf{u}+\\ &\quad d_2(\mathbf{x}_1,\mathbf{x}_2)\mathbf{w} \quad (26)\end{aligned}$$

with dim $(\mathbf{x}_2)=m$ and rank $g_2=m$ for all $\mathbf{x}_1$ and $\mathbf{x}_2$. Assume now that there exists a diffeomorphic map $\phi(x)$, $\phi=col(\phi_1,..\ \phi_{n-m})$ , $\phi(0)=0$ satisfying the system of partial differential equations

$$\frac{\partial\phi(\mathbf{x})}{\partial\mathbf{x}_1}g_1(\mathbf{x})+\frac{\partial\phi(\mathbf{x})}{\partial\mathbf{x}_2}g_2(\mathbf{x})=0,$$

or equivalently, that the Pfaffian system

$$d\mathbf{x}_1-g_1(\mathbf{x})g_2^{-1}(\mathbf{x})d\mathbf{x}_2=0 \quad (27)$$

is solvable. Necessary and sufficient conditions for the existence of a solution to (27) may be given by Frobenius Theorem (Louk'yanov, 1981).

The nonlinear transformation

$$\begin{pmatrix}\mathbf{x}'_1\\ \mathbf{x}'_2\end{pmatrix}=\Phi(\mathbf{x}_1,\mathbf{x}_2)=\begin{pmatrix}\phi(\mathbf{x}_1,\mathbf{x}_2)\\ \mathbf{x}_2\end{pmatrix} \quad (28)$$

reduces system (25) and (26) to the regular form

$$\begin{aligned}\mathbf{x}'_1 &= f'_1(\mathbf{x}'_1,\mathbf{x}'_2)+d'_1(\mathbf{x}'_1,\mathbf{x}'_2)\mathbf{w}\\ \mathbf{x}'_2 &= f'_2(\mathbf{x}'_1,\mathbf{x}'_2)+g_2(\mathbf{x}'_1,\mathbf{x}'_2)\mathbf{u}+d'_2(\mathbf{x}'_1,\mathbf{x}'_2)\mathbf{w}.\end{aligned} \quad (29)$$

where $f'_0$, $f'_1$, $f'_2$, $g'_2$, $d'_0$, $d'_1$ and $d'_2$ are calculated from 25) and (26) and (28).

Choosing now the sliding surface (13) as

$$\sigma=\mathbf{x}'_2+\sigma_1(\mathbf{x}'_1,\mathbf{w})=0 \quad (30)$$

where $\sigma_1(\mathbf{x}'_1,\mathbf{w})$ is a $\mathbf{C}^k$ ( $k\geq 2$) mapping with $\sigma_1(0,0)=0$, and setting $\mathbf{M}=g_2{}^{-1}$ and

$$\begin{aligned}\mathbf{u}_{eq} &= -g_2^{-1}\left[f'_2+\frac{\partial\sigma_1}{\partial\mathbf{x}'_1}f'_1+\right.\\ &\quad \left.\left(d'_2+\frac{\partial\sigma_1}{\partial\mathbf{x}'_1}d'_1\right)\mathbf{w}+\frac{\partial\sigma_1}{\partial\mathbf{w}}s\right]\end{aligned}$$

then the control (12) guarantees a sliding motion on $\sigma=0$. On this surface, the dynamics of the system is given by the reduced order equations

$$\begin{aligned}\mathbf{x}'_1 &= f'_1(\mathbf{x}'_1,\sigma_1)+d'_1(\mathbf{x}'_1,\sigma_1)\mathbf{w} \quad (31)\\ \mathbf{e} &= h'(\mathbf{x}'_1,\sigma_1)-q(\mathbf{w}) \quad (32)\end{aligned}$$

with $h'(\mathbf{x}'_1,\mathbf{x}'_2)=h(\Phi^{-1}(\mathbf{x}_1,\mathbf{x}_2))$. At this point, in order to guarantee that the reduced dynamics (31) is such that the error converges to zero, the following assumption is introduced.

**H3.** The linear approximation of (1) at $\mathbf{x}=0$ is controllable.

Now, representing the system (3) and (31) as

$$\begin{aligned}\mathbf{x}'_1 &= (\mathbf{A}_{11}+\mathbf{A}_{12}\Sigma_1)\mathbf{x}'_1+(\mathbf{A}_{12}L_1+\mathbf{D}_1)\mathbf{w}+\\ &\quad \varphi(\mathbf{x}'_1,\mathbf{w}) \quad (33)\\ \dot{\mathbf{w}} &= \mathbf{S}w+\psi(\mathbf{w}) \quad (34)\end{aligned}$$

where $\mathbf{A}_{11}=\frac{\partial f'_1}{\partial\mathbf{x}'_1}(0,0)$, $\mathbf{A}_{12}=\frac{\partial f'_1}{\partial\mathbf{x}'_2}(0,0)$, $\mathbf{D}_1=\mathbf{D}'_1(0,0)$, $\Sigma_1=\frac{\partial\sigma_1}{\partial\mathbf{x}'_1}(0,0)$, $L_1=\frac{\partial\sigma_1}{\partial\mathbf{w}}(0,0)$, $\mathbf{S}=\frac{\partial s}{\partial\mathbf{w}}(0)$ , and functions $\varphi(x'_1,w)$ and $\psi(\mathbf{w})$ vanishing at the origin with its first derivatives, it is possible to show that assumptions H1 and H5 guarantees the existence of a matrix $\Sigma_1$ such that matrix $(\mathbf{A}_{11}+\mathbf{A}_{12}\Sigma_1)$ is Hurwitz and a center manifold $\mathbf{x}_1=\pi_1(\mathbf{w})$ for system (33) and (34) satisfying the invariance equation

$$\begin{aligned}\frac{\partial\pi_1}{\partial\mathbf{w}}s(\mathbf{w}) &= f'_1(\pi_1(\mathbf{w}),\sigma_1(\pi_1(\mathbf{w}),\mathbf{w}))+\\ &\quad d'_1(\pi_1(\mathbf{w}),\sigma_1(\pi_1(\mathbf{w}),\mathbf{w}))\mathbf{w}. \quad (35)\end{aligned}$$

From this it is possible to deduce a condition for the solution sliding mode regulator problem for nonlinear system presented in the regular form (29).

**Proposition 2** *Under assumptions H1 and H3, if there exist $C^k$ ($k\geq 2$) mappings $c_1(\mathbf{w})$ and $\pi_1(\mathbf{w})$ with $c_1(0)=0$ and $\pi_1(0)=0$ which satisfy*

$$\begin{aligned}f'_1(\pi_1,c_1)+d'_1(\pi_1,c_1)\mathbf{w} &= \frac{\partial\pi_1}{\partial\mathbf{w}}s(\mathbf{w}) \quad (36)\\ h'(\pi_1,c_1)-q(\mathbf{w}) &= 0 \quad (37)\end{aligned}$$

*Then the SMRP is solvable.*

**Proof.** Proceeding along the previous discussion, setting

$$\sigma_1(\mathbf{x}'_1,\mathbf{w})=c_1(\mathbf{w})+\Sigma_1(\mathbf{x}'_1-\pi_1(\mathbf{w})),$$

then the Jacobian matrix of $f'_1(x'_1,\sigma_1(x'_1,0))$ is the matrix $(\mathbf{A}_{11}+\mathbf{A}_{12}\Sigma_1)$ which, by assumption H5, can be made stable by appropriate choice of $\Sigma_1$. Therefore, by a property of center manifolds, $\mathbf{x}_1(t)\rightarrow\pi_1(\mathbf{w}(t))$, and thus $\sigma_1(\pi_1(\mathbf{w}),\mathbf{w})=c_1(\mathbf{w})$ in the manifold $\mathbf{x}_1(t)=\pi_1(\mathbf{w}(t))$, and condition (36) reduces to condition (35), so, by continuity, if condition (37) holds, then the output tracking error converges to zero.

The later result provides an existence condition for derivating a sliding manifold on which the output tracking error is zeroed. However, in the general case, finding an exact nonlinear transformation (28) may represent a difficult task. In the following a class of nonlinear systems for which the nonlinear transformation can be calculated in an explicit form is considered.

Let us consider the nonlinear system (1) partitioned as

$$\begin{pmatrix} \dot{\mathbf{x}}_0 \\ \dot{\mathbf{x}}_1 \\ \dot{\mathbf{x}}_2 \end{pmatrix} = \begin{pmatrix} f_0 \\ f_1 \\ f_2 \end{pmatrix} + \begin{pmatrix} g_0 \\ g_1 \\ g_2 \end{pmatrix} \mathbf{u} + \begin{pmatrix} d_0 \\ d_1 \\ d_2 \end{pmatrix} \mathbf{w} \quad (38)$$

such that $x_0 = \mathbf{y}$, and rank $g_2 =$ rank $g = m$ for all $\mathbf{x}$. Assume also that matrix $g$ depends on vector $\mathbf{y}$ only, and that the system has relative degree $r = 2$ with respect to the output, i.e. $g_0(\mathbf{y}) = 0$. It is worth nothing the fact that many mechanical processes, among which the underactuated systems, have these properties. In this case the corresponding solution of the Pfaffian system

$$\begin{aligned} d\mathbf{x}_0 &= 0 \\ d\mathbf{x}_1 - g_1(\mathbf{x}_0)g_2^{-1}(\mathbf{x}_0)d\mathbf{x}_2 &= 0 \end{aligned}$$

can be found as

$$\phi(\mathbf{x}) = \begin{pmatrix} \mathbf{x}_0 \\ \mathbf{x}_1 - g_1(\mathbf{x}_0)g_2^{-1}(\mathbf{x}_0)\mathbf{x}_2 \end{pmatrix}.$$

Defining $\mathbf{z} = col(\ \mathbf{z}_0,\ \mathbf{z}_1)$, $\mathbf{z}_0 \in \Re^p$, $\mathbf{z}_1 \in \Re^{n-p-m}, \sigma \in \Re^m$, $\pi(\mathbf{w}) = col\,(\pi_0(\mathbf{w}), \pi_1(\mathbf{w}))$ and setting

$$\mathbf{z}_0 = \mathbf{x}_0 - \pi_0(\mathbf{w}),\ \mathbf{z}_1 = (\mathbf{x}_1 - g_1(\mathbf{y})g_2^{-1}(\mathbf{y})\mathbf{x}_2) - \pi_1(\mathbf{w})$$

and switching function

$$\sigma = \mathbf{x}_2 - \sigma_1(\mathbf{z}, \mathbf{w}), \quad (39)$$

where $\sigma_1 = c_1(\mathbf{w}) + K_0\mathbf{z}_0 + K_1\mathbf{z}_1$ and $\pi_0(\mathbf{w}) = q(\mathbf{w})$; $K_0$ and $K_1$ constant matrices, the the regular form

$$\begin{aligned} \dot{\mathbf{z}}_0 &= f_0'(\mathbf{z}, \pi(\mathbf{w}), \sigma) - \frac{\partial \pi_0(\mathbf{w})}{\partial \mathbf{w}} s(\mathbf{w}) + \\ & \quad d_0'(\mathbf{z}, \pi(\mathbf{w}), \sigma) w \quad (40) \\ \dot{\mathbf{z}}_1 &= f_1'(\mathbf{z}, \pi(\mathbf{w}), \sigma) - \frac{\partial \pi_1(\mathbf{w})}{\partial \mathbf{w}} s(\mathbf{w}) + \\ & \quad d_1'(\mathbf{z}, \pi(\mathbf{w}), \sigma) \mathbf{w} \quad (41) \\ \dot{\sigma} &= f_2'(\mathbf{z}, \pi(\mathbf{w}), \sigma) + g_2'(\mathbf{z}, \pi(\mathbf{w}))\mathbf{u} + \\ & \quad d_2'(\mathbf{z}, \pi(\mathbf{w}), \sigma, \mathbf{w}) \quad (42) \end{aligned}$$

is obtained, where $f_0'$, $f_1'$, $f_2'$, $g_2'$, $d_0'$, $d_1'$ and $d_2'$ are calculated from (38) and (39).

Setting $\mathbf{M} = g_2^{-1}$ and $\mathbf{u}_{eq} = -g_2^{-1}[f_2' + d_2']$, the control (12) guarantees a sliding mode on the $\sigma = 0$. The manifolds $\pi_1(\mathbf{w})$ and $c_1(\mathbf{w})$, are obtained by solving the equations

$$\begin{aligned} \frac{\partial \pi_0(\mathbf{w})}{\partial \mathbf{w}} s(\mathbf{w}) &= f_0'(0, \pi(\mathbf{w}), c_1(\mathbf{w})) + \\ & \quad d_0'(0, \pi(\mathbf{w}), c_1(\mathbf{w}))\mathbf{w} \\ \frac{\partial \pi_1(\mathbf{w})}{\partial \mathbf{w}} s(\mathbf{w}) &= f_1'(0, \pi(\mathbf{w}), c_1(\mathbf{w})) + \\ & \quad d_1'(0, \pi(\mathbf{w}), c_1(\mathbf{w}))\mathbf{w} \end{aligned}$$

and $K_0$ and $K_1$ are chosen such that the linear approximation of the sliding mode equation of $n-m$ order in the equilibrium point $\mathbf{z} = 0$, i.e.

$$\begin{pmatrix} \dot{\mathbf{z}}_0 \\ \dot{\mathbf{z}}_1 \end{pmatrix} = \begin{pmatrix} \mathbf{A}_{00} + \mathbf{B}_0 K_0 & \mathbf{A}_{01} + \mathbf{B}_0 K_1 \\ \mathbf{A}_{10} + \mathbf{B}_1 K_0 & \mathbf{A}_{11} + \mathbf{B}_1 K_1 \end{pmatrix} \begin{pmatrix} \mathbf{z}_0 \\ \mathbf{z}_1 \end{pmatrix}$$

where

$$\begin{pmatrix} \mathbf{A}_{00} & \mathbf{A}_{01} & \mathbf{B}_0 \\ \mathbf{A}_{10} & \mathbf{A}_{11} & \mathbf{B}_0 \end{pmatrix} = \begin{pmatrix} \frac{\partial f_0'}{\partial \mathbf{z}_0} & \frac{\partial f_0'}{\partial \mathbf{z}_1} & \frac{\partial f_0'}{\partial \sigma} \\ \frac{\partial f_1'}{\partial \mathbf{z}_0} & \frac{\partial f_1'}{\partial \mathbf{z}_1} & \frac{\partial f_1'}{\partial \partial \sigma} \end{pmatrix}$$

on the surface $\sigma = 0$ is asymptotically stable.

## 3. APPLICATION TO THE PENDUBOT

In this section, the proposed control scheme is applied to a model of an underactuated system, the well-known *Pendubot* (Figure 1), which can be described by the state space model

$$\begin{aligned} \dot{\mathbf{x}} &= \begin{pmatrix} x_2 \\ \frac{-\mathbf{D}_{12}(x_1)}{\mathbf{D}_{11}(x_1)\mathbf{D}_{22} - \mathbf{D}_{12}^2(\mathbf{x})} - q(\mathbf{x}) \\ x_4 \\ \frac{\mathbf{D}_{22}}{\mathbf{D}_{11}(x_1)\mathbf{D}_{22} - \mathbf{D}_{12}^2(\mathbf{x})} p(\mathbf{x}) \end{pmatrix} + \\ & \quad \begin{pmatrix} 0 \\ \frac{-\mathbf{D}_{12}}{\mathbf{D}_{11}(\mathbf{x}_1)\mathbf{D}_{22} - \mathbf{D}_{12}^2(\mathbf{x})} \\ 0 \\ \frac{\mathbf{D}_{22}}{\mathbf{D}_{11}(\mathbf{x}_1)\mathbf{D}_{22} - \mathbf{D}_{12}^2(\mathbf{x})} \end{pmatrix} \mathbf{u} \quad (43) \\ y &= x_1 \end{aligned}$$

(see Ramos *et al* 1997) for the definition of the terms involved in the model and their numerical values),

In order to apply the control scheme described in the previous section, suppose that the reference signal is produced by the exosystem $\dot{w}_1 = \alpha w_2, \dot{w}_2 = -\alpha w_1$ with $\alpha = 1$ and $w(0) =$col$(w_1(0), w_2(0))$ , and $e = x_1 - w_2$.

choosing the transformation $x_1' = x_1$, $x_2' = x_2 - g_2(x_1)g_4^{-1}(x_1)x_4$, $x_3' = x_3$ and $x_4' = x_4$, where $g_2(x_1)g_4^{-1}(x_1) = D_{12}(x_1)D_{22}^{-1}(x_1)$, system (43) is transformed into the regular form

$$\dot{\mathbf{x}} = \begin{pmatrix} x_2' - D_{12}D_{22}^{-1}x_4' \\ -(q+G_2)D_{22}^{-1} + \mathbf{C}_2(\mathbf{x}')x_4' \\ x_4' \\ \frac{D_{22}}{D_{11}D_{22}-D_{12}^2}p_1 \end{pmatrix} + \begin{pmatrix} 0 \\ 0 \\ 0 \\ \frac{D_{22}}{D_{11}D_{22}-D_{12}^2} \end{pmatrix}\mathbf{u}$$

Now, setting $\pi = Col(\pi_1, \pi_2, \pi_3)$ and defining $z_1 = \mathbf{x}_1' - \pi_1(\mathbf{w})$, $z_2 = x_2' - \pi_2(\mathbf{w})$, $z_3 = x_3' - \pi_3(\mathbf{w})$ and

$$\sigma = x_4 - \sigma_1(\mathbf{z}, \mathbf{w}), \quad \sigma_1 = c_1(\mathbf{w}) - k_1 z_1 - k_2 z_2 - k_3 z_3,$$

with $k_1, k_2$ and $k_3$ are constant parameters, after some tedious calculation, an approximate solution (Castillo 1992) can be found as

$$\begin{aligned} \pi_1(\mathbf{w}) &= w_2 \\ \pi_2(\mathbf{w}) &= -w_1 + c_1(\mathbf{w}) \\ \pi_3(\mathbf{w}) &= -0.974377w_2 + 0.0145446w_1^2 w_2 - \\ &\quad 0.00921993w_2^3 \\ c_1(\mathbf{w}) &= 0.974374w_1 - 0.0145446w_1^3 - \\ &\quad 0.05674899w_1^2 w_2. \end{aligned}$$

Finally, the control

$$\begin{aligned} \mathbf{u} &= \mathbf{u}_{eq} - kg_4^{-1}\text{signo}\sigma, \; k = 10 \\ \mathbf{u}_{eq} &= -g_4^{-1}\left(f_4' + k_1 f_1' + k_2 f_2' + k_3 f_3' - \frac{\partial c_1(\mathbf{w})}{\partial \mathbf{w}}\right), \end{aligned}$$

is chosen to maintain sliding mode motion on $\sigma = 0$. To stabilize the linear approximation of the sliding mode equation

$$\begin{aligned} \dot{z}_1 &= 1.5961k_1 z_1 + (1 + 1.5961k_2)z_2 + 1.5961k_3 z_3 \\ \dot{z}_2 &= 21.6707z_1 + 21.6707z_3 \\ \dot{z}_3 &= -k_1 z_1 - k_2 z_2 - k_3 z_3, \end{aligned}$$

the set of parameters $[k_1, k_2, k_3] = [-34.8401, -7.4834, -40.6080]$ were used. In all the simulations, the initial conditions of the system were chosen near the equilibrium point $x_1 = \frac{\pi}{2}$ and $x_3 = 0$ while variations in the nominal values of $m_i$ and $l_i$ were introduced. The simulation results (not shown here for space reasons) allows to assure that the controller guarantees small output tracking error, despite variation on the systems parameters.

## CONCLUSSIONS

In this work it have been shown that by choosing appropiately a discontinuous control law, it is possible to induce a center manifold on which the output tracking error is zeroed. The sliding regulator problem is introduced and conditions for the existence of a solution for the linear and nonlinear case as well are derived. For a particular class of nonlinear systems, an explicit solution to the problem may be obtained. Simulations performed on a model of the pendubot system show the performance of the proposed structure.

## REFERENCES


Castillo-Toledo B. (1990). Almost tracking through singular points via the nonlinear regulator theory. IV Latin American Congress on Automatic Control, Mexico, Vol.2, pp. pp. 590-594.

Davison E. J. and Goldenberg A. (1975), The robust control of a general servomechanism problem: the servo compensator, *Automatica*, Vol. 11, pp.461-471.

El-Chesawi, O.M.E., Zinober, A.S.I., Billings, S.A. 1983, Analysis and design of variable structure systems using a geometric approach. *International Journal of Control* 38, pp. 657-671.

Francis, B.A. (1977). The linear multivariable regulator problem. *SIAM J. Control Optimiz.*, Vol. 15, pp. 486-505.

Isidori A., Byrnes C.I.(1990). Output regulation of nonlinear systems, *IEEE Trans. Aut. Control*, Vol 35, No.2, pp. 131-140.

Luk'yanov A.G. and Utkin V.I. (1981). Methods for reducing dynamic systems to regular form, *Automation and Remote Control*, Vol. 42, No.4, (P.1), pp. 413-420.

Ramos L. E., B. Castillo-Toledo and J. Alvarez, Nonlinear regulation of an underactuated system, *Procceedings of the 1997 IEEE International Conference on Robotics and Automation*, Albuquerque, New Mexico-April 1997

Sebakhy O. A. and Wonham W. M., (1976). Design procedure for multivariable regulator, *Automatica*, Vol. 12, pp. 467-478.

Utkin, V.I. (1981). Sliding modes in control and optimization (in Russian), Nauka. Moscow.

Utkin, V.I. and Young K.-K.D. (1978). Methods for constructing discontinuity planes in multidimensional variable-structure systems. Automation and Remote Control. Vol.39, No.10. (P.1) pp.1466-1470.

# Backstepping Variable Structure Control of Nonlinear Systems with Unmatched Uncertainties

**Li, Jun  Xu, Demin*  Ren, Zhang**

*College of Marine Engineering, Northwestern Polytechnical Univ.,*
*Xi'an, P. R. China, 710072*
**Tel.: +86(29)8492824  Fax: +86(29)8494000*
**e-mail: npuxdm@pub.xa-online.sn.cn*

Abstract: In this paper, the application of variable structure control to process control problems is considered. Variable structure control is presented via the backstepping method as a natural tool for robust design of states feedback controllers for minimum phase nonlinear systems with unmatched un-certainties. An example is presented to illustrate some of the issues involved in the design of the variable structure robust controllers. *Copyright © 1999 IFAC*

Key Word: backstepping, variable structure control, nonlinear systems, uncertainty

## 1.INTRODUCTION

In the last decade several approaches have been developed for nonlinear controller design. Some of the most important results are sliding mode or variable structure control (Slotine and Coetsee, 1986), and adaptive recursive backstepping method (Kanellakopoulos and Kokotovic and Morse, 1991; Krstic and Kanellakopoulos and Kokotovic, 1995). These early results have been extended to output tracking and stabilization of nonlinear systems with parametric uncertainties and unmodelled dynamics, and have usually been carried out from a practical perspective based upon either adaptive or robust control schemes.

For VSC to be used, the uncertain system must satisfy a matching condition. The matching condition assumes that uncertain terms in the state equation are reached directly by inputs. But in practice many physical systems do not satisfy this condition. The adaptive recursive backstepping approach provides a systematic framework for the design of regulation strategies suitable for a large class of state feedback linearization nonlinear systems. For this method to be used, the uncertainties must be constant parameter values and the controlled linearizable plant belongs to the class of systems transformable into the parametric-strict feedback form.

The purpose of this paper is to show how the combination of recursive backstepping approach and variable structure control provides an effective tool for robust nonlinear process control design for minimum phase systems with unmatched non-parameter uncertainties. Here we consider the output tracking problem of this class of systems with the assumption of full state measurement.

## 2.PROBLEM FORMULATION

Consider the single-input single-output nonlinear systems with uncertainties

$$\begin{aligned} \dot{x}_i &= x_{i+1} + \Delta x_i, \quad i = 1,2,\ldots,n-1 \\ \dot{x}_n &= f(x,t) + \Delta f(x,p,t) + \\ &\quad [g(x,t) + \Delta g(x,p,t)]u + d(x,t) \\ y &= x_1 \end{aligned} \tag{1}$$

where $x \in R^n$ is the state, $u$、$y \in R^1$ are scalar input and scalar output respectively, $f(x,t)$ and $g(x,t)$ are known sufficiently smooth functions, and $f(0,0) = 0$, $p \in P \subset R^l$ a compact set that represents parametric uncertainty in the model. $\Delta x_i$ $(i = 1,\cdots,n-1)$, $\Delta f(x,p,t)$ and $\Delta g(x,p,t)$ are bounded uncertainties. $d(x,t)$ is bounded external disturbance.

The object is to design a feedback controller such that the system output $y(t)$ asymptotically tracks a desired trajectory $y_d(t)$. We also assume that $y_d(t)$ and its derivatives up to $n$ are bounded and are sufficiently smooth functions of $t$.

**Assume 1** There exist known smooth function set $a_i(x_1,\ldots,x_i)$, such that

$$|\Delta x_i| \le a_i(x_1,\ldots,x_i), \quad i = 1,2,\ldots,n-1$$

**Assume 2** $|\Delta f(x,p,t)| \le \bar{f}(x,t)$,
$|d(x,t)| \le \bar{d}(x,t)$, $\forall x \in R^n, t \in R, p \in P$

**Assume 3**

$$g^L(x,t) \le \Delta g(x,p,t) \le g^U(x,t),$$

$\forall x \in R^n, t \in R, p \in P$, and

$$g(x,t) + g^L(x,t) > \alpha$$

where $\alpha$ is an arbitrary positive number.

## 3.BACKSTEPPING VSC DESIGN

We define $n$ error variables using the virtual control $x_i, i = 1,2,\ldots,n-1$ and its desired value $x_{id}, i = 1,2,\ldots,n-1$

$$\begin{aligned} s_1 &= x_1 - x_{1d} \\ s_2 &= x_2 - x_{2d} \\ &\vdots \\ s_n &= x_n - x_{nd} \end{aligned} \tag{2}$$

where $x_{1d} = y_d$, and $x_{id}$ $(i = 2,\ldots,n-1)$ is to be computed.

***Step* 1:**
Differentiation of $s_1$ gives

$$\begin{aligned} \dot{s}_1 &= x_2 + \Delta x_1 - \dot{x}_{1d} \\ &= s_2 + x_{2d} - \dot{x}_{1d} + \Delta x_1 \end{aligned} \tag{3}$$

Define Lyapunov function as $V_1 = \frac{1}{2}s_1^2$, and the derivative of $V_1$ along (3) is computed as

$$\dot{V}_1 = s_1(s_2 + x_{2d} - \dot{x}_{1d} + \Delta x_1) \tag{4}$$

Since $\Delta x_1$ is unknown, we introduce nonlinear damping term "$-\frac{s_1 a_1^2}{2\varepsilon}$" to overcome the influence of $\Delta x_1$, the desired value is

$$x_{2d} = -ks_1 + \dot{x}_{1d} - \frac{s_1 a_1^2}{2\varepsilon} \tag{5}$$

where $k, \varepsilon$ are positive number. Putting (5) into (4) yields

$$\dot{V}_1 = -ks_1^2 + s_1 s_2 + s_1 \Delta x_1 - \frac{s_1^2 a_1^2}{2\varepsilon} \tag{6}$$

From Young's Inequality $x^2 + y^2 \ge 2xy$, we can get

$$\frac{s_1^2 a_1^2}{2\varepsilon} + \frac{\varepsilon}{2} \ge |s_1| a_1 \ge s_1 \Delta x_1 \tag{7}$$

as a result of the above

$$\dot{V}_1 \le -2kV_1 + s_1 s_2 + \frac{\varepsilon}{2} \tag{8}$$

***Step* 2:**
Differentiation of $s_2$ gives

$$\dot{s}_2 = x_3 + \Delta x_2 - \dot{x}_{2d} \\ = s_3 + x_{3d} - \dot{x}_{2d} + \Delta x_2 \qquad (9)$$

Because $x_{2d}$ is a function of $x_1$, the differentiation of desired value $x_{2d}$ will introduce uncertain terms. Therefore $\dot{x}_{2d}$ has both certain and uncertain terms. By defining them as $\bar{\dot{x}}_{2d}$ and $\Delta\dot{x}_{2d}$ respectively, $\dot{x}_{2d}$ is expressed as

$$\dot{x}_{2d} = \bar{\dot{x}}_{2d} - \Delta\dot{x}_{2d} \qquad (10)$$

where

$$\bar{\dot{x}}_{2d} = \ddot{x}_{1d} + \left(k + \frac{a_1^2}{2\varepsilon}\right)\dot{x}_{1d} - \left(k + \frac{a_1^2}{2\varepsilon} + \frac{s_1 a_1}{\varepsilon}\frac{\partial a_1}{\partial x_1}\right)x_2 \qquad (11)$$

$$\Delta\dot{x}_{2d} = \left(k + \frac{a_1^2}{2\varepsilon} + \frac{s_1 a_1}{\varepsilon}\frac{\partial a_1}{\partial x_1}\right)\Delta x_1$$

Since $\Delta x_1$ is bounded by $a_1(x_1)$, $\Delta\dot{x}_{2d}$ can be bounded by a smooth function $\beta_2(x_1)$ such that

$$\left|\Delta\dot{x}_{2d}\right| \le \beta_2(x_1) \qquad (12)$$

Define Lyapunov function as $V_2 = V_1 + \frac{1}{2}s_2^2$, we have

$$\dot{V}_2 = \dot{V}_1 + s_2\left(s_3 + x_{3d} - \bar{\dot{x}}_{2d} + \Delta\gamma_2\right) \qquad (13)$$

where $\Delta\gamma_2 = \Delta\dot{x}_{2d} + \Delta x_2$, and if mark $\bar{a}_2(x_1,x_2) = \beta_2(x_1) + a_2(x_1,x_2)$, then

$$\Delta\gamma_2 \le \bar{a}_2(x_1,x_2) \qquad (14)$$

Putting (8) into (13) yields

$$\dot{V}_2 \le -ks_1^2 + s_1 s_2 + \frac{\varepsilon}{2} + s_2(s_3 + x_{3d} - \bar{\dot{x}}_{2d} + \Delta\gamma_2) \qquad (15)$$

As we have done in step 1, the desired value $x_{3d}$ is

$$x_{3d} = -ks_2 - s_1 + \bar{\dot{x}}_{2d} - \frac{s_2\bar{a}_2^2}{2\varepsilon} \qquad (16)$$

From Young's Inequality, we can get

$$\frac{s_2^2\bar{a}_2^2}{2\varepsilon} + \frac{\varepsilon}{2} \ge |s_2|\bar{a}_2 \ge s_2\Delta\gamma_2 \qquad (17)$$

And putting (16) into (15) yields

$$\dot{V}_2 \le -2kV_2 + s_2 s_3 + \varepsilon \qquad (18)$$

The steps discussed above can be continued to the ($n$-1)th step.

***Step n:***

At this final step we adopt the variable structure control to overcome the influence of $\Delta g(x,p,t)$. Differentiation of $s_n$ gives

$$\dot{s}_n = \dot{x}_n - \dot{x}_{nd} \qquad (19)$$

where $\dot{x}_{nd}$ has both certain and uncertain terms. By defining them as $\bar{\dot{x}}_{nd}$ and $\Delta\dot{x}_{nd}$ respectively, $\dot{x}_{nd}$ is expressed as

$$\dot{x}_{nd} = \bar{\dot{x}}_{nd} - \Delta\dot{x}_{nd} \qquad (20)$$

$\Delta\dot{x}_{nd}$ can be bounded by a smooth function $\beta_n(x_1,\ldots,x_{n-1})$ such that

$$\left|\Delta\dot{x}_{nd}\right| \le \beta_n(x_1,\ldots,x_{n-1}) \qquad (21)$$

so the differentiation of $s_n$ can be rewritten as

$$\dot{s}_n = \dot{x}_n - \bar{\dot{x}}_{nd} + \Delta\dot{x}_{nd} \\ = f(x,t) + g(x,t)u + \Delta f + \Delta gu + d(x,t) - \bar{\dot{x}}_{nd} + \Delta\dot{x}_{nd} \qquad (22)$$

Define Lyapunov function $V_n = V_{n-1} + \frac{1}{2}s_n^2$, we have

$$\dot{V}_n \le -2kV_{n-1} + s_{n-1}s_n + \frac{(n-1)\varepsilon}{2} + s_n(f(x,t) - \bar{\dot{x}}_{nd} + g(x,t)u + \Delta gu + \eta) \qquad (23)$$

where $\eta = \Delta f + d(x,t) + \Delta\dot{x}_{nd}$.

There exist

$$\bar{\eta} = \bar{f}(x,t) + \bar{d}(x,t) + \beta_n(x_1,\ldots,x_{n-1})$$

such that $|\eta| \le \bar{\eta}$.

(I) Case 1: $\Delta g(x,p,t) = 0$

Introduce nonlinear damping term $-\frac{s_n\bar{\eta}^2}{2\varepsilon}$ to overcome the influence of $\eta$, the control law is

$$u = \frac{1}{g(x,t)}[-ks_n - s_{n-1} - f(x,t) + \bar{\dot{x}}_{nd} - \frac{s_n\bar{\eta}^2}{2\varepsilon}] \qquad (24)$$

Since

$$\frac{s_n^2\bar{\eta}^2}{2\varepsilon} + \frac{\varepsilon}{2} \ge |s_n|\bar{\eta} \ge s_n\eta \qquad (25)$$

and putting (24) into (23) yields

$$\dot{V}_n \le -2kV_n + \frac{n\varepsilon}{2} \qquad (26)$$

then $\lim_{t\to\infty} V_n \le \frac{n\varepsilon}{4k}$ and $\lim_{t\to\infty} s_1 \le \sqrt{\frac{n\varepsilon}{2k}}$.

Based on the above development, we are in position to formulate our result

**Theorem 1** Consider the system (1) with the assumption 1 ~ 3, in the case of $\Delta g(x,p,t) = 0$, the control law (24) can ensure the tracking error is bounded, and it is less then $\sqrt{\frac{n\varepsilon}{2k}}$.

The proof of the theorem 1 is outlined above.

(II) Case 2: $\Delta g(x,p,t) \ne 0$

In this case, the control law is

$$u = u_c + u_v \tag{27}$$

where

$$u_c = \frac{1}{g(x,t)}[-ks_n - s_{n-1} - f(x,t) + \bar{\ddot{x}}_{nd} - \frac{s_n \bar{\eta}^2}{2\varepsilon}] \tag{28}$$

$$u_v = \tau(x,t)v \tag{29}$$

$$v = \begin{cases} v_1 = -g^U u_c, & s_n > 0 \\ v_2 = -g^L u_c, & s_n < 0 \\ v_0, v_0 \in [v_1,\ v_2], & s_n = 0 \end{cases} \tag{30}$$

$$\tau(x,t) = \frac{1}{g(x,t) + \zeta(x,t)} \tag{31}$$

$$\zeta(x,t) = \begin{cases} g^U(x,t), & s_n v > 0 \\ g^L(x,t), & s_n v < 0 \end{cases} \tag{32}$$

**Theorem 2** Consider the system (1) with the assumption 1 ~ 3, in the case of $\Delta g(x,p,t) \ne 0$, the control law (27)~(32) can ensure the tracking error is bounded, and it is less then $\sqrt{\frac{n\varepsilon}{2k}}$.

**Proof:** Define $V_n$ is a Lyapunov function of system (1), we have

$$\begin{aligned}\dot{V}_n &\le -2kV_{n-1} + s_{n-1}s_n + \frac{(n-1)\varepsilon}{2} + \\ &\quad s_n\left(f(x,t) - \bar{\ddot{x}}_{nd} + g(x,t)u + \eta + \Delta g u\right) \\ &= -2kV_{n-1} + s_{n-1}s_n + \frac{(n-1)\varepsilon}{2} + \\ &\quad s_n\left(f(x,t) - \bar{\ddot{x}}_{nd} + g(x,t)u_c + \eta\right) \\ &\quad + s_n\left(\Delta g u_c + (g(x,t) + \Delta g)u_v\right) \\ &\le -2kV_n + \frac{n\varepsilon}{2} + s_n \Delta g u_c + \\ &\quad s_n v(g(x,t) + \Delta g)\tau \end{aligned} \tag{33}$$

Putting (31) and (32) into (33) yields

$$\dot{V}_n \le -2kV_n + \frac{n\varepsilon}{2} + s_n \Delta g u_c + s_n v \tag{34}$$

Apply (30), we can get

$$\max\{s_n[\Delta g u_c + v]\}\begin{cases} \le 0, s_n \ne 0 \\ = 0, s_n = 0 \end{cases} \tag{35}$$

then

$$\dot{V}_n \le -2kV_n + \frac{n\varepsilon}{2} \tag{36}$$

so $\lim_{t\to\infty} V_n \le \frac{n\varepsilon}{4k}$ and $\lim_{t\to\infty} s_1 \le \sqrt{\frac{n\varepsilon}{2k}}$ □

## 4.NUMERICAL SIMULATION

As an illustrative example, we consider the following system

$$\begin{aligned} \dot{x}_1 &= x_2 + \Delta x_1 \\ \dot{x}_2 &= x_1 \sin x_2 + x_2 + \alpha(t) x_1 \sin x_2 \\ &\quad + (g + \Delta g)u + d(t) \\ y &= x_1 \end{aligned} \tag{37}$$

where

$$\Delta g = 0.5\sin(5t),$$
$$\Delta x_1 = 0.1\sin t, \alpha(t) = \sin(4t),$$
$$d(t) = 2\cos(t)$$

are uncertainties. And the bounds are

$$\bar{f} = |x_1|, \bar{d} = 2.5, a_1 = 0.12,$$
$$g^U = 0.7, g^L = -0.7$$

The desired trajectory is $y_d = \sin(t)$.

Choose $\beta_2 = [k + \frac{a_1^2}{2\varepsilon}]a_1, k = 1$, and $\varepsilon = 0.05$.

The initial conditions is $x(0) = 1$ and $\dot{x}(0) = 0$. We adopt the boundary method to reduce the chattering, and the boundary width is $B = 0.05$.

The simulation results are plotted in figure 1.

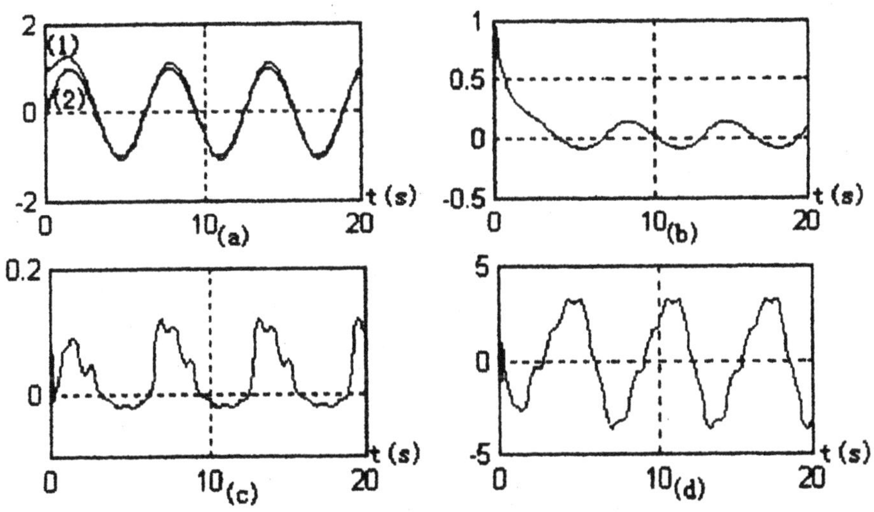


Fig. 1. Simulation result (a)output curve(curve 1: real output; curve 2: desired output) (b)tracking error $s_1$ (c) $s_2$ (d)control

It is clear from figure 1 that the tracking error of closed system is convergent and the control effort oscillates accompany the input signal. The controller is stable. The tracking error is decided by the constant $\varepsilon$ and $k$. For example, if increase $\varepsilon$ or decrease $k$, the tracking error will decrease, but the control value will increase.

## 5.Conclusion

In this paper we presented a new controller for nonlinear systems with unmatched uncertainties, which is based on the combination of the variable structure control and recursive backstepping approach. The key idea is applying the backstepping method to design the controller and introducing the suitable nonlinear damping to overcome the influence of uncertainties at each step. The proposed algorithm is an extension of conventional variable structure control that requires the uncertainties to satisfy the matching condition.

## REFERENCES

Kanellakopoulos, I., Kokotovic, P., and Morse, A. S. (1991). Systematic Design of Adaptive Controllerds for Feedback Linearizable Systems, In: *IEEE Trans. on Automatic Control*, **vol. 36**, pp. 1241-1253.

Krstic, M., Kanellakopoulos, I., and Kokotovic, P. (1995). *Nonlinear and Adaptive Control Design*, New York, Wiley.

Rios-Bolivar, M., Zinober, A. S. I., and Sira-Ramirez(1997). Dynamical Adaptive Sliding Mode Output Tracking Control of a Class of Nonlinear systems, In: *Int. J. of Robust and Nonlinear Control*, **vol. 7**, pp. 387-405.

Slotine, J. J. E. and Coetsee, J. A. (1986). Adaptive Sliding Synthesis for Nonlinear Systems, In: *Int. J. of Control*, **vol. 43(6)**. pp.1631-1651.

Won, M., and Hedrick, J. K. (1996). Multiple-surface sliding control of a class of uncertain nonlinear systems, In: *Int. J. of Control*, **vol. 64, no. 4**, pp. 693-706.

F-2c-23-1
14th Triennial World Congress, Beijing, P.R. China

# GLOBAL $H_\infty$ CONTROL FOR A CLASS OF INTERCONNECTED NONLINEAR SYSTEMS

**Shoulie Xie,** * **Lihua Xie** * **and Wei Lin** **,[1]

* *School of Electrical & Electronic Engineering, Nanyang Technological University, Singapore 639798*

** *Department of Electrical Engineering and Computer Science, Case Western Reserve University, Cleveland, OH 44106, USA*

Abstract: This paper is concerned with a global $\mathcal{H}_\infty$ control problem for a class of interconnected nonlinear systems. We first consider a fairly general class of large-scale nonlinear systems with strong nonlinear interconnections. It is shown that the decentralized $\mathcal{H}_\infty$ control problem for the system can be converted into centralized $\mathcal{H}_\infty$ control problems associated with a set of auxiliary systems whose solutions are related to the Hamilton-Jacobi-Isaacs (HJI) inequalities. Realizing that finding a global solution of the HJI inequality is usually impossible, we then concentrate on a globally *inverse* $\mathcal{H}_\infty$ control problem for a class of interconnected nonlinear systems which are transformable to interconnected systems with lower triangular structure. We show that the decentralized control laws and the weights can be constructed explicitly using a two-way (backward and forward) recursive design technique extended from (Isidori and Lin, 1998). *Copyright © 1999 IFAC*

Keywords: Interconnected systems, nonlinear systems, decentralized control, $\mathcal{H}_\infty$ control, inverse Lyapunov approach.

## 1. INTRODUCTION

Large-scale systems such as power systems and digital communication networks are composed of a set of small interconnected subsystems. It is generally impossible to incorporate many feedback loops into the controller design and is also too costly even if they can be implemented. These difficulties motivate the development of decentralized control theory where each subsystem is controlled independently on the basis of its own performance criterion and locally available information. The decentralized stabilization problem for interconnected linear systems with uncertainties satisfying the so-called strict matching conditions has been investigated in (Chen *et al.*, 1991; Gavel and Siljak, 1989; Shi and Singh, 1992; Siljak, 1991) and references therein. The interconnections among subsystems treated in these papers are mostly bounded by first-order polynomials of state. It was pointed out in (Shi and Singh, 1992) that a decentralized control system based on the first-order bounded interconnections may become unstable when the interconnections are of higher order. Recently, following the development of centralized control of nonlinear systems, a decentralized adaptive stabilization for a class of large-scale interconnected nonlinear systems was proposed in (Jain and Khorrami, 1997) where the strict matching condition was relaxed and higher-order polynomial interconnections among subsystems were considered.

[1] Wei Lin's research is supported in part by the U.S. NSF under Grant DMS-9634395, and by the Wareen E. Rupp Endowed Professorship.

In the past few years, the decentralized $\mathcal{H}_\infty$ control problem for linear systems has been considered in (Paz, 1993; Wang *et al.*, 1993), to name but a few. In spite of significant advance in centralized $\mathcal{H}_\infty$ control for nonlinear systems; see, for instance, (Ball *et al.*, 1993; Isidori and Kang, 1995; van der Schaft, 1991; van der Schaft, 1992) and references therein, few results on decentralized $\mathcal{H}_\infty$ control of large-scale interconnected nonlinear systems are available in the literature. Further, note that all these results on $\mathcal{H}_\infty$ control of nonlinear systems require solving the Hamilton-Jacobi-Isaacs (HJI) partial differential equations, which imposes a formidable difficulty. This is the main reason why nonlinear $\mathcal{H}_\infty$ control theory has not yet penetrated into many practical applications. Very recently, some interesting results have been obtained on global disturbance attenuation and almost disturbance decoupling, for a class of nonlinear systems with lower triangular structure (Isidori, 1995; Marino *et al.*, 1994). A significant feature of these results is that solving the HJI equations or inequalities is not required. Their basic limitation is, however, the lack of consideration for the *cost* needed to implement the desired control action, because the cost functions in the above-mentioned papers do not impose any penalty on control efforts. This issue has been addressed in (Isidori and Lin, 1998), where a global $\mathcal{L}_2$-gain design methodology was developed for minimum-phase nonlinear systems in the lower triangular form.

Due to the work (van der Schaft, 1991; van der Schaft, 1992; Lin and Xie, 1998), the relationship between an $\mathcal{L}_2$-gain of a nonlinear system and an $\mathcal{L}_2$-gain of its linearized system has become quite transparent. It is well known now that if the $\mathcal{H}_\infty$ control problem for the linearized system is solvable, one can find trivially a *local* solution to the $\mathcal{H}_\infty$ control problem of the orginal nonlinear system. Quite interestingly, the work (Isidori and Lin, 1998) suggests that in some special cases, for example, the linearizable systems or minimum-phase nonlinear systems with triangular structure, the solution to the problem of disturbance attenuation for the linearized system suffices to determine a feedback law that solves the *global* disturbance attenuation problem with internal stability, for the corresponding nonlinear system (Isidori and Lin, 1998). Note that in (Isidori and Lin, 1998), since weighting function $r(x)$ is fixed a *posteriori*, only the problem of *inverse* global $L_2$-gain analysis is addressed, i.e. determine a weighting function $r(x)$ and a globally stabilizing feedback law $\alpha(x)$ (which is constructed by starting from the solution of a strict Riccati inequality) that solve the problem of global $\mathcal{H}_\infty$ disturbance attenuation.

In this paper we shall investigate a global decentralized $\mathcal{H}_\infty$ control problem via state feedback control, for a class of large-scale interconnected nonlinear systems. First, we consider a rather general interconnected nonlinear system with strong nonlinear interconnections. We show how the global decentralized $\mathcal{H}_\infty$ control problem of the system can be converted into the centralized $\mathcal{H}_\infty$ control problems for a set of auxiliary nonlinear systems without interconnections. It is well known that solutions to the later problems are related to the HJI equations. This result extends the decentralized $\mathcal{H}_\infty$ control problem for linear interconnected systems to the nonlinear case. Observe, however, that so far there is no effective algorithm available for solving the HJI equations globally. To avoid this difficulty, we then focus our attention on a decentralized global *inverse* $\mathcal{H}_\infty$ control problem for a class of interconnected nonlinear systems which are transformable to interconnected systems with lower triangular structure. More specifically, we are interested in seeking a set of decentralized state feedback control laws as well as state-dependent weights of the control inputs such that the associated global $\mathcal{H}_\infty$ control problem is solvable. The main tools used in this paper consist of casting the decentralized global $\mathcal{H}_\infty$ control problem into centralized $\mathcal{H}_\infty$ control once and the two-way (backward and forward) recursive design technique developed in (Isidori and Lin, 1998) for feedback linearizable systems. In this paper we show how this technique can be extended to solve the decentralized global inverse $\mathcal{H}_\infty$ control problem, for interconnected nonlinear systems.

## 2. DECENTRALIZED $\mathcal{H}_\infty$ CONTROL OF INTERCONNECTED NONLINEAR SYSTEMS

Consider a large-scale nonlinear system composed of $N$ interconnected subsystems of the form:

$$\begin{aligned} \dot{x}_i &= A_i(x_i) + B_i(x_i)u_i + p_i(x_i)\omega_i + h_i(x) \\ y_i &= C_i(x_i) \\ z_i &= (y_i^T y_i + u_i^T r_i(x_i) u_i)^{\frac{1}{2}}, \qquad i = 1, 2, \cdots, N, \end{aligned} \tag{2.1}$$

where $x_i \in \Re^{n_i}$ is the state of $i$th subsystem, $i = 1, 2, \cdots, N$, $x = [x_1^T \ \cdots \ x_N^T]^T$ is the state of the overall interconnected system, $u_i \in \Re^{p_i}$, $\omega_i \in \Re^{r_i}$ and $z_i \in \Re$ are the control input, the disturbance input and the penalty output, respectively. $A_i(x_i)$, $B_i(x_i)$, $C_i(x_i)$, $p_i(x_i)$ and $r_i(x_i)$ are smooth with appropriate dimensions and satisfy $A_i(0) = 0$, $C_i(0) = 0$, $h_i(0) = 0$ and $r_i(x_i) \geq 0$ for any $x_i \in \Re^{n_i}$.

*Assumption 2.1.* The nonlinear interconnections $h_i(x) = [h_{i1}(x) \ \ h_{i2}(x) \ \ \cdots \ \ h_{in_i}(x)]^T$ in (2.1) are bounded by

$$|h_{ij}(x)| \le \eta_{ij}(x_i) \sum_{l=1}^{N} \zeta_{ijl}(x_l) \qquad (2.2)$$

where $\eta_{ij}(x_i)$, $\zeta_{ijl}(x_l)$, $1 \le j \le n_i$, $1 \le i,\ l \le N$, are non-negative continuous functions with $\zeta_{ijl}(0) = 0$.

*Remark 2.1.* Note that the interconnections $h_i(x)$, $i = 1, 2, \cdots, N$, in Assumption 2.1 are very general which include many types of interconnections considered in the existing literature as special cases, for example, the interconnections bounded by linear (first-order) polynomials (Chen *et al.*, 1991; Gavel and Siljak, 1989; Gong *et al.*, 1996) and higher-order polynomials (Shi and Singh, 1992). Furthermore, unlike the work in (Chen *et al.*, 1991; Gavel and Siljak, 1989; Gong *et al.*, 1996; Shi and Singh, 1992), no matching conditions are imposed in Assumption 2.1.

In this section, we deal with the *global decentralized $\mathcal{H}_\infty$ control* problem for the system (2.1) satisfying Assumption 2.1. More precisely, *given scalars $\gamma_i > 0$, $i = 1, 2, \cdots, N$, we are concerned with the design of decentralized control laws, $u_i = u_i(x_i)$, $i = 1, 2, \cdots, N$, for each subsystem such that the overall closed-loop interconnected system (2.1) is globally asymptotically stable and the $\mathcal{L}_2$-gain from the disturbance $\omega = [\omega_1^T \ \cdots \ \omega_N^T]^T$ to the controlled output $z = [z_1^T \ \cdots \ z_N^T]^T$ is less than $\gamma = [\gamma_1 \ \cdots \ \gamma_N]^T$ in the sense that*

$$\sum_{i=1}^{N} \int_0^\infty z_i^T z_i dt < \sum_{i=1}^{N} \gamma_i^2 \int_0^\infty \omega_i^T \omega_i dt + \delta(x_0) \qquad (2.3)$$

*for all $\omega_i \in \mathcal{L}_2[0, \infty)$, where $\delta(x_0)$ is a real-valued function of the initial state $x_0 = [x_1^T(0) \ \cdots \ x_N^T(0)]^T$ satisfying $\delta(0) = 0$.*

*Remark 2.2.* In the above problem statement, $\gamma_i$, $i = 1, 2, \cdots, N$, can be regarded as the prespecified level of $\mathcal{H}_\infty$ disturbance attenuation for each subsystem. When $\gamma_i = \gamma_0$, $i = 1, 2, \cdots, N$, (2.3) becomes

$$||z||_2^2 < \gamma_0^2 ||\omega||_2^2 + \delta(x_0) \qquad (2.4)$$

for all $\omega_i \in \mathcal{L}_2[0\ \infty)$, where $z = [z_1 \ z_2 \ \cdots \ z_N]^T$ and $\omega = [\omega_1^T \ \cdots \ \omega_N^T]^T$. In this case, a standard decentralized $\mathcal{H}_\infty$ control problem is recovered.

Before presenting a result on decentralized nonlinear $\mathcal{H}_\infty$ control, we recall the definition of global $\mathcal{H}_\infty$ disturbance attenuation for nonlinear systems. To this end, consider a nonlinear system of the form:

$$\begin{aligned} \dot{x} &= f(x) + g(x)u + p(x)\omega \\ y &= h(x) \qquad (2.5) \\ z &= \left(y^T y + u^T r(x) u\right)^{\frac{1}{2}} \end{aligned}$$

where $\omega \in \Re^r, u \in \Re^p$ and $z \in \Re$ are the disturbance input, the control input and the penalty output, respectively, $f(0) = 0$, $h(0) = 0$ and $r(x) \ge 0$ for all $x \in \Re^n$.

*Definition 2.1.* Given a real number $\gamma > 0$, the system (2.5) is said to have global $\mathcal{H}_\infty$ disturbance attenuation $\gamma$ if there exists a feedback control law $u = \alpha(x)$ with $\alpha(0) = 0$ such that for some positive definite and proper function $V(x)$, the HJI inequality

$$\frac{\partial V(x)}{\partial x}(f(x) + g(x)\alpha(x)) + \frac{1}{4\gamma^2}\left(\frac{\partial V(x)}{\partial x}p(x)\right)^2 + h^T(x)h(x) + \alpha^T(x)r(x)\alpha(x) < 0 \qquad (2.6)$$

is satisfied for all nonzero $x$.

*Remark 2.3.* It is well known that if (2.6) holds, then the feedback law $u = \alpha(x)$ globally asymptotically stabilizes the equilibrium $x = 0$ of the system (2.5) when $\omega = 0$, and it renders the $\mathcal{L}_2$-gain from $\omega$ to $z$ less than $\gamma$, see, e.g. (van der Schaft, 1991; van der Schaft, 1992) for details.

Associated with the interconnected system (2.1), we introduce the following auxiliary system:

$$\begin{aligned} \dot{x}_i &= A_i(x_i) + B_i(x_i)u_i + \left[p_i(x_i) \quad \beta^{\frac{1}{2}}\gamma_i\eta_i(x_i)\right]\tilde{\omega}_i \\ \tilde{y}_i &= \begin{bmatrix} C_i(x_i) \\ \beta_i^{-\frac{1}{2}}(d_i(x_i))^{\frac{1}{2}} \end{bmatrix} \qquad (2.7) \\ \tilde{z}_i &= (\tilde{y}_i^T\tilde{y}_i + u_i^T r_i(x_i)u_i)^{\frac{1}{2}}, \qquad i = 1, 2, \cdots, N, \end{aligned}$$

where $x_i$ is the state, $\tilde{\omega}_i$ is the disturbance input, $u_i$ is the control input, $\tilde{z}_i$ is the penalty output, $A_i(x_i)$, $B_i(x_i)$, $C_i(x_i)$, $p_i(x_i)$ and $r_i(x_i)$ are the same as in the system (2.1), $\beta_i$, $i = 1, 2, \cdots, N$, are some positive scalars and $\beta = \sum_{i=1}^{N} \beta_i$. Moreover,

$$\begin{aligned} \eta_i(x_i) &= diag\{\eta_{i1}(x_i), \cdots, \eta_{in_i}(x_i)\}, \\ \zeta_{il}(x_l) &= [\zeta_{i1l}(x_l) \ \cdots \ \zeta_{in_il}(x_l)]^T, \\ d_i(x_i) &= \sum_{l=1}^{N} \zeta_{li}^T \zeta_{li}. \end{aligned}$$

The main result of this section is the following theorem that shows that to solve the global decentralized $\mathcal{H}_\infty$ control problem for the system (2.1), it suffices to solve the problem for the auxiliary system (2.1).

*Theorem 2.1.* Consider the interconnected system (2.1) satisfying Assumption 2.1. Given $\gamma_i > 0, i = 1, 2, \cdots, N$, suppose that there exist state feedback control laws $u_i = \alpha_i(x_i)$ with $\alpha_i(0) = 0$ such that the system (2.7) has global $\mathcal{H}_\infty$ disturbance attenuation $\gamma_i$ from the disturbance $\tilde{\omega}_i$ to the penalty output $\tilde{z}_i$. Then, the decentralized control laws $u_i = \alpha_i(x_i)$ render the system (2.1) globally asymptotically stable with the $\mathcal{L}_2$-gain from the disturbance $\omega = [\omega_1^T \ \cdots \ \omega_N^T]^T$ to the controlled output $z = [z_1^T \ \cdots \ z_N^T]^T$ less than $\gamma = [\gamma_1 \ \cdots \ \gamma_N]^T$ for all $\omega_i \in \mathcal{L}_2[0, \infty)$, $i = 1, 2, \cdots, N$.

*Proof:* Suppose that the system (2.7) has global $\mathcal{H}_\infty$ disturbance attenuation $\gamma_i$. By Definition 2.1, there exist feedback control laws $u_i = \alpha_i(x_i)$ with $\alpha_i(0) = 0$, such that for some positive definite and proper function $V_i(x_i)$, the HJI inequalities

$$\frac{\partial V_i(x_i)}{\partial x_i}(A_i(x_i) + B_i(x_i)\alpha(x_i)) + \frac{1}{4\gamma_i^2}\left(\frac{\partial V_i(x_i)}{\partial x_i}p_i(x_i)\right)^2 + \frac{1}{4}\beta\frac{\partial V_i(x_i)}{\partial x_i}\eta_i(x_i)\eta_i^T(x_i)\left(\frac{\partial V_i(x_i)}{\partial x_i}\right)^T + C_i^T(x_i)C_i(x_i) + \beta_i^{-1}d_i(x_i) + \alpha_i^T(x_i)r_i(x_i)\alpha_i(x_i) < 0, \qquad (2.8)$$

are satisfied $\forall x_i \neq 0$, $i = 1, 2, \cdots, N$.

Define $V(x) = \sum_{i=1}^{N} V(x_i)$. By employing (2.8), it is then easy to show that along the state trajectory of the interconnected system (2.1),

$$\sum_{i=1}^{N}\left(\dot{V}_i - \gamma_i^2\omega_i^T\omega_i + z_i^T z_i\right) < 0$$

Therefore, the asymptotic stability of the interconnected system when $\omega \equiv 0$ is obvious. The $\mathcal{H}_\infty$ performance also follows by integrating the above inequality. □

Theorem 2.1 demonstrates that the decentralized $\mathcal{H}_\infty$ control problem for interconnected nonlinear systems can be cast into the associated centralized $\mathcal{H}_\infty$ control problems whose solutions are related to the HJI equations. This result extends the decentralized $\mathcal{H}_\infty$ control of interconnected linear systems (Wang *et al.*, 1993) to the nonlinear case. Since solving an HJI equation globally remains challenging, as illustrated in (van der Schaft, 1992; Lin and Xie, 1998), it is straightforward to prove, via the first approximation argument, that the solution to the decentralized $\mathcal{H}_\infty$ control problem of the linearized interconnected system of (2.1), also gives a *local* solution of the decentralized $\mathcal{H}_\infty$ control problem for the nonlinear system (2.1).

## 3. GLOBAL INVERSE $\mathcal{H}_\infty$ CONTROL OF INTERCONNECTED NONLINEAR SYSTEMS

In the previous section we have shown, theoretically speaking, that a global solution to the problem of decentralized $\mathcal{H}_\infty$ nonlinear control can be obtained if the global solutions of the HJI inequalities (2.8) can be found. Note, however, that there are still lack of effective methodologies of solving the HJI inequalities or equations currently. Moreover, it is usually impossible to solve the HJI inequality of the form (2.8) *globally,* for analytical and topological reasons related to the nature of a partial differential inequality or equation. Keeping these facts in mind, we pursue in this section a less ambitious design goal. More specifically, we will study the problem of global *inverse* $\mathcal{H}_\infty$ control rather than the *regular* global $\mathcal{H}_\infty$ control problem. Our aim is to seek not only a set of decentralized feedback control laws but also weighting functions such that the associated $\mathcal{H}_\infty$ problem is solvable *globally,* for a class of interconnected nonlinear systems. The global inverse $\mathcal{H}_\infty$ control result below owes a great deal of inspiration from the recent work (Isidori and Lin, 1998), and is indeed a generalization of the result in (Isidori and Lin, 1998).

Consider a class of interconnected nonlinear systems which are transformable to interconnected nonlinear systems of the form (Jain and Khorrami, 1997):

$$\begin{aligned}
\dot{x}_{i1} &= x_{i2} + p_{i1}(x_{i1})\omega_i + \phi_{i1}(x_{i1}; X_{i1}) \\
&\ \ \vdots \\
\dot{x}_{i,n_i-1} &= x_{i,n_i} + p_{i,n_i-1}(\bar{x}_{i,n_i-1})\omega_i \\
&\quad + \phi_{i,n_i-1}(\bar{x}_{i,n_i-1}; X_{i1}) \\
\dot{x}_{i,n_i} &= u_i + p_{i,n_i}(\bar{x}_{i,n_i})\omega_i + \phi_{i,n_i}(\bar{x}_{i,n_i}; X_{i1}) \\
y_i &= x_{i1}
\end{aligned} \qquad (3.1)$$

where $u_i \in \Re$ is the local control input, $\bar{x}_{ij} = [x_{i1} \ \ x_{i2} \ \ \cdots \ \ x_{ij}]^T$, $i = 1, 2, \cdots, N$, $j = 1, 2, \cdots, n_i$, $X_{i1} = [x_{11} \ x_{21} \ \cdots \ x_{i-1,1} \ x_{i+1,1} \ \cdots \ x_{N1}]^T$, $p_{i1}(x_{i1}), \cdots, p_{in_i}(\bar{x}_{in_i})$ and $\phi_{i1}(x_{i1}; X_{i1})$, $\cdots$, $\phi_{in_i}(\bar{x}_{i,n_i}; X_{i1})$ are smooth with $\phi_{ij}(0; 0) = 0$, $i = 1, \cdots, N$, $j = 1, \cdots, n_i$.

For convenience, we consider $n_i = n$, for $1 \le i \le N$, and represent the system in a compact form:

$$\begin{aligned}
\dot{x}_i &= A_i x_i + f_i(x_i) + p_i(x_i)\omega_i + B_i u_i + h_i(x_i; X_{i1}) \\
y_i &= C_i x_i
\end{aligned} \qquad (3.2)$$

where $x_i = [x_{i1} \ x_{i2} \ \cdots \ x_{in}]^T = \bar{x}_{in}$ and

$$A_i = \begin{bmatrix} 0 & 1 & 0 & \cdots & 0 \\ 0 & 0 & 1 & \cdots & 0 \\ \vdots & \vdots & \vdots & \ddots & \vdots \\ 0 & 0 & 0 & \cdots & 1 \\ 0 & 0 & 0 & \cdots & 0 \end{bmatrix}_{n\times n}, \quad B_i = \begin{bmatrix} 0 \\ 0 \\ \vdots \\ 0 \\ 1 \end{bmatrix}_{n\times 1},$$

$$C_i = [1 \ \ 0 \ \ \cdots \ \ 0]_{1\times n},$$

$$f_i(x_i) = \begin{bmatrix} \phi_{i1}(x_{i1}; 0) \\ \phi_{i2}(\bar{x}_{i2}; 0) \\ \vdots \\ \phi_{in}(\bar{x}_{in}; 0) \end{bmatrix}, \quad p_i(x_i) = \begin{bmatrix} p_{i1}(x_{i1}) \\ p_{i2}(\bar{x}_{i2}) \\ \vdots \\ p_{in}(\bar{x}_{in}) \end{bmatrix},$$

$$h_i(x_i, X_{i1}) = \begin{bmatrix} \phi_{i1}(x_{i1}; X_{i1}) - \phi_{i1}(x_{i1}; 0) \\ \phi_{i2}(\bar{x}_{i2}; X_{i1}) - \phi_{i2}(\bar{x}_{i2}; 0) \\ \vdots \\ \phi_{in}(\bar{x}_{in}; X_{i1}) - \phi_{in}(\bar{x}_{in}; 0) \end{bmatrix}$$

*Assumption 3.1.* The nonlinear interconnections $h_i(x_i; X_{i1})$, $i = 1, 2, \cdots, N$ in (3.2) are bounded by strong nonlinearities in $X_{i1}$:

$$|\phi_{ij}(\bar{x}_{ij}; X_{i1}) - \phi_{ij}(\bar{x}_{ij}; 0)| \leq \sum_{l=1, l\neq i}^{N} |\eta_{ij}(\bar{x}_{ij})||\zeta_{ijl}(x_{l1})| \quad (3.3)$$

where $\eta_{ij}(\bar{x}_{ij})$ and $\zeta_{ijl}(x_{l1})$, $1 \leq j \leq n$, $1 \leq i$, $l \leq N$ are smooth functions with $\zeta_{ijl}(0) = 0$.

Note that the Assumption 3.1 represents a fairly general form of interconnections which include those in (Chen *et al.*, 1991; Gavel and Siljak, 1989; Gong *et al.*, 1996; Shi and Singh, 1992) as special cases.

In the sequel, we shall focus on the *globally inverse* $\mathcal{H}_\infty$ *control* problem: *given some real numbers* $\gamma_i > 0$, $i = 1, 2 \cdots, N$, *find decentralized feedback laws* $u_i = \alpha_i(x_i)$, *with* $\alpha_i(0) = 0$, *and some non-negative continuous functions* $r_i(x_i)$ *such that the overall closed-loop system is internally asymptotically stable with the* $\mathcal{L}_2$*-gain between the disturbance input* $\omega = [\omega_1^T\ \omega_2^T\ \cdots\ \omega_N^T]^T$ *and the controlled output* $z = [z_1\ z_2\ \cdots\ z_N]^T$, *where* $z_i = (y_i^2 + r_i(x_i)u_i^2)^{\frac{1}{2}}$, *less than* $\gamma = [\gamma_1\ \cdots\ \gamma_N]^T$ *in the sense that*

$$\sum_{i=1}^{N} \int_0^\infty z_i^T z_i dt < \sum_{i=1}^{N} \gamma_i^2 \int_0^\infty \omega_i^T \omega_i dt + \delta(x_0)$$

*for all* $\omega_i \in \mathcal{L}_2[0\ \infty)$ *and all admissible nonlinear interconnections, where* $\delta(x_0)$ *is a function of the initial state* $x_0$ *satisfying* $\delta(0) = 0$.

In view of Assumption 3.1 and noting that $\zeta_{ijl}(x_{l1})$, $i = 1, 2, \cdots, N$, $j = 1, 2, \cdots, n$, $l = 1, 2, \cdots, N$, are smooth with $\zeta_{ijl}(0) = 0$, there exist smooth functions $\bar{\zeta}_{ijl}(x_{l1})$ such that $\zeta_{ijl}(x_{l1}) = \bar{\zeta}_{ijl}(x_{l1})x_{l1}$. Taking into consideration of Theorem 2.1, we introduce the following auxiliary systems associated with (3.2) satisfying Assumption 3.1:

$$\dot{x}_i = A_i x_i + f_i(x_i) + B_i u_i + \left[p_i(x_i)\ \ \beta^{\frac{1}{2}}\gamma_i\eta_i(x_i)\right]\tilde{\omega}_i$$
$$\tilde{y}_i = \begin{bmatrix} C_i x_i \\ \beta_i^{-\frac{1}{2}}(\bar{d}_i(x_{i1}))^{\frac{1}{2}} C_i x_i \end{bmatrix} \quad (3.4)$$

where $x_i$, $i = 1, 2, \cdots, N$, is the state, $\tilde{\omega}_i$ is the disturbance input, $u_i$ is the control input, $A_i(x_i)$, $B_i(x_i)$, $C_i(x_i)$, $f_i(x_i)$ and $p_i(x_i)$ are the same as in the system (3.2), and $\beta_i$, $i = 1, 2, \cdots, N$ are some positive scalars, $\beta = \sum_{l=1}^{N} \beta_l$, and

$$\eta_i = diag\{\eta_{i1}(x_{i1}), \eta_{i2}(\bar{x}_{i2}), \cdots, \eta_{in}(\bar{x}_{in})\}$$

$$\bar{\zeta}_{il}(x_{l1}) = \begin{cases} [\bar{\zeta}_{i1l}(x_{l1})\ \bar{\zeta}_{i2l}(x_{l1})\ \cdots\ \bar{\zeta}_{inl}(x_{l1})]^T & \text{if } l \neq i \\ [0\ 0\ \cdots\ 0]^T & \text{if } l = i \end{cases};$$

$$\bar{d}_i(x_{i1}) = \sum_{l=1, l\neq i}^{N} \bar{\zeta}_{li}^T(x_{i1})\bar{\zeta}_{li}(x_{i1}) = \sum_{l=1}^{N} \bar{\zeta}_{li}^T(x_{i1})\bar{\zeta}_{li}(x_{i1})$$

Let $P_i$ denote the value of $p_i(x_i)$ at $x_i = 0$. Since $f_i(x_i)$ is smooth with $f_i(0) = 0$, it can be rewritten as:

$$f_i(x_i) = \begin{bmatrix} \phi_{i11}(x_{i1}) & 0 & \cdots & 0 \\ \phi_{i21}(\bar{x}_{i2}) & \phi_{i22}(\bar{x}_{i2}) & \cdots & 0 \\ \vdots & \vdots & \ddots & \vdots \\ \phi_{in1}(\bar{x}_{in}) & \phi_{in2}(\bar{x}_{in}) & \cdots & \phi_{inn}(\bar{x}_{in}) \end{bmatrix} \begin{bmatrix} x_{i1} \\ x_{i2} \\ \vdots \\ x_{in} \end{bmatrix} \triangleq \Phi_{if}(x_i)x_i$$

where all the functions are smooth. It can be observed that $F_i \triangleq \Phi_{if}(0) = \frac{\partial f_i}{\partial x_i}|_{x_i=0}$ and is a lower triangular matrix.

The linearized system of (3.4) at $x_i = 0$ is

$$\dot{x}_i = \bar{A}_i x_i + B_i u_i + \left[P_i\ \ \beta^{\frac{1}{2}}\gamma_i\Gamma_i\right]\tilde{\omega}_i$$
$$\tilde{y}_i = \begin{bmatrix} C_i x_i \\ \beta_i^{-\frac{1}{2}}\tau_i^{\frac{1}{2}} C_i x_i \end{bmatrix} \quad (3.5)$$

where $\bar{A}_i = A_i + F_i$, $\Gamma_i = \eta_i(0)$ and $\tau_i = \bar{d}_i(0)$.

A solution to the global inverse $\mathcal{H}_\infty$ control problem for the interconnected nonlinear system (3.2) is given in the following theorem. Due to the space limitation, the proof of this main result is omitted. For the details, please refer to the full version of the paper (Xie *et al.*, 1998).

*Theorem 3.1.* Consider the interconnected system (3.2) satisfying Assumption 3.1 and let $R_i > 0$, $i = 1, 2, \cdots, N$, be some given weighting factors. Suppose that for each $i$, there exists a linear feedback law $u_i = K_i x_i$ for (3.5) which internally stabilizes the system and renders its $\mathcal{L}_2$-gain, between the disturbance input $\tilde{\omega}_i$ and the penalty output $\tilde{z}_i = (\tilde{y}_i^T\tilde{y}_i + R_i u_i^2)^{\frac{1}{2}}$, less than a prescribed number $\gamma_i > 0$. Then, there exist weighting factors $r_i(x_i)$, continuously depending on $x_i$ and satisfying $r_i(0) = R_i$ and $0 < r_i(x_i) \leq R_i$, and smooth decentralized feedback control laws $\alpha_i(x_i)$, $i = 1, 2, \cdots, N$, for (3.2) which globally stabilize the interconnected system (3.2) and render its $\mathcal{L}_2$-gain, between the disturbance input $\omega = [\omega_1^T\ \omega_2^T\ \cdots\ \omega_N^T]^T$ and the penalty output $z = [z_1\ z_2\ \cdots\ z_N]^T$, where $z_i = (y_i^2 + r_i(x_i)u_i^2)^{\frac{1}{2}}$, less than $\gamma = [\gamma_1\ \cdots\ \gamma_N]^T$ for all $\omega_i \in \mathcal{L}_2[0, \infty)$ and all admissible nonlinear interconnections.

*Remark 3.1.* Theorem 3.1 presents a solution to the decentralized global inverse $\mathcal{H}_\infty$ control problem based on an explicit construction of the weighting factors, the positive definite and proper Lyapunov functions and the decentralized control laws satisfying the HJI inequalities (2.8). The key point of our approach in overcoming the strong nonlinear interconnections is to cast the decentralized $\mathcal{H}_\infty$ control problem into the associated centralized $\mathcal{H}_\infty$ control which is then solved by extending the result in (Isidori and Lin, 1998). It should be observed that the cast system of (3.4) is multi-disturbance-input and multi-output one and has a more general structure than the system considered in (Isidori and Lin, 1998) which is a single-input and single-output system with $f_i(x_i) \equiv 0$.

*Remark 3.2.* It is clear from Theorem 3.1 that at the equilibrium point $x_i = 0, i = 1, 2, \cdots, N$ the constructed Lyapunov functions, the decentralized nonlinear control laws and the control weighting factors reduce to those associated with the decentralized $\mathcal{H}_\infty$ control of the linearized interconnected system. Therefore, our result in Theorem 3.1 provides an important link between the linear (local) decentralized $\mathcal{H}_\infty$ control of the linearized interconnected system (the interconnected nonlinear system) and the global $\mathcal{H}_\infty$ control of the interconnected nonlinear system.

## 4. CONCLUSION

The problem of decentralized global $\mathcal{H}_\infty$ control was studied in this paper. It has been shown that the decentralized nonlinear control problem for a large class of interconnected systems can be cast into centralized $\mathcal{H}_\infty$ control problems, and its solvability can be given in terms of the global solutions of a set of HJI inequalities. Realizing that it is usually impossible to solve the first order partial differential HJI inequality globally, we have addressed here a decentralized global *inverse* $\mathcal{H}_\infty$ control problem instead of the regular $\mathcal{H}_\infty$ control problem. A global *inverse* solution was constructed explicitly, using the two-way (backward and forward) recursive induction argument initially proposed in (Isidori and Lin, 1998).

## 5. REFERENCES


Ball, J. A., J. W. Helton and M. L. Walker (1993). $\mathcal{H}_\infty$ control for nonlinear systems with output feedback. *IEEE Trans. Automat. Contr.* **38**, 549–559.

Chen, Y. H., G. Leitmann and Z. K. Xiong (1991). Robust control design for interconnected systems with time-varying uncertainties. *Int. J. Contr.* **54**, 1457–1477.

Gavel, D. T. and D. D. Siljak (1989). Decentralized adaptive control: Structural conditions for stability. *IEEE Trans. Automat. Contr.* **34**(4), 413–426.

Gong, Z., C. Wen and D. P. Mital (1996). Decentralized robust controller design for a class of interconnected uncertain systems with unknown bound of uncertainty. *IEEE Trans. Automat. Contr.* **41**(6), 850–854.

Isidori, A. (1995). *Nonlinear Control Systems.* third ed.. Springer-Verlag. New York.

Isidori, A. and W. Kang (1995). $\mathcal{H}_\infty$ control via measurement feedback for general nonlinear systems. *IEEE Trans. Automat. Contr.* **40**, 466–472.

Isidori, A. and W. Lin (1998). Global $\mathcal{L}_2$-gain state feedback design for a class of nonlinear systems. Systems & Control Letters, Vol. 34, 295-302.

Jain, S. and F. Khorrami (1997). Decentralized adaptive control of a class of large-scale interconnected systems. *IEEE Trans. Automat. Contr.* **42**(2), 136–154.

Lin, W. and L. Xie (1998). A link between $\mathcal{H}_\infty$ control of a discrete-time nonlinear system and its linearization. *Int. J. Contr.* **69**(2), 301–314.

Marino, R., W. Respondek, A. J. van der Schaft and P. Tomei (1994). Nonlinear $\mathcal{H}_\infty$ almost disturbance decoupling. *Systems & Control Letters* **23**, 159–168.

Paz, R. A. (1993). Decentralized $\mathcal{H}_\infty$ control. In: *Proc. Amer. Contr. Conf.*. San Franciso, California. pp. 2381–2385.

Shi, L. and S. K. Singh (1992). Decentralized adaptive controller design for large-scale systems with higher order uncertainties. *IEEE Trans. Automat. Contr.* **37**(8), 1106–1118.

Siljak, D. D. (1991). *Decentralized Control of Complex Systems.* Academic Press. New York.

van der Schaft, A. J. (1991). On a state-space approach to nonlinaer $\mathcal{H}_\infty$ control. *Systems & Control Letters* **16**, 1–8.

van der Schaft, A. J. (1992). $\mathcal{L}_2$-gain analysis of nonlinear systems and nonlinear $\mathcal{H}_\infty$ control. *IEEE Trans. Automat. Contr.* **37**, 770–784.

Wang, Y., C. E. de Souza and L. Xie (1993). Decentralized output feedback control of interconnected uncertain systems. In: *Proc. the 2nd European Contr. Conf.*. Groningen, Netherland.

Xie, S., L. Xie and W. Lin (1998). Global $\mathcal{H}_\infty$ control for a class of interconnected nonlinear systems. Technical Report, School of Electrical and Electronic Engineering, Nanyang Technological University.

14th Triennial World Congress, Beijing, P.R. China

F-2c-23-2

# FURTHER RESULTS ON NONINTERACTING CONTROL WITH INTERNAL STABILITY OF NONLINEAR SYSTEMS

**X. Xia** *

* *Department of Electrical and Electronic Engineering, University of Pretoria, Pretoria 0002, South Africa. Fax: +27 (12) 362 5000, Email:* xxia@postino.up.ac.za.

Abstract: The parameterization approach is taken further to deal with the noninteracting control problem with internal stability of nonlinear systems. First a general conclusion and some sufficient conditions for the solvability of the problem are given. A complete solution is then provided to the noninteracting control problem with internal stability for a restricted class of nonlinear systems. *Copyright © 1999 IFAC*

Keywords: nonlinear systems, non-interacting control, stability, controllability distribution

## 1. INTRODUCTION

There has been considerable interest in the problem of noninteracting control with internal stability of nonlinear systems (Xia and Gao, 1993; Battilotti, 1994). According to the partition of the outputs, the nonlinear noninteracting control problem can be divided into two cases: the Morgan's problem (the one-one decoupling problem), and the block decoupling problem. At present, there are relatively complete solutions to the nonlinear stable Morgan's problem under both (nondegenerate) static state feedbacks and (nondegenerate) dynamic state feedbacks, under some mild regularity conditions (Battilotti and Dayawansa, 1991). For the general case of the stable block decoupling problem, the research results are limited (Grizzle and Isidori, 1989; Battilotti, 1992; Xia, 1993; Xia, 1994). In (Grizzle and Isidori, 1989), Grizzle and Isidori extended their results in (Isidori and Grizzle, 1988), and obtained a necessary condition for the stable block decoupling via nondegenerate static state feedbacks, that is, the so-called $P^*$-dynamics is asymptotically stable. A parameterization approach was taken in (Xia, 1993) to deal with the static block decoupling problem with exponential internal stability. Examples and counterexamples are demonstrated in (Xia, 1994). In (Battilotti, 1992), a sufficient condition was given for the nonlinear noninteracting control with stability via a dynamic state feedback.

The parameterization approach (Xia, 1993) has been shown to be a useful method to deal with problems of structural design with internal stability (Xia, 1994; Xia, 1996). This approach is further explored in this paper to continue studying the noninteracting control problem with internal stability of nonlinear systems. A general result on the solvability of the problem is first given. Conditions are then found for the uniqueness of the set of compatible controllability distributions that can arise as solutions to the noninteracting control problem. Sufficient conditions for the solvability of the noninteracting control problem with stability of this class of systems are thus given. Finally, a complete solution is provided to the problem for a restricted class of nonlinear systems.

## 2. A GENERAL ANALYSIS

Consider a nonlinear control system of the following form:

$$\dot{x} = f(x) + \sum_{i=1}^{m} g_i(x)u_i, \qquad (1)$$
$$y_j = h_j(x), j = 1, \ldots, \mu,$$

where $x \in N \subset I\!R^n$, $y_j \in I\!R^{\nu_j}$, $\sum_{j=1}^{\mu} \nu_j = p$, $f, g_1, \ldots, g_m$ are smooth vector fields, and $h_1, \ldots, h_\mu$ are smooth mappings.

Without loss of generality, it is assumed that $x_0 = 0$ is an equilibrium of $f(x)$, *i.e.*, $f(0) = 0$, and only the local case is considered.

Let $K_i = \ker dh_i$, $g(x) = (g_1(x), \ldots, g_m(x))$, $G(x) = \text{span}\{g(x)\}$, and assume that $\dim G = m$. For $i = 1, \ldots, \mu$, denote as $P_i^*$ the maximal controllability distribution of the system (1) contained in $K_i$, and $P^* = \bigcap_{i=1}^{m} P_i^*$. Denote as $O^*$ the maximal controllability distribution of the system (1) contained in $P^*$, and as $P_0 =< f, g \mid G >$ the strong accessibility distribution of the system (1). Under some regularity conditions, all these distributions exist, and are computable through standard algorithms (Sussmann and Jurdjevic, 1972).

It is well-known (Grizzle and Isidori, 1989) that a necessary and sufficient condition for the solvability of the noninteracting control problem is the existence of a set of controllability distributions $\{P_1, \ldots, P_\mu\}$ such that

$$P_i \subseteq K_i$$
$$G \cap P_i + \bigcap_{j \neq i}(G \cap P_j) = G, \qquad (2)$$

for $i = 1, \ldots, \mu$. For simplicity, assume that any set of controllability distributions satisfying (2) also satisfy the aforementioned regularity condition, and is referred to as a solution to the noninteracting control problem.

For the convenience of the discussion, assume that the set of controllability distribution $\{P_1, \ldots, P_\mu\}$ are given in the following form:

$$P_i = \ker dT_i(x). \qquad (3)$$

This amounts to solving $\mu$ partial differential equations whose solvability is guaranteed by the Frobenius theorem.

**Theorem 1.** The set of controllability distributions given by (3) is a solution to the noninteracting control problem if and only if

$$\text{rank} L_g T = \sum_{i=1}^{\mu} \text{rank} L_g T_i, \qquad (4)$$

where $T$ represents a matrix with rows $T_i$.

Under condition (4), the following conclusions hold:

i) Let $P = \cap_{i=1}^{\mu} P_i$, $O$ be the maximal controllability distribution contained in $P$. There exist smooth functions $T_{\mu+1}(x)$ and $T_{\mu+2}(x)$, together with $T_1(x), \ldots, T_\mu(x)$, constituting a local coordinate transformation, such that $O = \ker\{dT_1, \ldots, dT_{\mu+1}\}$. Let $u = \alpha_0(x) + \beta_0(x)v$ be the compatible friend (Xia, 1993) of $\{P_1, \ldots, P_\mu\}$, then the composite system writes in the new coordinates as, with obvious abuse of notations,

$$\begin{aligned}
\dot{x}_i &= \tilde{f}_i(x_i) + \tilde{g}_{ii}(x_i)v_i, i = 1, \ldots, \mu, \\
\dot{x}_{\mu+1} &= \tilde{f}_{\mu+1}(x_1, \ldots, x_\mu, x_{\mu+1}) \\
&\quad + \sum_{j=1}^{\mu} \tilde{g}_{\mu+1,j}(x_1, \ldots, x_\mu, x_{\mu+1})v_j \qquad (5) \\
\dot{x}_{\mu+2} &= \tilde{f}(x) + \sum_{j=1}^{\mu+1} \tilde{g}_{\mu+2,j}(x)v_j \\
y_i &= h_i(x_i), \quad \text{for} \quad i = 1, \ldots, \mu.
\end{aligned}$$

ii) Denote as $F(P_1, \ldots, P_\mu)$ the compatible friend set of $\{P_1, \ldots, P_\mu\}$, then $(\alpha, \beta) \in F(P_1, \ldots, P_\mu)$ if and only if $(\alpha, \beta)$ takes the following form

$$\alpha = \alpha_0 + \beta_0 \begin{bmatrix} \xi_1(x_1) \\ \vdots \\ \xi_\mu(x_\mu) \\ \xi(x) \end{bmatrix}, \beta = \beta_0 \begin{bmatrix} \eta_1(x_1) \\ \vdots \\ \eta_\mu(x_\mu) \\ \eta(x) \end{bmatrix}. (6)$$

Proof: Equation (4) is proved in (Xia, 1993), Lemma 6. i) can be proved along similar lines as in (Grizzle and Isidori, 1989). For a proof of ii), refer to (Xia, 1993), Theorem 3.

In (Grizzle and Isidori, 1989), it was proved (in Theorem 3.4) that the dynamics

$$\dot{x}_{\mu+1} = \tilde{f}_{\mu+1}(0, \ldots, 0, x_{\mu+1})$$

does not depend on the choice of $(\alpha, \beta) \in F(P_1, \ldots, P_\mu)$, whose asymptotic stability thus constitutes a necessity for the solvability of the stable noninteracting control problem.

Necessary and sufficient conditions are given in the following theorem.

**Theorem 2.** Suppose $\{P_1, \ldots, P_\mu\}$ is a solution to the noninteracting control problem of the system (1). Without loss of generality, the system (1) is assumed to be in the form of (5). Then there exist a pair $(\alpha, \beta) \in F(P_1, \ldots, P_\mu)$ such that the stable noninteracting control problem is solvable if and only if

i) Subsystem

$$\dot{x}_i = \tilde{f}_i(x_i) + \tilde{g}_{ii}(x_i)v_i \qquad (7)$$

is stabilizable by a static state feedback, for each $i = 1, \ldots, \mu$;

ii) Subsystem

$$\dot{x}_{\mu+1} = \tilde{f}(0, \ldots, 0, x_{\mu+1})$$

is asymptotically stable;

iii) Subsystem

$$\begin{aligned}
\dot{x}_{\mu+2} &= \tilde{f}_{\mu+2}(0, \ldots, 0, x_{\mu+2}) \\
&\quad + \tilde{g}_{\mu+2,\mu+1}(0, \ldots, 0, x_{\mu+2})v_{\mu+1} \qquad (8)
\end{aligned}$$

is stabilizable by a static state feedback.

Proof: (sufficiency) For each $i = 1, \ldots, \mu$, let $v_i = \alpha_i(x_i)$ be the stabilizing feedbacks in condition i), and let $v_{\mu+1} = \alpha_{\mu+1}(x_{\mu+2})$ be the stabilizing feedback in condition iii). From (6), the pair defined by the following feedback

$$v = \begin{bmatrix} \alpha_1(x_1) \\ \vdots \\ \alpha_\mu(x_\mu) \\ \alpha_{\mu+1}(x_{\mu+2}) \end{bmatrix} + w,$$

belong to $F(P_1, \ldots, P_\mu)$. The drift term of the modified dynamics corresponding to this feedback is

$$\begin{aligned} \bar{\tilde{f}}_i &= \tilde{f}_i(x_i) + \tilde{g}_{ii}(x_i)\alpha_i(x_i) \\ \bar{\tilde{f}}_{\mu+1} &= \tilde{f}_{\mu+1}(x_1, \ldots, x_\mu, x_{\mu+1}) \\ &\quad + \sum_{i=1}^{\mu} \tilde{g}_{\mu+1,i}(x_1, \ldots, x_\mu, x_{\mu+1})\alpha_i(x_i) \quad (9) \\ \bar{\tilde{f}}_{\mu+2} &= \tilde{f}_{\mu+2}(x) + \sum_{i=1}^{\mu} \tilde{g}_{\mu+2,i}(x)\alpha_i(x_i) \\ &\quad + \tilde{g}_{\mu+2,\mu+1}(x)\alpha_{\mu+1}(x_{\mu+2}) \end{aligned}$$

Since $\alpha_i(0) = 0$, for $i = 1, \ldots, \mu$, one has

$$\begin{aligned} \bar{\tilde{f}}_{\mu+1}(0, \ldots, 0, x_{\mu+1}) &= \tilde{f}_{\mu+1}(0, \ldots, 0, x_{\mu+1}), \\ \bar{\tilde{f}}_{\mu+2}(0, \ldots, 0, x_{\mu+2}) &= \tilde{f}_{\mu+2}(0, \ldots, 0, x_{\mu+2}) \\ &\quad + \tilde{g}_{\mu+2,\mu+1}(0, \ldots, 0, x_{\mu+2})\alpha_{\mu+1}(x_{\mu+2}), \end{aligned}$$

so (9) is in a low triangular form. From (Vidyasagar, 1980), the system (9) is asymptotically stable.

(necessity) From (6), the free term of the modified dynamics of the system (5) corresponding to any feedback pair belonging to $F(P_1, \ldots, P_\mu)$ takes the form of (9), with only $\alpha_{\mu+1}(x_{\mu+2})$ being replaced by $\alpha_{\mu+1}(x)$. If (9) is asymptotically stable, then directly from the stability concepts, the conditions in the theorem hold.

Even if the stability or stabilizability of the subsystems in Theorem 2 are assumed to be known, the conditions are still not checkable, since there is usually not a unique solution to the noninteracting control problem.

**Lemma 1.** A solution $\{P_1, \ldots, P_\mu\}$ to the noninteracting control problem is unique if and only if $P_i = P_i^*$, for $i = 1, \ldots, \mu$, and $O^* = 0$.

Proof: Sufficiency is proved in (Grizzle and Isidori, 1989). For necessity, note first,

$$O^* \subset \cap_{i=1}^{\mu} P_i^*. \quad (10)$$

Take $(\alpha, \beta) \in F(P_1^*, \ldots, P_\mu^*)$, apply Theorem 1 to obtain a decoupled decomposition for the closed loop system corresponding to $\{P_1^*, \ldots, P_\mu^*\}$. But due to (10), in characterizing the coordinate transformation $X = T(x)$, one can further decompose $T_i$ such that

$$\begin{aligned} P_0 &= \ker dT_{\mu+3}, \\ P_i^* &= \ker\{dT_{\mu+3}, dT_i\}, \\ O^* &= \ker\{dT_1, \ldots, dT_\mu, dT_{\mu+1}, dT_{\mu+3}\}. \end{aligned}$$

The corresponding decomposition can be written as

$$\begin{aligned} \dot{x}_i &= \tilde{f}_i(x_i, x_{\mu+3}) + \tilde{g}_{ii}(x_i, x_{\mu+3})v_i, i = 1, \ldots, \mu \\ \dot{x}_{\mu+1} &= \tilde{f}_{\mu+1}(x_1, \ldots, x_\mu, x_{\mu+1}, x_{\mu+3}) \\ &\quad + \sum_{j=1}^{\mu} \tilde{g}_{\mu+1,j}(x_1, \ldots, x_\mu, x_{\mu+1}, x_{\mu+3})v_j \\ \dot{x}_{\mu+2} &= \tilde{f}(x) + \sum_{j=1}^{\mu+1} \tilde{g}_{\mu+2,j}(x)v_j \\ \dot{x}_{\mu+3} &= \tilde{f}_{\mu+3}(x_{\mu+3}), \\ y_i &= h_i(x_i, x_{\mu+3}), \quad \text{for} \quad i = 1, \ldots, \mu. \end{aligned}$$

If $O^* \neq 0$, then $O^* \cap G \neq 0$. There is an element (a column) $\tilde{g}^j_{\mu+2,\mu+1}(x) \neq 0$ in $\tilde{g}_{\mu+2,\mu+1}(x)$. Suppose, without loss of generality, that the first component of $\tilde{g}^j_{\mu+2,\mu+1}(x)$ is different from zero. Denote the first component of $x_{\mu+2}$ as $\bar{x}_{\mu+2}$, *i.e.*, decompose $x_{\mu+2} = (\bar{x}_{\mu+2}, \hat{x}_{\mu+2})$, where $\bar{x}_{\mu+2} \in \mathbb{R}$. Define the distribution

$$P_1 = \ker\{dx_1, d\bar{x}_{\mu+2}, dx_{\mu+3}\},$$

then, by (Xia, 1993), Lemma 2.1, it is easy to verify that $\{P_1, P_2^*, \ldots, P_\mu^*\}$ satisfies (2), that is, it is a solution to the noninteracting control problem. But by the construction of $P_1$, $P_1^* \cap G \neq P_1 \cap G$. Thus $P_1 \neq P_1^*$, and the solution is not unique.

In summarizing, one has the following theorem.

**Theorem 3.** Suppose the noninteracting control problem of the system (1) is locally solvable at $x_0$, and the maximal solution $\{P_1^*, \ldots, P_\mu^*\}$ is regular. If

- the fixed dynamics of $\{P_1^*, \ldots, P_\mu^*\}$ is asymptotically stable,
- the subsystems (7) and (8) corresponding to $\{P_1^*, \ldots, P_\mu^*\}$ are asymptotically stabilizable by static state feedbacks,

then the problem of noninteracting control with internal stability is locally solvable at $x_0$ for the system (1).

If $O^* = 0$, the above two conditions are also necessary. For linear controllable systems, ii) is automatically satisfied, and i) is also necessary for the solvability of the problem.

The above assertions about linear systems are proved in (Grizzle and Isidori, 1989). In the linear case, these results also say that a stabilizing and decoupling feedback can be chosen from $F(P_1^*, \ldots, P_\mu^*)$ when the problem is solvable. In (Xia, 1994), an example was given to show that it is not the case for nonlinear systems. In the next section, a nontrivial class of systems will be identified for which a complete solution will be found for the solvability of noninteracting control problem with internal stability.

## 3. A COMPLETE SOLUTION FOR A CLASS OF SYSTEMS

Consider a nonlinear system with three inputs and two outputs,

$$\dot{x} = f(x) + g_1(x)u_1 + g_2(x)u_2 + g_3(x)u_3,$$
$$y_1 = h_1(x),$$
$$y_2 = h_2(x),$$

where $x \in \mathbb{R}^n$, $f, g_i$, and $h_j$ are all smooth.

A few assumption will be made to specify the class of systems.

(H1) $\{g_1(x), g_2(x), g_3(x)\}$ is involutive;

(H2) the relative degrees $r_1$ and $r_2$ of the outputs $y_1$ and $y_2$, respectively, exist and such that the decoupling matrix defined by

$$\begin{bmatrix} L_{g_1}L_f^{r_1}h_1(x) & L_{g_2}L_f^{r_1}h_1(x) & L_{g_3}L_f^{r_1}h_1(x) \\ L_{g_1}L_f^{r_1}h_2(x) & L_{g_2}L_f^{r_1}h_2(x) & L_{g_3}L_f^{r_1}h_2(x) \end{bmatrix}$$

is of full row rank for all $x$ in the neighborhood of the equilibrium $x = 0$ of the system;

It is well-known that, under the above two assumptions, the system is decouplable by a static state feedback (Xia and Gao, 1993), and the closed-loop system can be written as, with abuse of notations,

$$\begin{aligned}
\dot{x}_{1,i} &= x_{1,i+1}, i = 1, \ldots, r_1 - 1, \qquad & y_1 = x_{11} \\
\dot{x}_{1,r_1} &= u_1 \\
\dot{x}_{2,i} &= x_{2,i+1}, i = 1, \ldots, r_2 - 1, \qquad & y_2 = x_{21} \\
\dot{x}_{2,r_2} &= u_2 \\
\dot{x}_3 &= f(x_1, x_2, x_3, x_4) \\
\dot{x}_4 &= u_3
\end{aligned} \tag{11}$$

in which $x_1 = (x_{1,1}, \ldots, x_{1,r_1})$ and $x_2 = (x_{2,1}, \ldots, x_{2,r_2})$.

Denote as $\Delta_1^*$ and $\Delta_2^*$ the maximal controlled invariant distributions contained in $\ker dh_1$ and $\ker dh_2$, respectively. Then by the above assumptions,

$$\Delta_1^* = \ker\{dx_{11}, \ldots, dx_{1r_1}\}$$
$$\Delta_2^* = \ker\{dx_{21}, \ldots, dx_{2r_2}\}$$

And denote as $P_1^*$ and $P_2^*$ the maximal controllability distributions contained in $\ker dh_1$ and $\ker dh_2$, respectively. Then one has, for $i = 1, 2$, $P_i^* \subseteq \Delta_i^*$.

The third assumption is the regularity condition assumed in the previous section for the general case.

(H3) Any set of two controllability distributions $(P_1, P_2)$ satisfying

$$P_i \subseteq \ker dh_i,$$

$$P_1 \cap G + P_2 \cap G = G, \tag{12}$$

also satisfy the regularity condition, *i.e.*, the distributions $G, P_1, P_2, P_1 \cap G$, $P_2 \cap G$ and $O$, the maximal controllability distribution contained in $P_1 \cap P_2$, are nonsingular in a neighborhood of $x_0 = 0$.

First of all, from (H3), one can easily verify that $P_1^* \subset \Delta_1^*$, *i.e.*, $P_1^*$ is properly contained in $\Delta_1^*$ only if $\frac{\partial f}{\partial x_4} = 0$ and $\frac{\partial f}{\partial x_2} = 0$ and that $P_2^* \subset \Delta_2^*$ only if $\frac{\partial f}{\partial x_4} = 0$ and $\frac{\partial f}{\partial x_1} = 0$, in a neighborhood of $x_0 = 0$. So one has the following four situations:

A) $P_1^* = \Delta_1^*$ and $P_2^* = \Delta_2^*$;

B) $P_1^* \subset \Delta_1^*$ and $P_2^* = \Delta_2^*$;

C) $P_1^* = \Delta_1^*$ and $P_2^* \subset \Delta_2^*$;

D) $P_1^* \subset \Delta_1^*$ and $P_2^* \subset \Delta_2^*$;

Note that $X \subset Y$ means $X$ is properly contained in $Y$.

Each of the above four situations is to be discussed.

D) is the simplest case, since in this case one has

$$\frac{\partial f}{\partial x_1} = 0, \frac{\partial f}{\partial x_2} = 0, \frac{\partial f}{\partial x_4} = 0,$$

the system is actually not strongly accessible (Sussmann and Jurdjevic, 1972). Stable decoupling is possible if and only if the inaccessible dynamics defined by

$$\dot{x}_3 = f(x_3)$$

is asymptotically stable.

Now, consider the case A).

A set of two controllability distributions $P_1$ and $P_2$ contained in $\ker dh_1$ and $\ker dh_2$, respectively, can arise as a solution to the noninteracting control problem if and only if (12) holds. Since $P_1 \subseteq P_1^*$, $P_2 \subseteq P_2^*$, $\dim G = 3$ and by the property of controllability distributions, there can only be the following three cases:

A1) $P_1 = P_1^*$ and $P_2 = P_2^*$;

A2) $P_1 = P_1^*$ and $P_2 \cap G = \text{span}\{g_1 + k(x)g_3\}$ with $k(x)$ being a smooth function;

A3) $P_2 = P_2^*$ and $P_1 \cap G = \text{span}\{g_2 + k(x)g_3\}$ with $k(x)$ being a smooth function;

For A1), since $P = P_1 \cap P_2 = \text{span}\{\frac{\partial}{\partial x_3}, \frac{\partial}{\partial x_4}\}$, the maximal controllability distribution $O$ contained in $P$, which by assumption (H3) is nonsingular, can be either 2 dimensional or 1 dimensional.

When $\dim O = 2$, then $O = P$. By Theorem 2, the system is stably decouplable if and only if

$$\begin{aligned} \dot{x}_3 &= f(0, 0, x_3, x_4), \\ \dot{x}_4 &= u_3, \end{aligned} \tag{13}$$

is feedback stabilizable;

When $\dim O = 1$, then the fixed mode of $(P_1, P_2)$ is

$$\dot{x}_3 = f(0,0,x_3,0)$$

whose stability then constitutes, by Theorem 2, a necessary and sufficient condition for stable decoupling of the system.

For A2), since $P_2 \subset P_2^*$, one can assume

$$P_2 = \ker\{dx_{21}, \ldots, dx_{2r_2}, dT_1, \ldots, dT_s\}$$

for some smooth functions of $x$, $T_1, \ldots, T_s$, such that $\{dx_{21}, \ldots, dx_{2r_2}, dT_1, \ldots, dT_s\}$ is an independent set.

One can prove that either $s = 1$ or $s = 2$.

In fact, due to $P_1 = \ker\{dx_{11}, \ldots, dx_{1r_1}\}$, the equation (12) and the assumption of $P_2 \cap G = \text{span}\{g_1 + k(x)g_3\}$, one knows that there is at least one $T_i$ such that $\frac{\partial T_i}{\partial x_4} \neq 0$. Without loss of generality, assume that $\frac{\partial T_1}{\partial x_4} \neq 0$. Then by Theorem 4.2 (Xia, 1993), a compatible friend of $(P_1, P_2)$ can be computed by

$$\begin{aligned} v_1 &= u_1, \\ v_2 &= u_2, \\ v_3 &= L_f T_1 + L_g T_1\, u, \end{aligned}$$

which, because $L_{g_3} T_1 = \frac{\partial T_1}{\partial x_4} \neq 0$, can be denoted as

$$\begin{aligned} u_1 &= v_1, \\ u_2 &= v_2, \\ u_3 &= \alpha(x) + \beta(x) v, \end{aligned} \tag{14}$$

By definition,

$$P_2 = < \tilde{f}, \tilde{g} \mid \text{span}\{g_1 + k(x)g_3\} >,$$

in which $(\tilde{f}, \tilde{g})$ is the modified dynamics corresponding to the feedback (14). It can be calculated that

$$\begin{aligned} &\text{ad}_{\tilde{f}}^{r_1-k}(g_1 + k(x)g_3) \\ &= (0 \cdots 0\, 1\, 0 \cdots 0 * *)^T \end{aligned} \tag{15}$$

for $k = 1, \ldots, r_1$, in which the 1 is at the $k$-th position, and $*$ could be nonzero. Thus, $\dim P_2 \geq r_1$, implying $\dim P_2^\perp \leq n - r_1 = r_2 + 2$. So, $s \leq 2$. $s > 0$, because $P_2$ is properly contained in $P_2^*$.

- If $s = 2$, then actually

$$P_2 = \ker\{dx_{21}, \ldots, dx_{2r_2}, dx_3, dx_4\}.$$

By regularity assumptions on $P_2$, this can be the case only if

$$\frac{\partial f}{\partial x_1} = 0.$$

Since $P_1 \cap P_2 = 0$, by Theorem 2, stable decoupling is possible if and only if the subsystem

$$\begin{aligned} \dot{x}_{2,i} &= x_{2,i+1}, i = 1, \ldots, r_2 - 1, \\ \dot{x}_{2,r_2} &= u_2 \\ \dot{x}_3 &= f(x_2, x_3, x_4) \\ \dot{x}_4 &= u_3 \end{aligned} \tag{16}$$

is feedback stabilizable;

If $s = 1$, then performing a coordinate transformation $z_i = x_i$, $i = 1, 2, 3$, $z_4 = T_1(x)$, it is then easy to see that the fixed mode of $(P_1, P_2)$ is

$$\dot{x}_3 = f(0,0,x_3,0), \tag{17}$$

whose stability constitutes the necessary and sufficient condition for stable decoupling of the system.

For case A3), one can similarly obtain that stable decoupling is possible if and only if (17) is asymptotically stable or $\frac{\partial f}{\partial x_2} = 0$ and the subsystem

$$\begin{aligned} \dot{x}_{1,i} &= x_{1,i+1}, i = 1, \ldots, r_1 - 1, \\ \dot{x}_{1,r_1} &= u_1 \\ \dot{x}_3 &= f(x_1, x_3, x_4) \\ \dot{x}_4 &= u_3 \end{aligned} \tag{18}$$

is feedback stabilizable.

Next, consider case B). In this case, $\frac{\partial f}{\partial x_2} = 0$, $\frac{\partial f}{\partial x_4} = 0$, and

$$P_1^* = \ker\{dx_1, dx_3\} = \text{span}\{\frac{\partial}{\partial x_2}, \frac{\partial}{\partial x_4}\},$$

$$P_2^* = \ker\{dx_2\} = \text{span}\{\frac{\partial}{\partial x_1}, \frac{\partial}{\partial x_3}, \frac{\partial}{\partial x_4}\}.$$

Analogous to case A), there are only three cases for a set of two controllability distributions $(P_1, P_2)$ to be a solution to the noninteracting control problem:

B1) $P_1 = P_1^*$ and $P_2 = P_2^*$;

B2) $P_1 = P_1^*$ and $P_2 \cap G = \text{span}\{g_1 + k(x)g_3\}$ with $k(x)$ being a smooth function;

B3) $P_2 = P_2^*$ and $P_1 \cap G = \text{span}\{g_2 + k(x)g_3\}$ with $k(x)$ being a smooth function;

By Theorem 2, $(P_1, P_2)$ in B1) is a solution to the noninteracting control problem with internal stability if and only if

$$\begin{aligned} \dot{x}_{1,i} &= x_{1,i+1}, i = 1, \ldots, r_1 - 1 \\ \dot{x}_{1,r_1} &= u_1 \\ \dot{x}_3 &= f(x_1, x_3) \end{aligned} \tag{19}$$

is feedback stabilizable.

In the case of B2), one can prove, along similar lines as in the proof of A2), that $P_2$ takes the form

$$P_2 = \ker\{dx_1, dT\}$$

for a smooth function $T$ of $x$ such that $\frac{\partial T}{\partial x_4} \neq 0$. Note that the reasoning in the proof of A2) for the case of $s = 2$ is not valid because otherwise one would have $\frac{\partial f}{\partial x_1} = 0$, contradicting to the fact that $P_2 = \ker dx_2$, so there is only one $T$.

Since $P_1 \cap P_2 = 0$, there is no fixed mode for $(P_1, P_2)$, again by Theorem 2, $(P_1, P_2)$ is a solution to the noninteracting control problem with stability if and only if the subsystem (19) is stabilizable.

For case B3), one can prove, along similar lines as in the proof of A2), that $P_1$ takes the form

$$P_1 = \ker\{dx_2, dx_3, dT\}$$

in which $T$ is a smooth function of $x$ such that $\frac{\partial T}{\partial x_4} \neq 0$.

Since $P_1 \cap P_2 = 0$, there is no fixed mode for $(P_1, P_2)$, once again by Theorem 2, $(P_1, P_2)$ is a solution to the noninteracting control problem with stability if and only if the subsystem

$$\begin{aligned} \dot{x}_{1,i} &= x_{1,i+1}, i = 1, \ldots, r_1 - 1, \\ \dot{x}_{1,r_1} &= u_1 \\ \dot{x}_3 &= f(x_1, x_3) \\ \dot{x}_4 &= u_3 \end{aligned} \tag{20}$$

is feedback stabilizable.

The analysis of situation C) is *symmetric* to situation B), a necessary and sufficient condition for the existence of a solution to the noninteracting control problem with stability is that (remembering that $\frac{\partial f}{\partial x_1} = 0$ and $\frac{\partial f}{\partial x_4} = 0$)

$$\begin{aligned} \dot{x}_{2,i} &= x_{2,i+1}, i = 1, \ldots, r_2 - 1 \\ \dot{x}_{2,r_2} &= u_2 \\ \dot{x}_3 &= f(x_2, x_3) \end{aligned} \tag{21}$$

is feedback stabilizable, or

$$\begin{aligned} \dot{x}_{2,i} &= x_{2,i+1}, i = 1, \ldots, r_2 - 1 \\ \dot{x}_{2,r_2} &= u_2 \\ \dot{x}_3 &= f(x_2, x_3) \\ \dot{x}_4 &= u_3 \end{aligned} \tag{22}$$

is feedback stabilizable.

In summary, for the system (11), a complete solution to the noninteracting control problem with stability is provided by the following theorem.

**Theorem 4.** Under assumption (H1), (H2) and (H3), the noninteracting control problem with stability is solvable for the system (11) if and only if one of the following conditions holds:

1) The subsystem (13) is feedback stabilizable;

2) $\frac{\partial f}{\partial x_1} = 0$ and the subsystem (16) is feedback stabilizable;

3) $\frac{\partial f}{\partial x_2} = 0$ and the subsystem (18) is feedback stabilizable;

## 4. CONCLUSIONS

Following previous work (Xia, 1993; Xia, 1994; Xia, 1996), the parameterization approach has been taken further to deal with the noninteracting control problem with internal stability of nonlinear systems. A general result on the noninteracting control problem with internal stability of nonlinear systems has been given. Conditions have been found for the uniqueness of the set of compatible controllability distributions that can arise as solutions to the noninteracting control problem. Sufficient conditions have thus been presented for the solvability of the problem of this class of systems. Finally, a complete solution has been provided to the problem for a restricted class of nonlinear systems. It can be noted that all the examples and counterexamples in (Xia, 1994) fit into some categories in Theorem 4.

**Acknowledgment** The author would like to thank Professor I. K. Craig for his valuable comments during the preparation of this paper.

## 5. REFERENCES

Battilotti, S. (1992). A sufficient condition for nonlinear noninteracting control with stability via dynamic state feedback: block-partitioned outputs. *International Journal of Control* **55**, 1141 – 1160.

Battilotti, S. (1994). *Noninteracting Control with Stability for Nonlinear Systems.* Springer-Verlag. Berlin.

Battilotti, S. and W. P. Dayawansa (1991). Noninteracting control with stability for a class of nonlinear systems. *Systems and Control Letters* **19**, 327–338.

Grizzle, J. W. and A. Isidori (1989). Block noninteracting control with stability via static state feedback. *Mathematics of Control, Signals and Systems* **2**, 315–914.

Isidori, A. and J. W. Grizzle (1988). Fixed modes and nonlinear noninteracting control with stability. *IEEE Transactions on Automatic Control* **AC-33**, 907–914.

Sussmann, H. J. and V. Jurdjevic (1972). Controllability of nonlinear systems. *Journal of Differential Equations* **12**, 95–116.

Vidyasagar, M. (1980). Decomposition techniques for large-scale systems with nonadditive interactions, stability and stabilizability. *IEEE Transactions on Automatic Control* **AC-25**, 773 – 779.

Xia, X. (1993). Parameterization of decoupling control laws for affine nonlinear systems. *IEEE Transactions on Automatic Control* **AC-38**, 916 – 928.

Xia, X. (1994). Examples and counterexamples in stable noninteracting control of nonlinear systems. *Acta Automatica Sinica* **20**, 186 – 190.

Xia, X. (1996). A parameterization approach to disturbance decoupling problem with stability of nonlinear systems. *Automatica* **32**, 607 - 610.

Xia, X. and W. Gao (1993). *Nonlinear Systems Control and Decoupling.* Academic Press. Beijing,(in Chinese).

# Nonlinear $\mathcal{H}_\infty$ Control: An Enhanced Quasi-LPV Approach

**Yun Huang and Ali Jadbabaie**
**Control and Dynamical Systems**
**California Institute of Technology, Pasadena, CA 91125,** *USA*
**{yunhuang, alij}@cds.caltech.edu**

**Abstract**

Being formed into a state-dependent linear representation, an input-affine nonlinear system can be considered as a quasi linear parameter varying plant where the parameters are actually functions of the state variables. Recognizing that a critical element in this process is the choice of linear state-dependent representation, a technique is proposed that incorporates this design freedom into the control selection process in an optimal fashion. It is shown that the $\mathcal{L}_2$-gain analysis problem can be recast as an infinite dimensional convex optimization problem which can then be solved by gridding the parameter space. An iterative scheme similar to the D-K iteration in $\mu$ synthesis is proposed to solve the synthesis problem. Numerical examples are presented to illustrate the effectiveness of this method. *Copyright © 1999 IFAC*
**Keywords:** Nonlinear $\mathcal{H}_\infty$ control, Linear Parameter Varying (LPV) methods, Linear Matrix Inequalities (LMI), Frozen Riccati Equations.

## 1. Introduction

In the past decade, the nonlinear $\mathcal{H}_\infty$ control problem has been studied extensively in the state-space framework. The control objective is to attenuate the $\mathcal{L}_2$-gain from the exogenous inputs to the controlled outputs. Most tools developed [5, 7, 9] rely on the solution to the Hamilton-Jacobi Inequality (HJI). Unfortunately, there is no efficient algorithm to solve the HJE/HJI, which is a first order partial differential equation/inequality, for systems with more than a few states. Based on the well-developed linear theory, a few alternative approaches [2, 8] were developed to handle optimal regulation and $\mathcal{H}_\infty$ control problems for a class of input-affine nonlinear systems which can be put into the state-dependent linear like structure. Since the corresponding HJE/HJI is reduced to some algebraic equation or inequality, these techniques are computationally more appealing.

The Frozen Riccati Equation (FRE) approach, initially proposed by Cloutier *et al* [2], is the simplest among them in terms of computation complexity and implementation. It works well for many examples in spite of the lack of a sound theoretical justification to guarantee even mere stability. On the other hand, the LPV technique [10] synthesizes controllers that guarantee the performance when the underlying parameters have bounded rate variation. When applied to nonlinear systems, the varying parameters become functions of the states instead of "free" variables, which can cause severe conservatism. Our recent research [6] revealed that different linear representations used for the FRE design may lead to quite different performance levels for the same nonlinear plant. It has also been proven that there always exists such a representation that the FRE recovers the optimal control under some assumptions on the value function. Expecting the similar effect on other schemes that employ the same representation of nonlinear systems, we propose a quasi-LPV strategy that incorporates this design freedom into the control selection process.

In this paper, an infinite dimensional Linear Matrix Inequality (LMI) condition is derived to compute the best achievable $\mathcal{L}_2$-gain over all the linear representations of the dynamics and parameter-dependent scaling functions. More specific, the problem of $\mathcal{L}_2$-gain analysis can be formulated as an infinite dimensional convex optimization problem which can be solved by gridding the parameter space and using the recently developed interior-point methods. Unfortunately, the problem of state feedback synthesis is not convex in both design variables. The synthesis problem can be cast as an infinite dimensional Bilinear Matrix Inequality (BMI) problem. Because of the non-convexity of the condition in both design variables, there is no guarantee in finding the global optimum. However, analogous to the "$D-K$ iteration" for $\mu$ synthesis [11], an iterative procedure is presented that attempts to find the best achievable performance. A practical computation scheme is proposed along with the examples demonstrating that the $\mathcal{L}_2$-gain of the closed loop system can be greatly improved when the choice of the quasi-LPV representation for the nonlinear dynamics is considered as one of the design variables. It is our hope that the introduction of this extra design freedom in forming the quasi-LPV model can be viewed as an enhancement to the existing LPV methodology, and will reduce the conservatism to a minimum. Even though we will concentrate solely on state feedback $\mathcal{H}_\infty$ control in this paper, the same strategy can be applied to other control objectives as well.

The notation used is fairly standard. We denote $\mathbf{R}^+$ as the set of non-negative real numbers, and $\mathcal{C}^1(\mathbf{R}^m, \mathbf{R}^n)$ as the set of continuously differentiable functions that map $\mathbf{R}^m$ to $\mathbf{R}^n$. The Euclidean norm of a vector is denoted by $\|\cdot\|$, i.e., $\|z\|^2 = z^T z$ for $z \in \mathbf{R}^k$. The notation $\mathcal{L}_2$ will be used for bounded energy signals, i.e., $\mathcal{L}_2(0,T) = \{w : \int_0^T \|w(t)\|^2 dt < \infty\}$.

## 2. Preliminary LPV Results

We first review some results for LPV systems with bounded parameter variation rates. (See [10] and the references therein for details.)

Define the allowable parameter set as

$$\mathcal{F}_{\mathcal{P}}^{\nu} := \{\rho \in \mathcal{C}^1(\mathbf{R}^+, \mathbf{R}^m) : \rho(t) \in \mathcal{P},\ |\dot{\rho}_i| \leq \nu_i,\ i = 1, \cdots, m\}$$

where $\mathcal{P} \subset \mathbf{R}^m$ is a compact set, and $\nu_i \geq 0$.

### 2.1. Analysis Test

The LPV system is governed by

$$\begin{bmatrix} \dot{x}(t) \\ z(t) \end{bmatrix} = \begin{bmatrix} A(\rho(t)) & B(\rho(t)) \\ C(\rho(t)) & 0 \end{bmatrix} \begin{bmatrix} x(t) \\ w(t) \end{bmatrix} \quad (1)$$

where $\rho \in \mathcal{F}_{\mathcal{P}}^{\nu}$, and $A$, $B$, $C$ and $D$ are continuous functions of proper dimensions. The next lemma gives a sufficient condition for the system to have an $\mathcal{L}_2$-gain bounded by $\gamma$.

**Lemma 2.1** *If there exists a $C^1$ positive definite function $P : \mathbf{R}^m \to \mathbf{R}^{n\times n}$ such that*

$$\begin{bmatrix} \mathcal{M}(\rho) & P(\rho)B(\rho) \\ B^T(\rho)P(\rho) & -\gamma^2 I \end{bmatrix} < 0 \tag{2}$$

$$\mathcal{M}(\rho) = \sum_{i=1}^{m} \pm \nu_i \frac{\partial P}{\partial \rho_i} + P(\rho)A(\rho) + A^T(\rho)P(\rho) + C^T(\rho)C(\rho)$$

*for all $\rho \in \mathcal{P}$, then the LPV system (1) has $\mathcal{L}_2$-gain $\leq \gamma$ for all the admissible parameter trajectories.*

Note that (2) actually represents $2^m$ inequalities since $\sum_{i=1}^{m} \pm(\cdot)$ indicates that every combination of $+(\cdot)$ and $-(\cdot)$ should be included.

The proof of the lemma is omitted here. The key point in the proof is the use of a parameter-dependent Lyapunov function $V(x,t) = x^T(t)W(\rho(t))x(t)$. While the quadratic Lyapunov functions are used for the linear time-invariant systems, this modification captures the varying nature of the LPV plant.

### 2.2. State Feedback Synthesis

Next we consider the state-feedback synthesis problem where $u$ is the control input:

$$\begin{bmatrix} \dot{x}(t) \\ z_1(t) \\ z_2(t) \end{bmatrix} = \begin{bmatrix} A(\rho(t)) & B_1(\rho(t)) & B_2(\rho(t)) \\ C_1(\rho(t)) & 0 & 0 \\ C_2(\rho(t)) & 0 & I \end{bmatrix} \begin{bmatrix} x(t) \\ w(t) \\ u(t) \end{bmatrix} \tag{3}$$

For simplicity of the formulas, the standard assumptions are made in the above model. The feed-through term from $w$ to $z$ is assumed to be zero and the $D_{22}$ term where the control $u$ enters is scaled to $I$. More complicated results can be derived if these assumptions are relaxed.

We want to find a continuous function $F(\rho)$ such that the closed loop system has $\mathcal{L}_2$ gain less than $\gamma$ under the state feedback law $u(t) = F(\rho(t))x(t)$.

**Lemma 2.2** *If there exists a continuous differentiable matrix function $X(\rho) > 0$ for all $\rho \in \mathcal{P}$ that satisfies*

$$\begin{bmatrix} \mathcal{M}(\rho) & X(\rho)C_1^T(\rho) & B_1(\rho) \\ C_1(\rho)X(\rho) & -I & 0 \\ B_1^T(\rho) & 0 & -\gamma^2 I \end{bmatrix} < 0 \tag{4}$$

$$\mathcal{M}(\rho) = -\sum_{i=1}^{m} \pm \nu_i \frac{\partial X}{\partial \rho_i} + \hat{A}(\rho)X(\rho) + X(\rho)\hat{A}(\rho)^T - B_2(\rho)B_2^T(\rho)$$

*where $\hat{A}(\rho) = A(\rho) - B_2(\rho)C_2(\rho)$, then under the state feedback law*

$$u(t) = -(B_2(\rho(t))X^{-1}(\rho(t)) + C_2(\rho(t)))x(t) \tag{5}$$

*the closed loop system (3) has $\mathcal{L}_2$ gain $\leq \gamma$ for all the allowable parameter trajectories $\rho(t) \in \mathcal{F}_{\mathcal{P}}^{\nu}$.*

The above result is a natural generalization of the linear $\mathcal{H}_\infty$ theory to LPV plants. Instead of a quadratic Lyapunov function for the linear time invariant plant, a parameter dependent Lyapunov function $V(x,t) = x^T(t)X^{-1}(\rho(t))x(t)$ is assumed. As a result, we have to solve the parameter-dependent LMI (4), which is an infinite dimensional convex optimization problem.

### 2.3. Computation Considerations

An *ad hoc* computation scheme can be used to solve the state feedback problem stated in Lemma 2.2. First choose a set of $C^1$ functions $\{f_i(\rho)\}_{i=1}^{M}$ as a basis for $X(\rho)$, i.e.

$$X(\rho) = \sum_{i=1}^{M} f_i(\rho)X_i \tag{6}$$

where $X_i \in \mathcal{S}^{n\times n}$ is the coefficient matrix for $f_i(\rho)$. If $X(\rho)$ in (4) is substituted with (6), the $cHf$ state feedback problem then becomes the following optimization:

$$\min_{\{X_i\}_{i=1}^{M}} \gamma^2$$

subject to

$$\begin{bmatrix} \mathcal{M}(\rho) & \sum_{j=1}^{M} f_j(\rho)X_jC_1^T(\rho) & B_1(\rho) \\ C_1(\rho)\sum_{j=1}^{M} f_j(\rho)X_j & -I & 0 \\ B_1^T(\rho) & 0 & -\gamma^2 I \end{bmatrix} < 0 \tag{7}$$

$$\sum_{j=1}^{M} f_j(\rho)X_j > 0 \tag{8}$$

where

$$\begin{aligned} \mathcal{M}(\rho) &= -\sum_{i=1}^{m} \pm \nu_i \Big(\sum_{j=1}^{M} \frac{\partial f_j}{\partial \rho_i} X_j\Big) \\ &\quad + \sum_{j=1}^{M} f_j(\rho)(\hat{A}(\rho)X_j + X_j\hat{A}(\rho)^T) - B_2(\rho)B_2^T(\rho) \end{aligned}$$

Note that both (7) and (8) are LMI's in terms of the matrix variables $\{X_i\}_{i=1}^{M}$ when the parameter $\rho$ is fixed. To solve this infinite dimensional convex optimization problem, we can grid the parameter set $\mathcal{P}$ by $L$ points $\{\rho_k\}_{k=1}^{L}$ in each dimension. Then evaluate the above LMI's at these points. Since (7) consists of $2^m$ constraints, a total of $(2^m + 1)L^m$ affine matrix inequalities in terms of $M$ matrix variables $\{X_i\}$ have to be solved. An approximation of the density of the gridding points that guarantees the global solvability of the LMI's is given in [10].

This computation scheme has a few obvious limitations. Associated with gridding is the "curse of dimensionality". Fortunately, the number of varying parameters is usually not more than 3. Another significant problem is the lack of guidance in choosing the basis functions for $X(\rho)$. Despite these drawbacks, it provides us a tool to construct control laws for LPV systems that theoretically guarantees the performance for parameter trajectories with a priori variation rate.

## 3. Main Results

Gain scheduling has been a common practice to deal with nonlinear plants in many engineering applications. Traditionally, the designs are based on linear-invariant approximations of the nonlinear plant at several operation points. The recent development of the LPV techniques provide a systematic way to design controllers that schedule on varying parameters. Treating the nonlinear dynamics as the state-dependent linear system, we can directly apply the LPV technique to a large class of nonlinear systems. Based on the observation that the quasi-LPV model is not uniquely determined by the nonlinear dynamics, we propose a scheme that incorporates the freedom we have in choosing the LPV representation into the optimization problem for both $\mathcal{L}_2$-gain analysis and synthesis.

### 3.1. Quasi-LPV Model of Input-Affine Systems

We first establish the parameterization of all the state-dependent linear representations of some nonlinear dynamics.

**Lemma 3.1** *Suppose a function* $f : \mathbf{R}^n \to \mathbf{R}^n$ *is continuously differentiable with* $f(0) = 0$. *Then a continuous matrix-valued function* $A_o(x) : \mathbf{R}^n \to \mathbf{R}^{n\times n}$ *can alway be found such that* $f(x) = A_o(x)x$. *Moreover, all the possible* $A(x)$ *satisfying* $f(x) = A(x)x$ *can be parameterized as*

$$A(x) = A_o(x) + N(x)$$

*where* $N : \mathbf{R}^n \to \mathbf{R}^{n\times n}$ *satisfies* $N(x)x = 0$.

Note that $A(x)$ in the above lemma is not the linearization of $f(x)$. In fact, we have an infinite number of possible matrix representations $A(x)$ for a given $f(x)$. Following the lemma, the input-affine nonlinear dynamics can be formed into

$$\dot{x} = A(x)x + g(x)w,$$

which looks like a linear system except that the state matrices depend on the state variables instead of being constant as in the linear case. To keep the number of varying parameters in the state matrices $A$ and $g$ to a minimum, the dependence of the state matrices $A(x)$ and $g(x)$ on the state variables will be changed to $\rho(x) \in C^1(\mathbf{R}^n, \mathbf{R}^m)$ with $m \leq n$, i.e. $A(x)$ and $g(x)$ become $A(\rho(x))$ and $g(\rho(x))$ respectively. For instance, $\rho(x)$ can simply represent part of the state variables.

The following quasi-LPV representation of the nonlinear input-affine system will be used for the synthesis problem:

$$\begin{bmatrix} \dot{x} \\ z_1 \\ z_2 \end{bmatrix} = \begin{bmatrix} A(\rho(x)) & B_1(\rho(x)) & B_2(\rho(x)) \\ C_1(\rho(x)) & 0 & 0 \\ C_2(\rho(x)) & 0 & I \end{bmatrix} \begin{bmatrix} x \\ w \\ u \end{bmatrix} \qquad (9)$$

This is different from the true LPV model (3) only in that the underlying parameters $\rho_i$'s are now functions of the state variables, i.e $\rho = \rho(x)$. Not really varying freely in time, the trajectory of the parameters is actually determined by the behavior of the closed-loop system. Aiming at attenuating the effect of the exogenous inputs for all the allowable parameter trajectories, the LPV treatment will inevitably introduce some conservatism due to the connection between the parameters and the states.

One of the difficulties when applying Lemma 2.2 is to determine the bound of the variation rate of $\rho$ since it is no longer known a priori to us. One way to bypass this problem is to restrict $X$ in Lemma 2.2 to be constant, i.e only the quadratic storage functions are considered. Since the variation rate $\nu$ plays no role in this case, the $\mathcal{L}_2$-gain is guaranteed for all state trajectories. More often, however, it is too conservative to restrict our search to quadratic functions. A more practical approach is to use the best estimate of the operating range of the chosen varying parameters, and an approximation on the parameter variation rates. These assumptions should be verified by simulations after the controller is designed.

As stated in Lemma 3.1, we have infinite number of choices when constructing the quasi-LPV model from the nonlinear dynamics. However, they are not equal in terms of the achievable performance when the LPV technique is applied. Here is an illustrative example.

**Example 3.2**

$$\begin{cases} \dot{x}_1 &= x_2 & +(x_2-1)u \\ \dot{x}_2 &= x_1+x_2 & +w-(x_1+2)u \\ z &= [x_1\ x_2\ u]^T \end{cases}$$

Though the dynamics are linear without control, it is not any easier than any other nonlinear problem if the HJI has to be solved. Since this example was created using the converse Hamilton-Jacobi method [3], we know that a closed loop $\mathcal{L}_2$ gain of 1 can be achieved by state feedback $u = x_1 + 2x_2$, and $V(x) = x_1^2 + x_2^2$ serves as a storage function. However, if the obvious choice of $A = \begin{bmatrix} 0 & 1 \\ 1 & 1 \end{bmatrix}$ is used to form the quasi-LPV model, no finite $\mathcal{L}_2$ gain can be achieved by state feedback since the pair $[A(x)\ B_2(x)]$ is not controllable on two lines in the state space: $x_2 = \frac{1}{2}(1+\sqrt{5})x_1 + 2 + \sqrt{5}$ and $x_2 = \frac{1}{2}(1-\sqrt{5})x_1 + 2 - \sqrt{5}$, and $A$ is simply unstable. Fortunately, not every choice of $A(x)$ has this problem. If we are lucky enough to use this one:

$$A(x) = \begin{bmatrix} 0 & 1 \\ 1 & 1 \end{bmatrix} + \begin{bmatrix} \frac{1}{2}x_2 - 1 \\ -\frac{1}{2}x_1 - 2 \end{bmatrix} \begin{bmatrix} x_2 & -x_1 \end{bmatrix}$$

then it is easy to check that (4) is satisfied with any $\gamma > 1$ and $X(x) = \begin{bmatrix} 1 & 0 \\ 0 & 1 \end{bmatrix}$.

This example demonstrates that the differences in performance levels achieved by LPV controllers can be huge due to different choices of $A(x)$. In our recent research on the frozen Riccati approach to the nonlinear optimal regulation problem, we were able to show that there always exists a representation $f(x) = A(x)x$ such that the solution to the frozen Riccati equation recovers the optimal control, provided that the optimal value function has a certain form [6]. This means if the right $A(x)$ is used, then one only needs to solve the algebraic Riccati equation at any instant to get the optimal control. The bad news is that we cannot find the optimal $A(x)$ without figuring out the optimal value function. For the nonlinear $\mathcal{H}_\infty$ problem, we don't have rigorous theoretical justification of the existence of such optimal $A(x)$. However, the difference in the achievable performance delivered by different $A(x)$ under the quasi-LPV framework is too big to ignore. It is usually the case that the obvious choice does not lead to the best performance. In the next section, we will propose a systematic way to gear the original choice of $A(x)$ toward those that can result in the best performance.

### 3.2. $\mathcal{L}_2$-Gain Analysis

The following theorem is a direct application of the non-uniqueness of the quasi-LPV model. Consider the following model

$$\begin{bmatrix} \dot{x}(t) \\ z(t) \end{bmatrix} = \begin{bmatrix} A(\rho(t)) & B(\rho(t)) \\ C(\rho(t)) & 0 \end{bmatrix} \begin{bmatrix} x(t) \\ w(t) \end{bmatrix} \tag{10}$$

Assume the underlying parameter $\rho$ varies in the allowable set

$$\mathcal{F}_{\mathcal{P}}^{\nu} := \{\rho \in \mathcal{C}^1(\mathbf{R}^+, \mathbf{R}^m) : \rho \in \mathcal{P},\ \underline{\nu}_i(\rho) \le \dot{\rho}_i \le \overline{\nu}_i(\rho),\ i = 1, \cdots, m\} \tag{11}$$

where $\mathcal{P} \subset \mathbf{R}^m$ is a compact set. Note that the bounds on the rate variation of the parameter are allowed to vary with the parameter value.

**Theorem 3.3** *If there exists a positive $P(\rho(x)) > 0$ and a $W(\rho(x))$ with $W(\rho(x))x = 0$ that satisfy the following matrix inequality:*

$$\begin{bmatrix} \mathcal{M}(\rho) & P(\rho)B(\rho) \\ B^T(\rho)P(\rho) & -\gamma^2 I \end{bmatrix} < 0 \tag{12}$$

$$\mathcal{M}(\rho) = \sum_{i=1}^{m} \underline{\overline{\nu}}_i(\rho)\frac{\partial P}{\partial \rho_i} + P(\rho)A(\rho) + A^T(\rho)P(\rho) + W(\rho) + W^T(\rho) + C^T(\rho)C(\rho) \tag{13}$$

*for all $\rho \in \mathcal{P}$, then the nonlinear system (10) has $\mathcal{L}_2$-gain less than $\gamma$.*

The notation $\sum_{i=1}^{m} \underline{\overline{\nu}}_i(\rho)$ in (12) means that every combination of $\overline{\nu}_i(\rho)$ and $\underline{\nu}_i(\rho)$ should be included in the inequality. For instanece, when $m = 2$, $\overline{\nu}_1(\rho) + \overline{\nu}_2(\rho)$, $\overline{\nu}_1(\rho) + \underline{\nu}_2(\rho)$, $\underline{\nu}_1(\rho) + \overline{\nu}_2(\rho)$ and $\underline{\nu}_1(\rho) + \underline{\nu}_2(\rho)$ should be checked individually. In other words, (12) actually represents $2^m$ inequalities.

**Proof.*** Let

$$N(\rho) = P^{-1}(\rho)W(\rho).$$

Substituting $W(\rho)$ with $P(\rho)N(\rho)$, (13) becomes

$$\begin{aligned} \mathcal{M}(\rho) &= \sum_{i=1}^{m} \underline{\overline{\nu}}_i(\rho)\frac{\partial P}{\partial \rho_i} + P(\rho)(A(\rho) + N(\rho)) \\ &\quad + (A(\rho) + N(\rho))^T P(\rho) + C^T(\rho)C(\rho) \end{aligned}$$

Since $W(\rho(x))x = 0$, we have $N(\rho(x))x = 0$ from (3.2). Therefore (12) is just the $\mathcal{L}_2$ analysis constraints of the same nonlinear system for a different quasi-LPV representation $\hat{A}(\rho) = A(\rho) + N(\rho)$. □

Recognizing the freedom we have in choosing $A(\rho)$ for a nonlinear plant, we introduce an extra design variable $W(\rho)$ to the standard LPV analysis matrix inequalities (2). The interesting point here is the way $W(\rho)$ entering the LMI, which preserves the convexity of the constraint in both design variables $P(\rho)$ and $W(\rho)$. The same computational considerations previously reviewed can be directly applied. The only extra requirement is $W(\rho(x))x = 0$, which can be enforced by choosing a set of basis functions that guarantee it.

The purpose of introducing this extra design variable $W(\rho)$ is to reduce the conservatism of using the LPV method to handle the nonlinear system. As we pointed out earlier, the achievable performance level calculated by the LPV method is greatly dependent on the choice of $A(\rho)$. Most often, however, it is not at all trivial to pick a good $A(x)$. The free design variable $W(\rho)$ provides a systematic way to make the selection process tractable.

**Example 3.4** *Consider the TORA example [1] which was introduced as a benchmark for nonlinear control design. The non-dimensional equations of motion are given by*

$$\dot{x} = f(x) + B_1(x)w + B_2(x)u$$

*where*

$$f(x) = \begin{bmatrix} x_2 \\ \frac{-x_1 + \epsilon x_4^2 \sin x_3}{1 - \epsilon^2 \cos^2 x_3} \\ x_4 \\ \frac{\epsilon \cos x_3 (x_1 - \epsilon x_4^2 \sin x_3)}{1 - \epsilon^2 \cos^2 x_3} \end{bmatrix},$$

$$B_1(x) = \begin{bmatrix} 0 \\ \frac{1}{1 - \epsilon^2 \cos^2 x_3} \\ 0 \\ \frac{-\epsilon \cos x_3}{1 - \epsilon^2 \cos^2 x_3} \end{bmatrix},\quad B_2(x) = \begin{bmatrix} 0 \\ \frac{-\epsilon \cos x_3}{1 - \epsilon^2 \cos^2 x_3} \\ 0 \\ \frac{1}{1 - \epsilon^2 \cos^2 x_3} \end{bmatrix}.$$

*and $\epsilon = 0.3$ is the parameter. It can be shown that the constant state feedback controller $u = -Kx$ with*

$$K = [-0.2380\ 0.2248\ 1.0000\ 1.4638].$$

*is stabilizing for $x_3 \in [-\pi/2, \pi/2]$ when $w = 0$. Our objective is to compute the $\mathcal{L}_2$-gain of the closed system from the disturbance $w$ to the regulated output given by*

$$z = [0.5x_1\ \ x_3\ \ u]^T.$$

A natural choice of the quasi-LPV model is

$$A(\rho) = \begin{bmatrix} 0 & 1 & 0 & 0 \\ -\frac{1}{1 - \epsilon^2 \cos^2 x_3} & 0 & 0 & \frac{\epsilon x_4 \sin x_3}{1 - \epsilon^2 \cos^2 x_3} \\ 0 & 0 & 0 & 1 \\ \frac{\epsilon \cos x_3}{1 - \epsilon^2 \cos^2 x_3} & 0 & 0 & \frac{\epsilon^2 x_4 \cos x_3 \sin x_3}{1 - \epsilon^2 \cos^2 x_3} \end{bmatrix}$$

where $\rho = [x_3\ \ x_4]$ is used as the varying parameter. Restricting the operating range of $(x_3, x_4)$ to be $\mathcal{P} = [-\frac{\pi}{2}, \frac{\pi}{2}] \times [-0.5, 0.5]$, we grid $\mathcal{P}$ uniformly with 9 points on $x_3$ and 7 points on $x_4$. The range of $x_4$ is decided based on the simulation of the closed-loop system with the given state feedback. Since $x_4$ is the derivative of $x_3$, we are able to use the exact variation rate of $x_3$, which should incur no conservatism at all. Also from the simulation, a rate bound of 0.5 is tentatively set for $x_4$. Different basis functions for $P(\rho)$ are used for comparison:

$$\beta_1 = \{1, \frac{2}{\pi}x_3, (3(\frac{2}{\pi}x_3)^2 - 1)/2\}$$

$$\beta_2 = \{1, \tfrac{2}{\pi}x_3, (3(\tfrac{2}{\pi}x_3)^2 - 1)/2, (5(\tfrac{2}{\pi}x_3)^3 - 3(\tfrac{2}{\pi}x_3))/2, 35(\tfrac{2}{\pi}x_3)^4 - 30(\tfrac{2}{\pi}x_3)^2 + 3)/2\}$$

$$\beta_3 = \{1, \frac{2}{\pi}x_3, (3(\frac{2}{\pi}x_3)^2 - 1)/2, 2x_4, (3(2x_4)^2 - 1)/2, x_3x_4\}$$

| Method | Constant $P$ | Parameter-dependent $P$ | | |
|---|---|---|---|---|
| | | $\beta_1$ | $\beta_2$ | $\beta_3$ |
| LPV w/o $W(\rho)$ | not feasible | 73.6 | 73.2 | 71.4 |
| LPV w/ $W(\rho)$ | 66.4 | 8.2 | 8.0 | 7.4 |

**Table 1:** $\mathcal{L}_2$-gain of the closed-loop system of Example 3.4 with the given state feedback law

Note that the above basis are Legendre polynomials on the operating regions. The basis in $\beta_1$ and $\beta_2$ are functions of only $x_3$, thus the LPV algorithm should yield a guaranteed bound on the $\mathcal{L}_2$-gain since the exact variation rate is used for $x_3$.

To impose the constraint of $W(\rho(x))x = 0$, we assume that $W(\rho(x))$ is of the following form

$$W(\rho(x)) = V(\rho(x))[0\ 0\ x_4\ \ -x_3]$$

where $F(\rho)$ is a $4 \times 1$ vector function. Picking a basis of $\{g_i(\rho)\} = \{1,\ x_3,\ x_4\}$ for $V(\rho)$, we have

$$W(\rho(x)) = \sum_{i=1}^{3} g_i(\rho)V_i[0\ 0\ x_4\ \ -x_3].$$

with $V_i \in \mathbf{R}^4$.

Table 1 lists the $\mathcal{L}_2$-gain computed using both the standard and new LPV algorithms. The LMI's (12) from the standard LPV method on the gridding points are not feasible for constant $P$. With the extra variable $W(\rho)$ in the LMI's, we obtain significantly less conservative numbers. Also notice that allowing $P$ to be parameter-dependent greatly reduces conservatism. When $\beta_1$ is used, the coefficient matrices $Z_i$ returned by the optimization are as follows:

$$Z_1 = \begin{bmatrix} 0 \\ 0 \\ 0 \\ 0 \end{bmatrix},\ Z_2 = \begin{bmatrix} 2.70 \\ -1.85 \\ -0.68 \\ 1.12 \end{bmatrix},\ Z_3 = \begin{bmatrix} -0.11 \\ 1.48 \\ -0.32 \\ -0.08 \end{bmatrix}.$$

Similar results are obtained for other basis functions.

As we know, the $\mathcal{H}_\infty$ norm of the linearized system at the origin provides a lower bound for the best global $\mathcal{L}_2$-gain for the nonlinear system. Not to our surprise, the linearized closed-loop system with the given state feedback law indeed has a lower $\mathcal{H}_\infty$ norm of 6.6 at the origin. In fact, the best achievable $\mathcal{H}_\infty$ norm of the original system near zero is 4.6.

### 3.3. State Feedback $\mathcal{H}_\infty$ Synthesis: $X - N$ Iteration

Consider the quasi-LPV model (9). Assume the underlying parameter $\rho$ varies in the allowable set $\mathcal{F}_\mathcal{P}^\nu$ (11). The next theorem is based on Lemma 2.2 and Lemma 3.1.

**Theorem 3.5** *If we can find an $X(\rho(x)) > 0$ and an $N(\rho(x))$ with $N(\rho(x))x = 0$ that satisfy the following matrix inequality:*

$$\begin{bmatrix} \mathcal{M}(\rho) & X(\rho)C_1^T(\rho) & B_1(\rho) \\ C_1(\rho)X(\rho) & -I & 0 \\ B_1^T(\rho) & 0 & -\gamma^2 I \end{bmatrix} < 0 \qquad (14)$$

$$\begin{aligned} \mathcal{M}(\rho) &= -\sum_{i=1}^{m} \underline{\nu}_i(\rho)\frac{\partial X}{\partial \rho_i} + (\hat{A}(\rho) + N(\rho))X(\rho) \\ &\quad + X(\rho)(\hat{A}(\rho) + N(\rho))^T - B_2(\rho)B_2^T(\rho) \end{aligned}$$

*with $\hat{A}(\rho) = A(\rho) - B_2(\rho)C_2(\rho)$ for all $\rho \in \mathcal{P}$, then the closed loop system (9) has $\mathcal{L}_2$-gain $\leq \gamma$ under the state feedback law:*

$$u(x) = -(B_2(\rho(x))X^{-1}(\rho(x)) + C_2(\rho(x)))x. \qquad (15)$$

Compared to the standard LPV result stated in Lemma 2.2, a new matrix-valued function $N(\rho(x))$ is added to take advantage of the non-uniqueness of the quasi-LPV model of the nonlinear dynamics. The best $\gamma$ level associated with any particular $A(\rho)$ from Lemma 2.2 serves as an upper bound of the best achievable $\mathcal{L}_2$-gain of the closed-loop nonlinear system.

Note that the constraint (14) is not affine in $X(\rho)$ and $N(\rho)$ in contrast to the analysis result. For any value of $\rho$, (14) is a bilinear matrix inequality (BMI) in $X$ and $N$. Because the BMI can be reformulated as a nonconvex programming problem, no tractable algorithm is believed to exist for a general BMI. However, there are computational procedures suggested for solving the BMI via a global optimization [4], though they exhibit exponential growth.

Even though minimizing $\gamma$ over both $X$ and $N$ under the BMI constraint (14) is not a convex optimization, when we fix $X$ or $N$ in (14) and try to minimize $\gamma$ over the other, it is a convex problem. Similar to the "$D - K$" iteration for $\mu$ synthesis, an "$X - N$" iteration is proposed as a practical scheme to achieve the best closed loop performance:

1. Find a quasi-LPV representation (9) for the nonlinear open loop system.
2. Choose the varying parameters $\rho$, the range of the parameter $\mathcal{P}$ and the variation rate bound $\nu$.
3. Calculate the best achievable $\gamma$ level for the current quasi-LPV model (Lemma 2.2). If the performance is satisfactory, stop.
4. Relax $\gamma$ slightly, solve (4) for a feasible $X(\rho) > 0$.
5. Calculate the best $\gamma$ over all $N(\rho)$ with $X(\rho)$ fixed from the previous step (Theorem 3.5). If the new $\gamma$ level is satisfactory or unchanged, stop.
6. Relax $\gamma$ slightly, solve (14) with $X(\rho)$ from step 4 for a feasible $N(\rho)$.
7. Replace $A(\rho)$ with $A(\rho) + N(\rho)$. Go to step 3.

Note that this strategy is not guaranteed to converge to the global optimum because it is not convex in both $X$ and $N$. At step 4 and 6, we try to avoid using $X(\rho)$ or $N(\rho)$ associated with the minimum $\gamma$ at the previous step, because it usually does not lead to the best overall $\gamma$ level. From our experience, $X(\rho)$ and $N(\rho)$ that correspond to a relaxed $\gamma$ are more likely to bring $\gamma$ further down at the next step.

The practical computation scheme described in section 2.3, which involves gridding the parameter space $\mathcal{P}$ and employing basis functions for $X(\rho)$, can be used to solve LMI (4) in steps 3 and 4. At steps 5 and 6, the BMI (14) is reduced to an LMI in $N(\rho)$ since $X(\rho)$ is fixed. Selecting a set of basis functions for $N(\rho)$ that reflect the requirement of $N(\rho(x))x = 0$, we can get an approximate solution to (14) by solving the collection of the LMI's evaluated at the grid points of $\mathcal{P}$.

| LPV method | Best $\gamma$ (synthesis) | X | Best $\gamma$ (analysis) |
|---|---|---|---|
| Standard | 2.4 | $\begin{bmatrix} 0.95 & 0.00 \\ 0.00 & 0.09 \end{bmatrix}$ | 1.7 |
| Enhanced | 1.0 | $\begin{bmatrix} 0.99 & 0.00 \\ 0.00 & 0.95 \end{bmatrix}$ | 1.0 |

**Table 2:** Performance comparison of the standard and the enhanced LPV synthesis for Example 3.6

**Example 3.6**

$$\begin{cases} \dot{x}_1 &= x_1 + 2x_1x_2^2 + w_1 + 2u \\ \dot{x}_2 &= 4x_1^2 - x_2 + w_2 + 2x_1u \\ z &= [x_1\ x_2\ u]^T \end{cases}$$

The control object is to find a state feedback that minimizes the $\mathcal{L}_2$-gain of the closed loop system. Again, this example was created so that we know that the closed-loop $\mathcal{L}_2$-gain of 1 can be achieved by the state feedback $u = -2x_1 - 2x_1x_2$ and $V(x) = x_1^2 + x_2^2$ serves as a storage function.

We begin with an arbitrary choice of $A(x)$:

$$A(x) = \begin{bmatrix} 1 & 2x_1x_2 \\ 4x_1 & -1 \end{bmatrix}$$

and other state space matrices needed in model (9):

$$B_1 = \begin{bmatrix} 1 & 0 \\ 0 & 1 \end{bmatrix}; \quad B_2 = \begin{bmatrix} 2 \\ 2x_1 \end{bmatrix}$$

$$C_1 = \begin{bmatrix} 1 & 0 \\ 0 & 1 \end{bmatrix}; \quad C_2 = \begin{bmatrix} 0 & 0 \end{bmatrix}$$

Same as the in the previous example, $\rho = [x_1\ x_2]^T$ is used as the varying parameter. $\mathcal{P} := [-10, 10] \times [-10, 10]$ is uniformly gridded. To avoid the problem caused by the rate variation estimation, we only consider the quadratic storage functions, i.e., $X$ is restricted to be constant. Notice in the two dimensional case that all the $N(x)$ satisfying $N(x)x = 0$ can be parameterized as $N(x) = \Phi(x)[x_2 \quad -x_1]$ where $\Phi(x) : \mathbf{R}^2 \to \mathbf{R}^{2\times 1}$. A set of basis functions $\{1,\ x_1,\ x_2\}$ for $\Phi(x)$ is chosen for this example.

The computation results are shown in Table 2. With the initial choice of $A(\rho)$, the best $\gamma$ level over all the positive definite scaling functions $X$ obtained by the standard LPV scheme is 2.4, though the $\mathcal{L}_2$-gain analysis (Theorem 3.3) of the closed-loop system gives a better bound of 1.7. However, starting from the same initial $A(\rho)$, $\gamma$ is reduced to 1.0 after a single "$X - N$ iteration."

## 4. Conclusions

It is shown through examples that the performance of the quasi-LPV controller is greatly dependent upon the choice of the model. The standard LPV analysis and state feedback synthesis tests are enhanced by incorporating a new design variable which represents the parameterization of the quasi-LPV model. Under this new framework, the $\mathcal{L}_2$-gain analysis is reformulated as a minimization problem over an affine matrix inequality constraint, and an approximate solution can be obtained by a tractable computation which involves solving a collection of LMI's. In contrast, the state feedback synthesis problem appears to be a "bilinear" constraint which is not jointly convex in both unknown variables. However, an iterative algorithm is proposed which divides it into two convex optimization problems. Although not guaranteed to converge to the global optimum, its effectiveness is demonstrated by a numerical example.

Even though conservatism is reduced by the enhanced quasi-LPV scheme, the optimal achievable performance still may not be obtained. In order to limit the computational load, the number of underlying parameters can be much less than the number of states. Consequently, the search space of $A(x)$ is rather limited in comparison to the complete parameterization.

## 5. Acknowledgments

The authors would like to thank Andy Packard for providing the LPV software, and John Doyle for his inspiration.

### References

[1] R.T. Bupp, D.S. Bernstein, and V.T. Coppola. A benchmark problem for nonlinear control design. In *Proc. IEEE Conf. Decision and Control*, New Orleans, Louisiana, 1995.

[2] J.R. Cloutier, C.N. D'Souza, and C.P.Mracek Marcek. Nonlinear regulation and nonlinear $H_\infty$ control via the state-dependent riccati equation technique. In *Proc. 1st Internat. Conf. on Nonl. Problems in Aviation and Aeraspace*, Daytona Beach, FL, 1996.

[3] J.C. Doyle. Nonlinear games: Examples and counterexamples. In *Proc. IEEE Conf. Decision and Control*, Kobe, Japan, 1996.

[4] K.C. Goh, M.G. Safonov, and G.P. Papvassilopoulos. A global optimization approach for the BMI problem. In *Proc. IEEE Conf. Decision and Control*, pages 2009–2014, Lake Buena Vista, Florida, 1994.

[5] J.W. Helton and W. Zhan. An inequality governing nonlinear $\mathcal{H}_\infty$ control. *Systems and Control Letters*, vol.22(3):157–165, 1994.

[6] Y. Huang and W.M. Lu. Nonlinear optimal control: Alternatives to Hamilton-Jacobi equation. In *Proc. IEEE Conf. Decision and Control*, Kobe, Japan, 1996.

[7] W.M. Lu. Control of uncertain systems. Ph.D dissertation, California Institute of Technology, 1995.

[8] W.M. Lu and J.C. Doyle. Robustness analysis and synthesis of nonlinear uncertain systems. *IEEE Trans. Automatic Control*, vol.40(9):1668–1675, 1996.

[9] A.J. Van der Schaft. Nonlinear state space $h_\infty$ control theory. In H. J. Trentelman and Willems J. C., editors, *Perspectives in Control.* Eds. Birkhauser, 1993.

[10] Fen Wu, Hua Xin Yang, Andy Packard, and Greg Becker. Induced $L_2$-Norm control for lpv system with bounded parameter variation. *Int. Journal of Nolinear and Robust Control*, vol.6:983–998, 1996.

[11] P.M. Young. Robustness with parametric and dynamic uncertainty. Ph.D dissertation, California Institute of Technology, 1993.

14th Triennial World Congress, Beijing, P.R. China

F-2c-23-4

# A NECESSARY AND SUFFICIENT CONDITION FOR THE EXISTENCE OF THE STABILIZING SOLUTION OF THE HAMILTON-JACOBI EQUATION IN NONLINEAR CONTROL THEORY

**Noboru Sakamoto** *

* *Department of Aerospace Engineering, Nagoya University, Japan*
sakamoto@nuae.nagoya-u.ac.jp

Abstract: A necessary and sufficient condition for the existence of the stabilizing solution of the Hamilton-Jacobi equation arising from nonlinear control theory is presented. First, a method of solving the Hamilton-Jacobi equation is outlined using contact and symplectic geometries. It is shown that to solve the Hamilton-Jacobi equation all one has to do is solve ordinary differential equations and/or integrate a completely integrable Pfaffian system. Next, a necessary and sufficient condition of the existence of the stabilizing solution is given in terms of the flow of the Hamiltonian vector field associated with the Hamilton-Jacobi equation. 

Keywords: Hamilton-Jacobi equation, nonlinear control, symplectic geometry

## 1. INTRODUCTION

Since the pioneering work by van der Schaft (1991, 1992), a large number of research has been reported on nonlinear $H^\infty$ control theory (Imura *et al.*, 1995; Isidori and Astolfi, 1992). The solutions of nonlinear $H^\infty$ problems and their derivation are quite natural generalizations of those in linear $H^\infty$ control theory. Thus, it seems that nonlinear $H^\infty$ theory has been developed almost as much as linear $H^\infty$ theory. There are, however, very few applications of nonlinear $H^\infty$ theory that have been reported while linear $H^\infty$ theory is now being commonly used by industrial engineers. One of the biggest reasons for this is that although solution of a nonlinear $H^\infty$ problem is expressed in terms of the solution of the Hamilton-Jacobi equations, the method of solving the Hamilton-Jacobi equation has not been developed. It is also well-known that fundamental control and system theories such as classical optimal control and dissipativity lead the Hamilton-Jacobi equation in nonlinear systems (Willems, 1972; Hill and Moylan, 1976).

The research that considers exact solutions of the Hamilton-Jacobi equation is the work by van der Schaft (1991, 1992). There, it is proved that if the Riccati equation constructed from the linear approximation of a Hamilton-Jacobi equation can be solved, then there exists, locally, a solution of the Hamilton-Jacobi equation. Although this is a natural and practical result, it is not satisfactory for nonlinear control theory in the sense that it only asserts the local existence of the solution and cannot handle truly nonlinear systems such as the systems where center manifolds exist. It should be noted, however, that the mathematical method such as symplectic geometry and Hamiltonian mechanics employed in van der Schaft (1991, 1992) is very important because one can capture nonlinear phenomenan very well in that framework. After van der Schaft (1991, 1992), attention has been paid not to the Hamilton-Jacobi equation itself, which represents the essential difficulty of nonlinearity, but to the derivation of the Hamilton-Jacobi equation, which is rather similar to that of the Riccati equation in linear $H^\infty$ theory.

In this paper, we take the approach in van der Schaft (1991, 1992) and develop it further using contact and symplectic geometries. The Hamilton-Jacobi equation arising from nonlinear control theory is a partial differential equation of the first order. Theory of partial differential equations of the first order was developed together with the research in analytical mechanics such as contact and symplectic geometries (Abraham and Marsden, 1978; Carathéodory, 1965; Libermann and Marle, 1987; Oshima and Komatsu, 1977). The special structure of the Hamilton-Jacobi equation in nonlinear control theory allows analysis by symplectic geometry. It is shown in Sakamoto and Hamada (1998) that solving the Hamilton-Jacobi equation can be reduced to solving an ordinary differential equation. More specifically, to solve the Hamilton-Jacobi equation all one has to do is solve an ordinary differential equation and/or integrate a completely integrable Pfaffian system. This method can deal with all $C^\infty$ Hamilton-Jacobi equations including the ones whose linear approximations (Riccati equations) fail to have solutions (see Example 4). The symplectic geometrical notion of the stabilizing solution is also introduced in Sakamoto and Hamada (1998), which gives a nice geometrical insight for understanding the structure of the stabilizing solution. The purpose of the present paper is, in addition to reviewing the results of Sakamoto and Hamada (1998), to give a necessary and sufficient condition for the existence of the stabilizing solution of the Hamilton-Jacobi equation.

The paper is organized as follows. In Section 2 we overview the theory of partial differential equations of the first order. In Section 3 attention is paid to the Hamilton-Jacobi equation using its special structure and symplectic geometry. In Section 4 theory of the Riccati equation is outlined in symplectic geometry to see the correspondence to theory of the Hamilton-Jacobi equation in Section 3. In Section 5 a necessary and sufficient condition for the stabilizing solution of the Hamilton-Jacobi equation is presented. Here, basic facts from symplectic geometry and Hamiltonian mechanics are used. The condition is expressed in terms of flows of ordinary differential equations. Essentially, this also means that solving the Hamilton-Jacobi equation can be reduced to solving ordinary differential equations.

## 2. THEORY OF PARTIAL DIFFERENTIAL EQUATIONS OF THE FIRST ORDER

Let us consider a general partial differential equation of the first order of the form

$$\text{(PD)} \quad F(p_1, \cdots, p_n, z, x_1, \cdots, x_n) = 0$$

where $F$ is a real valued $C^\infty$ function, $x_1, \cdots, x_n$ are independent variables, $z$ is an unknown function, and $p_j = \partial z/\partial x_j$, $j = 1, \cdots, n$. Let $X$ be an $n$ dimensional space for $x = (x_1, \cdots, x_n)$, $N$ be an $n+1$ dimensional space for $(z, x)$, and $M$ be a $2n+1$ dimensional space for $(p, z, x)$. Points of the cotangent bundle $T^*N$ of $N$ are expressed as $(\xi_0, \cdots, \xi_n; z, x) = (z, x; \xi_0 dz + \sum_{i=1}^n \xi_i dx_i)$ and the contangential sphere bundle $S^*N$ of $N$ is defined by $(T^*N - N)/\mathbb{R}_+$, each fiber is the quotient space of fibers of $T^*N$ relative to the equivalent relation induced by the action of positive real numbers $\mathbb{R}_+$. There exists the canonical 1-form $\omega = dz - p_1 dx_1 - \cdots - p_n dx_n$ on $S^*N$, which defines the contact structure of $S^*N$. $M$ is an open submanifold $S^*N_+$ of the contact manifold $S^*N$ corresponding to $\xi_0 > 0$ and therefore has the contact structure by $\omega$.

Let $V$ be a submanifold of $M$ of codimension 1 defined by $F = 0$. Define $Z = \{(z, x) \in N \mid z = z(x)\}$ and let $\Lambda_Z$ be its conormal sphere bundle in $M$ defined

$$\begin{aligned}\Lambda_Z &= S_Z^* N_+ \\ &= S_Z^* N \bigcap S^* N_+ \\ &= \{(p, z, x) \in M \mid z = z(x), \\ &\qquad p_i = \partial z/\partial x_i(x), i = 1, \cdots, n\},\end{aligned}$$

then $z(x)$ is a solution of (PD) if and only if $\Lambda_Z \subset V$, where the conormal sphere bundle of $Z$ is defined by $S_Z^*N = (T_Z^*N - Z)/\mathbb{R}_+$ from the conormal bundle $T_Z^*N = \bigcup_{p \in Z} (T_p Z)^\perp$ of $Z$ (elements of $T_p^*N$ that annihilate $T_p Z$). A submanifold $\Lambda$ of the contact manifold $S^*N$ is said to be a Lagrangian submanifold if $\dim \Lambda = n$ and $\varphi|_\Lambda = \psi|_\Lambda = 0$, $\varphi, \psi \in \mathscr{F}(M)$ (the set of all $C^\infty$ functions defined on $M$) $\Rightarrow [\varphi, \phi]|_\Lambda = 0$, where $[\cdot, \cdot]$ is the Lagrangian bracket defined

$$\begin{aligned}[\varphi, \psi] = \sum_{i=1}^n \left( \frac{\partial \varphi}{\partial x_i} + p_i \frac{\partial \varphi}{\partial z} \right) \frac{\partial \psi}{\partial p_i} \\ - \sum_{i=1}^n \left( \frac{\partial \psi}{\partial x_i} + p_i \frac{\partial \psi}{\partial z} \right) \frac{\partial \varphi}{\partial p_i}.\end{aligned}$$

The conormal sphere bundle $S_Z^*N$ is a typical Lagrangian submanifold of $S^*N$.

Let $\pi : M \to N$, $\pi' : M \to X$ be natural projections. Then, it is easily seen that $\pi'|_{\Lambda_Z} : \Lambda_Z \to X$ is a diffeomorphism. Conversely, from contact geometry (Oshima and Komatsu, 1977) , it is known that if an Lagrangian submanifold $\Lambda$ of $M$ is such that $\pi'|_\Lambda : \Lambda \to X$ is a diffeomorphism, then there exists a function $z = z(x)$ such that locally $\Lambda$ is expressed as a conormal sphere bundle; $\Lambda = S^*_{\{z=z(x)\}} N_+$. Therefore, finding a solution $z = z(x)$ of (PD) is equivalent to finding an Lagrangian submanifold $\Lambda$ of $M$ contained in $V$ such that $\pi'|_{\Lambda_Z} : \Lambda_Z \to X$ is a diffeomorphism.

To find a Lagrangian submanifold with the above properties, it is necessary and sufficient (Oshima and Komatsu, 1977) to find functions $F_2, \cdots, F_{n+1}$ such that

$$[F_i, F_j]|_V = 0, \quad i, j = 1, \cdots, n+1 \qquad (1)$$

$$\left|\frac{\partial(F_1, \cdots, F_{n+1})}{\partial(p_1, \cdots, p_n, z)}\right|(q) \neq 0, \qquad (2)$$

where $F_1 = F$. The submanifold defined by $F = 0$, $F_2 = c_2, \cdots, F_{n+1} = c_{n+1}$ is a Lagrangian submanifold, and eliminating $p_1, \cdots, p_n$ from these equations by the implicit function theorem yields a complete solution (ones including arbitrary constants) of (PD). Let $X_F = [F, \cdot]$. $X_F$ is a differential operator on $\mathscr{F}(M)$ and therefore a vector field on $M$. It is readily seen that condition (1) is closely related to a linear partial differential equation on $\mathscr{F}(M)$

$$L_{X_F} v = X_F \cdot v = 0.$$

An integral curve of the vector field $X_F$ is a solution of the ordinary differential equation

$$\begin{cases} \dfrac{dx_i}{dt} = \dfrac{\partial F}{\partial p_i} \\ \dfrac{dp_i}{dt} = -\dfrac{\partial F}{\partial x_i} - p_i \dfrac{\partial F}{\partial z} & (i = 1, \cdots, n) \quad (3) \\ \dfrac{dz}{dt} = \displaystyle\sum_{i=1}^{n} p_i \dfrac{\partial F}{\partial p_i}. \end{cases}$$

The 1-dimensional distribution generated by $X_F$ is given by a Pfaffian system (the Lagrange-Charpit system)

$$\frac{dp_1}{-\frac{\partial F}{\partial x_1} - p_1 \frac{\partial F}{\partial z}} = \cdots = \frac{dp_n}{\frac{-\partial F}{\partial x_n} - p_n \frac{\partial F}{\partial z}}$$
$$= \frac{dx_1}{\frac{\partial F}{\partial p_1}} = \cdots = \frac{dx_n}{\frac{\partial F}{\partial p_n}} = \frac{dz}{\sum_{i=1}^{n} p_i \frac{\partial F}{\partial p_i}}.$$

## 3. THE HAMILTON-JACOBI EQUATION

In this section we investigate the Hamilton-Jacobi equation by applying the general theory of partial differntial equations of the first order in the previous section.

The Hamilton-Jacobi equation arising from nonlinear control theory has the form

$$\text{(HJ)} \quad H(p, x) = p^T f(x) - \frac{1}{2} p^T R(x) p + Q(x) = 0,$$

where $f : \mathbb{R}^n \to \mathbb{R}^n$, $R : \mathbb{R}^n \to \mathbb{R}^{n \times n}$, $Q : \mathbb{R}^n \to \mathbb{R}$, and $p_j = \partial z / \partial x_j$, $j = 1, \cdots, n$ for an unknown function $z$ of $x$. We assume that $f$, $R$, and $Q$ are all $C^\infty$ with $R(x)$ symmetric for all $x$ and $f(0) = 0$, $Q(0) = 0$.

A manifold is said to be symplectic if it is of even dimension, $2n$, and on a neighborhood of every point, there exists a nondegenerate closed 2-form $\theta$ called canonical form, that is, $\theta^n$ vanishes nowhere on the neighborhood and $d\theta = 0$. The cotangent bundle $T^*X$ of $X$ has a natural symplectic structure and the canonical form is $\theta = \sum_{i=1}^{n} dp_i \wedge dx_i$.

Noting that (HJ) does not depend on $z$, it is seen that finding a Lagrangian submanifold of $S^*N_+$ contained in $V = \{(p, z, x) | H(p, x) = 0\}$ on which $\pi'$ is a diffeomorphism is equivalent to finding a Lagrangian submanifold of the symplectic manifold $T^*X$ contained in $V$ on which $\pi'$ is a diffeomorphism. A submanifold $\Lambda$ of a $2n$ dimensional symplectic manifold is a Lagrangian submanifold if $\dim \Lambda = n$ and $\theta|_\Lambda = 0$. If, in addition, $T^*X$ has a vector space structure and $\Lambda$ is a subspace of $T^*X$, $\Lambda$ is called a Lagrangian subspace (Libermann and Marle, 1987).

Let $H_1 = H$. Since one can find functions $H_2, \cdots, H_n$ that do not depend on $z$ and satisfy

$$[H_i, H_j]|_V = 0, \quad i, j = 1, \cdots, n \qquad (4)$$

$$\left|\frac{\partial(H_1, \cdots, H_n)}{\partial(p_1, \cdots, p_n)}\right|(q) \neq 0, \qquad (5)$$

the submanifold defined by $H = 0$, $H_2 = c_2, \cdots, H_n = c_n$ is a Lagrangian submanifold contained in $V$. Thus, a complete solution is obtained by solving these equations about $p_1, \cdots, p_n$ and integrating a completely integrable Pfaffian equation

$$dz = p_1(x, c) dx_1 + \cdots + p_n(x, c) dx_n.$$

For (HJ), the ordinary differential equation (3) becomes

$$\begin{cases} \dfrac{dx_i}{dt} = \dfrac{\partial H}{\partial p_i} \\ \dfrac{dp_i}{dt} = -\dfrac{\partial H}{\partial x_i} \end{cases} \quad (i = 1, \cdots, n) \qquad (6)$$

since the $2n+1$-th equation of (3) can be solved independently by integration. It should be noted that this equation gives integral curves of the Hamiltonian vector field $X_H = \{H, \cdot\}$, where $\{\cdot, \cdot\}$ is the Poisson bracket defined by

$$\{\varphi, \psi\} = \sum_{i=1}^{n} \left( \frac{\partial \varphi}{\partial p_i} \frac{\partial \psi}{\partial x_i} - \frac{\partial \varphi}{\partial x_i} \frac{\partial \psi}{\partial p_i} \right)$$

for $\varphi, \psi \in \mathscr{F}(M)$. (6) is called the Hamilton's canonical equation in the classical mechanics (Abraham and Marsden, 1978). To find $H_2, \cdots, H_n$ that satisfy (4), it suffices to solve a partial differential equation

$$L_{X_H} v = X_H \cdot v = 0 \qquad (7)$$

with respect to $v \in \mathscr{F}(M)$. (7) is a linear equation and therefore can be reduced to a system of ordinary differential equations. A complete solution can also be obtained as follows. Find $n-1$ integrals $H_2, \cdots, H_n$ of the Lagrange-Charpit system

$$\frac{dp_1}{-\partial F/\partial x_1} = \cdots = \frac{dp_n}{-\partial F/\partial x_n} = \frac{dx_1}{\partial F/\partial p_1} = \cdots = \frac{dx_n}{\partial F/\partial p_n},$$

satisfying (5), solve $H = 0$, $H_2 = c_2, \cdots, H_n = c_n$ with respect to $p_1, \cdots, p_n$, and integrate the Pfaffian equation

$$dz = p_1(x,c)dx_1 + \cdots + p_n(x,c)dx_n.$$

In general, solution of a partial differential equation is not unique. In control theory the notion of stabilizing solution of (HJ) is of great interest. The following is the definition of the stabilizing solution in symplectic geometry.

*Definition 1.* A locally (globally) stabilizing solution of (HJ) is a Lagrangian submanifold $\Lambda \subset V$ of $T^*X$ such that $\pi'|_\Lambda : \Lambda \to X$ is a diffeomorphism and $X_H|_\Lambda$ is a locally (globally) asymptotically stable vector field.

*Remark 2.* As is well-known, Hamiltonian vector fields can not be asymptotically stable (Abraham and Marsden, 1978).

*Proposition 3.* (Sakamoto and Hamada, 1998) A sufficient condition for $X_H|_\Lambda$ to be a vector field on $\Lambda$ is that $\Lambda$ is a solution of (HJ).

*Example 4.* Let us consider the following equation

$$-2p_1x_1^3 + 2p_2x_2 - p_2^2 + x_1^4 = 0.$$

This is a Hamilton-Jacobi equation for $f$, $R$, and $Q$ of the form

$$f(x) = \begin{pmatrix} -x_1^3 \\ x_2 \end{pmatrix}, \quad R(x) = \begin{pmatrix} 0 & 0 \\ 0 & 1 \end{pmatrix}, \quad Q(x) = x_1^4.$$

Following the procedure explained in this section yields a stabilizing solution

$$z = \frac{1}{4}x_1^2 + x_2^2,$$

while the linearization argument by van der Schaft (1991, 1992) gives no information since the couple $(\partial f/\partial x(0) = \left(\begin{smallmatrix} 0 & 0 \\ 0 & 1 \end{smallmatrix}\right), R)$ is not stabilizable.

## 4. THE RICCATI EQUATION

In this section we show how the theory of the Riccati equation can be incorporated into the theory of the Hamilton-Jacobi equation.

Algebraic Riccati equations of the form

$$\text{(RIC)} \quad PA + A^TP - PRP + Q = 0$$

appear in a large number of linear control theory literature, where $A$ is an $n \times n$ square matrix and $R$, $Q$ are $n \times n$ symmetric matrices.

$$\text{Ham} = \begin{pmatrix} A & -R \\ -Q & -A^T \end{pmatrix}$$

is called the Hamiltonian matrix of (RIC). It is well-known (Francis, 1986) that a necessary and sufficient condition for (RIC) to have a stabilizing solution is that (i) Ham has no eigenvalues on the imaginary axis and (ii) the generalized eigenspace $E_-$ for stable eigenvalues of Ham satisfies

$$E_- \oplus \text{Im}\begin{bmatrix} 0 \\ I \end{bmatrix} = \mathbb{R}^{2n}. \tag{8}$$

The corresponding Hamilton-Jacobi equation to (RIC) is

$$\text{(HJ)}' \quad H'(p,x) = p^TAx - \frac{1}{2}p^TRp + \frac{1}{2}x^TQx = 0.$$

It is obvious that $z = (1/2)x^TPx$ is a solution of (HJ)' for a symmetric matrix $P$ if and only if $P$ is a solution of (RIC). It is not, however, clear how the results in the previous subsection relates to the theory of Riccati equations. For example, we may have to show that $V'$ defined by the equation (HJ)' which is a quadric hypersurface of $T^*X$ contains the Lagrangian submanifold corresponding to the solution $z = (1/2)x^TPx$ which is, in this case, a linear subspace of $T^*X$.

*Definition 5.* A solution $\Lambda$ of the Hamilton-Jacobi equation (HJ) is said to be a linear solution if $\Lambda$ is a Lagrangian subspace of $T^*X$.

*Theorem 6.* (Sakamoto and Hamada, 1998) Let $V'$ be the quadric hypersurface defined by $H' = 0$. If Ham has no eigenvalues on the imaginary axis, $V'$ contains $n$ dimensional Lagrangian subspaces. Moreover, if the generalized eigenspace $E_-$ corresponding to $n$ stable eigenvalues satisfies (8), there exists a Lagrangian subspace $\Lambda_-$ such that $\pi'|_{\Lambda_-}$ is a linear isomorphism onto $X$ and $X_{H'}|_\Lambda$ is asymptotically stable (existence of a linear stabilizing solution).

When the second part of the theorem 6 holds, $\Lambda_-$ is given by

$$\Lambda_- = \{ (p,x) \in T^*X \mid p = Px, \ x \in X \}$$

for a symmetric matrix $P$ which is a solution of (RIC). The Hamiltonian vector field $X_{H'}$ for $H'$ is

$$\begin{cases} \dfrac{dx}{dt} = Ax - Rp \\ \dfrac{dp}{dt} = -Qx - A^Tp \end{cases} \tag{9}$$

and the restriction $X_{H'}|_{\Lambda_-}$ makes sense as a vector field of $\Lambda_-$ because the left hand side of the second equation of (9) is $(PA - PRP)x$, the right hand side is $(-Q-A^TP)x$, and $P$ is a solution of (RIC). This is a proof of proposition 3 for linear systems. The differential equation for $X_{H'}|_{\Lambda_-}$ is

$$\frac{dx}{dt} = (A - RP)x.$$

This is asymptotically stable if and only if $\Lambda_-$ is a stabilizing solution of (HJ)'.

## 5. A NECESSARY AND SUFFICIENT CONDITION FOR THE EXISTENCE OF THE STABILIZING SOLUTION

In the previous section, we showed that the theory of the Riccati equation of linear control theory can be regarded as a special case of the theory of the Hamilton-Jacobi equation. In this section, we attempt to generalize the condition for the existence of the stabilizing solution of the Riccati equation to the Hamilton-Jacobi equation. The condition is expressed in terms of the flow of the Hamltonian vector field associated with the Hamilton-Jacobi equation, and therefore it can be considered as a restatement that solving the Hamilton-Jacobi equation is reduced to solving ordinary differential equations.

The key tool in this section is a fundamental result from Hamiltonian mechanics that the flow of a Hamiltonian vector field is a symplectic transform (Abraham and Marsden, 1978). More specifically, let $F_t$ be the flow of a Hamiltonian vector field on a symplectic manifold $(M, \theta)$. Then, for the symplectic form $\theta$, $F_t^*\theta = \theta$ and therefore,

$$DF_t(p,x)^T \cdot J \cdot DF_t(p,x) = J \tag{10}$$

for all $(p,x) \in M$ and $t$ for which the flow is defined, where $J = \begin{pmatrix} 0 & I_n \\ -I_n & 0 \end{pmatrix}$.

*Theorem 7.* Let $F_t$ be the flow of the Hamiltonian vector field $X_H$ associated with (HJ). A locally stabilizing solution of (HJ) exists on a neighborhood of $0 \in X$ if and only if there exist a neighborhood of $0 \in \mathbb{R}^n$, an immersion $g$ into $T^*X$ defined on the neighborhood with $g(0) = (0,0)$ and $Dg_2(0) : \mathbb{R}^n \to T_0^*X$ being an isomorphism, and a locally asymptotically stable vector field $X_A$ on $\mathbb{R}^n$ with 0 as its equilibrium such that for the flow $\tilde{F}_t$ of $X_A$

$$F_t \circ g = g \circ \tilde{F}_t, \quad t \geqslant 0 \tag{11}$$

on a neighborhood of $0 \in \mathbb{R}^n$. Moreover, when these conditions are satisfied the stabilizing solution $z(x)$ is given by the completely integrable Pfaffian equation

$$dz = g_{11} \circ g_2^{-1}(x)dx_1 + \cdots + g_{1n} \circ g_2^{-1}(x)dx_n, \tag{12}$$

where $g(u) = (g_1(u), g_2(u)) = (g_{11}(u), \cdots, g_{1n}(u), g_{21}(u), \cdots, g_{2n}(u))$, $g_1(u) \in T^*_{g_2(u)}X$, $g_2(u) \in X$.

Suppose, in addition, that $g_2$ is a diffeomorphism on $\mathbb{R}^n$, $X_A$ is a globally asymptotically stable vector field, and (11) holds on $\mathbb{R}^n$. Then, the stabilizing solution obtained from (12) is a global one.

*Outline of the proof:* We only show the sufficiency. First, note that the ordinary differential equation (6) that defines the flow $F_t$ of $X_H$ is

$$\begin{cases} \dfrac{dx}{dt} = f(x) - R(x)p \\ \dfrac{dp}{dt} = -\left(\dfrac{\partial f}{\partial x}\right)^T(x)p + \dfrac{1}{2}\left(\dfrac{\partial Rp}{\partial x}\right)^T(x)p \\ \qquad - \left(\dfrac{\partial Q}{\partial x}\right)^T(x). \end{cases} \tag{13}$$

$F_t$, $\tilde{F}_t$ and their derivatives $DF_t$, $D\tilde{F}_t$ satisfy the following differential equations.

$$\begin{cases} \dfrac{d}{dt}F_t(p,x) = X_H(F_t(p,x)) \\ \dfrac{d}{dt}\tilde{F}_t(u) = X_A(\tilde{F}_t(u)) \end{cases} \tag{14}$$

$$\begin{cases} \dfrac{d}{dt}DF_t(p,x) = DX_H(F_t(p,x))\, DF_t(p,x) \\ \dfrac{d}{dt}D\tilde{F}_t(u) = DX_A(\tilde{F}_t(u))\, D\tilde{F}_t(u) \end{cases} \tag{15}$$

Taking the derivative of (11), we have

$$DF_t(g(u))\, Dg(u) = Dg(\tilde{F}_t(u))\, D\tilde{F}_t(u). \tag{16}$$

Pre-multiply (16) by $Dg(u)^T DF_t(g(u))^T J$ and use (10) to get

$$\begin{aligned} &Dg(u)^T J\, Dg(u) \\ &= D\tilde{F}_t(u)^T\, Dg(\tilde{F}_t(u))^T J\, Dg(\tilde{F}_t(u))\, D\tilde{F}_t(u). \end{aligned}$$

Now note that the second linear dynamical system of (15) is asymptotically stable since $X_A$ is a locally asymptotically stable vector field. Therefore, there exists a neighborhood $U'$ of 0 in $\mathbb{R}^n$ such that $D\tilde{F}_t(u) \to 0$ as $t \to \infty$ for all $u \in U'$. Thus, we have

$$Dg(u)^T J\, Dg(u) = 0 \text{ for all } u \in U',$$

which is equivalent to

$$Dg_1(u)^T Dg_2(u) = Dg_2(u)^T Dg_1(u), \quad u \in U'. \tag{17}$$

Differentiating $F_t(g(u)) = g(\tilde{F}_t(u))$, $u \in U'$ with respect to $t$ at $t = 0$ and letting $x = g_2(u)$ yield

$$\begin{aligned} &X_H(g_1 \circ g_2^{-1}(x), x) \\ &= \begin{pmatrix} Dg_1(u)Dg_2^{-1}(x) \\ I \end{pmatrix} Dg_2(u)X_A(u) \end{aligned} \tag{18}$$

where we have used (14). Let $p = g_1 \circ g_2^{-1}$. $p$ is defined on the open set $U = g_1(U')$ of $X$, and $Dp(x) = Dg_1(u)Dg_2^{-1}(x)$, $x \in U$. Pre-multiply (18) by $[-I\ Dp(x)]$ and use (13) to get

$$\begin{aligned} &Dp(x)f(x) - Dp(x)R(x)p(x) + \left(\frac{\partial f}{\partial x}\right)^T(x)p(x) \\ &- \frac{1}{2}\left(\frac{\partial R(x)p}{\partial x}\right)^T(x)p(x) + \left(\frac{\partial Q}{\partial x}\right)^T(x) = 0, \end{aligned}$$

which is equivalent to

$$\frac{\partial^T}{\partial x}\left\{p(x)^T f(x) - \frac{1}{2}p(x)^T R(x)p(x) + Q(x)\right\} = 0.$$

But, because $f(0) = 0$, $Q(0) = 0$, and $p(0) = g_1 \circ g_2^{-1}(0) = g_1(0) = 0$, we obtain

$$p(x)^T f(x) - \frac{1}{2}p(x)^T R(x)p(x) + Q(x) = 0$$

for all $x \in U$. This is precisely the Hamilton-Jacobi equation. From (17), it follows that

$$\begin{aligned} Dp(x)^T &= (Dg_1(u)Dg_2(u)^{-1})^T \\ &= (Dg_2(u)^{-1})^T Dg_1(u)^T Dg_2(u)Dg_2(u)^{-1} \\ &= (Dg_2(u)^{-1})^T Dg_2(u)^T Dg_1(u)Dg_2(u)^{-1} \\ &= Dg_1(u)Dg_2(u)^{-1} \\ &= Dp(x), \end{aligned}$$

which implies the complete integrability of (12) because $g_{1j} \circ g_2^{-1} = p_j$, $j = 1, \cdots, n$. The Lagrangian submanifold $\Lambda_-$ corresponding to this solution is locally expressed

$$\Lambda_- = \{\,(p(x), x) \mid x \in U\,\} \subset T^*X.$$

From (18), the flow of the restriction $X_H|_{\Lambda_-}$ is given by

$$\begin{aligned} \frac{dx}{dt} &= f(x) - R(x)p(x) \\ &= Dg_2(u)X_A(u) \\ &= Dg_2(g_2^{-1}(x))X_A(g_2^{-1}(x)) \\ &= ((g_2)_* X_A)(x) \end{aligned}$$

This is locally asymptotically stable because the right hand side is the push-forward of the locally asymptotically stable vector field $X_A$ by the diffeomorphism $g_2 : U' \subset \mathbb{R}^n \to U \subset X$. ∎

*Remark 8.* It can be shown, in the linear case, that the condition (11) in the theorem is equivalent to requiring the eigenequation

$$\mathrm{Ham}\ T = TH_-,$$

where $H_-$ is a stable $n \times n$ real matrix whose eigenvalues are the stable eigenvalues of Ham and $T$ is a $2n \times n$ real matrix whose columns span the stable generalized eigenspace of Ham. Therefore, theorem 7 is a natural nonlinear enhancement of the existence theorem of the stabilizing solution of the Riccati equation.

## 6. CONCLUDING REMARKS

It has been shown using symplectic and contact geometries that the Hamilton-Jacobi equation in nonlinear control theory is solvable by rendering it to ordinary differential equations. Unlike the results in van der Schaft (1991, 1992), the method does not require assumption on linear approximations and gives an explicit procedure to find the solution. Furthermore, a necessary and sufficient condition for the existence of the stabilizing solution of the Hamilton-Jacobi equation has been presented. The condition is expressed using flows of ordinary differential equations. This can be considered as a restatement that the Hamilton-Jacobi equation can be reduced to ordinary differential equations.

The geometrical method employed in van der Schaft (1991, 1992) and this paper gives a regularity result of viscosity solution of the Hamilton-Jacobi equation (Soravia, 1997). Therefore, as long as one wishes to seek the stabilizing solution for sufficiently smooth nonlinear affine systems, the method of viscosity solutions is not necessary.

## 7. REFERENCES


Abraham, R. A. and J. E. Marsden (1978). *Foundations of Mechanics, 2nd ed.* Addison-Wesley.

Bardi, M. and I. Capuzzo-Dolcetta (1997). *Optimal Control and Viscosity Solutions of Hamilton-Jacobi-Bellman equations.* Birkhauser.

Carathéodory, C. (1965). *Calculus of Variations and Partial Differential Equations of the First Order, Part I, Partial Differential Equations of the First Order.* Holden-Day.

Francis, B. A. (1986). *A Course in $H_\infty$ Control Theory.* Springer.

Hill, D. J. and P. J. Moylan (1976). The stability of nonlinear dissipative systems. *IEEE Trans. Automat. Contr.* **21**, 708–711.

Imura, J., H. Maeda, T. Sugie and T. Yoshikawa (1995). Robust stabilization of nonlinear systems by $H_\infty$ state feedback. *Syst. Contr. Lett.* **24**, 103–114.

Isidori, A. and A. Astolfi (1992). Disturbance attenuation and $H_\infty$-control via measurement feedback in nonlinear systems. *IEEE Trans. Automat. Contr.* **37**, 1283–1293.

Libermann, P. and C.-M. Marle (1987). *Symplectic Geometry and Analytical Mechanics.* D. Reidel.

Oshima, T. and H. Komatsu (1977). *Partial Differential Equations of the First Order.* Iwanami. Tokyo (Japanese).

Sakamoto, N. and K. Hamada (1998). Symplectic/contact geometry and the Hamilton-Jacobi equation arising from nonlinear $H^\infty$ control theory. *presented at MTNS 98.*

Soravia, P. (1997). Nonlinear $H_\infty$ control (Appendix B in Bardi and Capuzzo-Dolcetta (1997)).

Willems, J. C. (1972). Dissipative dynamical systems-part I, II. *Arch. Rational Mechanics and Analysis* **45**, 321–393.

F-2c-23-5
14th Triennial World Congress, Beijing, P.R. China

# $H_\infty$ CONTROL, STABILIZATION AND INPUT OUTPUT STABILITY OF NONLINEAR SYSTEMS BASED ON HOMOGENEOUS TECHNIQUES [1]

**Yiguang Hong**

*Institute of Systems Science, Chinese Academy of Sciences, Beijing 100080, China. E-Mail: yghong@iss03.iss.ac.cn*

Abstract: In this paper, new relationships are studied between $H_\infty$ control and stabilization of (homogeneous) nonlinear control systems, and between $L_p$ stability and Lyapunov stability of (homogeneous) nonlinear systems. In the first part, by virtue of Hamilton-Jacobi-Isaacs (HJI) inequality, nonlinear $H_\infty$ control problems of stabilizable homogeneous systems and the systems that can be approximated by homogeneous ones are discussed for their designs. One of the main results is that, if the considered system, without exogenous input signals, can be stabilized via homogeneous feedback, then its $H_\infty$ control problem can be solved with continuous (but maybe nonsmooth) feedback laws. In the second part, the problem of input-output stability (IOS) is considered with relations among the lowest homogeneity degrees of the terms of considered systems, and both global and local results are obtained in different cases. *Copyright © 1999 IFAC*

Keywords: $H_\infty$ control, homogeneous systems, $L_p$ stability, nonsmooth feedback.

## Notations

| | | |
|---|---|---|
| $C^1$ | : | set of functions or vector fields, which are continuously differentiable. |
| $\|\cdot\|$ | : | Euclidean norm of vectors; or induced norm for matrices. |
| $o(\cdot)$ | : | $\lim_{\cdot \to 0} \frac{o(\cdot)}{\cdot} = 0.$ |
| $\mathcal{D}V$ | : | shorter form for $\frac{\partial V}{\partial x}$; $\mathcal{D}V^T$ for the transpose of $\mathcal{D}V$. |
| $\|\cdot\|_p$ | : | $L_p$ norm, $p \geq 1$. |
| $\Delta_\epsilon^r$ | : | dilation: $\Delta_\epsilon^r x = (\epsilon^{r_1} x_1, ..., \epsilon^{r_n} x_n)^T$, $r = (r_1, ..., r_n), r_i > 0$ |
| $S$ | : | $= \{x \in R^n : \Gamma(x) = 1\}$. |

## 1. INTRODUCTION

Many problems of homogeneous systems, with solid application backgrounds, have been studied, such as small time local controllability and stabilization (Celikovsky, *et al.*, 1997; Kawski, 1989), optimal control (Hermes, 1996), and input-to-state stability (Ryan, 1995). Homogeneous properties, in some sense, like linearity, can give rise to global results, and sometimes,the analysis and design for homogeneous systems are much simpler than the global analyses and designs for general nonlinear systems. Moreover, Brockett (1983) gave a necessary condition for smooth stabilizing feedback laws, which shows the limitations of smooth techniques, but homogeneous systems can be used as good research objects to employ nonsmooth feedback controls in their syntheses. M'Coskey, *et al.* (1997) and Rui, *et al.* (1997) offered some practical examples modeled from mechanical systems. In addition, depending on ho-

[1] This work was supported by the National Key Project, Project 863, and NSF of China.

mogeneous techniques, many nonlinear systems, which can be approximated by homogeneous nonlinear systems, can also be studied locally (Hermes, 1991).

Also, in recent years, the design and synthesis of nonlinear systems have aroused wide research interest and many useful results have been obtained. Nonlinear $H_\infty$ control is one of the important problems, concerning disturbance attenuation, model matching, robust control and so on.

Here the concept of nonlinear $H_\infty$ control problem is introduced briefly (referring to van der Schaft (1992), Isidori, *et al.* (1995) for details). Consider an affine nonlinear control system

$$\begin{cases} \dot{x} = f(x) + G_1(x)u + G_2(x)w, & w \in R^s \\ z = h(x) + d(x)u, & x \in R^n, z \in R^l, u \in R^m \end{cases} \tag{1}$$

where $f, g, h, d \in C^1$ and $f(0) = 0, h(0) = 0, G_1(0) \neq 0$. Here $w$ denotes exogenous input variables, including all disturbance signals; $u$ denotes the control signal; $z$ denotes the (penalty) output variable; $x$ denotes the state variable of the system. In addition, $d^T(x)d(x)$ is of full rank, which is a basic assumption in the study of $H_\infty$ control to guarantee that the (local) saddle solution exists uniquely.

**Definition 1.1.** Nonlinear $H_\infty$ control problem of (1.1) can be solved via state feedback if, for any $T > 0$, $w(\cdot) \in L_2(R^+, R^s), ||w(\cdot)|| < \infty$, we can find, if existing, a constant $\lambda^* > 0$ as small as possible, and for $\lambda > \lambda^*$, there exists a state feedback $u = u(x)$ such that

1) when the initial condition $x(0) = 0$, we have,

$$\int_0^T z^T(t)z(t)\, dt \leq \lambda^2 \int_0^T w^T(t)w(t)dt; \tag{2}$$

2) when $w(t) \equiv 0$, the closed loop system is asymptotically stable.

With the formulation based on differential game theory, nonlinear $H_\infty$ control problem changes to a problem of solving Hamilton-Jacobi-Isaacs (HJI) equations or inequalities. Unfortunately, in most cases, it is very difficult to construct concrete control laws for nonlinear $H_\infty$ designs because it is hard to analyze and solve the induced HJI equations or inequalities, and even to check the existence of their smooth solutions. Therefore, it is usually assumed beforehand that the equations or inequalities have smooth solutions in research. However, further efforts are made. On the one hand, analysis with viscosity solutions (Soravia, 1996; Hong, *et al.*, 1997), to generalize (theoretically) the meaning of solutions, is needed on the problem. On the other hand, to provide practical $H_\infty$ controllers for nonlinear systems, specific systems are studied. For example, the problem is studied for a particular class of homogeneous systems with a smooth feedback law in a form quite similar to the one of linear systems (Hong and Li, 1998).

As well known, the problem of $H_\infty$ control is related to input-output properties and $L_2$ gain problem. In fact, the problems of input-output stability (IOS) and input-to-state stability (ISS) have drawn much attention for many years and fruitful work was done (Desoer, *et al.*, 1975; Sontag, 1995; Vidyasagar, *et al.*, 1982). Especially, a brief survey about the research topic was reported in Sontag (1995). To formulate our problems, consider a nonlinear system

$$\begin{cases} \dot{x} = f(x) + G(x)w, & x \in R^n \\ z = h(x) \quad z \in R^l, & w \in R^s \end{cases} \tag{3}$$

with $f(0) = 0, h(0) = 0$, and the following concepts are necessary in the sequel.

**Definition 1.2.** System (3) is said to be $L_p$-stable or have a finite $L_p$ gain ($p \geq 1$) if there exists a finite number $\lambda > 0$ such that when $w(\cdot) \in L_p(R^+, R^s), ||w(\cdot)|| < \infty$, we have $z(\cdot) \in L_p(R^+, R^l)$, and

$$||z||_p \leq \lambda ||w||_p + C$$

where constant $C$ depends only on $x(0)$. Moreover, system (3) is said to be small signal $L_p$-stable if there exist positive constants $\lambda$, $r_*$, and $\varepsilon$ such that when $||x(0)|| \leq r_*$, $w(\cdot) \in L_p(R^+, R^s)$ and $||w(\cdot)|| \leq \varepsilon$, we have $z(\cdot) \in L_p(R^+, R^l)$, and

$$||z||_p \leq \lambda ||w||_p + C$$

where $C$ depends only on $x(0)$ and $|| \cdot ||_p$ denotes the norm on $L_p(R^+, R^l)$ or $L_p(R^+, R^s)$ as appropriate.

**Remark 1.3.** Let $[0, T^*)$ be the maximal existence interval of solution $x(t)$ of system (1), or (3), with initial condition $x(0)$ and given $(u(t), w(t))$ of (1), or $w(t)$ of (3). Then the following discussions for system (1) or (3) are held when $t \in [0, T^*)$, without any further emphasis.

In this paper, the above problems are considered based on homogeneous techniques. The paper is organized as follows. Some preliminary results of homogeneous systems are given at first in Section 2. Then nonlinear $H_\infty$ control is studied in Section 3 and 4. The studied nonlinear systems and the designed $H_\infty$ controls may not be differentiable at the origin, much different from the smooth controls proposed in Hong and Li (1998), whose dilation vector satisfies $r_1 = r_2 = \ldots = r_n$. Relations between the stabilizing control and $H_\infty$ control of homogeneous systems are shown. Moreover, the systems that can be approximated by homogeneous ones are also studied with the same technique and corresponding local results are presented. After that, the result about $L_p$ stability of homogeneous systems is given with a relation of the homogeneity degrees. At last, $L_p$ stability of a class of generalized systems is obtained in a local version.

Due to space limitation, here we only provide the results.

## 2. PRELIMINARY RESULTS FOR STABILIZATION AND $H_\infty$ CONTROL

Consider a class of homogeneous systems

$$\begin{cases} \dot{x} = f(x) + G_1(x)u + G_2(x)w, & , w \in R^t \\ z = h(x) + Du, & x \in R^n, z \in R^l, u \in R^m \end{cases} \quad (4)$$

where $f, G_1, G_2, h$ are continuous and $C^1$ except the origin and $f(0) = 0, h(0) = 0, G_1(0) \neq 0$. In addition, $f(x) = (f_1(x), ..., f_n(x))^T$ is a homogeneous vector field of degree $k \geq 0$ with dilation $\Delta_\epsilon^r$, $(r = (r_1, ..., r_n)$, and without loss of generality, assume $1 = r_n \leq r_{n-1} \leq ... \leq r_1)$, that is, for any $\epsilon > 0$,

$$f_i(\Delta_\epsilon^r x) = f_i(\epsilon^{r_1}x_1, ..., \epsilon^{r_n}x_n) = \epsilon^{k+r_i} f_i(x),$$

$G_j(x) = (G_{j1}^T(x), ..., G_{jn}^T(x))^T$ $(j = 1, 2)$ are also homogeneous of degree $s$, namely

$$G_{ji}(\Delta_\epsilon^r x) = G_{ji}(\epsilon^{r_1}x_1, ..., \epsilon^{r_n}x_n) = \epsilon^{s+r_i} G_{ji}(x),$$

and $s + r_n = 0$, which is to keep $G_1(0) \neq 0$. Moreover, $h(x)$ is a continuous homogeneous function of degree $q = k + r_n > 0$, $h(\Delta_\epsilon^r x) = \epsilon^q h(x)$, and $D$ is a constant matrix with the usual assumption $D^T D$ is of full rank. Without loss of generality, we assume $D^T D = \theta^2 I$ with a given constant $\theta > 0$, and $h^T(x)D = 0$.

**Definition 2.1.** A continuous map $\Gamma : R^n \to R$ is called a homogeneous norm with respect to dilation $\Delta_\epsilon^r$, if

1). $\Gamma(x) \geq 0$, and $\Gamma(x) = 0$ if and only if $x = 0$;

2). $\Gamma(\Delta_\epsilon^r x) = \epsilon \Gamma(x)$, $\quad \epsilon > 0$.

Suppose system (4), when $w \equiv 0$, is stabilized via a homogeneous asymptotically stabilizing control law $u(x)$, whose homogeneity degree can be determined easily as $k + r_n$, that is, $u(\Delta_\epsilon^r x) = \epsilon^{k+r_n} u(x)$.

It is easy to know that the analysis about homogeneous systems can be limited on 'unit sphere' $S$ because for any $x \in R^n$, there is a positive number $\epsilon$, such that $x = \Delta_\epsilon^r y$ and $y \in S$. With this, a stabilization result is given at first.

**Lemma 2.2.** Suppose that, when $w \equiv 0$, system (4) can be stabilized using a homogeneous stabilizing control law, then there are a homogeneous Lyapunov function $V$ of degree $p = k + 2r_n$ with dilation $\Delta_\epsilon^r$, and a constant $\alpha^0$, such that, when $\alpha > \alpha^0$, the feedback law in the form of

$$u^*(x) = \begin{cases} -\alpha G_1^T(x)\mathcal{D}V^T(x), & x \neq 0 \\ 0, & x = 0 \end{cases} \quad (5)$$

can stabilize the system.

**Remark 2.3.** Up to now, there have been different methods to construct (homogeneous) stabilizing feedback laws for homogeneous systems, such as Coron, *et al.* (1991) and Kawski (1989). Among them, Hermes (1996) proposed a homogeneous stabilizing control in the form of $u(x) = -\alpha[\mathcal{D}V_0(x)]^{1/(\sigma-1)}$, where $\sigma > 1$ is the homogeneity degree of $V_0$ and large enough such that $V_0$ is smooth, and Hermes (1996) also proved that the feedback law could be viewed as the optimal control of a cost functional in the form of

$$J = \int_0^\infty [K(u(\tau))^\sigma + h_\sigma^+(x(\tau))]d\tau, \quad K > 0$$

where $h_\sigma^+$ is positive definite homogeneous function of degree $\sigma$. However, controller (5) is built on the basis of $V$, quite different from the above one, $V_0$, because $V$ may not be differentiable at the origin.

The following lemma shows one of the key steps for the design of $H_\infty$ control of homogeneous systems. To do this, a further division is made on $S$, after the one used in Lemma 2.2.

**Lemma 2.4.** If, when $w \equiv 0$, system (4) can be stabilized using a homogeneous stabilizing control law, then for any continuous homogeneous function $\phi$ of degree $k + r_n$ with the same dilation, $\Delta_\epsilon^r$, there are a positive definite continuous homogeneous function $V \in C^1(R^n/\{0\})$ of degree $k + 2r_n$ and positive constants $\gamma_1, \gamma_2 > 0$, such that at least one of the following inequalities can be satisfied, when $x \neq 0$:

$$\begin{cases} \mathcal{D}Vf(x) + \frac{1}{\gamma_1}\|\phi(x)\|^2 \leq 0, \\ -\|\mathcal{D}VG_1\| + \frac{1}{\gamma_2}\|\phi(x)\| \leq 0 \end{cases}$$

Before further analysis with the help of differential game theory, a Hamiltonian function is defined for system (4)

$$H(V, x, u, w) = \mathcal{D}V(f + G_1u + G_2w) + z^T z - \lambda^2 w^T w$$

when $x \neq 0$, and it has a saddle point that can be written as $(u^*(x, \mathcal{D}V), w^*(x, \mathcal{D}V))$ with

$$u^* = -\frac{1}{2\theta^2}G_1^T\mathcal{D}V^T, w^* = \frac{1}{2\lambda^2}G_2^T\mathcal{D}V^T, \quad x \neq 0$$

## 3. HOMOGENEOUS $H_\infty$ CONTROL

In this section, our main purpose is to show that, if homogeneous systems like (4), when $w(t) \equiv 0$, can be stabilized via homogeneous feedback, then its $H_\infty$ control problem can be solved globally. Moreover, the $H_\infty$ control problem of a class of systems that can be approximated by stabilizable homogeneous systems can be solved locally using homogeneous feedback.

First of all, we discuss (4) in the case of $G_1 = G_2$. In fact, the results in this case can be extended easily to the case when the matching condition holds, that is, $G_2 = G_1 X$, where the matrix $X$ may not be of full rank.

**Lemma 3.1**. With the assumptions above, when $\lambda > \lambda^* = \theta$, if $\alpha > 0$ is large enough, then $V_\alpha = \alpha V$ is a solution of the following HJI inequality, when $x \neq 0$,

$$H(V_\alpha, x, u^*, w^*) = \alpha \mathcal{D}Vf - \frac{\alpha^2}{4\theta^2}\mathcal{D}VG_1G_1^T\mathcal{D}V^T$$

$$+\frac{\alpha^2}{4\lambda^2}\mathcal{D}VG_2G_2^T\mathcal{D}V^T + h^Th \leq 0. \qquad (6)$$

To make a bridge between (6) and following main results, another important lemma is needed.

**Lemma 3.2**. $V(x(t))$ defined as above for system (4) with $u = u^*(x(t))$ is absolutely continuous in any compact subintervals of the existence interval of the solution $x(t)$.

Then one of the main results is the following theorem.

**Theorem 3.3**. If system (4) can be stabilized by a homogeneous feedback law and $\lambda > \lambda^* = \theta$, then the $H_\infty$ control problem of (4) is solvable and its control law can be given in the form of (5).

Just as mentioned in the beginning of the section, it is easy to analyze the $H_\infty$ control problem of nonlinear systems in the case of $G_2 = G_1 X$ where $X$ is a matrix, according to above discussions.

At last, a general result can be obtained for (2.1), where $G_1(x)$ has no direct relation with $G_2(x)$.

**Theorem 3.4**. If system (4) can be stabilized using homogeneous feedback, then there is a positive constant $\lambda^*$, selected large enough, such that for any $\lambda > \lambda^*$, the $H_\infty$ control problem of system (4) can be solved.

**Remark 3.5**. Hermes (1995) proved that, if a single-input homogeneous system is stabilizable (but it may not be stabilized via homogeneous feedback), then it admits a dynamic homogeneous asymptotically stabilizing feedback. Therefore, when the system is disturbed by $w$ in the form of (4), its $H_\infty$ control can be realized by a dynamic homogeneous feedback.

As van der Schaft (1992) discussed the $H_\infty$ control problem of nonlinear systems and their linear approximation systems, the following discuses briefly about some relations between nonlinear systems and their homogeneous approximation systems. Consider

$$\begin{cases} \dot{x} = f(x) + \delta f(x) + (G_1(x) + \delta G_1(x))u \\ +(G_2(x) + \delta G_2(x))w, \\ z = h(x) + \delta h(x) + (D + \delta D(x))u \end{cases} \qquad (7)$$

where $f(x), h(x), G_1, G_2, D$ are defined as before; $\delta f_i(x) = o(\Gamma(x)^{k+r_i}), \delta h(x) = o(\Gamma(x)^q), \delta G_j(x) = (g_{j1}, ..., g_{jn})^T, g_{ji}(x) = o(\Gamma(x)^{s+r_i}), \quad j = 1, 2,$ and $\delta D(x) = o(1)$. In other words, $\delta f, \delta G_1, \delta G_2, \delta h$ and $\delta D$ are 'higher degree terms' of $f, G_1, G_2, h$ and $D$, in accordance with homogeneous norm $\Gamma(x)$, respectively. Thus, system (7) can be approximated by the systems in the form of (4), or in other words, system (4) may be called the homogeneous approximation system of (7). For some $\lambda > \lambda^*$, system (7) has a local solution to its $H_\infty$ control problem if the $H_\infty$ control problem of (4) has a solution and $||w||$ is small enough to keep the state in its valid neighborhood of $x = 0$. That means the robustness of (4) against 'higher degree perturbations'.

## 4. EXAMPLES AND FURTHER DISCUSSIONS

The following lemma is helpful in the theoretic analysis.

**Lemma 4.1**. When $a, b, y$ are all positive numbers, the following inequality holds:

$$ya \leq y^{1+b} + a^{1+\frac{1}{b}}$$

Let us look at several examples.

**Example 4.2**. Consider a nonlinear system

$$\begin{cases} \dot{x}_1 = x_1 - 14x_2^3 \\ \dot{x}_2 = u + w \\ z = \begin{bmatrix} x_2 \\ u \end{bmatrix} \end{cases} \qquad (8)$$

Namely, in the form of (4) $G_1 = G_2 = (0, 1)^T, f(x) = (x_1 - 14x_2^3, 0)^T, h(x) = x_2$ and $r = (3, 1), k = 0, s = -1, q = 1, \theta = 1$ (i.e., $\lambda^* = 1$ according to Lemma 3.1).

As well known, the stabilization problem of the class of homogeneous systems in the form like $\dot{x}_1 = x_1 - 14x_2^3, \dot{x}_2 = u$ has been studied in many papers like Coron et al (1991) and Kawski (1989) via various nonsmooth controls. Here is a way to consider both its stabilization problem and $H_\infty$ control problem.

The homogeneous norm for (8) can be taken in the form of $\Gamma(x) = (x_1^2 + x_2^6)^{1/6}$. According to Lemma 4.1, a function, of homogeneity degree 2, taken as

$$V = 4(3x_1^{\frac{4}{3}} - 2x_1x_2 + 4x_2^4)^{\frac{1}{2}}$$

is positive definite and $C^1$ when $x \neq 0$ and can be viewed as a candidate Lyapunov function.

Therefore, from Theorem 3.3, we know that the problems of stabilization and $H_\infty$ control can be solved with

$$u^* = \begin{cases} \frac{\alpha 8(x_1 - 8x_2^3)}{V}, & x \neq 0 \\ 0 & x = 0 \end{cases}$$

Q.E.D.

**Example 4.3.** Consider

$$\begin{cases} \dot{x}_1 = x_1 - 8x_3^3 + |x_2|^{\frac{3}{2}} sgn(x_2) + x_3^2 w \\ \dot{x}_2 = x_3 u_1 + x_1 x_3 \\ \dot{x}_3 = u_2 + x_2 \\ z = (x_1 + \frac{x_3}{2}\ u_1\ u_2)^T \end{cases} \quad (9)$$

It is easy to know that (9) is not a homogeneous system, but it can be approximated by a homogeneous one with dilation $(3,2,1)$:

$$\begin{cases} \dot{x}_1 = x_1 - 8x_3^3 + |x_2|^{\frac{3}{2}} sgn(x_2) + x_3^2 w \\ \dot{x}_2 = x_3 u_1 \\ \dot{x}_3 = u_2 \\ z = (\frac{x_3}{2}\ u_1\ u_2)^T \end{cases} \quad (10)$$

where $k = 0, s = -1$ and $sgn(\cdot)$ denotes the sign function. To solve the $H_\infty$ control problem of (9), consider (10). By Lemma 4.1,

$$V = (3x_1^{\frac{8}{3}} - x_1^{\frac{7}{3}} x_3 + 2x_3^8 - x_1 |x_2|^{\frac{5}{2}} sgn(x_2) + 4x_2^4)^{\frac{1}{4}}$$

is a positive definite homogeneous function. After analysis, a $H_\infty$ control law can be taken in the form:

$$u_1^* = \begin{cases} -\dfrac{\alpha x_3}{2}\dfrac{\partial V}{\partial x_2}, & x \neq 0 \\ 0, & x = 0 \end{cases};$$

$$u_2^* = \begin{cases} -\dfrac{\alpha}{2}\dfrac{\partial V}{\partial x_3}, & x \neq 0 \\ 0, & x = 0 \end{cases}$$

Q.E.D.

**Remark 4.4.** Lemma 3.1 and Theorem 3.4, to show the existence of a solution of the $H_\infty$ control problem, also suggest a way to estimate $\alpha$. The estimation directly depends on the choice of $V$ and division $E$ and $\bar{E}_0$ (Example 4.2, when the matching condition holds), and even the value of $\alpha/\lambda$ (Example 4.3). In the above examples, we use a very conservative estimate way to get $\alpha$. In fact, a suitable value of $\alpha$ can be selected much smaller than the one given in the above if we start with the original HJI inequality and analyze it more carefully.

**Example 4.5.** The work can be extened, by the help of stabilization analysis in Celikovsky, *et al.* (1997) and Coron, *et al.* (1991), to a class of single-input homogeneous systems that can be transformed into a class of triangular forms.

**Example 4.6.** In the case of $r_1 = r_2 = ... = r_n = 1$, under suitable conditions given in Hong and Li (1998), feedback laws for the $H_\infty$ control problem can be written in the form of $u^* = -\alpha(x^T Px)^{k/2} Px$

## 5. 5. $L_P$ STABILITY AND SMALL SIGNAL $L_P$ STABILITY

In this section, homogeneous properties are employed to study input-output stability in $L_p$ sense. Consider a nonlinear homogeneous system with $w$ as its input variables:

$$\begin{cases} \dot{x} = f(x) + G(x)w, & x \in R^n, w \in R^s \\ z = h(x) \quad f(0) = 0, h(0) = 0, & z \in R^l \end{cases} \quad (11)$$

where $f(x), G(x), h(x)$ are continuous, $C^1(R^n/\{0\})$ and homogeneous of degree $k \geq 0, s \geq -r_n, q > 0$, respectively, with dilation $\Delta_\epsilon^r, r_1 \geq ... \geq r_n = 1$ (so $s \geq -1$).

Further on, consider the free system corresponding to (11)

$$\dot{x} = f(x), \quad x \in R^n \quad (12)$$

With the results in Rosier (1992), the requirements about the smoothness of $f$ and the uniqueness of the solution of the free system are not necessary.

The main result about $L_p$ stability can be obtained then.

**Theorem 5.1.** If $q + s = k$, and $\dot{x} = f(x)$ is asymptotically stable, then (11) is $L_p$-stable when $p \geq \frac{k+1}{q}$.

**Remark 5.2.** As mentioned in Section 1, IOS and ISS have been studied widely for many years. Among them, ISS of homogeneous systems in the form of $\dot{x} = f(x, w)$ with

$$f(\epsilon x, \epsilon w) = \epsilon^\mu f(x, w), \quad \mu \geq 1, \epsilon > 0 \quad (13)$$

was studied with a nice result: if the origin is an asymptotically stable equilibrium of the unforced system $\dot{x} = f(x, 0)$, then (13) is input-to-state stable in $L_p$ sense for all $p \geq \mu$. In fact, if $h(x)$ is smooth ($q \geq 1$), then we can get a differential equation $\dot{z} = \hat{f}(z, w)$ from $z = h(x)$, which is in the form of (13) with $\mu = k + 1$, so Theorem 5.1 is consistent with the one in Ryan (1995). In addition, if $q = k + 1, s = -1$, like the conditions about system (4), then the system is $L_2$ stable.

Furthermore, a local result is given for a generalized class of systems, which is in the form

$$\begin{cases} \dot{x} = f(x) + \delta f(x) + [G(x) + \delta G(x)]w \\ z = h(x) + \delta h(x) \end{cases} \quad (14)$$

where $f$, $G$, and $h$ are homogeneous of degree $k \geq 0$, $s \geq -r_n = -1$, and $q > 0$, resepectively, and $\delta f = (\delta f_1, ..., \delta f_n)^T$, $f_i(x) = o(\Gamma(x)^{k+r_i})$, $i = 1, ..., n$, $\delta G = (\delta G_1, ..., \delta G_n)^T$, $\delta G_i(x) = o(\Gamma(x)^{s+r_i})$, $i = 1, ..., n$, $\delta h(x) = o(\Gamma(x)^q)$. Namely, $\delta f(x)$, $\delta h$, and $\delta G$ denote 'higher degree' terms corresponding to $f, h$, and $G$, respectively. In other words, (14) can be approximated by a system in the form of (11), or (11) can be viewed as the homogeneous approximation system of (14). Then the result of the above theorem holds locally.

**Theorem 5.3.** If $q + s \geq k$, and $\dot{x} = f(x)$ is asymptotically stable, then system (14) is small signal $L_p$-stable when $p \geq \max\{1, \frac{k+1}{q}\}$.

**Remark 5.4.** With the assumptions like exponential stability of the considered system when $w \equiv 0$ and some Lipschitz conditions with respect to $x$ and $w$, Vidyasagar and Vannelli (1982) gave a result of small signal input-output stability. In our problem, $k$ (or $s$) may be larger than 0 (or $-1$), and $q$ may be smaller than 1.

## 6. CONCLUSIONS

In the paper, $H_\infty$ control and $L_p$ stability of (stabilizable or stable) homogeneous systems are taken into consideration. If the systems can be stabilized via homogenous (maybe nonsmooth) feedback, then its $H_\infty$ control problem can be solved by homogeneous feedback controls. In the research, nonsmooth analysis on the considered systems is essential to obtain the results. Global results are obtained for homogeneous systems, while local results are got for the systems that can be approximated by homogeneous systems. Notice that the global results are given without additional assumptions like zero-state observability and some dissipative properties used, for instance, in van der Schaft (1992). In fact, the results are consistent with the ones of linear systems and Hong and Li (1998). In addition, the input-output stability, globally or locally, in $L_p$ sense of nonlinear systems is discussed in light of homogeneous techniques, Lyapunov function theory and ISS ideas.

## REFERENCES


Brockett, R. (1983). Asymptotic stability and feedback stabilization. In *Differential Geometric Control Theory*, **27**, (Brockett, Millman, Sussmann, (Ed)), 181-191, Birkhäuser, Boston.

Celikovsky, S. and H. Nijmeijer (1997). On the relation between local controllability and stabilizability for a class of nonlinear systems. *IEEE Trans. on Automat. Contr.* **42**, 90-94.

Coron, J.M. and L. Praly (1991). Adding an integrator for the stabilization problem. *Syst. Contr. Lett.*, **17**, 89-104.

Desoer, C. and M. Vidyasagar (1975). *Feedback Systems: Input-Output Properties.* Academic Press, New York.

Hermes, H. (1991). Homogeneous coordinates and continuous asymptotically stabilizing feedback controls. In *Differential Equations Stability and Control*, (S.Elaydi (Ed)), 249-260. Marcel Dekkere, New York.

Hermes, H. (1995). Homogeneous feedback controls for homogeneous systems. *Syst. Contr. Lett.*, **24**, 7-11.

Hermes, H. (1996). Resonance, stabilizing feedback controls, and regularity of viscosity solutions of Hamilton-Jacobi-Bellman equations. *Mathematics of Control, Signals, and Systems*, **9**, 59-72.

Hong, Y., and H. Li (1998). Nonlinear $H_\infty$ control and its related problem of homogeneous systems. *Int. J. of Control*, **71** 79-92.

Hong, Y., S. Yung, S. Mei and H. Qin (1997). Viscosity solution to nonlinear $H_\infty$ control. *Chinese Science Bulletin*, **42**, 890-893.

Isidori, A. and W. Kang (1995). $H_\infty$ control via measure feedback for general nonlinear systems. *IEEE Trans. Automat. Contr.*, **40**, 466-471.

Kawski, M. (1989). Stabilization of nonlinear systems in the plane. *Syst. Contr. Lett.*, **12**, 169-175.

M'Closkey, R.T. and R. Murray (1997). Exponential stabilization of driftless nonlinear control systems using homogeneous feedback. *IEEE Trans. Automat. Contr.*, **42**, 614-628.

Rosier, L. (1992). Homogeneous Lyapunov function for homogeneous continuous vector field. *Syst. Contr. Lett.*, **19**, 467-473.

Rui, C., M. Reyhanoglu, I. Kolmanovosky, S. Cho, and H. McClamroch (1997). Nonsmooth stabilization of an underactuated unstable two degrees of freedom mechanical system. *Proc. of CDC*, 3998-4003.

Ryan, E. P. (1995). Universal stabilization of a class of nonlinear systems with homogeneous vector fields. *Syst. Contr. Lett.*, **26**, 177-184.

Sontag, E.D. (1995). State-space and i/o stability for nonlinear systems. In *Feedback Control, Nonlinear Systems, and Complexity*, (B.A. Francis and A.R. Tannenbaum, (Ed)), 215-235. Lecture Notes in Control Information Sciences, Springer, Berlin and New York.

Soravia, P. (1996). $H_\infty$ control of nonlinear systems: differential games and viscosity solutions. *SIAM J. Contr. Optim.*, **35**, 1071-1097.

van der Schaft, A. J. (1992). $L_2$ gain analysis of nonlinear systems and nonlinear $H_\infty$ control. *IEEE Trans. Automat. Contr.*, **37**, 770-784.

Vidyasagar, M. and A. Vannelli (1982). New relationships between input-output and Lyapunov stability. *IEEE Trans. Automat. Contr.* **27**, 481-483.

# CONTROL OF NONLINEAR SYSTEMS VIA FEEDBACK LINEARIZATION AND CONSTRAINED MODEL PREDICTIVE CONTROL

**Won-Kee Son* Oh-Kyu Kwon***

* *School of electrical and computer engr., Inha Univ., Incheon 402-751, Korea, E-mail: okkwon95@dragon.inha.ac.kr*

Abstract: This paper combines the constrained model predictive control with the feedback linearization to solve the control problem of a nonlinear systems with the constrained input. This combined approach consists of two step: Firstly, the linearized model is obtained through feedback linearization. And secondly, constrained model predictive controller taken a consideration into constraints on input is designed using the linearized model. Effectiveness of the combined approach and performances of tracking for nonlinear system subject to constraint on input are shown via simulation, where a comparison is done between two case of unconstrained input and constrained input in feedback linearization. As a test nonlinear model, a flexible single link robot system is used. *Copyright © 1999 IFAC*

Keywords: Nonlinear systems, Feedback linearization, Constrained Model Predictive Control, Single link Robot system

## 1. INTRODUCTION

In many common control problems involving nonlinear systems, it is possible to construct a feedback control law which will transform the original nonlinear plant into a closed loop system which is exactly linear. Once a nonlinear plant is linearized by feedback, all of the tools developed for the theory of linear systems can be directly applied to control the new closed loop. This method allows to design an exact controller for the nonlinear systems affine in the control. Another method in the controlling nonlinear systems is gain scheduling which is a technique that can extend the validity of the linearization approach to a range of operating points. But this method has a basic limitation that the controller is guaranteed to work only in the neighborhood of an equilibrium point.

The goal of the current paper is to design a control system for nonlinear systems with the constrained input. In all physical systems there are limitations to the changes that can be made to the manipulated variables. Therefore a control algorithm should be designed to have the ability to account for such limitations. To fulfill overall operational requirements, it is therefore necessary to explicitly consider all constraints in the formulation of the control cost function. The CMBPC (Constrained Model-Based Predictive Control) (Wilkinson *et al.*, 1990), which is known to be numerically stable and to have the robust control performance is adopted, as a control technique which accounts for such constraints in this paper. This control law is a kind of GPC (Generalized Predictive Control). Therefore, to control nonlinear dynamic system with constraints, a combined approach which combines the feedback linearization and constrained model predictive control is proposed in this paper.

Another goal of this paper is to apply the combined approach to a flexible single link robot system proposed by Spong and Vidyasagar (1989). The model is nonlinear with single input and single output. In the simulation, it is shown that the combined approach works well even under some constraints on input variable in the flexible robot system.

The layout of the paper is as follows: In Section 2, the feedback linearization is reviewed for nonlinear systems. In Section 3, model predictive control is also reviewed for systems subject to constraints. A combined approach scheme is then proposed in Section 4, which combines the feedback linearization and the constrained model predictive control presented in Section 2 and 3, respectively. Finally, the combined approach is applied to flexible link robot system in the constrained input through some simulations in Section 5 to exemplify the tracking performance of the scheme. Conclusions are summarized in Section 6.

## 2. FEEDBACK LINEARIZATION

Consider systems represented by the state equations

$$\dot{x} = f(x) + g(x)u \tag{1}$$

with $f$ and $g$ being smooth vector fields. The problem of input-state linearization is solved in two steps. First, one finds a state transformation $z = z(x)$ and an input transformation $u = u(x, v)$ so that the nonlinear system dynamics is transformed into an equivalent *linear time invariant* dynamics, in the familiar form $\dot{z} = Az + bv$. Second, one uses standard linear techniques to design $v$. Note that systems in the form (1) are said to be *linear in control or affine.* It is useful to point out that if a nonlinear system has the form

$$\dot{x} = f(x) + g(x)w[u + \phi(x)]$$

with $w$ being an invertible scalar function and $\phi$ being an arbitrary functional, a simple variable substitution $v = w[u + \phi(x)]$ puts the dynamics into the form of (1). One can design a control law for $v$ and then $u$ by inverting $w$, *i.e.*, $u = w^{-1}(v) - \phi(x)$. Given a scalar function $h(x)$ and a vector field $f(x)$, we define a new scalar function $L_f h$, called the Lie derivative of $h$ with respect to $f$

$$L_f h = \nabla_h f = \frac{\partial h}{\partial x} f. \tag{2}$$

A single-input nonlinear system in the form (1), with $f(x)$ and $g(x)$ being smooth vector fields on $R^n$, is said to be input-state linearizable if there exits a region $\Omega$ in $R^n$, a diffeomorphism $\phi : \Omega \to R^n$, and a nonlinear feedback control law

$$u = \alpha(x) + \beta(x)v$$

such that the new state variables $z = \phi(x)$ and the new input $v$ satisfy a linear time-invariant relation

$$\dot{z} = Az + bv.$$

To simplify notations, we will often use $z$ to denote not only the transformed state, but the diffeomorphism $\phi$ itself, *i.e.*, write

$$z = z(x).$$

In order to check whether or not nonlinear systems can be transformed into a linear and controllable system via state feedback and coordinates transformation, we test the condition of following theorem.

*Theorem 1.* [Slotine, 1991] The nonlinear system (1), with $f(x)$ and $g(x)$ being smooth vector fields, is input-state linearizable if, there exists a region $\Omega$ such that the following conditions hold:

1. The vector fields $\{g, ad_f g, \ldots, ad_f^{n-1} g\}$ are linearly independent in $\Omega$.
2. The set $\{g, ad_f g, \ldots, ad_f^{n-2} g\}$ is involutive in $\Omega$. ∎

Based on the previous discussion, the input-state linearization of a nonlinear system can be performed through the following steps :

1) Construct the vector fields $g, ad_f g, \ldots, ad_f^{n-1} g$ for the given system.
2) Check whether the controllability and involutivity conditions are satisfied.
3) If both are satisfied, find the first state $z_1$ from equation $L_g z_1 = L_{ad_f g} z_1 = \ldots = L_{ad_f^{n-2} g} z_1 = 0$.
4) Compute the state transformation $z(x) = [z_1 \; L_f z_1 \ldots L_f^{n-1} z_1]^T$ and the input transformation, with

$$\alpha(x) = -\frac{L_f^n z_1}{L_g L_f^{n-1} z_1}, \quad \beta(x) = \frac{1}{L_g L_f^{n-1} z_1}$$

The closed loop system under the above control law is represented in the block diagram in Fig. 1.

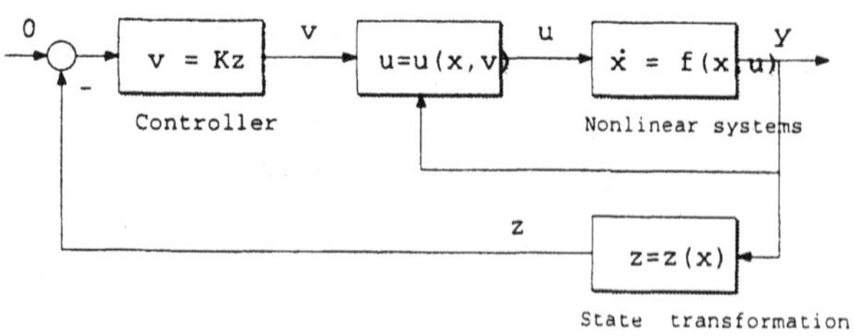


Fig. 1. Input-state linearization.

## 3. CONSTRAINED MODEL BASED PREDICTIVE CONTROL

The system model under consideration is described by the linear discrete-time state-space equation as follows:

$$x(k+1) = Ax(k) + B\Delta u(k) + Gw(k)$$
$$y(k) = Cx(k) + Rv(k), \qquad (3)$$

where the $n$-dimensional vector $x$ denotes the system states, $u$ the control input signals, and $y$ the controlled signals. The matrices $A$, $B$, $C$, $G$, $R$ contain the model parameters. The discrete noise signals $w$ and $v$ are modeled as zero-mean, white, normal, and stationary random processes. The difference operator $\Delta$ is defined as $q - 1$, where $q$ denotes the forward shift operator on the discrete time scale. The best j-step ahead predictor is obtained from expected value of the j-step ahead output signal with stochastic noise signal $v$ and disturbance $w$:

$$\hat{y}(k+j) = E[y(k+j)]$$
$$= CA^jE[x(k)] + \sum_{i=0}^{j-1} CA^{j-i-1}B\Delta u(k+i)$$
$$+\sum_{i=0}^{j-1} CA^{j-i-1}GE[w(k+i)] + RE[v(k+j)]. (4)$$

From the assumption $E[v] = 0$ and $E[w] = 0$, if $\hat{x}$ is substituted for the expected value of $x$, then (4) is simplified as follows:

$$\hat{y}(k+j) = CA^j\hat{x}(k) + \sum_{i=0}^{j-1} CA^{j-i-1}B\Delta u(k+i). (5)$$

In (5), the estimation of the state vector $\hat{x}$ is found from a stationary Kalman filter which contains the Kalman filter gain matrix $L$ to adapt the estimated model states to the measured outputs:

$$\bar{x}(k) = A\hat{x}(k-1) + B\Delta u(k-1)$$
$$\hat{x}(k) = \bar{x}(k) + L[y(k) - C\bar{x}(k)]. \qquad (6)$$

Here the Kalman filter is designed to minimize the variance of the error between the real state and the estimated state

$$e(k) = \hat{x}(k) - x(k). \qquad (7)$$

The control law for model-based predictive control should minimize the following cost function in systems with noise and disturbance:

$$J = \sum_{j=N_1}^{N_2} [r(k+j) - \hat{y}(k+j)]^2$$
$$+\lambda \sum_{j=0}^{N_u-1} [\Delta u(k+j)]^2. \qquad (8)$$

In the cost function (8), $N_1$ denotes the lower cost horizon, $N_2$ the upper cost horizon, $N_u$ the control horizon, $\lambda$ the weighting factor for control increments, and $r$ the reference. The cost function (8) can be expressed in vector notation as follows:

$$J = (r - \hat{y})^T(r - \hat{y}) + \lambda\Delta u^T\Delta u, \qquad (9)$$

where

$$r = [r(k+N_1)\dots r(k+N_2)]^T$$
$$\hat{y} = [\hat{y}(k+N_1)\dots\hat{y}(k+N_2)]^T$$
$$\Delta u = [\Delta u(k)\dots\Delta u(k+N_u-1)]^T.$$

The predicted output vector can be calculated by the following matrices:

$$\hat{y} = F\hat{x}(k) + H\Delta u, \qquad (10)$$

where

$$F = \begin{bmatrix} CA^{N_1} \\ \vdots \\ CA^{N_2} \end{bmatrix}, \quad H = \begin{bmatrix} h_{N_1,1} & \dots & h_{N_1,N_u} \\ \vdots & \ddots & \vdots \\ h_{N_2,1} & \dots & h_{N_2,N_u} \end{bmatrix},$$

$$h_{j,i} = \begin{cases} CA^{j-i}B, & j \geq i \\ 0, & j < i \end{cases}.$$

The free response $f$ is defined as the predicted output when the control input remains constant from now on:

$$f = F\hat{x}(k). \qquad (11)$$

Substituting (10) and (11) into the cost function (9) yield a quadratic minimization problem

$$J(\Delta u) = \Delta u^T[H^TH + \lambda I]\Delta u - 2(r-f)^TH\Delta u$$
$$+(r-f)^T(r-f). \qquad (12)$$

The well-known LS (Least Squares) solution solves this minimization problem as follows:

$$\Delta u = (H^TH + \lambda I)^{-1}H^T(r-f). \qquad (13)$$

Just the first element of the future control increments is a actually needed for the control law with the controller gain vector

$$u(k) = u(k-1) + K(r-f)$$
$$K = (I\ 0\ \cdots\ 0)(H^TH + \lambda I)^{-1}H^T. \quad (14)$$

A block diagram of model-based predictive controller with Kalman filtering in multivariable nonlinear systems is shown in Fig.2. The minimization problem of the cost function with constraints is reformulated as solving QP (Quadratic Programming) problem. The algorithm uses SVD (Singular Value Decomposition) to firstly determine the unconstrained minimum of the cost function. If the specified constraints are violated, then LDP (Least Distance Programming) incorporating NNLS (Non-Negative Least Squares) is applied to find an admissible solution.

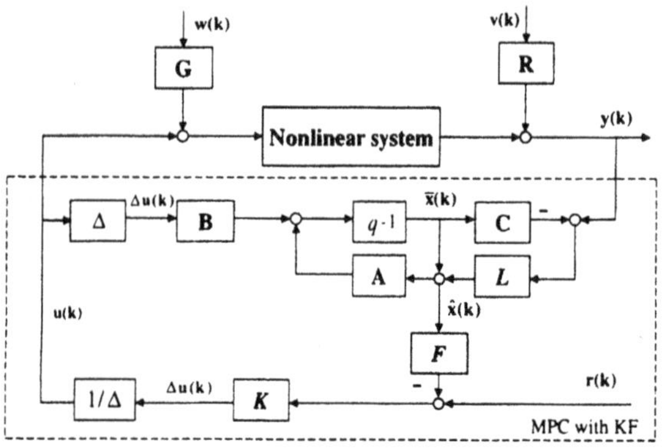


Fig. 2. Block diagram of MBPC.

Inspection of the cost function shows that the optimization problem being considered is Least Squares with linear Inequality constraints(LSI):

$$Min \ \| \Gamma \Delta u - b \|^2 \quad subject\ to \ \ \Im \Delta u \geq h \tag{15}$$

with

$$\Gamma = \begin{bmatrix} H \\ \lambda^{\frac{1}{2}} I \end{bmatrix}, \quad b = \begin{bmatrix} r - f \\ 0 \end{bmatrix}.$$

The matrix $\Im$ contains dynamic information about the constraints while $h$ gives the limiting values for the constraints. In solving a constrained quadratic cost function, QP technique based on a SVD approach is adopted. In solving the LSI problem (15), the unconstrained solution is first determined using SVD. The SVD of $\Gamma$ is as follows:

$$\Gamma = U \Sigma V^T, \tag{16}$$

where $U$ and $V$ are orthogonal matrices, and $\Sigma$ is a diagonal matrix of singular values. Using (16) the minimum norm result of the problem (15) is given as follows:

$$\Delta u = (\Gamma^T \Gamma)^{-1} \Gamma^T b = V \Sigma^{-1} U^T b.$$

If this result does not satisfy the specified constraints, the LSI problem is converted into an LDP problem to find the solution that satisfies the constraint set. Consider the partitioned SVD of $\Gamma$:

$$\Sigma = \begin{bmatrix} S & 0 \\ 0 & 0 \end{bmatrix}, \quad S = diag[s_1, \ \cdots \ , s_r]$$

$$\Gamma = U \begin{bmatrix} S & 0 \\ 0 & 0 \end{bmatrix} V^T = \begin{bmatrix} U_1 : U_2 \end{bmatrix} \begin{bmatrix} S \\ 0 \end{bmatrix} V^T. \tag{17}$$

By a transformation of variables, viz. $\Delta u = S\beta$, the LSI objective function can be rewritten as follows:

$$\phi(\Delta u) = \| b - \Gamma \Delta u \|^2 = \left\| \begin{bmatrix} U_1^T b \\ U_2^T b \end{bmatrix} - \begin{bmatrix} S\beta \\ 0 \end{bmatrix} \right\|^2$$

$$= \| \bar{b}_1 - S\beta \|^2 + \| \bar{b}_2 \|^2$$
$$= \| z \|^2 + \| \bar{b}_2 \|^2, \tag{18}$$

where $\bar{b}_i = U_i^T b$, $z = \bar{b}_i - S\beta$. Except for the constant $\| \bar{b}_2 \|^2$, which corresponds to the absolute minimum value of the unconstrained objective function, the original LSI problem is then equivalent to the following LDP formulation:

$$min \ \| z \|, \quad subject\ to \ \bar{\Im} z \geq \bar{h}. \tag{19}$$

where $\bar{\Im} = \Im V S^{-1}$ and $\bar{h} = h - \bar{\Im}\bar{b}_1$. In solving the LDP representation, $d$ is first computed in order to solve the non-negative least squares form:

$$min \ \| Qd \ - \ \gamma \|, \qquad subject\ to \ \ d \geq 0, \tag{20}$$

where

$$Q = \begin{bmatrix} \bar{\Im}^T \\ \bar{h}^T \end{bmatrix} \quad and \ \ \gamma = [0, \ \ldots \ , 0, 1]^T.$$

The residual vector

$$\alpha = Qd - \gamma, \quad \gamma \in R^{(n+1) \times 1} \tag{21}$$

is next computed and the LDP solution is obtained as

$$z_i = -\frac{\alpha_i}{\alpha_{n+1}}, \qquad i \ = \ 1, \ \cdots \ , n. \tag{22}$$

Finally, the solution of the original LSI problem is then recovered from

$$\Delta u = V S^{-1} (z + \bar{b}_1). \tag{23}$$

Calculated control moves are sent to the system only when a solution satisfying all specified constraints has been found.

## 4. A COMBINED APPROACH

In this section, using a feedback loop which linearizes the nonlinear plant with hard input constraint and such an obtained linearied model for controller design, the control problem is reformulated as a linear one involved in quadratic programming. The configuration of combined approach is shown in Fig.3 in presence of constraints on input. The internal loop is concerned only with linearizing the plant and tracking its own reference input $v$. The external loop produces a control output $v$ as a function of the external reference $y_r$ and of the state $z$ of the linearized model. A brief description of this combined approach follows. Consider the nonlinear systems described in (24)

$$\dot{x} = f(x) + g(x)u, \quad y = h(x) \tag{24}$$

with constraints on input

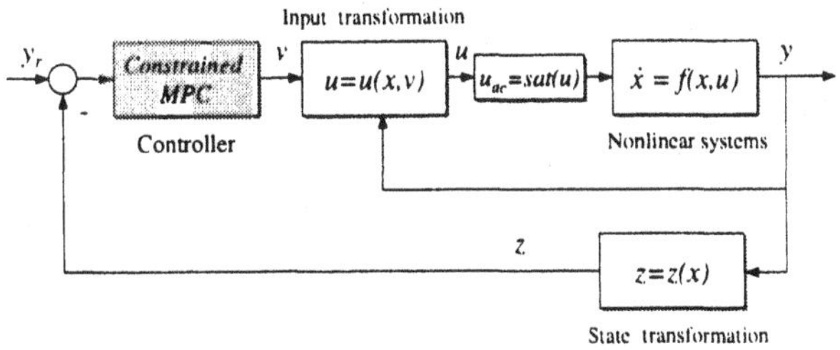


Fig. 3. Configuration of MPC in presence of constraint.

$$u = u(x,v) = \alpha(x) + \beta(x)v, -u_{lc} \le u \le u_{uc}, \quad (25)$$

where $f$ and $g$ are smooth vector fields and $f(0) = 0$, and $-u_{lc}, u_{uc}$ are lower and upper bound, respectively. Then, we can define a new control variable $v$ as an external reference input from (25)

$$v = \bar{\alpha}(x) + \bar{\beta}(x)u, \quad (26)$$

where

$$\bar{\alpha}(x) = -\frac{\alpha(x)}{\beta(x)} = L_f^n h(x)$$
$$\bar{\beta}(x) = \frac{1}{\beta(x)} = L_g L_f^{n-1} h(x).$$

Introducing a coordinate transformation, we can define a new linearized system as follows.

$$\dot{z} = Az + Bv, \quad y = Cz. \quad (27)$$

Note that (27) hold the linear properties only if the constraints conditions on its input are satisfied

$$v = \bar{\nu}(x, -u_{lc}, u_{uc}), \quad -v_{lc} \le v \le v_{uc}, \quad (28)$$

where the lower and upper bounds are given by

$$-v_{lc} = \bar{\alpha}(x) - u_{lc}\bar{\beta}(x)$$
$$v_{uc} = \bar{\alpha}(x) + u_{uc}\bar{\beta}(x). \quad (29)$$

Using the state transformation, the new constraints can be expressed using the state variables of the linearized system

$$v = \nu(z, -u_{lc}, u_{uc}), \quad (30)$$

where the lower and upper bounds are given by

$$-v_{lc} = \tilde{\alpha}(z) - u_{lc}\tilde{\beta}(z)$$
$$v_{uc} = \tilde{\alpha}(z) + u_{uc}\tilde{\beta}(z).$$

It is clear that, due to the feedback linearization, original hard constraints $(-u_{lc}, u_{uc})$ are mapped into the MPC constraints on $v$ that are in general *nonlinear and state dependent*. To avoid reaching these input constraints and to improve control performance, the MPC can be applied to the new linearized system given by (27).

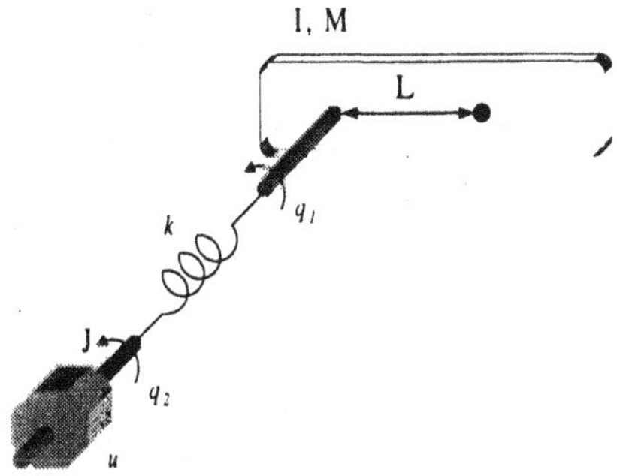


Fig. 4. A flexible joint robot mechanism.

## 5. SIMULATION

Consider the equation motion of a single-link flexible-joint robot[Slotine, 1991]

$$I\ddot{q}_1 + MgL\sin q_1 + k(q_1 - q_2) = 0$$
$$J\ddot{q}_2 - k(q_1 - q_2) = u.$$

Let's choose $[q_1 \ \dot{q}_1 \ q_2 \ \dot{q}_2]^T$ as a state. Then, we can find vector fields $f(x)$ and $g(x)$ can be written

$$f = \begin{bmatrix} x_2 \\ -\frac{1}{I}(MgL\sin x_1 + k(x_1 - x_3)) \\ x_4 \\ \frac{k}{J}(x_1 - x_3) \end{bmatrix}, g = \begin{bmatrix} 0 \\ 0 \\ 0 \\ \frac{1}{J} \end{bmatrix}.$$

The controllability matrix is obtained by simple computation

$$[g \ ad_f g \ ad_f^2 g \ ad_f^3 g] = \begin{bmatrix} 0 & 0 & 0 & -\frac{k}{IJ} \\ 0 & 0 & \frac{k}{IJ} & 0 \\ 0 & -\frac{1}{J} & 0 & \frac{k}{J^2} \\ \frac{1}{J} & 0 & -\frac{k}{J^2} & 0 \end{bmatrix}. \quad (31)$$

(31) has rank 4. Furthermore, since the vector fields $\{g, \ ad_f g, \ ad_f^2 g, \ ad_f^3 g\}$ are constant, they form an involutive set. Therefore, the system is input-state linearizable. Let us find out the state transformation $z = z(x)$. From given the expression of the controllability matrix in (31) and the controllability matrix is constant, the first component $z_1$ of the new state vector $\mathbf{z}$ should be satisfy

$$\frac{\partial z_1}{\partial x_1} \ne 0 \quad \frac{\partial z_1}{\partial x_2} = 0 \quad \frac{\partial z_1}{\partial x_3} = 0 \quad \frac{\partial z_1}{\partial x_4} = 0.$$

Thus, $z_1$ must be a function $x_1$ only. The simplest solution to the above equation is

$$z_1 = x_1.$$

The other states can be obtained from $z_1$. Also, we can obtain the input transformation $u$ as follows:

$$\begin{aligned} u = \alpha(x) + \beta(x)v &= -\frac{\nabla_{z_4} f}{\nabla_{z_4} g} + \frac{1}{\nabla_{z_4} g} v \\ &= \frac{J}{k}\left[v - MgL\sin x_1\left(x_2^2 + \frac{MgL}{I}\cos x_1 + \frac{k}{I}\right)\right. \\ &\quad \left. -k(x_1 - x_3)\left(\frac{k}{I} + \frac{k}{J} + \frac{MgL}{I}\cos x_1\right)\right]. \end{aligned} \quad (32)$$

As a result of the above state and input transformations, we obtain the following set of linear equations $\dot{\mathbf{z}} = A\mathbf{z} + bv$

$$\dot{z}_1 = z_2, \quad \dot{z}_2 = z_3, \quad \dot{z}_3 = z_4, \quad \dot{z}_4 = v. \quad (33)$$

In this simulation, a combined scheme is applied to flexible single link robot system with constrained input or not. The linearized model discretized in sampling time 0.01[*sec*] to design model predictive controller. Step input used as reference input and only amplitude constraint is considered. As shown in section 4., if we assume that the physical system input is limited to $-u_{lc} \leq u \leq u_{uc}$, the lower and upper bound for the new input $v$ of the linearized system can be given by (29) and (34).

$$\begin{aligned} \bar{\alpha}(x) = L_f^n h(x) &= \nabla z_4 f \\ &= \frac{MgL}{I} x_2^2 \sin x_1 + \frac{1}{I^2}(MgL\cos x_1 + k) \\ &\quad (MgL\sin x_1 + k(x_1 - x_3)) + \frac{k^2}{IJ}(x_1 - x), \\ \bar{\beta}(x) = L_g L_f^{n-1} &= \nabla z_4 g = \frac{k}{IJ}. \end{aligned} \quad (34)$$

Tracking performance of controller considered constraint on input taking change of saturation limit value is shown in Fig.5. As shown, for reducing saturation limit value rising time is more long and even unstable.

## 6. CONCLUSIONS

In this paper a combined control scheme combining the constrained model predictive control with feedback linearization is presented, and its performance is analyzed via some simulation applied to flexible single link robot system, which has shown that the controller presented works well even under some actuator saturation in robot system. The simulation also has compared the performance of combined scheme in the case of constrained input with that of the case with unconstrained input. However, the control method adopted in the current paper has not taken the parameter uncertainty into consideration, and so it requires further research to develop robust control methods.

## 7. REFERENCES


Isidori, A. (1989). *Nonlinear Control Systems : An Introduction.* Springer Verlag.

Khalil, H. (1996). *Nonlinear Systems.* Prentice Hall, second edition.

Nevistic, V., and L. Del Re (1994). Feasible suboptimal model predictive control for linear plants with state dependent constraints. *Proc. of American Control Conference*, Baltimore, 2862-2866.

Slotine, J-J. E., and W. Li (1991). *Applied Nonlinear Control.* Prentice Hall, Englewood Cliffs, New Jersey.

Spong, M. W., and M. Vidyasagar (1989). *Robot Dynamics and Control* Wiley, New York.

Vidyasagar, M. (1993). *Nonlinear Systems Analysis* Prentice Hall, Englewood Cliffs, New Jersey, second edition.

Wilkinson, D. J., A.J. Morris and M.T. Tham (1990). Constrained multivariable predictive control (a comparison with QDMC). *Proc. of the American Control Conference*, San Diego, California, U.S.A, 1620-1625.


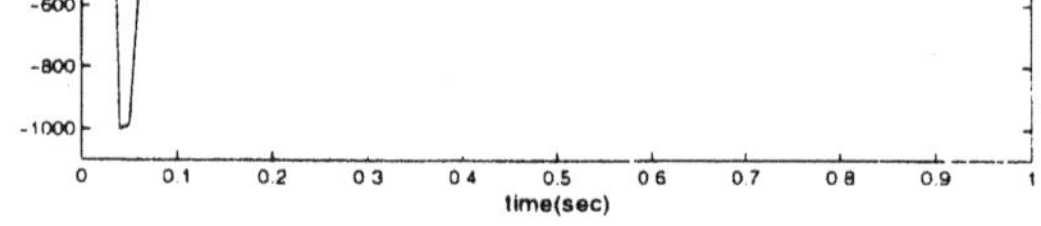


(a) Input torque

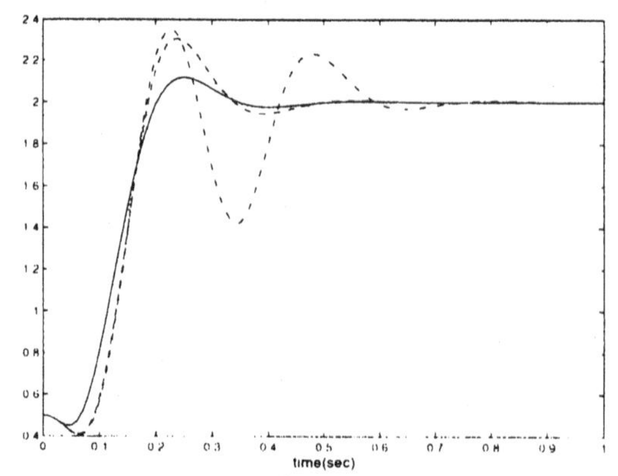


(b) Angle of link

Fig. 5. Tracking results of constrained controller taking change of saturation limit value

14th Triennial World Congress, Beijing, P.R. China



# Infinite Dimensional Optimal Control Theory

Jiongmin Yong

Department of Mathematics and Institute of Mathematical Finance
Fudan University, Shanghai 200433, China
email: *jyong@ms.fudan.edu.cn*

Abstract: One of the milestones in modern optimal control theory is the Pontryagin Maximum Principle, which was firstly established by Pontryagin and his colleagues in later 50s for finite dimensional optimal control problems. In this survey paper, we revisit some extensions of the maximum principle in infinite dimensional spaces, developed via the infinite dimensional spike variation technique. 

Keywords: Optimal control, distributed parameter systems, maximum principle.

## 1. INTRODUCTION

Infinite dimensional systems can be used to describe many phenomena in the real world. As is well-known, heat conduct, property of elastic-plastic material, behavior of fluid dynamics, population dynamics, diffusion-reaction processes,..., etc. are all within this area. Thus, controlling such systems is of great theoretical and practical importance.

It is known by now that *Pontryagin's maximum principle* ([15]), *Bellman's dynamic programming method* ([1]) and *Kalman's linear quadratic optimal control theory* ([9]) are three milestones of modern (finite dimensional) optimal control theory (see Fleming [7], for extensive discussion). The study of optimal control theory for infinite dimensional systems can be traced back to the beginning of 60's. A main goal of such a theory is to establish the infinite dimensional version of the above-mentioned three fundamental theories. In the past about 40 years, many mathematicians and control theorists have made great contributions in this research area. The purpose of this paper is mainly to briefly survey some works related to the Pontryagin's maximum principle (for infinite dimensional systems), done by a group of mathematicians (including the present author) from Fudan University and their collaborators.

The first infinite dimensional version of Pontryagin maximum principle was proved by A. G. Butkovsky ([2]) for integral equation systems and by G. L. Kharatishvile ([10]) for ordinary delay equation systems in 1961. In 1963, Yu. V. Egorov ([5]) constructed an example showing that the maximum principle does not necessarily hold for infinite dimensional systems with state constraints. Because of this counterexample, in 60s and 70s, many people left the area and switched to the discussion of time optimal control problem for linear infinite dimensional systems with convex control domains.

This work is supported in part by the NSFC under grant 79790130, the National Distinguished Youth Science Foundation of China under grant 19725106, the Chinese Education Ministry Science Foundation under grant 97024607, and the Li Foundation.

In 1981, X. Li and Y. Yao ([11]) firstly studied the time optimal control problem for infinite dimensional linear systems with general control domain. They proved that the closure of the reachable set of a linear systems is convex (although the reachable set itself is not necessarily convex). By this fact, they proved the maximum principle without assuming the convexity of the control domain $U$. This is the starting point of *this group* in developing infinite dimensional optimal control theory.

In 1985, X. Li and Y. Yao ([12]) discussed the optimal control problem for an evolution systems in Banach space having terminal state constraint, with a Lagrange type cost functional and general control domain. By assuming the finite codimensionality condition on the terminal state constraint set, together with some other minor conditions, they proved the maximum principle. The vector measure theory was successfully adopted and the corresponding spike variation technique was introduced.

At the same time, H. O. Fattorini ([6]) used the Ekeland variational principle to proved the maximum principle in Hilbert spaces under the condition that the reachable set of the variational system is finite codimensional.

In 1989, X. Li and J. Yong proved the maximum principle for the evolutionary control system in some Banach space $X$ which admits an equivalent norm under which the dual $X^*$ is strictly convex. This is the case if $X$ is either separable, or reflexive. The state constraint allows the initial and terminal states to be combined constrained. This includes the periodic optimal control. (That paper was appeared in 1991). In proving this result, Li and Yong unified and improved the methods of Li-Yao and Fattorini. We now call the method the *infinite dimensional spike variation technique*. This work is fundamentally influential to the later works of this group related to the maximum principle for optimal control of infinite dimensional systems.

Since 1990, this author and his collaborators extended the *infinite dimensional spike variational technique* to treat many other problems: problems with nonsmooth state equation and/or cost functional, problems with mixed state and control constraints, elliptic problems, variational inequalities, minimax problem, infinite dimensional Volterra-Stieltjes evolution equations, etc. (see [16–24,3,4]).

In the rest of this paper, we are going to look at the infinite dimensional maximum principle and the spike variational technique.

## 2. MAXIMUM PRINCIPLE

We start with the following basic assumption:

(H1) Let $X$ be a Banach space, $S \subseteq X \times X$ be a closed and convex set, $U$ be a separable metric space and $T > 0$ be a constant. Let $e^{At}$ be a $C_0$-semigroup on $X$ with generator $A : \mathcal{D}(A) \subseteq X \to X$.

(H2) Let $f : [0,T] \times X \times U \to X$ and $f^0 : [0,T] \times X \times U \to \mathbb{R}$ such that $f(t,y,u)$ and $f^0(t,y,u)$ are strongly measurable in $t \in [0,T]$, continuously Fréchet differentiable in $y \in X$ with $f(t,\cdot,\cdot)$, $f_y(t,\cdot,\cdot)$, $f^0(t,\cdot,\cdot)$ and $f^0_y(t,\cdot,\cdot)$ being continuous. Moreover, for some constant $L > 0$,

$$\begin{aligned} &|f_y(t,y,u)|,\ |f^0_y(t,y,u)|,\\ (2.1)\qquad &|f(t,0,u)|,\ |f^0(t,0,u)| \le L,\\ &\forall (t,y,u) \in [0,T] \times X \times U. \end{aligned}$$

We consider the following evolution equation

$$\begin{aligned} &y(t) = e^{At}y(0)\\ (2.2)\qquad &+\int_0^t e^{A(t-s)}f(s,y(s),u(s))ds,\quad t \in [0,T], \end{aligned}$$

where $u(\cdot)$ is called the *control*, which belongs to the following set:

$$\mathcal{U}[0,T] \equiv \{u : [0,T] \to U \mid u(\cdot) \text{ measurable }\}.$$

The differential form of (2.2) reads

$$(2.3)\qquad \dot{y}(t) = Ay(t) + f(t,y(t),u(t)),\quad t \in [0,T].$$

It is clear that under (H2), for any $(y(0),u(\cdot)) \in X \times \mathcal{U}[0,T]$, (2.2) admits a unique solution $y(\cdot) \triangleq y(\cdot\,;y(0),u(\cdot)) \in C([0,T];X)$. Any $(y(\cdot),u(\cdot)) \in C([0,T];X) \times \mathcal{U}[0,T]$ satisfying (2.2) is called a *feasible pair*. The set of all feasible pairs is denoted by $\mathcal{A}$.By (2.1),we see that for any $(y(\cdot),u(\cdot)) \in \mathcal{A}$, $f^0(\cdot,y(\cdot),u(\cdot)) \in L^1(0,T)$. Thus, we can define

$$(2.4)\qquad J(y(\cdot),u(\cdot)) = \int_0^T f^0(s,y(s),u(s))ds.$$

We point out that assuming condition (2.1) is only for the simplicity of presentation and it can be slightly relaxed.

Next, we let $\mathcal{A}_{ad}$ be the set of all feasible pairs $(y(\cdot),u(\cdot)) \in \mathcal{A}$, such that the following *state constraint* is satisfied:

$$(2.5)\qquad (y(0),y(T)) \in S.$$

Any $(y(\cdot),u(\cdot)) \in \mathcal{A}_{ad}$ is called an *admissible pair*. Now, we can state our optimal control problem as follows:

**Problem (C).** Find $(\bar{y}(\cdot),\bar{u}(\cdot)) \in \mathcal{A}_{ad}$, such that

$$J(\bar{y}(\cdot),\bar{u}(\cdot)) = \inf_{\mathcal{A}_{ad}} J(y(\cdot),u(\cdot)). \tag{2.6}$$

If such a pair exists, we call it an *optimal pair* and refer to $\bar{y}(\cdot)$ and $\bar{u}(\cdot)$ as an *optimal trajectory* and *control*, respectively.

Now, let us recall the result in finite dimensions. For such a case, we may take $X = \mathbb{R}^n$ and $A = 0$. Then, system (2.2) is the same as

$$\dot{y}(t) = f(t,y(t),u(t)), \quad \text{a.e. } t \in [0,T], \tag{2.7}$$

The following result is classical, which is called the *Pontryagin's Maximum Principle.*

**Theorem 2.1.** *Let (H1)-(H2) hold with $X = \mathbb{R}^n$ and $A = 0$. Let $(\bar{y}(\cdot),\bar{u}(\cdot))$ be an optimal pair for Problem (C). Then there exists a nontrivial pair $(\psi^0,\psi(\cdot)) \in \mathbb{R} \times C([0,T];\mathbb{R}^n)$, i.e., $(\psi^0,\psi(\cdot)) \neq 0$, satisfying*

$$\psi^0 \le 0, \tag{2.8}$$

$$\begin{aligned}\dot{\psi}(t) &= -f_y(t,\bar{y}(t),\bar{u}(t))^*\psi(t)\\ &\quad + \psi^0 f_y^0(t,\bar{y}(t),\bar{u}(t)), \quad \text{a.e. } t \in [0,T].\end{aligned} \tag{2.9}$$

$$\begin{aligned}&\langle\psi(0),x_0-\bar{y}(0)\rangle - \langle\psi(T),x_1-\bar{y}(T)\rangle \le 0,\\ &\qquad\qquad \forall (x_0,x_1) \in S,\end{aligned} \tag{2.10}$$

$$\begin{aligned}&\max_{u\in U} H(t,\bar{y}(t),u,\psi^0,\psi(t))\\ &= H(t,\bar{y}(t),\bar{u}(t),\psi^0,\psi(t)),\\ &\qquad\qquad \text{a.e. } t \in [0,T],\end{aligned} \tag{2.11}$$

*where*

$$\begin{aligned}&H(t,y,u,\psi^0,\psi)\\ &= \psi^0 f^0(t,y,u) + \langle \psi, f(t,y,u)\rangle,\\ &\forall (t,y,u,\psi^0,\psi) \in [0,T]\times\mathbb{R}^n\times U\times\mathbb{R}\times\mathbb{R}^n.\end{aligned} \tag{2.12}$$

In the above, we refer to $\psi(\cdot)$ as the *costate* or the *adjoint variable.* Thus, (2.9) is called the *adjoint system* along the optimal pair $(\bar{y}(\cdot),\bar{u}(\cdot))$. Also, we call (2.10) the *transversality condition* and (2.11) the *maximum condition.* The function $H$ defined by (2.12) is called the *Hamiltonian.*

**Remark 2.2.** It is clear that by taking $(\psi^0,\psi(\cdot)) = 0$, (2.8)–(2.11) trivially hold. Thus, it is crucial to require the pair $(\psi^0,\psi(\cdot))$ be nontrivial.

For infinite dimensional case, Theorem 2.1 is not necessarily true if we do not make some further assumptions. To see this, let us present the following example.

**Example 2.3.** Let $X$ be a Hilbert space of infinite dimension. Let $A = A^* : \mathcal{D}(A) \subseteq X \to X$ be a self-adjoint operator with the following properties: There exist sequences $\{\varphi_k\}_{k\ge1} \subseteq X$ and $\{\lambda_k\}_{k\ge1} \subseteq \mathbb{R}$ with the properties that $\{\varphi_k\}_{k\ge1}$ forms an orthonormal basis of $X$ and $\{e^{\lambda_k\cdot}\}_{k\ge1}$ forms a basis of $L^2(0,1)$, such that

$$\begin{cases} 0<\lambda_1\le\lambda_2\le\cdots, & \lim_{k\to\infty}\lambda_k = +\infty,\\ A\varphi_k = -\lambda_k\varphi_k, & k\ge1.\end{cases} \tag{2.13}$$

This can be easily achieved. Now, we let

$$\begin{cases} b\in X, \quad b_k \equiv \langle b,\varphi_k\rangle \neq 0, \quad k\ge1;\\ a\in X, \quad c\in\mathbb{R}, \quad c\neq\langle b,A^{-1}a\rangle;\\ q = (e^A - I)A^{-1}b; \quad U = [-2,2].\end{cases} \tag{2.14}$$

Consider the following system:

$$\dot{y}(t) = Ay(t) + bu(t), \qquad t\in[0,1], \tag{2.15}$$

with $u(\cdot)\in\mathcal{U} \equiv \{u:[0,1]\to U \mid u(\cdot)\text{measurable}\}$. Our constraint for the endpoints of the state is

$$y(0) = 0, \qquad y(1) = q, \tag{2.16}$$

i.e., $S = \{0\}\times\{q\}$ and the cost functional is given by

$$J(u(\cdot)) = \int_0^1 \Big[\langle a,y(t)\rangle + cu(t)\Big]dt \tag{2.17}$$

Solutions to system (2.15) are understood as mild solutions. Thus, for given $u(\cdot)\in\mathcal{U}$ and the initial state $y(0) = 0$, we have

$$y(t) = \int_0^t e^{A(t-s)}bu(s)ds, \qquad t\in[0,1]. \tag{2.18}$$

If we let $\bar{u}(t)\equiv1$, then the corresponding trajectory, denoted by $\bar{y}(\cdot)\equiv y(\cdot\,;\bar{u}(\cdot))$, satisfies

$$\bar{y}(1) = \int_0^1 e^{As}bds = (e^A - I)A^{-1}b = q. \tag{2.19}$$

Thus, $(\bar{y}(\cdot),\bar{u}(\cdot))$ is admissible. If $\hat{u}(\cdot)\in\mathcal{U}$ is such that $(y(\cdot\,;\hat{u}(\cdot)),\hat{u}(\cdot))$ is also admissible, then,

$$\begin{aligned}0 &= y(1;\hat{u}(\cdot)) - \bar{y}(1)\\ &= \sum_{k\ge1}\Big\{\int_0^1 e^{-\lambda_k(1-s)}(\hat{u}(s)-\bar{u}(s))ds\Big\}b_k\varphi_k\\ &= \sum_{k\ge1}\Big\{\int_0^1 e^{\lambda_k s}(\hat{u}(s)-\bar{u}(s))ds\Big\}e^{-\lambda_k}b_k\varphi_k.\end{aligned} \tag{2.20}$$

Hence,

$$\int_0^1 e^{\lambda_k s}(\hat{u}(s)-\bar{u}(s))ds = 0, \qquad k\ge1. \tag{2.21}$$

Since $\{e^{\lambda_k\cdot}\}_{k\ge1}$ forms a basis of $L^2(0,1)$, it follows from (2.21) that $\hat{u}(s) = \bar{u}(s)$, a.e. $s\in[0,1]$.

This means that $(\bar{y}(\cdot), \bar{u}(\cdot))$ is the only admissible pair and hence it is optimal. Now, suppose the pair $(\bar{y}(\cdot), \bar{u}(\cdot))$ satisfies the Pontryagin's Maximum Principle; that is there exists a pair $(\psi^0, \psi(\cdot)) \neq 0$, with $\psi^0 \leq 0$ and

$$\begin{aligned} \psi(t) &= e^{A(1-t)}\psi(1) + \psi^0 \int_t^1 e^{A(s-t)} a ds \\ &= e^{A(1-t)}\psi(1) + \psi^0 (e^{A(1-t)} - I)A^{-1}a, \end{aligned} \tag{2.22}$$

such that

$$\begin{aligned} &\max_{|u|\leq 2} H(\bar{y}(t), u, \psi^0, \psi(t)) \\ &= H(\bar{y}(t), \bar{u}(t), \psi^0, \psi(t)), \text{ a.e. } t \in [0,1], \end{aligned} \tag{2.23}$$

where

$$\begin{aligned} &H(y, u, \psi^0, \psi) \\ &= \psi^0(\langle a, y\rangle + cu) + \langle \psi, b\rangle u, \\ &\forall (y, u, \psi^0, \psi) \in X \times U \times \mathbb{R} \times X^*. \end{aligned} \tag{2.24}$$

Then, (2.23) implies (note $\bar{u}(t) \equiv 1$)

$$\begin{aligned} &\max_{|u|\leq 2} [u(\psi^0 c + \langle \psi(t), b\rangle)] \\ &= \psi^0 c + \langle \psi(t), b\rangle, \text{ a.e. } t \in [0,1]. \end{aligned} \tag{2.25}$$

In order this to be true, we must have (note the continuity of $\psi(\cdot)$)

$$\psi^0 c + \langle \psi(t), b\rangle \equiv 0, \qquad t \in [0,1]. \tag{2.26}$$

Combining the above with (2.22), we have

$$\begin{aligned} 0 &= \psi^0 c + \langle \psi(t), b\rangle \\ &= \psi^0 c + \langle \psi(1), e^{A(1-t)} b\rangle \\ &\quad + \psi^0 \langle (e^{A(1-t)} - I)A^{-1}a, b\rangle \\ &= \psi^0 [c - \langle A^{-1}a, b\rangle] \\ &\quad + \sum_{k\geq 1} \langle \psi(1) + \psi^0 A^{-1}a, \varphi_k\rangle e^{-\lambda_k(1-t)} b_k. \end{aligned} \tag{2.27}$$

Thus, it follows that

$$\begin{cases} \psi^0[c - \langle A^{-1}a, b\rangle] = 0, \\ \langle \psi(1) + \psi^0 A^{-1}a, \varphi_k\rangle = 0, \qquad k \geq 1, \end{cases} \tag{2.28}$$

which gives

$$\begin{cases} \psi^0[c - \langle A^{-1}a, b\rangle] = 0, \\ \psi(1) + \psi^0 A^{-1}a = 0. \end{cases} \tag{2.29}$$

By (2.14), we derive from (2.29) that

$$\psi^0 = 0, \quad \psi(1) = 0. \tag{2.30}$$

But, this gives the pair $(\psi^0, \psi(\cdot)) = 0$, which leads to a contradiction. □

The above example shows that, in general, for infinite dimensional control problem, the usual Pontryagin's type maximum principle does not necessarily hold. To ensure the Pontryagin's Maximum Principle to hold, we need some additional assumptions. Now, we again let $(\bar{y}(\cdot), \bar{u}(\cdot)) \in \mathcal{A}_{ad}$ be a fixed optimal pair. For any $u(\cdot) \in \mathcal{U}[0,T]$, we introduce the *variational system* of the state equation along the optimal pair $(\bar{y}(\cdot), \bar{u}(\cdot))$ (under control $u(\cdot)$):

$$\begin{aligned} \xi(t) &= e^{At} y_0 \\ &+ \int_0^t e^{A(t-s)} f_y(s, \bar{y}(s), \bar{u}(s))\xi(s) ds \\ &+ \int_0^t e^{A(t-s)} [f(s, \bar{y}(s), u(s)) \\ &\quad - f(s, \bar{y}(s), \bar{u}(s))] ds, \\ &\qquad t \in [0,T], u(\cdot) \in \mathcal{U}[0,T]. \end{aligned} \tag{2.31}$$

Clearly, under (H1)–(H2), for any $(y_0, u(\cdot)) \in X \times \mathcal{U}[0,T]$, (2.31) admits a unique solution $\xi(\cdot) \triangleq \xi(\cdot\,; y_0, u(\cdot)) \in C([0,T]; X)$. The differential form of (2.31) is

$$\begin{cases} \dot{\xi}(t) = [A + f_y(t, \bar{y}(t), \bar{u}(t))]\xi(t) \\ \qquad + f(t, \bar{y}(t), u(t)) - f(t, \bar{y}(t), \bar{u}(t)), \\ \xi(0) = y_0. \end{cases} \tag{2.32}$$

Define

$$\mathcal{R} = \{\xi(T; 0, u(\cdot)) \in X \mid u(\cdot) \in \mathcal{U}[0,T]\}, \tag{2.33}$$

$$Q = \{y_1 - \xi(T; y_0, 0) \mid (y_0, y_1) \in S\}. \tag{2.34}$$

The set $\mathcal{R}$ is called the *reachable set* of the variational system (2.31). For simplicity, we will call $\mathcal{R}$ the reachable set associated with $(\bar{y}(\cdot), \bar{u}(\cdot))$ hereafter. The set $Q$ is referred to as the *modified endpoint constraint set*. Following definition is necessary in sequel.

**Definition 2.4.** Let $Z$ be a Banach space. A subset $S$ of $Z$ is said to be *finite codimensional* in $Z$, denoted by $\operatorname{codim}_Z S < \infty$, if there exists a point $z_0 \in \overline{\text{co}}\, S$, such that $\text{span}\{S - z_0\} \triangleq$ the closed subspace spanned by $\{z - z_0 \mid z \in S\}$ is a finite codimensional subspace of $Z$ and $\overline{\text{co}}(S - z_0)$ has a nonempty interior in this subspace.

Our maximum principle for the optimal controls in infinite dimensions is as follows.

**Theorem 2.5.** *Let $X$ be a Banach space. Let (H1)–(H2) hold. Let $(\bar{y}(\cdot), \bar{u}(\cdot))$ be an optimal pair of Problem (C) with the associated reachable set $\mathcal{R}$ and the modified endpoint constraint set $Q$. Suppose further that the following (H3) and (H4) hold:*

(H3) $\mathcal{R} - Q \equiv \{r - q \mid r \in \mathcal{R}, q \in Q\}$ is finite codimensional in $X$.

(H4) The dual $X^*$ of $X$ is strictly convex.

Then there exists a nontrivial pair $(\psi^0, \psi(\cdot)) \in \mathbb{R} \times C([0,T];\mathbb{R}^n)$, i.e., $(\psi^0, \psi(\cdot)) \neq 0$, satisfying

$$\psi^0 \leq 0, \tag{2.35}$$

$$\begin{aligned}\psi(t) &= e^{A^*(T-t)}\psi(T) \\ &+ \int_t^T e^{A^*(s-t)} f_y(s,\bar{y}(s),\bar{u}(s))^* \psi(s)ds \\ &+ \int_t^T e^{A^*(s-t)} \psi^0 f_y^0(s,\bar{y}(s),\bar{u}(s))ds, \\ &\qquad \text{a.e. } t \in [0,T],\end{aligned} \tag{2.36}$$

$$\begin{aligned}&\langle \psi(0), x_0 - \bar{y}(0)\rangle - \langle \psi(T), x_1 - \bar{y}(T)\rangle \leq 0, \\ &\qquad \forall (x_0, x_1) \in S,\end{aligned} \tag{2.37}$$

$$\begin{aligned}&\max_{u \in U} H(t,\bar{y}(t),u,\psi^0,\psi(t)) \\ &= H(t,\bar{y}(t),\bar{u}(t),\psi^0,\psi(t)), \\ &\qquad \text{a.e. } t \in [0,T],\end{aligned} \tag{2.38}$$

where

$$\begin{aligned}&H(t,y,u,\psi^0,\psi) \\ &= \psi^0 f^0(t,y,u) + \langle \psi, f(t,y,u)\rangle, \\ &\forall (t,y,u,\psi^0,\psi) \in [0,T] \times X \times U \times \mathbb{R} \times X^*.\end{aligned} \tag{2.39}$$

The differential form of (2.36) is the following:

$$\begin{aligned}\dot{\psi}(t) &= -A^*\psi(t) - f_y(t,\bar{y}(t),\bar{u}(t))^*\psi(t) \\ &+ \psi^0 f_y^0(t,\bar{y}(t),\bar{u}(t)), \quad \text{a.e. } t \in [0,T].\end{aligned} \tag{2.40}$$

**Remark 2.6.** (H4) is actually very general. In the case $X$ being separable or reflexive, one can find an equivalent norm under which (H4) holds. Thus, if our state space is one of $L^p(\Omega)$, $W^{m,p}(\Omega)$ $(1 \leq p < \infty)$, $C(\overline{\Omega})$, $C([-r,0];\mathbb{R}^n)$, then, (H4) can be assumed. These cases are mostly common in applications.

**Remark 2.7.** (H3) is crucial for the nontriviality of the pair $(\psi^0, \psi(\cdot))$. We can prove that if one of $\mathcal{R}$ and $Q$ is finite co-dimensional in $X$, so is $\mathcal{R} - Q$. Also, we will see some interesting cases covered by this condition.

## 3. SPIKE VARIATIONAL TECHNIQUE

In this section, we give the main idea of the proof of Theorem 2.5. The main technique is referred to as the *spike variational technique*. Let us first list some main lemmas needed in the proof of the maximum principle. The following result, due to I. Ekeland, which will be used for treating the state constraint.

**Lemma 3.1.** *Let $(V,d)$ be a complete metric space and let $F : V \to (-\infty, +\infty]$ be a proper (i.e., $F \not\equiv +\infty$), lower semi-continuous function bounded from below. Let $\varepsilon > 0$ and $v_0 \in V$ be such that*

$$F(v_0) \leq \inf_{v \in V} F(v) + \varepsilon. \tag{3.1}$$

*Then, there exists a $v_\varepsilon \in V$, such that*

$$F(v_\varepsilon) \leq F(v_0), \qquad d(v_\varepsilon, v_0) \leq \sqrt{\varepsilon}, \tag{3.2}$$

*and for all $v \in V$,*

$$-\sqrt{\varepsilon} d(v, v_\varepsilon) \leq F(v) - F(v_\varepsilon). \tag{3.3}$$

The Ekeland variational principle will be applied to the cost functional whose domain will be $X \times \mathcal{U}[0,T]$. Thus, we need to introduce the following metric on $\mathcal{U}[0,T]$:

$$\begin{aligned}&\bar{d}(u(\cdot), v(\cdot)) = \big|\{t \in [0,T] \mid u(t) \neq v(t)\}\big|, \\ &\qquad \forall u(\cdot), v(\cdot) \in \mathcal{U}[0,T].\end{aligned} \tag{3.4}$$

Then $(\mathcal{U}[0,T], \bar{d})$ is a metric space. From Lemma 3.1, we see that it will be good if this space id complete. This is the case actually:

**Lemma 3.2.** *Let $U$ be a measurable space. Then, $(\mathcal{U}[0,T], \bar{d})$ is a complete metric space.*

It should be pointed out that we only require $U$ be a measurable space. In particular, the above result applies to the case that $U$ is a metric space (not necessarily complete) with the $\sigma$-field generated by all open sets. Also, $[0,T]$ can be replaced by any measure space with a non-atomic measure; in particular, we can replace $[0,T]$ by any domain $\Omega \subset \mathbb{R}^n$ with the Lebesgue measure.

As we said in Remark 2.7, the finite codimensionality of the set $\mathcal{R} - Q$ in $X$ guarantees the nontriviality of $(\psi^0, \psi(\cdot))$. This will be achieved essentially through the following result.

**Lemma 3.3.** *Let $Q$ be finite codimensional in $X$. Let $\{f_n\}_{n \geq 1} \subseteq X^*$ with*

$$|f_n|_{X^*} \geq \delta > 0, \qquad f_n \xrightarrow{*} f \in X^*, \tag{3.5}$$

*and*

$$\langle f_n, x\rangle \geq -\varepsilon_n, \qquad \forall x \in Q, \quad n \geq 1, \tag{3.6}$$

*where $\varepsilon_n \to 0$. Then, $f \neq 0$.*

We note that the control domain $U$ is just a metric space and does not necessarily have any algebraic structure. Thus, we cannot talk about the convexity of $U$. Consequently, in the derivation of optimality conditions, the control variation is restricted to be a very special type. Let us describe it as follows: Let $u \in \mathcal{U}[0,T]$ be a feasible control. Pick any other $v \in \mathcal{U}[0,T]$ and a measurable set $E \subseteq [0,T]$, we define $u_E = v$ on the set $E$ and $u_E = u$ on the set $[0,T] \setminus E$. That is, we only change the values of $u$ on the set $E$. Clearly, $u_E \in \mathcal{U}[0,T]$. We call $u_E$ is a *spike perturbation* of $u$ (sometimes, it is also called a "*needlelike*" perturbation). Let us recall that for a finite measure space $(\Omega, \mathcal{F}, \mu)$, we say that $\mu$ is *non-atomic* if for any $E \in \mathcal{F}$ and any $\rho \in (0,1)$, there exists a set $F \in \mathcal{F}$ with $F \subseteq E$ and $\mu(F) = \rho\mu(E)$. The following result is crucial in making variation of controls. It tells us how we can "approximate" the function 1 by "oscillatory" function $\frac{1}{\rho}\chi_E$ for any given $\rho \in (0,1)$.

**Lemma 3.4.** *Let $X$ be a Banach space and $(\Omega, \mathcal{F}, \mu)$ be a finite measure space with $\mu$ being non-atomic. For any $\rho \in (0,1)$, let $\mathcal{E}_\rho = \{E \in \mathcal{F} \mid \mu(E) = \rho\mu(\Omega)\}$. Then, for any $h \in L^1(\Omega; X)$,*

$$\inf_{E \in \mathcal{E}_\rho} \Big| \int_\Omega (\frac{1}{\rho}\chi_E - 1) h d\mu \Big| = 0. \tag{3.7}$$

Finally, for $Q$ be a convex and closed subset in some Banach space $Z$, we define

$$d_Q(z) = \inf_{z' \in Q} |z - z'|_Z. \tag{3.8}$$

We call $d_Q$ the *distance function* (to the set $Q$). This function will play an important role in construction of penalty function. It is clear that $d_Q(\cdot)$ is convex function, and Lipschitz continuous with the Lipschitz constant being 1. We define the *subdifferential* of the function $d_Q(\cdot)$ as follows:

$$\begin{aligned} \partial d_Q(z) = \{\zeta \in Z^* \mid \\ d_Q(z') - d_Q(z) \ge \langle \zeta, z' - z \rangle, \ \forall z' \in Z\}. \end{aligned} \tag{3.9}$$

The following lemma collects interesting properties of $\partial d_Q(z)$.

**Lemma 3.5.** *Let $Q$ be a convex and closed subset in some Banach space $Z$. Then,*

(i) *The set $\partial d_Q(z)$ is nonempty, convex and weak*-compact in $Z^*$. For any $z, \xi \in Z$, the limit*

$$\begin{aligned} \lim_{\rho \downarrow 0} \frac{d_Q(z + \rho\xi) - d_Q(z)}{\rho} &\triangleq d_Q^0(z;\xi) \\ &= \max\{\langle \zeta, \xi \rangle \mid \zeta \in \partial d_Q(z)\}, \end{aligned} \tag{3.10}$$

*exists, which is positively homogeneous and subadditive in $\xi$. Moreover, $\partial d_Q(z)$ is can be represented by*

$$\begin{aligned} \partial d_Q(z) = \{\zeta \in Z^* \mid \\ \langle \zeta, \xi \rangle \le d_Q^0(z;\xi), \ \forall \xi \in Z \}. \end{aligned} \tag{3.11}$$

*Further, if $z_\alpha \in Z$ and $\zeta_\alpha \in Z^*$ are two nets with*

$$z_\alpha \xrightarrow{s} z, \qquad \zeta_\alpha \xrightarrow{*} \zeta, \qquad \zeta_\alpha \in \partial d_Q(z_\alpha), \tag{3.12}$$

*then, $\zeta \in \partial d_Q(z)$. Consequently, the map $z \mapsto \partial d_Q(z)$ is psuedo-continuous on $Z$.*

(ii) *For any $z \notin Q$,*

$$|\zeta|_{Z^*} = 1, \qquad \forall \zeta \in \partial d_Q(z). \tag{3.13}$$

*In particular, if $Z^*$ is strictly convex, then, for any $z \notin Q$, the set $\partial d_Q(z)$ consists of one point.*

We now briefly sketch the proof of the maximum principle stated in §2 (Theorem 2.5). The main steps are the following: (i) Define a penalty functional, via which we will transform the original problem with endpoint constraint to another one called approximate problem, for convenience, which has no endpoint constraint. (ii) Apply Ekeland Variational Principle to find an optimal pair for the approximate problem and this pair turns out to be close to the original optimal pair. (iii) Use spike variation technique to derive the (first order) necessary conditions for the optimal pair of approximate problem. (iv) Pass to the limit to obtain the necessary conditions for the original optimal pair, the maximum principle.

Let us first introduce some notation. For all $(x, u(\cdot)), (\widehat{x}, \widehat{u}(\cdot)) \in X \times \mathcal{U}[0,T]$, we define

$$\begin{aligned} &\widetilde{d}((x, u(\cdot)), (\widehat{x}, \widehat{u}(\cdot))) \\ &\quad = |x - \widehat{x}|_X + \bar{d}(u(\cdot), \widehat{u}(\cdot)), \end{aligned} \tag{3.14}$$

where $\bar{d}$ is the distance defined by (3.4). Then, by Lemma 3.2, we can show that $(X \times \mathcal{U}[0,T], \widetilde{d})$ is a complete metric space. Next, we know that under (H1)–(H2), for any $(y_0, u(\cdot)) \in X \times \mathcal{U}[0,T]$, (2.2) admits a unique solution $y(\cdot)$ with $y(0) = y_0$ and the cost functional (2.4) is also uniquely determined by $(y_0, u(\cdot))$. In the sequel, we denote the unique solution of (2.2) with $y(0) = y_0$ by $y(\cdot\,; y_0, u(\cdot))$ and the corresponding cost functional by $J(y_0, u(\cdot))$. Now, let $(\bar{y}(\cdot), \bar{u}(\cdot)) \in \mathcal{A}_{ad}$ be the optimal pair given in Theorem 2.5, for which we want to derive the necessary conditions. We denote $\bar{y}(0) = \bar{y}_0$. Without loss of generality, we may assume that $J(\bar{y}_0, \bar{u}(\cdot)) = 0$; otherwise, we may consider the optimal control problem with the cost functional being $J(y_0, u(\cdot)) - J(\bar{y}_0, \bar{u}(\cdot))$.

*Step 1.* Introduction of the penalty functional.

For any $\varepsilon > 0$, we define

$$\begin{aligned} J_\varepsilon(y_0, u(\cdot)) = \Big\{ &d_S^2(y_0, y(T; y_0, u(\cdot))) \\ &+ \big[(J(y_0, u(\cdot)) + \varepsilon)^+\big]^2 \Big\}^{1/2}, \\ &\forall (y_0, u(\cdot)) \in X \times \mathcal{U}[0,T], \end{aligned} \tag{3.15}$$

where

$$d_S(y_0, y_1) = d((y_0, y_1), S)$$

(3.16) $$\equiv \inf_{(x_0,x_1)\in S} \left\{|y_0 - x_0|^2 + |y_1 - x_1|^2\right\}^{1/2},$$

$$\forall (y_0, y_1) \in X \times X.$$

We can easily prove that functional $J_\varepsilon(y_0, u(\cdot))$ is continuous on the space $(X \times \mathcal{U}[0,T], \tilde{d})$.

*Step 2.* Application of Ekeland Variational Principle.

By the definition of $J_\varepsilon(y_0, u(\cdot))$, we see that

$$\begin{cases} J_\varepsilon(y_0, u(\cdot)) > 0, \quad \forall (y_0, u(\cdot)) \in X \times \mathcal{U}[0,T], \\ J\varepsilon(\bar{y}_0, \bar{u}(\cdot)) = \varepsilon \leq \inf\limits_{X \times \mathcal{U}[0,T]} J_\varepsilon(y_0, u(\cdot)) + \varepsilon. \end{cases}$$

Thus, by Lemma 2.1, there exists a pair $(y_0^\varepsilon, u^\varepsilon(\cdot)) \in X \times \mathcal{U}[0,T]$, such that

(3.17) $$\tilde{d}((x_0^\varepsilon, u^\varepsilon(\cdot)), (\bar{x}_0, \bar{u}(\cdot))) \leq \sqrt{\varepsilon},$$

(3.18) $$J_\varepsilon(y_0, u(\cdot)) + \sqrt{\varepsilon}\tilde{d}((y_0, u(\cdot)), (y_0^\varepsilon, u^\varepsilon(\cdot))) \geq J_\varepsilon(y_0^\varepsilon, u^\varepsilon(\cdot)), \ \forall (y_0, u(\cdot)) \in X \times \mathcal{U}[0,T].$$

The above implies that if we let $y^\varepsilon(\cdot) = y(\cdot\,; y_0^\varepsilon, u^\varepsilon(\cdot))$, then $(y^\varepsilon(\cdot), u^\varepsilon(\cdot))$ is an optimal pair for the problem with the state equation being (2.2) and the cost functional being the left hand side of (3.18).

*Step 3.* Derivation of the necessary conditions for $(y^\varepsilon(\cdot), u^\varepsilon(\cdot))$.

We fix any $(\eta, v(\cdot)) \in B_1(0) \times \mathcal{U}[0,T]$ and $\rho \in (0,1]$. For $E_\rho \subset [0,T]$, with $|E_\rho| = \rho T$, define

(3.19) $$u_\rho^\varepsilon(t) = \begin{cases} u^\varepsilon(t), & t \in [0,T] \setminus E_\rho\,, \\ v(t), & t \in E_\rho\,. \end{cases}$$

Then $u_\rho^\varepsilon(\cdot) \in \mathcal{U}[0,T]$ and

(3.20) $$\bar{d}(u_\rho^\varepsilon(\cdot), u^\varepsilon(\cdot)) \leq |E_\rho| = \rho T.$$

We let $y_\rho^\varepsilon(\cdot) = y(\cdot\,; y_0^\varepsilon + \rho\eta, u_\rho^\varepsilon(\cdot))$ and recall $y^\varepsilon(\cdot) = y(\cdot\,; y_0^\varepsilon, u^\varepsilon(\cdot))$. The following lemma gives a "Taylor expansion" of $y_\rho^\varepsilon(\cdot)$ and $J(y_0^\varepsilon + \rho\eta, u_\rho^\varepsilon(\cdot))$ with respect to $\rho$ at $\rho = 0$. The proof can be obtained by using Lemma 3.4. This result is very important in deriving the necessary conditions for the pair $(y^\varepsilon(\cdot), u^\varepsilon(\cdot))$.

**Lemma 3.6.** *There exists a family $E_\rho \subseteq [0,T]$ satisfying $|E_\rho| = \rho T$, so that*

(3.21) $$\sup_{t\in[0,T]} |y_\rho^\varepsilon(t) - y^\varepsilon(t) - \rho z_\varepsilon(t)| = o(\rho),$$

(3.22) $$J(y_0^\varepsilon + \rho\eta, u_\rho^\varepsilon(\cdot)) = J(y_0^\varepsilon, u^\varepsilon(\cdot)) + \rho z_\varepsilon^0 + o(\rho),$$

*where*

(3.23)
$$\begin{cases} z_\varepsilon(t) = e^{At}\eta \\ \quad + \displaystyle\int_0^t e^{A(t-s)} f_y(s, y^\varepsilon(s), u^\varepsilon(s)) z_\varepsilon(s)\,ds \\ \quad + \displaystyle\int_0^t e^{A(t-s)} \big[f(s, y^\varepsilon(s), v(s)) \\ \quad - f(s, y^\varepsilon(s), u^\varepsilon(s))\big]\,ds, \quad t \in [0,T], \\ z_\varepsilon^0 = \displaystyle\int_0^T f_y^0(s, y^\varepsilon(s), u^\varepsilon(s)) z_\varepsilon(s)\,ds \\ \quad + \displaystyle\int_0^T \big[f^0(s, y^\varepsilon(s), v(s)) \\ \quad - f^0(s, y^\varepsilon(s), u^\varepsilon(s))\big]\,ds. \end{cases}$$

From the above, we see that $J_\varepsilon(y_0^\varepsilon + \rho\eta, u_\rho^\varepsilon(\cdot)) = J_\varepsilon(y_0^\varepsilon, u^\varepsilon(\cdot)) + o(1)$, as $\rho \to 0$. Furthermore,

(3.24)
$$\begin{aligned} & -\sqrt{\varepsilon}(|\eta| + T) \\ & \leq \frac{1}{\rho}\Big(J_\varepsilon(y_0^\varepsilon + \rho\eta, u_\rho^\varepsilon(\cdot)) - J_\varepsilon(y_0^\varepsilon, u^\varepsilon(\cdot))\Big) \\ & = \frac{1}{2J_\varepsilon(y_0^\varepsilon, u^\varepsilon(\cdot)) + o(1)}\Big\{\frac{1}{\rho}\big[d_S^2(y_0^\varepsilon + \rho\eta, y_\rho^\varepsilon(T)) \\ & \quad - d_S^2(y_0^\varepsilon, y^\varepsilon(T))\big] \\ & \quad + \frac{1}{\rho}\{[(J(y_0^\varepsilon + \rho\eta, u_\rho^\varepsilon(\cdot)) + \varepsilon)^+]^2 \\ & \quad - [(J(y_0^\varepsilon, u^\varepsilon(\cdot)) + \varepsilon)^+]^2\}\Big\}. \end{aligned}$$

Let $\rho \to 0$, by Lemma 3.5, we get

(3.25) $$-\sqrt{\varepsilon}(|\eta| + T) \leq \langle \bar{\varphi}_\varepsilon, \eta \rangle + \langle \bar{\psi}_\varepsilon, z_\varepsilon(T) \rangle + \bar{\psi}_\varepsilon^0 z_\varepsilon^0(T),$$

where

(3.26)
$$\begin{cases} (\bar{\varphi}_\varepsilon, \bar{\psi}_\varepsilon) = \dfrac{d_S(y_0^\varepsilon, y^\varepsilon(T))}{J_\varepsilon(y_0^\varepsilon, u^\varepsilon(\cdot))}(a_\varepsilon, b_\varepsilon), \\ \bar{\psi}_\varepsilon^0 = \dfrac{(J(y_0^\varepsilon, u^\varepsilon(\cdot)) + \varepsilon)^+}{J_\varepsilon(y_0^\varepsilon, u^\varepsilon(\cdot))}, \end{cases}$$

with $(a_\varepsilon, b_\varepsilon) \in \partial d_S(y_0^\varepsilon, y^\varepsilon(T))$ and

(3.27) $$|\bar{\varphi}_\varepsilon|_{X^*}^2 + |\bar{\psi}_\varepsilon|_{X^*}^2 + (\bar{\psi}_\varepsilon^0)^2 = 1.$$

On the other hand, by (3.9), we have

(3.28) $$\langle \bar{\varphi}_\varepsilon, x_0 - y_0^\varepsilon \rangle + \langle \bar{\psi}_\varepsilon, x_1 - y^\varepsilon(T) \rangle \leq 0, \quad \forall (x_0, x_1) \in S.$$

Conditions (3.26)–(3.28) can be regarded as necessary conditions for the pair $(y^\varepsilon(\cdot), u^\varepsilon(\cdot))$. We may proceed further to derive an approximate maximum principle. However, we prefer to pass to the limit first and derive the maximum principle directly for the original optimal pair $(\bar{y}(\cdot),\bar{u}(\cdot))$.

*Step 4.* Pass to the limit.

By (H3) and Lemma 3.3, we can find a subsequence (still denoted by itself), such that

$$(\bar{\varphi}_\varepsilon, \bar{\psi}_\varepsilon, \bar{\psi}^0_\varepsilon) \xrightarrow{*} (\bar{\varphi}, \bar{\psi}, \bar{\psi}^0) \neq 0. \tag{3.29}$$

Then, we can obtain

$$\begin{aligned} &\langle \bar{\varphi}, \eta - (x_0 - \bar{y}_0) \rangle \\ &+ \langle \bar{\psi}, z(T) - (x_1 - \bar{y}(T)) \rangle + \bar{\psi}^0 z^0 \geq 0, \\ &\forall (x_0, x_1) \in S, v(\cdot) \in \mathcal{U}[0,T], \eta \in B_1(0). \end{aligned} \tag{3.30}$$

Next, let $\psi(\cdot)$ be the unique solution of (2.36) with

$$\psi^0 = -\bar{\psi}^0 \leq 0, \quad \psi(T) = -\bar{\psi}. \tag{3.31}$$

By a duality relation,

$$\begin{aligned} &\langle \psi(T), z(T) \rangle - \langle \psi(0), \eta \rangle + \psi^0 z^0 \\ &= \int_0^T \big[ H(t, \bar{y}(t), v(t), \psi^0, \psi(t)) \\ &\quad - H(t, \bar{y}(t), \bar{u}(t), \psi^0, \psi(t)) \big] dt, \end{aligned} \tag{3.32}$$

for all $\eta \in B_1(0)$ and $v(\cdot) \in \mathcal{U}[0,T]$. Hence, by setting $\eta = 0$ and $(x_0, x_1) = (\bar{y}_0, \bar{y}(T))$ in (3.30), we obtain

$$\begin{aligned} &\int_0^T \big[ H(t, \bar{y}(t), v(t), \psi^0, \psi(t)) \\ &\quad - H(t, \bar{y}(t), \bar{u}(t), \psi^0, \psi(t)) \big] dt \leq 0, \\ &\forall v(\cdot) \in \mathcal{U}[0,T]. \end{aligned} \tag{3.33}$$

Then, all the conclsion follows easily.

The following result gives the cases when $\psi^0 \neq 0$.

**Proposition 3.7.** *Let (H1)-(H4) hold and $(\bar{y}(\cdot), \bar{u}(\cdot))$ be an optimal pair with $\bar{y}_0 = \bar{y}(0)$. Then, $\psi^0 \neq 0$ under any of the following conditions:*

(i) $\bar{y}(T) - G(T,0)\bar{y}_0 \in \operatorname{Int}(Q - \mathcal{R})$*; in particular,* $0 \in \operatorname{Int}\mathcal{R}$ *or* $Q = X$*;*

(ii) *Suppose $S = Q_0 \times Q_1$ with $Q_0, Q_1 \subset X$ closed and convex. Let $\bar{y}(T) \in \operatorname{Int} Q_1$; in particular, $Q_1 = X$;*

(iii) *Suppose $S = Q_0 \times Q_1$ with $Q_0, Q_1 \subset X$ closed and convex. Let $\bar{y}_0 \in \operatorname{Int} Q_0$ and $G(T,0)$ be injective; in particular, $Q_0 = X$ and $G(T,0)^{-1} \in \mathcal{L}(X)$.*

(iv) *Let*

$$\begin{aligned} &\Big\{ \bigcup_{\lambda \geq 0} \lambda \partial d_{Q-\mathcal{R}}(\bar{y}(T) - G(T,0)\bar{y}_0) \Big\} \\ &\bigcap \Big\{ \psi \in X^* \mid \begin{pmatrix} G^*(T,0) \\ I \end{pmatrix} \psi \\ &\in \bigcup_{\lambda \geq 0} \lambda \partial d_S(y_0, \bar{y}(T)) \Big\} = \{0\}. \end{aligned} \tag{3.34}$$

We have seen that in many cases, $\psi^0 \neq 0$. The case that $y_0$ is fixed and $y(T)$ is free is the most classical one. This case amounts to say that $Q_1 = X$, which is contained in (ii). It is not hard for the readers to identify the cases listed above.

## 4. APPLICATIONS

Let us point out some important special cases covered by Theorem 2.5.

In what follows, let us assume that (H1)-(H2) and (H4) hold, and let $G(t,s)$ be the evolution operator generated by $A + f_y(t, \bar{y}(t), \bar{u}(t))$ with $(\bar{y}(\cdot), \bar{u}(\cdot))$ being some optimal pair. We consider the different type endpoint constraints.

*1. The control problem with fixed end points.*

In this case, the constraint set of the problem is $S = \{(y_0, y_1)\}$ and the endpoint constraint takes the following form:

$$y(0) = y_0, \qquad y(T) = y_1. \tag{4.1}$$

Thus, the modified state constraint set is $Q = \{x_1 - G(T,0)x_0\}$, which is a singleton. Hence, in order the maximum principle holds, we need the reachable set $\mathcal{R}$ associated with $(\bar{y}(\cdot), \bar{u}(\cdot))$ to be finite codimensional in $X$. This is the case if $e^{At}$ is a $C_0$ group on $X$, $f(t,y,u) = f_1(t,y) + u$, with $U$ being a ball in $X$. This result is essentially due to Fattorini.

*2. The control problem with a terminal state constraint.*

Let $x_0 \in X$ and $Q_1 \subseteq X$ be convex and closed. We consider the problem with the endpoint constraints:

$$y(0) = x_0, \qquad y(T) \in Q_1. \tag{4.2}$$

Thus, the constraint set is $S = \{x_0\} \times Q_1$ and the modified constraint set is given by $Q = Q_1 - G(T,0)x_0$. Clearly, this set is finite codimensional in $X$ if and only if $Q_1$ is so. Hence, if $Q_1$ is finite codimensional in $X$, then, maximum principle holds. This result is due to Li and Yao.

*3. The control problem with separated endpoint constraints.*

Let $Q_0, Q_1 \subseteq X$ be convex and closed sets. We consider the optimal control problem with the endpoint constraints:

$$y(0) \in Q_0, \qquad y(T) \in Q_1, \tag{4.3}$$

Then, the constraint set can be written as $S = Q_0 \times Q_1$ and the modified constraint set is given by $Q = Q_1 - G(T,0)Q_0$. Now, we know that if $Q_1$ or $G(T,0)Q_0$ is finite codimensional in $X$, so is

$Q$. In this case, we have the maximum principle. In the case $e^{At}$ is a $C_0$ group, which is the case for the wave equation without damping, $G(T,0)$ is invertible; and we need only to have the finite codimensionality of either $Q_0$ or $Q_1$.

Next, let us discuss an interesting case — The optimal periodic control problem. To study such a problem, let us first assume (H1)-(H2) and (H4). In addition, we make the following periodic conditions:

(H5) For any $(t,x,u) \in [0,\infty) \times X \times U$,

$$(4.4) \qquad \begin{cases} f(t+T,x,u) = f(t,x,u), \\ f^0(t+T,x,u) = f^0(t,x,u). \end{cases}$$

It is required that the state satisfies the following periodic condition:

$$(4.5) \qquad y(0) = y(T).$$

Thus, in the present case, the state constraint set is $S = \{(x,x) \mid x \in X\}$. This is the "diagonal" set in $X \times X$. Clearly, this set is *not* finite codimensional in $X \times X$ (if $\dim X = \infty$). The modified constraint set is given by $Q = (I - G(T,0))X$. Now the question is when this set is finite codimensional in $X$? We can show that when $e^{At}$ is a compact semigroup, $Q$ is finite co-dimensional. Thus, we have the following:

**Proposition 4.1.** *Let (H1)-(H2) and (H4)-(H5) hold. Let $e^{At}$ be a compact semigroup. Let $(\bar{y},\bar{u})$ be an optimal pair for the optimal periodic control problem. Then, there exists a pair $(\psi^0,\psi(\cdot)) \neq 0$, such that (2.35)-(2.38) hold and*

$$(4.6) \qquad \psi(0) = \psi(T).$$

*Moreover, if $\mathcal{N}(I - G^*(T,0)) = \{0\}$, $\psi^0 = -1$.*

For some concrete examples, see [14].

## 5. ELLIPTIC PROBLEMS

We now briefly discuss the elliptic problems. For simiplicity, we consider the following system:

$$(5.1) \qquad \begin{cases} -\Delta y(x) = f(x,y(x),u(x)), & \text{in } \Omega, \\ y\big|_{\partial\Omega} = 0. \end{cases}$$

The cost functional is given by

$$(5.2) \qquad J(y,u) = \int_\Omega f^0(x,y(x),u(x))dx,$$

with $(y,u)$ being a pair of state and control satisfying (5.1). We make the following assumptions.

(S1) $\Omega \subset \mathbb{R}^n$ is a bounded region with $\partial\Omega$ being $C^{1,\gamma}$, for some $\gamma > 0$ and $U$ is a separable metric space.

(S2) The function $f : \Omega \times \mathbb{R} \times U \to \mathbb{R}$ has the following properties: $f(\cdot\,,y,u)$ is measurable on $\Omega$, $f(x,\cdot\,,u)$ is in $C^1(\mathbb{R})$ with $f(x,\cdot\,,\cdot)$ and $f_y(x,\cdot\,,\cdot)$ being continuous on $\mathbb{R} \times U$. Moreover,

$$(5.3) \qquad f_y(x,y,u) \le 0, \quad \forall (x,y,u) \in \Omega \times \mathbb{R} \times U,$$

and for any $R > 0$, there exists a constant $M_R > 0$, such that

$$(5.4) \qquad \begin{aligned} &|f(x,y,u)| + |f_y(x,y,u)| \le M_R, \\ &\qquad \forall (x,u) \in \Omega \times U, |y| \le R. \end{aligned}$$

(S3) The function $f^0 : \Omega \times \mathbb{R} \times U \to \mathbb{R}$ satisfies (S2) except (5.3).

Next, we set

$$\mathcal{U} = \{u : \Omega \to U \mid u \text{ is measurable }\}.$$

Any element $u \in \mathcal{U}$ is referred to as a control. The following result is basic.

**Proposition 5.1.** *Let (S1)-(S2) hold. Then, for any $p \in [1,\infty)$ and any $u \in \mathcal{U}$, there exists a unique $y \equiv y(\cdot\,;u) \in W_0^{1,p}(\Omega) \bigcap L^\infty(\Omega)$ solving (5.1). Furthermore, there exists a constant $C_p > 0$, independent of $u \in \mathcal{U}$, such that*

$$(5.5) \qquad \|y(\cdot\,;u)\|_{W_0^{1,p}(\Omega)\cap L^\infty(\Omega)} \le C_p, \qquad \forall u \in \mathcal{U}.$$

If we let $p > n$, by Sobolev imbedding theorem, the solution $y$ of (5.1) is actually in $C^\beta(\bar{\Omega})$ for some $\beta \in (0,1)$. In what follows, we fix a $p \in [1,\infty)$. Any pair $(y,u) \in W_0^{1,p}(\Omega) \times \mathcal{U}$ satisfying (5.1) is called a *feasible pair* and we refer to the corresponding $y$ and $u$ as a *feasible state* and a *feasible control*, respectively. We let $\mathcal{A}$ be the set of all feasible pairs. Clearly, under (S1)-(S2), $\mathcal{U}$ coincides with the set of all feasible controls and for each feasible control $u \in \mathcal{U}$, there corresponds a unique feasible state and the cost functional (5.2) is well-defined. Therefore, under (S1)-(S3), we can write $J(y,u)$ as $J(u)$ without any ambiguity. We introduce one more assumption:

(S4) $\mathcal{Y}$ is a Banach space with strict convex dual $\mathcal{Y}^*$, $F : W_0^{1,p}(\Omega) \to \mathcal{Y}$ is continuously Fréchet differentiable and $Q \subset \mathcal{Y}$ is closed and convex.

We see that under (S1)-(S3), for any $u \in \mathcal{U}$, the solution $y(\cdot\,;u)$ of (5.1) is in $W_0^{1,p}(\Omega)$. Thus, state constraint of the following type makes sense:

$$(5.6) \qquad F(y) \in Q.$$

We will present many examples of the state constraints covered by (5.6). Now, we let $\mathcal{A}_{ad}$ be the set of all pair $(y,u) \in \mathcal{A}$, such that the state constraint (5.6) is satisfied. Any $(y,u) \in \mathcal{A}_{ad}$ is called an *admissible pair*. Our optimal control problem can be stated as follows:

**Problem (SD).** Find an admissible pair $(\bar{y}, \bar{u}) \in \mathcal{A}_{ad}$, such that

$$J(\bar{y}, \bar{u}) = \inf_{\mathcal{A}_{ad}} J(y, u). \tag{5.7}$$

We refer to such a pair $(\bar{y}, \bar{u})$, if it exists, an *optimal pair* and refer to $\bar{y}$ and $\bar{u}$ as an *optimal state* and *optimal control*, respectively.

Next, we let $(\bar{y}, \bar{u}) \in \mathcal{A}_{ad}$ be an optimal pair of Problem (SD). For any $u \in \mathcal{U}$, we let $z = z(\cdot\,; u) \in W_0^{1,p}(\Omega)$ be the unique solution of the following problem:

(5.8)
$$\begin{cases} Az(x) = f_y(x, \bar{y}(x), \bar{u}(x))z(x) \\ \qquad + f(x, \bar{y}(x), u(x)) - f(x, \bar{y}(x), \bar{u}(x)), \quad \text{in } \Omega, \\ z\big|_{\partial\Omega} = 0. \end{cases}$$

This system is referred to as the *variational system* along the pair $(\bar{y}, \bar{u})$. We define

$$\mathcal{R} = \{z(\cdot\,; u) \mid u \in \mathcal{U}\}. \tag{5.9}$$

This set is called the *reachable set* of the variational system (1.10). Clearly $\mathcal{R} \subseteq W_0^{1,p}(\Omega)$. We define the Hamiltonian as follows:

(5.10)
$$H(x, y, u, \psi^0, \psi) = \psi^0 f^0(x, y, u) + \psi f(x, y, u),$$
$$\forall (x, y, u, \psi^0, \psi) \in \Omega \times \mathbb{R} \times U \times \mathbb{R} \times \mathbb{R}.$$

The following is the first order necessary conditions for optimal pairs of our Problem (SD).

**Theorem 5.2.** *Let (S1)-(S4) hold. Let $(\bar{y}, \bar{u}) \in \mathcal{A}_{ad}$ be an optimal pair of Problem (SD). Let $F'(\bar{y})\mathcal{R} - Q$ be finite codimensional in $\mathcal{Y}$. Then, there exists $(\psi^0, \psi, \varphi) \in [-1, 0] \times W_0^{1,p'}(\Omega) \times \mathcal{Y}^*$, such that $(\psi^0, \varphi) \neq 0$,*

$$\langle \varphi, \eta - F(\bar{y}) \rangle_{\mathcal{Y}^*, \mathcal{Y}} \leq 0, \qquad \forall \eta \in Q. \tag{5.11}$$

(5.12)
$$\begin{cases} A\psi(x) = f_y(x, \bar{y}(x), \bar{u}(x))\psi(x) \\ \qquad + \psi^0 f_y^0(x, \bar{y}(x), \bar{u}(x)) - F'(\bar{y}(\cdot))^*\varphi, \text{ in } \Omega, \\ \psi\big|_{\partial\Omega} = 0. \end{cases}$$

(5.13)
$$\max_{u \in U} H(x, \bar{y}(x), u, \psi^0, \psi(x)),$$
$$= H(x, \bar{y}(x), \bar{u}(x), \psi^0, \psi(x)), \quad \text{a.e. } x \in \Omega.$$

*In addition, if $F'(\bar{y})^*$ is injective, then,*

$$(\psi^0, \psi) \neq 0. \tag{5.14}$$

We refer to (5.11), (5.12) and (5.13) as the *transversality condition*, the *adjoint system* (along the given optimal pair) and the *maximum condition*, respectively.

Let us make some remarks on the above result. First of all, by Proposition 1.1, we know that $\bar{y} \in L^\infty(\Omega)$.Thus,by (S2) and (S3),$f_y(\cdot, \bar{y}(\cdot), \bar{u}(\cdot))$, $f_y^0(\cdot\,, \bar{y}(\cdot), \bar{u}(\cdot)) \in L^\infty(\Omega)$. On the other hand, by (S5), $F'(\bar{y}) : W_0^{1,p}(\Omega) \to \mathcal{Y}$. Thus, $F'(\bar{y})^* : \mathcal{Y}^* \to W^{-1,p'}(\Omega)$ $(p' = \frac{p}{p-1})$. Consequently, $F'(\bar{y})^*\varphi \in W^{-1,p'}(\Omega)$. Thus, (5.12) admits a unique solution $\psi(\cdot) \in W_0^{1,p'}(\Omega)$ (if $\psi^0$ is specified).

Let us now look at various kind of state constraints covered by (5.6).

1) Let $Q = \mathcal{Y}$. This corresponds to the case that there is no state constraint. In such a case, by the transversality condition (5.11), we have $\varphi = 0$. This is also the case if the state constraint is *inactive*, i.e., the set $Q$ has nonempty interior in which the optimal state $\bar{y}$ lies.

2) Let $\mathcal{Y} = C(\overline{\Omega})$. We note that under usual norm of continuous functions, the dual $\mathcal{Y}^*$ is not strictly convex. But, since $C(\overline{\Omega})$ is separable, we can find an equivalent norm $|\cdot|_0$ on $C(\overline{\Omega})$, under which the dual $(C(\overline{\Omega}), |\cdot|_0)^*$ is strictly convex. It is also clear that under this equivalent norm, we still have the Riesz Representation Theorem: $C(\overline{\Omega})^* = \mathcal{M}(\overline{\Omega}) \equiv \{$ all the regular Borel measures on $\overline{\Omega}\}$. We consider the following type state constraint:

$$h(x, y(x)) \leq 0, \qquad \forall x \in \Omega, \tag{5.15}$$

where $h \in C(\overline{\Omega} \times \mathbb{R})$ satisfies the following: For each $x \in \overline{\Omega}$, $h(x, \cdot)$ is $C^1$ with $h_y \in C(\overline{\Omega} \times \mathbb{R})$. Moreover,

$$h(x, 0) < 0, \qquad \forall x \in \partial\Omega. \tag{5.16}$$

We define

(5.17)
$$\begin{cases} F(y)(x) = h(x, y(x)), \; x \in \overline{\Omega}, y \in C(\overline{\Omega}); \\ Q = \{\eta \in C(\overline{\Omega}) \mid \eta(x) \leq 0, \; \forall x \in \overline{\Omega}\}. \end{cases}$$

Then, (5.15) is of form (5.6). We claim that $Q$ has a nonempty interior in $\mathcal{Y}$. In fact, by taking $y_0(x) \equiv 2$, we see that the ball in $\mathcal{Y}$ centered at $y_0$ with radius 1 is contained in $Q$. This implies that $Q$ is codimension 0 in $\mathcal{Y}$. Let us look at the property that the corresponding $\varphi$ possesses. By the transversality condition (5.11), we have

$$\int_\Omega (\eta(x) - h(x, \bar{y}(x)))d\varphi(x) \leq 0, \; \forall \eta \in Q. \tag{5.18}$$

Denote $\Omega_0 = \{x \in \Omega \mid h(x, \bar{y}(x)) = 0\}$. This set is called the *active* set, meaning that the state constraint is active. The complement $\Omega \setminus \Omega_0$ is called the *inactive* set. Since $\bar{y}$ and $h$ are continuous, $\Omega \setminus \Omega_0$ is open. Also, by (5.16), we see that $\Omega_0 \bigcap \partial\Omega = \phi$. Now, for any open set $E \subset \Omega \setminus \Omega_0$, we let $\zeta \in C_0^\infty(E)$. Then, by the definition of $E$, for all $\varepsilon > 0$ small enough, $\eta_\varepsilon \equiv h(\cdot\,, \bar{y}(\cdot)) + \varepsilon\zeta(\cdot) \in Q$. Thus, taking $\eta = \eta_\varepsilon$ in (5.18), we obtain

$$(5.19) \qquad \int_E \zeta(x) d\varphi(x) \leq 0, \qquad \forall \zeta \in C_0^\infty(E).$$

This shows that the support supp $\varphi$ of $\varphi$ is disjoint with $E$. Since $E$ is an arbitrary open set in $\Omega \setminus \Omega_0$, we arrive at

$$(5.20) \qquad \operatorname{supp} \varphi \subset \Omega_0.$$

Finally, from (5.18), when $x \in \Omega_0$, the integrand is nonpositive, thus $\varphi$ is a nonnegative valued measure supported on $\Omega_0$.

3) Let $\mathcal{Y} = \mathbb{R}^m$ and $p > n$. Let $x_1, \cdots, x_m \in \Omega$. Define $F : W_0^{1,p}(\Omega) \to \mathbb{R}^m$ by the following:

$$(5.21) \qquad F(y(\cdot)) = (y(x_1), \cdots, y(x_m)), \; \forall y(\cdot) \in W_0^{1,p}(\Omega).$$

Clearly, $F$ a is linear bounded operator. Thus it is Fréchet differentiable. We let $Q \subset \mathbb{R}^m$ be a (nonempty) convex and closed set. Since, $\mathbb{R}^m$ is finite dimensional, $Q$ is of course finite codimensional in $\mathcal{Y}$. In the present case, our state constraint has the following form:

$$(5.22) \qquad (y(x_1), \cdots, y(x_m)) \in Q.$$

See [24] for some extensive discussions.

## REFERENCES


[1] R. Bellman, *Dynamic Programming*, Princeton Univ. Press, New Iersey, 1957.

[2] A. G. Butkovsky, *Maximum principle of optimal control for distributed parameter systems, Avto. & Tele.*, 22 (1961), 1288–1301 (in Russian).

[3] E. Casas, L. A. Fernández, and J. Yong, *Optimal control problems of quasilinear parabolic equations, J. Royal Soc. Edinburgh*, 125A (1995), 545–565.

[4] E. Casas and J. Yong, *Maximum principle for state-constrained optimal control problems governed by quasilinear elliptic equations, Diff. Int. Eqs.*, 8 (1995), 1–18.

[5] Yu. V. Egorov, *Optimal control in Banach spaces, Dokl. Acad. Nauk SSSR*, 150 (1963), 241-244.

[6] H. O. Fattorini, *A unified theory of necessary conditions for nonlinear nonconvex control systems, Appl. Math. & Optim.*, 15 (1987), 141–185.

[7] W. H. Fleming, *Future Directions in Control Theory — A Mathematical Perspective,* SIAM Philadelphia, 1988.

[8] B. Hu and J. Yong, *Pontryagin maximum principle for semilinear and quasilinear parabolic equations with pointwise state constraints,SIAM J.Control & Optim.*, 33 (1995), 1875–1880.

[9] R. E. Kalman, *Contributions to the theory of optimal control, Bol. Soc. Math. Mexicana,* 5 (1960), 102–119.

[10] G. L. Kharatishvili, *The maximum principle in the theory of optimal processes with a delay, Soviet Math. Dokl.*, 2 (1961), 28–32.

[11] X.Li and Y.Yao, *Time optimal control of distributed parameter systems, Scientia Sinica,* 24 (1981), 455–465.

[12] X. Li and Y. Yao, *Maximum principle of distributed parameter systems with time lags, Lecture Notes in Control & Inform. Sci.*, Vol.75, Springer-Verlag, 1985, 410–427.

[13] X. Li and J. Yong, *Necessary conditions of optimal control for distributed parameter systems, SIAM J. Control & Optim.*, 29 (1991), 895–908.

[14] X. Li and J. Yong, *Optimal Control Theory for Infinite Dimensional Systems,* Birkäuser, Boston, 1995.

[15] L. S. Pontryagin, V. G. Boltyanskii, R. V. Gamkrelidze, and E. F. Mischenko, *Mathematical Theory of Optimal Processes,* Wiley, New York, 1962.

[16] J. Yong, *Maximum principle of the optimal controls for a nonsmooth semilinear evolution system, Lecture Notes in Control & Inform. Sci.*, Springer-Verlag, 144 (1990), 559–569.

[17] J. Yong, *Optimal control for distributed parameter systems with mixed constraints, Colloquium Math.*, 60/61 (1990), 35–48.

[18] J. Yong, *Time optimal control for semilinear distributed parameter systems—existence theory and necessary conditions,Kodai Math. J.*, 14 (1991), 239–253.

[19] J. Yong, *Existence theory of optimal control for distributed parameter systems, Kodai Math. J.*, 15 (1992), 193–220.

[20] J. Yong, *Pontryagin maximum principle for semilinear second order elliptic partial differential equations and variational inequalities with state constraints*, Diff. Int. Eqs., 5 (1992), 1307–1334.

[21] J. Yong, *Infinite dimensional Volterra-Stieljes evolution equations and related optimal control problems*, SIAM J. Control & Optim., 31 (1993), 539–568.

[22] J. Yong, *Optimal control for nonlinear abstract evolution systems*, Diff. Int. Eqs., 6 (1993), 1145–1159.

[23] J. Yong, *A minimax control problem for second order elliptic PDEs*, Kodai Math. J., 16 (1993), 469–486.

[24] J. Yong and P. Zhang, *Necessary conditions of optimal impulse controls for distributed parameter systems*, Bull.Austral.Math.Soc., 45 (1992), 305–326.

14th Triennial World Congress, Beijing, P.R. China

# THE LINEAR QUADRATIC OPTIMAL CONTROL IN HILBERT SPACES WITH UNBOUNDED CONTROLS

Xunjing Li

Department of Mathematics
Fudan University
Shanghai 200433, China
Fax:86-21-65646073
Email:xjli@ms.fudan.edu.cn

Abstract. The LQR control problems for finite dimensional systems has been very popular as the most basic part of modern control theory and system theory. The purpose of this paper is to give a brief survey on the works done by the people from Fudan Group related to LQR control problems for infinite dimensional systems, especially, on the works to unbounded control and indefinite performance. *Copyright © 1999 IFAC*

Keywords. LQR control, Boundary value problem, Optimal control, Performance functions, Algebraic Riccati equations, (Fredholm) integral equations, Frequence domains.

## 1. INTRODUCTION

The linear quadratic regulator control ( LQR, for short ) problems in finite dimensional system was first studied by Bellman-Glisksberg-Gross in 1958. In 1960, Kalman founded the relation between the LQR problems and optimal linear state feedback control. This result was called Kalman LQR control theory, which has been very popular as the most basic part of modern control theory and system theory.

Infinite dimensional systems include the state equations such as integro-differential equations, partial differential equation, abstract evolution equations and functional differential equation. In the mid-1960s, Lions discussed the LQR control problem for partial differential equations. The general LQR control problem for evolution equation with bounded controls was investigated by Lukes-Russell in 1969, by Curtain-Pritchard in 1976, by Balakrishnan in 1977 and other authors. The most distinguished feature of LQR control problem in infinite dimensional systems is that the control operator may be unbounded. The LQR control problem with unbounded controls was studied by Lions, Curtain & Pritchard and Balakrishnan in 1970s. In 1981, J.L.Lions studied the LQR control problem for parabolic equation in $\mathbf{R}^n$ ($n \leq 3$) with pointwise controls by introducing some suitable function spaces. Since 1980s, Lasiecka-Triggiani (1991,1992) etc. had systematically studied the LQR control problems

This work is partially supported by the National Key Project of China, the National Nature Science Foundation of China, NSF of the Chinese State Education Commission and Lab. of Math. for Nonlinear Sciences at Fudan University.

for parabolic, hyperbolic, and other equations with boundary controls and pointwise controls. Riccati equation is an important tool in studying the LQR control problems, which can give the state feedback of optimal control. Flandoli and Da Prato, Ichikawa as well as Bensoussan, Da Prato, Delfour and Mitter had done a direct study of the Riccati equation. However, all the above works are deal with the standard LQR control problem, in which the performance functions has nonnegative coefficients.

The study of LQR control problems with indefinite performance functions comes from theoretical and practical needs. In LQ differetial game problems, the performance functions appear to be indefinite in general. The performance functions could have negative coefficients in certain LQR control problems relating to network analysis. The LQR control problems with indefinite performance functions are also related to the analysis of the stability of nonlinear system and other problems.

For the LQR control problem with indefinite performance function and bounded control, You (1983) obtained the synthesis for the optimal control via the solution to Fredholm integral equation and the solution to Riccati integral equation. Chen (1985) directly discussed an equivalence between two types of integral equations in some sense. Yao, Yao-You and Chen discussed the LQR control problem for parabolic equations with pointwise controls in the domain or on the boundary. They reduced the problem to the minimization of densely defined quadratic functional on some Hilbert space. Li-Yong (1995) developed the results of You (1983) and Chen (1985) into the LQR control problem with indefinite performance function and unbounded control in Hilbert space including Neumann boundary control or pointwise control (for space dimension $n = 1$) for a parabolic equation.

Recently, Wu-Li (1997) generalized and improved the results of You (1983), Chen (1985), Li-Yong (1995) and Lasiecka-Triggiani (1991, 1992) etc. into the finite horizon LQR problems with unbounded controls and indefinite performance functions in Hilbert space, which covers Dirichlet, Neumann boundary control and pointwise control (for $n \leq 3$) for a parabolic equation. Wu-Li (1998) also established the necessary and sufficient frequency theorem or so called positive real lemma, and derived the synthesis for optimal control of the corresponding infinite horizon LQR control problem, which does not require the assumption for the exact controllability of the control system and thus generalized the result of Yakubovich and other authors into parabolic equation.

## 2. LQR CONTROL PROBLEMS FOR FINITE HORIZON

Let $X$ and $U$ be real Hilbert spaces, $-A$ generate an analytic semigroup $e^{-At}$ on $X$ and $A^\gamma$ be defined for all $\gamma \in \mathbf{R}$. Denote $F^*$ be the adjoint operators of $F$, which will be clear in context. Consider the abstract control system

$$y(t) = e^{-At}x + \int_0^t A^\alpha e^{-A(t-s)} Bu(s)\, ds, \tag{1}$$

where $\alpha \in [0,1)$ is constant, $x \in X$ and $B \in \mathcal{L}(U, X)$. In the case of parabolic equation on a bounded domain $\Omega \subset R^n$, the relevant values of $\alpha$ are as follows: $\alpha = \frac{3}{4} + \varepsilon$, $\forall \varepsilon > 0$, for Dirichlet boundary control with $U = L^2(\partial\Omega)$, $X = L^2(\Omega)$; $\alpha = \frac{1}{4} + \varepsilon$, $\forall\varepsilon > 0$ for Neumann boundary control with $U = L^2(\partial\Omega)$, $X = L^2(\Omega)$; and $\frac{n}{4} < \alpha < 1$, for pointwise control with $U = H^{-2\alpha}(\Omega)$, $X = L^2(\Omega)$.

Fix $T \in (0, \infty)$. If $u(\cdot) \in \mathcal{U} \triangleq L^2(0,T;U)$, then $y(\cdot) \in \mathcal{X} \triangleq L^2(0,T;X)$ (see Lasiecka-Triggiani (1991, 1992) and Li-Yong (1995)). Define $L \in \mathcal{L}(\mathcal{U};\mathcal{X})$ and $L_T$ as follows:

$$\begin{cases} (Lu)(t) \triangleq \int_0^t A^\alpha e^{-A(t-s)} Bu(s)\, ds, \\ \qquad \text{for } t \in [0,T], \quad u(\cdot) \in \mathcal{U}; \\ L_T u \triangleq \int_0^T A^\alpha e^{-A(T-s)} Bu(s)\, ds, \\ \qquad \text{for} \quad u(\cdot) \in D(L_T); \\ D(L_T) \triangleq \{u(\cdot) \in \mathcal{U} \mid \\ \qquad \int_0^T A^\alpha e^{-A(T-s)} Bu(s)\, ds \in X\}. \end{cases} \tag{2}$$

$$J(x; u(\cdot)) = \langle Q_1 y(T), y(T)\rangle + \int_0^T \{\langle Qy(t), y(t)\rangle + \langle Ru(t), u(t)\rangle\}\, dt, \tag{3}$$

where $u(\cdot) \in D(L_T)$, $y(\cdot)$ is given by (1) and $y(T) \triangleq e^{-AT}x + L_T u(\cdot)$; Hereafter, it is assumed throughout this paper that $Q$, $Q_1 \in \mathcal{L}(X)$ are self-adjoint but allowed to be indefinite and not necessarily smoothing in the sense that, $(A^*)^\beta Q_1$ or $(A^*)^\beta Q \in \mathcal{L}(X)$ for some nonnegative $\beta > \alpha - \frac{1}{2}$ respectively; $R \in \mathcal{L}(U)$ and $\langle Ru, u\rangle \geq \delta \|u\|_U^2$, $\forall u \in U$ for some $\delta > 0$.

**$LQR$ Control Problem** *For given $x \in X$, find a $u(\cdot) \in D(L_T)$ so that the performance function (3) is minimized.*

Wu-Li (1997) proved the equivalence between the solvability of LQR control problem and that of the corresponding two-point boundary value problem. In general, a state trajectory is not necessarily continuous. After establishing some singular inequalities,Wu-Li (1997) applies them to prove that any optimal control and its state trajectory are continuous. This results does not require any assumptions such as the uniqueness of optimal control and the solvability of LQR control problem at other initial state condition. Then Wu-Li (1997) obtain the equivalence between the solvability of LQR control problem and that of Fredholm integral equation in some sense and derive the synthesis for the optimal control via the solution to Fredholm integral equation. Finally, Wu-Li (1997) proved

**Proposition 1** Suppose that

$$Q_1 = Q_0 + (A^*)^{-\beta} E A^{-\beta}$$

with $Q_0 \geq 0$ or $Q_0 \leq 0$, $E^* = E \in \mathcal{L}(X)$, $Q_0^{\frac{1}{2}} L_1(t)$ or $(-Q_0)^{\frac{1}{2}} L_1(t)$ is closed respectively, and $\beta > \alpha - \frac{1}{2}$ nonnegative. Moreover, for some absolute constant $\delta > 0$, it holds that

$$\begin{aligned} &\langle (R + L^* Q L) u(\cdot), u(\cdot)\rangle \\ &\quad + \langle Q_1 L_T u(\cdot), L_T u(\cdot)\rangle \\ &\quad \geq \delta \|u(\cdot)\|_{\mathcal{U}}^2, \quad \forall u(\cdot) \in D(L_T). \end{aligned} \tag{4}$$

Then $LQ$ Problem is uniquely solvable at any $x \in X$ with the optimal pair $(\bar{u}(\cdot), \bar{y}(\cdot))$ The Riccati integral equation

$$\begin{aligned} P(t) &= e^{-A^*(T-t)} Q_1 e^{-A(T-t)} \\ &+ \int_t^T e^{-A^*(\sigma-t)} Q e^{-A(\sigma-t)}\, d\sigma \\ &- \int_t^T e^{-A^*(\sigma-t)} [(A^*)^\alpha P(\sigma)]^* B \\ &\quad R^{-1} B^* (A^*)^\alpha P(\sigma) e^{-A(\sigma-t)}\, d\sigma, \end{aligned} \tag{5}$$

has a solution $P(\cdot) \in C([0,T);\ \mathcal{L}(X)) \triangleq \{F : [0,T) \mapsto \mathcal{L}(X) | F(\cdot)x \in C([0,T);X),\ \forall x \in X\}$ in the strong sense with $P^*(t) = P(t)$, $(A^*)^\gamma P(\cdot) \in C([0,T);\mathcal{L}(X))$ for $\gamma \in [0,1)$ and for some constant $C(\gamma) > 0$ and absolute constant $C > 0$

$$\langle P(0)x, x\rangle = J(x;\bar{u}(\cdot)), \quad \forall\, x \in X; \tag{6}$$

$$\lim_{t \to T} P(t)x = Q_1 x \qquad \forall x \in X; \tag{7}$$

$$\begin{aligned} &P(t) = -(A^*)^{-\alpha} H(t,t); \\ &\bar{u}(t) = -R^{-1} B^* (A^*)^\alpha P(t) \bar{y}(t), \\ &\qquad t \in [0,T); \end{aligned} \tag{8}$$

$$\begin{aligned} &\|(A^*)^\gamma P(t)\| \leq \frac{C(\gamma)}{|T-t|^\gamma}, \\ &\|P(t)\| \leq C, \ \ \forall\, t \in [0,T). \end{aligned} \tag{9}$$

Lasiecka-Triggiani (1991, 1992) and Li-Yong (1995) derived (7) in the weak sense. Li-Yong (1995) obtained (7) under the additional assumption $Q_1 \geq 0$ and $Q_1^{\frac{1}{2}}$ is compact.

## 3. LQR CONTROL PROBLEMS FOR INFINITE HORIZON

Let $X$ and $U$ be Hilbert spaces. With the dynamics (1), define

(10)

$$\begin{aligned} J(x;u(\cdot)) = \int_0^\infty &\{\langle Qy(t), y(t)\rangle \\ &+ 2\mathrm{Re}\langle Sy(t), u(t)\rangle + \langle Ru(t), u(t)\rangle\} dt, \end{aligned}$$

where $S \in \mathcal{L}(X,U)$. With the dynamics (1), a control $u(\cdot)$ is called admissible at an initial state condition $x \in X$ if $u(\cdot) \in L^2(0,\infty;U)$ and $y(\cdot) \in L^2(0,\infty;X)$.

**Problem $LQR$** *For given $x \in X$, find an admissible $u(\cdot) \in L^2(0,\infty;U)$ so that the performance function (10) is minimized.*

It has been proved that the exponential stabilizability of (1) is equivalent to the existence of admissible controls at any $x \in X$, which is called stability condition by Yakubovich. For the simplicity, Wu-Li (1998) assume that $e^{-At}$ is exponentially stable.

$$\begin{aligned} &[Lu(\cdot)](t) \triangleq \int_0^t A^\alpha e^{-A(t-s)} Bu(s)\, ds, \\ &\qquad \forall u(\cdot) \in L^2(0,\infty;U). \end{aligned} \tag{11}$$

$$\begin{aligned} &\widetilde{\Phi}(\omega) \triangleq R+ \\ &B^*(A^*)^\alpha(-i\omega + A^*)^{-1} Q A^\alpha (i\omega + A)^{-1} B \\ &\quad + SA^\alpha(i\omega + A)^{-1} B \\ &\quad + B^*(A^*)^\alpha(-i\omega + A^*)^{-1} S^*, \\ &\qquad \forall \omega \in \mathbf{R}, \end{aligned} \tag{12}$$

where $\widetilde{\Phi}(\cdot)$ is called the frequency characteristic assciated with Problem $LQ$.

Wu-Li (1998) proved the equivalence between the solvability of Problem $LQR$ and that of a corresponding two-point boundary value problem. Similarly, Wu-Li (1998) establish some singular inequalities and prove that any optimal control and its state trajectory are continuous. Finally, Wu-Li (1998) established

**Proposition 2** The following statements are equivalent:

**(I)** Problem $LQ$ is uniquely solvable at any $x \in X$ with the optimal pair $(\bar{u}(\cdot), \bar{y}(\cdot))$;

**(II)** The algebraic Riccati equation

$$\begin{aligned} &-\langle Ax, Px\rangle - \langle Px, Ax\rangle + \langle Qx, x\rangle \\ &-\langle [S + B^*(A^*)^\alpha P]^* \\ &R^{-1}[S + B^*(A^*)^\alpha P]x, x\rangle = 0, \\ &\forall\ x \in D(A), \end{aligned} \tag{13}$$

admits a solution $P$ satisfying $(A^*)^\alpha P \in \mathcal{L}(X)$ and

$$\widetilde{A} \triangleq -A^\alpha\{A^{1-\alpha} + BR^{-1}[S + B^*(A^*)^\alpha P]\} \tag{14}$$

with

$$\begin{aligned} D(\widetilde{A}) = \{&x \in D(A^{1-\alpha}) | \{A^{1-\alpha} \\ &+ BR^{-1}[S + B^*(A^*)^\alpha P]\}x \in D(A^\alpha)\} \end{aligned} \tag{15}$$

generates an exponentially stable and analytic semigroup. In this case, the state feedback optimal control of Problem $LQ$ is given by

$$\begin{aligned} \bar{u}(t) = -R^{-1}[S + B^*(A^*)^\alpha P]\bar{y}(t), \\ \text{for}\quad t \geq 0; \end{aligned} \tag{16}$$

**(III)** There exists some constant $\sigma > 0$ such that

$$\begin{aligned} &\langle (R + L^*QL + SL + L^*S^*)u(\cdot), u(\cdot)\rangle \\ &\geq \sigma\langle u(\cdot), u(\cdot)\rangle, \\ &\forall\ u(\cdot) \in L^2(0, \infty; U); \end{aligned} \tag{17}$$

**(IV)** There exists some constant $\sigma > 0$ such that

$$\begin{aligned} \langle \widetilde{\Phi}(\omega)u, u\rangle \geq \sigma\langle u, u\rangle, \\ \forall\ u \in U,\quad \omega \in \mathbf{R}. \end{aligned} \tag{18}$$

## REFERENCES


Chen,S. (1985), *Riccati equations arising in infinite dimensional optimal control problem, Control Theory Appl.* **2**, 64–72. (in Chinese)

Lasiecka,I. & Triggiani,R.(1991), *Differential and Algebraic Riccati Equations with Application to Boundary/ Point Control Problems: Continuous Theory and Approximation Theory,* LNCS Vol. 164, Springer-Verlag, New York.

Lasiecka,I. & Triggiani,R.(1992), *Riccati differential equations with unbounded coefficients and non-smooth terminal condition—The case of analytic semigroups, SIAM J. Math. Anal.* **23**, 449–481.

Li,X. & Yong,J.(1995), *Optimal Control Theory for Infinite Dimensional Systems,* Systems & Control: Foundations & Applications, Birkhäuser, Boston Basel Berlin.

Wu,H. & Li,X.(1997), *A linear quadratic problem with unbounded control in Hilbert spaces,* preprint.

Wu,H. & Li,X.(1998), *An infinite horizon linear quadratic problem with unbounded control in Hilbert space,* preprint.

You,Y.(1983), *Optimal control for linear systems with quadratic indefinite criterion on Hilbert spaces, Chin. Ann. Math. Ser.B,* **4**, 21–32.

14th Triennial World Congress, Beijing, P.R. China

F-2d-01-3

# WHAT SYSTEM IS OPTIMAL FOR GIVEN CONTROL LAW AND CRITERION ?

**Katsuhisa Furuta** [*,1] **Yongcai Xu** [*]

[*] *Department of Mechanical and Environmental Informatics, Tokyo Institute of Technology, 2-12-1, Oh-okayama, Meguro-ku, Tokyo 152, Japan.*
*E-mail: furuta@mei.titech.ac.jp Fax: +81-3-5734-3487*

Abstract: This paper presents Optimal System Problem, that is what system is optimal when both control law and control criterion are given. Unlike optimal control and the inverse optimal control which have been well researched, Optimal System Problem has not been considered untill now. In this paper, only the case that the criterion is quadratic and the input is single is considered. A necessary and sufficient condition of existence of optimal system which makes closed-loop system stable by given control feedback control law is presented. Also a computation method to get an existing optimal system is discussed. At the last part of this paper, an example is presented to compare optimal control, inverse optimal control and optimal system problem. 

Keywords: Optimal Control; Inverse Optimal Control; Optimal System

## 1. INTRODUCTION

In the conventional optimal control problem,the optimal control law is designed so as to minimize the criterion function for a given system. (Kalman, 1964) considered the problem inversely, and solved the problem of what quadratic criterion is optimized for the given stabilizing control law to the controlled system , while (Suga, 1967) considered more general criterion for the inverse problem. The similar inverse problem has also been considered by (Lenz *et al.*, 1988) for $H_\infty$ optimal control. In such so called inverse optimal control, the criterion has been considered for the given control law and controlled system.

When there exist the controller and criterion, there may be a problem of designing system which makes the closed loop system optimal for the criterion by using the controller. This problem has not been considered, but may be useful in the design of systems which requires high performance with the given control system.

This paper simply presents the system which is optimal for the given quadratic criterion function and the control law in the case of the single input. The approach can be extended to the multivariable case easily. An example for comparison of optimal control, inverse problem in optimal control and optimal system is also presented.

[1] This research is supported by COE, Grant-in-Aid for Scientific Research (09CE2004).

## 2. PRELIMINARIES

This paper considers single input linear time invariant system, whose dynamical equation is denoted by

$$\dot{\mathbf{x}} = \mathbf{A}\mathbf{x} + \mathbf{b}u \quad \mathbf{x}(0) = \mathbf{x}_0 \tag{1}$$

where $\mathbf{A} \in \mathcal{R}^{n\times n}$, $\mathbf{b} \in \mathcal{R}^n$, $\mathbf{x} \in \mathcal{R}^n$ and $(\mathbf{A},\mathbf{b})$ is assumed to be controllable.

The criterion is given by

$$J = \frac{1}{2}\int_0^\infty (\mathbf{x}^T\mathbf{Q}\mathbf{x} + ru^2)dt \qquad (2)$$

where $\mathbf{Q} \geq 0$, $r$ is a positive scalar.

The optimal control law is given by 3, see (Furuta *et al.*, 1988).

$$u = \mathbf{k}\mathbf{x} = -r^{-1}\mathbf{b}^T\mathbf{P}\mathbf{x} \qquad (3)$$

where $\mathbf{P}$ is the positive definite solution of the Algebraic Riccati Equation 4:

$$\mathbf{A}^T\mathbf{P} + \mathbf{P}\mathbf{A} + \mathbf{Q} - \mathbf{P}\mathbf{b}r^{-1}\mathbf{b}^T\mathbf{P} = 0 \qquad (4)$$

This paper considers the problem of determining $\mathbf{A}$ and $\mathbf{b}$ when $\mathbf{Q}$, $r$ in the criterion function 2 and the control law $\mathbf{k}$ are given. This problem is said to be Optimal System (OS) Problem .

Without loss of generality, the following assumptions are taken:

- $r = 1$
- the given control law is represented by $\mathbf{f}$ defined as $\mathbf{f} = [f_0 \ldots f_{n-1}]^T := \mathbf{P}\mathbf{b}$ instead of $\mathbf{k}$ for convenience.
- $\mathbf{Q}$ and $\mathbf{f}$ are real.
- $(\mathbf{A}, \mathbf{b})$ is in controller canonical form, i.e.

$$\mathbf{A} = \begin{bmatrix} 0 & 1 & 0 & \ldots & 0 \\ 0 & 0 & 1 & \ldots & 0 \\ \vdots & \vdots & \vdots & \ddots & \vdots \\ 0 & 0 & 0 & \ldots & 1 \\ -\alpha_0 & -\alpha_1 & \ldots & \ldots & -\alpha_{n-1} \end{bmatrix}, \ \mathbf{b} = \begin{bmatrix} 0 \\ \vdots \\ 0 \\ 1 \end{bmatrix}$$

Hence the Riccati Equation 4 becomes

$$\mathbf{A}^T\mathbf{P} + \mathbf{P}\mathbf{A} + \mathbf{Q} - \mathbf{f}\mathbf{f}^T = 0 \qquad (5)$$

## 3. MAIN RESULTS

It is well known that solution of the Algebraic Riccati Equation is not unique, see (Zhou *et al.*, 1996) chapter 13. But if $(\mathbf{A}, \mathbf{b}, \mathbf{Q}, r)$ satisfy some conditions, then there exists a positive definite solution, furthermore this positive solution is unique.

As compare , however, if $(\mathbf{A}, \mathbf{f})$ is taken in controller canonical form as above, normally the solution for $\mathbf{A}$ and $\mathbf{P}$ is unique, which is shown in the following theorem.

*Theorem 1.* Let $\mathbf{L} = [l_{ij}]$, $\mathbf{T} = [t_{ij}] = \mathbf{Q} - \mathbf{f}\mathbf{f}^T$ and $\mathbf{h} = [h_1 \ldots h_n]^T$

where

$$l_{ij} = \begin{cases} 0 & \text{if } \nu < 1 \text{ or } \nu > n \\ 2f_{\nu-1}(-1)^{\nu-1} & \text{if } 1 \leq \nu \leq n \end{cases} \qquad (6)$$

$$\nu = 2i - j \quad \text{for } i, j = 1, \ldots, n$$

$$h_i = \sum_{j=max(1,2i-n)}^{j=min(n,2i-1)} t_{j(2i-j)}(-1)^{j+1} - 2(-1)^n f_{2i-2-n}\mu(2i-2-n) \qquad (7)$$
$$\text{for } i = 1, \ldots, n$$

$$\mu(j) = \begin{cases} 0 & \text{if j} < 0 \\ 1 & \text{if } j \geq 0 \end{cases}$$

Then the equation 5 has a unique solution if and only if the matrix $\mathbf{L}$ is nonsingular, furthermore the $n$ parameters of $\mathbf{A}$ is given by

$$[\alpha_0 \ldots \alpha_{n-1}]^T = \mathbf{L}^{-1}\mathbf{h} \qquad (8)$$

If $\mathbf{A}$ is determined, then $\mathbf{P}$ is uniquely determined to satisfy both equation 5 and control law $\mathbf{f} = \mathbf{P}\mathbf{b}$.

**Proof:** Add and subtract $s\mathbf{P}$ to equation 5 to get

$$\mathbf{P}(s\mathbf{I} - \mathbf{A}) + (-s\mathbf{I} - \mathbf{A}^T)\mathbf{P} + \mathbf{f}\mathbf{f}^T - \mathbf{Q} = 0$$

Now multiply the above equation from the left by $\mathbf{b}^T(-s\mathbf{I} - \mathbf{A}^T)^{-1}$ and from the right by $(s\mathbf{I} - \mathbf{A})^{-1}\mathbf{b}$ to get

$$\mathbf{b}^T(-s\mathbf{I} - \mathbf{A}^T)^{-1}\mathbf{f} + \mathbf{f}^T(s\mathbf{I} - \mathbf{A})^{-1}\mathbf{b} - \mathbf{b}^T(-s\mathbf{I} - \mathbf{A}^T)^{-1}\mathbf{T}(s\mathbf{I} - \mathbf{A})^{-1}\mathbf{b} = 0 \qquad (9)$$

Let

$$\Delta(s) = det(s\mathbf{I} - \mathbf{A}) = s^n + \alpha_{n-1}s^{n-1} + \ldots + \alpha_1 s + \alpha_0$$

Then

$$det(-s\mathbf{I} - \mathbf{A}^T) = det(-s\mathbf{I} - \mathbf{A}) = \Delta(-s)$$

note that

$$(s\mathbf{I} - \mathbf{A})^{-1}\mathbf{b} = [1 \ s \ \ldots \ s^{n-1}]^T$$

Substitute these into equation 9,multiply both sides by $\Delta(s)\Delta(-s)$, to get

$$\Delta(s)\sum_{i=0}^{i=n-1} f_i(-1)^i s^i + \Delta(-s)\sum_{i=0}^{i=n-1} f_i s^i = [1 \ -s \ \ldots \ (-s)^{n-1}]\mathbf{T}[1 \ s \ \ldots \ s^{n-1}]^T \qquad (10)$$

It is obviously that all those coefficients of odd-th power to $s$ is 0. And since the above equation holds for all $s$ that satisfies $\Delta(s) \neq 0$, all the coefficients of $s^{2i}(i = 0, 1, \ldots, n-1)$ must be equal. Hence there are total $n$ equations, since the number of parameters of $\mathbf{A}$ to be determined is $n$, it is reasonable that the solution for $\mathbf{A}$ is unique. By simple calculation, easily get

$$\begin{bmatrix} 2f_0 & 0 & \cdots & 0 \\ \vdots & \vdots & \ddots & \vdots \\ 0 & 0 & \cdots & (-1)^n 2f_n \end{bmatrix} \begin{bmatrix} \alpha_0 \\ \vdots \\ \alpha_{n-1} \end{bmatrix} = \begin{bmatrix} t_{11} \\ \vdots \\ t_{nn}(-1)^{n+1} - 2(-1)^n f_{n-2} \end{bmatrix} \quad (11)$$

or

$$\mathbf{L}[\alpha_0 \ldots \alpha_{n-1}]^T = \mathbf{h}$$

Therefore, the solution for $\mathbf{A}$ exists and be unique if and only if $\mathbf{L}$ is nonsingular. Furthermore, if $\mathbf{L}$ is nonsingular, the solution for $\mathbf{A}$ is given by equation 8

Now we prove $\mathbf{P}$ is uniquely determined when $\mathbf{A}$ and $\mathbf{f}$ are given. Let

$$\mathbf{P} = [\mathbf{p_1} \ldots \mathbf{p_n}]$$

$$\mathbf{T} = \mathbf{Q} - \mathbf{ff}^T = [\mathbf{t_1} \ldots \mathbf{t_n}]$$

where $\mathbf{p_i}$ is the $i_{th}$ column of $\mathbf{P}$. $\mathbf{t_i}$ is the $i_{th}$ column of $\mathbf{T}$.

Then

$$\mathbf{PA} = [\mathbf{p_1} \ldots \mathbf{p_n}] \begin{bmatrix} 0 & 1 & 0 & \ldots & 0 \\ 0 & 0 & 1 & \ldots & 0 \\ \vdots & \vdots & \vdots & \ddots & \vdots \\ 0 & 0 & 0 & \ldots & 1 \\ -\alpha_0 & -\alpha_1 & \ldots & \ldots & -\alpha_{n-1} \end{bmatrix}$$
$$= [-\alpha_0\mathbf{p_n}\ (\mathbf{p_1} - \alpha_1\mathbf{p_n}) \ldots (\mathbf{p_{n-1}} - \alpha_{n-1}\mathbf{p_n})]$$

$$\mathbf{A}^T\mathbf{P} = [\mathbf{A}^T\mathbf{p_1}\ \mathbf{A}^T\mathbf{p_2} \ldots \mathbf{A}^T\mathbf{p_n}]$$

Hence from equation 5, it is easy to get

$$\mathbf{A}^T\mathbf{p_{j+1}} + \mathbf{p_j} - \alpha_j\mathbf{p_n} + \mathbf{t_{j+1}} = 0 \quad \text{for } j = 1, \ldots, n-1$$

or

$$\mathbf{p}_j = -\mathbf{t_{j+1}} - \mathbf{A}^T\mathbf{p_{j+1}} + \alpha_j\mathbf{p_n} \quad (12)$$
$$\text{for } j = 1, \ldots, n-1$$

Since $\mathbf{p_n} = \mathbf{Pb} = \mathbf{f}$ is given, then $\mathbf{p_{n-1}}, \ldots, \mathbf{p_2}, \mathbf{p_1}$ can be uniquely determined in turn by the equation above .

□

*Remark 2.* If $\mathbf{L}$ is singular, then $\mathbf{A}$ is not unique if $\mathbf{h} \in Im(L)$ , otherwise $\mathbf{A}$ satisfying equation 5 does not exist.

This can be illustrated by the following Example.

**Example 1:** To consider a second order system with

$$\mathbf{Q} = \begin{bmatrix} q_1 & 0 \\ 0 & q_2 \end{bmatrix}$$

where $q_1 \geq 0$, $q_2 \geq 0$

It is easy to see that

$$\mathbf{L} = \begin{bmatrix} 2f_0 & 0 \\ 0 & -2f_1 \end{bmatrix} \qquad \mathbf{h} = \begin{bmatrix} q_1 - f_0^2 \\ -q_2 + f_1^2 - 2f_0 \end{bmatrix}$$

Either $f_0 = 0$ or $f_1 = 0$, $L$ is singular , now suppose $f_0 = 0$, i.e. $\mathbf{f} = [0\ f_1]^T$

**Case1:** $q_1 \neq 0$, then $\mathbf{h}$ is not in the set of $Im(\mathbf{L})$, hence such $\mathbf{A}$ that satisfies equation 5 does not exist.

**Case2:** $q_1 = 0$, then $\mathbf{h} \in Im(\mathbf{L})$ and a solution for $\mathbf{A}$ is not unique, in fact

$$\mathbf{A} = \begin{bmatrix} 0 & 1 \\ -\alpha_0 & \dfrac{q_2 - f_1^2}{2f_1} \end{bmatrix} \quad \mathbf{P} = \begin{bmatrix} \alpha_0 f_1 & 0 \\ 0 & f_1 \end{bmatrix}$$

for any $\alpha_0 \in \mathcal{R}$

are all solutions of equation 5.

*Theorem 1* presents the condition for $\mathbf{A}$ being existing and also a method to get $\mathbf{A}$ and $\mathbf{P}$, however it does not guarantee the closed-loop system by state feedback $\mathbf{u} = -\mathbf{f}^T\mathbf{x}$ be stable. Hence the solution may not be practical or has no meaning in practice if the closed-loop system $\mathbf{A}_f := \mathbf{A} - \mathbf{bf}^T$ is not stable. Although there may exist a unique positive definite solution of $\mathbf{P}$ for Riccati Equation 4 for some $(\mathbf{A}, \mathbf{b}, \mathbf{Q}, r)$, and $\mathbf{P} > 0$ implies $\mathbf{A}_f$ be stable, note that a solution of Riccati equation is not unique, and the solution of $\mathbf{P}$ for OS Problem 5 must satisfy $\mathbf{Pb} = \mathbf{f}$, therefore it may not be the positive definite one, in other words, $\mathbf{A}_f$ may be unstable.

Example 2 illustrates this statement.

**Example 2 :** To consider a second order system with

$$\mathbf{Q} = \begin{bmatrix} 2 & 0 \\ 0 & 4 \end{bmatrix} \quad \mathbf{f} = \begin{bmatrix} -1 \\ 1 \end{bmatrix}$$

Then from *Theorem 1*, it is easy to see that a unique solution for OS Problem exists:

$$\mathbf{A} = \begin{bmatrix} 0 & 1 \\ 0.5 & -0.5 \end{bmatrix} \quad \mathbf{P} = \begin{bmatrix} -2 & -1 \\ -1 & 1 \end{bmatrix}$$

the closed-loop system by state feedback $\mathbf{u} = -\mathbf{f}^T\mathbf{x}$ is

$$\mathbf{A_f} = \begin{bmatrix} 0 & 1 \\ 1.5 & -1.5 \end{bmatrix}$$

Obviously the closed-loop system is not stable, neither is $\mathbf{P}$ positive definite. Note that for such given $(\mathbf{A}, \mathbf{b}, \mathbf{Q}, r)$, there exists a unique positive definite $\mathbf{P}$ satisfying equation 4, which is given by

$$\mathbf{P} = \begin{bmatrix} 4.5584 & 2.0000 \\ 2.0000 & 2.3723 \end{bmatrix}$$

However it does not satisfy $\mathbf{Pb} = \mathbf{f} = [-1\ 1]^T$, hence can not be OS Problem solution.

Before investigating the conditions for closed-loop system $\mathbf{A_f}$ to be stable , the relationship between the closed-loop stability and $\mathbf{P}$ is investigated first.

*Lemma 3.* For an open loop system described by equation 1, if the closed-loop system by state feedback $\mathbf{u} = -\mathbf{f}^T\mathbf{x}$ is stable, i.e. $\mathbf{A_f} = \mathbf{A} - \mathbf{bf}^T$ is stable, then $\mathbf{P}$ satisfying equation 5 must be positive semi-definite , i.e. $\mathbf{P} \geq 0$ .

**Proof:** Proof for this Lemma is easy, from 5, to get

$$\begin{aligned}(\mathbf{A} - \mathbf{bf}^T)^T\mathbf{P} + \mathbf{P}(\mathbf{A} - \mathbf{bf}^T) \\ = -\mathbf{Q} + \mathbf{ff}^T - 2\mathbf{Pbf}^T\end{aligned}$$

note $\mathbf{f} = \mathbf{Pb}$, the equation above is equivalent to

$$\mathbf{A_f}^T\mathbf{P} + \mathbf{PA_f} = -(\mathbf{Q} + \mathbf{ff}^T) \tag{13}$$

if $\mathbf{A_f}$ is stable, then the solution for the above Lyapunov equation is given by

$$\mathbf{P} = \int_0^\infty e^{\mathbf{A_f}^T t}(\mathbf{Q} + \mathbf{ff}^T)e^{\mathbf{A_f}t}dt$$

since

$$\mathbf{Q} \geq 0$$

then

$$\mathbf{Q} + \mathbf{ff}^T \geq 0$$

hence

$$\mathbf{P} \geq 0$$

□

*Remark 4. Lemma 3* shows $\mathbf{P} \geq 0$ is necessary for $\mathbf{A_f}$ to be stable, however it is not sufficient. That is $\mathbf{P} \geq 0$ does not imply $\mathbf{A}_f$ stable. This is clearly illustrated by Example 3.

**Example 3:** Still a second order system is considered with

$$\mathbf{Q} = \begin{bmatrix} 1 & -1 \\ -1 & 1 \end{bmatrix} \qquad \mathbf{f} = \begin{bmatrix} -k_0 \\ k_0 \end{bmatrix}$$

Then the closed-loop system and $\mathbf{P}$ can be computed:

$$\mathbf{A}_f = \begin{bmatrix} 0 & 1 \\ \frac{1+k_0^2}{2k_0} & -\frac{(1-k_0)^2}{2k_0} \end{bmatrix} \qquad \mathbf{P} = \begin{bmatrix} k_0 & -k_0 \\ -k_0 & k_0 \end{bmatrix}$$

It shows clearly that $\mathbf{P} \geq 0$ for all $k_0 > 0$ , but $\mathbf{A_f}$ is unstable for any $k_0 \neq 0$ .

*Remark 5. Lemma 3* presents a necessary condition for the closed-loop system to stable, Example 3 shows shows such condition is not sufficient. An sufficient condition is that the corresponding $\mathbf{P}$ in equation 5 is positive definite. See (Zhou *et al.*, 1996) for reference.

The state feedback closed-loop system to be stable is essential in practice, so such question will be asked that under which condition the closed-loop system is stable or OS Problem solution is practical.

*Theorem 6.* If there exist a unique solution for $\mathbf{A}$ of equation 5, i.e. $\mathbf{L}$ defined in 6 is nonsingular, then $\mathbf{A}_f = \mathbf{A} - \mathbf{bf}^T$ is stable if and only if

(1) $f_{n-1} > 0$
(2) $n(s) := f_{n-1}s^{n-1} + ... + f_1 s + f_0$ has no roots on the right half complex plane.

**Proof:** Let

$$G(s) = -\mathbf{f}^T(s\mathbf{I} - \mathbf{A})^{-1}\mathbf{b} \tag{14}$$

$$\Phi(s) = -\mathbf{f}^T(s\mathbf{I} - \mathbf{A} + \mathbf{bf}^T)\mathbf{b} \tag{15}$$

$$\begin{aligned} d(s) :&= det(s\mathbf{I} - \mathbf{A} + \mathbf{bf}^T) \\ &= s^n + (\alpha_{n-1} + f_{n-1})s^{n-1} + ... \\ &\quad +(\alpha_1 + f_1)s + (\alpha_0 + f_0) \end{aligned} \tag{16}$$

Then

$$\begin{aligned}\Phi(s) &= \frac{-f_{n-1}s^{n-1} - ... - f_1 s - f_0}{s^n + (\alpha_{n-1} + f_{n-1})s^{n-1} + ... + (\alpha_0 + f_0)} \\ &= \frac{-n(s)}{d(s)}\end{aligned} \tag{17}$$

From equation 9, it is easy to get

$$\begin{aligned}&- G(j\omega)^* - G(j\omega) + G(j\omega)^*G(j\omega) \\ &- \mathbf{b}^T(-j\omega\mathbf{I} - \mathbf{A}^T)^{-1}\mathbf{Q}(j\omega\mathbf{I} - \mathbf{A})^{-1}\mathbf{b} = 0\end{aligned}$$

Since

$$\mathbf{Q} \geq 0$$

then

$$\mathbf{b}^T(-j\omega\mathbf{I} - \mathbf{A}^T)^{-1}\mathbf{Q}(j\omega\mathbf{I} - \mathbf{A})^{-1}\mathbf{b} \geq 0$$

Therefore

$$(1 - G(j\omega)^*)(1 - G(j\omega)) \geq 1 \qquad (18)$$

i.e.

$$(1 + \mathbf{b}^T(-j\omega\mathbf{I} - \mathbf{A}^T)^{-1}\mathbf{f})(1 + \mathbf{f}^T(j\omega\mathbf{I} - \mathbf{A})^{-1}\mathbf{b}) \geq 1$$

Take the inverse of the above inequality, get

$$(1 + \Phi(j\omega)^*)(1 + \Phi(j\omega)) \leq 1 \qquad (19)$$

The above inequality shows that the Nyquist diagram of $\Phi(s)$ stays in the unit disk centered at (-1,0).

Now suppose the two conditions in *Theorem 6* are satisfied.

Since $\mathbf{L}$ is nonsingular, from 11 it is easy to see that $f_0 \neq 0$, then as $\omega$ varies from $0_+$ to $+\infty$, the Nyquist diagram of $\Phi(j\omega)$ starts at the left real axis, and since $f_{n-1} > 0$, as $\omega$ varies from $+\infty$ to $-\infty$, the Nyquist diagram of $\Phi(j\omega)$ still stays on the left of the origin point $(0,0)$, i.e. the Nyquist diagram of $\Phi(j\omega)$ does not circle the origin point $(0,0)$ (refer to Fig.1 ). By the Nyquist Stable Criterion, $n(s)$ and $d(s)$ must have the same number of roots that lie on the right half complex plane. By supposition 2, $n(s)$ has no roots on the right half complex plane, so neither does $d(s)$, i.e. $\mathbf{A} - \mathbf{b}\mathbf{f}^T$ is stable. The proof for sufficiency is completed.

Now consider the proof of necessity, suppose $\mathbf{A_f} = \mathbf{A} - \mathbf{b}\mathbf{f}^T$ is stable.

By *Lemma 3*, $\mathbf{P} \geq 0$ from which it is easy to see that $f_{n-1} \geq 0$. Furthermore, $\mathbf{L}$ is nonsingular, by 11 $f_{n-1} \neq 0$ , hence $f_{n-1} > 0$.

From the proof of sufficiency, by $f_{n-1} > 0$ and equation 19, it is concluded that $n(s)$ and $d(s)$ have the same number of roots that lie on the right half complex plane. $\mathbf{A_f} = \mathbf{A} - \mathbf{b}\mathbf{f}^T$ is stable means that $d(s)$ has no roots on the right half complex plane, therefore $n(s)$ has no roots on the right half complex plane,either.
□

*Theorem 6* gives a sufficient and necessary condition for the obtained optimal system to yield a stable closed-loop when being applied by the given control law. While *Theorem 1* presents a condition for the existence of optimal system and a method to get it. All these are based on $(\mathbf{A}, \mathbf{b})$ in controller canonical form, hence normally the optimal system is unique if it exists. The following example gives an comparison between optimal control, inverse optimal control and optimal system, also the case that $(\mathbf{A}, \mathbf{b})$ is not in controller canonical form is discussed, which shows optimal system is not unique if without controller canonical form assumption.

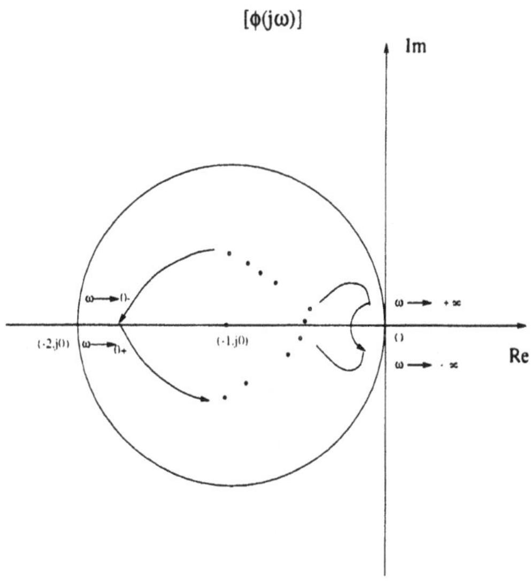


Fig. 1. the Nyquist Diagram of $\Phi(j\omega)$

**Example 4:** For sake of simplicity, still a single input second order system is considered.

First, consider optimal control problem, a system is described as 1, where

$$\mathbf{A} = \begin{bmatrix} 0 & 1 \\ 0 & 0 \end{bmatrix} \quad \mathbf{b} = \begin{bmatrix} 0 \\ 1 \end{bmatrix} \qquad (20)$$

The objective function is given by

$$J = \frac{1}{2}\int_0^\infty (x_1^2 + x_2^2 + u^2)dt$$

or

$$\mathbf{Q} = \begin{bmatrix} 1 & 0 \\ 0 & 1 \end{bmatrix} \quad r = 1 \qquad (21)$$

It is easy to get the optimal feedback control law 22 which minimizes criterion function 21 and stabilizes this given system.

$$u = -[1\ \sqrt{3}]\begin{bmatrix} x_1 \\ x_2 \end{bmatrix} \qquad (22)$$

or

$$\mathbf{f} = [1\ \sqrt{3}]^T$$

the corresponding matrix $\mathbf{P}$ is

$$\mathbf{P} = \begin{bmatrix} \sqrt{3} & 1 \\ 1 & \sqrt{3} \end{bmatrix} \qquad (23)$$

Note that if an optimal control does exist, it is unique, however, inverse problem is not unique, usually it is a class of criterion function that is optimized by the given control law as shown in (Kalman, 1964). Now consider the inverse problem.

The system and control law is given by 20 and 22, respectively. What kind of criterion function is optimized? If the criterion function is quadratic, then it is given by 24, 25 gives corresponding **P**.

$$\mathbf{Q} = \begin{bmatrix} 1 & \frac{1}{2}\alpha \\ \frac{1}{2}\alpha & 1 \end{bmatrix} \quad r = 1 \tag{24}$$

$$\mathbf{P} = \begin{bmatrix} \sqrt{3} - \frac{1}{2}\alpha & \\ 1 & \sqrt{3} \end{bmatrix} \tag{25}$$

where $\alpha$ is an arbitrary real scalar.

It is clearly shown that the criterion function is not unique, but a class set. Such a class of criteria can be verified to yield the same unique optimal control law 22. Furthermore, if the criterion function is not restricted to be quadratic, then more general criteria shown in 26 and 27 are also available, see (Suga, 1967) for details.

$$J = \int_0^\infty [\ (1 - e^{-t})x_1^2 + 2e^{-t}x_1x_2 + x_2^2 + u^2]dt \tag{26}$$

$$J = \int_0^\infty (\ x_2^2 - 2tx_1x_2 + u^2)dt \tag{27}$$

Finally, consider the optimal system problem dealt in this paper.

Let the given criterion and control law be 24 and 22, respectively. It is easy to verify that all conditions in *Theorem 1* and *Theorem 6* are satisfied, hence a unique optimal system in controller canonical form can be obtained by the method presented in *Theorem 1*, it is the same as 20.

Now consider more general case that **A** is not in controller canonical form, as stated before, $n$ parameter of **A** can be determined from equation 5, therefore suppose $\mathbf{A}, \mathbf{b}$ in the form as shown in 28 and 20, respectively.

$$\mathbf{A} = \begin{bmatrix} a_1 & a_2 \\ -\alpha_0 & -\alpha_1 \end{bmatrix} \tag{28}$$

where $a_1$ and $a_2$ are given

To make problem simple, the given criterion is the same as 21 instead of 24. By some computation, parameters of **A** and **P** are obtained as shown in 29 and 30, respectively.

$$\begin{cases} \alpha_0 = a_1(\frac{1}{\sqrt{3}} + \frac{4}{2\sqrt{3}a_2 - 6a_1}) \\ \alpha_1 = \frac{a_2 - 1}{\sqrt{3}} \end{cases} \tag{29}$$

$$\mathbf{P} = \begin{bmatrix} p_{11} & 1 \\ 1 & \sqrt{3} \end{bmatrix} \tag{30}$$

where $p_{11} = \frac{1}{\sqrt{3}} + \frac{4}{2\sqrt{3}a_2 - 6a_1}$

$$a_2 \neq \sqrt{3}a_1 \tag{31}$$

Relationship 31 is made to ensure the problem nonsingular, if it does not hold, treatment is similar to that in *Remark 2*.

Note that the optimal system obtained above does not ensure a stable closed-loop system, as stated in *Remark 5*, to ensure the closed-loop system stable, $\mathbf{P} > 0$ is required, which means

$$a_2 > \sqrt{3}a_1 \tag{32}$$

If 32 does hold, then system 28—29 and criterion 21 also yields the same stabilizing optimal control law 22.

Optimal system in case of **A** being in general form for second order system is studied, study of $n$ dimensional system will be discussed in later publications.

## 4. CONCLUSION

This paper has presented a method to compute the optimal system parameters for given control law and criterion. Also both the condition for existence of solution and conditions for closed-loop system to be stable are presented. At present, we focus only on the quadratic criterion and what we have done is preliminary. There are a lot of work should be done subject to so called Optimal System Problem. In the future, the $H_\infty$ case may be considered. We expect more applications of our results.

## 5. REFERENCES

Furuta, Katsuhisa, Akira Sano and Derek Atherton (1988). *State Variable Methods in Automatic Control.* John Wiley and Sons. Chichester, New York.

Kalman, R.E. (1964). When is a linear system optimal?. *Trans. ASME Ser.D.J.Basic Eng.* **86**, 51–60.

Lenz, Kathryn E., Pramod P.Khargonekar and John C.Doyle (1988). When is a controller $h_\infty$ optimal?. *Math. Control Signals Systems* **1**, 107–122.

Suga, Ichiro (1967). Some inverse problem in optimal control. *Instrument and Automatic Control, Journal of SICE, Japan* **6**, 549–557.

Zhou, Kemin, John C. Doyle and Keith Glover (1996). *Robust and Optimal Control.* Prentice Hall. New Jersey.

# Some Qualitative Properties of Certain $C_0$ Semigroups Arising in Elastic Systems with Various Dampings

Zhuangyi Liu

Department of Mathematics and Statistics
University of Minnesota, Duluth, MN 55812, USA

Jiongmin Yong

Department of Mathematics and Institute of Mathematical Finance
Fudan University, Shanghai 200433, China
email: *jyong@ms.fudan.edu.cn*

Abstract: This paper presents some results concerning various qualitative properties of the semigroup for centain elastic systems with differenti dampings. Among others, we obtain conditions under which the semigroup is analytic, differentiable, exponentially stable, etc. *Copyright © 1999 IFAC*

Keywords: Qualitative analysis, distributed parameter systems, partial differential equations.

## 1. INTRODUCTION

Let $H_0$ and $H_1$ be two Hilbert spaces, $-A_0 : \mathcal{D}(A_0) \subseteq H_0 \to H_0$ and $-A_1 : \mathcal{D}(A_1) \subseteq H_1 \to H_1$ be two densely defined closed linear operators generating $C_0$ contraction semigroups $e^{-A_0 t}$ and $e^{-A_1 t}$ on $H_0$ and $H_1$, respectively. Let $B : \mathcal{D}(B) \subseteq H_1 \to H_0$ and $C : \mathcal{D}(C) \subseteq H_0 \to H_1$ be other two densely defined and closed linear operators. Consider the following system of coupled linear evolution equations on $\mathcal{H} \triangleq H_0 \times H_1$:

$$(1.1) \quad \begin{cases} \dot{x}(t) = -A_0 x(t) + By(t), & x(0) = x_0, \\ \dot{y}(t) = Cx(t) - A_1 y(t), & y(0) = y_0. \end{cases}$$

This work is supported in part by the NSFC under grant 79790130, the National Distinguished Youth Science Foundation of China under grant 19725106, the Chinese Education Ministry Science Foundation under grant 97024607, and the Li Foundation.

Many elastic systems with various dampings can be put into such an abstract form. We keep in mind that an important special case of the above is that $C = -B^*$ (with identifications $H_0^* = H_0$ and $H_1^* = H_1$). Now, if we define

$$(1.2) \quad \mathcal{A} = \begin{pmatrix} -A_0 & B \\ C & -A_1 \end{pmatrix} : \mathcal{D}(\mathcal{A}) \subseteq \mathcal{H} \to \mathcal{H},$$

with $\mathcal{D}(\mathcal{A}) = [\mathcal{D}(A_0) \bigcap \mathcal{D}(C)] \times [\mathcal{D}(A_1) \bigcap \mathcal{D}(B)]$, then (1.1) can be rewritten as follows:

$$(1.3) \qquad \dot{z}(t) = \mathcal{A}z(t), \qquad z(0) = z_0,$$

with $z(\cdot) = (x(\cdot), y(\cdot))$ and $z_0 = (x_0, y_0)$. We raise the following natural questions:

(i) When $\mathcal{A}$ generates a $C_0$ semigroup?

(ii) If $\mathcal{A}$ generates a $C_0$ semigroup $e^{\mathcal{A}t}$, when it has the spectrum determining growth property (SDGP, for short)? and when it is strongly stable, exponentially stable, differentiable, of Gevrey class, or analytic?

The above two questions correspond to the well-posedness of (1.1), the asymptotic behavior and some regularity of the solutions to (1.1), respectively. In most applications, under very mild conditions, $\mathcal{A}$ does generate a $C_0$ semigroup $e^{\mathcal{A}t}$ on $\mathcal{H}$. Then the second question becomes very interesting. As it is known, for example, the exponential stability is very crucial for further studying optimal regulator problem associated with (1.1) in infinite time horizon; or, if we regard (1.1) as the closed-loop system of some control system, the exponential stability might be one of the goals that one wants to achieve. On the other hand, differentiability, of Gevrey class, and analyticity of $e^{\mathcal{A}t}$ will give us some further regularity information about the solutions to (1.1), which will be very useful analysis and/or design of the corresponding physical systems described by (1.1).

We point out that in the case

$$\begin{aligned} &\|e^{\mathcal{A}t}\| \le M, \quad \forall t \ge 0, \\ &\rho(\mathcal{A}) \supset i\mathbb{R} \triangleq \{i\beta \mid \beta \in \mathbb{R}\}, \end{aligned} \tag{1.4}$$

which implies the strong stability of $e^{\mathcal{A}t}$, one has the following for $e^{\mathcal{A}t}$:

$$\begin{aligned} &\text{analytic} \Rightarrow \text{of Gevrey class } \delta > 1 \\ &\Rightarrow \text{differentiable} \Rightarrow \text{SDGP} \\ &\Rightarrow \text{exponentially stable.} \end{aligned} \tag{1.5}$$

Also, it is always true that the exponential stability implies the strong stability.

## 2. Qualitative Properties

We assume the following:

$$\begin{cases} e^{-A_0 t} \text{ and } e^{-A_1 t} \text{ are contraction} \\ \text{semigroups on } H_0 \text{ and } H_1, \text{ resp.,} \\ B : \mathcal{D}(B) \subseteq H_1 \to H_0 \text{ and} \\ C : \mathcal{D}(C) \subseteq H_0 \to H_1 \text{ are closed,} \\ \overline{\mathcal{D}(A_0) \bigcap \mathcal{D}(C)} = H_0, \\ \overline{\mathcal{D}(A_1) \bigcap \mathcal{D}(B)} = H_1, \\ \operatorname{Re}\{ \langle A_0 x, x \rangle + \langle A_1 y, y \rangle \\ \quad - \langle By, x \rangle - \langle Cx, y \rangle \} \ge 0, \\ \forall x \in \mathcal{D}(A_0) \bigcap \mathcal{D}(C), \\ y \in \mathcal{D}(A_1) \bigcap \mathcal{D}(B). \end{cases} \tag{2.1}$$

By Lumer-Phillips Theorem, we see that both $-A_0$ and $-A_1$ are *dissipative* and

$$\begin{cases} (\lambda + A_0)^{-1} \in \mathcal{L}(H_0), \\ (\lambda + A_1)^{-1} \in \mathcal{L}(H_1), \end{cases} \quad \forall \operatorname{Re} \lambda > 0. \tag{2.2}$$

Furthermore, by (2.1), we have that the operator $\mathcal{A}$ defined by (1.2) is densely defined on $\mathcal{H}$ and it is dissipative:

$$\begin{aligned} &\operatorname{Re} \langle \mathcal{A} \begin{pmatrix} x \\ y \end{pmatrix}, \begin{pmatrix} x \\ y \end{pmatrix} \rangle \le 0, \\ &\forall \begin{pmatrix} x \\ y \end{pmatrix} \in \mathcal{D}(\mathcal{A}). \end{aligned} \tag{2.3}$$

Thus, if $\mathcal{A}$ generates a $C_0$ semigroup $e^{\mathcal{A}t}$ on $\mathcal{H}$, it must be a contraction semigroup. Note that in the case $C = -B^*$, the first condition in (2.1) implies the last in (2.1).

Now, suppose $\mathcal{A}$ generates a contraction semigroup $e^{\mathcal{A}t}$ on $\mathcal{H}$. We would like to look at some qualitative properties of this semigroup. Needless to say, the properties of $e^{\mathcal{A}t}$ not only depend on $A_0$ and $A_1$, but also depend on the *coupling operators* $B$ and $C$. In what follows, we assume the following which is true in many applications:

$$\begin{aligned} &e^{-A_1 t} \text{ is analytic and} \\ &\quad \text{exponentially stable.} \end{aligned} \tag{2.4}$$

Thus, $e^{-A_1 t}$ is reasonably "good". One would like to know whether $e^{\mathcal{A}t}$ is also reasonably "good" if $e^{-A_0 t}$ happens to be not very "good". In this context, we will see that the coupling operators $B$ and $C$ play very interesting and subtle roles.

Note that under (2.4), the fractional power $A_1^b$ (as well as $(A_1^*)^b = (A_1^b)^*$) is well-defined for all $b \in \mathbb{R}$, in particular, $A_1^{-1}$ exists. Concerning the coupling operators, we suppose that for some $b, c \ge 0$, it holds

$$\mathcal{D}(A_1^b) \subseteq \mathcal{D}(B), \quad \mathcal{D}([A_1^c]^*) \subseteq \mathcal{D}(C^*). \tag{2.5}$$

We say that (1.1) is *weakly coupled* if $b, c \le 1$ and $b + c < 2$; and is *strongly coupled* if $b, c \ge 1$. If neither the above happens, we say that (1.1) is *mixedly coupled*, which will not be considered in this paper. Note, however, that sometimes, by changing the underline spaces properly, one can transform a mixedly coupled case to either a weakly or strongly coupled case. Roughly speaking, weakly coupled case corresponds to the case that the "orders" of coupling operators $B$ and $C$ are lower than that of $A_1$. The strongly coupled case is the other way around.

The following two theorems are concerned with the properties of $e^{\mathcal{A}t}$ for the weakly coupled case. We first state some sufficient conditions for $e^{\mathcal{A}t}$ having particular properties.

**Theorem 2.1.** *Let (2.1) and (2.4) hold. Let $0 \le b, c \le 1$ such that (2.5) holds, and $e^{\mathcal{A}t}$ be a contraction semigroup.*

(i) $e^{\mathcal{A}t}$ *is exponentially stable if* $i\mathbb{R} \subset \rho(\mathcal{A})$, $b + c < 1$, *and one of the following holds:*

(2.6) $$\overline{\lim_{|\beta|\to\infty}} \|(i\beta + A_0)^{-1}\| < \infty;$$

(2.7) $$\overline{\lim_{|\beta|\to\infty}} \|(i\beta + A_0 - BA_1^{-1}C)^{-1}\| < \infty.$$

(ii) $e^{\mathcal{A}t}$ *is differentiable if* $b+c \le 1$ *and one of the following holds:*

(2.8) $$\overline{\lim_{|\beta|\to\infty}} \log(1 + |\beta|)\|(i\beta + A_0)^{-1}\| = 0;$$

(2.9) $$\overline{\lim_{|\beta|\to\infty}} \{ \log(1 + |\beta|) \cdot \|(i\beta + A_0 - BA_1^{-1}C)^{-1}\| = 0.$$

(iii) *Let* $\mu \in (0,1)$. $e^{\mathcal{A}t}$ *is of Gevrey class* $\delta > 1/\mu$ *if one of the following holds:*

(2.10) $$b + c \le 1, \quad \overline{\lim_{|\beta|\to\infty}} |\beta|^{\mu}\|(i\beta + A_0)^{-1}\| < \infty;$$

(2.11) $$b + c < 1 + \mu, \quad \overline{\lim_{|\beta|\to\infty}} |\beta|^{\mu}\|(i\beta + A_0 - BA_1^{-1}C)^{-1}\| < \infty.$$

(iv) $e^{\mathcal{A}t}$ *is analytic if one of the following holds:*

(2.12) $$\begin{cases} b + c \le 1, \\ \overline{\lim_{|\beta|\to\infty}} |\beta|\|(i\beta + A_0)^{-1}\| < \infty, \\ \text{i.e., } e^{-A_0 t} \text{ is analytic;} \end{cases}$$

(2.13) $$\begin{cases} b + c < 2, \\ \overline{\lim_{|\beta|\to\infty}} |\beta|\|(i\beta + A_0 - BA_1^{-1}C)^{-1}\| < \infty, \\ \text{i.e., } e^{(-A_0+BA_1^{-1}C)t} \text{ is analytic.} \end{cases}$$

In the above, when assuming (2.6), we have assumed that for $|\beta|$ large enough, $i\beta \in \rho(-A_0)$. It is not necessary to assume that $i\mathbb{R} \subset \rho(-A_0)$ (thus, it allows $A_0 = 0$). The similar remark applies to (2.7)-(2.13), as well as the following (2.15) and (2.16). Next, we look at some necessary conditions for $e^{\mathcal{A}t}$ having some particular properties.

**Theorem 2.2.** *Let (2.1) and (2.4) hold. Let* $0 \le b, c \le 1$ *such that (2.5) holds, and* $e^{\mathcal{A}t}$ *and* $e^{(-A_0+BA_1^{-1}C)t}$ *be contraction semigroups.*

(i) *Let* $e^{\mathcal{A}t}$ *be exponentially stable and* $b + c < 1$. *Then*

(2.14)
$$i\mathbb{R} \subset \rho(-A_0) \Rightarrow e^{-A_0 t} \text{ is exponentially stable,}$$
$$i\mathbb{R} \subset \rho(-A_0 + BA_1^{-1}C) \Rightarrow e^{(-A_0+BA_1^{-1}C)t} \text{ is exponentially stable.}$$

(ii) *Let* $e^{\mathcal{A}t}$ *be differentiable and* $b + c \le 1$. *Then both* $e^{-A_0 t}$ *and* $e^{(-A_0+BA_1^{-1}C)t}$ *have the SDGP. Moreover, if*

(2.15) $$\lim_{|\beta|\to\infty} \log(1 + |\beta|)\|(i\beta - \mathcal{A})^{-1}\| = 0,$$

*then both* $e^{-A_0 t}$ *and* $e^{(-A_0+BA_1^{-1}C)t}$ *are differentiable.*

(iii) *Let* $\mu \in (0,1)$ *such that*

(2.16) $$\overline{\lim_{|\beta|\to\infty}} |\beta|^{\mu}\|(i\beta - \mathcal{A})^{-1}\| < \infty.$$

*Then, in the case* $b + c \le 1$ *(resp.* $b + c < 1 + \mu$*),* $e^{-A_0 t}$ *(resp.* $e^{(-A_0+BA_1^{-1}C)t}$*) is of Gevrey class* $\delta > 1/\mu$.

(iv) *Let* $e^{\mathcal{A}t}$ *be analytic. Then, in the case* $b+c < 2$ *(resp.* $b + c \le 1$*),* $e^{(-A_0+BA_1^{-1}C)t}$ *(resp.* $e^{-A_0 t}$*) is analytic.*

Let us make some comments on the above results. First of all, for the case $b + c \le 1$ (resp. $b + c < 1$ and $i\mathbb{R} \subset \rho(\mathcal{A}) \cap \rho(-A_0)$), $e^{\mathcal{A}t}$ and $e^{-A_0 t}$ are both analytic (resp. exponentially stable) or neither. In particular, if $e^{-A_0 t}$ is not analytic (resp. exponentially stable), neither is $e^{\mathcal{A}t}$. This is the case if, say, $A_0$ is unbounded and skew-symmetric. Thus, roughly speaking, if the coupling is too weak, the property of $e^{\mathcal{A}t}$ will only be as "good" as $e^{-A_0 t}$. On the other hand, if the coupling is reasonably strong (i.e., $b+c > 1$), then due to the nicer property of $e^{-A_1 t}$, the property of $e^{\mathcal{A}t}$ could be better than $e^{-A_0 t}$, provided $e^{-[A_0 - BA_1^{-1}C]t}$ is better. We note that the case $b + c = 1$ for the exponential stability of $e^{\mathcal{A}t}$ and the case $b + c = 2$ for the analyticity of $e^{\mathcal{A}t}$ are not included in the above. These cases turn out to be critical and will be included in the following results.

For strongly coupled case, we have corresponding results, under some additional (compatible) conditions. We omit the details here and we refer the readers to Liu and Yong (1999) for details.

REFERENCES

Liu, Z., and Yong, J.(1999), Qualitative properties of certain $C_0$ semigroups arising in elastic systems with various dampings, *Adv. Diff. Eqs.*, to appear.

# OPTIMIZATION OF BOUNDARY AND STARTING CONTROLS IN MULTI-DIMENSIONAL HYPERBOLIC SYSTEMS

**Alexander V. Arguchintsev**

*Irkutsk State University, Mathematical Department*
*1, K.Marx Street, Irkutsk, 664003. Russia.*
*E-mail: arguch@math.isu.ru*

Abstract: An optimization problem of initial-boundary conditions for first-order multi-dimensional semilinear hyperbolic systems is considered in this paper. The special feature of the general optimization problem for boundary conditions is non-validity of the classic optimality condition of Pontryagin's type. The idea of special 'inner' variations of admissible smooth controls is used for deriving a new optimality condition. The numerical algorithm based on the optimality condition is given. Some possible applications are discussed. *Copyright ©1999 IFAC*

Keywords: Optimal control, distributed-parameter systems, optimality, convergence of numerical methods.

## 1. INTRODUCTION

An optimization problem of initial-boundary conditions for first-order multi-dimensional semilinear hyperbolic systems is considered in this paper. The classic second-order hyperbolic equations and the systems with orthogonal characteristic families (Goursat-Darboux type) can be reduced to considered systems. The problems of such kind arise in modeling a number of phenomena of aero- and hydrodynamics, chemical technology, population dynamics, transfer processes (Chan and Guo, 1989; Markus and Mizel, 1980; Ostrovsky and Volin, 1975).

In (Arguchintsev and Vasiliev, 1996a, 1996b) necessary optimality conditions and successive approximations algorithms were derived for discontinuous and continuously differentiable distributed controls in the right-hand side of hyperbolic operators. Extension of these results to multi-dimensional hyperbolic systems is proved to be a non-trivial operation. In (Choo *et al.*, 1981) a differential (linearized) maximum principle was derived for boundary controls.

The special feature of the general optimization problem for boundary conditions is non-validity of the classic optimality condition of Pontryagin's type. For instance, a counter example was constructed in (Wolfersdorf, 1980) for systems with two orthogonal characteristic families and distributed control as a function of only one independent variable.

The idea of special 'inner' variations of admissible smooth controls was used in (Arguchintsev and Vasiliev, 1996b) for optimal control problems in hyperbolic systems with one spatial variable. For such systems it is possible to use a characteristics method for deriving optimality conditions. Inner variations can be applied for controls satisfying either restrictions of inclusion type or integral restrictions. Such unusual class of admissible controls appeared while researching inverse problems of mathematical physics. It turns out that the suggested approach is efficient for solving inverse problems of optimal control. The unknown parameters of controlled dynamic system can be considered as new smooth controls.

The object of this paper is to apply the most general approach for arbitrary first-order hyperbolic systems. A scheme of numerical method based on the optimality condition is given. Some possible applications to solving inverse problems are discussed.

## 2. PROBLEM STATEMENT

Let in some domain $P = S \times T$, $S = \{s \in E^m : a_i \leq s_i \leq b_i, i = 1, 2, \ldots, m,\}$, $T = [t_0, t_1]$ a controlled process be given by a system of semilinear hyperbolic equations:

$$\frac{\partial x}{\partial t} + \sum_{i=1}^{m} \mathbf{A}_i(s,t)\frac{\partial x}{\partial s_i} = f(x,s,t). \tag{1}$$

Here $x = x(s,t)$ is the $n$-component state function, $\mathbf{A}_i = \mathbf{A}_i(s,t)$ are symmetric matrices with continuously differentiable elements and eigenfunctions of a constant sign in $P$. The initial-boundary conditions are

$$\begin{aligned} &x(s,t_0) = p(u(s),s), \quad s \in S; \\ &\mathbf{B}^-(s,t)x = g(v,s,t), \quad (s,t) \in \Gamma, \end{aligned} \tag{2}$$

where $\Gamma$ - a lateral boundary of $P$, a special matrix-function $\mathbf{B}^-$ is constructed on the base of the matrix

$$\mathbf{B} = \tau \mathbf{I} + \sum_{i=1}^{m} \xi_i \mathbf{A}_i,$$

$\mathbf{I}$ - the unit matrix, $(\xi_1, \xi_2, \ldots, \xi_m, \tau)$ - the unit vector of the external normal to the boundary of $P$, functions $p(u(s),s), g(s,t)$ are supposed to be continuously differentiable with respect to their arguments.

The admissible controls are $r$-dimensional continuously differentiable vector-functions $u = u(s)$ and $k$-dimensional vector-functions $v = v(s,t)$ satisfying

$$\begin{aligned} &u(s) \in U \subset E^r, \quad s \in S, \\ &v(s,t) \in V \subset E^k, \quad (s,t) \in \Gamma. \end{aligned} \tag{3}$$

Admissible controls $u = u(s)$, $v = v(s,t)$ generate smooth solution $x = x(s,t)$ of the problem (1)-(2).

The objective of the control for the process described by (1)-(3) is

$$\mathcal{J}(u) = \int_S \varphi(x(s,t_1),s)ds + \iint_P F(x,s,t)dtds \rightarrow min. \tag{4}$$

## 3. AN OPTIMALITY CONDITION

Consider inner variations of controls. Let $u = u(s)$ be an admissible control. Give the variated control by the formula

$$u_\varepsilon(s) = u(s + \varepsilon\delta(s)) \tag{5}$$

for the admissible functions (3).

Here $\delta(s)$ is a continuously differentiable vector-function, $a_i \leq s_i + \delta_i(s) \leq b_i, s \in S, j = 1, 2, \ldots, n$. It is evident that control function (5) is also admissible.

Then we consider the objective functional increment formula for two admissible controls:

$$\Delta_\varepsilon \mathcal{J}(u) = -\int_S < \psi(s,t_0), \Delta p(u(s),s) > ds + \int_S o_\varphi(||\Delta x(s,t_1)||)\,ds - \int_P o_H(||\Delta x(s,t)||)\,dsdt. \tag{6}$$

Here $< \cdot, \cdot >$ is a designation of a scalar product in $E^n$, $o(\varepsilon)/\varepsilon \rightarrow 0$ as $\varepsilon \rightarrow 0$.

The function $\psi = \psi(s,t), \psi(s,t) \in E^n$ satisfies the conjugate problem

$$\frac{\partial \psi}{\partial t} + \sum_{i=1}^{m} \frac{\partial(\mathbf{A}_i \psi)}{\partial s_i} = -\frac{\partial H(\psi,x,s,t)}{\partial x},$$

$$\psi(s,t_1) = -\frac{\partial \varphi(x(s,t_1),s)}{\partial x}, \quad s \in S,$$

$$\mathbf{B}^+\psi(s,t) = 0, \quad (s,t) \in \Gamma, \quad \mathbf{B}^+ = \mathbf{B} - \mathbf{B}^-;$$

Gamiltonian H is

$$H(\psi,x,s,t) = < \psi(s,t), f(x,s,t) > -F(x,s,t).$$

Let

$$h(u(s),s) = < \psi(s,t_0), p(u(s),s) >,$$
$$\omega([u],s) = < h_u(u,s), \dot{u}(s) > .$$

Then it follows from (6):

$$\Delta_\varepsilon \mathcal{J}(u) = -\varepsilon \int_S \delta(s)\omega([u],s)ds + o(\varepsilon), \tag{7}$$

where the remainder term in (7) is estimated on the base ofnonclassical variant of energy inequality (Arguchintsev and Vasiliev, 1996a, 1996b).

By the increment formula (7) we have the following theorem.

**Theorem** *Consider the problem (1)-(4). For an optimal control $u^*$ the following necessary optimality condition holds:*

$$\mu(u^*) = \int_S \omega([u^*], s)ds = 0. \tag{8}$$

**Remark** *To simplify our presentation we formulated a new necessary optimality condition only for starting control ($t = t_0$). Analogous result can be formulated for boundary controls.*

This optimality condition allows to construct convergent iterative procedures.

Let an admissible control $u^0$ be given and $u^k$ be calculated by the method. Construct an admissible function $\delta(s)$ which has the coinciding sign with $\omega([u^k], s)$. The ways of constructing such functions were considered in (Arguchintsev and Vasiliev, 1996b). Then we determine a one-parametric collection of controls $u^k_\varepsilon$ (see (5)) and solve a problem

$$\varepsilon_k = argmin_{\varepsilon \in [0,1]} \mathcal{J}(u^k_\varepsilon).$$

The next approximation is given by the formula

$$u^{k+1} = u^k_{\varepsilon_k}, k = 0, 1, 2, \ldots.$$

The corresponding convergence theorem to the necessary optimality condition (8) is formulated by the same way as in (Arguchintsev and Vasiliev, 1996b). The given iterative process is relaxative and convergent in the sense $\mu(u^k) \to 0$ as $k \to \infty$ under additional standard conditions (Arguchintsev and Vasiliev, 1996b). Main of these conditions are boundedness from below of a functional and Lipschitz inequality for partial derivatives with respect to x from right-hand sides of a system and objective functions.

## 4. APPLICATIONS

The described results were applied to solving two types of inverse problems - tsunami wave problem and inverse optimal control problem.

Consider the problem of determination of a bottom shift or initial wave parameters which generates a wave of the given characteristics. This inverse problem of mathematical physics can be formulated in terms of optimal control theory. We study the system of shallow water equations (Voltsynger, 1985; Arguchintsev and Terletsky, 1990)

$$\eta_t + hv_y + hw_z = -\gamma_t - h_y v - h_z w,$$

$$v_t + g\eta_y = 0, \; w_t + g\eta_z = 0, \tag{9}$$

$$\eta(s, t_0) = u_1(s); v(s, t_0) = u_2(s); w(s, t_0) = u_3(s).$$

Here $t$ is a time variable, $t \in [t_0, t_1]$, $y$ and $z$ are the spatial variables, $\eta$ - wave profile, $v$ and $w$ are the velocities, $h$ - depth, $\gamma$ - bottom shift, $g$ - acceleration of free fall.

Substitution:

$$\xi = \eta\sqrt{g/h}$$

transforms system (9) to the form (1) with a symmetric matrix of coefficients. The problem is in determination of initial parameters $u_1, u_2, u_3$ , which generates the wave of the given characteristics $\eta(s, t_1), v(s, t_1), w(s, t_1)$. It is essential that the initial wave parameters are continuously differentiable functions.

This problem can be considered as an optimal control problem by the initial conditions of the hyperbolic system (9) with smooth admissible controls and a quadratic cost functional. The problem has been solved by the gradient procedures and the inner variations method. The inner variation method was especially efficient for additional integral restrictions for control functions.

It turns out that the inner variations methods can be successfully applied for solving inverse problems of optimal control. These problems are in determination of functional parameters for which the given controlled process is optimal. The method is used as an auxiliary procedure for solving inverse problems. The unknown functional parameters can be considered as new smooth controls. It is important to note that gradient procedures in the case of smooth admissible controls are not rigorous justified, because an auxiliary controls in these procedures can be nonsmooth functions. The first results were obtained for one first-order hyperbolic equation with starting control.

This work was supported in part by the Russian Foundation for Basic Research (Project 98-01-00400) and the Educational Ministry of Russian Federation (Grant Centre - St.Petersburg State University).

## REFERENCES


Arguchintsev, A.V. and O.V. Vasiliev (1996a). Iterative processes of the maximum principle and their modifications in distributed parameter systems. *Differential Equations.* 1996. **32**, pp. 797-803.

Arguchintsev, A.V. and O.V. Vasiliev (1996b). Optimization methods for discontinuous and smooth controls in semi-linear hyperbolic systems. In: *13-th Triennial World Congress of IFAC. Preprints.* **D**, 1996, pp. 339-343. Elsevier Science Ltd, San Francisco.

Arguchintsev, A.V. and V.A. Terletsky (1990). On solving the inverse tsunami problem withing the framework of two-dimensional model by optimal control methods. *Issledovaniya tsunami*, **4**, pp. 52-57.

Chan, W.L. and B.Z. Guo (1989). Optimal birth control of population dynamics. *J. Math. Anal. Appl.*, **144**, pp. 532-552.

Choo, K.G., K.L. Teo and Z.S. Wu (1981-1982). On an optimal control problem involving first order hyperbolic systems with boundary controls. *Numer. Funct. Anal. and Optim.*, **4**, pp. 171-190.

Markus, M. and M. Mizel (1980). Semilinear hereditary hyperbolic systems with nonlocal boundary conditions. *J. Math. Anal. Appl.*, **76**, pp.440-475.

Ostrovsky, G.M. and Y.M. Volin (1975). *Modelling complex chemical technology systems*, Chimiya, Moscow (in Russian).

Voltsynger, N.E. (1985). *Long waves on shallow water*, Gydrometeoizdat, Leningrad (in Russian).

Wolfersdorf, L. (1980). A counter example to the maximum principle of Pontryagin for a class of distributed parameter systems. *Z. Angew. Math. Mech.*. **6**, p. 204.

# OPTIMAL SYNTHESIS AND STABILIZATION OF TIME-DELAY DYNAMICAL SYSTEMS

**R. Gabasov**

*Department of Applied Mathematics*
*Belarus State University*
*Scorina av. 4, 220050 Minsk, Belarus*
*kirill@nsys.minsk.by*

**F.M. Kirillova, N.V.Balashevich**

*Institute of Mathematics*
*National Academy of Sciences*
*Surganov str. 11, 220072 Minsk, Belarus*
*kirill@nsys.minsk.by*

Abstract: A new approach to the solution of the classical optimal synthesis problem for linear time-delay dynamical systems is under consideration. Positional solutions of optimal control problems are constructed by solving special algebraic equations over the optimal Lagrange vector and switching points of the optimal open-loop control at each current moment of a control process. A stabilization problem for two types of dynamical systems with delay (delay in feedback loop, delays in control) is investigated. Algorithms of constructing bounded stabilizing feedbacks based on positional solutions of special auxiliary optimal control problems are justified. Two schemes of realization of stabilizing feedbacks are proposed. *Copyright ©1999 IFAC*

Keywords: linear system, delay, optimal synthesis, stabilization

## 1. INTRODUCTION

In the optimal control theory it is generally recognized that the synthesis of dynamical systems is an ultimate goal of the theory. The problem of optimal synthesis was formulated just in the beginning of 50's (Tsien, 1954; Feldbaum, 1963; Song, 1963) and was solved for linear two-dimensional time-optimal problems. Attempts of developing the first results undertaken during the next forty years didn't lead to significant success. The optimal synthesis problem was successfully solved only for the linear-quadratic Kalman-Lyotov problem which does not contain geometric constraints on control. Taking into account such constraints is a principal element of the transfer from the classical stage of the optimal control theory to the modern one. Geometric constraints on controls are the most commonly encountered nonlinearity in applied problems (Thoma, 1973). So every actual optimal control problem contains such constraints and it is precisely they don't allow to construct the optimal feedback in an explicit form.

In the late 80's the authors of the paper suggested a new approach to the solution of the classical optimal synthesis problem. From positions of this approach it was impossible to solve the optimal synthesis problem in the beginning of 50's because of the lack of necessary mathematical methods as well as because of inadequate development of computer technology. New approach is oriented on constructing a realization of optimal feedback during each particular control process in real time. Also it was shown that this approach can be successfully used for solving a variety of problems which initial statements does not contain extremal elements. Among these problems the stabilization problem of dynamical systems (Krasovskii, 1965; Yong, 1987) is one of the most important for both the theory and applications.

In the paper the optimal synthesis problem for dynamical systems with delays is considered. An algorithm of operating an optimal controller which constructs a realization of optimal feedback in real time is described. Then the method of constructing a positional solution to the special optimal control problem is used to generate bounded stabilizing feedbacks for dynamical systems with delays in feedback loop and with delays in control.

## 2. OPTIMAL CONTROLLER FOR TIME-DELAY SYSTEM

In the class of piecewise continuous functions $u(t)$, $t \in T = [0, t^*]$, consider a linear optimal control problem

$$J(u) = h_0' x(t^*) \longrightarrow \max,$$
$$\dot{x}(t) = Ax(t) + A_1 x(t - \theta_x) + bu(t) + $$
$$+ b_1 u(t - \theta_u), \quad t \in T,$$
$$x(t) = x_0(t), \quad t \in T_{0x} = [-\theta_x, 0], \qquad (1)$$
$$u(t) = u_0(t), \quad t \in T_{0u} = [-\theta_u, 0],$$
$$Hx(t^*) = g, \quad |u(t)| \le 1, \quad t \in T,$$

where $x \in R^n$, $u \in R$, $\theta_x > 0$, $\theta_u > 0$, $H \in R^{m \times n}$, $\operatorname{rank} H = m$, $x_0(\cdot) = (x_0(t), t \in T_{0x}) \in C[T_{0x}]$; $u_0(\cdot) = (u_0(t), t \in T_{0u})$ belongs to the class of piecewise continuous functions $KC[T_{0u}]$.

Imbed problem (1) into the family of problems

$$J_\tau(u) = h_0' x(t^*) \longrightarrow \max,$$
$$\dot{x}(t) = Ax(t) + A_1 x(t - \theta_x) + bu(t) + $$
$$+ b_1 u(t - \theta_u), \quad t \in T_\tau = [\tau, t^*],$$
$$x(t) = æ(t), \quad t \in T_{\tau x} = [\tau - \theta_x, \tau], \qquad (2)$$
$$u(t) = v(t), \quad t \in T_{\tau u} = [\tau - \theta_u, \tau],$$
$$Hx(t^*) = g, \quad |u(t)| \le 1, \quad t \in T_\tau,$$

depending on a moment $\tau$ and functions $æ_\tau(\cdot) \in C[T_{\tau x}]$, $v_\tau(\cdot) \in KC[T_{\tau u}]$.

Denote by $u^0(t|\tau, æ_\tau(\cdot), v_\tau(\cdot))$, $x^0(t|\tau, æ_\tau(\cdot), v_\tau(\cdot))$, $t \in T_\tau$, an optimal open-loop control and the corresponding optimal trajectory of (2).

A piecewise continuous functional

$$u^0(\tau, æ_\tau(\cdot), v_\tau(\cdot)) = u^0(0|\tau, æ_\tau(\cdot), v_\tau(\cdot)), \qquad (3)$$
$$æ_\tau(\cdot) \in C[T_{\tau x}], \quad v_\tau(\cdot) \in KC[T_{\tau u}], \quad \tau \in T,$$

is said to be an optimal feedback.

Usually the use of feedback controls is caused by unknown disturbances acting to the system in the control process. Suppose that the actual behaviour of system (2) closed by feedback (3) is described by the equation

$$\dot{x}(t) = Ax(t) + A_1 x(t - \theta_x) + bu^0(t, x_t(\cdot), u_t^0(\cdot)) +$$
$$+ b_1 u^0(t - \theta_u, x_{t-\theta_u}(\cdot), u^0_{t-\theta_u}(\cdot)) + w(t), \qquad (4)$$
$$t \in T, \quad x(t) = x_0(t), \quad t \in T_{0x},$$
$$u(t) = u_0(t), \quad t \in T_{0u},$$

where $w(t)$, $t \in T$, is an unknown piecewise continuous $n$-vector-function of disturbances, $u_t^0(\cdot) = (u^0(s), t - \theta_u \le s \le t)$ is a control fed to the input of system (4) on the interval $[t - \theta_u, t[$.

Denote by $x^*(t)$, $t \in T$, a trajectory of system (4) corresponding to a disturbance $w^*(t)$, $t \in T$, realized in some particular control process.

The function $u^*(t) = u^0(t, x_t^*(\cdot), u_t^*(\cdot))$, $t \in T$, $u^*(t) = u_0(t)$, $t \in T_{0u}$, is a realizing control in the process under consideration.

A device able to calculate the control $u^*(t)$, $t \in T$, in real time is called an optimal controller (Gabasov, *et al.*, 1992, 1995; Gao, and Li, 1997).

An algorithm of operating an optimal controller is based on the solution to special algebraic equations. In order to construct these equations let us rewrite (2) with the use of the Cauchy formula

$$\int_\tau^{t^*} c(t)u(t)dt \to \max, \qquad (5)$$

$$\int_\tau^{t^*} h(t)u(t)dt = g - z(\tau), \quad |u(t)| \le 1, \quad t \in T_\tau.$$

Here

$$c(t) = h_0'[F(t^*, t)b + F(t^*, t+\theta_u)b_1], \ t \in [\tau, t^* - \theta_u],$$
$$c(t) = h_0' F(t^*, t)b, \ t \in ]t^* - \theta_u, t^*],$$
$$h(t) = H[F(t^*, t)b + F(t^*, t+\theta_u)b_1], \ t \in [\tau, t^* - \theta_u],$$
$$h(t) = HF(t^*, t)b, \ t \in ]t^* - \theta_u, t^*],$$
$$z(\tau) = HF(t^*, \tau)æ(\tau) +$$
$$+ \int_{\tau - \theta_x}^{\tau} HF(t^*, t + \theta_x)A_1 æ(t)dt +$$
$$+ \int_{\tau - \theta_u}^{\tau} HF(t^*, t + \theta_u)b_1 v(t)dt,$$

$F(s, t)$, $s \in T$, $0 \le t \le s$, is a solution to equation

$$\partial F(s, t)/\partial t = -F(s, t)A - F(s, t + \theta_x)A_1,$$
$$F(s, s) = E, \quad F(s, t) \equiv 0, \ t > s.$$

According to the Pontryagin maximum principle the optimal open-loop control $u^0(t|\tau, z(\tau))$, $t \in T_\tau$, has the form

$$u^0(t|\tau, z(\tau)) = \operatorname{sign} \Delta_\tau^0(t), \ t \in T_\tau \qquad (6)$$

where

$$\Delta_\tau^0(t) = \psi^{0\prime}_\tau(t)b + \psi^{0\prime}_\tau(t + \theta_u)b_1, t \in [\tau, t^* - \theta_u],$$
$$\Delta_\tau^0(t) = \psi^{0\prime}_\tau(t)b, t \in ]t^* - \theta_u, t^*],$$

is a cocontrol, $\psi_\tau^0(t)$, $t \in T_\tau$, is a solution to the conjugate system

$$\dot{\psi}(t) = -A'\psi(t) - A_1'\psi(t + \theta_x), \ t \in T,$$
$$\psi(t^*) = h_0 - H'y(\tau), \ \psi(t) \equiv 0, \ t > t^*,$$

$y(\tau)$ is the optimal Lagrange vector. Thus the optimal control $u^0(t|\tau, z(\tau))$, $t \in T_\tau$, is completely defined by $y(\tau)$ and moments where $\Delta_\tau^0(t)$, $t \in T_\tau$, changes the sign. The set

$$t_1(\tau), \ldots, t_{p(\tau)}(\tau);\ y(\tau) \qquad (7)$$

where $\Delta_\tau^0(t_i(\tau)) = 0$, $i = \overline{1, p(\tau)}$, is called defining elements of the optimal control. From terminal constrains of (5) and condition (6) it is easy to show that elements (7) satisfy the equations

$$f(\tau; t_i(\tau), i = \overline{1, p(\tau)}; z(\tau)) = 0, \qquad (8)$$

$$q_j(t_i(\tau), i = \overline{1, p}; y(\tau)) = 0,\ j = \overline{1, p(\tau)};$$

where

$$f(\tau; t_i, i = \overline{1, p}; z) = \sum_{i=0}^{p} \int_{t_i}^{t_{i+1}} k_i h(t) dt + z - g,$$

$$q_j(t_i, i = \overline{1, p}; y) = c(t_j) - y' h(t_j),\ j = \overline{1, p};$$

$$k_i = \begin{cases} \operatorname{sign} \Delta_\tau^0(t), & t \in [t_i, t_{i+1}[\cup[\tau, t^* - \theta_u], \\ \operatorname{sign} \Delta_\tau^1(t), & t \in [t_i, t_{i+1}[\cup[t^* - \theta_u, t^*], \end{cases}$$

$t_0 = \tau$, $t_{p+1} = t^*$, $\delta_\tau^1(t) = \psi_\tau'(t) b$, $t \in T_\tau$.

Equations (8) are called the defining equations of the optimal controller. Under rather general conditions the Jacobi matrix of equations (8) is nonsingular on the optimal control with respect to variables (7). So the solution to the defining equations can be constructed by the Newton method if a good enough initial approximation is known. The algorithm of operating the optimal controller consists of the three procedures: 1) the starting procedure, 2) the procedure of constructing the defining elements on regular intervals, 3) the procedure of changing the structure (the set of the defining equations and variables).

The first procedure constructs the initial values $t_1(0), \ldots, t_{p(0)}(0)$; $y(0)$ of elements (7). For this purpose, the optimal open-loop control $u^0(t|0, x_0(\cdot), u_0(\cdot))$, $t \in T$, of (1) is constructed before the beginning of the control process, based on a priori information. In doing so there are no limitations on the time of calculation. The switching points and the Lagrange vector of the constructed solution set vector (7).

The second procedure is used on intervals of control where the number of defining equations (8) does not change. The solution $t_i(\tau)$, $i = \overline{1, p(\tau)}$, $y(\tau)$ is used as an initial approximation for constructing a solution $t_i(\tau + h)$, $i = \overline{1, p(\tau)}$, $y(\tau + h)$ to the defining equations at the moment $\tau + h$. The number $h > 0$ is chosen in such a way, that it would take no more than three iterations to obtain the solution by the Newton method with given accuracy. This allows to evaluate the amount of calculations. If the time the computer device takes to carry out this work does not exceed $h$ units one may say that the optimal controller using this device realizes the optimal feedback in real time. The appropriate estimates are easy to obtain from the properties of the Newton method.

The third procedure is intended for revealing moments for changing the structure of the defining equations and for calculating initial values for the defining elements on new intervals of constancy of the structure. Details are presented in (Al'sevich, *et al.*, 1992).

## 3. STABILIZATION OF DYNAMIC SYSTEMS WITH DELAYS IN FEEDBACK LOOP

Stabilization of dynamical systems is the fundamental problem of both the theory of stability and the theory of optimal control. In (Balashevich, *et al.*, 1994) an approach based on constructing optimal controllers for special control problems is proposed for stabilizing dynamical systems without delays. In the case of controlling remote objects delays arise in the exchange of states and control signals. For control systems with delays in feedback loop two types of feedback are considered below. The first type is an immediate generalization of the feedback for control systems without delays and connects with finite-dimensional imbedding. The second type is defined by imbedding the initial problem into the family of problems depending on a functional element.

Consider a linear control system

$$\dot{x} = Ax + bu,\ x(0) = x_0,\ (x \in R^n,\ \ u \in R). \quad (9)$$

Assume that system (9) is controlled from a remote center. A signal $x$ reaches the control center with a delay $s_1 \geq 0$ and a control signal $u$ reaches the system with a delay $s_2 \geq 0$. Following the basic idea of the approach (Balashevich, *et al.*, 1994), to construct the stabilizing control for (9) we use the problem of optimal damping the dynamical system by piecewise continuous controls on a finite interval $T = [0, \theta]$, $\theta > s = s_1 + s_2$,

$$\rho = \max_{t \in T} |u(t)| \longrightarrow \min,$$

$$\dot{x} = Ax + bu, \quad x(0) = x^*(\tau), \quad x(\theta) = 0 \quad (10)$$

where $x^*(\tau)$ is a state of (9) at a moment $\tau$.

A control $u^0(t|x^*(\tau))$, $t \in T$, is said to be an optimal open-loop damping control if it gets the minimal intensity $\rho$ and ensures the damping of system (9) in the time $\theta$: $x(\theta) = 0$.

Under the optimal control $u^0(t|x_0)$, $t \in T$, at the moment $\theta$ the trajectory of system (9) reaches the equilibrium $x = 0$ and will stay there when $u(t) \equiv 0$, $t > \theta$. But this way does not take into account the delays $s_1$, $s_2$ and corresponds to ideal conditions. In the majority of real situations the motion of the system is described by the equation

$$\dot{x} = Ax + bu + w(t), \quad x(0) = x_0, \qquad (11)$$

where $w(t)$, $t \geq 0$, is an unknown piecewise continuous function of disturbances. So the necessity of constructing positional solutions to (10) arises.

The general scheme of the positional control consists in the following. On the interval $[0, s[$ no signal sent by system (11) to the control center at the moment $t = 0$ can return back, so the optimal open-loop control $u^0(t|x_0)$, $t \in [0, s[$, is fed to (11). Realized under this control and a disturbance $w^*(t)$, $t \in [0, s[$, the states $x^*(t)$, $t \in [0, s[$, are measured and sent to the control center. From the moment $s_1$ the control center according to the received information $x^*(t)$, $t \geq 0$, produces the control $u^*(t + s)$, $t \geq 0$. These signals begin to reach system (11) from the moment $t = s$.

Thus the realization of the stabilizing control is reduced to constructing a feedback. The approach under consideration is oriented on the optimal feedback. Two algorithms of constructing the optimal feedback depending on the way of accounting the delays $s_1$, $s_2$ are possible.

*3.1 Optimal stabilizing feedback of the first type*

Optimal feedback control of the first type is a generalization of the classical concept for the case of delays in the feedback loop. To construct this feedback we imbed problem (10) into the family

$$\rho = \max_{t \in T} |u(t)| \longrightarrow \min,$$

$$\dot{x} = Ax + bu, \quad x(0) = z, \quad x(\theta) = 0 \qquad (12)$$

depending on an $n$-vector $z$.

Let $u^0(t|z)$, $t \in T$, be an optimal open-loop control of problem (10), $Z$ be the set of initial states $z \in R^n$ for which there exist solutions to problem (12). If system (9) is controllable then $Z = R^n$.

The function

$$u^0(z) = u^0(s|z), \quad z \in Z, \qquad (13)$$

is called the optimal stabilizing feedback of the first type. Feedback (13) takes into account the delay in the feedback loop.

In the particular control process it is sufficient to know only the values $u^*(t) = u^0(x^*(t - s))$, $t \geq s$, of optimal feedback (13) along the realizing trajectory $x^*(t)$, $t \geq 0$, but not the whole function $u^0(z)$, $z \in Z$.

A device able to generate the fuction $u^*(t)$, $t \geq s$, during each particular control process in real time is called a stabilizer.

Algorithm of operating a stabilizer is based on the solution of defining equations. Main items of the algorithm are similar to that described above.

*3.2 Optimal stabilizing feedback of the second type*

Let us introduce a piecewise continuous function $v(\cdot) = (v(t), \ t \in [0, s[)$ and imbed problem (10) into the family of problems

$$\rho = \max_{t \in [s,\theta]} |u(t)| \longrightarrow \min,$$

$$\dot{x} = Ax + bu, \quad x(0) = z, \quad x(\theta) = 0, \qquad (14)$$
$$u(t) = v(t), \ t \in [0, s[,$$

depending on an $n$-vector $z$ and a function $v(\cdot)$.

Denote by $u^0(t|z, v(\cdot))$, $t \in [s, \theta]$, the optimal open-loop control of problem (14), by $Z \times V$ a set of pairs $(z, v(\cdot))$ for which there exist solutions to problem (14). In the case of controllable system problem (14) has a solution for any pair $(z, v(\cdot))$.

The functional

$$u^0(z, v(\cdot)) = u^0(s|z, v(\cdot)), \ (z, v(\cdot)) \in Z \times V,$$

is said to be an optimal stabilizing feedback of the second type.

For calculating the value $u^*(\tau)$, $\tau \geq s$, of the realization of the optimal stabilizing feedback the stabilizer uses the information on the state $x^*(\tau - s)$ and on the produced control $u^*(t)$, $t \in [\tau - s, \tau[$. On the base of this data the vector

$$\bar{x}(\tau) = F(s)x^*(\tau - s) + \int_{\tau - s}^{\tau} F(\tau - t)bu^*(t)dt$$

is calculated and the system of defining equations

$$\sum_{i=0}^{p(\tau)} \int_{t_i(\tau)}^{t_{i+1}(\tau)} F(\theta - s - t)bdtk_i(\tau)\rho(\tau) +$$
$$+ F(\theta - s)\bar{x}(\tau) = 0;$$
$$-y'(\tau)F(\theta - s - t_i(\tau))b = 0, \ i = \overline{1, p(\tau)}; \qquad (15)$$
$$y'(\tau)F(\theta - s)\bar{x}(\tau) = \rho(\tau)$$

is solved. Here $t_0(\tau) = s$; $k_i(\tau) = \text{sign}\,\Delta^0_\tau(t_i(\tau) + 0)$, $i = \overline{0, p(\tau)}$; $\Delta^0_\tau(t) = -y'(\tau)F(\theta - s - t)b$, $t \in [s, \theta]$. Equations (15) are solved for the elements $\{t_i(\tau), \ i = \overline{1, p(\tau)}; \ y(\tau), \ \rho(\tau)\}$ defining the open-loop control $u^0(t|x^*(\tau - s), v^*_\tau(\cdot))$, $t \in [s, \theta]$. The scheme of solving the defining equations is analogous to that described above. Having solved system (15) we set

$$u^*(\tau) = k_0(\tau)\rho(\tau).$$

It can be shown that in spite of essensially more complicated type of feedback its realization has the same complexity as the realization of the stablizing feedback for systems without delay. The minimal intensity $\rho(z)$ in problem (14) can be considered (Balashevich, *et al.*, 1994) as the Lyapunov function for proving the asymptotic stability of closed system when $w(t) \equiv 0, t \geq 0$. Its total derivative along the trajectory of this system

$$d\rho/d\tau = -|\Delta_\tau^0(\theta - s)|\rho(\tau) \leq 0.$$

As $\Delta_\tau^0(\theta - s) \not\equiv 0$, so $\rho(x(\tau)) \to 0$ for $\tau \to \infty$. It follows from the definition of the function $\rho$ that also $x(\tau) \to 0, \tau \to \infty$.

## 4. STABILIZATION OF LINEAR SYSTEMS WITH DELAY IN CONTROL

Consider a dynamical system with delays in control

$$\dot{x}(t) = Ax(t) + \sum_{j=0}^{l} b_j u(t - h_j), \; x(0) = x_0, \quad (16)$$

$$u(t) = u_0(t), \; t \in [-h, 0[,$$

where $x(t) \in R^n$, $u(t) \in R$, $0 = h_0 < h_1 < \ldots < h_l = h$.

Suppose that system (16) is relatively controllable. As accessible controls piecewise continuous functions $u(t)$, $t \geq 0$, with quantization period $\nu$ are used: $u(t) = u^k$, $t \in [(k-1)\nu, k\nu[$, $k = 1, 2, \ldots$; $h_j = m_j\nu$, $m_j$ is a natural number, $j = \overline{1, l}$.

While $u(t) \equiv 0$, $t \geq 0$, system (16) is unstable and the problem consists in stabilizing the system by bounded controls.

Let $G$ be some neighborhood of the point $x = 0$, $D$ be the set of piecewise continuous functions $v(\cdot) = (v(t), t \in [-h, 0[)$; $V = \{v(\cdot)\colon \|v(t)\| \leq L, t \in [-h, 0[\} \subset D$ be some neighborhood of the function $v(t) \equiv 0$, $t \in [-h, 0[$.

A functional $u(x, v(\cdot))$, $x \in G$, $v(\cdot) \in V$, is said to be a bounded stabilizing feedback for system (16) in $\Phi = G \times V$ if 1) $u(0, 0) = 0$; 2) $|u(x, v(\cdot))| \leq L$, $x \in G$, $v(\cdot) \in V$; 3) the closed system

$$\dot{x}(t) = Ax(t) + \sum_{j=0}^{k-1} b_j u(x(t - h_j), v_{t-h_j}(\cdot)) +$$

$$+ \sum_{j=k}^{l} b_j u_0(t - h_j), \; t \in [h_k, h_{k+1}[, \; k = \overline{1, l-1};$$

$$\dot{x}(t) = Ax(t) + \sum_{j=0}^{l} b_j u(x(t - h_j), v_{t-h_j}(\cdot)), \quad (17)$$

$$t \geq h, \quad x(0) = x_0,$$

has a unique continuous solution $x(t)$, $t \geq 0$, for every $x_0 \in G$, $u_0(\cdot) \in V$, where $v_t(\cdot) = (v(s) = u^*(t + s), s \in [-h, 0[)$, $u^*(t)$ is a control applied to the system at a moment $t$; $u^*(t) = u_0(t)$, $t \in [-h, 0[$; 4) the trivial solution $x(t) \equiv 0$, $t \geq 0$, $u_0(t) = 0$, $t \in [-h, 0[$, of system (17) is asymptotically stable in $\Phi$.

The problem of stabilizing dynamical systems by bounded controls consists not only in constructing a functional $u(x, v(\cdot))$ for which system (17) is asymptotically stable in some domain $\Phi$ where $|u(x, v(\cdot))| \leq L$ but also it is such that for a given $L$ the domain $\Phi$ of attraction of the asymptotically stable solution $x(t) \equiv 0$, $t \geq 0$, $u_0(t) \equiv 0$, $t \in [-h, 0[$, to be the biggest, i.e. to be a little different from the set of controllability of system (17). If the problem is understood not so, it is solved simply because an arbitrary stabilizing feedback $u(x, v(\cdot))$ will be the bounded stabilizing feedback in any domain $\Phi = \{(x, v(\cdot)) : |u(x, v(\cdot))| \leq L\}$ while the trajectory of the closed system doesn't leave the set $G$.

Constructing a bounded stabilizing feedback in an explicit form is an extraordinary difficult problem. As in the previous consideration one can show that in a particular control process the whole function $u(x, v(\cdot))$, $x \in G$, $v(\cdot) \in V$, is not required, all one need to do is to know its values $u^*(t) = u(x^*(t), v_t^*(\cdot))$, $t \geq 0$, along the curve $(x^*(t), v_t^*(\cdot)), t \geq 0$. Here $x^*(t), t \geq 0$, is a realized trajectory of system (17) with an initial vector $x_0^*$ and an initial control $u_0^*(t)$, $t \in [-h, 0[$; $v_t^*(\cdot)$ is a set of values $u^*(t + s)$, $s \in [-h, 0[$, applied to system (17) on the interval $[t - h, t[$. Moreover it is enough to know the value $u^*(\tau) = u(x^*(\tau), v_\tau^*(\cdot))$ only at a current moment $\tau$.

A function $u^*(t) = u^*(k\nu)$, $t \in [k\nu, (k+1)\nu[$, $k = 0, 1, \ldots$, is said to be a discrete realization of a bounded stabilizing feedback and any device which for a chosen $\nu$ is able to calculate values $u^*(k\nu)$, $k = 0, 1, \ldots$, during a concrete process in real time is said to be a discrete stabilizer.

Thus the problem of stabilizing dynamical systems with delays in control is reduced to constructing an algorithm of operating a stabilizer. This algorithm is based on an algorithm of operating an optimal controller for an auxiliary problem of optimal control (Gabasov, *et al.*, 1995).

Choose a parameter of the method $\theta = N\nu$ and introduce an auxiliary (accompanying) optimal control problem:

$$L(z, v(\cdot)) = \min L, \; \dot{x}(t) = Ax(t) + \sum_{j=0}^{l} b_j u(t - h_j),$$

$$x(0) = z, x(\theta) = 0, \; u(t) = v(t), t \in [-h, 0[, \quad (18)$$

$$u(t) \leq L, t \in [0, \theta - h[, \; u(t) \equiv 0, t \in [\theta - h, \theta[.$$

A functional $u^0(t|z, v(\cdot))$, $t \in [0, \theta - h[$, is said to be an optimal open-loop control of problem (18) if the corresponding trajectory $x^0(t|z, v(\cdot))$, $t \in [0, \theta]$, satisfies the constraint $x^0(\theta|z, v(\cdot)) = 0$, and the control criterion of (18) reaches the minimal value.

The optimal feedback is defined by the equality

$$u^0(z, v(\cdot)) = u^0(0|z, v(\cdot)), z \in R^n, v(\cdot) \in D. \quad (19)$$

Introduce the set $\Phi_\theta = \{(x, v(\cdot)), x \in R^n, v(\cdot) \in D : |u^0(x, v(\cdot))| \leq L\}$.

Let $S_L$ be the set of controllability of system (16) by bounded controls, i.e. the set containing all pairs $(x_0, u_0(\cdot))$, for which it is possible to transfer system (16) to the position $(x = 0,\ u(t) = 0,\ t \in [t^* - h, t^*[)$ in a finite time $t^*$ by bounded piecewice continuous controls $|u(t)| \leq L$, $t \in [0, t^* - h[$. For any $\varepsilon > 0$ one can indicate such a number $\theta < \infty$ that the $\varepsilon$-neighborhood $[\Phi_\theta]_\varepsilon$ of the set $\Phi_\theta$ contains $S_L$.

The functional (19) is a bounded stabilizing feedback for system (16) in $\Phi_\theta$. Its properties 1) $u^0(0,0) = 0$; 2) $u^0(x, v(\cdot)) \in U_L$, $(x, v(\cdot)) \in \Phi_\theta$, are evident. One can show that equation (17) has the Filippov solution. For proving the asymptotic stability of closed system (17) $(u(x(t), v_t(\cdot)) = u^0(x(t), v_t(\cdot)))$, the Lyapunov function method is used. The function of minimal intensity of control $L(z, v(\cdot))$, $(z, v(\cdot)) \in \Phi_\theta$, is considered as the Lyapunov function.

The general scheme of operating the intensity stabilizer consists in the following.

Suppose that the stabilizer has operated at moments $0, \nu, \ldots, l\nu$. At the moment $\tau = l\nu$ to generate the control $u^*(\tau)$ it uses the value of the optimal open-loop control $u^0(0|z, v(\cdot))$ of problem (18) for $z = x^*(\tau)$, $v(t) = u^*(\tau + t)$, $t \in [-h, 0[$.

Denote $\xi_0 = 1/L$, $\xi_k = u^k/L$, $i = \overline{1, K}$, where $K = N - m_l$. The problem under consideration is equivalent to the linear programming problem with $K + 1$ variables and $n$ general constraints

$$\xi_0 \longrightarrow \max, \ (F(N\nu)x^*(\tau) + \quad (20)$$

$$+ \sum_{j=0}^{l} \int_0^{h_j} F(N\nu - t) b_j u^*(\tau - h_j + t) dt) \xi_0 +$$

$$+ \sum_{k=1}^{K} \int_{(k-1)\nu}^{k\nu} \sum_{j=0}^{l} F(N\nu - t - h_j) b_j dt \xi_k = 0,$$

$$\xi_0 \geq 0, \ |\xi_k| \leq 1, \ k = \overline{1, K}.$$

To generate the control $u^*(\tau + \nu)$ the stabilizer has to know the solution to problem (18) for the initial position $(z = x^*(\tau + \nu),\ v(t) = u^*(\tau + \nu + t),\ t \in [-h, 0[)$. This problem differs from problem (20) only by the column corresponding to $\xi_0$ and this difference is the smaller the smaller $\nu$. In this situation the natural and the most effective method of solving problem (20) is the dual method of linear programming (Gabasov, 1994) which for solving problem (18) at the moment $\tau + \nu$ takes as initial approximation the optimal support $K^0(\tau)$ (Gabasov, 1994) of problem (20). If the time it takes for calculating the control $u^*(\tau + \nu)$ does not exceed $\nu$ then one can say that the stabilizer constructs a realization of a bounded stabilizing feedback in real time.

## 5. ACKNOWLEDGEMENTS

The research has been supported by Grant F96-011 from Byelorussian Republican Foundation for Basic Research and by the State Programme for Basic Research of Belarus.

## REFERENCES

Al'sevich V.V., R. Gabasov and F.M. Kirillova (1992). An optimal regulator for linear systems with delay with finite-dimensional terminal constraints. *Russian Acad. Sci. Dokl. Math.*, **45**, No. 3, 503–508.

Balashevich N.V., R. Gabasov and F.M. Kirillova (1994). An optimal damper for dynamical systems. *Automation and Remote Control*, **55**, No. 5, 615–622.

Feldbaum, A.A. (1963). *Foundations of Theory of Optimal Automatic Systems.* GIFML, Moscow.

Gabasov, R. (1994). Adaptive method of solving linear programming problems. *Preprint Series of University of Karlsruhe*, Institute for Statistics and Mathematics, Germany, Karlsruhe.

Gabasov, R., F.M. Kirillova and O.I. Kostyukova (1992). Constructing of optimal controls of feedback type in a linear problem. *Soviet Math. Dokl.*, **44**, No. 2, 608–613.

Gabasov, R., F.M. Kirillova and S. V. Prischepova (1995). *Optimal Feedback Control.* Lecture Notes in Control and Information Sciences. (Thoma M. ed.) Springer. **207**.

Gao Xuedong and Li Zongyuan (1997). Synthesis of optimal feedback controls for piecewise linear system. *Control Theory Appl.*, **14**, No. 4, 473–480.

Krasovskii N.N. (1965). On the stabilization of dynamic systems by supplementary forces. *Differential Equations*, **1**, No. 1, 1–9.

Song Jian (1963). Method of system synthesis with general criterion of optimality. *Automatica Sinica*, **1**, No. 1, 86–97.

Thoma M. (1973). *Theorie linearer regelsysteme.* Vieweg, Braun-Sscweig.

Tsien, H.S. (1954). *Engineering Cybernetics.* Mc-Graw-Hill, N.Y.-Toronto-London.

Yong Jiongmin (1987). Stabilization of linear systems by time-delay feedback controls. *Quarterly of Applied Mathematics*, **45**, No. 2, 377–388.

# REALIZATION OF OPTIMAL OUTPUT FEEDBACKS

**R. Gabasov**

*Department of Applied Mathematics*
*Belarus State University*
*Scorina av. 4, 220050 Minsk, Belarus*
*kirill@nsys.minsk.by*

**F.M. Kirillova, N.V.Balashevich**

*Institute of Mathematics*
*National Academy of Sciences*
*Surganov str. 11, 220072 Minsk, Belarus*
*kirill@nsys.minsk.by*

Abstract: Optimal synthesis problem for dynamical systems under uncertainty is under consideration. A nonstochastic model of uncertainty is used around which guaranteed results are obtained. Under incomplete and inexact information, a problem of optimal observation is introduced. Algorithms of operating optimal estimators which are able to construct estimates of available information required for operating the optimal output controller in real time are described. *Copyright ©1999 IFAC*

Keywords: output feedback, estimator, controller, realization

## 1. INTRODUCTION

The classical statement of the optimal synthesis problem (Tsien, 1954; Feldbaum, 1963; Song, 1963) assume that at each current moment $\tau \in T$ of control process a complete and exact information on the current state $x^*(\tau)$ of the control system is available. It was possible for this assumption to be accepted in the 50s as the order of systems being studied was not very high. But today this assumption is too restrictive because available information on states of modern complex control systems is, as a rule, incomplete and inexact. Now one comes to a decision based on the readings of a measurer able to measure only some output signals of the system with limited accuracy.

That is why the output feedbacks defined not on system states but on its available output signals are urgent for the modern theory of control. The problem of constructing optimal output feedbacks has been much studied in the framework of the Kalman-Lyotov stochastic problem. Below, only nonstochastic optimal control problems are considered which are treated from the point of view of the theory of guaranteed optimization (Schweppe, 1973; Krasovskii, 1977; Chernousko, 1988; Kurzhanski, 1989). The results presented use and develop the previous ideas and principles published by the authors.

## 2. PROBLEM STATEMENT

In the class of piecewise continuous functions $u(t)$, $t \in T = [0, t^*]$, let us consider a linear optimal control problem

$$J(u) = h_0'x(t^*) \longrightarrow \max, \ \dot{x} = Ax + bu + d\mu,$$
$$x(0) = x_0 \in X_0 = \{x \in R^n : Dx = \delta,$$
$$d_* \leq x \leq d^*\}, \ \text{rank}D = l < n, \qquad (1)$$
$$x(t^*) \in X^* = \{x \in R^n : h_i'x \geq g_i, i = \overline{1,m}\},$$
$$|u(t)| \leq 1, \ t \in T,$$
$$y(t) = c'x(t) + \xi(t), \ \xi_* \leq \xi(t) \leq \xi^*.$$

The problem has the following sense. The initial state of the optimized system is not known exactly. A priori information on the initial state is exhausted by the inclusion $x_0 \in X_0$. By analogy with the theory of filtration we say that the set $X_0$ is a priori distribution of the initial state of the control system, omitting the adjective "probability". In the control process current states are not available, only the output signal $c'x(t)$, $t \in T$, of the system is measured and the measurer does it with the error $\xi(t)$, $t \in T$, so at each moment

of the control only a number $y(t)$ is at our disposal. As an error of measurement, any piecewise continuous function $\xi(t)$, $t \in T$, satisfying the inequalities $\xi_* \leq \xi(t) \leq \xi^*$, $t \in T$, may be realized. A disturbance $\mu(t)$, $t \in T$, is an unknown bounded piecewise continuous function. Depending on a degree of uncertainty in the function $\mu(t)$, $t \in T$, three problems can be stated:
1) optimal control problem without identification of disturbances;
2) optimal control problem with identification of parametric disturbances;
3) optimal control problem with partial identification of parametric disturbances.
At each moment $\tau \in T$ based on the written signal $y_\tau(\cdot) = (y(t),\ t \in T_\tau = [0, \tau])$, the optimal controller has to produce controls $u^0(\tau, y_\tau(\cdot))$, $|u^0(\tau, y_\tau(\cdot))| \leq 1$, $\tau \in T$, which at the moment $t = t^*$ steer the system to the terminal set $X^*$ with assurance and give the maximal guaranteed value for the control criterion $h_0'x(t^*)$.

Let us illustate our approach to solving optimal control problems under uncertainty by the simplest problem in which an uncertaity is only in an initial state i.e. instead of system (1) the system

$$\dot{x} = Ax + bu \tag{2}$$

is considered.

For more precise definitions let us consider a particular control process at any current moment $\tau \in T$. Denote by $y_\tau^*(\cdot) = (y^*(t),\ t \in T_\tau)$ a signal of the measurer written by that time, by $u_\tau^*(\cdot)$ a control used. As the information, available at the moment $\tau$, contains, in general case, additional knowledge of the initial state $x_0^*$ realized in the process, so let us use this knowledge to decrease a priori uncertainty. A vector $x^* \in X_0$ is called an initial state compatible with the functions $u_\tau^*(\cdot)$, $y_\tau^*(\cdot)$ if the corresponding trajectory of (2) together with some possible function of errors $\xi^*(t)$, $t \in T_\tau$, generates the signal $c'x^*(t) + \xi^*(t)$, $t \in T_\tau$, coinciding with the written signal $y^*(t)$, $t \in T_\tau$: $y^*(t) \equiv c'x^*(t) + \xi^*(t)$, $t \in T_\tau$.

Denote by $\hat{X}_0(\tau)$ the set of all vectors $x^* \in X_0$ compatible with $u_\tau^*(\cdot)$, $y_\tau^*(\cdot)$ and call it $\tau$-a-posteriori distribution of the initial state. The set

$$\hat{X}_\tau(\tau) = \{x : x = F(\tau)x_0 + \int_0^\tau F(\tau - s)bu^*(t)dt,$$
$$x_0 \in \hat{X}_0(\tau)\}$$

where $\dot{F} = AF$, $F(0) = E$, is said to be $\tau$-a-posteriori distribution of $x(\tau)$.

A piecewise continuous function $u_\tau(\cdot) = (u(t), t \in T(\tau) = [\tau, t^*))$, $|u(t)| \leq 1$, $t \in T(\tau)$, is said to be an admissible (guaranteeing) control if at the moment $t = t^*$ it steers all states from the set $\hat{X}_\tau(\tau)$ to the terminal set $X^*$. Let

$$X(t^*|u_\tau(\cdot)) = \{x : x = F(t^* - \tau)x_\tau +$$
$$+ \int_\tau^{t^*} F(t^* - s)bu(s)ds,\ x_\tau \in \hat{X}_\tau(\tau)\}.$$

Then the admissibility of the control $u_\tau(\cdot)$ can be written briefly: $X(t^*|u_\tau(\cdot)) \subset X^*$. As a control criterion, the functional $J(u_\tau(\cdot)) = \min h_0'x$, $x \in X(t^*|u_\tau(\cdot))$, is considered.

An admissible control $u_\tau^0(\cdot) = \big(u^0(t|\tau, \hat{X}_\tau(\tau)), t \in T(\tau)\big)$, is called an optimal (guaranteeing) open-loop control if it satisfies the equality

$$J(u_\tau^0(\cdot)) = \max J(u_\tau(\cdot)).$$

The optimal (guaranteeing) output feedback is defined by the equality

$$u^0(\tau, y_\tau(\cdot)) = u^0(\tau|\tau, \hat{X}_\tau(\tau)), \quad y_\tau(\cdot) \in Y_\tau,\ \tau \in T,$$

where $Y_\tau$ is the set of all signals of measurer for which there exists $u^0(t|\tau, \hat{X}_\tau(\tau))$, $t \in T(\tau)$.

The function $u^*(\tau) = u^0(\tau, y_\tau^*(\cdot))$, $\tau \in T$, is said to be a realization of the optimal output feedback along the realized output signal $y_\tau^*(\cdot)$. Any device able to calculate values $u^*(\tau)$, $\tau \in T$, during each particular control process in real time is called an optimal controller.

Following (Gabasov, *et al.*, 1993) it can be shown that for calculating $u^*(\tau)$ it is sufficient to solve the determined optimal control problem

$$h_0'x(t^*) \longrightarrow \max, \quad \dot{x} = Ax + bu, \quad x(\tau) = 0, \tag{3}$$
$$h_i'x(t^*) \geq g_i(\tau), i = \overline{1,m},\ |u(t)| \leq 1, t \in T(\tau),$$

in real time where $g_i(\tau) = g_i - \alpha_i(\tau)$, $i = \overline{1,m}$, are obtained by solving $m$ linear problems of optimal observation

$$\alpha_i(\tau) = \min_{x \in \hat{X}_\tau(\tau)} h_i'F(t^* - \tau)x, \quad i = \overline{1,m}, \tag{4}$$

in real time.

## 3. OPTIMAL OBSERVATION PROBLEM. OPTIMAL CONTINUOUS ESTIMATOR

In "pure form" optimal observation problem used for solving optimal control problems under uncertainty consists in the following.

On the interval $T = [0, t^*]$ the behavior of the dynamical system is described by the equation

$$\dot{x} = Ax \quad (x \in R^n). \tag{5}$$

The initial state of system (5) is not known exactly. It is only known that the vector $x_0$ belongs to the set $X_0$ (1). To refine the actual initial state $x_0^*$ realized in a particular process the behavior of (5) is observed and readings $y(t)$, $t \in T$, are obtained.

The measurer at each moment $t \in T$ can measure the given linear combination $c'x(t) = \sum_{j=1}^{n} c_j x_j(t)$ of the components $x_1(t), \dots, x_n(t)$ of the current state $x(t)$ of system (5) with an error $\xi(t)$. Suppose that as the function $\xi(t)$, $t \in T$, an arbitrary piecewise continuous function satisfying the inequalities $\xi_* \le \xi(t) \le \xi^*$, $t \in T$, may be realized. The information inherent in the signal $y(t)$, $t \in T$, allows to decrease a priori uncertainty of the realized vector $x_0^*$.

A set $\hat{X}_0$ is called a posteriori distribution of the initial state of (5) if it consists of those and only those elements of $X_0$ which together with admissible functions $\xi(t)$, $t \in T$, can produce the written signal $y(t)$, $t \in T$.

The analytical description of the set $\hat{X}_0$ has the form: a vector $x \in \hat{X}_0$ if and only if it satisfies

$$\xi_* \le y(t) - c'F(t)x \le \xi^*, \ t \in T; \qquad (6)$$

$$Dx = \delta, \ d_* \le x \le d^*.$$

The description of $\hat{X}_0$ contains a continuum of inequalities, thus, its structure may be very complex for effective tabulation. Luckily, in many applied problems the set $\hat{X}_0$ is not used entirely, all one has to do is to know some of its estimates (numerical characteristics). Thus, in order to solve (1), the estimates of the form

$$p'x \longrightarrow \max, \quad x \in \hat{X}_0, \qquad (7)$$

are necessary, i.e. the extension of the set $\hat{X}_0$ in the direction of the vector $p \in R^n$.

Problem (7) is called a linear optimal observation problem (Gabasov, and Kirillova, 1990). In view of (6) this problem is a linear semi-infinite extremal problem. Similar problems occur in various applications and have a sufficiently developed theory (Kortanek, and Hettich, 1993).

To define positional solutions to problem (7) let us imbed it into the family of problems

$$p'x \longrightarrow \max, \quad x \in \hat{X}_0(\tau), \qquad (8)$$

where $\hat{X}_0(\tau)$ is a posteriori distribution of an initial state of (5) corresponding to $y_\tau(\cdot)$.

A vector $x_\tau^0 = x^0(\tau, y_\tau(\cdot)) \in \hat{X}_0(\tau)$, which gives the criterion of (8) the maximum value, is called an optimal open-loop solution to problem (8).

Let $Y_\tau$ be a set of signals $y_\tau(\cdot)$ which can be realized on $[0, \tau]$ for any initial states $x_0 \in X_0$ and any measuring errors $\xi_* \le \xi(t) \le \xi^*$, $t \in T_\tau$. The functional $x^0(\tau, y_\tau(\cdot))$, $y_\tau(\cdot) \in Y_\tau$, $\tau \in T$, is said to be a positional solution to (7).

As in the case of an optimal synthesis problem, consider a particular observation process in which an output signal $y^*(t)$, $t \in T$ is realized. It is clear that in this process values of the functional $x^0(\tau, y_\tau(\cdot))$, $y_\tau(\cdot) \in Y(\tau)$, $\tau \in T$, are necessary only along the curve $y^*(t)$, $t \in T$.

The function $x^*(\tau) = x^0(\tau, y_\tau^*(\cdot))$, $y_\tau^*(\cdot) \in Y_\tau$, $\tau \in T$, is said to be a realization of the positional solution to the optimal observation problem and any device able to calculate its values in real time is called an optimal continuous estimator.

Describe an algorithm of operating the optimal estimator.

Suppose that the estimator has been constructed and operated on the interval $T_\tau$. Using the written signal $y^*(t)$, $t \in T_\tau$, at the moment $\tau$ the estimator has to solve problem (8). In detailed notation problem (8) has the form

$$p'x \longrightarrow \max, \ \alpha_*(t) \le a'(t)x \le \alpha^*(t), \ t \in T_\tau,$$

$$Dx = \delta, \quad d_* \le x \le d^*, \qquad (9)$$

where $\alpha_*(t) = \xi_* - y^*(t)$, $\alpha^*(t) = \xi^* - y^*(t)$, $a'(t) = (a_j(t), j \in J) = -c'F(t)$, $J = \{1, 2, \dots, n\}$.

Let $\hat{x}_\tau$ be an optimal feasible point of problem (9). Introduce the sets where constraints of problem (9) are active:

$$T_a(\tau) = T_a^+(\tau) \cup T_a^-(\tau),$$

$$T_a^+(\tau) = \{t \in T_\tau : a'(t)\hat{x}_\tau = \alpha^*(t)\},$$

$$T_a^-(\tau) = \{t \in T_\tau : a'(t)\hat{x}_\tau = \alpha_*(t)\},$$

$$J^+(\tau) = \{j \in J : \hat{x}_{\tau j} = d_j^*\},$$

$$J^-(\tau) = \{j \in J : \hat{x}_{\tau j} = d_{*j}\}.$$

The optimal feasible point $\hat{x}_\tau$ satisfies

$$\Delta_j(\tau) \ge 0 \text{ if } j \in J^-(\tau); \ \Delta_j(\tau) \le 0 \text{ if } j \in J^+(\tau);$$

$$\Delta_j(\tau) = 0 \text{ if } j \in J^-(\tau),$$

where $\Delta_j(\tau) = \sum_{t \in T_a(\tau)} a_j(t)\eta(t) + \eta' D_j + p_j$, $\eta_\tau = (\eta(t), t \in T_a(\tau); \eta)$ is an optimal dual feasible point of problem (9):

$$\eta(t) \ge 0, t \in T_a^+(\tau); \ \eta(t) \le 0, t \in T_a^-(\tau); \ \eta \in R^l.$$

A pair $\{\hat{x}_\tau, \eta_\tau\}$ is called a positional solution to problem (9).

Introduce a set $J_0(\tau) = \{j \in J : \Delta_j(\tau) = 0\}$. A solution to problem (9) is called nonsingular if 1) $\text{rank}\Big(D_j'; \ a_j(t), t \in T_a(\tau); \ \dot{a}_j(t), t \in T_a(\tau) \setminus \{0, \tau\}; \ j \in J_0(\tau)\Big) = |J_0(\tau)|$; 2) $\text{rank}\Big(D_j'; \ a_j(t), t \in T_a(\tau); \ j \in J_0(\tau)\Big) = |T_a(\tau)| + l$; 3) $d_*j < \hat{x}_{\tau j} < d_j^*$, $j \in J_0(\tau)$; 4) $\eta(t) \ne 0$, $t \in T_a(\tau)$.

Here $J_0(\tau) = \{j \in J : \Delta_j(\tau) = 0\}$. Suppose that $\{0\} \notin T_a(\tau)$, $\tau > 0$.

Let $\{\hat{x}_{\tau_*}, \eta_{\tau_*}\}$ be a nonsingular solution to (9) at $\tau = \tau_*$. Describe rules of constructing solution to problem (9) for $\tau \in ]\tau_*, t^*]$. Two situations are possible: 1) $\tau_* \in T_a(\tau_*)$; 2) $\tau_* \notin T_a(\tau_*)$.

Consider the case 1). Assume $a'(\tau_*)\hat{x}_{\tau_*} = \alpha_*(\tau_*)$, $\dot{a}'(\tau_*)\hat{x}_{\tau_*} \neq \dot{\alpha}_*(\tau_*)$.

As is shown in (Gabasov, *et al.*, 1993), the set

$$æ(\tau) = \{\hat{x}_\tau, \eta_\tau, T_a(\tau)\}, \tag{10}$$

consisting of the optimal primal and dual feasible points of problem (9) and the active moments, is sought from the equations

$$\hat{x}_{\tau j} = d_j^*, j \in J^-;\ \hat{x}_{\tau j} = d_{*j}, j \in J^+;\ D\hat{x}_\tau = \delta,$$

$$a'(t_i(\tau))\hat{x}_\tau = \alpha_i(t_i(\tau)),\ i = \overline{1,q}; \tag{11}$$

$$\dot{a}'(t_i(\tau))\hat{x}_\tau = \dot{\alpha}_i(t_i(\tau)),\ i = \overline{1,q-1};\ t_q(\tau) = \tau,$$

$$\Delta_j(t_i(\tau)) = 0,\ j \in J_0.$$

Here $q = |T_a(\tau)|$, $J_0 = J_0(\tau_*)$, $I_0 = \{1, \ldots, q\}$, $I_0^+ = \{i \in I_0 :\ t_i(\tau) \in T_a^+(\tau)\}$, $I_0^- = I_0 \backslash I_0^+$, $J^+ = (J \backslash J_0) \cap J^+(\tau)$, $J^- = (J \backslash J_0) \cap J^-(\tau)$, $\alpha_i(t) = \alpha^*(t)$ for $i \in I_0^+$; $\alpha_i(t) = \alpha_*(t)$ for $i \in I_0^-$. Equations (11) are said to be the defining equations of the optimal estimator. The set

$$\{q,\ J_0,\ J^+,\ J^-,\ I_0^+,\ I_0^-\}$$

is called a structure of the defining equations.

The algorithm of operating the optimal estimator consists of three procedures:
1) the starting procedure;
2) the procedure of constructing defining elements (10) on intervals with constant structure;
3) the procedure of changing the structure.
Details of these procedures are analogous to that described in (Gabasov, *et al.*, 1993).

Consider the case 2). For $\tau \in T^+(\tau_*)$ elements (10) has the form

$$\hat{x}_\tau = \hat{x}_{\tau_*},\ \eta_\tau = \eta_{\tau_*},\ T_a(\tau) = T_a(\tau_*). \tag{12}$$

A solution to problem (9) is constructed according to rule (12) till a moment $\bar{\tau} \in [\tau_*, t^0]$ where $a'(\bar{\tau})\hat{x}_{\bar{\tau}} = \alpha_*(\bar{\tau})$, $\dot{a}'(\bar{\tau})\hat{x}_{\bar{\tau}} \neq \dot{\alpha}_*(\bar{\tau})$ or $a'(\bar{\tau})\hat{x}_{\bar{\tau}} = \alpha^*(\bar{\tau})$, $\dot{a}'(\bar{\tau})\hat{x}_{\bar{\tau}} \neq \dot{\alpha}^*(\bar{\tau})$.

For $\tau \in T^+(\bar{\tau})$ a solution to problem (9) is constructed according to the case 1.

## 4. OPTIMAL DISCRETE ESTIMATOR

To introduce a discrete estimator let us change some conditions of observation. The object of observation remains as before (see (5)). Formally equation (1) of the measurer also remains. But this time, we suppose that on the interval $[0, t^*]$ measurements are written at discrete moments $t_k = k\nu$, $k = 0, 1, \ldots, N$, $\nu = t^*/N$. Errors of measurements are numbers $\xi_k$, $k = 0, 1, \ldots, N$, satisfying the inequalities $\xi_* \leq \xi_k \leq \xi^*$, $k = 0, 1, \ldots, N$. Formally the notion of a posteriori distribution $\hat{X}_0$ of an initial state remains but its analytical form is radically changed: a vector $x_0 \in \hat{X}_0$ if and only if it satisfies the finite set of inequalities

$$\xi_* \leq y(k\nu) - c'F(k\nu)x_0 \leq \xi^*, \quad k = \overline{0, N},$$

$$Dx_0 = \delta, \quad d_* \leq x_0 \leq d^*.$$

Of course now, the structure of the set $\hat{X}_0$ is also essentially more complex than of the set $X_0$. But this structure allows one to solve optimal observation problems considerably easier and without any demand on analytical properties of available signals $y(t)$, $t \in T$. This is very important for solving applied problems.

Now, the optimal observation problem takes the form

$$p'x \longrightarrow \max,$$

$$\xi_* \leq y(k\nu) - c'F(k\nu)x \leq \xi^*,\ k = \overline{0, N}, \tag{13}$$

$$Dx = \delta,\ d_* \leq x \leq d^*,$$

i.e. it is a linear programming problem. The optimal feasible point $x^0$ of (13) presents an open-loop solution to the discrete optimal observation problem.

To introduce a positional solution to the optimal observation problem let us imbed problem (13) into a family of problems (8) where as $\tau$ discrete moments $k\nu$, $k = \overline{0, N}$, are considered, the sets $\hat{X}_0(\tau)$ are appropriately changed $\hat{X}_0(\tau) = \{x :\ \xi_* \leq y(i\nu) - c'F(i\nu)x \leq \xi^*,\ i = \overline{0, k},\ Dx = \delta,\ d_* \leq x \leq d^*\}$, $y_\tau(\cdot) = \{y(i\nu),\ i = \overline{0, k}\}$.

Let $x_\tau^0 = x^0(\tau, y_\tau(\cdot)) \in \hat{X}_0(\tau)$ be an optimal feasible point of (8) at $\tau = k\nu$. The functional

$$x^0(\tau, y_\tau(\cdot)), \tag{14}$$

$$y_\tau(\cdot) \in Y_\tau,\ \tau \in T_\nu = [0, \nu, \ldots, N\nu],$$

is said to be a positional solution to the discrete optimal observation problem. In spite of simplification of the optimal observation problem in discrete case, it is very difficult to construct functional (14). So let us follow the approach described above for the continuous optimal observation problem. Suppose that functional (14) is constructed. Use it in a particular observation process when the signal $y^*(k\nu)$, $k = \overline{0, N}$, is written. This signal corresponds to the estimates $\alpha^*(k\nu) = p'x^0(k\nu, y_{k\nu}^*(\cdot))$ of a posteriori distribution $\hat{X}_0(k\nu)$. It is clear that in order to obtain these estimates not the whole functional (14) is used, but it is sufficient to know its values along the sequence $y^*(k\nu)$, $k = \overline{0,N}$. The function $x^*(k\nu) = x^0(k\nu, y^*_{k\nu}(\cdot))$, $y^*_{k\nu}(\cdot) \in Y(k\nu)$, $k = \overline{0,N}$, is said to be a discrete realization of a positional solution to the optimal observation problem and a device able to calculate its values in real time is called an optimal discrete estimator.

To describe an algorithm of operating a discrete estimator let us study a situation at an arbitrary moment $\tau = k\nu$. According to the definition, in order to calculate $\alpha^*(\tau)$, the optimal estimator needs to know a solution to the linear programming problem

$$p'x \longrightarrow \max, \tag{15}$$

$$\xi_* - y^*(i\nu) \leq -c'F(i\nu)x \leq \xi^* - y^*(i\nu),\ i = \overline{0,k},$$
$$Dx = \delta, \quad d_* \leq x \leq d^*.$$

At the previous moment $(k-1)\nu$ the estimator dealt with the problem

$$p'x \longrightarrow \max, \tag{16}$$

$$\xi_* - y^*(i\nu) \leq -c'F(i\nu)x \leq \xi^* - y^*(i\nu), i = \overline{0,k-1},$$
$$Dx = \delta, \quad d_* \leq x \leq d^*.$$

As problems (15) and (16) differ only by one constraint, so knowing the solution to (16), a very few iterations would be sufficient to construct a solution $x^0(k\nu, y^*_{k\nu}(\cdot))$ to (15) with the help of the dual method of linear programming (Gabasov, 1994). If the time that takes to do this job does not exceed $\nu$ then one may say that the positional solution to optimal observation problem is realized in real time.

## 5. REALIZATION OF OPTIMAL OUTPUT FEEDBACK

As it was noted in §2, calculation of the realization of the optimal output feedback $u^*(\tau)$, $\tau \in T$, is performed by solving problems (3), (4) in real time.

According to the Pontryagin maximum principle the optimal open-loop control $u^0(t|\tau)$, $t \in T(\tau)$, of problem (3) has the form

$$u^0(t|\tau) = \operatorname{sign} \Delta^0_\tau(t),\ t \in T(\tau)$$

where $\Delta^0_\tau(t) = \psi'_\tau(t)b$, $t \in T(\tau)$, is a cocontrol, $\psi_\tau(t)$, $t \in T(\tau)$, is a solution to the adjoint system

$$\dot{\psi} = -A'\psi, \quad \psi(t^*) = h_0 - \sum_{i \in I(\tau)} h_i\nu_i(\tau),$$

$\nu(\tau) = (\nu_i(\tau),\ \ i \in I(\tau))$ is the Lagrange optimal vector, $I(\tau) = \{i \in I : h'_i x^0_\tau(t^*) = g_i(\tau)\}$, $I = \{1, 2, \ldots m\}$, $x^0_\tau(t)$, $t \in T(\tau)$, is the optimal trajectory of problem (3).

Thus, the defining elements of problem (3) are the switching points of the optimal control and the Lagrange vector

$$æ(\tau) = \big(t_1(\tau), \ldots, t_p(\tau);\ \ \nu(\tau)\big)\big). \tag{17}$$

Elements (17) satisfy the equations

$$h'_i x(t^*) = g_i(\tau), \quad i \in I(\tau),$$
$$\Delta^0_\tau(t_k|\tau) = 0, \quad k = \overline{1,p(\tau)}.$$

The algorithm of operating the optimal controller is analogous to one described in (Gabasov, *et al.*, 1992).

The values $g_i(\tau) = g_i - \alpha_i(\tau)$, $i = \overline{1,m}$, used in (3), are calculated by solving problems (4).

Consider one ($i$-th) of problems (4). With the help of the Cauchy formula it can be written as

$$\alpha_i(\tau) = h'_i F(t^*)x + \int_0^\tau h'_i F(t^* - t)bu^*(t)dt \to \min_x,$$

$$\alpha_*(t) \leq a'(t)x \leq \alpha^*(t),\ t \in T_\tau, \tag{18}$$
$$Dx = \delta,\ d_* \leq x \leq d^*,$$

where $\alpha_*(t) = \xi_* - y^*(t) + \int_0^t h'_i F(t-s)bu^*(s)ds$, $\alpha^*(t) = \xi^* - y^*(t) + \int_0^t h'_i F(t-s)bu^*(s)ds$, $t \in T_\tau$, $a'(t) = -h'_i F(t)$.

According to the scheme described in §3 the optimal estimator at a current moment $\tau \in T$ constructs a solution $x^0(\tau, y^*_\tau(\cdot))$ to problem (18) and calculates an estimate $\alpha_i(\tau) = h'_i F(t^*)x^0(\tau, y^*_\tau(\cdot)) + \int_0^\tau h'_i F(t^* - t)bu^*(t)dt$.

All $m$ problems (18) are solved in parallel based on the known values of the output signal $y^*_\tau(\cdot)$ and the realization of the optimal feedback $u^*(t)$, $t \in T_\tau$. Constructed values $\alpha_i(\tau)$, $i = \overline{1,m}$, are used by the optimal controller for calculating the realization $u^*(\tau)$.

In the case of constructing a discrete realization of the optimal output feedback $u^*(k\nu) = u^0(k\nu, y^*_{k\nu}(\cdot))$, $k = \overline{0, N-1}$, the optimal controller at each current moment $\tau = k\nu$ uses a solution to the linear programming problem

$$h'_0 \sum_{j=1}^{N-k} \int_{(k+j-1)\nu}^{(k+j)\nu} F(t^* - t)bdtu_j \longrightarrow \max,$$

$$h'_i \sum_{j=1}^{N-k} \int_{(k+j-1)\nu}^{(k+j)\nu} F(t^* - t)bdtu_j \geq g_i(k\nu), \tag{19}$$

$$i = \overline{1,m}, \quad |u_j| \leq 1,\ j = \overline{1, N-k},$$

which is equivalent to problem (3).

The solution $(u_j^0,\ j = \overline{1, N-k})$ to problem (19) is constructed by the dual method (see §4). The value $u^*(k\nu) = u_1^0$ is fed into the input of (2).

Optimal observation problems (4) for the moment $\tau = k\nu$ in the discrete case take the form

$$\alpha_i(k\nu) = h_i' F(t^*)x +$$

$$+\sum_{j=0}^{k-1} \int_{j\nu}^{(j+1)\nu} h_i' F(t^* - t) b dt u^*(j\nu) \longrightarrow \min_x,$$

$$\alpha_*(j\nu) \le a'(j\nu)x \le \alpha^*(j\nu),\ j = \overline{0,k}, \qquad (20)$$

$$Dx = \delta,\ d_* \le x \le d^*,$$

where $\alpha_*(j\nu) = \xi_* - y^*(j\nu) + \sum_{r=0}^{j-1} \int_{r\nu}^{(r+1)\nu} h_i' F(j\nu - s) b ds u^*(r\nu)$, $\alpha^*(j\nu) = \xi^* - y^*(j\nu) + \sum_{r=0}^{j-1} \int_{r\nu}^{(r+1)\nu} h_i' F(j\nu - s) b ds u^*(r\nu)$, $a'(j\nu) = -h_i' F(j\nu)$.

According to the scheme described in §4 the optimal estimator at a current moment $\tau \in T$ constructs a solution $x^0(k\nu, y_{k\nu}^*(\cdot))$ to problem (20) and calculates an estimate $\alpha_i(k\nu) = h_i' F(t^*) x^0(k\nu, y_{k\nu}^*(\cdot)) + \sum_{j=0}^{k-1} \int_{j\nu}^{(j+1)\nu} h_i' F(t^* - t) b dt u^*(j\nu)$.

As in the case of continuous realization, problems (20) are solved in parallel by $m$ estimators and constructed values $\alpha_i(k\nu)$, $k = \overline{0, N-1}$, are used by the optimal controller for calculating the realization $u^*(k\nu)$.

## 6. EXAMPLE

Operation of proposed optimal feedbacks is illustrated by the problem of minimization of fuel consumption in steering a two-mass mechanical system in the vicinity of the equilibrium state. The mathematical model of the problem has the form

$$\int_0^{20} u(t)dt \longrightarrow \min,\ \dot{x}_1 = x_3,\ \dot{x}_2 = x_4,$$

$$\dot{x}_3 = -x_1 + x_2,\ \dot{x}_4 = 0.1x_1 - 1.02x_2 + w,$$

$$x(0) \in X_0,\ x(20) \le 0.$$

The set $X_0$ is described by the relations

$$x_1 = 0,\ x_2 = 0,\ 0.8 \le x_3 \le 1.1,\ 0.4 \le x_4 \le 0.6.$$

An initial state is the vector $x(0) = (0, 0, 1, 0.5)$. The measurer $y = x_1 + \xi$, $|\xi(t)| \le 0.2$, is used.

Let a discrete realization of the optimal feedback is constructed. The parameter $\nu$ is put to be equal to 0.5, the realized error function has the form $\xi^*(t) = 0.2 \sin 3t$, $t \in T$. The system is under the disturbance of the form $w(t) = w_1\omega_1(t) + w_2\omega_2(t) + w_3\omega_3(t)$, $t \in T$: $w^*(t) = 0.15 \sin 0.5t + 0.1 \sin 2t + 0.05 \sin 5t$, $t \in T$.

In the case when a disturbance does not affect the system the value of the control criterion is 2.6082, in the case of identification of parametric disturbances the values of the control criterion is 3.1619. In the ideal conditions of the open-loop control of the system without uncertainty the value of the control criterion is equal to 2.6014.

## 7. ACKNOWLEDGEMENTS

The research has been supported by Grant F96–011 from Byelorussian Republican Foundation for Basic Research and by the State Programme for Basic Research of Belarus.

## REFERENCES


Chernousko, F.L. (1988). *Estimating of Phase State of Dynamic Systems.* Nauka, Moscow.

Feldbaum, A.A. (1963). *Foundations of Theory of Optimal Automatic Systems.* GIFML, Moscow.

Gabasov, R. (1994). Adaptive method of solving linear programming problems. *Preprint Series of University of Karlsruhe,* Institute for Statistics and Mathematics, Germany, Karlsruhe.

Gabasov, R. and F.M. Kirillova (1990). Adjoint problems of control, observation and identification. *Doklady Academii Nauk BSSR,* **34**, 777–780.

Gabasov, R., F.M. Kirillova and O.I. Kostyukova (1992). Constructing of optimal controls of feedback type in a linear problem. *Soviet Math. Dokl.,* **44**, No. 2, 608–613.

Gabasov, R., F.M. Kirillova and O.I. Kostyukova (1993). An algorithm of the optimization in real time a partially defined linear control system. *Automat. Remote Control,* **54**, No. 4, 555–563.

Gabasov, R., F.M. Kirillova and S. V. Prischepova (1995). *Optimal Feedback Control.* Lecture Notes in Control and Information Sciences. (Thoma M. Ed.) Springer-Verlag, **207**, London.

Kortanek, K. and R. Hettich (1993). Semi-infinite programming: Theory, methods and applications. *SIAM Review,* **35**, 380–429.

Krasovskii, N.N. (1977). *Control by Dynamic System.* Nauka, Moscow.

Kurzhanski, A.B. (1989). Identification — a theory of guaranteed estimates. In: *From Data to Model* (J.C. Willems, Ed.), pp. 135–214. Springer-Verlag, Berlin.

Schweppe, F.C. (1973). *Uncertain Dynamic Systems.* Englewood Cliffs, Prentice Hall.

Song Jian (1963). Method of system synthesis with general criterion of optimality. *Automatica Sinica,* **1**, No. 1, 86–97.

Tsien, H.S. (1954). *Engineering Cybernetics.* Mc-Graw-Hill, N.Y.-Toronto-London.

# TIME REVERSIBILITY AND SPECTRAL COMPLETENESS VIA EXACT OBSERVABILITY

**LIU, Kangsheng**[1]

*Department of Applied Mathematics, Zhejiang University*
*Hangzhou, 310027, China. Email: ksl@math.zju.edu.cn*

Abstract: Consider a linear system $\dot{Z}(t) = AZ(t), t > 0$ in a Hilbert space $H$. When $H$ is finite-dimensional, every state trajetory of the system has a unique backward-time continuation and the set of all finite linear combinations of the generalized eigenvectors of $A$ is dense in $H$. For infinite-dimensional system, however, both time reversibility and spectral completeness can be absent. This paper reveals the two properties in several elastic systems with boundary damping by the observability method (Liu and Russell, 1998). *Copyright © 1999 IFAC*

Keywords: distributed-parameter systems, reversibility, eigenvectors, eigenmode analysis, observability, vibration, damping

## 1. INTRODUCTION

Consider linear system

$$\begin{cases} \dfrac{dZ}{dt} = AZ(t),\ t > 0, \\ Z(0) = Z_0 \end{cases} \tag{1}$$

where $A$ is the infinitesimal generator of a $C_0$ semigroup $e^{tA}$ of bounded linear operators on a Hilbert space $H$ of infinite dimension. The system (1) is said to be reversible in time if $e^{tA}$ can be embedded in a $C_0$ group. A nonzero vector $y \in \mathcal{D}(A)$ is called a generalized eigenvector of $A$, corresponding to the eigenvalue $\lambda \in \sigma_p(A)$, if $(\lambda I - A)^m y = 0$ for some positive integer $m$. Denote by $\mathrm{Sp}(A)$ the set of all finite linear combinations of the generalized eigenvectors. The system (1) is said to have spectral completeness if $\mathrm{Sp}(A)$ is dense in $H$. This paper is devoted to the time reversibility and the spectral completeness of several elastic systems with boundary damping.

When $A$ is defined on a finite-dimensional space, the time reversibility and the spectral completeness always hold. In infinite-dimensional space, however, both properties can be absent. For example, let

$$H = \{y \in H^1(0,1) |\ y(0) = 0\} \times L^2(0,1), \tag{2}$$

$$A(y, v) = (v, y''), \tag{3}$$

$$\mathcal{D}(A) = \{(y, v) \in H |\ y \in H^2(0,1),\ v \in H^1(0,1), v(0) = y'(1) + v(1) = 0\} \tag{4}$$

[1]Supported partially by the National Key Project of China and by the NSFC grant 69874034.

where $H^1(0,1)$ is the Sobolev space of order 1 (Adams, 1975), $y' = \frac{\partial y}{\partial x}$. Then $e^{tA} = 0$ for $t > 2$ and $\sigma(A) = \emptyset$, see (Chen and Zhou, 1993). The operator $A$ defined by (2)-(4) is associated with a string vibrating equation with boundary damping.

Recently, several characterizations of $C_0$ groups on a Hilbert space have been given by Liu (1999), Liu and Russell (1998):

PROPOSITION 1. *Let $A$ be the infinitesimal generator of a $C_0$ semigroup $e^{tA}$ on a Hilbert space $H$. Then the following statements are equivalent:*

(a) *$e^{tA}$ can be embedded in a $C_0$ group on $H$.*
(b) *There exist real number $\alpha$, bounded positive definite self-adjoint operator $P$ and skew adjoint operator $A_0$ with $\mathcal{D}(A_0) \subset H$ such that*
$$A = P^{\frac{1}{2}}(A_0 + \alpha I - \frac{1}{2}P^{-1})P^{-\frac{1}{2}}.$$
(c) *The system $(A, I)$ is exactly controllable and there exists a sequence $\{\lambda_n\}$ in$\mathbb{C}\backslash\sigma_p(A)$ such that $Re\,\lambda_n \to -\infty$.*
(d) *There exists $\gamma_0 > 0$ such that*
$$\{\lambda \in \mathbb{C} \mid Re\,\lambda \geq \gamma_0\} \subset \rho(-A),$$
$$\sup\{\|(\lambda + A)^{-1}\| \mid Re\,\lambda \geq \gamma_0\} < +\infty$$
*and*
$$\lim_{Re\,\lambda \to +\infty} \|(\lambda + A)^{-1}x\| = 0 \qquad \forall x \in H.$$

Note that $e^{tA}$ extends to a $C_0$ group if and only if $e^{tA^*}$ does. Thus, when $A$ has the compact resolvent, the condition (c) can be replaced by the observability inequality
$$\int_0^T \|e^{tA}y\|^2 dt \geq \delta\|y\|^2 \quad \forall\, y \in H. \tag{5}$$

This provides a method for establishing time reversibility of some infinite-dimensional linear systems. Moreover, the condition (b) enables us to prove (Liu and Russell, 1998):

PROPOSITION 2. *The infinitesimal generator $A$ of a $C_0$ group on a Hilbert space has a complete system of generalized eigenvectors if for some $\lambda_0$, $(\lambda_0 I - A)^{-1}$ is a compact operator of finite order.*

A compact operator $\mathcal{L}$ is said to be of finite order if the sequence of eigenvalues (counting multiplicities) of $\mathcal{L}^*\mathcal{L}$ is in $l^q$ for some $1 \leq q < \infty$ (see Gokhberg and Krein, 1969). This condition holds generally for elastic systems with boundary damping (Liu and Russell, 1998). Therefore, in order to obtain both time reversibility and spectral completeness, It suffices to verify the observability inequality (5). That is so-called *time reversibility and spectral completeness via exact observability.* In the next sections, by verifying (5) the two properties in several elastic systems with boundary damping are revealed.

## 2. EULER-BERNOULLI BEAM

Consider the Euler-Bernoulli equation of nonhomogeneous beam with boundary damping:
$$\begin{cases} \rho(x)y_{tt} + (p(x)y'')'' = 0, \; 0 < x < L \\ y(0,t) = y'(0,t) = 0, \qquad t > 0 \\ p(L)y''(L,t) + \alpha_1 y_t'(L,t) = 0 \\ (py'')'(L,t) - \alpha_2 y_t(L,t) = 0 \\ y(x,0) = y_0(x), \; y_t(x,0) = v_0(x) \end{cases} \tag{6}$$

where $\alpha_1, \alpha_2 \geq 0$, $0 < \rho(\cdot), p(\cdot) \in C^1[0,1]$. The energy of the beam at time $t$ is
$$E(t) = \frac{1}{2}\int_0^1 \left[p(x)|w''|^2 + \rho(x)|w_t|^2\right] dx.$$

The beam is clamped at $x = 0$. The boundary conditions at $x = L$ describe a control scheme used for the boundary stabilization of beam. The boundary bending moment and shear force apply at $x = L$ with angle velocity and velocity feedbacks so that the energy of beam decays in time; see (Chen et al. 1987a,b). Let
$$V = \{v \in H^2(0,L) |\; v(0) = v(1) = 0\},$$
$$\|v\|_V = \left(\int_0^L p(x)|v''|^2 dx\right)^{\frac{1}{2}}.$$

Let
$$H = V \times L^2_\rho(0,1)$$

equipped with the norm
$$\|(y,v)\|_H = \left[\int_0^L \left(p(x)|y''|^2 + \rho(x)|v|^2\right) dx\right]^{\frac{1}{2}}.$$

Then, $V$ and $H$ are (complex) Hilbert spaces. Define in $H$
$$A(y,v) = (v\,,\; -\frac{1}{\rho(x)}(p(x)y'')''), \tag{7}$$

$$\mathcal{D}(A) = \{(y,v)|\ y,v \in V,\ p(x)y'' \in H^2(0,L),\ (-p(1)y''(1),(py'')'(1)) = \mathrm{diag}(\alpha_1,\alpha_2)(v'(1),v(1))\}. \quad (8)$$

Then, the system (6) can be written as the abstract evolution equation (1) in $H$, where $Z = (y,v), v = y_t, Z_0 = (y_0,v_0)$. It is clear that $A$ generates a $C_0$ semigroup, $e^{tA}$, of contractions on $H$, and $A^{-1}$ is a compact operator. The exponential stability of $e^{tA}$ is proved by Chen et al. (1987a,b) for constant coefficient case and by Liu, K. and Liu, Z. (1998) for variable coefficient case. In this section, the following theorem will be proved.

THEOREM 1. *The system (6) has both time reversibility and spectral completeness.*

*Proof.* It suffices to verify (5) for $Z_0 \in \mathcal{D}(A)$. Set $(y(\cdot,t),v(\cdot,t)) = e^{tA}Z_0$. For $Z_0 = (y_0,v_0) \in \mathcal{D}(A)$, It holds that

$$y \in C^2([0,\infty);L^2_\rho(0,L)) \cap C^1([0,\infty);V),$$

$$y_t = v,\ \eta(t) = y_t'(L,t)$$

and $y$ satisfies (6). Thus,

$$2E(t) = \|e^{tA}Z_0\|^2_H.$$

It may be assumed without loss of generality that $y(x,t)$ is real-valued. It follows from multiplying the first equation of (6) by $y_t$ and integrating by parts that

$$E(0) - E(t) = \int_0^t \left[\alpha_1 y_t'(L,s)^2 + \alpha_2 y_t(L,s)^2\right] ds.$$

Therefore,

$$TE(0) - \int_0^T E(t)dt = \int_0^T (T-t)\left[\alpha_1 y_t'(L,t)^2 + \alpha_2 y_t(L,t)^2\right] dt. \quad (9)$$

Multiply the first equation of (6) by $(T-t)xy'$ and integrate in $x$ and $t$,

$$\int_0^T\int_0^L [\rho y_{tt} + (py'')''](T-t)xy'dxdt = 0.$$

Then integrate by parts:

$$\int_0^T\int_0^L \rho y_{tt}(T-t)xy'dxdt = -\int_0^L Tx\rho y_0'v_0dx - \frac{1}{2}\int_0^T L\rho(L)(T-t)y_t(L,t)^2dt + \int_0^T\int_0^L [x\rho y'y_t + \frac{T-t}{2}(x\rho)'y_t^2]dxdt,$$

$$\int_0^L (py'')''xy'dx = L\alpha_2 y_t(L,t)y'(L,t) + \int_0^L [\frac{3p}{2} - xp'](y'')^2dx - p(L)y''(L,t)[y'(L,t) + \frac{L}{2}y''(L,t)].$$

Thus,

$$\int_0^T (T-t)L\left[\rho(L)y_t(L,t)^2 + p(L)y''(L,t)^2\right] dt \le 2\int_0^T (T-t)y'(L,t)L\alpha_2 y_t(L,t)dt - 2\int_0^T (T-t)y'(L,t)p(L)y''(L,t)dt + C_1\int_0^T E(t)dt - 2T\int_0^L x\rho y_0'v_0dx \quad (10)$$

where $C_1$ is some positive number independent of $Z_0$. Since

$$\int_0^T (T-t)y'(L,t)y''(L,t)dt \le \epsilon\int_0^T (T-t)y''(L,t)^2dt + \frac{LT}{\epsilon}\int_0^T\int_0^L (y'')^2dxdt$$

for all $\epsilon > 0$, (10) implies

$$\frac{1}{2}\int_0^T (T-t)L\left[\rho(L)y_t(L,t)^2 + p(L)y''(L,t)^2\right] dt \le C_2\int_0^T E(t)dt - 2T\int_0^L x\rho y_0'v_0dx \quad (11)$$

Combination of (9), (11) and the boundary condition at $x = L$ yields

$$TE(0) \le C_T\int_0^T E(t)dt - 2T\int_0^L x\rho y_0'v_0dx$$

for some $C_T > 0$ independent of $Z_0$. Application of a standard compactness argument yields the desired estimate (5). To verify that the compact operator $A^{-1}$ is of finite order, define in $L^2_\rho(0,L)$

$$A_0v = \frac{1}{\rho}(pv'')'',$$

$$\mathcal{D}(A_0) = \{v \in V|\ pv'' \in H^2(0,L),\ (pv'')(L) = (pv'')'(L) = 0\}.$$

Then $A_0$ is a positive definite self-adjoint operator with $A_0^{-1}$ being a compact operator of finite order. Let

$$L = \begin{pmatrix} 0 & -A_0^{-1} \\ I & 0 \end{pmatrix}.$$

Then $L$ is a compact operator on $H$ of finite order. It is easy to prove that $A^{-1}-L$ has the finite rank. Thus $A^{-1}$ is a compact operator of finite order.

## 3. EXAMPLES IN HIGH DIMENSION

Consider the following wave and plate equations with boundary damping

$$\begin{cases} w_{tt} - \Delta w = 0 \text{ in } \Omega \times I\!R^+ \\ \dfrac{\partial w}{\partial \nu} + \alpha(x) w_t = 0 \text{ on } \partial\Omega \times I\!R^+, \\ w(x,0) = \xi(x),\ w_t(x,0) = \eta(x),\ x \in \Omega \end{cases} \tag{12}$$

$$\begin{cases} w_{tt} + \Delta^2 w = 0 \text{ in } \Omega \times I\!R^+ \\ y = 0 \quad \text{on} \quad \partial\Omega \times I\!R^+ \\ \dfrac{\partial w}{\partial \nu} = 0 \text{ on } \Gamma_0 \times I\!R^+ \\ \Delta w = -\dfrac{\partial w_t}{\partial \nu} \text{ on } \Gamma_1 \times I\!R^+ \\ w(x,0) = \xi(x),\ w_t(x,0) = \eta(x),\ x \in \Omega \end{cases} \tag{13}$$

where $\Omega$ is a bounded open subset in $I\!R^n$ $(n \geq 2)$ with the boundary $\partial\Omega$ smooth enough, $\nu$ is the unit normal vector of $\partial\Omega$ pointing towards the exterior of $\Omega$, $\alpha(x) \geq 0$ is a smooth function on $\partial\Omega$, $\overline{\Gamma_0} \cup \overline{\Gamma_1} = \partial\Omega$, $\Gamma_0 \cap \Gamma_1 = \emptyset$. Majda (1976) prove that the system (12) has no time reversibility if $\alpha(x_0) \neq 0$ for some $x_0 \in \Omega$. He (Majda,1975) also shows by examples that the spectral completeness of (12) can be absent. However, for the plate equation (13) the two properties hold.

THEOREM 2. *If* $\overline{\Gamma_0} \cap \overline{\Gamma_1} = \emptyset$, *then the system (13) has both time reversibility and spectral completeness.*

*Sketch of the Proof.*

Let

$$V = \{v \in H^2(\Omega) |\ v = 0 \text{ on } \partial\Omega,\ \frac{\partial v}{\partial \nu} = 0 \text{ on } \Gamma_0\},$$

$$H = V \times L^2(\Omega), \quad A(w,v) = (v, -\Delta^2 w),$$

$$\mathcal{D}(A) = \{(w,v) \mid w \in H^4(\Omega),\ w,v \in V,\ w = 0 \text{ on } \partial\Omega,\ \frac{\partial w}{\partial \nu} = 0 \text{ on } \Gamma_0,\ \Delta w = -\frac{\partial v}{\partial \nu} \text{ on } \Gamma_1\}.$$

Then $A$ generates a $C_0$ semigroup $S(t)$ of contractions on $H$. For $(\xi,\eta) \in D(A)$, set $(w(t), w_1(t)) = S(t)(\xi,\eta)$. Then the function $w(\cdot)$ is in

$$C([0,\infty); H^4(\Omega)) \cap C^1([0,\infty); V) \cap C^2([0,\infty); W).$$

It satisfies $w_t(\cdot) = w_1(\cdot)$ and (13). Let

$$E(t) = \frac{1}{2}\|S(t)(\xi,\eta)\|^2_{\mathcal{H}} = \frac{1}{2}\int_\Omega (|\Delta w|^2 + |w_t|^2) dx.$$

It suffices to prove the observability inequality

$$\int_0^T E(t)\,dt \geq \delta E(0) \tag{14}$$

for some $T, \delta > 0$ and all real-valued $(\xi,\eta) \in \mathcal{D}(A)$. It follows from multiplying the first equation of (13) by $w_t$ and integrating by parts that

$$TE(0) - \int_0^T E(t)dt = \int_0^T (T-t) \int_{\Gamma_1} \left(\frac{\partial w_t}{\partial \nu}\right)^2 d\sigma dt.$$

Choose $h \in C^1(\overline{\Omega}; I\!R^n)$ such that $h = 0$ on $\Gamma_0$ and $h = \nu$ on $\Gamma_1$. Then multiply the first equation of (13)$(u = 0)$ by $(T-t)h \cdot \nabla w$ to get

$$\int_0^T (T-t) \int_{\Gamma_1} \left[\left(\frac{\partial w_t}{\partial \nu}\right)^2 + \frac{1}{2}\text{div}h \frac{\partial w_t}{\partial \nu}\frac{\partial w}{\partial \nu}\right] d\sigma dt \leq -\int_\Omega \eta h \cdot \nabla\xi dx + C_T \int_0^T E(t)dt.$$

where $C_T$ is some positive number independent of $(\xi,\eta)$. Now, the observability inequality (14) follows from a standard compactness argument.

## 4. RAYLEIGH BEAM

Consider the following Rayleigh beam equation with boundary damping:

$$\begin{cases} y_{tt} - \gamma y''_{tt} + y'''' = 0, \text{ on } (0,1) \times I\!R^+ \\ y(0,t) = y'(0,t) = 0 \\ \gamma y'_{tt}(1,t) - y'''(1,t) = 0 \\ y''(1,t) = -k y'_t(1,t) \\ y(x,0) = \xi(x),\ y_t(x,0) = \eta(x) \end{cases} \tag{15}$$

where $\gamma$, $k > 0$ are constant. It is known that the waves in the Rayleigh beam go at finite speed, like in a string. In this section, time reversibility and spectral completeness of (15) are showed for $k \neq \sqrt{\gamma}$. It is not clear whether the solution to (15) disappears if $k = \sqrt{\gamma}$.

THEOREM 3. *When $k \neq \sqrt{\gamma}$, the system (15) has both time reversibility and spectral completeness.*

*Sketch of the Proof.*

The energy of the Rayleigh beam at time $t$ is

$$E(t) = \frac{1}{2}\int_0^1 (|y''|^2 + \gamma|y_t'|^2 + |y_t|^2)dx.$$

It follows from multiplying the first equation of (15) by $y_t$ and integrating by parts that

$$TE(0) - \int_0^t E(t)dt = k\int_0^T (T-t)y_t'(1,t)^2dt. \tag{16}$$

Multiply the first equation of (15) by $(T-t)xy'$. Then integrate by parts again:

$$\begin{aligned}
&(\frac{\gamma}{2}+\frac{k^2}{2})\int_0^T (T-t)y_t'(1,t)^2dt + T\int_0^1 x\xi'\eta dx \\
&+\gamma T\int_0^1 \xi'\eta'\,dx + \gamma T\int_0^1 x\xi''\eta'dx \\
&\le k\int_0^T (T-t)y'(1,t)y_t'(1,t)dt + C_T\int_0^T E(t)dt
\end{aligned} \tag{17}$$

where $C_T$ is some positive number. (16) and (17) imply

$$\begin{aligned}
TE(0) \le \frac{2k}{\gamma+k^2}\Big[&T\sqrt{\gamma}E(0) \\
&- T\int_0^1 x\xi'\eta dx - \gamma T\int_0^1 \xi'\eta' dx \\
&+ k\int_0^T (T-t)y'(1,t)y_t'(1,t)dt\Big] \\
&+ (1+C_T)\int_0^T E(t)dt.
\end{aligned}$$

Thus, a standard compactness argument gives the desired estimate.

## 5. CONCLUSIONS

The time reversibility and the spectral completeness hold generally in elastic systems with boundary damping in which the speed of wave propagation is infinite. In the case when the speed of wave propagation is finite, the two properties can be absent even in one-dimensional space. For the wave equation with boundary damping in high-dimensional space, the time reversibility is absent in general. The observability method is quite efficient for revealing the two properties.

## REFERENCES


Adams, R. A. (1975), *Sobolev Spaces*, Academic Press.

Chen, G., M. C., Delfour, A. M., Krall, and G., Payre (1987a), Modeling, stabilization and control of serially connected beams, *SIAM J. Cont. Opt.*, **25**, pp. 526-546.

Chen, G., S.G. Krantz, D.W. Ma and C.E. Wayne (1987b), The Euler-Bernoulli beam equation with boundary energy dissipation, in *Operator Methods for Optimal Control Problems*, S.J. Lee ed., Lect. Notes in Pure and Appl. Math Ser., pp. 67-96, Marcel Dekker, Inc., New York.

Chen, G. and J. Zhou (1993), *Vibration and Damping in Distributed Systems* Vol. **I**, CRC Press, Inc.

Gokhberg, I. C. and M. G. Krein (1969), *Introduction to the Theory of Linear Nonselfadjoint Operators*, AMS, Providence.

Liu, K. (1999), A characterization of strongly continuous groups of linear operators on a Hilbert space, *Bulletin of London Math. Society*, to appear.

Liu, K. and Z., Liu (1998), Boundary stabilization of nonhomogeneous beam by frequency domain multiplier mothed, submitted.

Liu, K. and D. L. Russell (1998), Exact controllability, time reversibility and spectral completeness of linear systems in Hilbert spaces, submitted.

Majda, A. (1976), The location of the spectrum for the dissipative acoustic operator, *Indiana Univ. Math. J.*, **25**, pp. 973-987.

Majda, A. (1975), Disappearing solutions for the dissipative wave equation, *Indiana Univ. Math. J.*, **24**, pp. 1119-1133.

14th Triennial World Congress, Beijing, P.R. China



# EXPONENTIAL STABILIZATION OF TIMOSHENKO BEAM WITH LOCALLY DISTRIBUTED FEEDBACK*

DONG-HUA SHI DE-XING FENG

*Laboratory of Systems and Control, Institute of Systems Science*
*Chinese Academy of Sciences, Beijing 100080, China*
*Fax: (8610)62587343, E-mail: dxfeng@iss03.iss.ac.cn*

**Abstract** The problem of exponential stabilization of Timoshenko beam with locally distributed controls is investigated in this paper. In the case that the beam is non-uniform and the two wave speeds are different, first it is shown, under some locally distributed controls, that the vibration of the beam decays exponentially by using Huang's theorem and Birkhoff's asymptotic expansion method. Then explicit asymptotic locations of eigenfrequencies are also derived. When the two wave speeds are same, by using Chen's theorem, the exponential decay is established as well for a uniform beam. 

**Keywords** non-uniform Timoshenko beam, locally distributed feedback control, Birkhoff's asymptotic expansion method, exponential stabilization.

## 1. INTRODUCTION

The main objective of this paper is to investigate the locally distributed feedback stabilization of a beam . It is well known that if the cross-section dimensions of the beam is not negligible compared with its length, then it is necessary to consider the effect of the rotatory inertia and, moreover, if the deflection due to shear is not negligible either, the dynamic model of the beam , which is called Timoshenko one, must be described as follows:

$$\begin{cases} \rho\dfrac{\partial^2 w}{\partial t^2} - \dfrac{\partial}{\partial x}\left(K\dfrac{\partial w}{\partial x}\right) + \dfrac{\partial}{\partial x}\left(K\varphi\right) + u_1(x,t) = 0, \\ I_\rho\dfrac{\partial^2 \varphi}{\partial t^2} - \dfrac{\partial}{\partial x}\left(EI\dfrac{\partial \varphi}{\partial x}\right) + K\left(\varphi - \dfrac{\partial w}{\partial x}\right) + u_2(x,t) = 0, \\ 0 < x < \ell,\ t > 0. \end{cases} \tag{1}$$

Here a non-uniform beam of length $\ell$ moves in $w$-$x$ plane, $\rho(x)$ is the mass density, $w(x,t)$ the deflection of the beam from its equilibrium and $\varphi(x,t)$ the total rotatory angle of the beam at $x$, for whose precise physical meaning, see (Timoshenko, 1955; Russell, 1986). $I_\rho(x)$ and $EI(x)$ are the mass moment of inertia and rigidity coefficient of the cross section, respectively, and $K(x)$ is the shear modulus of elasticity. $u_1(x,t)$ and $u_2(x,t)$ are locally distributed controls, i.e. there exists a subinterval $[\alpha,\beta] \subset [0,\ell]$, such that

$$u_1(x,t) \equiv 0,\ \ u_2(x,t) \equiv 0, \qquad x \notin [\alpha,\beta], \quad t \geq 0.$$

In the case of the cantilever configuration, the appropriate boundary conditions should be described as follows:

$$\begin{cases} w(0,t) = 0,\ \ \varphi(0,t) = 0, \\ K(\ell)[\varphi(\ell,t) - w'(\ell,t)] = \mu_1(t), \\ -EI(\ell)\varphi'(\ell,t) = \mu_2(t), \end{cases} \tag{2}$$

where $\mu_1(t)$ and $\mu_2(t)$ are the applied lateral force and moment at $x = \ell$ respectively. To avoid lengthy cacculations which are not necessary for our purpose, one can assume that $\mu_1(t) = \mu_2(t) \equiv 0$. Here and below

* This research was supported by the National Key Project of China

the prime and the dot always denote derivatives with respect to space and time variables, respectively.

In recent years, the control problem of large space flexible structures has attracted much more attention of many authors, for example, see (Chen et al, 1991; Chen et al, 1987a; Wang and Chen, 1991). In (kim and Renardy, 1987), the boundary feedback control of Timoshenko beam is considered. In (Chen et al, 1987b), the locally distributed control of Euler-Bernoulli beam is also studied. The purpose of this paper is to show the exponential stabilization of Timoshenko beam vibration by using locally distributed feedback controls. As is shown above, the vibration of a Timoshenko beam is discribed by two coupled wave equations with variable coeffcient. After tremendous work, one can realize that it is hard to seek right mutipliers especially for the case with nonhomogenous boundary conditions, so the methods used in (Ho, 1990; Liu, 1997) are not suitable here. On the other hand, it is very difficult to know whether or not the eigenvalues of the system satisfies the gap conditions, which is necessary in usual sufficient condition for the stabilization of a elastic vibration. Thus those methods in (Lagnese, 1983; Russell, 1978) are not efficient enough to study our problem.

Huang's theorem (see (Huang,1985)) is a sufficient and necessary condition about exponential stability of an infinite dimensional system. However, it is difficult to verify for Timoshenko beam system with varible coefficient because no explicit formula of resolvent is available.(Many authors used this condition for the system with constant coefficient, see (Liu et al, 1989). When using Birkhoff's asymptotic expansion method introduced by (Birkhoff and Langer, 1923), one can yield a successful treatment. Using Birkhoff's method to estimate the norm of the resolvent operator herein is more complex and difficult than it does in (Wang and Chen, 1991), where the same method is used to estimate the locations of eigenfrequences. Moreover, the form of the asymptotic solutions also looks rather different from that in (Wang and Chen, 1991), because the first-order equation with a large parameter is a non-homogenous one . In Section 2, some locally distributed feedback controls are proposed, and the well-posedness of the corresponding closed loop system is given via the semigroup theory. In Section 3, In the case that the wave speeds are different( in general , this condition is satisfied. see (Russell, 1986), it is shown that the vibration of the beam decays exponentially by virtue of Huang's theorem and Birkhorff's asymptotic expansion method. In Section 4, asymptotic estimates of eigenvalues are derived by using Rouche's theorem. Finally in Section 5, while two wave speeds are equal and the beam is uniform, the exponential stability of the closed loop system is also proved under certain locally distributed controls.

## 2. WELL-POSEDNESS OF THE CLOSED LOOP SYSTEM

One proposes the following locally distributed feedback controls:

$$\begin{cases} u_1(x,t) = \rho(x)b_1(x)\dot{w}(x,t), \\ u_2(x,t) = I_\rho(x)b_2(x)\dot{\varphi}(x,t), \end{cases} \tag{3}$$

where $b_j(x) \in C^1[0,\ell]$ for $j = 1, 2$, and

$$\begin{cases} b_j(x) \equiv 0, \ x \notin [\alpha,\beta] \\ b_j(x) \geq 0, \ x \in [\alpha,\beta] \\ b_j(x) \geq \gamma > 0, \ x \in [c,d] \subset [\alpha,\beta]. \end{cases} \tag{4}$$

Then the closed loop system corresponding to (1) and (2) becomes

$$\begin{cases} \rho\ddot{w} - (Kw')' + (K\varphi)' + \rho b_1 \dot{w}(x,t) = 0, \\ I_\rho\ddot{\varphi} - (EI\varphi')' + K(\varphi - w') + I_\rho b_2\dot{\varphi}(x,t) = 0, \\ w(0,t) = 0, \ \varphi(0,t) = 0, \\ K(\ell)[\varphi(\ell,t) - w'(\ell,t)] = 0, \ EI(\ell)\varphi'(\ell,t) = 0, \end{cases} \tag{5}$$

Now one incorporates the closed loop system (5) into a certain function space. To this end, one define the product Hilbert space $\mathcal{H} = V_0^1 \times L_\rho^2(0,\ell) \times V_0^1 \times L_{I_\rho}^2(0,\ell)$, where

$$V_0^k = \left\{\varphi \in H^k(0,\ell) \mid \varphi(0) = 0\right\}, \quad k = 1, 2,$$

and $H^k(0,\ell)$ is the usual Sobolev space of order $k$. The inner product in $\mathcal{H}$ is defined as follows:

$$\left(Y_1, Y_2\right)_{\mathcal{H}} = \int_0^\ell K(\varphi_1 - w_1')(\varphi_2 - w_2')\,dx + \int_0^\ell EI\varphi_1'\varphi_2'\,dx + \int_0^\ell \rho z_1 z_2\,dx + \int_0^\ell I_\rho\psi_1\psi_2\,dx,$$

where $Y_k = [w_k, z_k, \varphi_k, \psi_k]^T \in \mathcal{H}$ for $k = 1, 2$. One defines linear operators $\mathcal{A}$ and $\mathcal{B}$ in $\mathcal{H}$:

$$\mathcal{A}\begin{bmatrix} w \\ z \\ \varphi \\ \psi \end{bmatrix} = \begin{bmatrix} z \\ -(K(\varphi - w'))'/\rho \\ \psi \\ (EI\varphi')'/I_\rho - K(\varphi - w')/I_\rho \end{bmatrix}, \quad \begin{bmatrix} w \\ z \\ \varphi \\ \psi \end{bmatrix} \in \mathcal{D}(\mathcal{A}),$$

$$\mathcal{D}(\mathcal{A}) = \Big\{ [w,z,\varphi,\psi]^T \in \mathcal{H} \mid w,\varphi \in V_0^2, \ z,\psi \in V_0^1, \ \varphi(\ell) = w'(\ell), \ \varphi'(\ell) = 0 \Big\}.$$

$$\mathcal{B}\begin{bmatrix} w \\ z \\ \varphi \\ \psi \end{bmatrix} = \begin{bmatrix} 0 \\ -b_1 z \\ 0 \\ -b_2\psi \end{bmatrix}, \quad \begin{bmatrix} w \\ z \\ \varphi \\ \psi \end{bmatrix} \in \mathcal{D}(\mathcal{B}) = \mathcal{H}.$$

Let $\mathcal{A}_1 = \mathcal{A} + \mathcal{B}$. Then one can write the closed loop system (5) as the following linear evolution equation in $\mathcal{H}$:

$$\frac{dY(t)}{dt} = \mathcal{A}_1 Y(t), \tag{6}$$

where $Y(t) = [w(\cdot,t), \dot{w}(\cdot,t), \varphi(\cdot,t), \dot{\varphi}(\cdot,t)]^\tau$. The energy corresponding to the solution to the closed loop system (6) is

$$E(t) = \frac{1}{2}\|Y(t)\|^2_{\mathcal{H}} = \frac{1}{2}\Big[\int_0^\ell EI|\varphi'|^2 dx + \int_0^\ell K|\varphi - w'|^2 dx + \int_0^\ell \rho|\dot{w}|^2 dx + \int_0^\ell I_\rho|\dot{\varphi}|^2 dx\Big],$$

where $Y(t) = [w(\cdot,t), \dot{w}(\cdot,t), \varphi(\cdot,t), \dot{\varphi}(\cdot,t)]^\tau$ is the solution to (6). It is easy to verify the following lemmas:

**Lemma 2.1** The linear operator $\mathcal{A}$ defined above is skew adjoint in the Hilbert space $\mathcal{H}$.

**Lemma 2.2** $\mathcal{A}_1$ has compact resolvent. The following is the main result in this section.

**Theorem 2.3** For any initial data $Y_0 \in \mathcal{H}$, (6) has a unique weak solution $Y(t)$ such that $Y(\cdot) \in C([0,\infty),\mathcal{H})$. Moreover, if $Y_0 \in \mathcal{D}(\mathcal{A})$, then $Y(t)$ is the unique strong solution to (6) such that $Y(\cdot) \in C^1([0,\infty),\mathcal{H}) \cap C([0,\infty),\mathcal{D}(\mathcal{A}))$.

## 3.EXPONENTIAL STABILITY OF THE CLOSED LOOP SYSTEM(I)

In general the wave speeds $\sqrt{K(x)/\rho(x)}$ and $\sqrt{EI(x)/I_\rho(x)}$ are different.(see Russell, 1986) For simplicity, one denotes by $\rho_1(x)$ and $\rho_2(x)$ the reciprocals of two speeds $\sqrt{\rho(x)/K(x)}$ and $\sqrt{I_\rho(x)/EI(x)}$ respectively.

To use Birkhoff's asymptotic expansion method, throughout this section and the next section, one assumes that $\rho(x), K(x), EI(x), I_\rho(x) \in C^1[0,\ell]$ and

$$\rho(x), K(x), EI(x), I_\rho(x) \geq \gamma_1 > 0, \quad x \in [0,\ell], \tag{7}$$

$$\rho_1(x) \neq \rho_2(x), \qquad \forall x \in [0,\ell]. \tag{8}$$

In (Huang, 1985), it is shown that for a uniformly bounded $C_0$-semigroup $T(t)$ on a Hilbert space $\mathcal{H}$, $T(t)$ is exponentially stable iff

$$\{i\omega \mid \omega \in I\!R\} \subset \rho(\mathcal{A}), \tag{9}$$

and

$$K_1 \triangleq \sup\{\|(i\omega - \mathcal{A})^{-1}\| \mid \omega \in I\!R\} < \infty, \tag{10}$$

where $\mathcal{A}$ is the infinitesimal generator of $T(t)$. Note that the semigroup $T(t)$ generated by $\mathcal{A}_1$ is uniformly bounded in terms of Theorem 2.3. Therefore, for the exponential stability of the semigroup $T(t)$ generated by $\mathcal{A}_1$, it is enough to show that (9) and (10) hold. First one has the following lemma:

**Lemma 3.1** The imaginary axis is a subset of the resolvent set of $\mathcal{A}_1$, $\rho(\mathcal{A}_1)$, i.e. $\{i\omega \mid \omega \in I\!R\} \subset \rho(\mathcal{A}_1)$.

Next one proceeds to show that (10) holds for $\mathcal{A}_1$. To do this, one shall use Birkhoff's asymptotic expansion method. First one needs to transform the resolvent equation into a first order system with a large parameter.

For any $Y_1 = [w_1, z_1, \varphi_1, \psi_1]^\tau \in \mathcal{H}$, let $Y = [w, z, \varphi, \psi]^\tau \in \mathcal{D}(\mathcal{A})$ such that $\lambda Y - \mathcal{A}_1 Y = Y_1$, i.e.

$$\begin{cases} \lambda w - z = w_1, \\ (\lambda + b_1)z + (K(\varphi - w'))'/\rho = z_1, \\ \lambda\varphi - \psi = \varphi_1, \\ (\lambda + b_2)\psi - (EI\varphi')'/I_\rho + K(\varphi - w')/I_\rho = \psi_1, \\ w(0) = z(0) = \varphi(0) = \psi(0) = \varphi'(\ell) = 0, \\ \varphi(\ell) = w'(\ell). \end{cases} \tag{11}$$

Eliminating $z$ and $\psi$ in (11) and setting $u = [w, \varphi, w'/\lambda, \varphi'/\lambda]^\tau$, then (11) is equivalent to

$$\begin{cases} u' = \lambda\begin{bmatrix} 0 & I_2 \\ A^2 & 0 \end{bmatrix}u + \begin{bmatrix} 0 & 0 \\ B & D \end{bmatrix}u + \lambda^{-1}\begin{bmatrix} 0 & 0 \\ C & 0 \end{bmatrix}u \\ \quad + \begin{bmatrix} 0 \\ f_1 \end{bmatrix} + \lambda^{-1}\begin{bmatrix} 0 \\ g_1 \end{bmatrix}, \\ B_1 u(0) = 0, \qquad B_2 u(\ell) = 0, \end{cases} \tag{12}$$

where $I_2$ is $2\times 2$ unit matrix and $A, B, C, D, f_1, g_1, B_1, B_2$ are defined as follows:

$$A = \begin{bmatrix} \rho_1 & 0 \\ 0 & \rho_2 \end{bmatrix}, \quad B = \begin{bmatrix} \rho_1^2 b_1 & 0 \\ 0 & \rho_2^2 b_2 \end{bmatrix},$$

$$C = \begin{bmatrix} 0 & \frac{K'}{K} \\ 0 & \frac{K}{EI} \end{bmatrix}, \quad D = \begin{bmatrix} -\frac{K'}{K} & 1 \\ -\frac{K}{EI} & -\frac{EI'}{EI} \end{bmatrix},$$

$$f_1 = \begin{bmatrix} -\rho_1^2 w_1 \\ -\rho_2^2 \varphi_1 \end{bmatrix}, \quad g_1 = \begin{bmatrix} -\rho_1^2(z_1 + b_1 w_1) \\ -\rho_2^2(\psi_1 + b_2\varphi_1) \end{bmatrix},$$

$$B_1 = \begin{bmatrix} 1 & 0 & 0 & 0 \\ 0 & 1 & 0 & 0 \end{bmatrix}, \quad B_2 = \begin{bmatrix} 0 & 1 & -\lambda & 0 \\ 0 & 0 & 0 & 1 \end{bmatrix}.$$

To diagonalize the dominant coefficient matrix $\begin{bmatrix} 0 & I_2 \\ A^2 & 0 \end{bmatrix}$, let $u = Qv$ or $v = Q^{-1}u$, where

$$Q = \frac{1}{2}\begin{bmatrix} A^{-1} & -A^{-1} \\ I_2 & I_2 \end{bmatrix}, \quad Q^{-1} = \begin{bmatrix} A & I_2 \\ -A & I_2 \end{bmatrix}$$

then (12) becomes

$$\begin{cases} v' = \lambda M v + \widetilde{B} v + \lambda^{-1}\widetilde{C} v + F_1 + \lambda^{-1} G_1 \\ \widetilde{B}_1 v(0) = 0, \quad \widetilde{B}_2 v(\ell) = 0 \end{cases} \tag{13}$$

where

$$M = \begin{bmatrix} A & 0 \\ 0 & -A \end{bmatrix}, \; \widetilde{B} = -Q^{-1}Q' + Q^{-1}\begin{bmatrix} 0 & 0 \\ B & D \end{bmatrix} Q,$$
$$\widetilde{C} = Q^{-1}\begin{bmatrix} 0 & 0 \\ C & 0 \end{bmatrix} Q, \quad F_1 = \begin{bmatrix} f_1 \\ f_1 \end{bmatrix}, \quad G_1 = \begin{bmatrix} g_1 \\ g_1 \end{bmatrix},$$
$$\widetilde{B}_1 = \begin{bmatrix} \rho_1^{-1}(0) & 0 & -\rho_1^{-1}(0) & 0 \\ 0 & \rho_2^{-1}(0) & 0 & -\rho_2^{-1}(0) \end{bmatrix},$$
$$\widetilde{B}_2 = \begin{bmatrix} -\lambda & \rho_2^{-1}(\ell) & -\lambda & -\rho_2^{-1}(\ell) \\ 0 & 1 & 0 & 1 \end{bmatrix}.$$

The solution of (13) can be written as

$$v(x) = T_c(x)\Theta + v_p(x), \tag{14}$$

where $\Theta = [\theta_1, \theta_2, \theta_3, \theta_4]^\tau$ is a constant vector to be determined later, $T_c(x)$ is the fundamental solution matrix of the homogeneous equation associated with (13), and $v_p(x)$ is a particular solution of non-homogeneous equation (13), which satisfy

$$T_c' = (\lambda M + \widetilde{B} + \lambda^{-1}\widetilde{C})T_c, \tag{15}$$
$$v_p(x) = T_c(x)\textstyle\int_0^x T_c^{-1}(s)(F_1(s) + \lambda^{-1}G_1(s))\,ds. \tag{16}$$

Next, applying the Birkhoff's arguments in (Birkhoff and Langer, 1923) to (15) with the procedure similar to that in (Wang and Chen, 1991), one gets for $\lambda \in G^- = \{\lambda \mid \operatorname{Re}\lambda \le 0, |\lambda| > N\}$ with $N$ given large enough,

$$T_c(x) = \left(\exp(\mu_j(x))\{\delta_{kj} + \xi_{kj}(x,\lambda)/\lambda\}\right)_{4\times4}, \tag{17}$$
$$T_c^{-1}(x) = \left(\exp(-\mu_k(x))\{\delta_{kj} + \eta_{kj}(x,\lambda)/\lambda\}\right)_{4\times4}, \tag{18}$$

where $\mu_j(x) = \int_0^x \widetilde{b}_j(s)\,ds + \lambda\int_0^x m_j(s)\,ds$, $\widetilde{b}_j(s)$ and $m_j(s)$ are the (j,j)-entry of matrix $\widetilde{B}$ and $M$, respectively. A simple calculation gives

$$\widetilde{b}_1 = \frac{1}{2}\left(\rho_1 b_1 - \frac{K'}{K} + \frac{\rho_1'}{\rho_1}\right),$$
$$\widetilde{b}_2 = \frac{1}{2}\left(\rho_2 b_2 - \frac{EI'}{EI} + \frac{\rho_2'}{\rho_2}\right),$$
$$\widetilde{b}_3 = \frac{1}{2}\left(-\rho_1 b_1 - \frac{K'}{K} + \frac{\rho_1'}{\rho_1}\right),$$
$$\widetilde{b}_4 = \frac{1}{2}\left(-\rho_2 b_2 - \frac{EI'}{EI} + \frac{\rho_2'}{\rho_2}\right).$$

From (Birkhoff and Langer, 1923), one knows that $\xi_{kj}(x,\lambda)$ and $\eta_{kj}(x,\lambda)$ for $\lambda \in G^-$ are analytic and bounded uniformly for $\lambda \in G^-$ and $x \in [0,\ell]$.

Denote

$$v_p(x) = [p_1(x), p_2(x), p_3(x), p_4(x)]^\tau$$

and

$$v_c(x) = T_c(x)\Theta = [r_1(x), r_2(x), r_3(x), r_4(x)]^\tau,$$

it can be easily seen that for $k, j = 1, \ldots, 4$

$$p_k(x) = \exp(\mu_k(x))q_k(x) + \lambda^{-1}\sum_{j=1}^4 \xi_{kj}(x,\lambda)\exp(\mu_j(x))q_j(x), \tag{19}$$
$$q_j(x) = \int_0^x \exp(-\mu_j(s))\Big[h_j(s) + \lambda^{-1}\sum_{n=1}^4 \eta_{jn}(s,\lambda)h_n(s)\Big]\,ds, \tag{20}$$
$$r_k(x) = \exp(\mu_k(x))\theta_k + \lambda^{-1}\sum_{j=1}^4 \xi_{kj}(x,\lambda)\exp(\mu_j(s))\theta_j, \tag{21}$$

where

$$h_1 = h_3 = -\rho_1^2\left(w_1 + \frac{b_1 w_1 + z_1}{\lambda}\right),$$
$$h_2 = h_4 = -\rho_2^2\left(\varphi_1 + \frac{b_2\varphi_1 + \psi_1}{\lambda}\right).$$

After a tremendous but straightforward caculation, one can obtain the following theorem, whose proof is omitted here due to space limitation.

**Theorem 3.2** With the previous assumptions and notations, the closed loop system (5) is exponentially stable, more precisely, the $C_0$-semigroup $T(t)$ generated by $\mathcal{A}_1$ is exponentially stable.

## 4. ASYMPTOTIC DISTRIBUTION OF EIGENVALUES OF THE SYSTEM

In this section, under the assumptions (7) and (8), the well-known Rouche's throrem will be used to investigate the asymptotic behavior of the eigenvalues of the system (5). The method used here is similar to that in (Wang and Chen, 1991). By virtue of the arguement of the last section with $Y_1 = 0$, $\lambda \in G^-$ is an eigenvalue of $\mathcal{A}_1$ if and only if

$$\det\left(\widehat{E} + \lambda^{-1}\widehat{E}_1\right) = 0, \tag{22}$$

where $\widehat{E}, \widehat{E}_1$ are defined as follows:

$$\widehat{E} = \begin{bmatrix} \widetilde{B}_1 \\ \widehat{B}_2(\exp(\mu_j(\ell))\delta_{kj})_{4\times 4} \end{bmatrix},$$

$$\widehat{E}_1 = \begin{bmatrix} \widetilde{B}_1(\xi_{kj}(0,\lambda))_{4\times 4} \\ \widehat{B}_2(\exp(\mu_j(\ell))\xi_{kj}(\ell,\lambda))_{4\times 4} \end{bmatrix},$$

$$\widehat{B}_2 = \begin{bmatrix} -1 & \dfrac{1}{\lambda\rho_2(\ell)} & -1 & -\dfrac{1}{\lambda\rho_2(\ell)} \\ 0 & 1 & 0 & 1 \end{bmatrix},$$

In terms of the properties of the determinants, one obtains

$$\det\left(\widehat{E} + \lambda^{-1}\widehat{E}_1\right) = \det\widehat{E} + \lambda^{-1}E_0(\lambda). \tag{23}$$

Since $\xi_{kj}(x,\lambda), 1 \le k,j \le 4$ for $\lambda \in G^-$ are analytic and bounded uniformly for $\lambda \in G^-$ and $x \in [0,\ell]$, $E_0(\lambda)$ is analytic and uniformly bounded for $\lambda \in G^-$.

A simple caculation yields that $\det\widehat{E} = 0$ is equivalent to

$$\exp(\mu_1(\ell)) + \exp(\mu_3(\ell) = 0, \tag{24}$$

or

$$\exp(\mu_2(\ell)) + \exp(\mu_4(\ell) = 0. \tag{25}$$

The roots of (24) and (25) are

$$\lambda_{jn} = -\frac{\int_0^\ell \rho_j b_j\, ds}{2\int_0^\ell \rho_j\, ds} + i\frac{(2n+1)\pi}{2\int_0^\ell \rho_j\, ds},$$
$$j = 1,2,\ n = 0, \pm 1, \pm 2, \cdots.$$

Now, for large positive integer $n$, define two rectangles as follows:

$$\Lambda_{jn} = \left\{\lambda \in G^- \,\middle|\, |\operatorname{Re}(\lambda - \lambda_{jn})| = 1/\sqrt{n},\right.$$
$$\left.|\operatorname{Im}(\lambda - \lambda_{jn})| = 1/\sqrt{n}\right\},\ j = 1,2. \tag{26}$$

To apply Rouche's theorem, it needs to estimate $\det\widehat{E}$ on $\Lambda_{jn}$ for $j = 1,2$. In the following discussion, one assumes that

$$\int_0^\ell \rho_2 b_2\, ds\left(\int_0^\ell \rho_2\, ds\right)^{-1} \neq \int_0^\ell \rho_1 b_1\, ds\left(\int_0^\ell \rho_1\, ds\right)^{-1}. \tag{27}$$

Then by a straight estimation, one knows that there exists $n_6 > 0$, such that

$$|\lambda^{-1}E_0(\lambda)| < |\det\widehat{E}|, \quad \forall\ \lambda \in \Lambda_{jn},\ j = 1,2,\ n > n_6. \tag{28}$$

Therefore, it follows from Rouche's theorem that the following theorem holds:

**Theorem 4.1** Assume that $\lambda$ is an eigenvalue of the system (5). If $|\lambda|$ is sufficiently large and the condition (27) holds, then there exists some $n$ such that

$$|\lambda - \lambda_{jn}| < \frac{1}{\sqrt{n}}. \quad j = 1 \text{ or } 2,$$

where $n$ is dependent only on $\lambda$. Therefore, $\lambda$ satisfies

$$\lambda = -\frac{\int_0^\ell \rho_j b_j\, ds}{2\int_0^\ell \rho_j\, ds} + i\frac{(2n+1)\pi}{2\int_0^\ell \rho_j\, ds} + \mathcal{O}\left(\frac{1}{\sqrt{n}}\right),$$
$$j = 1,2,\ n \text{ large}.$$

**Remark 4.2** If $\dfrac{\int_0^\ell \rho_1 b_1\, ds}{\int_0^\ell \rho_1 ds} = \dfrac{\int_0^\ell \rho_2 b_2\, ds}{\int_0^\ell \rho_2 ds}$ and $\dfrac{\int_0^\ell \rho_1 ds}{\int_0^\ell \rho_2 ds}$ is an irrational number, then by a well-known result of number theory (see Theorem 7.9 in (Stark, 1979)),

$$\inf_{n\neq m} |\lambda_{1n} - \lambda_{2m}| = \inf_{n\neq m}\left|\frac{i(2n+1)\pi}{2\int_0^\ell \rho_1 ds} - \frac{i(2m+1)\pi}{2\int_0^\ell \rho_2 ds}\right| = 0,$$

which means that there is no gap between the zeros of $\det\widehat{E}$. In this case, it is hard to obtain the explict asymptotic expressions like in Theorem 4.1 for the eigenvalues by the Rouche's theorem.

## 5. EXPONENTIAL STABILITY OF THE CLOSED LOOP SYSTEM(II)

In Section 3, the condition that $\rho_1(x) \neq \rho_2(x)$ for all $x \in [0,\ell]$ plays an essential role to the applicability of Birkhoff's asymptotic expansion method. For the exponential stabilization under the case that the wave speeds are equal ($\rho_1 = \rho_2$), in order to avoid lengthy discussion, one considers only the uniform beam, i.e., one assumes that $\rho, K, I_\rho, EI$, are positive constants. First, one needs two lemmas.

**Lemma 5.1** For the uniform beam, assume that $\mathcal{A}\Psi_n = i\omega_n\Psi_n$, where $\Psi_n = [w_n, z_n, \varphi_n, \psi_n]^\tau \in \mathcal{D}(\mathcal{A})$, $||\Psi_n||_{\mathcal{H}} = 1$ and $\omega_n^2 \to \infty$ when $n \to \infty$, then there exists $\delta > 0$, such that

$$\liminf_{n\to\infty} \int_\alpha^\beta \rho|z_n|^2\, dx + \int_\alpha^\beta I_\rho|\psi_n|^2\, dx \ge \delta.$$

An immediate consequence of Lemma 2.1 and Lemma 2.2 is that $\mathcal{A}$ has pure point spectrum, i.e. $\sigma(\mathcal{A}) = \{i\omega_n \mid n = 1, 2, \ldots\}$, and that for every $n$, the dimension of the eigensubspace corresponding to the eigenvalue $i\omega_n$ is finite.

**Lemma 5.2** For a uniform beam, with the above notations, one has $\omega_n = \mathcal{O}(n)$.

By using the result in (Chen et al,1991), one is now able to prove the exponential stability of the closed loop system.

**Theorem 5.3** For a uniform beam, the closed loop system (1) (2) (3) is exponentially stable under the condition that $\rho_1 = \rho_2$.

## REFERENCES


Birkhoff, G..D. and Langer, R. E.(1923). The boundary problems and developments associated with a system of ordinary linear differential equations of the first order, *Proc. Amer. Acad. Arts Sci.*, **2**, 51–128.

Chen, G. et al.(1987a). Modeling, stabilization and control of serially connected beams, *SIAM J. Control & Optimization*, **25**, 526–546.

Chen, G. et al.(1987b). The Euler-Bernoulli beam equation with boundary energy dissipation, *Operator Methods for Optimal Control Problems*, S. J. Lee ed., Lecture Notes in Pure and Applied Mathematics Series, 67–96, Marcel Dekker Inc., New York.

Chen, G. et al.(1991). Exponential decay of energy of evolution equations with locally distributed damping, *SIAM J. Appl. Math.*, **51**, pp.266–301.

Courant, R. and Hilbert, D.(1965). *Methods of Mathematical Physics*, Vol.I, Interscience, New York.

Desch, W. et al.(1987). Boundary stabilization of an Euler-Bernoulli beam with viscoelastic damping, *Proc. IEEE Conf. on Decision and Control*, Los Angeles, CA, 1792–1795.

Ho, L. F.(1990). Exact controllability of the one-dimensional wave equation with locally distributed control, *SIAM J. Control & Optimization*, **28**, 733–748.

Huang, F. L.(1985) Characteristic conditions for exponential stability of linear dynamical system in Hilbert spaces, *Ann. Diff. Eqs.*, **1**, 43–53.

Kim, J. H. and Renardy, Y.(1987). Boundary control of the Timoshenko beam, *SIAM J. Control & Optimization*, **25**, 1417–1429.

Lagnese, J.(1983). Control of wave processes with distributed controls supported on a subregion, *SIAM J. Control & Optimization*, **21**, 68–85.

Liu, K. S.(1997). Locally distributed control and damping for the conservative systems, *SIAM J. Control & Optimization*, **35**, 1574–1590.

Liu, K. S., Huang, F. L. and Chen G.(1989). Exponential stability analysis of a long chain of coupled vibrating strings with dissipative linkage, *SIAM J. Appl. Math.*, **49**, 1694–1707.

Pazy, A.(1983). *Semigroups of Linear Operators and Applications to Partial Diffrential Equations*, Springer-Verlag, New York.

Russell, D. L.(1978). Controllability and Stabilizability theory for linear partial differential equations, recent progress and open questions, *SIAM Review*, **20**, 639–740.

Russell, D. L.(1986) Mathematical models for the elastic beam and their control-theoretic implications, in *Semigroups, Theory and Applications*, Vol.2, H.Brezis, H. G. Crandall and F. Kapell, Eds., Essex, England: Longman, 177–217.

Stark, Harold M.(1979). *An Introduction to Number Theory*, The MIT Press, Cambrige, Massachusetts, 1979.

Timoshenko, S.(1955). *Vibration Problems in Engineering*, Van Nostand, New York.

Wang, H. K. and Chen, G.(1991). Asymptotic locations of eigenfrequencies of Euler-Bernoulli beam with nonhomogeneous structural and viscous damping coefficients, *SIAM J. Control & Optimization*, **29**, 347–367.

F-2d-02-5
14th Triennial World Congress, Beijing, P.R. China

# ANALYSIS AND SYNTHESIS OF OPTIMAL AND ROBUSTLY OPTIMAL CONTROL SYSTEMS UNDER BOUNDED DISTURBANCES

**Vsevolod M. Kuntsevich** * **Alexei V. Kuntsevich** **,[1]

* *Space Research Institute, 40 Prospekt Akademika Glushkova, 252022 Kiev, Ukraine, E-mail:* vmkun@d305.icyb.kiev.ua
** *Institute for Mathematics, University of Graz, Heinrichstr.,36, A-8010 Graz, Austria, E-mail:* alex@bedvgm.kfunigraz.ac.at

Abstract: Analysis of dynamic (closed-loop control) systems under bounded disturbances by calculating minimal invariant sets is discussed. It is suggested to use the radius of a minimal invariant set as a particular numerical measure of the stability degree of a system and apply such a criterion for optimal linear control synthesis by means of minimizing this radius. It has been proven that the radius of a minimal invariant set reaches its minimum at a nilpotent parameter matrix of a system. The solution to the optimal control synthesis problem can be found, therefore, by calculating the linear feedback parameters providing the nilpotent matrix of the closed-loop system. In the case of scalar controls, the problem has got an analytical solution. *Copyright © 1999 IFAC*

Keywords: control synthesis, optimal control, invariant, sets, bounded disturbances, discrete systems, set values.

## 1. INTRODUCTION

Set-valued models have been widely applied to system identification and observation, robust and adaptive control synthesis. Set-membership identification for parameters and perturbations, respectively worst-case analysis, has become a regular tool in modern control theory and requires completely different methods compared to average case analysis. Nowadays, the so-called hard bound approach, which uses only a priori given bounds on disturbances, unknown system parameters and model errors, attracts more and more mathematicians and engineers.

During the last decade, the use of hard bound models (set-valued estimates) has been widening rapidly, particularly in identification (see Milanese et al. (1996),Walter and Pronzato (1997)). In contrary, considerably less efforts have been paid to apply the hard-bound approach to automatic control synthesis. The reason for this goes back to the necessity of calculating minimal invariant sets. As soon as an invariant set has been determined, one obtains a measure of the dynamic precision of a system, which allows us to estimate the maximal deviation of solutions from a nominal solution and can be used the same way as a mean square deviation in stochastic dynamics. Actually, the existence of a bounded invariant set (the property of the ultimate boundedness or dissipativity) is one of the most significant group properties of that class of closed-loop control systems (dynamic plants) determined by noise and/or parameters within the given bounds.

Apparently, the problem of calculating an invariant set for a plant under bounded disturbances has been posed for the first time by Bulgakov (1946). Further, Yoshizawa (1966) and Barbashin

[1] Supported by the Special Research Center F-003 and the JF-NBA project No. 6243, Austria

(1970) have been using the apparatus of Lyapunov functions for calculating stable limit cycles. They succeeded to obtain some significant results, particularly for nonlinear systems. However, these results revealed some serious disadvantages of the approach. In fact, the final success essentially depends on the choice of a particular Lyapunov function.

The early results on calculating minimal invariant sets by use of an exact algebraic formula first proved in Kuntsevich and Pshenitchnyi (1996a) can be found in Kuntsevich and Pshenitchnyi (1996b, 1997).

Here the results by Kuntsevich and Pshenitchnyi (1996a, 1996b, 1997) are extended towards synthesis of optimal and robustly optimal control systems. With the aim to apply the apparatus of minimal invariant sets to the control synthesis problem, the radius of a minimal invariant set is used as a criterion. The key theorem claims that the radius of a minimal invariant set reaches its minimum at the nilpotent parameter matrix. Hence the problem of optimal control synthesis takes the form of calculating the feedback parameters providing the nilpotent matrix of the closed-loop control system. The solution to this problem for the case of scalar controls is calculated analytically.

In view of the fact that our main goal is to develop computational methods, we present in the following the main concepts for discrete-time systems.

## 2. ANALYSIS OF DYNAMIC SYSTEMS UNDER BOUNDED DISTURBANCES

### 2.1 *Calculating the minimal invariant set of a linear disturbed system*

Consider the class of linear control systems given by the equation

$$X_{n+1} = \boldsymbol{A}X_n + \boldsymbol{B}U_n + F_n, \quad n = 0, 1, \dots, \quad (1)$$

where $X_n$ and $U_n$ are a state $m$-dimensional vector, resp. a control $p$-dimensional vector, at the $n$th discrete time instant, $\boldsymbol{A}$ and $\boldsymbol{B}$ are the matrices of the respective dimensions and $F_n$ is an $m$-dimensional vector of bounded disturbances:

$$F_n \in \mathfrak{F}, \quad n = 0, 1, \dots, \quad (2)$$

$\mathfrak{F}$ is a convex compact set.

When a linear control feedback is used,

$$U_n = \boldsymbol{C}X_n, \quad (3)$$

the equation of a closed-loop system takes the form

$$X_{n+1} = \boldsymbol{L}X_n + F_n, \quad n = 0, 1, \dots, \quad (4)$$

where $\boldsymbol{L} = \boldsymbol{A} + \boldsymbol{BC}$.

Independent from other matrix properties which might be needed later, $\boldsymbol{L}$ is assumed to be a Schur-Cohn matrix, i.e., all eigenvalues are in the open unit circle:

$$|\lambda_i(\boldsymbol{L})| \le q < 1, \quad i = 1, \dots, m. \quad (5)$$

We assume a system (1) given by matrices $\boldsymbol{A}$ and $\boldsymbol{B}$ is controllable according to Kalman's criterion. Hence the set of matrices $\boldsymbol{C}$ providing the fulfilment of the condition (5) is not empty.

Introduce the set

$$\mathfrak{M}(\boldsymbol{L}, \mathfrak{F}) = \mathfrak{F} + \boldsymbol{L}\mathfrak{F} + \boldsymbol{L}^2\mathfrak{F} + \dots, \quad (6)$$

where the sum of sets is a Minkowski sum. Recall, set $\mathfrak{M}$ is called an invariant set for the system (1-2), if for $X \in \mathfrak{M}$, the condition $\boldsymbol{L}X + F \in \mathfrak{M}$ is fulfilled with any $F \in \mathfrak{F}$, i.e., $\boldsymbol{L}\mathfrak{M} + \mathfrak{F} \subseteq \mathfrak{M}$.

*Definition 1.* A compact set is called a minimal invariant set of a system, if it is invariant and does not contain an invariant proper subset.

The following two theorems have been proven by Kuntsevich and Pshenitchnyi (1997).

*Theorem 1.* Under the assumption (5), the set $\mathfrak{M}(\boldsymbol{L}, \mathfrak{F})$ given by formula (6) is a compact set in $\mathbb{R}^m$.

*Theorem 2.* $\mathfrak{M}(\boldsymbol{L}, \mathfrak{F})$ given by formula (6) is a minimal invariant set for a system (4). If $X_0 \notin \mathfrak{M}(\boldsymbol{L}, \mathfrak{F})$, then for any $\varepsilon > 0$, all the states of system (4) are inside the $\varepsilon$-neighborhood of $\mathfrak{M}(\boldsymbol{L}, \mathfrak{F})$ starting at a particular discrete-time moment which depends on $\varepsilon$.

Theorems 1 and 2 determine a convex minimal invariant set as soon as $\mathfrak{F}$ is a convex set.

Note that if a matrix $\boldsymbol{A}$ is nilpotent, the series (6) is finite.

### 2.2 *Invariant sets of linear nonstationary systems*

Calculating minimal invariant sets of linear nonstationary dynamic (closed-loop control) systems under bounded disturbances are discussed here by recollecting the results by Kuntsevich and Pshenitchnyi (1996b, 1997).

Assume the class of dynamic systems is given by the equation

$$X_{n+1} = \boldsymbol{A}_nX_n + \boldsymbol{B}U_n + F_n, \quad n = 0, 1, \dots,$$

where all the notations are the same as in (1), except $\boldsymbol{A}_n$ is a matrix of time-varying parameters:

$$\boldsymbol{A}_n = \mathring{\boldsymbol{A}} + \delta\boldsymbol{A}_n, \quad n = 0, 1, \dots .$$

Here $\overset{\circ}{\boldsymbol{A}}$ is a constant matrix and

$$\delta \boldsymbol{A}_n \in \delta\mathfrak{A} = \{\delta \boldsymbol{A} : \ \|\delta \boldsymbol{A}\| \leq \sigma\}, \quad n = 0, 1, \dots .$$

Assume linear feedback (3) for this class of systems and consider the closed-loop control system:

$$X_{n+1} = \boldsymbol{L}_n X_n + F_n, \quad n = 0, 1, \dots , \qquad (7)$$

where

$$\boldsymbol{L}_n = \overset{\circ}{\boldsymbol{A}} + \boldsymbol{CB} + \delta\boldsymbol{A}_n, \quad n = 0, 1, \dots .$$

Again, assume $\boldsymbol{L} = \overset{\circ}{\boldsymbol{A}} + \boldsymbol{CB}$ is a Schur-Cohn matrix (see (5)). Introduce the $\varepsilon$-neighborhood of the set $\mathfrak{F}$:

$$\mathfrak{F}_\varepsilon = \mathfrak{F} + \varepsilon\mathfrak{S},$$

where $\mathfrak{S}$ is a unit solid sphere with the center at the origin, and the set

$$\mathfrak{M}(\boldsymbol{L}, \mathfrak{F}_\varepsilon) = \mathfrak{M}(\boldsymbol{L}, \mathfrak{F}) + \varepsilon\mathfrak{S} + \varepsilon\boldsymbol{L}\mathfrak{S} + \varepsilon\boldsymbol{L}^2\mathfrak{S} + \dots . \qquad (8)$$

Kuntsevich and Pshenitchnyi (1997) proved that (8) is a minimal invariant set for the class of nonstationary systems (7) for any $\varepsilon$ satisfying the inequality

$$\sigma \leq \frac{1-q}{\alpha} \frac{\varepsilon}{d(\mathfrak{F}) + \varepsilon},$$

where $d(\mathfrak{F})$ is the diameter of the set $\mathfrak{F}$ and $\alpha$ is a constant providing the fulfilment of the inequality $\boldsymbol{L}^n \leq \alpha q^n$. The latter is correct for a matrix satisfying the condition (5) and a constant $\alpha > 1$.

### 2.3 *Invariant sets of dynamic systems with bounded nonlinear parts*

The following class of systems have been analized by Kuntsevich and Pshenitchnyi (1996a).

$$X_{n+1} = \boldsymbol{L}X_n + \Phi(X_n), \quad n = 0, 1, \dots , \qquad (9)$$

where $\boldsymbol{L}$ meets the condition (5) and $\Phi(\cdot)$ is a continuous vector function $\mathbb{R}^m \to \mathbb{R}^m$, therewith

$$\mathfrak{F} = \mathrm{cl}\, \Phi(\mathbb{R}^m) \qquad (10)$$

is a compact set in $\mathbb{R}^m$. Here $\mathrm{cl}(\cdot)$ means closing a set.

It is straightforward that the systems (9,10) and (4,2) at equivalent by means of having the same trajectories. Hence $\mathfrak{M}(\boldsymbol{L}, \mathfrak{F})$ calculated by formula (6) is a minimal invariant set of system (9,10) by virtue of Theorem 2.

Analysis of nonautonomous systems with bounded nonlinear parts leads to the similar result (see Kuntsevich and Pshenitchnyi (1996a)).

### 2.4 *Invariant sets of special type nonlinear systems*

Consider the class of systems given by the equation

$$X_{n+1} = \Phi(X_n) + F_n, \quad n = 0, 1, \dots , \qquad (11)$$

where $\Phi(\cdot)$ is a nonlinear vector function $\mathbb{R}^m \to \mathbb{R}^m$ satisfying the inequality

$$\|\Phi(X)\| \leq \mu\|X\|, \quad \mu < 1, \qquad (12)$$

and the disturbances $F_n$ are bounded by (2).

Under the condition (12), a system (11) has a bounded invariant set. However, one cannot calculate this set explicitly in contrary to the case of linear systems (4). On the other hand, since one has got an upper bound for the values of $\Phi(X_n)$, it becomes possible to estimate the radius of a minimal invariant set,

$$r(\mathfrak{M}) = \max\{\|X\| : \ X \in \mathfrak{M}\},$$

by the following iterative procedure.

Assume $\mathfrak{M}_0 = \mathfrak{F}$, $r(\mathfrak{M}_0) = r(\mathfrak{F}) = \rho$.

Calculate $\mathfrak{M}_1 = \Phi(\mathfrak{F}) + \mathfrak{F}$. The radius of $\mathfrak{M}_1$ is obviously bounded:

$$r(\mathfrak{M}_1) \leq r(\Phi(\mathfrak{F})) + \rho \leq \rho + \mu\rho.$$

After estimating the radius $r(\mathfrak{M}_k)$ by its upper bound $\rho\sum_{j=0,\dots,k}\mu^j$, estimating $r(\mathfrak{M}_{k+1})$ is straightforward.

*Theorem 3.* The radius of the minimal invariant set $\mathfrak{M}(\Phi, \mathfrak{F})$ for the system (11,12) is upper bounded by

$$r(\mathfrak{M}(\Phi, \mathfrak{F})) \leq \rho\sum_{k=0}^{\infty}\mu^k = \frac{\rho}{1-\mu}.$$

### 2.5 *Minimum radius invariant set*

Hereinafter, the radius of a minimal invariant set (or an estimate for this set) is used as a control performance criterion.

A minimal invariant set, which has minimal radius among all other invariant sets, will be called a *minimum radius invariant set.*

Consider a control system (4,2) with

$$\boldsymbol{L} = \begin{Vmatrix} 0 & 1 & \dots & 0 \\ \vdots & & \ddots & \\ 0 & 0 & \dots & 1 \\ l_1 & l_2 & \dots & l_m \end{Vmatrix}, \qquad (13)$$

meaning $\boldsymbol{L}$ takes the standard Frobenius form.

Think of $\boldsymbol{L}$ as the sum

$$\boldsymbol{L} = \boldsymbol{L}_0 + \boldsymbol{L}_m$$

of a nilpotent (with the nilpotency index $m$) matrix

$$L_0 = \begin{Vmatrix} 0 & 1 & \dots & 0 \\ \vdots & & \ddots & \\ 0 & 0 & \dots & 1 \\ 0 & 0 & \dots & 0 \end{Vmatrix}$$

and a matrix

$$L_m = \begin{Vmatrix} & \mathbf{0} & \\ \dots & \dots & \dots \\ l_1 & \dots & l_m \end{Vmatrix}.$$

*Theorem 4.* $\mathfrak{M}(L_0, \mathfrak{F})$ is a minimum radius invariant set of system (4,2,13).

Theorem 4 claims that minimum of the radius is reached at the nilpotent parameter matrix $L = L_0$.

**PROOF.** Consider the sum of the first two items in formula (6):

$$\mathfrak{M}(L, \mathfrak{F})_1 = \mathfrak{F} + L\mathfrak{F} = (\mathfrak{F} + L_0\mathfrak{F}) + L_m\mathfrak{F}.$$

Obviously, $L_m\mathfrak{F} = \{0\}$, if $\forall F \in \mathfrak{F}\ \sum_{i=1,\dots,m} l_i f_i = 0$. The latter is true, iff $a_i = 0$, $i = 1, \dots, m$. The sum of nonempty sets is larger than each of them, i.e., $\mathfrak{M}(L_0, \mathfrak{F})_1 \subset \mathfrak{M}(L, \mathfrak{F})_1$. Therefore, $r\,(\mathfrak{M}(L_0, \mathfrak{F})_1) < r\,(\mathfrak{M}(L, \mathfrak{F})_1)$.

Similar considerations for the consecutive sets $\mathfrak{M}(L, \mathfrak{F})_k$, the sum of $k+1$ items in formula (6), lead to the same result. By induction, one obtains

$$r\,(\mathfrak{M}(L_0, \mathfrak{F})_\infty) < r\,(\mathfrak{M}(L, \mathfrak{F})_\infty).$$

The use of Theorem 4 will be shown in the next section.

## 3. SYNTHESIS OF CONTROL SYSTEMS UNDER BOUNDED DISTURBANCES

### 3.1 *Optimal control synthesis with known parameter matrix*

By virtue of Theorem 4, the synthesis problem is reduced to calculating a feedback matrix $C_{opt}$ providing the needed property of a matrix $L$. Such matrix $C_{opt}$ always exists, if system (1) is controllable according to Kalman's criterion.

Calculating a feedback matrix is the easiest for scalar controls. In this case, it is appropriate to consider matrices $A$ and $B$ of the standard form

$$A = \begin{Vmatrix} 0 & \vdots & I \\ \dots & \dots & \dots \\ & A_m^T & \end{Vmatrix}, \quad B = \begin{Vmatrix} 0 \\ \vdots \\ b_m \end{Vmatrix}$$

without loss of generality. Obviously, matrix $L = A + BC$ is of the standard form (13).

Theorem 4 claims the minimum radius invariant set at $L = L_0$, i.e.,

$$L_m = A_m + b_m C = 0.$$

Hence

$$C_{opt} = -\frac{1}{b_m} A_m.$$

For vector controls, the synthesis problem cannot be solved that elegantly and easily it is a particular case of a modal control synthesis problem (see Kuntsevich and Kuntsevich (1994, 1996) for instance). The solution of the synthesis problem in this case certainly cannot be given in explicit form and will require solving rather complex minimax problems.

### 3.2 *Synthesis of optimal control systems with uncertain parameters*

*Remark.* Synthesizing a control, which provides a minimum radius invariant set for a closed-loop system, one obviously has to check first the robust stabilizability of the given class of systems. Unfortunately, the robust stabilizability problem has not been solved yet. That is why the robust stabilizability of the class of systems under consideration is assumed here a priori.

Here, the consideration is extended to the class of systems with uncertain parameters assuming *the coefficients of $L$ may take their values in a given compact convex set $\mathfrak{L}$.*

*Definition 2.* Assuming Schur-Cohn condition (5) is fulfilled for every matrix $L \in \mathfrak{L}$, a minimal invariant set of the class of systems (4) with uncertain parameters is the union of the minimal invariant sets calculated for every particular matrix $L$ from the set $\mathfrak{L}$:

$$\mathfrak{M}(\mathfrak{L}, \mathfrak{F}) = \bigcup_{L \in \mathfrak{L}} \mathfrak{M}(L, \mathfrak{F}).$$

In particular, if

$$\mathfrak{L} = \operatorname*{conv}_{i=1,\dots,N} \{L_i\},$$

meaning $\mathfrak{L}$ is a polytope with $N$ vertices, calculating a minimal invariant set becomes less difficult. The following result was first presented by Kuntsevich (1996).

*Theorem 5.* If the condition (5) is fulfilled for every $L \in \mathfrak{L}$, then

$$\bar{\mathfrak{M}}(\mathfrak{L}, \mathfrak{F}) = \operatorname{conv}\{\mathfrak{M}(L_i, \mathfrak{F}) : i = 1, \dots, N\}$$

is i) an invariant compact convex set for system (4) with uncertain parameter matrix $L \in \mathfrak{L}$ and ii) a non-reducible upper bound convex estimate for the minimal invariant set of the class of systems.

The most significant corollary from the above is obvious:

*Corollary 1.* Under the conditions of Theorem 5, the radius of a convex estimate $\bar{\mathfrak{M}}(\mathfrak{L},\mathfrak{F})$ equals the radius of a minimal invariant set $\mathfrak{M}(\mathfrak{L},\mathfrak{F})$.

The radius $r(\bar{\mathfrak{M}}(\mathfrak{L},\mathfrak{F}))$ can be calculated directly by evaluating the norms of the vertices. However, the use of such a minimization criterion is limited to low dimensions $m$, since one has to pay tremendous computational efforts for higher dimensions.

Alternatively, consider the class of systems with scalar control and uncertain parameters $A_m \in \mathfrak{A}_m$ and $b_m \in \mathfrak{b}_m$, where

$$\mathfrak{A}_m = \underset{i=1,\dots,N}{\text{conv}}\{A_m^i\}$$

is a polytope and

$$\mathfrak{b}_m = \{b_m : \underline{b}_m \le b_m \le \bar{b}_m\},$$

is an interval, $A_m^i$, $i = 1,\dots,N$, are the vertices of a given polytope $\mathfrak{A}_m$ and $\underline{b}_m, \bar{b}_m$ are given constants.

Let us point to the features of a nilpotent matrix except those making use of eigen values. Consider the $m$-th degree of a matrix $\boldsymbol{L}$ of the standard form (13)

$$\boldsymbol{L}^m = \left\|\begin{matrix} L_m^T \\ \dots\dots \\ G(L_m)\end{matrix}\right\|, \quad G(0) = 0. \tag{14}$$

It is straightforward that $\boldsymbol{L}^m = 0$, iff $L_m = 0$. If $L_m = \varepsilon L_m^0$ and $\varepsilon \to 0$, $\boldsymbol{L}^m \to 0$ at a degree of $\varepsilon^{m-1}$. This is a strong reason for the use of a measure of deviation of the vector $L_m$ from the null vector as a parametric synthesis criterion. This requires solution of the minimax problem

$$\min_{C} \max_{\substack{A_m \in \mathfrak{A}_m \\ b_m \in \mathfrak{b}_m}} \{\|A_m + b_m C\|\}. \tag{15}$$

It is worth to note that the optimization problem (15) appears to be quite the same as considered in Kuntsevich and Kuntsevich (1994), where another performance criterion was used.

Exploring the way the synthesis problem has been solved above analytically for a system with known parameter matrix, calculate the center $\overset{\circ}{A}_m$ of the set $\mathfrak{A}_m$ as the center of the minimal upper bound solid sphere containing the polytope and the center $\overset{\circ}{b}_m = 0.5(\underline{b}_m + \bar{b}_m)$ of the interval set $\mathfrak{b}$ (here, we assume $0 \notin \mathfrak{b}$). Now, represent the given parameter estimates in the centralized form

$$\mathfrak{A}_m = \overset{\circ}{A}_m + \delta\mathfrak{A}_m \text{ and } \mathfrak{b}_m = \overset{\circ}{b}_m + \delta\mathfrak{b}_m, \tag{16}$$

where $\delta\mathfrak{A}_m = \text{conv}\{A_m^i - \overset{\circ}{A}_m : i = 1,\dots,N\}$.

The following theorem gives the solution for the problem.

*Theorem 6.* Under the conditions of Theorem 4 and for the given set-valued parameter estimates (16), the feedback vector

$$C_{opt} = -\overset{\circ}{A}_m/\overset{\circ}{b}_m$$

is a minimizer for the problem (15).

**PROOF.** Further in the proof, the index $m$ will be omitted. Introduce the notation

$$\varphi(\delta A, \delta b, C) = \|\overset{\circ}{A} + \delta A + (\overset{\circ}{b} + \delta b)C\| \tag{17}$$

for the objective function of the minimax problem (15). Here $\delta A \in \delta\mathfrak{A}$ and $\delta b \in \delta\mathfrak{b}$. Assume $C_{opt} = -\overset{\circ}{A}/\overset{\circ}{b}$ and substitute $C$ by $C_{opt}$ in (17) to obtain

$$\varphi(\delta A, \delta b, C_{opt}) = \left\|\delta A - \frac{\delta b}{\overset{\circ}{b}}\overset{\circ}{A}\right\|. \tag{18}$$

Equivalently,

$$\varphi(\delta A, \delta b, C_{opt}) = \varphi(\delta L) = \|\delta L\| \tag{19}$$

when introducing the notation $\delta L = \delta A - \delta b \overset{\circ}{b}{}^{-1}\overset{\circ}{A}$, where $\delta L$ takes its values on the set $\delta\mathfrak{L} = \delta\mathfrak{A} - \delta\mathfrak{b}\overset{\circ}{b}{}^{-1}\overset{\circ}{A}$.

Set $\delta\mathfrak{L}$ is a convex compact set, since the sets $\delta\mathfrak{A}$ and $\delta\mathfrak{b}$ are convex and compact. Furthermore, these two sets contain the origin, hence $\delta\mathfrak{L}$ contains the origin too. This important property will be used later in the proof.

Consider a parameter vector $C = C_{opt} + C_1$, where $C_1$ is an arbitrary increment vector, with the aim to prove that

$$\varphi(\delta A, \delta b, C_{opt}) \le \varphi(\delta A, \delta b, C),$$

i.e., $C_{opt}$ is a minimizer for the problem (15).

Substitute $C$ by $C_{opt} + C_1$ and obtain

$$\varphi(\delta A, \delta b, C) = \|\delta A - \delta b \overset{\circ}{b}{}^{-1}\overset{\circ}{A} + (\overset{\circ}{b} + \delta b)C_1\|.$$

Next, take into account the notations accepted in (19) and rewrite the above in the form:

$$\varphi(\delta A, \delta b, C) = \varphi(\delta L, b, C_1) = \|\delta L + bC_1\|. \tag{20}$$

Now, recall that $b$ is strictly negative or strictly positive on $\mathfrak{b}$ ($0 \notin \mathfrak{b}$) and $\delta\mathfrak{L}$ contains the origin. Hence for any $C_1$, one can find a vector $\delta L_\star \in \delta\mathfrak{L}$ such that

$$\|\delta L_\star + bC_1\| \ge \|\delta L_\star\|.$$

Theorem 6 generalizes the corresponding result published by Kuntsevich and Kuntsevich (1997) to the case of a convex polytope $\mathfrak{A}_m$.

Obviously, the robustly optimal control synthesis problem should have been considered at a parameter matrix belonging to a Schur-Cohn set.

However, it seems almost impossible to overcome the major difficulty of applying this constraint, because there is no constructive (in view of computations) description of a set of stable matrices. Hence, one has to check the robust stability of the synthesized control system after calculating $C_{opt}$. If the robust stability conditions are not fulfilled, the only way is to run a set-membership estimating procedure anew with the aim to reduce an uncertainty in parameters and then recalculate $C_{opt}$.

## 4. CONCLUSION

By accepting the hard-bound concept, one necessarily has to reformulate anew all major analysis and control synthesis problems and also one faces new unique problems having nothing in common with those met when using stochastic models.

The purpose of this paper is finding an easily calculated measure of a system reaction to bounded disturbances. The presented methods of analysis and synthesis make use of the performance criterion, which is minimizing the radius of a minimal invariant set of a dynamic (closed-loop control) system. These methods are applied to linear and some types of nonlinear systems. The class of nonlinear models under consideration is restricted to those with strictly bounded nonlinear parts. It is very important to extend this class.

A more difficult and commonly faced problem of the robust stabilizability remains yet unsolved. However, the existing control synthesis methods proved to be applicable (and sometimes very efficient) to a wide class of real life control systems, where the hard-bound concept is the only acceptable.

## References


E.A. Barbashin. *Introduction to the theory of stability*. Wolters-Noordhoff, Groningen, 1970.

B.V. Bulgakov. On the accumulation of disturbances in linear oscillatory systems with constant parameters. *Doklady Acad. Sci. URSS*, 51:343 345, 1946. (in Russian).

A.V. Kuntsevich. Set-membership identification for robust control. In *CESA'96 IMACS Multiconference (Lille-France, 1996)*, volume 2, pages 1168 1172, Lille, 1996.

A.V. Kuntsevich and V.M. Kuntsevich. Linear adaptive control for nonstationary systems under bounded noise. *System and Control Letters*, 31:33 40, 1997.

V.M. Kuntsevich and A.V. Kuntsevich. Modal control synthesis under uncertainty. In *IFAC Simposium on Robust Control (Rio de Janeiro, 1994)*, pages 124 129, Rio-de-Janeiro, 1994.

V.M. Kuntsevich and A.V. Kuntsevich. Robust stability analysis and control synthesis under uncertainty. *Systems Science*, 22(3):17 28, 1996.

V.M. Kuntsevich and B.N. Pshenitchnyi. Invariant and stationary sets of nonlinear discrete systems under bounded disturbances. *Problemy upravleniya i informatiki (Journal of Automation and Information Sciences)*, (1-2):35 45, 1996b. in Russian.

V.M. Kuntsevich and B.N. Pshenitchnyi. Minimal invariant sets of dynamic systems with bounded disturbances. *Cybernetics and Systems Analysis*, 32(1):58 64, 1996a. Translated from *Kibernetika i Sistemnyi Analiz*, No. 1, pp.74-81, 1996.

V.M. Kuntsevich and B.N. Pshenitchnyi. Minimal invariant sets of discrete systems with stable linear parts. *Problemy upravleniya i informatiki (Journal of Automation and Information Sciences)*, (1):81 91, 1997. in Russian.

M. Milanese, J. Norton, H. Piet-Lahanier, and E. Walter, editors. *Bounding approaches to system identification*. Plenum Press, New-York and London, 1996.

E. Walter and L. Pronzato. *Identification of Parametric Models from Experimental Data*. Springer-Verlag, Berlin Heidelberg, 1997.

T. Yoshizawa. *Stability theory by Liapunov's second method*. Number 9 in Publ. of the Math. Soc. of Japan. The Mathematical Society of Japan, 1966.

# THE OPTIMAL STABILIZATION OF PLASMA UNDER UNCERTAINTY CONDITIONS

**Vyacheslav F. Gubarev Oleg S. Yakovlev**

*Space Research Institute of Ukrainian National Academy of Sciences and National Space Agency of Ukraine*
*40, Acad. Glushkov Prsp., 252022 Kiev-22, Ukraine*
*Fax: 380+44+266-4124, E-mail: gvf@d310.icyb.kiev.ua*

Abstract: The model of magnetohydrodynamic plasma equilibrium under uncertainty conditions is offered. Initially the processes in plasma was described by the complicated system of partial differential and ordinary equations that may be qualified as system with distributed parameters. Then using real physical assumptions and hypothesis the model was reduced to the system of ordinary differential equations, which contains vector of control and vector of perturbation. Original approach for plasma stabilization problem based on the method of analytical synthesis of the admissible set of nonlinear feedback is proposed. The admissible range of the controller parameters is obtained and optimization method that permits to evaluate optimal solution of stabilization problem in admissible area is developed. 

Keywords: Nonlinear control, Uncertain dynamic systems, Modeling errors, Robust control, Stabilization

## 1. INTRODUCTION

In the modern and perspective thermonuclear installations based on the tokamak the functions that implement the systems of automatic control and stabilization are permanently extended. But for stable plasma confinement it is especially important to adjust and to stabilize the parameters which are responsible for configuration of plasma and magnetic surfaces (Hofmann and Jardin, 1990), (Hofmann *et al.*, 1992). In a new generation of research tokamaks with parameters close to thermonuclear and in experimental reactors on their base there are the tendency to complicating of the magnetic configuration mainly due to utilize in them the poloidal divertor. Then for the mathematical description of control processes in such devices it is necessary to use at least two-dimensional equations of the plasma equilibrium instead of their simplified substitution as it was earlier done for tokamaks with plasma and magnetic surface shape close to circle. There are also other reasons for the development of a new approach to problems of plasma control in thermonuclear installations and reactors.

For the description of magnetic and plasma configurations in toroidal tokamaks are used the Maxwell equations for quasistationary electromagnetic fields described in the cylindrical coordinate system $(r, \varphi, z)$ with an axis $z$ to be coincided with a torus axis. In axisymmetrical configurations the physical parameters do not depend from $\varphi$ and two-dimensional description defined for any axial cross- section of a torus, passing through an axis $z$ is valid. In problems connected to a magnetic configuration the basic characteristic is a poloidal flux $\psi(r, z)$ of a magnetic field which is penetrating through toroidal contour with coordinates $(r, z)$. Then the surfaces $\psi(r, z) = \text{const}$ determine a system of magnetic surfaces of tokamak playing a key role in confinement and termoisolation of plasma. The system of the closed magnetic surfaces inside the vacuum chamber determines plasma area situated in removal from the walls of the chamber. Extreme surface of these enclosed each other closed surfaces define the shape and position of plasma in the chamber during its formation, heating and confinement. In tokamaks with poloidal divertor it is accepted to consider as the plasma boundary a separatrix magnetic surface which divides the closed and disconnected magnetic surfaces. This separatrix obligatory contains singular lines where magnetic surfaces are branched (× – points in cross-section).

The main goal of the magnetic and plasma configuration control system consists in adjustment and stabilization of the position and the shape of this extreme magnetic surface including some of its characteristic outside the plasma if this surface is a separatrix one.

Note that above-stated characteristics of a configuration are determined by equations of magnetohydrodynamic equilibrium ensuring balance of all acting forces in each point of area occupied by plasma. The serious mathematical problems arise when one is trying to solve control tasks on the base of this equilibrium equations directly. In reference (Gubarev, 1996) it is offered the approach which considerably simplifies both statement and solving of control problems. It is proposed to extend the set of the possible equilibrium states limited by the mentioned equations of equilibrium. For this it is recommend to refuse from global equilibrium conditions for all plasma area $S_p$ and to remain these conditions only for finite number of points which lie on controlled magnetic surface $\psi(r_b, z_b) = \text{const}$.

In this case the positions of $n - 1$ specially selected points on this surface are adjusted and the position of the other part of the bounded surface due to equilibrium condition are always remained smooth curve connecting this points.

Magnetic configuration are formed by magnetic field created by the poloidal plasma current and the currents in external poloidal coils. Then besides equilibrium equations it is necessary to attract circuit equations that describe the processes in electromagnetically coupling system, including plasma circuit as passive contour. So the coupled system namely plasma plus external poloidal field coils with the controlled power supply sources is a dynamic system. In this system takes place also plasma position variation. By means of current control in the poloidal field coils it is possible to realize such regimes that plasma shape evolves according to a preprogramming scenario, irrespective of changes in the plasma parameters. Magnetic measurements outside the plasma boundary are usually used for the determination of the selected points positions to be controlled. The positions of these points are tightly depended on the poloidal coil currents. So the desired plasma position and shape may be provided if coil currents are changed in accordance with the law calculated on the base of magnetic measurements. In other words the plasma control system may be realized as servomechanism with respect to coil currents. Such approach are regarded and studied in this paper.

## 2. PROBLEM FORMULATION

The realization in principle the approach proposed above will be considered for simplified case. Electromagnetic poloidal system consists of two coils: 1) inductor for plasma current induce and control, 2) coil for plasma position control that corresponds one selected point on plasma surface. It implies that $n = 2$ but magnetic configuration are nevertheless remained complicated one. So the momentum approach developed in (Gubarev, 1992) was used for deriving suitable for optimal control design model. This approach allows to transform two-dimensional system with distributed parameters into system of ordinary differential equations containing uncertainty.

The resultant expression of this equations for tokamak with simple poloidal system consisting of two coils: inductor and plasma position control coil may be written in dimensionless form

$$
\begin{aligned}
&\frac{d}{dt}\left[(1+l_1)j_p\right] + \lambda_1 \frac{d}{dt}\left[(1+l_2)j_1\right] \\
&+\frac{d}{dt}(l_3 j_2) + \frac{j_p}{\tau_p} = 0, \\
&\lambda_2 \frac{d}{dt}\left[(1+l_2)j_p\right] + \frac{d}{dt}(l_4 j_2) + \frac{dj_1}{dt} + \frac{j_1}{\tau_1} = v_1 (1) \\
&\frac{d}{dt}(l_3 j_p) + \frac{d}{dt}(l_4 j_2) + \frac{dj_2}{dt} + \frac{j_2}{\tau_2} = v_2 .
\end{aligned}
$$

Here we use the following notations: $j_p$ is a plasma current; $j_1$ is an inductor current; $j_2$ is a plasma position coil current; $v_1$ is a control that corresponds adjustable inductor voltage; $v_2$ is a control that corresponds plasma position control coil voltage; $\lambda_1, \lambda_2, \tau_1, \tau_2$ are the plasma and coil constants, $\tau_p(t)$ — a positive time-variance function; $l_1, l_2, l_3, l_4$ — small undetermined values.

Original approach for relevant plasma stabilization problem based on the method of analytical synthesis of the admissible set of nonlinear feedback is proposed. This method permits first to obtain analytically the set of controllers guaranteeing asymptotic stability. Then on this set the class of controllers that provide the robust stability with respect to any perturbed model can be computed using the same methodology. At last among robust controllers the optimal nonlinear controller is determined.

Nonlinear control for plasma boundary stabilization problem are developed because in comparison with linear control nonlinear one allow to optimize simultaneously several contradictory criteria. In particularly, for problem under consideration due to nonlinear control it is possible to provide fast attenuation of large deviation in system and at the same time to minimize the system response on the bounded errors in measurements and on unmodeled effects.

It is practically impossible to reach a given closed loop performance in system with linear control only.

In other words it is proposed to define the control with amplification factors depending on the state distance from the stationary point. This are very important for plasma position and shape control to be under consideration in this paper because it is extremely desired to return highly disturbed plasma to initial position and to avoid sliding mode control with rapid switching near zero.

It is also worth mentioning that proposed control design methodology is well suited for state-space uncertain dynamic systems with identified low order parametric stationary nominal model and an identification error giving a measure of the model perturbation. Using numerical simulation of the plasma behavior when its parameters are disturbed it is possible to evaluate a tight unmodeled dynamics bound. In addition the dynamics not modeled by this nominal model may be corresponded with nonstationary and even nonlinear perturbation system. The method that allows to obtain the set of asymptotically stable feedback laws based on Sylvester inequalities guarantees the desired dynamic properties of the closed loop system for these cases also.

## 3. CONTROL DESIGN

In case $l_1 = l_2 = l_3 = l_4 = 0$ the equation system (1) is broken on two subsystems

$$\frac{dj_p}{dt} + \lambda_1 \frac{dj_1}{dt} + \frac{j_p}{\tau_p} = 0,$$
$$\lambda_2 \frac{dj_p}{dt} + \frac{dj_1}{dt} + \frac{j_1}{\tau_1} = v_1; \quad (2)$$

$$\frac{dj_2}{dt} + \frac{j_2}{\tau_2} = v_2. \quad (3)$$

The system (2), (3) we define as nominal. Desired nominal evaluated mode of this system described as

$$j_p = j_{pn},\ j_1 = j_{1n},\ j_2 = j_{2n};$$
$$v_1 = v_{1n},\ v_2 = v_{2n}. \quad (4)$$

In practice the fast changes values $l_1$, $l_2$, $l_3$, $l_4$ result in occurrence of initial deviation of the system (2), (3), and slow changes these values in occurrence of unknown but bounded disturbances $w$.

Deviations of the variables (4) from nominal evaluated mode are designated as

$$x_1 = j_p - j_{pn},\ x_2 = j_1 - j_{1n},\ x_3 = j_2 - j_{2n};$$
$$u_1 = v_1 - v_{1n},\ u_2 = v_2 - v_{2n}. \quad (5)$$

Equations of disturbance motion in accordance with the expressions (2) - (5) may be written in form

$$\frac{dx_1}{dt} = \frac{1}{\lambda_1\lambda_2 - 1}\left[\frac{x_1}{\tau_p} - \frac{\lambda_1 x_2}{\tau_1} + \lambda_1 u_1\right] + h_1 w,$$
$$\frac{dx_2}{dt} = \frac{1}{\lambda_1\lambda_2 - 1}\left[-\frac{\lambda_2 x_1}{\tau_p} + \frac{x_2}{\tau_1} - u_1\right] + h_2 w; \quad (6)$$

$$\frac{dx_3}{dt} = -\frac{x_3}{\tau_2} + u_2 + h_3 w. \quad (7)$$

We have that initial deviation vector $x_0 \in \Omega_0$, unknown disturbances $w \in W$, $|W| < N$.

The equation (7) is robust stable with a control law

$$u_2 = -\lambda_3 x_3, \quad (8)$$

$\lambda_3$ is a large constant.

The equations (6) may be written in standard form

$$\frac{dx}{dt} = A(t)x + B(t)u + H(t)w, \quad (9)$$

where $x \in \mathcal{R}^n$ is a deviation vector of the system, $u$ is a control, $w \in W$ is a disturbance. It is required to find the set of admissible feedback laws $u(t, x_1, x_2, \ldots, x_n)$ ensuring asymptotic stability of the equilibrium position $x = 0$ of the system when $w = 0$ and to choose parameters of this set with aim to optimize the scalar convolution of two partial criteria

$$I = I_1 + \alpha I_2, \quad (10)$$

where $I_1 = \|x\|_\infty$; $I_2 = T_2$ for which $\|x\| < R$, $t > T_2$, $w \in W$; $\alpha$ is a weight coefficient.

The set of admissible feedback laws is sought in the form

$$u = Cx, \quad (11)$$

the Lyapunov function is defined as positive definite quadratic form $V = x^T D(t)x$ and is used to construct the set of matrix $C$ ensuring asymptotic stability.

Coefficients of the Lyapunov function have to satisfy Sylvester inequalities

$$\mathcal{D}_r(t) > \varepsilon, \quad r = \overline{1,n} \tag{12}$$

where $\mathcal{D}_r$ — main minors of the matrix $\mathcal{D}$, $\varepsilon$ — a small positive value.

If we differentiate the Lyapunov function with respect to $t$ along the motion paths (9) taking into account (11) then

$$\frac{dV}{dt} = x^T L^1(A,B,C,D)x + M(x,w), \quad L^1 = \left\|l^1_{ij}(t)\right\|^n_1, \tag{13}$$

where

$$l^1_{ij} = g^1_{ij} + p_i c_j + p_j c_i, \quad g^1_{ij} = \sum_{r=1}^{n} d_{ri}a_{rj} + d_{rj}a_{ri} + \frac{dd_{ij}}{dt}, \quad p_i = \sum_{r=1}^{n} d_{ri}b_r. \tag{14}$$

It necessary to obtain the set of matrices $C$ which deliver negative definite matrix $L^1$. For this purpose we solve the Sylvester inequalities (Yakovlev *et al.*, 1991)

$$L^1_r(-1)^r > \varepsilon, \quad r = \overline{1,n}$$

with respect to elements of matrix $C$ and obtain

$$c_1 = -\frac{1}{2p_1}(g^1_{11} + k_1),$$
$$c_r = -\frac{1}{(p_1)^2}(g^1_{r1}p_1 - \frac{g^1_{11}p_r}{2} + \frac{p_r k_1}{2}) - \frac{1}{(p_1)^2}\sum_{m=0}^{r-2}\frac{L^2_r\begin{bmatrix}1 & r-m & \dots r-1\\ 1 & r-m+1 \dots & r\end{bmatrix}}{L^2_{r-m}\begin{bmatrix}1\\1\end{bmatrix}}k_{r-m}$$
$$\times\left(-\frac{k_1 L^2_{r-m}\begin{bmatrix}1\\1\end{bmatrix}}{L^2_{r-m-1}\begin{bmatrix}1\\1\end{bmatrix}}\prod_{\lambda=2}^{r-m-1}(1-k^2_\lambda)\right)^{\frac{1}{2}}(-1)^{r+m}, \quad r = \overline{2,n}. \tag{15}$$

The numbers in square brackets designate the deleted lines and columns indexes.

The elements of the symmetric matrix $L^2$ are expressed in form

$$l^2_{ij} = g^1_{ij}(p_1)^2 - g^1_{1j}p_1p_i - g^1_{i1}p_1p_j + g^1_{11}p_ip_j, \; i,j \geq 2;$$
$$l^2_{i1} = g^1_{i1}p_1 - g^1_{i1}p_i \tag{16}$$
$$-p_1p_ic_1 + (p_1)^2c_i, \; i \geq 2;$$
$$l^2_{11} = l^1_{11}.$$

The factors $k_{sr}$, $s \leq r$ are piecewise continuous functions satisfying to restrictions

$$k_{ss}(t,x) > \alpha(x), \quad |k_{sr}(t,x)| < 1 - \alpha(x), \quad s = \overline{1,m}, \; r = \overline{1,n}, \tag{17}$$

$\alpha(x)$ — positive definite function accepted small values.

Then the set of admissible feedback laws $U$ is determined by the expressions (11), (14)–(17). Lyapunov function coefficients have to satisfy conditions which guarantee reality of right sides of expression (15)

$$G^2_r\begin{bmatrix}1\\1\end{bmatrix}(-1)^{r+1} > \varepsilon, \quad r = \overline{2,n}. \tag{18}$$

The Lyapunov function derivative in the closed system (9), (11), (15) may be written in form

$$\frac{dV}{dt} = -\frac{1}{(p_1)^2}\Bigg\{-k_1\left(\sum_{i=1}^{n}p_ix_i\right)^2 + \sum_{i,j=2}^{n}l^2_{ij}x_ix_j - 2\sum_{i=1}^{n}p_ix_i\sum_{r=2}^{n}\sum_{m=2}^{r}\frac{L^2_{r+m}\begin{bmatrix}1 & r & \dots r+m-1\\ 1 & r+1 \dots & r+m\end{bmatrix}}{L^2_r\begin{bmatrix}1\\1\end{bmatrix}}$$
$$\times k_r\left(-k_1\frac{L^2_r\begin{bmatrix}1\\1\end{bmatrix}}{L^2_{r-1}\begin{bmatrix}1\\1\end{bmatrix}}\prod_{\lambda=2}^{r-1}(1-k^2_\lambda)\right)^{\frac{1}{2}} \times(-1)^r x_{r+m}\Bigg\} + M(x,w). \tag{19}$$

The term in figured brackets in right side of expression (19) is the negative definite quadratic form but the function $M(x,w)$ is a linear with respect to variable $x$. Then it is exist a finite value $\|x\|_\infty$ outside which, the Lyapunov function derivative is negative for any values $k_r(t,x)$, $r = \overline{1,n}$ and coefficients $d_{ij}$, satisfying its restrictions.

The expression (19) makes it possible to analyze affect of the factors $k_r(t,x)$ on Lyapunov function damping rate and to choose its structures.

The maximum distance from boundaries of stability of the closed system is achieved for the average factor values

$$k_r(t,x) \equiv 0, \quad r = \overline{1,n}. \tag{20}$$

Hence the robust subset of admissible feedback laws $U_1 \subset U$ is determined by the expressions (11), (14)–(16), (20).

For the equations (6) the robust subset $U_1$ become form $U_1 = \{u : \ u = c_1x_1 + c_2x_2\}$,

$$c_1 = -\frac{1}{2p_1}(g_{11}^1 + k_1),$$
$$c_2 = -\frac{1}{(p_1)^2}(g_{21}^1 p_1 - \frac{g_{11}^1 p_2}{2} + \frac{p_2 k_1}{2}), \quad (21)$$
$$k_1(t,x) > \alpha(x).$$

The coefficients of the equations (21) is determined by the expression (14), where

$$a_{11} = \frac{1}{\tau_p(\lambda_1\lambda_2 - 1)};$$
$$a_{12} = -\frac{\lambda_1}{\tau_1(\lambda_1\lambda_2 - 1)};$$
$$a_{21} = -\frac{\lambda_2}{\tau_p(\lambda_1\lambda_2 - 1)};$$
$$a_{22} = \frac{1}{\tau_1(\lambda_1\lambda_2 - 1)};$$
$$b_1 = \frac{\lambda_1}{\lambda_1\lambda_2 - 1};$$
$$b_2 = -\frac{1}{\lambda_1\lambda_2 - 1}.$$

## 4. PARAMETER OPTIMIZATION

The factor $k_1$ may be chosen or to be a constant or to be a nonlinear function. The structure of function $k_1(t,x)$ is picked with aim to agree the demands of partial criteria.

By means of the coefficients $d_{ij}$ changing in accordance with the restrictions (12), (18), and the factor $k1$ changing in accordance with the restrictions (17) it is possible to impact performance of the transient processes in closed system (2), (4), (21) without loosing asymptotic stability. The concrete values of these variables are determined in process of numerical optimization of the criteria (10).

## 5. REFERENCES


Gubarev, V.F. (1992). Moment models for problems of control in physical systems. *Journal of Automation and Information* **25**(2), 23–30.

Gubarev, V.F. (1996). Control of plasma parameters in tokamaks with complicated magnetic configuration. *Journal of Automation and Information* **28**(1-2), 93–108.

Hofmann F., S.C. Jardin (1990). Plasma shape and position control in highly elongated tokamaks. *Nuclear Fusion* **30**(10), 2013–2020.

Hofmann, F., N. Pomphrey and S.C. Jardin (1992). Application of a new algorithm to plasma shape control in BPX. *Nuclear Fusion* **32**(5), 897–902.

Yakovlev, O.S., V.A. Majorov and V.V. Kazakov (1991). An ergatic control system with incomplete information. *Cybernetics and Computing Technology* **92**, 30–34.

# MULTIDISCIPLINARY OPTIMIZATION WITH EVOLUTIONARY COMPUTING FOR CONTROL DESIGN

**Wael Khatib Valceres Silva Andrew Chipperfield Peter Fleming**

*w.khatib@shef.ac.uk*
*Department of Automatic Control and Systems Engineering,*
*University Of Sheffield, Mappin Street, Sheffield S1 3JD, UK*

Abstract: Multidisciplinary optimization (MDO) is concerned with complex systems exhibiting challenges in terms of organization and scale. Complex multivariable control design can benefit from advances in MDO architectures. An example of model-based PI control design of a gas turbine engine is presented. Three MDO architectures including collaborative optimization are applied to this example. The optimization is carried out through a multiobjective genetic algorithm framework. Implementation and results provided suggest viability of such techniques for complex control design problems and potential improvements over existing techniques. *Copyright © 1999 IFAC*

Keywords: Efficient algorithms, Gas turbines, Genetic algorithms, Optimization, PI controllers.

## 1. INTRODUCTION

Multidisciplinary optimization (MDO) is concerned with the design of complex systems that are made up of a number of interacting or coupled systems. MDO research tends to be focused in the aircraft industry. The design of an airplane requires the bringing together of resources representing aerodynamics, structures, performance, control, metallurgy, systems and other disciplines in order to produce an optimal design. The main challenges faced in MDO design problems are computational cost and organizational complexity (Sobieszczanski-Sobieski and Haftka, 1996). The complexity of design optimization depends on problem size, the complexity of the pertinent disciplines and the nature of the constraints and the objectives. Should there be more than one discipline controlling the same design variables for a particular objective, then the problem becomes far more complex growing at a super linear rate compared with an aggregation of disparate disciplines. This is mainly due to the effect of coupling between the variables. Organizational complexity is due to the fact that the various disciplines traditionally reflect different departments, teams, analysis methods, schools of thought, software and hardware platforms, standards, etc. The organizational challenge in MDO is for an efficient exchange of data, systems integration and other aspects of communication.

Evolutionary computing refers to computer-based problem solving systems that use evolutionary algorithms (EAs). EAs generally use computational models that exploit mechanisms based broadly on the neo-Darwinian theory of evolution. The main techniques used in EAs include: genetic algorithms (GAs), evolutionary programming (EP), evolution strategies (ES) and genetic programming (GP). EAs are amenable to parallelization and can help reduce the computational cost. These algorithms are stochastic in nature and can usually start an optimization process without much *a priori* knowledge. No derivative information is required as in the traditional gradient based methods and this helps EAs deal with difficult search spaces characterized by multimodal disjoint feasible areas.

Most design problems including MDO ideally require the treatment of multiple objectives. These objectives are often conflicting or competing. An elegant and powerful method for dealing with multiple objectives is the concept of Pareto optimality. Using this approach, the designer is no longer searching for a single optimum, rather a compromise satisfying the various objectives. The collection of compromise solutions is referred to as the non-dominated set. Within this set, attempted improvement in one objective will result in degradation in one or more of the others. EAs are amenable to multiobjective optimization (MO). This is because an EA works on a population of solutions instead of the traditional single point search. The search with this population can help achieve a faster and more comprehensive mapping of the trade-off hyper surface.

An overview of the application of EAs to MDO can be found in Khatib and Fleming (1997). The work described here employs a GA with an implementation of MO as proposed by Fonseca and Fleming (1993). This multiobjective GA (MOGA) presents a simple but powerful tool for design optimization were preferences can be articulated progressively by the decision maker as more insight into the problem in hand is gained.

## 2. MDO ARCHITECTURES

Various approaches for formulating MDO problems have appeared over the years on an *ad hoc* basis. Balling and Sobieszczanski-Sobieski (1996) introduce a consistent method for classifying the various approaches using compact and consistent notation. For practical purposes, the various approaches can be grouped into three main categories:

- Single-level approaches
- Collaborative optimization
- Concurrent sub-space optimization

A detailed discussion of these methods can be found in Khatib and Fleming (1998). Single-level approaches address a design problem as one whole unit. Concurrent sub-space optimization relies on recursive iteration loops and can be less efficient. The collaborative optimization (CO) approach works through decomposition of complex large-scale problems into smaller sub-problems or elements. Each element proceeds with its own optimization using separate decision variables. The outputs of the various disciplines can be pooled into a shared resource area to be redistributed or observed by a system designer or optimizer to ensure a viable overall design. The decomposition boundaries depend on the physical organization of the problem, the available resources and/or the mathematical limits. This gives rise to soft and hard boundaries. The evolutionary collaborative architecture proposed by Khatib and Fleming (1998) allows the various disciplines to progress simultaneously in search of optimal designs. This approach draws on a seemingly good match between the two elements of CO and EAs. They share amenability to parallelization and this is also reflected in existing practices in industry.

## 3. GAS TURBINE ENGINES CONTROL

Gas turbine engines are highly nonlinear plants that have multiple inputs and outputs. The operating conditions span extremes of temperature, pressure and load conditions. The engine performance requirements cover a wide flight envelope that includes a continuum of set points of altitude and speed in terms of the Mach number. These requirements add to the complexity of designing suitable controllers that can achieve high performance levels while maintaining stability and safe operation with minimum overall cost. A successful design of an engine controller can be achieved as part of, say, an overall MDO design of an aircraft. Such designs are still not attempted on a full scale in industry due to the limitations of available resources for MDO. However, MDO techniques can be applied to the design process for the controller *per se*. To demonstrate this, we present an example for designing a PI control strategy for a gas turbine engine that comprises two PI controllers. A nonlinear SIMULINK model for a SPEY gas turbine engine is used to design and evaluate controller performance.

The engine model has three inputs: fuel flow, exhaust nozzle area and inlet guide vane angle. Sensors provided from engine outputs include: high pressure (HP) spool speed, low pressure (LP) spool speed, engine and fan pressure ratios and Mach number. These measurements can be used to provide various pairings of input-output for closed-loop control. Important engine parameters such as thrust and surge margin cannot be measured directly. Such parameters are controlled implicitly through other related measurable values such as pressure ratios. Chipperfield and Fleming (1996) used a linear model to design a PI control strategy for this engine. Silva and Fleming (1998) used a non-linear model for control configuration design and also for robust $H\infty$ controller design. Using findings from these studies, a closed-loop control strategy is chosen for parametric optimization of a designed PI controller structure for a particular operating point to demonstrate the use of MDO architectures for control design (Fig. 1). Two PI controllers are implemented for controlling thrust and LP surge margin. Because these two variables cannot be measured directly, they are controlled implicitly through the engine pressure ratio and the bypass duct Mach number respectively. The third input, inlet guide vane angle, is gain scheduled against the measured output values of the HP spool speed.

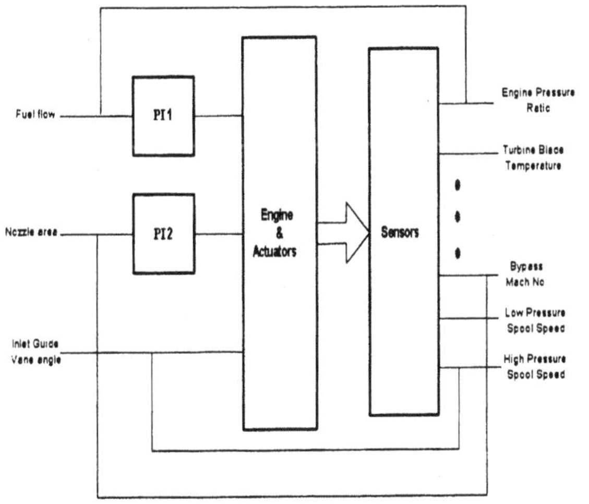


**Fig. 1.** Gas Turbine (Spey) Control Structure

The design is expected to satisfy the following objectives and constraints:

a) minimum thrust (performance requirement).
b) maximum turbine blade temperature (physical limit).
c) minimum LP surge margin (stability limit).
d) thrust rise and settling time for speed of response performance.
e) steady state error limits for HP/LP spool speed.
f) steady state error limit for exhaust nozzle area.
g) limits on overshoot/undershoot for transient regions of HP/LP spool speeds.
h) other engine limits for spool speeds and exhaust nozzle area.

## 1. PI MOGA DESIGN ARCHITECTURES

This work employs one single-level and two collaborative design frameworks using MOGA. The single-level MOGA addresses the two controllers and all the objectives simultaneously. Each individual in the MOGA population is a chromosome made up of the four controller parameters in a Gray-coded binary representation. The two CO implementations are different in the way the design problem is decomposed. The first example, CO I, has two optimization sub-problems (Fig. 2): one dealing with the proportional gains in both controllers, and the other dealing with the integral gains. The objectives for each of these sub-problems reflect potential scope of influence. The proportional gains are designed to optimize all the objectives and constraints of the problem simultaneously, while the integral gains are considered for optimization of thrust and turbine blade temperature (TBT) dynamics (objectives b & d). The second example, CO II, follows the physical boundaries by having each of two sub-problems dealing with one of the two PI controllers. The pertinent objectives also reflect the level of influence: PI1 design optimizes thrust and spool speed dynamics (objectives d and g), steady state errors for spool speeds (objectives e) whilst minimizing TBT (objective b). PI2 design optimizes the nozzle area variations (objective f) while maintaining LP surge margin and TBT within safe limits (objectives b and c). Note that TBT is common to both controllers. This suggests coupling in the system and can help the two design sub-problems achieve a satisfactory overall result.

In terms of the underlying GA engine in the MOGA framework, the two main operators that guide the optimization process are selection and crossover. The selection pressure which favors the fitter individuals is based on how well each of the individuals performs in terms of Pareto optimality. Grouping the proportional and integral terms for both controllers together for each individual suggests that an integral term, say, might be promoted in the population based on influence of a proportional term in the same individual. This form of nepotism can be inefficient in computing terms and might even lead to less optimal solutions.

Because of the parallel nature of MOGA, the two sub-problems in CO I and CO II run concurrently. Each problem has its own population of solutions. Each population experiences selection and crossover driven by its own pertinent objectives. The two populations do not share or individuals through migration. To achieve a consistent overall design, each population presents its proposed solutions to a shared data pool which performs the engine simulations. Results of the simulations are fed back into the pool and each separate MOGA scans for its relevant outputs to drive its own GA operators.

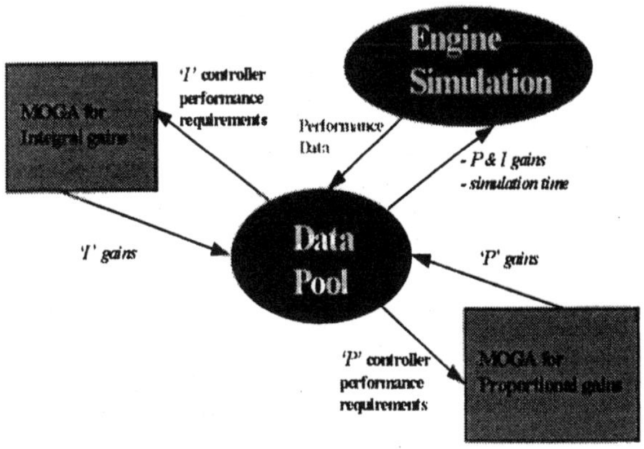


**Fig. 2.** MOGA CO I architecture

The control response is measured after a step input in thrust demand, corresponding to changes from 85% to 90% of high pressure spool speed. Using two populations of solutions simply increases the requirements to compute the MOGA operators for ranking, selection, crossover and mutation. These costs are negligible when compared to the cost of engine simulations. The number of simulations required is the same as for the single-level optimization case. Each of the three MOGAs employs 40 individuals and is evolved for 50 generations.

## 5. RESULTS

Each of the three MOGAs achieves a set of nondominated solutions (21, 12 and 12 controller sets for the single-level, CO I and CO II respectively). Most of these controllers exhibit similar performance characteristics. The gross thrust response (normalised) is presented in Fig. 3 - 5 for the three design architectures.

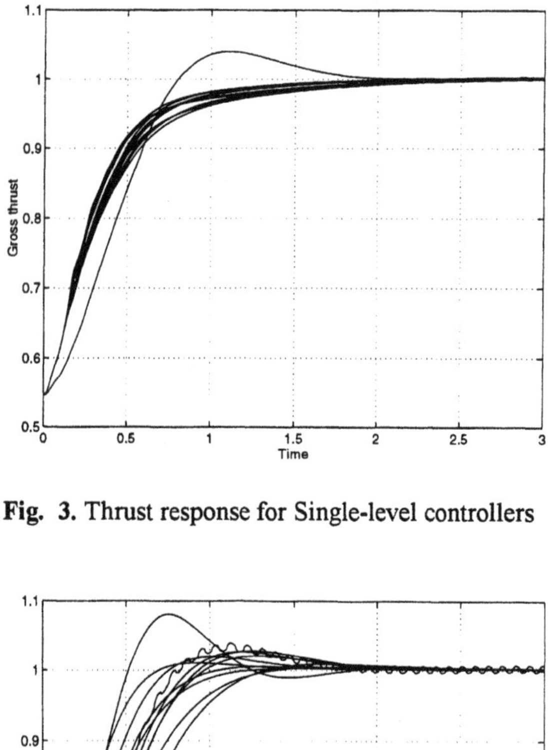


**Fig. 3.** Thrust response for Single-level controllers

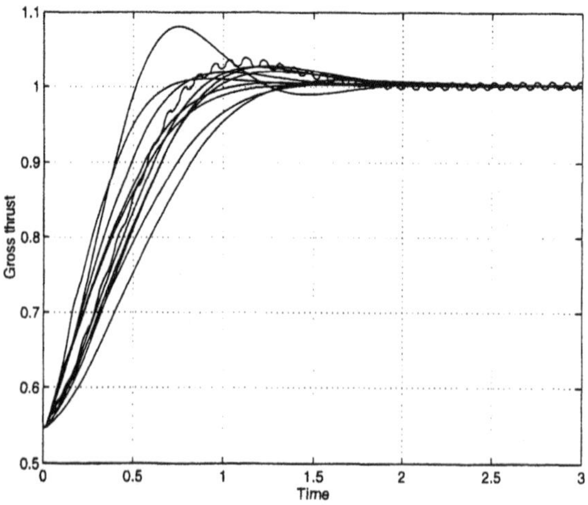


**Fig. 4.** Thrust response for CO I controllers

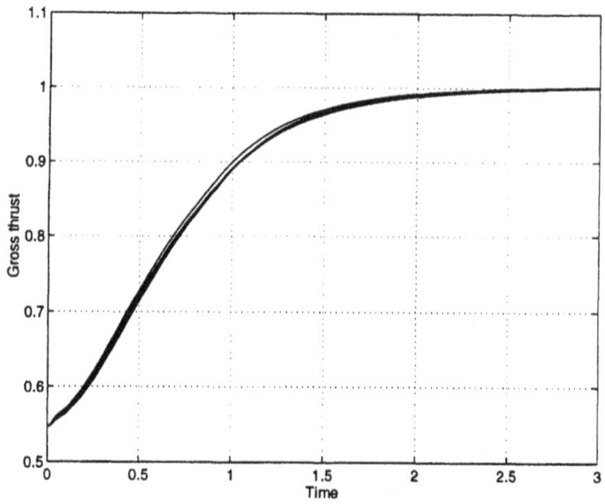


**Fig. 5.** Thrust response for CO II controllers

To study the control performance in more detail, further filtering is applied to choose one controller set for analysis from each MOGA. This is done by choosing the best controller that satisfies objectives *a* - *d* only. Again, this is done in a multiobjective manner. The performance graphs refer to the final selected controller set for each case.

All the performance graphs indicate that the CO I controller outperforms the other two. CO II perfomance indicates slower, though stable, responses when compared to the other two. The superiority of CO I was found to hold for all the other nondominated controllers in each case.

Fig. 6 shows how the thrust response is faster achieving the desired steady state value with little or no errors after nearly 1 second. Similar performance is observed in the HP spool speed response (Fig. 9). Fig. 7 shows how the overshoot in the TBT reposnse is less for CO I compared with the single-level case. The response for LPSM indicates better transient behaviour for CO I. The steady state value of the surge margin is also higher for CO I indicating increased robustness.

Thus, decomposing the problem into the two different types of controller (P and I) as in CO I gives better results than CO II where the split is along actual controller strucures corresonding to different input-output combinations. This decomposition also outperforms the classical single-level approach. The engine is one whole unit made up of various interacting sub-units. The single level approach can allow some perfomance dynamics, pertinent only to a subset of the controller parameters, to influence other controller parameters. This is implying inefficient or even erroneuous optimisation paths. The control performance issues are affected more by the type of a controller (P or I) than by the physical location of a controller (PI1 or

PI2). Hence, CO I appears to offer a better decomposition scheme than CO II. This allows the relevant perfomance metrics to be allocated to each optimization strand in a more efficient and appropriate way.

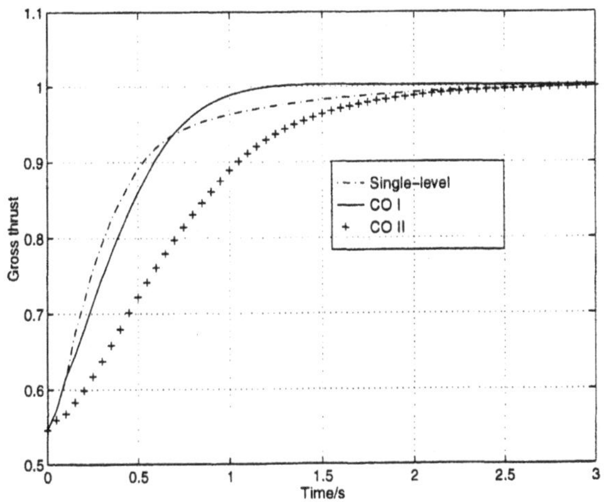


**Fig. 6.** Thrust response for all architectures

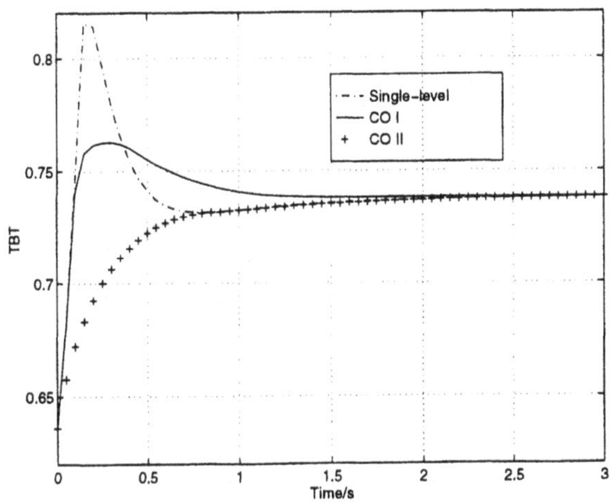


**Fig. 7.** TBT response for all architectures

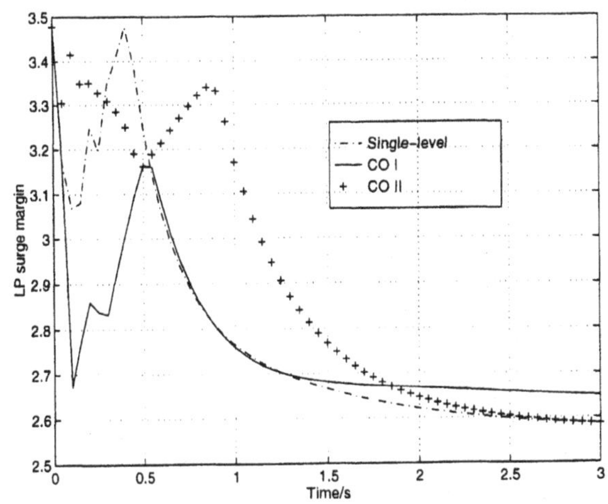


**Fig. 8.** LPSM response for all architectures

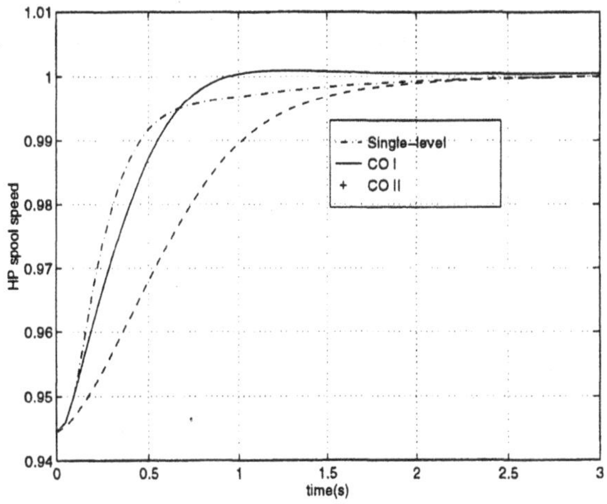


**Fig. 9.** HP spool speed response for all architectures

# 6. CONCLUSIONS

The field of MDO is traditionally based in the aircraft industry. It is used to tackle some of the more difficult design problems involving interacting systems with strong coupling. Complex systems tend to have strong coupling between the various elements and are usually multidisciplinary in nature.

The complex optimization scenarios in some control design problems can benefit from applying MDO techniques. MOGAs have already been applied successfully to various control applications. Both MOGA and CO MDO architectures lend themselves well to parallelization and thus to finding solutions to problems with multiple conflicting objectives. Application of these techniques to a non-linear multivariable control strategy for a gas turbine engine illustrates the viability of the proposed approach.

The design methodolgy using collaborative optimisation decomposes a large-scale problem along convenient and efficient boundaries. This allows for improvements in terms of computational costs, problem set-up and organisational issues. The final overall designs can also offer improvements over traditional single-level all-in-one appraoches.

These techniques may be applied to other control problems where the underlying MDO challenges are present in the control scenario under study. These challenges are envisaged to be present in complex contrrol scenarios such as multivariable control

## Acknowledgements

The work of the first author is supported by an industrial CASE award by the Engineering and Physical Sciences Research Council (EPSRC, UK) and British Aerospace PLC. The support of Dr Phil Greenway of Sowerby Research Centre (BAe) is acknowledged. The work of the second author is supported by CAPES, 'Coordenação de Aperfeiçoamento de Pessoal de Nível Superior' and FUNREI, 'Fundação de Ensino Superior de São João del Rei', Brazil. The authors also wish to thank DERA Pyestock for the Spey engine model.

## References

Balling, R.J. and J. Sobieszczanski-Sobieski (1996). Optimization of Coupled Systems: A Critical Overview of Approaches. *AIAA Journal*, **34(1)**, pp. 6-17.

Chipperfield, A.J. and P.J. Fleming (1996). Systems integration using evolutionary algorithms. *International Conference on Control'96 - IEE Conference Publication n° 427*, **1**, pp. 705-710.

Fonseca, C.M. and P.J. Fleming (1993). *Genetic Algorithms for Multiobjective* Optimization: Formulation, Discussion and Generalisation In *Genetic Algorithms: Proceedings of the Fifth International Conference* (S. Forrest, ed.), San Mateo, CA: Morgan Kaufmann.

Khatib W. and P.J. Fleming (1998). Evolutionary Computing Applied to MDO Test Problems. *7th AIAA/USAF/NASA/ISSMO Symposium on Multidisciplinary Analysis and Optimzation*, St. Louis, MO.

Khatib W. and P.J. Fleming (1997). Evolutionary computing for multidisciplinary optimisation. *Proc 2nd IEE/IEEE International Conference on Genetic Algorithms in Engineering Systems: Innovations and Applications GALESIA'97*, Glasgow, pp 7-12.

Silva, V.V.R. and P.J. Fleming (1998). Multiobjective non-linear robust control design for gas turbine engines using evolutionary computing. *In Proceedings of ' Mathematical Theory Networks Systems Conference'* - MTNS'98, July, 'Universidade de Padova', Italy.

Sobieszczanski-Sobieski, J. and R.T. Haftka (1996). Multidisciplinary Aerospace Design Optimisation: Survey of Recent Developments. *AIAA* Paper 96-0711, Reno, NV.

# OPTIMAL FINITE-PRECISION PID CONTROLLER STRUCTURES USING GENETIC ALGORITHMS

**J.F. Whidborne* R.S.H. Istepanian****

* *Dept of Mechanical Engineering, King's College London, Strand, London WC2R 2LS, U.K.* email:james.whidborne@kcl.ac.uk, fax:+44-(0)171-873-5071
** *Dept of Electrical and Computer Engineering, Ryerson Polytechnic University, 350 Victoria St, Toronto M5B 2K3, Canada.*

Abstract: An approach to optimizing the structure of a finite-precision PID controller implementation by means of genetic algorithms is presented. The approach is multi-objective, and provides a set of solutions that are (near) Pareto optimal with respect to minimal memory requirements and to minimal change in closed loop poles. The method is applied to the PID controller structure for a steel rolling mill control system.


Keywords: Genetic algorithms, digital controller structures, multi-objective optimization, PID controllers, finite word-length.

## 1. INTRODUCTION

It is well known that controller implementations with fixed-point arithmetic offer the advantages of speed, memory space, cost and simplicity compared to floating-point arithmetic (Masten and Panahi, 1997). Thus, to date, fixed-point processors are still the dominant architecture in many modern digital control engineering applications such as automotive, consumer, military and medical applications. However, a closed-loop control system will suffer a performance degradation and may even become unstable when the designed infinite-precision controller is implemented with a fixed-point digital processor due to the finite precision of the parameter representation resulting from the Finite Word-Length (FWL). This so-called FWL effect is in fact strongly dependent upon the parameterization of the controller. Thus, over the years, many results have been reported in the literature dealing with FWL implementation and their relevant parameterization issues, for example, Knowles and Edwards (1965), Morony (1983), Williamson (1991) and Gevers and Li (1993). Most of these approaches are concerned with finding parameterizations of the digital controller that are insensitive to small uncertainties in the parameters. That is, given an initial realization of the digital controller, find an equivalent controller realization such that closed loop stability and performance is maintained if the controller parameters are perturbed.

In this paper, a more direct approach is taken; that is, find an FWL controller that is an approximation to the originally designed controller such that the closed loop stability and performance degradation and the FWL memory requirement are simultaneously minimized. The approach is based on the generation of near-equivalent finite word-length controller representations by means of the solution to a linear system similarity completion problem, and using a Genetic Algorithm (GA) to find the (near) optimal FWL approximation. GA techniques require the solution space to be encoded in binary strings. Since controllers implemented with FWL are coded in binary, a one-to-one relationship between the genotype and the phenotype within the GA can be defined. GA's have been used to design FWL digital filters, for example, Xu and Daley (1995), but the authors are not aware of any work on using GA's for closed loop FWL controller design. GA's also allow the optimal structure of the solution to be found (Man *et al.*, 1997; Tang *et al.*, 1996; Dakev *et al.*, 1997), this means that the word-length does not need to be defined *a priori*, but the GA will select the best from a predefined set.

In this paper, a multi-objective genetic algorithm is used. This allows the designer to trade-off FWL implementation stability and performance against the memory and other requirements of the implementation. For reasons of simplicity, in this paper, the performance and stability are assessed simply by the maximal change in closed loop poles, and only the parameter memory storage requirement is considered. However, the method is entirely generic, and any set of stability/performance and implementation measures could be used. The developed approach is applied to the problem of the implementation of a PID controller for a steel rolling mill problem (Hori, 1996) which was also solved using a hill-climbing approach by Istepanian *et al.* (1998).

The paper is organized as follows. In the next section, some of the concepts underpinning the method are introduced, namely, multi-objective optimization, multi-objective genetic algorithms, FWL representation. The method also requires a linear system similarity completion problem to be solved, this also presented in the next section. In Section 3, details of the developed method are presented. The steel rolling mill problem is presented in Section 4, and in the final section, some conclusions are drawn.

## 2. PRELIMINARY CONCEPTS AND THEORY

### 2.1 *Multi-objective optimization*

The majority of engineering design problems are multi-objective, in that there are several conflicting design aims which need to be simultaneously achieved. If these design aims are expressed quantitatively as a set of $n$ design objective functions $\phi_i(p) : i = 1 \ldots n$, where $p$ denotes the design parameters chosen by the designer, the design problem could be formulated as a multi-objective optimization problem:

$$\min_{p \in \mathcal{P}} \{\phi_i(p), \text{ for } i = 1 \ldots n\}. \quad (1)$$

where $\mathcal{P}$ denotes the set of possible design parameters $p$. In most cases, the objective functions are in conflict, so the reduction of one objective function leads to the increase in another. Subsequently, the result of the multi-objective optimization is known as a Pareto-optimal solution (Pareto, 1906). A Pareto-optimal solution has the property that it is not possible to reduce any of the objective functions without increasing at least one of the other objective functions.

A point $p^* \in \mathcal{P}$ is defined as being Pareto-optimal if and only if there exists no other point $p \in \mathcal{P}$ such that $\phi_i(p) \leq \phi_i(p^*) \forall i = 1, \ldots, n$ and $\phi_j(p) < \phi_j(p^*)$ for at least one $j$.

### 2.2 *Genetic algorithms*

Genetic algorithms are search procedures based on the evolutionary process in nature. The idea is that the GA operates a population of individuals, each individual representing a potential solution to the problem, and applies the principle of survival of the fittest on the population, so that the individuals evolve towards better solutions to the problem.

The individuals are given a chromosomal representation, which corresponds to the genotype of an individual in nature. Three operations can be performed on individuals in the population, selection, cross-over and mutation. These correspond to the selection of individuals in nature for breeding, where the fitter members of a population breed and so pass-on their genetic material. The cross-over corresponds to the combination of genes by mating, and mutation to genetic mutation in nature. The selection is weighted so that the 'fittest' individuals are more likely to be selected for cross-over, the fitness being a function of the function which is being minimized. By means of these operations, the population will evolve towards a near-optimal solution.

GA's are very-well suited for multi-objective optimization problems. The approach used here is the Multi-Objective Genetic Algorithm (MOGA) (Fonseca and Fleming, 1993*a*; Fonseca and Fleming, 1993*b*), which is an extension on an idea by Goldberg (1989). This formulation maintains the genuine multi-objective nature of the problem. The idea behind the MOGA is to develop a population of Pareto-optimal or near Pareto-optimal solutions. This is achieved by finding a set of solutions which are non-dominated. An individual $j$ with a set of objective functions $\phi^j = (\phi^j_1, \ldots, \phi^j_n)$ is said to be non-dominated if for a population of $N$ individuals, there are no other individuals $k = 1, \ldots, N, k \neq j$ such that $\phi^k_i \leq \phi^j_i \forall i = 1, \ldots, n$ and $\phi^k_i < \phi^j_i$ for at least one $i$. With the MOGA, non-dominated individuals are given the greatest fitness, and individuals that are dominated by many other individuals are given a small fitness. Using this mechanism, the population evolves towards a set of non-dominated, near Pareto-optimal individuals. Details of this mechanism are given in Fonseca and Fleming (1993*a*).

In addition to finding a set of near Pareto-optimal individuals, it is desirable that the sample of the whole Pareto-optimal set given by the set of non-dominated individuals is fairly uniform. A commonly used mechanism to ensure this is known as fitness sharing (Fonseca and Fleming, 1993*a*), and works by reducing the fitness of individuals that are genetically close to each other. However, as will be seen in Section 3.3, not all the bits of a candidate solution bit string are necessarily active. Thus, 2 individuals may have the same genotype, but different gene strings. Thus it could become very complicated to measure the difference between 2 genotypes in order to implement fitness sharing, so for the sake of simplicity in this paper, multiple genotypes are simply removed from the population.

### 2.3 *FWL representation*

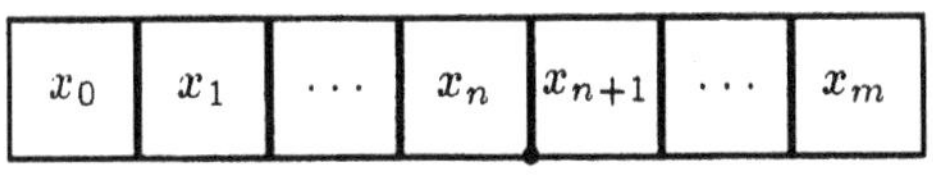


Fig. 1. FWL fixed-point representation

A typical 2's complement FWL fixed-point representation of a number $q(x)$ is shown in Figure 1, is given by a $m+n+1$ binary string $x$ where $x = [x_0, x_1 \ldots, x_m, x_{m+1}, \ldots, x_{m+n}]$, $x_i \in \{0,1\}$, $m \in \{0,1,2,\ldots\}$, $n \in \{0,1,2,\ldots\}$ and the value $q(x)$ is given by

$$q(x) = -x_0 2^m + \sum_{i=1}^{m} x_i 2^i + \sum_{i=m+1}^{m+n} x_i 2^{m-i}. \quad (2)$$

The set of possible values which can be taken by an FWL variable represented by a $m+n+1$ binary string $x$ is defined as $\mathcal{Q}_{n,m}$ given by $\mathcal{Q}_{n,m} = \{q : q = q(x), x_i \in \{0,1\} \forall i\}$. The set of $p \times q$ matrices with elements from $\mathcal{Q}_{n,m}$ is denoted by $\mathcal{Q}^{p\times q}_{n,m}$.

### 2.4 *A linear system similarity completion problem*

*Theorem 1.* Given a 2 state SISO LTI system $F(z)$, where

$$F \stackrel{s}{=} \left[\begin{array}{c|c} A & B \\ \hline C & D \end{array}\right] \quad (3)$$

$$\stackrel{s}{=} \left[\begin{array}{cc|c} a_{11} & a_{12} & b_1 \\ a_{21} & a_{22} & b_2 \\ \hline c_1 & c_2 & d \end{array}\right], \quad (4)$$

then (provided certain rank and singularity conditions are satisfied) given $(\tilde{a}_{ii}, \tilde{a}_{jk} \neq 0, \tilde{b}_l, \tilde{c}_m)$; $i,j,k,l,m \in \{1,2\}, j \neq k$, there exists an equivalent state-space representation $\tilde{F}$ where

$$\tilde{F} \stackrel{s}{=} \left[\begin{array}{c|c} \tilde{A} & \tilde{B} \\ \hline \tilde{C} & \tilde{D} \end{array}\right] \quad (5)$$

$$\stackrel{s}{=} \left[\begin{array}{cc|c} \tilde{a}_{11} & \tilde{a}_{12} & \tilde{b}_1 \\ \tilde{a}_{21} & \tilde{a}_{22} & \tilde{b}_2 \\ \hline \tilde{c}_1 & \tilde{c}_2 & d \end{array}\right], \quad (6)$$

such that $\tilde{F}(z) = F(z)$ where

$$F(z) = C\,[zI - A]^{-1}\, B + D \quad (7)$$

and

$$\tilde{F}(z) = CT\left[zI - TAT^{-1}\right]^{-1} T^{-1}B + D, \quad (8)$$

where

$$T = \begin{bmatrix} t_{11} & t_{12} \\ t_{21} & t_{22} \end{bmatrix} \quad (9)$$

is non-singular.

**Proof:** Proof is by construction.

**Construction of $\tilde{A}$.** From (8),

$$\tilde{A} = TAT^{-1}, \quad (10)$$

so by a property of similar matrices (Nering, 1990, p.115), $\operatorname{tr}\tilde{A} = \operatorname{tr}A$. Hence, given $\tilde{a}_{ii}, i \in \{1,2\}$, then $\tilde{a}_{nn}, n \in \{1,2\}, n \neq i$ is given by $\tilde{a}_{nn} = a_{11} + a_{22} - \tilde{a}_{ii}$. Similarly, from the property of similar matrices that $\det\tilde{A} = \det A$ (by definition), given $\tilde{a}_{jk} \neq 0, j,k \in \{1,2\}, k \neq j$, then $\tilde{a}_{kj}$ is given by $\tilde{a}_{kj} = (\tilde{a}_{11}\tilde{a}_{22} - \det A)/\tilde{a}_{jk}$.

**Construction of $T$, $\tilde{B}$ and $\tilde{C}$.** From (10), $AT - T\tilde{A} = 0$, which gives (Horn and Johnson, 1991, p. 255)

$$\left[I \otimes A - \tilde{A}^T \otimes I\right] \begin{bmatrix} t_{11} \\ t_{21} \\ t_{12} \\ t_{22} \end{bmatrix} = \begin{bmatrix} 0 \\ 0 \\ 0 \\ 0 \end{bmatrix}, \quad (11)$$

where $[I \otimes A - \tilde{A}^T \otimes I]$ is rank 2.

Now, from (8), $TB = \tilde{B}$ and $CT = \tilde{C}$. Hence, given $\tilde{b}_l, l \in \{1,2\}$,

$$\begin{bmatrix} b_1 & b_2 \end{bmatrix} \begin{bmatrix} t_{l1} \\ t_{l2} \end{bmatrix} = \tilde{b}_l \quad (12)$$

and given $\tilde{c}_m, m \in \{1,2\}$,

$$\begin{bmatrix} c_1 & c_2 \end{bmatrix} \begin{bmatrix} t_{1m} \\ t_{2m} \end{bmatrix} = \tilde{c}_m. \quad (13)$$

From (12) and (13), we can construct a $Y \in \mathbb{R}^{4\times 4}$ where Y is rank 2, and

$$Y \begin{bmatrix} t_{11} \\ t_{21} \\ t_{12} \\ t_{22} \end{bmatrix} = z, \qquad (14)$$

where $z \in \mathbb{R}^4$ is a column vector with elements of $\tilde{b}_l$, $\tilde{c}_m$ and two zero elements.

If $[I \otimes A - \tilde{A}^T \otimes I] + Y$ is non-singular rank 4, we can calculate $T$ and hence calculate $\tilde{B}$ and $\tilde{C}$.

**Construction of $\tilde{D}$.** From (8), $\tilde{d} = d$. ∎∎

## 3. MOGA FOR OPTIMAL FWL CONTROLLER STRUCTURES

### 3.1 *Procedure outline*

The problem can be summarized as follows. Given a discrete time nominal plant $G(z)$ and designed PID (or other second order) controller $K(z)$, find an FWL controller $K_q(z)$ and state-space parameterization such that the difference between the closed loop system and the original closed loop system and the memory requirements are simultaneously minimized.

In order to solve the above problem using the MOGA, two performance indices need to be defined. The first index is a closed loop measure of the implementation accuracy of the FWL controller compared to the original controller, and the second index is a measure of the total number of bits required to store the FWL controller parameters. These indices are defined in Section 3.4.

The main elements of the approach are now summarized.

i) Randomly generate partially filled FWL parameterizations of the 2 state controller e.g.

$$\tilde{K} \stackrel{s}{=} \left[\begin{array}{cc|c} q_1 & ? & q_2 \\ q_3 & ? & ? \\ \hline ? & q_4 & D_K \end{array}\right] \qquad (15)$$

where $q_i \in \mathcal{Q}_{n,m}, i = 1, \ldots, 4$, i.e. each $q_i$ is FWL.

ii) Solve a linear system similarity completion problem such that $\tilde{K}(z) = K(z)$.

iii) Obtain $K_q \approx \tilde{K}$ by rounding the non-FWL parameters in $\tilde{K}$ so that they are FWL, i.e.

$$K_q \stackrel{s}{=} \left[\begin{array}{cc|c} q_1 & q_5 & q_2 \\ q_3 & q_6 & q_7 \\ \hline q_8 & q_4 & q_9 \end{array}\right] \qquad (16)$$

where $q_i \in \mathcal{Q}_{n,m}, i = 1, \ldots, 9$.

iv) Calculate the closed loop implementation index and the memory requirement index for $K_q$.

v) Use the MOGA to evolve a set of near Pareto-optimal solutions.

### 3.2 *FWL effect measure*

A measure of the difference between the FWL implemented closed loop system and the original closed loop system is defined in this section based on the difference between the closed loop pole positions of the systems.

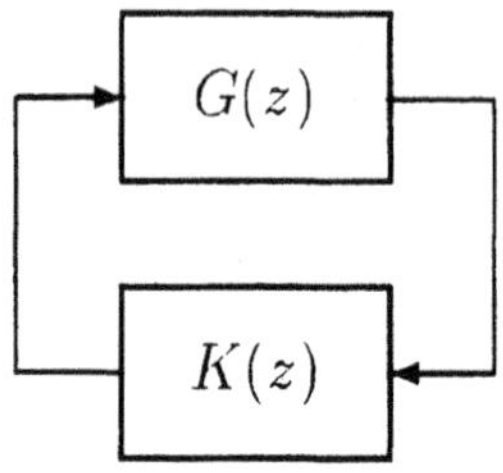


Fig. 2. Discrete time feedback system

Consider the discrete time system shown in Figure 2. Let $(A_g, B_g, C_g, D_g)$ be a state-space description of the SISO plant $G(z) = C_g(zI - A_g)^{-1}B_g + D_g$, where $A_g \in \mathbb{R}^{m\times m}$, $B_g \in \mathbb{R}^{m\times 1}$, $C_g \in \mathbb{R}^{1\times m}$ and $D_g \in \mathbb{R}$. Let $(A_k, B_k, C_k, D_k)$ be a state-space description of the 2 state controller, $K(z) = C_k(zI - A_k)^{-1}B_k + D_k$, where $A_k \in \mathbb{R}^{2\times 2}$, $B_k \in \mathbb{R}^{2\times 1}$, $C_k \in \mathbb{R}^{1\times 2}$ and $D_k \in \mathbb{R}$. In this paper, $(A_k, B_k, C_k, D_k)$ is also called a realization of $K(z)$. The realizations of $K(z)$ are not unique. In fact, if $(A_k^0, B_k^0, C_k^0, D_k^0)$ is a realization of $K(z)$, then so is $(T^{-1}A_k^0 T, T^{-1}B_k^0, C_k^0 T, D_k^0)$ for any non-singular similarity transformation $T$. The transition matrix of the closed loop system is $\bar{A} =$

$$\begin{bmatrix} A_g + B_g D_k (I - D_g D_k)^{-1} C_g & B_g (I - D_k D_g)^{-1} C_k \\ B_k (I - D_g D_k)^{-1} C_g & A_k + B_k (I - D_k D_g)^{-1} D_g C_k \end{bmatrix}.$$

When $(A_k, B_k, C_k, D_k)$ is implemented with a finite word-length device, the elements are rounded to the realization $K_q = (A_q, B_q, C_q, D_q)$, that is $K(z)$ is perturbed to $K_q(z)$ where $K_q(z) = C_q(zI - A_q)^{-1}B_q + D_q$ and where $A_q \in \mathcal{Q}_{n,m}^{2\times 2}$, $B_q \in \mathcal{Q}_{n,m}^{2\times 1}$, $C_q \in \mathcal{Q}_{n,m}^{1\times 2}$ and $D_q \in \mathcal{Q}_{n,m}$. The transition matrix of the FWL implemented closed loop system is thus $\bar{A}_q =$

$$\begin{bmatrix} A_g + B_g D_q (I - D_g D_q)^{-1} C_g & B_g (I - D_q D_g)^{-1} C_q \\ B_q (I - D_g D_q)^{-1} C_g & A_q + B_q (I - D_q D_g)^{-1} D_g C_q \end{bmatrix}.$$

A closed loop measure, $\eta$, of the accuracy of the FWL implementation is taken as

$$\eta(K_q) \triangleq \max_{i \in \{1,\ldots,m+2\}} \left| \lambda_i - \lambda_i^Q \right| \qquad (17)$$

where $\{\lambda_i : i = 1, \ldots, m+2\}$ represents the set of all eigenvalues of $\bar{A}$ and $\{\lambda_i^Q : i = 1, \ldots, m+2\}$ the corresponding set of all eigenvalues of $\bar{A}_q$.

### 3.3 *Encoding of solution space*

In order to generate partially filled parameterizations of $K$ such as (15), the genotype of each individual consists of a 75 bit binary string $[x_1, \ldots, x_{75}], x_i \in \{0,1\}$. The bit length of the integer part of the parameters' representation $m \in 0, \ldots, 7$ is represented by $[x_1, \ldots, x_3]$, and is given by $m = \sum_{i=1}^{3} x_i 2^i$. The bit length of the fractional part of the parameters' representation $n \in 0, \ldots, 15$ s represented by $[x_4, \ldots, x_7]$, and is given by $n = \max(\sum_{i=1}^{4} x_i 2^{i-3}, 15 - m)$. The parameters $(i, j, l, m) \in \{1,2\}$ are represented by $[x_8, \ldots, x_{11}]$ and given by $[x_8 + 1, \ldots, x_{11} + 1]$ respectively. The four $m + n + 1$ word-length parameters $q_i, i = 1, \ldots, 4$, where $(q_1, q_2, q_3, q_4) = (a_{ii}, a_{jk}, b_l, c_m)$ respectively, are represented by $(x_{12+16(i-1)}, \ldots, x_{12+m+n+16(i-1)})$. Thus not all the bits in $x$ are necessarily active. The values of $q_i$ are calculated by (2).

### 3.4 *MOGA implementation*

The remaining parameters of an equivalent controller to $K(z)$ are calculated using Theorem 1, an FWL controller $K_q(x)$ is calculated by a simple rounding operation on the remaining parameters.

Two objective functions, $\phi_1(x), \phi_2(2)$ are defined. A measure of implementation accuracy (if the implemented controller and system is stable) is defined as

$$\phi_1 = \begin{cases} \eta(K_q) & \text{if } \left|\lambda_i^K\right| \leq 1 \forall i \\ & \text{and } \left|\lambda_i^Q\right| < 1 \forall i, \\ \max_i \left|\lambda_i^K\right| & \text{if } \left|\lambda_i^K\right| > 1 \exists i, \\ \max_i \left|\lambda_i^Q\right| & \text{if } \left|\lambda_i^K\right| \leq 1 \forall i \\ & \text{and } \left|\lambda_i^Q\right| \geq 1 \exists i \end{cases} \qquad (18)$$

where $\{\lambda_i^K : i = 1, 2\}$ represents the pair of eigenvalues of $A_q$ and $\{\lambda_i^Q : i = 1, \ldots, m + 2\}$ represents the set of all eigenvalues $\bar{A}_q$. An implementation memory function, $\phi_2$, is defined as the total number of bits used to implement $K_q$. This function is calculated bearing in mind that parameters of $K_q$ that are $1, 0, -1$ or that are a power of 2 require less memory requirement than the $m + n + 1$ bits from (2).

## 4. APPLICATION TO STEEL ROLLING MILL SYSTEM

The proposed approach is applied to a PID controller of steel rolling mill (Hori, 1996).

The continuous time linearized ideal two-mass inertia model of the system plant, $G(s)$, is given as

$$A_g = \begin{bmatrix} 0 & -9763.7203 & 0 \\ 1 & 0 & -1 \\ 0 & 13424.9859 & 0 \end{bmatrix},$$

$$B_g = \begin{bmatrix} 249.03 \\ 0 \\ 0 \end{bmatrix}, C_g = \begin{bmatrix} 1 & 0 & 0 \end{bmatrix}.$$

The plant, $G(s)$, is discretized with sampling period of 0.001 seconds.

A PID vibration suppression and disturbance rejection controller is designed as

$$K(s) = \frac{0.00269s}{0.001s + 1} - 0.435 - \frac{14.26}{s}.$$

The bilinear transform,

$$s = \frac{2}{h}\frac{1 - z^{-1}}{1 + z^{-1}},$$

gives the digital PID controller as

$$K(z) = -\frac{0.01426}{z - 1} - \frac{1.1956}{z - 0.3333} + 1.3512.$$

The initial realization $K^0$ is set to

$$A_k^0 = \begin{bmatrix} 1 & 0 \\ 0 & 0.3333 \end{bmatrix}, B_k^0 = \begin{bmatrix} -1 \\ -1 \end{bmatrix},$$

$$C_k^0 = \begin{bmatrix} 0.01426 & 1.1956 \end{bmatrix}, D_k^0 = 1.3512.$$

The MOGA is implemented in MATLAB using the GA Toolbox (Chipperfield *et al.*, 1994). An elitism strategy is used whereby all non-dominated individuals propagate through to the next population. Selection is performed using stochastic universal sampling with a fitness determined by the number of dominating individuals. Single point cross-over is used with a probability of 0.7. Each bit has a probability of mutation of 0.00933.

The MOGA is run with a population of 120. After 400 generations (which takes about 1 hour on a 450 MHz Pentium II), a set of non-dominated solutions is obtained. This set is shown in Figure 3. The figure also shows the stable dominated solutions along with FWL implementations of the initial realization, $K^0$, for various word-lengths. In addition, the diagonal similarity balanced realization of $K(z)$ is calculated using the **ssbal** function from the MATLAB Control System Toolbox (MATLAB, 1998, pp 9/200–201), and FWL implementations of the realization for various word-lengths are shown. Note that the axis for $\phi_1$ is shown to log scale for clarity. The figure clearly shows that improved FWL state-space realizations for the controller can be obtained using the approach.

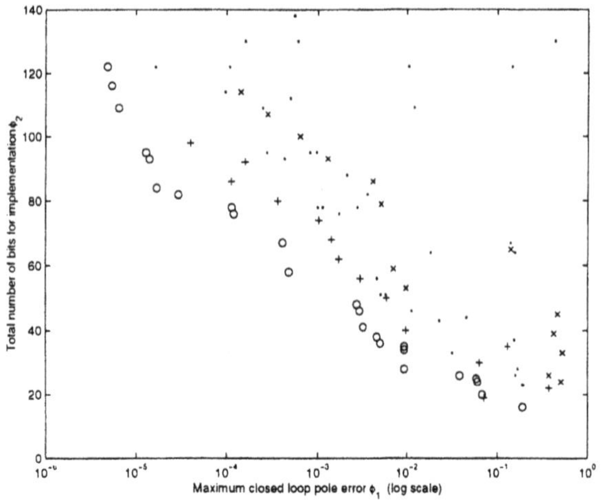


Fig. 3. Solution set showing trade-off with MOGA non-dominated set (o), MOGA dominated set (·), $K_q^0$ realizations (×) and similarity balanced realizations (+)

## 5. CONCLUSION

A GA-based method to determine optimal FWL structures for PID digital controllers has been developed. The method is illustrated by a practical example. The method exploits the fact that the implementation of FWL controllers is by means of binary numbers, as is the representation in genetic algorithms.

The method requires the solution of a linear system similarity completion problem. A solution to the linear system similarity completion problem for 2 state SISO systems has been presented. This solution would need to be extended to general linear systems for the methodology to be extended to higher order controllers. Otherwise, the method is entirely generic, and any computable set of stability and performance measures could be included. In addition, other implementation measures could be included, such as computation time, as well as measures of other important quantization effects, namely quantization noise and scaling requirements.

## 6. REFERENCES


Chipperfield, A., P.J. Fleming, H. Pohlheim and C.M. Fonseca (1994). *Genetic Algorithm Toolbox: User's Guide.* Dept. Automatic Control and Systems Engineering, University of Sheffield, U.K.

Dakev, N.V., J.F. Whidborne, A.J. Chipperfield and P.J. Fleming (1997). $H_\infty$ design of an EMS control system for a maglev vehicle using evolutionary algorithms. *Proc. IMechE, Part I: J of Sys and Contr.*, **311**(4), 345–355.

Fonseca, C.M. and P.J. Fleming (1993*a*). Genetic algorithms for multiobjective optimization: formulation, discussion and generalization. In: *Genetic Algorithms: Proceeding of the Fifth International Conference.* San Mateo, CA. pp. 416–423.

Fonseca, C.M. and P.J. Fleming (1993*b*). Multiobjective genetic algorithms. In: *IEE Colloquium on Genetic Algorithms for Control Systems Engineering*, number 1993/130. London, U.K. pp. 6/1–6/5.

Gevers, M. and G. Li (1993). *Parametrizations in Control, Estimations and Filtering Problems: Accuracy Aspects.* Springer-Verlag, Berlin.

Goldberg, D.E. (1989). *Genetic Algorithms in Search, Optimization and Machine Learning.* Addison-Wesley, Reading, MA.

Hori, Y. (1996). A review of torsional vibration control methods and a proposal of disturbance observer based new techniques. In: *Proc. 13th IFAC World Congress.* San Fransisco, CA. pp. 7–13.

Horn, R.A. and C.R. Johnson (1991). *Topics in Matrix Analysis.* Cambridge University Press, Cambridge, U.K.

Istepanian, R.H., J. Wu, J.F. Whidborne and J. Chu (1998). Maximizing lower bound stability measure of finite precision PID controller realizations by nonlinear programming. In: *Proc. 1998 Amer. Contr. Conf.* Philadelphia, PA. pp. 2596–2600.

Knowles, J.B. and R. Edwards (1965). Effect of a finite-word-length computer in a sampled-data feedback system. *Proc. IEE,* **112**(6), 1197–1207.

Man, K.F., K.S. Tang, S. Kwong and W.A. Halang (1997). *Genetic Algorithms for Control and Signal Processing.* Springer, London, U.K.

Masten, M.K. and I Panahi (1997). Digital signal processors for modern control systems. *Control Engg Practice,* **5**(4), 449–458.

MATLAB (1998). *Control System Toolbox: User's Guide, version 4.* The MathWorks, Inc., Natick, MA.

Morony, P. (1983). *Issues in the Implementation of Digital Feedback Compensators.* MIT Press, Cambridge, MA.

Nering, E.D. (1990). *Linear Algebra and Matrix Theory,* 2 ed. John Wiley, New York.

Pareto, V. (1906). *Manuale di Economia Politica.* Societa Editrice Libraria. Milan, Italy.

Tang, K.S., K.F. Man and D.-W. Gu (1996). Structured genetic algorithm for robust $H^\infty$ control system design. *IEEE Trans. Industrial Electronics,* **43**(15), 575–582.

Williamson, D. (1991). *Digital Control and Implementation: Finite Wordlength Considerations.* Prentice Hall, Englewood Cliffs, NJ.

Xu, D.J. and M.L. Daley (1995). Design of optimal digital filter using a parallel genetic algorithm. *IEEE Trans. Circuits Syst.: II,* **42**(10), 673–675.

14th Triennial World Congress, Beijing, P.R. China

F-2d-03-4

# A COOPERATIVE/COMPETITIVE CO-EVOLUTIONARY LEARNING APPROACH TO SORTING NETWORK PROBLEMS

**Ming Chang   Kazuhiro Ohkura   Kanji Ueda** *,[1]

* *Faculty of Engineering, Kobe University*
*Rokkodai, Naka-Ku, 657, Japan*
*email: {chang, ohkura, ueda}@mi-2.mech.kobe-u.ac.jp*
*Phone: +81-78-803-1119   Fax: +81-78-803-1131*

Abstract: A combination of cooperative co-evolutionary learning and competitive co-evolutionary learning is described and the performance of this method is examined on the sorting network problems. A sorting network is divided according to its parallelly performable steps. The parallel steps are encoded by *cooperatively* co-evolving populations. Each population corresponds to one parallel step. A complete sorting network is constructed by individuals from each population. Simultaneously, a population of test data evolving *competitively* against the cooperative populations is adopted in order to utilize the *Red Queen Effect* which accelerates an evolutionary process so that the emergence of function specialization in the cooperative co-evolutionary populations occurs quickly. Several computer simulations are conducted to investigate the optimization performance and analyze co-evolving behaviors.


Keywords: Co-evolutionary Learning, Cooperate, Compete, Sorting Network Problems.

## 1. INTRODUCTION

Major *Evolutionary Algorithms* (EAs) evolve a population of competing individuals using fitness-biased selection, random mating, and a gene-level representation of individuals together with certain simple genetic operators for modeling inheritance of traits. In most case, the individuals in the population have the same structure, *e.g.* in the case of *canonical* Genetic Algorithms (cGA), each individual in population is represented by bit-strings with the same length. Individuals are selected for mating according to their fitness which evaluated by a user-predefined optimization criterion. These EAs have been successfully applied to a wide variety of problems. At the same time, difficulties have also arised. For example, when EAs were used to evolve complex structures with potentially unrestricted capabilities, as in the evolution of animats or mechanisms for cognition structures, a search space of an infinite degree of freedom is facing, and the invariant genotype representation of individuals in traditional EAs will not be suitable for the purpose of evolving more and more complex structures. Furthermore, the user-predefined criteria may also be problematic because in some cases the criteria that evaluate individuals were not known precisely *a priori*.

Harvey [1,2] has proposed *Species Adaptation Genetic Algorithms* (SAGA) to obtain complex structures, where genotypes with variable length are used. Since genotype structures are variable, it may get more and more intricate during the evolution process. The resulting complex structures

[1] Partially supported by the financial support through "Methodology for Emergent Synthesis" project (project number: 96P00702) in Research for the Future Program of the Japan Society for the Promotion of Science (JSPS)

in this case are considered as extensions toward *outside* *i.e.*, toward the infinite search space. On the other hand, under the hypothesis that " to evolve more and more complex structures, explicit notions of modularity need to be introduced in order to provide reasonable opportunities for complex solutions to evolve in the form of interacting co-adapted subcomponents", De Jong *et al* [10,11] have proposed *Cooperative Coevolutionary Algorithms* (CCGAs), in which the complex structures are considered as the emergence of modularity, hierarchies, and so on, within obtained structures that are *inside* the search space.

In the present paper, an extended *co-evolutionary* approach is proposed in order to evolve complex structures. *Co-Evolutionary Algorithms* have several populations interacting with one another as well with their environment, *Relative Fitness Function*[3] is employed between them. Co-evolution can give rise to the *Red Queen Effect*, since interacting populations alter each other's fitness landscape.

The proposed approach combine the *Cooperative Co-evolutionary Learning* [10–13] with *Competitive Co-evolutionary Learning* [5,8], and its performance is examined on the sorting network problems.

The next section gives a brief introduction of the sorting network problems. Cooperative and competitive co-evolutionary learning is explained in Section 3 and 4 respectively. In Section 5 the proposed approach to solving sorting network problems is explained in detail, and the simulation results are shown in Section 6.

## 2. THE SORTING NETWORK PROBLEM

According to [6], a *comparator network* with $n$ inputs can be expressed as an ordered tuple $(L_1, L_2, ..., L_d, n)$. Each $L_k$ is called a *level*, and is a set of ordered pairs $\langle i,j \rangle$ where $1 \le i < j \le n$. Each $\langle i,j \rangle \in L_k$ is called a *comparator*, and is stated to be at *level* $k$. The values $i$ and $j$ are called *channels*. Every channel is involved in at most one comparison at each level, that is, $|\{\langle i,j \rangle \in L_k | i < j \le n\}| + |\{\langle j,i \rangle \in L_k | 1 \le j < i\}| \le 1$, for all $1 \le i \le n$ and $1 \le k \le d$.

Comparator networks are used to sort an input $x = \langle x_1, ..., x_n \rangle$ of $n$ distinct values as follows. Let $C = (L_1, L_2, ..., L_d, n)$ be a comparator network. Define the *value* of $C$ in channel $i$ at level $k$ on input $x$ to be $V(i,k)$, where $V(i,0) = x_i$ for $1 \le i \le n$, and for $k > 0$,

$$V(i,k) = \begin{cases} min(V(i,k-1), V(j,k-1)) \\ \qquad \text{if}\langle i,j \rangle \in L_k, \\ max((V(i,k-1), V(j,k-1)) \\ \qquad \text{if}\langle j,i \rangle \in L_k, \\ V(i,k-1) \qquad \text{otherwise} \end{cases}$$

The *output* of $C$ on input $x$ is defined to be $\langle V(1,d), V(2,d), ..., V(n,d) \rangle$. If the output of $C$ is in ascending order for all input $x$, then $C$ is called a *sorting network*. Each level of a comparator network consists of a set of comparisons which may be performed in parallel. The number of levels is called as the *depth* of the network.

The sorting network problem is to find an efficient network which correctly sorts any sequence of inputs. It has been proved that a sorting network which can sort any {0,1} sequence will correctly sort any sequence that consists of arbitrary numbers correctly.

There are two different measures for sorting networks [7,9]:

- its **depth:** the number of parallel steps.
- its **length:** the total number of comparators.

The current upper and lower bounds of the *depth* and the *length* of the most efficient network of $n$-input ($n \le 16$) are shown in [7].

## 3. COOPERATIVE CO-EVOLUTIONARY LEARNING

In cooperative co-evolutionary learning, individual is only a rule, like in the Michigan approach. The whole population is a complete learning system. Individuals collectively make up a complete solution for the problem, and the fitness of a individual is computed by estimating how well it *cooperates* with other individuals to produce good solutions. One of the most difficult problems faced by this approach is the credit assignment problem, *i.e.*, the problem of assigning credits to individual rules based on the payoff of the complete system. There has not been a satisfactory solution to this problem so far.

## 4. COMPETITIVE CO-EVOLUTIONARY LEARNING

Most competitive co-evolutionary learning use two genetically distinct populations, and individuals in the populations constitute a complete learning system. Because a *competitive fitness function* [14] is used between the populations, fitness landscape for each population can be altered by changes in the other population. This sort of mechanism can be found in a *host-parasite* system or a *predator-prey* system, *i.e.*, the fitness of a individual in a *host* population is based on direct competition with individual(s) from the *parasite* population. Success of one population is perceived by the other population as its failure, which must be overcame in order to maintain one's chances of survival. This typically results in *arm races* in which the complexities of both populations increase stepwise. The key to a successful co-evolutionary learning is an *arm race* between opposed populations. A number of biologists have suggested that this co-evolutionary *arm race* is a prime source of evolutionary innovations and adaptations[4].

## 5. PROPOSED APPROACH

A sorting network can be divided into several layered parallel steps in which the order of comparison-swaps do not affect the performance of the network. Each population in cooperative co-evolutionary learning corresponds to one of parallel steps. Individuals in each population contains all information about a parallel step in the network, and collectively construct a complete sorting network with individuals of the other populations. The constructed sorting network is evaluated by how well it sorts the input test sequences. If it correctly sorted a test sequence, the individual will get a bonus, if it fails, the test sequence will get a bonus. Fitness of each individual is further altered according to the number of comparators in it so that the selection favors the shorter network. As a result, the more compact network can be found.

There is also a population that consists of test sequence, initialized by randomly generated bit strings. Individuals in the test population are evaluated by how difficult it is to be correctly sorted by the sorting networks. The selection process for the test population favors the test sequences which the network failed to sort correctly. This *competitive fitness function* between the networks and the test sequences will induce the *Red Queen Effect*, and the evolutionary rate of both population will be accelerated.

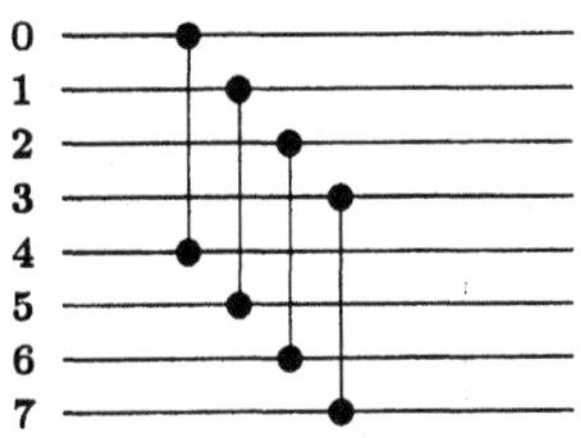


Fig. 1. A parallel step in a 8-input sorting network.

### 5.1 *Individual Representation*

Figure 1 gives a convenient graphical representation of a parallel step of a 8-input sorting network. Each horizontal line represents an input for the sorting network assigned the index number from 0 to 7. Each connection between two nodes represents a *comparator* which compares the two elements and exchange them if one on the upper line is larger than one on the lower line. The four comparators in Figure 1 are indexed as *a, b, c, d* from left to right, respectively. The input of the sorting network is on the left of the representation. Elements in the output sequence are sorted and the largest element is on the bottom line.

| comparator : | *a* | *b* | *c* | *d* |
|---|---|---|---|---|
| regulator gene : | 1 | 1 | 1 | 1 |
| input index : | 0 | 1 | 2 | 3 |
| input index : | 4 | 5 | 6 | 7 |

Fig. 2. A genotype for a parallel step.

Figure 2 shows the genotype of the parallel step in the above sorting network. Genotype of each comparator is represented by three numbers. For example, genotype of comparator *a* is represented by 1 0 4. The second and third numbers, 0 and 4, represent the two line indexes in comparator *a*, the first number is *regulator gene* that controls *phenotypic expression* of the genotype of comparator *a*. "1" means the genotype will be expressed while "0" means it will not.

### 5.2 *Genetic Operator*

Three kinds of mutation operators are defined as following.

5.2.1. *Mutation of Regulator Gene* : $M_r$ Figure 3 shows mutation in a regulator gene in the genotype. The $M_r$ operator simple flip the value of the regulator gene from 1 to 0, or *vice versa*. Figure 4 express the phenotype after the $M_r$ occured. Thus, the *length* of sorting networks can be controlled by $M_r$.

| | | | | | | | | |
|---|---|---|---|---|---|---|---|---|
| 1 | *1̲* | 1 | 1 | | 1 | *0̲* | 1 | 1 |
| 0 | 1 | 2 | 3 | ⟶ | 0 | 1 | 2 | 3 |
| 4 | 5 | 6 | 7 | | 4 | 5 | 6 | 7 |

Fig. 3. Mutation of regulator gene.

5.2.2. *Transversion Operator:* $M_v$ Figure 5 shows the operation of transversion operator in the genotype. The $M_v$ operator randomly swaps input indexes in the same rank in the genotype presentation. Figure 6 shows phenotype after $M_v$ occured.

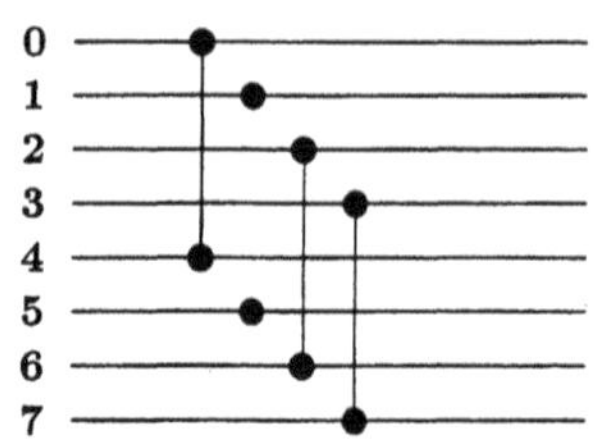


Fig. 4. Phenotype after $M_r$ occured in the original genotype.

| | | | | | | | | |
|---|---|---|---|---|---|---|---|---|
| 1 | 1 | 1 | 1 | | 1 | 1 | 1 | 1 |
| 0 | 1 | 2 | 3 | → | 0 | 1 | 2 | 3 |
| 4 | 5 | *6* | *7* | | 4 | 5 | *7* | *6* |

Fig. 5. Operation of Tranversion.

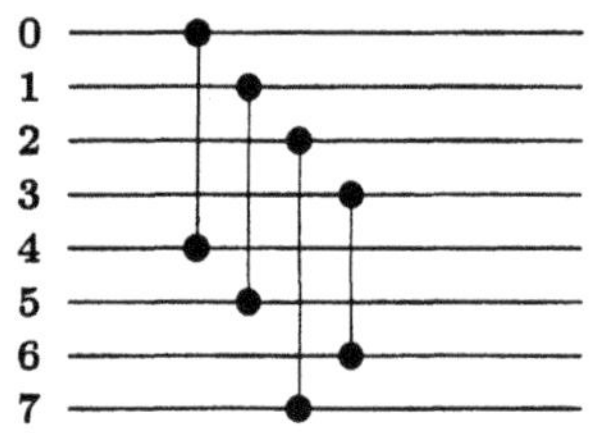


Fig. 6. Phenotype after $M_v$ occurred.

5.2.3. *Transition Operator :* $M_i$ Figure 7 shows the operation of transition operator in the genotype. The $M_i$ operator swaps input index of the same genotypes of comparators. Figure 8 shows that $M_i$ itself does not introduce a new phenotype. But the combination of $M_i$ and $M_v$ can introduce a new phenotype that can not be generated using only $M_v$.

| | | | | | | | | |
|---|---|---|---|---|---|---|---|---|
| 1 | 1 | 1 | 1 | | 1 | 1 | 1 | 1 |
| 0 | 1 | *2* | 3 | → | 0 | 1 | *6* | 3 |
| 4 | 5 | *6* | 7 | | 4 | 5 | *2* | 7 |

Fig. 7. Operation of Transition.

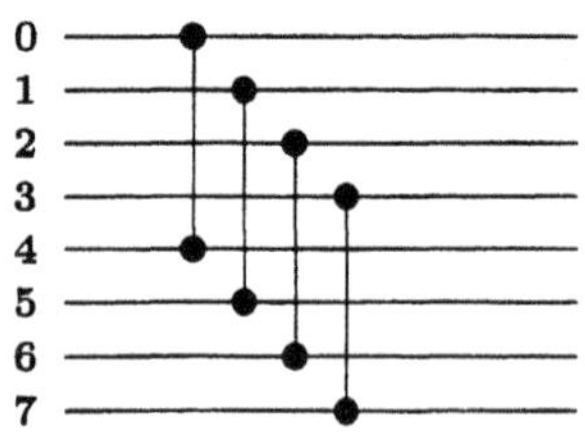


Fig. 8. $M_i$ itself does not introduce a new phenotype.

### 5.3 *A Cooperative/Competitive Approach*

The pseudo-code given in figure 9 describes the basic procedure of the proposed approach. At first, the test sequences population (Compop($gen$)) and the populations for each cooperating parallel step (Coopop($gen$)) are initialized. Since a complete sorting network is constructed by assembling individuals from each cooperative populations, one can assign fitness to a particular individual of a particular cooperative population by assembling it along with selected individuals of the other cooperative populations to form one or more sorting networks whose fitness are computed by competing with the test sequences population. The obtained sorting networks are evaluated by using the test sequences population. The pay-off for the sorting network is 1 (0) if it (un)correctly sorted one test sequence in the population, and the pay-off for the test sequence is 0 (1). The resulted fitness of the formed sorting networks are used to assign fitness to the individual components being evaluated. That is, the fitness of a particular individual of a particular cooperative population is calculated by estimating how well it "cooperates" with other populations to produce good solutions, while the fitness of individual of the competitive population is calculated by estimating how well it "compete" against the formed sorting networks.

```
gen = 0
initialize(Compop(gen));
initialize(Coopop(gen));
for each cooperative population s do begin
  evaluate(Coopop_s(gen),Compop(gen));
  end
while termination condition = false do begin
  gen = gen + 1;
  reproduce(Compop(gen),Compop(gen - 1));
  for each cooperative population s do begin
  reproduce(Coopop_s(gen),Coopop_s(gen - 1));
  evaluate(Coopop_s(gen),Compop(gen));
  end
end
```

Fig. 9. Algorithm of the proposed approach.

The initial fitness of each individual in the cooperative populations is calculated by combining it with randomly selected individuals from each of the other populations to construct a network and testing the resulting network against the test sequences population. The fitness of the network is assigned to the individual. After this, each cooperative population is coevolved in a round-robin fashion using a *canonical* GA. The fitness of an individual in the cooperative population is obtained by combining it with the best current individuals of the remaining population and testing the resulting network against the test sequences population. Then the fitness of the network is assigned to the individual. After all cooperative populations are updated, the test sequences population will be updated, then a new generation begin.

## 6. EXPERIMENTAL RESULTS

The Algorithm in figure 9 has been tested on 8-input sorting network problem. For 8-input sort-

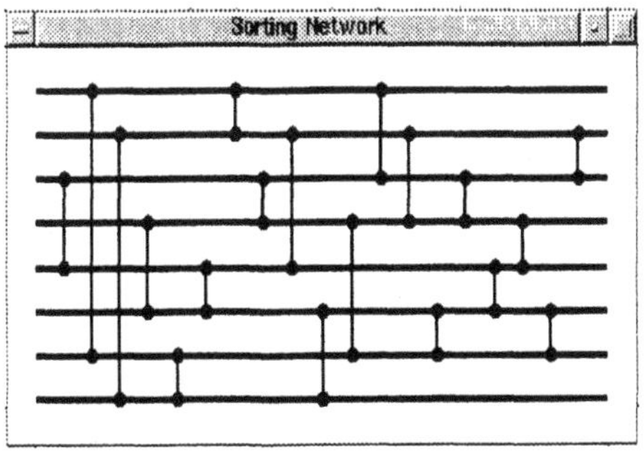


Fig. 10. A 8-input 19-comparators sorting network using 6 parallel steps.

ing network, the possible input test sequences is $2^8$, it is a relatively small number, so all the possible test sequences are included in the test sequences population and remained unshanged through the evolution process. This means the comparative learning is excluded. Under the conditions, the sorting network as good as the best known is found. The parameters and the found sorting network are shown in figure 11 and figure 10. Tournament selection is employed in cooperative population. A tournament consists of two players, and $P_T$ shows the possibility of the winner being copied to the loser.

| | |
|---|---|
| depth: | 6 |
| cooperative population size : | 10 |
| $M_R$ : | 0.001 |
| $M_I$ : | 0.4 |
| $M_V$ : | 0.8 |
| $P_T$ : | 0.8 |
| test sequences population size : | 256 |

Fig. 11. Parameters of the approach

The algorithm has also been tested for the test sequences population size of 200. In this case, the best sorting network has not been found. The experimental results show that in this case, the balance between the evolutionary rates of the competing populations affect the performance greatly.

The 16-input sorting network problem has been challenged by the proposed algorithms. But no sorting network that can sort correctly all the test sequences found. A network with 62 comparators that correctly sorted about 95% of all the test sequences has been found after 3 hours on SUN workstation, and no more progress on the found network is obtained for further runs.

## 7. CONCLUSION & FURTHER WORK

In the proposed approach, each of the cooperative population is coevolved in a round-robin fashion. This sequential version of the algorithm could be called more accurately as quasi-coevolution. A fully coevolutionary version is also possible in which each population only occasionally communicates with the other populations.

The credit assignment used here is very simple, that evaluated individual in the cooperative population by combining it with the current best individuals in the remaining population. This procedure may work well on problem whose variables (cooperative populations) are reasonably independent, but have difficulties with problems with interacting variables-the sorting network problems is just this kind of problems.

In the further work, the 16-input sorting network problem will be challenged continuously, and a new credit assignment will be applied where individuals from the population which corresponds to the first parallel step of the sorting network will be evaluated by how well it sorts the inputed original test sequences. The individuals from population that corresponds to the second parallel step will be evaluated by how well it sorts the test sequences that has been sorted by the individuals in the first population, and so on. Since each the population focuses only on the tests that has been sorted by the previous populations, the *functional specialization* will occur in this case.

## 8. REFERENCES


[1] Inman Harvey, "Species adaptation genetic algorithms: The basis for a continuing SAGA," in F.J. Varela and P.Bourgine, editors, Toward a Practice of Autonomous Systems: Proceedings of the First European Conference on Artificial Life, 346-354, MIT Press/Bradford Books, Cambridge, MA, 1992.

[2] I. Harvey. "Evolutionary robotics and SAGA: the case for hill crawling and tournament selection. in C. Langton, editor, Artifical Life III, Santa Fe Institute Studies in the Sciences of Complexity, *Proc. Vol. XVI,* 299-326. Addison Wesley, 1993.

[3] Peter. J. Angeline, "Competitive Fitness," The Handbook of Evolutionary Computation, C4.3: 1-2, Bäck T., Fogel D., Michalewicz Z. (eds.). Oxford University Press, 1997.

[4] R. Dawkins, "The Blind Watchmaker," Longman, Essex, 1986.

[5] W. D. Hillis, "Co-evolving parasites improve simulated evolution as an optimization procedure," Physical D, 42 : 228-234, 1990.

[6] I. Parberry, "A Computer-assisted optimal depth lower bound for nine-input sorting networks," *Mathematical Systems Theory,* 24 : 101-116, 1991.

[7] H. Juillé, "Evolution of Non-Deterministic Incremental Algorithms as a New Approach for Search in State Spaces," in L. J. Eshelman, editor, Proceedings of the Sixth Interna-

tional Conference on Genetic Algorithms and their Applications *(ICGA'95)*, Pittsburgh, PA, July 1995.

[8] C. D. Rosin and R. K. Belew, "Methods for Competitive Co-evolution : Finding opponents worth beating," in L. J. Eshelman, (ed.), Proceeding of the Sixth International Conference on Genetic Algorithms and their Applications *(ICGA'95)*, Pittsburgh, PA, July 1995.

[9] Hugues Juillé, "Incremental Co-evolution of Organisms : A New Approach for Optimization and Discovery of Strategies," in Proceedings of the Third European Conference in Artificial Life, pp 246-260, Granada, Spain, June 1995.

[10] Mitchell A. Potter and Kenneth A. De Jong, "A Cooperative Coevolutionary Approach to Function Optimization," in the Third Parallel Problem Solving From Nature, Jerusalem, Isrusal. pp 249-257, Springer Verlag, 1997.

[11] Kenneth A. De Jong and Mitchell A. Potter, "Evolving Complex Structures via Cooperative Coevolution," in the Fourth Annual Conference on Evolutionary Programming, San Diego, CA. March 1-3,1995.

[12] Mitchell A. Potter and Kenneth A. De Jong, "Evolving Neural Networks with Collaborative Species," in Proceedings of the 1995 Summer Computer Simulation Conference, Ottawa, Ontario, July 24-26, 1995.

[13] Mitchell A. Potter, Kenneth A. De Jong, John J. Grefenstette, "A Coevolutionary approach to Learning Sequential Decision Rules," in L. J. Eshelman, editor, Proceedings of the Sixth International Conference on Genetic Algorithms and their Applications *(ICGA'95)*, Pittsburgh, PA, July 1995.

[14] Jan Paredis, "Coevolutionary Algorithms," The Handbook of Evolutionary Computation, C7.4 : 1-21, Bäck T., Fogel D., Michalewicz Z. (eds.). Oxford University Press, 1997.

14th Triennial World Congress, Beijing, P.R. China



F-2d-03-6

# GENETIC DRIVER FOR CAR RACING

**K S Tang, K F Man and S Kwong**

*City University of Hong Kong*
*Tat Chee Avenue, Kowloon, Hong Kong*
*email: eekstang@cityu.edu.hk*

**Abstract:** Genetic Algorithm is adopted to design the driver for a racing simulation, called Robot Auto Racing Simulation. By receiving the car's local situations, the car can be driven by the simple rules with control parameters optimized by Genetic Algorithm. The competition results demonstrate that it is the best amongst the existing control methods. *Copyright © 1999 IFAC*

**Keywords:** genetic algorithms, parameter optimization, robot

## 1. ROBOT AUTO RACING SIMULATION

The Robot Auto Racing Simulation (RARS) (RARS ftp-site, 1995) is a simulation of auto racing in which the cars are driven by robots. The RARS can link with maximum six racing cars while each one, with identical physical characteristic, can be driven by its own control program.

Two parameters are supplied to the simulator for steering and controlling the throttle of the racing car. The first parameter is the wheel attack angle, $\alpha$, which is the deviation between the angle that the car is pointing and the velocity vector. The maximum rate-of-change of $\alpha$ is 183 deg/sec so that the driver can steer and orient his vehicle appropriately.

The next control parameter $v_b$ is the speed of the bottom of the wheel relative to the car. It is set to be larger than the car's speed for acceleration, and vice verse. If the car is coasting at zero power, $v_b$ is equal to the speed of the car. If $v_b$ is held constant for a while, the car's speed will approach $v_b$, unless it is cornering, in which case $v_b$ may greatly exceed the speed. The positive direction of $v_b$ is limited by the full power available in the system, which is assumed to be a constant.

By obtaining the values of $\alpha$ and $v_b$ from the robot, the simulator can update the state of the car. The new information will then be passed to the robot driver for next iteration. The duration of each update is 0.25s. Figure 1 depicts the information flow between the robot and the simulator.

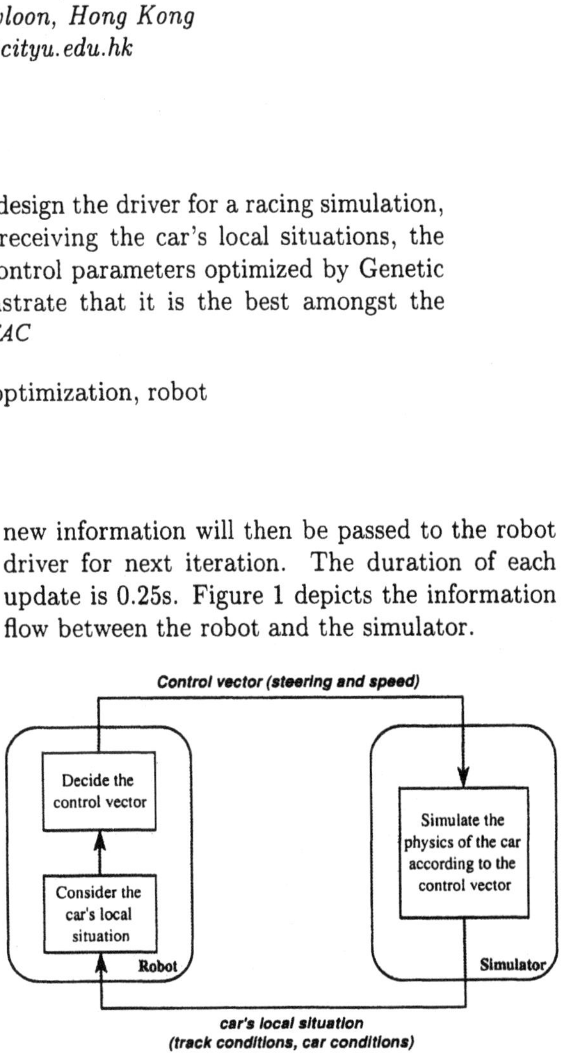


Fig. 1. Information flow between robot and simulator

There are totally 17 different racing tracks in RARS. Car racing in each track can be observed with a simple bird's eye view as shown in Fig. 2.

## 2. BASIC DRIVING CONCEPTS

In order to control the racing car, some basic driving concepts must be implemented in the robot.

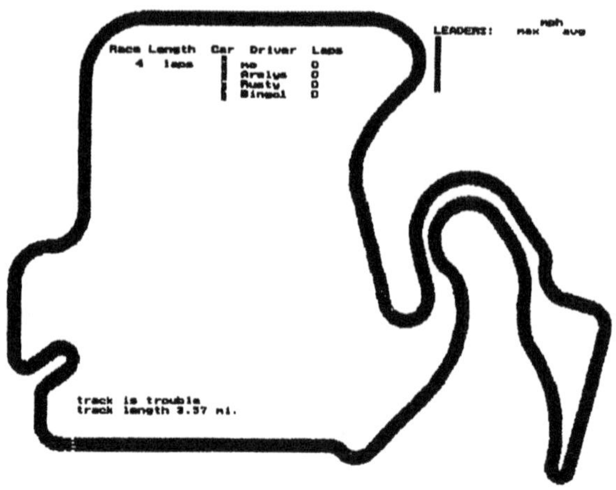

Fig. 2. A bird's eye view of a racing track

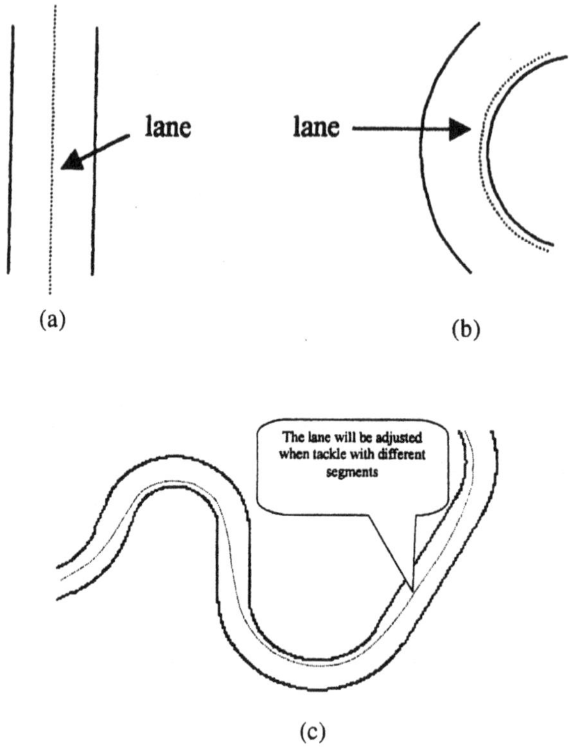


Fig. 3. Different road segments and their corresponding desired path (a) straight lane (b) curved lane (c) combination of straight and curved lane

## 2.1. *Steering*

From the servo-mechanism in steering, the steering gain and the steering damp must be considered. The wheel attack angle of the racing car, $\alpha$, is determined by the following equation:

$$\alpha = G_s \frac{d_l}{w_l} - D_s \frac{v_p}{v} \qquad (1)$$

where $G_s$ is the servo gain constant; $D_s$ is the servo damping constant; $v$ is the current speed; $v_p$ is the speed component perpendicular to the track; and $w_l$ and $d_l$ are the width of the lane and the distance between the car and the racing side wall, respectively.

## 2.2. *Desired racing path*

For safety point of view, a car should be maintained in the mid of the lane for maximum clearance. However, such scheme is not preferable in racing since this path may not be shortest. Referring to the professional racing car drivers, the desired path can be defined as in Fig. 3 for different racing segments. When the lane is a straight one, the racing car should be kept at the mid of the lane. While the car is turning around a curved segment, it should be driven near the inner wall for shorter path.

Fig. 3(c) shows a more general case in which straight lane and curved lane are combined. In such case, an interpolation is performed on the attack angle, as shown in Fig. 4. It is the appropriate angle to direct the car from current lane to the desired lane. The new wheel attack angle is computed as

$$\alpha_i = \frac{x}{x+y}\alpha_x + \frac{y}{x+y}\alpha_y \qquad (2)$$

where $\alpha_x$ and $\alpha_y$ are the wheel attack angle of the current segment and the next segment, respectively.

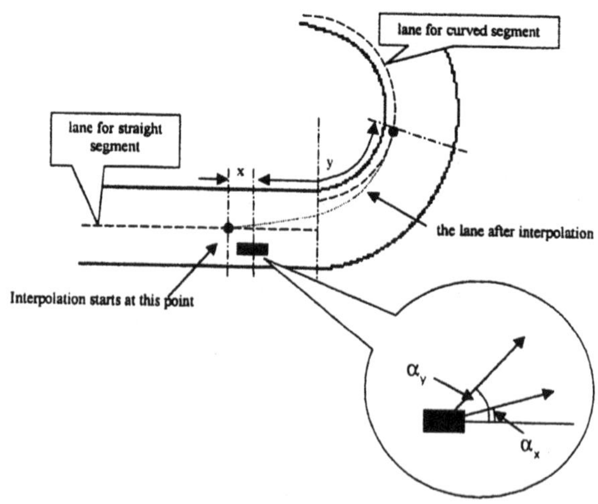


Fig. 4. Interpolation of the wheel attack angle

### 2.3. *Cornering speed*

In a circular path, the sliding problem must be prevented, otherwise, the racing car will not run on the desired path. Sliding occurs while the centripetal force larger than a particular value. Therefore, there exists a maximum allowable speed, known as cornering speed, for each corner such that no sliding is occurred. It is computed by

$$v_c = \sqrt{F_c \cdot (r + L)} \qquad (3)$$

where $v_c$ is the cornering speed, $F_c$ is the centripetal force, $r$ and $L$ is the radius and the length of the lane, respectively.

### 2.4. *Braking distance*

Braking distance is applied when the car runs from a high speed segment to a low-speed segment. For example, the car is entering a curved segment from a straight road. In order to meet the cornering speed in the curved segment, a brake must be applied. The distance that the car has to apply the brake is called the braking distance. It is computed by

$$D_b = \frac{(v_c^2 - v_t^2)}{2 \cdot d} \qquad (4)$$

where $D_b$ is the braking distance, $d$ is the deceleration of the car, $v_c$ and $v_t$ are the current speed and the targeted speed of the car, respectively.

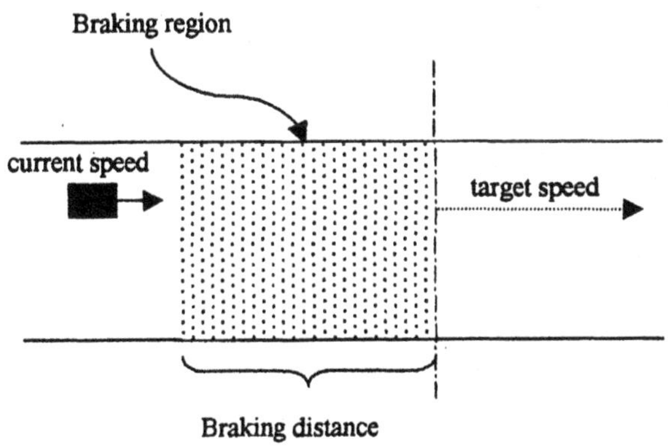


Fig. 5. Braking distance

### 2.5. *Damage index*

To emulate the real situation, the damage from either collisions or off-road adventures is accumulated in a damage index. Such index is checked after every collision. If it exceeds 30000, the car is forced to be out of the race.

## 3. GENETIC DRIVER

As explained in Sect. 2, the purpose of the robot driver is to keep the racing car in the desired path. The model of the robot driver is depicted in Fig. 6. It consists of a number of racing rules and the control parameters.

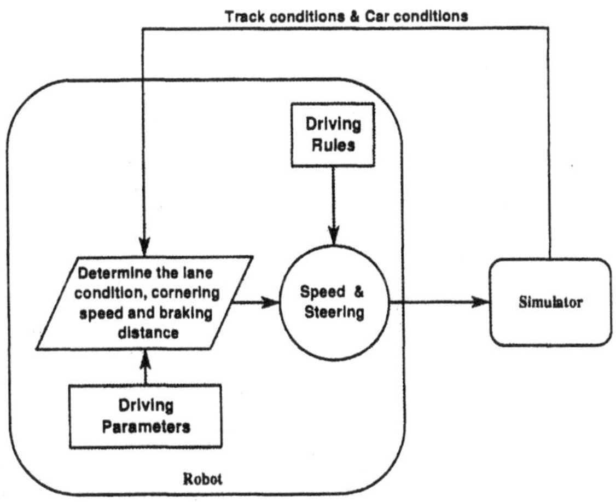


Fig. 6. Robot driver model

### 3.1. *Driving rules*

Decisions are to be made for different situations during the race. It is governed by the following driving rules, emulating the human behavior.

Fig. 7. Driving rules

### 3.2. *Driving parameters*

In order to compute the values of $\alpha$ and $v_b$, a number of parameters are required. They are, namely, steering gain, steering damp, wall fraction, wheel attack angles interpolation starting point and ending point, short segment factor, sharp segment factor, decelerating factor and centripetal force. The detail descriptions of these parameters are as below:

1. steering gain ($G_s$)– This parameter indicates the servo gain constant, as shown in Eqn. 1.
2. steering damp ($D_s$) – This is another parameter for the purpose of steering which prevents the oscillation of the car.
3. wall fraction – As mentioned in Sect. 2.2, the racing car should be maintained near the inner wall of the track for turning. This parameter, in the range of 0 to 1, indicates the optimal distance away from the wall. For example, if it is equal to 0.2 and the width of the track is 100 units, the desired lane should be 20 units away from the wall.
4. wheel attack angles interpolation starting point and ending point – They are in the range from 0 to 1, indicating the starting point and ending point of interpolation of the wheel attack angles, as shown in Fig. 4.
5. short segment factor – If the ratio of the segment length to track width is less than this factor, this segment is regarded as a short segment.
6. sharp segment factor – If the ratio of the segment radius to segment length is less than this factor, this segment is regarded as a sharp segment.
7. decelerating factor ($d$) – It is the constant required in Eqn. 4 for the calculation of the braking distance.
8. centripetal force ($F_c$) – It is an essential parameter, as shown in Eqn. 3, to compute the cornering speed.

### 3.3. *Genetic algorithm*

Genetic algorithm (GA) is applied to optimize the parameters so as to obtain an optimal genetic driver for car racing. The basic principles of GA were first proposed by Holland (Holland, 1975). Thereafter, a series of literatures (Man, *et al.*, 1997; Michalewicz, 1996) and reports (Man, *et al.*, 1996; Srinivas and Patnaik, 1994) have become available. GA is inspired by the mechanism of natural selection where stronger individuals would likely be the winners in a competing environment. A direct analogy of such natural evolution is here used.

The chromosome is encoded in a real number string. The lap time in racing is considered as the fitness function. The fitness value is obtained by running a lap in a track without other competitor. Such positive value reflects the degree of "goodness" of the driver.

Through out a genetic evolution, offspring are generated by the selected parent via the genetic operations, crossover and mutation. These offspring are then undergone the same objective function evaluation process. The offspring will replace the old chromosomes in the population if they are outperforming. Figure 8 depicts a general flow of GA cycle.

```
Genetic Algorithm ()
{
    // start with an initial time
    t := 0;
    // initialize a usually random population of individuals
    init_population P (t);
    // evaluate fitness of all initial individuals of population
    evaluate P (t);
    // evolution cycle
    while not terminated do
        // increase the time counter
        t := t + 1;
        // select a sub-population for offspring production
        P' := select_parents P (t);
        // recombine the "genes" of selected parents
        recombine P' (t);
        // perturb the mated population stochastically
        mutate P' (t);
        // evaluate its new fitness
        evaluate P' (t);
        // select the survivors from actual fitness
        P := survive P,P' (t);
    od
}
```

Fig. 8. GA cycle

Arithmetical crossover and non-uniform mutation (Michalewicz, 1996) are adopted in the Genetic Driver since they are effective in handling constraints for numerical problems.

Consider two chromosomes $s_1 = [a_1, a_2, \cdots, a_m]$ and $s_2 = [b_1, b_2, \cdots, b_m]$, the resulting offspring after arithmetical crossover are

$$s_1' = \beta \cdot s_1 + (1 - \beta) \cdot s_2$$
$$s_2' = \beta \cdot s_2 + (1 - \beta) \cdot s_1$$

where $\beta \in (0, 1)$

The mutating gene $a_k$ is modified as follows:

$$a_k = \begin{cases} a_k + \Delta(u_{(k)} - a_k) & \text{if a random digit is 0} \\ a_k - \Delta(a_k - l_{(k)}) & \text{if a random digit is 1} \end{cases}$$

where $\Delta(x)$ is a random value from 0 to $x$, $u_{(k)}$ and $l_{(k)}$ are the upper bound and the lower bound of the k-th gene.

## 4. RESULTS

The Genetic Driver joins the races in different tracks competing with four other drivers, including a fuzzy driver also designed by the authors. In the fuzzy driver, the driving rules are replaced with a fuzzy logic controller which has been fine-tuned manually. The competition results are tabulated in Table 1.

Table 1 Competition Results

| | | DRIVERS | | | | |
|---|---|---|---|---|---|---|
| TRACKS | | Genetic | Fuzzy | Arelys | Rusty | Bingol |
| Brazil | speed | 67.26 | 67.36 | 65.25 | — | 65.72 |
| | damage | 0 | 0 | 0 | out† | 0 |
| | rank | 2 | 1 | 4 | — | 3 |
| Clkwis | speed | 61.28 | 60.69 | 58.19 | 59.85 | 57.59 |
| | damage | 0 | 0 | 0 | 0 | 0 |
| | rank | 1 | 2 | 4 | 3 | 5 |
| Dreiw | speed | 75.98 | 77.00 | — | 67.85 | — |
| | damage | 17301 | 0 | out | 0 | out |
| | rank | 2 | 1 | — | 3 | — |
| Fourmile | speed | 78.68 | 76.24 | — | 86.64 | 76.27 |
| | damage | 0 | 18325 | out | 0 | 0 |
| | rank | 2 | 4 | — | 1 | 3 |
| Indy500 | speed | 113.98 | 115.24 | 117.77 | — | 116.82 |
| | damage | 0 | 0 | 0 | out | 0 |
| | rank | 4 | 3 | 1 | — | 2 |
| Kevin | speed | 95.10 | 91.72 | 95.47 | — | 95.26 |
| | damage | 0 | 0 | 0 | out | 0 |
| | rank | 3 | 4 | 1 | — | 2 |
| Oval2 | speed | 63.43 | 67.36 | 67.90 | 67.96 | 67.90 |
| | damage | 0 | 0 | 0 | 0 | 0 |
| | rank | 5 | 4 | 2 | 1 | 2 |
| Ramdu | speed | 63.73 | 69.75 | 67.85 | — | 64.23 |
| | damage | 0 | 0 | 0 | out | 0 |
| | rank | 1 | 2 | 3 | — | 4 |
| Speed2 | speed | 91.09 | 86.29 | 88.66 | 97.40 | 87.09 |
| | damage | 0 | 0 | 0 | 11074 | 0 |
| | rank | 2 | 5 | 3 | 1 | 4 |
| Rars | speed | 72.27 | 69.78 | 67.85 | — | 64.23 |
| | damage | 0 | 0 | 0 | out | 0 |
| | rank | 1 | 2 | 3 | — | 4 |
| Stef2 | speed | 48.17 | 46.91 | 45.52 | 49.26 | 42.71 |
| | damage | 0 | 0 | 0 | 0 | 0 |
| | rank | 2 | 3 | 4 | 1 | 5 |
| Tight2 | speed | 60.11 | 59.28 | 57.16 | 58.20 | 57.69 |
| | damage | 0 | 0 | 0 | 0 | 0 |
| | rank | 1 | 2 | 5 | 3 | 4 |
| Timin | speed | 76.95 | 76.05 | 75.55 | 75.83 | 76.22 |
| | damage | 0 | 0 | 0 | 23265 | 0 |
| | rank | 1 | 3 | 5 | 4 | 2 |
| Trackfil | speed | 61.07 | 62.38 | 60.39 | 65.11 | 59.21 |
| | damage | 23909 | 0 | 0 | 0 | 0 |
| | rank | 3 | 2 | 4 | 1 | 5 |
| Trouble | speed | 68.54 | 67.69 | — | 65.86 | 64.29 |
| | damage | 0 | 0 | out | 18965 | 0 |
| | rank | 1 | 2 | — | 3 | 4 |
| Weird | speed | 75.99 | 76.99 | — | — | — |
| | damage | 0 | 0 | out | out | out |
| | rank | 2 | 1 | — | — | — |
| Zandvort | speed | 79.86 | 76.01 | 76.71 | — | 74.94 |
| | damage | 0 | 0 | 0 | out | 0 |
| | rank | 1 | 3 | 2 | — | 4 |

† The car is out of the race.

## 5. CONCLUSION

The use of GA in parameter optimization provides a better design in the robot driver. The result of racing is encouraged and such approach provides a potential solution for the auto-driving car. The Genetic Driver is the champion or the first runner-up in most of the races. Moreover, it (and also the Fuzzy Driver) can tackle with different kinds of track segments and finish all the races.

## REFERENCES


Holland, J.H. (1975). *Adaptation in natural and artificial systems*, MIT Press.

Man, K.F., K.S. Tang and S. Kwong, (1996). Genetic algorithms: concepts and applications, *IEEE Trans. Industrial Electronics*, **43(5)**, pp. 519–534.

Man, K.F., K.S. Tang, S. Kwong and W.A. Halang, (1997). *Genetic algorithms for control and signal processing*, Springer-Verlag London, ISBN 3-540-76101-2.

Michalewicz, Z. (1996). *Genetic Algorithms + Data Structures = Evolution Program*, 3rd Ed., Springer-Verlag.

RARS ftp–site (1995). available via ftp://ftp.ijs.com/rars

Srinivas, M. and L. M. Patnaik, (1994). Genetic algorithms: A survey, *Computer*, pp. 17–26, June.

14th Triennial World Congress, Beijing, P.R. China



# GLOBAL OPTIMALITY FOR GENERAL MULTIPLE LINEAR QUADRATIC CONTROL IN DISCRETE TIME SYSTEMS

**Fu-cai QIAN**

*School of Automation and Information Engineering, Xi'an University of Technology, Xi'an.710048, China*

**Bai-wu WAN**

*Systems Engineering Institute, Xi'an Jiaotong University, Xi'an 710049, China*
*Email: wanbw@xjtu.edu.cn*

**Duan LI**

*Department of Systems Engineering and Engineering Management, Chinese University of Hong Kong, Hong Kong*
*Email: dli@se.cuhk.edu.hk*

Abstract: This paper considers a class of the optimal control problem that are often difficult to solve directly due to large dimension, nonlinearity, nonseparability and nonconvexity of the problem. An iterative double loop algorithm is proposed. The inner loop solves the separable parametric optimization problems, the outer loop overcomes the nonseparability of the objective function. This algorithms can search out the global optimal solution. *Copyright © 1999 IFAC*

Keywords: Multiobjective optimization, optimal control, global optimization, Linear quadratic control, Discrete systems.

## 1. INTRODUCTION

Due to its wide application and its mathematical elegance in tractability, the linear quadratic control formulation and its solution have played a significant role in development of the modern control theory. The use of multiple objective is well justified by many realworld control problems (Li, et al., 1988; Makila, 1989; Li, 1990a; Li, 1990b), Various multiobjective linear quadratic control schemes have thus been motivated and investigated. There are two major research directions: i) characterization of the set of noninferior solutions for multiobjective linear quadratic control, and ii) search for a specific noninferior solution for multiple linear quadratic control, for example, the minimax solution, the ideal point, and the hierarchical ordering. The selection of the final implementing control is left unsolved when the first research direction is pursued, while the solution methodologies develop in the second research direction are confined to specific selection criteria.

A general class of multiple linear quadratic control problems in discrete time systems is studied in this paper, where the overall objective function is a nonlinear function of multiple quadratic performance indices. In (Li, and Burger, 1994), a multilevel solution procedure was developed, but the converging point at which this solution scheme

terminated could be a local optimal solution. The main research result in this paper is to attain the global optimality of search algorithm.
The organization of this paper is as follows. In section 2, the mathematical model of general multiple linear quadratic control in discrete time systems is given. The convexification on the overall objective function is performed in section 3. A multilevel algorithm which can search out the global optimal solution is developed in section 4. An illustrative example is presented in section 5.

## 2. MATHEMATICAL MODEL

The dynamics of a discrete time control system are assumed to be governed by a system of first-order linear difference equations with a known initial state $x(0)$

$$x(t+1) = A(t)x(t) + B(t)u(t) \quad t=0,1,\cdots,T-1 \quad (1)$$

where $t$ is the time index, $T$ is the terminal time, $x(t) \in R^n$ is the state vector, $u(t) \in R^p$ is the control vector, and $A(t)$ and $B(t)$ are matrices of appropriate dimensions.

Multiple performance indices $J_i$ ($i$=1,2, ⋯ ,$k$) are used to measure the system performance and each performance index to be minimized is assumed to be of a quadratic form

$$J_i = \frac{1}{2}x'(T)S_i x(T) + \frac{1}{2}\sum_{t=0}^{T-1}[x'(t)Q_i(t)x(t) + u'(t)R_i(t)u(t)] \quad (2)$$

where $S_i$ and $Q_i(t)$ ($i$=1,2,⋯ ,$k$) are positive semidefinite symmetric $n \times n$ matrices, and $R_i(t)$ ($i$=1,2,⋯,$k$) are positive definite symmetric $p \times p$ matrices.

The overall objective function, $J$, is a second-order differentiable nonlinear function of the $k$ quadratic performance indices

$$J = \varphi(J_1, J_2, \cdots, J_k) \quad (3)$$

It is assumed that $J \in R^+ / \{0\}$ and is a strictly increasing function of $J_i$, i,e.,

$$\frac{\partial J}{\partial J_i} > 0 \quad (i=1,2,\cdots,k) \quad (4)$$

The overall objective function $J$ reflects the decision maker's disutility function over the system. The interpretation of (4) is that improving each individual performance index leads to an improvement of the overall performance measure.

The general multiple linear quadratic control problem in discrete time systems is formulated now as

$$\min J = \varphi(J_1, J_2, \cdots, J_k) \quad (5a)$$

$$\text{s.t. } x(t+1) = A(t)x(t) + B(t)u(t)$$

$$t=0,1,\cdots,T-1 \quad (5b)$$

Problem(5) is difficult to be solved directly. An efficient solution scheme will be proposed in this paper to search a global optimal solution to (5).

## 3. CONVEXIFICATION OF THE OVERALL OBJECTIVE FUNCTION

In order to achieve the global optimality, first convexification on the overall objective function will be performed.

Raising the power to p with $p \geq 1$ on both sides of (3) yields $J^p = \varphi^p(J_1, J_2, \cdots J_k)$ (6)

Define $\varphi^p[(J_1^{-p})^{-\frac{1}{p}}, (J_2^{-p})^{-\frac{1}{p}}, \cdots, (J_k^{-p})^{-\frac{1}{p}}] = \hat{\varphi}(J_1^{-p}, J_2^{-p}, \cdots J_k^{-p})$

and $\hat{J} = J^p$

Equation (6) can be rewritten in the following compact form $\hat{J} = \hat{\varphi}(J_1^{-p}, J_2^{-p}, \cdots, J_k^{-p})$

It can be easily checked that the partial derivative of $\hat{J}$ with respect to $J_i$ is negative

$$\frac{\partial \hat{J}}{\partial J_i^{-p}} = \frac{\partial J^p}{\partial J_i^{-p}} = \frac{\partial J^p}{\partial J}\frac{\partial J}{\partial J_i}\frac{\partial J_i}{\partial J_i^{-p}} = -J^{p-1}J_i^{p+1}\frac{\partial J}{\partial J_i} < 0$$

**Theorem 1** (Li, and Burger, 1994) There exists a finite positive number $q_1$ such that function $\hat{J}$ becomes convex with respect to $J_1^{-p}, J_2^{-p}, \cdots, J_k^{-p}$ when $p > q_1$.

By performing the p-th power transformation, the following equivalent problem to problem(5) can be formed

$$\min \hat{J} = \hat{\varphi}(J_1^{-p}, J_2^{-p}, \cdots, J_k^{-p}) \quad (8a)$$

$$\text{s.t. } x(t+1) = A(t)x(t) + B(t)u(t)$$

$$t=0,1,\cdots,T-1 \quad (8b)$$

The set of optimal solution to problem (8) is the same as the set of optimal solution to problem(5). A prominent feature which is essential for the global optimality is that the overall objective function in (8) $\hat{J}$, is a convex function of $J_1^{-p}, J_2^{-p}, \cdots J_k^{-p}$.

## 4. MULTILEVEL SOLUTION SCHEME

Construct the parameter optimization problem as follows.

$$\min r(y) \quad (9a)$$

$$\text{s.t. } w_i J_i(u,x) \le y \quad i=1,2,\cdots,k \tag{9b}$$

$$x(t+1) = A(t)x(t) + B(t)u(t)$$

$$t=0,1,\cdots,T-1 \tag{9c}$$

where $y \in R$ is auxiliary variable, $w = \{1, w_2, w_3, \cdots, w_k\}$ is parameter vector, $r(y)$ is any strictly increasing and second-order differentiable function of $y$.

**Theorem 2** The global optimal solution of problem (8) is in the set of solutions of parameter problems (9).

Proof. Let $(u^*, x^*)$ be the optimal solution of problem (8), take $w_i^* = \dfrac{J_1(u^*,x^*)}{J_i(u^*,x^*)}$ $i=2,3,\cdots,k$

$$y^* = J_1(u^*, x^*)$$

obviously, $(u^*, x^*, y^*)$ satisfies the constrain conditions of problem (9).

Suppose that $(u^*, x^*, y^*)$ is not the optimal solution of problem (9), then there exists $(\hat{u}, \hat{x}, \hat{y})$, such that $r(\hat{y}) < r(y^*)$, since $r(y)$ is strictly increasing, $\hat{y} < y^*$.

Thus $\quad w_i^* J_i(\hat{u}, \hat{x}) \le \hat{y} < y^* = J_1(u^*, x^*)$

Namely, $J_i(\hat{u}, \hat{x}) < J_i(u^*, x^*)$

By $\hat{\varphi}$ is strictly decreasing of $J_i^{-p}$, so

$$\hat{\varphi}(u^*, x^*) > \hat{\varphi}(\hat{u}, \hat{x})$$

This is a contradiction to the assumption that $(u^*, x^*)$ is the optimal solution of problem (8). Thus, $(u^*, x^*, y^*)$ is the optimal solution of auxiliary problem (9)

**Lemma** Assume that the optimal solution $(u^*, x^*, y^*)$ of problem (9) is a regular point. In addition, $(u^*, x^*, y^*)$, together with the associated Kuhn-Tucker multiplies $\lambda_i$ ($i=1,2,\cdots,k$), satisfies the second-order sufficient conditions of problem (9). Assume further that all active inequalities in problem (9) are nondegenerate. Then

$$\frac{\partial r(y^*)}{\partial w_i} = \lambda_i^* J_i(u^*, x^*) \quad i=2,3,\cdots,k$$

Proof. Problem (9) can be rewritten as

$$\min r(y) \tag{10a}$$

$$\text{s.t. } -\frac{y}{J_i} \le -w_i \quad i=1,2,\cdots,k \tag{10b}$$

$$x(t+1) = A(t)x(t) + B(t)u(t) \tag{10c}$$

The Lagrange function of problem (10) is

$$L_1(u,x,y,\sigma,\mu) = r(y) + \sum_{i=1}^{k} \sigma_i[-\frac{y}{J_i} + w_i]$$
$$+\sum_{t=0}^{T-1} \mu_t[A(t)x(t) + B(t)u(t) - x(t+1)] \tag{11}$$

The Lagrange function of Problem (9) is

$$L_2(u,x,y,\lambda,\gamma) = r(y) + \sum_{i=1}^{k} \lambda_i[w_i J_i - y]]$$
$$+\sum_{t=0}^{T-1} \gamma_t[A(t)x(t) + B(t)u(t) - x(t+1)] \tag{12}$$

The Lagrange function (11) at the optimal point satisfies the following first-order necessary conditions

$$\frac{\partial L_1}{\partial y} = \frac{\partial r}{\partial y} - \sum_{i=1}^{k} \frac{\sigma_i}{J_i} = 0 \tag{13a}$$

$$\frac{\partial L_1}{\partial \mu} = A(t)x(t) + B(t)u(t) - x(t+1) = 0 \tag{13b}$$

$$\frac{\partial L_1}{\partial x(t)} = 0 \tag{13c}$$

$$\frac{\partial L_1}{\partial u(t)} = 0 \tag{13d}$$

$$\frac{\partial L_1}{\partial \sigma} \le 0 \tag{13e}$$

$$\sigma_i[-\frac{y}{J_i} + w_i] = 0 \quad \sigma_i > 0 \tag{13f}$$

The Lagrange function (12) at the optimal point satisfies the following first-order necessary conditions

$$\frac{\partial L_2}{\partial y} = \frac{\partial r}{\partial y} - \sum_{i=1}^{k} \lambda_i = 0 \tag{14a}$$

$$\frac{\partial L_2}{\partial \gamma} = A(t)x(t) + B(t)u(t) - x(t+1) = 0 \tag{14b}$$

$$\frac{\partial L_2}{\partial x(t)} = 0 \tag{14c}$$

$$\frac{\partial L_2}{\partial u(t)} = 0 \tag{14d}$$

$$\frac{\partial L_2}{\partial \lambda} \le 0 \tag{14e}$$

$$\lambda_i[w_i J_i - y] = 0 \quad \lambda_i > 0 \tag{14f}$$

Comparing the two sets of necessary conditions in Eqs. (13) and (14), it is evident that $(u^*, x^*, y^*, \lambda^*, \gamma^*)$ solves Eq. (13) if and only if $(u^*, x^*, y^*, \sigma^*, \mu^*)$ solves Eq. (14), where ($\sigma_i^* = \lambda_i^* J_i^*$, $\mu^* = \gamma^*$). Note here that if the ith

constraint in Eq. (9b) is not binding, both $\lambda_i^*$ and $\sigma_i^*$ are equal to zero, and if the ith constraint in Eq. (8b) is binding, $y^*$ is equal to $w_i J_i^*$ and both $\lambda_i$ and $\sigma_i$ are nonzero due to the nondegenerate assumption.
From the sensitivity theorem,

$$\frac{\partial r(y^*)}{\partial(-w_i)} = -\sigma_i^*, \text{ namely,}$$

$$\frac{\partial r(y^*)}{\partial w_i} = \sigma_i^* \quad (i=2,3,\cdots,k)$$

Thus, further $$\frac{\partial \varphi(y^*)}{\partial w_i} = \lambda_i J_i^*$$

Optimal solution of problem (9) depends on the assigned value of $w$. In abstract, the optimal solution of problem (9) can be parameterized by the weighting vector $w$.

$$u^* = u(w),\ x^* = x(w),\ y^* = y(w) \quad (15a)$$

$$\lambda^* = \lambda(w),\ J_i^* = J_i(w) \quad (i=1,2,\cdots,k) \quad (15b)$$

substituting above those functions into overall objective function $\hat{J}$, then $\hat{J}^* = \hat{J}^*(w)$, since $\hat{J}$ is convex function of $J_1^{-p}, J_2^{-p}, \cdots, J_k^{-p}$, the point at which satisfies

$$\frac{\partial \hat{J}(w)}{\partial J_i^{-p}} = 0 \quad i=1,2,\cdots,k \quad (16)$$

is the global optimal solution.

**Theorem 3** If, for a given parameter $w$, the optimal Kuhn-Tucker multiplier associated with the ith constraint in Eq. (9b) is strictly positive at the optimal solution, $\{x^*, y^*, \lambda^*\}$, of problem (8), then the following equality

$$\frac{\partial J_i^*}{\partial w_j} = \frac{1}{w_i}\{\frac{\lambda_j^* J_j^*}{\frac{dr(y^*)}{dy}} - \delta(i-j)J_i^*\} \quad i=1,2,\cdots,k$$

$j=2,3,\cdots,k$

where $\delta(\cdot)$ is the Kronecker function, $\delta(0)=1$ and $\delta(x)=0$ for all $x \neq 0$.

Proof. If the optimal Kuhn-Tucker multiplier associated with the ith constraint in Eq. (9b) is strictly positive at $\{u^*, x^*, y^*, \lambda^*\}$, then

$$w_i J_i^* - y^*(w) = 0 \quad (17)$$

Taking derivative with respect to $w_j$ on both sides of Eq. (17) yields

$$\delta(i-j)J_i^* + w_i \frac{\partial J_i^*}{\partial w_j} - \frac{\partial y^*}{\partial w_j} = 0$$

from

$$\frac{\partial y^*}{\partial w_j} = \frac{\partial y^*}{\partial r(y^*)}\frac{\partial r(y^*)}{\partial w_j} = \frac{\partial r}{\partial w_j}\Big/\frac{dr}{dy} = \frac{\lambda_j^* J_j^*}{\frac{dr(y^*)}{dy}}$$

So $$\frac{\partial J_i^*}{\partial w_j} = \frac{1}{w_i}\{\frac{\lambda_j^* J_j^*}{\frac{dr(y^*)}{dy}} - \delta(i-j)J_i^*\}$$

$i=1,2,\cdots,k \quad j=2,3,\cdots,k$

if all the optimal Kuhn-Tucker multipliers associated with the constraints in Eq. (9b) are strictly positive at $\{u^*, x^*, y^*, \lambda^*\}$, then

$$\frac{\partial \hat{J}^*}{\partial w_j} = \sum_{i=1}^{k} \frac{\partial \hat{J}}{\partial J_i^{-p}} \frac{\partial J_i^{-p}}{\partial J_i} \frac{\partial J_i}{\partial w_j} = -p\sum_{i=1}^{k} J_i^{-p-1} \frac{\partial \hat{J}}{\partial J_i^{-p}} \frac{\partial J_i}{\partial w_j}$$

The dual function of problem (9) is

$$H(\lambda) = \min_{u,x,y,\gamma} L(u,x,y,\gamma;\lambda)$$

since $r(y)$ and $J_i$ are convex function of $y, u, x$, problem (9) is equivalent to

$$\max_{\lambda>0} \min_{u,x,y} L(u,x,y;\lambda,\gamma) \quad (18)$$

problem (18) can be solved by two-level algorithm. For given $\lambda, w$, the first level solves

$$\min_{u,x,y} L(u,x,y;\lambda,\gamma)$$

namely, the first level solves the following two subproblems.

subproblem 1: $\min r(y) - \sum_{i=1}^{k} \lambda_i y$

s.t. $y > 0$

subproblem 2: $\min \sum_{i=1}^{k} w_i \lambda_i J_i$

s.t. $x(t+1) = A(t)x(t) + B(t)u(t)$

subproblem 1 is a simple mathematical programming problem; subproblem 2 is an optimal control problem, it can be solved by dynamic programming.

The second level solves the following dual problem

$$\max_{\lambda} H(\lambda)$$

The updating formulation of $\lambda$ is

$$\lambda^{s+1} = \lambda^s + \alpha(w_i J_i - y)$$

where $s$ is the iteration number, $\alpha$ is step size.
$\hat{J}$ is a convex function of $J_1^{-p}, J_2^{-p}, \cdots, J_k^{-p}$, thus, the point $w$ at which such that

$$\frac{\partial \hat{J}}{\partial J_i^{-p}}(w) = 0 \quad (i=1,2,\cdots,k) \quad (19)$$

correspond the global optimal solution.

Let $w^s = (1, w_2^s, \cdots, w_k^s)$ be the value of parameter $w$ at iteration $s$, the left side of (19) is expanded at point $w^s$, then

$$\frac{\partial \hat{J}}{\partial J_i^{-p}}(w^s) + \sum_{r=2}^{k}\sum_{j=1}^{k}\left( \frac{\partial^2 \hat{J}}{\partial J_i^{-p} \partial J_j^{-p}} \cdot \frac{\partial J_j^{-p}}{\partial J_j} \cdot \frac{\partial J_j}{\partial w_r}(w_r - w_r^s) \right) = 0 \quad i=1,2,\cdots,k \tag{20}$$

Let $\nabla \hat{J} = \left[ \frac{\partial \hat{J}}{\partial J_1^{-p}}, \frac{\partial \hat{J}}{\partial J_2^{-p}}, \cdots, \frac{\partial \hat{J}}{\partial J_k^{-p}} \right]^T$

$$\hat{H} = \begin{bmatrix} \frac{\partial^2 \hat{J}}{\partial (J_1^{-p})^2} & \frac{\partial^2 \hat{J}}{\partial J_1^{-p} \partial J_2^{-p}} & \cdots & \frac{\partial^2 \hat{J}}{\partial J_1^{-p} \partial J_k^{-p}} \\ \frac{\partial^2 \hat{J}}{\partial J_2^{-p} \partial J_1^{-p}} & \frac{\partial^2 \hat{J}}{\partial (J_2^{-p})^2} & \cdots & \frac{\partial^2 \hat{J}}{\partial J_2^{-p} \partial J_k^{-p}} \\ \vdots & \vdots & \vdots & \vdots \\ \frac{\partial^2 \hat{J}}{\partial J_k^{-p} \partial J_1^{-p}} & \frac{\partial^2 \hat{J}}{\partial J_k^{-p} \partial J_2^{-p}} & \cdots & \frac{\partial^2 \hat{J}}{\partial (J_k^{-p})^2} \end{bmatrix}$$

by (20),

$$\nabla \hat{J} - p\hat{H} \cdot \sum_{r=2}^{k} \begin{bmatrix} \frac{1}{J_1^{p+1}} \frac{\partial J_1}{\partial w_r} \\ \vdots \\ \frac{1}{J_k^{p+1}} \frac{\partial J_k}{\partial w_r} \end{bmatrix} (w_r - w_r^s) = 0 \tag{21}$$

let $\Lambda = diag(\frac{1}{J_1^{p+1}}, \frac{1}{J_2^{p+1}}, \cdots, \frac{1}{J_k^{p+1}})$

$$d_r = \left[ \frac{\partial J_1}{\partial w_r}, \frac{\partial J_2}{\partial w_r}, \cdots, \frac{\partial J_k}{\partial w_r} \right]^T$$

$$D = [d_2, d_3, \cdots, d_k]$$

then

$$\nabla \hat{J} - p\hat{H} \cdot \Lambda \cdot D \cdot (w - w^s) = 0 \tag{22}$$

From $\hat{J}$ is convex function of $J_1^{-p}, J_2^{-p}, \cdots, J_k^{-p}$, so its Hessian $\hat{H}$ is positive definite. Thus.

$$D \cdot (w - w^s) = \frac{1}{p} \cdot \Lambda^{-1} \cdot \hat{H}^{-1} \cdot \nabla \hat{J} \tag{23}$$

Formulation (23) is a set of linear equations, thus, the gradient-type algorithm can be applied to solve (23). In this way

$$w^{s+1} = w^s + (D^T D)^{-1} \cdot D^T \cdot \Lambda^{-1} \cdot \hat{H}^{-1} \cdot \nabla \hat{J}(w^s)$$

let $E = (D^T D)^{-1} \cdot D^T \cdot \Lambda^{-1} \cdot \hat{H}^{-1}$

then $w^{s+1} = w^s + \beta \cdot E \cdot \nabla \hat{J}(w^s)$

$$r=2,3,4,\cdots,k \tag{24}$$

Since $w = (1, w_2, \cdots, w_k)$ has $k$-1 independent variables, the surface $\{ J_1(w), J_2(w), \cdots, J_k(w) \}$ is $k$-1 dimension in performance index space, it shows that the rank of matrix $D$ is $k$-1, so $D^T D$ is invertible.

The overall multilevel method is now summarized as follow.

step1 Select the initial parameter vector $w_0$, choose a small number $\varepsilon$ for the error tolerance in stopping condition, and set iteration number $s$=0.

step2 For a given $w^s$, problem (9) is solved by two-level algorithm, its solution is $(u^s, x^s, y^s, \lambda^s)$.

Remark Assume at iteration $s$, some inequalities in (9b) are not binding when a solution $(u^s, x^s, y^s, \lambda^s)$ is generated at step 2. The value of $w^s$ needs to be reassessed and problem (9) needs to be solved again in order to calculate $\frac{\partial J_i^*}{\partial w_j}$. In the situation where $J_1(u^s, x^s) = y^s$, solution $(u^s, x^s, y^s)$ remains feasible and optimal in (9) with all constraints in (9b) binding when the inactive constraints in (9b) are modified to

$$\frac{y^s}{J_i(u^s, x^s)} J_i(u, x) \le y$$

if $\quad w_i J_i(u^s, x^s) < y^s$

In the situations where $J_1(u^s, x^s) < y^s$, solution $(u^s, x^s, y^s)$ remains feasible and optimal in (9) with all constraints in (9b) binding when the inactive constraints in (9b) are modified to

$$\frac{y^s}{J_1(u^s, x^s)} J_1(u, x) \le y \tag{25}$$

$$\frac{y^s}{J_i(u^s, x^s)} J_i(u, x) \le y,$$

$$w_i J_i(u^s, x^s) < y^s$$

the following reassessment scheme can be developed for both situations where either $J_1(u^s, x^s) = y^s$ or $J_1(u^s, x^s) < y^s$

$$w_i^{s+1} = \frac{J_1(u^s, x^s)}{J_i(u^s, x^s)} \tag{26}$$

step3 Check if a norm of $\left\| \nabla \hat{J}(w) \right\| < \varepsilon$, if yes, the search terminates. Otherwise, go to step4.

step4 Updating $w$ using Eq.( 24) or (26), set $s$=$s$+1 and go back to step2.

## 5. EXAMPLE

$$\min J = J_1^2 \cdot J_2 + J_2^{0.5}$$

$$s.t.\ x(t+1) = \begin{bmatrix} 1/(t+1)^2 & 1 \\ -1/6 & -t/(t+1) \end{bmatrix} x(t)$$

$$+ \begin{bmatrix} (t+2)/(t+1) \\ (t+4)/(t+2) \end{bmatrix} u(t) \qquad t=0,1,2,3,4$$

$$x(0) = [1,2]'$$

$$J_i = \frac{1}{2}\left\{ \sum_{t=0}^{5} x'(t) \begin{bmatrix} 0.5 + \sqrt[(t+i)]{t+i} & 0.1 \cdot i \\ 0.1 \cdot i & 0.1 + 0.1 \cdot (t+1)^i \end{bmatrix} x(t) + 0.1 \cdot \sum_{t=0}^{4} (t+1)^i u^2(t) \right\}$$

$J$ is nonconvex with respect to $J_1$, $J_2$ and this problem has two optimal solutions. The value of the objective function $J$ associated with the local solution is 199.34, the value of the objective function $J$ associated with the global solution is 164.96. The initial value of $w_2$ is set to 10 and the step size $\beta$ is 0.02, $\varepsilon$=0.01, the local optimal solution by the Li's method (Li, and Burger, 1994) is obtained after algorithm iterating 23 times. Using the approach in this paper, where take $p$=2, the global optimal solution is obtained after iterating 63 times.

By Li's method, the values of control and state in each time are as follows:
$u(0)$=-0.57322; $u(1)$=0.03542; $u(2)$=0.0016;
$u(3)$=-0.00159; $u(4)$=-0.00221;

$x_1(1)$=1.85356; $x_1(2)$=-0.7792; $x_1(3)$=0.30126;
$x_1(4)$=-0.10266; $x_1(5)$=0.02879;

$x_2(1)$=-1.31311; $x_2(2)$=0.39739; $x_2(3)$=-0.12526;
$x_2(4)$=0.03801; $x_2(5)$=-0.01478.

By the method in this paper, the values of control and state in each time are as follows:
$u(0)$=-0.72883; $u(1)$=0.17120; $u(2)$=-0.06707;
$u(3)$=0.04314; $u(4)$=-0.05374;

$x_1(1)$=1.54234; $x_1(2)$=-0.98193; $x_1(3)$=0.6419;
$x_1(4)$=-0.40317; $x_1(5)$=0.2457

$x_2(1)$=-1.62433; $x_2(2)$=0.84042; $x_2(3)$=-0.49722;
$x_2(4)$=0.32632; $x_2(5)$=-0.26552

## 6. CONCLUSION

A threelevel iterative method is proposed in this paper for a class of nonseparable dynamical optimization problem. The proposed approach tackles difficult and complex optimization problems through embedding, separation, decomposition and coordination. The resulting auxiliary parametric optimization problems at the inner loop are of a minimax type and often have a sample structure that is mathematically tractable by existing efficient solution algorithms. The outer loop can overcome the nonseparablility of the objective function.

## REFERENCES


Li D. (1990a). Multiple objectives and nonseperability in stochastic dynamic programming, *Int. J. Systems Science*, Vol 21, pp589-607.

Li D. (1990b). On the minimax solution of multiple lienar-quaratic problems, *IEEE Trans. on Automatic Control*, Vol 35, pp1153-1156.

Li D. (1995). Zero duality gap for a class of non-convex optimization problems, *Journal of optimization Theory and Applications*, Vol. 85.

Li D. and J. A. Burger (1994). Multilevel dynamic programming for general multiple linear-quadratic control in discrete-time systems, *Computers Math. Applic.*, Vol. 27, No. 9/10, pp59-66.

Li D. and Y. Y. Haimes (1988). Decomposition technique in multiobjective discrete-time dynamic problems, In: *Control and Dynamic Systems*, Vol. 28, Academic Press.

Makila P. M. (1989). on multiple criteria stationary linear quadratic control, *IEEE Trans. on Automatic Control*, Vol 34, pp1311-1313.

# LQG optimal control of constrained discrete time systems

**Andrew E.B. Lim**
Department of Systems Engineering and Engineering Management
Chinese University of Hong Kong
Shatin, New Territories
Hong Kong
Email: andrew@ee.mu.oz.au

**Abstract**

In this paper, we study the discrete time LQG optimal control problem, with convex quadratic constraints. We show that the Separation Theorem does not hold. However, a generalization of this result which we call the Quasi-Separation Theorem applies. The optimal control is obtained by solving a convex, finite-dimensional optimization problem that is equivalent to solving a sequence of unconstrained LQG control problems.


**Index Terms**– LQG optimal control; quadratic constraints; Separation Theorems; discrete time systems; optimization theory; duality theory.

## 1 Introduction

In this paper, we study the discrete time LQG optimal control problem, with convex quadratic constraints. In many applications, performance specifications must be considered explicitly in order for a useful solution to be found. Furthermore, many performance criteria are naturally expressed in the form of quadratic cost functionals, and quadratic constraints [4, 11]. Hence, the LQG control with quadratic constraints is a useful framework in which to study many problems where there are many distinct performance criteria that must be satisfied concurrently.

In this paper, we focus on two issues: How can we calculate the optimal control, and is there a Separation Theorem. We show that the Separation Theorem does not hold in the case of quadratic constraints: The optimal control depends on the solution of the filtering Riccati equation. However, we show that this dependence is superficial, and hence the label of quasi-separation. The optimal control can be determined by solving a convex finite-dimensional (dual) optimization problem. Interestingly, solving this finite-dimensional problem is equivalent to solving a sequence of unconstrained LQG control problem.

## 2 Deterministic case

Assume $N < \infty$. Let $l_2^N(\mathbf{R}^n)$ denote the Hilbert space of sequences $(x(0), \cdots, x(N))$ with $x(k) \in \mathbf{R}^n$ such that $\|x(k)\| < \infty$ for every $k$, with inner product

$$\langle x,\ y\rangle = \sum_{k=0}^{N} x'(k) \cdot y(k)$$

Let $\xi \in \mathbf{R}^n$, $A(k) \in \mathbf{R}^{n\times n}$ and $B(k) \in \mathbf{R}^{n\times m}$, $k = 0, \cdots, N-1$ be given. For every $u(\cdot) \in l_2^{N-1}(\mathbf{R}^m)$, let $x(\cdot) \in l_2^N(\mathbf{R}^n)$ be the solution of the linear system

$$x(k+1) = A(k)\,x(k) + B(k)\,u(k), \quad x(0) = \xi \quad (1)$$

We refer to such a pair $(x, u) \in l_2^N(\mathbf{R}^n) \times l_2^{N-1}(\mathbf{R}^m)$ as an admissible pair. Let $z = (\bar{z}, 0) \in l_2^N(\mathbf{R}^n) \times l_2^{N-1}(\mathbf{R}^m)$ where $\bar{z}$ is the solution of the difference equation (1) with $u(k) = 0$. Define the set

$$\mathcal{Y} = \{(x,u) \in l_2^N(\mathbf{R}^n) \times l_2^{N-1}(\mathbf{R}^m) \mid x(k+1) = A(k)\,x(k) + B(k)\,u(k), x(0) = 0\} \quad (2)$$

It follows then that $z + \mathcal{Y}$ is the set of all admissible pairs for the linear system (1). Moreover, $\mathcal{Y}$ is a closed subspace of $l_2^N(\mathbf{R}^n) \times l_2^{N-1}(\mathbf{R}^m)$ and therefore, $z + \mathcal{Y}$ is an affine subspace of $l_2^N(\mathbf{R}^n) \times l_2^{N-1}(\mathbf{R}^m)$.

Define the functionals $f_i : l_2^N(\mathbf{R}^n) \times l_2^{N-1}(\mathbf{R}^m) \to \mathbf{R}$, $i = 0, \cdots, r$ by

$f_i(x, u) =$

$$\frac{1}{2}\sum_{k=0}^{N-1}(x'(k)\,Q_i(k)\,x(k)+u'(k)\,R_i(k)\,u(k))$$
$$+\frac{1}{2}x'(N)\,Q_i(N)\,x(N) \tag{3}$$

where $Q_i(k) \in \mathbf{R}^{n\times n}$ are positive semi-definite and $R_i(k) \in \mathbf{R}^{m\times m}$ are strictly positive definite. Let $c_i \in \mathbf{R}$ $i = 1, \cdots r$ be given constants. The deterministic LQ optimal control problem subject to quadratic constraints can be stated as follows:

$$\begin{cases} f_0(x,u) \to \min \\ f_i(x,u) \le c_i, \quad i = 1,\cdots,r \\ (x,u) \in z + \mathcal{Y} \end{cases} \tag{4}$$

We make the following assumption:

**Assumption 2.1** *There exists $(x,u) \in l_2^N(\mathbf{R}^n) \times l_2^{N-1}(\mathbf{R}^m)$ which is feasible for (4).*

We have the following result on the existence of an optimal solution $(x^*, u^*)$ for (4).

**Theorem 2.1** *Suppose that Assumption 2.1 is true. Then there exists a unique optimal admissible pair $(x^*, u^*)$ for the problem (4).*

For every $\lambda = (\lambda_1, \cdots, \lambda_r) \ge 0$ and $\zeta = (x, u) \in z + \mathcal{Y}$, define the Lagrangian

$$L(\zeta,\lambda) = f_0(\zeta) + \sum_{i=1}^{r}\lambda_i\,(f_i(\zeta,\lambda) - c_i) \tag{5}$$

For convenience, we denote:

$$\begin{aligned} Q(k,\lambda) &= Q_0(k) + \sum_{i=1}^{r}\lambda_i Q_i(k) \\ R(k,\lambda) &= R_0(k) + \sum_{i=1}^{r}\lambda_i R_i(k) \end{aligned}$$

The following assumption is needed for us to use results from duality theory.

**Assumption 2.2** *For every $\lambda_i \ge 0$, $i = 1,\cdots,r$ (not all equal to zero), there exists $(x,u) \in z + \mathcal{Y}$ such that*

$$\sum_{i=1}^{r}\lambda_i(f_i(x,u) - c_i) \quad < \quad 0$$

**Theorem 2.2** *Let $L(\zeta(\lambda),\lambda)$ be given by (5). Then there exists a $\lambda^* \ge 0$ which is optimal for*

$$\max_{\lambda\ge 0}\left\{\frac{1}{2}\xi'\cdot S(0,\lambda)\cdot\xi - \lambda' c\right\} \tag{6}$$

$$\begin{aligned} S(k) &= Q(k,\lambda) + A'(k)S(k+1)A(k) \\ &\quad - A'(k)S(k+1)B(k) \\ &\quad \times (R(k,\lambda) + B'(k)S(k+1)B(k))^{-1} \\ &\quad \times B'(k)S(k+1)A(k) \\ S(N) &= Q_i(N,\lambda) \end{aligned} \tag{7}$$

*Furthermore, the optimal control $u^* \in l_2^{N-1}(\mathbf{R}^m)$ of (4) exists and is given by*

$$\begin{aligned} u^*(k) &= \\ &-[R(k,\lambda^*) + B'(k)\,S(k+1,\lambda^*)\,B(k)]^{-1} \\ &\times B'(k)\,S(k+1,\lambda^*)\,A(k)\,x(k) \end{aligned} \tag{8}$$

Note that the optimal control is obtained by solving the finite-dimensional optimization problem (6)-(7) for $\lambda^*$. Moreover, (6)-(7) is convex, since the 'primal' problem (4) is convex. Hence, (6)-(7) can be solved using numerical algorithms for finite-dimensional optimization problems if the gradient of the cost functional (6) with respect to $\lambda$ can be calculated. In Section 5, we show how these gradients can be calculated.

## 3 Full observation case

Let $(\Omega, \mathcal{F}, P)$ be a probability space and $x_0 \sim N(\xi, P_0)$ and $w(k) \sim N(0, \hat{Q}(k))$ for $k = 1, \cdots N$ be normally distributed $\mathbf{R}^n$-valued random variables defined on $(\Omega, \mathcal{F}, P)$. Let $L_2^N(\Omega, \mathcal{F}, P; \mathbf{R}^n)$ be the space of all $\mathbf{R}^n$-valued processes $x = (x(0), \cdots, x(N))$ on $(\Omega, \mathcal{F}, P)$ such that $E(x'(k)\cdot x(k)) < \infty$. Let $\mathcal{F}_k = \sigma\{x_0, w(1), \cdots, w(N)\}$ be the $\sigma$-algebra generated by $x_0, w(1), \cdots, w(k)$. The class of admissible controls $\mathcal{U}$ is the set of $\mathbf{R}^m$-valued processes $u = (u(0), \cdots, u(N-1)) \in L_2^{N-1}(\Omega, \mathcal{F}, P; \mathbf{R}^m)$ such that $u(k)$ is measurable with respect to $\mathcal{F}_k$. Let $\mathcal{X}$ denote the set of processes $x = (x(0), \cdots, x(N)) \in L_2^N(\Omega, \mathcal{F}, P; \mathbf{R}^n)$ such that $x(k)$ is measurable with respect to $\mathcal{F}_k$. For every $u \in \mathcal{U}$, let $x \in \mathcal{X}$ be the solution of the stochastic difference equation

$$\begin{aligned} x(k+1) &= A(k)\,x(k) + B(k)\,u(k) + w(k+1) \\ x(0) &= x_0 \end{aligned} \tag{9}$$

where $A(k)$ and $B(k)$ are as in (1). Let

$$\begin{aligned} \mathcal{Y} = \{(x,u) \in \mathcal{X}\times\mathcal{U}\,|\, \\ x(k+1) = A(k)\,x(k) + B(k)\,u(k), \\ x(0) = 0\} \end{aligned} \tag{10}$$

and $z = (\bar{z}, 0) \in \mathcal{X} \times \mathcal{U}$ where $\bar{z}$ is the solution of (9) with $u(k) = 0$. Then $z + \mathcal{Y}$ is a closed affine subspace of $\mathcal{X} \times \mathcal{U}$ and is the set of all solutions of (9). Define $f_i : \mathcal{X} \times \mathcal{U} \to \mathbf{R}$, $i = 0, \cdots, r$ by

$$f_i(x,u) = \frac{1}{2} \sum_{k=0}^{N-1} (x'(k)\, Q_i(k)\, x(k) + u'(k)\, R_i(k)\, u(k)) + \frac{1}{2} x'(N)\, Q_i(N)\, x(N) \quad (11)$$

where $Q_i(k)$ and $R_i(k)$ are as in (11). In this section, we consider the problem

$$\begin{cases} f_0(x,u) \to \min \\ f_i(x,u) \le c_i, \quad i = 1, \cdots, r \\ (x,u) \in z + \mathcal{Y} \end{cases} \quad (12)$$

Analogous to Assumption 2.1, we shall assume the following:

**Assumption 3.1** *There exists $(x,u) \in \mathcal{X} \times \mathcal{U}$ which is feasible for (12).*

The following result for the existance of an optimal solution $(x^*, u^*)$ for (12) follows from Assumption (3.1). It can be proved in exactly the same manner as Theorem 2.1.

**Theorem 3.1**
*Suppose that Assumption 3.1 holds. Then there exists a unique optimal solution $(x^*, u^*)$ of (12).*

Let $\zeta = (x, u)$. We define, as in the deterministic case, the Lagrangian $L(\zeta, \lambda)$ by (5) where now, $f_i(\zeta)$ are given by (11). To solve (12) we need the following assumption:

**Assumption 3.2** *For every $\lambda_i \ge 0$, $i = 1, \cdots, r$ (not all equal to zero), there exists $(x,u) \in z + \mathcal{Y}$ such that*

$$\sum_{i=1}^{r} \lambda_i (f_i(x,u) - c_i) \quad < \quad 0$$

Under Assumption 3.2, we have the following result:

**Theorem 3.2** *Then there exists a $\lambda^* \ge 0$ which is optimal for the problem*

$$\begin{cases} \max_{\lambda \ge 0} \{ \frac{1}{2} \xi' \cdot S(0,\lambda) \cdot \xi \\ \quad + \frac{1}{2} \sum_{k=0}^{N} tr\{\hat{Q}(k)\, S(k,\lambda)\} - \lambda' c \} \\ \text{subject to: } (7) \end{cases} \quad (13)$$

*Furthermore, the optimal control for (12) is*

$$u(k) = -[R(k,\lambda^*) + B'(k)\, S(k+1,\lambda^*)\, B(k)]^{-1} \times B'(k)\, S(k+1,\lambda^*)\, A(k)\, x(k) \quad (14)$$

Unlike the unconstrained [1, 12] and the linearly constrained [8] cases, certainty equivalence does not hold because (6)-(7) and (13) do not generally have the same optimal solution. The problem (13) (as in the deterministic case (6)-(7)) is a convex, finite-dimensional optimization problems over $\lambda \in \mathbf{R}^r$ and can be solved using gradient-type optimization algorithms.

## 4 Partially observed case

Let $x_0 \sim N(\xi, P_0)$, $w(k) \sim N(0, \hat{Q}(k))$ and $v(k) \sim N(0, \hat{R}(k))$ be independent Gaussian random variables defined on $(\Omega, \mathcal{F}, P)$. We shall assume that $\hat{R}(k) \in \mathbf{R}^{p \times p}$ are strictly positive definite. Consider the following partially observed discrete time system on $(\Omega, \mathcal{F}, P)$:

$$\begin{aligned} x(k+1) &= A(k)\, x(k) + B(k)\, u(k) + w(k+1), \\ x(0) &= x_0 \quad (15) \\ y(k) &= C(k)\, x(k) + v(k) \quad (16) \end{aligned}$$

The class of admissible controls $\mathcal{U}_{ad}$ is defined as follows: Let $\bar{\mathcal{F}}_k = \sigma\{y(0), \cdots, y(k)\}$ be the $\sigma$-algebra generated by $y(0), \cdots, y(k)$. Then $\mathcal{U}_{ad}$ is the set of processes $u = (u(0), \cdots, u(N-1))$ on $(\Omega, \mathcal{F}, P)$ such that $u(k)$ is measurable with respect to $\bar{\mathcal{F}}_k$; that is, $u(k)$ is a function of at most $y(0), \cdots, y(k)$.

Let
$f_i : L_2^N(\Omega, \mathcal{F}, P; \mathbf{R}^n) \times L_2^{N-1}(\Omega, \mathcal{F}, P; \mathbf{R}^m) \to \mathbf{R}$, $i = 0, \cdots, r$ be given by (11). The partially observed, LQG optimal control problem with integral quadratic constraints is defined as follows:

$$\begin{cases} \min f_0(x,u) \\ f_i(x,u) \le c_i, \quad i = 1, \cdots, r \\ (x,u) \text{ satisfies } (15), u \in \mathcal{U}_{ad} \end{cases} \quad (17)$$

As with the full observation case, we need the following assumptions:

**Assumption 4.1** *There exists $(x,u)$ such that $u \in \mathcal{U}_{ad}$, $x$ satisfies (15) and $f_i(x,u) \le c_i$ for $i = 1, \cdots, r$.*

**Assumption 4.2** *For every $\lambda_i \ge 0$, $i = 1, \cdots, r$ (not all zero), there exists $(x,u)$ satisfying $u \in \mathcal{U}_{ad}$ and (15) such that*

$$\sum_{i=1}^{r} \lambda_i (f_i(x,u) - c_i) \quad < \quad 0$$

Using standard techniques, we can transform the partial observation problem (17) into a full observation problem of the form (12). To summarize this, we need the following basic results from

Kalman filtering theory [2]. Denoting $\hat{x}(k) = E[x(k)|y(0), \cdots, y(k-1)] = E[x(k)|\bar{\mathcal{F}}_{k-1}]$, it is well known that the process $\hat{x}(0), \cdots, \hat{x}(N)$ is given by the Kalman filter:

$$\begin{aligned} \hat{x}(k+1) &= A(k)\,\hat{x}(k) + B(k)\,u(k) - \hat{K}(k)\nu(k), \\ x(0) &= \xi \end{aligned} \tag{18}$$

where

$$\begin{aligned} \hat{K}(k) &= A(k)\Sigma(k)C'(k) \\ &\quad \times(\hat{R}(k+1) + C(k)\Sigma(k)C'(k))^{-1} \\ \nu(k) &= y(k) - C(k)\,\hat{x}(k) \end{aligned}$$

and

$$\begin{aligned} \Sigma(k+1) &= \hat{Q}(k+1) + A(k)\,\Sigma(k)\,A'(k) \\ &\quad -A(k)\,\Sigma(k)C'(k) \\ &\quad \times[\hat{R}(k+1) + C(k)\Sigma(k)C'(k)]^{-1} \\ &\quad \times C(k)\Sigma(k)\,A'(k) \\ \Sigma(0) &= P_0 \end{aligned} \tag{19}$$

The result we call the Quasi-Separation theorem can be stated as follows:

**Theorem 4.1 (Quasi-Separation Theorem)** *There exists $\lambda^* \geq 0$ which is optimal for the problem:*

$$\left\{ \begin{array}{l} \max_{\lambda\geq 0} \{\frac{1}{2}\,\xi' S(0,\lambda)\xi \\ +\frac{1}{2}\sum_{k=0}^{N} tr\,\{\Gamma(k)\,S(k,\lambda) \\ +Q(k,\lambda)\,\Sigma(k)\} - \lambda' c\} \\ \quad \text{Subject to: (7)} \end{array} \right. \tag{20}$$

*where $\Gamma(k) = \hat{Q}(k) + A(k-1)\,\Sigma(k-1)\,A'(k-1) - \Sigma(k)$. Furthermore the optimal control for (17) exists and is given by*

$$\begin{aligned} u^*(k) &= \\ &-[R(k,\lambda^*) + B'(k)\,S(k+1,\lambda^*)\,B(k)]^{-1} \\ &\times B'(k)\,S(k+1,\lambda^*)\,A(k)\,\hat{x}(k) \end{aligned} \tag{21}$$

*where $\hat{x}(k)$ is the process given by the Kalman filter (18).*

The reader should note that certainty equivalence does not hold, and the Separation Theorem does not hold in the sense of [12]. To see this, the reader should observe (20). It is clear that the solution $\lambda^*$ of (20) dependends on $\Sigma(k)$ associated with the filtering problem (18)-(19). Therefore, the filtering and control problems are not independent. However, this dependence of $\lambda^*$ (and hence the solution of the control problem) on $\Sigma(k)$ adds no complication. Since $\Sigma(k)$ is independent of $\lambda$, it needs to be calculated only once, and the optimization problem (20) may be solved with no further re-calculation of $\Sigma(k)$ (and hence, with no further reference to the filtering problem). It is in this sense that the control and filtering problems are independent, and hence our naming of Theorem 4.1 a Quasi-Separation Theorem.

## 5 Optimal parameter selection problems and a separation theorem for gradient calculations

In view of Theorems 2.2, 3.2 and 4.1, the optimal control for constrained LQG problems is obtained by solving the finite dimensional optimization problems (6)-(7), (13) and (20). In the literature, these finite dimensional optimization problems are known as optimal parameter selection problems [10]. They can be solved as mathematical programming problems (using efficient gradient-type optimization algorithms) so long as the value of the cost functional and gradient of the cost functional can be calculated for any given $\lambda$.

We show that calculating the gradient of the cost functional (6)-(7) is equivalent to solving an unconstrained LQ control problem. In the full observation and partial observation problems, we show that a similar result holds. Furthermore, by virtue of the Separation Theorem for unconstrained LQG, it follows that there is a Separation Theorem associated with calculating the gradient.

The gradients for (6)-(7), (13) and (20) are derived in [5]. For the sake of space constraints, we simply state the results.

**Theorem 5.1** *Let $\lambda \in \mathbf{R}^r$, $\lambda \geq 0$ and*

$$J(\lambda) = \frac{1}{2}\xi' \cdot S(0,\lambda) \cdot \xi - \lambda' c$$

*The gradient of the cost functional $J(\lambda)$ evaluated at $\lambda$ is*

$$\frac{dJ(\lambda)}{d\lambda} = \left[\frac{\partial J(\lambda)}{\partial \lambda_1}, \ldots, \frac{\partial J(\lambda)}{\partial \lambda_r}\right]$$

$$\begin{aligned} \frac{\partial J(\lambda)}{\partial \lambda_i} &= \\ &\frac{1}{2}\sum_{k=0}^{N-1}[\beta'(k)\,Q_i(k)\,\beta(k) + \eta'(k)\,R_i(k)\,\eta(k)] \\ &+\frac{1}{2}\beta'(N)\,Q_i(N)\,\beta(N) - c_i \end{aligned} \tag{22}$$

*where $\beta(k)$ is the solution of the equation*

$$\begin{aligned} \beta(k+1) &= A(k)\,\beta(k) + B(k)\,\eta(k), \\ \beta(0) &= \xi \end{aligned} \tag{23}$$

*with $\eta(k)$ given by*

$$\eta(k) = -[R(k,\lambda) + B'(k)\,S(k+1,\lambda)\,B(k)]^{-1} \times B'(k)\,S(k+1,\lambda)\,A(k)\,\beta(k) \qquad (24)$$

As stated before, the problem of calculating the gradient of $J(\lambda)$ is equivalent to solving an unconstrained LQ optiimal control problem, and evaluating the value of the constraints with this optimal control.

Similarly, we have the following result for the dual problem (13) for the full observation problem (12).

**Theorem 5.2** *Let $\lambda \in \mathbf{R}^r$, $\lambda \geq 0$ be given. Then the cost functional of the problem (13) is of the form*

$$J(\lambda) = \frac{1}{2}\xi' \cdot S(0,\lambda) \cdot \xi + \frac{1}{2}\sum_{k=0}^{N} tr\{\hat{Q}(k)\,S(k,\lambda)\} - \lambda' c \qquad (25)$$

*The gradient of $J(\lambda)$ evaluated at $\lambda$ is*

$$\frac{dJ(\lambda)}{d\lambda} = \left[\frac{\partial J(\lambda)}{\partial \lambda_1}, \cdots, \frac{\partial J(\lambda)}{\partial \lambda_r}\right]$$

$$\frac{\partial J(\lambda)}{\partial \lambda_i} = E\left\{\frac{1}{2}\sum_{k=0}^{N-1}[\beta'(k)\,Q_i(k)\,\beta(k) + \eta'(k)\,R_i(k)\,\eta(k)] + \frac{1}{2}\beta'(N)\,Q_i(N)\,\beta(N)\right\} - c_i \qquad (26)$$

*where*

$$\beta(k+1) = A(k)\,\beta(k) + B(k)\,\eta(k) + w(k+1), \quad \beta(0) = x_0 \qquad (27)$$

*with $\eta(k)$ given by*

$$\eta(k) = -[R(k,\lambda) + B'(k)\,S(k+1,\lambda)\,B(k)]^{-1} \times B'(k)\,S(k+1,\lambda)\,A(k)\,\beta(k) \qquad (28)$$

As in the deterministic case, the problem of calculating the gradient is equivalent to solving an unconstrained full observation LQG control problem, and evaluating the constraints with this optimal control.

The partially observed problem (17) is solved by transforming it into the full observation problem (12). It can be shown [5] that the gradient of the cost functional of (20) is obtained by solving an unconstrained, partially observed LQG problem. For this reason, there is a Separation Theorem for calculating the gradient.

**Theorem 5.3 (Gradient Separation Theorem)** *Let $\lambda \in \mathbf{R}^r$, $\lambda \geq 0$ be given. Then the cost functional of the problem (20) is*

$$J(\lambda) = \frac{1}{2}\xi' S(0,\lambda)\xi + \frac{1}{2}\sum_{k=0}^{N} tr\,\{\Gamma(k)\,S(k,\lambda) + Q(k,\lambda)\,\Sigma(k)\} - \lambda' c \qquad (29)$$

*where $\Gamma(t) = \hat{Q}(k) + A(k-1)\,\Sigma(k-1)\,A'(k-1) - \Sigma(k)$. The gradient of $J(\lambda)$ evaluated at $\lambda$ is*

$$\frac{dJ(\lambda)}{d\lambda} = \left[\frac{\partial J(\lambda)}{\partial \lambda_1}, \cdots, \frac{\partial J(\lambda)}{\partial \lambda_r}\right]$$

$$\frac{\partial J(\lambda)}{\partial \lambda_i} = E\left\{\frac{1}{2}\sum_{k=0}^{N-1}\left[\hat{\beta}'(k)\,Q_i(k)\,\hat{\beta}(k) + \eta'(k)\,R_i(k)\,\eta(k)\right] + \frac{1}{2}\hat{\beta}'(N)\,Q_i(N)\,\hat{\beta}(N)\right\} - c_i \qquad (30)$$

*where*

$$\hat{\beta}(k+1) = A(k)\,\hat{\beta}(k) + B(k)\,\eta(k) - \hat{K}(k)\nu(k), \quad \hat{\beta}(0) = \xi \qquad (31)$$

$$\eta(k) = -[R(k,\lambda) + B'(k)\,S(k+1,\lambda)\,B(k)]^{-1} \times B'(k)\,S(k+1,\lambda)\,A(k)\,\hat{\beta}(k) \qquad (32)$$

*$\nu(0), \cdots, \nu(N-1)$ is the innovations process given by $\nu(k) = y(k) - C(k)\,\hat{\beta}(k)$ and $y(0), \cdots, y(N-1)$ is the output of the system of equations:*

$$\beta(k+1) = A(k)\beta(k) + B(t)\eta(k) + w(k+1),$$

$$\beta(0) = x_0 \qquad (33)$$

$$y(k) = C(k)\beta(k) + v(k+1) \qquad (34)$$

*where $x_0$, $w(0), \cdots, w(N-1)$ and $v(0), \cdots, v(N-1)$ are independent Gaussian random variables which satisfy the conditions stated in Section 4.*

Theorem 5.2 shows that calculating the gradient of (25) is equivalent to solving a full observation, unconstrained LQG control problem. These gradients are used in numerical optimization algorithms to find the optimal solution of the problem defined by (25). These algorithms are iterative schemes which update the current estimate of the optimal solution by making use of gradient information.

Hence at each iteration, the gradient must be calculated and for this reason, the problem of calculating the optimal control is equivalent to solving a sequence of unconstrained, full observation LQG control problems. Similarly, it can be seen from Theorem 5.3 that calculating the gradient of (29) is equivalent to solving an unconstrained, partially observed LQG control problem. Therefore, solving the partially observed LQG control problem with quadratic constraints is equivalent to solving a sequence of unconstrained partially observed LQG control problems. By virtue of the Separation Theorem for the unconstrained LGQ control, one may view the concept of the Quasi-Separation Theorem in this way.

On the other hand, when it comes to applications, Theorems 5.2 and 5.3 are not satisfactory in their current form because the gradient is expressed in terms of an expectation. The following result shows how one can calculate the gradients (26) and (30).

**Corollary 5.1** *Let $\lambda \in \mathbf{R}^r$, $\lambda \geq 0$ be given. The gradients (26) and (30) are given by*

$$\begin{aligned} &\frac{\partial J(\lambda)}{\partial \lambda_i} \\ &= \frac{1}{2}\sum_{k=0}^{N-1} tr[Q_i(k) + G'(k)\,R_i(k)\,G(k)]\,\Lambda(k) \\ &\quad -\frac{1}{2}tr\{Q_i(N)\,\Lambda(N)\} - c_i \qquad (35) \end{aligned}$$

$$\begin{aligned} \Lambda(k+1) &= \bar{A}(k)\,\Lambda(k)\,\bar{A}'(k) + \Omega(k+1), \\ \Lambda(0) &= \Lambda_0 \qquad (36) \end{aligned}$$

*where*

$$\begin{aligned} G(k) &= -[R(k,\lambda) + B'(k)\,S(k+1,\lambda)\,B(k)]^{-1} \\ &\quad \times B'(k)\,S(k+1,\lambda)\,A(k) \\ \bar{A}(k) &= A(k) + B(k)\,G(k) \end{aligned}$$

*with $\Omega(k) = \hat{Q}(k)$, $\Lambda_0 = \Sigma_0 + \xi\,\xi'$ for the cost functional (25), and $\Omega(k) = \Gamma(k)$, $\Lambda_0 = \xi\,\xi'$ for the cost functional (29).*

## References


[1] B.D.O. Anderson and J.B. Moore. *Optimal Control: Linear Quadratic Methods* (Prentice Hall, New Jersey, 1989).

[2] B.D.O. Anderson and J.B. Moore. *Optimal Filtering* (Prentice-Hall, Englewood Cliffs, NJ 1979).

[3] M. Athans (Ed.). Special issue on the linear-quadratic-Gaussian estimation and control problem. *IEEE Trans. Aut. Contr.*, **16**, pp 527–847, (1971).

[4] G. Kreisselmeier and R. Steinhauser. Application of vector performance optimization to a robust control loop design for a fighter aircraft. *Internat. J. Contr.*, **37**, pp 251 - 284, (1983).

[5] A.E.B. Lim. LQG optimal control of constrained discrete time systems: Separation theorems and quasi-separation theorems, (preprint).

[6] A.E.B. Lim and J.B. Moore. A quasi-separation theorem for LQG optimal control with IQ constraints, *Syst. & Contr. Lett.*, **32**, pp 21 - 33, (1997).

[7] A.E.B. Lim, J.B. Moore. Separation theorem for linearly constrained LQG optimal control - continuous time case. *Proc.* 35$^{th}$ *Conference on Decision and Control.* pp 4152 - 4157 (1996).

[8] A.E.B. Lim, J.B. Moore and L. Faybusovich. Separation theorem for linearly constrained LQG optimal control. *Systems & Control Letters* **28** (1996) 227-235.

[9] D.G. Luenberger. *Optimization by Vector Space Methods* (John Wiley, New York, 1968).

[10] K.L. Teo, C.J. Goh and K.H. Wong. *A Unified Computational Approach to Optimal Control Problems.* (Longman Scientific and Technical, London, 1991).

[11] H.T. Toivonen and P.M. Makila. Computer-aided design procedure for multiobjective control problems. *Internat. J. Contr.*, **49**, pp 655 - 666, (1989).

[12] W.M. Wonham. On the separation theorem of stochastic control. *SIAM J. Control*, no 6, 1968, pp 312-326.

14th Triennial World Congress, Beijing, P.R. China





# LQ OPTIMAL CONTROL FOR MULTIRATE SAMPLED DATA SYSTEMS

**J. Tornero, J. Salt, and P. Albertos**

*Department of Systems Engineering and Control*
*Universidad Politécnica de Valencia, Valencia, Spain.*
*P.O.Box 22012 E-46071. Tel. 34-96-3879570 Fax 34-96-3879579*
*e-mail:* {jtornero,julian,pedro}@isa.upv.es

**Abstract**: In this paper a new approach to model a MIMO multirate sampled data system allowing an easy computation of the solution of a Linear Quadratic Regulator problem is presented. The process model is split into a time-invariant-single-rate part, involving the basic process dynamics, and a periodic one reflecting the external sampling strategy. Also, the cost index is easily translated into an equivalent one, incorporating the sampling schema, and the optimal solution is found by solving a sequence of recurrent Riccati equations. The approach is validated by means of a simple but general example, being also applied to a typical example. *Copyright © 1999 IFAC*

**Keywords**: Sampled data systems, linear quadratic regulator, optimal multirate control.

## 1. INTRODUCTION

In the last years, the integrated communication and control applied to complex dynamical systems like advanced aircraft, spacecraft, autonomous manufacturing plants or complex industrial processes, require advanced control schemas where it is unrealistic to assume a unique sampling period and synchronicity in the change of all the system variables. In general, in any distributed control systems where there are several processors and communication channels, this assumption is no more valid.

This contribution deals with linear time-invariant MIMO systems. A multirate sampling scheme is considered for the input and output vectors, as shown in Fig. 1. Although the sampling schema can be totally random and/or asynchronous, it is generally accepted that there is a sampling/updating periodicity for each signal, ($TN_i$, $TM_j$), and there exist a least common multiple (l.c.m.) $T_0$, the <u>frame</u> period, and a greatest common divisor T, the <u>basic</u> period, such as $T_0 = NT$, with $N = \text{l.c.m.}(N_i, M_j)$.

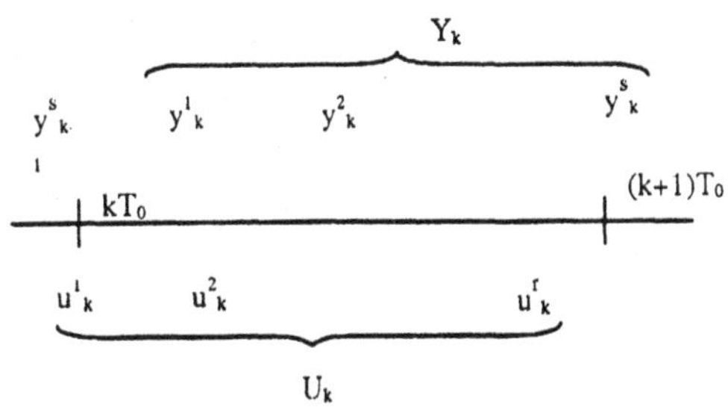


Fig.1. Non-uniform sampling.

This problem has been extensively treated in the last two decades and it is possible to find many contributions dealing with modelling and analysis, as well as controller design, for this kind of processes. One of the approaches to treat the modelling step is to assume an enlarged MIMO system, with $N$x$m$ inputs and $N$x$p$ outputs (Araki and Yamamoto, 1986), (Berg, *et al.*, 1988), (Godbout, *et al.*, 1990), (Albertos, *et al.*, 1996). In these approaches, a discrete time-invariant state equation, for the frame period $T_0$, with extended state, input and output vectors, is used. Similar to this one is the Vector Switch Decomposition (Kranc, 1957), later on implemented by (Thompson, 1986, 1988).

Another viewpoint is taken in (Meyer and Burrus, 1975) and (Albertos, 1990). The state is not extended and block vectors are defined for the input and output signals. Finally, another popular technique is denoted as the lifting technique, developed and applied in (Bamieh, *et al.*, 1991), (Voulgaris and Bamieh, 1992). In (Meyer, 1990) the equivalence among different techniques in multirate digital control is proved.

The approach in this work follows the original idea proposed by (Kalman and Bertram, 1959), based on a continuous time process model, time-events, and their interaction. This was the viewpoint initially used in (Tornero and Edmunds, 1983), (Tornero, 1985), later on also followed by (Longhi, 1994).

A simplification for the cost function calculation in the LQ problem with mutirate schemes, is proposed. The application of the multirate schema to LQ control has been considered in many of the above referred contributions. Another notes of interest could be (Glasson, 1983), usually cited as the first contribution in this field, and (Apostolakis and Jordan, 1991), (AlRahmani and Franklin, 1992) and (Meyer, 1992).

The paper is organised in five sections. The control problem is formally stated in the next section. The multirate SD model of the plant is described in section 3. The multirate optimal regulator design and the solution for the specific case we have considered are presente din section 4. Some illustrative examples are outlined in section 5. The paper ends with some conclusions.

## 2. PROBLEM STATEMENT

Consider the linear time-invariant continuous-time plant described by

$$\begin{aligned} \dot{x}(t) &= A_c \cdot x(t) + B_c \cdot u(t) \\ y(t) &= C_c \cdot x(t) \end{aligned} \tag{1}$$

where $t \in Z^+, x \in R^n$ is the state, $u \in R^m$ is the control input vector and $y \in R^p$ is the output vector. The single-rate sampled-data (**SD**) system, at the base period $T$, is expressed as follows

$$x((k+1)T) = A_{SR}(T) \cdot x(kT) + B_{SR}(T) \cdot u(kT) \tag{2}$$

where

$$A_{SR}(T) = \exp(A_c \cdot T), \quad B_{SR}(T) = \int_0^T \exp(A_c \cdot \tau) \cdot B_c \cdot d\tau$$

These matrices can be also obtained in a more compact form, as follows

$$\begin{bmatrix} A_{SR}(T) & B_{SR}(T) \\ 0 & I \end{bmatrix} = \exp\begin{bmatrix} A_c \cdot T & B_c \\ 0 & 0 \end{bmatrix} \tag{3}$$

To control this system, a discrete time quadratic optimisation index, $J_{SR}$, is defined by:

$$J_{SR} = \frac{1}{2} \sum_{k=0}^{\infty} \left\{ x(k)^T Q_{SR}(T) x(k) + u(k)^T R_{SR}(T) u(k) \right\} \tag{4}$$

This index can be also written in a more compact form as

$$J_{SR} = \frac{1}{2} \sum_{k=0}^{\infty} \begin{bmatrix} x(k)^T & u(k)^T \end{bmatrix} \cdot \begin{bmatrix} Q_{SR}(T) & 0 \\ 0 & R_{SR}(T) \end{bmatrix} \cdot \begin{bmatrix} x(k) \\ u(k) \end{bmatrix} =$$

$$= \frac{1}{2} \sum_{k=0}^{\infty} \begin{bmatrix} x(k)^T & u(k)^T \end{bmatrix} \cdot \overline{Q}_{SR}(T) \cdot \begin{bmatrix} x(k) \\ u(k) \end{bmatrix} \tag{5}$$

The state feedback optimal control

$$u(k) = K.x(k) \tag{6}$$

is given by the solution of the well known discrete time algebraic Riccati equation.

Now, let us introduce the multirate updating schema. It is assumed that the state is available at any instant of time, but the $i$-th component $u_i(t), i = 1,2,\ldots m$ of the input vector $u(t)$ can be only updated every $T_i$ time instants. Let us denote by $T$, the base period, the maximum common divisor of the updating periods, such that $T_i = N_i T$, $N_i \in Z^+$. Similarly, let us denote by $T_o$, the frame period, the l.c.m. of them, that is, $T_o = r_i T_i$. The whole system is $T_o$-periodic.

The goal is to find the optimal control law, w.r.t. the cost index (4), for the system (2), with the multirate control input updating scheme previously defined, that is, when not all the input components are updated at the same time. For that purpose, a multirate model is derived and the cost index is accordingly modified.

## 3. MULTIRATE SAMPLED DATA PLANT MODEL

Pursuing the ideas in (Tornero, 1985) and (Longhi, 1994), the multirate SD system can be expressed by two discrete time systems: one related to the plant and another one related to the input-hold mechanism,

$$\begin{aligned} x((k+1)T) &= A_{SR}(T)x(kT) + B_{SR}(T)(I-\Delta(k))v(kT) + \\ &\quad + B_{SR}(T)\Delta(k)u(kT) \\ v((k+1)T) &= (I-\Delta(k))v(kT) + \Delta(k)u(kT) \end{aligned} \tag{7}$$

where $v(\bullet) \in R^m$ is a new state vector associated to the input-hold mechanism, and $\Delta(\bullet)$ is an $N$-periodic $m$x$m$ matrix given by

$$\begin{aligned} &\Delta(k) \equiv diag\{\delta_i(k), i = 1,2,\ldots,m\} \\ &\delta_i(k) \equiv \begin{cases} 1, & k = j\cdot N_i, \\ 0 & k \neq j\cdot N_i, \end{cases} \quad j \in Z^+ \end{aligned} \tag{8}$$

This multirate model can be expressed in a more compact form, by

$$\bar{x}((k+1)\cdot T) = A_{MR}(k,T)\cdot \bar{x}(kT) + B_{MR}(k,T)\cdot u(kT) \tag{9}$$

where $\bar{x}(kT)^T \equiv \left[x(kT)^T \quad v(kT)^T\right] \in R^{n+m}$ is the enlarged state vector and the $N$-periodic matrices $A_{MR}(\bullet)$, $B_{MR}(\bullet)$ are given by

$$\begin{aligned} A_{MR}(k,T) &= \begin{bmatrix} A_{SR}(T) & B_{SR}(T)\cdot(I-\Delta(k)) \\ 0 & I-\Delta(k) \end{bmatrix} \\ B_{MR}(k,T) &= \begin{bmatrix} B_{SR}(T)\Delta(k) \\ \Delta(k) \end{bmatrix} \end{aligned} \tag{10}$$

Eq. (9) can be written as follows

$$[\bar{x}]_{(k+1)T} = [A_{MR}(k,T) \quad B_{MR}(k,T)]\cdot \begin{bmatrix} \bar{x} \\ u \end{bmatrix}_{kT} \tag{11}$$

providing an easy way to compute it. In fact, these multirate matrices can be easily obtained as the product of two independent matrices:

i) $H_{SRD}(T) \in R^{(n+m)x(n+m)}$
related to the single rate dynamic matrices, and

ii) $H_{MRS}(k) \in R^{(n+m)x(2n+m)}$
related to the multirate sampling matrix

This yields as state equation

$$\begin{bmatrix} x \\ v \end{bmatrix}_{(k+1)T} = H_{SRD}(T)\cdot H_{MRS}(k)\cdot \begin{bmatrix} x \\ v \\ u \end{bmatrix}_{kT} =$$

$$= \begin{bmatrix} A_{SR}(T) & B_{SR}(T) \\ 0 & I \end{bmatrix}\cdot \begin{bmatrix} I & 0 & 0 \\ 0 & I-\Delta(k) & \Delta(k) \end{bmatrix}\cdot \begin{bmatrix} x \\ v \\ u \end{bmatrix}_{kT}$$

## 4. MULTIRATE LQR DESIGN

A multirate discrete time quadratic optimisation index $J_{MR}$, equivalent to $J_{SR}$, can be deduced. Together with the multirate process model (11), it will determine an optimal control design problem. Similar to (5), this index should be written as

$$J_{MR} = \frac{1}{2}\sum_0^{\infty}\left[x(k)^T \quad v(k)^T \quad u(k)^T\right]\cdot \bar{Q}_{MR}(k,T)\cdot \begin{bmatrix} x(k) \\ v(k) \\ u(k) \end{bmatrix} \tag{12}$$

Taking into account that the system input is either updated or held, according to the multirate updating schema, the original index can be rewritten as:

$$\begin{aligned} J_{SR} &= \frac{1}{2}\sum_{k=0}^{\infty}\left\{x(k)^T\cdot Q_{SR}(T)\cdot x(k) + u(k)^T\cdot R_{SR}(T)\cdot u(k)\right\} = \\ &= \frac{1}{2}\sum_{k=0}^{\infty}\{x(k)^T\cdot Q_{SR}(T)\cdot x(k) + u(k)^T\Delta^T(k)R_{SR}(T)\Delta(k)\cdot u(k) + \\ &\quad + v(k)^T[I-\Delta(k)]^T R_{SR}(T)[I-\Delta(k)]\cdot v(k)\} \end{aligned}$$

This gives the following matrix equivalence

$$\begin{aligned} \bar{Q}_{MR}(k,T) &= \begin{bmatrix} I & 0 & 0 \\ 0 & I-\Delta(k) & \Delta(k) \end{bmatrix}^T \begin{bmatrix} Q_{SR}(T) & 0 \\ 0 & R_{SR}(T) \end{bmatrix} \\ &\quad \begin{bmatrix} I & 0 & 0 \\ 0 & I-\Delta(k) & \Delta(k) \end{bmatrix}; \\ \bar{Q}_{MR}(k,T) &= \begin{bmatrix} Q_{MR}(k,T) & M_{MR}(k,T) \\ M_{MR}(k,T) & R_{MR}(k,T) \end{bmatrix} \end{aligned} \tag{19}$$

Summarising, the multirate SD LQR problem can be expressed as

$$J_{MR} = \frac{1}{2}\sum_{k=0}^{\infty}\left\{\begin{array}{l}\bar{x}(k)^T \cdot Q_{MR}(k,T)\cdot \bar{x}(k)+ \\ +2\cdot \bar{x}(k)^T \cdot M_{MR}(k,T)\cdot u(k)+ \\ +u(k)^T \cdot R_{MR}(k,T)\cdot u(k)\end{array}\right\}$$

subject to

$$\bar{x}((k+1)\cdot T) = A_{MR}(k,T)\cdot \bar{x}(kT) + B_{MR}(k,T)\cdot u(kT)$$

where $\bar{x}(kT)^T \equiv [x(kT) \;\; v(kT)] \in R^{n+m}$ is the enlarged state vector and the $N$-periodic matrices $A_{MR}(\bullet)$, $B_{MR}(\bullet)$ are given by (10)-(11).

## 5. ILLUSTRATIVE EXAMPLES

The application of this multirate SD model to a multivariable system implies the generation of a periodic single rate system model at a base period where some inputs are not considered at certain instants of time. A single rate SD system can also be considered in order to illustrate how the model is generated, as well as the design of the optimal linear quadratic regulator.

### *5.1 Multirate computation of a single rate control.*

Let us assume (1) to be a scalar process. The single rate SD system (4) is expressed by

$$x((k+1)T) = a\cdot x(kT) + b\cdot u(kT)$$

where $u((k+i)T) = u(kT)$, $i = 0,1,\ldots,N-1$. If the input is updated only every $N$ base periods, the $N$-periodic single rate system model is given by

$$[A_{MR}(k) \;\; B_{MR}(k)] = H_{SRD}\cdot H_{MRS}(k) = = \begin{bmatrix} a & b \\ 0 & 1\end{bmatrix}\cdot\begin{bmatrix} 1 & 0 & 0 \\ 0 & 1-\delta(k) & \delta(k)\end{bmatrix}$$

$$\begin{array}{ll} x((k+1)T) = & ax(kT) + \;\; b(1-\delta(k))v(kT) + \;\; b\delta(k)u(kT) \\ v((k+1)T) = & \qquad (1-\delta(k))v(kT) + \;\; \delta(k)u(kT)\end{array}$$

where

$$\delta(k) \equiv \begin{cases} 1, & k = j\cdot N, \\ 0 & k \neq j\cdot N,\end{cases} \quad j\in Z^+$$

If a discrete quadratic optimisation index

$$J_{SR} = \frac{1}{2}\sum_{k=0}^{\infty}\left\{x(k)^T \cdot q\cdot x(k) + u(k)^T \cdot r\cdot u(k)\right\}$$

is defined for the initial $T$-periodic system, the multirate SD LQR problem can be expressed as

$$J_{MR} = \frac{1}{2}\sum_{k=0}^{\infty}\left\{\begin{array}{l}\bar{x}(k)^T \cdot Q_{MR}(k,T)\cdot \bar{x}(k)+ \\ +2\cdot \bar{x}(k)^T \cdot M_{MR}(k,T)\cdot u(k)+ \\ +u(k)^T \cdot R_{MR}(k,T)\cdot u(k)\end{array}\right\}$$

where $\bar{x}(kT)^T \equiv \left[x(kT)^T \;\; v(kT)^T\right] \in R^2$

and

$$\bar{Q}_{MR}(k,T) = \begin{bmatrix} Q_{MR}(k,T) & M_{MR}(k,T) \\ M_{MR}^T(k,T) & R_{MR}(k,T)\end{bmatrix}$$

$$= H_{MRS}^T(k)\cdot \bar{Q}_{SR}\cdot H_{MRS}(k) =$$

$$= \begin{bmatrix} I & 0 & 0 \\ 0 & I-\delta(k) & \delta(k)\end{bmatrix}^T \cdot \begin{bmatrix} q & 0 \\ 0 & r\end{bmatrix}\cdot\begin{bmatrix} I & 0 & 0 \\ 0 & I-\delta(k) & \delta(k)\end{bmatrix}$$

To get the solution implies to solve the two Riccati-like equations in a given sequence. First,

$$\begin{bmatrix} p_{11} & p_{12} \\ p_{12} & p_{22}\end{bmatrix}_{N\cdot k} = \begin{bmatrix} q & 0 \\ 0 & 0\end{bmatrix} +$$

$$\begin{bmatrix} a & 0 \\ 0 & 0\end{bmatrix}\begin{bmatrix} p_{11} & p_{12} \\ p_{12} & p_{22}\end{bmatrix}_{N\cdot k+1}\left[I + \begin{bmatrix} b \\ 1\end{bmatrix}\frac{1}{r}[b \;\; 1]\begin{bmatrix} p_{11} & p_{12} \\ p_{12} & p_{22}\end{bmatrix}_{N\cdot k+1}\right]^{-1}\begin{bmatrix} a & 0 \\ 0 & 0\end{bmatrix}$$

and then, for $i = 0,\ldots,N-2$, (i.e., $N$-1 times) the following equation,

$$\begin{bmatrix} p_{11} & p_{12} \\ p_{12} & p_{22}\end{bmatrix}_{N\cdot k-i-1} = \begin{bmatrix} q & 0 \\ 0 & r\end{bmatrix} + \begin{bmatrix} a & 0 \\ b & 1\end{bmatrix}\begin{bmatrix} p_{11} & p_{12} \\ p_{12} & p_{22}\end{bmatrix}_{N\cdot k-i}\begin{bmatrix} a & b \\ 0 & 1\end{bmatrix}$$

For the particular numerical case where $a = 0.5$, $b = 0.5$, $q = 1$, $r = 1$ with $N$=2 the infinite time solution gives the following Riccati matrix and Kalman gain

$$\begin{bmatrix} p_{11} & p_{12} \\ p_{12} & p_{22}\end{bmatrix}_{\infty} = \begin{bmatrix} 1.31216 & 0.31216 \\ 0.31216 & 1.31216\end{bmatrix}$$

$$[K_1 \;\; K_2]_{\infty} = [0.163978 \;\; 0.0]$$

which shows that the state feedback associated to the input is null when there is no updating.

An equivalent result should be obtained using the original system, if appropriately treated. For that purpose, the conventional system is defined by

$$x(2(k+1)) = a^2\cdot x(2k) + b\cdot(1+a)\cdot u(2k)$$

and the equivalent discrete linear quadratic optimisation index is

$$J = \frac{1}{2}\sum_{k=0}^{\infty}\left\{\begin{array}{l} x(2k)^T\left(q\cdot\left(1+a^2\right)\right)\cdot x(2k)+ \\ +2\cdot x(2k)^T(a\cdot q\cdot b)\cdot u(2k)+ \\ +u(2k)^T\left(2r+b^2 q\right)\cdot u(2k)\end{array}\right\}$$

A Riccati solution $p = 1.24866$ and the Kalman gain $K = 0.163978$, which are in agreement with those computed for the multirate model, are obtained.

### 5.2 *Multirate optimal control*

Consider the weakly coupled multivariable plant proposed in (Araki and Yamamoto, 1986) and (Godbout, *et al.*, 1990). The CT state-space model is given by:

$$A_c = \begin{bmatrix} -2.5 & 0 & 0 \\ 0 & -2 & 0 \\ 0 & 0 & 1 \end{bmatrix}; B_c = \begin{bmatrix} 2.5 & 0 \\ 10 & -1.2 \\ 5/6 & 1 \end{bmatrix};$$

The inputs, $u_1$ and $u_2$, are updated every $T_1=0.1$sec and $T_2=0.15$ sec, respectively. The state is available at any basic sampling period, that is, $T=0.1$sec, being $T_0=0.3$ sec. In the cost index (4), the same weight has been assigned to both state and control variables.

In this case, there are four different updating conditions, $\{0,0.1,0.15,0.2\}$, denoted by $i=0,1,2,3$, repeated with period $T_0$. The corresponding feedback matrices are computed by solving a set of four Riccati-like equations. In a compact form, the control law is:

$$u(i) = K_i x + H_i u(i-1)$$

where $K_i$ is the state feedback matrix and $H_i$ reflects the holding effect. In this case, the results are:

1) t=0

$$K_0 = \begin{bmatrix} 0.1343 & 0.6147 & 0.1130 \\ 0.0272 & -0.1029 & 0.4091 \end{bmatrix}; H_0 = \begin{bmatrix} 0 & 0 \\ 0 & 0 \end{bmatrix}$$

Both inputs are updated.

2) t=0.1 *sec*

$$K_1 = \begin{bmatrix} 0.1325 & 0.6201 & 0.0875 \\ 0 & 0 & 0 \end{bmatrix}; H_1 = \begin{bmatrix} 0 & -0.0657 \\ 0 & 1 \end{bmatrix}$$

$u_1$ is updated, but $u_2$ is taken as an unchanged state variable.

3) t=0.15 *sec*

$$K_2 = \begin{bmatrix} 0 & 0 & 0 \\ 0.0155 & -0.1570 & 0.3983 \end{bmatrix}; H_2 = \begin{bmatrix} 1 & 0 \\ -0.1074 & 0 \end{bmatrix}$$

4) t=0.2 *sec*

$$K_3 = \begin{bmatrix} 0.1328 & 0.6195 & 0.0914 \\ 0 & 0 & 0 \end{bmatrix}; H_3 = \begin{bmatrix} 0 & -0.0544 \\ 0 & 1 \end{bmatrix}$$

As expected, for the sampling instants $T_0$+{.,.,*0.15*,.} the first raw of $K$ is null, and for $T_0$+{.,0.1,.,*0.2*} the second raw of $K$ also vanishes.

The infinite time Riccati matrix is given by

$$P = \begin{bmatrix} 4.4268 & -0.4662 & -0.1625 & 0.2042 & 0.0211 \\ -0.4662 & 2.9427 & -0.2989 & 0.9467 & -0.1379 \\ -0.1625 & -0.2989 & 8.7367 & 0.1518 & 0.4155 \\ 0.2042 & 0.9467 & 0.1518 & 1.5315 & -0.0520 \\ 0.0211 & -0.1379 & 0.4155 & -0.0520 & 1.0300 \end{bmatrix}$$

The system response, from the initial condition $x^T(0) = [0.5 \ -0.4 \ 0.3]$, is shown in Fig. 2. The multirate control input is plotted in Fig. 3.

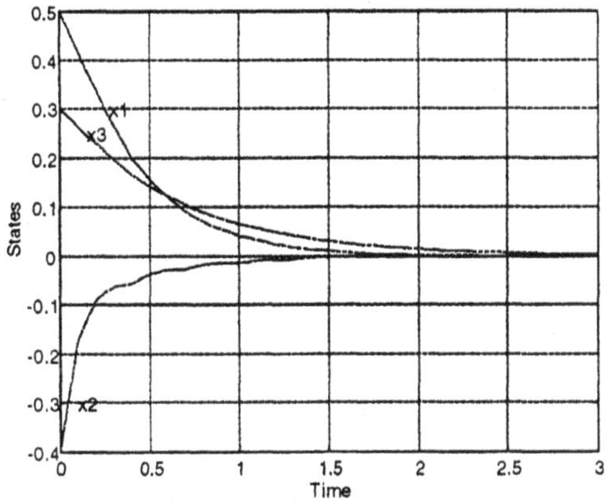


Fig. 2. State evolution.

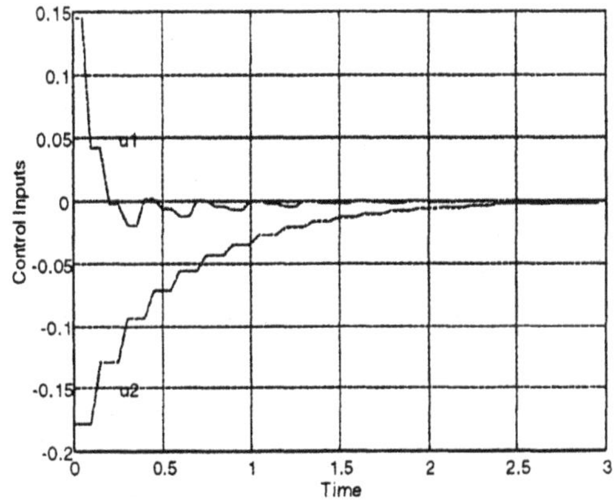


Fig. 3 Multirate control input.

## 6. CONCLUSIONS

In this paper, a compact model to represent discrete time systems with a multirate updating schema is presented. The proposed model enables an easy computation of the involved matrices from the single rate system matrices and a sampling matrix including the features of the updating schema. This allows a great simplicity to change either the system parameters or the sampling events, in an independent way.

This model has been also shown extremely useful to compute the LQR for SD systems with the multirate policy. The initial single rate performance index matrices are transformed into multirate index matrices, by using the same approach, that is, by means of the same sampling matrix.

The result is a periodic state feedback control law which is computed by sequentially solving a set of Riccati-like equations.

A simple single-rate example shows the computational procedure and an illustrative example shows the applicability of the approach.

## REFERENCES


Albertos, P. (1990) Block Multirate Input-Output Model for sampled data Control Systems. *IEEE Trans. on Aut. Control.* Vol. **AC-35**, No 9, pp 1085-1088.

Albertos P., Salt J. ,Tornero J. (1996). Dual-rate adaptive control. *Automatica,* Vol. **32**, No 7, pp 1027-1030.

Al-Rahmani H. M., G. F. Franklin. (1992) Multirate Control: a new approach. *Automatica.* Vol. **28,** No.1, pp 35-44.

Apostolakis I.S., Jordan D. (1991). A time invariant approach to multirate optimal regulator design. *Int. J. of Control,* **53**, 1233-1254.

Araki M, Yamamoto K. (1986). Multivariable multirate sampled data systems: state-space description, transfer characteristics, and Nyquist criterion. *IEEE Trans. on Aut. Control,* **31**, 145-154.

Araki, M. (1993). Recent development in digital control theory. *Proc. 12th IFAC World Congr.,* Vol.**9**, pp. 951-960.

Bamieh B., Francis B., Tannenbaum A (1991). A Lifting Thecnique for Linear Periodic Systems with Aplications to Sampled - Data Control. *Systems and Control Letters,* vol **17**, pp 79-88.

Berg M.C., Amit N., Powell J.D. (1988). Multirate Digital Control System Design. *IEEE Trans. on Aut. Control,* Vol. **33**, No 12., pp 1139-1150.

Glasson D. P. (1983). Development and Applications of Multirate Digital Control.*Control System Magazine,* 2-8

Godbout L.F., Jordan D., Apostolakis I.S. (1990). Closed-loop model for general multirate digital control systems. *IEE Proceedings,* Part-D. **137**, 326-336.

Kalman R.E., Bertram J.E. (1959). A unified approach to the theory of samplig systems. *Journal Franklin Inst.,* vol. 267, pp 405-436, 1959.

Kranc G.M. (1957). Compensation of an error-sampled system by a multirate controller. *Trans. AIEE,* vol. **6**, M.6, part II, pp. 149-155, 1957.

Longhi, S. (1994). Structural Properties of Multirate Sampled Systems. *IEEE Trans. on Aut. Control,* Vol **AC-39**, No 3, pp 692-696.

Meyer R.A., Burrus C.S. (1975). A unified analysis of multirate and periodically time-invarying digital filters. *IEEE Trans. Circuits Syst.* **CAS-22**, pp.162-168.

Meyer D.G. (1990). A parametrization of stabilizing controllers for multirate sampled data systems. *IEEE Trans. on Aut. Control.* Vol. **AC-35**, No.2, pp 233-236.

Meyer D.G. (1992). Cost translation and a lifting approach to the multirate LQG problem. *IEEE Trans. on Aut. Control.* Vol. **AC-37**, No.9, pp 1411-1415.

Thompson P. M. (1986). Gain and phase margin of multirate sampled data feedback systems. *Int. J. Control,* **44**, 833-846.

Thompson P.M. (1988). Program CC Version 4: Reference Manual. *Hawthorne, Systems Technology, Inc.,* **II**, 566-585.

Tornero, J., Edmunds, J. (1983) Irregular but periodic Sampling. *Control System Centre Report n° 584/1983 University of Manchester (UMIST)*

Tornero, J. (1985) Non-conventional sampled data Systems Modelling. *Control System Centre Report n° 640/1985 University of Manchester (UMIST)*

Voulgaris P., Bamieh B (1992). Optimal $H^\infty$ and $H^2$ Control of Hybrid Multirate Systems. *Systems and Control Letters,* vol. **20**, pp 249-261.

14th Triennial World Congress, Beijing, P.R. China


# OPTIMAL CONTROL PROBLEM FOR A DISCRETE DISTRIBUTED SYSTEM WITH FIXED FINAL STATE

**L.Chraïbi**[1] **J.Karrakchou**[1*] **A.Ouansafi**[2] **M.Rachik**[3]

[1] *Ecole Mohammadia d'ingénieurs, B.P 765, Rabat- Morocco*
*E-mail: j.karrak@emi.ac.ma.*
[2] *Faculté des Sciences, Dép. Math & Info, BP 1014, Rabat- Morocco.*
*E-mail: ouansafi@emi.ac.ma*
[3] *Faculté des Sciences Benmsik Casa II, Casablanca Morocco*

Abstract: The problem considered is that of minimizing a quadratic cost functional for discrete distributed systems with a fixed initial and final states. It is shown that under suitable controllability assumptions, there is a closed relationship between this problem and the problem of exact controllability with the minimization of a variable time energy criterion. Results using HUM approach ta treat the exact controllability for analoguous systems are then extended to the variable case and applied to give an explicit expression of the optimal control and the optimal cost. Copyright © 1999 IFAC.

Key words: Controllability, Discrete-time, Dynamic systems, Feedback control, Optimal control, Quadratic control, Reachability.

## 1. INTRODUCTION

The paper is concerned with the problem of linear quadratic optimal control of discrete distributed systems subject to initial and final state. The system considered is described by the difference equation

$$\begin{cases} x_{i+1} = \Phi x_i + D u_i \quad ; i \in \{0, .., N-1\} \\ x_0 \in X, \end{cases} \tag{1}$$

where $x_i \in X$ , $\Phi : X \longrightarrow X$ and $D : U \longrightarrow X$ are bounded operators. $X$ and $U$ are Hilbert spaces. For a desired state $d \in X$, consider $\Sigma_P$ the set of admissible controls given by

$$\Sigma_P = \left\{ u = (u_0, .., u_{N-1}) \in U^N : x_N^u = d \right\}, \tag{2}$$

where $x_N^u$ is the state of system (1) at stage $N$ corresponding to the control $u$.

The control problem with a fixed final state, denoted by $(P)$, is the following:
Find $u_P = (u_0^P, u_1^P, \ldots, u_{N-1}^P)$ in $\Sigma_P$ minimizing the cost functional

$$J(u) = \sum_{i=0}^{N-1} \left\{ \langle Q x_i, x_i \rangle + \langle R u_i, u_i \rangle \right\}, \tag{3}$$

where $Q$ and $R$ are self-adjoints non negative with $\langle Ru, u\rangle \geq \delta \left\|u\right\|^2$, $\delta > 0$.
The continuous case of such problems has been studied by Curtain (1984), and the optimal control is expressed, in the case of approximately controllable systems, as the weak limit of a sequence of approximating controls. These results were complemented by Emirsajlow (1989), where both open-loop and feedback descriptions of the optimal control were presented under a sufficient controllability condition.
In the discrete case, Lee, *et al.* (1972), has considered the problem with free final conditions and the optimal control is characterized by a Riccati equation, in both finite and infinite horizon case.

* Supported in part by the CNCPRST-Morocco.

The present work shows that by introducing a feedback law, the resolution of problem $(P)$ can be obtained by the resolution of some controllability problem with the minimum energy. Then techniques used in (Karrakchou, *et al.*, 1995) and (Chraïbi, *et al.*, 1998) for solving the problem of exact controllability by the H.U.M approach (Hilbert Uniqueness Method ) are extended to the variable case and applied to give an explicit expression of the solution of the problem.

In what follows $\sigma_\alpha^\beta$ denotes the finite family of Z defined by: $\sigma_\alpha^\beta = \{\alpha, \alpha+1, \cdots, \beta-1, \beta\}$.

If $Y$ is a Hilbert space then $l^2(\sigma_\alpha^\beta, Y) = \left\{(x_i)_{i\in\sigma_\alpha^\beta}, x_i \in Y\right\}$ is a Hilbert space with usual addition, scalar multiplication and with an inner product defined by:

$$\langle x, y\rangle_{l^2(\sigma_\alpha^\beta, Y)} = \sum_{i\in\sigma_\alpha^\beta} \langle x_i, y_i\rangle. \quad (4)$$

Let $H_P$ be the reachability operator of the system (1) defined by

$$H_P : l^2(\sigma_0^{N-1}, U) \longrightarrow X, \quad H_P(u) = \sum_{k=0}^{N-1} \Phi^{N-k-1} Du_k. \quad (5)$$

If $V_P$ denotes the set of reachable states,

$$V_P = range H_P,$$

a result of existence and uniqueness of the problem $(P)$ is given by the following proposition:

**Proposition 1** *The problem $(P)$ admits a unique solution if and only if $d - \Phi^N x_0 \in V_P$.*

**Proof.** $d - \Phi^N x_0 \in V_P$ means that there exists a control $u$ in $l^2(\sigma_0^{N-1}, U)$ such that

$$d - \Phi^N x_0 = H_P(u) = \sum_{k=0}^{N-1} \Phi^{N-k-1} Du_k, \quad (6)$$

therefore

$$d = \Phi^N x_0 + \sum_{k=0}^{N-1} \Phi^{N-k-1} Du_k = x_N^u, \quad (7)$$

hence $u \in \Sigma_P$ and $\Sigma_P \neq \emptyset$. Moreover $\Sigma_P$ is closed and convex then the functional attains a unique minimum $u = (u_0^P, \ldots, u_{N-1}^P) \in \Sigma_P$. Conversely, if $(P)$ admits a unique solution $u$, the corresponding response at the step $N$ is $d$, that means

$$x_N^u = d = \Phi^N x_0 + H_P(u), \quad (8)$$

hence $d - \Phi^N x_0 \in range H_P = V_p$. ■

## 2. RELATIONSHIP BETWEEN THE PROBLEM (P) AND A MINIMUM ENERGY PROBLEM

First, let recall some results concerning the unconstrained problem (1)-(3) presented by Lee, *et al.* (1972).

**Proposition 2** *The optimal control of the problem (1)-(3) with a free-terminal state has a feedback form and is given by*

$$u_i = -R^{-1}D^*K_{i+1}(I + DR^{-1}D^*K_{i+1})^{-1}\Phi x_i, \quad i \in \sigma_0^{N-1}, \quad (9)$$

*where $(x_i)_{i\in\sigma_0^{N-1}}$ is the optimal state and $\{K_i, i \in \sigma_0^{N-1}\}$ is a family of bounded self-adjoint non negative operators, unique solution of the discrete Riccati equation:*

$$\begin{cases} K_i = \Phi^*K_{i+1}(I + DR^{-1}D^*K_{i+1})^{-1}\Phi + Q \\ \qquad i = N-1, N-2, \ldots, 0, \\ K_N = 0. \end{cases} \quad (10)$$

Now consider the control system (1) starting at time $i$ with an initial state $h \in X$,

$$\begin{cases} x_{j+1} = \Phi x_j + Du_j, \; j \in \sigma_i^{N-1}, \\ x_i = h. \end{cases} \quad (11)$$

Let $v \in l^2(\sigma_i^{N-1}, U)$ then for the control

$$u_j = v_j - R^{-1}D^*K_{j+1}(I + DR^{-1}D^*K_{j+1})^{-1}\Phi x_j \quad j \in \sigma_i^{N-1}, \quad (12)$$

the last difference equation becomes

$$x_{j+1} = (I - DR^{-1}D^*K_{j+1}(I + DR^{-1}D^*K_{j+1})^{-1})\Phi x_j + Dv_j, \quad j \in \sigma_i^{N-1}, \quad (13)$$

then a factorization by $(I + DR^{-1}D^*K_{j+1})^{-1}$ yields to

$$x_{j+1} = (I + DR^{-1}D^*K_{j+1})^{-1}\Phi x_j + Dv_j, \quad j \in \sigma_i^{N-1}. \quad (14)$$

Taking $B(j) = (I + DR^{-1}D^*K_{j+1})^{-1}\Phi$ for $j \in \sigma_0^{N-1}$, the difference equation

$$\begin{cases} x_{j+1} = B(j)x_j + Dv_j, \; j \in \sigma_i^{N-1}, \\ x_i = h; \end{cases} \quad (15)$$

can be considered as a perturbed system of equation (11) by

$$\left\{-DR^{-1}D^*K_{j+1}B(j)\right\}_{j\in\sigma_i^{N-1}}.$$

**Proposition 3** *If the system (11) is excited by a control law u given by the relation (12), with* $v = (v_j)_{j\in\sigma_i^{N-1}} \in l^2(\sigma_i^{N-1}, U)$, *the following equality holds for* $i \in \sigma_0^{N-1}$,

$$\begin{aligned}\langle K_i h, h\rangle &= \sum_{j=i}^{N-1} \langle x_j, (Q+ \\ &\quad B^*(j)K_{j+1}DR^{-1}D^*K_{j+1}B(j))\,x_j\rangle \\ &- 2\sum_{j=i}^{N-1} \langle D^*K_{j+1}B(j)x_j, v_j\rangle \\ &- \sum_{j=i}^{N-1} \langle K_{j+1}Dv_j, Dv_j\rangle .\end{aligned} \tag{16}$$

**Proof.** Using (12), the right member of the identity (16) may be written as

$$\begin{aligned}\text{R.M} &= \sum_{j=i}^{N-1} \langle x_j, Qx_j\rangle \\ &+ \sum_{j=i}^{N-1} \langle B^*(j)K_{j+1}D\,(v_j - u_j)\,, x_j\rangle \\ &- 2\sum_{j=i}^{N-1} \langle D^*K_{j+1}B(j)x_j, v_j\rangle \\ &- \sum_{j=i}^{N-1} \langle K_{j+1}Dv_j, Dv_j\rangle ,\end{aligned} \tag{17}$$

that yields to

$$\begin{aligned}\text{R.M} &= \sum_{j=i}^{N-1} \langle x_j, Qx_j\rangle \\ &- \sum_{j=i}^{N-1} \langle B^*(j)K_{j+1}Dv_j, x_j\rangle \\ &- \sum_{j=i}^{N-1} \langle B^*(j)K_{j+1}Du_j, x_j\rangle \\ &- \sum_{j=i}^{N-1} \langle K_{j+1}Dv_j, Dv_j\rangle ,\end{aligned} \tag{18}$$

by substitution of (11) and (15)

$$\begin{aligned}\text{R.M} &= \sum_{j=i}^{N-1} \langle x_j, (Q+\Phi^*K_{j+1}B(j))\,x_j\rangle \\ &- 2\sum_{j=i}^{N-1} \langle B(j)x_j, K_{j+1}x_{j+1}\rangle \\ &+ \sum_{j=i}^{N-1} \langle K_{j+1}B(j)x_j, B(j)x_j\rangle \\ &- \sum_{j=i}^{N-1} \langle K_{j+1}Dv_j, Dv_j\rangle ,\end{aligned} \tag{19}$$

using the Riccati equation (10) the last equality becomes

$$\begin{aligned}\text{R.M} &= \sum_{j=i}^{N-1} \{\langle x_j, K_jx_j\rangle - \langle x_{j+1}, K_{j+1}x_{j+1}\rangle\} \\ &= \langle h, K_ih\rangle . \blacksquare\end{aligned}$$

In the sequel the relation (12) and the controlled system (15) will be considered for $i = 0$ and $h = x_0$; so (15) may be viewed as a perturbed system of (1) by $\left\{-DR^{-1}D^*K_{j+1}B(j)\right\}_{j\in\sigma_0^{N-1}}$.
Consider now the reachability operator corresponding to system (15) and given by

$$H_E : l^2(\sigma_0^{N-1}, U) \longrightarrow X$$

$$H_E(v) = \sum_{j=0}^{N-1} \left(\prod_{k=0}^{N-j-2} B(N-k-1)\right) Dv_j. \tag{20}$$

where for a given finite sequence of operators $(A(k))_k$ in a Hilbert space, the operator $\prod_{k=i}^{j} A(k)$ denotes

$$\prod_{k=i}^{j} A(k) = \begin{cases} A(i).A(i+1).\cdots.A(j-1).A(j) \ , & \text{if } i \le j, \\ I & \text{if } j = i-1, \\ 0 & \text{if } j < i-1. \end{cases}$$

It is obvious to see the following result

**Proposition 4**

$$rangeH_P = rangeH_E = V_P.$$

**Proof.** Given $y \in rangeH_E$ , that means there is a control $v$ such that $y = H_E(v)$. Taking a control $u$ such that

$$u_i = v_i - F(i)\sum_{j=0}^{i-1} \left(\prod_{k=0}^{i-j-2} B(i-k-1)\right) Dv_j, \quad i \in \sigma_0^{N-1}. \tag{21}$$

with $F(i) = R^{-1}D^*K_{i+1}B(i)$, it is showed that $H_P(u) = H_E(v) = y$.
Conversely the control system (1) is the perturbation of (15) by $\left\{DR^{-1}D^*K_{j+1}B(j)\right\}_{j\in\sigma_0^{N-1}}$, thus the reverse inclusion is established. ■

Let $\Sigma_E$ be the set of admissible controls described by

$$\Sigma_E = \left\{v \in l^2(\sigma_0^{N-1}, U) : x_N^v = d\right\}, \tag{22}$$

where $x_N^v$ is the final state corresponding to the control $v$ and satisfying the perturbed equation (15). Consider the following minimum energy problem $(E)$ stated as follows:

Find $v_E = (v_0^E, \ldots, v_{N-1}^E) \in \Sigma_E$ minimizing the performance

$$W(v) = \sum_{i=0}^{N-1} \langle R(i)v_i, v_i\rangle , \tag{23}$$

where $R(i) = R + D^*K_{i+1}D, \quad i \in \sigma_0^{N-1}$.

For the same arguments as in the proposition 1,

the problem $(E)$ admits a unique solution if and only if

$$d-\left(\prod_{j=0}^{N-1} B(N-j-1)\right) x_0 \in rangeH_E = V_P. \tag{24}$$

Moreover the following equivalence is established.

**Proposition 5** *The two following assertions are equivalent*

$$i)\quad d-\Phi^N x_0 \in V_P$$
$$ii)\quad d-\left(\prod_{j=0}^{N-1} B(N-j-1)\right) x_0 \in V_P.$$

**Proof.** $d-\Phi^N x_0 \in V_P$ means that there exists a control $u$ such that

$$d = x_N^u = x_N^v,$$

where $u$ and $v$ are related by (12). Hence

$$x_N^v = d = \left(\prod_{j=0}^{N-1} B(N-j-1)\right) x_0 + H_E(v), \tag{25}$$

That means

$$d-\left(\prod_{j=0}^{N-1} B(N-j-1)\right) x_0 \in rangeH_E = V_P. \blacksquare$$

The fundamental result of this section is stated as follows

**Theorem 6** *Assuming that $d-\Phi^N x_0 \in V_P$, the solution of the problem $(P)$ is given by*

$$u_i^P = v_i^E - R^{-1}D^* K_{i+1}B(i)x_i^E, \quad i \in \sigma_0^{N-1}, \tag{26}$$

*where $v_E = (v_0^E, \ldots, v_{N-1}^E)$ is the optimal control corresponding to the problem $(E)$ and $x_E$ is the optimal response. Moreover the optimal cost satisfies*

$$J(u_p) = W(v_E) + \langle K_0 x_0, x_0\rangle. \tag{27}$$

**Proof.** Since $d-\Phi^N x_0 \in V_P$, the last proposition assures the existence of $v_E$. In the other part for an arbitrary control $v \in \sum_E$ consider the control given by (12) and apply the identity (16) for $i = 0$ and $h = x_0$, that yields to

$$\begin{aligned}\langle K_0 x_0, x_0\rangle &= \sum_{j=0}^{N-1} \langle Q x_j, x_j\rangle \\ &+ \sum_{j=0}^{N-1} \langle R(R^{-1}D^* K_{j+1}B(j)x_j)\,, \\ &\quad R^{-1}D^* K_{j+1}B(j)x_j\rangle \\ &- 2\sum_{j=0}^{N-1} \langle R(R^{-1}D^* K_{j+1}B(j)x_j), v_j\rangle \\ &- \sum_{j=0}^{N-1} \langle D^* K_{j+1} D v_j, v_j\rangle.\end{aligned} \tag{28}$$

Observing that $R^{-1}D^* K_{j+1}B(j)x_j = v_j - u_j$ for $j \in \sigma_0^{N-1}$, the equality (28) becomes

$$\begin{aligned}\langle K_0 x_0, x_0\rangle &= \sum_{j=0}^{N-1} \{\langle Q x_j, x_j\rangle + \langle R u_j, u_j\rangle\} \\ &- \sum_{j=0}^{N-1} \langle (R + D^* K_{j+1} D) v_j, v_j\rangle.\end{aligned} \tag{29}$$

Thus

$$\langle K_0 x_0, x_0\rangle = J(u) - W(v)\ , \text{ for every } v \in \Sigma_E. \tag{30}$$

Since the last relation holds true for the solution of the problem $(E)$ ,

$$J(u) - J(u_p) = W(v) - W(v_E) \geq 0; \tag{31}$$

that ends the proof. ■

**Remark 1** *As a consequence of the above theorem, it suffices to resolve the exact controllability problem $(E)$ to find the control $u_P$.*

## 3. RESOLUTION OF THE PROBLEM

### *3.1 Minimum energy problem*

In this subsection the minimum energy problem of a discrete controlled system is examined when, in a general way, the dynamic of the system and the energy operator are variable. The H.U.M approach is used and all the results found in (Karrakchou, *et al.*, 1995) have a straightforward extension. Indeed consider the discrete controlled system

$$x_{i+1} = \Phi(i)x_i + Dv_i \quad ; i \in \sigma_0^{N-1}; \qquad x_0 \in X, \tag{32}$$

where $\Phi(i) : X \longrightarrow X$ are bounded. The control problem is to find $v = (v_0, \ldots, v_{N-1}) \in \Sigma_E = \left\{v \in l^2(\sigma_0^{N-1}, U) : x_N^v = d\right\}$ minimizing the energy criterion

$$W(v) = \sum_{i=0}^{N-1} \langle R(i)v_i, v_i\rangle, \tag{33}$$

where $x = (x_i)_{i\in\sigma_0^N}$ is solution of the difference equation (32), and $R(i) : U \longrightarrow U, i \in \sigma_0^{N-1}$, are self-adjoint non negative with $\langle R(i)u, u\rangle \geq \alpha \|u\|^2, \alpha > 0$.

Consider the reachability operator

$$H_R : l^2(\sigma_0^{N-1}, U) \longrightarrow X$$
$$H_R(v) = \sum_{j=0}^{N-1} \left(\prod_{k=0}^{N-j-2} \Phi(N-k-1)\right) DR^{-\frac{1}{2}}(j)v_j. \tag{34}$$

and define over the space $X$ the scalar product

$$\langle x, y\rangle_{X_E} = \langle H_R^* x, H_R^* y\rangle_{l^2(\sigma_0^{N-1}, U)}. \tag{35}$$

Assuming that the system (32) is approximately controllable, the seminorm $\|\cdot\|_{X_E}$ obtained from the above scalar product defines a norm on $X$. Then $X_E$ denotes the completion space of $X$ for the norm $\|\cdot\|_{X_E}$. Let $\Lambda_R$ be the bounded selfadjoint operator defined on $X$ by

$$\Lambda_R : X \to X : \Lambda_R\, x = H_R H_R^* x, \qquad (36)$$

so $\Lambda_R$ has an extension to an isomorphism denoted also by $\Lambda_R$ and defined from $X_E$ to its dual $X_E'$.

**Proposition 7** *Suppose that the control system (32) is approximately controllable.*
*if* $d - \left(\prod_{j=0}^{N-1} \Phi(N-j-1)\right) x_0 \in X_E'$ *, the optimal control of the problem (32)-(33) is given by*

$$v_j = R^{-1}(j) D^* \left(\prod_{k=j+1}^{N-1} \Phi^*(k)\right) g \ , \ j \in \sigma_0^{N-1}, \qquad (37)$$

*where* $g$ *is solution of the algebraic equation*

$$\Lambda_R g = d - \left(\prod_{j=0}^{N-1} \Phi(N-j-1)\right) x_0. \qquad (38)$$

*The minimum energy is*

$$W(v) = \|g\|_{X_E}^2 \ . \qquad (39)$$

**Proof.** The proof is based essentielly on techniques and tools used in (Karrakchou, *et al.*, 1995) and (Chraïbi, *et al.*, 1998) ■

*3.2 Main result*

From the above proposition, an explicit solution to the problem $(P)$ may be given,

**Theorem 8** *Suppose that (1) is approximately controllable and* $d - \Phi^N x_0 \in V_P$. *Then the optimal control of the problem (P) is given, for* $i \in \sigma_0^{N-1}$*, by*

$$\begin{aligned} u_i^P &= (R + D^* K_{i+1} D)^{-1} D^* \left(\prod_{k=i+1}^{N-1} B^*(k)\right) g \\ &- R^{-1} D^* K_{i+1} B(i) x_i^E, \end{aligned} \qquad (40)$$

*where* $g$ *is the unique solution of the equation*

$$\Lambda_R g = d - \left(\prod_{j=0}^{N-1} B(N-j-1)\right) x_0, \qquad (41)$$

*with* $\Lambda_R = H_E \overline{R}^{-1} H_E^*$ *, and*

$$\begin{aligned} \overline{R} &: l^2(\sigma_0^{N-1}, U) \to l^2(\sigma_0^{N-1}, U) \\ u &\mapsto (R(0)u_0, \ldots, R(N-1)u_{N-1}). \end{aligned} \qquad (42)$$

*Moreover the optimal cost is*

$$J(u_p) = \|g\|_{X_N}^2 + \langle K_0 x_0, x_0 \rangle \ . \qquad (43)$$

**Proof.** From the proposition 4 it is clear that if (1) is approximately controllable, the system (11) is it. Moreover since $d - \Phi^N x_0 \in V_P$, the problem $(E)$ satisfies the hypothesis of the proposition 6 and $v_E$ can be expressed by (37) and (38); by substitution in (26) (theorem 1) the expression (40) is obtained ■

## 4. CONCLUSION

This paper examined the linear quadratic optimal control of distributed discrete systems with final equality constraints. A closed relationship between the original problem and a minimum energy one was combined with the H.U.M approach to give an explicit expression of the optimal control. The numerical simulation and the inequality constraints case are under investigation.

## REFERENCES


Chraïbi, L., J.Karrakchou, A.Ouansafi and M.Rachik.(1998). Exact controllability for distributed systems with a discrete delayed control . *CIMASI'98-October 27-29, Casablanca-Morocco.*

Curtain, R. F.(1984). Linear-quadratic control problem with fixed endpoints in infinite dimensions. *J.Optimiz.Theor. Applics.* **44**, 55-74.

Emirsajlow, Z. (1989). A feedback for an infinite dimensional linear quadratic control problem with a fixed terminal state. *IMA J.Math. and Inf.* ,**6**, 97-117.

Karrakchou, J., M.Rachik and A.Namir (1995). Une nouvelle approche pour la contrôlabilité des systèmes à retard . *IFAC Conference on Systems Structure and Control. Nantes-France.*

Lee, K.Y., S.Chow and R.O Barr (1972). On the control of discrete-time distributed parameter systems. *SIAM J.Cont.,vol.***10,2,** 361-376.

14th Triennial World Congress, Beijing, P.R. China





# SINGULAR OPTIMAL CONTROL PROBLEM OF LINEAR SINGULAR SYSTEMS WITH LINEAR-QUADRATIC COST*

Chen Yilin Ma Shuping Cheng Zhaolin

*Department of Mathematics, Shandong University, Jinan 250100, P.R. China.*
*E-mail:Chengzha@jn-public.sd.cninfo.net*

**Abstract:** This paper investigates the singular linear-quadratic optimal control problem of a kind of linear time-invariant singular systems containing impulse mode. The relation between this problem and the linear-quadratic optimal control problem of standard state-space linear systems is established by using an equivalent transformation based on the Weierstrass canonical form of singular systems. The existence and uniqueness of the solution, the method to compute the optimal control and corresponding state trajectory, and the feedback synthesis of the optimal control of this problem is also given. *Copyright © 1999 IFAC*

**Keywords:** Linear-quadratic cost; optimal control; singular control; linear singular systems; Weierstrass canonical form.

## 1. INTRODUCTION

This paper considers the optimal control problem of the linear singular systems

$$E\dot{x} = Ax + Bu, \quad y = Cx \tag{1}$$

with initial condition

$$Ex(0^-) = \alpha_0 \tag{2}$$

and quadratic cost

$$\begin{aligned} J(u, x(0^-)) &= x^\tau(T)E^\tau SEx(T) \\ &+ \int_0^T (y^\tau(t)y(t) + u^\tau(t)Ru(t))dt \end{aligned} \tag{3}$$

where $E, A, S \in R^{n*n}$, $B \in R^{n*\tau}$, $C \in R^{m*n}$, $R \in R^{r*\tau}$, $E, A, B, C, S$ and $R$ are constant matrices, $S \geq 0, R \geq 0, E$ is singular. The pencil $sE - A$ is regular, i.e. det $(sE - A) \not\equiv 0$. Furthermore, we assume that the system (1) contains impulse mode, which is $E, A$ satisfying

$$degreedet(sE - A) < rankE. \tag{4}$$

For relating simplicity, denoting above linear-quadratic cost problem as problem $P$, solving $P$ is finding $u^*$ and $x^*(0^-)$ such that

$$J(u^*, x^*(0^-)) = \min_{\substack{u \in U \\ x(0^-) \in \Gamma(0^-)}} J(u, x(0^-)), \tag{5}$$

where $U$ is an admissible control set, $\Gamma(0^-) = \{x(0^-) : Ex(0^-) = \alpha_0\}$ is an initial manifold.

In the last two decades, there have been many excellent works about research for the problem $P$, such as Pandolfi(1981), Cobb(1983), Bender and Laub(1987) , Cheng and his cooperator(1988), Katayama and Minamino(1992). Now it can be said that the linear-quadratic cost optimal control theory of linear time-invariant singular systems for the case of $R > 0$ has been mature, and it has

* Research was supported by the National Nature Sciences Fundation of PRC and the Nature Sciences Fundation of Shandong Province, PRC.

been reduced to solving nonsingular linear-quadratic cost problem of standard state-space systems with the order of rank$E$, but for the case of $R \geq 0$, it is still an open problem. In fact, since singular systems contain impulse mode, the singular linear-quadratic cost optimal control problem of singular systems is more complicated than that of standard state-space linear systems, and there doesn't exist corresponding relation between both of them. An interesting problem is whether the singular linear-quadratic cost optimal control problem of linear singular systems will can be reduced to the nonsingular linear-quadratic cost problem of standard state-space linear systems or not. This paper tries to answer this problem.

## 2. PRELIMINARIES

Considering the system (1), notice that the pencil $sE - A$ is regular, there exist following transformations:

$$\left\{\begin{array}{l} PEQ = \begin{bmatrix} I_{n_1} & & \\ & J & \\ & & 0 \end{bmatrix}, \\ PAQ = \begin{bmatrix} A_1 & & \\ & I_{n_2} & \\ & & I_{n_3} \end{bmatrix}, \\ PB = [B_1^T, B_2^T, B_3^T]^T, \quad CQ = [C_1, C_2, C_3], \end{array}\right. \tag{6}$$

where $P, Q \in R^{n*n}$ are nonsingular matrices, $A_1 \in R^{n_1*n_1}$, $I_{n_i} \in R^{n_i*n_i}$ is a unit matrix, $B_i \in R^{n_i*r}$, $C_i \in R^{m*n_i}$, $i = 1, 2, 3$, $J \in R^{n_2*n_2}$ is a nilpotent Jordan matrix with the nolpotent index $\nu$, and the order of each Jordan block of $J$ is not less than 2. Denoting

$$x = Q\tilde{x}, \quad \tilde{x} = [x_1^T, x_2^T, x_3^T]^T, \tag{7}$$

in which $x_i \in R^{n_i}, i = 1, 2, 3$, then the following system

$$\left\{\begin{array}{ll} \dot{x}_1 & = A_1 x_1 + B_1 u, \quad J\dot{x}_2 = x_2 + B_2 u, \\ 0 & = x_3 + B_3 u, \\ y & = C_1 x_1 + C_2 x_2 + C_3 x_3 \end{array}\right. \tag{8}$$

is called the Weierstrass canonical form of system (1). Obviously, according to above transformations, the initial condition (2) transforms as

$$x_1(0) = \alpha_{01}, \quad Jx_2(0^-) = \alpha_{02}, \tag{9}$$

in which

$$\alpha_{01} = [I_{n_1}, 0, 0]P\alpha_0, \quad \alpha_{02} = [0, I_{n_2}, 0]P\alpha_0, \tag{10}$$

the cost (3) transforms as

$$\begin{aligned} J(u, x(0^-)) & = J_1(u, x_2(0^-)) \\ & = [x_1^T(T), x_2^T(T)]S_1[x_1^T(T), x_2^T(T)]^T \\ & \quad + \int_0^T (y^T(t)y(t) + u^T(t)Ru(t))dt, \end{aligned} \tag{11}$$

where

$$S_1 = \begin{bmatrix} I_{n_1} & 0 & 0 \\ 0 & J^T & 0 \end{bmatrix} (P^{-1})^T S P^{-1} \begin{bmatrix} I_{n_1} & 0 \\ 0 & J \\ 0 & 0 \end{bmatrix}. \tag{12}$$

Denoting the optimal control problem of system (8) with initial condition (9) and cost (11) as $P_1$, obviously, solving $P$ is equal to solving $P_1$, which is finding $u^*$ and $x_2^*(0^-)$ such that

$$J_1(u^*, x_2^*(0^-)) = \min_{\substack{u \in U \\ x_2(0^-) \in \Gamma_2(0^-)}} J_1(u, x_2(0^-)), \tag{13}$$

where $\Gamma_2(0^-) = \{x_2(0^-) : Jx_2(0^-) = \alpha_{02}\}$ is an initial manifold. Notice that the state trajectory of system (8) is:

$$\left\{\begin{array}{l} x_1(t) = e^{A_1 t}x_1(0) + \int_0^t e^{A_1(t-\tau)}B_1 u(\tau)d\tau, \\ x_2(t) = -\sum_{i=0}^{\nu-1} J^i B_2 u^{(i)}(t) - (\sum_{i=0}^{\nu-2} \delta^{(i)}(t)J^{i+1}) \\ \qquad (x_2(0^-) + \sum_{i=0}^{\nu-1} J^i B_2 u^{(i)}(0)), \\ x_3(t) = -B_3 u(t), \end{array}\right. \tag{14}$$

where $u^{(i)}(t)$ and $\delta^{(i)}(t)$ are the $i$th-order derivative of $u(t)$ and Dirac-$\delta$ function, i.e. $\delta(t)$, respectively. To assure that there is no impulsive behaviour in the system state trajectory, the initial values of control $u(t)$ and its derivatives must and only must satisfy the following condition:

$$Jx_2(0^-) = -\sum_{i=0}^{\nu-1} J^{i+1}B_2 u^{(i)}(0). \tag{15}$$

In accordance with the above view, this paper provides the admissible control set as

$$U = \{u : u \textit{ has up to the } \nu - 1th \textit{ order continuous derivative, and } u \textit{ satisfies } (15)\}. \tag{16}$$

Obviously, $U$ is defined by $Jx_2(0^-)$, but for any given $Jx_2(0^-)$, it is not clear whether $U$ is nonempty or not. For this we have:

*Lemma 1.* For any given $Jx_2(0^-), U$ is nonempty, if and only if:

$$rank[B_2, JB_2, \cdots, J^{\nu-1}B_2] = n_2. \tag{17}$$

*Lemma 2 (see Dai, 1989).* System (1) is impulse controllable, if and only if the equation (17) holds, or equivalently, matrix $[J, B_2]$ has full row rank.

*Lemma 3.* Matrix $[B_2, JB_2, \cdots, J^{\nu-1}B_2]$ has full column rank, if and only if the matrix $J^{\nu-1}B_2$ has full column rank.

*Lemma 4.* If system (1) is impulse controllable, and $J^{\nu-1}B_2$ is of full column rank, then

1) $r = \sigma$, i.e. the number of inputs of system (1) is equal to the number of the Jordan blocks of $J$; 2) each Jordan block of $J$ has the same order. Otherwise, if 1), 2) hold, and system (1) is impulse controllable, then $J^{\nu-1}B_2$ has full column rank.

*Lemma 5 (see Dai, 1989).* System (1) is impulse observable, if and only if

$$rank \begin{bmatrix} C_2 \\ C_2 J \\ \vdots \\ C_2 J^{\nu-1} \end{bmatrix} = n_2, \tag{18}$$

or equivalently, the matrix $[J^{\tau}, C_2^{\tau}]^{\tau}$ is of full column rank.

*Lemma 6.* If system (1) is impulse controllable, and $J^{\nu-1}B_2$ is of full column rank, then system (1) is impulse observable if and only if matrix $C_2 J^{\nu-1} B_2$ is of full column rank.

*Remark.* The state trajectory of singular system (1) contains impulsive behaviour or not, which is determined by the consistent condition (15) of initial values of both the control and the state trajectory. Lemma 1, 3 point out the necessary and sufficient condition of which the equation (15) has and only has a solution for any initial value $Jx_2(0^-)$,. Lemma 4 points out the relation between above necessary and sufficient condition and the geometric structure of coefficient matrices of the system. Lemma 6 gives a new equivalent description about impulse observability of singular systems according to above necessary and sufficient condition.

## 3. CASE OF FINITE TIME HORIZON

This section discusses the solution of problem $P$, in above section it has been shown that solving problem $P$ is equal to solving problem $P_1$. It is different to previous literature, in this section, the weighting matrix $R$ is allowed to be positive semi-definite. Denote

$$z(t) = \begin{bmatrix} u^{(\nu-2)}(t) \\ \vdots \\ u(t) \end{bmatrix}, \quad v(t) = u^{(\nu-1)}(t), \tag{19}$$

where $u(t)$ is an admissible control of problem $P$, $u(t)$ satisfies consistent condition (15). Using a dynamic model to describe $z(t)$, obviously, $z(t)$ satisfies

$$\dot{z}(t) = Dz(t) + Fv(t), \tag{20}$$

$$Hz(0) = -\alpha_{02}, \tag{21}$$

where

$$D = \begin{bmatrix} 0 & & & \\ I_r & \ddots & & \\ \vdots & \ddots & \ddots & \\ 0 & \cdots & I_r & 0 \end{bmatrix}, \quad F = \begin{bmatrix} I_r \\ 0 \\ \vdots \\ 0 \end{bmatrix},$$
$$H = [J^{\nu-1}B_2, \cdots, JB_2], \tag{22}$$

and $D \in R^{(\nu-1)r*(\nu-1)r}$, $F \in R^{(\nu-1)r*r}$, $I_r \in R^{r*r}$ is a unit matrix, $\alpha_{02}$ is defined by (9)-(10). By lemma 1 and lemma 3, it is clear that for any $\alpha_{02}$ the equation (21) has a unique solution $z(0)$ iif system (1) is impulse controllable, and $J^{\nu-1}B_2$ has full column rank. This leads us to stipulate the above characteristics for discussed system. Obviously, combining (19) and (14) yields

$$\begin{bmatrix} x_1(t) \\ x_2(t) \\ x_3(t) \\ u(t) \end{bmatrix} = \begin{bmatrix} I_{n1} & 0 & 0 \\ 0 & T_1 & -J^{\nu-1}B_2 \\ 0 & T_2 & 0 \\ 0 & T_3 & 0 \end{bmatrix} \begin{bmatrix} x_1(t) \\ z(t) \\ v(t) \end{bmatrix} \tag{23}$$

in which

$$\begin{aligned} T_1 &= -[J^{\nu-2}B_2, \cdots, B_2], \\ T_2 &= -[0, \cdots, 0, B_3], \\ T_3 &= [0, \cdots, 0, I_r] \end{aligned} \tag{24}$$

and $T_2 \in R^{n_3*(\nu-1)r}$, $T_3 \in R^{r*(\nu-1)r}$. Denoting the solution of (21) as

$$z(0) = h \tag{25}$$

According to (23), (21) and (25), rewrite the system (8), the initial condition (9) and the cost (11), we have

$$\begin{bmatrix} \dot{x}_1 \\ \dot{z} \end{bmatrix} = \begin{bmatrix} A_1 & G \\ 0 & D \end{bmatrix} \begin{bmatrix} x_1 \\ z \end{bmatrix} + Lv, \tag{26}$$

$$\begin{bmatrix} x_1(0) \\ z(0) \end{bmatrix} = \begin{bmatrix} \alpha_{01} \\ h \end{bmatrix}, \tag{27}$$

$$\begin{aligned} &J_1(u, x_2(0^-)) = J_2(v) \\ &= [x_1^{\tau}(T), z^{\tau}(T)] S_2 [x_1^{\tau}(T), z^{\tau}(T)]^{\tau} \\ &+ \int_0^T [x_1^{\tau}(t), z^{\tau}(t), v^{\tau}(t)] \tilde{Q} [x_1^{\tau}(t), z^{\tau}(t), v^{\tau}(t)]^{\tau} dt, \end{aligned} \tag{28}$$

where

$$\left\{ \begin{aligned} G &= [0, \cdots, 0, B_1], \quad L = [0, F^{\tau}]^{\tau}, \\ S_2 &= \begin{bmatrix} I_{n_1} & 0 & 0 \\ 0 & (JT_1)^{\tau} & 0 \end{bmatrix} (P^{-1})^{\tau} \\ &\quad SP^{-1} \begin{bmatrix} I_{n1} & 0 \\ 0 & JT_1 \\ 0 & 0 \end{bmatrix}, \\ \tilde{Q} &= \begin{bmatrix} Q_0 & Q_1 \\ Q_1^{\tau} & Q_2 \end{bmatrix}, \\ Q_0 &= \begin{bmatrix} C_1^{\tau}C_1 & C_1^{\tau}(C_2T_1 + C_3T_2) \\ (C_2T_1 + C_3T_2)^{\tau}C_1 + T_3^{\tau}RT_3 & (C_2T_1 + C_3T_2)^{\tau}(C_2T_1 + C_3T_2) \end{bmatrix}, \\ Q_1 &= -\begin{bmatrix} C_1^{\tau} \\ (C_2T_1 + C_3T_2)^{\tau} \end{bmatrix} C_2J^{\nu-1}B_2, \\ Q_2 &= (C_2J^{\nu-1}B_2)^{\tau} C_2J^{\nu-1}B_2 \end{aligned} \right. \tag{29}$$

and $G \in R^{n_1*(\nu-1)r}, L \in R^{(n_1+(\nu-1)r)*r}$. Denote the optimal control problem of system (26) with initial condition (27) and cost (28) as $P_2$, and denote the admissible control set as $V = \{v : v \text{ is a continuous function}\}$, obviously, $P_2$ is the optimal control problem of standard state-space linear systems. we analyzing the equivalence of problem $P_1$ to $P_2$.

*Definition 1.* The pair $(u(t), x(t))$ is said to be an admissible control-trajectory pair of the problem $P$ if $u(t)$ is an admissible control of $P$ and $x(t)$ is a state trajectory of system (1) drived by $u(t)$.

*Definition 2.* Support that there are two optimal control problems $P_k, k = 1,2$. Let $\theta_k$ be any addmissible control-trajectory pairs of $P_k$, $J_k$ be the cost of $P_k$, $S_k$ be the set formed by all of addmissible control-trajectory pairs of $P_k$. We say that $P_1$ is equivalent to $P_2$ if:
1) there exists a bijective mapping $F$ of $S_1$ onto $S_2$;
2) for any $\theta_1 \in S_1$, if $\theta_2 \in S_2$ is the image of $\theta_1$ by $F$, then

$$J_1(\theta_1) = J_2(\theta_2).$$

*Theorem 1.* Assume that the system (1) is impulse controllable, impulse observable, and $J^{\nu-1}B_2$ has full column rank, then problem $P$ is equal to $P_2$, and $P_2$ is a nonsingular linear-quadratic cost problem. Furthermore, the following proposition is estabilished: 1) $P$ has a unique solution; 2) The optimal control can be synthesized as linear feedback of the state; 3) The state trajectory of the optimal closed-loop system has no impulsive behaviour.

*Proof.* Obviously $P$ is equivalent to $P_1$. Next we prove that $P_1$ is equivalent to $P_2$. Denote admissible control-trajectory pair of $P_1$ and $P_2$ as $S_1$ and $S_2$ respectively, noticing that the transform of $S_2$ to $S_1$ described by (23) is surjective, and the transform matrix is of full column rank, we can see that the transform is both injective and surjective ( or bijective mapping of $S_2$ to $S_1$). According to definition 2, $P_1$ is equivalent to $P_2$, hence $P$ is equivalent to $P_2$. It has pointed out that $P_2$ is linear-quadratic cost problem of standard state-space linear system (26) with the initial condition (27) and cost (28). Since the system (1) is impulse controllable and impulse observable, and the matrix $J^{\nu-1}B_2$ is of full column rank, by lemma 7 we know that the submatrix $Q_2$ of weighting matrix $\tilde{Q}$ in cost (28) is nonsingular, therefore $P_2$ is a nonsingular linear-quadratic cost problem. Testing the state dimensions of system (26), by lemma 4, it is clear that the state dimensions of system (26) is of

$$rank E = n_1 + (\nu - 1)r \tag{30}$$

Solving $P_2$ is well known, next we solve $P$ by solving $P_2$:

1)Denoting the optimal control-trajectory pair of $P_2$ as $(v^*(t), [x_1^{*\tau}(t), z^{*\tau}(t)]^\tau)$, by the optimal control theory (see Anderson and Moore, 1971), we have

$$v^*(t) = -Q_2^{-1}(Q_1^\tau + L^\tau\Phi(t))[x_1^{*\tau}(t), z^{*\tau}(t)]^\tau \tag{31}$$

where $\Phi(t)$ satisfies Riccati differential equation:

$$\begin{aligned}\dot{\Phi}(t) = & -\Phi(t)(\begin{bmatrix} A_1 & G \\ 0 & D \end{bmatrix} - LQ_2^{-1}Q_1^\tau) \\ & -(\begin{bmatrix} A_1^\tau & 0 \\ G^\tau & D^\tau \end{bmatrix} - Q_1Q_2^{-1}L^\tau)\Phi(t) \\ & +\Phi(t)LQ_2^{-1}L^\tau\Phi(t) - (Q_0 - Q_1Q_2^{-1}Q_1^\tau)\end{aligned} \tag{32}$$

with the terminal condition

$$\Phi(T) = S_2, \tag{33}$$

$[x_1^{*\tau}(t), z^{*\tau}(t)]^\tau$ satisfies:

$$\begin{bmatrix} \dot{x}_1(t) \\ \dot{z}(t) \end{bmatrix} = (\begin{bmatrix} A_1 & G \\ 0 & D \end{bmatrix} - LQ_2^{-1}(Q_1^\tau + L^\tau\Phi(t))) \begin{bmatrix} x_1(t) \\ z(t) \end{bmatrix} \tag{34}$$

with the initial condition (25), and the optimal cost is

$$J_2(v^*) = [\alpha_{01}^\tau, h^\tau]\Phi(0)[\alpha_{01}^\tau, h^\tau]^\tau \tag{35}$$

where $A_1, G, D, L, Q_i, i = 0,1,2, S_2, \alpha_{01}$ and $h$ are given as (22), (25)-(29) respectively.

2)Denoting the optimal control-trajectory pair of $P$ as $(u^*(t), x^*(t))$, by (7) and (23) we have

$$\begin{bmatrix} x^*(t) \\ u^*(t) \end{bmatrix} = \begin{bmatrix} Q & 0 \\ 0 & I_r \end{bmatrix} \begin{bmatrix} I_{n_1} & 0 & 0 \\ 0 & T_1 & -J^{\nu-1}B_2 \\ 0 & T_2 & 0 \\ 0 & T_3 & 0 \end{bmatrix} \begin{bmatrix} x_1^*(t) \\ z^*(t) \\ v^*(t) \end{bmatrix}, \tag{36}$$

and the optimal cost is

$$J(u^*, x^*(0^-)) = J_1(u^*, x_2^*(0^-)) = J_2(v^*). \tag{37}$$

3)Optimal closed-loop synthesis. Denoting

$$M(t) = \begin{bmatrix} I_{n_1} & 0 \\ 0 & T_1 \\ 0 & T_2 \end{bmatrix} + \begin{bmatrix} 0 \\ J^{\nu-1}B_2 \\ 0 \end{bmatrix} Q_2^{-1}(Q_1^\tau + L^\tau\Phi(t)), \tag{38}$$

it is easy to prove that the following lemma holds.

*Lemma 7.* If system (1) is impulse controllable and impulse observable, and $J^{\nu-1}B_2$ is of full column rank, then for any $t \in [0,T]$, $M(t)$ has full column rank.

Using lemma 7 and (36), (31),(24) we have

$$u^*(t) = T_0[M^T(t)M(t)]^{-1}M^T(t)Q^{-1}x^*(t) \quad (39)$$

where

$$T_0 = [0, \cdots, 0, I_r], \quad (40)$$

and $T_0 \in R^{r*(n_1+(\nu-1)r)}$. Denoting

$$K(t) = T_0[M^T(t)M(t)]^{-1}M^T(t)Q^{-1}, \quad (41)$$

then the optimal closed-loop system of $P$ is obtained, that is

$$E\dot{x}^*(t) = (A + BK(t))x^*(t) \quad (42)$$

with the initial value

$$x^*(0) = QM(0)\begin{bmatrix} \alpha_{01} \\ h \end{bmatrix}. \quad (43)$$

It is easy to know that

$$Ex^*(0) = \alpha_0, \quad (44)$$

i.e., $x^*(0)$ belongs to the initial manifold $\Gamma(0^-)$.

*Example 1.* Solving the following linear-quadratic optimal cost control problem in the cases of both $r = 1$ and $r = 0$:

$$\begin{bmatrix} 1 & 0 \\ 0 & 0 \end{bmatrix}\begin{bmatrix} \dot{x}_1 \\ \dot{x}_2 \end{bmatrix} = \begin{bmatrix} 0 & 1 \\ 1 & 0 \end{bmatrix}\begin{bmatrix} x_1 \\ x_2 \end{bmatrix} + \begin{bmatrix} 0 \\ 1 \end{bmatrix}u,$$

$$\begin{bmatrix} 1 & 0 \\ 0 & 0 \end{bmatrix}\begin{bmatrix} x_1(0^-) \\ x_2(0^-) \end{bmatrix} = \begin{bmatrix} 1 \\ 0 \end{bmatrix},$$

$$J = \frac{1}{2}x_1^2(1) + \frac{1}{2}\int_0^1 (x_2^2(t) + ru^2(t))dt.$$

Testing the above system, it is evident that the system satisfies all conditions demanded by theorem 1 whether $r = 1$ or $r = 0$. Hence we can solve it by using the method shown in the proof of theorem 1.

## 4. CASE OF INFINITE-TIME HORIZON

This section is to discuss the case of $T = \infty$, that is, to discuss the following optimal control problem of system (1) with initial condition (2) and linear-quadratic cost

$$J_0(u, x(0^-)) = \int_0^\infty (y^T(t)y(t) + u^T(t)Ru(t))dt \quad (45)$$

where $R$ is stipulated for $R \geq 0$. Denoting this problem as $P_0$, and the set of admissible control-trajectory pair of $P_0$ as $U$. Assume that $U$ is defined as (16), and system (1) is impulse controllable, impulse observable, and $J^{\nu-1}B_2$ is of full column rank. By (23) and definition 2, it is evident that the problem $P_0$ is equivalent to the following optimal control problem of system (26) with initial condition (27) and linear-quadratic cost

$$\tilde{J}_0(v) = \int_0^\infty [x_1^T(t), z^T(t), v^T(t)]\tilde{Q} [x_1^T(t), z^T(t), v^T(t)]^T dt \quad (46)$$

Where $\tilde{Q}$ is defined as (29), and the submatrix $Q_2$ of $\tilde{Q}$ is nonsingular, the set of admissible control-trajectory pairs is defined as $V = \{v : v \ is \ continous \ function\}$. Denote the latter as $\tilde{P}_0$, Obviously, $\tilde{P}_0$ is nonsingular linear-quadratic cost problem of standard state-space linear systems. It is well known that existence and uniquness of the solution to $\tilde{P}_0$ are dependent on the stabilizability of system (26), using PBH criterion to observe system (26), we obtain that the stabilizability of system (26) is equivalent to the R-stabilizability of system (1). Thus we have:

*Theorem 2.* If system (1) is impulse controllable, impulse observable, R-stabilizable, and $J^{\nu-1}B_2$ is of full column rank, then problem $P_0$ is equivalent to a nonsingular linear-quadratic cost problem of stand state-space linear systems with state dimensions equalled to rank$E$, and: 1) The solution of $P_0$ is unique; 2) The optimal control can be synthesized as a constant gain state feedback; 3) The state trajectory of the system drived by optimal the optimal control doesn't contain any impulse behaviour.

For computation of the optimal control-trajectory pair, the feedback synthesis of optimal control, the feedback gain matrix, and the optimal cost of problem $P_0$ can use the formulas (31), (34)-(36), and (38)-(43). It is only needed to use these formulas to replace $\Phi(t)$ with $\Phi$, where $\Phi$ is a positive semi-definite solution of the following Riccati algebra equation:

$$(\begin{bmatrix} A_1^T & 0 \\ G^T & D^T \end{bmatrix} - Q_1Q_2^{-1}L^T)\Phi + Q_0 - Q_1Q_2^{-1}Q_1^T$$

$$+\Phi(\begin{bmatrix} A_1 & G \\ 0 & D \end{bmatrix} - LQ_2^{-1}Q_1^T) - \Phi LQ_2^{-1}L^T\Phi = 0 \quad (47)$$

Notice that the conditions of theorem 2 is not sufficient for the asymptotic stability of optimal closed-loop system of problem $P_0$, however, for the case of infinite-time horizon, it can not be neglected to assure the asymptotic stability of the optimal closed-loop system. Next we try to give the sufficient condition to assure the asymptotic stability of the optimal closed-loop system. Denote

$$A_0 = \begin{bmatrix} A_1 & G \\ 0 & D \end{bmatrix} - LQ_2^{-1}Q_1^{\tau},\ \tilde{Q}_0 = Q_0 - Q_1Q_2^{-1}Q_1^{\tau}. \quad (48)$$

By the equivelance of both $P_0$ and $\tilde{P}_0$, and (36), it is clear that the asymptotic stability of optimal closed-loop system of $P_0$ is identified with that of $\tilde{P}_0$, i.e.,the asymptotic stability of optimal closed-loop system of $P_0$ is determined by detectability of the matrix pair $(A_0, \tilde{Q}_0)$. Analysing the pair $(A_0, \tilde{Q}_0)$, under the following assumptions

$$span(C_1) \subseteq span(C_2J^{\nu-1}B_2). \quad (49)$$

$$W^{\tau}(s)Q_3W(s) + R > 0,\ \forall s, s \in C, Res \geq 0 \quad (50)$$

where

$$W(s) = C_2(sJ - I_{n_2})^{-1}B_2 - C_3B_3 \quad (51)$$

is the transfer matrix of the fast subsystem of system (1). by complicated computing, Using the PBH criterion, we can obtain that the pair $(A_0, \tilde{Q}_0)$ is detectable if and only if the system (1) is R- detectable. So we have:

*Theorem 3.* If system (1) is impulse controllable, impulse observable, R-stabilizable, R-detectable, $J^{\nu-1}B_2$ is of full column rank, and (49), (50) hold, then the optimal closed-loop system of $P_0$ is asymptotically stable.

## 5. CONCLUSIONS

This paper has discussed the singular optimal control problem of linear singular systems with linear-quadratic cost. Based on the Weierstrass canonical form of singular systems and the idear of partitioning optimal control problems into equivalence classes, this paper has pointed out that

1) If the system is impulse controllable, impulse observable, and the matrix $J^{\nu-1}B_2$ is of full column rank, then the singular linear-quadratic cost control problem of linear singular systems is equivalent to the nonsingular linear-quadratic cost problem of standard state-space linear systems, and the state dimensions of corresponding linear system are reduced to $rankE$.

2) If the system is controllable and the matrix $J^{\nu-1}B_2$ is of full column rank, but the system is not impulse observable, then the singular linear-quadratic cost optimal control problem of linear singular systems is equivalent to the singular linear-quadratic cost optimal control problem of standard state-space linear systems, and the state dimensions of corresponding linear system are reduced to $rankE$.

For computation of singular linear-quadratic cost optimal control problem of singular systems, this paper has focused on discussing the kind of singular systems which has impulse controllability, impulse observability, and $J^{\nu-1}B_2$ has full column rank. In both the case of finite-time horizon and the case of infinite-time horizon, the following results are given, which include the existence and uniqueness of the solution, the computing method of the solution, and the feedback synthesis of the optimal control. In particular, for the case of infinite-time horizon, the asymptotic stability of the optimal closed-loop system is discussed and some sufficient condition and necessary condition are proposed. Furthermore, the relation between the asymptotic stability of the optimal colsed-loop system and the transfer matrix of the fast subsystem of singular system is also revealed.

## REFERENCES


Anderson, B.D.O. and J.B. Moore (1971). *Linear Optimal Control.* Prentice-Hall, New Jersey.

Bender, D. J. and A. J. Laub (1987). The linear linear-quadratic optimal regulator for descriptor systems. *IEEE Trans. Aut. Contr.* **AC-32,** 672-688.

Cheng, Z., H. Hong, and J. Zhang (1988). The optimal regulation of generalized state-space systems with linear-quadratic cost. *Automatica, 24,* 707-710.

Clements, D. J. and B. D. O. Anderson (1978). *Singular Optimal Control: The Linear-Quadratic Problem.* Springer-Verlag, New York.

Cobb, D.(1983). Descriptor variable systems and optimal state regulation. *IEEE Trans. Auto. Contr.,* **AC-28**, 601-611.

Dai, L. (1989). *Singular Control Systems.* Springer-Verlag, New York.

Katayama, T. and K. Minamino (1992). Linear quadratic regulator and spectral factorization for continuous-time descriptor systems. Proc. 31st IEEE Conf. Decision and Control, Tucson, Arizona,USA, December, 967-972.

Pandolfi, T. (1981). On the regulator problem for linear degenerate control systems. *J. Optimiz. Theory Applic.,* **33** , 241-254.

# THE OPTIMAL CONTROL OF PROCESSING SYSTEMS BY ECONOMICAL CRITERIA AS APPLIED TO DISTILLATION

**Krivosheev,V.P., Torgashov,A.Y.**

*Computer Systems and Technologies Department, Vladivostok State University of Economics and Service, 41 Gogolya street, Vladivostok, 690600,Russia,*
*Andrei.Torgashov@vvsu.ru.*

Abstract: This paper considers a solution of the task of optimal control of a complex processing systems by economical criteria such as cost, profit and output of final product. At the first stage, the optimal values of controlled process parameters are defined by using step by step steady-state optimization method designed by the authors and taking into account features of economical criteria. At the second stage, a maintenance of suggested parameters is implied by using digital minimum variance controller. *Copyright © 1999 IFAC*

Keywords: Optimization, Steady-state, Process control, Optimal control, Minimum variance control.

## 1. INTRODUCTION

The economical effect of production is the main index of operating processing system (PS). Therefore, present paper seeks to develop optimal digital control system which has economical criterion instead of the objective functional. Authors propose the approach for achievement of the stated problem, which divide the initial statement of problem into two subproblems: steady-state and dynamic optimization. This division is justified by complexity of solving problem of optimization in initial statement by such methods as calculus of variations, dynamic programming in continuous form, principle of maximum as well as ,in practice, sometimes it is difficult to obtain the adequate mathematical description of the PS.

The paper reveals essence and aspects of steady-state and dynamic optimizations consistently demonstrating necessary conditions for the corresponding subproblems. These conditions are then applied for synthesis of optimal control actions realized on digital devices fairly simply. Successful application of the method proposed is illustrated by example.

## 2. STATEMENT OF PROBLEM AND METHODS OF ITS SOLUTION

Let PS consist of $N$ basic processing plants. Hence, the general form of statement of problem can be presented by the following mathematical notation

$$J^*(\mathbf{x}_{b0}(t)) = \min \int_0^t F(\mathbf{x}_{b0}(t), \mathbf{u}_1(t), \dots, \mathbf{u}_N(t), \mathbf{y}_1(t), \dots, \mathbf{y}_N(t))dt, \quad (1)$$

$$\mathbf{u}_1(t), \dots, \mathbf{u}_N(t) \in \Omega$$

$$\mathbf{x}_b(0) = \mathbf{x}_{b0}^0; \mathbf{u}_i(0) = \mathbf{u}_i^0; \mathbf{y}_i(0) = \mathbf{y}_i^0; \mathbf{y}_i(t_k) = \mathbf{y}_i^k;$$

$$V_i \frac{\partial y_{is}}{\partial l_s} + \frac{\partial y_{is}}{\partial t} = f_{is}(\mathbf{x}_{bi}(t), \mathbf{u}_i(t), \mathbf{y}_i(t)); \quad (2)$$

$$x_{i\mu} = y_{i-1,\mu}; \mu = 1, \dots, l; \quad (3)$$

$$y_{i\mu} \geq y_{i,\mu}^{zd}; \mu = 1, \dots, \sigma; \quad (4)$$

$$y_{i\xi} \le y_{i\xi}^{zd}; \xi = k,...,\chi; \quad (5)$$

$$u_{ij}^{\min} \le u_{ij} \le u_{ij}^{\max}; j = 1,...,r, \quad (6)$$

where $\mathbf{x}_{b0}(t)$ - vector of disturbances; $\mathbf{u}_i(t)$ - vector of manipulated variables; $\mathbf{y}_i(t)$ - vector of output; $V_i$ - linear speed of substance flow; $i=1,...,N$; $s=1,...,n$. The criterion of optimization, as a functional (1), characterizes effectiveness of PS when PS is changed over from one state to other under the influence of disturbances and manipulated variables. The restriction (2) represents mathematical description of process taking place in the $i$-th element of PS. The restriction (3) defines interconnection between input parameters of following plant and output parameters of previous plant. Restrictions (4), (5) reflect the requirements of process rule to output and intermediate magnitudes of parameters of PS. Restrictions (6) define a capability of variations of manipulated variables.

The given problem (1)-(6) is difficult to solve since it has large dimension and restrictions. Therefore, that problem falls into two subordinate subproblems. Each subproblem has its own objective. The aim of the first subproplem is the steady-state optimization with a single criterion of optimization for overall PS. The aim of the second subproblem is dynamic optimization. It consists of the shift of plant into optimal steady-state over minimum interval of time.

## 3.STEADY-STATE OPTIMIZATION

The well-known methods solving the problem of steady-state optimization with large dimension such as dynamic programming (Bellman, 1961) and discrete principle of maximum (Boyarinov and Kafarov, 1975) are not universal. Their application is limited by dimension of input vector. The authors have developed a step by step method of steady-state optimization which has no restrictions on dimension of input vector. The method is based on a module optimization principle at the first step and correction of derived values of variable parameters for achieving optimal magnitude of overall criterion of PS at the second step. Let us consider the algorithm of optimization problem solution for element of PS as module with particular criterion of optimization. The element of PS means the equipment in which processing takes place, for example, distillation column, absorber, extractor, reactor, and etc.. These modules are defined by the designation of each of the element of PS.

The following modules can be used for distillation columns: purity maximization of the initial mixture separation into fractions; maximization of extracting particular product of desired quality; minimization of consumption of power used for obtaining the product of the desired quality. The modules for chemical reactors can be such as: selectivity maximization under the given degree of conversion (with uncomplete conversion); desired product output maximization (with complete conversion).

To define the necessary conditions of extremum for economical criteria of PS a selection factor $\eta$ can be introduced. This factor characterizes the specific selection of the product or semi-product of the desired purity with the mainstream flow in reference to feed supplied to a separate element of PS (EPS)

$$\eta_i = F_i Z_{ib} / (F_{i-1} Z_{Nb}^{zd}), \quad (7)$$

where $F_i$, $F_{i-1}$ - final and feed flow rate of $i$-th EPS, respectively; $Z_{ib}$ - concentration of intermediate or basic product in the final flow of $i$-th EPS; $Z_{Nb}^{zd}$ - given maximum permissible concentration of objective component in the marketable product. The block diagram below can be considered as a series of connected elements along mainstream flow (Fig.1).

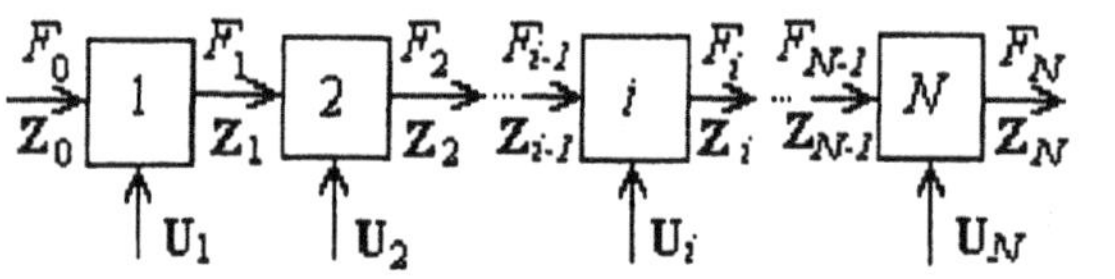


Fig.1. The block diagram of series connection between EPS.

The criterion of optimization can be represented as

$$Q = Q(F_N, E), \quad (8)$$

where $F_N$ - flow rate of marketable product; $E$- production expenditure of $F_N$. For example, the problem of optimization is maximization (8), i.e.

$$Q^*(F_0, \mathbf{Z}_0) = \max_{\mathbf{U}_1,...,\mathbf{U}_N \in \Omega} Q(F_N, E), \quad (9)$$

where $\mathbf{Z}_0 = \{Z_{01},...,Z_{0m}\}$ - vector of flow rate of feed in the first EPS; $\mathbf{U}_i = \{U_{i1},...,U_{ir}\}$ - vector of manipulated variables for $i$-th EPS. $\Omega$ is defined by the following conditions

$$F_i = \psi_i(F_{i-1}, \mathbf{Z}_{i-1}, \mathbf{U}_i), \quad (10)$$

$$\mathbf{Z}_i = \phi_i(F_{i-1}, \mathbf{Z}_{i-1}, \mathbf{U}_i), \quad (11)$$

$$Z_{i\xi} = Z_{i\xi}^{zd}, \quad (12)$$

$$U_{i\mu}^{\min} \le U_{i\mu} \le U_{i\mu}^{\max}, \quad (13)$$

where $i=1,...,N$; $\xi$- some from $s=1,...,m$ or sum from some $s$; $j= 1,...,r$.

The condition (12) determines the restrictions for the quality of products, namely, its performance results in satisfying requirements on concentrations of all other components in output flows. Using relationship (7), $F_N$ can be represented as

$$F_N = F_0 \prod_{i=1}^{N} \eta_i (Z_{Nb}^{zd} / Z_{ib}) . \qquad (14)$$

The Lagrangian function can be applied as follows:

$$\overline{Q} = Q(F_N, E) + \sum_{i=1}^{N} \lambda_{i\xi} (Z_{i\xi} - Z_{i\xi}^{zd}), \qquad (13)$$

and the necessary conditions will be stated for extremum of function (15) assuming that $U_{i\mu}$ affects on parameters of $i$-th step as

$$\frac{\partial \overline{Q}}{\partial U_{i\mu}} = \frac{\partial \overline{Q}}{\partial E} \cdot \frac{\partial E}{\partial U_{i\mu}} + \frac{\partial \overline{Q}}{\partial \eta_i} \cdot \frac{\partial \eta_i}{\partial U_{i\mu}} + \frac{\partial \overline{Q}}{\partial \eta_i} \cdot \frac{\partial \eta_i}{\partial Z_{ib}} \times \frac{\partial Z_{ib}}{\partial U_{i\mu}} + \frac{\partial \overline{Q}}{\partial Z_{ib}} \cdot \frac{\partial Z_{ib}}{\partial U_{i\mu}} + \lambda_{i\xi} \frac{\partial Z_{i\xi}}{\partial U_{i\mu}} = 0. \qquad (16)$$

Thus, the desired manipulated variables solving the set of equations (16) under conditions (12) can be obtained. The particular economical criteria such as cost, profit and output of final product under the given cost must be considered. The cost is represented by

$$C = E / F_N , \qquad (17)$$

where $E = P_0 F_0 + \sum_{i=1}^{N} \sum_{\nu=1}^{S} P_{\nu i} q_{\nu i} - \sum_{i=1}^{N} \sum_{\lambda=1}^{L} P_{\lambda i} f_{\lambda i} + K$ ; $P_0$ - price of the unit of raw material; $P_{\nu i}$ -price of the unit of energy flow $q_{\nu i}$ ; $P_{\lambda i}$ - price of the unit of secondary product $f_{\lambda i}$; $K$ - direct component of expenditure in production of marketable product $F_N$. The values $q_{\nu i}$, $f_{\lambda i}$ can be considered as manipulated variables

$$\mathbf{U}_i = \{q_{1i}, \ldots, q_{Si}, f_{1i}, \ldots, f_{Li}\} . \qquad (18)$$

Taking account of (14), the (17) can be written as:

$$C = E \Big/ F_0 \prod_{i=1}^{N} \eta_i \frac{Z_{Nb}^{zd}}{Z_{ib}} . \qquad (19)$$

Let us suppose that condition (12) takes place in general case for any $i$. Then, the Lagrangian function can be formulated as

$$\overline{C} = E \Big/ F_0 \prod_{i=1}^{N} \eta_i \frac{Z_{Nb}^{zd}}{Z_{ib}} + \sum_{i=1}^{N} \lambda_{i\xi} (Z_{i\xi} - Z_{i\xi}^{zd}). \qquad (20)$$

The necessary conditions of extremum for the function (20) can be derived assuming that $U_{i\mu}$ influences only the parameters of the $i$-th EPS

$$\frac{\partial \overline{C}}{\partial U_{i\mu}} = \frac{\partial E}{\partial U_{i\mu}} \Big/ F_0 \prod_{i=1}^{N} \eta_i \frac{Z_{Nb}^{zd}}{Z_{ib}} - \frac{\partial \eta_i}{\partial U_{i\mu}} \times \frac{E}{\eta_i} \Big/ F_0 \prod_{i=1}^{N} \eta_i \frac{Z_{Nb}^{zd}}{Z_{ib}} + \lambda_{i\xi} \frac{\partial Z_{i\xi}}{\partial U_{i\mu}} = 0 \qquad (21)$$

The conditions (21) can be written in view of (14) as

$$\frac{\partial E}{\partial U_{i\mu}} \Big/ F_N - \frac{\partial \eta_i}{\partial U_{i\mu}} \cdot \frac{E}{\eta_i} \Big/ F_N + \lambda_{i\xi} \frac{\partial Z_{i\xi}}{\partial U_{i\mu}} = 0 \qquad (22)$$

or

$$\frac{\partial E}{\partial U_{i\mu}} = \frac{E}{\eta_i} \cdot \frac{\partial \eta_i}{\partial U_{i\mu}} - \lambda_{i\xi} F_N \frac{\partial Z_{i\xi}}{\partial U_{i\mu}} . \qquad (23)$$

Passing to the increments and multiplying all terms of (22) by $\Delta U_{i\mu} / E$, the following result can be obtained

$$\frac{\Delta E}{E} = \frac{\Delta \eta_i}{\eta_i} - \lambda_{i\xi} \frac{F_N}{E} \Delta Z_{i\xi} . \qquad (24)$$

The profit is considered as

$$R = (P - C) F_N . \qquad (25)$$

Substituting $F_N$ into (25) from (14) and $C$ from (17)

$$R = P \cdot F_0 \cdot \prod_{i=1}^{N} \eta_i (Z_{Nb}^{zd} / Z_{ib}) - E . \qquad (26)$$

Forming the Lagrangian function as

$$\overline{R} = PF_0 \prod_{i=1}^{N} \eta_i \frac{Z_{Nb}^{zd}}{Z_{ib}} - E + \sum_{i=1}^{N} \lambda_{i\xi} (Z_{i\xi} - Z_{i\xi}^{zd}), \qquad (27)$$

the necessary conditions for extremum of function (27) can be written as

$$\frac{\partial \overline{R}}{\partial U_{i\mu}} = \frac{PF_0}{\eta_i} \times \prod_{k=1}^{N} (\eta_k Z_{Nb}^{zd} / Z_{ib}) (\partial \eta_i / \partial U_{i\mu}) - \partial E / \partial U_{i\mu} + \lambda_{i\xi} (\partial Z_{i\xi} / \partial U_{i\mu}) = 0. \qquad (28)$$

Then, conditions (28) may be rewritten in view of (14) as

$$\frac{PF_N}{\eta_i} \cdot \frac{\partial \eta_i}{\partial U_{i\mu}} - \frac{\partial E}{\partial U_{i\mu}} + \lambda_{i\xi} \frac{\partial Z_{i\xi}}{\partial U_{i\mu}} = 0. \qquad (29)$$

Passing to the increments and multiplying all terms of (22) by $\Delta U_{i\mu} / (PF_N)$ the following equation has been obtained

$$\Delta E / (PF_N) = \Delta \eta_i / \eta_i + \lambda_{i\xi} \Delta Z_{i\xi} / (PF_N) . \qquad (30)$$

$F_N$ is considered as criterion of optimization

$$F_N = F_0 \prod_{i=1}^{N} \eta_i (Z_{Nb} / Z_{ib}) , \qquad (31)$$

$$C = C^{zd}. \tag{32}$$

Forming the Lagrangian function

$$\overline{F}_N = F_0 \cdot \prod_{i=1}^{N} \eta_i (Z_{Nb}^{zd} / Z_{ib}) + \\ + \sum_{i=1}^{N} \lambda_{i\xi} (Z_{i\xi} - Z_{i\xi}^{zd}) + \lambda (E / F_N - C^{zd}) \tag{33}$$

the necessary conditions of extremum for function (33) have form

$$\frac{\partial \overline{F}_N}{\partial U_{i\mu}} = \frac{F_N}{\eta_i} \cdot \frac{\partial \eta_i}{\partial U_{i\mu}} + \lambda_{i\xi} \frac{\partial Z_{i\xi}}{\partial U_{i\mu}} + \\ + \lambda \left( F_N \cdot \frac{\partial E}{\partial U_{i\mu}} - \frac{EF_N}{\eta_i} \cdot \frac{\partial \eta_i}{\partial U_{i\mu}} \right) \Big/ F_N^2 = 0 \tag{34}$$

Passing to the increments and multiplying all terms of (34) by $\Delta U_{i\mu} / F_N$ the following result cam be obtained

$$\frac{\Delta E_i}{E} = \frac{\Delta \eta_i}{\eta} \left( 1 - \frac{F_N^2}{\lambda \cdot E} \right) - \frac{\lambda_{i\xi}}{\lambda} \cdot \frac{F_N}{E} \Delta Z_{i\xi}. \tag{35}$$

Let us consider the calculating aspects of the method distinguishing two stages. At the 1-st stage, each EPS is assigned by module and added by additional conditions (22), (29), (34) for cost, profit and output, respectively with due account of restrictions (12), (32) depending on general criterion for overall PS. Then beginning with the 1-st EPS, the problems of optimization for conditional extremum are solved by selected modules with corresponding conditions (22), (29), (34) for all EPS under given $F_0$, $Z_0$. If the analytical solution is impossible for the problem of optimization for each of EPS, the optimization is performed by using numerical methods for search extremum of generalized particular criteria in the form of penalty function as it was shown by Boyarinov and Kafarov (1975). In this case, the selected module is added by the components corresponding to only one of the necessary conditions (24), (30), (35) of extremum which is general for criterion PS, namely :

1) if general criterion is cost, then

$$\overline{H}_{pi} = H_{pi} \pm \alpha_i (\Delta E_i / E - \Delta \eta_i / \eta_i)^2 \pm \\ \pm \beta_{i\xi} (Z_{i\xi} - Z_{i\xi}^{zd})^2 ; \tag{36}$$

2) if general criterion is profit, then

$$\overline{H}_{pi} = H_{pi} \pm \alpha_i (\Delta E_i / (PF_N) - \Delta \eta_i / \eta_i)^2 \pm \\ \pm \beta_{i\xi} (Z_{i\xi} - Z_{i\xi}^{zd})^2 ; \tag{37}$$

3) if general criterion is output under given cost, then

$$\overline{H}_{pi} = H_{pi} \pm \alpha_i (\Delta \eta_i / \eta_i - \Delta Z_{ib} / Z_{ib}) \pm \\ \pm \beta_{i\xi} (Z_{i\xi} - Z_{i\xi}^{zd})^2 \pm \gamma (E / F_N - C^{zd}) . \tag{38}$$

The multipliers $\alpha$, $\beta$, $\gamma$ are coefficients of the penalty in expressions (36)-(38). The plus sign in the conditions (36)-(38) is assumed if the particular criterion $H_{pi}$ must be minimized. The minus sign can be assumed in the opposite case.

The second stage of optimization is necessary by virtue of the assumption which made when deriving the necessary conditions (24), (30), (35): the power consumption is assumed to be changed only at the $i$-th EPS when $U_{i\mu}$ is varied. The second stage incorporates the estimate of sensitivity of the criterion of optimization PS after the first stage as well as correction of manipulated variables $U_{i\mu}$ obtained at the fist stage. The sensitivity will be insignificant if the following condition

$$\left| \Delta Q(F_0, \mathbf{Z}_0, \mathbf{U}_1^*, \ldots, \mathbf{U}_N^*) / \Delta U_{i\mu} \right| < \varepsilon_{i\mu} \tag{39}$$

takes place under varying the $U_{i\mu}$ in domain of the feasible manipulated variables. The value $\varepsilon_{i\mu}$ can be assumed from condition

$$\left| \Delta \overline{U}_{i\mu} \partial Q(F_0^L, \mathbf{Z}_0^L, \mathbf{U}_1^L, \ldots, \mathbf{U}_N^L) / \partial U_{i\mu} \right| = \varepsilon_{i\mu}, \tag{40}$$

where $\partial Q(F_0^L, \mathbf{Z}_0^L, \mathbf{U}_1^L, \ldots, \mathbf{U}_N^L) / \partial U_{i\mu}$ is taken with respect to the start point of local optimization for the $i$-th EPS as

$$\Delta \overline{U}_{i\mu} = S_{U_{i\mu}} A_{U_{i\mu}} / 100 , \tag{41}$$

where $S_{U_{i\mu}}$ - accuracy class of the transducer for measurement $U_{i\mu}$; $A_{U_{i\mu}}$ - range of measurement of the parameter $U_{i\mu}$. If the condition (39) is no satisfied for any $U_{i\mu}$ then a correction of manipulated variables is implemented, for example, by the methods of directed search of extremum for criterion $Q$.

## 4.DYNAMIC OPTIMIZATION

By a dynamic optimization is meant a shift of the plant (in that case - EPS) from one steady-state to another over the minimum time. It is assumed that the each EPS has ARMA model in the neighborhood of operating point. As it was shown by Isermann (1981) and Puzirev (1992) the minimum variance controllers had advantages in comparison with other kinds of control devices under conditions when stochastic disturbances take place. The synthesis of the multivariable minimum variance controller was considered by Isermann (1981). The application of these controllers is basic innovation in achieving extremums of particular criteria (36)-(38) during solving stated the problem.

## 5.EXAMPLE

The distillation plant (Fig.2) is taken for illustration of synthesizing optimal digital control system. The following notations are adopted in Fig.2: $l_1$, $l_2$- number of the light key component in K-1 and K-2, respectively; $x_{Di}$, $x_{wi}$ - concentration of the light key component in distillate and bottom product; $F_2$- flow rate of final product. Moreover, the following restrictions

$$\sum_{i=1}^{l1} x_{D1}^{i} \geq \sum_{i=1}^{l1} x_{D1}^{izd},\ \sum_{i=1}^{l2} x_{D2}^{i} \geq \sum_{i=1}^{l2} x_{D2}^{izd},\ \sum_{i=1}^{l2} x_{w2}^{i} \leq \sum_{i=1}^{l2} x_{w2}^{izd}$$

are imposed on compositions of products, where *zd* denotes the given concentration of component.

Suppose that the economical criterion is given as cost (13). The following modules can be specified for PS as: for K-1–maximization of extracting particular product of desired quality; for K-2–minimization of consumption of power used for obtaining the product of the desired quality. Then, the steady-state optimization is implemented for each column in accordance with the above mentioned equations (18), (36). The results are depicted in the figure Fig.3 in which index «*» denotes the optimal steady-state and «*b*» denotes the basic steady-state.

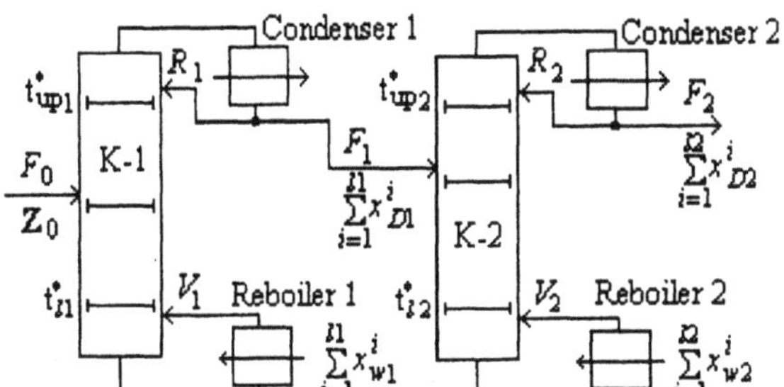


Fig.2. Considered PS.

The optimal regime parameters of columns are final result of steady-state optimization. These parameters are temperatures in check points (top and bottom part of column) of K-1 and K-2.

The aim of dynamic optimization is the maintenance of the parameters $t^{*}_{up1}, t^{*}_{l1}$ and $t^{*}_{up2}, t^{*}_{l2}$ (temperatures in the top and bottom part of columns K-1 and K-2). The synthesis of optimal control law is given only for K-1 because it may be found similarly for K-2. The distillation column K-1 has two controlled parameters – $t^{*}_{up1}, t^{*}_{l1}$ and two manipulated variables – $R_1$, $V_1$ (reflux, vapor boilup, respectively).

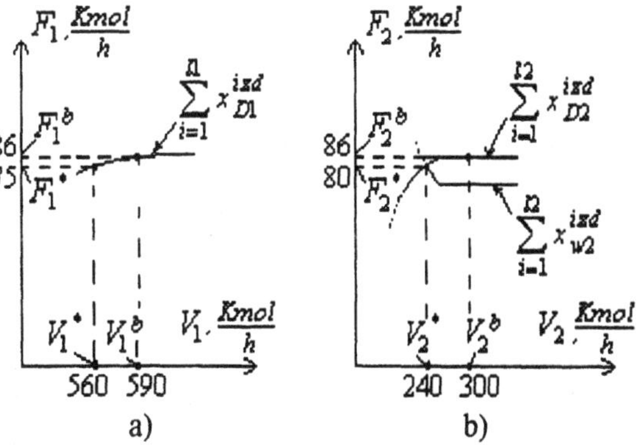


Fig. 3. The results of steady-state optimization: a)- for K-1; b)- for K-2.

Using the identification procedure the following model for K-1 can be derived as

$$\mathbf{y}(t)+\begin{pmatrix} a_1^1 & 0 \\ 0 & a_1^2 \end{pmatrix}\mathbf{y}(t-1)+\begin{pmatrix} a_2^1 & 0 \\ 0 & a_2^2 \end{pmatrix}\mathbf{y}(t-2)=$$
$$=\begin{pmatrix} b_{11}^0 & b_{12}^0 \\ b_{21}^0 & b_{22}^0 \end{pmatrix}\mathbf{u}(t)+\begin{pmatrix} b_{11}^1 & b_{12}^1 \\ b_{21}^1 & b_{22}^1 \end{pmatrix}\mathbf{u}(t-1)+$$
$$+\begin{pmatrix} b_{11}^2 & b_{12}^2 \\ b_{21}^2 & b_{22}^2 \end{pmatrix}\mathbf{u}(t-2)+\xi(t)+\begin{pmatrix} c_1 & 0 \\ 0 & c_2 \end{pmatrix}\xi(t-1)$$

where $a_1^1$=-2.88; $a_2^1$=2.76; $a_3^1$=-0.8835; $a_1^2$=-2.88; $a_2^2$=2.765; $a_3^2$=-0.8838; $b_{11}^0$=-0.01932; $b_{11}^1$=0.0367; $b_{11}^2$=-0.017; $b_{12}^0$=0.0025; $b_{12}^1$=-0.0057; $b_{12}^2$=0.0025; $b_{21}^0$=-0.008; $b_{21}^1$=0.014; $b_{21}^2$=-0.0068; $b_{22}^0$=0.003293; $b_{22}^1$=-0.006521; $b_{22}^2$=0.003228; $c_1$=-0.2; $c_2$=-0.8.

The minimum variance strategy for K-1 is represented by following equation

$$\begin{pmatrix} u_1(t) \\ u_2(t) \end{pmatrix}=\frac{1}{f_{11}f_{22}-f_{12}f_{22}}\times$$
$$\times\begin{pmatrix} f_{22}\left[y_1^d(1+c_1q^{-1})-y_1(f_0^1+f_1^1q^{-1}+f_2^1q^{-2})\right]- \\ f_{11}\left[y_2^d(1+c_2q^{-1})-y_2(f_0^2+f_1^2q^{-1}+f_2^2q^{-2})\right]- \\ -f_{12}\left[y_2^d(1+c_2q^{-1})-y_2(f_0^2+f_1^2q^{-1}+f_2^2q^{-2})\right] \\ -f_{21}\left[y_1^d(1+c_1q^{-1})-y_1(f_0^1+f_1^1q^{-1}+f_2^1q^{-2})\right] \end{pmatrix},$$

where $f_0^1$=2.68; $f_1^1$=-2.76; $f_2^1$=0.8835; $f_0^2$=2.081; $f_1^2$=-2.76; $f_2^2$=0.883; $u_1(t)=R_1(t)$; $u_2(t)=V_1(t)$; $y_1(t)=t_{up1}(t)$; $y_2(t)=t_{l1}(t)$; $y_1^d(t)=t^{*}_{up1}(t)$;

$f_{11}=b_{11}^0+b_{11}^1q^{-1}+b_{11}^2q^{-2}$; $f_{12}=b_{12}^0+b_{12}^1q^{-1}+b_{12}^2q^{-2}$;

$$f_{21} = b_{21}^0 + b_{21}^1 q^{-1} + b_{21}^2 q^{-2}; \quad f_{22} = b_{22}^0 + b_{22}^1 q^{-1} + b_{22}^2 q^{-2};$$
$$y_2^d(t) = t_{l1}^*(t).$$

The examination of transient processes shows that multivariable minimum variance controller provides a high quality of control within present cross channels and extends the area of step-by-step method of steady-state optimization which was developed by the authors.

## 6.CONCLUSION

It was shown that the application of the proposed approach improved the efficiency of the controlled process by means of stabilization of calculated optimal values of parameters PS with using minimum-variance controller. This is illustrated by example of the 10% of energy consumption decrease in the distillation processes using the steady-state and dynamic optimizations.

## REFERENCES

Bellman R. (1961). *Adaptive control processes: a guided tour*, Princeton University Press, Princeton, New Jersey.

Boyarinov, A.I. and V.V. Kafarov (1975). *The methods of optimization in chemical technology*, Himiya, Moscow.

Isermann, R. (1981). *Digital control systems.* Springer-Verlag, Berlin.

Puzirev, V.A. (1992). *Microprocessor-based self-tuning controllers*, Inergoatomizdat, Moscow.

14th Triennial World Congress, Beijing, P.R. China



# OPTIMAL INVENTORY-PRODUCTION CONTROL PROBLEM WITH STOCHASTIC DEMAND

**E.K. Boukas** *,[1] **P. Shi A. Andijani R.K. Agarwal**

* *To whom correspondance should be addressed:*
*Professor, Mechanical Engineering Department,*
*École Polytechnique de Montréal,*
*P.O. Box 6079, Succ. "Centre-ville", Montréal (Qc) Canada,*
*H3C 3A7. Email: boukas@anas.meca.polymtl.ca.*

Abstract: This paper deals with the inventory-production control problem where the produced items are supposed to be deteriorating with a rate that depends on the stochastic demand rate. The inventory-production control problem is formulated as a jump linear quadratic control problem. The optimal policy that solves the optimal control problem is obtained in terms of a set of coupled Riccati equations. The guaranteed cost control problem is also investigated. 

Keywords: Production systems, Inventory control, Stochastic process, Markov process, Jump linear quadratic control, Guaranteed cost control problem.

## 1. INTRODUCTION

In this paper, we study the problem of continuous-time inventory-production control with deteriorating rate of the items, which has been a subject being extensively investigated for the last ten years. For more information regarding the contributions, we refer the reader to Haiping and Wang (1990) and Hamid (1989) and the references therein.

The goal of this paper is to determine the optimal production policy that reduces the cost of the facility under consideration. Under some reasonable assumptions, we give the closed form solution of the optimal policy of the optimization problem in terms of a set of coupled Riccati equations. The uncertain model of the inventory control problem is also proposed where the uncertainties on parameters are supposed to be of the norm bounded type. The guaranteed cost inventory control problem is also investigated.

[1] Research of this author was supported by the Natural Sciences and Engineering Research Council of Canada under grants OGP0036444.

The paper is organized as follows. In section 2, the problem is stated and the assumptions are formulated. In the first part of section 3, we will focus our attention on the case where the demand rate is modeled by a continuous-time stochastic Markov process with finite state space and the deteriorating rate is constant and independent on the demand rate. In the second part of section 3, we will consider the case where the demand is stochastic but the deteriorating rate is depending on the demand rate. In a similar way, the control policy will be established. In section 4, the guaranteed cost control inventory-production problem is investigated.

## 2. PROBLEM STATEMENT

Let us assume that we have a production system that produces one item. Let the production rate of this facility be denoted by $u(t) \in \mathbb{R}_+$. Let $\nu(t)$ be a continuous-time Markov process with finite state space denoted by $\mathcal{S} = \{1, 2, \ldots, n_s\}$. Let the demand rate of the produced item be a stochastic demand rate and denote it by $d(\nu(t)) \in D \subset \mathbb{R}_+$.

For each value of the process $\nu(t)$ at time $t$, $t \in [t_n, t_{n+1}]$, where $t_n$ is the instant of the $n$th jump of the process $\nu(t)$, the demand rate will take a constant value, denoted by $d(\nu(t))$. Let us assume that the deterioration rate depends of the demand rate $d(\nu(t))$, i.e. $\gamma(d(\nu(t)))$. For instance, let $d_1$ and $d_2$ be two levels of the demand rate, with $d_2 > d_1$, then $\gamma(d_2) < \gamma(d_1)$. The evolution of the stochastic process $\{\nu(t), t \geq 0\}$ that determines the demand rate is then described by the following probability transitions:

$$P[\nu(t+h) = \beta | \nu(t) = \alpha] = \begin{cases} q_{\alpha\beta} h + o(h), & \text{if } \alpha \neq \beta \\ 1 + q_{\alpha\alpha} h + o(h), & \text{otherwise} \end{cases} \quad (1)$$

with $q_{\alpha\beta} \geq 0$ for all $\alpha \neq \beta$ and $q_{\alpha\alpha} = -\sum_{\beta \in \mathcal{S}} q_{\alpha\beta}$ for all $\alpha \in \mathcal{S}$, and $\lim_{h\to 0} \frac{o(h)}{h} = 0$.

The differential equation that describes the evolution of the inventory of our facility is therefore given by:

$$\dot{x}(t) = -\gamma(d(\nu(t))) I_{\{x(t)\geq 0\}} x(t) + u(t) - d(\nu(t)), x(0) = x_0, \nu(0) = \alpha \quad (2)$$

where $I_{\{P(x)\}}$ is the index function defined as follows:

$$I_{\{P(x)\}} = \begin{cases} 1 & \text{if } P(x) \text{ is true} \\ 0 & \text{otherwise} \end{cases}$$

and $\gamma(d(\alpha))$ is constant when $\nu(t) = \alpha$, $x(0) = x_0$ and $\nu(0) = \alpha$ are respectively the initial values of the state and the mode at time $t = 0$.

Our objective in this paper is to seek an admissible feedback control law that minimizes the following cost function:

$$J_u(x_0, \alpha, 0) = \mathsf{E}\left[\int_0^T [c_1 x^2(t) + c_2 u^2(t)] \, dt | x(0) = x_0, \nu(0) = \alpha\right] \quad (3)$$

where $c_1$ ($c_1 \neq 0$) is the inventory holding cost when $x(t)$ is positive or the shortage cost when $x(t)$ is negative, $c_2$ ($c_2 \neq 0$) is the production unit cost and $\mathsf{E}$ stands for the mathematical expectation operator.

## 3. INVENTORY-PRODUCTION CONTROL PROBLEM

Let us assume that the deteriorating rate $\gamma(\nu(t))$, in Eq. (2) is constant and does not depend on the demand rate $d(\nu(t))$. The demand rate, $d(\nu(t))$ is the same as in section 2.

*Definition 3.1.* A control $u(\cdot) = \{u(t) : t \geq 0\}$ is said to be *admissible* if: (i) $u(\cdot)$ is adapted to the $\sigma$-algebra generated by the random process $\nu(\cdot)$, denoted as $\sigma\{\xi(s) : 0 \leq s \leq t\}$, and (ii) $u(t) \in \mathbb{R}_+$ for all $t \geq 0$.

Let $\mathcal{U}$ denote the set of all admissible controls of our control problem.

*Definition 3.2.* A measurable function $u(x(t), \nu(t))$ is an admissible *feedback* control, if (i) for any given initial $x$ and $\alpha$ of the continuous state and the mode, the following equations have an unique solution $x(\cdot)$:

$$\dot{x}(t) = -\gamma(d(\nu(t)) I_{\{x(t)\geq 0\}} x(t) + u(x(t), \nu(t)) - d, \; x(0) = x \quad (4)$$

and (ii) $u(\cdot) = u(x(\cdot), \nu(\cdot)) \in \mathcal{U}$.

It can be shown that the HJB equation is given by:

$$\frac{\partial v}{\partial t}(x(t), \nu(t), t) + \min_{u(t)} \Big[ [-\gamma I_{\{x(t)\geq 0\}} x(t) + u(t) - d(\nu(t))]^\top \frac{\partial v}{\partial x}(x(t), \nu(t), t) + \sum_{\beta \in \mathcal{S}} q_{\nu(t)\beta} v(x(t), \beta, t) + c_1 x^2(t) + c_2 u^2(t) \Big] = 0 \quad (5)$$

Since the control is unconstrained, it results that the optimal control law is given by:

$$u^\star(t) = -\frac{1}{2c_2} \frac{\partial v}{\partial x}(x(t), \nu(t), t) \quad (6)$$

Hence, substituting this control, $u^\star(t)$, into the HJB equation, we have the following first-order partial differential equation for the value function:

$$\frac{\partial v}{\partial t}(x(t), \nu(t), t) - \frac{1}{4c_2}\left[\frac{\partial v}{\partial x}(x(t), \nu(t), t)\right]^2 + [-\gamma I_{\{x(t)\geq 0\}} x(t) - d(\nu(t))] \frac{\partial v}{\partial x}(x(t), \nu(t), t) + \sum_{\beta \in \mathcal{S}} q_{\nu(t)\beta} v(x(t), \beta, t) + c_1 x^2(t) = 0 \quad (7)$$

To solve this partial differential equation, we try the following expression for the value function $v(x(t), \nu(t), t)$:

$$v(x(t), \nu(t), t) = p(\nu(t), t) x^2(t) + q(\nu(t), t) x(t) + r(\nu(t), t) \quad (8)$$

where $p(\nu(t),t)$, $q(\nu(t),t)$ and $r(\nu(t),t)$ are time-varying positive real functions depending on $\nu(t)$ and $t$. They remain to be determined.

Based on this expression, and after computing the expressions of $\frac{\partial v}{\partial t}(x(t),\nu(t),t)$ and $\frac{\partial v}{\partial x}(x(t),\nu(t),t)$ and substituting them into (7), and using the fact that the corresponding HJB equation is equal to 0 for all $x(t)$ in $\mathbb{R}$, one obtains:

$$\begin{cases} \begin{cases} \dot{p}(\alpha,t) - 4\gamma I_{\{x(t)\geq 0\}}p(\alpha,t) - \frac{3}{c_2}p^2(\alpha,t) \\ \quad + \sum_{\beta\in\mathcal{S}} q_{\alpha\beta}p(\beta,t) + c_1 = 0, \\ p(\alpha,T) = 0 \end{cases} \\ \begin{cases} \dot{q}(\alpha,t) - 2\gamma I_{\{x(t)\geq 0\}}q(\alpha,t) \\ \quad -\frac{3}{c_2}p(\alpha,t)q(\alpha,t) + \sum_{\beta\in\mathcal{S}} q_{\alpha\beta}q(\beta,t) \\ \quad -4d(\alpha)p(\alpha,t) = 0, \\ q(\alpha,T) = 0 \end{cases} \\ \begin{cases} \dot{r}(\alpha,t) - \frac{3}{4c_2}q^2(\alpha,t) - 2d(\alpha)q(\alpha,t) \\ \quad + \sum_{\beta\in\mathcal{S}} q_{\alpha\beta}r(\beta,t) = 0, \\ r(\alpha,T) = 0 \end{cases} \end{cases} \tag{9}$$

for all $\alpha \in \mathcal{S}$

Therefore, if there exists a set of positive real functions $p(\alpha,t)$ and $q(\alpha,t)$ satisfying the coupled Riccati equations (9), then (7) has a set of solutions $v(x(t),\nu(t),t)$. Consequently, there exists an optimal control law of the form (6) that solves our stochastic optimization problem (3). To summarize the above, we have the following theorem.

*Theorem 3.1.* If the set of coupled Riccati equations (9) has positive real function, $p(\alpha,t)$ and $q(\alpha,t)$, for all $\alpha \in \mathcal{S}$, then the linear quadratic control problem of (1)-(3) has an optimal solution given by the following expression:

$$u(t) = -\frac{1}{2c_2}\left[p(\alpha,t)x(t) + q(\alpha,t)\right], \ \forall\alpha \in \mathcal{S} \tag{10}$$

To get the corresponding optimality conditions for our problems, let us now apply this result to our case in which:

$$f(x(t),u(t),\nu(t)) = -\gamma(d(\nu(t)))I_{\{x(t)\geq 0\}}x(t) + u(t) - d(\nu(t)) \tag{11}$$

$$g(x(t),u(t),\nu(t)) = c_1x^2(t) + c_2u^2(t) \tag{12}$$

For this problem, it can be shown that the optimality conditions are given by:

$$\frac{\partial v}{\partial t}(x(t),\nu(t),t) + \min_{u(t)}\Big[[-\gamma(d(\nu(t)))I_{\{x(t)\geq 0\}}x(t) + u(t) - d(\nu(t))]^{\top}\frac{\partial v}{\partial x}(x(t),\nu(t),t) + \sum_{\beta\in\mathcal{S}} q_{\nu(t)\beta}v(x(t),\beta,t) + c_1x^2(t) + c_2u^2(t)\Big] = 0 \tag{13}$$

Since the control is unconstrained, it results that the optimal control law is given by:

$$u^{\star}(t) = -\frac{1}{2c_2}\frac{\partial v}{\partial x}(x(t),\nu(t),t) \tag{14}$$

Hence, substituting this control, $u^{\star}(t)$, into the HJB equation, we have the following first-order partial differential equation for the value function:

$$\frac{\partial v}{\partial t}(x(t),\nu(t),t) - \frac{1}{4c_2}\left[\frac{\partial v}{\partial x}(x(t),\nu(t),t)\right]^2 + \left[-\gamma(\nu(t))x(t)I_{\{x(t)\geq 0\}} - d(\nu(t))\right]\frac{\partial v}{\partial x}(x(t),\nu(t),t) + \sum_{\beta\in\mathcal{S}} q_{\nu(t)\beta}v(x(t),\beta,t) + c_1x^2(t) = 0 \tag{15}$$

To solve this partial differential equation, we try the same expression for the value function $v(x(t),\nu(t),t)$ as we did in the previous and use the same steps. Then based on Eq. (15), one obtains:

$$\begin{cases} \begin{cases} \dot{p}(\alpha,t) - 4\gamma(d(\alpha))I_{\{x(t)\geq 0\}}p(\alpha,t) \\ \quad -\frac{3}{c_2}p^2(\alpha,t) + \sum_{\beta\in\mathcal{S}} q_{\alpha\beta}p(\beta,t) + c_1 = 0, \\ p(\alpha,T) = 0 \end{cases} \\ \begin{cases} \dot{q}(\alpha,t) - 2\gamma(d(\alpha))I_{\{x(t)\geq 0\}}q(\alpha,t) \\ \quad -\frac{3}{c_2}p(\alpha,t)q(\alpha,t) + \sum_{\beta\in\mathcal{S}} q_{\alpha\beta}q(\beta,t) \\ \quad -4d(\alpha)p(\alpha,t) = 0, \\ q(\alpha,T) = 0 \end{cases} \\ \begin{cases} \dot{r}(\alpha,t) - \frac{3}{4c_2}q^2(\alpha,t) - 2d(\alpha)q(\alpha,t) \\ \quad + \sum_{\beta\in\mathcal{S}} q_{\alpha\beta}r(\beta,t) = 0 \\ r(\alpha,T) = 0 \end{cases} \end{cases} \tag{16}$$

for all $\alpha \in \mathcal{S}$.

Therefore, if there exists a set of positive real functions $p(\alpha,t)$ and $q(\alpha,t)$ for $\alpha \in \mathcal{S}$, satisfying the coupled Riccati equations (16), then (15) has a set of solutions $v(x(t),\nu(t),t)$. Consequently, there exists an optimal control law of the form (10) to solve our stochastic optimization problem (3). To summarize the above, we have the following theorem.

*Theorem 3.2.* If the set of coupled Riccati equations (16) have positive real functions, $p(\alpha,t)$ and $q(\alpha,t)$, for all $\alpha \in \mathcal{S}$, then the linear quadratic control problem of (1)-(3) has an optimal solution given by the following expression:

$$u(t) = -\frac{1}{2c_2}\left[p(\alpha,t)x(t) + q(\alpha,t)\right], \ \alpha \in \mathcal{S}. \tag{17}$$

*Remark 3.1.* Since in practice the production rate, is all the time positive, the optimal production rate, $u^*(t)$, is then given by:

$$u^*(t) = \max\left(0, -\frac{1}{2c_2}\frac{\partial v}{\partial x}(x(t), \nu(t), t)\right)$$

□

Let us now deal with the infinite horizon optimization problem by letting $T$ goes to infinite. In this case we need more assumptions to guarantee the existence of the optimal solution. These assumptions can be reduced to the stochastic stabilizability and the stochastic detectability. For more information regarding these concepts, we refer the reader to Ji and Chizeck (1990) and the references therein. Before introducing the definitions of stochastic stability and stochastic stabilizability, let us put our optimization problem in more general formulation with multiple states, multiple modes and multiple production rates. This formulation is given by the following system of differential equations:

$$\dot{\mathbf{x}}(t) = \mathbf{A}(\nu(t))\mathbf{x}(t) + \mathbf{B}(\nu(t))\left[\mathbf{u}(t) - \mathbf{d}(\nu(t))\right], \quad \mathbf{x}(0) = \mathbf{x}_0, \nu(0) = \alpha \tag{18}$$

where $\mathbf{x}(t) \in \mathbb{R}^n$ is state, $\mathbf{u}(t) \in \mathbb{R}^n_+$ is control input, $\mathbf{d}(\nu(t)) \in \mathbb{R}^n_+$ is the demand vector being a function of $\nu(t)$ which is the same as in Section 2. $\mathbf{A}(\nu(t))$ and $\mathbf{B}(\nu(t))$ are matrix functions of $\nu(t)$ with the form

$$\mathbf{A}(\nu(t)) = \begin{cases} \begin{bmatrix} -\gamma_1(\nu(t)) & 0 & 0 & \cdots & 0 \\ 0 & -\gamma_2(\nu(t)) & 0 & \cdots & 0 \\ \vdots & \vdots & \vdots & \ddots & \vdots \\ 0 & 0 & 0 & \cdots & -\gamma_n(\nu(t)) \end{bmatrix} & \text{if } x \geq 0 \\ 0 & \text{otherwise.} \end{cases}$$

$$\mathbf{B}(\nu(t)) = \begin{bmatrix} 1 & 0 & 0 & \cdots & 0 \\ 0 & 1 & 0 & \cdots & 0 \\ \vdots & \vdots & \vdots & \ddots & \vdots \\ 0 & 0 & 0 & \cdots & 1 \end{bmatrix} = \mathbf{I}_n$$

where $\mathbf{I}_n$ is the identity matrix of dimension $n$.

Associated with system (18), we consider the following cost function

$$J_{\mathbf{u}}(\mathbf{x}_0, \alpha, 0) = \mathsf{E}\left[\int_0^\infty [\mathbf{x}^\top \mathbf{R}_1(\alpha)\mathbf{x}(t) + \mathbf{u}^\top(t) \mathbf{R}_2(\alpha)\mathbf{u}(t)]\, dt | \mathbf{x}(0) = \mathbf{x}_0, \nu(0) = \alpha\right], \tag{19}$$

where $\mathbf{R}_1(\alpha)$ and $\mathbf{R}_2(\alpha)$, $\alpha \in \mathcal{S}$ are symmetric and positive definite matrices. There are diagonal matrices.

Let $\mathbf{x}(t, \mathbf{x}_0, \nu(0))$ denote the trajectory of the state $\mathbf{x}(t)$ (stock level at time $t$) from the initial state $(\mathbf{x}_0, \nu(0))$. We introduce the following stochastic stability and stochastic stabilizability concepts for continuous-time jump linear systems.

*Definition 3.3.* For system (18) with $\mathbf{u}(t) \equiv 0$ and $\mathbf{d}(\nu(t)) \equiv 0$, for all $\nu(t) \in \mathcal{S}$, the equilibrium point 0 is *stochastically stable*, if for every initial state $(\mathbf{x}_0, \nu_0)$

$$\int_0^{+\infty} \mathsf{E}\left\{\|\mathbf{x}(t, \mathbf{x}_0, \nu_0)\|^2\right\} dt < +\infty.$$

*Definition 3.4.* We say the system (18) is *stochastically stabilizable*, if for every initial state $(\mathbf{x}_0, \nu(0))$, there exists a linear feedback control law $\mathbf{u}(t) = \mathbf{L}(\nu(t))\mathbf{x}(t) + \mathbf{d}(\nu(t))$, $\nu(t) = \alpha$, such that the closed-loop system

$$\dot{\mathbf{x}}(t) = [\mathbf{A}(\nu(t)) + \mathbf{L}(\nu(t))]\, \mathbf{x}(t)$$

is stochastically stable.

If we assume that the system is stochastically stabilizable and stochastically detectable, then we can prove that the optimal control law exists and it is given by solving a set of coupled Riccati equations similar to the one we give for the case of one item (see Ji and Chizeck, 1990).

## 4. GUARANTEED COST CONTROL

The class of uncertain systems under consideration is described by the state equations

$$\begin{aligned} \dot{\mathbf{x}}(t) &= [\mathbf{A}(\nu(t)) + \Delta\mathbf{A}(\nu(t))]\, \mathbf{x}(t) \\ &\quad + [\mathbf{B}(\nu(t)) + \Delta\mathbf{B}(\nu(t))]\, (\mathbf{u}(t) - \mathbf{d}(\nu(t))), \\ &\quad \mathbf{x}(0) = \mathbf{x}_0, \nu(0) = \alpha \end{aligned} \tag{20}$$

where $\mathbf{x}(t) \in \mathbb{R}^n$ is state, $\mathbf{u}(t) \in \mathbb{R}^n$ is control input, $\mathbf{d}(\nu(t)) \in \mathbb{R}^n$ is another input being a function of $\nu(t)$ which is the same as in Section 2. $\mathbf{A}(\nu(t))$ and $\mathbf{B}(\nu(t))$ are matrix functions of $\nu(t)$ with the same forms as in the previous section, $\Delta\mathbf{A}(\nu(t))$ and $\Delta\mathbf{B}(\nu(t))$ are unknown matrices which represent time-varying parameter uncertainties and assumed to belong to certain bounded compact sets being the form:

$$\Delta\mathbf{A}(\nu(t)) = \begin{cases} \begin{bmatrix} \Delta\gamma_1(\nu(t)) & 0 & 0 \cdots & 0 \\ 0 & \Delta\gamma_2(\nu(t)) & 0 \cdots & 0 \\ \vdots & \vdots & \vdots \ddots & \vdots \\ 0 & 0 & 0 \cdots & \Delta\gamma_n(\nu(t)) \end{bmatrix} \\ \quad \text{if } \mathbf{x} \geq 0 \\ \mathbf{0} \quad \text{otherwise} \end{cases}$$

$$\Delta\mathbf{B}(\nu(t)) = \begin{bmatrix} \Delta b_1(\nu(t)) & 0 & 0 \cdots & 0 \\ 0 & \Delta b_2(\nu(t)) & 0 \cdots & 0 \\ \vdots & \vdots & \vdots \ddots & \vdots \\ 0 & 0 & 0 \cdots & \Delta b_n(\nu(t)) \end{bmatrix}.$$

The admissible parameter uncertainties are assumed to be of the following forms

$$\begin{aligned} \Delta\mathbf{A}(\nu(t)) &= \mathbf{H}(\nu(t))\mathbf{F}(\nu(t),t)\mathbf{E}_1(\nu(t)), \\ \Delta\mathbf{B}(\nu(t)) &= \mathbf{H}(\nu(t))\mathbf{F}(\nu(t),t)\mathbf{E}_2(\nu(t)) \end{aligned} \quad (21)$$

where $\mathbf{H}(\nu(t)), \mathbf{E}_1(\nu(t))$ and $\mathbf{E}_2(\nu(t))$, for any $\nu(t) = \alpha$, $\alpha \in \mathcal{S}$, are known constant matrices of appropriate dimensions and $\mathbf{F}(\nu(t),t)$ is an unknown time-varying matrix satisfying

$$\mathbf{F}^\top(\nu(t),t)\mathbf{F}(\nu(t),t) \leq \mathbf{I},\ \forall \nu(t) \in \mathcal{S}. \quad (22)$$

Associated with system (20), we consider the following cost function

$$J_{\mathbf{u}}(\mathbf{x}_0,\alpha,0) = \mathsf{E}\Bigg[\int_0^\infty [\mathbf{x}^\top \mathbf{R}_1(\alpha)\mathbf{x}(t) + \mathbf{u}^\top(t)$$

$$\mathbf{R}_2(\alpha)\mathbf{u}(t)]\,dt|\mathbf{x}(0) = \mathbf{x}_0, \nu(0) = \alpha\Bigg], \quad (23)$$

where $\mathbf{R}_1(\alpha)$ and $\mathbf{R}_2(\alpha)$, $\alpha \in \mathcal{S}$ are symmetric and positive definite matrices.

Motivated by the results in Feng *et al.* (1992) and Petersen (1987), we also introduce the following quadratic stability concept for uncertain jumping system (20).

*Definition 4.1.* The system (20) (with $\mathbf{u}(t) \equiv 0$ and $\mathbf{d}(\nu(t)) \equiv 0$) is said to be stochastically quadratically stable if and only if there exists a set of symmetric matrices $\{\mathbf{P}(\alpha) > 0,\ \alpha \in \mathcal{S}\}$ satisfying the following coupled matrix inequalities:

$$\Big[\mathbf{A}(\nu(t)) + \mathbf{H}(\nu(t))\mathbf{F}(\nu(t),t)\mathbf{E}_1(\nu(t))\Big]^\top$$
$$\mathbf{P}(\alpha) + \mathbf{P}(\alpha)\Big[\mathbf{A}(\nu(t)) + \mathbf{H}(\nu(t))\mathbf{F}(\nu(t),t)$$
$$\mathbf{E}_1(\nu(t))\Big] + \sum_{\beta \in \mathcal{S}} q_{\alpha\beta}\mathbf{P}(\beta) < 0$$

for all admissible uncertainties $\mathbf{F}(\nu(t),t)$ in (21)-(22).

The following result is from Costa and Boukas (1997).

*Proposition 4.1.* If system (20) (with $\mathbf{u}(t) \equiv 0$ and $\mathbf{d}(\nu(t)) \equiv 0$) is stochastically quadratically stable, then it is stochastically stable.

*Lemma 4.1.* Given matrices $\mathbf{Q},\mathbf{H},\mathbf{E}$ and $\mathbf{R}$ of appropriate dimensions and with $\mathbf{Q}$ and $\mathbf{R}$ symmetric and $\mathbf{R} > 0$, then

$$\mathbf{Q} + \mathbf{H}FE + \mathbf{E}^\top\mathbf{F}^\top\mathbf{H}^\top < 0$$

for all $\mathbf{F}$ satisfying $\mathbf{F}^\top\mathbf{F} \leq \mathbf{R}$, if and only if there exists some $\varepsilon > 0$ such that

$$\mathbf{Q} + \varepsilon^2\mathbf{H}H^\top + \varepsilon^{-2}\mathbf{E}^\top\mathbf{R}E < 0.$$

*Lemma 4.2.* Given matrices $\mathbf{H},\mathbf{E}$ and $\mathbf{R}$ of appropriate dimensions with $\mathbf{R} > 0$ symmetric, then there exists some $\varepsilon > 0$ such that

$$\mathbf{H}FE + \mathbf{E}^\top\mathbf{F}^\top\mathbf{H}^\top \leq \varepsilon^2\mathbf{H}H^\top + \varepsilon^{-2}\mathbf{E}^\top\mathbf{R}E$$

for all $\mathbf{F}$ satisfying $\mathbf{F}^\top\mathbf{F} \leq \mathbf{R}$.

The following result can be derived immediately by using Lemmas 4.1 and 4.2.

*Theorem 4.1.* The system (20) (with $\mathbf{u}(t) \equiv 0$ and $\mathbf{d}(\nu(t)) \equiv 0$) is stochastically quadratically stable if and only if there exist a sequence of $\{\varepsilon_\alpha > 0,\ \alpha \in \mathcal{S}\}$ and a set of symmetric and positive definite matrices $\{\mathbf{P}(\alpha),\ \alpha \in \mathcal{S}\}$ such that the following inequality

$$\mathbf{A}^\top(\alpha)\mathbf{P}(\alpha) + \mathbf{P}(\alpha)\mathbf{A}(\alpha) + \sum_{\beta \in \mathcal{S}} q_{\alpha\beta}\mathbf{P}(\beta)$$
$$+\varepsilon_\alpha^2\mathbf{P}(\alpha)\mathbf{H}(\alpha)\mathbf{H}^\top(\alpha)\mathbf{P}(\alpha)$$
$$+\varepsilon_\alpha^{-2}\mathbf{E}^\top(\alpha)\mathbf{E}(\alpha) < 0$$

holds for all $\alpha \in \mathcal{S}$.

*Definition 4.2.* A control law $\mathbf{u}(t) = \mathbf{L}(\alpha)\mathbf{x}(t) + \mathbf{d}(\alpha)$, $\alpha \in \mathcal{S}$ is said to define a *stochastic quadratic guaranteed cost control* with associated cost matrix $\mathbf{P}(\alpha) > 0, \alpha \in \mathcal{S}$ for the system (20) and cost function (23) if

$$\mathbf{x}^\top\left[\mathbf{R}_1 + \mathbf{L}^\top(\alpha)\mathbf{R}_2\mathbf{L}(\alpha)\right]\mathbf{x} + 2\mathbf{x}^\top\mathbf{P}(\alpha)\Big[\mathbf{A}(\alpha)$$
$$+\mathbf{H}_1(\alpha)\mathbf{F}(\alpha)(\mathbf{E}_1(\alpha) + \mathbf{E}_2(\alpha)\mathbf{L}(\alpha))$$
$$+\mathbf{B}(\alpha)\mathbf{L}(\alpha)\Big]\mathbf{x} + \mathbf{x}^\top\left(\sum_{\beta \in \mathcal{S}} q_{\alpha\beta}\mathbf{P}(\beta)\right)$$
$$x < 0,\ \alpha \in \mathcal{S} \quad (24)$$

for all non-zero $\mathbf{x} \in \mathbb{R}^n$ and all matrices $\mathbf{F}(\nu(t),t)$ satisfying (22).

Now we are ready to present our main results in this section.

*Theorem 4.2.* Consider the system (20) with cost function (23) and suppose the control law $\mathbf{u}(t) = \mathbf{L}(\alpha)\mathbf{x}(t) + \mathbf{d}(\alpha),\ \alpha \in \mathcal{S}$ is a stochastic quadratic guaranteed cost control with cost matrix $\mathbf{P}(\alpha) > 0,\ \alpha \in \mathcal{S}$. Then the closed loop uncertain system

$$\dot{\mathbf{x}}(t) = \big[\mathbf{A}(\alpha) + \mathbf{H}(\alpha)\mathbf{F}(\alpha,t)(\mathbf{E}_1(\alpha) + \mathbf{E}_2(\alpha)\mathbf{L}(\alpha)) + \mathbf{B}(\alpha)\mathbf{L}(\alpha)\big]\mathbf{x}(t) \quad (25)$$

is stochastically quadratically stable for all $\mathbf{F}(\alpha)$ satisfying $\mathbf{F}^\top(\alpha)\mathbf{F}(\alpha) \leq \mathbf{I}$, and for all $\alpha \in \mathcal{S}$. Furthermore, the corresponding value of the cost function (23) satisfies the bound

$$J_{\mathbf{u}}(\mathbf{x}_0, \alpha, 0) \leq \mathsf{E}\left\{\mathbf{x}_0^\top \mathbf{P}(\alpha)\mathbf{x}_0\right\},\ \forall\, \mathbf{F}^\top(\alpha)\mathbf{F}(\alpha) \leq \mathbf{I}.$$

Conversely, if there exists a control law $\mathbf{u}(t) = \mathbf{L}(\alpha)\mathbf{x}(t) + \mathbf{d}(\alpha),\ \alpha \in \mathcal{S}$ such that the resulting closed loop system (25) is stochastically quadratically stable, then this control law is a stochastic quadratic guaranteed cost control with some cost matrix $\bar{\mathbf{P}}(\alpha) > 0,\ \alpha \in \mathcal{S}$.

*Theorem 4.3.* Suppose there exists a sequence of $\{\varepsilon_\alpha > 0,\ \alpha \in \mathcal{S}\}$ such that the set of coupled Riccati equations

$$\begin{aligned}
&\Big[\mathbf{A}(\alpha) - \mathbf{B}(\alpha)\left(\varepsilon_\alpha \mathbf{R}_2(\alpha) + \mathbf{E}_2^\top(\alpha)\mathbf{E}_2(\alpha)\right)^{-1} \\
&\mathbf{E}_2(\alpha)\mathbf{E}_1(\alpha)\Big]^\top \mathbf{P}(\alpha) + \mathbf{P}(\alpha)\Big[\mathbf{A}(\alpha) - \mathbf{B}(\alpha) \\
&\left(\varepsilon_\alpha \mathbf{R}_2(\alpha) + \mathbf{E}_2^\top(\alpha)\mathbf{E}_2(\alpha)\right)^{-1}\mathbf{E}_2(\alpha)\mathbf{E}_1(\alpha)\Big] \\
&+\sum_{\beta\in\mathcal{S}} q_{\alpha\beta}\mathbf{P}(\beta) + \varepsilon_\alpha \mathbf{P}(\alpha)\mathbf{H}(\alpha)\mathbf{H}^\top(\alpha)\mathbf{P}(\alpha) \\
&-\varepsilon_\alpha \mathbf{P}(\alpha)\mathbf{B}(\alpha)\left(\varepsilon_\alpha \mathbf{R}_2(\alpha) + \mathbf{E}_2^\top(\alpha)\mathbf{E}_2(\alpha)\right)^{-1} \\
&\mathbf{B}^\top(\alpha)\mathbf{P}(\alpha) + \frac{1}{\varepsilon_\alpha}\mathbf{E}_1^\top(\alpha) \\
&\Big[\mathbf{I} - \mathbf{E}_2(\alpha)\left(\varepsilon_\alpha \mathbf{R}_2(\alpha) + \mathbf{E}_2^\top(\alpha)\mathbf{E}_2(\alpha)\right)^{-1}\mathbf{E}_2^\top(\alpha)\Big] \\
&\mathbf{E}_1(\alpha) + \mathbf{R}_1(\alpha) = 0 \qquad (26)
\end{aligned}$$

has a set of symmetric and positive definite solution $\{\mathbf{P}(\alpha),\ \alpha \in \mathcal{S}\}$. Then, given any sequence of $\{\delta_\alpha > 0,\ \alpha \in \mathcal{S}\}$, there exists a set of symmetric and positive definite solution $\{\bar{\mathbf{P}}(\alpha),\ \alpha \in \mathcal{S}\}$ such that $\mathbf{P}(\alpha) < \bar{\mathbf{P}}(\alpha) < \mathbf{P}(\alpha) + \delta_\alpha \mathbf{I},\ \alpha \in \mathcal{S}$ and the control law

$$\begin{aligned}
\mathbf{u}(t) = &-\left(\varepsilon_\alpha \mathbf{R}_2(\alpha) + \mathbf{E}_2^\top(\alpha)\mathbf{E}_2(\alpha)\right)^{-1} \\
&\left(\varepsilon_\alpha \mathbf{B}^\top(\alpha)\mathbf{P}(\alpha) + \mathbf{E}_2^\top(\alpha)\mathbf{E}_1(\alpha)\right) \\
&\mathbf{x}(t) + \mathbf{d}(\alpha) \qquad (27)
\end{aligned}$$

is a stochastic quadratic guaranteed cost control for the system (20) with cost matrix $\bar{\mathbf{P}}(\alpha)$, $\alpha \in \mathcal{S}$.

*Lemma 4.3.* Suppose that (26) admits a set of symmetric and positive definite solutions $\{\mathbf{P}(\alpha),\ \alpha \in \mathcal{S}\}$ for $\varepsilon_\alpha = \varepsilon_\alpha^1 > 0,\ \alpha \in \mathcal{S}$. Then, for any given $\varepsilon_\alpha \in (0, \varepsilon_\alpha^1)$, (26) has a set of symmetric and positive definite solutions $\{\mathbf{P}_\alpha^0,\ \alpha \in \mathcal{S}\}$ as well.

## 5. CONCLUSION

In this paper, we dealt with the inventory-production control problem where the random process representing the demand is modeled as a Markov chain with constant jump rates. The optimal policy which minimizes the cost of facilities is determined. It has been shown that the above problem can be solved if a set of coupled differential Riccati equations has symmetric and positive definite solutions. The guaranteed cost inventory-production has also been investigated when the deteriorating rates for the items are uncertain.

## References


Boukas, E.K. (1993). Control of systems with controlled jump Markov disturbances. *Control Theory and Advanced Technology,* **9**(2), 577–595.

Costa, O.L.V. and E.K. Boukas (1997). *Necessary and sufficient condition for robust stability and stabilizability of continuous-time linear systems with Markovian jumps.* Preprint.

Feng, X., K.A. Loparo, Y. Ji and H.J. Chizeck (1992). Stochastic stability properties of jump linear systems. *IEEE Trans. Automat. Control,* **37**(1), 38–53.

Hamid, B. (1989). Replenishment schedule for deteriorating items with time proportional demand. *Journal of Operations Research Society,* **40**(1), 75–81.

Ji, Y. and H.J. Chizeck (1990). Controllability, stabilizability, and continuous-time Markovian jump linear quadratic control. *IEEE Trans. Automat. Control,* **35**(7), 777–788.

Petersen, I.R. (1995). Guaranteed cost LQG control of uncertain linear systems. *IEE Proceedings-D,* **142**(2), 95–102.

# 2D MODEL-FOLLOWING SERVO SYSTEM AND ITS APPLICATIONS TO LINEAR MULTIPASS PROCESSES

**Minoru Yamada* Li Xu** Osami Saito*****

* *Gifu National College of Technology, shinsei-cho, Gifu, 501-0495 Japan, e-mail:myamada@gifu-nct.ac.jp*
** *Asahi University, hozumi, Gifu, 501-0296 Japan, e-mail:xuli@alice.asahi-u.ac.jp*
*** *Chiba University, Inageku, Chibashi, 263-8522 Japan, e-mail:saito@cute.te.chiba-u.ac.jp*

Abstract: This paper gives, in view of the feature of practical 2D (2-dimensional) systems, a formulation of the 2D model-following servo(MFS) problem for the case where one of the independent variables of the considered 2D systems is unbounded and the other one is bounded. Its application to the design of non-unit memory linear multipass processes is shown. A numerical example for metal rolling process is also given to verify the effectiveness. *Copyright © 1999 IFAC*

Keywords: Model-following control, Servo, Two-dimensional systems, LQR control method, Controllability

## 1. INTRODUCTION

In practical situations, the input and output signals of an $n$D (multidimensional) system are usually unbounded in, at most, one dimension (Agathoklis and Bruton, 1983). It is well known, for instance, that iterative learning control systems (ILCS) and linear multipass processes (LMP) possess in fact two kind of dynamics: the one along the time direction on each iteration or pass, and the one with respect to the iteration index, and thus can be characterized by 2D models (Geng *et al.*, 1990). A common feature of these systems is that, though the iterative index is not subject to any boundary restriction, the iterated passes are always bounded in finite time intervals. For control of such systems, the conditions for conventional stability and stabilization problems obtained without considering this practical feature may be over restrictive (Agathoklis and Bruton, 1983). In view of these facts, $n$D control problems in the practical sense that not more than two of the independent variables of the considered system can take infinite values simultaneously, have recently been investigated (Agathoklis and Bruton, 1983; Xu *et al.*, 1997).

The results on 2D tracking control obtained in Xu *et al.* (1994), however, are mainly based on the consideration of steady-state performance of the resultant control systems. Therefore, no guarantee is provided in general for the transient performance. Optimal control and model-following control would be naturally considered as an effective strategy to improve the system transient performance (Furuta, 1987; M. Yamada, 1996). For practical applications, it would be more reasonable and practicable to consider 2D optimal control in the practical sense of allowing only one variable unbounded.

The purpose of this paper is to consider the 2D model-following servo(MFS) problem for the case that one of the two variables of the considered 2D systems is unbounded, and the other one is bounded. This problem is applied to the design of non-unit memory LMP.

## 2. FORMULATION OF 2D MODEL-FOLLOWING SERVO PROBLEM

In this section, a formulation for the 2D MFS problem is given. In particular, the problem will be considered on the support $D_N \triangleq \{(i,j) \mid i=0, 1,\ldots, j=0,1,\ldots,N\}$ , where the variable $j$ is bounded while the variable $i$ is unbounded.

Consider a 2D plant described by the Fornasini-Marchesini (F-M) state space model with unmeasurable constant disturbances $\boldsymbol{d} \in \mathbb{R}^n$ and $\boldsymbol{w} \in \mathbb{R}^p$:

$$\begin{aligned}\boldsymbol{x}(i+1,j+1) &= \boldsymbol{A}_1\boldsymbol{x}(i,j+1) + \boldsymbol{A}_2\boldsymbol{x}(i+1,j) \\ &\quad + \boldsymbol{B}_1\boldsymbol{u}(i,j+1) + \boldsymbol{B}_2\boldsymbol{u}(i+1,j) + \boldsymbol{d} \\ \boldsymbol{y}(i,j) &= \boldsymbol{C}_1\boldsymbol{x}(i,j) + \boldsymbol{D}_1\boldsymbol{u}(i,j) + \boldsymbol{w} \quad (1)\end{aligned}$$

with boundary conditions $\boldsymbol{x}(i,0) = \boldsymbol{x}_0$, $i = 0,1,\ldots,\boldsymbol{x}(0,j) = \boldsymbol{x}_j^0$, $j=1,2,\ldots,N$ where $\boldsymbol{x} \in \mathbb{R}^n$, $\boldsymbol{u} \in \mathbb{R}^l$ and $\boldsymbol{y} \in \mathbb{R}^p$ are the local state, input and output vectors respectively, and $\boldsymbol{A}_1$, $\boldsymbol{A}_2$, $\boldsymbol{B}_1$, $\boldsymbol{B}_2$, $\boldsymbol{C}_1$ and $\boldsymbol{D}_1$ are real matrices of appropriate dimensions. It is assumed that $p \le l$.

Similarly, let the 2D model system be specified by

$$\begin{aligned}\boldsymbol{x}_m(i+1,j+1) &= \\ &\boldsymbol{A}_{m1}\boldsymbol{x}_m(i,j+1) + \boldsymbol{A}_{m2}\boldsymbol{x}_m(i+1,j) \\ &+ \boldsymbol{B}_{m1}\boldsymbol{r}(i,j+1) + \boldsymbol{B}_{m2}\boldsymbol{r}(i+1,j) \\ \boldsymbol{y}_m(i,j) &= \boldsymbol{C}_{m1}\boldsymbol{x}_m(i,j) \quad (2)\end{aligned}$$

with boundary conditions $\boldsymbol{x}_m(i,0) = \boldsymbol{x}_{m0}$, $i = 0,1,\ldots$, $\boldsymbol{x}_m(0,j) = \boldsymbol{x}_{mj}^0$, $j=1,2,\ldots,N$ where $\boldsymbol{x}_m \in \mathbb{R}^{n_m}$, $\boldsymbol{r} \in \mathbb{R}^{l_m}$ and $\boldsymbol{y}_m \in \mathbb{R}^p$ are the local state, input and output vectors respectively. The model system is assumed asymptotically stable on $D_N$. Namely, letting $\bar{U} \triangleq \{z \in \mathbb{C} \mid |z| \le 1\}$ and $z$ be the backward shift, the asymptotic stability of (2) on $D_N$ means $\det(\boldsymbol{I} - z\boldsymbol{A}_{m1}) \neq 0, \ \forall z \in \bar{U}$.

Further, let

$$\boldsymbol{r}(i,j) = \begin{cases} \bar{\boldsymbol{r}}(j) & , (i,j) \in D_N \\ 0 & , (i,j) \notin D_N \end{cases} . \quad (3)$$

Then, $\boldsymbol{r}(i,j)$ can be viewed as a kind of step signal in the direction of $i$ on $D_N$.

*Problem 1.* Given the plant (1), the model system (2) and the input $\bar{\boldsymbol{r}}(j)$, $j=0,1,\ldots,N$, determine a control input $\boldsymbol{u}(i,j)$ that makes the tracking error

$$\boldsymbol{e}(i,j) = \boldsymbol{y}(i,j) - \boldsymbol{y}_m(i,j) \quad (4)$$

small for $(i,j) \in D_N$ and $\lim_{i\to\infty} \boldsymbol{e}(i,j) = 0$.

This problem will be simply referred to as a 2D MFS problem. It is noted that, when $\boldsymbol{y}(i,j)$ converges to (the step response) $\boldsymbol{y}_m(i,j)$ for all $j$ as $i \to \infty$, $\boldsymbol{u}(i,j)$ also approaches some steady-state value in the direction of $i$. Therefore, it is reasonable to propose the following quadratic performance index

$$J = \sum_{i=0}^{\infty}\sum_{j=0}^{N} \|\boldsymbol{e}(i,j)\|_{\boldsymbol{Q}}^2 + \sum_{i=1}^{\infty}\sum_{j=0}^{N} \|\boldsymbol{u}(i+1,j) - \boldsymbol{u}(i,j)\|_{\boldsymbol{R}}^2 \quad (5)$$

where $\boldsymbol{Q}$ and $\boldsymbol{R}$ are symmetric positive definite matrices. The 2D MFS problem can now be interpreted as selecting $\boldsymbol{u}(i,j)$ to minimize $J$, which will be referred to as *Problem 1'* in the sequel for convenience.

## 3. TRANSFORMATION TO 1D CASE AND CONTROLLER DESIGN

The following equivalent description for the system (1) is obtained by calculating $\boldsymbol{x}(i+1,1)$, $\boldsymbol{x}(i+1,2),\ldots,\boldsymbol{x}(i+1,j)$ from (1).

$$\begin{aligned}&\boldsymbol{x}(i+1,j) \\ &= \boldsymbol{A}_2^j\boldsymbol{x}_0 + \sum_{k=1}^{j}\boldsymbol{A}_2^{j-k}\boldsymbol{A}_1\boldsymbol{x}(i,k) + \sum_{k=1}^{j}\boldsymbol{A}_2^{j-k}\boldsymbol{B}_1\boldsymbol{u}(i,k) \\ &\quad + \sum_{k=0}^{j-1}\boldsymbol{A}_2^{j-k-1}\boldsymbol{B}_2\boldsymbol{u}(i+1,k) + \sum_{k=0}^{j-1}\boldsymbol{A}_2^k\boldsymbol{d} \\ &\qquad i = 0,1,2,\ldots, \qquad j=1,2,\ldots,N. \quad (6)\end{aligned}$$

Define the vectors

$$\begin{aligned}\mathcal{X}(i) &= \left[\boldsymbol{x}^T(i,1)\ \boldsymbol{x}^T(i,2)\cdots\boldsymbol{x}^T(i,N)\right]^T, \\ \mathcal{U}(i) &= \left[\boldsymbol{u}^T(i,0)\ \boldsymbol{u}^T(i,1)\cdots\boldsymbol{u}^T(i,N)\right]^T, \\ \mathcal{Y}(i) &= \left[\boldsymbol{y}^T(i,0)\ \boldsymbol{y}^T(i,1)\cdots\boldsymbol{y}^T(i,N)\right]^T. \quad (7)\end{aligned}$$

(6) can then be represented in the 1D form:

$$\begin{aligned}\mathcal{X}(i+1) &= \tilde{\boldsymbol{A}}\mathcal{X}(i) + \tilde{\boldsymbol{B}}_1\mathcal{U}(i) + \tilde{\boldsymbol{B}}_2\mathcal{U}(i+1) \\ &\quad + \tilde{\boldsymbol{F}}_1\boldsymbol{x}_0 + \tilde{\boldsymbol{F}}_1\boldsymbol{d} \\ \mathcal{Y}(i) &= \tilde{\boldsymbol{C}}\mathcal{X}(i) + \tilde{\boldsymbol{D}}\mathcal{U}(i) + \tilde{\boldsymbol{H}}_1\boldsymbol{x}_0 + \tilde{\boldsymbol{H}}_2\boldsymbol{w} \quad (8)\end{aligned}$$

with the initial condition $\mathcal{X}(0) = \left[\boldsymbol{x}_1^{0T}\cdots\boldsymbol{x}_N^{0T}\right]^T$ where

$$\tilde{\boldsymbol{A}} = \begin{bmatrix} \boldsymbol{A}_1 & 0 & \ldots & 0 \\ \boldsymbol{A}_2\boldsymbol{A}_1 & \boldsymbol{A}_1 & \ldots & 0 \\ \vdots & \vdots & \ddots & \vdots \\ \boldsymbol{A}_2^{N-1}\boldsymbol{A}_1 & \boldsymbol{A}_2^{N-2}\boldsymbol{A}_1 & \ldots & \boldsymbol{A}_1 \end{bmatrix} \in \mathbb{R}^{nN\times nN},$$

$$\tilde{\boldsymbol{B}}_1 = \begin{bmatrix} 0 & \boldsymbol{B}_1 & 0 & \ldots & 0 \\ 0 & \boldsymbol{A}_2\boldsymbol{B}_1 & \boldsymbol{B}_1 & \ldots & 0 \\ \vdots & \vdots & \vdots & \ddots & \vdots \\ 0 & \boldsymbol{A}_2^{N-1}\boldsymbol{B}_1 & \boldsymbol{A}_2^{N-2}\boldsymbol{B}_1 & \ldots & \boldsymbol{B}_1 \end{bmatrix}$$

$$\in \mathbb{R}^{nN\times l(N+1)},$$

$$\tilde{B}_2 = \begin{bmatrix} B_2 & 0 & \dots & 0 & 0 \\ A_2B_2 & B_2 & \dots & 0 & 0 \\ \vdots & \vdots & \ddots & \vdots & \vdots \\ A_2^{N-1}B_2 & A_2^{N-2}B_2 & \dots & B_2 & 0 \end{bmatrix} \in \mathbb{R}^{nN\times l(N+1)},$$

$$\tilde{F}_1 = \left[A_2^T \; A_2^{2T} \cdots A_2^{NT}\right]^T \in \mathbb{R}^{nN\times n},$$

$$\tilde{F}_2 = \left[I_n \; (I_n + A_2)^T \cdots \sum_{k=0}^{N-1} A_2^{kT}\right]^T \in \mathbb{R}^{nN\times n},$$

$$\tilde{C} = \begin{bmatrix} 0 & \cdots & 0 \\ C_1 & & 0 \\ & \ddots & \\ 0 & & C_1 \end{bmatrix} \in \mathbb{R}^{p(N+1)\times nN},$$

$$\tilde{D} = \text{block diag}(\overbrace{D_1, \dots, D_1}^{N+1}) \in \mathbb{R}^{p(N+1)\times l(N+1)},$$

$$\tilde{H}_1 = \left[C_1^T \; 0 \cdots 0\right]^T \in \mathbb{R}^{p(N+1)\times n},$$

$$\tilde{H}_2 = I_{p(N+1)}.$$

Note that $I_k$ denotes a $k \times k$ identity matrix.

In the same manner, the model system (2) is transformed to

$$\begin{aligned} \mathcal{X}_m(i+1) &= \tilde{A}_m\mathcal{X}_m(i) + (\tilde{B}_{m1} + \tilde{B}_{m2})\bar{r} + \tilde{F}_m x_{m0} \\ \mathcal{Y}_m(i) &= \tilde{C}_m\mathcal{X}_m(i) + \tilde{H}_m x_{m0} \end{aligned} \tag{9}$$

where $\bar{r} = \left[\bar{r}^T(0)\; \bar{r}^T(1) \cdots \bar{r}^T(N)\right]^T$, and $\tilde{A}_m$, $\tilde{B}_{m1}$, $\tilde{B}_{m2}$, $\tilde{C}_m$, $\tilde{F}_m$ and $\tilde{H}_m$ are defined similarly as $\tilde{A}$, $\tilde{B}_1$, $\tilde{B}_2$, $\tilde{C}$, $\tilde{F}$ and $\tilde{H}$ respectively. Further, from (4) we have

$$\begin{aligned} \mathcal{E}(i) &\triangleq \left[e^T(i,0) \cdots e^T(i,N)\right]^T \\ &= \mathcal{Y}(i) - \mathcal{Y}_m(i) \\ &= \tilde{C}\mathcal{X}(i) - \tilde{C}_m\mathcal{X}_m(i) + \tilde{D}\mathcal{U}(i) + \tilde{H}_1 x_0 \\ &\quad + \tilde{H}_2 w - \tilde{H}_m x_{m0}. \end{aligned} \tag{10}$$

In view of (8), (9) and (10), the following augmented system is obtained

$$\begin{aligned} \mathcal{X}_a(i+1) &= A_a\mathcal{X}_a(i) + B_a\mathcal{U}_a(i) \\ \mathcal{E}(i-1) &= C_a\mathcal{X}_a(i) \end{aligned} \tag{11}$$

where

$$\mathcal{X}_a(i) = \begin{bmatrix} \mathcal{U}(i) - \mathcal{U}(i-1) \\ \mathcal{X}(i) - \mathcal{X}(i-1) \\ \mathcal{E}(i-1) \\ \hline \mathcal{X}_m(i) - \mathcal{X}_m(i-1) \end{bmatrix} \triangleq \begin{bmatrix} \mathcal{X}_{a1}(i) \\ \mathcal{X}_{a2}(i) \end{bmatrix},$$

$$\mathcal{U}_a(i) = \mathcal{U}(i+1) - \mathcal{U}(i),$$

$$A_a = \left[\begin{array}{cccc|c} 0 & 0 & 0 & & 0 \\ \tilde{B}_1 & \tilde{A} & 0 & & 0 \\ \tilde{D} & \tilde{C} & I_{p(N+1)} & & -\tilde{C}_m \\ \hline 0 & 0 & 0 & & \tilde{A}_m \end{array}\right] \triangleq \begin{bmatrix} A_{a1} & A_{a2} \\ 0 & \tilde{A}_m \end{bmatrix} \in \mathbb{R}^{\{n_a+l+p\}\times\{n_a+l+p\}},$$

$$B_a = \begin{bmatrix} I \\ \tilde{B}_2 \\ 0 \\ \hline 0 \end{bmatrix} \triangleq \begin{bmatrix} B_{a1} \\ 0 \end{bmatrix} \in \mathbb{R}^{\{n_a+l+p\}\times l(N+1)},$$

$$C_a = \left[0 \; 0 \; I_{p(N+1)} \,|\, 0\right] \triangleq \left[C_{a1} \; 0\right] \in \mathbb{R}^{p(N+1)\times\{n_a+l+p\}},$$

where $n_a = (n+l+p+n_m)N$. On the other hand, the performance index (5) can be expressed as

$$J = \sum_{i=1}^{\infty} \left\{ \|C_a\mathcal{X}_a(i)\|^2_{Q_a} + \|\mathcal{U}_a(i)\|^2_{R_a} \right\} \tag{12}$$

where

$$Q_a \triangleq \text{block diag}(Q, \dots, Q) \in \mathbb{R}^{p(N+1)\times p(N+1)},$$

$$R_a \triangleq \text{block diag}(R, \dots, R) \in \mathbb{R}^{l(N+1)\times l(N+1)}.$$

*Problem 1'* has thus been transformed to the 1D LQR problem: find the control sequence $\mathcal{U}_a(i)$, $i \geq 1$ that minimize the quadratic performance index (12) subject to (11).

*Theorem 2.* If $(A_{a1}, B_{a1}, C_{a1})$ is stabilizable and detectable (in the 1D sense), *Problem 1'* is solvable and the optimal control sequence $\mathcal{U}(k)$ is given by the equation

$$\begin{aligned} \mathcal{U}(i) &= \mathcal{U}(i-1) - K_1\mathcal{X}_{a1}(i-1) - K_2\mathcal{X}_{a2}(i-1) \\ &= \begin{cases} -K_{11}\mathcal{U}(0) - K_{12}\mathcal{X}(0) - K_2\mathcal{X}_m(0), & i = 1 \\ -K_{11}\mathcal{U}(i-1) - K_{12}\mathcal{X}(i-1) \\ \quad -K_{13}\sum_{k=0}^{i-2}\mathcal{E}(k) - K_2\mathcal{X}_m(i-1), & i = 2, 3, \dots \end{cases} \end{aligned} \tag{13}$$

where

$$K_1 = \left(R_a + B_{a1}^T P_1 B_{a1}\right)^{-1} B_{a1}^T P_1 A_{a1} \triangleq [K_{11} \; K_{12} \; K_{13}] \tag{14}$$

$$K_2 = \left(R_a + B_{a1}^T P_1 B_{a1}\right)^{-1} B_{a1}^T \left(P_1 A_{a2} + P_{12}\tilde{A}_m\right) \tag{15}$$

with $P_1$ being a symmetric positive semidefinite solution of (16), and $P_{12}$ determined by (17).

$$P_1 = A_{a1}^T P_1 A_{a1} + C_{a1}^T Q_a C_{a1} - A_{a1}^T P_1 B_{a1} \left( R_a + B_{a1}^T P_1 B_{a1} \right)^{-1} B_{a1}^T P_1 A_{a1} \tag{16}$$

$$P_{12} = A_{a1}^T P_1 A_{a2} + A_{a1}^T P_{12} \tilde{A}_m - A_{a1}^T P_1 B_{a1} \left( R_a + B_{a1}^T P_1 B_{a1} \right)^{-1} B_{a1}^T \times \left( P_1 A_{a2} + P_{12} \tilde{A}_m \right) \tag{17}$$

**Proof** The proof can be shown in the similar way as M. Yamada (1996). □

Some further results can be given with respect to the system of (8).

*Lemma 3.* (a) $(A_{a1}, B_{a1})$ is stabilizable if and only if $\text{rank}[I - \tilde{A}z \; \tilde{B}_1 z + \tilde{B}_2] = nN$, $\forall z \in \bar{U}$ and

$$\text{rank} \begin{bmatrix} \tilde{A} - I & \tilde{B}_1 + \tilde{B}_2 \\ \tilde{C} & \tilde{D} \end{bmatrix} = (n+p)N + p. \tag{18}$$

(b) $(C_{a1}, A_{a1})$ is detectable if and only if $(\tilde{C}, \tilde{A})$ is detectable.

**Proof** The proof can be shown by the definitions of $A_{a1}$, $B_{a1}$, $C_{a1}$. (For space limitation, the details are omitted.) □

## 4. SOLVABILITY CONDITIONS RELATED TO THE 2D PLANT

It has been shown that, if $(A_{a1}, B_{a1}, C_{a1})$ is stabilizable and detectable in the 1D sense, there exists an optimal control sequence for the 2D MFS problem. It is natural to ask, then, what these conditions mean for the original 2D system. The purpose of this section is to show the relation between the conditions obtained for the 1D system $(A_{a1}, B_{a1}, C_{a1})$ and some properties directly related to 2D system (1). In this section, we will call the 2D system (1) practically stabilizable and detectable in the direction of $i$, if $(A_1, B_1, C_1)$ is stabilizable and detectable in the 1D sense.

*Lemma 4.* If 2D system (1) is practically stabilizable in the direction of $i$, i.e., $(A_1, B_1)$ is stabilizable in the 1D sense, $\text{rank}[I - \tilde{A}z \; \tilde{B}_1 z + \tilde{B}_2] = nN$, $\forall z \in \bar{U}$.

**Proof** By the definitions of $\tilde{A}$, $\tilde{B}_1$, $\tilde{B}_2$, $\tilde{C}$,

$$\text{rank}[I - \tilde{A}z \; \tilde{B}_1 z + \tilde{B}_2] \geq N \, \text{rank}[I - A_1 z \; B_1 z]$$

In view of the assumption that $(A_1, B_1)$ is stabilizable, i.e. $\text{rank}[I - A_1 z \; B_1 z] = n$, $\forall z \in \bar{U}$, and the fact that the number of the rows of $[I - \tilde{A}z \; \tilde{B}_1 z + \tilde{B}_2]$ is $nN$, we see that $\text{rank}[I - \tilde{A}z \; \tilde{B}_1 z + \tilde{B}_2] = nN$, $\forall z \in \bar{U}$. □

It should be noted that the condition of *Lemma 4* is only a sufficient one.

*Lemma 5.* If 2D system (1) is practically stable in the direction $i$, and further

$$\text{rank} \begin{bmatrix} A_1 - I & B_1 \\ C_1 & D_1 \end{bmatrix} = n + p \tag{19}$$

and

$$\text{rank}\, D_1 = p \tag{20}$$

hold, $(A_{a1}, B_{a1})$ is stabilizable in the 1D sense.

**Proof** By the definitions of $\tilde{A}$, $\tilde{B}_1$, $\tilde{B}_2$, $\tilde{C}$, it is easy to see that the rank condition (18) holds true if (19) and (20) are satisfied. Then the claim follows from *Lemmas 3* and *4*. □

*Lemma 6.* $(\tilde{C}, \tilde{A})$ or equivalently $(C_{a1}, A_{a1})$, is detectable if and only if 2D system (1) is practically detectable in the direction of $i$, i.e., $(C_1, A_1)$ is detectable.

**Proof** The sufficiency can be shown similarly as the proof of *Lemma 4*, and the necessity can be easily seen by representing $[\tilde{C}^T \; (I - \tilde{A}z)^T]^T$ in terms of $C_1$ and $A_1$. □

Summarizing *Lemmas 5, 6* and *Theorem 2*, the following condition for solvability of the 2D MFS problem which is directly related to 2D system (1) is obtained.

*Theorem 7.* If 2D system (1) is practically stabilizable and detectable in the direction of $i$ and the rank conditions (19) and (20) hold, *Problem 1'* is solvable and the optimal control input is given by (13).

## 5. APPLICATION TO DESIGN OF LINEAR MULTIPASS PROCESSES

In this section, it is shown that the method proposed in the previous section for the 2D MFS problem can be applied to control of non-unit memory LMP (Rogers and Owens, 1992).

Consider the discrete non-unit memory LMP represented by the state space model (Rogers and Owens, 1992):

$$X_i(j+1) = \mathbf{\Phi} X_i(j) + \mathbf{\Gamma} U_i(j) + \sum_{k=1}^{M} \mathbf{\Gamma}_{k-1} Y_{i-k}(j)$$

$$Y_i(j) = \mathbf{\Lambda} X_i(j) + \mathbf{\Theta}_0 U_i(j) + \sum_{k=1}^{M} \mathbf{\Theta}_k Y_{i-k}(j)$$

$$X_i(0) = \boldsymbol{x}_0,\ i = 0, 1, \ldots,\ j = 0, 1, \ldots, N \quad (21)$$

where $X_i(j) \in \mathbb{R}^{n_s}$, $U_i(j) \in \mathbb{R}^{l_s}$ and $Y_i(j) \in \mathbb{R}^{p_s}$ are the state, input and output vectors on the $i$-th pass respectively, $\boldsymbol{\Phi}$, $\boldsymbol{\Gamma}$, $\boldsymbol{\Lambda}$, $\boldsymbol{\Gamma}_j$, $\boldsymbol{\Theta}_j$ are constant matrices with suitable size, and the pass length $N$ is a finite constant. It has been known (Xu *et al.*, 1997) that the non-unit memory LMP of (21) can be represented by 2D F-M model as follows.

$$\begin{aligned}\xi(i+1, j+1) &= \hat{\boldsymbol{A}}_1\xi(i, j+1) + \hat{\boldsymbol{A}}_2\xi(i+1, j)\\ &\quad + \hat{\boldsymbol{B}}_1 U(i, j+1) + \hat{\boldsymbol{B}}_2 U(i+1, j)\\ Y(i,j) &= \hat{C}_1\xi(i,j) + \hat{\boldsymbol{D}}_1 U(i,j)\\ \xi(i, 0) &= \xi_0,\\ &\qquad i = 0, 1, \ldots,\ j = 0, 1, \ldots, N \quad (22)\end{aligned}$$

where

$$\xi(i,j) = \left[X_i^T(j)\ Y_{i-1}^T(j) \cdots Y_{i-M}^T(j)\right]^T \in \mathbb{R}^{n_s + p_s M}$$

$$U(i,j) = U_i(j) \in \mathbb{R}^{l_s},\ Y(i,j) = Y_i(j) \in \mathbb{R}^{p_s}$$

$$\hat{\boldsymbol{A}}_1 = \begin{bmatrix} 0 & 0 & 0 & \cdots & 0 & 0 \\ \boldsymbol{\Lambda} & \boldsymbol{\Theta}_1 & \boldsymbol{\Theta}_2 & \cdots & \boldsymbol{\Theta}_{M-1} & \boldsymbol{\Theta}_M \\ 0 & \boldsymbol{I}_{p_s} & 0 & \cdots & 0 & 0 \\ 0 & 0 & \boldsymbol{I}_{p_s} & \cdots & 0 & 0 \\ \vdots & \vdots & \vdots & \ddots & \vdots & \vdots \\ 0 & 0 & 0 & \cdots & \boldsymbol{I}_{p_s} & 0 \end{bmatrix},$$

$$\hat{\boldsymbol{A}}_2 = \begin{bmatrix} \boldsymbol{\Phi}\ \boldsymbol{\Gamma}_0\ \boldsymbol{\Gamma}_1 \cdots \boldsymbol{\Gamma}_{M-1} \\ 0 \end{bmatrix}$$

$$\hat{\boldsymbol{B}}_1 = \left[0\ \boldsymbol{\Theta}_0^T\ 0 \cdots 0\right]^T, \hat{\boldsymbol{B}}_2 = \left[\boldsymbol{\Gamma}^T\ 0 \cdots 0\right]^T$$

$$\hat{\boldsymbol{C}}_1 = \left[\boldsymbol{\Lambda}\ \boldsymbol{\Theta}_1\ \boldsymbol{\Theta}_2 \cdots \boldsymbol{\Theta}_M\right],\ \hat{\boldsymbol{D}}_1 = \boldsymbol{\Theta}_0.$$

It is now possible to apply the proposed approach to design a MFS system for LMP described by (22) ($\boldsymbol{d} = 0$, $\boldsymbol{w} = 0$ in (1)). According to *Theorem 7*, the MFS problem for (22) is solvable if (22) is practically stabilizable and detectable in the direction of $i$.

*Lemma 8.* $(\hat{\boldsymbol{A}}_1, \hat{\boldsymbol{B}}_1)$ is stabilizable in the 1D sense if and only if

$$\text{rank}\left[\boldsymbol{I}_{p_s} - \sum_{k=1}^{M}\boldsymbol{\Theta}_k z^k\ \ \boldsymbol{\Theta}_0 z\right] = p_s, \forall z \in \bar{U}. (23)$$

**Proof** $(\hat{\boldsymbol{A}}_1, \hat{\boldsymbol{B}}_1)$ is stabilizable in the 1D sense if and only if

$$\text{rank}\left[\boldsymbol{I} - \hat{\boldsymbol{A}}_1 z\ \ \hat{\boldsymbol{B}}_1 z\right] = n_s + p_s M,\ \forall z \in \bar{U}.$$

Then the claim follows from the definitions of $\hat{\boldsymbol{A}}_1$ and $\hat{\boldsymbol{B}}_1$ and the fact that the number of the rows of $[\boldsymbol{I} - \hat{\boldsymbol{A}}_1 z\ \hat{\boldsymbol{B}}_1 z]$ is $n_s + p_s M$. □

*Lemma 9.* $(\hat{\boldsymbol{C}}_1, \hat{\boldsymbol{A}}_1)$ is always detectable.

**Proof** The claim follows immediately from the fact that the rank condition

$$\text{rank}\begin{bmatrix} \hat{\boldsymbol{C}}_1 \\ \boldsymbol{I} - \hat{\boldsymbol{A}}_1 z \end{bmatrix} = n_s + p_s M,\ \forall z \in \mathbb{C} \quad (24)$$

always holds true. □

It can also be shown that

$$\text{rank}\begin{bmatrix} \hat{\boldsymbol{A}}_1 - \boldsymbol{I} & \hat{\boldsymbol{B}}_1 \\ \hat{\boldsymbol{C}}_1 & \hat{\boldsymbol{D}}_1 \end{bmatrix} = \text{rank}\,\boldsymbol{\Theta}_0 + n_s + p_s M. \quad (25)$$

Therefore the left-hand side of (25) is full rank if and only if $\text{rank}\,\boldsymbol{\Theta}_0 = p_s$, because the number of the rows is $n_s + p_s(M+1)$. Further, the rank conditions (20) and (23) also hold true whenever $\text{rank}\,\boldsymbol{\Theta}_0 = p_s$. Thus from *Theorem 7* and *Lemmas 8, 9* we have the conclusion that if $\text{rank}\,\boldsymbol{\Theta}_0 = p_s$, the controller for LMP described (21) can be designed by using the MFS approach.

## 6. NUMERICAL EXAMPLE FOR METAL ROLLING PROCESS

Consider the metal rolling process shown in Fig. 1 which can be viewed as a typical industry example of LMP (Rogers and Owens, 1992). The simplified model of this multipass metal rolling process is described by the equation (Edwards and Owens, 1982; Foda and Agathoklis, 1992) :

$$y_i(t) = \frac{\lambda}{\lambda + Mp^2}\left\{\left(1 + \frac{Mp^2}{\lambda_1}\right)y_{i-1}(t) - \frac{1}{\lambda_2}F_m\right\} (26)$$

where $p$ denotes the differentiation operator $d/dt$, $y_i(t)$ is the $i$th actual roll-gap thickness, $F_m$ is the force developed by the motor, $M$ is the lumped mass of the roll-gap adjusting mechanism, $\lambda_1$ is the stiffness of the adjusting mechanism spring, $\lambda_2$ is the hardness of the metal strip and $\lambda = \lambda_1\lambda_2/(\lambda_1 + \lambda_2)$ is the composite stiffness of the metal strip and the roll mechanism.

By replacing differentiation by backward difference, denoting the sampling period by $T$, and defining $t = jT$ and $y_i(j) = y_i(jT)$, a LMP model for (26) can be obtained as follows.

$$\begin{aligned} X_i(j+1) &= \boldsymbol{\Phi}X_i(j) + \boldsymbol{\Gamma}U_i(j) + \boldsymbol{\Gamma}_0 y_{i-1}(j) \\ y_i(j) &= \boldsymbol{\Lambda}X_i(j) + \boldsymbol{\Theta}_0 U_i(j) + \boldsymbol{\Theta}_1 y_{i-1}(j) \end{aligned} \quad (27)$$

where

$$X_i(j) = \left[\,y_i(j-1)\ y_i(j-2)\ y_{i-1}(j-1)\ y_{i-1}(j-2)\,\right]^T,$$

$$U_i(j) = F_m,$$

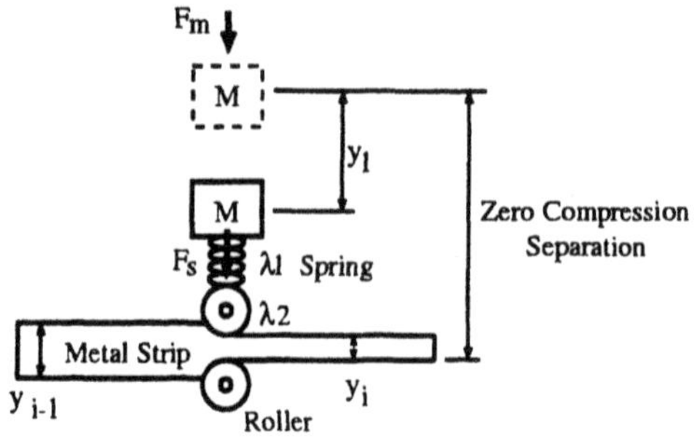


Fig. 1. Metal rolling process

$$\mathbf{\Phi} = \begin{bmatrix} a_1 & a_2 & a_4 & a_5 \\ 1 & 0 & 0 & 0 \\ 0 & 0 & 0 & 0 \\ 0 & 0 & 1 & 0 \end{bmatrix}, \mathbf{\Gamma} = \begin{bmatrix} b \\ 0 \\ 0 \\ 0 \end{bmatrix}, \mathbf{\Gamma}_0 = \begin{bmatrix} a_3 \\ 0 \\ 1 \\ 0 \end{bmatrix},$$

$$\mathbf{\Lambda} = \begin{bmatrix} a_1 & a_2 & a_4 & a_5 \end{bmatrix}, \ \mathbf{\Theta}_0 = b, \ \mathbf{\Theta}_1 = a_3$$

$$a_1 = \frac{2M}{S}, \ a_2 = -\frac{M}{S},$$

$$a_3 = \frac{\lambda}{S}\left(T^2 + \frac{M}{\lambda_1}\right), a_4 = -\frac{2\lambda M}{\lambda_1 S},$$

$$a_5 = \frac{\lambda M}{\lambda_1 S}, \ b = -\frac{\lambda T^2}{\lambda_2 S}, \ S = \lambda T^2 + M.$$

Due to the results of section 5, a MFS system for (27) can be constructed if $b \neq 0$.

Suppose

$$\lambda_1 = 600, \ \lambda_2 = 2000, \ M = 100, \ T = 0.1$$

$$\boldsymbol{A}_{m1} = \begin{bmatrix} 0.6 & 0.08 \\ 1 & 0 \end{bmatrix}, \ \boldsymbol{A}_{m2} = \begin{bmatrix} 0 & 0 \\ 0 & 0 \end{bmatrix},$$

$$\boldsymbol{B}_{m1} = \begin{bmatrix} 1 \\ 0 \end{bmatrix}, \boldsymbol{B}_{m2} = \begin{bmatrix} 0 \\ 0 \end{bmatrix}, \boldsymbol{C}_{m1} = \begin{bmatrix} 0.32 & 0 \end{bmatrix},$$

$$X(0) = 0, X_m(0) = [-1/3.2 \ \ 0]^T,$$

$$\tilde{\boldsymbol{r}}(j) = -1, \ j = 0, 1, \ldots, 10 \qquad (28)$$

Further, let the performance index be $N = 10$, $\boldsymbol{R} = 1 \times 10^7$, $\boldsymbol{R} = 0.1$. The response of the model system to $\tilde{\boldsymbol{r}}(j)$ and the corresponding outputs of the plant due to the obtained optimal control are shown in Fig. 2 and Fig. 3 respectively.

## 7. CONCLUSIONS

In this paper, the 2D MFS problem has been formulated for the case where one of the independent variables of the system is unbounded and the other one is finitely bounded. The results of this paper can also be shown for Roesser model in a similar way.

## 8. REFERENCES

Agathoklis, P. and L. Bruton (1983). Practical-BIBO stability of n-dimensional discrete systems. *IEE Proc. G. Electron. Circuits & Systems* **130**(6), 236–242.

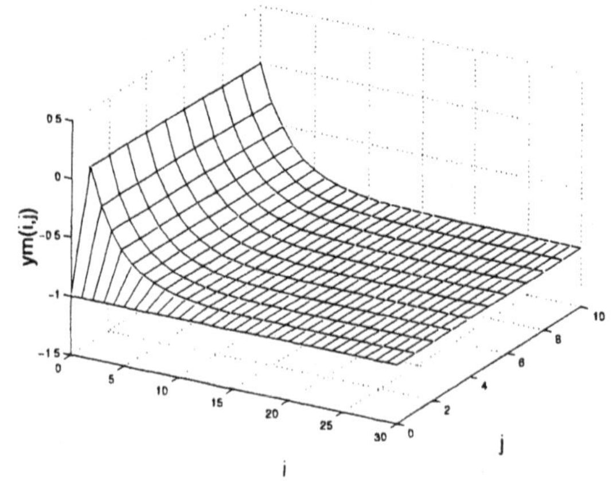


Fig. 2. The output $y_m(i,j)$ of the model system (28)

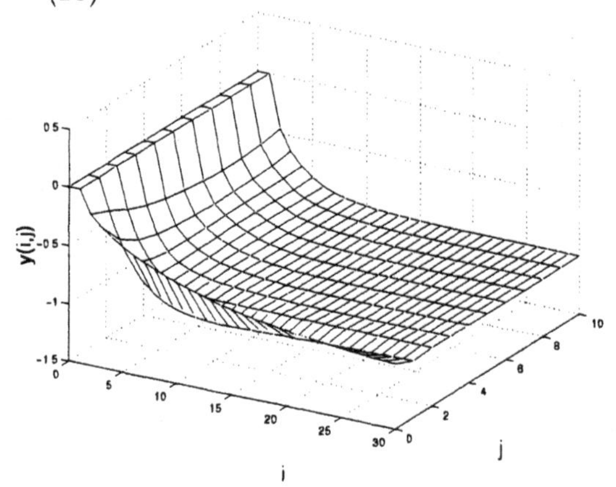


Fig. 3. The output $y(i,j)$ of the plant (26) ($\boldsymbol{Q} = 1 \times 10^7$, $\boldsymbol{R} = 0.1$)

Edwards, j. and D. Owens (1982). *Analysis and Control of Multipass Processes.* Research Studies Press. Chichester.

Foda, S. and P. Agathoklis (1992). Control of the Metal Rolling Process: A Multidimensional System Approach. *J. Franklin Inst.* **329**(2), 317–332.

Furuta, K. (1987). Alternative robust servo-control system and its digital control. *Int. J. Control* **45**(1), 183–194.

Geng, Z., R. Carroll and J. Xie (1990). Two-dimensional model and algorithm analysis for a class of iterative learning control systems. *Int. J. Control* **52**(4), 833–862.

M. Yamada, Li Xu, O. Saito (1996). 2D Model-Following Servo System. In: *Proceedings of IFAC'96 Vol. D.* san francisco. pp. 291–296.

Rogers, E. and D. H. Owens (1992). *Stability Analysis for Linear Repetitive Processes.* Springer Verlag. Berlin.

Xu, L., M. Yamada, O. Saito and K. Abe (1994). Practical Tracking Problem of $n$D Systems and Its Applications to Iterative Control Systems. In: *Proc. of First Asian Control Conference, Vol. 3.*

Xu, L., O. Saito and K. Abe (1997). $n$D Control Systems in a Practical Sense. *Applied Mathematics and Computer Science* **7**(4), 907–941.

# OPTIMAL CONTROL SOLUTIONS FOR THE BALL-AND-BEAM PROBLEM

**Mark W. Palmer**
*Marquette University*
*Department of Electrical and Computer Engineering*
*P.O. Box 1881*
*Milwaukee, WI 53202-1881*
*Email: palmer@spiff.eece.mu.edu, USA*

**Dr. Xin Feng**
*Marquette University*
*Department of Electrical and Computer Engineering, USA*

Abstract: The ball-and-beam problem is a commonly recurring nonlinear control application used by authors to demonstrate the effectiveness of various control methods. We present theory and methods for finding optimal fixed final time solutions to the ball-and-beam problem. Optimal control theory is applied to the ball-and-beam state equations and cost function to obtain a system of eight first-order differential equations. This system is solved numerically using the FORTRAN 77 routine COLNEW to obtain the optimal solution. The optimal solution provides an objective benchmark to which other controls can be compared. *Copyright © 1999 IFAC*

Keywords: optimal control, nonlinear control, boundary-value problem, numerical solutions

## 1. INTRODUCTION

The ball-and-beam problem is a commonly recurring control application in the engineering literature. It was recommended as early as 1984 as a nontrivial control application by Astrom and Wittenmark, and has been used since then as a test for various control techniques. (For examples, see Li, *et. al.* (1995), Wang (1998), Sastry and Tomlin).

Considering that the ball-and-beam application has emerged as a benchmark test for evaluating control effectiveness, the authors suggest that it can be helpful to know the "ideal," or optimal control for a given set of evaluation standards. Once the optimal control is found, it can serve as an objective baseline against which other proposed control solutions are rated.

In this paper, optimal control theory is applied to the ball-and-beam system, obtaining solutions for the fixed final time optimal control case. Section 2 shows how to derive the boundary-value problem given the dynamics of the ball-and-beam system and optimization criteria. Section 3 describes how the resulting boundary-value problem can be solved numerically with public domain FORTRAN 77 code, with results shown in Section 4.

## 2. THEORY

The ball-and-beam experiment is depicted in Figure 1. The beam rotates in a vertical plane in response to input torque applied at the fulcrum. The ball is free to roll along the beam, but it is required that the ball roll without slipping or losing contact with the beam. State is described by the vector $\mathbf{x} = \begin{bmatrix} r & \dot{r} & \theta & \dot{\theta} \end{bmatrix}^T$, with

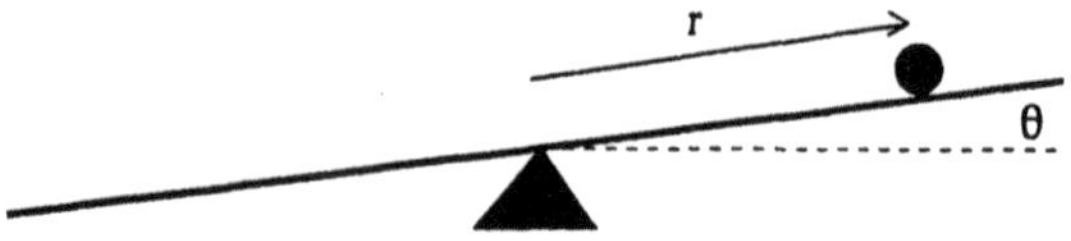


Figure 1: The ball and beam system.

the positive senses of $r$ and $\theta$ shown in Figure 1. Positive torque is the counterclockwise direction, which increases $\theta$. Typically, the objective is to manipulate the input in such a way that the ball reaches a stable state at the origin ( $\mathbf{x} = \begin{bmatrix} 0 & 0 & 0 & 0 \end{bmatrix}^T$ ).

The dynamics of the ball-and-beam system can be written as

$$\dot{\mathbf{x}} = \begin{bmatrix} x_2 \\ \alpha(x_1 x_4^2 - \beta\sin(x_3)) \\ x_4 \\ 0 \end{bmatrix} + \begin{bmatrix} 0 \\ 0 \\ 0 \\ 1 \end{bmatrix} u, \quad (1)$$

where $u$ is the input and $\beta$ is the acceleration due to gravity. The factor $\alpha$ describes the mass and moment of inertia of the ball (Hauser, *et. al.*, 1992):

$$\alpha = \frac{M}{\frac{I}{R^2} + M}. \quad (2)$$

In (2), $I$ is the moment of inertia of the ball, $M$ is the mass of the ball, and $R$ is the ball's radius.

In the general continuous-time optimization problem, a performance index, or "cost" $J$ is associated with the system

$$J(t_0) = \phi(x(T),T) + \int_{t_0}^{T} L(x,u,t)\,dt, \quad (3)$$

where $L$ is a function of the trajectory and $\phi$ is the penalty cost applied to deviation from the targeted final state. For our example, no deviation is allowed at the final time from the stationary point $\mathbf{x} = 0$, so $\phi(x(T),T) = 0$. We choose $L$ to be

$$L(x,u,t) = x_1^2(t) + x_2^2(t) + x_3^2(t) + x_4^2(t) + u^2(t), \quad (4)$$

so that total state deviation from zero and control magnitude, integrated over time, is to be minimized. We want to reach the final state at T=6 seconds from the starting point $\mathbf{x}(0) = \begin{bmatrix} 2 & 0 & 0 & 0 \end{bmatrix}^T$. The goal in this optimization problem is to find the input *u(t)* over the time $0 \le t \le 6$ that drives the ball-and-beam system from the starting point to $\mathbf{x}(T) = \begin{bmatrix} 0 & 0 & 0 & 0 \end{bmatrix}^T$, while minimizing the cost function (3).

To obtain the optimal controller, start by writing the Hamiltonian equation:

$$H(x,u,t) = L(x,u,t) + \lambda^T f(x,u,t) \quad (5)$$

$$\begin{aligned} H = {} & x_1^2(t) + x_2^2(t) + x_3^2(t) + x_4^2(t) + u^2(t) + \\ & \lambda_1 x_2 + \lambda_2 \alpha x_1 x_4^2 - \lambda_2 \alpha\beta\sin(x_3) + \\ & \lambda_3 x_4 + \lambda_4 u \end{aligned} \quad (6)$$

This introduces the costates, written as $\lambda$ in the notation used by Lewis and Syrmos (1995). The costate equations are found using

$$\dot{\lambda} = -\frac{\partial H}{\partial x}, \quad (7)$$

which gives

$$\begin{aligned} \dot{\lambda}_1 &= -\alpha x_4^2 \lambda_2 - 2x_1 \\ \dot{\lambda}_2 &= -\lambda_1 - 2x_2 \\ \dot{\lambda}_3 &= \alpha\beta\cos(x_3)\lambda_2 - 2x_3 \\ \dot{\lambda}_4 &= -2\alpha x_1 x_4 \lambda_2 - \lambda_3 - 2x_4 \end{aligned} . \quad (8)$$

The system of equations in (8) when combined with (1) gives eight equations in nine variables. The stationarity condition

$$\begin{aligned} \frac{\partial H}{\partial u} &= 0 \\ 2u + \lambda_4 &= 0, \end{aligned} \quad (9)$$

allows us to reexpress $u$ in terms of the state and costate

$$u = -\frac{1}{2}\lambda_4, \quad (10)$$

so the final system has eight first-order differential equations in eight variables.

$$\begin{aligned} \dot{x}_1 &= x_2 \\ \dot{x}_2 &= \alpha x_1 x_4^2 - \alpha\beta\sin(x_3) \\ \dot{x}_3 &= x_4 \\ \dot{x}_4 &= -\frac{1}{2}\lambda_4 \\ \dot{\lambda}_1 &= -\alpha x_4^2 \lambda_2 - 2x_1 \\ \dot{\lambda}_2 &= -\lambda_1 - 2x_2 \\ \dot{\lambda}_3 &= \alpha\beta\cos(x_3)\lambda_2 - 2x_3 \\ \dot{\lambda}_4 &= -2\alpha x_1 x_4 \lambda_2 - \lambda_3 - 2x_4 \end{aligned} . \quad (11)$$

Since the initial and final states are each four conditions, a total of eight boundary conditions have been specified, and the boundary-value problem is complete.

## 3. METHODS

Due to the instability in the system of differential equations shown in (11), initial-value based methods such as shooting methods can be unreliable, computationally intensive, and may fail to give results. Instead of these methods, the use of the FORTRAN 77 boundary-value solver COLNEW, an updated version of COLSYS written by U. Ascher and G. Bader, is suggested (Ascher, *et. al.*, 1981). This software is available online from *http://gams.nist.gov*.

In the notation of COLNEW, $x$ replaces $t$ as the independent variable, and $u$ will replace $x$ as the state variable. The differential equations should be written and ordered according to

$$u_n^{(m_n)}(x) = F_n(x;\mathbf{z}(\mathbf{u})) \qquad (12)$$
$$n = 1,\ldots,d$$
$$a < x < b\,.$$

The subscript $n$ stands for the "differential equation number $n$." There are a total of $d$ differential equations in the problem. The equations are ordered in such a way that the order $m_n$ of the equations is nondecreasing. In the case of the optimal ball-and-beam problem, there are $d$=8 equations all of order $m_n$=1. The variables $a$ and $b$ are limits of the range of the independent variable, 0 and 6.

Since all the equations are first order, no state derivatives appear on the right-hand side of the equations. This is not true of all problems that COLNEW encounters, so one further transformation of variables is made which orders all state variables and their derivatives into a new vector, **z**. The $u$ elements have a fixed ordering in the **z** vector:

$$\mathbf{z}(\mathbf{u}) = (u_1, u_1', \ldots, u_1^{(m_1-1)}, u_2, \ldots, u_d, \ldots, u_d^{(m_d-1)})\,. \qquad (13)$$

But again, since all of the equations in this problem are first order, this just becomes

$$\mathbf{z}(\mathbf{u}) = (u_1, u_2, u_3, u_4, u_5, u_6, u_7, u_8)\,. \qquad (14)$$

The final format of the system of differential equations is

$$\begin{aligned}
F_1 &= z_2 \\
F_2 &= \alpha z_1 z_4^2 - \alpha\beta \sin(z_3) \\
F_3 &= z_4 \\
F_4 &= -\frac{1}{2} z_8 \\
F_5 &= -\alpha z_4^2 z_6 - 2z_1 \\
F_6 &= -z_5 - 2z_2 \\
F_7 &= \alpha\beta \cos(z_3) z_6 - 2z_3 \\
F_8 &= -2\alpha z_1 z_4 z_6 - z_7 - 2z_4
\end{aligned} \qquad . \qquad (15)$$

For the boundary conditions, the following format applies:

$$g_j(\varsigma_j;\mathbf{z}(\mathbf{u})) = 0 \qquad j = 1,\ldots,m^* \qquad (16)$$

where $m^*$ is

$$m^* = \sum_{n=1}^{d} m_n\,, \qquad (17)$$

which for this problem equals eight. Boundary conditions must be ordered so that the locations of the conditions are nondecreasing from $a$ to $b$:

$$a \le \varsigma_1 \le \varsigma_2 \le \cdots \le \varsigma_{m^*} \le b\,. \qquad (18)$$

The conditions are written as functions of **z** set equal to zero. The $\zeta$ values and $g$ expressions which completely describe the eight boundary conditions are

$$\varsigma = [0 \quad 0 \quad 0 \quad 0 \quad 6 \quad 6 \quad 6 \quad 6]^T \qquad (19)$$

$$\mathbf{g} = [z_1 - 2 \quad z_2 \quad z_3 \quad z_4 \quad z_1 \quad z_2 \quad z_3 \quad z_4]^T\,.$$

Lastly, COLNEW requires that the Jacobians of the system of equations (15) and the boundary conditions (19) be available.

The FORTRAN front-end code for setting up this problem for COLNEW is available from the authors via email. The code returns state and costate for the optimal solution trajectory by calling the routine APPSLN (included in the COLNEW routines) after the COLNEW routine has completed. To obtain an entire solution trajectory, simply call APPSLN from a loop that varies X from 0 to 6.

## 4. RESULTS

For parameters $\alpha$=0.7143 and $\beta$=9.81, the initial state-costate vector returned by COLNEW for this problem is z(0) = [2 0 0 0 8.525772371443 6.960732795714 -22.53455679375 -4.002366083995]$^T$. Plotting ball position, velocity, beam angle and control input u from

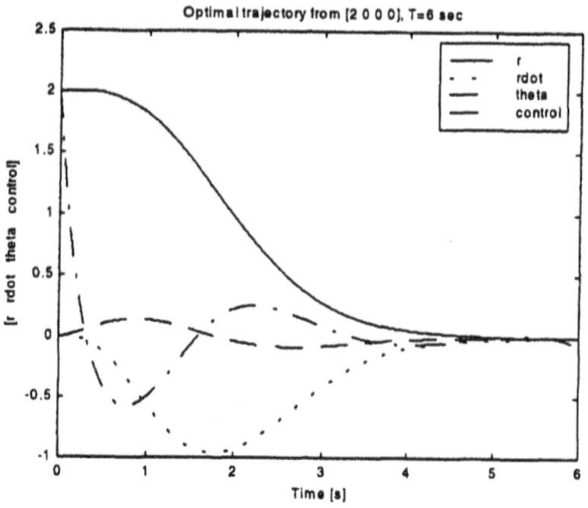


Figure 2: Selected state trajectories and control for the optimal ball-and-beam problem.

time 0 to 6 seconds gives the results shown in Figure 2. Recall that the control $u$ is computed from the costate using (10).

It is possible to use one initial state-costate vector from APPSLN as the initial condition in an ordinary differential equation solver to get the trajectories shown in Figure 2. However, care must be taken to keep the errors small. Accumulated error will depend on the precision of the initial state-costate vector, the error tolerance used in the ODE solver, and the amount of time over which a solution is computed. In this example where T is only 6 seconds, the error is not noticeable so long as the initial state-costate and solver computations are accurate to about nine digits. However, for problems where T is greater than 12 seconds, error growth will be noticeable even using full double-precision accuracy. In all cases, it is important to look at the final computed state to verify that the solver has reached the expected target.

## 5. CONCLUSION

This paper presented the computation of optimal control and solution trajectories for the fixed final time case of the ball-and-beam problem. Optimal control theory was applied to the ball-and-beam equations to formulate a system of eight first-order differential equations with eight boundary values. The FORTRAN boundary-value solver COLNEW solved the system. Optimal control solutions to benchmark control problems such as the ball-and-beam provide objective performance ideals that various control strategies can be rated against.

## REFERENCES


Ascher, U., J. Christiansen, R. D. Russell. Collocation Software for Boundary-Value ODEs. *ACM Transactions on Mathematical Software*, **7**, no. 2, 209-222.

Astrom, K. J. and B. Wittenmark. *Computer Controlled Systems, Theory and Design*. Prentice Hall, Englewood Cliffs, NJ, 1984.

Hauser, J., S. Sastry, P. Kokotovic. Nonlinear control via approximate input-output linearization: The ball and beam example. *IEEE Transactions on Automatic Control*, **37**, no. 3, March 1992, 392-398.

Lewis, F. L. and V. L. Syrmos. *Optimal Control (2nd Edition)*, J. Wiley, New York, 1995, 134.

Li, H. H., N. Godfrey, and Y. Ji. A Beam-and-Ball Controller Prototype. *IEEE Micro*, **15**, no. 6, 64.

Sastry, S. S. and C. J. Tomlin. "Switching through Singularities," Department of Electrical Engineering and Computer Sciences, University of California, Berkeley.

Wang, L. X. Stable and Optimal Fuzzy Control of Linear Systems. *IEEE Transactions on Fuzzy Systems*, **6**, no. 1, 137-143.

# OPTIMAL CCONTROL STRATEGIES FOR ACTIVE STEERING OF RAILWAY VEHICLES

**T.X.Mei and R.M.Goodall**

*Department of Electronic and Electrical Engineering*
*Loughborough University, Loughborough, LEICS. LE11 3RQ, UK*
*Email: T.X.Mei@lboro.ac.uk*

Abstract: This paper presents a study of controls for actively steered wheelsets on railway vehicles. The paper explores the potential benefits from this form of active control and studies what could be achieved when modern control techniques are used on the vehicles via mechatronic components. Simulation models are developed for vehicles with both solid axle and independently rotating wheelsets where a mechanically simplified configuration is used, leading to reduced weight and cost. Optimal controllers are developed and applied to the vehicles with the different wheelsets. Parameter uncertainty and robustness of controllers are also investigated in the paper. *Copyright © 1999 IFAC*

Keyword: Active Control, Optimal Control, Robustness, Modelling and Simulation

## 1. INTRODUCTION

It has now been recognised by the railway industry that future railway vehicles must be more cost-effective and energy-efficient with improved ride quality and higher travel speed. This may only be achieved through an extensive use of advanced electronic control, embedded within the vehicle system from the earliest stages of the design process. As the first phase of a comprehensive study for Mechatronic solutions for future railway vehicles, this paper investigates controls for actively-steered railway wheelsets.

Over the last two decades there have been varying levels of interest in the use of active suspensions for railway vehicles (Goodall, 1997). Significant use however has only happened during the 1990s with the advent of tilting trains in service and it seems likely that more general use of active control of the secondary suspensions between the bogie and the vehicle body (i.e. to give improved ride quality) will progressively increase. However, although there has been a small amount of experimental work, active control of the primary suspension between the wheels and the bogie remains a largely theoretical possibility (Aknin et al, 1991; Leo, 1985; Wickens, 1993).

In this paper, the possibility to use advanced control technology and mechatronic concepts to replace the traditional mechanical wheelset and use 'steer-by-wire' is investigated. It seeks both to stabilise the system and to improve the performance of curving without compromising the ride quality. The paper uses a mechanically simple 2-axle vehicle to study the active control of two types of wheelset: a normal solid-axle wheelset, and a wheelset with an axle but having independently rotating wheels (often referred as IRW). Each type has quite distinct control options because of their particular characteristics/limitations. Optimal controls are developed for the control of active steering. The aim of this paper to explore what could be achieved with active control on wheelset steering and to lay a foundation for further studies.

## 2. MODELLING OF A 2-AXLE VEHICLE

Figure 1 shows a plan view of the 2-axle vehicle, which mainly consists of a vehicle body and two

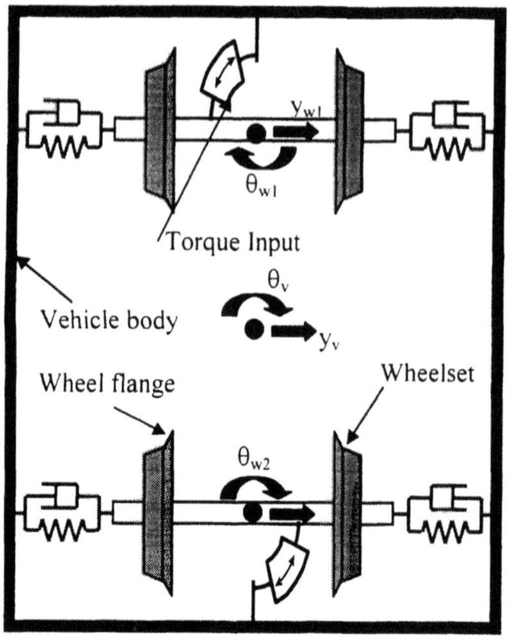


Figure 1. Plan View of a 2-Axle vehicle

wheelsets. Bogies are removed and the wheelsets are mounted onto the vehicle body via the springs and dampers in the lateral direction, the values of which have been chosen to give a satisfactory ride quality on the vehicle body purely through passive means. No longitudinal springs or dampers are shown, although in practice some form of longitudinal connection is needed to transmit traction and braking forces from the wheels to the vehicle body. The torque input shown in the diagram can be from passive components to stabilise the system or from an actuator. This vehicle configuration is mechanically simpler and lighter, which has drawn considerable attention by the railway industry. The wheelsets can be either the solid-axle or independently-rotating ones.

The conventional wheelset for the railway vehicle is composed of two coned (or otherwise profiled) wheels rigidly fixed to a common axle and therefore constrained to rotate at the same angular velocity. This arrangement has the advantages of natural centring and curving, but when without any mechanical connections it also exhibits a sustained kinematic oscillation in the lateral plane commonly referred as the "*wheelset hunting*". This is overcome on conventional railway vehicles using springs connected from the wheelset to the bogie or the body of the vehicle. However this added stiffness degrades the ability of the wheelset to curve and results in severe wear of the wheels and rails. It has always been a difficult trade-off for design engineers to compromise between the curving, the dynamic performance and the stability.

A model of the vehicle with two solid-axle wheelsets can be represented by the following equation, derived from wheelset models given by (Li and Goodall, 1998). Note that all state variables are related to the local track references and all symbols and parameters are given in the Appendix.

$$\mathbf{X} = \mathbf{A} \cdot \mathbf{X} + \mathbf{B} \cdot \mathbf{u} + \mathbf{T} \cdot \mathbf{w} \qquad (1)$$

where **X**, **u** and **w** are the state variables, torque input and track curving input (curvature and cant angle) respectively:-

$$\mathbf{X} = \begin{bmatrix} \dot{y}_{w1} & y_{w1} & \dot{\theta}_{w1} & \theta_{w1} & \dot{y}_{w2} & y_{w2} & \dot{\theta}_{w2} & \theta_{w2} & \dot{y}_v & y_v & \dot{\theta}_v & \theta_v \end{bmatrix}^T$$

$$\mathbf{u} = \begin{bmatrix} T_{w1} & T_{w2} \end{bmatrix}^T$$

$$\mathbf{w} = \begin{bmatrix} 1/R_1 & \theta_{c1} & 1/R_2 & \theta_{c2} \end{bmatrix}^T$$

$$\mathbf{A} = \begin{bmatrix} a_{11} & a_{12} & 0 & a_{14} & 0 & 0 & 0 & 0 & a_{19} & a_{1a} & a_{1b} & a_{1c} \\ 1 & 0 & 0 & 0 & 0 & 0 & 0 & 0 & 0 & 0 & 0 & 0 \\ 0 & a_{32} & a_{33} & 0 & 0 & 0 & 0 & 0 & 0 & 0 & 0 & 0 \\ 0 & 0 & 1 & 0 & 0 & 0 & 0 & 0 & 0 & 0 & 0 & 0 \\ 0 & 0 & 0 & 0 & a_{55} & a_{56} & 0 & a_{58} & a_{59} & a_{5a} & a_{5b} & a_{5c} \\ 0 & 0 & 0 & 0 & 1 & 0 & 0 & 0 & 0 & 0 & 0 & 0 \\ 0 & 0 & 0 & 0 & 0 & a_{76} & a_{77} & 0 & 0 & 0 & 0 & 0 \\ 0 & 0 & 0 & 0 & 0 & 0 & 1 & 0 & 0 & 0 & 0 & 0 \\ a_{91} & a_{92} & 0 & 0 & a_{95} & a_{96} & 0 & 0 & a_{99} & a_{9a} & 0 & 0 \\ 0 & 0 & 0 & 0 & 0 & 0 & 0 & 0 & 1 & 0 & 0 & 0 \\ a_{b1} & a_{b2} & 0 & 0 & a_{b5} & a_{b6} & 0 & 0 & 0 & 0 & a_{bb} & a_{bc} \\ 0 & 0 & 0 & 0 & 0 & 0 & 0 & 0 & 0 & 0 & 1 & 0 \end{bmatrix}$$

$$\mathbf{B} = \begin{bmatrix} 0 & 0 & 1 & 0 & 0 & 0 & 0 & 0 & 0 & 0 & -1 & 0 \\ 0 & 0 & 0 & 0 & 0 & 0 & 1 & 0 & 0 & 0 & -1 & 0 \end{bmatrix}^T$$

N.B. **T** is a 12x4 matrix, many elements of which are zero. The non-zero elements $t_{ij}$ are defined in the appendix.

This is of course a linearised model, whereas the real system has profound non-linearities associated with the wheel flanges (see Fig. 1), which contact the rail when the lateral movement available on the coned section is exceeded. However the overall aim is to avoid flange contact in normal running because this is what causes high wear on the wheel and rail; whereas this aim is difficult for a conventional (passive) vehicle design, with active control it is achievable and the use of a linearised model is therefore justified. Analysis of the model shows that, when the torque input is not used in any way, there are two modes given in Table 1 with positive real parts which represent the dynamic instabilities of the two wheelsets. The third and forth modes are body lateral and yaw, and the last two pairs are higher frequency well-damped modes associated with the wheelsets. Adding a yaw stiffness between the vehicle body and each wheelset can stabilise the wheelset, however this will worsen the curving performance.

Table 1. Open loop Eigenvalues of solid axle vehicle

| Eigenvalue | Damping(%) | Freq. (Hz) |
|---|---|---|
| 6.83 ± 42.38I | -15.9 | 6.83 |
| 6.79 ± 42.39I | -15.8 | 6.83 |
| -4.03 ± 4.56I | 66.2 | 0.97 |
| -3.71 ± 4.51I | 63.5 | 0.93 |
| -277 ± 22.4I | 99.7 | 44.2 |
| -277 ± 22.4I | 99.7 | 44.2 |

Table 2. Open loop Eigenvalues of IRW vehicle

| Eigenvalue | Damping(%) | Freq. (Hz) |
|---|---|---|
| 2.21 ± 28.2I | -7.82 | 4.5 |
| 2.17 ± 28.2I | -7.69 | 4.49 |
| -4.04 ± 4.55i | 66.4 | 0.97 |
| -3.72 ± 4.51i | 63.7 | 0.93 |
| -313 ± 0.0I | 100 | 49.8 |
| -569 ± 0.0I | 100 | 90.5 |
| 0 | | 0 |
| 0 | | 0 |

For the independently rotating wheelset, the wheels fitted onto an axle can rotate independently from each other and there is an extra degree of freedom for each wheelset. However the natural mechanical steering action is lost as the two wheels on a wheelset are uncoupled. The model of the vehicle with IRW can be obtained by adding two extra states (Satou, *et al.*, 1991], i.e. relative rotating velocities between the two wheels of the leading and trailing wheelset, to the model above. Again the model is formulated using the local track references.

$$\begin{bmatrix}\dot{\mathbf{X}}\\ \dot{\phi}_{w1}\\ \dot{\phi}_{w2}\end{bmatrix} = \begin{bmatrix}\mathbf{A} & \mathbf{A}_{12}\\ \mathbf{A}_{21} & \mathbf{A}_{22}\end{bmatrix}\cdot\begin{bmatrix}\mathbf{X}\\ \dot{\phi}_{w1}\\ \dot{\phi}_{w2}\end{bmatrix} + \begin{bmatrix}\mathbf{B}\\ 0\quad 0\\ 0\quad 0\end{bmatrix}\cdot\mathbf{u} + \begin{bmatrix}\mathbf{T}\\ t_a\quad 0\\ 0\quad t_b\end{bmatrix}\cdot\mathbf{w} \qquad (2)$$

where

$$\mathbf{A}_{12} = \begin{bmatrix}0 & 0 & a_{3d} & 0 & 0 & 0 & 0 & 0 & 0 & 0 & 0 & 0\\ 0 & 0 & 0 & 0 & 0 & 0 & a_{7c} & 0 & 0 & 0 & 0 & 0\end{bmatrix}^T$$

$$\mathbf{A}_{21} = \begin{bmatrix}0 & a_{d2} & a_{d3} & 0 & 0 & 0 & 0 & 0 & 0 & 0 & 0 & 0\\ 0 & 0 & 0 & 0 & 0 & a_{e6} & a_{e7} & 0 & 0 & 0 & 0 & 0\end{bmatrix}$$

$$\mathbf{A}_{22} = \begin{bmatrix}a_{dd} & 0\\ 0 & a_{ee}\end{bmatrix}$$

It is important to realise that a wheelset with independently rotating wheels still exhibits an unstable quasi-kinematic mode. Table 2 gives the eigenvalues of the model. The two zero eigenvalues reflect the indeterminacy of the wheelsets associated with the loss of steering action.

## 3. OPTIMAL CONTROL

A general diagram of the control structure is shown in Figure 2. There are several issues that must be addressed when any control is designed for the active steering of railway vehicles. Firstly the controller must stabilise the system and the closed loop system should be robust against variation of operational parameters such as wheel conicity and creepage coefficients. Secondly the controller should not interfere with the natural curving action of the solid axle wheelsets. The vehicle with the independently rotating wheelsets must be controlled to follow the track. Finally the controller should not compromise the ride quality.

### *3.1. Design of Optimal Controllers*

The primary purpose of the controller is steering and so four state variables, the lateral and yaw displacements of the two wheelsets ($y_{w1}, y_{w2}, \theta_{w1}, \theta_{w2}$) are used in the cost function for designing the optimal controller:

$$J = \int(\mathbf{y}^T\cdot\mathbf{Q}\cdot\mathbf{y} + \mathbf{u}^T\cdot\mathbf{R}\cdot\mathbf{u})dt \qquad (3)$$

where

$$\mathbf{y} = [y_{w1}\quad \theta_{w1}\quad y_{w2}\quad \theta_{w2}]^T$$

Similar cost functions can be readily derived for the design of the optimal controls for the vehicle with the independently rotating wheelset. It is assumed at this initial stage that all state variables are available for the control feedback and a state observer will later be designed to estimate the variables that are impractical to measure. Also the properties of actuators required to implement the control actions are neglected at the stage.

Selection of proper weighting factors (**Q**, **R**) is very important to ensure a good design of the optimal controls. This is particularly true for control design of the solid axle vehicle. Adequate values of the

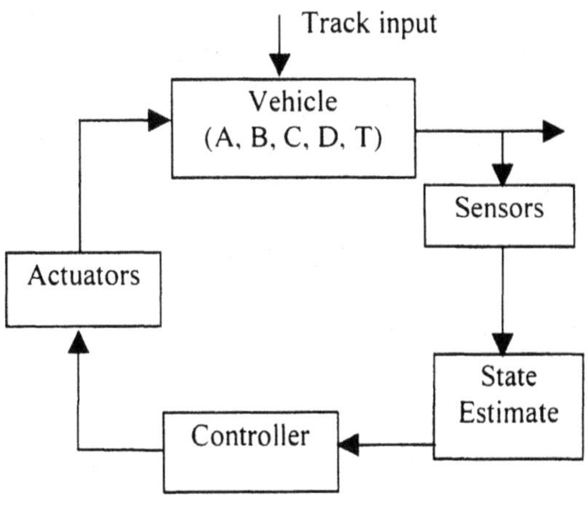


Figure 2. Diagram of the control structure

weighting factors are required, such that the controller does not interfere with the natural curving action of the solid axle wheelset. This is less critical for the IRW vehicles because of the extra degree of freedom of the vehicle, but extra control is required to guide the vehicles on the railway track. In the design **R** is fixed at $\mathbf{R}=10^{-12} * \mathbf{I}_{2x2}$ to maintain a good tracking performance and **Q** is to be adjusted.

### *3.2. Results of Solid Axle Vehicle*

The railway track profile used in all simulations is shown in Figure 3. A curved track with radius of 1000 *m* connected to straight tracks on either end via a transition is selected to study the control performance on curves. A time delay is set to reflect the difference between the leading and trailing wheelsets. The curved track is canted inwards to give a zero cant deficiency, i.e. zero lateral acceleration at the running speed of 60 m/s.

A weighting factor $\mathbf{Q} = \mathbf{diag}[0.5\ 50\ 0.5\ 50]$ has been found to be suitable for the solid axle vehicle model. The values are selected based on the engineering insight of the requirement and some on-line tuning.

Figure 4 presents the simulation result of the vehicle controlled by the optimal controller. The yaw displacement, which is also the angle of attack, is magnified by 10 times in the figure.

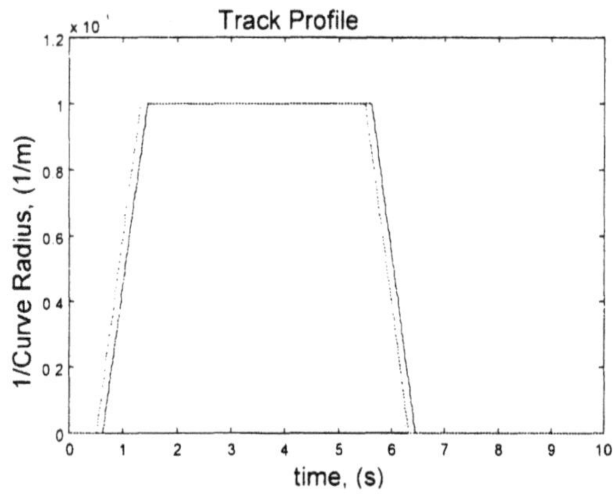


Figure 3. Track profile used in the study

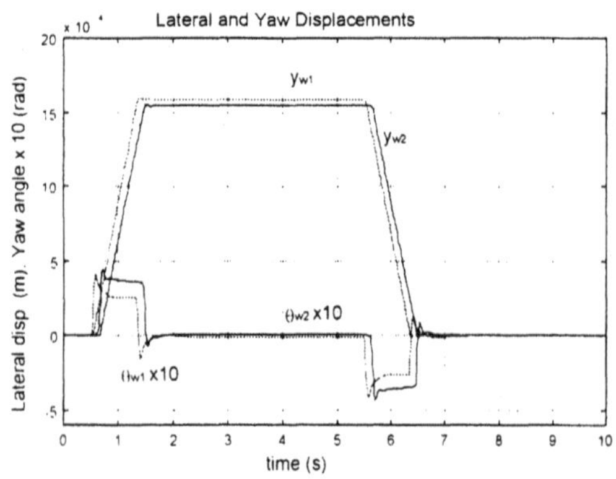


Figure 4. Solid axle vehicle, wheelset lateral displacement and yaw angle

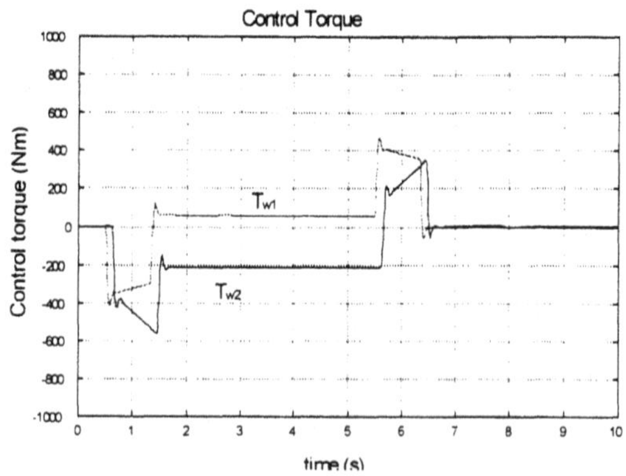


Figure 5. Solid axle vehicle, control torque

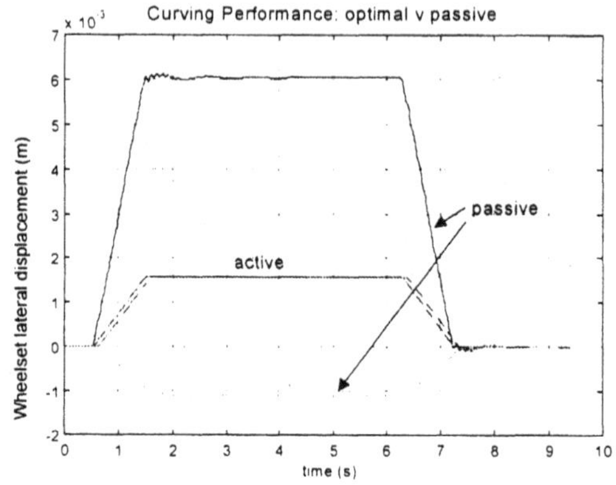


Figure 6 Comparison of Curving Performance

The lateral displacement of the wheelset required for pure rolling on the steady curve is 1.6 *mm* under the given condition and quasi-static value of the angle of attack should be zero as there is no cant deficiency. Both are largely achieved by the optimal control as shown in Figure 4, although there are small deviations. Those deviations are also reflected in Figure 5 by the non-zero quasi-static torque causing extra creepage on the constant curve. The control torque required on the curve transition is about 400-500*Nm*.

If a different Q value is used, the vehicle will behave undesirably. The lateral displacement on the constant curve will be forced by the controller away from that required for the natural curving action, which increases the angle of attack unnecessarily, and there is a large increase in the control torques under this condition (values 10 times as big are possible).

Figure 6 demonstrates a significant improvement of the curving performance of the actively controlled wheelset, compared with the passive vehicle. The natural curving action is achieved for the actively steered wheelset, while the two wheelsets of the passive vehicle behave differently and they deviate from the ideal curving position in the opposite directions. This large deviation is created by the yaw stiffness used to stablise the passive vehicle.

It is possible to add body yaw acceleration into the cost function in order to improve the ride quality on the vehicle body. Figure 7 compares the yaw acceleration of the vehicle body with the controller without this extra term, which shows an apparent improvement. However it also worsens the lateral displacement of the wheelset (by 0.2*mm*) and the angle of attack on the curve (by 0.003*mrad*), and in general it has proved better to concentrate upon optimising the curving performance.

### *3.3. Results of IRW Vehicle*

The natural mechanical steering action is lost when the two wheels on a wheelset are uncoupled, and so the primary purpose of the active control must be to steer the wheelset, for which various approaches are available. One obvious option is to steer the wheelset to keep it central within the flangeway clearance. Because there is no natural curving action from the IRW wheelset, the controller design can be treated as a simple regulator. The value of $\mathbf{Q}_3 = 50*\mathbf{I}_{4x4}$ has been chosen in the design of the optimal controllers for the IRW vehicle, with the same value of **R** as for the solid axle case.

Figure 7 gives the simulation results of the IRW vehicle with the optimal control. A much better performance is achieved compared with the solid axle vehicle. The quasi-static lateral displacement is only 0.5mm and the angle of attack is zero on the constant curve. Even on the curve transition the improvement on the angle of attack is obvious compared to the performance shown in Figure 4. It is worth noting that the control torque required to steer and control the IRW wheelset is much smaller, just over 20*Nm* as shown in Figure 8. This is because the extra degree of freedom of the uncoupled wheelset enable the two wheels to rotate freely and hence the controller would not have to 'fight' against the creepage forces as can happen for the solid axle vehicle.

Similarly a fifth element can be added to the weighting factor matrix **Q** in the controller design for the IRW vehicle to improve the ride quality. Simulation results show that improvement can be achieved on the ride quality, but again the lateral and yaw displacements of the wheelset are slightly affected adversely.

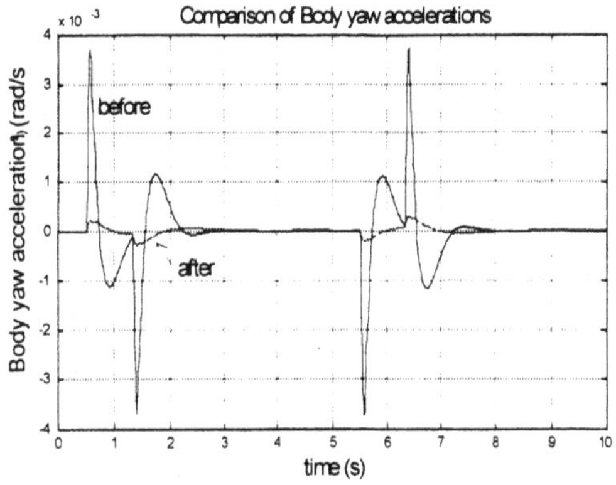


Figure 7. Improved body acceleration

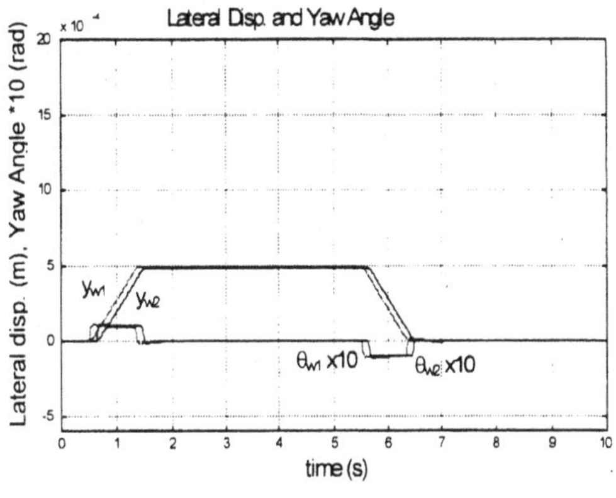


Figure 8. IRW vehicle, wheelset lateral displacement and yaw angle

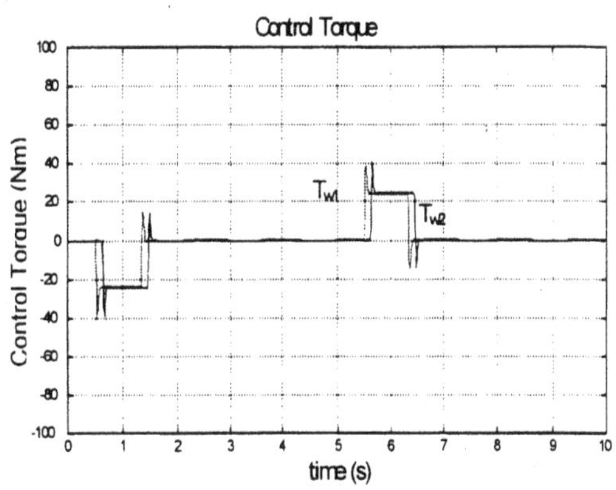


Figure 9. IRW vehicle, control torque

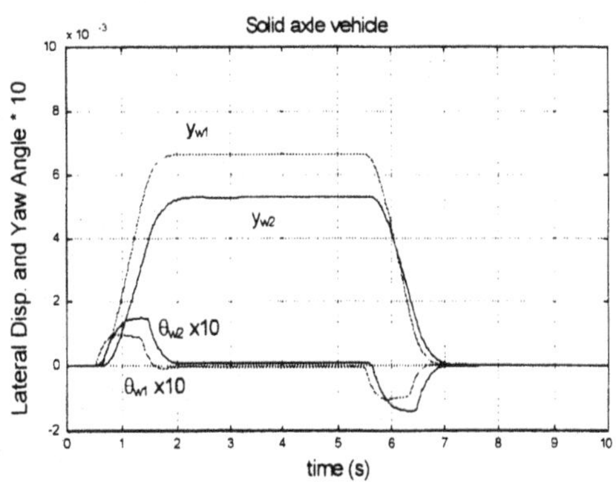


Figure 10. Solid axle vehicle with parameter Variations (worst case)

### *3.4. Parameter Uncertainty and Robustness*

Railway vehicles are subject to variation of working conditions, which may substantially change its dynamic properties. Two significant factors are the wheelset conicity and creepage coefficients between the wheels and the track. The former is due to wear of wheels and the latter is largely due to different track and weather conditions etc. The wheelset conicity can vary between 0.05 and 0.4 and the creepage between 5*MN* and 10*MN*. Therefore it is vital that any controller likely to be used must be robust enough to accommodate those parameter uncertainties. In the optimal control design this can be achieved by carefully selecting the weighting factors in the cost function. A close examination of the closed loop eigenvalues has found that the optimal controllers for both solid axle and IRW vehicles are stable with reasonable amount of margins when the two parameters vary from one extreme value to another. Figure 8 gives the simulation results under the worst variation where the wheelset conicity is 0.05 and the creepage coefficient is 5*MN*. Also the quasi-static lateral displacement of the solid axle wheelset required for the natural curving is now increased to 6.4*mm* because of the much lower conicity value.

## CONCLUSION AND FURTHER WORK

This paper has studied the optimal control for active steering/control of a 2-axle railway vehicle with the solid axle wheelset and independently rotating wheelset. It has demonstrated that a much improved curving performance can be achieved by applying active control to the vehicle primary suspension and the stability is also guaranteed. Optimal control works well with the solid axle wheelsets and even better with IRW, but careful selection of the weighting factor is needed. This is particularly true for the solid axle wheelset to reduce the interference with the natural curving action.

The requirement of full state feedback is obviously a drawback of the optimal control and it is neither practical nor economical to have all the state variables measured. A feasible solution as seen in many other applications is to design a state estimator to provide all the required feedback signals and this is being studied in a closely related research project (Li, and Goodall, 1999). Alternatively it is desirable to develop other control structures that do not require the full state feedback. There are various options available and intuitively formulated controls are the natural starting point, e.g. for the solid axle vehicle an extra lateral force proportional to the yaw velocity and for the IRW vehicles a yaw damping may be used to stabilise the systems. Studies of alternative control methods are currently being carried out and results should be available soon.

The property of possible actuators to be used to implement the developed controllers and its effect on the system is to be investigated. Finally a more comprehensive study on the ride quality is to be carried out for all controls, where the track roughness will have to be considered.

## APPENDIX. SYMBOLS AND PARAMETERS

$y_{w1}, y_{w2}, y_v$ -- Lateral displacement of leading, trailing wheelset and vehicle body

$\theta_{w1}, \theta_{w2}, \theta_v$ --Yaw angle of leading, trailing wheelset and vehicle body

$\phi_{w1}, \phi_{w2}$ -- Relative rotating angle between two wheels of leading and trailing wheelset

$m_v, m_w$ -- Vehicle mass (13,500 *kg*) and Wheelset mass (1250 *kg*)

$I_v$ , $I_w$ -- Vehicle yaw inertia (170,000 $kg\ m^2$) and Wheelset yaw inertia (700 $kg\ m^2$)

$I_{wr}$ -- Wheel rotational inertia (100 $kg\ m^2$)

$L_g$ -- Half gauge of wheelset (0.7 *m*)

$L_v$ -- Half space of the vehicle (3.7 *m*)

$r_0$ -- Wheel radius (0.45 *m*)

$\lambda$ -- Wheel conicity (0.2)

$K_s$ -- Lateral stiffness per wheelset (230 *kN/m*)

$C_s$ -- Lateral damping per wheelset (50 *kN s/m*)

$f_{11}, f_{22}$ -- Longitudinal and lateral creepages (10 *MN*)

$R_1$, $R_2$ -- Track curve radius at two wheelsets

$\theta_{c1}, \theta_{c2}$ -- Track cant angles at the two wheelsets

$T_{w1}, T_{w2}$ -- Controlled torque for two wheelsets

$V_s$ -- Vehicle travel speed (60 *m/s*)

$g$ -- Gravity (9.8 m/s$^2$)

$a_{11} = a_{55} = -(2f_{22}/V_s + C_s) / m_w$ , $a_{12} = a_{56} = -K_s / m_w$ ,

$a_{14} = a_{58} = 2f_{22} / m_w$ , $a_{19} = a_{59} = C_s / m_w$ ,

$a_{1a} = a_{5a} = K_s / m_w$ , $a_{1b} = - a_{5b} = L_v C_s / m_w$ ,

$a_{1c} = - a_{5c} = L_v K_s / m_w$ , $a_{32} = a_{-6} = -2\lambda f_{11} L_g/(r_0 * I_w)$

$a_{33} = a_{--} = 2f_{22}L_g^2 / (V_s * I_w)$, $a_{91} = a_{95} = C_s / m_v$ ,

$a_{92} = a_{96} = K_s / m_v$ , $a_{99} = -2C_s / m_v$ , $a_{9a} = -2K_s / m_v$

$a_{b1} = -a_{b5} = L_v C_s / I_v$ , $a_{b2} = -a_{b6} = L_v K_s / I_v$ ,

$a_{bb} = -2L_v^2 C_s / I_v$ , $a_{bc} = -2\ L_v^2 K_s / I_v$ ,

$a_{3d} = a_{-e} = - 2r_0 f_{11} L_g/(V_s I_w)$,

$a_{d2} = a_{e6} = - r_0 f_{11} L_g/(V_s I_{wr})$ , $a_{d3} = a_{e-} = - \lambda f_{11} / I_{wr}$ ,

$a_{dd} = a_{ee} = - r_0^2 f_{11} /(V_s I_{wr})$,

$t_a = t_b = r_0 f_{11} L_g / I_{wr}$

$t_{11} = t_{53} = V_s^2$, $t_{12} = t_{54} = g$,

$t_{31} = t_{-3} = 2f_{11} L_g^2 / I_w$

## REFERENCE


Aknin, P., Ayasse, J.B., and Devallez , A. (1991) Active steering of railway wheelsets, *12th IAVSD Conference*, Lyon, France

Goodall, R.M.(1997), Active railway suspensions: implementation status and technological trends, *Vehicle System Dynamics*, **28(1997)**, pp. 87-117

Leo, R. (1985) Creep-controlled wheelsets for high speed service – theory and test results, *Proc. 8th Int. Wheelset Congress*, **Vol. 1.2,** Madrid, Spain, pp1-16.

Li, H. and Goodall. R.M. (1998) Modelling and Analysis of a railway wheelset for active control, *UK Control 98*, Swansea, UK

Li, H. and Goodall, R.M. (1999) "State estimation for active steering of railway vehicles", *IFAC99 Congress*, Beijing, China.

Satou, E. and Miyamoto, M.(1991), Dynamics of a bogie with independently rotating wheels", *12th IAVSD Symposium*, France.

Wickens,A.H. (1993) Dynamic stability of articulated and steered railway vehicles guided by lateral displacement feedback, *13th IAVSD Symposium*, Chengdu, China.

14th Triennial World Congress, Beijing, P.R. China

F-2d-05-6

# FORCING PULSATILE BEHAVIOUR IN SEMICONDUCTOR LASERS BY PERIODIC CONTROL

**Paola Abbati, Sergio Bittanti, Fabio Previdi**

*Dipartimento di Elettronica e Informazione - Politecnico di Milano*[1]
*via G. Ponzio 34/5, 20133 Milano (ITALY)*
*tel. +39-2-23993650 - fax +39-2-23993412*
*e-mail: previdi@elet.polimi.it*

Abstract: In this paper, we apply optimal control theory for the operation of a standard edge-emitter semiconductor laser so as to produce good quality pulsing output signals. This is achieved by using a sinusoidal injection current whose characteristics (frequency and amplitude) are suitably tuned. The performances obtained from this optimal control strategy are compared with those achievable with a novel self-pulsing device recently designed. *Copyright © 1999 IFAC*

Keywords: Periodic control, Hopf-bifurcation, variational methods, self-pulsing semiconductor laser, optical nonlinearities.

## 1. INTRODUCTION

Pulsing semiconductor lasers have received notable attention in the last years especially for their applications in optical clocks, optical synchronisation and Compact Disk devices. Standard Edge-Emitter Semiconductor Lasers - EESL- (operating in Continuos Wave -CW- mode) cannot generate pulsatile output in their normal configuration. To achieve a pulsing optical output there are basically two alternatives: either to modify the geometric characteristics of the device cavity (leading e.g. to the so called Saturable Absorber Semiconductor Laser -SASL- (Pauli, 1978; Ueno and Lang, 1985; Yamada, 1993) or to magnify the Kerr effect through a suitable modification of the active material (Kerr Effect Semiconductor Laser - KESL-) (Abbati, *et al.*, 1998a).

In this paper we explore a third possibility, having the advantage of avoiding physical and structural modifications of the device, namely that of feeding a standard EESL with a sinusoidal injection current in place of a constant one. To achieve this aim a proper choice of the input current frequency is a fundamental design task. For, the possibility of disposing of a model-based design tool would be of most usefulness since it would allow for a suitable choice of the input characteristics without an over-effort in experimental work. This tool is provided by periodic control theory where the subject of designing periodic inputs for the appropriate operation of dynamical systems has been widely explored (see the original papers by Bittanti *et al.* (1973), Berstein and Gilbert (1980), Gilbert (1977), Zhang *et al.* (1996) and the recent survey by Bittanti and Colaneri (1998)).

In the following, we apply this theory to the standard model of a single-mode edge emitter semiconductor laser and we propose a periodic operation in place of the traditional steady-state one. The so obtained pulsing optical output is compared with that achievable by the magnification of the Kerr effect recently studied in the paper by Abbati *et al.* (1998b). The paper is organised as follows: in Sect. 2, the model of the standard EESL is introduced. Then the basic rationale of the frequency domain approach to periodic control is outlined and used for the performance optimization of the EESL (Sect. 3). As already said another possibility to obtain optical pulses is to resort to KESL in self-oscillating mode. In Sect. 4 it is explained how to modify the model of Sect. 2 in order to incorporate the Kerr effect, so achieving the mathematical model of the KESL.

[1]Paper supported by MURST project "Identification and control of industrial systems" and by CNR-CESTIA.

Finally we compare the characteristics of the pulsatile signals obtained following the two approaches.

## 2. A MODEL FOR SINGLE MODE EDGE-EMITTER SEMICONDUCTOR LASERS

The EESL model relies on the theory of a two-level system interacting with a single-mode electromagnetic field (Svelto, 1989). "Single-mode" means that the laser produces a strictly monochromatic optical output. The basic variables entering the equations are:

- $x_{ph}(\cdot)$ is the photon density inside the cavity of the laser;
- $x_c(\cdot)$ is the carrier density inside the cavity of the laser;
- $u(\cdot)$ is the normalized injection current, namely the quotient of the injection current I(.) over the so called threshold current $I_{th}$, i.e., $u(t) = I(t)/I_{th}$.

The injection current creates an inversion of population, i.e. it moves a great amount of carriers in the higher energy level; the more the carriers in this higher level the more is the rate of generation of photons in the cavity and the rate of carrier density is subject to a correspondent decrease. Taking also into account the natural decay effect due to the spontaneous lifetime decay the basic model for a EESL is described by the equations (Svelto, 1989; Iga, 1994):

$$\dot{x}_{ph}(t) = \frac{4\chi^2}{\gamma_\perp} x_c(t) x_{ph}(t) - 2k\, x_{ph}(t) + \beta\, \gamma_{//} x_c(t) \qquad (1.a)$$

$$\dot{x}_c(t) = -\frac{4\chi^2}{\gamma_\perp} x_c(t) x_{ph}(t) - \gamma_{//} x_c(t) + \frac{I_{th}}{e} u(t) \qquad (1.b)$$

The last term in Eq. (1.a) is related to the spontaneous emission process (often neglected in the literature). The parameters appearing in these equations have the following meaning:

- $k$ is related to the characteristic relaxation time of the electromagnetic field amplitude; in particular, $2k = 1/\tau_{ph}$, where $\tau_{ph}$ is the photon lifetime in the resonator;
- $\gamma_{//} = 1/\tau_s$, where $\tau_s$ is the carrier interband lifetime;
- $\gamma_\perp$ is related to the relaxation time of the polarization of the electromagnetic field;
- $\chi$ describes the coupling between the electromagnetic field and the active material and can be linked to the slope of the gain-current characteristic;
- $\beta$ determines the significance of the spontaneous emission process;
- $e$ is the electron charge.

The parameters have been chosen according to the values of a standard semiconductor stripe (Pauli, 1978; Ueno and Lang, 1985; Yamada, 1993; Avrutin, 1993) and are listed in Table 1. Eqs. (1) is a nonlinear dynamical system with two state variables. The optical output of the device is the number of outgoing photons:

$$y(t) = \alpha x_{ph}(t) \qquad (2)$$

where $\alpha$ is a geometric constant taking into account the cavity volume, so that $y(t)$ is the optical output in term of photon number.

Table 1 Parameter values used in numerical simulations.

| PHYSICAL QUANTITY | SYMBOL | VALUE |
|---|---|---|
| Stimulated-emission coefficient | $4\chi^2/\gamma_\perp$ | $10^4\ s^{-1}$ |
| Electromagnetic field relaxation coefficient | $2k$ | $(2\ 10^{-12}\ s)^{-1}$ |
| Population inversion relaxation coefficient | $\gamma_{//}$ | $(2\ 10^{-9}\ s)^{-1}$ |
| Spontaneous-emission coefficient | $\beta$ | $10^{-3}$ |
| Cavity volume | $\alpha$ | $10^{-10}\ cm^3$ |

Note that these parameters are typical for GaAs lasers emitting a (monochromatic) radiation around a wavelength of about 850 nm. A typical step response of this system is depicted in Fig. 1.

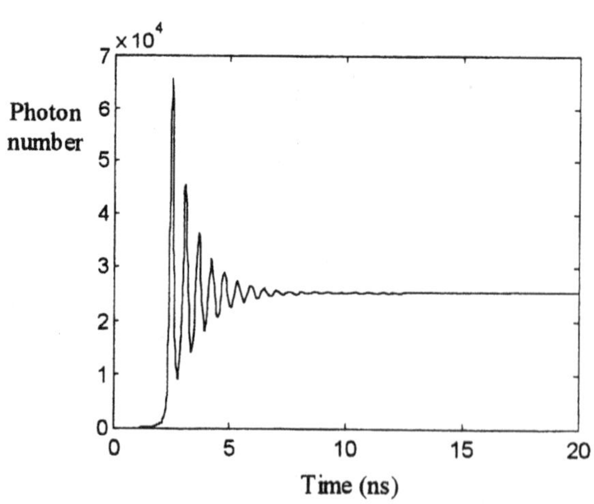


Fig. 1. The typical response of the system described by Eqs. (1)-(2) to a step input current.

As can be seen after an initial quiescent period (up to a time $\overline{T}$, where $\overline{T} = 2\ ns$ in the considered case), damping oscillations occur; this is the celebrated relaxation oscillation effect widely studied since the early days of the laser era (Svelto, 1989; Iga, 1994). The initial apparent delay can be explained as follows: initially, when all state variables are set to zero, some time interval is necessary before the external injection current can produce a significant increase in the value of the carrier density $x_c(\cdot)$ according to Eq. (1.b); only after this time interval $x_c(\cdot)$ can exercise its forcing effect on $x_{ph}(\cdot)$ as in Eq. (1.a).

*Remark 1.* To avoid possible misunderstandings, it is worthy pointing out that the signal y(t) represent the intensity of the laser beam, the light wavelength (i.e., its colour) being fixed (and not included in the model above).

## 3. PERIODIC CONTROL OF A STANDARD EESL

As already pointed out, EESL devices are usually operated at constant injection current. In order to explore its possibility of generating pulsatile light intensity, we consider the case of sinusoidal injection current. The objective of our study is to see if a nice pulsatile behaviour of the optical output can be achieved in this condition under appropriate choice of the sinusoid frequency. Optical pulses are necessary in telecommunication systems (where they work as optical clocks for optical synchronisation purposes) and they are used for reading optical disks. In these applications good quality of the pulses in terms of shape and frequency precision are fundamental tasks. To this purpose write Eqs. (1) and (2) in the general form

$$\begin{cases} \dot{x} = f(x,u) \\ y = h(x) \end{cases} \tag{3}$$

where $u(t) \in \Re^1$, $x(t) \in \Re^2$, $y(t) \in \Re^1$, $u(\cdot)$ belongs to the set of the piecewise continuous functions. Under the action of a sinusoidal input of period $\tau$, suppose that a periodic regime is imposed, and consider the problem of maximizing the performance index

$$J(u) \equiv \frac{1}{\tau}\int_0^\tau g(y,u)dt, \quad \tau > 0 \tag{4}$$

which is an average performance of the system in periodic regime. The simplest class of input signals is that of constant ones ($u(t) = \bar{u}$). Correspondingly the system operates in a steady-state condition: $x(t) = \bar{x}$, where $\bar{x}$ is such that $f(\bar{x},\bar{u}) = 0$, and the output takes the value $y(t) = \bar{y} = h(\bar{x})$. The problem becomes then the optimal programming problem of maximizing $J = \bar{J} = g(h(\bar{x}),\bar{u})$ under the constraint $f(\bar{x},\bar{u}) = 0$.

A slightly more complex problem arises if the family of the considered signals is enlarged to the class of periodic functions, in the hope of achieving a performance index better than the one associated to optimal steady-state conditions. A sensible way to face this problem is to consider as input signal the one obtained by superimposing to $\bar{u}$ a periodic variation $\hat{\delta u}(\cdot)$, and see whether

$$J[\bar{u} + \hat{\delta u}(\cdot)] > J[\bar{u}] \equiv \bar{J} \tag{5}$$

Now, under weak variations, if $\delta u(t) = U\exp(j\omega t) + U^*\exp(-j\omega t)$, where $U^*$ denotes the conjugate of $U$, the perturbed performance index can be given the form (Bittanti *et al.*, 1973):

$$J[\bar{u} + \delta u(\cdot)] = J[\bar{u}] + \frac{1}{2}U^*\Pi(\omega)U \tag{6}$$

where the so called $\Pi$-function $\Pi(\omega)$ is given by:

$$\Pi(\omega) = \breve{p}|\breve{G}(j\omega)|^2 + 2\breve{q}\,\mathrm{Re}\,\breve{G}(j\omega) + \breve{r} \tag{7}$$

In Eq. (7):

- $\breve{G}(s)$ is the transfer function of the linearized system obtained from the nonlinear Eqs. (3) around the operating point $(\bar{x},\bar{u})$;
- $\breve{p} = g_{yy}(\bar{y},\bar{u})$;
- $\breve{q} = g_{yu}(\bar{y},\bar{u})$;
- $\breve{r} = g_{uu}(\bar{y},\bar{u})$.

In view of expression (6), the steady state performance $J[\bar{u}]$ can be improved by cycling if $\delta u(\cdot)$ is a sinusoidal signal of frequency $\omega$ for which $\Pi(\omega) > 0$. Moreover, assuming that the set of frequencies such that $\Pi(\omega) > 0$ is not empty, the value $\hat{\omega} = \arg\max_{\omega} \Pi(\omega)$ corresponds to the optimal periodic weak variation $\hat{\delta u}(\cdot)$ to the constant input signal. As performance index we have considered:

$$J = \frac{1}{\tau}\int_0^\tau \{y^2(t) - \mu u^2(t)\}dt \tag{8}$$

which expresses the objective of maximizing the produced optical power without an excessive intensity in the injection current which can damage the device. The value of the parameter $\mu$ is chosen according to dimensional considerations. The optimal steady-state is then obtained as $\max\{\bar{y}^2 - \mu\bar{u}^2\}$ under the constraint $f(\bar{x},\bar{u}) = 0$, $\bar{y} = h(\bar{x})$. It turns out that, for the parameters given in Table 1, and assuming $\mu = 1.7 \cdot 10^9$, the optimal steady state is given by $\bar{u} = 3.125$, a (normalized) current intensity belonging to the typical range of values for injection currents used in applications.

The linearized system (A,B,C,D) around the stationary operating point $(\bar{u},\bar{x})$, is the following one (in which we have re-introduced the typical relaxation times $\tau_{ph}$ and $\tau_s$ instead of the primitive parameters appearing in Eqs. (1)):

$$A=\begin{bmatrix} \dfrac{\bar{u}+1-w}{2\tau_{ph}(1-\beta)}-\dfrac{1}{\tau_{ph}} & \dfrac{\bar{u}-1+w}{2\tau_s}+\dfrac{\beta}{\tau_s} \\ -\dfrac{\bar{u}+1-w}{2\tau_{ph}(1-\beta)} & -\dfrac{\bar{u}-1+w}{2\tau_s}-\dfrac{1}{\tau_s} \end{bmatrix}$$

$$B=\begin{bmatrix} 0 \\ \dfrac{\gamma_\perp}{4\chi^2\tau_s\tau_{ph}} \end{bmatrix},\ C=\begin{bmatrix}1 & 0\end{bmatrix},\ D=0 \qquad (9)$$

where: $w=\sqrt{(\bar{u}-1)^2+4\bar{u}\beta}$;

As for the associated $\Pi$-function, its diagram is depicted in Fig. 2. From this diagram it can be seen that it is indeed possible to improve performance by cycling and that an appropriate value of the sinusoid frequency is $\hat{\omega}=2.303\cdot10^{10}\ rad\,s^{-1}$.

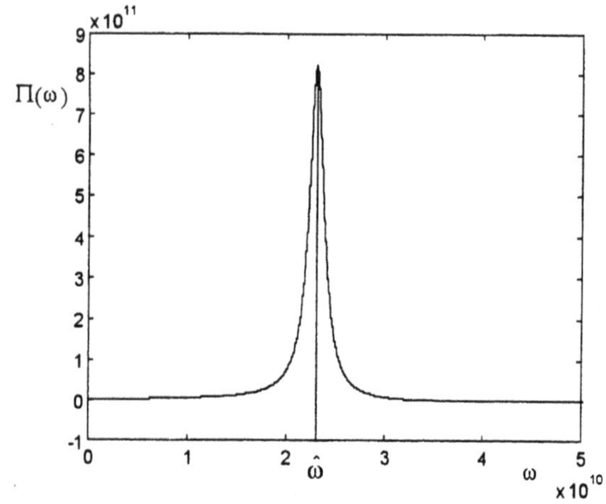


Fig. 2. The function $\Pi(\omega)$.

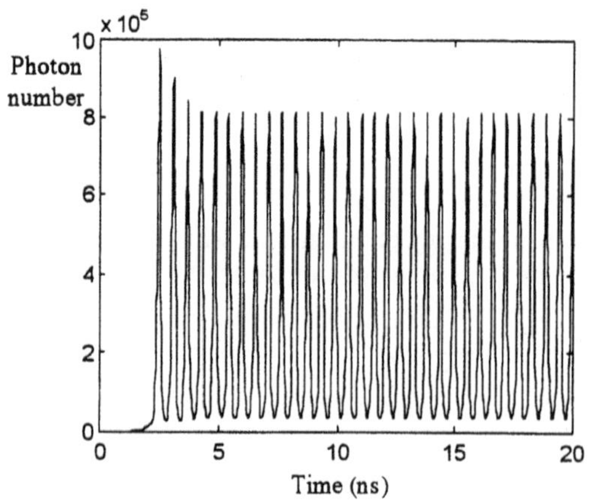


Fig. 3. Optical output obtained from the EESL of Eqs. (1) with periodic control.

In Fig. 3 it is shown the system output when a periodic control $\hat{\delta u}(t)=\varepsilon\, sin(\hat{\omega}t)$ with $\varepsilon=0.25$ is superimposed to the input $\bar{u}=3.125$. In Fig. 4 it is shown the performance index as a function of the amplitude of the periodic control.

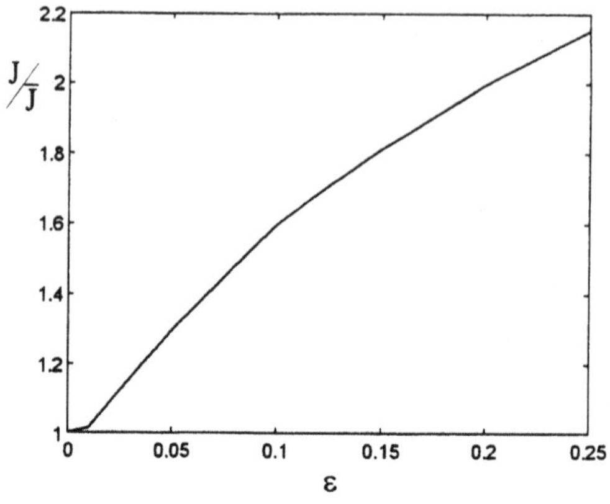


Fig. 4. Periodic performance index (normalized to the stationary one) for increasing periodic control amplitude.

*Remark 2.* There exists an equivalent way to assess the existence of a proper periodic control for a system, known as the circle criterion. Consider the following region in the complex plane:

$$\mathcal{L}=\{s:\ |s-c|-\rho>0\}$$

where: $s$ is the complex variable; $c=\dfrac{\breve{q}}{\breve{p}}+j0$;

$$\rho=\sqrt{\left(\frac{\breve{q}}{\breve{p}}\right)^2-\frac{\breve{r}}{\breve{p}}}\ \text{ with } \breve{p}\neq0.$$

The considered steady state can be improved by cycling if $\delta u(\cdot)$ is a sinusoidal signal of frequency $\omega$ for which $\breve{G}(j\omega)\in\mathbb{C}\setminus\mathcal{L}$, i.e. the whole complex plane except the region $\mathcal{L}$ (Fig. 5).

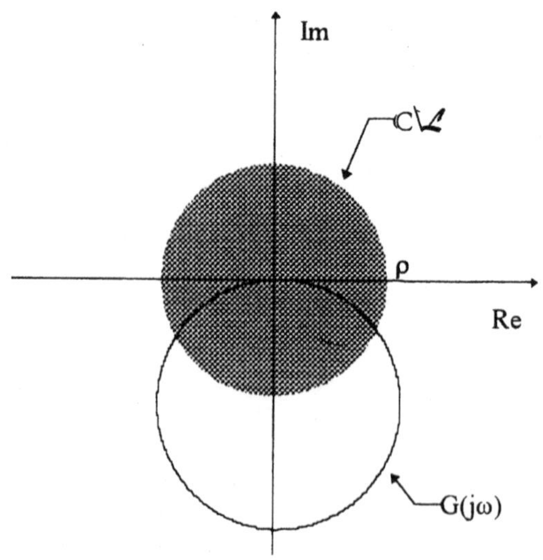


Fig. 5. The circle criterion.

## 4. COMPARISON BETWEEN PERIODICALLY OPERATED EESL AND KERR EFFECT SELF-PULSING LASERS

In this section the performances of a periodic controlled EESL will be compared with those of a Kerr Effect self-pulsing laser (KESL). The following Eqs. (10) describe the carrier and photon dynamics:

$$\dot{x}_{ph}(t)=\frac{4\chi^2}{\gamma_\perp}x_c(t)x_{ph}(t)-2k\left(1-\eta x_{ph}(t)\right)x_{ph}(t)+ +\beta\gamma_{//}x_c(t) \quad (10.a)$$

$$\dot{x}_c(t)=-\frac{4\chi^2}{\gamma_\perp}x_c(t)x_{ph}(t)-\gamma_{//}x_c(t)+\frac{I_{th}}{e}u(t) \quad (10.b)$$

This model differs from that of a standard EESL described by Eqs. (1) for the term $\eta\, x_{ph}^2(t)$ which takes into account the optical nonlinearity in the active material, experimentally linked to Kerr effect (Abbati *et al.*, 1998b; Butcher and Cotter, 1990). This term originates in the following physical process: the active region of the laser is subject to an absorbing mechanism driven by the electromagnetic field intensity in the laser cavity, i.e. by the photon number $x_{ph}$, as $x_{ph}\propto|E|^2$, where $|E|$ is the field amplitude. This mechanism leads to a pulsing output according to the following rationale:

- The number of outgoing photons is reduced, which is equivalent to an increase in photon lifetime $\tau_{ph}$. This phenomenon can be taken into account through a modification of the term associated for photon removal due to the cavity losses in the rate equation model. It is therefore possible to write $\tau_{ph}=\hat{\tau}_{ph}\left(1-\eta x_{ph}\right)^{-1}$, where $\hat{\tau}_{ph}$ is the non perturbed photon-lifetime. Note that $\eta$ is the parameter that takes into account the Kerr effect. The value of such a parameter can be chosen by the designer by a suitable selection of the optical properties of the active material in the cavity.
- The longer lifetime of the photons inside the cavity greatly increases the stimulated emission process (i.e. the fundamental phenomenon of laser light emission) and consequently a great increase in photon emission can be observed.
- Over a certain threshold the absorption mechanism is no more effective and the rate of stimulated emission drops.

In particular the absorption mechanism, identified as the responsible for pulsation appearance, can be related to the third-order susceptibility $\chi^{(3)}$ of the material (Abbati *et al.*, 1998b). In fact, the photon lifetime increment can be also seen as the consequence of a refractive index increment, induced by the absorption process and controlled by the optical field intensity in the laser cavity. Finally, the parameter $\eta$ can then be linked to the nonlinear refractive index $n_2$ of the material of the active region. The system described in Eqs. (10) is again a nonlinear dynamical system with two state variables. This system has the feature of presenting a Hopf bifurcation depending on parameter $\eta$. For small values of $\eta$ the system shows a single stable equilibrium point; for high values of $\eta$ the system output has a cyclic behaviour. Fig. 6 shows which kind of output behaviour is obtained as a function of the (constant) control variable and the parameter $\eta$.

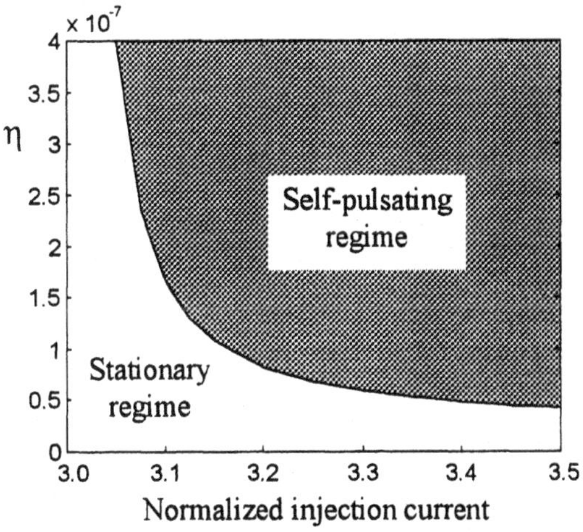


Fig. 6. Bifurcation diagram for Eqs. (10).

Fig. 7 shows the different behaviours of the optical output below and above the bifurcation point when a constant injection current is used; it is evident that a self-pulsing regime can be established. The value of the term $\eta x_{ph}$ at which self-pulsing regime establishes ranges between $3\cdot10^{-2}$ and $10^{-3}$ according to the active material characteristics and the injection current intensity.

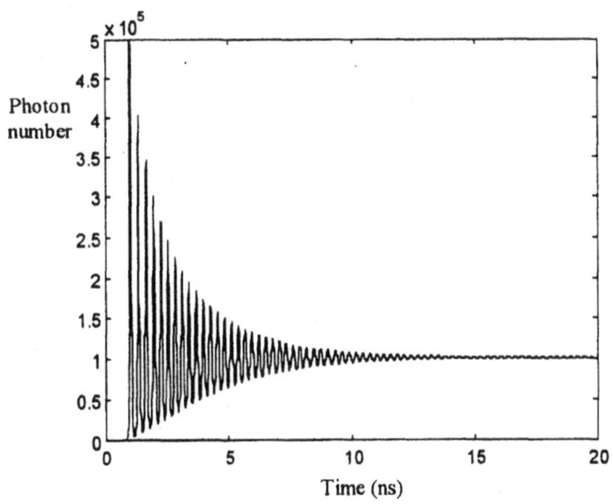


Fig. 7.a. Photon dynamics with $\eta=2\cdot10^{-7}$.

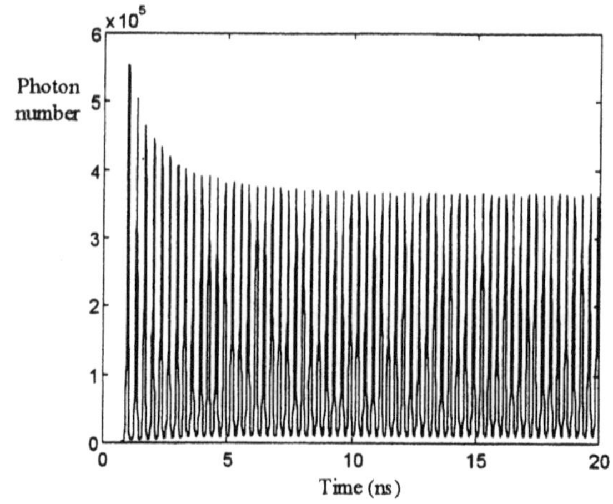


Fig. 7.b. Photon dynamics with $\eta = 1.5 \cdot 10^{-7}$.

In Fig. 8 a comparison between the output from a periodic controlled EESL and the output from a KESL is shown. It can be seen that the pulse intensity obtained using periodic control are more intense than that of the KESL. Moreover, the difficult in the choice of the frequency of the periodic input has been overcome by the use of the Π-function in the periodic control approach. Finally, the periodic controlled pulsatile behaviour has been obtained without the use of a too much intense injection current; this is an advantage since the coupling of the high optical power density in the cavity (due to the photon pulses) and of high injection current densities in the cavity can lead to irreversible damage of the device, highly probable in KESL devices due to the presence of the Kerr effect field intensity enhancement which can produce high photon concentrations in localized regions of the cavity.

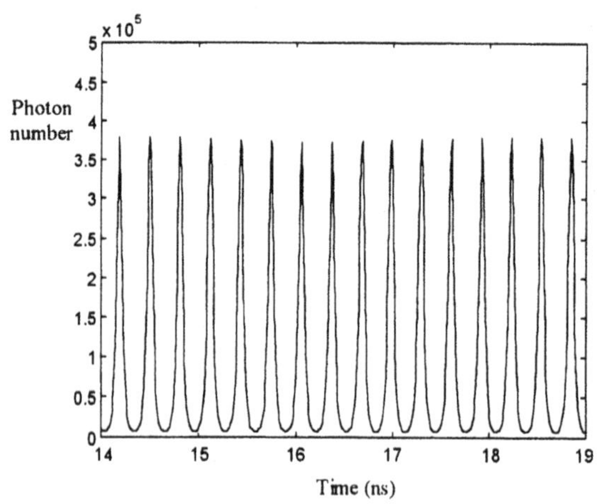


Fig. 8. Comparison of the optical output from a KESL device (solid) and a periodic controlled EESL device (dotted).

## 5. CONCLUDING REMARKS

Periodic control has been used to deal with a variety of applications. In this paper, we apply it to a new field, that of optoelectronics. Precisely, we show that better performances are achievable by forcing a pulsatile behaviour in a standard edge-emitter semiconductor laser. These performances are even better than those which can be obtained, at the price of structural and physical modifications, in self-pulsing laser devices.

## REFERENCES


Abbati P., F. Previdi and M. Milani (1998a). Optical feedback control of self-pulsations in semiconductor lasers. Presented at *13th U.K. National Quantum Electronic Conference*, Cardiff, Wales (UK).

Abbati P., F. Previdi and M. Milani (1998b). Optical feedback sustained self-pulsations in semiconductor lasers. *Nuovo Cimento D*, in press.

Avrutin E.A. (1993). Analysis of spontaneous emission and noise in self-pulsing laser diodes. *Proc. IEEE*, **140**, 16-20.

Berstein D.S. and E.G. Gilbert (1980). Optimal periodic control: the π-test revisited. *IEEE Trans. Automatic Control*, **25**, 673-684.

Bittanti S. and P. Colaneri (1998). Periodic Control. In: *Encyclopedia of Electrical and Electronic Engineering* (J. Webster Ed.). J. Wiley.

Bittanti S., G. Fronza and G.O. Guardabassi (1973). Periodic Control: A Frequency Domain Approach. *IEEE Trans. Automatic Control*, **18**, 33-38.

Butcher P.N. and D. Cotter (1990). *The elements of nonlinear optics*. Cambridge University Press.

Gilbert E.G. (1977). Optimal periodic control: a general theory of necessary conditions. *SIAM Jour. Control and Opt.*, **15**, 717-746.

Iga K. (1994). Fundamental of laser optics. Plenum Press, New York.

Pauli, T. L. (1978). Saturable absorption effects in the self-pulsing (AlGa)As junction laser. *Appl. Phys. Lett.*, **34**, 652-655.

Svelto O. (1989). *Principle of Lasers*. Plenum Press, New York.

Ueno M. and R. Lang (1985). Conditions for self-sustained pulsation and bistability in semiconductor laser. *J. Appl. Phys.*, **58**, 1689-1692.

Yamada M. (1993). A theorethical analysis of self-sustained pulsation phenomena in narrow-stripe semiconductor lasers. *IEEE J. Quantum Electron.*, **29**, 1330-1336.

Zhang C., J. Zhang and K. Furuta (1996). Performance analysis of periodically time-varying controllers. *Proc. of the IFAC World Congress*, San Francisco (CA), 207-212.

14th Triennial World Congress, Beijing, P.R. China





# ROBUST MAXIMUM PRINCIPLE FOR MINIMAX BOLZA PROBLEM WITH TERMINAL SET

**V.G.Boltyanski * A.S.Poznyak ****

* *CIMAT, Guanajuato,AP 402, Mexico 36000*
***CINVESTAV-IPN, Seccion de Control Automatico, AP 14 740, CP 07000, Mexico D.F., Mexico*
** *CINVESTAV-IPN, Seccion de Control Automatico, Av.IPN 2508, A.P.14-740, Mexico D.F., 07000, Mexico, e-mail: apoznyak@ctrl.cinvestav.mx*

Abstract: This paper presents a version of Robust Maximum Principle applied to the Minimax Bolza problem. We deal with the cost function containing a terminal term as well as integral one.A fixed horizon and terminal set are considered. We construct the optimal control strategies for the class of uncertain systems given by an ordinary differential equation with unknown parameters from a given finite set. This problem belongs to the class of minimax optimization problems. The proof is based on the recent results obtained for Minimax Mayer Problem (Boltyansky and Poznyak, 1999). Illustrative examples conclude this study. 

Keywords: Maximum principle, optimization, minimax robust control, Boltz and Mayer problems.

## 1. INTRODUCTION

During last two decades, the minimax control problem, dealing with different classes of nonlinear systems, received great attention from many researchers because of its theoretical and practical importance. Basically, the results of this area were based on the traditions of two classical approaches: *Maximum Principle* (Pontryagin *et all*, 1969 ) and *Dynamic Programming Method* (Bellman, 1956). In the case of a complete model description, both of them can be directly applied to construct the optimal control.

Faced with any sort of uncertainties (parametric type, unmodelled dynamics, external perturbations and etc.) these results cannot be applied. The main way to obtain a solution suitable for a class of given models is to formulate a corresponding *minimax control problem*, where maximization is taken over a set of uncertainty and minimization is taken over control strategies within a given set. Several approaches are effective in this situation: the *game-theoretic approach* (Basar and Bernhard, 1991), *Lagrange Multipliers Approach* (Kurjanskii, 1977 ) and *frequency domain methods* (Zames, 1981), which dealt with to minimize the norm of the transfer function between the disturbance inputs and the performance output, the minimax controller design is formulated as $H^{\infty}$- *optimization problem.* As it was shown in (Basar and Bernhard, 1991) this specific problem can be successfully solved in time domain, leading to rapprochement with dynamic game theory and to the establishment of a relation with *risk - sensitivity quadratic stochastic control.* The paper (Ming *et all*, 1991) presents a control design method for continuous - time plants whose uncertain parameters in the output matrix are only known to lie within an ellipsoidal set. An algorithm for minimax control which at every iteration minimizes approximately the defined Hamiltonian is described in (Pytlak, 1990). In (Didinsky and

Basar, 1994), using *"the cost - to - come" method*, the authors show that the original problem with incomplete information can be converted into an equivalent full information minimax control problem of higher dimension, which can be solved using Dynamic Programming Approach. Minimax control of a class of dynamic systems with mixed uncertainties is investigated in (Basar, 1994). A continuous deterministic uncertainty which affects the system dynamics, and a discrete stochastic uncertainties leading to jumps in the system structure at random times are studied. The solution presents a finite dimensional compensator using two finite sets of partial differential equations. Robust controller for linear time-varying systems given by stochastic differential equation was studied in (Poznyak and Taksar, 1996). The solution is based on the stochastic *Lyaponov-like analysis* with the martingale technique implementation. Another problems dealing with discrete time models of deterministic and/or stochastic nature and their corresponding solutions are discussed in (Blom and Everdij, 1993), (Bernhard, 1994) and (Ugrinovskii, 1997). The paper (Demianov V.F. and T.K. Vinogradova, 1973) pretends to formulate Maximum Principle for Minimax Lagrange problem with a fixed horizon and with parametric uncertainty within a given compact, but on our opinion, it contains mismatch result which has been given without any proof. A comprehensive survey of different parameter space methods for robust control design can be found in (Siljak, 1989).

The purpose of this paper is to explore the possibilities of the Maximum Principle approach for the class of minimax control problems dealing with construction of the optimal control strategies for uncertain systems given by a system of ordinary differential equations with unknown parameters from a given finite set. The problem under consideration belongs to the class of optimization problems of minimax type. A version of *Robust Maximum Principle* applied to the Minimax Bolza problem with terminal set is presented. In this study we deal with the cost function containing a terminal term as well as integral one. A fixed horizon is considered. The proof is based on the recent results obtained for Minimax Mayer Problem (Boltyansky and Poznyak, 1999) and uses, so named, the *Tent Method* (Boltyansky, 1975) to formulate the necessary conditions of optimality in Hamiltonian form. The main result deals with finite parametric uncertain sets involved into the description of a model. Illustrative examples conclude this study.

## 2. MINIMAX PROBLEM

Consider the controlled plant

$$\dot{x} = f^{\alpha}(x,u) \qquad (1)$$

where $x = \left(x^1,...,x^n\right)^T \in R^n$ is its state vector, $u = \left(u^1,...,u^r\right)^T \in R^r$ is the control that may run over a given control region $U \subset R^r$, and $\alpha$ is a parameter that may run over a given parametric set $\mathcal{A}$. In this paper we assume that the parametric set $\mathcal{A}$ is finite. On the right-hand side

$$f^{\alpha}(x,u) = \left(f^{\alpha,1}(x,u),...,f^{\alpha,n}(x,u)\right)^T \in R^n$$

we impose the usual restrictions: continuity with respect to the collection of the arguments $x, u$ and differentiability (or Liptshitz condition) with respect to $x$. One more restriction will be formulated below.

**Definition 1.** *A function $u(t)$ $t_0 \le t \le t_1$ is said to be an admissible control if it is piecewise continuous and $u(t) \in U$ for all $t \in [t_0, t_1]$.*

We assume for convenience that every admissible control is right-continuous, i.e.,

$$u(t) = u(t+0) \text{ for } t_0 \le t < t_1$$

and moreover, $u(t)$ is continuous at the terminal moment, i.e., $u(t_1) = u(t_1 - 0)$. We fix also an initial point $x_0 = (x_0^1,...,x_0^n)^T \in R^n$. For a given admissible control $u(t)$, $t_0 \le t \le t_1$, consider the corresponding solution $x^{\alpha}(t) = \left(x^{\alpha,1}(t),...,x^{\alpha,n}(t)\right)^T$ of (1) with the initial condition $x^{\alpha}(t_0) = x_0$. We suppose that for any admissible control $u(t)$, $t_0 \le t \le t_1$, all solutions $x^{\alpha}(t), \alpha \in \mathcal{A}$ are defined on the whole segment $[t_0, t_1]$ (this is an additional restriction on the right-hand side of (1)). In the space $R^n$, let us define the terminal set $\mathcal{M}$ given by the inequality

$$g(x) \le 0 \qquad (2)$$

where $g(x)$ is a smooth real function of $x \in R^n$. For a given admissible control $u(t)$, $t_0 \le t \le t_1$, we are interested in the corresponding trajectory starting from the initial point $x_0$. But we do not know the realized value of $\alpha \in \mathcal{A}$. Therefore we have to consider the *family* of trajectories $x^{\alpha}(t)$ with insufficient information about the trajectory which is realized.

**Definition 2.** *We say that the control $u(t)$, $t_0 \le t \le t_1$, realizes the terminal condition (2) if for every $\alpha \in A$ the corresponding trajectory $x^{\alpha}(t)$ satisfies the inclusion $x^{\alpha}(t_1) \in M$.*

As the cost function, we select function containing the integral term as well as terminal one:

$$h^{\alpha} := h_0(x^{\alpha}(t_1)) + \int_{t=t_0}^{t=t_1} f^{n+1}\left(x^{\alpha}(t), u(t), t\right) dt$$

The end time-point $t_1$ is not fixed and $x^{\alpha}(t) \in R^n$. Analogously, since the realized value of the parameter $\alpha$ is unknown, we define the *worst (highest) cost*

$$F = \max_{\alpha \in \mathcal{A}} h^{\alpha} \tag{3}$$

The function $F$ depends only on the considered admissible control $u(t)$, $t_0 \le t \le t_1$.

**Definition 3.** *We say that the control is robust optimal if (i) it realizes the terminal condition, (ii) it realizes the minimal worst (highest) cost $F$ (among all admissible controls which satisfy the terminal condition).*

Thus the *Robust Optimization Problem* consists of finding the control action $u(t)$, $t_0 \le t \le t_1$, which realizes

$$\min_{u(t)} F = \min_{u(t)} \max_{\alpha \in \mathcal{A}} h^{\alpha} \tag{4}$$

where the minimum is taken over all admissible control strategies satisfying the terminal condition. It is the *Minimax Bolza Problem.* Now we introduce $(n+1)$-dimensional space $R^{n+1}$ of the variables $x = (x_1, ..., x_n, x_{n+1})$ where the first $n$ coordinates satisfy (1) and the component $x_{n+1}$ is given by

$$x^{\alpha,n+1}(t) := \int_{\tau=t_0}^{t} f^{n+1}\left(x^{\alpha}(\tau), u(\tau), \tau\right) d\tau$$

or, in the differential form,

$$\dot{x}^{\alpha,n+1}(t) = f^{n+1}\left(x^{\alpha}(t), u(t), t\right) dt \tag{5}$$

with the initial condition $x^{\alpha,n+1}(t_0) = 0$. As a result, in the space $R^{n+1}$ we obtain the Robust Optimization Problem in Mayer form discussed in (Boltyansky and Poznyak, 1999) with the cost function

$$h^{\alpha} = h_0(x^{\alpha}(t_1)) + x^{\alpha,n+1}(t_1) \tag{6}$$

where the function $h_0(x^{\alpha})$ does not depend on the last coordinate $x^{\alpha,n+1}$, i.e., $\frac{\partial}{\partial x^{\alpha,n+1}} h_0(x^{\alpha}) = 0$. The Mayer Problem with the cost function (6) is equivalent to the Optimization Problem in Bolza form. Let

$$\bar{x}^{\alpha}(t) = \left(x^{\alpha,1}(t), ..., x^{\alpha,n}(t), x^{\alpha,n+1}(t)\right) \in R^{n+1}$$

be a solution of the system (1), (5). We also introduce for any $\alpha \in \mathcal{A}$ the following cojugate variables

$$\psi_{\alpha}(t) = (\psi_{\alpha,1}(t), ..., \psi_{\alpha,n}(t), \psi_{\alpha,n+1}(t)) \in R^{n+1}$$

satisfying

$$\dot{\psi}_{\alpha,i} = -\sum_{k=1}^{n+1} \frac{\partial f^{\alpha,k}(x^{\alpha}(t), u(t))}{\partial x^{\alpha,i}} \psi_{\alpha,k} \tag{7}$$

with the terminal condition

$$\psi_{\alpha,j}(t_1) = b_{\alpha,j},\ t_0 \le t \le t_1,\ \alpha \in \mathcal{A},\ j = 1, ..., n+1$$

Let $\bar{\psi}_{\diamond} = (\psi_{\alpha,i}) \in R_{\diamond}$ be a covariant vector and

$$\bar{f}^{\diamond}(\bar{x}^{\diamond}, u) = \left(f^{\alpha,k}\right), \quad \bar{x}^{\diamond} = \left(x^{\alpha,k}\right)$$

Introduce the Hamiltonian function

$$\mathcal{H}^{\diamond}(\bar{\psi}_{\diamond}, \bar{x}^{\diamond}, u) = \langle \bar{\psi}_{\diamond}, \bar{f}^{\diamond}(\bar{x}^{\diamond}, u) \rangle =$$

$$= \sum_{\alpha \in \mathcal{A}} \left\langle \bar{\psi}_{\alpha}, \bar{f}^{\alpha}(x^{\alpha}, u) \right\rangle = \sum_{\alpha \in \mathcal{A}} \sum_{i=1}^{n} \psi_{\alpha,i} f^{\alpha,i}(x^{\alpha}, u) \tag{8}$$

Remark that $\mathcal{H}^{\diamond}(\bar{\psi}_{\diamond}, \bar{x}^{\diamond}, u)$ is the sum of "usual" Hamiltonian functions:

$$\mathcal{H}^{\diamond}(\bar{\psi}_{\diamond}, \bar{x}^{\diamond}, u) = \sum_{\alpha \in \mathcal{A}} \left\langle \bar{\psi}_{\alpha}, \bar{f}^{\alpha}(x^{\alpha}, u) \right\rangle$$

The function (8) allows to rewrite the conjugate equations (7) for the plant (1) in the following vector form:

$$\frac{d}{dt} \bar{\psi}_{\diamond} = -\frac{\partial \mathcal{H}^{\diamond}(\bar{\psi}_{\diamond}, \bar{x}^{\diamond}(t), u(t))}{\partial \bar{x}^{\diamond}} \tag{9}$$

Let now $b_{\diamond} = (b_{\alpha,i}) \in \bar{R}_{\diamond}$ be a covariant vector. Denote by $\psi_{\diamond}(t)$ the solution of the equation (9) with the terminal condition $\psi_{\diamond}(t_1) = b_{\diamond}$. We say that the control $u(t)$, $t_0 \le t \le t_1$, satisfies the maximum condition with respect to $x^{\diamond}(t), \psi_{\diamond}(t)$ if

$$u(t) = \arg\max_{u \in U} \mathcal{H}^{\diamond}(\psi_{\diamond}(t), x^{\diamond}(t), u) \quad \forall t \in [t_0, t_1],$$

i.e.,

$$\mathcal{H}^{\diamond}(\psi_{\diamond}(t), x^{\diamond}(t), u(t)) \ge \mathcal{H}^{\diamond}(\psi_{\diamond}(t), x^{\diamond}(t), u) \quad \forall u \in U,\ t \in [t_0, t_1]. \tag{10}$$

## 3. ROBUST MAXIMUM PRINCIPLE

Applying the Theorem 1 from (Boltyansky and Poznyak, 1999) to this optimization problem, we derive the following result.

**Theorem** *(MaximumPrinciple for Bolza problem with a terminal set). Let $u(t)$ $(t \in [t_0, t_1])$ be an admissable **Robust** control and $x^{\alpha}(t)$ be the corresponding solution of (1) with the initial condition $x^{\alpha}(t_0) \in x_0$ $(\alpha \in \mathcal{A})$. The parametric uncertainty set $\mathcal{A}$ is assumed to be finite. Assume also that the terminal condition is satisfied:*

$$x^{\alpha}(t_1) \in \mathcal{M} \text{ for all } \alpha \in \mathcal{A}.$$

*For robust optimality of a control action $u(t)$, $t_0 \le t \le t_1$ it is necessary that there exist a vector $b_{\diamond} \in \bar{R}_{\diamond}$ and nonnegative real functions $\mu(\alpha)$, $\nu(\alpha)$ defined on $\mathcal{A}$ such that the following conditions are satisfied: (i) **(the maximality condition)**: denote by $\bar{\psi}_{\diamond}(t)$, $t_0 \le t \le t_1$ the solution of the equation (7) with the terminal condition; then the* *robust optimal control $u(t)$, $t_0 \le t \le t_1$ satisfies the maximality condition (10); moreover*

$$\mathcal{H}^\circ(\psi_\circ(t), x^\circ(t), u(t)) = 0 \quad \forall t_0 \le t \le t_1 ;$$

*(ii)* ***(complementary slackness conditions):*** *for every $\alpha \in \mathcal{A}$, either the equality $h^\alpha = F^0$ holds, or $\mu(\alpha) = 0$, i.e.,*

$$\mu(\alpha) \cdot \left(h^\alpha - F^0\right) = 0;$$

*moreover, for every $\alpha \in \mathcal{A}$, either the equality $g(x^\alpha(t_1)) = 0$ holds, or $\nu(\alpha) = 0$, i.e.,*

$$\nu(\alpha) \cdot g(x^\alpha(t_1)) = 0;$$

*(iii)* ***(transversality condition):*** *for every $\alpha \in \mathcal{A}$, the equalities*

$$\psi_\alpha(t_1) + \mu(\alpha) \text{ grad } h_0(x^\alpha(t_1)) +$$

$$\nu(\alpha) \text{ grad } g(x^\alpha(t_1)) = 0$$

*and*

$$\psi_{\alpha, n+1}(t_1) + \mu(\alpha) = 0$$

*hold; (iv) (nontriviality condition): there exists $\alpha \in \mathcal{A}$ such that either $\psi_\alpha(t_1) \neq 0$, or at least one of the numbers $\mu(\alpha), \nu(\alpha)$ is distinct from zero.*

## 4. ILLUSTRATIVE EXAMPLE

To illustrate the approach suggested in this paper, let us consider the family of the controlled objects given by

$$\dot{x} = -\lambda x + b^\alpha u + w^\alpha, \quad \alpha \in \mathcal{A} = \{\alpha_1, ..., \alpha_q\}$$

where $x \in R^1$ is scalar variable, $\lambda > 0$, and $b^\alpha, w^\alpha$ are real constants, satisfying

$$w^\alpha + |b^\alpha| < 0 \tag{11}$$

for every $\alpha \in \mathcal{A}$ and $x(0) \in \mathcal{X}_0$. In spite of the simplicity, this model contains all features specific for uncertain systems. Indeed, the set $\mathcal{W} = \{w^\alpha\}$ characterizes an external uncertainty, the set $\mathcal{B} = \{b^\alpha\}$ describes an internal uncertainty and the set $\mathcal{X}_0$ characterizes an initial condition uncertainty. If at least one of these sets contains more than unique element, we deal with uncertain model. We will consider the case when

$$u \in U = \{u : |u| \le 1\}, \tag{12}$$

the set $\mathcal{M}$ of final states is defined by the inequality

$$g(x) = x \le 0 \tag{13}$$

and the set $\mathcal{X}_0$ contains unique element $\mathcal{X}_0 = \{x_0\}$, $x_0 = 1$. The problem is to find an admissible control $u$ (defined within the interval $t \in [0, t_1]$ which is unknown) such that the worst penalty function given by

$$F^0 = \max_{\alpha \in \mathcal{A}} \left(x^\alpha(t_1)\right)^2 \tag{14}$$

be minimal. This is the minimax Mayer problem. Remark that, by (??), (12), we have

$$\dot{x} < -\delta \ \forall x > 0, \ \alpha \in \mathcal{A}, \ -1 \le u \le 1,$$

where $\delta$ is a positive constant. Consequently, starting at time $t_0 = 0$ from the initial point $x_0 = 1$, each trajectory $x^\alpha(t)$ arrives to the terminal set $\mathcal{M}$ in a time $t_1 < \delta^{-1}$ and remains in $\mathcal{M}$. In other words, any control $u(t)$, $0 \le t \le t_1$, satisfies the terminal condition, if $t_1$ is close enough to $\delta^{-1}$. Assume that a control $u(t)$, $0 \le t \le t_1$, is robust optimal and hence the conclusion of Theorem 1 holds. Directly applying Theorem, we deduce:

$$\mathcal{H}^\circ = \sum_{\alpha \in \mathcal{A}} \psi_\alpha(-\lambda x^\alpha + b^\alpha u + w^\alpha),$$

(we remark that $n = 1$ and hence we may omit the index $i$ in the right-hand side). Furthermore, $\dot{\psi}_\alpha = \lambda \psi_\alpha$ $(\alpha \in \mathcal{A})$ and hence $\psi_\alpha = c_\alpha e^{\lambda t}$ where $c_\alpha = \text{const}$. Thus

$$\mathcal{H}^\circ = -\lambda \sum_{\alpha \in \mathcal{A}} c_\alpha x^\alpha(t) e^{\lambda t} + \sum_{\alpha \in \mathcal{A}} c_\alpha b^\alpha e^{\lambda t} u(t) + \sum_{\alpha \in \mathcal{A}} c_\alpha w^\alpha e^{\lambda t} \equiv 0. \tag{15}$$

Assume first that $\sum_{\alpha \in \mathcal{A}} c_\alpha b^\alpha \neq 0$. Then, by the maximum condition $(i)$, we have

$$u = \arg \max_{u \in U} \mathcal{H}^\circ = \text{sign}\left(\sum_{\alpha \in \mathcal{A}} c_\alpha b^\alpha\right) = \pm 1. \tag{16}$$

**a).** Consider more detailed the case $u \equiv 1$. By (11) and using the initial condition $x^\alpha(0) = 1$, we obtain

$$x^\alpha(t) = \frac{b^\alpha + w^\alpha}{\lambda} + \left(1 - \frac{b^\alpha + w^\alpha}{\lambda}\right) e^{-\lambda t}.$$

Consequently, by (15),

$$\mathcal{H}^\circ = -\sum_{\alpha \in \mathcal{A}} c_\alpha(\lambda - b^\alpha - w^\alpha) = 0,$$

i.e.,

$$\sum_{\alpha \in \mathcal{A}} \beta^\alpha c_\alpha = 0 \tag{17}$$

where the coefficients $\beta^\alpha := \lambda - b^\alpha - w^\alpha$, $\alpha \in \mathcal{A}$ are positive, by (11).

**b).** For $u \equiv -1$, by an analogous reasoning, we obtain the equality (17) with $\beta^\alpha := \lambda + b^\alpha - w^\alpha > 0$, $\alpha \in \mathcal{A}$. If even $\sum_{\alpha \in \mathcal{A}} c_\alpha b^\alpha = 0$, then for $t = 0$ we obtain from (15)

$$-\lambda \sum_{\alpha \in \mathcal{A}} c_\alpha + \sum_{\alpha \in \mathcal{A}} c_\alpha w^\alpha = 0$$

(since $x^\alpha(0) = 1$ for every $\alpha \in \mathcal{A}$), i.e., we again obtain the equality (17) with positive coefficients $\beta^\alpha := \lambda - w^\alpha > 0$, $\alpha \in \mathcal{A}$. (cf. (11)). Thus at any rate we have the equality (17) with some positive coefficients.

Let now take into consideration the conditions $(ii)$, $(iii)$, $(iv)$ of Theorem 1. Admit that all values $x^{\alpha}(t_1)$ are distinct from zero, i.e.,$x^{\alpha}(t_1) < 0$, $\alpha \in \mathcal{A}$. From $(ii)$ we obtain $\nu(\alpha) = 0$, $\alpha \in \mathcal{A}$, and hence, by $(iii)$,

$$\psi_{\alpha}(t_1) + \mu(\alpha) \cdot \text{ grad } f^0(x^{\alpha}(t_1)) = 0, \ \alpha \in \mathcal{A}.$$

This means, by $(iv)$, that at least one of the numbers $\mu(\alpha)$, $\alpha \in \mathcal{A}$, is distinct from zero. Multiplying the last equality by $\beta^{\alpha}$ and summing, we obtain, by (17)

$$\sum_{\alpha \in \mathcal{A}} \beta^{\alpha} \mu(\alpha) \cdot \text{ grad } f^0(x^{\alpha}(t_1)) = 0,$$

contradicting the ineqialities

$$\beta^{\alpha} > 0, \ \mu(\alpha) \geq 0, \ \text{grad } f^0(x^{\alpha}(t_1)) < 0,$$

and taking into account that at least one of $\mu(\alpha)$ being positive. This contradiction shows that at least one of the values $x^{\alpha}(t_1)$ has to be equal to zero. This is the main conclusion of the robust maximum principle for the controlled object, and we use it in the numerical example given below.

*Example 1.* We now consider a particular case of the controlled object with $\lambda = 1$ and suppose that $q = 4$, i.e., $\mathcal{A} = \{1, 2, 3, 4\}$, where the coefficients $b^{\alpha}, w^{\alpha}$ have the following values:

$$b^1 = b^2 = 1, \ b^3 = b^4 = -1,$$
$$w^1 = w^3 = -1.9, \ w^2 = w^4 = -2.1.$$

in other words, we consider the controlled object

$$\dot{x} = -x - 2 \pm u \pm 0.1$$

with all combinations of signs depending on $\alpha \in \mathcal{A}$. By the terminal condition, it is necessary that $x^{\alpha}(t_1) \leq 0 \ \ \forall \ \ \alpha \in \mathcal{A}$ and for the robust optimality at least one of the values $x^{\alpha}(t_1)$ being equal to zero. We have $\dot{x}^2 - \dot{x}^1 = -(x^2 - x^1) - 0.2$, and hence (since $x^1(0) = x^2(0) = 1$) we obtain

$$x^2(t) - x^1(t) = -0.2\left(1 - e^t\right). \qquad (18)$$

Similarly,$x^4(t) - x^3(t) = -0.2\left(1 - e^t\right)$.Consequently

$$x^2(t) < x^1(t), \quad x^4(t) < x^3(t) \ \forall t > 0. \qquad (19)$$

This means that neither of the equalities $x^2(t_1) = 0$, $x^4(t_1) = 0$ is possible. Thus at least one of the values $x^1(t_1), x^3(t_1)$ is equal to zero, say, $x^1(t_1) = 0$ (the case $x^3(t_1) = 0$ is analogous). Thus $x^1(t_1) = 0$, $x^3(t_1) \leq 0$. Assume that $x^3(t_1) < 0$, and consequently, $x^4(t_1) < 0$. As a result, the values $\alpha = 2, 3, 4$ are $g$-inactive. Hence, by the condition $(ii)$ of Theorem, $\nu(2) = \nu(3) = \nu(4) = 0$.Moreover, since

$$x^2(t_1) = -0.2\left(1 - e^{-t_1}\right),$$
$$x^4(t_1) = x^3(t_1) - 0.2\left(1 - e^{-t_1}\right)$$

(cf. (18) and (19)), the value $x^4(t_1)$ is the least from all $x^{\alpha}(t_1)$, $\alpha \in \mathcal{A}$. Consequently, $F^0 = f^0(x^{\alpha}(t_1))$ only for $\alpha = 4$, i.e., the values $\alpha = 1, 2, 3$ are $f^0$-inactive. This means (by the condition $(ii)$ of Theorem) that $\mu(1) = \mu(2) = \mu(3) = 0$. Now, by $(iii)$, we have

$$\psi_1(t_1) + \nu(1) = 0, \quad \psi_2(t_1) = \psi_3(t_1) = 0,$$
$$\psi_4(t_1) + \mu(4) \cdot \text{ grad } f^0(x^4(t_1)) = 0.$$

Consequently,

$$\psi_1(t_1) \leq 0, \quad \psi_2(t_1) = \psi_3(t_1) = 0, \quad \psi_4(t_1) \geq 0.$$

This means that the equality $\psi_1(t_1) - \psi_4(t_1) = 0$ is impossible, otherwise we should have $\psi_1(t_1) = \psi_4(t_1) = 0$ and $\nu(1) = \mu(4) = 0$, contradicting the condition $(iv)$ of Theorem. Thus, the sum

$$\sum_{\alpha} \psi_{\alpha}(t_1) b^{\alpha} = \psi_1(t_1) - \psi_4(t_1)$$

is negative, i.e., $u \equiv -1$. But evidently, the control $u \equiv -1$ must be removed, since by the terminal condition and $x^1(t_1) = 0$, we have $x^3(t_1) > 0$. Hence, the assumption $x^3(t_1) < 0$ is contradictory, i.e., $x^1(t_1) = x^3(t_1) = 0$. Consequently, the variable $y = x^1 - x^3$ satisfies the condition $y(0) = y(t_1) = 0$. Since, $\dot{y} = -y + 2u(t)$, we obtain

$$y(t) = \left(2 \int_0^t u(t)\, e^{\tau}\, d\tau\right) e^{-t}.$$

Now the equality $y(t_1) = 0$ implies

$$\int_0^{t_1} u(t) e^t \, dt = 0. \qquad (20)$$

Furthermore, by (20) and $x^1(0) = 1$, we have

$$x^1(t) = \left(\int_0^t (u(t) - 1.9) e^t \, dt + 1\right) e^{-t} =$$
$$\left(-1.9 \int_0^t e^t \, dt + 1\right) e^{-t} = -1.9 + 2.9 e^{-t},$$

and consequently (by $x^1(t_1) = 0$) we obtain $t_1 = \ln \frac{29}{19}$. Now from (18) and (19) we find

$$x^1(t_1) = x^3(t_1) = 0,$$
$$x^2(t_1) = x^4(t_1) = -\frac{2}{29}, \quad F^0 = \left(\frac{2}{29}\right)^2. \qquad (21)$$

Thus, if a control $u(t)$, $0 \leq t \leq t_1$, is robust optimal, then $t_1 = \frac{29}{19}$ and (20) holds. Evidently, the condition (20) is sufficient, too. Indeed, it implies $t_1 = \ln \frac{29}{19}$ and consequently (21) holds. It is easily shown that in this case there are infinitely many robust optimal controls. Indeed, denote by $C^{\diamond}$ the set of all piecewise continuous functions defined on the segment $S = [0, \ln \frac{29}{19}]$ with the norm $\| \ u \ \| = \max_{t \in S} |u(t)|$. Then $C^{\diamond}$ is an infinite-dimensional normed space (it is not a Banach space, since it is not complete). The left-hand side of (20) is a bounded linear functional on $C^{\diamond}$, and consequently (20) defines an infinitely-dimensional subspace $L$ of $C^{\diamond}$. Thus every function $u \in L$ with $\| \ u \ \| \leq 1$ is an admissible, robust optimal control. This shows the range of the set of robust optimal controls.

Surely, there are examples, in which robust optimal controls are unique.

*Example 2.* Consider the same controlled object with $\mathcal{A} = \{1, 2\}$, $\lambda = b^1 = b^2 = 1$, $w^1 = -2, w^2 = -3$. Suppose that $x_0$, $\mathcal{M}$, and $f^0(x)$ are as in Example 1. Repeating the calculation given above, we can prove that a control $u(t)$, $0 \leq t \leq t_0$, is robust optimal if and only if $t_1 = \ln \frac{4}{3}$ and $u(t) \equiv -1$.

## 5. CONCLUSION

The main point is that **Hamilton function used in this Robust Maximum Principle is shown to be equal to the sum of the standard Hamiltonians corresponding to a fixed value of uncertain parameter**. Families of state and conjugated variables differential equations together with transversality and complementary slackness conditions are demonstrated to form a closed system of equations, sufficient to construct a corresponding robust control. This methodology can be successfully applied to the solution of the analogous problem but with parametric compact set of uncertainties. The next studies will realize this approach.

## 6. REFERENCES


Basar T. and P. Bernhard. $H^\infty$ Optimal Control and Related Minimax Desigh Problems: A Dynamic Game Approach. Boston, MA: Birkhauser.

Basar T. Minimax Control of Switching Systems Under Sampling. *Proceedings of the 33-rd Conference on Decision and Control.* Lake Buena vista, Florida. December, pp. 716 - 721.

Bellman R. Dynamic Programming and Modern Control Theory. Academic Press. NY.

Bernhard P. Minimax Versus Stochastic Partial Information Control. *Proceedings of the 33-rd Conference on Decision and Control.* Lake Buena vista, Florida. December, pp. 2572 - 2577.

Blom H. and Everdij M. Embedding Adaptive JLQG into LQ Martingale with a Complete Observable Stochastic Control Matrix. *Proceedings of the 32-nd Conference on Decision and Control.* San Antonio, Texas. December, pp. 849 - 854.

Boltyansky V.G.. The Method of Tents in the Theory of extremal Problems. *Uspehi Mat. Nauk*, v.30, No.3, pp.3-65.

Boltyansky V.G. and A.S.Poznyak. Robust Maximum Principle in Minimax Control. *Int. J. of Control* (to be published).

Didinsky G. and T.Basar. Minimax Adaptive Control of Uncertain Plants. *Proceedings of the 33-rd Conference on Decision and Control.* Lake Buena vista, Florida. December, pp. 2839 - 2844.

Dubovitzkii A.Ya. and A.A.Milyutin, *Necessary Conditions of Weak Extremum in General problem of Optimal Control.* Nauka. Moscow (in Russian).

Demianov V.F. and T.K. Vinogradova, About Minimax Principle in Optimal Control Problems. *Doklady Academii Nauk USSR*, vol.213, 3, pp.512-514.

Krasovskii N.N. Game Problem on Movements Correction. *Applied Mathematic and Mechanics.* No.33, issue 3 (in Russian).

Kurjanskii A.B. *Control and Observations in Uncertain Conditions.* Nauka, Moscow (in Russian).

Ming K. Lau, Boyd S., Koshut R.L. and G.F.Franklin. Robust Control Desigh for Ellipsoidal Plant Set. *Proceedings of the 30-th Conference on Decision and Control.* Brighton, England. December, pp. 291 - 296.

Pontryagin L.S., Boltyansky V.G., Gamkrelidze R.V. and E.F. Mishenko. *Mathematical Theory of Optimal Processes.* Nauka, Moscow (in Russian).

Poznyak A.S. and M.I.Taksar, Robust Control of Linear Stochastc Systems with Fully Observable State. *Applicationes Mathematicae*, v.24, No. 1, pp.35-46.

Pytlak R. Strong Variation Algoritm for Minimax Control: Parallel Approach. *Proceedings of the 29-th Conference on Decision and Control.* Honolulu, Hawall. December, pp. 2469 - 2474.

Siljak D.D. Parameter Space Methods for Robust Control Design: A Guided Tour. *IEEE Transactions on Automatic Control*, v. AC-34, No.7, pp.674- 688.

Ugrinovskii V.A. and I.R.Petersen. Finite Horizon Minimax Optimal Control of Stochastic Partially Observed Time Varying Uncertain Systems, *Proceedings of the 36-rd Conference on Decision and Control.* San Diego, December, CL, p.2265.

Zames J. Feedback and Optimality sensitivity: Model Reference Transformation, Multiplicative Seminorms and Approximate Inverses. *IEEE Transactions on Automatic Control*, v. AC-26, pp.301- 320.

14th Triennial World Congress, Beijing, P.R. China

# STABILITY PROPERTIES OF AN ITERATIVE OPTIMAL CONTROL ALGORITHM

**P D Roberts**

*Control Engineering Research Centre, City University, London, UK*
*email: p.d.roberts@city.ac.uk*

Abstract: The local stability of a technique for iterative solution of continuous nonlinear optimal control problems is considered. The algorithm is based on dynamic integrated system optimisation and parameter estimation, DISOPE, and it is shown that it's local dynamic behaviour can be described in terms of a linear unit-memory repetitive process where the initial conditions change from iteration to iteration. The resulting stability theorem requires the solution of a complex eigenvalue problem. The nature of the solution of this problem is investigated where significant insight is achieved through the use of Maple V symbolic computation. The theoretical results are verified using a MATLAB based graphical user interface specifically developed to analyse these problems.


Keywords: optimal control, iterative methods, stability analysis, computational methods.

## 1. INTRODUCTION

The solution of a nonlinear dynamic optimal control problem often requires an algorithm which updates a trial solution from iteration to iteration. In such algorithms, it is important to ensure stability so that the iterations converge in a satisfactory manner. In this paper, particular emphasis is given to an algorithm, known as dynamic integrated system optimisation and parameter estimation, DISOPE, (Roberts, 1993), for the solution of continuous nonlinear optimal control problems subject to model-reality differences. Recent research (e.g. Roberts, 1996) has shown that two-dimensional system theory, expressed in the form of an unit memory repetitive process (Rogers and Owens, 1992) can be employed to analyse the local convergence and stability properties of the DISOPE algorithm. However, the analytic investigation of the resulting stability theorem is computationally difficult because of the intractability of the resulting eigenvalue type equations resulting from initial conditions which change from iteration to iteration. In this paper, symbolic computation together with numerical analysis employing a MATLAB graphical user interface, GUI, is used to study the nature of the eigenvalue solution patterns.

## 2. DYNAMIC INTEGRATED SYSTEM OPTIMISATION AND PARAMETER ESTIMATION (DISOPE)

Dynamic integrated system optimisation and parameter estimation, DISOPE, is a technique for solving optimal control problems where there are differences in structure and parameter values between reality and the model employed in the computations. The main property of the procedure is that by iterating on appropriately modified model based problems the correct optimal solution is achieved in spite of the model-reality differences. Algorithms have been developed in both continuous and discrete time for a general non-linear optimal control problem with terminal weighting, mixed free and specified terminal conditions, and bounded controls (Roberts, 1993; Becerra, 1994). Algorithms for hybrid systems are also available (Roberts and Becerra, 1998). In this paper, however, in order to focus on the stability and convergence aspects of the algorithm, we will consider the simpler situation of a free end point continuous time problem without terminal weighting and without control constraints.

Consider the optimal control problem denoted as the real optimal problem ROP:

$$\min_{\mathbf{u}(t),\mathbf{x}(t)} \int_0^T L^*(\mathbf{x}(t),\mathbf{u}(t))dt \qquad (1)$$
$$\text{s.t. } \dot{\mathbf{x}}(t) = \mathbf{f}^*(\mathbf{x}(t),\mathbf{u}(t)),\ \mathbf{x}(0) = \mathbf{x}_o$$

where $T$ is the final time, $\mathbf{x}(t) \in \Re^n$, $\mathbf{u}(t) \in \Re^m$, $L^*: \Re^n \times \Re^m \to \Re$ is the real performance measure function and $f^*: \Re^n \times \Re^m \to \Re^n$ represents reality.

Because of modelling uncertainties and system complexity, instead of solving ROP directly, we consider the following, linear quadratic model based optimal control problem (MOP):

$$\min_{\mathbf{u}(t),\mathbf{x}(t)} \int_0^T \{\tfrac{1}{2}(\mathbf{x}(t)^T\mathbf{Q}\mathbf{x}(t) + \mathbf{u}(t)^T\mathbf{R}\mathbf{u}(t)) + \gamma(t)\}dt \qquad (2)$$
$$\text{s.t. } \dot{\mathbf{x}}(t) = \mathbf{A}\mathbf{x}(t) + \mathbf{B}\mathbf{u}(t) + \alpha(t),\ \mathbf{x}(0) = \mathbf{x}_o$$

where $\mathbf{Q} \in \Re^{n\times n} \geq 0$, $\mathbf{R} \in \Re^{m\times m} > 0$, $\mathbf{A} \in \Re^{n\times n}, \mathbf{B} \in \Re^{n\times m}$; and $\alpha(t) \in \Re^n$ and $\gamma(t) \in \Re$ are continuous parameters.

By expanding MOP to include additional equality constraints to match model with reality and then examining the overall optimality conditions, it can be shown that the solution of ROP can be found by iterating on the modified model based problem MMOP (Roberts , 1993):

$$\begin{aligned}\min_{\mathbf{u}(t),\mathbf{x}(t)} \int_0^T \Big\{ & \tfrac{1}{2}(\mathbf{x}(t)^T\mathbf{Q}\mathbf{x}(t) + \mathbf{u}(t)^T\mathbf{R}\mathbf{u}(t)) + \gamma_i(t) \\ & - \lambda_i(t)^T\mathbf{u}(t) - \beta_i(t)^T\mathbf{x}(t) \\ & + \tfrac{1}{2}r_1\|\mathbf{u}(t) - \mathbf{u}_i(t)\|^2 + \tfrac{1}{2}r_2\|\mathbf{x}(t) - \mathbf{x}_i(t)\|^2 \Big\}dt \end{aligned} \qquad (3)$$
$$\text{s.t. } \dot{\mathbf{x}}(t) = \mathbf{A}\mathbf{x}(t) + \mathbf{B}\mathbf{u}(t) + \alpha_i(t),\ \mathbf{x}(0) = \mathbf{x}_o$$

under specified parameters $\alpha_i(k)$ and $\gamma_i(k)$, and specified modifiers $\lambda_i(k)$ and $\beta_i(k)$, which are updated between iterations using the relationships:

$$\alpha_i(t) = \mathbf{f}^*(\mathbf{x}_i(t),\mathbf{u}_i(t)) - \mathbf{A}\mathbf{x}_i(t) - \mathbf{B}\mathbf{u}_i(t)$$
$$\gamma_i(t) = L^*(.) - \tfrac{1}{2}(\mathbf{x}_i(t)^T\mathbf{Q}\mathbf{x}_i(t) - \mathbf{u}_i(t)^T\mathbf{R}\mathbf{u}_i(t)) \qquad (4)$$

$$\lambda_i(t) = -\left[\frac{\partial \mathbf{f}^*(.)}{\partial \mathbf{u}} - \mathbf{B}\right]^T \mathbf{p}_i(t) - \left[\nabla_u L^*(.) - \mathbf{R}\mathbf{u}_i(t)\right]$$
$$\beta_i(t) = -\left[\frac{\partial \mathbf{f}^*(.)}{\partial \mathbf{x}} - \mathbf{A}\right]^T \mathbf{p}_i(t) - \left[\nabla_x L^*(.) - \mathbf{Q}\mathbf{x}_i(t)\right] \qquad (5)$$

where $i$ is the iteration number and $\mathbf{p}_i(t)$ is the co-state variable obtained from the solution of MMOP. The terms proportional to scalars $r_1$ and $r_2$ are introduced in MMOP to provide convexification augmentation with the aim of improving convergence. In the implementation of DISOPE convergence and stability are also regulated by using the relaxation scheme:

$$\mathbf{u}_{i+1}(t) = \mathbf{u}_i(t) + k_u(\hat{\mathbf{u}}_i(t) - \mathbf{u}_i(t))$$
$$\mathbf{x}_{i+1}(t) = \mathbf{x}_i(t) + k_x(\hat{\mathbf{x}}_i(t) - \mathbf{x}_i(t)) \qquad (6)$$
$$\mathbf{p}_{i+1}(t) = \mathbf{p}_i(t) + k_p(\hat{\mathbf{p}}_i(t) - \mathbf{p}_i(t))$$

where $k_u$, $k_x$ and $k_p$ are scalar gain parameters, and, $\hat{\mathbf{u}}_i(k)$, $\hat{\mathbf{x}}_i(k)$ and $\hat{\mathbf{p}}_i(k)$ are obtained from standard optimality conditions (Lewis and Syrmos, 1995), producing:

$$\hat{\mathbf{u}}_i(t) = \overline{\mathbf{R}}^{-1}[-\mathbf{B}^T\hat{\mathbf{p}}_i(t) + \overline{\lambda}_i(t)]$$
$$\tfrac{d}{dt}\hat{\mathbf{x}}_i(t) = \mathbf{A}\hat{\mathbf{x}}_i(t) + \mathbf{B}\hat{\mathbf{u}}_i(t) + \alpha_i(t),\ \hat{\mathbf{x}}_i(0) = \mathbf{x}_o \qquad (7)$$
$$\tfrac{d}{dt}\hat{\mathbf{p}}_i(t) = -\overline{\mathbf{Q}}\hat{\mathbf{x}}_i(t) - \mathbf{A}^T\hat{\mathbf{p}}_i(t) + \overline{\beta}_i(t)],\ \hat{\mathbf{p}}_i(T) = 0$$

where $\overline{\mathbf{R}} = \mathbf{R} + r_1\mathbf{I}_m$, $\overline{\mathbf{Q}} = \mathbf{Q} + r_2\mathbf{I}_n$, $\overline{\lambda}(t) = \lambda(k) + r_1\mathbf{u}(t)$ and $\overline{\beta}(t) = \beta(k) + r_2\mathbf{x}(t)$.

### *2.1 Linear DISOPE and it's Unit-Memory Linear Repetitive Process Representation*

In order to gain insight into local convergence and stability properties of the technique, we consider the special situation where the real optimal control problems is also linear with a quadratic performance index. In the problem ROP defined by Equation (1), let:

$$L^*(\mathbf{x}(t),\mathbf{u}(t)) = \tfrac{1}{2}(\mathbf{x}(t)^T\mathbf{Q}^*\mathbf{x}(t) + \mathbf{u}(t)\mathbf{R}^*\mathbf{u}(t))$$
$$\mathbf{f}^*(\mathbf{x}(t),\mathbf{u}(t)) = \mathbf{A}^*\mathbf{x}(t) + \mathbf{B}^*\mathbf{u}(t) \qquad (8)$$

Then from Equations (4) and (5) the model parameter and modifier vectors become:

$$\overline{\lambda}_i(t) = -[\mathbf{B}^* - \mathbf{B}]^T\mathbf{p}_i(t) - [\mathbf{R}^* - \overline{\mathbf{R}}]\mathbf{u}_i(t)$$
$$\overline{\beta}_i(t) = -[\mathbf{A}^* - \mathbf{A}]^T\mathbf{p}_i(t) - [\mathbf{Q}^* - \overline{\mathbf{Q}}]x_i(t) \qquad (9)$$
$$\alpha_i(t) = [\mathbf{A}^* - \mathbf{A}]\mathbf{x}_i(t) + [\mathbf{B}^* - \mathbf{B}]\mathbf{u}_i(t)$$

Note that it is not necessary to calculate $\gamma_i(t)$.

Manipulation of Equations (6), (7) and (9), to eliminate $\alpha_i(t)$, $\overline{\lambda}_i(t)$ and $\overline{\beta}_i(t)$, enables us to write the algorithm in the form of the unit-memory linear repetitive process (Rogers and Owens, 1992):

$$\tfrac{d}{dt}\mathbf{X}_i(t) = \mathbf{A}_o\mathbf{X}_i(t) + \mathbf{B}_o\mathbf{Y}_i(t)$$
$$\mathbf{Y}_{i+1}(t) = \mathbf{C}\mathbf{X}_i(t) + \mathbf{D}_1\mathbf{Y}_i(t) \qquad (10)$$

where $\mathbf{X}_i(t) = \begin{bmatrix}\hat{\mathbf{x}}_i(t)\\ \hat{\mathbf{p}}_i(t)\end{bmatrix}$; $\mathbf{Y}_i(t) = \begin{bmatrix}\mathbf{u}_i(t)\\ \mathbf{x}_i(t)\\ \mathbf{p}_i(t)\end{bmatrix}$ and

$$\mathbf{A}_o = \begin{bmatrix}\mathbf{A} & -\mathbf{B}\overline{\mathbf{R}}^{-1}\mathbf{B}^T\\ -\overline{\mathbf{Q}} & -\mathbf{A}^T\end{bmatrix} \quad \mathbf{C} = \begin{bmatrix}\mathbf{O}_{m,n} & -k_u\overline{\mathbf{R}}^{-1}\mathbf{B}^T\\ k_x\mathbf{I}_n & \mathbf{O}_n\\ \mathbf{O}_n & k_p\mathbf{I}_n\end{bmatrix}$$

$$\mathbf{B}_o = \begin{bmatrix}(\mathbf{B}^* - \mathbf{B}\overline{\mathbf{R}}^{-1}\mathbf{R}^*) & (\mathbf{A}^* - \mathbf{A}) & \mathbf{B}\overline{\mathbf{R}}^{-1}(\mathbf{B} - \mathbf{B}^*)^T\\ \mathbf{O}_{n,m} & (\overline{\mathbf{Q}} - \mathbf{Q}^*) & (\mathbf{A} - \mathbf{A}^*)^T\end{bmatrix}$$

$$\mathbf{D}_1 = \begin{bmatrix} (\mathbf{I}_m - k_u\overline{\mathbf{R}}^{-1}\mathbf{R}^*) & \mathbf{O}_{m,n} & k_u\overline{\mathbf{R}}^{-1}(\mathbf{B}-\mathbf{B}^*)^T \\ \mathbf{O}_{n,m} & (1-k_x)\mathbf{I}_n & \mathbf{O}_n \\ \mathbf{O}_{n,m} & \mathbf{O}_n & (1-k_p)\mathbf{I}_n \end{bmatrix}$$

with initial condition:

$$\hat{\mathbf{p}}_i(0) = -[\phi_{22}(T,0)]^{-1}\{\phi_{21}(T,0)x_o - \int_0^T \phi_2(T,\tau)\mathbf{B}_o\mathbf{Y}_i(\tau)d\tau\} \quad (11)$$

where:

$$e^{\mathbf{A}_o(T-\tau)} = \begin{bmatrix} \phi_1(T,\tau) \\ \phi_2(T,\tau) \end{bmatrix} = \begin{bmatrix} \phi_{11}(T,\tau) & \phi_{12}(T,\tau) \\ \phi_{21}(T,\tau) & \phi_{22}(T,\tau) \end{bmatrix} \quad (12)$$

## 3. STABILITY THEOREM

Without loss of generality, in order to analyse the stability properties of the above system, we will consider the situation where the initial condition $\mathbf{x}_0 = \mathbf{0}$. Then, in the limit as $i \to \infty$, $\mathbf{X}_i(t) = \mathbf{0}$, $\mathbf{Y}_i(t) = \mathbf{0}$, $t \in [0,T]$; and the iteration dependent initial condition $\mathbf{X}_i(0)$ is given from Equation (11) as:

$$\mathbf{X}_i(0) = \int_0^T \mathbf{W}(T,t)\,\mathbf{Y}_i(t)dt \quad (13)$$

where:

$$\mathbf{W}(T,t) = \begin{bmatrix} \mathbf{O}_{n,2n+m} \\ -\phi_{22}(T,0)^{-1}\phi_2(T,t)\mathbf{B}_0 \end{bmatrix} \quad (14)$$

The following stability theorem can be stated (Roberts, 1994):

*Theorem* 1: A sufficient and necessary condition for asymptotic stability of the continuous unit memory system represented by Equations (10) with initial condition (13) is that all solutions for $\lambda$ in the equation:

$$\left|\mathbf{I}_{2n} - \int_0^T \mathbf{W}(T,\tau)\mathbf{F}(\lambda)\exp\{(\mathbf{A}_o + \mathbf{B}_o\mathbf{F}(\lambda))\tau\}d\tau\right| = 0 \quad (15)$$

with $\mathbf{F}(\lambda) = [\lambda\mathbf{I}_{2n+m} - \mathbf{D}_1]^{-1}\mathbf{C}$, lie in the unit circle in the complex plane. *Proof*: see (Roberts, 1994)

## 4. COMPUTATIONAL SOLUTION

Finding the solutions for $\lambda$ in Equation (15) and testing the stability conditions of Theorem 1 is not a trivial problem. One technique is to compute the associated determinant values at all points in a grid over a selected region of the complex plane but noting that the equation becomes ill-conditioned at values of $\lambda$ close to the eigenvalues of $\mathbf{D}_1$. In particular, since the interest is in stability, we wish to find the particular $\lambda$ solution which is furthest from the origin. If this solution is outside the unit circle, then the corresponding iterative DISOPE algorithm is unstable with divergent iterations, if inside then it is stable with convergent iterations Canonical transformations can be used to assist the solution of Equation (15) as follows.

We require to compute:

$$D(\lambda) = \left|\mathbf{I}_{2n} - \int_0^T \mathbf{W}(T,\tau)\mathbf{F}(\lambda)\exp\{(\mathbf{A}_o + \mathbf{B}_o\mathbf{F}(\lambda))\tau\}d\tau\right| \quad (16)$$

where $\mathbf{W}(T,\tau)$ and $\mathbf{F}(\lambda)$ are defined in association with Equation (12).

Assuming distinct eigenvalues of $\mathbf{A}_0$, using Jordan form canonical transformation we can write:

$$e^{\mathbf{A}_o(T-\tau)} = \begin{bmatrix} \hat{\mathbf{U}} \\ \overline{\mathbf{U}} \end{bmatrix} e^{\mathbf{P}(T-\tau)}\mathbf{V} \quad (17)$$

where $\mathrm{P} = \mathrm{diag}(\rho_i)$ is the eigenvalue matrix of $\mathbf{A}_0$ and $\mathbf{U}$ and $\mathbf{V}$ are the corresponding left and right eigenvectors (with $\mathbf{U}$ partitioned as shown). Then:

$$\mathbf{W}(T,\tau) = \begin{bmatrix} \mathbf{O}_{n,2n+m} \\ \sum_{k=1}^{2n}\mathbf{E}_k e^{\rho_k(T-\tau)} \end{bmatrix} \quad (18)$$

where $\mathbf{E}_k = [\phi_{22}(T,0)]^{-1}\overline{\mathbf{U}}_k\mathbf{V}_k^T\mathbf{B}_o$.

Also, we can write:

$$\mathbf{F}(\lambda)\exp\{(\mathbf{A}_o + \mathbf{B}_o\mathbf{F}(\lambda))\tau\} = \tilde{\mathbf{U}}e^{\tilde{\mathrm{P}}\tau}\tilde{\mathbf{V}} = \sum_{k=1}^{2n}\tilde{\mathbf{F}}_k e^{\tilde{\rho}_k\tau} \quad (19)$$

where $\tilde{\mathrm{P}} = \mathrm{diag}(\tilde{\rho}_k)$ is the eigenvalue matrix of $\mathbf{A}_0+\mathbf{B}_0\mathbf{F}(\lambda)$, $\tilde{\mathbf{U}}$ and $\tilde{\mathbf{V}}$ are the corresponding left and right eigenvectors, and $\tilde{\mathbf{F}}_k = \mathbf{F}(\lambda)\tilde{\mathbf{U}}_k\tilde{\mathbf{V}}_k^T$. Hence:

$$\mathrm{D}(\lambda) = \left|\mathbf{I}_{2n} - \int_0^T \sum_{j=1}^{2n}\sum_{k=1}^{2n} \begin{bmatrix} \mathrm{O}_{n,2n} \\ \mathbf{E}_j\tilde{\mathbf{F}}_k \end{bmatrix} e^{\rho_j T} e^{(\tilde{\rho}_k - \rho_j)\tau} d\tau\right| \quad (20)$$

That is:

$$\mathrm{D}(\lambda) = \left|\mathbf{I}_n - \sum_{j=1}^{2n}\sum_{k=1}^{2n}\mathbf{G}_{jk}\frac{e^{\tilde{\rho}_k T} - e^{\rho_j T}}{\tilde{\rho}_k - \rho_j}\right| \quad (21)$$

where $\mathbf{G}_{jk}$ is the $n$+1,$n$ partition of $\mathbf{E}_j\tilde{\mathbf{F}}_k$.

The MATLAB 3 - dimensional visual graphics facilities and the 2 - dimensional contour facility, together with the MATLAB optimisation toolbox, can be employed to search for zeros of the determinant $\mathrm{D}(\lambda)$ relative to the unit circle. Searching for intersections of the zero real and zero imaginary contours has been found to be a particularly effective technique. To aid this work a comprehensive user-interaction MATLAB graphical user interface, GUI. has been developed (Roberts, 1998). When this GUI is employed to investigate an optimal control example with four states and three control inputs (Roberts,1994) the zero real and zero imaginary contours shown in Figure 1 are obtained with $k_u = k_x = k_p = 1$, $r_1 = r_2 = 0$. The figure shows the contours in the upper half complex plane in relation to the unit circle. It is observed that although there are solutions for $\lambda$ within the unit circle there are also solutions outside along the negative real axis. This clearly indicates instability and the DISOPE iterations will diverge with these

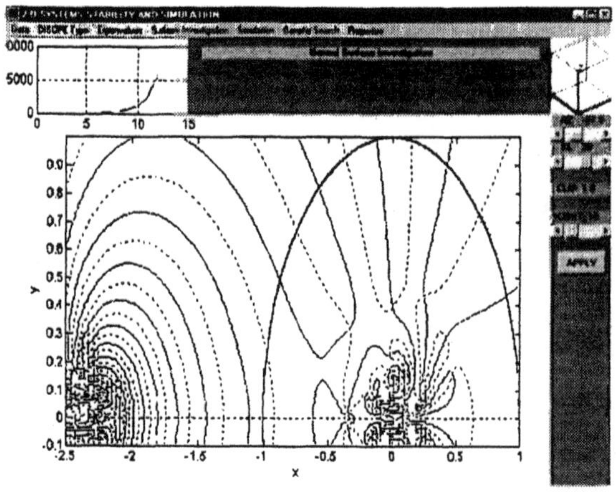


Fig. 1. MATLAB GUI illustration of zero real and imaginary contours

particular relaxation gain and convexification augmentation parameters. This is in fact demonstrated by the graph shown in the top left corner of Figure 1 which shows the progression of response error norm during a previous simulation of the iterations. Figure 1 also indicates that the $\lambda$ solutions crowd together so that there are many solutions in small regions of the complex $\lambda$ plane, which, for this particular example, occur at selected eigenvalues of $\mathbf{D}_1$. Extensive investigations using the GUI appear to indicate that there are an infinite number of $\lambda$ solutions occurring in infinitesimally small regions.

## 5. SYMBOLIC COMPUTATION FOR THE SCALAR CASE

In order to gain insight into the nature of the $\lambda$ solutions of Equation (15) consider the simplest possible situation of a single state and single control input optimal control problem. In this situation $n = 1$, $m = 1$ and $\mathbf{A}$, $\mathbf{B}$, $\mathbf{Q}$, $\mathbf{R}$ and $\mathbf{A}^*$, $\mathbf{B}^*$, $\mathbf{Q}^*$, $\mathbf{R}^*$ all become scalar and the analysis is simplified. However, even for this apparently trivial case full evaluation of Equation (15) using Equation (21) soon becomes unmanageable and, for the author, it was found that progress was possible only using symbolic computation, in this particular investigation with Maple V (Maple, 1996).

For the scalar situation Equation (21) becomes:

$$\mathrm{D}(\lambda) = 1 - \sum_{j=1}^{2}\sum_{k=1}^{2} \mathrm{G}_{jk} \frac{e^{\tilde{\rho}_k T} - e^{\rho_j T}}{\tilde{\rho}_k - \rho_j} \tag{22}$$

where $\mathrm{G}_{jk}$, which is defined in terms of eigenvectors of $\mathbf{A}_0$ and $\mathbf{A}_0 + \mathbf{B}_0\mathbf{F}(\lambda)$ (see Equations (18) and (19)), is now scalar. Also, $\mathbf{A}_0 + \mathbf{B}_0\mathbf{F}(\lambda)$ becomes:

$$\mathbf{A}_0 + \mathbf{B}_0\mathbf{F}(\lambda) =$$

$$\begin{bmatrix} \dfrac{\mathrm{A}\lambda - \mathrm{A} + k_x\mathrm{A}^*}{\lambda - 1 + k_x} & -\dfrac{(\mathrm{B}\lambda - \mathrm{B} + k_p\mathrm{B}^*)(\mathrm{B}\lambda - \mathrm{B} + k_u\mathrm{B}^*)}{(\overline{\mathrm{R}}\lambda - \overline{\mathrm{R}} + k_u\mathrm{R}^*)(\lambda - 1 + k_p)} \\ -\dfrac{(\overline{\mathrm{Q}}\lambda - \overline{\mathrm{Q}} + k_x\mathrm{Q}^*)}{\lambda - 1 + k_x} & -\dfrac{\mathrm{A}\lambda - \mathrm{A} + k_p\mathrm{A}^*}{\lambda - 1 + k_p} \end{bmatrix} \tag{23}$$

Because $\mathbf{A}_0$ is symplectic we can write $\rho_1 = \rho$, $\rho_2 = -\rho$, $\mathrm{E}_1 = \mathrm{E}(\rho)$, $\mathrm{E}_2 = \mathrm{E}(-\rho)$. In general, unless $k_x = k_p$, $\mathbf{A}_0 + \mathbf{B}_0\mathbf{F}(\lambda)$ is not symplectic. However, we can still write $\tilde{\rho}_1 = \sigma(\lambda) + \beta(\lambda)$, $\tilde{\rho}_2 = \sigma(\lambda) - \beta(\lambda)$, giving $\mathrm{G}_{11} = g(\rho,\sigma,\beta)$, $\mathrm{G}_{12} = g(\rho,\sigma,-\beta)$, $\mathrm{G}_{21} = g(-\rho,\sigma,\beta)$, $\mathrm{G}_{22} = g(-\rho,\sigma,-\beta)$ where:

$$\rho = {}_{+}\sqrt{\frac{\overline{\mathrm{R}}\mathrm{A}^2 + \overline{\mathrm{Q}}\mathrm{B}^2}{\overline{\mathrm{R}}}} \tag{24}$$

$$\sigma(\lambda) = -\frac{1}{2}\frac{\mathrm{A}\lambda - \mathrm{A} + k_p\mathrm{A}^*}{\lambda - 1 + k_p} + \frac{1}{2}\frac{\mathrm{A}\lambda - \mathrm{A} + k_x\mathrm{A}^*}{\lambda - 1 + k_x} \tag{25}$$

$$\beta(\lambda) = {}_{+}\frac{1}{2}\left(\left(\frac{\mathrm{A}\lambda - \mathrm{A} + k_p\mathrm{A}^*}{\lambda - 1 + k_p} + \frac{\mathrm{A}\lambda - \mathrm{A} + k_x\mathrm{A}^*}{\lambda - 1 + k_x}\right)^2 \right. \tag{26}$$

$$\left. +4\frac{(\mathrm{B}\lambda - \mathrm{B} + k_p\mathrm{B}^*)(\mathrm{B}\lambda - \mathrm{B} + k_u\mathrm{B}^*)(\overline{\mathrm{Q}}\lambda - \overline{\mathrm{Q}} + k_x\mathrm{Q}^*)}{(\overline{\mathrm{R}}\lambda - \overline{\mathrm{R}} + k_u\mathrm{R}^*)(\lambda - 1 + k_p)(\lambda - 1 + k_x)}\right)^{\frac{1}{2}}$$

and, after more symbolic computation we find:

$$g(\rho,\sigma,\beta) =$$

$$\frac{1}{4}\frac{\big(d_0 + d_1(\sigma + \beta)\big)\big(c_0 + c_1\rho + (c_2 + c_3\rho)(\sigma - \beta)\big)}{\big(\rho\cosh(T\rho) - \mathrm{A}\sinh(T\rho)\big)\beta d_2} \tag{27}$$

with

$$c_0(\lambda) =$$

$$\begin{aligned} &-k_u(\lambda - 1 + k_x)\overline{\mathrm{Q}}(\mathrm{B}^*\overline{\mathrm{R}} - \mathrm{B}\mathrm{R}^*)(\mathrm{B}\lambda - \mathrm{B} + k_p\mathrm{B}^*)\tilde{\mathrm{Q}} \\ &-k_x(\lambda - 1 + k_p)\overline{\mathrm{R}}(\mathrm{A}^*\overline{\mathrm{Q}} - \mathrm{A}\mathrm{Q}^*)(\mathrm{A}\lambda - \mathrm{A} + k_x\mathrm{A}^*)\tilde{\mathrm{R}} \\ &-k_p(\lambda - 1 + k_x)\big(\overline{\mathrm{Q}}\mathrm{B}(\mathrm{B}^* - \mathrm{B}) + \overline{\mathrm{R}}\mathrm{A}(\mathrm{A}^* - \mathrm{A})\big)\tilde{\mathrm{Q}}\tilde{\mathrm{R}} \end{aligned} \tag{28}$$

$$c_1(\lambda) =$$

$$\begin{aligned} &-k_x(\lambda - 1 + k_p)\overline{\mathrm{R}}(\mathrm{Q}^* - \overline{\mathrm{Q}})(\mathrm{A}\lambda - \mathrm{A} + k_x\mathrm{A}^*)\tilde{\mathrm{R}} \\ &+k_p(\lambda - 1 + k_x)\overline{\mathrm{R}}(\mathrm{A}^* - \mathrm{A})\tilde{\mathrm{Q}}\tilde{\mathrm{R}} \end{aligned} \tag{29}$$

$$c_2(\lambda) = k_x(\lambda - 1 + k_x)(\lambda - 1 + k_p)\overline{\mathrm{R}}(\overline{\mathrm{Q}}\mathrm{A}^* - \mathrm{Q}^*\mathrm{A})\tilde{\mathrm{R}} \tag{30}$$

$$c_3(\lambda) = k_x(\lambda - 1 + k_x)(\lambda - 1 + k_p)\overline{\mathrm{R}}(\mathrm{Q}^* - \overline{\mathrm{Q}})\tilde{\mathrm{R}} \tag{31}$$

$$d_0(\lambda) = -(\mathrm{A}\lambda - \mathrm{A} + k_x\mathrm{A}^*) \tag{32}$$

$$d_1(\lambda) = \lambda - 1 + k_x \tag{33}$$

$$d_2(\lambda) = k_x(\lambda - 1 + k_x)^2(\lambda - 1 + k_p)\overline{\mathrm{R}}\tilde{\mathrm{R}}\tilde{\mathrm{Q}} \tag{34}$$

where $\tilde{\mathrm{Q}} = \overline{\mathrm{Q}}\lambda - \overline{\mathrm{Q}} + k_x\mathrm{Q}^*$, $\tilde{\mathrm{R}} = \overline{\mathrm{R}}\lambda - \overline{\mathrm{R}} + k_u\mathrm{R}^*$.

Equation (22) can be manipulated further to provide:

$$\mathrm{D}(\lambda) =$$

$$\begin{aligned} &1 + a_c(\rho,\sigma,\beta)\cosh(T\rho) + a_s(\rho,\sigma,\beta)\sinh(T\rho) \\ &+b_c(\rho,\sigma,\beta)\cosh(T\beta) + b_s(\rho,\sigma,\beta)\sinh(T\beta) \end{aligned} \tag{35}$$

where:

$$\begin{aligned} a_c(\rho,\sigma,\beta) = {} & \frac{g(-\rho,\sigma,\beta)}{\rho + \sigma + \beta} + \frac{g(-\rho,\sigma,-\beta)}{\rho + \sigma - \beta} \\ & -\frac{g(\rho,\sigma,\beta)}{\rho - \sigma - \beta} - \frac{g(\rho,\sigma,-\beta)}{\rho - \sigma + \beta} \end{aligned} \tag{36}$$

$$a_s(\rho,\sigma,\beta) = -\frac{g(-\rho,\sigma,\beta)}{\rho+\sigma+\beta} - \frac{g(-\rho,\sigma,-\beta)}{\rho+\sigma-\beta} - \frac{g(\rho,\sigma,\beta)}{\rho-\sigma-\beta} - \frac{g(\rho,\sigma,-\beta)}{\rho-\sigma+\beta} \quad (37)$$

$$b_c(\rho,\sigma,\beta) = \left(\frac{g(\rho,\sigma,\beta)}{\rho-\sigma-\beta} - \frac{g(-\rho,\sigma,\beta)}{\rho+\sigma+\beta} + \frac{g(\rho,\sigma,-\beta)}{\rho-\sigma+\beta} - \frac{g(-\rho,\sigma,-\beta)}{\rho+\sigma-\beta}\right)e^{T\sigma} \quad (38)$$

$$b_s(\rho,\sigma,\beta) = \left(\frac{g(\rho,\sigma,\beta)}{\rho-\sigma-\beta} - \frac{g(-\rho,\sigma,\beta)}{\rho+\sigma+\beta} - \frac{g(\rho,\sigma,-\beta)}{\rho-\sigma+\beta} + \frac{g(-\rho,\sigma,-\beta)}{\rho+\sigma-\beta}\right)e^{T\sigma} \quad (39)$$

*5.1 Case $k_p = k_x$, $Q^* = Q$, $\overline{Q} = Q$, $A = A^*$*

For simplicity we will now consider the special situation when $k_p = k_x$, for which $\mathbf{A}_o + \mathbf{B}_o\mathbf{F}(\lambda)$ is now symplectic, and the only model-reality difference occurs in the control parameters (i.e. $B^* \neq B$) and control performance function weighting ($R^* \neq R$). Then Equations (23) to (34) provide $\sigma = 0$ and:

$$\mathbf{A}_o + \mathbf{B}_o\mathbf{F}(\lambda) = \begin{bmatrix} A & -\dfrac{(B\lambda - B + k_x B^*)(B\lambda - B + k_u B^*)}{(\overline{R}\lambda - \overline{R} + k_u R^*)(\lambda - 1 + k_x)} \\ -Q & -A \end{bmatrix} \quad (40)$$

$$\beta(\lambda) = +\sqrt{A^2 + \frac{Q(B\lambda - B + k_x B^*)(B\lambda - B + k_u B^*)}{(\overline{R}\lambda - \overline{R} + k_u R^*)(\lambda - 1 + k_x)}} \quad (41)$$

$$c_o(\lambda) = -(\lambda - 1 + k_x)^2 Q^2 \big(k_u(B^*\overline{R} - BR^*)(B\lambda - B + k_x B^*) + k_x B(B^* - B)(\overline{R}\lambda - \overline{R} + k_u R^*)\big) \quad (42)$$

$$c_1(\lambda) = c_2(\lambda) = c_3(\lambda) = 0 \quad (43)$$

$$d_o(\lambda) = -A(\lambda - 1 + k_x) = -Ad_1(\lambda) \quad (44)$$

$$d_2(\lambda) = k_x(\lambda - 1 + k_x)^4 Q\overline{R}(\overline{R}\lambda - \overline{R} + k_u R^*) \quad (45)$$

giving, from Equation (27)

$$g(\rho,\sigma,\beta) = \frac{c_o d_1}{4d_2}\frac{\beta - A}{(\rho\cosh(T\rho) - A\sinh(T\rho))\beta} \quad (46)$$

We now find from Equations (36), (37) and (46), and then substitution using Equations (24), (41), to (45):

$$1 + a_c(\rho,\sigma,\beta)\cosh(T\rho) + a_s(\rho,\sigma,\beta)\sinh(T\rho) = 1 - \frac{c_o d_1}{d_2(\rho^2 - \beta^2)} = 0 \quad (47)$$

Similarly, we find:

$$b_c(\rho,\sigma,\beta) = \frac{\rho}{(\rho\cosh(T\rho) - A\sinh(T\rho))} \quad (48)$$

$$b_s(\rho,\sigma,\beta) = -\frac{\rho A}{(\rho\cosh(T\rho) - A\sinh(T\rho))\beta} = -\frac{A}{\beta}b_c(\rho,\sigma,\beta) \quad (49)$$

Hence, from Equation (35):

$$D(\lambda) = b_c(\rho,\sigma,\beta)\left(\cosh(T\beta) - \frac{A\sinh(T\beta)}{\beta}\right) \quad (50)$$

We require to determine the values of $\lambda$ for which $D(\lambda) = 0$, in which $\beta = \beta(\lambda)$. It is clearly seen now that these zero solutions occur when:

$$\cosh(T\beta) - \frac{A\sinh(T\beta)}{\beta} = 0 \quad (51)$$

*Solution Pattern* We will now investigate the $\lambda$ solutions of Equation (51). First we note that real solutions are possible only when $A \geq 0$. It is also readily seen that for large terminal time $T$ these real solutions occur when $\beta \cong \pm A$ Then from equation (41), the corresponding approximate real $\lambda$ solutions are:

$$\lambda \cong 1 - \frac{B^*}{B}k_u, \quad \lambda \cong 1 - \frac{B^*}{B}k_x \quad (52)$$

giving the following necessary conditions for stability for the relaxation gains $k_x$ and $k_u$ when $A \geq 0$ and $T \to \infty$:

$$0 \leq k_u \leq 2\frac{B}{B^*}, \quad 0 \leq k_x \leq 2\frac{B}{B^*} \quad (53)$$

requiring B and $B^*$ to be of the same sign.

To investigate the complex solutions of Equation (51) we substitute $\beta = \sigma + j\omega$, note that solutions occur only when $\sigma = 0$ and, hence, determine an infinite set of complex solutions $\beta = j\omega$ given by:

$$\cos(T\omega) - \frac{A\sin(T\omega)}{\omega} = 0 \quad (54)$$

Looking at Equation (41) we see that the bound $\omega \to \infty$ occurs when:

$$\lambda = 1 - \frac{R^*}{\overline{R}}k_u, \quad \lambda = 1 - k_x \quad (55)$$

which are eigenvalues of the matrix $\mathbf{D}_1$ in Equation (10). Hence, further necessary conditions for stability are

$$0 \leq k_u \leq 2\frac{\overline{R}}{R^*}, \quad 0 \leq k_x \leq 2 \quad (56)$$

At this point it is worth remembering that, for the situation where the initial conditions are fixed, the sufficient and necessary condition for the stability of the unit memory repetitive linear process represented by Equation (10), with $\mathbf{X}_i(0)$ constant, is that the

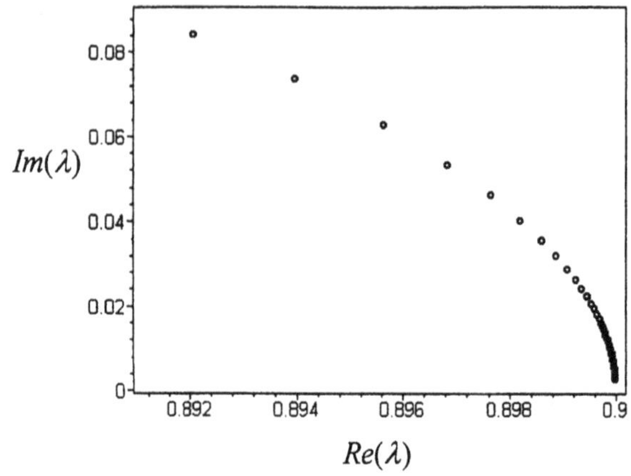


Fig. 2. Complex $\lambda$ solutions for scalar example

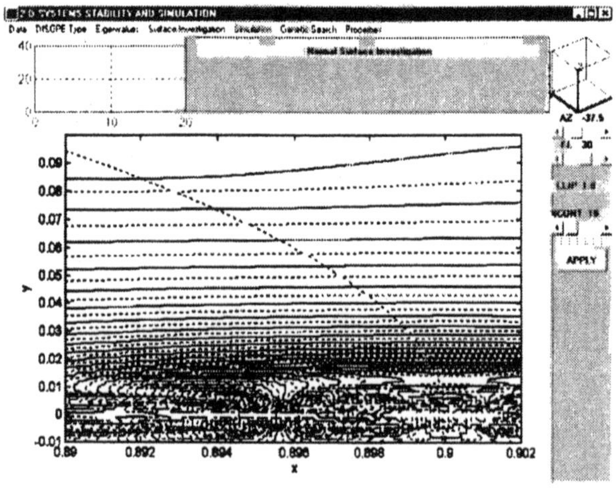

Fig. 3. MATLAB GUI result for scalar example.

eigenvalues of matrix $\mathbf{D}_1$ should all lie within the complex plane unit circle (Rogers and Owens, 1992). The full infinite set of complex solutions will occur between the first solution of Equation (54) and Equation (55) crowding towards Equation (55), This is illustrated in Figure 2 which shows the first 50 numerical complex solutions for $\lambda$ when A = 1, $T$ = 10, $k_u = 0.1$, $k_x = 1$, $R^* = \overline{R} = 1$, $B^* = 1$, B = 0.1, Q = 1. This result is also verified by the MATLAB GUI output shown in Figure 3. For stability of the system with the varying initial conditions as described, it is sufficient and necessary that the whole of this infinite set of complex solutions, together with any real solution of Equation(51), lie within the unit circle. This particular example satisfies this condition.

## 6. CONCLUSIONS

Stability properties of a unit memory repetitive process with initial conditions which are updated from iteration to iteration have been investigated. This process represents the local convergence behaviour of a particular algorithm for solving nonlinear continuous optimal control problems subject to model reality differences. A related stability theorem which requires the solution of a complex eigenvalue problem solution has been considered and insight has been achieved through symbolic and numerical computation. These computations have illustrated that the number of eigenvalue type solutions are infinite and crowd together in the complex plane. These results have impact on the study of local convergence and stability properties of iterative techniques for solving nonlinear optimal control problems. Although, so far, significant symbolic results have only been obtained by investigation of a simple first-order single control problem, where even so the analysis is by no means trivial, it is expected that the insight achieved can be extended to higher order systems. Research is continuing in an attempt to verify this.

## 7. ACKNOWLEDGEMENT

This work is supported by the Engineering and Physical Sciences Research Council (Grant GR/L64478)

## 8. REFERENCES

Becerra V.M. 1994, *Development and applications of novel optimal control algorithms*, PhD Thesis, City University, UK.

Lewis F.L and Syrmos , 1995, *Optimal Control*, Wiley, New York, USA.

*Maple V.* 1997, Waterloo Maple Inc.

Roberts P.D., 1993, An algorithm for optimal control of nonlinear systems with model reality differences, *12th IFAC World Congress on Automatic Control*, Sydney, Australia, **8**, 407-412.

Roberts P.D., 1994, Unit memory repetitive processes aspects of iterative optimal control algorithms, *Proceedings of 33rd Conference on Decision and Control*, Lake Buena Vista, Florida, USA, December 1994, pp. 1394-1399.

Roberts, P.D., 1996, Computing the stability of iterative optimal control algorithms through the use of two-dimensional system theory, *Proceedings of UKACC International Conference on CONTROL'96*, Exeter, UK, Institution of Electrical Engineers Conference Publication No. 427, pp 981-986.

Roberts, P.D. and Becerra, V.M. 1998, Optimal control of non-linear systems under model-reality differences - a hybrid approach, *4th International Conference on Optimization: Techniques and Applications, ICOTA'98*, July 1 – 3, Perth, Australia.

Roberts, P.D., 1998, A MATLAB graphical user interface for investigating the local stability of iterative optimal control algorithms, *UKACC International Conference on CONTROL'98*, 1 – 4 September, Swansea, UK.

Rogers E. and Owens D.H., 1992, *Stability Analysis for Linear Repetitive Processes*, Springer Verlag, Berlin, Germany.

14th Triennial World Congress, Beijing, P.R. China

# OPTIMAL REGULATOR FOR NEUTRAL CONTROL SYSTEMS WITH MULTI-DELAYS

Cun-Chen Gao[†], Xiao-Hua Liu, Rui-Na Tang and Hui-Chao Zou
Department of Mathematics and Computer Science
Yantai Teachers University, Yantai, 264025, P.R.CHINA

[†] *Email ccgao@mail.ytnc.edu.cn*

**Abstract** With optimal regulator problem for neutral control systems with multi-delays is considered. For neutral system with multi-delays, Riccati type partial differential equations are given. Optimal regulator is structured by using the solutions of the Riccati type partial differential equations. Some closed-loop properties of the optimal regulator are also obtained. *Copyright © 1999 IFAC*

**Keywords:** neutral, control systems, delay, optimal regulators, Riccati equations

## 1. INTRODUCTION

ptimal regulator problem for retarded control system s well-studied; it is known that optimal regulator for tarded control system has some same good closed-op properties as that for lumped parameter systems .Uchida). For neutral systems, a few researches about e optimal control are presented (R.Datko; M.Sendanla; Ito; A.J.Prichard; N.Abe). R. Datko has demonstrated at the optimal control satisfied a linear feedback law. . Sendaula has suggested that the finite time interval timal control law was characterized by the solutions of nultaneous differential equations. K. Ito has treated utral system as an evolution equations and discussed timal regulator problem with operator Riccati equans. A. J. Prichard and D. Salamon have established general semigroup framework for solving optimal regator problem for distributed parameter systems with bounded input and output operators. N. Abe has dissed optimal regulator problem in a finite time interval r the linear neutral system with singular-delay; as one these examples, they treat a neutral system.

the previous researches stated above, it has been left me problems. For example, does the feedback control w have an integral kernel description? If so, what form es the kernel have? In this paper we solve optimal regator problem in a concrete form by using Riccati type rtial differential equations (PDEs) and derive the optial regulator of the integral kernel form for neutral conol system with multi-delays. Some closed-loop propers of the optimal regulator similar to lumped parameter system and retarded systems are discussed.

The finite time interval problem is represented and solved in section 2, Riccati type PDEs are given, the existence theorem of solution of the Riccati type PDEs are also established by applying the characteristic surface method (K.Uchida). In section 3, stationary Riccati type PDEs are formulated. In section 4, stabilizability and detectability for neutral systems with multi-delays and solving the optimal regulator problem are defined. In section 5, some closed-loop properties of the optimal regulator are obtained.

## 2. OPTIMAL REGULATOR IN FINITE TIME INTERVAL

We consider that the linear neutral control system with multi-delays

$$\dot{x}(t) = A_0x(t)+\sum_{i=1}^{N}[A_ix(t-\sigma_i)+A_{-i}\dot{x}(t-\sigma_i)]+Bu(t) \quad (1)$$

where $A_0, A_i, A_{-i}(i = 1,2,\dots N)$ and $B$ are $n\times n, n\times n, n\times n$ and $n\times m$ constant matrices, respectively. Initial condition $x(t) = \phi(t), -\sigma \le t \le 0, \sigma = \max_{1\le i\le N}\{\sigma_i\}$ belongs to $C^1([-\sigma,0],R^n)$ such that $\phi(0) = \sum_{i=1}^{N}[A_i\phi + A_{-i}\dot{\phi}]$. The constraint guarantees solution $x(t) \in C^1$. If $\phi$ belongs to

$C^1$ without the constraint, then the solution $x(t)$ is general discontinuous at points $t = n\sigma_i, n = 0,1,2,\cdots; i = 1,\cdots,N$.

Examples of linear neutral system are lossless transmission lines (R.Bellman) or repetitive control systems (Y.S.Qin).

Considering optimal regulator problem in a finite time interval for the linear neutral system (1). Suppose that the finite time cost functional be given as the following

$$J(t) = \int_0^T [x^{\top}(t)Qx(t) + u^{\top}(t)Ru(t)]dt \tag{2}$$

where $Q = C^{\top}C, C \in R^{n\times r}$ is a constant matrix and $R \in R^{m\times m}$ is a positive definite constant matrix. The '$\top$' denotes transpose. Consider the problem of finding a continuous control $u(t)$ in the finite time interval which minimizes the value of the finite time cost functional (2) subject to the system (1).

*Theorem 1* Suppose that $P_0(t), P_1(t,\eta)$ and $P_2(t,\xi,\eta), 0 \le t \le T, -\sigma \le \xi \le 0, -\sigma \le \eta \le 0$ be a $C^1$–solution of the following Riccati type PDEs

$$\frac{d}{dt}P_0(t) = -A_0^{\top}\Phi_0(t) - \Phi_0^{\top}(t)A_0 + \Phi_0^{\top}(t)BR^{-1}B^{\top}\Phi_0(t) - (\sum_{i=1}^{N} A_i)^{\top}P_1^{\top}(t,0) - P_1(t,0)\sum_{i=1}^{N} A_i - Q \quad (3-a)$$

$$\frac{\partial}{\partial t}P_1(t,\eta) = \frac{\partial}{\partial\eta}P_1(t,\eta) - A_0^{\top}\Phi_1(t,\eta) + \Phi_0^{\top}(t)BR^{-1}B^{\top}\Phi_1(t,\eta) - \sum_{i=1}^{N} A_iP_2(t,0,\eta) \quad (3-b)$$

$$\frac{\partial}{\partial t}P_2(t,\xi,\eta) = \frac{\partial}{\partial\xi}P_2(t,\xi,\eta) - \frac{\partial}{\partial\eta}P_2(t,\xi,\eta) - \Phi_1^{\top}(t,\xi)BR^{-1}B^{\top}\Phi_1(t,\eta) \quad (3-c)$$

where $\Phi_0(t) = P_0(t) + (\sum_{i=1}^{N} A_{-i})^{\top}P_1^{\top}(t,0)$ and $\Phi_1(t,\eta) = P_1(t,\eta) + (\sum_{i=1}^{N} A_{-i})^{\top}P_2(t,0,\eta)$, with the terminal conditions

$$P_0(T) = 0, P_1(T) = 0, P_2(T,\xi,\eta) = 0, \quad -\sigma \le \xi \le 0, -\sigma \le \eta \le 0 \tag{4}$$

the symmetric conditions

$$P_0^{\top}(t) = P_0(t), P_2^{\top}(t,\xi,\eta) = P_2(t,\xi,\eta) \tag{5}$$

and the boundary conditions

$$\Phi_0(t) - P_1(t,-\sigma) = 0,\ \Phi_1(t,\eta) - P_2(t,-\sigma,-\sigma) = 0 \tag{6}$$

Then the optimal control law $u^*(t)$ for cost functional (2) can be written as

$$u^*(t) = -R^{-1}B^{\top}\{\Phi_0(t)x(t) + \int_{-\sigma}^{0} \Phi_1(t,\eta)\sum_{i=1}^{N}[A_ix(t+\eta) + A_{-i}\dot{x}(t+\eta)]d\eta\} \tag{7}$$

and the optimal cost $J^*(T)$ is represented as

$$J^*(T) = \phi^{\top}(0)P_0(0)\phi(0) + 2\phi^{\top}(0)\int_{-\sigma}^{0} P_1(0,\eta)\sum_{i=1}^{N}[A_i\phi(\eta) + A_{-i}\dot{\phi}(\eta)]d\eta + \int_{-\sigma}^{0}\int_{-\sigma}^{0}\sum_{i=1}^{N}[A_i\phi(\xi) + A_{-i}\dot{\phi}(\xi)]^{\top}P_2(0,\xi,\eta)\sum_{i=1}^{N}[A_i\phi(\eta) + A_{-i}\dot{\phi}(\eta)]d\xi d\eta \tag{8}$$

where $\phi$ is the initial function.

*Proof* Let $x(t), 0 \le t \le T$ be an unique solution for the initial condition $\phi$ and control law $u(t)$ and $P_0(t), P_1(t,\eta)$ and $P_2(t,\xi,\eta)$ be the solution of (3)-(6). Defining a functional $V(x)$ is that

$$V(x) = x^{\top}(t)P_0(t)x(t) + 2x^{\top}(t)\int_{-\sigma}^{0} P_1(t,\eta)\sum_{i=1}^{N}[A_ix(t+\eta) + A_{-i}\dot{x}(t+\eta)]d\eta + \int_{-\sigma}^{0}\int_{-\sigma}^{0}\sum_{i=1}^{N}[A_ix(t+\xi) + A_{-i}\dot{x}(t+\xi)]^{\top}P_2(t,\xi,\eta) \sum_{i=1}^{N}[A_ix(t+\eta) + A_{-i}\dot{x}(t+\eta)]d\xi d\eta \tag{9}$$

Now, we consider the following identity

$$V(x(T)) - V(x(0)) = \int_{-\sigma}^{0} \dot{V}(x(t))dt \tag{10}$$

Applying Riccati type PDE (3), boundary condition (6) and the terminal conditions $V(x(t)) = 0$, to the right-hand side of the above identity, we can obtain

$$J(T) = V(x(0)) + \int_0^T [u^*(t) - u(t)]^{\top}R[u^*(t) - u(t)]dt \tag{11}$$

where $u^*(t)$ is given by (7). It is clear that for any conditions control inputs the integration term of the right-hand side of (11) is nonnegative. It is also clear that for $u^*(t)$ the integral term is zero. Since the first term of the right-hand side does not depend on control input, $u^*(t)$ is the unique optimal control law, and optimal cost is given by $V(x(t))$ that is (8). The proof of Theorem 1 is completed.

Now, we present global existence theorem of the Riccati type PDEs.

*eorem 2* If $\|\sum_{i=1}^{N} A_{-i}\|_\infty < 1$ hold, then there exists an que $C^1-$ solution of Riccati type PDEs (3)-(6) in the on $R(T) = \{(t,\xi,\eta) : 0 \le t \le T, -\sigma \le \xi \le 0, -\sigma \le$ $0\}$, where $\|W\|_\infty = \max_i \sum_j w_{ij}$.

*of* Suppose that $J^0(T)$ denote the cost functional (2) the zero input ($u(t) \equiv 0$), then we have that $J^*(T) \le$ $T$). The boundedness of the triplet $P_0(t), P_1(t,\eta)$ and $t,\xi,\eta)$ follows from this inequality and (8). Next, we sider a set of the characteristic curves which are solu-ns $\frac{d}{dt}\xi(t) = -1, \frac{d}{dt}\eta(t) = -1$, that is

$$C = \{(t,\xi,\eta) : \xi = \xi(t), \eta = \eta(t)\}.$$

ng each curve, we see that $P_1(t,\cdot)$ and $P_2(t,\cdot,\cdot)$ sat- ODEs. The following statements are equivalent: the cati type PDEs (3) hold dt all points $(t,\xi,\eta)$ in the ion $R(T)$ and the OEDs hold on all the characteris- curves in $C$ on $R(T)$. If $\|\sum_{i=1}^{N} A_{-i}\|_\infty < 1$ holds, then can prove that the mapping $F$, defined by integrating a) and ODEs along characteristic curves, is the con-ction mapping. The fact means from the fixed point orem and the boundedness that there exists an unique ution in the region $R(T)$. The proof of the theorem is npleted.

e difference between the existence theorem for neutral tems and retarded systems is that the neutral cases uire an additional norm condition $\|\sum_{i=1}^{N} A_{-i}\|_\infty < 1$. te that the norm condition is a sufficient condition for stence of Riccati type PDEs.

## . STATIONARY SOLUTION OF RICATI PARTIAL DIFFERENTIAL EQUA-ONS

this section we consider the existence theorem for sta-nary solution of the Riccati type PDEs. By using solution, we construct the optimal regulator. The tionary solution is the triplet $P_0, P_1(\eta)$ and $P_2(\xi,\eta)$, $\le \xi \le 0, -\sigma \le \eta \le 0$ that satisfies the following cati type PDEs.

$$A_0^\top \Phi_0 + \Phi_0^\top A_0 + (\sum_{i=1}^{N} A_i)^\top P_1^\top(0) - P_1(0)\sum_{i=1}^{N} A_i$$
$$-\Phi_0^\top BR^{-1}B^\top \Phi_0 + Q = 0 \quad \text{(12-a)}$$

$$\frac{\partial}{\partial\eta}P_1(\eta) - A_0^\top \Phi_1(\eta) + (\sum_{i=1}^{N} A_i)^\top P_2(0,\eta)$$
$$-\Phi_0^\top BR^{-1}B^\top \Phi_1(\eta) = 0 \quad \text{(12-b)}$$

$$\frac{\partial}{\partial\xi}P_2(\xi,\eta) - \frac{\partial}{\partial\eta}P_2(\xi,\eta) - \Phi_1^\top(\xi)BR^{-1}B^\top \Phi_1(\eta) = 0 \quad \text{(12-c)}$$

where $Q = C^\top C$, $\Phi_0 = P_0 + (\sum_{i=1}^{N} A_{-i})^\top P_1^\top(0)$ and $\Phi_1(\eta) = P_1(\eta) + (\sum_{i=1}^{N} A_{-i})^\top P_2(0,\eta)$ with the symmetric conditions

$$P_0^\top = P_0, P_2^\top(\xi,\eta) = P_2(\xi,\eta) \quad (13)$$

and the boundary conditions

$$\Phi_0 - P_1(-\sigma) = 0, \; \Phi_1(\eta) - P_2(-\sigma,-\sigma) = 0 \quad (14)$$

*Theorem 3* Suppose that $\|\sum_{i=1}^{N} A_{-i}\|_\infty < 1$. If there exists a positive constant $k$ such that

$$\max\{\max_i P_0^{ii}(t), \max_{-\sigma\le\eta\le 0} \max_i P_2^{ii}(t,\eta,\eta)\} \le k \quad (15)$$

for $t > 0$, where superscript $ii$ denotes $(i,i)$-element of the matrix, then there is a $C-$ solution of the stationary Riccati equations in the interval $-\sigma \le \xi \le 0$, $-\sigma \le \eta \le 0$.

Before turning to the proof of the theorem, we state a lemma for the properties of the nonstationary Riccati type PDEs.

*Lemma 1* The solution of the Riccati type PDEs (3)-(6) have the following some properties

- $(i)$ $P_0^{ii}(0) \ge 0$
- $(ii)$ $P_2^{ii}(0,\eta,\eta) \ge 0$
- $(iii)$ $P_0^{ij}(0) \le \frac{1}{2}(P_0^{ii}(0) + P_0^{jj}(0))$
- $(iv)$ $P_2^{ij}(0,\xi,\eta) \le \frac{1}{2}(P_2^{ii}(0,\eta,\eta) + P_2^{jj}(0,\xi,\xi))$
- $(v)$ $P_1^{ij}(0,\eta) \le \frac{1}{2}(P_0^{ii}(0) + P_2^{jj}(0,\eta,\eta))$

This lemma is easily proved by using the relation $J(T) \ge 0$ and choosing properly initial functions.

*Proof of Theorem 3* Let $u_1$ and $u_2$ denote the optimal control for terminal condition $T_1$ and $T_2 (T_1 \le T_2)$, respectively. Then the relation $J(u_2;\phi,T_2) \ge J(u_1;\phi,T_1)$ holds. From Lemma 1 $(i)$, $P_0^{ii}(T_1) \le P_0^{ii}(T_2)$ and the boundedness assumption (15), there exists the limit $P_0^{ii}$ such that

$$\lim_{T\to\infty} P_0^{ii}(t) = P_0^{ii} \quad (16)$$

Furthermore, from Lemma 1 $(iii)$, we can obtain that

$$\lim_{T\to\infty} P_0^{ij}(t) = P_0^{ij} \quad (17)$$

If $T_2 > T_1$, then $P_2^{ii}(T_1,\eta,\eta) \le P_2^{ii}(T_2,\eta,\eta)$ and there exist the limit for fixed $\eta$ such that

$$\lim_{T\to\infty} P_2^{ii}(t,\eta,\eta) = P_2^{ii}(\eta,\eta) \quad (18)$$

We show that (18) is uniformly convergence with respect to $\eta$. Let $C^{ii}$ denote the characteristic curve defined as follows

$$C^{ii} = \{(\xi,\eta,t^i) : \xi = \eta, -\sigma \le \eta \le 0, t^i = t^i(\eta;\sigma_i), \sigma_i \ge 0\}$$

$$\frac{dt^i}{d\eta} = -1, -\sigma \le \eta \le 0, t^i(0;\sigma) = \sigma \quad (19)$$

We consider $P_2^{ii}(t,\xi,\eta)$ along this characteristic curve. $P_2^{ii}(t,\xi,\eta)$ obeys the ODE. Therefore, by using Ascoli-Arzela theorem, we obtain a sequence $\{\sigma_k\}$ such that

$$\lim_{\sigma_k\to\infty} P_2^{ii}(0,\eta,\eta;t_i(\eta,\sigma_k)) = \widetilde{P}_2^{ii}(\eta,\eta) \tag{20}$$

uniformly in $\eta$, $-\sigma \leq \eta \leq 0$, for some continuous function $\widetilde{P}_2^{ii}(\eta,\eta)$. From the uniqueness of the monotonicity convergent sequence, we have $\widetilde{P}_2^{ii}(\eta,\eta) = P_2^{ii}(\eta,\eta)$. Furthermore, from Lemma 1 $(iv),(v)$ and assumption (15), we show that non-diagonal terms of $|P_2(T_1,\xi,\eta)-P_2(T_2,\xi,\eta)|$ and $|P_1(T_1,\eta)-P_1(T_2,\eta)|$ are not greater than its diagonal term. As mentioned above, the convergence

$$\lim_{T\to\infty} P_1^{ij}(t,\eta) = P_1^{ij}(\eta), \tag{21}$$

$$\lim_{T\to\infty} P_2^{ij}(t,\xi,\eta) = P_2^{ij}(\xi,\eta), \tag{22}$$

hold uniformly in $(\xi,\eta)$, $-\sigma \leq \xi \leq 0$, $-\sigma \leq \eta \leq 0$, where $P_1^{ij}(\eta)$ and $P_2^{ij}(\xi,\eta)$ are continuous functions. The proof of Theorem 3 is completed.

# 4. OPTIMAL REGULATOR

In this section, we deal with optimal regulator for neutral system with multi-delays under assumption of $\|\sum_{i=1}^{N} A_{-i}\| < \infty$ in the infinite time interval. We define stabilizability and detectability for neutral system with multi-delays, and we show that if system (1) is stabilizable and detectable, then there exists the solution of the stationary Riccati equations. We show finally that the optimal closed-loop system is stable.

*Definition 1* (stabilizability) Suppose that control input is given by the following linear state feedback form

$$u(t) = K_0x(t) + \int_{-\sigma}^{0} K_1(\eta)[\sum_{i=1}^{N}(A_ix(t+\eta)+A_{-i}\dot{x}(t+\eta))]d\eta \tag{23}$$

where $(K_0,K_1) \in \mathbf{R}^{n\times m} \times L^2([-\sigma,0];\mathbf{R}^{n\times m})$. For the initial function $\phi$, if the unique solution $x(t)$ satisfies the relation

$$\lim_{T\to\infty} \int_{Q}^{\sigma} x^{\top}(t)x(t)dt < \infty \tag{24}$$

then the system is said to be $L^2$-stable. Furthermore, the system is stabilizable if there exists $(K_0,K_1)$ for given initial function, such that the closed-loop system is $L^2$-stable.

*Definition 2* (detectable) The system is detectable if the adjoint system is stabilizable.

$$\dot{y}(t) = A_0^{\top}y(t) + \sum_{i=1}^{N}[A_i^{\top}y(t+\sigma) + A_{-i}^{\top}\dot{y}(t+\sigma)] + C^{\top}u(t) \tag{25}$$

It is known that neutral root chains in the right half plane can not be shifted into the left half plane by linear stable feedback control of the form (23)(D,A.O'ccnnor). Since $\max|\lambda(M)| \leq \|M\|_\infty$, the assumption $\|\sum_{i=1}^{N} A_{-i}\|_\infty < 1$ (i.e. $\max|\lambda(\sum_{i=1}^{N})| < 1$) guarantees that neutral root chains lie in the left half plane, that is, there exist only finite number of poles in the closed right half plane. Unless the assumption $\max|\lambda(\sum_{i=1}^{N})| < 1$ holds, the stabilizability problem has been left unsolved until now.

We now state an optimal regulator problem in the infinite time interval and derive the solution of the optimal regulator problem. Let the infinite time cost functional be given as

$$J(\infty) = \int_0^{\infty} [x^{\top}(t)C^{\top}Cx(t) + u^{\top}(t)Ru(t)]dt. \tag{26}$$

Our problem is to find a control input which minimizes the cost functional (26) for the system (1).

*Theorem 4* Suppose that $\|\sum_{i=1}^{N} A_{-i}\|_\infty < 1$ and the system (1) is stabilizable and detectable. There exists an unique optimal control law given as

$$\begin{aligned} u(t) = &-R^{-1}B^{\top}\{\Phi_0x(t) \\ &+ \int_{-\sigma}^{0} \Phi(\eta)\sum_{i=1}^{N}[A_ix(t+\eta) + A_{-i}\dot{x}(t+\eta)]d\eta\} \end{aligned} \tag{27}$$

where $\Phi_0$ and $\Phi_1(\eta)$ are defined by the solution of the stationary Riccati equations (12)-(14). The optimal cost $J(\infty)$ is represented as

$$\begin{aligned} J(\infty) = &\ \phi^{\top}(0)P_0\phi(0) \\ &+2\phi^{\top}(0)\int_{-\sigma}^{0} P_1(\eta)\sum_{i=1}^{N}[A_i\phi(\eta) + A_{-i}\dot{\phi}(\eta)]d\eta \\ &+\int_{-\sigma}^{0}\int_{-\sigma}^{0}\sum_{i=1}^{N}[A_i\phi(\xi) + A_{-i}\dot{\phi}(\xi)]^{\top}P_2(\xi,\eta) \\ &\sum_{i=1}^{N}[A_i\phi(\eta) + A_{-i}\dot{\phi}(\eta)]d\xi d\eta \end{aligned} \tag{28}$$

*Proof* In the same way as in the proof of Theorem 1, we can prove the optimality. Since the system (1) is stabilizable, there exist a feedback gain $(K_0,K_1)$ such that the closed-loop system is $L^2$-stable. Let $\widetilde{x}(t)$ and $\widetilde{u}(t)$ denote the corresponding trajectory and input, respectively. Then we have

$$J(\sigma) \leq \int_0^{\sigma} [\widetilde{x}^{\top}(t)Q\widetilde{x}(t) + \widetilde{u}^{\top}(t)R\widetilde{u}(t)]dt. \tag{29}$$

ce the closed-loop system is $L^2$-stable, $J(\tau)$ is unded and we consider that (15) of Theorem 3 is satis- for suitable choice of initial function. Therefore from eorem 3 there exists the solution of the stationary Ric- i equation. J. Zabczyk shown that the semigroup gen- ted by the optimal closed-loop system is exponential ble by applying an operator Lyapunov equation and assumption of detectability. We introduce the Lya- ov type PDEs. Using similar procedure to that of J. czyk to the Riccati type PDEs, $L^2$-stability of the sed-loop system is proved under the assumption of de- tability. At last, we prove that the uniqueness. Now, suppose that there are two solutions of the stationary cati equations. Defining a function as the following

$$\begin{aligned} V^{12}(t) &= x^{1\top}(t)U_0x^2(t) \\ &+x^{1\top}(t)\int_{-\sigma}^{0} U_1(\eta)\sum_{i=1}^{N}[A_ix^2(t+\eta)+A_{-i}\dot{x}^2(t+\eta)]d\eta \\ &+\int_{-\sigma}^{0}\sum_{i=1}^{N}[A_ix^1(t+\xi)+A_{-i}\dot{x}^1(t+\xi)]^\top \\ &U_1^\top(\xi)d\xi x^2(t)+\int_{-\sigma}^{0}\sum_{i=1}^{N}[A_ix^1(t+\xi)+A_{-i}\dot{x}^1(t+\xi)]^\top \\ &U_2(\xi,\eta)\sum_{i=1}^{N}[A_ix^2(t+\eta)+A_{-i}\dot{x}^2(t+\eta)]d\xi d\eta \end{aligned} \tag{30}$$

ere $U_i, i=0,1,2$ are difference of two solutions and $i=1,2$ denote the trajectory of two closed-loops. lculating $dV^{12}(t)/dt$, we obtain $dV^{12}(t)/dt=0$. Since closed-loop systems are $L^2$-stable, $\lim_{t\to\infty}V^{12}(t)=0$ ds. It means that $U_0=0, U_1(\eta)=0$ and $U_2(\xi,\eta)=0$, $\le\xi\le 0, -\sigma\le\eta\le 0$. Therefore, the uniqueness is ved.

## . CLOSED-LOOP PROPERTIES

study in this section some properties of the optimal ulator analogous to that of the finite-dimensional op- al regulators, that is circle condition and the relation ween Hamiltonian matrix and closed-loop poles.

ltiplying (12-b) by $\sum_{i=1}^{N}(A_ie^{s\eta}+sA_{-i}e^{s\eta})$ from the right, egrating both sides with respect to $\eta$ over the interval $\sigma,0]$ by parts by using the boundary condition (14), we tain

$$\begin{aligned} &-P_1(0)\sum_{i=1}^{N}(A_i+sA_{-i})+\Phi_0^\top\sum_{i=1}^{N}(A_ie^{-s\sigma}+sA_{-i}e^{-s\sigma}) \\ &+(sI+A_0^\top-\Phi_0^\top BR_{-1}B^\top)\int_{-\sigma}^{0}\Phi_1(\eta)\sum_{i=1}^{N}(A_ie^{s\eta} \\ &+sA_{-i}e^{s\eta})d\eta+\sum_{i=1}^{N}(A_i^\top-sA_{-i}^\top)\int_{-\sigma}^{0}P_2(0,\eta) \\ &\sum_{i=1}^{N}(A_ie^{s\eta}+sA_{-i}e^{s\eta})d\eta=0 \end{aligned} \tag{31}$$

Multiplying transposed version of (12-b) in which the independent variable is denoted by $\xi$ instead $\eta$ by $\sum_{i=1}^{N}(A_i^\top e^{-s\xi}-sA_{-i}^\top e^{-s\xi})$ from the left, and calculating the same integral operation as above, we get that

$$\begin{aligned} &-\sum_{i=1}^{N}(A_i-sA_{-i})^\top P_1(0)+\sum_{i=1}^{N}(A_ie^{s\sigma}+sA_{-i}e^{s\sigma})^\top\Phi_0 \\ &-\int_{-\sigma}^{0}\sum_{i=1}^{N}(A_ie^{-s\xi}-sA_{-i}e^{-s\xi})^\top\Phi_1^\top(\xi)(sI-A_0^\top \\ &+BR_{-1}B^\top\Phi_0)d\xi+\int_{-\sigma}^{0}\sum_{i=1}^{N}(A_ie^{-s\xi}-sA_{-i}e^{-s\xi})^\top \\ &P_2(\xi,0)d\xi\sum_{i=1}^{N}(A_i+sA_{-i})=0 \end{aligned} \tag{32}$$

Furthermore, multiplying both sides of (12-c) by $\sum_{i=1}^{N}(A_ie^{s\eta}+sA_{-i}e^{s\eta})$ from the right, and by $\sum_{i=1}^{N}(A_i^\top e^{-s\xi}-sA_{-i}^\top e^{-s\xi})$ from the left, integrating them with respect to $\xi$ and $\eta$ over the same interval $[-\sigma,0]$ by using the second boundary condition (14), we can obtain that

$$\begin{aligned} &-2\sum_{i=1}^{N}(A_i-sA_{-i})^\top\int_{-\sigma}^{0}P_2(0,\eta)\sum_{i=1}^{N}(A_ie^{s\eta}+sA_{-i}e^{s\eta})d\eta \\ &+2\sum_{i=1}^{N}(A_ie^{-s\sigma}-sA_{-i}e^{-s\sigma})^\top\int_{-\sigma}^{0}\Phi_1(\eta) \\ &\sum_{i=1}^{N}(A_ie^{s\eta}+sA_{-i}e^{s\eta})-\int_{-\sigma}^{0}\int_{-\sigma}^{0}\sum_{i=1}^{N}(A_ie^{-s\xi}-sA_{-i}e^{-s\xi})^\top \\ &\Phi_1^\top(\xi)BR^{-1}B^\top\Phi_1(\eta)\sum_{i=1}^{N}(A_ie^{s\eta}+sA_{-i}e^{s\eta})d\xi d\eta=0 \end{aligned} \tag{33}$$

Adding together for the equations (31), (32), (33) and Riccati equation (12-a), that is $(12-a)+(31)+(32)+(33)$, and rearranging terms, we have

$$\begin{aligned} &F^\top(-s)\Delta(s)+\Delta^\top(-s)F(s) \\ &+F^\top(-s)BR^{-1}B^\top F(s)-Q=0 \end{aligned} \tag{34}$$

where

$$F(s)=\Phi_0+\int_{-\sigma}^{0}\Phi_1(\eta)\sum_{i=1}^{N}(A_ie^{s\eta}+sA_{-i}e^{s\eta})d\eta \tag{35}$$

$$\Delta(s)=sI-A_0-\sum_{i=1}^{N}(A_ie^{-s\sigma}+sA_{-i}e^{-s\sigma}) \tag{36}$$

Then, by using a standard successive manipulation for finite dimensional Riccati equation into (34), it can be rewritten as the following form

$$\begin{aligned} &[I+R^{-1}B^\top F(-s)\Delta^{-1}(-s)B]^\top \\ &R[I+R^{-1}B^\top F(-s)\Delta^{-1}(-s)B] \\ &= R+B^\top\Delta^{-1\top}(-s)Q\Delta^{-1}(s)B \end{aligned} \tag{37}$$

It can be called *Kalman Equation*. Set $s=j\omega$ in (36) and note that the matrix $B^\top\Delta^{-1\top}(-j\omega)Q\Delta^{-1}(j\omega)B$ is positive semi-definite for all real $\omega$. Thus we have

$$[I + R^{-1}B^{\top}F(-j\omega)\Delta^{-1}(-j\omega)B]^{*}R \quad [I + R^{-1}B^{\top}F(-j\omega)\Delta^{-1}(-j\omega)B] \geq R \tag{38}$$

for all real $\omega$. This inequality is called *Circle Condition.*

Circle Condition plays an important role about robustness in finite dimensional systems and retarded system[1,14]. Therefore, it may be true that optimal regulator for neutral systems with multi-delays has some good closed-loop properties as same as the above.

At last, we study the relation between the Hamiltonian matrix and the optimal closed-loop poles. Defining Hamiltonian matrix is that

$$H(s) = \begin{bmatrix} A_0 + \sum_{i=1}^{N}(e^{-s\sigma}A_i + se^{-s\sigma}A_{-i}) & -BR^{-1}B^{\top} \\ -Q & -A_0^{\top} - \sum_{i=1}^{N}(e^{s\sigma}A_i^{\top} + se^{s\sigma}A_{-i}^{\top}) \end{bmatrix}$$

Calculating $\det[sI - H(s)]$ with the Kalman equation (36), we have

$$\det[sI - H(s)] = (-1)^n \det[\Delta^{\top}(-s) + F^{\top}(-s)BR^{-1}B^{\top}]\det[\Delta(s) + F(s)BR^{-1}B^{\top}] \tag{39}$$

Since closed-loop system is $L^2$-stable, closed-loop poles lie in the left-half plane. The characteristic equation is given as $\det[\Delta(s) + BR^{-1}B^{\top}F(s)]$, the formula (39) ensures that the optimal closed-loop poles coincide with the eigenvalues associated with the Hamiltonian matrix $H(s)$ in the left-half complex plane.

## 6. CONCLUSION

It has been shown that the infinite time interval optimal regulator for neutral control system with multi-delays is constructed by using the stationary solution of the Riccati type PDEs on the assumption that no neutral chain lies in the closed-right half plane. It has been also shown that the optimal control law is given by the state feedback from and that the closed-loop system is $L^2$-stable and has the properties similar to those of finite dimensional control systems and retarded control system with multi-delays. It is necessary furthermore to investigate whether optimal regulator problem is well-posed or not, when there exist infinite number of poles in the closed right half plane.

It is known that there are two different definition of state, one is $(\phi(0), \sum_{i=1}^{N}(A_i + A_{-i})\phi)$ and the other is $(\phi(0) - \sum_{i=1}^{N} A_{-i}\phi(-\sigma), \phi)$. In this paper we have used the first one as the definition of state. Even if we use the second one, it is possible to study optimal regulator problem similarly. Since the limited space, we omit the discussion.

## REFERENCES


Uchida, K., E.Shimemura and N.Abe (1987). Circle condition and stability margin of the optimal regulator for systems with delays. *Int.J.Control*, **46**, 1203-1204.

Datko,R. (1974). Neutral autonomous functional equations with quadratic cost. *SIAM J. Control*, **12**, 70-82

Sendanla, M. (1976). On the optimal control theory for neutral systems. *IEEE Proc. ISCAS*, 408-411

Ito, K. and T.J.Tarn (1985). A linear quadratic optimal control for neutral systems. *Nonlinear Analysis, Theory, Methods & Applications*, **9**, 699-727

Prichard, A.J. and D. Salamon (1987). The linear quadratic control problem for infinite dimensional systems with unbounded input and output operators. *SIAM J. Control and Optimization*, **25**, 121-144

Abe, N., K. Uchida and E. Shimemura (1990). Optimal regulator for neutral systems. *11th IFAC World Congress on Automatic Control in the Serviceof Mankind*, **2**, 119-124, Tallinn, Estonia, USSR.

Hale, J. (1977). Theory and Application on Functional Differential Equations. Chapter 1; 12, Springer Verlag, New York.

Bellman, R., and K.L. Cooke (1963). Differential-Difference Equations. Chapter 1-4, Academic Press, New York.

Qin, Y.S., Y.Q.Liu, L Wang and Z. X. Zheng (1989). Motion Stability of Dynamic System with Delays (Second Press). 1-21, Academic Press, Beijing.

O'ccnnor, D.A., and T.J.Tarn (1983). On stabilization by state feedback for neutral differential difference equations. *IEEE AC*, **28** 615-618

Zabczyk, J. (1976). Remark on the algebraic Riccati equation in Hilbert space. *Appl. Math. & Optim.*, **2**, 251-258

Gao, C.C., and F.Y.Zheng (1994). The control representative on a kind of control systems determined by partial differential equation of the hyperbilic type. *Advances in Modelling & Analysis, C*, **40(3)**, 1-11

Liu, Y.Q., and C. C. Gao (1992a). On stabilization for the multi-groups and the multi-delays neutral type linear control systems. *Intern. '92 AMSE Conference on Modelling, Simulation & Control*, **5**, 530-575, Hefei, P.R.China.

Liu, Y.Q., and C. C. Gao (1993b). On the robust stability of interval coefficients neutral type linear constant control systems with multi-group multi-delays (1). *Proc. '93 Am. Contr. Conf.*, 1037-1040

# ON THE GENERALIZED EIGENSPACE APPROACH FOR SOLVING RICCATI EQUATIONS *

**Li Qiu** *

* *Department of Electrical & Electronic Engineering*
*Hong Kong University of Science & Technology*
*Clear Water Bay, Kowloon, Hong Kong*
*Tel: 852-2358-7067, Fax: 852-2358-1485*
*Email: eeqiu@ee.ust.hk*

Abstract: In this paper, we show that a general discrete time Riccati equation can be solved by finding a deflating subspace of a $2n \times 2n$ matrix pencil. In many applications, the data in the Riccati equation is formed from the squares of the raw physical data. In this case, we show that continuous-time and discrete-time Riccati equations can be solved without squaring the raw data by finding deflating subspaces of larger matrix pencils. Examples show that the solutions without squaring the data have numerical advantages over the existing solutions of Riccati equations using the generalized eigenspace approach. *Copyright © 1999 IFAC*

Keywords: Riccati equations, numerical methods, optimal control, matrix equations

## 1. INTRODUCTION

For a matrix pair $(A, B) \in \mathbb{C}^{n\times n} \times \mathbb{C}^{n\times m}$ and a Hermitian matrix

$$\begin{bmatrix} Q & S \\ S^* & R \end{bmatrix} \in \mathbb{C}^{(n+m)\times(n+m)},$$

the associated continuous time algebraic Riccati equation (CARE) is of the form (assuming $R$ to be nonsingular)

$$A^*X + XA + Q - (XB + S)R^{-1}(B^*X + S^*) = 0 \quad (1)$$

and the associated discrete time algebraic Riccati equation (DARE) is of the form (assuming $\mathrm{Ker}R \cap \mathrm{Ker}B = \{0\}$)

$$A^*XA - X + Q - (A^*XB + S)(B^*XB + R)^{-1} \times (B^*XA + S^*) = 0. \quad (2)$$

A Hermitian solution $X$ of CARE (1) is said to be stabilizing if all eigenvalues of $A + BF$, where $F = -R^{-1}(B^*X + S^*)$ is the corresponding state feedback gain, have negative real parts. A Hermitian solution $X$ of DARE (2) is said to be stabilizing if all eigenvalues of $A + BF$, where $F = -(B^*XB + R)^{-1}(B^*XA + S^*)$ is the corresponding state feedback gain, are inside the unit circle. The stabilizing solution, if exists, is unique.

The solution of ARE has been extensively studied. Two recent collections of works (Bittanti *et al.*, 1991; Patel *et al.*, 1994) contain good survey papers on this topic. There are several ways to find the stabilizing solution of an ARE. Among them the so-called Schur method has gained popularity in recent years.

The Schur method for CARE (1) is based on the fact that the graph of the unique stabilizing solution of (1), if exists, is equal to the invariant subspace of the $2n \times 2n$ Hamiltonian matrix

$$\begin{bmatrix} A - BR^{-1}S^* & -BR^{-1}B^* \\ -Q + SR^{-1}S^* & -A^* + SR^{-1}B^* \end{bmatrix} \quad (3)$$

corresponding to the eigenvalues with negative real parts. This fact, apart from its theoretic interest, has been the basis of the general purpose

* This research is supported by the Hong Kong Research Grants Council and Sino Software Research Institute.

Riccati solvers in, such as, MATLAB Control System Toolbox.

The Schur method for DARE (2) leads to generalized eigenproblems. If $R$ is nonsingular, the graph of the unique stabilizing solution of (2), if exists, is equal to the deflating subspace of $2n \times 2n$ symplectic matrix pencil

$$\left( \begin{bmatrix} A - BR^{-1}S^* & 0 \\ -Q + SR^{-1}S^* & I \end{bmatrix}, \begin{bmatrix} I & BR^{-1}B^* \\ 0 & A^* - SR^{-1}B^* \end{bmatrix} \right) \quad (4)$$

corresponding to the eigenvalues inside the unit circle (Pappas *et al.*, 1980; Emami-Naeini and Franklin, 1980). However, applications where $R$ is singular occur frequently. To cope with this, it is proposed in the literature (Emami-Naeini and Franklin, 1980; Van Dooren, 1981) and implemented in MATLAB Control System Toolbox to find the stabilizing solution using the following $(2n+m) \times (2n+m)$ matrix pencil

$$\left( \begin{bmatrix} A & 0 & B \\ Q & -I & S \\ S^* & 0 & R \end{bmatrix}, \begin{bmatrix} I & 0 & 0 \\ 0 & -A^* & 0 \\ 0 & -B^* & 0 \end{bmatrix} \right). \quad (5)$$

In this case, the graph of the stabilizing solution, if exists, is equal to the space spanned by the first $2n$ coordinates of the deflating subspace of (5) corresponding to the eigenvalues inside the unit circle. This although pragmatically overcomes the difficulty brought about by the singular or near singular $R$, it leaves some disappointment: the use of $(2n+m) \times (2n+m)$ pencil is theoretically unpleasant (lack of symmetry) and numerically undesirable (presence of bigger matrices). In this paper, we show that it is possible to solve (2) using a $2n \times 2n$ matrix pencil.

For a given Hermitian matrix

$$\begin{bmatrix} Q & S \\ S^* & R \end{bmatrix},$$

it is always possible to carry out factorization

$$\begin{bmatrix} Q & S \\ S^* & R \end{bmatrix} = \begin{bmatrix} C^* \\ D^* \end{bmatrix} J \begin{bmatrix} C & D \end{bmatrix} \quad (6)$$

where $J$ is a signature matrix such as

$$J = \begin{bmatrix} I & 0 \\ 0 & -I \end{bmatrix}$$

or some other nonsingular Hermitian matrix. In many applications, such as $\mathcal{H}_2$ and $\mathcal{H}_\infty$ control and various factorizations of rational matrices, the raw data for a Riccati equation is given in terms of $(A, B, C, D, J)$ instead of $(A, B, Q, R, S)$. In this case, CARE (1) becomes

$$A^*X + XA + C^*JC - (XB + C^*JD) \times (D^*JD)^{-1}(B^*X + D^*JC) = 0, \quad (7)$$

and DARE (2) becomes

$$A^*XA - X + C^*JC - (A^*XB + C^*JD) \times (B^*XB + D^*JD)^{-1}(B^*XA + D^*JC) = 0. \quad (8)$$

Although matrices of the form $C^*JC$ and $DJD^*$ appear in the equation, it is desirable to avoid this kind of squares in the numerical solution due to finite wordlength reasons. As soon as the matrix squares are formed, the number of accurate digits is lost by half. No matter what method in solving AREs (1) and (2) is used to solve AREs (7) and (8), it is impossible to recover such loss of accuracy. The readers may wish to examine Examples 1-2 in Section 5 for better appreciation of this point. In this paper, we will show that AREs (7) and (8) can be solved via $(2n+m+p) \times (2n+m+p)$ matrix pencils without forming the matrix squares explicitly. Examples show that the proposed new method does behave better numerically.

An immediate criticism of this proposed method is the involvement of matrices of large dimensions. This is indeed a new problem created in addressing an old problem. On one hand we would like to argue that the computer wordlength is a more precious resource than the computing time. On the other hand, we will show that the $(2n+m+p) \times (2n+m+p)$ matrix pencil is always sparse and structured, so iteration method can be used to find the required eigenspace.

## 2. SOLUTION OF THE DISCRETE TIME ARE VIA $2N \times 2N$ PENCIL

Since $\mathrm{Ker}B \cap \mathrm{Ker}R = \{0\}$, there exists symmetric matrix $X_0$ such that $B^*X_0B + R$ is nonsingular. Let

$$\begin{bmatrix} Q_0 & S_0 \\ S_0^* & R_0 \end{bmatrix} = \begin{bmatrix} A^*X_0A - X_0 & A^*X_0B \\ B^*X_0A & B^*X_0B \end{bmatrix} + \begin{bmatrix} Q & S \\ S^* & R \end{bmatrix}$$

and let $\tilde{X}$ be the stabilizing solution of the following Riccati equation

$$A^*XA - X + Q_0 - (A^*XB + S_0)(B^*XB + R_0)^{-1} \times (B^*XA + S_0^*) = 0.$$

Then it is easy to see that $X = \tilde{X} + X_0$ is a stabilizing solution of (2). Hence the following theorem follows immediately.

*Theorem 1.* Let $X_0$ be any Hermitian matrix such that $B^*X_0B + R$ is nonsingular and

$$\begin{bmatrix} Q_0 & S_0 \\ S_0^* & R_0 \end{bmatrix} = \begin{bmatrix} A^*X_0A - X_0 & A^*X_0B \\ B^*X_0A & B^*X_0B \end{bmatrix} + \begin{bmatrix} Q & S \\ S^* & R \end{bmatrix}.$$

Let $\begin{bmatrix} V_1 \\ V_2 \end{bmatrix}$ span the deflating subspace of

$$\left( \begin{bmatrix} A - BR_0^{-1}S_0^* & 0 \\ -Q + S_0R_0^{-1}S_0^* & I \end{bmatrix}, \begin{bmatrix} I & BR_0^{-1}B^* \\ 0 & A^* - S_0R_0^{-1}B^* \end{bmatrix} \right) \quad (9)$$

corresponding to the stable eigenvalues. Then (2) has a stabilizing solution if and only if $V_1$ is invertible. In this case, the stabilizing solution is given by $X = X_0 + V_2V_1^{-1}$.

One may prefer a matrix pencil which directively leads to $X$ without the extra addition $\tilde{X} + X_0$. This is possible.

*Theorem 2.* Let $X_0$ be any Hermitian matrix such that $R_0 := B'X_0B + R$ is nonsingular and $\begin{bmatrix} U_1 \\ U_2 \end{bmatrix}$ span the deflating subspace of

$$\left( \begin{bmatrix} A - BR_0^{-1}S_0^* & 0 \\ -Q + SR_0^{-1}S_0^* & I \end{bmatrix}, \begin{bmatrix} I - BR_0^{-1}B^* & BR_0^{-1}B^* \\ SR_0^{-1}B^*X_0 & A^* - SR_0^{-1}B^* \end{bmatrix} \right) \quad (10)$$

corresponding to the stable eigenvalues. Then (2) has a stabilizing solution if and only if $U_1$ is invertible. In this case, the stabilizing solution is given by $X = U_2U_1^{-1}$.

## 3. SOLUTION WITHOUT SQUARE-UP

When (6) holds, CARE (1) becomes CARE (7). It is desirable to solve CARE (7) without forming $Q, S, R$.

Consider matrix pencil

$$\left( \begin{bmatrix} A & B & 0 & 0 \\ C & D & 0 & -J^{-1} \\ 0 & 0 & -A^* & -C^* \\ 0 & 0 & -B^* & -D^* \end{bmatrix}, \begin{bmatrix} I & 0 & 0 & 0 \\ 0 & 0 & 0 & 0 \\ 0 & 0 & I & 0 \\ 0 & 0 & 0 & 0 \end{bmatrix} \right). \quad (11)$$

*Proposition 1.* The matrix pencil (11) has $p + m$ eigenvalues at infinity and the rest of the $2n$ eigenvalues are finite and symmetric to the imaginary axis.

Let $\mathcal{V}$ be the deflating subspace of (11) corresponding to the finite eigenvalues with negative real parts and let a basis of $\mathcal{V}$ be given by the columns of $\begin{bmatrix} V_1 \\ V_2 \\ V_3 \\ V_4 \end{bmatrix}$, partitioned consistently with that of (11).

*Theorem 3.* CARE (7) has a stabilizing solution if and only if $V_1$ is invertible. In this case, the unique stabilizing solution of (7) is given by $X = V_3V_1^{-1}$ and the corresponding feedback gain is given by $F = V_2V_1^{-1}$.

The DARE (8) can also be solved without squares. Consider matrix pencil

$$\left( \begin{bmatrix} A & B & 0 & 0 \\ C & D & 0 & 0 \\ 0 & 0 & I & 0 \\ 0 & 0 & 0 & 0 \end{bmatrix}, \begin{bmatrix} I & 0 & 0 & 0 \\ 0 & 0 & 0 & J^{-1} \\ 0 & 0 & A^* & C^* \\ 0 & 0 & B^* & D^* \end{bmatrix} \right). \quad (12)$$

*Proposition 2.* The matrix pencil (12) has at least $p$ eigenvalues at 0, at least $m$ at $\infty$, and the rest $2n$ eigenvalues are symmetric to the unit circle.

Let $\mathcal{U}$ be the deflating subspace of (12) corresponding to the eigenvalues inside the unit circle. Then it is easy to see that span $\begin{bmatrix} 0 \\ 0 \\ 0 \\ I \end{bmatrix} \subset \mathcal{U}$. Let $\mathcal{V}$ be any complement of span $\begin{bmatrix} 0 \\ 0 \\ 0 \\ I \end{bmatrix}$ in $\mathcal{U}$ and let a basis of $\mathcal{V}$ be given by the columns of $\begin{bmatrix} V_1 \\ V_2 \\ V_3 \\ V_4 \end{bmatrix}$.

*Theorem 4.* The Riccati equation (8) has a stabilizing solution if and only if $V_1$ is invertible. In this case, the unique stabilizing solution of (8) is given by $X = V_3V_1^{-1}$ and the corresponding state feedback gain is given by $F = V_2V_1^{-1}$.

## 4. THE POWER METHOD

An immediate criticism of the proposed use of $(2n+m+p) \times (2n+m+p)$ pencils in solving ARE is the involvement of matrices of large dimensions. This is indeed a new problem created in addressing an old problem. On one hand we would like to argue that the computer wordlength is a more precious resource than the computing time. On the other hand, we see that the $(2n + m + p) \times (2n + m + p)$ matrix pencil is always sparse and structured. It is hopeful that the sparseness and the structure can be utilized to reduce the computational complexity. In this section, we present a crude method based on power iteration to find the required eigenspaces of the matrix pencils involved.

Let us first consider the problem of finding the eigenspace corresponding to the $m$ dominant eigenvalues of a general regular $n \times n$ pair $(M, N)$. First let us assume that $N$ is nonsingular. In this case, the eigenvalues and corresponding eigenspaces are identical to those of $N^{-1}M$. The following algorithm, called block or subspace power iteration, is well known for the eigenspace of the dominant eigenvalues (Saad, 1992).

*Algorithm 1.*

**Start.** Choose an initial $m$ dimensional subspace $\mathcal{V}_0$.

**Iteration.** Until convergence, compute $\mathcal{V}_{i+1} = N^{-1}M\mathcal{V}_i$.

For generic initial subspace, this algorithm will converge to the required eigenspace if $|\lambda_m(M, N)| > |\lambda_{m+1}(M, N)|$.

Now let us consider the matrix pencil (11). The eigenspace that we are interested in is the one corresponding to the eigenvalues on the left hand side of the complex plane. We need to transform the problem so that these eigenvalues become dominant ones. Observe that the map $(M_0, N_0) \to (M_0+\alpha N_0, M_0-\alpha N_0)$ where $\alpha > 0$ maps all eigenvalues in the left hand side of the complex plane to the outside of the unit circle. This technique is called shifting. After shifting, matrix pencil (11) becomes

$$(M, N) = \left( \begin{bmatrix} A-\alpha I & B & 0 & 0 \\ C & D & 0 & -J^{-1} \\ 0 & 0 & -A^*-\alpha I & -C^* \\ 0 & 0 & -B^* & -D^* \end{bmatrix}, \begin{bmatrix} A+\alpha I & B & 0 & 0 \\ C & D & 0 & -J^{-1} \\ 0 & 0 & -A^*+\alpha I & -C^* \\ 0 & 0 & -B^* & -D^* \end{bmatrix} \right).$$

Since the pair $(M, N)$ is still sparse and structured, both $M\mathcal{V}$ and $N^{-1}\mathcal{V}$ can be computed efficiently without forming $(M, N)$ explicitly.

For the pencil (12), we are interested in the eigenspace corresponding to the eigenvalues inside the unit circle. Apparently, shifting can also be used to transform these eigenvalues to the dominant ones. The map $(M_0, N_0) \to (N_0 - \alpha M_0, M_0 - \alpha N_0)$ for $-1 < \alpha < 1$ does the job. In this case, it amounts to consider the following matrix pair.

$$(M, N) = \left( \begin{bmatrix} I-\alpha A & -\alpha B & 0 & 0 \\ -\alpha C & -\alpha D & 0 & J^{-1} \\ 0 & 0 & A^*-\alpha I & C^* \\ 0 & 0 & B^* & D^* \end{bmatrix}, \begin{bmatrix} A-\alpha I & B & 0 & 0 \\ C & D & 0 & -\alpha J^{-1} \\ 0 & 0 & I-\alpha A^* & -\alpha C^* \\ 0 & 0 & -\alpha B^* & -\alpha D^* \end{bmatrix} \right). \quad (13)$$

The sparseness of the matrix pair (13) is maximized if $\alpha = 0$ is chosen, but it results in a singular $N$. Experiment shows that Algorithm 1 still works if $N^{-1}\mathcal{V}$ is interpreted as the preimage of $\mathcal{V}$ under $N$. This leads the following conjecture.

*Conjecture 1.* Let Algorithm 1 be applied to a regular matrix pencil $(M, N)$. If $|\lambda_m(M, N)| > |\lambda_{m+1}(M, N)|$, then for generic $\mathcal{V}_0$, $\mathcal{V}_i$ converges to $\mathcal{V}$.

The effort in proving this conjecture is underway.

The convergence of the power iteration depends on $\frac{|\lambda_{m+1}(M,N)|}{|\lambda_m(M,N)|}$. Examples show that the convergence can be very slow for simple problems. It is an interesting problem how to take advantage of the sparseness and structure of the matrix pencils to get a uniformly better algorithm than the unstructured QZ algorithm which is the most commonly used one to find the required eigenspaces.

## 5. NUMERICAL EXAMPLES

Two MATLAB .m files, caresys.m and daresys.m, are written which implement the $(2n + m + p) \times (2n + m + p)$ pencil methods in Section 3 for CARE (7) and DARE (8) respectively. Both caresys.m and daresys.m take as input variables $(A, B, C, D, J)$ and *option* which determines if the eigenspace is computed by the QZ method or the power method. These programs are compared with care.m and dare.m in the MATLAB Control Systems Toolbox. The computation is done in a SPARCstation 10 using MATLAB version 5.0.

Due to time limitation, we are only able to finish the numerical experiment for the discrete-time case. The numerical experiment for continuous-time case is underway and will be added to the paper in the next stage.

Traditional measure of the quality of an approximate solution $\hat{X}$ is $\|R(\hat{X})\|$ where $R(X)$ is the left hand side of the ARE. When the exact solution is unknown, this is probably is the only choice. However, how to compute $R(\hat{X})$ becomes a new nontrivial numerical problem we choose not to address in this paper. The computation of $R(\hat{X})$ is directly related to the computation of

$$F = -(D^*JD)^{-1}(B^*X + D^*JC)$$

in the continuous time case and

$$F = -(B^*XB + D^*JD)^{-1}(B^*XA + D^*JC)$$

in the discrete time case since

$$R(\hat{X}) = A^*X + XA + C^*JC + (XB + C^*JD)F$$

and

$$R(\hat{X}) \\ = A^*XA - X + C^*JC + (A^*XB + C^*JD)F$$

respectively. In applications, it is $F$, not $X$, that is needed. In the following examples, the exact solutions are all known. Hence we use $\|\hat{X} - X\|$ as the measure of the quality of solutions.

*Example 2.* Suppose

$$\left[\begin{array}{c|c} A & B \\ \hline C & D \end{array}\right] = \left[\begin{array}{cc|cc} 0 & 0 & 1 & 1 \\ 0 & 0 & 1 & 1+\epsilon \\ \hline 1 & 1 & 0 & 0 \\ 1 & 1+\epsilon & 0 & 0 \\ 0 & 0 & 1 & 0 \\ 0 & 0 & 0 & 1 \end{array}\right]$$

and $J = I$. It is easy to verify that for nonzero $\epsilon > 0$ the unique stabilizing solution to CARE (7) is $X = I$. However, since

$$Q = C^*C = \begin{bmatrix} 2 & 2+\epsilon \\ 2+\epsilon & 2+2\epsilon+\epsilon^2 \end{bmatrix}$$

is deemed rank one if $\epsilon^2$ is less than the machine epsilon, the matrix

$$\begin{bmatrix} A - BR^{-1}S^* & -BR^{-1}B^* \\ -Q + SR^{-1}S^* & -A^* + SR^{-1}B^* \end{bmatrix}$$

is deemed to have eigenvalues on the imaginary axis. In fact, care.m in the MATLAB Control System Toolbox fails when $\epsilon \leq 10^{-8}$.

*Example 3.* Suppose

$$\left[\begin{array}{c|c} A & B \\ \hline C & D \end{array}\right] = \left[\begin{array}{cc|cc} 1 & 0 & 1 & 0 \\ 0 & 1 & 0 & 1 \\ \hline 1 & 1 & 0 & 0 \\ 1 & 1+\epsilon & 0 & 0 \end{array}\right]$$

and $J = I$. It is easy to verify that for nonzero $\epsilon > 0$ the unique stabilizing solution to DARE (8) is

$$X = C^*C = \begin{bmatrix} 2 & 2+\epsilon \\ 2+\epsilon & 2+2\epsilon+\epsilon^2 \end{bmatrix}.$$

However, since

$$Q = C^*C = \begin{bmatrix} 2 & 2+\epsilon \\ 2+\epsilon & 2+2\epsilon+\epsilon^2 \end{bmatrix}$$

is deemed rank one if $\epsilon^2$ is less than the machine epsilon, the matrix pencil

$$\left( \begin{bmatrix} A & 0 & B \\ Q & -I & S \\ S^* & 0 & R \end{bmatrix}, \begin{bmatrix} I & 0 & 0 \\ 0 & -A^* & 0 \\ 0 & -B^* & 0 \end{bmatrix} \right)$$

is deemed to have generalized eigenvalues on the imaginary axis. Indeed, dare.m in the MATLAB Control System Toolbox fails for this example when $\epsilon \leq 10^{-8}$.

However, daresys.m works well for $\epsilon \geq 10^{-15}$, especially when the power method option is used. The results of daresys.m for different $\epsilon$ is summerized in Table 1. It is seen that the power method works much better than the QZ method in this example. The reason is probably because the error of MATLAB's internal QZ program, which can not be controlled by an user, is too big. The power method also converges very fast. This is due to the fact that the generalized eigenvalues of the matrix pencil involved has eigenvalues at infinity and zero only. A final observation is that $F$ is much more sensitive than $X$ to computation errors. The reason for this is not clear.

The above examples are not pathological cases. In the so-called low-gain and high-gain design of feedback control, the extreme values of the system data occur in quite a natural way.

*Example 4.* A program daresym.m, which implements the $2n \times 2n$ pencil method for DARE, is written. Programs daresym.m, daresys.m, and dare.m are tested using the benchmark examples in (Benner *et al.*, 1996). For most of the examples (examples 1-11), they all work almost equally well. Example 12 has the following data:

$$A = \begin{bmatrix} 0 & \alpha \\ 0 & 0 \end{bmatrix}; B = \begin{bmatrix} 0 \\ 1 \end{bmatrix}; C = \begin{bmatrix} 1 & 0 \\ 0 & 1 \\ 0 & 0 \end{bmatrix}; D = \begin{bmatrix} 0 \\ 0 \\ 1 \end{bmatrix}.$$

The exact solution is given by

$$X = \begin{bmatrix} 1 & 0 \\ 0 & 1+\alpha^2 \end{bmatrix}.$$

Table 2 gives the computation results for several different values of $\alpha$. The relative residue means

$$\|R(\hat{X})\|_F / \|\hat{X}\|_F.$$

The data in Example 13 is given in the following way:

$$A_0 = \text{diag}(0,1,3), V = I - \frac{2}{3}vv', v' = \begin{bmatrix} 1 & 1 & 1 \end{bmatrix}.$$

$$A = VA_0V, B = I, C = \begin{bmatrix} \sqrt{\alpha}I \\ 0 \end{bmatrix}, D = \begin{bmatrix} 0 \\ \sqrt{\alpha}I \end{bmatrix}.$$

The exact solution is

$$X = V\text{diag}\left(\alpha, \alpha\frac{1+\sqrt{5}}{2}, \alpha\frac{9+\sqrt{85}}{2}\right)V.$$

Table 3 shows the computation results. When $\alpha = 10^{14}$, both dare and daresym give the warning message "May not find symmetric solution; spectrum too near unit circle.", but daresys works perfectly. The relative error means

$$\|\hat{X} - X\|_F / \|X\|_F.$$

The relative residue means

$$\|R(\hat{X})\|_F / \|\hat{X}\|_F.$$

## 6. CONCLUSION

In this paper, we first proved that a general regular DARE can be solved by using a $2n \times 2n$ matrix pencil. This result is presented mainly to give a parallel development to the CARE theory. It also indicates that there is not much reason to use the $(2n+m) \times (2n+m)$ matrix pencil (5) in solving the DARE (2).

The main point of this paper is that square of data should be avoided in numerical computation. We have shown that in the case when the raw data for the Riccati equation is given in terms of $(A, B, C, D, J)$ instead of $(A, B, Q, R, S)$, it is possible to compute the solution of Riccati equations without squaring the data even though the squares appear in the equation. The study carried out here is preliminary. It is of interest to see if there are other methods without square-up with better numerical properties and how we can improve the numerical properties of the methods developed in this paper. Further study in the following directions is underway:

(1) Carry out further numerical experiment to compare the ARE solvers based on the matrix pencils (11) and (12) and those based on the traditional matrix pencils with data square-up.

| | | daresys (QZ) | daresys (power) |
|---|---|---|---|
| $\epsilon = 10^{-8}$ | $\|\hat{X} - X\|_F$ | $1.95 \times 10^{-15}$ | $1.54 \times 10^{-15}$ |
| | $\|\hat{F} - F\|_F$ | $5.44 \times 10^{-16}$ | $9.61 \times 10^{-16}$ |
| $\epsilon = 10^{-12}$ | $\|\hat{X} - X\|_F$ | $2.58 \times 10^{-9}$ | $8.82 \times 10^{-16}$ |
| | $\|\hat{F} - F\|_F$ | $1.32 \times 10^{8}$ | $1.05 \times 10^{-15}$ |
| $\epsilon = 10^{-14}$ | $\|\hat{X} - X\|_F$ | $2.78 \times 10^{-3}$ | $1.78 \times 10^{-15}$ |
| | $\|\hat{F} - F\|_F$ | $1.20 \times 10^{13}$ | $1.48 \times 10^{-15}$ |
| $\epsilon = 10^{-15}$ | $\|\hat{X} - X\|_F$ | $6.86 \times 10^{-2}$ | $9.93 \times 10^{-16}$ |
| | $\|\hat{F} - F\|_F$ | $4.32 \times 10^{14}$ | $1.04 \times 10^{-15}$ |

Table 1. For Example 2.

| | | dare | daresym | daresys |
|---|---|---|---|---|
| $\alpha = 10^6$ | $\hat{X} - X$ | $\begin{bmatrix} 0 & 0 \\ 0 & 0 \end{bmatrix}$ | $\begin{bmatrix} 0 & 0 \\ 0 & 2.2122 \times 10^7 \end{bmatrix}$ | $\begin{bmatrix} 0 & 0 \\ 0 & 0 \end{bmatrix}$ |
| | Relative residue | $4.8828 \times 10^{-16}$ | $2.2122 \times 10^{-5}$ | $1.2207 \times 10^{-16}$ |
| $\alpha = 10^{10}$ | $\hat{X} - X$ | $\begin{bmatrix} 0 & 0 \\ 0 & 8.2412 \times 10^6 \end{bmatrix}$ | Completely wrong | $\begin{bmatrix} 0 & 0 \\ 0 & 0 \end{bmatrix}$ |
| | Relative residue | $8.2416 \times 10^{-14}$ | Completely wrong | $1.0000 \times 10^{-20}$ |

Table 2. Example 12 in [3]

| | | dare | daresym | daresys |
|---|---|---|---|---|
| $\alpha = 10^6$ | Relative error | $2.5238 \times 10^{-11}$ | $5.2532 \times 10^{-15}$ | $1.3222 \times 10^{-15}$ |
| | relative residue | $4.8828 \times 10^{-16}$ | $4.7712 \times 10^{-15}$ | $1.1008 \times 10^{-15}$ |
| $\alpha = 10^{10}$ | Relative error | $1.4289 \times 10^{-7}$ | $4.2733 \times 10^{-7}$ | $1.0192 \times 10^{-14}$ |
| | Relative residue | $1.3024 \times 10^{-7}$ | $3.8965 \times 10^{-7}$ | $9.8596 \times 10^{-15}$ |
| $\alpha = 10^{14}$ | Relative error | $4.1923 \times 10^{-4}$ | $7.2512 \times 10^{-4}$ | $2.3031 \times 10^{-15}$ |
| | Relative residue | $3.7490 \times 10^{-4}$ | $6.5812 \times 10^{-4}$ | $2.6729 \times 10^{-15}$ |

Table 3. Example 13 in [3]

(2) What can be said about the structure of the matrix pencils (11) and (12)? How can we use the structure and sparseness to simplify computation and to increase accuracy? Is there a structure preserving QZ decomposition which can be used to find the eigenspace needed? For structure preserving Schur decomposition with application in solving AREs, see (Paige and Van Loan, 1981; Byers, 1986; Bunse-Gerstner and Faβbender, 1997).

(3) Extend our results to possibly accommodate descriptor systems (with an $E$ matrix in the appropriate places in the AREs) or singular AREs (singular $R$ in continuous-time case and $\text{Ker}B \cap \text{Ker}R \neq \{0\}$ in discrete-time case). See (Arnold, III and Laub, 1984; Ionescu *et al.*, 1997) for motivation and existing literature.

## 7. REFERENCES


Arnold, III, M. F. and A. J. Laub (1984). Generailized eigenproblem algorithms and software for algebraic Riccati equations. *Proceedings of IEEE* **72**, 1746–1754.

Benner, P., A. J. Laub and V. Mehrmann (1996). A collection of benchmark examples for the numerical solution of algbraic Riccati equation II: discrete-time case. preprint.

Bittanti, S., Laub, A. J. and Willems, J. C., Eds.) (1991). *The Riccati Equation.* Springer-Verlag. Berlin.

Bunse-Gerstner, A. and H. Faβbender (1997). A Jacobi-like method for solving algebraic Riccati equations on parallel computers. *IEEE Trans. Automat. Contr.* **42**, 1071–1084.

Byers, R. (1986). A Hamiltonian QR algorithm. *SIAM J. Sci. Stat. Comput.* **7**, 212–229.

Emami-Naeini, A. and G. F. Franklin (1980). Comments on 'On the numerical solution of the discrete-time algebraic Riccati equation'. *IEEE Trans. Automat. Contr.* **25**, 1015–1016.

Ionescu, V., C. Oara and M. Weiss (1997). General matrix pencil techniques for the solution of algebraic Riccati equations: a unified approach. *IEEE Trans. Automat. Contr.* **42**, 1085–1097.

Paige, C. and C. Van Loan (1981). A Schur decomposition for Hamiltonian matrices. *Linear Algebra and its Applications* **41**, 11–32.

Pappas, T., A. J. Laub and N. R. Sandell Jr. (1980). On the numerical solution of discrete-time algebraic Riccati equation. *IEEE Trans. Automat. Contr.* **25**, 631–641.

Patel, R. V., Laub, A. J. and Van Dooren, P. M., Eds.) (1994). *Numerical Linear Algebra Techniques for Systems and Control.* IEEE Press.

Saad, Y. (1992). *Numerical Methods for Large Eigenvalue Problems.* Manchester University Press. Manchester, UK.

Van Dooren, P. (1981). A generalized eigenvalue approach for solving Riccati equations. *SIAM J. Sci. Stat. Comput.* **2**, 121–135.

# LOCAL LEVEL OPTIMAL CONTROL OF LINEAR DISCRETE STOCHASTIC SYSTEMS

**Myo-Taeg Lim and Beom-Soo Kim** *

* *School of Electrical Engineering, Korea University Seoul 136-701, KOREA, mlim@eeserver.korea.ac.kr*

Abstract: In this paper the algebraic regulator and filter Riccati equations of discrete-time control systems are decomposed into reduced-order continuous-time algebraic Riccati equations. That is, the exact solution of the global discrete algebraic Riccati equation is found in terms of the reduced-order nonsymmetric continuous-time algebraic Riccati equations. In addition, the optimal global Kalman filter is decomposed into local optimal reduced-order filters both driven by the system measurements and the system optimal control input. Using the separation principle the corresponding LQG optimal problem is obtained in terms of the local subproblems.


Keywords: Discrete-time systems, Optimal control, Kalman filters, Order reduction, Algebraic Riccati equations

## 1. INTRODUCTION

In this paper, we present a closed-loop decomposition technique that guarantees complete decomposition of the optimal global filters and regulators in terms of local ones and distribution of all required off-line and on-line computations. In the regulation problem (optimal linear-quadratic control problem), we show how to decompose exactly the discrete-time algebraic Riccati equation into two reduced-order continuous-time algebraic Riccati equations. Note that the reduced-order continuous-time algebraic Riccati equations are nonsymmetric. In the filtering problem, in addition of using duality between filter and regulator to solve the discrete-time filter algebraic Riccati equation in terms of the reduced-order continuous-time algebraic Riccati equations, we have obtained completely independent local optimal Kalman filters driven by system measurements and system optimal control. In the last part of this paper we use the separation principle to solve the linear-quadratic Gaussian control problem of discrete systems.

## 2. DECOMPOSITION OF LINEAR-QUADRATIC CONTROL PROBLEM

By applying the proposed method, the discrete algebraic Riccati equation will be completely and exactly decomposed into two reduced-order continuous-time local algebraic Riccati equations. This decomposition allows us to design the linear controllers for subsystems completely independently of each other and thus, to achieve the complete and exact separation for the linear-quadratic regulator problem.

Consider a time-invariant discrete system partitioned into two subsystems

$$\begin{aligned} x_1(k+1) &= A_1 x_1(k) + A_2 x_2(k) + B_1 u(k) \\ x_2(k+1) &= A_3 x_1(k) + A_4 x_2(k) + B_2 u(k) \\ x_1(0) &= x_{10},\ x_2(0) = x_{20} \end{aligned} \quad (1)$$

with state variables $x_1 \in R^{n_1}$ and $x_2 \in R^{n_2}$, and control inputs $u \in R^m$. The performance criterion of the corresponding linear-quadratic control problem is defined by

$$J = \frac{1}{2} \sum_{k=0}^{\infty} \left[ x(k)^T Q x(k) + u(k)^T R u(k) \right] \quad (2)$$

where

$$x(k) = \begin{bmatrix} x_1(k) \\ x_2(k) \end{bmatrix}, Q = \begin{bmatrix} Q_1 & Q_2 \\ Q_3 & Q_4 \end{bmatrix} \geq 0, R > 0 \quad (3)$$

It is well known that the solution of the above optimal regulation problem is given by (Dorato and Levis, 1971; Kwakernaark and Sivan, 1972)

$$\begin{aligned} u(k) &= -R^{-1} B^T p(k+1) \\ &= -\ R + B^T P B\ ^{-1} B^T P A x(k) \end{aligned} \quad (4)$$

where $p(k)$ is a costate variable and $P$ is the positive semidefinite stabilizing solution of the discrete algebraic Riccati equation (Dorato and Levis, 1971; Kwakernaark and Sivan, 1972; Lewis, 1986)

$$P = Q + A^T P [I + SP]^{-1} A \tag{5}$$
$$= Q + A^T PA - A^T PB \left[R + B^T PB\right]^{-1} B^T PA$$

The Hamiltonian form of (1) and (2) can be written as the forward recursion (Lewis, 1986)

$$\begin{bmatrix} x(k+1) \\ p(k+1) \end{bmatrix} = \mathbf{H} \begin{bmatrix} x(k) \\ p(k) \end{bmatrix} \tag{6}$$

with

$$\mathbf{H} = \begin{bmatrix} A + BR^{-1}B^T A^{-T} Q & -BR^{-1}B^T A^{-T} \\ -A^{-T} Q & A^{-T} \end{bmatrix} \tag{7}$$

where $\mathbf{H}$ is the symplectic matrix which has the property that the eigenvalues of $\mathbf{H}$ can be grouped into two disjoint subsets $\Gamma_1$ and $\Gamma_2$, such that for every $\lambda_c \in \Gamma_1$ there exists $\lambda_d \in \Gamma_2$, which satisfies $\lambda_c \times \lambda_d = 1$, and we can choose either $\Gamma_1$ or $\Gamma_2$ to contain only the stable eigenvalues (Salgado *et al.*, 1988).

For partitioned discrete system (1) the corresponding matrices in (4)-(7) are given by

$$A = \begin{bmatrix} A_1 & A_2 \\ A_3 & A_4 \end{bmatrix}, \quad B = \begin{bmatrix} B_1 \\ B_2 \end{bmatrix},$$
$$S = BR^{-1}B^T = \begin{bmatrix} S_1 & S_2 \\ S_3 & S_4 \end{bmatrix}$$
$$S_1 = B_1 R^{-1} B_1^T,$$
$$S_2 = S_3^T = B_1 R^{-1} B_2^T, \quad S_4 = B_2 R^{-1} B_2^T \tag{8}$$

Partitioning costate variables $p(k)$ as $p(k) = \left[p_1^T(k) \; p_2^T(k)\right]^T$ with $p_1(k) \in R^{n_1}$ and $p_2(k) \in R^{n_2}$, we get

$$\begin{bmatrix} x_1(k+1) \\ x_2(k+1) \\ p_1(k+1) \\ p_2(k+1) \end{bmatrix} = \mathbf{H} \begin{bmatrix} x_1(k) \\ x_2(k) \\ p_1(k) \\ p_2(k) \end{bmatrix}$$
$$= \begin{bmatrix} \overline{A_1} & \overline{A_2} & \overline{S_1} & \overline{S_2} \\ \overline{A_3} & \overline{A_4} & \overline{S_3} & \overline{S_4} \\ \overline{Q_1} & \overline{Q_2} & \overline{A_{11}^T} & \overline{A_{21}^T} \\ \overline{Q_3} & \overline{Q_4} & \overline{A_{12}^T} & \overline{A_{22}^T} \end{bmatrix} \begin{bmatrix} x_1(k) \\ x_2(k) \\ p_1(k) \\ p_2(k) \end{bmatrix} \tag{9}$$

Note that in the following there is no need for the analytical expressions for matrices with "bar". These matrices only have to be formed by a computer in the process of calculations.

Interchanging second and third rows in (9) by grouping subsystem states and costate variables we get

$$\begin{bmatrix} x_1(k+1) \\ p_1(k+1) \\ x_2(k+1) \\ p_2(k+1) \end{bmatrix} = \begin{bmatrix} T_1 & T_2 \\ T_3 & T_4 \end{bmatrix} \begin{bmatrix} x_1(k) \\ p_1(k) \\ x_2(k) \\ p_2(k) \end{bmatrix} \tag{10}$$

where

$$T_1 = \begin{bmatrix} \overline{A_1} & \overline{S_1} \\ \overline{Q_1} & \overline{A_{11}^T} \end{bmatrix}, T_2 = \begin{bmatrix} \overline{A_2} & \overline{S_2} \\ \overline{Q_2} & \overline{A_{21}^T} \end{bmatrix}$$
$$T_3 = \begin{bmatrix} \overline{A_3} & \overline{S_3} \\ \overline{Q_3} & \overline{A_{12}^T} \end{bmatrix}, T_4 = \begin{bmatrix} \overline{A_4} & \overline{S_4} \\ \overline{Q_4} & \overline{A_{22}^T} \end{bmatrix} \tag{11}$$

Introducing the notation

$$U(k) = \begin{bmatrix} x_1(k) \\ p_1(k) \end{bmatrix}, \quad V(k) = \begin{bmatrix} x_2(k) \\ p_2(k) \end{bmatrix} \tag{12}$$

we have the discrete system under new notation

$$U(k+1) = T_1 U(k) + T_2 V(k)$$
$$V(k+1) = T_3 U(k) + T_4 V(k) \tag{13}$$

Applying the following transformation

$$\mathbf{T_1} = \begin{bmatrix} I - HL & -H \\ L & I \end{bmatrix}, \mathbf{T_1^{-1}} = \begin{bmatrix} I & H \\ -L & I - LH \end{bmatrix}$$
$$\begin{bmatrix} \eta(k) \\ \xi(k) \end{bmatrix} = \mathbf{T_1} \begin{bmatrix} U(k) \\ V(k) \end{bmatrix} \tag{14}$$

to (13) produces two completely decoupled subsystems

$$\begin{bmatrix} \eta_1(k+1) \\ \eta_2(k+1) \end{bmatrix} = \eta(k+1) = (T_1 - T_2 L)\eta(k) \tag{15}$$

$$\begin{bmatrix} \xi_1(k+1) \\ \xi_2(k+1) \end{bmatrix} = \xi(k+1) = (T_4 + LT_2)\xi(k) \tag{16}$$

where $L$ and $H$ satisfy

$$-L + T_4 L - T_3 - L(T_1 - T_2 L) = 0 \tag{17}$$

$$-H(T_4 + LT_2) + T_2 + (T_1 - T_2 L) H = 0 \tag{18}$$

Solutions of nonlinear generalized algebraic Riccati equations (17) are discussed in (Medanic, 1982). Its solution can be obtained by using either well known eigenvector method (Bittanti *et al.*, 1991) or by the algorithm developed in (Avramovic *et al.*, 1980). The second transformation for $H$ represents a linear Sylvester type algebraic equation and it can be easily solved after matrix $L$ has been found.

The rearrangement and modification of variables in (9) is done by using the permutation matrix $E_1$ of the form

$$\begin{bmatrix} x_1(k) \\ p_1(k) \\ x_2(k) \\ p_2(k) \end{bmatrix} = \begin{bmatrix} I_{n_1} & 0 & 0 & 0 \\ 0 & 0 & I_{n_1} & 0 \\ 0 & I_{n_2} & 0 & 0 \\ 0 & 0 & 0 & I_{n_2} \end{bmatrix} \begin{bmatrix} x_1(k) \\ x_2(k) \\ p_1(k) \\ p_2(k) \end{bmatrix}$$

$$= E_1 \begin{bmatrix} x(k) \\ p(k) \end{bmatrix} \tag{19}$$

From (12), (14)-(16), and (19), we obtain the relationship between the original coordinates and the new ones

$$\begin{bmatrix} \eta_1(k) \\ \xi_1(k) \\ \eta_2(k) \\ \xi_2(k) \end{bmatrix} = E_1^T \mathbf{T}_1 E_1 \begin{bmatrix} x(k) \\ p(k) \end{bmatrix} = \Pi \begin{bmatrix} x(k) \\ p(k) \end{bmatrix}$$
$$= \begin{bmatrix} \Pi_1 & \Pi_2 \\ \Pi_3 & \Pi_4 \end{bmatrix} \begin{bmatrix} x(k) \\ p(k) \end{bmatrix} \tag{20}$$

Since $p(k) = Px(k)$, where $P$ satisfies the discrete algebraic Riccati equation (5), it follows from (20) that

$$\begin{bmatrix} \eta_1(k) \\ \xi_1(k) \end{bmatrix} = (\Pi_1 + \Pi_2 P)x(k)$$
$$\begin{bmatrix} \eta_2(k) \\ \xi_2(k) \end{bmatrix} = (\Pi_3 + \Pi_4 P)x(k) \tag{21}$$

In the original coordinates, the required optimal solution has a closed-loop nature. We have the same characteristic for the new systems (15) and (16), that is,

$$\begin{bmatrix} \eta_2(k) \\ \xi_2(k) \end{bmatrix} = \begin{bmatrix} P_1 & 0 \\ 0 & P_2 \end{bmatrix} \begin{bmatrix} \eta_1(k) \\ \xi_1(k) \end{bmatrix} \tag{22}$$

Then (21) and (22) yield

$$\begin{bmatrix} P_1 & 0 \\ 0 & P_2 \end{bmatrix} = (\Pi_3 + \Pi_4 P)(\Pi_1 + \Pi_2 P)^{-1} \tag{23}$$

Following the same logic, we can find $P$ reversely by introducing

$$E_1^{-1} \mathbf{T}_1^{-1} E_1 = \Omega = \begin{bmatrix} \Omega_1 & \Omega_2 \\ \Omega_3 & \Omega_4 \end{bmatrix} \tag{24}$$

and it yields

$$P = \left(\Omega_3 + \Omega_4 \begin{bmatrix} P_1 & 0 \\ 0 & P_2 \end{bmatrix}\right) \left(\Omega_1 + \Omega_2 \begin{bmatrix} P_1 & 0 \\ 0 & P_2 \end{bmatrix}\right)^{-1} \tag{25}$$

Partitioning (15) and (16) as

$$\begin{bmatrix} \eta_1(k+1) \\ \eta_2(k+1) \end{bmatrix} = \begin{bmatrix} a_1 & a_2 \\ a_3 & a_4 \end{bmatrix} \begin{bmatrix} \eta_1(k) \\ \eta_2(k) \end{bmatrix}$$
$$= (T_1 - T_2 L) \begin{bmatrix} \eta_1(k) \\ \eta_2(k) \end{bmatrix} \tag{26}$$

$$\begin{bmatrix} \xi_1(k+1) \\ \xi_2(k+1) \end{bmatrix} = \begin{bmatrix} b_1 & b_2 \\ b_3 & b_4 \end{bmatrix} \begin{bmatrix} \xi_1(k) \\ \xi_2(k) \end{bmatrix}$$
$$= (T_4 + LT_2) \begin{bmatrix} \xi_1(k) \\ \xi_2(k) \end{bmatrix} \tag{27}$$

and using (22) yield to two reduced-order nonsymmetric algebraic Riccati equations

$$P_1 a_1 - a_4 P_1 - a_3 + P_1 a_2 P_1 = 0 \tag{28}$$

$$P_2 b_1 - b_4 P_2 - b_3 + P_2 b_2 P_2 = 0 \tag{29}$$

It is very interesting that the algebraic Riccati equation of discrete-time control systems is completely and exactly decomposed into two reduced-order nonsymmetric continuous-time algebraic Riccati equations (28)-(29). Note that the nonsymmetric algebraic Riccati equations have been studied in (Medanic, 1982). An efficient algorithm for solving the general nonsymmetric algebraic Riccati equation is derived in (Avramovic *et al.*, 1980).

Local subsystems in the new coordinates are, respectively, given by

$$\eta_1(k+1) = (a_1 + a_2 P_1)\, \eta_1(k) \tag{30}$$

$$\xi_1(k+1) = (b_1 + b_2 P_2)\, \xi_1(k) \tag{31}$$

## 3. DECOMPOSITION OF THE KALMAN FILTER FOR LINEAR DISCRETE SYSTEMS

In this section the optimal filter gain will be completely determined in terms of the reduced-order continuous-time algebraic Riccati equations, based on the duality property between the optimal filter and regulator.

Consider the linear discrete-time invariant stochastic system

$$x_1(k+1) = A_1 x_1(k) + A_2 x_2(k) + G_1 w_1(k)$$
$$x_2(k+1) = A_3 x_1(k) + A_4 x_2(k) + G_2 w_1(k) \tag{32}$$

with the corresponding measurements

$$y(k) = C_1 x_1(k) + C_2 x_2(k) + w_2(k) \tag{33}$$

$w_1(k)$ and $w_2(k)$ are zero-mean stationary, white Gaussian noise stochastic processes with intensities $W_1 > 0$ and $W_2 > 0$, respectively, and $y \in \mathrm{R}^l$ are the system measurements. In the following $A_i, G_j, C_j,\ i = 1,2,3,4;\ j = 1,2$, are constant matrices.

The optimal Kalman filter, driven by the innovation process, is given by, (Kwakernaark and Sivan, 1972)

$$\hat{x}(k+1) = A\hat{x}(k) + K[y(k) - C\hat{x}(k)] \tag{34}$$

where

$$A = \begin{bmatrix} A_1 & A_2 \\ A_3 & A_4 \end{bmatrix}, C = [C_1 \ C_2],$$

$$G=\begin{bmatrix}G_1\\G_2\end{bmatrix},\quad K=\begin{bmatrix}K_1\\K_2\end{bmatrix} \tag{35}$$

The filter gain $K$ is obtained from

$$K = AP_FC^T\left(W_2 + CP_FC^T\right)^{-1} \tag{36}$$

where $P_F$ is the positive semidefinite stabilizing solution of the discrete-time filter algebraic Riccati equation given by

$$P_F = AP_FA^T - AP_FC^T\left(W_2 + CP_FC^T\right)^{-1} CP_FA^T + GW_1G^T \tag{37}$$

Using the decomposition procedure proposed from the previous section and the duality property between the optimal filter and regulator, we propose a decomposition scheme such that the local filters are completely decoupled and both of them are driven by the system measurements.

The results of interest which can be deduced from Section 2 are summarized in the form of the following lemma.

**Lemma 1.** *Consider the optimal closed-loop linear discrete system*

$$\begin{aligned}x_1(k+1)&=(A_1 - B_1F_1)x_1(k) + (A_2 - B_1F_2)x_2(k)\\ x_2(k+1)&=(A_3 - B_2F_1)x_1(k) + (A_4 - B_2F_2)x_2(k)\end{aligned} \tag{38}$$

*then there exists a nonsingular transformation* $\mathbf{T}$

$$\begin{bmatrix}\theta_1(k)\\ \theta_2(k)\end{bmatrix} = \mathbf{T}\begin{bmatrix}x_1(k)\\ x_2(k)\end{bmatrix} \tag{39}$$

*such that*

$$\begin{aligned}\theta_1(k+1) &= (a_1 + a_2P_1)\theta_1(k)\\ \theta_2(k+1) &= (b_1 + b_2P_2)\theta_2(k)\end{aligned} \tag{40}$$

*where $P_1$ and $P_2$ are the unique solutions of the exact local algebraic Riccati equations derived in the previous section, that is*

$$\begin{aligned}P_1a_1 - a_4P_1 - a_3 + P_1a_2P_1 &= 0\\ P_2b_1 - b_4P_2 - b_3 + P_2b_2P_2 &= 0\end{aligned} \tag{41}$$

*Matrices $a_i$, $b_i$, $i = 1,2,3,4$, can be found from (26)-(27). The nonsingular transformation $\mathbf{T}$ is given by*

$$\mathbf{T} = (\Pi_1 + \Pi_2P) \tag{42}$$

*Even more, the global solution $P$ can be obtained from the reduced-order local algebraic Riccati equations, that is*

$$P=\left(\Omega_3+\Omega_4\begin{bmatrix}P_1&0\\0&P_2\end{bmatrix}\right)\left(\Omega_1+\Omega_2\begin{bmatrix}P_1&0\\0&P_2\end{bmatrix}\right)^{-1} \tag{43}$$

*Known matrices $\Omega_i$, $i = 1,2,3,4$ and $\Pi_1$, $\Pi_2$ are given in terms of the transformation matrices in Section 2.*

The desired decomposition of the Kalman filter (26) will be obtained by using a duality between the optimal filter and regulator, and the same decomposition method developed in Section 2. Consider the optimal closed-loop Kalman filter (26) driven by the system measurements, that is

$$\begin{aligned}\hat{x}_1(k+1) &= (A_1 - K_1C_1)\hat{x}_1(k)\\ &\quad+(A_2 - K_1C_2)\hat{x}_2(k) + K_1y(k)\\ \hat{x}_2(k+1) &= (A_3 - K_2C_1)\hat{x}_1(k)\\ &\quad+(A_4 - K_2C_2)\hat{x}_2(k) + K_2y(k)\end{aligned} \tag{44}$$

By using (32) and duality between the optimal filter and regulator, that is

$$\begin{aligned}&A \to A^T,\ Q \to GW_1G^T,\ B \to C^T\\ &BR^{-1}B^T \to C^TW_2^{-1}C\end{aligned} \tag{45}$$

the filter "state-costate equation" can be defined as

$$\begin{bmatrix}x(k+1)\\p(k+1)\end{bmatrix}=\overline{\mathbf{H}}\begin{bmatrix}x(k)\\p(k)\end{bmatrix}=\begin{bmatrix}\overline{H}_{11}&\overline{H}_{12}\\ \overline{H}_{21}&\overline{H}_{22}\end{bmatrix}\begin{bmatrix}x(k)\\p(k)\end{bmatrix} \tag{46}$$

where

$$\begin{aligned}\overline{H}_{11} &= A^T + C^TW_2^{-1}CA^{-1}GW_1G^T\\ \overline{H}_{12} &= -C^TW_2^{-1}CA^{-1}\\ \overline{H}_{21} &= -A^{-1}GW_1G^T\\ \overline{H}_{22} &= A^{-1}\end{aligned} \tag{47}$$

Partitioning $p(k)$ as $p(k) = \left[p_1^T(k)\ p_2^T(k)\right]^T$ with $p_1(k) \in R^{n_1}$ and $p_2(k) \in R^{n_2}$, (46) can be rewritten as follows

$$\begin{bmatrix}x_1(k+1)\\x_2(k+1)\\p_1(k+1)\\p_2(k+1)\end{bmatrix}=\begin{bmatrix}\overline{A_1^T}&\overline{A_3^T}&\overline{S_1}&\overline{S_2}\\ \overline{A_2^T}&\overline{A_4^T}&\overline{S_3}&\overline{S_4}\\ \overline{Q_1}&\overline{Q_2}&\overline{A_{11}}&\overline{A_{12}}\\ \overline{Q_3}&\overline{Q_4}&\overline{A_{21}}&\overline{A_{22}}\end{bmatrix}\begin{bmatrix}x_1(k)\\x_2(k)\\p_1(k)\\p_2(k)\end{bmatrix} \tag{48}$$

Interchanging the second and third rows in (48) and introducing $x(k) = \left[x_1^T(k)\ x_2^T(k)\right]^T$ yield

$$\begin{bmatrix}x_1(k+1)\\p_1(k+1)\\x_2(k+1)\\p_2(k+1)\end{bmatrix}=\begin{bmatrix}T_1&T_2\\T_3&T_4\end{bmatrix}\begin{bmatrix}x_1(k)\\p_1(k)\\x_2(k)\\p_2(k)\end{bmatrix} \tag{49}$$

where

$$\begin{aligned}T_1 &= \begin{bmatrix}\overline{A_1^T}&\overline{S_1}\\ \overline{Q_1}&\overline{A_{11}}\end{bmatrix},\quad T_2 = \begin{bmatrix}\overline{A_3^T}&\overline{S_2}\\ \overline{Q_2}&\overline{A_{12}}\end{bmatrix}\\ T_3 &= \begin{bmatrix}\overline{A_2^T}&\overline{S_3}\\ \overline{Q_3}&\overline{A_{21}}\end{bmatrix},\quad T_4 = \begin{bmatrix}\overline{A_4^T}&\overline{S_4}\\ \overline{Q_4}&\overline{A_{22}}\end{bmatrix}\end{aligned} \tag{50}$$

Decomposition of the above discrete system can be achieved by applying transformation to (14), which yields two completely decoupled subsystems

$$\begin{bmatrix} \eta_1(k+1) \\ \eta_2(k+1) \end{bmatrix} = \begin{bmatrix} a_{1F} & a_{2F} \\ a_{3F} & a_{4F} \end{bmatrix} \begin{bmatrix} \eta_1(k) \\ \eta_2(k) \end{bmatrix} = (T_1 - T_2 L) \begin{bmatrix} \eta_1(k) \\ \eta_2(k) \end{bmatrix} \quad (51)$$

$$\begin{bmatrix} \xi_1(k+1) \\ \xi_2(k+1) \end{bmatrix} = \begin{bmatrix} b_{1F} & b_{2F} \\ b_{3F} & b_{4F} \end{bmatrix} \begin{bmatrix} \xi_1(k) \\ \xi_2(k) \end{bmatrix} = (T_4 + L T_2) \begin{bmatrix} \xi_1(k) \\ \xi_2(k) \end{bmatrix} \quad (52)$$

Note that the decoupling transformation has the form of (14) with $H$ and $L$ matrices obtained from (17)-(18) with $T_i$'s taken from (50). By duality and Lemma 1 the following reduced-order nonsymmetric algebraic Riccati equations exist

$$P_{F1}a_{1F} - a_{4F}P_{F1} - a_{3F} + P_{F1}a_{2F}P_{F1} = 0 \quad (53)$$

$$P_{F2}b_{1F} - b_{4F}P_{F2} - b_{3F} + P_{F2}b_{2F}P_{F2} = 0 \quad (54)$$

By using the permutation matrix $E_1$ we can define

$$\Pi_F = \begin{bmatrix} \Pi_{1F} & \Pi_{2F} \\ \Pi_{3F} & \Pi_{4F} \end{bmatrix} = E_1^T \begin{bmatrix} I - HL & -H \\ L & I \end{bmatrix} E_1 \quad (55)$$

Then, the desired transformation is given by

$$\mathbf{T_2} = (\Pi_{1F} + \Pi_{2F}P_F) \quad (56)$$

The transformation $\mathbf{T_2}$ applied to the filter variables (44) as

$$\begin{bmatrix} \hat{\eta}_s \\ \hat{\eta}_f \end{bmatrix} = \mathbf{T_2^{-T}} \begin{bmatrix} \hat{x}_1 \\ \hat{x}_2 \end{bmatrix} \quad (57)$$

produces

$$\begin{bmatrix} \hat{\eta}_{ss1}(k+1) \\ \hat{\eta}_{ss2}(k+1) \end{bmatrix} = \mathbf{T_2^{-T}} \begin{bmatrix} A_1 - K_1C_1 & A_2 - K_1C_2 \\ A_3 - K_2C_1 & A_4 - K_2C_2 \end{bmatrix} \mathbf{T_2^T} \begin{bmatrix} \hat{\eta}_{ss1}(k) \\ \hat{\eta}_{ss2}(k) \end{bmatrix} + \mathbf{T_2^{-T}} \begin{bmatrix} K_1 \\ K_2 \end{bmatrix} y(k) \quad (58)$$

such that the complete closed-loop decomposition is achieved, that is

$$\hat{\eta}_{ss1}(k+1) = (a_{1F} + a_{2F}P_{F1})^T \hat{\eta}_s(k) + K_{ss1}y(k)$$
$$\hat{\eta}_{ss2}(k+1) = (b_{1F} + b_{2F}P_{F2})^T \hat{\eta}_f(k) + K_{ss2}y(k) \quad (59)$$

where

$$\begin{bmatrix} K_{ss1} \\ K_{ss2} \end{bmatrix} = \mathbf{T_2^{-T}} \begin{bmatrix} K_1 \\ K_2 \end{bmatrix} \quad (60)$$

It is important to point out that the matrix $P_F$ in (56) can be obtained in terms of $P_{F1}$ and $P_{F2}$ instead of $P_1$ and $P_2$ by using (43), and $\Omega_{1F}$, $\Omega_{2F}$, $\Omega_{3F}$, $\Omega_{4F}$ obtained from

$$\Omega_F = \begin{bmatrix} \Omega_{1F} & \Omega_{2F} \\ \Omega_{3F} & \Omega_{4F} \end{bmatrix} = E_1^{-1} \begin{bmatrix} I & H \\ -L & I - LH \end{bmatrix} E_1 \quad (61)$$

## 4. LINEAR-QUADRATIC GAUSSIAN OPTIMAL CONTROL PROBLEM

Consider the discrete linear stochastic control system represented in the partitioned form by

$$x_1(k+1) = A_1x_1(k) + A_2x_2(k) + B_1u(k) + G_1w_1(k)$$
$$x_2(k+1) = A_3x_1(k) + A_4x_2(k) + B_2u(k) + G_2w_1(k)$$
$$y(k) = C_1x_1(k) + C_2x_2(k) + w_2(k) \quad (62)$$

with the performance criterion

$$J = \frac{1}{2} E\left\{ \sum_{k=0}^{\infty} \left[ z^T(k)z(k) + u^T(k)Ru(k) \right] \right\}, R > 0 \quad (63)$$

where $x_i \in R^{n_i}$, $i = 1, 2$, comprise state vectors, $u \in R^m$ is the control input, $y \in R^l$ is the observed output, $w_1 \in R^r$ and $w_2 \in R^l$ are independent zero-mean stationary Gaussian mutually uncorrelated white noise processes with intensities $W_1 > 0$ and $W_2 > 0$, respectively, and $z \in R^s$ is the controlled output given by

$$z(k) = D_1x_1(k) + D_2x_2(k) \quad (64)$$

All matrices are of appropriate dimensions and assumed to be constant.

The optimal control law of the system (62) with performance criterion (63) is given by (Kwakernaark and Sivan, 1972)

$$u(k) = -F\hat{x}(k) \quad (65)$$

with the time-invariant filter

$$\hat{x}(k+1) = A\hat{x}(k) + Bu(k) + K[y(k) - C\hat{x}(k)] \quad (66)$$

where

$$A = \begin{bmatrix} A_1 & A_2 \\ A_3 & A_4 \end{bmatrix}, \quad B = \begin{bmatrix} B_1 \\ B_2 \end{bmatrix}, \quad C = [C_1 \; C_2], \quad K = \begin{bmatrix} K_1 \\ K_2 \end{bmatrix} \quad (67)$$

The regulator gain $F$ and filter gain $K$ are obtained from

$$F = \left(R + B^TPB\right)^{-1} B^TPA \quad (68)$$

$$K = AP_FC^T \left(W_2 + CP_FC^T\right)^{-1} \quad (69)$$

where $P$ and $P_F$ are positive semidefinite stabilizing solutions of the discrete-time algebraic regulator and filter Riccati equations (Dorato and Levis, 1971), respectively, given by

$$P = D^T D + A^T PA - A^T PB(R + B^T PB)^{-1} B^T PA \quad (70)$$

$$P_F = AP_F A^T - AP_F C^T (W_2 + CP_F C^T)^{-1} CP_F A^T + GW_1 G^T \quad (71)$$

where

$$D = [D_1 \ \ D_2], \ \ G = \begin{bmatrix} G_1 \\ G_2 \end{bmatrix} \quad (72)$$

The exact decomposition method of the discrete algebraic regulator and filter Riccati equations produces two sets of two reduced-order nonsymmetric algebraic Riccati equations, that is, for the regulator

$$P_1 a_1 - a_4 P_1 - a_3 + P_1 a_2 P_1 = 0 \quad (73)$$

$$P_2 b_1 - b_4 P_2 - b_3 + P_2 b_2 P_2 = 0 \quad (74)$$

and for the filter

$$P_{F1} a_{1F} - a_{4F} P_{F1} - a_{3F} + P_{F1} a_{2F} P_{F1} = 0 \quad (75)$$

$$P_{F2} b_{1F} - b_{4F} P_{F2} - b_{3F} + P_{F2} b_{2F} P_{F2} = 0 \quad (76)$$

where the unknown coefficients are obtained from previous sections.

It has been shown in the previous section that the optimal global Kalman filter, based on the exact decomposition technique, is decomposed into local optimal filters both driven by the system measurements. These two filters can be implemented independently and in parallel. They are given by

$$\hat{\eta}_{ss1}(k+1) = (a_{1F} + a_{2F} P_{F1})^T \hat{\eta}_{ss1}(k) + K_{ss1} y(k) + B_{ss1} u(k)$$

$$\hat{\eta}_{ss2}(k+1) = (b_{1F} + b_{2F} P_{F2})^T \hat{\eta}_{ss2}(k) + K_{ss2} y(k) + B_{ss2} u(k) \quad (77)$$

where

$$\begin{bmatrix} B_{ss1} \\ B_{ss2} \end{bmatrix} = \mathbf{T_2^{-T}} B = (\Pi_{1F} + \Pi_{2F} P_F)^{-T} B \quad (78)$$

The optimal control in the new coordinates has been obtained as

$$u(k) = -F\hat{x}(k) = -F\mathbf{T_2^T} \begin{bmatrix} \hat{\eta}_{ss1}(k) \\ \hat{\eta}_{ss2}(k) \end{bmatrix} = -[F_{ss1} \quad F_{ss2}] \begin{bmatrix} \hat{\eta}_{ss1}(k) \\ \hat{\eta}_{ss2}(k) \end{bmatrix} \quad (79)$$

where $F_{ss1}$ and $F_{ss2}$ are obtained from

$$[F_{ss1} \ F_{ss2}] = F\mathbf{T_2^T} = R + B^T PB^{\ -1} B^T PA (\Pi_{1F} + \Pi_{2F} P_F)^T \quad (80)$$

The optimal value of $J$ is given by the very well-known form (Kwakernaark and Sivan, 1972)

$$J_{opt} = \frac{1}{2} tr \left[ D^T D P_F + PK \left( C P_F C^T + W_2 \right) K^T \right] \quad (81)$$

Note that $F$, $K$, $P$, and $P_F$ are obtained from (68)-(71), which can be determined in terms of the reduced-order solutions by taking advantage of (43) and (73)-(76).

## 5. CONCLUSIONS

An approach for the decomposition of the LQG optimal control of discrete systems has been proposed. The importance of the proposed method is in the fact that the global optimal control and filtering can be completely and exactly decomposed into local control and filtering problems, which reduces both off-line and on-line required computations and allows complete time-scale parallelism of the filtering and control tasks.

## 6. REFERENCES


Avramovic, B., P. Kokotovic, J. Winkelman and J. Chow (1980). Area decomposition for electromechanical models of power systems. *Automatica* **16**, 637–648.

Bittanti, S., A. Laub and J. C. Willems (1991). *The Riccati Equation.* Communications and Control Engineering Series. Springer-Verlag. Berlin.

Dorato, P. and A. Levis (1971). Optimal linear regulator: the discrete time case. *IEEE Trans. Automat. Contr.* **AC-16**, 613–620.

Kwakernaark, H. and R. Sivan (1972). *Linear Optimal Control Systems.* Wiley Interscience. New York.

Lewis, F. (1986). *Optimal Control.* Wiley. New York, NY.

Medanic, J. (1982). Geometric properties and invariant manifolds of the Riccati equation. *IEEE Trans. Automat. Contr.* **AC-27**, 670–677.

Salgado, M., R. Middleton and G. C. Goodwin (1988). Connection between continuous and discrete Riccati equation with applications to Kalman filtering. *Proc. IEE, Part D.* **135**, 28–34.

14th Triennial World Congress, Beijing, P.R. China

F-2d-06-6

# Applying Mathematical Control in Economic Systems

**Jinghao Zhu and Bingyong Tang**

(*Department of Applied Mathematics, Tongji university, 200092,Shanghai,*
*email address: jinghaok@online.sh.cn;*
*Baronbow institute, China textile University,200051, Shanghai,*
*email address: tangby@ctu.edu.cn*)

Abstract: An optimal control problem for a closed economy is discussed. In order to control the price level, we find a relationship between the government spending and the money stock..

The result in the continuous form can be used in the discrete case. 

Keywords:Optimal control; Lie derivative; Lie bracket; Nonlinear systems

## 1. An optimal control problem for a closed economy

As an example of nonlinear dynamic control system (see Nijmeijer 1990), we construct from the discrete form the following equations describing the evolution of a closed economy in continuous time.

$$\dot{Y} = C(Y) + I(Y,R,K) + G/P \qquad (1)$$

$$\dot{R} = L(Y,R) - M/P \qquad (2)$$

$$\dot{K} = I(Y,R,K) \qquad (3)$$

$$P = Q(Y,R,K). \qquad (4)$$

In this model the quantities have the following interpretation:

Y: real output
C: real private consumption
I: real private net investment
R: nominal interest rate
K: real capital stock
P: price level
L: real money demand
G: nominal government spending
M: nominal money stock.

We assume that

$$Q(Y,R,K) > 0, \qquad \frac{\partial I}{\partial Y} > 0. \qquad (5)$$

Equation (1) is a dynamic IS (Investments-Saving) equation and (2) is a dynamic LM (Loan-Money) equation. The capital accumulation is described via the dynamic Keynesian equation (3). To bring (1)—(3) into the form of a control system we interpret G and M as the controls of the system and define the price level P as the target variable which is to be controlled. Defining

$x = (Y, R, K)^T, u = (G, M)^T,$

$F_1(x,u) = C(Y) + I(Y,R,K) + G/P,$

$F_2(x,u) = L(Y,R) - M/P,$

$F_3(x,u) = I(Y,R,K),$

and $F(x,u) = (F_1, F_2, F_3)^T$, we can rewrite (1)—(4) into the following nonlinear control system

$$\dot{x} = F(x,u), \qquad (6)$$

$$P = Q(x). \qquad (7)$$

Here we assume that $F(x,u), Q(x)$ are smooth with respect to both $x$ and $u$. Then we consider the optimal control problem: Find $\hat{u}$ among the vector functions in $L_2([0,T], R^2)$ such that

$$\int_0^T \dot{P}_{\hat{u}}(t)dt = \min_u \int_0^T \dot{P}_u(t)dt$$

where $P_u(t) = Q(x(t))$ with $x(t)$ being the solution of (6) when the control takes $u$ and the initial state $x(0) = x_0$.

## 2. Pontryagin principle and Lie derivative

For given $u$ and the initial state $x$ we see that,
along the solution $x(t)$ of (6) as $x(0) = x$,

$$\dot{P}_u = L_{F(x,u)}Q(x(t)), \qquad (8)$$

where $L_{F(x,u)}Q(x)$ denotes the Lie derivative of $Q(x)$ with respect to the vector field $F(x,u)$ for the fixed quantity $u$. Defining

$$V(x) = \min_u \int_0^T L_{F(x,u)}Q(x)dt, \qquad (9)$$

and assuming $\hat{u}$ to be the optimal feedback controller, by the Pontryagin principle, (see Pontryagin,L.S. 1962, Russell, 1979) we have

$$\frac{\partial V}{\partial x}F(x,u) + L_{F(x,u)}Q(x) = 0. \qquad (10)$$

Therefore the optimal feedback control will satisfy the following Lie differential equation

$$L_{F(x,\hat{u})}(V(x) + Q(x)) = 0. \qquad (11)$$

About $V(x)$ with $R_x(T)$ denoting the reachable-set

$$\{x_u(T) | u \in U, \dot{x}_u(t) = F(x_u, u), x_u(0) = x\}$$

where $U$ denotes the set of all admissible controls, we have the following lemma.

**Lemma 1.** Under the above notations, we have

$$V(x) = \left[\min_{y \in R_x(T)} Q(y)\right] - Q(x). \qquad (12)$$

**Proof:** By the definition of $V(x)$ in (9), we have

$$V(x) = \min_u \int_0^T \dot{P}_u(t)dt = \min_u (P_u(T) - P_u(0))$$

$$= \min_u [Q(x_u(T)) - Q(x)]$$

$$= \left[\min_{y \in R_x(T)} Q(y)\right] - Q(x).$$

## 3. The condition of the optimal design

If $(\hat{G}, \hat{M})^T$ denotes the optimal feedback, (11)
implies that it will satisfies the following linear equation of components of the control

$$\frac{\partial(V+Q)}{\partial\ \ Y}\frac{\hat{G}}{Q}-\frac{\partial(V+Q)}{\partial\ \ R}\frac{\hat{M}}{Q}$$
$$=-\frac{\partial(V+Q)}{\partial\ \ Y}(C(Y)+I(Y,R,K))$$
$$-\frac{\partial(V+Q)}{\partial\ \ R}L(Y,R)-\frac{\partial(V+Q)}{\partial\ \ K}I(Y,R,K) \quad (13)$$

Since for the optimal control $\hat{u}$ by (11) we have

$$\int_0^T L_{F(x,\hat{u})}Q(x)dt=-\int_0^T L_{F(x,\hat{u})}V(x)dt$$
$$=V(x_0)-V(x(T)). \quad (14)$$

so the sufficient and necessary condition is

$$V(x(T))=0. \quad (15)$$

We may use Lie series to approximate $V(x)$ .( see Sussmann,H.J. 1983).

For the case in the closed economy system (1)—(3) we have the following result.

**Theorem 2.** The nonlinear control system (2)—(3) is locally accessible.

**Proof:** Defining

$$f_1(Y,R,K)=C(Y)+I(Y,R,K),$$
$$f_2(Y,R,K)=L(Y,R),$$
$$f_3(Y,R,K)=I(Y,R,K);$$
$$g_1(Y,R,K)=\left(\frac{1}{P},0,0\right)^T,$$
$$g_2(Y,R,K)=\left(0,\frac{-1}{P},0\right)^T,$$

and

$$f(Y,R,K)=(f_1,f_2,f_3)^T,$$
$$g(Y,R,K)=(g_1,g_2),$$

we can rewrite (1)—(3) into the following form,

with $x$ denoting $(Y,R,K)^T$ and $u$ denoting $(G,M)^T$,

$$\dot{x}=f(x)+g(x)u. \quad (16)$$

Since

$$(g_1,g_2,[f,g_1])$$
$$=\begin{pmatrix} 1/Q & 0 & L_f(1/Q)-(1/Q)\frac{\partial f_1}{\partial\ Y} \\ 0 & -1/Q & -\frac{1}{Q}\frac{\partial f_2}{\partial\ Y} \\ 0 & 0 & \frac{1}{Q}\frac{\partial f_3}{\partial\ Y} \end{pmatrix} \quad (17)$$

has rank 3, noting the assumption for the system (4), where $[f,g_1]$ denotes the Lie bracket of vector fields $f$ and $g_1$, by theorem 3.32 in [1], we see that the system (1)—(3) is locally accessible.

If we always consider the case in which $\dot{P}_u\geq 0$, then Lemma 1 implies $V(x)\equiv 0$. So in this case the sufficient and necessary condition for the optimal feedback (see (13)) is the following linear equation

$$\frac{\partial Q}{\partial Y}\frac{\hat{G}}{Q}-\frac{\partial Q}{\partial R}\frac{\hat{M}}{Q}$$
$$=-\frac{\partial Q}{\partial Y}(C(Y)+I(Y,R,K))-\frac{\partial Q}{\partial R}L(Y,R)$$
$$-\frac{\partial Q}{\partial K}I(Y,R,K). \quad (18)$$

Further we can consider the following problem: Under (18), in the optimal case, find the relationship between the nominal money stock and the government spending. On the other hand, for the discrete form of (1)—(3) we can use the difference equation in stead of (18).

## REFERENCES

Nijmeijer & Van der Schaft (1990). *Nonlinear Dynamic Control Systems,* Springer—Verag.

Pontryagin,L.S.(1962).*The mathematical theory of optimal process,* Wiley, New York

Russell,D.L.(1979). *Mathematics of finite dimensional control system.* Marcel Dekker, Inc., New York

Sussmann,H.J.(1983) Lie bracket and controllability: a sufficient condition for scalar-input systems, **Vol. 21, No. 5,** pp.686—713 SIAM J. Control and optimization

14th Triennial World Congress, Beijing, P.R. China



F-2d-07-1

# OPTIMAL PID CONTROLLERS INCLUDING ROLL OFF AND SCHMIDT PREDICTOR STRUCTURE

**Birgitta Kristiansson, Bengt Lennartson**

*Control Engineering Laboratory, Chalmers University of Technology*
*SE-412 96 Göteborg, Sweden*
*e-mail: bkn@ios.chalmers.se, bl@s2.chalmers.se*

Abstract: A new evaluation method for controllers recently proposed, has been somewhat modified to handle additional control goals for extended versions of PID controllers. To improve the roll off of the loop gain at high frequencies, an extra low pass filter has been included in the PID structure. When the filter is incorporated in the design, the system behavior does not deteriorate. For plants with significant time delay, a Schmidt predictor may be included. It is however shown that the more complicated predictor structure implies marginal improvements in terms of performance and robustness, compared to an ordinary PID, when both are optimized according to the same criterion. 

Keywords: PID control, performance, robustness, optimal control, autotuners, delay compensation

## 1. INTRODUCTION

When new control strategies and new controller design methods are introduced, achievable system properties like closed loop performance and robustness are usually evaluated and compared with corresponding results from well known control design schemes. Very often the PID controller designed by Ziegler-Nichols' (Z/N) method (Ziegler and Nichols, 1942) has been used as a reference. Unfortunately these evaluations and the conclusions drawn have not always been quite objective. The reason is that the influence on the performance and the robustness in one frequency range, due to changes made in another, has often been neglected. Recently a new evaluation method has been proposed (Lennartson and Kristiansson, 1997), which takes such influences into account. The criteria involved in the evaluation, which are essentially formulated as H-infinity norms, are also used as a basis for a control design procedure, where a low frequency criterion is minimized subject to constraints on mid and high frequency norms. In this paper this evaluation and design method is developed somewhat further.

The method has been used to achieve new design rules for standard PID controllers for a lot of different plant models. These rules are presented in (Lennartson and Kristiansson, 1997; Kristiansson and Lennartson, June1998; Dec 1998). Compared to standard tuning rules, it is shown that significant improvements of low frequency performance and closed loop bandwidth can be obtained, without loosing high frequency robustness and stability margins. Based on these experiences a set of auto-tuning rules derived.

Some extended versions of the PID controller are then examined. First, the consequences of introducing an extra high frequency roll off in the controller are considered. It is shown that if a low pass filter is incorporated during the optimization of the design, it will bring no deterioration to the low and mid frequency performance and robustness of the system. However, the ordinary PID-parameters are modified, which implies that one has to consider the additional roll off filter in the design.

For plants with considerable delay a Schmidt predictor may be included. It has been noticed that, when the controller is designed for a corresponding plant without delay, as is often recommended, the stability margins of the closed loop system may be significantly reduced, when the time delay and the corresponding predictor and time delay are added. But when the Schmidt predictor is included in the controller during the optimization of the delayed system, no such deterioration occurs. An interesting question is then how much the performance may be improved when the Schmidt predictor is included compared to an ordinary PID controller, when both controllers are optimized and evaluated according to the same criteria. The results obtained including the predictor are however not significantly better than the ones obtained without the predictor even when the delay is large. In this investigation a new restriction on the loop transfer function was introduced, in order to obtain a reasonable reference step response and not too high loop transfer at the transition between the mid and the high frequency regions.

## 2. EVALUATION METHOD

Consider the SISO-system in Figure 1, with a plant $G(s)$ controlled by a controller $K(s)$. The inputs are the reference signal, $r(t)$, the process disturbance, $v(t)$, and the measurement noise, $w(t)$. The outputs are the controlled output, $y(t)$, and the control signal, $u(t)$.

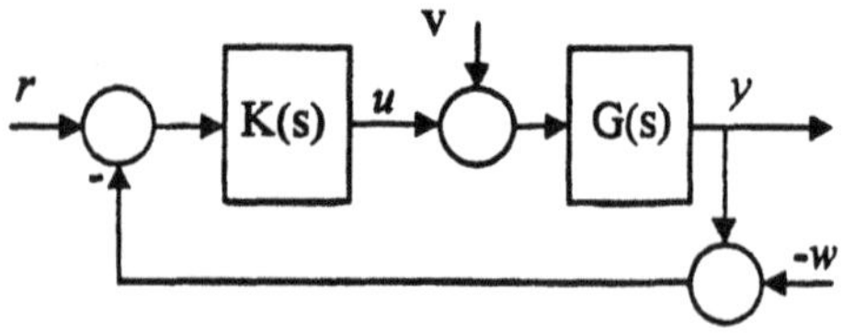


Figure 1: Closed loop system with plant $G(s)$ and controller $K(s)$

When integral action is included in the controller, its transfer function can be formulated as

$$K(s) = \frac{k_i}{s} + k_\infty K'(s) \qquad (1)$$

where $k_i$ is the integral gain and $k_\infty$ is defined such that $s^m K(s) \to k_\infty$ as $s \to \infty$. The parameter $m$ expresses the high frequency slope or roll off in the controller. Note that $s^m K'(s) \to 1 - s^{m-1} k_i/k_\infty$ as $s \to \infty$.

Let us now introduce the H-infinity criteria

$$J_{LF} = \| W_{LF} GS \|_\infty = \max_\omega | W_{LF}(j\omega) G(j\omega) S(j\omega) | \qquad (2)$$
$$J_{HF} = \| KS \|_\infty = \max_\omega | K(j\omega) S(j\omega) | \qquad (3)$$
$$J_{VHF} = \| W_{VHF} KS \|_\infty = \max_\omega | W_{VHF}(j\omega) K(j\omega) S(j\omega) | \qquad (4)$$

where the weighting functions $W_{LF}(s) = 1/s$ and $W_{VHF}(s) = s^m$. The transfer function $S(s)$ is the sensitivity function, that is $S(s) = [1 + G(s)K(s)]^{-1}$. Note that $W_{LF}(j\omega)G(j\omega)S(j\omega) \to 1/ki$, when $\omega \to 0$ and $W_{VHF}(j\omega)K(j\omega)S(j\omega) \to k_\infty$, when $\omega \to \infty$

Obviously, $J_{LF}$ can be used to characterize the low frequency (LF) range, and $J_{HF}$ characterizes the high frequency (HF) range where $KS$ normally has its maximum, which typically occurs at infinity when $m=0$ and at 5-10 times the bandwidth of the closed loop system when $m=1$. The very high frequency measure $J_{VHF} = J_{HF}$ when m=0. On the other hand, when $m \neq 0$, then $J_{VHF} / \omega^m$ touches the $|KS(j\omega)|$ curve from the right and it crosses the unit gain level at the frequency $\omega = J_{VHF}^{\ 1/m}$. Hence, by reducing $J_{VHF}$ the controller roll off occurs at lower frequencies, which generally is preferable.

To characterize the mid frequency (MF) range the generalized maximum sensitivity is used. It is now redefined compared to the definition in (Lennartson and Kristiansson, 1997) as

$$GM_S = max\,(\|S\|_\infty,\ \|T\|_\infty M_S/M_T,$$
$$max_\omega | W_L(GK + (2 - M_{Am})/2M_{Am}) |\ 2M_S) \qquad (5)$$

where $T$ is the complementary sensitivity function, $T(s) = 1 - S(s)$, $GK(s)$ is the loop transfer function, $W_L$ is a weighting function with unit gain above $\omega_\pi$, and zero gain below $\omega_\pi$, where $\omega_\pi$ is defined as the lowest frequency where the phase shift of the loop $GK(j\omega)$ is $-\pi$. In this paper the default values are $M_S$=1.7, $M_T$ =1.3 and $M_{Am}$= 3. A demand on $GM_S = M_S$ then corresponds to $\|S\|_\infty \leq M_S = 1.7$, $\|T\|_\infty \leq M_T = 1.3$, the amplitude margin $Am \geq 3$ and $|GK|_{\omega \geq \omega\pi}$ limited by a circle with radius $0.5$ and center at $(0.5 - 1/M_{Am} + j0)$. These default values guarantee a well damped reference step response, and good stability margins ($Am \geq 3.0$, and the phase margin $\varphi_m \geq 45°$). The limits on $|GK|$ introduced here are marked in Figure 5 in Section 4.

Åström and coworkers (1998) argue that the more simple criterion $\|S\|_\infty$ is a suitable robustness measure.

They do not introduce any explicit high frequency measure, and consider therefore the criterion $||S||_\infty$ as a tuning parameter, while we typically assume that $GM_S$ has a fixed value, and $J_{HF}$ is considered as the tuning parameter. In other words, we try to minimize the low frequency criterion ($J_{LF}$) by accepting a reasonable control gain ($J_{HF}$), while keeping the stability margin $GM_S$ fixed. In (Åström, et al, 1998) improved low frequency behavior is achieved by reducing the stability margin instead.

Our fixed measure of the stability margin ($GM_S$ = 1.7) motivates a more complicated criterion. Take for example PI control of an integrator process ($1/s$) or state feedback of a double integrator process ($1/s^2$). A demand $||S||_\infty$ = 1.7 then implies too low damping ($\zeta$= 0.3, $M_T$=1.95), while $GM_S$ = 1.7 means a well-damped system ($\zeta$ =0.67) due to the additional restriction on $||T||_\infty$. The criterion $J_{HF}$ (=3.1) is equal in the two cases.

The third restriction in (5) means that the amplitude margin $Am$ is limited, and above $\omega_\pi$ the loop gain is not allowed to increase too much. Note that, without this restriction on the loop gain above $\omega_\pi$, an optimized Schmidt predictor for a plant with long time delay, implies unacceptable loop gain at the transition between the mid and the high frequency regions. See further Section 4!

In controller design, independent of method, typically some parameters in a tuning vector $\rho$ has to be selected. Generally there is a trade-off between properties in different frequency ranges. Hence, when two different controllers are compared in one frequency region, it is necessary for a fair evaluation, to keep the properties in the other two frequency regions constant and equal. As indicated above our main goal is to compare and minimize the low frequency criterion $J_{LF}$. Hence, the following constrained optimization problem is considered in our evaluations and control design procedures:

$$\min_\rho J_{LF}(\rho) \qquad GM_S(\rho) = c_1 \quad J_{HF}(\rho) = c_2 \tag{6}$$

Here $c_1$ and $c_2$ are arbitrary constants. Finally, we note that fixed $GM_S$ and $J_{HF}$ values implies that modifications of the controller roll off ($J_{VHF}$) do not influence the low frequency performance $J_{LF}$. This will be demonstrated in the next section.

## 3. THE PID CONTROLLER

The well-known PID controller, designed by different methods, has been used to illustrate the evaluation method presented above. A traditional way to write the PID controller is

$$K(s) = K_{PID}(1 + \frac{1}{sT_i} + \frac{sT_d}{1 + sT_f}) \tag{7}$$

where the parameters are $K_{PID}$ = gain, $T_i$ = integration time, $T_d$= derivative time and $T_f$= filter time constant.

The HF behavior from reference signal to output, $G_{yr} = GK/(1+GK) \approx GK \; (GK<<1)$. Hence, the proper but not strictly proper PID controller (7) ($m=0$) implies that the corresponding roll off is determined exclusively by the plant. Corresponding transfer function to the control signal, , shows that there is no roll off at all from reference input and measurement noise to the control signal. If this is not acceptable, e.g. due to significant high frequency measurement noise, the PID controller (7) may be extended by a low pass filter. This filter may be applied only to the proportional and derivative parts of the controller, since the integral part already has a roll off. That gives the formulation:

$$K(s) = K_{PID}(\frac{1}{sT_i} + (1 + \frac{sT_d}{1 + sT_f})\frac{1}{(1 + s\gamma T_f)}) \tag{8}$$

where $\gamma$ is the ratio between the roll off filter constant and the derivative filter constant. Hence $\gamma$ = 0 brings back the classical PID controller (7). Let us introduce the parameters $a=T_i/T_d$, $b=T_d/T_f$ and the integral gain $k_i=K_{PID}/T_i$. For $\gamma = 0$ $(m=0)$ the high frequency gain $k_\infty = K_{PID}(1+b)$, and for $\gamma>0$ $(m=1)$ $k_\infty = k_i ab(1+b)/\gamma$.

Alternatively, the output from the 'text book' PID-controller ($T_f$=0 in (7)) may be filtered by a higher order filter. A second order filter gives the following modified PID controller

$$K(s) = K_{PID}(1 + \frac{1}{sT_i} + sT_d)\frac{1}{1 + 2\zeta T_f s + T_f^2 s^2} \tag{9}$$

This formulation has the same structure as (8) except that it also allows complex controller poles. Based on our experience so far it is reasonable to choose the damping factor around $\zeta$=0.7. It means that (9) also has only four tuning parameters as (7). This PID structure needs however to be further investigated.

The results accounted for in this paper is obtained by a controller of type (8). The optimization is made in two steps. First (step 1) $J_{LF}$ is minimized with $GM_S$ and $J_{HF}$ fixed and $\gamma = 0$ (no roll off filter at all). To show that the roll off does not influence the LF properties, $J_{LF}$ is then minimized again in a second step with $GM_S$ and $J_{HF}$ fixed as above but with different values of $\gamma$. Figure 2 illustrates that $J_{LF}$ is quite independent on $\gamma$.

For the sake of comparison, corresponding result is shown also for a sub-optimal PID controller (7), where the parameters $a$ $(T_i/T_d)$ and $b$ $(T_d/T_f)$ are not used for the optimization, but just given the standard values *4* and *10* respectively, often suggested in the literature (Ziegler and Nichols, 1942; Åström and Hägglund, 1984; Mantz and Tacconi, 1989; Åström and Hägglund, 1995).

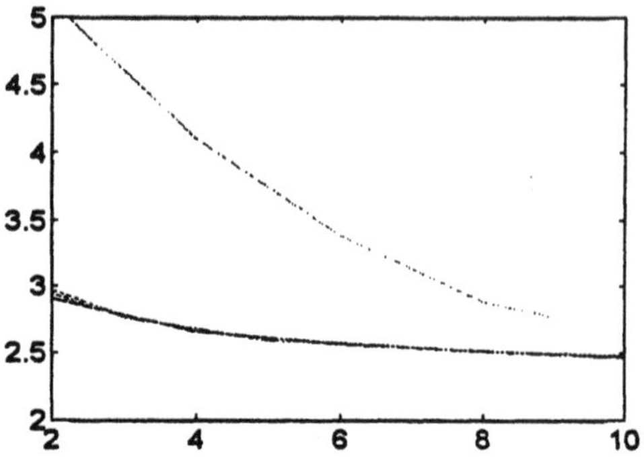


Figure 2: $J_{LF}$ as a function of $J_{HF}$ for $G=e^{-s}/(1+s)^3$ controlled by a modified PID controller (8) with $\gamma$ = 0, 0.3 and 1.0 respectively (the curves are coincident) max*KS*=6.0, $GM_S$=1.7 Corresponding results for a PID controller with standard values of the parameters *a* and *b* (*a* =4, *b*=10) and no roll off is also shown at the top.

In another investigation $J_{LF}$ and $J_{HF}$ (max*KS*) are fixed to the values given by step 1 ($GM_S$=1.7), $\gamma$ is fixed to *1* (multiple poles) and $J_{VHF}$ is minimized. This is a way to maximize the roll off without altering the LF and MF properties. As could be expected the optimal values of the parameters *a* and even more *b* are shifted when the roll off filter is included. Figure 3 illustrates this fact. The ratio *b* increases, and hence $T_f$ decreases, since the roll off filter then assists the derivative filter in keeping the desired $J_{HF}$ level. Note that the values of *a* are remarkably low!

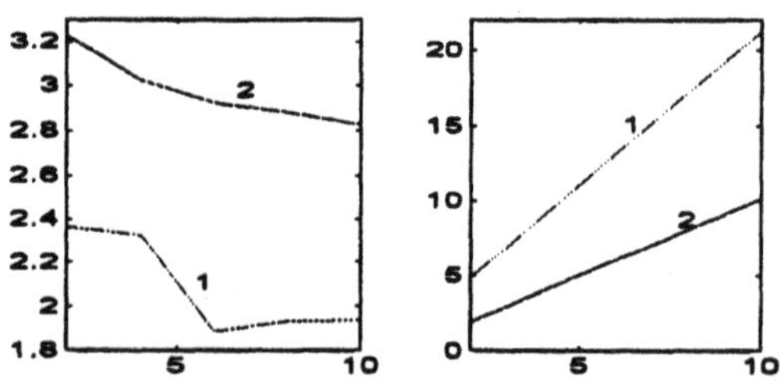


Figure3: The parameters *a* (left) and *b* (right) as functions of $J_{HF}$ for $G=e^{-s}/(1+s)^3$ controlled by an optimal PID controller with (1) and without (2) extra roll off ($\gamma$= 1)

When $J_{HF}$ is kept constant, the bandwidth of $G_{ur}$ (*KS*) as well as the maximum $J_{VHF}$ decrease when the roll off filter is included. The reduced bandwidth and $J_{VHF}$ correspond to a slightly decreased maximum of the control signal See Figure 4!

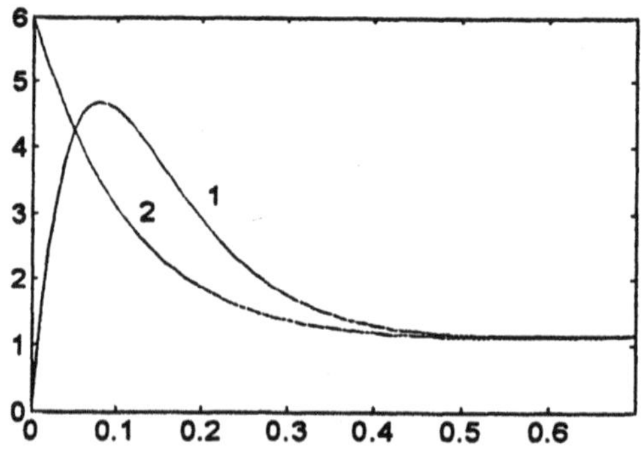


Figure 4: The control signal *u(t)* when the reference *r(t)* is a unit step, for $G = e^{-s}/(1+s)^3$ and an optimal PID controller with (1) and without (2) extra roll of ($\gamma$=1). $J_{HF}$= 6.0

During all our simulations we have seen regularities which have made it possible to formulate a set of rules for auto-tuning of PID controllers without extra roll off, when the plant is stable and non oscillating. Among the investigated plant models are transfer functions with short and long time delays, low and high order, minimum and non minimum phase, multiple poles and distributed poles. The rules are

$$k_\infty = (30\,\kappa^2 - 35\,\kappa + 12)/|G(j0)|$$
$$k_0 = (1.4\,\kappa^2 - 2\kappa + 0.95)\,\omega_\pi/|G(j0)|$$
$$a = 3;\ b = 2.5\,k_\infty|G(j\omega_\pi)| \qquad (10)$$

The variable $\kappa=|G(j\omega_\pi)|/|G(j0)$ ($\omega_\pi$ is the frequency where arg $(G(j\omega)) = -\pi$.) is a measure of the difficulty to control a process (Hang, et al, 1991). The higher $\kappa$-value, the more difficult it is to control. The rules in (10), coming from a further development of those presented in (Kristiansson and Lennartson, June 1998) have shown to give good results for all processes tested. In most cases they have led to $GM_s$-values in the range 1.7±5 % and bandwidths in the range 0.6-1.0 times $\omega_\pi$ for the actual process, with a tendency to smaller bandwidths for processes with higher $\kappa$-values. In fact the normalized bandwidth with good accuracy follows the formula $\omega_b/\omega_\pi = 1.05$-$0.73\kappa$. For plants with $\kappa < 0.05$ and $\kappa > 0.75$, the *GMs*-values reached are slightly higher.

The proposed set of rules is based on knowledge about the plant, which can be picked up from a step response ($|G(j0)|$) and a simple relay experiment ($\omega_\pi$, $|G(j\omega_\pi)|$) (Åström and Hägglund,1995).

## 4. STABLE NON OSCILLATING PLANTS WITH CONSIDERABLE TIME DELAY

When the plant suffers from a great delay, it is tempting to try to control it by a Schmidt predictor, which gives the controller the transfer function:

$$K(s) = \frac{K_0(s)}{1 + K_0(s)G_0(s)(1 - e^{-sL})} \quad (11)$$

where $L$ is the delay and $G_0(s)$ is the plant without delay. $K_0(s)$ is traditionally the controller designed for $G_0(s)$. A Schmidt predictor designed in that way is however not an optimal choice of controller. With the primary controller, $K_0(s)$, optimized for a corresponding plant without delay, the system will loose too much of its stability compared to an optimal PID without Schmidt predictor. But if the whole controller $K(s)$, with $K_0(s)$ written as in (7) is optimized with constrained $GM_S$, then the Schmidt predictor structure achieves the same stability margins as the original optimal PID solution.

In this paper the definition of $GM_S$ (5) is modified compared to the definition in (Lennartson and Kristiansson, 1997). The main reason is that for some controllers, e.g. the PID controller with a Schmidt predictor, the open loop gain may increase drastically for $\omega > \omega_\pi$ when $L$ is large. See Fig 5! Such undesirable behavior is prohibited by the modified $GM_S$ criterion. (5).

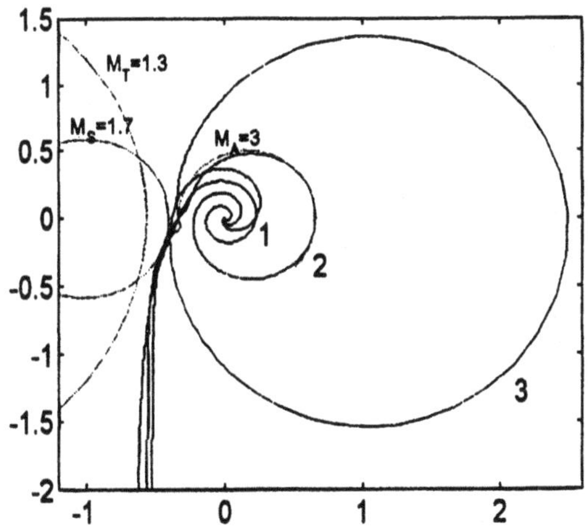


Figure 5: The Nyquist plot for the open loop when $G=e^{-s4}/(1+s)^3$ is controlled by a PID controller without (1) and with a Schmidt predictor, when there is a restriction on the loop gain for $\omega > \omega_\pi$ (2) and when there is only a restriction on the amplitude margin (3). $J_{HF}$ =7.0. The limiting circles for $GM_S$ are dotted.

Some profit can undoubtedly be achieved by including a Schmidt predictor in the controller. The $J_{LF}$ decreases and the bandwidth increases for fixed $J_{HF}$ and $GM_S$. The improvements are about 20 %. When the delay grows, the benefits are however to some degree eaten up by the effects of the restriction of the loop gain for $\omega > \omega_\pi$. See Fig 6 and Fig 7!

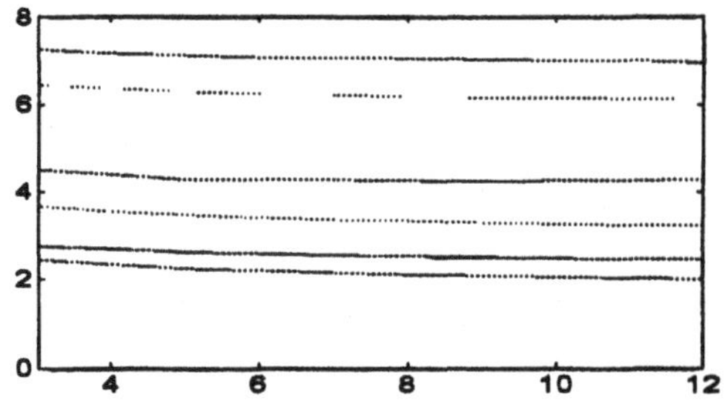

Figure 6: $J_{LF}$ as function of $J_{HF}$ for $G=e^{-sL}/(1+s)^3$ controlled by a PID controller with (Sch) and without (PID) a Schmidt predictor. The curves are from the bottom: Sch/L=1, PID/L=1, Sch/L=2, PID/L=2 , Sch/L=4, PID/L=4

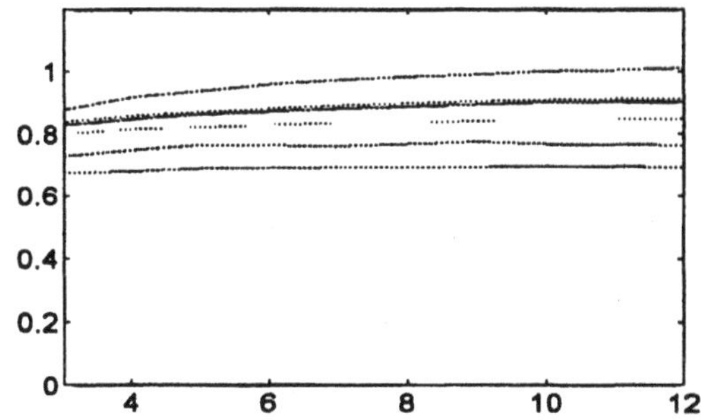

Figure 7: The normalized bandwidth $\omega_b/\omega_\pi$ as a function of $J_{HF}$ for $G=e^{-sL}/(1+s)^3$ controlled by a PID controller with (Sch) and without(PID) a Schmidt predictor. The curves are from the top Sch/L=1, Sch/L=2, PID/L=1, Sch/L=4, PID/L=2 PID/L=4

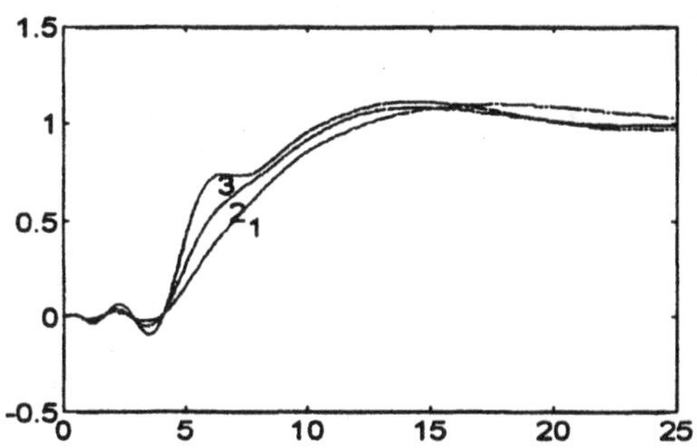


Figure 8: The output signal after a step in the reference signal for $G=e^{-s4}/(1+s)^3$ controlled by an optimal PID controller without (1) and with a Schmidt predictor when there is a restriction on the loop gain for $\omega > \omega_\pi$ (2) and when there is only a restriction on the amplitude margin (3). $J_{HF}$ =7.0

A registration of step responses shows that the benefits brought by the Schmidt predictor in the controller are marginally compared to a well tuned PID controller. Figure 8 shows the output signal and Figure 9 shows the control signal after a step in the reference signal. Both figures demonstrate a somewhat abnormal behavior of the system when there is no limitation on the loop gain for frequencies above $\omega_\pi$ (curve 3). The strange level away after 6-7 seconds that can be seen in the output, and the swinging of the control signal to even negative values, must be blamed on the weak damping of the pair of poles that occur at about 3 times $\omega_\pi$ when there is no limitation on the loop gain above $\omega_\pi$. When the restriction is introduced, the damping of these poles is increased by about 60 %. The small ripple at the beginning of the output signal is essentially an effect of the Padé approximation of the delay.

Even the output after a step disturbance has been studied, but after the filtering of the process there are no remarkable anomalies in this signal, and the three system responses (c f Fig 8) are hence quite equal.

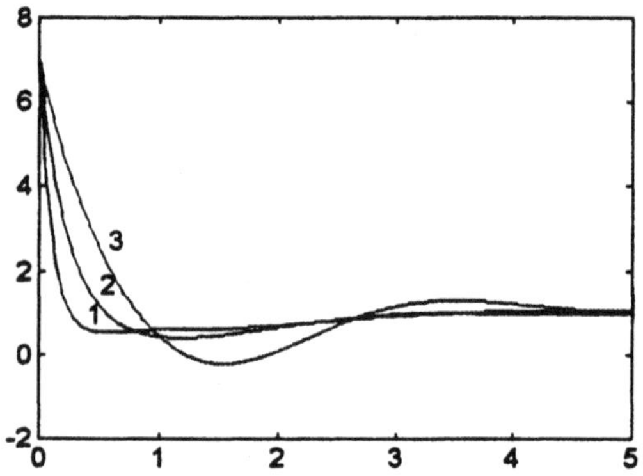


Figure 9: The control signal after a step in the reference signal for $G=e^{-s4}/(1+s)^3$ controlled by an optimal PID controller without (1) and with a Schmidt predictor when there is a restriction on the loop gain for $\omega > \omega_\pi$ (2) and when there is only a restriction on the amplitude margin (3). $J_{HF}=7.0$

## 5. CONCLUSIONS

In this paper we have presented some modifications of an earlier presented evaluation method for controllers. The performance and robustness criteria are slightly extended in order to investigate the significance of roll off in the controller. Furthermore, an additional restriction on the loop gain is introduced to prevent it from being too large at higher frequencies.

Design of PID controllers based on the method above is discussed. A set of autotuning rules is given. It is also shown that additional roll off by an extra low pass filter may well be added to the controller. If this roll off filter is included during the design process, it does not deteriorate the low and mid frequency system behavior. Note however that the controller parameters are modified, especially the filter time constant. Hence it is important to include the roll off filter in the design.

If the plant to be controlled has a considerable time delay, a Schmidt predictor may be included in the controller. However, the profits are shown to be small, compared to an ordinary PID controller, when the two control structures (PID with and without predictor) are optimized and evaluated by the same criteria.

## 6. REFERENCES

Hang, C.C., K. J. Åström and W. K. Ho. Refinements of the Ziegler-Nichols tuning formula, *IEE Proc. D*, Vol 138, No 2, March 1991

Kristiansson, B. and B. Lennartson. Robust design of PID controllers including auto-tuning rules, *Proc American Control Conference '98*, Phoenix, June 1998

Kristiansson, B. and B. Lennartson. Optimal PID controllers for unstable and resonant plants, *Proc 37th Conf on Decision and Control*, Tampa, Dec 1998

Lennartson, B. and B. Kristiansson. Pass band and high frequency robustness for PID control. *Proc 36th Conf on Decision and Control*, San Diego, Dec 1997

Mantz, R. J. and E. J. Tacconi. Complementary rules to Ziegler and Nichols rules for a regulating and tracking controller. *Int. J. Control*, 49:1465-1471, 1989

Ziegler, J.G. and N.B. Nichols. Optimum settings for automatic controllers. *Trans. ASME*, 64:759-768, 1942

Åström K. J. and T. Hägglund. A frequency domain method for automatic tuning of simple feedback loops. *Proc. 23rd Conf. On Decision and Control*, Las Vegas, NV, Dec 1984

Åström K.J. and T Hägglund. *PID Controllers: Theory, Design and Tuning*. Instrument Society of America, 1995

Åström, K.J. and H. Panagopoulos and T. Hagglund, Design of PI controllers based on non-convex optimization *Automatica* **Vol. 34**, p. 585-601, May, 1998.

*Acknowledgement:* Valuable comments given by professor Karl-Johan Åström and professor Tore Hägglund are very much appreciated.

# Analytical Design for Linear Continuous Integrator/Time Delay Systems

**Weidong Zhang [1], Xiaoming Xu [1], and Youxian Sun [2]**

1 *Department of Automation, Shanghai Jiaotong University, Shanghai 200030, P. R. China, Email:wdzhang@mail1.sjtu.edu.cn*
2 *National Laboratory of Industrial Control Technology, Institute of Industrial Process Control of Zhejiang University, Hangzhou 310027, P. R. China*

Abstract: Smith predictor has been proved to be a useful scheme for stable processes with time delay. In this paper an analytical design method of Smith predictor is presented for integrator/time delay processes based on the theory of complex function. The design procedure incorporates a two-stage procedure. In the first stage, the controller that stabilizes the closed loop system is parameterized. In the second stage, high enough order Pade approximation is introduced to approximate the time delay, and convert the design problem of system with time delay to that of system without time delay. The optimal Smith predictor is then derived analytically. Examples are given to illustrate the method.


Keywords: Optimal control, Smith predictor, integrator/time delay process, process control.

## 1 INTRODUCTION

A frequently encountered problem in process industries is that of controlling the process with a significant time delay. Dealing with this problem serves then as a starting point for the design of almost any process control system, regardless of its configuration.

The Smith predictor (Smith, 1959) is well known as an effective dead time compensator for stable processes with time delays and studied by many researchers (Laughlin *et al.*, 1987; Zhang and Sun, 1997; Zhang *et al.*, 1998a). This technique utilizes an inner feedback compensation loop based on a first order plus time delay model of the process, and theoretically allows arbitrary improvement of the closed loop dynamics. However, it has been shown earlier (Hang and Wong, 1979; Furukawa and Shimemura, 1983) that it cannot be used for the process having an integral mode since a constant load disturbance will result in a steady state error. From the viewpoint of modern control theory it is clear that the predictor has made the process unobservable to the controller. To overcome the problem, Watanabe and Ito (1981) proposed a modification, in which the single, ideally exact process model used in the conventional predictor was replaced by two, deliberately mismatched. Astrom *et al.* (1994) pointed out that the response of such a system tended to be slow. An alternative Smith predictor was accordingly developed. The advantage of the controller is that it decouples the disturbance response from the setpoint response. Zhang and Sun (1996) and Zhang *et al.* (1998b) extended the discussion, particularly with respect to the controller structure and tuning rules. Though the setpoint response and the disturbance response are improved in these methods, the Smith predictor is modified and the new structure becomes very complicated.

The design problem of Smith predictor for linear continuous integrator/dead time processes is studied in this paper. Based on the theory of complex function, an analytical design method is developed. Its main advantage is that the Smith predictor structure without any modification is used. The design procedure involves two steps. First, the controller that stabilizes the closed loop system is parameterized. Second, high enough order Pade approximation is introduced to approximate the time delay, and convert the design problem of system with time delay to that of system without time delay. The optimal Smith predictor is then derived analytically. The sufficient and necessary condition for testing robustness of the closed loop system is also provided.

## 2 PARAMETERIZATION OF CONTROLLER

A block diagram of Smith predictor is shown in Fig. 1, where *R(s)* is the controller, *G(s)* is the plant, $G_m(s)$ is the plant model, and $G_{mo}(s)$ is the delay free part of $G_m(s)$. In this paper, we consider a single input/single output process described by

$$G_m(s) = \frac{1}{Ts}e^{-\theta s},\ G_{mo}(s) = \frac{1}{Ts} \tag{1}$$

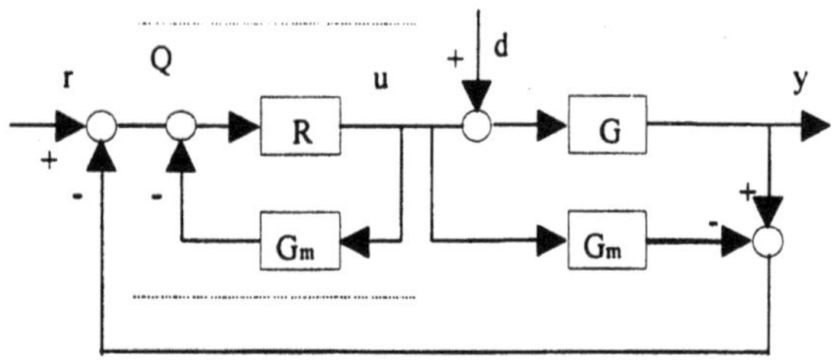


Fig. 1 Structure of Smith predictor

The structure of Smith predictor can be related to the well-known Youla parameterization through (Laughlin *et al.*, 1987; Zhang, 1998)

$$Q(s) = \frac{R(s)}{1 + G_{mo}(s)R(s)} \tag{2}$$

As we know, there always exist setpoint changes or disturbances in process industries. They are usually approximated by steps. As a basic closed loop specification the error is often required to vanish asymptotically for these inputs. Suppose that the model is exact, i.e. $G(s) = G_m(s)$. The transfer function that from the disturbance *d* to the system output *y* can be written as *G(s)(1-G(s)Q(s))*. It is easy to find that the system is of type II. In classic Smith predictor, the controller *R(s)* is either PI or PID controller. Such a system may not track the setpoint changes or reject disturbances asymptotically. A new design procedure will be developed in this paper, of which a remarkable characteristic is that the controller parameters and structure are derived simultaneously by an optimal performance criterion.

For *Q(s)* to guarantee the internal stability and exhibit no offset to step inputs it is necessary and sufficient that (Zhang *et al.*, 1998b; Doyle *et al.*, 1992)

$$\lim_{s\to 0} G(s)(1 - G(s)Q(s)) = 0 \tag{3}$$

or equivalently

$$\lim_{s\to 0}(1 - G(s)Q(s)) = 0 \tag{4}$$

$$\lim_{s\to 0}\frac{d}{ds}(1 - G(s)Q(s)) = 0 \tag{5}$$

The *Q(s)* that satisfies the first condition can be expressed as

$$Q(s) = s(T + TsQ_1(s))$$

where $Q_1(s)$ is stable. Combining it with the second condition yields

$$\lim_{s\to 0}\frac{d}{ds}\left[1 - \frac{1}{Ts}e^{-\theta s}(T + TsQ_1(s))\right]$$
$$= \lim_{s\to 0}\frac{d}{ds}\left[e^{-\theta s} + se^{-\theta s}Q_1(s)\right]$$
$$= \lim_{s\to 0}\begin{bmatrix}(-\theta)e^{-\theta s} + e^{-\theta s}Q_1(s) + \\ s\dfrac{d}{ds}(e^{-\theta s}Q_1(s))\end{bmatrix}$$
$$= \lim_{s\to 0}\begin{bmatrix}(-\theta)e^{-\theta s} + e^{-\theta s}Q_1(s) + \\ s(-\theta)e^{-\theta s}Q_1(s) + se^{-\theta s}\dfrac{d}{ds}Q_1(s)\end{bmatrix}$$

It is easy to see that $Q_1(0)=\theta$ satisfies the condition. Thus, *Q(s)* is given by an expression of

$$Q(s) = s(T + T\theta s + Ts^2Q_2(s)) \tag{6}$$

where $Q_1(s)$ is stable. The advantage of the procedure is that the controller parameterization of stable plant is utilized for unstable plant. As the controller parameterization based on the coprime factorization is very complicated, it will destroy the structure of Smith predictor.

## 3 ANALYTICAL SMITH PREDICTOR

The main difficulty in designing Smith predictor analytically is that the introduction of time delay leads to an irrational transfer function. It does not admit a finite-dimensional realization. However, many well-established design techniques, such as the characteristics locus method and the LQR (linear quadratic regulator), usually require a finite-dimensional representation of the system concerned. Hence, they can not cope with the control problem of irrational plants directly.

In order to utilize the existing theory, Pade approximation is introduced in this paper to approximate the time delay. The nature of Pade approximation has been studied by many papers, for example, Saff and Varga (1977) and Lam (1990), and will not be discussed here. Employ the $n$ order all-pass Pade approximation, we have

$$G_m(s) = \frac{1}{Ts}\frac{Q_{nn}(-\theta s)}{Q_{nn}(\theta s)} \tag{7}$$

where

$$Q_{nn}(\theta s) = \sum_{j=0}^{n} \frac{(2n-j)!n!}{(2n)!j!(n-j)!}(\theta s)^j$$

and $n$ is an integer large enough. Hence, the error between the Pade approximation and the time delay is small enough. The closed loop transfer function can be written as

$$T(s) = G(s)Q(s) \tag{8}$$

Then the optimal performance index can be expressed as $\min\|W(s)G(s)(1-T(s))\|_2$. Here $W(s)$ is the weighting function. If the model is exact, then we can take $W(s)=1/s$ (see, e.g., Doyle *et al.*, 1992). This leads to

$$\|W(s)G(s)(1-T(s))\|_2^2$$

$$= \left\| \frac{1}{s^2}\frac{Q_{nn}(-\theta s)}{TQ_{nn}(\theta s)} \left[ 1 - \frac{Q_{nn}(-\theta s)}{TsQ_{nn}(\theta s)} s(T + T\theta s + Ts^2 Q_2(s)) \right] \right\|_2^2$$

$$= \left\| \frac{1}{Ts^2}\left[ \frac{Q_{nn}(\theta s)}{Q_{nn}(-\theta s)} - (1 + \theta s + s^2 Q_2(s)) \right] \right\|_2^2$$

$$= \left\| \frac{Q_{nn}(\theta s)}{Ts^2 Q_{nn}(-\theta s)} - \frac{1+\theta s}{Ts^2} - \frac{Q_2(s)}{T} \right\|_2^2$$

It is known that $Q_{nn}(\theta s)$ is Hurwitz, i.e., all the zeros of all-pass Pade approximation are in the right half plane and all the poles are in the left half plane (Lam, 1990; Zhang, 1998). Then we obtain

$$\|W(s)G(s)(1-T(s))\|_2^2$$

$$= \left\| \frac{Q_{nn}(\theta s)}{Ts^2 Q_{nn}(-\theta s)} - \frac{1+\theta s}{Ts^2} \right\|_2^2 + \left\| \frac{Q_2(s)}{T} \right\|_2^2 \tag{9}$$

Therefore, the design problem of system with time delay is converted to that of system without time delay. Obviously, the right part is minimized when $Q_2(s)=0$. In this case, the optimal, generally improper $Q(s)$ is given by an expression of

$$Q_{im}(s) = Ts(1+\theta s) \tag{10}$$

An improper transfer function can not be realized in practice. This $Q_{im}(s)$ must be rolled-off at high frequency to get a proper suboptimal $Q(s)$. Now use the following low-pass filter to roll $Q_{im}(s)$ off

$$J(s) = \frac{3\lambda s + 1}{(\lambda s + 1)^3} \tag{11}$$

It follows that

$$Q(s) = Q_{im}(s)J(s) = \frac{Ts(1+\theta s)(3\lambda s + 1)}{(\lambda s + 1)^3} \tag{12}$$

As $\lambda$ tends to be zero, the system performance tends to be optimal. Elementary computations give

$$R(s) = T\frac{3\lambda\theta s^2 + (3\lambda + \theta)s + 1}{s(\lambda^3 s^2 + (3\lambda^2 - 3\lambda\theta)s - \theta)} \tag{13}$$

$R(s)$ is not a PID controller. This implies that the integrator/time delay process is very difficult to control.

For reason of clarity, as well as practical interest, this paper restricts attention to a first order plant. The design procedure developed here can be extended to high order system. However, the results will be very complicated.

## 4 NOMINAL PERFORMANCE AND ROBUSTNESS

Zhang and Sun (1997) and Zhang (1998) presented two methods for controller design, in which the time domain performance could be estimated quantitatively. In the design method of this paper, the quantitative time domain performance is also provided.

Notice that the closed loop transfer function is

$$T(s) = \frac{(1+\theta s)(3\lambda s+1)}{(\lambda s+1)^3}e^{-\theta s} \qquad (14)$$

Therefore, the system performance is only related to the ratio of $\lambda$ and $\theta$. The relationship is shown in Fig. 2. It is seen that a notable feature of the control system is that the overshoot always exists and tends to be a constant with the increase of $\lambda$. The responses of closed loop system with different parameters are shown in Fig. 3. Another feature exposed to us is that though there always exists overshoot, the system does not oscillate and the settling time is usually long.

The design of advanced control system is intrinsically a model-based activity. However, even the most detailed model is still an approximation of reality. The plant-model mismatch always exists. Hence, it is of great practical importance that the controller performs well even when the dynamic behavior of the real process differs from that described by its model. We will show how to test the robustness of the proposed scheme. Suppose that the profile of the plant-model mismatch is $\Delta(\omega)$. Based on the modern robust control theory developed recently (see, e.g., Doyle *et al.*, 1992), it is known that the closed loop system is robust stable if and only if

$$\left\|\Delta(\omega)T(j\omega)\right\|_\infty < 1 \qquad (15)$$

In other words, by tuning the parameter $\lambda$, the arbitrary robust stability can be obtained for the closed loop system. The robust performance can be tested by the sufficient and necessary condition as follows

$$\left\| |W(j\omega)S(j\omega)| + |\Delta(\omega)T(j\omega)| \right\|_\infty < 1 \qquad (16)$$

It is found that the relationship between the parameter $\lambda$ and the system performance is monotonous. The larger $\lambda$ is, the better the robust stability is and the worse the nominal performance is.

The test condition of robust performance is very complicated, of which a rigorous theoretical treatment is more of mathematical than of practical interest. In the context of process control, the $\lambda$ value between $1.0\theta$~$3.0\theta$ will yield good compromise between the stability and nominal performance. This is a rule of thumb.

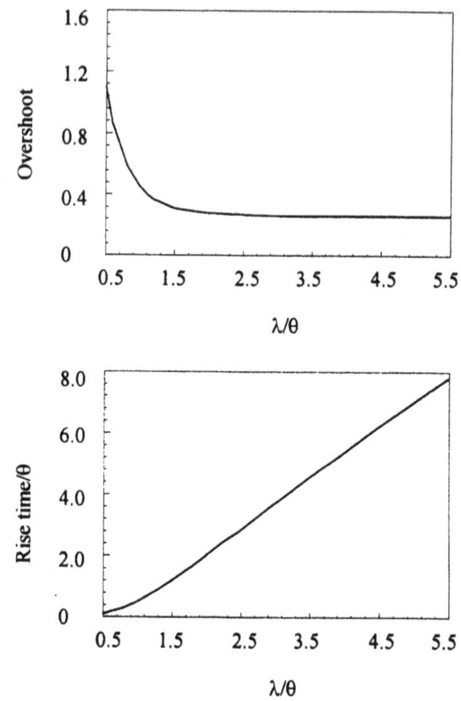


Fig. 2 Relationship between the controller parameter and system performance

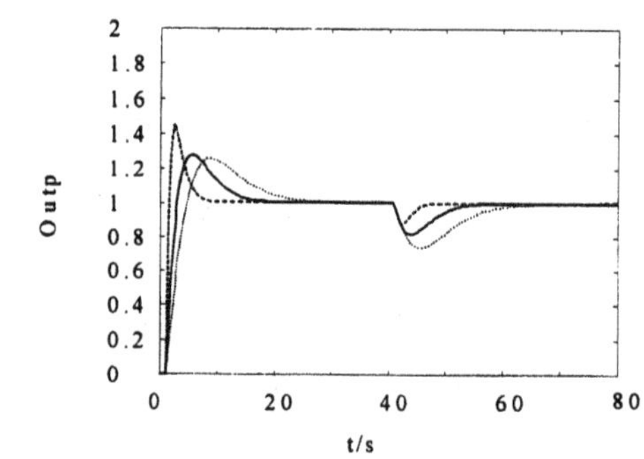


Fig. 3 Responses of the nominal systems
(Dotted line: $\lambda$=1.0$\theta$; solid line: $\lambda$=1.5$\theta$; dashed line: $\lambda$=2.0$\theta$)

*Example 1* Consider a middle-column temperature control system of high-purity distillation column. The transfer function model and parameters calculated from the step tests are

$$G_m(s) = \frac{1}{Ts}e^{-\theta s}$$

where $T$=100 and $\theta$=5.5. The parameters of conventional PID controller are P=13 and I=18.7. Let $\lambda$ be $1.1\theta$ for the new method. A unit step setpoint is introduced at time t=0. At time t=200, a load disturbance $d(s)$=-1/$s$ is introduced. The responses of closed loop system are shown in Fig. 4. It is seen that the system with PID controller has very large overshoot. The proposed method gives better response.

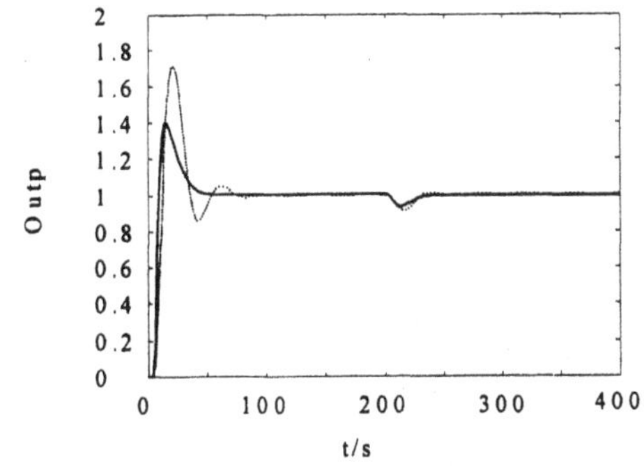


Fig. 4 Step responses of the closed loop systems (Dotted line: PID controller; solid line: the new method)

## 5 CONCLUSIONS

This paper devotes to the problem of controlling an integrator/time delay process. The controller parameterization is provided and an analytical design procedure is developed within the framework of standard Smith predictor. The range of applicability of this best known scheme is thus extended significantly.

The robust stability and robust performance are discussed and the sufficient and necessary test conditions are provided. The relationship between the performance of closed loop system and the controller parameter is investigated and utilized to specify the degree of controller detuning. It is shown that the nominal performance in time domain can be estimated quantitatively and adjusted monotonously by tuning the parameter $\lambda$. It is also found that the overshoot of closed loop system always exists even though the system tends to be optimal. This is an important characteristic in the control of integrator/time delay processes.

Morari and Zafriou (1989) pointed out that the internal model control structure could not be used for the control of unstable processes. The discussion of this paper shows that the structure can be used for the control of critical unstable processes.

## ACKNOWLEDGMENT

Project supported by National Natural Science of China (NSF69804007).

## REFERENCES

Astrom, K.J., C.C. Hang, and B.C. Lim (1994). A New Smith Predictor for Controlling a Process with an Integrator and Long Time Delay, *IEEE Trans. Automat. Contr.*, **39**, 343

Doyle, J.C., B.A. Francis, and A.R. Tannenbaum (1992). *Feedback Control Theory*, Macmillan Publishing Company, NY

Furukawa, F. and E. Shimemura (1983). Predictive Control for Systems with Time Delay, *Int. J. Control*, **37**, 399

Hang, C.C. and F.S. Wong (1979). Modified Smith Predictor for the Control of Process with Dead Time, *Proc. ISA Annual Conf.*, 33

Lam, J. (1990). Convergence of a Class of Pade Approximations for Delay Systems, *Int. J. Control*, **52**, 989

Laughlin, D.L., D.E. Rivera, and M. Morari (1987). Smith Predictor Design for Robust Performance, *Int. J. Control*, **46**, 477

Morari, M. And E. Zafriou (1989). *Robust Process Control*, Pretice Hall, NY

Saff, E.B. and R.S. Varga (1959). *Pade and Rational Approximation*, Academic Press, London, 1977

Smith, O.J.M.. A Controller to Overcome Dead Time, *ISA J.*, **6**, 28

Watanabe, K. and M. Ito (1981). A Process Model Control for Linear System with Delay, *IEEE Trans. Automat. Contr.*, **26**, 1261

Zhang, W. D. and Y. X. Sun (1996). Modified Smith Predictor for the Integrator/Time Delay Processes, *Ind. Eng. Chem. Res.*, **35**, 2796

Zhang,W.D., and Y.X. Sun (1997). A Class of Optimal Smith Predictor and Its Robust Tuning, *ACTA Automatica Sinica*, **23**(5), 660 (in chinese)

Zhang,W.D., Y.X. Sun, and X.M. Xu (1998a). Robust Digital Controller Design of Process with Dead Time-Some New Results, *IEE Proc.-Part D*, **145**(2),159

Zhang,W.D., Y.X. Sun, and X.M. Xu (1998b). Two Degree-of-Freedom Smith Predictor for Processes with Time Delay, *Automatica*, to appear in October

Zhang, W.D. (1998). *Analytical Design Methods for Process Control*, Postdoctoral Research Report, Shanghai Jiaotong University

14th Triennial World Congress, Beijing, P.R. China

# THE EQUIVALENCE OF TIME AND FREQUENCY-DOMAIN MULTIVARIABLE OPTIMAL SYSTEMS

Chih Min Lin* Young Fon Song* Paul C Austin+

*Department of Electrical Engineering, Yuan Ze University, Chung-Li ,Tao-Yuan 320, Taiwan, Republic of China
Tel : 886-3-4638800 ext 418
Fax : 886-3-4639355
E-mail : cml@saturn.yzu.edu.tw
+Department of Electrical and Electronic Engineering, University of Auckland, Private Box 92019, Auckland, New Zealand

Abstract: The equivalence of multivariable optimal systems designed in time-domain and frequency-domain is derived. Besides the equivalence of input signal and closed-loop transfer function matrix, the equivalence of loop-gain transfer function matrix is also described, such that multivariable optimal systems designed in both time- as well as frequency-domains also satisfy the return difference equality, i.e. both approaches have the same stability margins. In ensuing discussion, the plants employed for both time-domain and frequency-domain optimal systems are shown to be consistent, which have to contain minimum-phase zeros but can be stable or unstable. *Copyright © 1999 IFAC*

Keywords: multivariable systems, optimal control, stability analysis, time domain, frequency domain.

## 1. INTRODUCTION

The linear quadratic optimal control design has been widely investigated in past, both in time as well as frequency domains. In time-domain approach, the most important property is that the linear-quadratic-regulator (LQR) optimal control satisfies the return difference equality, so that it possesses the stability margins with a gain margin $(1/2, \infty)$ and phase margain $\geq 60°$ (Anderson and moore, 1990). As well known, the plants employed for LQR problem have to contain minimun-phase zeros but they can be stable or unstable. In frequency-domain approach, the parametric controller as presented by Youla *et al.* (1976) and Youla and Bongiorno (1985), achieve Wiener-Hopf linear quadratic optimal control and also ensure asymptotic stability for the closed-loop system. However, these studies of Wiener-Hopf optimal systems have not discussed the return difference equality property such as in LQR optimal systems. Wiener-Hopf optimal control system designs, wherein the feedback-loop satisfies the return difference equality, have been discussed by Lin and Shi (1996) and by Lin and Tu (1996).

For a single-input single-output (SISO) system, the equivalence of optimal systems designed in time- as well as frequency-domains has been partially solved by Hsu *et al.* (1994). However in their discussion, the equivalence was considered only for closed-loop transfer function and input signal. Whereas the equivalence of loop-gain transfer function was not considered. As a result, the feedback-loop of the optimal system, by frequency-domain approach, does not satisfy the return difference equality, i.e. the frequency-domain approach does not possess the stability margins as that possessed by time-domain LQR optimal systems. As well known, even the closed-loop transfer function is equivalent, if the stability margins are different, the systems can not be said to be equivalent completely. Moreover, in their discussion, the plants employed for designing the frequency-domain optimal system were required to be stable but they can contain non-minimum phase zeros. This is inconsistent with the time-domain LQR optimal systems, in which the plants employed have to contain minimum-phase zeros but they can be stable or unstable. Hence the equivalence of optimal system designed by time-domain and frequency-domain approaches as discussed by Hsu *et al.* [3] is not complete. In the present paper, a complete equivalence of optimal system designed in time-domain and frequency-domain has been derived. The systems here are the multi-input multi-output (MIMO) systems; and apart from presenting the equivalence of closed-loop transfer function matrix and input signal, the equivalence of loop-gain transfer function matrix is also satisfied, such that the optimal systems obtained by both time-domain and frequency-domain approaches can satisfy the return difference equality, i.e. they possess the same stability margins. Also, the plants employed in the present study are consistent for both time-domain and frequency-domain optimal systems. That is the plants employed have to contain minimum-phase zeros but they can be stable or unstable.

The notations used in this paper are as follows : $\bar{s}$ denotes the complex conjugate of s; $\Omega^T(s)$, $\Omega^*(s)$ and $\mathrm{tr}\Omega(s)$ denote the transpose, conjugate transpose and trace of $\Omega(s)$, respectively, and $\Omega_*(s)$ is the conjugate transpose of the transfer function matrix $\Omega(-\bar{s})$ or $\Omega_*(s)=\Omega^*(-\bar{s})$ which, for real rational matrices, reduces to $\Omega^T(-s)$. A transfer function matrix $\Psi(s)$ can be expressed as

$$\Psi(s)=\{\Psi(s)\}^+\{\Psi(s)\}^- \tag{1}$$

where $\{\Psi(s)\}^+$ and $\{\Psi(s)\}^-$ denote the spectral factorization of $\Psi(s)$ with $\{\Psi(s)\}^+$ containing all the poles and zeros in $Re(s)\leq 0$ while $\{\Psi(s)\}^-$ containing all the poles and zeros in $Re(s)\geq 0$. Also a transfer function matrix $\Lambda(s)$ may be fractionized as

$$\Lambda(s)=[\Lambda(s)]_+ + [\Lambda(s)]_- + [\Lambda(s)]_\infty \tag{2}$$

where $[\Lambda(s)]_+$ , $[\Lambda(s)]_-$ and $[\Lambda(s)]_\infty$ denote the part of the partial fraction expansion associated with all poles of $\Lambda(s)$ in $-\infty < R_e(s) \leq 0$, $0 < R_e(s) < \infty$ and at $s=\infty$, respectively.

## 2. PROBLEM FORMULATION

Consider a MIMO linear time-invariant model-following system shown in Fig. 1, where the plant is square and contain minimum-phase zeros, but it can be stable or unstabe, and assume its inverse exists. The problem is to minimize the system error $e(t)$ within the limit of power of the control signal $u(t)$. The cost function is hence defined as

$$J=\int_0^\infty e^T(t)e(t)dt+\int_0^\infty u^T(t)\Gamma^2 u(t)dt \tag{3}$$

where $\Gamma^2$ is a positive-definite real matrix.

Let $T(s)$ denote the system transfer function matrix from input signal $R(s)$ to output signal $Y_o(s)$, and $M_r(s)$ be the reference model. Following the Wiener-Hopf approaches presented by Chang (1961), Youla *et al.* (1976) and Youla and Bongiorno (1985), the optimal solution for minimizing $J$ may be given as

$$T(s)=Y^{-1}(s)\left[Y_*^{-1}(s)M_r(s)R(s)\right]_+ R^{-1}(s) \tag{4}$$

where

$$Y(s)=\left\{I+G_{p*}^{-1}(s)\Gamma^2 G_p^{-1}(s)\right\}^+ \tag{5}$$

Let

$$T(s)=T_c(s)G_f(s) \tag{6}$$

where

$$T_c(s)=G_p(s)(I+L(s))^{-1} \tag{7}$$

in which the loop-gain transfer function matrix

$$L(s)=G_c(s)G_p(s) \tag{8}$$

By appropriately setting

$$G_f(s)=\left[Y_*^{-1}(s)M_r(s)R(s)\right]_+ R^{-1}(s) \tag{9}$$

it reveals that

$$T_c(s)=Y^{-1}(s) \tag{10}$$

For the feedback-loop in Fig. 1 to satisfy the return difference equality, the loop-gain transfer function matrix $L(s)$ has to contain minimum-phase zeros (Anderson and Moore, 1990). Hence we assume the plant contains minimum-phase zeros. Using (5) and (10), for a given $G_p(s)$ and $\Gamma$, $Y(s)$ and $T_c(s)$ can be determined. Then the loop-gain transfer function matrix $L(s)$ and the controller $G_c(s)$ can be computed as

$$L(s) = G_c(s)G_p(s) = T_c^{-1}(s)G_p(s) - I \quad (11)$$

and

$$G_c(s) = L(s)G_p^{-1}(s) \quad (12)$$

And the prefilter $G_f(s)$ can be obtained from (9).

In the following, the time-domain approach LQR optimal system is reviewed. Consider a time-domain approach model-following control system as shown in Fig. 2, where the plant contains minimum-phase zeros. Assume its state-space representation is given as

$$\dot{X} = AX + Bu$$
$$Y_o = CX \quad (13)$$

and the feedback control law is given as

$$u = -KX = [-K_p \ \ -K_r]X$$
$$= [-K_p \ \ -K_m \ \ -K_c]X \quad (14)$$

where

$$X^T = [X_p^T \ \ X_r^T] = [X_p^T \ \ X_m^T \ \ X_c^T] \quad (15)$$

$$A = \begin{bmatrix} A_p & 0 \\ 0 & A_r \end{bmatrix}, \quad B = \begin{bmatrix} B_p \\ 0 \end{bmatrix},$$
$$A_r = \begin{bmatrix} A_m & B_m C_c \\ 0 & A_c \end{bmatrix}, \quad (16)$$

$$C = [-C_p \ \ C_m \ \ 0] \quad (17)$$

The problem is to determine the optimal feedback gain vector $K$, such that the following cost function is minimized.

$$V = \int_0^\infty [H^T(t)QH(t) + u^T(t)\Lambda u(t)]dt \quad (18)$$

where $H(t) = MX(t)$ is a linear combination of the states, $Q$ is a positive semidefinite matrix and $\Lambda$ is a positive definite real matrix. It can be shown that the cost functions described in (3) and (18) have the same physical meaning when we choose $M = C$ and $Q = qI$ ($q$ being a positive real number). Following the derivation given by Anderson and Moore (1990), the optimal feedback gain vector $K$ can be obtained as

$$K = \Lambda^{-1}B^T P = [K_p \ \ K_r] = [\Lambda^{-1}B_p^T P_{11} \ \ \Lambda^{-1}B_p^T P_{12}] \quad (19)$$

where

$$P = \begin{bmatrix} P_{11} & P_{12} \\ P_{12} & P_{22} \end{bmatrix} \quad (20)$$

which satisfies the following algebraic Riccati equation

$$A^T P + PA - \Lambda^{-1}PBB^T P + qC^T C = 0 \quad (21)$$

Further, the equation (21) can be decomposed into the following equations:

$$A_p^T P_{11} + P_{11}A_p - \Lambda^{-1}P_{11}B_p B_p^T P_{11} + qC_p^T C_p = 0 \quad (22)$$
$$A_p^T P_{12} + P_{12}A_r - \Lambda^{-1}P_{11}B_p B_p^T P_{12} - qC_p^T C_r = 0 \quad (23)$$
$$A_r^T P_{22} + P_{22}A_r - \Lambda^{-1}P_{12}^T B_p B_p^T P_{12} - qC_r^T C_r = 0 \quad (24)$$

Since $P_{11}$ can be obtained from (22), and $P_{12}$ can be calculated from (23), $K = [K_p \ \ K_r]$ can be computed from (19).

In the following, the conditions of equivalence between the optimal systems designed in time-domain and frequency-domain are proceeded.

## 3. THE EQUIVALENCE OF OPTIMAL SYSTEMS IN TIME AND FREQUENCY DOMAINS

The relationships between the input signal, reference signal and reference model of single-input single-output optimal systems in time-domain and frequency-domain have been discussed by Hsu *et al.* (1994). By extending their results to multi-input multi-output systems, it reveals the following lemma.

**Lemma 1**: For a time-domain approach optimal model-following system shown in Fig. 2, with $\Gamma^2 = q^{-1}\Lambda$, the input signal $R_i(s)$ can be described by a frequency-domain approach optimal method as

$$R_i(s) = K_r(sI - A_r)^{-1}X_r(0)$$
$$= -\Gamma^{-1}\left[Y_*^{-1}(s)M_r(s)R(s)\right]_+ \quad (25)$$

Using (9) and (25), it is obtained that

$$G_f(s) = -\Gamma K_r(sI - A_r)^{-1}X_r(0)R^{-1}(s) \quad (26)$$

Since (25) describes the relation between the reference signal $R(s)$ and the input signal $R_i(s)$, it represents the input signal equality.

Let $T_{om}(s)$ denote the overall transfer function matrix of the optimal system obtained by time-domain approach, the system output can be given as $Y_o(s) = T_{om}(s)R(s)$. Using (10) and (25), and referring to Fig. 2, it is obtained that

$$Y_o(s) = -Y^{-1}(s)\Gamma K_r(sI - A_r)^{-1}X_r(0)$$
$$= Y^{-1}(s)\left[Y_*^{-1}(s)M_r(s)R(s)\right]_+ \quad (27)$$

This implies that

$$T_{om}(s) = Y^{-1}(s)\left[Y_*^{-1}(s)M_r(s)R(s)\right]_+ R^{-1}(s) \quad (28)$$

This is the equivalence of closed-loop transfer function matrix.

In the discussion by Hsu *et al.* (1994), the loop controller $G_c(s)$ is placed in the forward path. In this case, the plants employed have to be stable but they can contain non-minimum phase zeros. Hence they can not discuss the equivalence of loop-gain transfer function matrix, i.e. they can not discuss the equivalence of stability margins. However, the stability margins are the most important properties of optimal systems. In the presenting study in this paper, the loop controller $G_c(s)$ is placed in the feedback path as shown in Fig. 1. In this case, the plants employed have to contain minimum-phase zeros but can be stable or unstable. And then the equivalence of loop-gain transfer function matrix can be shown in the following, so as to achieve complete equivalence of optimal systems designed in time-domain and frequency-domain.

For optimal control, by frequency-domain approach, the return difference equality of feedback-loop given in Fig. 1 is described as

$$\Gamma^2 + G_{p*}(s)G_p(s) = [I + G_c(s)G_p(s)]_*\Gamma^2[I + G_c(s)G_p(s))] \quad (29)$$

and for optimal control by time-domain approach, the return difference equality shown in Fig. 2 can be described as

$$\Lambda + [C_p(sI - A_p)^{-1}B_p]_*(qI)[C_p(sI - A_p)^{-1}B_p] = [I + K_p(sI - A_p)^{-1}B_p]_*\Lambda[I + K_p(sI - A_p)^{-1}B_p] \quad (30)$$

The equivalence of these two approaches can be obtained by setting $\Gamma^2 = q^{-1}\Lambda$ and $L(s) = G_c(s)G_p(s) = K_p(sI - A_p)^{-1}B_p$. Based on this equivalence, the loop-gain transfer function matrices, both in time-domain and frequency-domain optimal systems, can satisfy the return difference equality, such that they possess the same stability margins. Moreover, the plants employed for both optimal systems are also consistent.

## 4. EXAMPLE

Consider a model-following system with step input. The unstable plant is described as

$$Gp(s) = \frac{1}{(s-5.441)(s+7.651)}\begin{bmatrix} -17.25(s+1.766) & -1.58(s+8.179) \\ -0.17(s+101.33) & -0.25(s+7.127) \end{bmatrix} \quad (31)$$

Selecting $\Gamma^2 = I$, from (5), it is obtained that

$$Y(s) = \begin{bmatrix} -0.0618s - 1.5517 & 0.3907s - 0.2623 \\ 0.0420s + 4.1148 & -4.2657s - 8.2397 \end{bmatrix} \quad (32)$$

From (10), it can be solved that

$$T_c(s) = \frac{1}{(s^2 + 22.3706s + 56.0749)}\begin{bmatrix} -17.25s - 33.3205 & -1.58s + 1.0607 \\ -0.17s - 16.6400 & -0.25s - 6.2751 \end{bmatrix} \quad (33)$$

From (11), $L(s)$ is computed as

$$L(s) = \frac{1}{(s-5.441)(s+7.651)}\begin{bmatrix} 19.7549s + 93.4247 & 2.6202s + 20.5219 \\ 2.6202s + 16.5863 & 0.4056s + 3.1316 \end{bmatrix} \quad (34)$$

Solving (12), the feedback controller is obtained as

$$G_c(s) = \begin{bmatrix} -1.1111 & -3.4584 \\ -0.1449 & -0.7064 \end{bmatrix} \quad (35)$$

Assume that the reference model is given as

$$M_r(s) = \begin{bmatrix} \dfrac{9}{s^2 + 5.4s + 9} & 0 \\ 0 & \dfrac{0.64}{s^2 + 1.44s + 0.64} \end{bmatrix} \quad (36)$$

then, using (9), the prefilter $G_f(s)$ is obtained as

$$G_f(s) = \begin{bmatrix} \dfrac{0.0765s^2 + 0.3503s - 5.3480}{s^2 + 5.4s + 9} & \dfrac{-0.0165s^2 - 0.0701s - 0.1899}{s^2 + 1.44s + 0.64} \\ \dfrac{0.0304s^2 + 0.2217s + 0.1704}{s^2 + 5.4s + 9} & \dfrac{-0.0057s^2 - 0.0245s - 0.0716}{s^2 + 1.44s + 0.64} \end{bmatrix} \quad (37)$$

The previous results can be also obtained by the time-domain optimal control approach. Let the state-space representation of $G_p(s)$ be

$$\dot{X}_p = A_pX_p + B_pu$$
$$Y_o = C_pX_p \quad (38)$$

where

$$A_p = \begin{bmatrix} -16.9540 & -27.3418 \\ 7.6199 & 14.7440 \end{bmatrix},$$

$$B_p = \begin{bmatrix} 0.7697 & 0.0448 \\ -0.3968 & -0.0386 \end{bmatrix}$$

and

$$C_p = \begin{bmatrix} -3.2297 & 37.2109 \\ 7.7614 & 15.4844 \end{bmatrix} \quad (39)$$

Solving the Riccati equation, it is obtained that

$$K_p = \begin{bmatrix} -23.2532 & -94.8971 \\ -5.0156 & -16.3334 \end{bmatrix} \quad (40)$$

The feedback controller $G_c(s)$ can also be computed by $K_p(sI - A_p)^{-1}BG_p^{-1}(s)$, which is the same as the one obtained by the frequency-domain optimal control design given in (35). Considering previous reference model, it is given that

$$A_m = \begin{bmatrix} -6.0780 & 0 & 0 & -17.85 \\ 0 & -1.9258 & -1.8025 & 0 \\ 0 & 0.8741 & 0.4858 & 0 \\ 0.7351 & 0 & 0 & 0.6780 \end{bmatrix},$$

$$B_m = \begin{bmatrix} 0.7289 & 0 \\ 0 & 0.1805 \\ 0 & 0 \\ 0 & 0 \end{bmatrix}$$

and

$$C_m = \begin{bmatrix} 0 & 0 & 0 & 16.7970 \\ 0 & 0 & 6.7477 & 0 \end{bmatrix} \tag{41}$$

Then $A_r$ and $C_r$ can be written as

$$A_r = \begin{bmatrix} -6.078 & 0 & 0 & -17.85 & 0.7289 & 0 \\ 0 & -1.9258 & -1.8025 & 0 & 0 & 0.1085 \\ 0 & 0.8741 & 0.4858 & 0 & 0 & 0 \\ 0.7351 & 0 & 0 & 0.6780 & 0 & 0 \\ 0 & 0 & 0 & 0 & 0 & 0 \\ 0 & 0 & 0 & 0 & 0 & 0 \end{bmatrix}$$

and

$$C_r = \begin{bmatrix} 0 & 0 & 0 & 16.7970 & 0 & 0 \\ 0 & 0 & 6.7477 & 0 & 0 & 0 \end{bmatrix} \tag{42}$$

For the time-domain approach, using (19), (22) and (23), it is obtained that

$$K_r = \begin{bmatrix} 0.0863 & 0.4275 & 2.1286 & 11.3459 & 0.0765 & 0.0165 \\ -0.0791 & 0.1498 & 0.8 & 0.1194 & -0.0304 & 0.0057 \end{bmatrix} \tag{43}$$

Further, from (26), $G_f(s)$ can be computed as

$$G_f(s) = \begin{bmatrix} \dfrac{0.0765s^2 + 0.3503s - 5.3480}{s^2 + 5.4s + 9} & \dfrac{-0.0165s^2 - 0.0701s - 0.1899}{s^2 + 1.44s + 0.64} \\ \dfrac{0.0304s^2 + 0.2217s + 0.1704}{s^2 + 5.4s + 9} & \dfrac{-0.0057s^2 - 0.0245s - 0.0716}{s^2 + 1.44s + 0.64} \end{bmatrix} \tag{44}$$

This result is the same as that obtained by frequency-domain design approach. The state feedforward gain can also be computed by the frequency-domain design approach. From (27), it is obtained that

$$K_r = \begin{bmatrix} 0.0863 & 0.4275 & 2.1286 & 11.3459 & 0.0765 & 0.0165 \\ -0.0791 & 0.1498 & 0.8 & 0.1194 & -0.0304 & 0.0057 \end{bmatrix} \tag{45}$$

Again, this result is the same as that obtained by the time-domain design approach.

## 5. CONCLUSION

For multi-input multi-output systems, optimal model-following systems can be designed either in time-domain or frequency-domain. Besides showing the equivalence of closed-loop transfer function matrix and input signal, the equivalence of loop-gain transfer function matrix has also been shown, and the equivalence of these two methods is refined completely. In this study the plants employed are also consistent.

## REFERENCES


Anderson, B. D. O. and J. B. Moore (1990). *Optimal control- linear quadratic methods.* Prentice-Hall Inc., Englewood Cliffs, New Jersey.

Chang, S. S. L. (1961). *Synthesis of optimal control systems.* The Maple Press Company, York, P.A..

Hsu, Y. T., , T. P. Tsai and Y. T Juang,. (1994). The equivalence of model-following systems designed in the time domain and frequency domain. *IEEE Transactions on Automatic Control*, **39**, 1722-1726.

Lin C. M. and Z. R. Shi (1996). Quantitative performance robustness linear quadratic optimal system design. *Journal of Guidance, Control, and Dynamic,* **19**, 600-604.

Lin C. M. and M. H. Tu (1996). Multivariable model reference linear quadratic optimal systems. *Automatica,* **32**, 1061-1065.

Youla, D. C., , J. J. Bongiorno Jr. and , H. A Jabr (1976). Modern Wiener-Hopf design of optimal controllers, part II : the multivariable case. *IEEE Transactions on Automatic Control*, **21**, 319-338.

Youla, D. C. and J. J. Bongiorno Jr. (1985). A feedback theory of two-degree-of-freedom optimal Wiener-Hopf design. *IEEE Tranactions on Automatic Control* , **30**, 652-665, 1985.

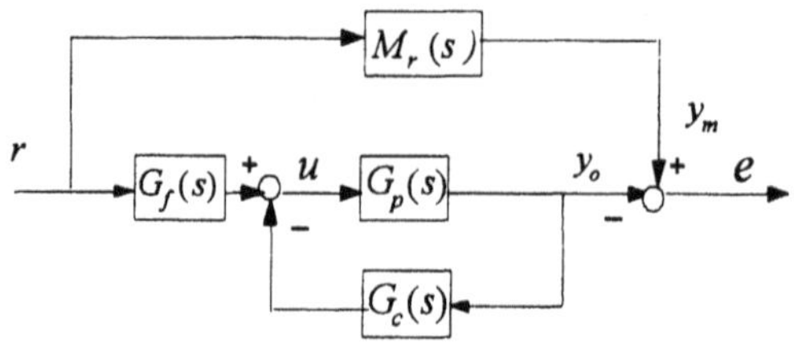


Fig. 1 The block diagram of a frequency-domain approach model-following system

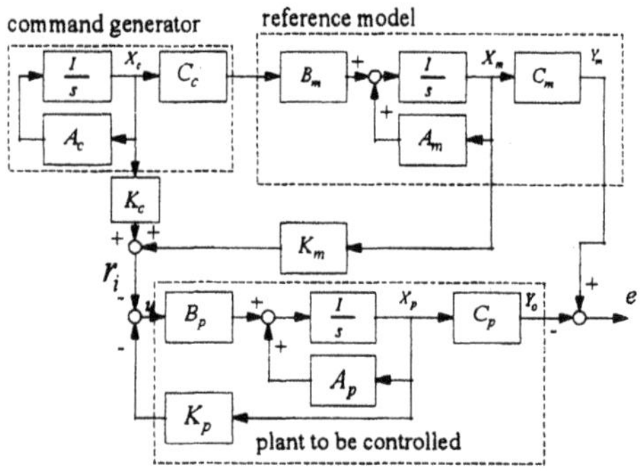


Fig. 2 The block diagram of a time-domain approach model-following system

F-2d-07-4
14th Triennial World Congress, Beijing, P.R. China

# QUADRATIC INDEX ANALYSIS IN PREDICTIVE CONTROL

J.S. Senent* M. Martínez* J. Sanchis*

* *Departamento de Ingeniería de Sistemas y Automática.*
*Universidad Politécnica de Valencia.*
*Camino de Vera, 14*
*E-46022 Valencia*
SPAIN
*e-mail: juanse@isa.upv.es*
*Fax : + 34 963879579*

Abstract: The formulation of the cost index has a remarkable influence in the performance of a predictive controller. In most of predictive control articles and industrial applications the only approximation used is quadratic. These indices produce optimal control actions, as they minimise the objective function, but at the same time, these actions can be suboptimal from the process control point of view. This problem is due to four factors: multiple objectives appear in a single index; it is difficult to find the weight factors; controller performance depends on prediction horizon (HP); and the most important, the index is not a process control oriented. This paper shows, with some simple examples, why these indices are not process control oriented and how to solve some of the problems mentioned using new index formulations. *Copyright © 1999 IFAC*

Keywords: Predictive control, Quadratic performance indices, Performance analysis

## 1. MOTIVATION AND GOALS

### 1.1 *Motivation*

In all the formulations of model predictive control (MPC) the process inputs, or manipulated variables (MV), are computed to optimise the future behaviour of the process. To carry out this idea three components are needed:

- A process model: to predict the future behaviour of the process variables or controlled variables (CV).
- An optimiser: to compute the MV.
- A performance index of the process: objective function used by the optimiser to compute the MV.

The main process model formulations are: impulse response models, step response models (Keyser *et al.*, 1988), transfer functions (Clarke *et al.*, 1987), linear state-space models (Camacho and Bordons, 1995), non-linear state-space models (Michalska and Mayne, 1993), neural networks (Zamarreno and Vega, 1997). As for the optimiser, the more frequently used are: quadratic programming (Chow and Clarke, 1994) non-linear programming (Chen and Allgöwer, 1997), dynamic programming (de Madrid, 1995), genetic algorithms (Martínez *et al.*, 1998) and simulated annealing (Senent *et al.*, 1998). But in the case of the performance index, the quadratic approximation is mainly used.

Is a quadratic index the best aproximation?

Most of the criticisms of this index come from the users of commercial programs that implement

MPC (Qin and Badgwell, 1996) and they can be summarised as follows:

> It becomes increasingly difficult to translate control requirements into relative weights for a single objective function... Even when a consistent set of relative weights can be found, care must be taken to avoid scaling problems.

Is the only possible formulation?

Non-linear MPC (NLMPC) (Michalska and Mayne, 1993) and (Chen and Allgöwer, 1997) makes possible the use of non-linear process models, non-linear constraints and non-quadratic performance indices. But it is difficult, in reference literature to find formulations other than quadratic.

### 1.2 *Goals*

The main goal is to describe the problems caused by the use of quadratic indices in MPC. First an analysis of the main criticisms of the index are stated and then several solutions for each problem are sketched.

This article is organised as follows: in section 2 the main criticisms of the quadratic index approximation are enumerated and explained using a set of simple examples. In section 3 several solutions to the above problems are described and finally a new index formulation is proposed.

## 2. QUADRATIC INDEX ANALYSIS IN MPC

### 2.1 *Index formulation*

In the most frequent formulation, quadratic indices penalise future errors and MV moves (it is also possible to penalise the increment of MV moves). Also, two weight matrices Q and R (academic denomination) are included to improve the robustness and performance of the control loop. The inverse of the Q elements are called *equal concern factors*, and the elements of R are called *move suppression factors* in the commercial programs.

The quadratic index used in most commercial MPC programs (Qin and Badgwell, 1996) is:

$$J = \sum_{j=1}^{HP} \| e(k+j) \|_Q^2 + \sum_{j=0}^{HC-1} \| u(k+j) \|_R^2 \qquad (1)$$

where $e(k+j)$ are the future errors of the process, $u(k+j)$ are future control actions, $HP$ is the length of the prediction horizon, $HC$ is the length of the control actions sequence and $Q$ y $R$ are the weighting matrices. Finally,

$$\| e(k+j) \|_Q^2 = e(k+j)^T Q e(k+j)$$

y

$$\| u(k+j) \|_R^2 = u(k+j)^T R u(k+j)$$

### 2.2 *Criticisms of the quadratic index*

Four criticisms of quadratic indices are found in the reference literature:

(1) **Multiple objectives in a single index:** Two kinds of contributions appear: those coming from the error penalisation and those from the control moves penalisation. But these two contributions are in conflict: aggressive control actions must be applied to obtain small errors, and when control actions are not aggressive then error can be high.

Some commercial MPC programs solve this problem by performing a two-step optimisation (Qin and Badgwell, 1996): first, errors are minimised and then, if several solutions to this optimisation problem exist, the solution with the smallest norm is chosen. Several solutions will exist only when dealing with fat plants (the number of MV is greater than the number of CV) or when the reference trajectory is a funnel.

(2) **Finding weighting factors:** The question is how to find the weight matrices Q and R in equation 1. Even when performing a two-step optimisation the factors to weigh errors or control actions are difficult to find. An economic criteria is used in these cases, but sometimes it is difficult to evaluate the economic cost produced by VC deviations or MV moves (Cifuentes, 1998). Besides the economic criteria, VC units are another reason that leads to weighing or normalising the terms in the index. For instance, one CV can be the temperature in a tank $°C$ and the MV can be the opening percentage %. But even when a consistent set of relative weights can be found, care must be taken to avoid scaling problems that lead to an ill-conditioned solution (Qin and Badgwell, 1996). An ill-conditioned problem arises when small errors produce large MV moves.

(3) **Prediction horizon dependence:** The performance of the control loop depends (once the Q and R matrices are found) on the length of HP. Such length fixes the weights in the index of the transient and steady-state response. And it is remarkable to remind that this parameter is tuned by trial and

error in most commercial programs (Qin and Badgwell, 1996) (*cut and try*).

(4) **It is not control process oriented**. It is not possible to define the closed loop behaviour, for example settling time or steady-state error. Once the Q and R matrices are defined the only tuning parameter is the HP. But when HP is tuned it is not possible to know which contribution -the transient response or the steady-state -will be more important.

In general, the weight of the transient response and the steady-state in the index depends on: the number of terms (the HP length); the weight factors; and the formulation of the terms (usually squared error).

To solve this problem some authors explicitly include the time in the index, for example (Yoon and Clarke, 1993):

$$J = \sum_{j=1}^{HP} \mu\beta^j \parallel e(k+j) \parallel_Q^2 + \sum_{j=0}^{HC-1} \rho\beta^j \parallel u(k+j) \parallel_R^2 \qquad (2)$$

where $\mu \geq 0$, $\rho \geq 0$ y $\beta > 1$ to weigh the steady-state or $\beta < 1$ to weigh the transient response.

However, more tuning parameters appear in this index (equation 2), and although some rules of thumb exist to tune them, they are only for first and second order linear models.

## 2.3 *Illustrative examples*

With the following simple examples we want to illustrate the problems related to points three and four above. It will be shown that the index can make the optimisation algorithm obtain optimal solutions, producing the minimum cost, but they are not the best from the process control point of view [1].

In all the examples the process model will be:

$$\frac{Y(s)}{U(s)} = G(s) = \frac{1}{s^2+s+1} \qquad (3)$$

with $y(0) = 0$, set-point $y_{SP} = 1$ and sample time $T = 1$ (time units).

[1] It is important to remember that most of the available optimisation algorithms (Quasi-Newton or Sequential Quadratic Programming) perform local searches in the solution space. They start evaluating the objective function in one point (or in a set of points) and with the results of the evaluation they compute the search direction where the solution is estimated to be (MathWorks Inc., 1996).

The performance index will be:

$$J_q = \sum_{j=1}^{HP} e^2(k+j) \qquad (4)$$

that comes from equation 1 with $Q = 1$ and $R = 0$. To simplify the analysis, the control actions have been excluded from the index ($R = 0$).

Finally, it is assumed that the operator objective, when the index is programmed, is to obtain a small steady-state error.

[Case 1]

If, in the instant $k$, the optimiser must choose between two possible sequences of control actions $u_1$ and $u_2$ (where $HP = 41$, and $HC = 2$):

$$u_1 = [3, 1, 1, \overset{40}{\smile}, 1] \quad \rightarrow \quad J(u_1)_q = 2.28$$
$$u_2 = [2, 1.1, 1.1, \overset{40}{\smile}, 1.1] \rightarrow \quad J(u_2)_q = 1.92$$

the optimiser would choose $u_2$. But if the evolution of the error is observed (figure 1) this sequence produces a greater steady-state error than the one produced by $u_1$ (although the transient response is better). The lower cost solution is different from the one that the operator would choose. This is because the steady-state behaviour of the process has not enough weight in the index. Therefore, the HP must be longer.

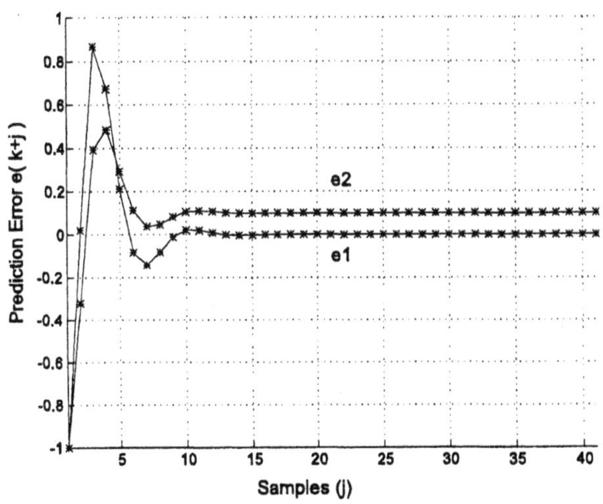


Fig. 1. Temporal evolution of future errors over the HP.

[Case 2]

The same control action sequences as in case 1 are considered, only the HP is changed ($HP = 81$). The sequences and cost index values are:

$$u_1 = [3, 1, 1, \overset{80}{\smile}, 1] \quad \rightarrow \quad J(u_1)_q = 2.28$$
$$u_2 = [2, 1.1, 1.1, \overset{80}{\smile}, 1.1] \rightarrow J(u_2)_q = 2.32$$

As the weight of the steady-state is greater than in case 1, the optimiser would choose the solution expected by the operator.

Due to the fact that the steady-state terms in the index are equal to 0.01 ($0.1^2$) the length of HP must be increased to get these results. One alternative to reduce the number of terms is to use indices without error deformations. For example:

$$J_a = \sum_{j=1}^{HP} |e(k+j)| \qquad (5)$$

The results for the two previous cases would be:

$$\text{Case 1: } J(u_1)_a = 3.15$$
$$J(u_2)_a = 5.98$$
$$\text{Case 2: } J(u_1)_a = 3.15$$
$$J(u_2)_a = 9.98$$

In both cases the minimum cost is found in the sequence whose trajectory finishes in the set-point (operator's objective). Therefore it is not necessary to increase the HP length to weigh the steady-state as was necessary when a squared error index was used. But it is possible that, even with these non-deforming indices, the optimiser chooses sequences different from the operator's point of view.

**[Case 3]**

In this case $HP = 21$ and the control action sequences and costs are:

$$u_1 = [8, 1, 1, \overset{20}{\smile}, 1] \rightarrow$$
$$J(u_1)_q = 20.48 \; J(u_1)_a = 9.96$$
$$u_2 = [0.5, 1.25, 1.25, \overset{20}{\smile}, 1.25] \rightarrow$$
$$J(u_2)_q = 3.11 \; J(u_2)_a = 6.92$$

With both indices the optimiser would choose the $u_2$ sequence, because it has the minimum cost. But only the trajectory generated by $u_1$ finishes in the set-point (operator's objective).

### 2.4 *A complete example*

In this case the two compared sequences will be obtained by different methods. On one hand, the $u_1$ sequence is the best solution computed through the minimisation of equation 4 (using two optimisation algorithms: a quasi-Newton method and the simplex method, both produce the same solution). On the other hand, the $u_2$ sequence is proposed by an expert operator. The results are (with $HP = 21$ and $HC = 2$, $y_{SP} = 5$):

$$u_1 = [7.95, 4.88, 4.88, \overset{20}{\smile}, 4.88] \rightarrow J(u_1)_q = 33.47$$
$$u_2 = [5, 5, 5, \overset{20}{\smile}, 5] \rightarrow J(u_2)_q = 37.58$$

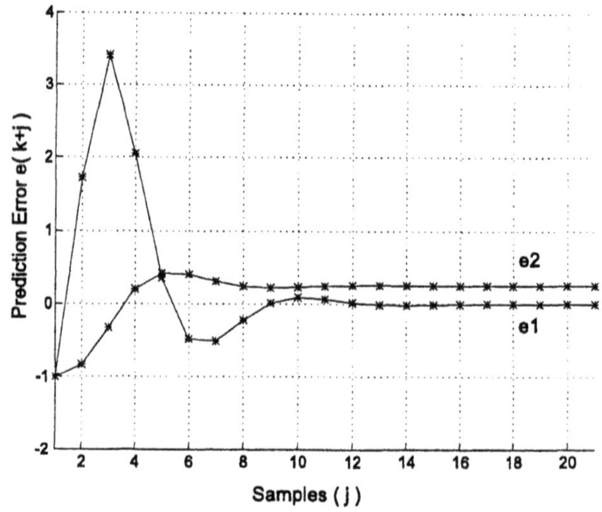


Fig. 2. Temporal evolution of the future errors over the HP.

| | $\sum e_j^2$ | $\sum \lvert e_j \rvert$ | operator |
|---|---|---|---|
| 1 | $u_2$ | $u_1$ | $u_1$ |
| 2 | $u_1$ | $u_1$ | $u_1$ |
| 3 | $u_2$ | $u_2$ | $u_1$ |

Table 1. Summary of the results of the three cases. Each column contains the solution using the three criteria: $\sum e_j^2$, $\sum |e_j|$, operator.

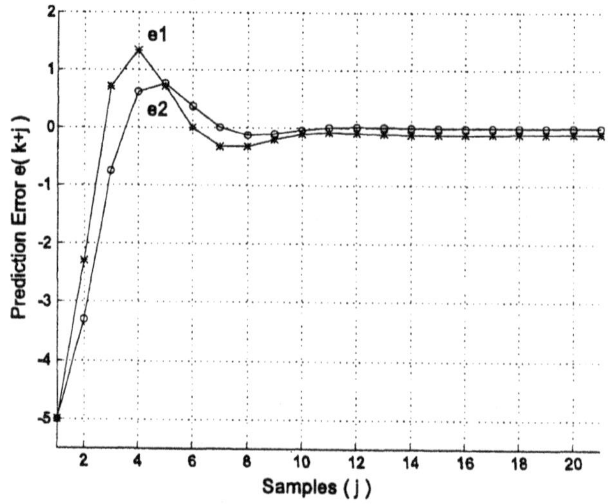


Fig. 3. Temporal evolution of future errors over the HP.

$J(u1)_q < J(u2)_q$ as $u_1$ is the optimal sequence, but the trajectory generated has a steady-state error and more overshoot that the one generated by the operator's sequence $u_2$. Therefore, $u_2$ is better from the process control point of view.

If instead of using the squared error the absolute value of the error appears in the index (equation 5) the results would be:

$$u_1 = [6.47, 4.99, 4.99, \overset{20}{\smile}, 4.99] \rightarrow J(u_1)_a = 10.34$$

$$u_2 = [5, 5, 5, \overset{20}{\smile}, 5] \rightarrow J(u_2)_a = 11.14$$

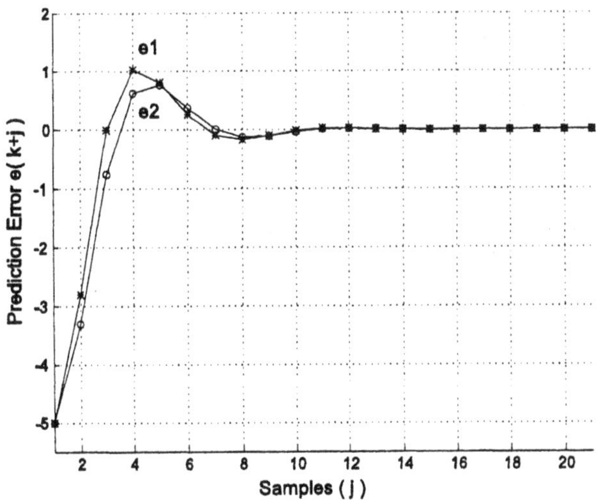


Fig. 4. Temporal evolution of future errors over the HP.

Now, the only difference is that the trajectory generated by $u_1$ has more overshoot than the one generated by $u_2$, as the steady-state error and settling time are practically the same.

Remarks: The weight in the index of the transient response and steady-state is not known and cannot be set before the simulation. Also, when using squared error indices some terms can have more or less weight depending on the error to be greater o lower than 1.

Absolute value indices need fewer terms (a shorter HP) than squared error indices to get the same specifications because there is no error deformation, so computing time is reduced. But those optimisation algorithms based on gradient information present numerical problems when dealing with indices like the one in equation 5. In these cases the computing time is longer than when dealing with indices like equation 1.

## 3. SOLUTIONS

After reviewing the problems or limitations of using these indices we will sketch some solutions for each:

(1) **Multiple objectives in a single index.** This problem will be solved, as in some commercial MPC programs, through the two-step optimisation procedure mentioned above.

(2) **Prediction horizon dependence.** Using indices with an infinite or quasi-infinite HP (Michalska and Mayne, 1993) and (Chen and Allgöwer, 1997) will solve the problem.

(3) **It is not control process oriented.** The terms in the index should be replaced by variables representing the performance of the CV, without HP dependence. As an example, the time one CV needs to achieve the set-point within some error limits (settling time) could be used, for instance set-point=$20^oC \pm 1^oC$. The index formulation could be:

$$J = \sum_{j=1}^{n} \alpha_j |st(j) - stsp(j)| \qquad (6)$$

where:

| | |
|---|---|
| $n$ | is the number of CV. |
| $\alpha_j$ | is the weight factor. |
| $st(j)$ | is the settling time predicted for CV $j$. |
| $stsp(j)$ | is the settling time desired for CV $j$. |

(4) **Finding weight factors.** The number of weight factors must be reduced or, at least, they should be directly understood by the operator. For example, the above index (equation 6) is a good candidate. The weight factors $\alpha_j$ are found easily as all the terms have the same units (time units).

## 4. CONCLUSIONS

The main problems using quadratic indices in predictive control are: they have multiple objectives in a single index; it is difficult to find the weight factors; the performance of the controller depends on the HP; and the most important, it is not control process oriented.

In this article it is shown, with four simple examples, how the last two problems can lead the optimiser to obtain suboptimal solutions from the process control point of view.

In these examples it is also shown that using squared error indices produces worse solutions than using absolute value indices (with the same HP).

Finally, in section 3 several solutions to the mentioned problems are described. Also, a new index is presented. This index has none of the problems mentioned above, but the computing time needed to obtain the optimal solution is greatly increased.

After the analysis of the problems, the next step will be the use of indices which recommendations in section 3, but avoiding:

- A huge computational time, and so allowing their use in real applications.
- Numerical problems due to the non-continuity or non-derivability of the objective function.

As an alternative to the current optimisation tools used in predictive control, we propose the heuristic optimisation tools (for example, simulated annealing or genetic algorithms) because they are numerically robust and the computing time is not very high when dealing with complex indices.

## 5. REFERENCES


Camacho, E.F. and C. Bordons (1995). *Model Predictive Control in the Process Industry.* Springer.

Chen, H. and F. Allgöwer (1997). A quasi-infinite horizon nonlinear model predictive control scheme with guaranteed stability. *Accepted for Automatica.*

Chow, C.M. and D.W. Clarke (1994). Actuator nonlinearities in predictive control. In: *Advances in Model-Based Predictive Control* (D. W. Clarke, Ed.). pp. 245–260. Oxford University Press.

Cifuentes, F. (1998). Developing multivariable control proyects in the REPSOL company. Technical report. Repsol Company.

Clarke, D.W., C. Mohtadi and P.S. Tuffs (1987). Generalized Predictive Control-Part I. *Automatica* **23**(2), 137–148.

de Madrid, A. Pérez (1995). Aplicación de técnicas de programación dinámica a control predictivo basado en modelos. PhD thesis. Universidad Nacional de Educación a Distancia. Facultad de Ciencias.

Keyser, R.M.C. De, G.A. Van de Velde and F.A.G. Dumortier (1988). A Comparative Study of Self-adaptive Long-range Predictive Control Methods. *Automatica* **24**(2), 149–163.

Martínez, M., J.S. Senent and F.X. Blasco (1998). Genelarized Predictive Control Using Genetic Algorithms (GAGPC). *Engineering Applications of Artificial Intelligence.*

MathWorks Inc. (1996). *Optimization Toolbox. User's Guide. Version 5.* The MathWorks Inc.

Michalska, H. and D.Q. Mayne (1993). Robust Receding Horizon Control of Constrained Nonlinear Systems. *IEEE Trans. on Automatic Control* **38**(11), 1623–1633.

Qin, S. Joe and T. A. Badgwell (1996). An Overview of Industrial Model Predictive Control Technology. Technical report. Dept. of Chemical Engineering, The University of Texas at Austin.

Senent, J.S., M.A. Martínez, X. Blasco and J. Sanchis (1998). MIMO predictive control of temperature and humidity inside a greenhouse using simulated annealing (SA) as optmizer of a multicriteria index. In: *Lecture Notes in Artificial Intelligence 1416.*. Vol. II. pp. 271–279. Springer.

Yoon, T.W. and David W. Clarke (1993). Receding-horizon predictive control with exponential weighting. *Int. J. Systems Sci.* **24**(9), 1745—1757.

Zamarreno, J.M. and P. Vega (1997). Identification and predictive control of a melter unit from the sugar industry using recurrent neural nets. *Artificial Intelligence in Engineering* (11), 365–373.

# SUBOPTIMALITY ANALYSIS OF RECEDING HORIZON PREDICTIVE CONTROL WITH TERMINAL CONSTRAINTS

**Xiaojun Geng*** and **Yugeng Xi****

*Institute of Automation, Shanghai Jiao Tong University,*
*Shanghai 200030, P. R. China*
*Email: *xjgeng@hotmail.com; **ygxi@mail.sjtu.edu.cn*

Abstract: In this paper, the suboptimality of predictive control with terminal constraints is analyzed. and compared with traditional optimal control. First, by studying the property of receding finite-horizon optimization, the upper bound of suboptimality of predictive control for nonlinear system is deduced. The result is then applied to linear system, and a quantitative measure of suboptimality is obtained. *Copyright© 1999 IFAC.*

Keywords: predictive control; optimal control; suboptimal systems; nonlinear systems.

## 1. INTRODUCTION

The optimal control theory has put forward the modern control theory. However, its application to the industry was limited because the industrial process is always time-varying and has uncertainty. Predictive control, developed in 70's, improved the optimal control by adopting on-line receding horizon optimization and feedback information from real process, and has proved successful to the complicated industrial processes.

Stemming from receding horizon concept with terminal constraint, Kwon and Pearson (1977, 1978) designed a stable feedback controller for linear discrete time-varying systems, and analyzed the relation between it and the standard optimal control strategy in terms of the properties of Riccati equations. Later, Mayne et. al. (1990a) showed that Receding Horizon Predictive Control with Terminal Constraints (CRHPC) can yield a stable closed-loop system when applied to continuous nonlinear system, and gave an implementable control algorithm (Mayne, et al., 1990b). Alamir and Bornard (1994) gave some stability results of CRHPC for nonlinear discrete-time systems. Meadows et. al (1995) further extended it to feedback linearizable systems. Later, in order to overcome feasibility difficulties resulting from the terminal constraints, some effective approaches to relax the terminal condition are proposed (see Rossiter, 1994; Rossiter and et. al, 1996). Those works are mainly based on General Predictive Control (GPC).

Freeman and Kokotovic (1996) pointed out that optimal control can guarantee the stability of system if an infinite-horizon optimization index is adopted, whereas with finite horizon optimization index the conclusion does not hold. However, based on receding finite horizon optimization, predictive control can stabilize system under certain conditions. CRHPC, constraining the state to a designed value in the terminal of predictive horizon, is a stable control strategy and has attracted many researchers. The study on CRHPC can be classified into two categories, one is based on system I/O model and required to track a expected output, the other is based on state space model and required to drive all the states to origin. Suppose that the system is observable, then the substance of above two problems is congruous. In order to highlight the approach in this paper, only the state-regulating problem is considered.

It's obvious that optimal control achieves global optimal because the whole control horizon is considered. However, in receding horizon control, at each stop only a finite horizon optimization is considered, so receding horizon control can't guarantee global optimality. Thus, the suboptimality of CRHPC deserves our attention.

In this paper, based on above stable strategies, the suboptimality of CRHPC for discrete nonlinear systems with respect to the traditional optimal control is analyzed. Section 2 gives the description of receding horizon predictive control with terminal constraints. In section 3, the formulation of suboptimality is presented, and the main result of this paper, that is, the upper bound of suboptimality is given. The result is applied to linear systems in section 4 and a quantitative measure for suboptimality is obtained. Finally, some concluding remarks are given in section 5.

## 2. RECEDING HORIZON PREDICTIVE CONTROL WITH TERMINAL ONSTRAINTS

Consider the general discrete nonlinear system described by

$$x(k+1) = f(x(k),u(k)) \tag{1}$$

where $x(0)$ is known, and $x(k) \in R^n, u(k) \in R^m$. $f$ is continuous in origin satisfying $f(0,0) = 0$.

At time k, given current state $x(k)$, let $\pi_N(k)$ represent a $N$ step control sequence:

$$\pi_N(k) = \{v_0(k), v_1(k), \cdots, v_{N-1}(k)\}$$
$$v_i(k) \in R^m$$

The state resulted from (1) with input sequence $\pi_N(k)$ is given as:

$$\{z_0(k), z_1(k), \cdots, z_N(k)\}, z_i(k) \in R^n$$

And the corresponding finite horizon optimization index is defined by

$$J_1(k) \overset{\Delta}{=} J(x(k), \pi_N(k)) = \sum_{j=0}^{N-1} L(z_j(k), v_j(k)) \tag{2}$$

The optimal control problem at $k$ then takes the form:

$$\begin{aligned} &\min_{\pi_N(k)} J_1(k) \\ &s.t. \quad z_{j+1}(k) = f(z_j(k), v_j(k)) \\ &\qquad z_0(k) = x(k), z_N(k) = 0 \end{aligned} \tag{3}$$

where $z_j(k), v_j(k)$ denote the predictive state and predictive control concerned in the course of current finite horizon optimization respectively. In addition, $L$ is continuous in (0,0) and has the following properties (Meadows et al. 1995):

a) $L(0,0)=0$
b) There exists a non-decreasing $\gamma : [0,\infty) \to [0,\infty)$, such that $\gamma(0) = 0$ and
$$0 < \gamma(\|x,u\|) \le L(x,u)$$

for all $(x,u) \ne (0,0)$. ($\|\cdot,\cdot\|$ represents a norm on $R^n \times R^m$). These lead to the following additional properties of $L$:

c) $L(x,u) > 0; \quad \forall (x,u) \ne (0,0)$
d) $L(x,u) \to 0 \quad \Rightarrow (x,u) \to 0$
e) $L(x,u) = 0 \quad \Rightarrow (x,u) = (0,0)$

Predictive control as a receding horizon optimization strategy is described as: at time $k$, obtain the optimal control sequence

$$\pi_N^*(k) = \{v_0^*(k), v_1^*(k), \cdots, v_{N-1}^*(k)\}$$

by solving the problem (3), and implement the current control move

$$u(k) = v_0^*(k) \tag{4}$$

to the system. In the next step $k+1$, repeat the above procedure based on the new initial state $x(k+1)$, and so on. The control strategy yields a closed-loop stable system. The stability results concerned can be found in Alamir et. al. (1994) and Meadows et. al (1995).

## 3. THE SUBOPTIMALITY OF CRHPC FOR NONLINEAR SYSTEMS

To analyze the suboptimality of CRHPC, the optimal control performance with infinite horizon is taken as a standard. Consider system (1), if we can obtain a control sequence $\{u(k), k = 0,1,\cdots\}$, such that the following performance index

$$J_0 = \sum_{k=0}^{\infty} L(x(k), u(k)) \quad (5)$$

exists and is minimized. Then the control sequence is optimal, where $L$ is congruous with that in section 2. Note that the performance index of infinite horizon optimal control satisfies

$$J_0^* = \sum_{k=0}^{\infty} L(x^*(k), u^*(k)) < \infty .$$

There exists $\lim_{k \to \infty} L(x^*(k), u^*(k)) = 0$ in terms of the property c) of $L$. Then

$$(x^*(k), u^*(k)) \to (0,0)$$

so the control system is stable.

In order to analyze the suboptimality of predictive control strategy above, it is necessary to study the relation between $J_0^*$ and the global performance index of predictive control $J_{10}$,

$$J_{10} = \sum_{k=0}^{\infty} L(x(k), u(k)) \quad (6)$$

where $x(k), u(k)$ is the real state and control value during the whole receding horizon predictive control. Starting from this point, the main results will be presented below.

Solving the finite-horizon optimization problem at time $k$, we can obtain $N$ step optimal control $\pi_N^*(k) = \{v_0^*(k),\ v_1^*(k), \cdots, v_{N-1}^*(k)\}$, and the corresponding state sequence described as $\{z_0^*(k), z_1^*(k), \cdots, z_N^*(k)\}$, where $z_N^*(k) = 0$, The optimal performance index in this step is defined as $J_1^*(k)$.

Based on receding optimization, only $v_0^*(k)$ as the current control is implemented, that is $u(k) = v_0^*(k)$. After that the state value at time $k+1$ is obtained as $x(k+1) = f(x(k), v_0^*(k))$, which is $z_1^*(k)$. By repeating the procedure above, we get $\pi_N^*(k+1)$ and $J_1^*(k+1)$, in addition, and implement $v_0^*(k+1)$ as $u(k+1)$, and so on. Consider a control sequence at time $k+1$ defined as follows:

$$\begin{aligned} &\pi_N(k+1) \\ &= \{v_0(k+1), \cdots, v_{N-2}(k+1), v_{N-1}(k+1)\} \\ &= \{v_1^*(k), \cdots, v_{N-1}^*(k), 0\} \end{aligned}$$

Let $\{z_1(k+1), \cdots z_N(k+1)\}$ denote the state resulted by $\pi_N(k+1)$. Clearly,

$$\begin{aligned} &z_1(k+1) = z_2^*(k), \\ &\ldots\ldots, \\ &z_{N-1}(k+1) = z_N^*(k) = 0, \\ &z_N(k+1) = f(z_{N-1}(k+1), \\ &z_N(k+1) = f(z_{N-1}(k+1), \\ &v_{N-1}(k+1)) = f(0,0) = 0 \end{aligned}$$

which satisfies the terminal constraints, so $\pi_N(k+1)$ is a feasible control sequence with respect to the optimization problem (3) at time $k+1$. The corresponding performance index is defined as $J_1(k+1)$. Comparing $J_1(k+1)$ with the optimal performance index in $k+1$, there exists

$$J_1(k+1) \geq J_1^*(k+1)$$

Note that:

$$\begin{aligned} &J_1(k+1) \\ &= \sum_{j=0}^{N-1} L(z_j(k+1), v_j(k+1)) \\ &= \sum_{j=0}^{N-2} L(z_{j+1}^*(k), v_{j+1}^*(k)) \\ &= \sum_{j=0}^{N-1} L(z_j^*(k), v_j^*(k)) - L(z_0^*(k), v_0^*(k)) \\ &= J_1^*(k) - L(x(k), u(k)) \end{aligned}$$

We have,

$$J_1^*(k) - J_1^*(k+1) \geq L(x(k), u(k)) \quad (7)$$

where the items on the left-hand represent the finite-horizon optimal performance index (2) in step $k$ and $k+1$ respectively, and that on the right-hand corresponds to the real value implemented.

Summing (7) from $k$=0 to $k$= $k^*$, then

$$J_1^*(0) - J_1^*(k^*) \geq \sum_{i=0}^{k^*-1} L(x(k), u(k)) \quad (8)$$

When $k^* \to \infty$, we get $\lim_{k^* \to \infty} J_1^*(k^*) = 0$ because the RHC controller is a stable control strategy (Alamir and Bornard , 1994; Meadows, et al. 1995). Thus

$$J_1^*(0) \geq \sum_{k=0}^{\infty} L(x(k),u(k)) \qquad (9)$$

where the term on the right-hand is just the global performance index of receding horizon predictive control. Then we have the following:

**Theorem 1** Considering CRHPC for discrete nonlinear system (1), if control law (4) is adopted, there exists a upper bound for global performance index (6), which is the finite-horizon optimal index of the first step optimization of (3).

**Remarks:** In practice, the controller is designed not only to guarantee the stability of closed-loop system, but also to meet some requirements of performance index. Considering linear systems with uncertainty, Petersen and Mcfarlane (1994) approached so-called optimal guaranteed cost control to guarantee the closed-loop index $J \leq x_0^T P x_0$, where $x_0$ is the initial state value and $P$ is a positive-definite matrix satisfying some certain conditions. This inequality is similar to our suboptimality relation. From this point of view, CRHPC also realizes a guaranteed cost control of state feedback actually, for its closed-loop cost, which is the global performance index, is smaller than a certain value. Moreover, this upper bound is dependent on optimization horizon, which will be explained in the following:

**Corollary 1** The suboptimality of CRHPC relates to the optimization horizon $N$. The greater $N$, the better the suboptimality.

**Proof** First, take the optimization horizon as $N$, we obtain the optimal control sequence through solving the first step finite horizon optimization, defined as $\pi_N^*(0) = \{v_0^*(0), \cdots, v_{N-1}^*(0)\}$, then the corresponding state resulted from it is $\{z_0^*(0), z_1^*(0), \cdots, z_N^*(0)\}$, and the optimal performance index is $J_{1N}^*(0)$ 。

Then take the optimization horizon as $M(M > N)$, and define the first step optimal performance index as $J_{1M}^*(0)$. Similar to the previous, the M-step control sequence at time 0 $\pi_M(0) = \{v_0^*(0), \cdots, v_{N-1}^*(0)\ ,0, \cdots, 0\ \}$ is a feasible control with respect to the terminal constraints and its performance index is defined as $J_{1M}$. Hence,

$$\begin{aligned} J_{1N}^* = J_{1M} &= \min \sum_{i=0}^{N-1} L(z_i, v_i) \geq J_{1M}^* \\ &= \min \sum_{i=0}^{M-1} L(z_i, v_i) \end{aligned}$$

It shows the greater the optimization horizon, the smaller the first step finite horizon optimal index, so the better the suboptimality of the CRHPC.

**Corollary 2** The CRHPC can approach the global optimal control performance arbitrarily.

**Proof** Let $J_{10} = \sum_{k=0}^{\infty} L(x(k),u(k))$. It follows from theorem 1 that

$$J_1^*(0) \geq J_{10} \geq J_0^* \qquad (10)$$

where $J_0^*$ denotes the global performance index of traditional optimal control. If $N \to \infty$, $J_1^*(0) \to J_0^*$, that means, for any small $\varepsilon > 0$, there exists a $N^*$ so that if $N > N^*$, $|J_1^*(0) - J_0^*| < \varepsilon$. Hence

$$|J_{10} - J_0^*| \leq |J_1^*(0) - J_0^*| < \varepsilon$$

So, $J_{10}$ can approach $J_0^*$ with arbitrary precision. They are equivalent when $N \to \infty$.

## 4. THE SUBOPTIMALITY OF CRHPC FOR LINEAR SYSTEM

Apply theorem 1 to the following linear discrete system

$$x(k+1) = Ax(k) + Bu(k) \qquad (11)$$

and $(A,B)$ is controllable. The finite horizon optimization problem defined as

$$\min J_1 = \min \sum_{i=0}^{N-1} \{z^T(i)Qz(i) + v^T(i)Rv(i)\}$$

$$\begin{aligned} s.t. \quad & z(0) = x(k) \\ & z(N) = 0 \\ & z(k+1) = Az(k) + Bv(k) \end{aligned}$$

On one hand, in terms of the knowledge of linear systems we know that

$$J_1^*(x_0) = x_0^T P(0) x_0$$

where $x_0 = x(0)$ and $P(0)$ is the solution of the following discrete-time Riccati equation

$$P(k) = A^T P(k+1)A + Q - A^T P(k+1)B(B^T P(k+1)B + R)^{-1} B^T P(k+1)A$$
$$P(N) = \infty \qquad (12)$$

Since $P(N) = \infty$ will lead to confusion in computation, we can derive a recursion for $P(t) = K(t)^{-1}$ to calculate $P(0)$ (Kwon and Pearson, 1978).

On the other hand, for infinite horizon optimal control, we obtain $J_0^*(x_0) = x_0^{\ T} P x_0 = \|x_0\|_P^2$, where $P$ is the solution of the algebra Riccati equation. It follows from the monotonicity of the solution of Riccati equation

$$\|x_0\|_P^2 < \|x_0\|_{P(0)}^2$$

From theorem 1, we obtain

$$J_{10} \leq J_1^*(x_0) = \|x_0\|_{P(0)}^2$$

To sum up, we have the following results:

**Theorem 2** Considering RHC with terminal constraints for linear discrete systems, there exists a upper bound $\|x_0\|_{P(0)}^2$ for global performance index corresponding to optimal control, which $x_0$ denotes the initial state value and $P(0)$ denotes the solution of discrete Riccati equation (12).

**Remarks:** According to the above analysis, the suboptimality of CRHPC for linear discrete systems can be measured by

$$\frac{J_{10}}{J_0^*(0)} = \frac{\|x_0\|_{P(0)}^2}{\|x_0\|_P^2} \qquad (13)$$

where $P$ denotes the solution of (12) with $P_k = P_{k+1} = P$. The result is identical with that of Kwon's obtained through analyzing the properties of Ricatti equation (Kwon and Pearson, 1978); The latter method is only suitable for linear system, whereas the results of this paper can be applied to a larger scope.

## ACKNOWLEDGEMENT

This paper was supported by the Chinese National Natural Science Foundation.

## 5. CONCLUSION

The control performance obtained by optimal control is optimal globally. However, predictive control strategy is suboptimal for it uses rolling optimization strategy with finite horizon index. In this paper, for general discrete nonlinear systems, we analyzed the property of RHC with terminal constraints and deduced the upper bound of suboptimality. In particular, we got a quantitative measure of suboptimality of CRHPC for linear systems. The result is congruous with that of Kwon's. Clearly, these results are valuable for predictive control theoretically.

## REFERENCES:


Alamir M. and Bornard G. (1994), On the Stability of Receding Horizon Control of Nonlinear Discrete-Time Systems, *Systems & Control Letters*, **Vol.23**, pp.291-296.

Edward S. Meadows, Michael A. H., and et al. (1995), Receding Horizon Control and Discontinuous State Feedback Stabilization, *Int. J. Control*, **Vol. 62**, pp. 1217-1299.

Freeman and P. V. Kokotovic (1996). Inverse optimality in robust stabilization, *SIMA J. Control and Optimization*, **Vol. 34**, pp. 1365-1391.

Kwon, W. H., and A. E. Pearson (1977), A Modified Quadratic Cost Problem and Feedback Stabilization of a Linear System, *IEEE Trans. Automat. Contr.*, **Vol. 22**, pp. 838-842.

Kwon, W. H., and A. E. Pearson (1978), On Feedback Stabilization of Time-varying Discrete Linear Systems, *IEEE Trans. Automat. Contr.*, **Vol. 23**, pp. 479-481.

Mayne, D. Q., and H. Michalska (1990a), Receding Horizon Control of Nonlinear Systems, *IEEE Trans. on Automa. Contr.*, **Vol. 35**, pp. 814-824.

Mayne, D. Q., and H. Michalska (1990b), An Implementable Receding Horizon Control for Stabilization of Nonlinear Systems, *the 29th IEEE Conference on Decision and Control*, pp.3396-3397.

Petersen, I. R., D. C. Mcfarlane (1994), Optimal Guaranteed Cost Control and Filtering for Uncertain Linear Systems, *IEEE Trans. on Automa. Contr.*, **Vol. 39**, pp. 1971-1977.

Rossiter (1994), GPC controllers with guaranteed stability and mean-level control of unstable plant, Proc. 33rd Conf. Decision Contr., pp. 3579-3580.

Rossiter, et al (1996), Infinite Horizon Stable Predictive Control, *IEEE Trans. on Automa. Contr.*, **Vol. 41**, pp. 1522-1527.

# ANALYSIS AND OPTIMAL CONTROL OF TIME-VARYING SYSTEM VIA WAVELETS

**Liu Xiangqian** **Zhang Lin**

Department of Automation , Tsinghua University , Beijing , 100084, P. R. China
liuxq@mail. au. tsinghua. edu. cn , zhlin@mail. au. tsinghua. edu. cn

**Abstract**: In this paper, Haar wavelet integral operational matrix is introduced and applied to analyze and optimal control linear time-varying system. The method converts the original problems to the solution of linear algebraic equations. Hence, computational difficulties are considerably reduced. Based on the property of time-frequency localization of Haar wavelet bases, the solution of a system includes both the frequency information and the time information. Other orthogonal functions do not have this property. Examples are given, and the results are shown to be very accurate and satisfactory. *Copyright © 1999 IFAC*

**Keywords:** Approximate analysis, Algebraic approaches, Boundary value problem, Time-varying systems, Optimal control,

## 1. INTRODUCTION

Orthogonal functions are frequently used by many researehers as a convenient and powerful tool for obtaining the approximate solutions of physical systmes(Mohan and Datta,1995; Lee, L.T and Tsay, Y.F.,1986; Wang, M.L. *et al.*, 1986;). The main feature of this technique is to convert a differential equation into an algebraic equation, and hence the solution of the control problem is either greatly reduced or much simplified accordingly. All previous orthogonal functions, however, are supported on the whole interval $a \le t \le b$. This kind of global support is evidently a drawback for certain analysis work, particularly systems involving abrupt variations or a local function vanishing outside a short interval of time or space. In recent years, Wavelets and wavelet-based analysis have found their way into many different fields of science and engineering (Chen,C.F. and Hsiao,C.H.,1997; Gu,J.S. and Jiang,W.S., 1996; Daubechies, 1992). It could be a possible tool for overcoming the difficulty in communication, physics and image processing.

In this paper, Haar wavelet integral operational matrix is introduced and then applied to the analysis and optimal control of linear time-varying systems. The extension is achieved through representing the product of two wavelet series in a new wavelet series. The method reduces the original problems to solve linear algebraic equations and avoids solving a system of coupled two-point boundary-value problems. Therefore computational difficulties are greatly reduced. Moreover, in the process of solving practical problems, the different resolution bases can be selected on the local time domain in order to understand more clearly system behavior on the local time domain. This advantage results from the property of Haar wavelet localization , and other orthogonal functions do not have this property because they are supported on the whole interval. Examples, which demonstrate the results, are given.

## 2. HAAR WAVELET INTEGRAL OPERATIONAL MATRIX

In this section the integral operational matrix $\mathbf{P}$ will be obtained, it is the basement of the following analysis. For any function $f(t) \in L^2(R)$,the discrete wavelet expansion of $f(t)$ is represented as

$$f(t) = \sum_{m,k} b_{m,k}\psi_{m,k}(t) = \mathbf{b}^T \boldsymbol{\theta}(t), \qquad (1)$$

where $\mathbf{b}$ is Haar wavelet coefficient vector, $\boldsymbol{\theta}(t)$ is a set of Haar wavelets $\psi_{m,k}(t)$ ($m,k \in Z$) which is chosen according to necessity. The two vectors are defined by

$$\mathbf{b} = [b_1, b_2, \cdots, b_m]^T, \quad (2)$$

$$\boldsymbol{\theta}(t) = [\psi_1, \psi_2, \cdots, \psi_m]^T, \quad (3)$$

where $\psi_i$ ($i = 1,2...,m$) is some $\psi_{m,k}(t)$.

Haar wavelet integral operational matrix $\mathbf{P}$ is defined as

$$\int_0^t \boldsymbol{\theta}(t)\mathrm{d}t = \mathbf{P}\,\boldsymbol{\theta}(t), \quad (4)$$

where the element $\mathbf{P}(i,j)$ of the integral operational matrix $\mathbf{P}$ is obtained by the same algorithm as the discrete wavelet expansion, i.e.

$$\mathbf{P}(i,j) = \left\langle \int_0^t \psi_i \mathrm{d}t, \psi_j \right\rangle. \quad (5)$$

Of course, if different bases are chosen, different integral operational matrices will be got. It is the advantage of wavelets. In other words, different bases can be chosen according to different interested intervals . Other orthogonal functions do not have this property.

## 3. ANALYSIS OF LINEAR TIME-VARYING SYSTEM

Consider a linear time-varying system characterized by:

$$\dot{\mathbf{x}}(t) = \mathbf{A}(t)\mathbf{x}(t) + \mathbf{B}(t)\mathbf{u}(t), \quad (6)$$

$$\mathbf{x}(0) = \mathbf{x}_0, \quad (7)$$

where the state variable $\mathbf{x}(t)$ and the input variable $\mathbf{u}(t)$ are respectively $n$-vector and $r$-vector (Note that $\mathbf{u}(t)$ is known when the system is analyzed). The time-varying coefficient matrices $\mathbf{A}(t)$ and $\mathbf{B}(t)$ are $n \times n$ and $n \times r$ matrices, respectively, $\mathbf{x}_0$ is initial vector.

Let Haar wavelet base approximations of $\mathbf{A}(t)$, $\mathbf{B}(t)$, $\mathbf{u}(t)$, $\mathbf{x}_0$, $\mathbf{x}(t)$ be respectively:

$$\mathbf{A}(t) = \sum_{i=1}^{m} \mathbf{A}_i \psi_i = [\mathbf{A}_1 \ \cdots \ \mathbf{A}_m] \begin{bmatrix} \psi_1 \mathbf{I} \\ \vdots \\ \psi_m \mathbf{I} \end{bmatrix}, \quad (8)$$

$$\mathbf{B}(t) = \sum_{i=1}^{m} \mathbf{B}_i \psi_i, \quad (9)$$

$$\mathbf{u}(t) = \mathbf{U}\,\boldsymbol{\theta}(t), \quad (10)$$

$$\mathbf{x}(t) = \mathbf{X}\,\boldsymbol{\theta}(t), \quad (11)$$

$$\mathbf{x}_0 = [\mathbf{k}, 0, \cdots, 0]\,\boldsymbol{\theta}(t), \quad (12)$$

where $\mathbf{U} = [\mathbf{U}_1 \ \mathbf{U}_2 \ \cdots \ \mathbf{U}_m] \in R^{r \times m}$,
$\mathbf{X} = [\mathbf{X}_1 \ \mathbf{X}_2 \ \cdots \ \mathbf{X}_m] \in R^{n \times m}$,

Then

$$\mathbf{A}(t)\mathbf{x}(t) = [\mathbf{A}_1 \ \cdots \ \mathbf{A}_m] \begin{bmatrix} \psi_1 \mathbf{I} \\ \vdots \\ \psi_m \mathbf{I} \end{bmatrix} [\mathbf{X}_1 \cdots \mathbf{X}_m] \begin{bmatrix} \psi_1 \\ \vdots \\ \psi_m \end{bmatrix}.$$

By decomposing $\psi_i \psi_j$ into Haar wavelet base approximations $\varphi_{ij}\,\boldsymbol{\theta}(t)$ yields

$$\mathbf{A}(t)\mathbf{x}(t) = [\mathbf{A}_1 \ \cdots \ \mathbf{A}_m] \begin{bmatrix} \mathbf{XH}_1 \\ \vdots \\ \mathbf{XH}_m \end{bmatrix} \boldsymbol{\theta}(t) = \sum_{i=1}^{m} \mathbf{A}_i \mathbf{XH}_i\,\boldsymbol{\theta}(t) \quad .(13)$$

where

$$\mathbf{H}_i = \begin{bmatrix} \varphi_{i1} \\ \vdots \\ \varphi_{im} \end{bmatrix} \quad .(14)$$

Similarly

$$\mathbf{B}(t)\mathbf{u}(t) = [\mathbf{B}_1 \ \cdots \ \mathbf{B}_m] \begin{bmatrix} \mathbf{UH}_1 \\ \vdots \\ \mathbf{UH}_m \end{bmatrix} \boldsymbol{\theta}(t) = \sum_{i=1}^{m} \mathbf{B}_i \mathbf{UH}_i\,\boldsymbol{\theta}(t) \quad .(15)$$

Integration of equation (6) gives

$$\mathbf{x}(t) - \mathbf{x}(0) = \int_0^t \mathbf{A}(t)\mathbf{x}(t)\mathrm{d}t + \int_0^t \mathbf{B}(t)\mathbf{u}(t)\mathrm{d}t \quad .(16)$$

Substituting equations (11), (12),(13) and (15) into equation (16) yields

$$\mathbf{X}\,\boldsymbol{\theta}(t) - [\mathbf{k}, 0 \cdots, 0]\,\boldsymbol{\theta}(t) = \int_0^t \sum_{i=1}^{m} \mathbf{A}_i \mathbf{XH}_i\,\boldsymbol{\theta}(t)\mathrm{d}t + \int_0^t \sum_{i=1}^{m} \mathbf{B}_i \mathbf{UH}_i\,\boldsymbol{\theta}(t)\mathrm{d}t$$

Substituting (4) into the above equation gets

$$\mathbf{X}\,\boldsymbol{\theta}(t) - [\mathbf{k}, 0, \cdots, 0]\,\boldsymbol{\theta}(t) = \sum_{i=1}^{m} \mathbf{A}_i \mathbf{XH}_i \mathbf{P}\,\boldsymbol{\theta}(t) + \sum_{i=1}^{m} \mathbf{B}_i \mathbf{UH}_i \mathbf{P}\,\boldsymbol{\theta}(t)$$

or

$$\mathbf{X} - \sum_{i=1}^{m} \mathbf{A}_i \mathbf{XH}_i \mathbf{P} = [\mathbf{k}, 0, \cdots, 0] + \sum_{i=1}^{m} \mathbf{B}_i \mathbf{UH}_i \mathbf{P} \ .$$

Therefore, the solution of equation (6) can be found by

$$\begin{bmatrix} \mathbf{X}_1 \\ \mathbf{X}_2 \\ \vdots \\ \mathbf{X}_m \end{bmatrix} = \left[ \mathbf{I} - \sum_{i=1}^{m} ((\mathbf{H}_i \mathbf{P})^T \otimes \mathbf{A}_i) \right]^{-1} \left( \begin{bmatrix} \mathbf{v}_1 \\ \mathbf{v}_2 \\ \vdots \\ \mathbf{v}_m \end{bmatrix} + \sum_{i=1}^{m} ((\mathbf{P}^T \mathbf{H}_i^T) \otimes \mathbf{B}_i) \begin{bmatrix} \mathbf{u}_0 \\ \mathbf{u}_1 \\ \vdots \\ \mathbf{u}_{m-1} \end{bmatrix} \right) \quad (17)$$

where $\mathbf{I}$ denotes the identity matrix, $\otimes$ denotes the Kronecker product, and $\mathbf{v}_i$ is the $i$-th column of matrix $\mathbf{V}$ which is defined by

$$\mathbf{V} = [\mathbf{k}, 0, \cdots, 0] \ .$$

Let $\bar{\mathbf{X}}$ donates the vector obtained by putting matrix $\mathbf{X}$ into one column, then

$$\bar{\mathbf{X}} = \mathbf{K}\bar{\mathbf{V}} + \mathbf{L}\bar{\mathbf{U}} \quad (18)$$

where

$$\mathbf{K} = \left[ \mathbf{I} - \sum_{i=1}^{m} ((\mathbf{H}_i \mathbf{P})^T \otimes \mathbf{A}_i) \right]^{-1} \quad (19)$$

$$\mathbf{L} = \left[ \mathbf{I} - \sum_{i=1}^{m} ((\mathbf{H}_i \mathbf{P})^T \otimes \mathbf{A}_i) \right]^{-1} \sum_{i=1}^{m} ((\mathbf{P}^T \mathbf{H}_i^T) \otimes \mathbf{B}_i) \quad (20)$$

Consequently, the solution of system (6) is

$$\mathbf{x}(t) = \mathbf{X}\,\boldsymbol{\theta}(t) = [\mathbf{X}_1 \cdots \mathbf{X}_m] \begin{bmatrix} \psi_1 \\ \vdots \\ \psi_m \end{bmatrix}. \quad (21)$$

## 4. OPTIMAL CONTROL OF LINEAR TIME-VARYING SYSTEMS

The control of the linear time-varying system of (6) with respect to a quadratic performance index

$$J = \frac{1}{2}\mathbf{x}^T(t_f)\mathbf{S}\mathbf{x}(t_f) + \frac{1}{2}\int_0^{t_f}[\mathbf{x}^T(t)\mathbf{Q}\mathbf{x}(t) + \mathbf{u}^T(t)\mathbf{R}\mathbf{u}(t)]dt \qquad (22)$$

is considered. The problem is to find the control vector $\mathbf{u}(t)$ to minimize the above cost functional: where $t_f$ is the final time, the matrices $\mathbf{S}$ and $\mathbf{Q}$ are symmetric positive semidefinite and matrix $\mathbf{R}$ is symmetric positive definite.

Let $\mathbf{Y} = \mathbf{X}^T = [\mathbf{Y}_1 \quad \mathbf{Y}_2 \quad \cdots \quad \mathbf{Y}_m]$, from equation (11) yield

$$\mathbf{x}(t_f) = \mathbf{Y}^T\boldsymbol{\theta}(t_f) \qquad (23)$$

The cost functional (22) can be rewritten as

$$J = \frac{1}{2}\boldsymbol{\theta}^T(t_f)\mathbf{Y}\mathbf{S}\mathbf{Y}^T\boldsymbol{\theta}(t_f) + \frac{1}{2}\int_0^{t_f}[\boldsymbol{\theta}^T(t)\mathbf{Y}\mathbf{Q}\mathbf{Y}^T\boldsymbol{\theta}(t) + \boldsymbol{\theta}^T(t)\mathbf{U}\mathbf{R}\mathbf{U}^T\boldsymbol{\theta}(t)]dt = \frac{1}{2}\mathrm{tr}(\mathbf{N}\mathbf{Y}\mathbf{S}\mathbf{Y}^T) + \frac{1}{2}[\mathrm{tr}(\mathbf{M}\mathbf{Y}\mathbf{Q}\mathbf{Y}^T) + \mathrm{tr}(\mathbf{M}\mathbf{U}\mathbf{R}\mathbf{U}^T)] \qquad (24)$$

where tr$(A)$ is the trace of matrix $A$, and

$$\mathbf{M} = \int_{t_0}^{t_f}\boldsymbol{\theta}(t)\boldsymbol{\theta}^T(t)dt, \qquad \mathbf{N} = \boldsymbol{\theta}(t_f)\boldsymbol{\theta}^T(t_f) \qquad (25)$$

Substituting equations (23),(25) and (18) into (24),the cost functional $J$ becomes a function of $u_{ij}$. Thus far, the optimization problem is reduced to that of finding $u_{ij}$ to minimize the cost functional $J$. From the necessary condition of optimization, then gives

$$\frac{\partial J}{\partial u_{ij}} = 0 \qquad i = 1,\cdots,r, j = 1,\cdots,m \qquad (26)$$

From (18) obtains

$$\frac{\partial x_{st}}{\partial u_{ij}} = \mathbf{e}^T_{(t-1)n+s}\mathbf{K}\mathbf{e}_{(j-1)n+i} = \frac{\partial y_{ts}}{\partial u_{ij}}, \ s = 1,\cdots,n, t = 1,\cdots,m \qquad (27)$$

where $\mathbf{e}_i$ is the vector that $i$-th element is 1 and other elements are zeros. Using formula

$$\frac{\partial}{\partial u_{ij}}\mathrm{tr}(\mathbf{M}\mathbf{U}^T\mathbf{R}\mathbf{U}) = \mathbf{e}_i^T\mathbf{R}\mathbf{U}\mathbf{M}\mathbf{e}_j - \frac{1}{2}\mathbf{e}_j^T\mathbf{M}\mathbf{e}_j\mathbf{e}_i^T\mathbf{U}\mathbf{e}_j\mathbf{e}_i^T\mathbf{R}\mathbf{e}_i \qquad (28)$$

then yields

$$\frac{\partial J}{\partial u_{ij}} = \frac{1}{2}\sum_{s=1}^{n}\sum_{t=1}^{m}\frac{\partial}{\partial y_{ts}}(\mathrm{tr}(\mathbf{S}\mathbf{Y}^T\mathbf{N}\mathbf{Y}) + \mathrm{tr}(\mathbf{Q}\mathbf{Y}^T\mathbf{M}\mathbf{Y}))\frac{\partial y_{ts}}{\partial u_{ij}} + \frac{1}{2}\frac{\partial}{\partial u_{ij}}\mathrm{tr}(\mathbf{R}\mathbf{U}^T\mathbf{M}\mathbf{U}) = \frac{1}{2}\sum_{s=1}^{n}\sum_{t=1}^{m}[\mathbf{e}_t^T(\mathbf{N}\mathbf{Y}\mathbf{S} + \mathbf{M}\mathbf{Y}\mathbf{Q})\mathbf{e}_s\frac{\partial x_{st}}{\partial u_{ij}} - (\frac{1}{2}\mathbf{e}_s^T\mathbf{S}\mathbf{e}_s\mathbf{e}_t^T\mathbf{Y}\mathbf{e}_s\mathbf{e}_t^T\mathbf{N}\mathbf{e}_t + \frac{1}{2}\mathbf{e}_s^T\mathbf{Q}\mathbf{e}_s\mathbf{e}_t^T\mathbf{Y}\mathbf{e}_s\mathbf{e}_t^T\mathbf{M}\mathbf{e}_t)\frac{\partial x_{st}}{\partial u_{ij}}] + \frac{1}{2}(\mathbf{e}_i^T\mathbf{R}\mathbf{U}\mathbf{M}\mathbf{e}_j - \frac{1}{2}\mathbf{e}_j^T\mathbf{R}\mathbf{e}_j\mathbf{e}_i^T\mathbf{U}\mathbf{e}_j\mathbf{e}_i^T\mathbf{M}\mathbf{e}_i)$$

$$\frac{\partial J}{\partial u_{ij}} = \frac{1}{2}\sum_{s=1}^{n}\sum_{t=1}^{m}\mathbf{e}_t^T(\mathbf{N}\mathbf{Y}\mathbf{S} + \mathbf{M}\mathbf{Y}\mathbf{Q})\mathbf{e}_s\mathbf{e}^T_{(t-1)n+s}\mathbf{K}\mathbf{e}_{(j-1)n+i} + \frac{1}{2}(\mathbf{e}_i^T\mathbf{R}\mathbf{U}\mathbf{M}\mathbf{e}_j - \frac{1}{2}\mathbf{e}_j^T\mathbf{R}\mathbf{e}_j\mathbf{e}_i^T\mathbf{U}\mathbf{e}_j\mathbf{e}_i^T\mathbf{M}\mathbf{e}_i) - \frac{1}{2}\sum_{s=1}^{n}\sum_{t=1}^{m}(\frac{1}{2}\mathbf{e}_s^T\mathbf{S}\mathbf{e}_s\mathbf{e}_t^T\mathbf{Y}\mathbf{e}_s\mathbf{e}_t^T\mathbf{N}\mathbf{e}_t + \frac{1}{2}\mathbf{e}_s^T\mathbf{Q}\mathbf{e}_s\mathbf{e}_t^T\mathbf{Y}\mathbf{e}_s\mathbf{e}_t^T\mathbf{M}\mathbf{e}_t)\mathbf{e}^T_{(t-1)n+s}\mathbf{K}\mathbf{e}_{(j-1)n+i} = 0$$

After putting the matrix into one column, then

$$\frac{1}{2}\sum_{s=1}^{n}\sum_{t=1}^{m}[(\mathbf{S}\mathbf{e}_{ts}\mathbf{e}^T_{(t-1)n+s}\mathbf{K}\mathbf{e}_{(j-1)n+i})^T \otimes (\mathbf{e}_t^T\mathbf{N}) + (\mathbf{Q}\mathbf{e}_s\mathbf{e}^T_{(t-1)n+s}\mathbf{K}\mathbf{e}_{(j-1)n+i})^T \otimes (\mathbf{e}_t^T\mathbf{M}) - \frac{1}{2}(\mathbf{e}_s\mathbf{e}_t^T\mathbf{M}\mathbf{e}_t\mathbf{e}^T_{(t-1)n+s}\mathbf{K}\mathbf{e}_{(j-1)n+i})^T \otimes (\mathbf{e}_s^T\mathbf{Q}\mathbf{e}_s\mathbf{e}_t^T) - \frac{1}{2}(\mathbf{e}_s\mathbf{e}_t^T\mathbf{N}\mathbf{e}_t\mathbf{e}^T_{(t-1)n+s}\mathbf{K}\mathbf{e}_{(j-1)n+i})^T \otimes (\mathbf{e}_s^T\mathbf{S}\mathbf{e}_s\mathbf{e}_t^T)]\bar{\mathbf{Y}} + \frac{1}{2}[(\mathbf{M}\mathbf{e}_j)^T \otimes (\mathbf{e}_i^T\mathbf{R}) - \frac{1}{2}(\mathbf{e}_j\mathbf{e}_i^T\mathbf{M}\mathbf{e}_i)^T \otimes (\mathbf{e}_j^T\mathbf{R}\mathbf{e}_j\mathbf{e}_i^T)]\bar{\mathbf{U}} = 0$$

Denote

$$\mathbf{E}_{ij} = \frac{1}{2}\sum_{s=1}^{n}\sum_{t=1}^{m}[(\mathbf{S}\mathbf{e}_{ts}\mathbf{e}^T_{(t-1)n+s}\mathbf{K}\mathbf{e}_{(j-1)n+i})^T \otimes (\mathbf{e}_t^T\mathbf{N}) + (\mathbf{Q}\mathbf{e}_s\mathbf{e}^T_{(t-1)n+s}\mathbf{K}\mathbf{e}_{(j-1)n+i})^T \otimes (\mathbf{e}_t^T\mathbf{M}) - \frac{1}{2}(\mathbf{e}_s\mathbf{e}_t^T\mathbf{M}\mathbf{e}_t\mathbf{e}^T_{(t-1)n+s} \cdot \mathbf{K}\mathbf{e}_{(j-1)n+i})^T \otimes (\mathbf{e}_s^T\mathbf{Q}\mathbf{e}_s\mathbf{e}_t^T) - \frac{1}{2}(\mathbf{e}_s\mathbf{e}_t^T\mathbf{N}\mathbf{e}_t\mathbf{e}^T_{(t-1)n+s}\mathbf{K}\mathbf{e}_{(j-1)n+i})^T \otimes (\mathbf{e}_s^T\mathbf{S}\mathbf{e}_s\mathbf{e}_t^T)]$$

$$\mathbf{F}_{ij} = \frac{1}{2}[(\mathbf{M}\mathbf{e}_j)^T \otimes (\mathbf{e}_i^T\mathbf{R}) - \frac{1}{2}(\mathbf{e}_j\mathbf{e}_i^T\mathbf{M}\mathbf{e}_i)^T \otimes (\mathbf{e}_j^T\mathbf{R}\mathbf{e}_j\mathbf{e}_i^T)],$$

hence yields

$$\mathbf{E}_{ij}\bar{\mathbf{Y}} + \mathbf{F}_{ij}\bar{\mathbf{U}} = 0 \qquad (29)$$

Let $i$ go from 1 to $r$, $j$ go from 1 to $m$, $mr$ equations will be obtained. Denote

$$\mathbf{E} = \begin{pmatrix} \mathbf{E}_{11} \\ \vdots \\ \mathbf{E}_{1m} \\ \mathbf{E}_{21} \\ \vdots \\ \mathbf{E}_{2m} \\ \vdots \\ \mathbf{E}_{r1} \\ \vdots \\ \mathbf{E}_{rm} \end{pmatrix}, \qquad \mathbf{F} = \begin{pmatrix} \mathbf{F}_{11} \\ \vdots \\ \mathbf{F}_{1m} \\ \mathbf{F}_{21} \\ \vdots \\ \mathbf{F}_{2m} \\ \vdots \\ \mathbf{F}_{r1} \\ \vdots \\ \mathbf{F}_{rm} \end{pmatrix} \qquad (30)$$

obtain $\quad \mathbf{E}\bar{\mathbf{Y}} + \mathbf{F}\bar{\mathbf{U}} = \mathbf{0} \qquad (31)$

Put (31) and (18) together and use $\mathbf{Y} = \mathbf{X}^T$ then $\mathbf{Y}, \mathbf{X}, \mathbf{U}$ can be obtained. Therefore the corresponding state $\mathbf{x}$ and performance index $J$ under the control $\mathbf{u}(t)$ can be obtained.

## 5. EXAMPLES

Example 1: Consider the following time-varying linear system:

$$\dot{\mathbf{x}}(t) = \begin{pmatrix} 0 & 0 \\ t & 0 \end{pmatrix}\mathbf{x}(t) + \begin{pmatrix} 1 \\ 1 \end{pmatrix}u(t),$$

where input u(t) is unit step and $\mathbf{x}(0) = (0 \quad \frac{2}{3})^T$.

According to the section 3, 16 bases are chosen then the solution of the system are obtained. It is shown by Fig.1(the top figure is the trajectory of x1 and the below figure is the trajectory of x2, the smooth curve is the exact solution and the un continuous line is the approximate solution). More bases are chosen and more accurate solution will be got. This example demonstrates the fast, flexible and convenient capabilities of the new method.

Example 2: Consider the system:

$$\dot{x}(t) = x(t) + u(t),$$
$$x(0) = 1, x(1) = 0, \quad (32)$$

and the cost functional

$$J = \frac{1}{2}\int_0^1 [x^2(t) + u^2(t)]dt \quad (33)$$

According to the section 4, 16 bases are chosen then the optimal control law satisfied the given conditions are obtained. It is shown by Fig 2(the smooth curve is the exact solution and the uncontinous line is the approximate solution, the top figure is the trajectory of system and the below one is the control law). This example shows the sharpness and effect of the approach.

## 6. CONCLUSIONS

In this paper the algorithm to get Haar wavelet integral operational matrix is introduced, and the analysis and optimal control of linear time-varying systems are studied. Because of the property of time-frequency localization of Haar wavelet bases, the solution of a system includes both the frequency information and the time information. This point is significant in control system analysis and synthesis. Furthermore, in order to study the system behavior in the local time domain very well, the different resolution bases can be selected in the local time domain.

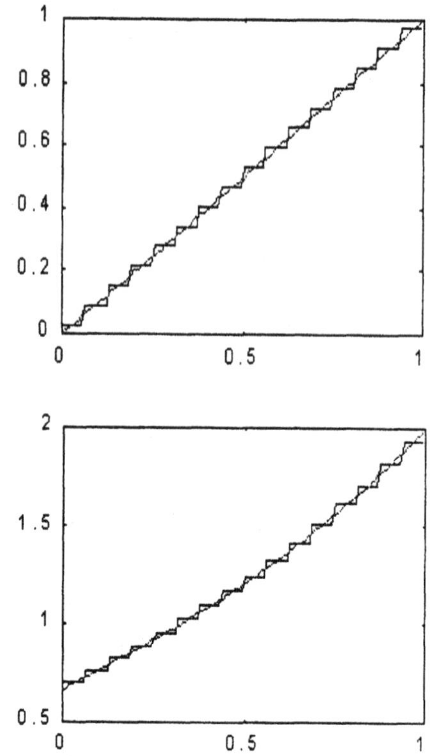


Fig.1 Solution of the system in example 1

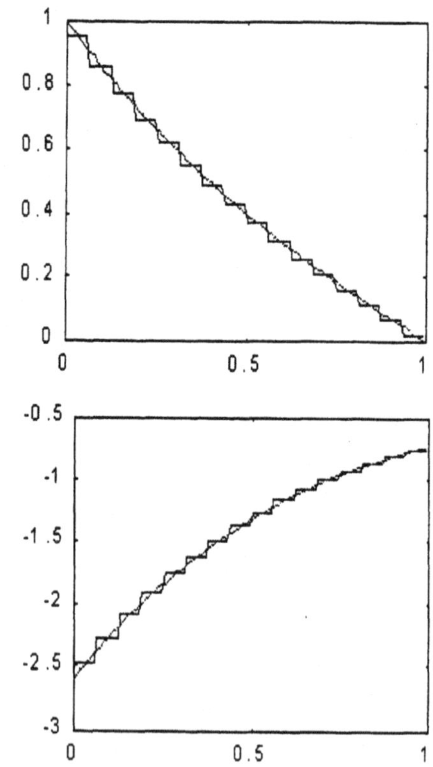


Fig.2 Optimal trajectory and control law of the example 2

## REFERENCES


Mohan, B.M., Datta, V.B. (1995), Analysis of linear time-invariant time-delay systems via orthogonal functions, *International Journal of System Science*, **26**, 91-111

Chen., C. F. and Hsiao, C. H., (1997), Haar wavelet Method for solving lumped and distributed-parameter systems, *IEE Proc-Control Theory Application*, **144**, 97-94

Daubechies Ingrid (1992), *Ten Lectures on Wavelets*, SIAM, Philadephia,

Gu, J.S. and Jiang, W.S. (1996), The Haar wavelets operational matrix of integration, *International Journal of System Science*, **27**, 623-628

Lee, L. T. and Tsay, Y. F. Tsay (1986). Analysis and optimal control of discrete linear time-varying systems via discrete general orthogonal polynomials, *International Journal of Control*, **44**,1427-1436

Wang, M. L., Chang, R. Y. and Yang ,S. Y.(1986), Analysis and optimal control of time-varying systems via generalized orthogonal polynomials, *International Journal of Control*, **44**,,895-910

14th Triennial World Congress, Beijing, P.R. China

F-2d-08-1

# OPTIMALITY OF TWO-DEGREE OF FREEDOM CONTROLLERS IN $\mathcal{H}_2$- AND $\mathcal{H}_\infty$-NORM SPACE, THEIR ROBUSTNESS AND MINIMAL SENSITIVITY

**L. Keviczky** and **Cs. Bányász**

*Computer and Automation Research Institute, Hungarian Academy of Sciences*
*H-1111 Budapest, Kende u 13-17, HUNGARY*
*Phone: +361-4665-435; Fax: +361-4667-503*
*e-mail: h10kev@ella.hu; banyasz@sztaki.hu*

Abstract: A special generic structure of two-degree of freedom controllers and a connected closed-loop system parametrization with optimal regulator design methods are discussed. The optimality is expressed by two simple serial compensators based on the concept of the nearest inverse and can be performed in $\mathcal{H}_2$- and $\mathcal{H}_\infty$-norm space. It is shown how these optimal filters are computed. The optimal sensitivity properties of this scheme are also analysed. *Copyright© 1999 IFAC*

Keywords: two-degree of freedom controller, combined identification and control, iterative controller refinement, optimal and robust control

## 1. INTRODUCTION

A wide class of control problems can be properly handled in the framework of the so-called two-degree of freedom feedforward/feedback closed-loop control systems, having many variants published since the classical works as, e.g., Horowitz (1963).

Keviczky and Bányász (1994) introduced a special scheme, which prooved to be good to handle the combined identification and control problems and provided a useful base for their further work connected to this area. The scheme (which was later named as a *generic* two-degree of freedom (*G2DF*) system) is based on a special (Keviczky-Bányász, or shortly *K-B*) parametrization, shown on Fig. 1.

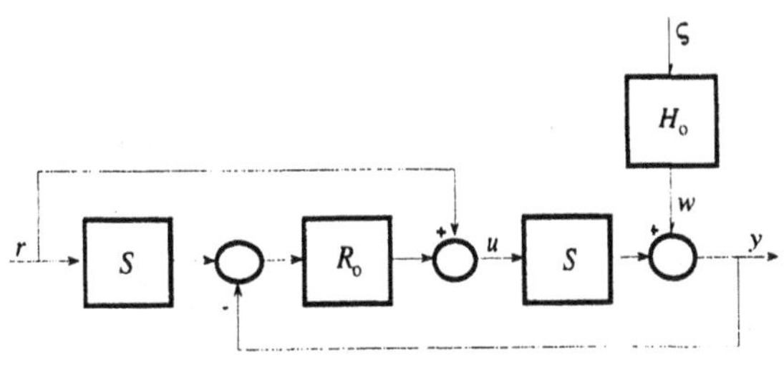


Fig. 1. The *K-B* parametrization of a *2DF* system

Here $S \in \mathcal{RH}_\infty$ and $H_o \in \mathcal{RH}_\infty$ are the true process and noise models, respectively. The ideal regulator $R_o \in \mathcal{RH}_\infty$ is assumed to fulfill an optimization criterion depending on $S$ and $H_o$. $R_o$ is an all realizable stabilizing (*ARS*) regulator iff. it has a special structure (Maciejowski, 1989)

$$R_o = Q_o / (1 - Q_o S) \tag{1}$$

where $Q_o \in \mathcal{RH}_\infty$ is the so-called Youla-parameter.

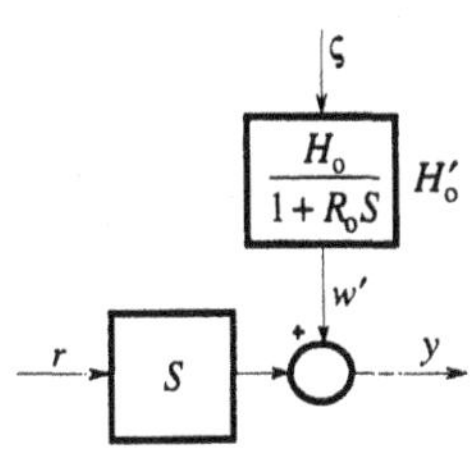


Fig. 2. The equivalent opened closed-loop using *K-B* parametrization

The most obvious advantage of this parametrization, that it opens the closed-loop as the equivalent Fig. 2 shows, which makes closed-loop identification (ID) easy. What we have to "pay" for this simplicity is that the true noise model $H_o$ virtually changes to an augmented model

$$H_o' = H_o(1 - Q_o S) \tag{2}$$

The *K-B* parametrization for closed-loop control is not so widely known as the celebrated Youla-(Kucera)-parametrization, however it is much closer to a control engineering view. Compare Fig.1 with the dual Youla-scheme embedded the true open-loop system in a very sophisticated way. (The negligible drawback of this parametrization is that it can only be applied for *2DF* controls.)

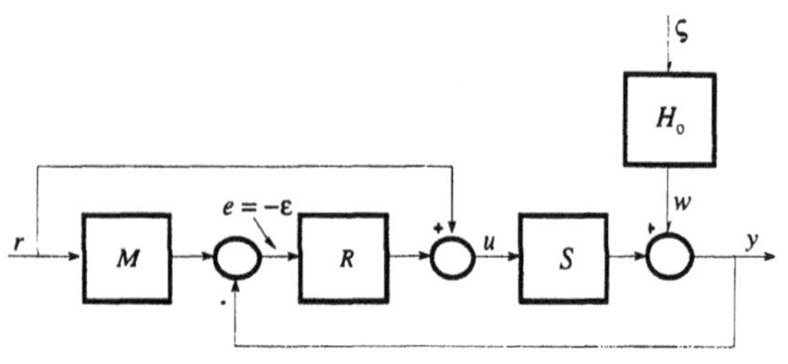


Fig. 3. The model-based *K-B* parametrization of a *2DF* system

An obvious advantage of the *K-B* parametrization is that it provides an easy way to handle model uncertainties. Fig. 3 shows the case, when only the model $M \in \mathcal{RH}_\infty$ of the process is available and an all realizable stabilizing regulator $R \in \mathcal{RH}_\infty$

$$R = Q/(1 - QS) \tag{3}$$

obtained with the same optimization principle as $R_o$ is applied. In this case the parametrization results in the following equivalent open-loop system

$$y = \frac{1}{1+Q\Delta} S\, r + H_o\left(1 - \frac{QS}{1+Q\Delta}\right)\varsigma = S' r + H' \varsigma \tag{4}$$

where $\Delta = S - M$ is the additive model uncertainty. Here $S'(\Delta \to 0) = S$ and $H'(\Delta \to 0) = H_o'$ for exact model matching. Observe that the negative of the control error equals to the model error of a parallel model identification scheme $(e = -\varepsilon)$, which is a special unique feature of the *K-B* scheme. After some straightforward computations the usual prediction error (PE), used in ID methods, is

$$\hat{\varepsilon}_{pr}(\Delta \to 0) = \frac{1}{\hat{H}'}\left\{Z_o \Delta r + \left(H_o' - \hat{H}'\right)\varsigma\right\} + \varsigma \tag{5}$$

where $Z_o$ is the sensitivity function

$$Z_o = 1/(1 + R_o S) = 1 - Q_o S \tag{6}$$

Investigating the covariance estimates the same formulas can be obtained for the $K$ - $B$ parametrization, as for the well known open-loop case (Ljung, 1987).

## 2. OPTIMAL *2DF* SYSTEMS: *A GENERIC SCHEME*

The model-based *K-B* parametrization of a *2DF* system provides an excellent way to handle the regulator design, too, applying an additional $P_r K_r$ serial compensator, because this parametrization virtually opens the closed-loop, as Fig. 2 showed. The obtained *generic (G2DF) optimal controller* structure is shown on Fig. 4.

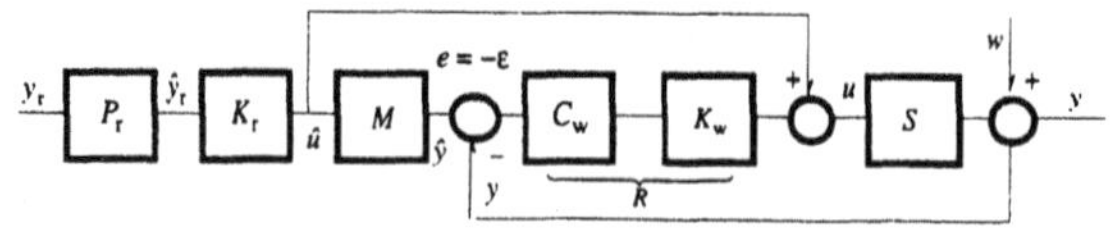


Fig. 4. *Generic (G2DF) scheme* for all cases

Here $y_r, u, y$ and $w$ are the reference, process input, output and disturbance signals in the closed-loop, where discrete-time representations are considered for computer controlled systems. $P_r$ and $K_r$ are the desired tracking (servo) reference model and a serial compensator, respectively. The ideal overall transfer functions of the closed-loop can be formulated as

$$y^o(k) = P_r y_r(k) + (1 - P_w)w(k) = y_t^o(k) + y_d^o(k) \tag{7}$$

So the desired tracking (servo) transfer function $y_t^o$ could be provided by the $P_r$ and the desired regulating (or disturbance rejection) behavior $y_d^o$ by the $P_w$ reference models. (If $P_w$ is the best $LS$ predictor of $w$, then the obtained regulator is the minimum variance (*MV*) regulator.) Assume that

$$S = S_+ S_- z^{-d} \quad \text{and} \quad M = M_+ M_- z^{-d} \tag{8}$$

are stable, $S_+$ and $M_+$ mean the inverse stable (*IS*), $S_-$ and $M_-$ the inverse unstable (*IU*) factors, respectively. Selecting the regulator as

$$R = C_w K_w = \frac{P_w G_w M_+^{-1}}{1 - P_w G_w M_- z^{-d}} \tag{9}$$

where

$$C_w = \frac{1}{1 - P_w G_w M_- z^{-d}}, \qquad K_w = P_w G_w M_+^{-1} \tag{10}$$

and using the serial compensator

$$K_r = G_r M_+^{-1} \tag{11}$$

provides the

$$\begin{aligned} y(k) &= P_r G_r S_- z^{-d} y_r(k) + \left(1 - P_w G_w S_- z^{-d}\right) w(k) \\ &= y_t(k) + y_d(k) \end{aligned} \tag{12}$$

reachable *"best"* overall transfer functions in case of the theoretically ideal exact model matching when $M = S$ and the *K-B* parametrization was formulated. Observe that introducing the Youla parameter

$$Q = K_w = P_w G_w M_+^{-1} \tag{13}$$

the regulator given by (9) has the $ARS$ structure required in (1) and (3).

## 3. OPTIMALITY IN $\mathcal{H}_2$ and $\mathcal{H}_\infty$-NORM SPACE

The most advantageous property of the *generic scheme* is that its optimization either in $\mathcal{H}_2$ or $\mathcal{H}_\infty$-norm space can be provided by the proper computation of $G_r$ and $G_w$. Because of the basic character of a two-degree of freedom control system the tracking and regulatory optimization can be performed independently, so $G_r = G_{r,j}$ and $G_w = G_{w,j}$ should not necessarily follow the same $(j = 2, \infty)$ optimization principle. We have proved (Bányász and Keviczky , 1998) that the optimal solutions can be obtained by serial compensators, for the optimal tracking paradigm

$$K_{r,j} = G_{r,j}\, S_+^{-1} \quad , \quad j = 2, \infty \tag{14}$$

canceling only the $IS$ factor of the process and attenuating the remaining $IU$ component by the optimal

$$G_{r,2} = \frac{X_2^r}{B_r S_-\left(z^{-1}\right)} = P_{rw} G_w \quad , \quad G_{r,\infty} = \frac{N}{B_r h(z)} \tag{15}$$

filters, depending on the norm $\mathcal{H}_j, (j = 2, \infty)$ applied and

$$P_{rw} = \frac{X_2^r}{B_r} \quad , \quad G_w = \frac{1}{S_-\left(z^{-1}\right)} \tag{16}$$

The above optimal serial compensator $K_{r,j}, (j = 2, \infty)$ can be considered as the <u>*nearest stable inverse*</u> of the process, where the stable and proper reference model is assumed in the form of

$$P_r = \frac{B_r}{A_r} = \frac{\tilde{B}_r z^{-m_-}}{A_r} \quad , \quad \tilde{B}_r(0) = b_{r0} \quad , \quad A_r(0) = 1 \tag{17}$$

with coprime $\tilde{B}_r$ and $A_r$ and $m_-$ is the number of unstable zeros in $S_-(z)$. The optimality in $\mathcal{H}_2$-norm space requires $X_2$ computed by the solution of the Diophantine equation (I)

$$S_-\left(z^{-1}\right) B_r = X_1^r A_r + X_2^r S_-(z) z^{-d} \tag{18}$$

Calculating the left side of (18)

$$S_-\left(z^{-1}\right) B_r = z^{m_-} \tilde{S}_-(z) \tilde{B}_r z^{-m_-} = \tilde{S}_-(z) \tilde{B} \tag{19}$$

it is easy to see that the $B_r = \tilde{B}_r z^{-m_-}$ form, required in (17) defining $P_r$, is necessary to the solvability. Here the notation $S_-\left(z^{-1}\right) = z^{m_-} \tilde{S}_-(z)$ is introduced, which virtually increases the time delay $d$ by $m_-$.

The optimality in $\mathcal{H}_\infty$-norm space requires $N$ and $h(z)$, where

$$N = \left[B_r h(z) - \mu A_r h\left(z^{-1}\right)\right] / S_- \tag{20}$$

and the division is with no residue. Furthermore $h(z)$ is a Hurwitz polynomial of degree at most $(m_- - 1)$, coming from the optimal solution lies in optimal interpolation theory and is known as the Navenlina-Pick problem. One can easily follow the application of this principle according to the paper of Wang and Chen (1988) for our discrete-time case. The closed-form optimal $W^o(z)$ minimizing $\|W(z)\|_\infty$ is of the all-pass form for $m_-$ unstable zeros

$$W^o(z) = \begin{cases} \mu \dfrac{h\left(z^{-1}\right)}{h(z)} & , \quad \text{if } m_- \geq 1 \\ 0 & , \quad \text{if } m_- = 0 \end{cases} \tag{21}$$

where the constant $\mu$ and the coefficients of $h(z)$ are real and are uniquely determined by the following interpolation constraints

$$W(z_i) = \left(P_r - X_c S_+ S_-\right)\Big|_{z_i} \quad , \quad i = 1, 2, \ldots, m_- - 1 \tag{22}$$

where $z_1, z_2, \ldots, z_{m_-}$ denote the distinct zeros of $S_-(z)$. (The multiplicity of zeros can be easily considered by additional differential interpolation constraints.)

The optimal solution for the optimal regulatory (or noise rejection) paradigm in $\mathcal{H}_2$-norm space is provided by the optimal serial compensator

$$K_w = P_w G_w\, S_+^{-1} = G_{w,2}\, S_+^{-1} = K_{w,2} \tag{23}$$

if the additive output noise acting on the process and/or on a closed-loop control system is assumed to be a disturbance with rational spectra and driven by

$$w(k) = H_o \varsigma(k) = \frac{E}{D} \varsigma(k) \tag{24}$$

where $H_o$ is stable and $IS$ with coprime $E$ and $D$, $\varsigma$ is a zero mean white noise sequence. Here $P_w$

$$P_w = \frac{X_2^w}{E} = P_{w,2} = P_w^o \tag{25}$$

is the $d$-step ahead realizable closed-loop $MV$ predictor of the output disturbance and $X_2^w$ is given by the solution of the Diophantine equation (II)

$$E S_-\left(z^{-1}\right) = X_1^w D + X_2^w S_- z^{-d} = F D + X_2^w S_- z^{-d} \tag{26}$$

The residual (in this case correlated) prediction error

$$\varepsilon_{\mathrm{w}}^{\mathrm{c}}(k+d \mid k)=F'\varsigma(k) \tag{27}$$

where

$$F'=X_1^{\mathrm{w}}/S_-\left(z^{-1}\right)=H_{\mathrm{o}}\left(1-P_{\mathrm{w}}G_{\mathrm{w}}S_-z^{-d}\right) \tag{28}$$

Please note that it is of course possible to optimize the *d-step ahead* prediction in $\mathcal{H}_\infty$-norm space, too, similarly to $K_{\mathrm{r},\infty}$, i.e., to compute $P_{\mathrm{w}}=P_{\mathrm{w},\infty}$ and $K_{\mathrm{w},\infty}$ instead of $P_{\mathrm{w}}=P_{\mathrm{w},2}$ and $K_{\mathrm{w}}=K_{\mathrm{w},2}$.

## 4. STABILITY AND SENSITIVITY OF THE *GENERIC SCHEME*

Denote the relative model errors by

$$\ell=\frac{\Delta}{M}=\frac{S-M}{M}\quad,\quad \ell_+=\frac{\Delta_+}{M_+}=\frac{S_+-M_+}{M_+}$$
$$\ell_-=\frac{\Delta_-}{M_-}=\frac{S_--M_-}{M_-} \tag{29}$$

It is easy to show that the characteristic equation of the closed-loop for small modeling error in the *generic scheme* is

$$\left(M_++P_{\mathrm{w}}\Delta_-+P_{\mathrm{w}}\Delta_+\right)\Big|_{(\Delta_-,\Delta_+)\to 0}=M_+=0 \tag{30}$$

Because the internal stability requires the stability of the true process this is an additional condition for the closed-loop stability.

The robust stability condition (Morari and Zafiriou, 1989) for the *generic scheme* gives a very interesting result

$$|\ell|\le\|\ell\|_\infty\le\frac{1}{\|P_{\mathrm{w}}\|_\infty} \tag{31}$$

i.e., it depends only on the $P_{\mathrm{w}}$ design requirement.

Investigate the sensitivity of the optimal regulator when the process is *IS*: so $G_{\mathrm{r}}=G_{\mathrm{w}}=1$, $M=M_+$ and $M_-=1$. Using (9) it is easy to obtain that

$$\ell_{\mathrm{R}}^{\mathrm{o}}=\frac{\Delta R}{R_{\mathrm{o}}}=\frac{R_{\mathrm{o}}-R}{R_{\mathrm{o}}}=-\ell \qquad \text{for } \forall\omega \tag{32}$$

This means that any identification method minimizing $\|\ell\|_\infty$ will provide the most robust regulator

$$\left\|\ell_{\mathrm{R}}^{\mathrm{o}}\right\|_\infty=\left\|\frac{\Delta R}{R_{\mathrm{o}}}\right\|_\infty=\left\|\frac{\Delta}{M}\right\|_\infty=\|\ell\|_\infty \tag{33}$$

This equation calculates only the uncertainty coming from the process model. To calculate the influence of the uncertainty in $P_{\mathrm{w}}$ the whole first order difference of $\Delta R$ should be formed as

$$\Delta R\cong\frac{\partial R}{\partial M}\Delta M+\frac{\partial R}{\Delta P_{\mathrm{w}}}\Delta P_{\mathrm{w}}=-RM^{-1}\Delta M$$
$$+RP_{\mathrm{w}}^{-1}/\left(1-P_{\mathrm{w}}\right)\Delta P_{\mathrm{w}} \tag{34}$$

Thus the relative uncertainty can be obtained by

$$\ell_{\mathrm{R}}=\frac{\Delta R}{R}\cong-\frac{\Delta M}{M}+\frac{1}{1-P_{\mathrm{w}}}\frac{\Delta P_{\mathrm{w}}}{P_{\mathrm{w}}}=-\ell+\frac{1}{1-P_{\mathrm{w}}}\ell_{\mathrm{w}} \tag{35}$$

Recent result of Gevers (1996) and Gevers, *et al.* (1997) proved that the most robust regulator, having the minimal sensitivity for process and noise model variations is given by

$$R_{\mathrm{opt}}=\frac{-F_1}{F_1M+F_2H} \tag{36}$$

where

$$F_1=\frac{\partial R}{\partial M}\quad\text{and}\quad F_2=\frac{\partial R}{\partial H} \tag{37}$$

Using (9) and its equivalent form

$$R=\frac{H-F'}{MF'} \tag{38}$$

obtained by (28), simple calculations give the sensitivity derivatives as

$$F_1=\frac{\partial R}{\partial M}=\frac{1}{M_+}\frac{\partial R}{\partial M_-}=-R\frac{1}{M} \tag{39}$$

$$F_2=\frac{\partial R}{\partial H}=\frac{\partial}{\partial H}\left(\frac{H-F'}{F'}\right)=\frac{1}{MF'} \tag{40}$$

Thus (36) results in the optimal regulator, providing the minimal controller variance as

$$R_{\mathrm{opt}}=\frac{-R\frac{1}{M}}{-R\frac{1}{M}M+\frac{1}{MF'}H}=\frac{R}{\frac{H}{F'}-RM}=R \tag{41}$$

which is the regulator of the *generic scheme* (9) optimized for output noise rejection in $\mathcal{H}_2$-norm space. This latter good feature adds another point to the long list of advantages of the *generic scheme*. Some others will also come in the next sections.

## 5. ON THE ROBUSTNESS OF THE *GENERIC SCHEME*

In this section the properties of a robustness measure for the *generic scheme* are discussed for inverse stable processes, when $M_+=M$, $M_-=1$ similarly to the process and $G_{\mathrm{w}}=1$. (The results are very similar for inverse unstable processes, however, the necessary formulas are much more complicated.) Let us denote the well-known robustness measure: the distance between the point $(-1+0j)$ and the Nyquist curve of the open-loop frequency characteristics $Y = RS$ by $\rho$

$$\rho = \rho(\omega, R) = |1 + RS| \tag{42}$$

(Note that in the most general case of $|Y(\omega = \infty)| = 0$ this measure falls into the range $0 \le \rho \le 1$.) Denote the worst (maximum) values with respect to $R$ by

$$\hat{\rho}_m = \max_R \left\{ \min_\omega [\rho(\omega, R)] \right\} = \max_R \left\{ \min_\omega (\rho) \right\} \tag{43}$$

Straightforward calculations give that

$$\rho_m(R) = \min_\omega |1 + RS| = \frac{1}{\left\| \dfrac{1 - P_w}{1 + P_w \ell} \right\|_\infty} \tag{44}$$

and the upper limits are easily obtained as

$$\frac{\rho_m(R)}{\hat{\rho}_m^o} \le \|1 + P_w \ell\|_\infty \le 1 + \|P_w\|_\infty \|\ell\|_\infty \tag{45}$$

where

$$\hat{\rho}_m^o = \hat{\rho}_m(\ell = 0) = \frac{1}{\|1 - P_w\|_\infty} = \min_\omega \left| \frac{1}{1 - P_w} \right| \tag{46}$$

so $\hat{\rho}_m^o$ is the theoretically best optimal (maximal) robustness measure obtained if $\|\ell\|_\infty \to 0$ (i.e., for exact model matching case) during an iterative controller refinement.

Because it is easy to construct lower limits, too, the following inequalities hold

$$\begin{aligned} 1 - \|P_w\|_\infty \|\ell\|_\infty \le \min_\omega |1 + P_w \ell| \le \frac{\rho_m(R)}{\hat{\rho}_m^o} \le \\ \le \|1 + P_w \ell\|_\infty \le 1 + \|P_w\|_\infty \|\ell\|_\infty \end{aligned} \tag{47}$$

which mean that any iterative combined identification and control scheme provides a convergence

$$\rho_m(R) \to \hat{\rho}_m^o = \max_{R \in \mathcal{R}} \left\{ \min_\omega [\rho(\omega, R)] \right\} \tag{48}$$

if $\|\ell\|_\infty \to 0$.

## 6. DESIGN EXAMPLE

*Example 1.* Let the discrete-time transfer function of a unity gain third order time delay $(d = 2)$ process

$$S = \frac{-0.4(1 - 1.4z^{-1})(1 - 0.75z^{-1})}{(1 - 1.5z^{-1} + 0.7z^{-2})(1 - 0.8z^{-1})} z^{-2} \tag{49}$$

with one $(m_- = 1)$ unstable zero at $z_1 = 1.4$. Be the tracking and regulatory reference models

$$P_r = \frac{0.8z^{-1}}{1 - 0.2z^{-1}} \quad , \quad P_w = \frac{0.2z^{-1}}{1 - 0.8z^{-1}} = \frac{B_w}{A_w} \tag{50}$$

The optimization discussed above requires the monic factorization:

$$S_-(z) = \frac{1 - 1.4z^{-1}}{-0.4} \quad , \quad S_-(z^{-1}) = \frac{1 - 1.4z}{-0.4} \tag{51}$$

$$S_+(z) = \frac{0.16(1 - 0.75z^{-1})}{(1 - 1.5z^{-1} + 0.7z^{-2})(1 - 0.8z^{-1})} \tag{52}$$

Calculate the $K_{r,2}$ optimal serial compensator first. The Diophantine equation (18) to be solved is now

$$\frac{1 - 1.4z}{-0.4} 0.8z^{-1} = X_1^r (1 - 0.2z^{-1}) + X_2^r \left( \frac{1 - 1.4z^{-1}}{-0.4} \right) z^{-2} \tag{53}$$

having the solutions

$$X_1^r = 2.8 + 0.36z^{-1} + 0.084z^{-2}, \quad X_2^r = 0.0048 \tag{54}$$

and from (15)

$$G_{r,2} = \frac{0.0048}{0.8z^{-1} \dfrac{1 - 1.4z}{-0.4}} = \frac{0.00171}{1 - 0.714z^{-1}} \tag{55}$$

So finally we get

$$K_{r,2} = \frac{0.0107(1 - 1.5z^{-1} + 0.7z^{-2})(1 - 0.8z^{-1})}{(1 - 0.714z^{-1})(1 - 0.75z^{-1})} \tag{56}$$

It is worth mentioning that the final serial term appearing in Fig. 4

$$P_r K_{r,2} = \frac{0.00857(1 - 1.5z^{-1} + 0.7z^{-2})(1 - 0.8z^{-1})z^{-1}}{(1 - 0.714z^{-1})(1 - 0.75z^{-1})(1 - 0.2z^{-1})} \tag{57}$$

is also realizable.

Calculate the $K_{r,\infty}$ optimal serial compensator next. Equations (20), (21) and (22) show that the $h(z)$ Hurwitz polynomial is a constant unity and $\mu = 2/3 = 0.6667$, furthermore $N = -\mu = -0.6667$. Thus from (15)

$$G_{r,\infty} = -0.8333z \tag{58}$$

and finally we get

$$K_{r,\infty} = \frac{-5.21(1 - 1.5z^{-1} + 0.7z^{-2})(1 - 0.8z^{-1})}{(1 - 0.75z^{-1})} \tag{59}$$

Because $P_w$ is given and not the transfer function $H$ generating $w$ the optimal regulator requires the computation of $G_w$ now from (16), i.e.

$$G_w = \frac{1}{\left(\dfrac{1-1.4z}{-0.4}\right)} = \frac{-0.4}{1-1.4z} = \frac{0.286}{1-0.714z^{-1}} \tag{60}$$

and the optimal regulator assuming $M = S$ is obtained from (9) as

$$\begin{aligned} R &= \frac{1}{1 - P_w G_w S_- z^{-d}} P_w G_w S_+^{-1} = \frac{B_w}{A_w S_-\left(z^{-1}\right) - B_w S_- z^{-d}} S_+^{-1} \\ &= \frac{1.429\left(1 - 1.5z^{-1} + 0.7z^{-2}\right)\left(1 - 0.8z^{-1}\right)z^{-2}}{\left(1 - 0.914z^{-1} + 0.143z^{-2} + 0.571z^{-4} - 0.8z^{-5}\right)\left(1 - 0.75z^{-1}\right)} \end{aligned} \tag{61}$$

## 7. CONCLUSIONS

This paper presents a unified way how the optimality of two-degree of freedom controllers can be handled in the control engineering practice. A *generic optimal controller* structure and regulator design method is shown, which can be easily optimized in $\mathcal{H}_2$- and $\mathcal{H}_\infty$-norm space if the regulator and two related serial compensators are selected in a special way. It is shown how these filters – depending on the reference models for the servo and the output disturbance rejection properties and the invariant process factors - should be selected by the solutions of Diophantine equations if optimality is important. It is proved that the optimal regulator of the *generic scheme* optimized in $\mathcal{H}_2$-norm space is the theoretical regulator providing minimal controller variance to the process and noise model variations, according to Gevers' condition. This procedure allows the control engineers to design high performance control systems providing both desired tracking and disturbance rejection behaviors reaching the theoretical ultimate limits.

The optimal sensitivity properties of this new scheme are also analysed. The paper clearly shows by explicit formulas how the accuracy of the identified model (and via the selected identification method) influences the optimality, the stability and even the sensitivity of the optimal controller.

Finally a design example is given to represent the introduced two-degree of freedom controller optimization technique.

## 8. REFERENCES


Gevers, M. (1996). Learning from identification to control design. *ESF COSY Workshop*, Valencia, pp. 1-60.

Gevers, M., L. Ljung and P.M.J. van den Hof (1997). Asymptotic variance expressions for closed-loop identification and their relevance in identification and control. *11th IFAC/IFORS Symp. SYSID'97*, Fukuoka, Japan, pp. 1449-1454.

Horowitz, I.M. (1963). *Synthesis of Feedback Systems*, Academic Press, New York.

Keviczky, L. and Cs. Bányász (1994). A new structure to design optimal control systems," *IFAC Workshop on New Trends in Design of Control Systems*, Smolenice, Slovakia, pp. 102-105.

Keviczky, L. (1995a). Combined identification and control: Another way, (Invited plenary paper.) *5th IFAC Symp. on Adaptive Systems in Control and Signal Processing ACASP'95*, Budapest, Hungary, pp. 13-30.

Keviczky, L. and Cs. Bányász (1995b). A new optimal regulator design method. *European Control Conference ECC'95*, Rome, Italy, pp. 3359-3364.

Bányász Cs. and L. Keviczky (1998). Designing high performance two-degree of freedom controllers. *IEEE Instrumentation & Measurement Technology Conference IMTC'98*, St. Paul, MN, USA, pp. 1359-1366.

Ljung, L. (1987). *System Identification*. Prentice Hall, Englewood Cliffs, N.J.

Maciejowski, J.M. (1989). *Multivariable Feedback Design*, Addison Wesley.

Morari, M. and E. Zafiriou (1989). *Robust Process Control*. Prentice-Hall International, Inc., London.

Vidyasagar, M. (1985). *Control System Synthesis: a Factorization Approach*. Cambridge, Mass.: MIT Press.

Wang. S. and B. Chen (1988). Optimal model matching control for time-delay systems. *Int. J. Control*, **47**, 3, pp. 883-894.


---

This work was supported by the Hungarian NSF (OTKA).

# RATIONAL $l_1$-SUBOPTIMAL FEEDBACK SYSTEM DESIGN [1]

Mark E. Halpern

*Dept of Electrical and Electronic Engineering*
*the University of Melbourne*
*Grattan St, Parkville, Vic 3052, Australia.*
*m.halpern@ee.mu.oz.au, r.evans@ee.mu.oz.au*

Abstract: A difficulty with the $l_1$ methodology is that optimal solutions can be of arbitrarily high order. In this paper, the use of a rational signal model is proposed for obtaining rational suboptimal solutions of low order to scalar $l_1$ problems. This model allows an arbitrary number of distinct positive closed-loop poles to be incorporated. Solving an interpolation constrained $l_1$ optimization problem then involves two finite linear programs and gives a solution which makes use of only the best out of all the prespecified poles. Using the signal model in a scalar two-block problem gives rational suboptimal solutions in a simple procedure involving four linear programs. 

Keywords: Discrete-time systems, optimal control, pole assignment, linear programming.

## 1 INTRODUCTION

The problem of optimal rejection of persistent bounded disturbances in feedback systems was posed and solved by Barabanov and Granichin (1984) and independently formulated by Vidyasagar (1986). A subsequent solution proposed by Dahleh and Pearson (1987) used linear programming and duality theory to solve an $l_1$ minimization problem subject to interpolation constraints. Dahleh and Pearson (1988) found a similar mathematical problem arises in the stability robustness context. It is well known that this one-block problem has a solution which has finite impulse response (FIR). It can be obtained by solving a linear program (LP) corresponding to the primal problem or another corresponding with the dual problem. A difficulty with the methodology is that the order of the optimal solution can be arbitrarily high, in contrast to the case with $l_2$ and $H_\infty$ optimisation problems where the solution order corresponds with the number of interpolation constraints. Another issue is the possible nonexistence of solutions to $l_1$ optimization problems containing interpolation on the stability boundary; see (Barabanov and Granichin, 1984; Vidyasagar, 1991). Controller implementation in the presence of such difficulties requires some relaxing of the requirement for $l_1$ optimality. One simply obtained suboptimal solution is the best FIR solution of given order, found by solving a truncated version of the primal or dual LP.

In an effort to obtain an improvement in terms of a smaller $l_1$ norm than is obtained using a reduced order FIR solution, a single real positive pole was incorporated by Halpern, *et al.* (1996) into the solution of the scalar problem, giving the solution a rational rather than polynomial transfer function. Both the primal and dual optimization problems remained finite linear programs. The approach could be applied directly to the problem of $l_1$ optimal design with one prespecified pole. Also, it was shown that, for a simple example, where the op-

[1]This work was supported by the Australian Research Council.

timum cost could be calculated analytically, simple calculus could be applied to the dual to find the exact pole location to minimise the $l_1$ norm. Other than in such simple cases, finding the best pole location required repeated solution of a LP, using a different pole value each time. The case with two poles, real and complex is developed by Halpern (1998). There again, results are largely restricted to pole placement. A different approach for the design of reduced order controllers in an $l_1$ context is by Blanchini and Sznaier (1997).

In this paper the approach used by Halpern, *et al.* (1996) is extended to allow the incorporation of more than one pole into the solution and simultaneously to aid in the selection of poles. A signal model is proposed, comprising an unknown FIR part together with an arbitrary number of real prespecified positive distinct poles under a restriction on the signs of their residues, which are otherwise unknown. Applying the $l_1$ optimization approach to this signal model yields, for a problem with $q$ interpolation constraints, a solution involving at most $q$ poles, giving an element of *optimal pole selection* to the approach. A consequence is that one can obtain good results without an informed selection of the poles; a large number of candidate poles can be incorporated into the problem and solving the two linear programs automatically chooses the best number of poles to include and rejects the others by setting their residues to zero.

In the $l_1$ context, the theory for multiblock problems is less complete, and obtaining the optimal solution can be rather difficult; they are usually obtained from a sequence of suboptimal solutions. There are several methods for obtaining suboptimal solutions. They generally involve obtaining upper and lower bounds on the optimal cost as the order of an approximating solution is increased in some manner. When the upper and lower bounds are close enough, the solution corresponding with the upper bound can be used, but thuis may involve a high order FIR approximation and these require high order controllers, even though the true optimum controllers could be of low order. The problem was first posed by by Dahleh and Pearson (1988); Staffans (1991) introduced the terminology, finitely many variables (FMV) to describe the truncated primal minimization and the term finitely many equations (FME) to describe a truncated dual problem; see also Staffans (1992) and used these pairs of solutions to find true oprimal solutions. A detailed exposition of the issues involved is by Diaz-Bobillo and Dahleh (1993) where the delay augmentation (DA) algorithm was introduced for the general multiblock problem. A related approach is by Casavola (1996). An alternative approach is by Barabanov and Sokolov (1994).

Finding the optimal solution to avoid controller order inflation requires determining which components of the solution are finitely supported. There are at least two difficulties with previous approaches. Firstly, there are only limited results indicating which types of problem do have FIR components in their optimal solutions. Secondly, identification of the FIR parts of a solution relies on observed behaviour of iterated solutions. Meanwhile, Staffans (1991) has shown with a simple example that the optimal solution can have very high order with miniscule performance improvement over a much lower order solution. This fact provides strong motivation for considering suboptimal low order solutions.

The method proposed in this paper is to use the signal formulation used in the earlier part of the paper; applied to the multiblock problem, it is in effect a modification of the FMV approach and is also along the lines of a suggestion by Diaz-Bobillo and Dahleh (1993) to incorporate the known poles into the solutions. Since it is not known a-priori which, if any, components of the optimal solution will be rational, but is expected they could have poles from the weighting functions, the solution is allowed to contain these poles, along with others if desired. Solving the primal linear program then selects the best of the poles from those specified. The approach gives automatic selection of which, if any, parts of the solution are FIR and readily allows all components of the solution to be rational.

## 2 PRELIMINARIES

The $z$-transform, $\hat{h}$ , of a sequence, $h = \{h_i\}_{i=0}^{\infty}$, is defined by $\hat{h} = \sum_{i=0}^{\infty} h_i z^i$. With this definition, a stable transfer function has all of its poles at values of $z : |z| > 1$ . The symbol $z$ also denotes the unit delay. The $l_1$ norm of a sequence $h$ is denoted $\|h\|_1$ and is given by $\|h\|_1 = \sum_{i=0}^{\infty} |h_i|$.

The scalar $l_1$ optimisation problem can be written (see Dahleh and Pearson 1987):

$$\nu = \min_{\phi \in l_1} \sum_{i=0}^{\infty} |\phi_i| \tag{1}$$

$$\text{subject to} \quad \hat{\phi}(a_j) = b_j, \quad j = 1, 2, \ldots, q, \tag{2}$$

where $a_1, \ldots, a_q$ are given distinct complex numbers, all satisfying $|a_i| < 1$ and the $b_j$'s are given complex numbers, not all zero. The $b_j$'s corresponding with real $z_j$'s are also real. Any complex $a_j$'s occur in conjugate pairs, as do the corresponding $b_j$'s. It is well known that the solution to (1,2) has an FIR $z$-transform, namely with the form

$$\hat{\phi}(z) = \sum_{i=0}^{N_0} \phi_i z^i$$

where $N_0$ is some integer greater than or equal to $q-1$. The value of $N_0$ can, in some cases be much larger than $q-1$ and it is this fact that motivates the work here; solutions with orders in the range $q-1,\ldots,N_0-1$ are sought, with $l_1$ norms as small as possible. One simple approach is to insist the solution has the form $\hat{\phi}(z) = \sum_{i=0}^{N} \phi_i z^i$ for some chosen $N \in \{q-1,\ldots,N_0-1\}$ and then to solve a linear program for the values of $\phi_0,\ldots,\phi_N$. In this paper an approach, using rational rather than polynomial solutions, is presented.

### 2.1 One real pole

One real pole was incorporated by Halpern, *et al.* (1996) into $\hat{\phi}(z)$ by insisting $\hat{\phi}(z)$ took the form

$$\hat{\phi}(z) = \sum_{i=0}^{m-1} \phi_i z^i + \frac{\phi_m z^m}{1+v_1 z}. \tag{3}$$

Here $m$ is chosen by the designer in the range $m \in \{q-1,\ldots,N_0-1\}$ to give a desired amount of overparametrisation. A positive stable pole has $v_1 \in (-1,0]$. The $l_1$ norm of $\phi$ of (3) is given by

$$\|\phi\|_1 = \sum_{i=0}^{m-1} |\phi_i| + \frac{|\phi_m|}{1+v_1} \tag{4}$$

For given $v_1, m$, the problem of minimising $\|\phi\|_1$ is

$$\nu = \min_{\phi_0,\ldots,\phi_m} \sum_{i=0}^{m-1} |\phi_i| + \frac{|\phi_m|}{1+v_1} \tag{5}$$

subject to interpolation constraints,

$$\sum_{i=0}^{m-1} \phi_i a_j^i + \frac{\phi_m a_j^m}{1+v_1 a_j} = b_j, \qquad j = 1,2,\ldots,q. \tag{6}$$

This can be written as a finite LP and solved numerically. Moreover it has the same form as (1,2) and is a weighted $l_1$ norm minimisation. This fact allows its dual problem to be simply obtained as shown by Halpern *et al.* (1996). For simple problems, using duality enables closed form expressions for the optimal pole location to be obtained.

## 3 RESULTS

### 3.1 Signal Model

Here $n$ real, distinct, stable positive poles fixed at $z = -1/v_1, -1/v_2 \ldots, -1/v_n$ are incorporated, by setting

$$\hat{\phi}(z) = \sum_{i=0}^{m-1} \phi_i z^i + z^m \sum_{i=1}^{n} \frac{r_i}{1+v_i z}, \tag{7}$$

where, for convenience, the $v_i'$s are ordered according to $-1 < v_1 < \cdots < v_n \leq 0$ and collected into a vector $\underline{v} = \{v_1,\ldots,v_n\}$.

Unfortunately, if $n > 1$, there is no single expression corresponding with (4) for $\|\phi\|_1$. For the two cases where all $r_i \geq 0$ or all $r_i \leq 0$ and only those,

$$\|\phi\|_1 = \sum_{i=0}^{m-1} |\phi_i| + \sum_{i=1}^{n} \frac{|r_i|}{1+v_i}. \tag{8}$$

Accordingly, in this paper, the signs of the residues are restricted to the two cases where all $r_i \geq 0$ or all $r_i \leq 0$. Halpern (1998) has shown that there is a considerable increase in problem complexity and a large number of different LP's may be needed if different signed residues are allowed, even with only two poles, i.e. $n = 2$.

### 3.2 One-block problem

Here the signal model (7) is incorporated into the solution of a scalar $l_1$ optimisation problem subject to $q$ interpolation constraints.

The problem of minimising $\|\phi\|_1$ of (8) subject to $q$ interpolation constraints and to nonnegativity (or nonpositivity) constraints on all $r_i$ can be written as a linear program which can be solved to find the values of $\phi_0,\ldots,\phi_{m-1}, r_1,\ldots,r_n$. The problem with all $r_i \geq 0$ is named LP1a, and that with all $r_i \leq 0$, LP1b. Problem LP1a is shown below.

LP1a:

$$\nu_{1a} = \min_{\substack{\{\phi_0,\ldots,\phi_{m-1},\\ r_1\geq 0,\ldots,r_n\geq 0\}}} \sum_{i=0}^{m-1} |\phi_i| + \sum_{i=1}^{n} \frac{|r_i|}{1+v_i} \tag{9}$$

subject to

$$\sum_{i=0}^{m-1} \phi_i a_j^i + a_j^m \sum_{i=1}^{n} \frac{r_i}{1+v_i a_j} = b_j, \tag{10}$$

where $j = 1,2,\ldots,q$.

Similarly $\nu_{1b}$ is defined as the optimal cost for LP1b, obtained by changing the constraints on all $r_i$ to $r_i \leq 0$. The final result is obtained after solving both LP1a and LP1b:

$$\nu_1 = \min\left[\nu_{1a}, \nu_{1b}\right]$$

and the optimal variables are those corresponding with the smaller of $\nu_{1a}, \nu_{1b}$.

A sufficient condition to guarantee the existence of a solution to both LP1a and LP1b is that the number of unsigned variables is greater than or equal to the number of equality constraints, that is

$$m \geq q. \tag{11}$$

Note that in some cases, solutions will exist to LP1a or LP1b with $m < q$. The existence of solutions to both LP1a and LP1b is not needed; indeed a low order solution violating (11) may be desirable; however the sign constraints on the $r_i$'s make it difficult to guarantee a-priori the existence of any solution unless (11) is satisfied.

It is easy to show that, like for the standard $l_1$ norm minimisation (where there are no restrictions on the signs of the variables), at most $q$ out of the variables $\phi_0, \ldots, \phi_{m-1}, r_1, \ldots, r_n$ can be nonzero in the solution of LP1a or LP1b. This means that if enough poles are incorporated so that $n > q$, then some of the prespecified poles will not appear in the solution since their residues will be zero. In other words, solving the LP1's automatically selects the number of poles and the best ones from all the $v_i$ to comprise the solution.

### 3.3 Mixed sensitivity minimization

The problem of minimizing the $l_1$ norm of mixed sensitivity was first posed by Dahleh and Pearson (1988). For this problem, a rational plant $P(z)$ with $n_p$ distinct unstable poles $a_1, \ldots, a_{n_p}$ and $n_z$ distinct nonminimum phase zeros $a_{n_p+1}, \ldots, a_{n_p+n_z}$ is given. Also given is a pair of stable rational weighting functions $W_1(z), W_2(z)$, which, to simplify the presentation, is taken to be all-pole, that is

$$W_1(z) = \frac{1}{D_1(z)},$$

$$W_2(z) = \frac{1}{D_2(z)}$$

where $D_1(z)$ is a polynomial of order $n_{d1}$ and $D_2(z)$ is a polynomial of order $n_{d2}$. The problem is to compute a suboptimal solution to the two-block row problem

$$\nu_2^{row} = \min_{\text{stabilizing } C(z)} \|\Phi_1 \Phi_2\|_1 \tag{12}$$

or, alternatively,a suboptimal solution to the two-block column problem

$$\nu_2^{col} = \min_{\text{stabilizing } C(z)} \left\| \begin{matrix} \Phi_1 \\ \Phi_2 \end{matrix} \right\|_1 \tag{13}$$

where $\Phi_1 = W_1 S$ and $\Phi_2 = W_2 T$ where

$$S(z) = \frac{1}{1+PC}$$

and

$$T(z) = \frac{PC}{1+PC}.$$

This problem can be written, see (Dahleh and Pearson, 1988; Staffans, 1991,1992)

$$\nu_2^{row} = \min_{\Phi_1, \Phi_2 \in l_1} \|\Phi_1 \Phi_2\|_1$$

or

$$\nu_2^{col} = \min_{\Phi_1, \Phi_2 \in l_1} \left\| \begin{matrix} \Phi_1 \\ \Phi_2 \end{matrix} \right\|_1$$

subject to $q = n_p + n_z$ interpolation constraints,

$$\hat{\Phi}_1(a_j) = \begin{cases} 0, & j = 1, \ldots, n_p, \\ W_1(a_j), & j = n_p+1, .., n_p+n_z \end{cases} \tag{14}$$

and to a set of $n_c$ convolution constraints obtained by equating like powers of $z$ in

$$W_1^{-1}(z)\hat{\Phi}_1(z) + W_2^{-1}(z)\hat{\Phi}_2(z) = 1. \tag{15}$$

Note that, as pointed out by Staffans (1991) the constraints (15) which are equivalent to $S+T = 1$, when taken together with the interpolation constraints (14) on $S$, imply $T$ satisfies corresponding required interpolation constraints.

To incorporate the signal model proposed in this paper, set

$$\hat{\Phi}_1(z) = \sum_{i=0}^{m-1} (\Phi_1)_i z^i + z^m \sum_{i=1}^{n} \frac{(r_1)_i}{1+v_i z} \tag{16}$$

and

$$\hat{\Phi}_2(z) = \sum_{i=0}^{p-1} (\Phi_2)_i z^i + z^p \sum_{i=1}^{n} \frac{(r_2)_i}{1+v_i z} \tag{17}$$

where $m, p, n, v_1, \ldots, v_n$ are prespecified and the variables are $(\Phi_1)_0, \ldots, (\Phi_1)_{m-1}, (r_1)_1, \ldots, (r_1)_n$, $(\Phi_2)_0, \ldots, (\Phi_2)_{p-1}$ and $(r_2)_1, \ldots, (r_2)_n$, all real scalars.

The interpolation constraints (14) are like those (10) for the one-block problem. To obtain the convolution constraints, (16, 17) are substituted into (15) to obtain, after multiplying both sides by all denominator terms, the following equation which is polynomial in $z$:

$$D_1(z)\left\{ \left[ \prod_{i=1}^{n}(1+v_i z) \sum_{i=0}^{m-1} (\Phi_1)_i z^i \right] + z^m \prod_{i=1}^{n}(1+v_i z) \sum_{i=1}^{n} \frac{(r_1)_i}{1+v_i z} \right\}$$

$$+ D_2(z)\left\{ \left[ \prod_{i=1}^{n}(1+v_i z) \sum_{i=0}^{p-1} (\Phi_2)_i z^i \right] \right.$$

$$+z^p \prod_{i=1}^{n}(1+v_i z)\sum_{i=1}^{n}\frac{(r_2)_i}{1+v_i z}\Bigg\}$$
$$= \prod_{i=1}^{n}(1+v_i z). \qquad (18)$$

Equating like powers of $z$ gives $n_c$ convolution constraints where

$$n_c = n - 1 + \max\left[n_{d1}+m, n_{d2}+p\right]. \qquad (19)$$

For the minimization of either (12) or (13), the optimal solution with structure constrained to the form (16, 17) is then obtained by solving four linear programs, and choosing the best solution. These linear programs are named LP2a–d and are obtained by choosing all possible sign restrictions on the residues, namely

LP2a: $\{(r_1)_i \geq 0, (r_2)_i \geq 0\}$,

LP2b: $\{(r_1)_i \geq 0, (r_2)_i \leq 0\}$,

LP2c: $\{(r_1)_i \leq 0, (r_2)_i \geq 0\}$,

LP2d: $\{(r_1)_i \leq 0, (r_2)_i \leq 0\}$.

As an example, problem LP2a for the row problem, cost function (12) is outlined below:

LP2a:

$$\nu_{2a}^{row} = \min_{\substack{\{(\Phi_1)_0,\ldots,(\Phi_1)_{m-1},\\ (\Phi_2)_0,\ldots,(\Phi_2)_{p-1},\\ (r_1)_1\geq 0,\ldots,(r_1)_n\geq 0,\\ (r_2)_1\geq 0,\ldots,(r_2)_n\geq 0\}}} \sum_{i=0}^{m-1}|(\Phi_1)_i| +$$
$$\sum_{i=1}^{n}\frac{|(r_1)_i|}{1+v_i} + \sum_{i=0}^{p-1}|(\Phi_2)_i| + \sum_{i=1}^{n}\frac{|(r_2)_i|}{1+v_i} \qquad (20)$$

subject to $q$ interpolation constraints (14) and $n_c$ convolution constraints from (18).

Like for the one-block problem, in order to guarantee the existence of solutions to all four problems, sufficient unsigned variables are needed, namely (assuming all $r_i = 0$),

$$m + p \geq q + \max\left[n_{d1}+m, n_{d2}+p\right]$$
$$\Rightarrow p \geq q + n_{d1}, \quad m \geq q + n_{d2}. \qquad (21)$$

## 4 EXAMPLE

### 4.1 1-block problem

The following simple problem is considered here:

$$\nu = \min_{\hat{\phi}\in l_1} \|\phi\|_1$$

subject to $\hat{\phi}(0.95) = 0.8$,
$\hat{\phi}(0.8) = 0.7$,
and $\hat{\phi}(0.6) = 0.9$.

The optimal solution is

$$\hat{\phi}(z) = 1.0636 - 0.7584z^3 + 0.8785z^{16} \qquad (22)$$

giving $\|\phi\|_1 = 2.700$. Any other solution satisfying the three given interpolation constraints must have a larger $l_1$ norm. With $N = 3$, the optimal truncated FIR solution is

$$\hat{\phi}(z) = 2.862 - 4.000z - 2.026z^3 \qquad (23)$$

giving $\|\phi\|_1 = 8.887$. Applying the approach in this paper, $m$ is set to $m = 3$ and $\underline{v}$ to $\underline{v} = \{-.9, -.8, -.7, -.6, -.5, -.4, -.3, -.2\}$. Solving LP1a and LP1b gives

$$\hat{\phi}(z) = 1.288 - 1.211z^2 + \frac{.1023z^3}{1-.9z},$$

giving $\|\phi\|_1 = 3.52$ with an $l_1$ norm considerably less than that of (23).

### 4.2 2-block problem

This subsection contains several examples which use the following plant and weighting functions:

$$P(z) = \frac{z(z-.5)}{(z-.1)(1-.5z)}$$

$$W_1(z) = \frac{0.02}{1-.2z}, W_2(z) = \frac{0.004\rho}{1-.6z}$$

where $\rho$ is a positive scalar. Clearly, $q = 3$ and $n_{d1} = n_{d2} = 1$. For all the following examples $\underline{v}$ is set to $\underline{v} = \{-.6, -.2\}$ so that $n = 2$. Satisfying (21) requires $m \geq 4$ and $p \geq 4$.

The first example considered here was posed by Dahleh and Pearson (1988) and Staffans (1991) of minimizing $\|\Phi_1\Phi_2\|_1$ with $\rho = 1$. Setting $m = p = 4$, and solving LP2a–d gives the exact optimal solution, found by Staffans (1991):

$$\hat{\Phi}_1(z) = \frac{1}{50} - \frac{113z}{450} + \frac{23z^2}{45},$$

$$\hat{\Phi}_2(z) = \frac{287z}{5625} - \frac{1531z^2}{18750} - \frac{8029z^3}{281250(1-3z/5)}.$$

The second example is that by Staffans (1992) and Diaz-Bobillo and Dahleh (1993) of minimizing

$$\nu = \left\|\begin{matrix}\Phi_1\\ \Phi_2\end{matrix}\right\|_1$$

this time with $\rho = 6$. Setting $m = p = 3$, (violating (21), but nonetheless allowing a solution to be found) and solving the LP2's gives $\nu = 1.04135$ with

$$\hat{\Phi}_1(z) = 0.02 - .258547z + .598851z^2 - \frac{.131163z^3}{1 - .2z}$$

$$\hat{\Phi}_2(z) = .31506z + .591639z^2 - \frac{.0538629z^3}{1 - .6z}.$$

The corresponding controller is of second order, namely:

$$C(z) = \frac{2.625 - 2.567z + 0.627z^2}{1 - 3.127z + 1.255z^2}.$$

Finally, using $m = p = 4$ gives $\nu = 1.00263$ with a third order controller as by Diaz-Bobillo and Dahleh (1993).

Since the approach is based on FMV, there is no need to order the components of the solution. Also, there is no need to determine which rows in the solution are active.

## 5 CONCLUSION

A scheme for incorporating an arbitrary number of prespecified real "candidate" closed-loop poles into the scalar $l_1$ norm optimisation framework has been given. For one-block problems, the approach can be used to obtain reduced order sub-optimal rational solutions, obtained by solving a pair of linear programs, which in effect select the best poles out of the prespecified set. For two-block problems, the method is a fairly natural one for allowing solutions with infinite support and allows rational suboptimal solutions to be obtained rather simply by solving four linear programs.

## 6 REFERENCES

Barabanov, A.E. and O.N. Granichin (1984). Optimal Controller with Linear Plant with Bounded Noise. Translated from Automatika i Telemechanica, **5**, pp. 39–46.

Barabanov, A.E. and A.A. Sokolov (1994). Geometrical Solution to $l_1$-Optimization Problem with Combined Conditions. In: *Proceedings 1st Asian Control Conference*, Tokyo, **3**, pp. 327–330.

Blanchini, F. and M. Sznaier (1997). A Convex Optimization Approach to Fixed-Order Controller Design for Disturbance Rejection in SISO Systems. In: *Proceedings 36th IEEE Conference on Decision and Control*, San Diego, CA, pp. 1540–1545.

Casavola, A. (1996). A Polynomial Approach to the $l_1$-Mixed Sensistivity Optimal Control Problem. *IEEE Trans. Automat. Contr.*, **41**, pp. 751–756.

Dahleh, M.A. and J.B. Pearson (1987). $l^1$-Optimal Feedback Controllers for MIMO Discrete-Time Systems. *IEEE Trans. Automat. Contr.*, **AC-32**, pp. 314–322.

Dahleh, M.A. and J.B. Pearson (1988). Optimal Rejection of Persistent Disturbances, Robust Stability, and Mixed Sensitivity Minimization. *IEEE Trans. Automat. Contr.*, **33**, pp. 722–731.

Diaz-Bobillo, I.J. and M.A. Dahleh (1993). Minimization of the Maximum Peak-to-Peak Gain: The General Multiblock Problem. *IEEE Trans. Automat. Contr.*, **38**, pp. 1459–1482.

Halpern, M.E., R.J. Evans and R.D. Hill (1996). Optimal Pole Placement Design for SISO Discrete-Time Systems. *IEEE Trans. Automat. Contr.*, **41**, pp. 1322–1326.

Halpern, M.E. (1998). Optimal Pole Placement in an $l_1$ Framework. In: *Proceedings of the 37th IEEE Conference on Decision and Control*, Tampa, FL, pp. 3311–3314.

Staffans, O.J. (1991). Mixed Sensitivity Minimization Problems with Rational $l^1$-Optimal Solutions. *J. Optimiz. Theory Appl.*, **70**, pp. 173–189.

Staffans, O.J. (1992). MIMO $l^1$-Optimization with a Scalar Control. *J. Optimiz. Theory Appl.*, **74**, pp. 545–564.

Vidyasagar, M. (1986). Optimal Rejection of Persistent Bounded Disturbances, *IEEE Trans. Automat. Contr.*, **AC-31**, pp. 527–534.

Vidyasagar, M. (1991). Further Results on the Optimal Rejection of Persistent Bounded Disturbances, *IEEE Trans. Automat. Contr.*, **36**, pp. 642–652.

14th Triennial World Congress, Beijing, P.R. China

# On Continuity Property of $\ell^1$ Optimization Design in General Case

**Sheng-Ping Li**
**Tian-You Chai**

*Department of Mechantronic Engineering,*
*Shantou University, Shantou, 515063, P. R. China*
*email: spli@mailserv.stu.edu.cn*
*Research Center of Automation,*
*Northeast University, Shenyan, 110006, P. R. China*
*email: tychai@mail.neu.edu.cn*

Abstract: In this paper we investigate the continuity property of $\ell^1$ optimization design in general case, extending the main result in (Dahleh *et al.* ,1990) where the continuity of $\ell^1$ optimization design is proved under some restrictive assumption. At first, in the general case, the existence of suboptimal solution of $\ell^1$ optimization problem is proved and a unified $\ell^1$ optimization design algorithm is presented .Then the continuity property of $\ell^1$ optimization design in general case is proved. This conclusion is the basis of the development of adaptive $\ell^1$ control theory. *Copyright © 1999 IFAC*

Keywords: optimization, control design, continuity, suboptimal control, adaptive control.

## 1. INTRODUCTION

The continuity property of $\ell^1$ optimization design with respect to the parameters of plant is of importance in studies of certain adaptive schemes of $\ell^1$ control which is also called adaptive $\ell^1$ control (Dahleh *et al.* ,1990; Voulgaris *et al.* , 1990). A standard procedure for the problem of solving a $\ell^1$ optimization design (Dahleh *et al.* ,1988; Khammash, 1996) is first to transfer it to a standard $\ell^1$ norm minimization problem which depends on the parameters of plant using parameterization formula (Vidyasagar, 1985) and then; applying duality (Lueberger, 1969) to reduce the norm minimization to linear programming problem under the assumption that no zeros and poles of the plant are present on unit circle. In studies of adaptive schemes of $\ell^1$ design, it appears desirable that $\ell^1$ optimization design approach should satisfy certain continuity property. A $\ell^1$ optimization design approach can be found in (Dahleh *et al.* ,1988) which is proved that satisfies certain continuity property (Dahleh *et al.*, 1990) ,under the assumption that the plant has no zeros and poles on the unit circle. It is noted that this assumption is very restrictive to practical plants. However, it is proved in (Vidyasagar, 1991) that there is no optimal solution to the $\ell^1$ norm optimization problem in the case that the plant has zeros or poles on the unit circle. therefore, the continuity property of $\ell^1$ design in general case remains to be established.

As mentioned above, in general the optimal solution of

$\ell^1$ norm optimization problem may not always be obtained. In this paper, a unified algorithm is addressed to solve the $\ell^1$ norm optimization problem. Applying this approach to the $\ell^1$ optimization problem, if no zero and pole is present on the unit circle one optimal solution can be obtained; otherwise, the suboptimal solution that arbitrarily approximates to optimal solution can only be obtained. then, main results that $\ell^1$ optimization design in general case has desirable continuity property with respect to parameters of the plant are established.

## 2. NOTATIONS AND PRELIMINARIES

The following notation and terminology are employed: x(i) represent the i th elements of the real vector x: ,

$$\|x\|_\infty \stackrel{def}{=} \max_i |x(i)|,$$

$$\|x\|_1 \stackrel{def}{=} \sum_{i=0}^{\infty} |x(i)|, \ell^1 \stackrel{def}{=} \{x: \|x\|_1 < \infty\},$$

its Z-transform is $\hat{x} \stackrel{def}{=} \sum_{i=0}^{\infty} x(i) z^i$,

$A \stackrel{def}{=} \{\hat{x}: \|x\|_1 < \infty\}$. Then the space A is isomerphic to $\ell^1$. Let $p_k$ be the truncation operator defined by $p_k x = \{x_0, x_1, \ldots, x_k, 0, 0, \ldots\}$.

As the standard $\ell^1$ optimization problem ,we will consider a feedback system as in figure 1where

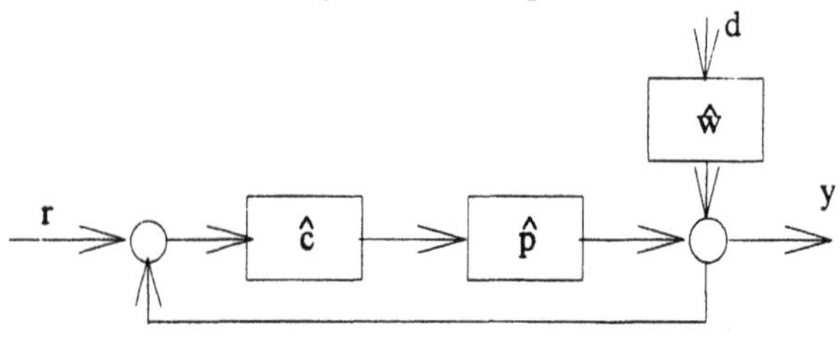


fig.1 feedback control system

r is referent input, d is external disturbance satisfying $\|d\|_\infty \le 1, \hat{w} \in A$ is weighting function, y is output. In this setting the design requirement is to find a compensator which internally stabilize the plant $\hat{p}$ and optimally reduces the effect of the disturbance on the output. This problem can be formulated as follows:

$$\mu = \inf_{\hat{c} \in S(\hat{p})} \left\| (1 + \hat{p}\hat{c})^{-1} \hat{w} \right\|_A,$$

where $S(\hat{p}) = \{\hat{c}: \hat{c} \text{ internally stabilize } \hat{p}\}$.

By using a parameterization of $S(\hat{p})$ (Vidyasagar, 1985), above optimal problem is equivalent to the following minimum distance problem in A:

$$\mu = \inf_{\hat{q} \in A} \left\| \hat{h} - \hat{q}\hat{v} \right\|_A, \qquad (\text{opt}_0)$$

here $\hat{h} = \hat{d}\hat{v}\hat{w}, \hat{v} = \hat{d}\hat{n}\hat{w} (\hat{p} = \hat{n}\hat{d}^{-1}, \hat{x}\hat{n} + \hat{y}\hat{d} = 1,$ $\hat{n}, \hat{d}, \hat{x}, \hat{y} \in A)$. By the similar procedure, other $\ell^1$ optimization design problem, such as robust stability and mixed sensitivity minimization, may be reduced to this form (Dahleh *et al.* ,1988).

## 3. AN UNIFIED SOLVING ALGORITHM FOR $(\text{opt}_0)$

Let $\hat{k} = \hat{q}\hat{v}$ ,the optimization design problem $(\text{opt}_0)$ that defines the $\ell^1$-optimal controller can be rewritten as

$$\mu = \inf_{\hat{k} \in S} \left\| \hat{h} - \hat{k} \right\|_A, \qquad (\text{opt})$$

where $S = \{\hat{k} \in A: \exists \hat{q} \in A \text{ satisfying } \hat{k} = \hat{q}\hat{v}\}$.

The existence of optimal solution of (opt) is determined by transmission zeros of $\hat{v}$ which corresponds to the zeros and poles of the plant. Furthermore, if $\hat{v}$ has no transmission zeros on the unit circle, then (opt) has an optimal solution to (opt); otherwise, it is no longer possible to compute optimal solution to (opt).it is only possible to compute suboptimal solution arbitrarily close to the optimal solution (Vidyasagar, 1991). In this subsection, we present an unified approach to tackle the $\ell^1$ optimization problem, despite that whether $\hat{v}$ has transmission zeros on unit circle or not.

Consider the following family of related finite dimensional problems which we refer to as truncated problems, and which are indexed by the non-negative integer $\delta$:

$$\mu_\delta = \min_{k \in S_\delta} \|h - k\|_1, \qquad (\text{opt}_\delta)$$

where

$$S_\delta = \{k: \hat{k} \in S, \hat{h} - \hat{k} \text{ is a polynomial of degree } \delta, \hat{h} \in A\}$$

Of course, the problem $(\text{opt}_\delta)$ always has a bounded objective function but the $S_\delta$ may be empty for the given problem (opt) for a given $\delta$ .In addition, if there exists $\delta^*$ such that $S_{\delta^*}$ is non-empty we can define a monotonically increasing integer sequence $\{\delta(i)\}_{i=0}^{\infty}$ for which, taking $\delta(0) \ge \delta^*$,the corresponding sequence of problems $\{opt_{\delta(i)}\}$ will have well defined infimal norms $\mu_{\delta(i)}$.The sequence $\mu_{\delta(i)}$ will clearly be monotonically non-increasing and, it is expected that $\lim_{i \to \infty} \mu_{\delta(i)} = \mu_{opt}$.Now, we are ready to state the main result of this subsection.

Theorem 1. There always exists $\delta^*$ such that $S_{\delta^*}$ is non-empty, and moreover, $\mu_\delta - \mu$ can be made arbitrarily small by taking $\delta$ sufficiently large (at least $\delta \geq \delta^*$).

Proof For problem(opt),there exists a $k^* \in S$ such that

$$\|h - k^*\| \leq \mu + \varepsilon/3, \tag{3.1}$$

for any $\varepsilon > 0$.Let $a_i$ $(i = 1,2,\ldots,n_z)$ is the zeros of $\hat{v}$ on the unit disk ,where $n_z$ is the number of such zeros. It is clear that $\hat{k}^*$ satisfies $\hat{k}^*(a_i)$=0 $(i = 1,2,\ldots,n_z)$. As a result, we can always construct a polynomial $\hat{k}^*_{a1}$ such that $\hat{k}^*_{a1}(a_i)$=0 $(i = 1,2, \ldots, n_z)$ and $\|k^*_{a1}\|_1$=1, and simultaneously satisfying $\hat{k}^* = \hat{k}^*_{a1}\hat{k}_2$ where $\hat{k}_2 \in A$. Using the fact that polynomial is dense in A, and since $\hat{h} \in A$ and $\hat{k}_2 \in A$, there respectively exists a polynomial approximation $\hat{h}_a$ of $\hat{h}$ and a polynomial approximation $\hat{k}_{a2}$ of $\hat{k}_2$ such that

$$\|h - h_a\|_1 < \varepsilon/3, \tag{3.2}$$

and

$$\|k_2 - k_{a2}\|_1 < \varepsilon/3, \tag{3.3}$$

for the given $\varepsilon$.Let $\hat{k}^*_a = \hat{k}^*_{a1}\hat{k}_{a2}$. It is obviously that $\hat{k}^*_a$ is a polynomial satisfying $\hat{k}^*_a(a_i) = 0$ $(i = 1,2, \ldots, n_z)$ it follows from (3.1), (3.2), (3.3) and $\|k^*_{a1}\|_1$=1 that

$$\|k^* - k^*_a\|_1 = \|k^*_{a1}(k_2 - k_{a2})\|_1$$
$$\leq \|k^*_a\|_1 \|k_2 - k_{a2}\|_1 < \varepsilon/3.$$

Using the triangle inequility, we have

$$\|h_a - k^*_a\|_1 = \|h_a - k^* + k^* - k^*_a\|_1 \leq \|h_a - k^*\|_1$$
$$+ \|k^* - k^*_a\|_1 < \|h_a - h + h - k^*\|_1 + \varepsilon/3$$
$$\leq \|h - k^*\| + \|h_a - h\| + \varepsilon/3 < \mu + \varepsilon.$$

Let $\delta^*$ denote the degree of polynomial $h_a - k^*_a$. Consequently, from above arguments we conclude that there exists $\delta^*$ such that $k^*_a \in S_{\delta^*}$ satisfying $\|h_a - k^*_a\|_1 < \mu + \varepsilon$,for the given $\varepsilon > 0$.Since $\varepsilon$ is arbitrarily given. This completes the proof.

Remark 1. The above theorem implies that any $\ell^1$ optimization problem $(opt)$ can be arbitrarily approximated by its corresponding truncated problem $(opt_\delta)$.therefore, the suboptimal solution of $(opt)$ can be found by solving the corresponding truncated problem $(opt_\delta)$ which there always exists optimal solution in A.

Remark 2. Let $\hat{b}_a = \hat{h}_a - \hat{k}_a$.When no unit circle zeros of $\hat{v}$ are present, the exact minimizer of $(opt)$ corresponds to a polynomial $\hat{b}_a$ and hence, by theorem 1, can be found by solving the corresponding truncated problem $(opt_\delta)$ for a suitably chosen $\delta$.

With above results and the use of a standard linear programming technique for handling the objective function (Mendlovitz, 1989) we see that $(opt_\delta)$ is equivalent to the following linear program in the coefficients $b_a(k)$ and the auxiliary variables $b^+_a(k), b^-_a(k)$ and $\lambda$:

$$\min \lambda \tag{$LP_\delta$}$$

s.t.
$$b^+_a(k) - b^-_a(k) = b_a(k),$$
$$b^+_a(k),\ b^-_a(k) \geq 0, k = 0,1,\ldots,\delta,$$
$$\sum_{k=0}^{\delta}[b^+_a(k) + b^-_a(k)] \leq \lambda,$$
$$h_a - b_a \in S_\delta,$$

where $S_\delta$ is the feasible set of $(opt_\delta)$.For any $\delta$, $(LP_\delta)$ has a finite number of variables and a finite number of constraints.

From above arguments, we present $\ell^1$ optimization design algorithm as following:

Step 1 Let $\delta = n_z - 1$.

Step 2 Solving $(LP_\delta)$,obtain $\hat{b}^*_a$ and $\mu_\delta$.

Step 3 Verify $|\mu_\delta - \mu_{\delta-1}|$ is sufficiently small or not. If not, let $\delta = \delta + 1$ and go to step 2.

Step 4 From $\hat{b}^*_a = (1 + \hat{p}\hat{c}^*)^{-1}\hat{w}$ ,obtain the controller

$$\hat{c}^* = \frac{w - \hat{b}^*_a}{\hat{p}\hat{b}^*_a}. \tag{3.4}$$

Remark 3. If the plant has no zeros and poles on the unit circle which implies that no unit circle zero of $\hat{v}$ are present, then the degree of $\hat{b}^*_a$ may stop increasing at certain $\delta$.

## 4. CONTINUITY PROPERTY OF $\ell^1$OPTIMIZATION DESIGN

This subsection is concerned with the continuity property of $\ell^1$ optimization design in general case. This property exhibits the type of dependence of the $\ell^1$ controller have on the system parameters, which is also the basic property used in adaptive $\ell^1$ control and control of slowly varying system.

Let $\hat{p}_i = \hat{n}_i \hat{d}_i^{-1}$ denote a sequence of plants, where $\hat{n}_i, \hat{d}_i \in A, \hat{d}_i(0) \neq 0; \hat{n}_i, \hat{d}_i$ are coprime. By complex analysis theory, both $\hat{n}_i, \hat{d}_i$ will have a finite number of zeros in the closed unit disk. It follows from $(opt_0)$ that $\hat{h}_i = \hat{d}_i \hat{y}_i \hat{w}$ and $\hat{v}_i = \hat{d}_i \hat{n}_i \hat{w}$, where $\hat{x}_i \hat{n}_i + \hat{y}_i \hat{d}_i = 1$. Now, let CF denote the set of all plants $\hat{p} = \hat{n}\hat{d}^{-1}, \hat{n}, \hat{d} \in A$, $\hat{d}(0) \neq 0$; $\hat{n}, \hat{d}$ are coprime. CF is equipped with the standard graph topology (Vidyasagar, 1985 ) induced over $\ell^\infty$. The topology is better understood from the notion of convergence: a sequence of plants $\hat{p}_i = \hat{n}_i \hat{d}_i^{-1} \in CF$ converges to a plant $\hat{p} = \hat{n}\hat{d}^{-1} \in CF$, if $\hat{n}_i$ converges to $\hat{n}$ and $\hat{d}_i$ converges to $\hat{d}$ in A. With above notion, we can obtain the following lemma that establishes the continuity of $\hat{h}_i$ and $\hat{v}_i$ with respect to $\hat{p}_i$.

Lemma 1. If $\hat{p}_i \to \hat{p}$ in CF, then $\hat{v}_i \to \hat{v}$, $\hat{h}_i \to \hat{h}$ in A.

Proof By Bezout equivalence, we have

$$(\hat{n}_i \quad \hat{d}_i)\begin{pmatrix}\hat{x}_i \\ \hat{y}_i\end{pmatrix} - (\hat{n} \quad \hat{d})\begin{pmatrix}\hat{x} \\ \hat{y}\end{pmatrix} = 0.$$

It follows from triangle inequivalence that

$$\left\| (\hat{n} \quad \hat{d})\left[\begin{pmatrix}\hat{x}_i \\ \hat{y}_i\end{pmatrix} - \begin{pmatrix}\hat{x} \\ \hat{y}\end{pmatrix}\right] \right\|$$

$$\leq \left\| (\hat{n}_i \quad \hat{d}_i) - (\hat{n} \quad \hat{d}) \right\| \left\| \begin{pmatrix}\hat{x}_i \\ \hat{y}_i\end{pmatrix} \right\|. \tag{4.1}$$

Since $\hat{p}_i \to \hat{p}$ in CF, then $\hat{n}_i \to \hat{n}$ and $\hat{d}_i \to \hat{d}$ in A. From (4.1)and Bezout equivalence, we have $\hat{x}_i \hat{n} + \hat{y}_i \hat{d} \to 1$.This means that $\hat{x}_i \to \hat{x}$ and $\hat{y}_i \to \hat{y}$.Since $\hat{h}_i = \hat{d}_i \hat{y}_i \hat{w}$ and $\hat{v}_i = \hat{d}_i \hat{n}_i \hat{w}$,lemma 1 follows.

Since $\hat{n}_i, \hat{d}_i$ have a finite number of zeros in the closed unit disk, so is $\hat{v}_i$.Thus, we may introduce following lemma.

Lemma 2. Let $\hat{v}_i \in A$ so that $\hat{v}_i$ converges to $\hat{v} \in A$,and $\hat{v}_i(z) \neq 0$ is analytic in an open disk containing the unit disk. Then, every zero $a_i$ of $\hat{v}$ on the unit disk is a limit point of zeros of $\hat{v}_i$.

Proof Similar to lemma 1 in (Dahleh *et al.* ,1990).

Lemma 3. Define $\mu_i$ by $\mu_i = \inf_{k_i \in S_i} \|h_i - k_i\|_1$, where $S_i = \{k_i \in \ell^1 : \exists q_i \in \ell^1 \text{satisfying } k_i = q_i v_i\}$. If $h_i \to h$ and $v_i \to v$ in $\ell^1$, then $\mu_i \to \mu$.

Proof For simplicity, suppose that $a_j, a_{ij}$ $(j = 1,2,\ldots n_z)$ are the distinct zeros of $\hat{v}$ and $\hat{v}_i$ that lie on the unit disk, respectly. By duality theorem (Lueberger , 1969), we have

$$\mu_i = \max_{\lambda \in L_i}\left\{\sum_{j=1}^{n_z}\left[\lambda_j \operatorname{Re}\hat{h}_i(a_{ij}) + \lambda_{n_z+j} \operatorname{Im}\hat{h}_i(a_{ij})\right]\right\},$$

where the set

$$L_i = \{\bar{\lambda} : \left|\sum_{j=1}^{n_z}\left[\lambda_j \operatorname{Re}(a_{ij}^k) + \lambda_{n_z+1} \operatorname{Im}(a_{ij}^k)\right]\right| \leq 1,$$

$$k = 0,1,2,\ldots\}.$$

For a given $\varepsilon > 0$, by the fact $h_k$ converges to h and lemma 1, there exists an integer $I_1$ such that $\hat{v}_i$ and $\hat{v}$ have the same number zeros on the unit disk, and

$$|\mu_i - \mu_i'| < \varepsilon, \text{for all } i > I_1, \tag{4.2}$$

where

$$\mu_i' = \max_{\lambda \in L_i}\left\{\sum_{j=1}^{n_z}\left[\lambda_j \operatorname{Re}\hat{h}(a_j) + \lambda_{n_z+j} \operatorname{Im}\hat{h}(a_j)\right]\right\}.$$

In addition, for the given $\varepsilon > 0$,there exists a $I_2$ such that for all $i > I_2$,we have

$$L_i \subseteq L_\varepsilon = \{\bar{\lambda} : \left|\sum_{j=1}^{n_z}\left[\lambda_j \operatorname{Re}(a_j^k) + \lambda_{n_z+1} \operatorname{Im}(a_j^k)\right]\right|$$

$$\leq 1 + \varepsilon, k = 0,1,2,\cdots\},$$

let $\bar{r} = \left\{\sum_{j=1}^{n_z}\left[\lambda_j \operatorname{Re}(a_j^k) + \lambda_{n_z+1} \operatorname{Im}(a_j^k)\right]\right\}_{k=0}^{\infty}$. This implies that

$$\mu_i' \leq \max_{\lambda \in L_\varepsilon}\left\{\sum_{j=1}^{n_z}\left[\lambda_j \operatorname{Re}\hat{h}(a_j) + \lambda_{n_z+j} \operatorname{Im}\hat{h}(a_j)\right]\right\} =$$

$$\max_{\|\bar{r}\|_\infty \leq 1+\varepsilon} \langle h, \bar{r} \rangle = (1+\varepsilon)\mu. \tag{4.3}$$

Combining (4.2) and (4.3),we have

$$\mu_i \leq (1+\varepsilon)\mu + \varepsilon, \text{ for all } i > \max(I_1, I_2). \tag{4.4}$$

With the same argument, we can show that there exists $I_3$ such that

$$\mu \leq (1+\varepsilon)\mu_i + \varepsilon, \text{ for all } i > I_3. \tag{4.5}$$

From (4.4) and (4.5),the proof is completed.

As mentioned in theorem 1, for a given $\ell^1$ optimization problem, there always exists a bounded integer $\delta$ such that $S_\delta$ is non-empty, and also suboptimal solution exists. Next, we prove a lemma concerning a unified bounded $\delta$ on the existence of suboptimal solution of any convergent sequence plants.

Lemma 4. Let $h_i \to h, v_i \to v$ in $\ell^1$,then there exists a unified bounded integer $\delta^*$ such that $S_{i\delta^*}$ (i=1,2,...)

is non-empty and suboptimal solution of $\mu_i$ exists for i=1,2,...

Proof For the given $\varepsilon > 0$, for (opt) there exists a $k^* \in S$ such that

$$\|h - k^*\|_1 \le \mu + \varepsilon/3, \quad (4.6)$$

where $\hat{k}^* = \hat{k}^*_{a1}\hat{k}^*_2$ satisfying $\hat{k}_{a1}(a_j) = 0$ $(j = 1,2,\ldots n_z)$ and $\|k_{a1}\|_1 = 1$ and $\hat{k}^*_2 \in A$.By the hypothesis that $h_i \to h, v_i \to v$ and lemma 2, following the same argument as in proof of theorem 1, there exists an integer $I_1$ and $\delta_0$ such that $\hat{v}_i$ and $\hat{v}$ have the same number of zeros that lie on the unit disk, and $\left\|\hat{k}^*_{ai} - k^*\right\|_1 < \varepsilon/3, \|h_{a1} - h\|_1 < \varepsilon/3$,for $i > I_1$.

here,$\hat{k}^*_{ai}$ and $\hat{h}_{a1}$ are polynomial of degree $\delta_0$, $\hat{k}^*_{a1} = \hat{k}^*_{ai}\hat{k}^*_{2a}$ and satisfying $\hat{k}_{a1i}(a_{ij}) = 0$ $(j = 1,2,\ldots n_z)$ and $\|k_{a1i}\|_1 = 1$, $\hat{k}^*_{2a}$ is the approximation polynomial of $\hat{k}^*_2$.Hence, $\hat{k}^*_{ai} \in S_{i\delta_0}$.

Now, let $\mu_{\delta_0 i} = \|h_{ai} - k^*_{ai}\|_1$. By triangle inequivality and (3.8), we have

$$|\mu_{\delta_0 i} - \mu| < \left|\|h_{ai} - k^*_{ai}\|_1 - \|h - k^*\|_1\right| + \varepsilon/3$$
$$\le \|h_{ai} - h\|_1 + \|k^*_{ai} - k^*\|_1 + \varepsilon/3 < \varepsilon, \text{for } i > I_1. \quad (4.7)$$

On the other hand, it follows from lemma 3, there exists $I_2$ such that

$$|\mu_i - \mu| < \varepsilon, \text{ for } i > I_2. \quad (4.8)$$

Combining (4.7) and (4.8), we conclude that

$$|\mu_{\delta^* i} - \mu_i| \le \left|\,|\mu_{\delta^* i} - \mu| + |\mu_i - \mu|\,\right| < 2\varepsilon, \text{ for } i > \max(I_1, I_2).$$

Finaly, a unified upper bound given by any $\delta^*$ equal to the largest of $\delta_i$, $1 \le i \le \max(I_1, I_2)$ and $\delta_0$. This completes the proof.

By lemma 4, there exists an integer $\delta^*$ such that for all i, the suboptimal solution $\hat{b}^*_{ai} = \hat{h}_{ai} - \hat{k}_{ai}$ are polynomials of degree at most $\delta^*$. With this result, we can obtain following lemma.

Lemma 5. If $h_i \to h, v_i \to v$ in $\ell^1$. Then $\mu_{\delta^* i} \to \mu_{\delta^*}$.

Proof For an arbitrarily given $\varepsilon$>0,it follows from the same argument of lemma 4 and triangle inequivality that there exists I such that

$$|\mu_{\delta^* i} - \mu_{\delta^*}| = \left|\|h_{ai} - k^*_{ai}\|_1 - \|h_a - k^*_a\|_1\right|$$
$$\le \|h_{ai} - h_a + k^*_{ai} - k^*_a\|_1 \le \|h_{ai} - h_a\|_1$$
$$+ \|k^*_{a2}(k^*_{a1i} - k^*_{a1})\|_1 < \varepsilon,$$

for i>I. The proof is completed.

Next, using lemma 1,lemma 4 and lemma 5, we will show the main result of this subsection.

Theorem 2. Let $\hat{b}^*_a$ denote the suboptimal solution of (opt) that satisfies $\mu_\delta = \|b^*_a\|_1 = \|h_a - k^*_a\|_1$ which is assumed to have a unique optimal solution. If $\hat{p}_i \to \hat{p}$ in A, then there exists an integer $\delta^*$ such that $\hat{b}^*_{ai}(i = 1,2,\ldots)$and $\hat{b}^*_a$ are polynomials of degree at most $\delta^*$ and $\hat{b}^*_{ai} \to \hat{b}^*_a$ in A.

Proof In lemma 1, it has been shown that if $\hat{p}_i \to \hat{p}$ then $h_i \to h$ and $v_i \to v$, and also by lemma 4 and lemma 5, it is concluded that there exists an integer $\delta^*$ such that suboptimal solution $\hat{b}^*_{ai}(i = 1,2,\ldots)$ and $\hat{b}^*_a$ are polynomials of degree at most $\delta^*$ and $\mu_{\delta^* i} \to \mu_{\delta^*}$,which means that $\left\{\|b^*_{ai}\|_1\right\}_{i=1}^{\infty}$ is a bounded sequence, and there exists a positive C such that $\|b^*_{ai}\|_1 < C$ for all $i$ .Since $\hat{b}^*_{ai}$ are polynomials of degree $\delta^*$ for all i, by compactness of the unit ball in$R^{\delta^*}$,we conclude that there exists a convergent subsequence of $\{b^*_{ai}\}$, and denote it by $\{b^*_{ai_k}\}$. Assume that $\lim_{k\to\infty} \hat{b}^*_{ai_k} = \bar{b}^*_a$. Since $b^*_{ai_k}$ are suboptimal solution, hence, $\hat{k}^*_{ai_k}(a_{i_k j}) = 0$ (i.e.$k^*_{ai_k} \in S_{\delta^* i_k}$) where $a_{i_k j}$ are the zeros of $\hat{v}_{i_k}$ that lie on unit disk. This means that $\hat{b}^*_{ai_k}(a_{i_k j}) = \hat{h}^*_{ai_k}(a_{i_k j})$. Hence,

$$\bar{b}^*_a(a_j) = \lim_{k\to\infty} \hat{b}^*_{ai_k}(a_{i_k j}) = \bar{h}_a(a_j).$$

It is implied that

$$\bar{k}^*_a(a_j) = 0 \text{(i.e. } \bar{k}^*_a \in S_{\delta^*}). \quad (4.9)$$

On the other hand, using lemma 5 we have

$$\lim_{k\to\infty} \|b^*_{ai_k}\|_1 = \|\bar{b}^*_a\|_1 = \mu_{\delta^*}. \quad (4.10)$$

Combination with (4.9)and(4.10), and by the uniqueness assumption, it follows that all the cluster points of $b^*_{ai_k}$ are the same and equal to $\bar{b}^*_a$. The theorem is proved.

Remark 4. From remark 2, the problem of obtaining $\hat{b}^*_a$ is equivalent to a standard linear programming problem $(LP_\delta)$.The nonuniqueness of the solution of $(LP_\delta)$is due to the fact that the object function of $(LP_\delta)$is parallel with one of the faces of the feasible set in $(LP_\delta)$. This situation is nongeneric because of the fact that any small perturbation in the plant parameters will remove this situation, and the unique solution will occur at a corner of the feasible set.

In the development of a comprehensive theory of adaptation, it is of importance to study the continuity property of feedback design. The following result claims that $\ell^1$ optimization design has desirable continuity. This is a consequence of theorem 2 and (3.4).

Theorem 3. If $\hat{p}_i$ converges to $\hat{p}$ in CF, denote their corresponding $\ell^1$ controllers by $\hat{c}_i$ and $\hat{c}$ respectively, then $\hat{c}_i$ converges to $\hat{c}$.

Proof If $\hat{p}_i$ converges to $\hat{p}$ in CF, then by theorem 2 we have $\hat{b}_{ai}^* \to \hat{b}_a^*$ in A. Rewrite the $\ell^1$ controller given by (3.4) as following

$$\hat{c}^* = \frac{w - \hat{b}_a^*}{\hat{p}\hat{b}_a^*}.$$

Since

$$\left\|\hat{p}_i \hat{b}_{ai}^* - \hat{p}\hat{b}_a^*\right\|_A \le \left\|\hat{p}_i - \hat{p}\right\|_A \left\|\hat{b}_{ai}^*\right\|_A + \left\|\hat{b}_{ai}^* - \hat{b}_a^*\right\|_A \left\|\hat{p}\right\|_A,$$

and

$$\left\|(\hat{w} - \hat{b}_{ai}^*) - (\hat{w} - \hat{b}_a^*)\right\|_A = \left\|\hat{b}_{ai}^* - \hat{b}_a^*\right\|_A.$$

By the notion of convergence in graph topology induced over $\ell^\infty$,it is clear that $\hat{p}_i \to \hat{p}$ and $\hat{b}_{ai}^* \to \hat{b}_a^*$ means $\hat{c}_i \to \hat{c}$.Theorem 3 follows.

## 5. CONCLUSION

In this paper, the existence of suboptimal solution to $\ell^1$ optimization in the general case is proved and, then an unified solving algorithm $\ell^1$ optimization problem in this case is presented.

Then , the continuity property of $\ell^1$ optimization design with respect to the parameters of plant is proved in general case, which may be employed to develop a novel theory of adaptive $\ell^1$ control. This conclusion extends the result in (Dahleh, *et al* . , 1990) , where the same conclusion is proved in the case that no zero and pole are present on the unit circle which is very restrictive in practical application.

## REFERENCES


Dahleh,M. and M.A.Dahleh (1990). Optimal rejection of persistent and bounded disturbances: continuity properties and adaption. *IEEE Trans. Automat. Contr.*, AC-35, 687-696.

Dahleh,M.A.and J.B.Pearson (1988).Optimal rejection of persistent disturbances, robust stability and mixed sensitivity minimization. *IEEE Trans. Automat. Contr.*, AC-33, 722-731.

Khammash, M.H.(1996). Synthesis of globally optimal controllers for robust performance to unstructured uncertainty. *IEEE Trans. Automat. Contr*, AC-41, 189-198.

Lueberger , D. G. (1969). *Optimization by Vector Space Methods.* New York: Wiley.

Mendlovitz, M.A. (1989). A simple solution to the $\ell^1$ optimization problem. *Systems Control Lett.* 461-463.

Vidyasagar, M. (1991). Further results on the optimal rejection of persistent bounded disturbances. *IEEE Trans. Automat.Contr.*, AC-36, 642-652.

Vidyasagar, M. (1985). *Control System Synthesis: A Fatorization Approach.* Cambridge, MA: M.I.T. Press.

Voulgaris,P.G.,M.A. Dahleh and L.S. Valavani (1992). On slowly varying systems: $\ell^\infty$ to $\ell^\infty$ performance and implication to robust adaptive control. *ACC*, 2, 1197-1201.

# LMI SOLUTION TO SINGULAR $\mathcal{H}_2$ SUBOPTIMAL CONTROL PROBLEMS

**Kang-Zhi Liu * Takashi Sato ***

* *Department of Electronics and Mechanical Engineering*
*Chiba University, 1-33 Yayoi-cho, Inage-ku, Chiba 263-8522, Japan*
*Fax: +81-43-290-3039 Email: liu@cute.te.chiba-u.ac.jp*

**Abstract.** In this paper, we derive an LMI solution to general $\mathcal{H}_2$ suboptimal control problems without any unnecessary assumptions. Necessary and sufficient conditions for the $\mathcal{H}_2$ norm of the closed loop transfer matrix to be bounded by a given number are given by 4 coupled LMIs. A suboptimal controller is also obtained. The obtained solution applies to both nonsingular and singular cases. *Copyright © 1999 IFAC*

**Keywords.** $\mathcal{H}_2$ control, suboptimal problem, singular problem, LMI

## 1. INTRODUCTION

It has been well recognized that as a performance index for control systems, $\mathcal{H}_2$ norm is more adequate than $\mathcal{H}_\infty$ and $\mathcal{L}_1$ norms. This problem has been known as LQG control problem and widely studied since 1960's. The purpose is to find a controller such that the closed loop system is internally stable and the $\mathcal{H}_2$ norm of certain closed loop transfer matrix is minimized. The modern version of this problem is treated in the generalized feedback systems of Fig. 1 under the name of $\mathcal{H}_2$ control problem. Refer to (Zhou *et al.*, 1995; Stoorvogel, 1992) and the references therein.

Subject to some nonsingularity assumption on the invariant zero of the generalized plant(Zhou *et al.*, 1995), an elegant 2-Riccati-equation solution has been derived. However, when the nonsingularity assumption is removed, which is the usual case in practice, the optimal $\mathcal{H}_2$ problem turns out to be very hard to solve. Stoorvogel(Stoorvogel, 1992) gave, by using a geometric approach, a formula for the infimum of $\mathcal{H}_2$ norm and a method for the construction of minimizing sequence of controllers when the infimum is attainable by a strictly proper controller. However, when the infimum of $\mathcal{H}_2$ norm is either attained by biproper controllers or not attainable at all, the problem remains open. To our knowledge, there has not been a complete solution to this optimal problem up to date.

Meanwhile, in the mixed $\mathcal{H}_2$ /$\mathcal{H}_\infty$ or $\mathcal{H}_2$ /$\mu$ norm optimization problems(Bernstein and Haddad, 1989; Hall and How, 1993; Khargonekar and Rotea, 1991; Zhou *et al.*, 1993; Stoorvogel, 1993), basically an upper bound of the $\mathcal{H}_2$ norm, rather than the $\mathcal{H}_2$ norm itself, is minimized. Further, some restrictive technical assumptions are also made in solving these problems.

In this paper, we consider the following suboptimal $\mathcal{H}_2$ problem instead of the optimal problem.

**$\gamma$-optimal $\mathcal{H}_2$ Control Problem** : *Consider the feedback system of Fig. 1 and let the closed loop transfer matrix from $w$ to $z$ be denoted by $H_{zw}(s)$. For a given $\gamma > 0$, find a controller such the closed loop system is internally stable and $||H_{zw}||_2 < \gamma$. Such a controller is called a $\gamma$-optimal $\mathcal{H}_2$ controller.*

This relaxation of optimality allows us to make full use of LMI approach(Iwasaki and Skelton, 1994; Gahinet and Apkarian, 1994) to derive $\mathcal{H}_2$ controller relatively easily. And this relaxation of optimality is reasonable in practice. Both the solvability condition and the formula of a $\gamma$-optimal $\mathcal{H}_2$ controller will be derived in this paper.

We remark that the infimum on the bound of $||H_{zw}||_2$ derived in (Stoorvogel, 1992) can be used to determine the lower bound for $\gamma$.

## 2. PRELIMINARIES

The notations used in this paper are fairly standard. The $\mathcal{H}_2$ norm $||H||_2$ of a stable transfer matrix $H(s)$ is defined by

$$||H||_2 = \left\{ \frac{1}{2\pi} \int_{-\infty}^{\infty} \text{Trace}\{H^*(j\omega)H(j\omega)\}d\omega \right\}^{1/2}. \quad (1)$$

Both $(A,\ B,\ C,\ D)$ and

$$\left[\begin{array}{c|c} A & B \\ \hline C & D \end{array}\right]$$

denote the state space realization for a transfer matrix $D + C(sI - A)^{-1}B$. Further, a linear fractional transformation $\mathcal{F}_\ell$ is defined as

$$\mathcal{F}_\ell(G, K) = G_{11} + G_{12}(I - KG_{22})^{-1}KG_{21}$$
$$: G \quad = \begin{bmatrix} G_{11} & G_{12} \\ G_{21} & G_{22} \end{bmatrix}.$$

### 2.1 *Closed loop system*

The system considered in this paper is described by the general feedback structure of Fig. 1 in which $G(s)$ is the generalized plant and $K(s)$ is a proper controller.

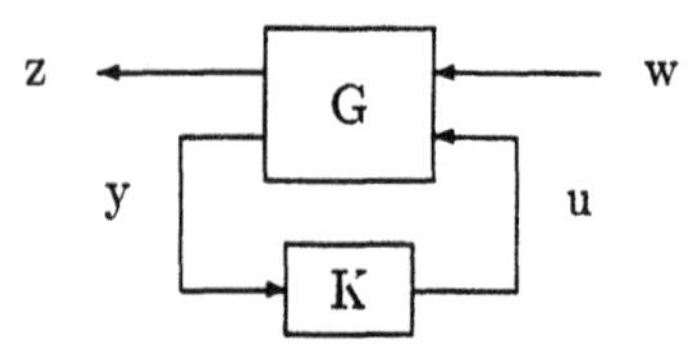


Fig. 1. Generalized feedback system

$w$ is the exogenous input, $u$ is the control input, $z$ is the controlled output and $y$ is the measured output. Let

$$G(s) = \begin{bmatrix} G_{11}(s) & G_{12}(s) \\ G_{21}(s) & G_{22}(s) \end{bmatrix}$$

$$= \left[\begin{array}{c|cc} A & B_1 & B_2 \\ \hline C_1 & D_{11} & D_{12} \\ C_2 & D_{21} & 0 \end{array}\right]$$

be a minimal realization of $G(s)$. The closed loop transfer matrix $H_{zw}$ from $w$ to $z$ is given by

$$H_{zw} = \mathcal{F}_\ell(G, K).$$

### 2.2 *Characterization of $\mathcal{H}_2$ norm*

The following lemma gives the state space characterizations of $\mathcal{H}_2$ norm which lays the foundation for LMI approach to $\mathcal{H}_2$ control problems.

**Lemma 1**

*(1) In order that $||C(sI - A)^{-1}B + D||_2 < \gamma$, $D = 0$ must hold.*

*(2) Assume $A$ is stable. Then there exists a unique $X \geq 0$ satisfying*

$$XA + A'X + C'C = 0 \quad (2)$$

*and the $\mathcal{H}_2$ norm is given by*

$$||C(sI - A)^{-1}B||_2^2 = \text{Trace}(B'XB). \quad (3)$$

*(3) The following statements are equivalent*

*(i) $A$ is stable and $||C(sI - A)^{-1}B||_2 < \gamma$.*

*(ii) There exists matrix $X = X' > 0$ such that*

$$XA + A'X + C'C < 0 \quad (4)$$
$$\text{Trace}(B'XB) < \gamma^2. \quad (5)$$

*(iii) There exist matrices $X = X'$, $W = W'$ such that*

$$\begin{bmatrix} XA + A'X & C' \\ C & -I \end{bmatrix} < 0 \quad (6)$$
$$\begin{bmatrix} W & B' \\ B & X^{-1} \end{bmatrix} > 0 \quad (7)$$
$$\text{Trace}W < \gamma^2 \quad (8)$$

**(Proof)** Refer to (Zhou *et al.*, 1995) for a proof of parts (1) and (2).

(3) (i) $\Rightarrow$ (ii): If (i) holds, there is a sufficiently small $\varepsilon > 0$ such that when $C$ is dilated to $C \rightarrow [C', \varepsilon^{1/2}I]'$

$$|| \begin{bmatrix} C \\ \varepsilon^{1/2}I \end{bmatrix} (sI - A)^{-1}B||_2 < \gamma$$

still holds. Applying part (2) to this new system, we have

$$XA + A'X + C'C = -\varepsilon I < 0 \quad (9)$$

$$\text{Trace}(B'XB) < \gamma^2$$

hold for some $X = X' \geq 0$. Suppose $X$ is singular, then there is a vector $v \neq 0$ satisfying $Xv = 0$. Multiplying $v'$ and $v$ to the left side and right side of Eq. (9) yields $v'C'Cv < 0$. This is a contradiction. Therefore, $X > 0$.

(ii) $\Rightarrow$ (i): According to Lyapunov stability theory, Eq. (4) has a solution $X > 0$ implies $A$ is stable. Let $\hat{C}$ be the matrix defined by

$$XA + A'X + C'C = -\hat{C}'\hat{C} < 0.$$

Application of part (2) gives

$$\left\|\begin{bmatrix} C \\ \hat{C} \end{bmatrix}(sI - A)^{-1}B\right\|_2 < \gamma.$$

Hence $\|C(sI - A)^{-1}B\|_2 < \gamma$.

(ii) $\Leftrightarrow$ (iii): The equivalence of Eq. (4) and Eq. (6) follows from Schur complement arguement. Moreover, Eq. (5) is equivalent to Eq. (8) and $W - B'XB > 0$. Finally, $W - B'XB > 0$ and $X > 0$ is equivalent to Eq. (7). □

The following lemma characterizes all solutions to a linear matrix equation. This lemma will be used to remove the constraint on the direct feedthrough term in the closed loop transfer matrix.

**Lemma 2** *(Boullion and Odell, 1971)*

*Consider the linear matrix equation*

$$D + AXB = 0.$$

*(1) This equation has a solution $X$ iff*

$$D - AA^{\dagger}DB^{\dagger}B = 0.$$

*(2) All solutions are parameterized by*

$$X = -A^{\dagger}DB^{\dagger} + S - A^{\dagger}ASBB^{\dagger}$$

*in which $S$ is an arbitrary matrix with compatible size.*

## 3. LMI SOLUTION

It is assumed that:

**A1** $(A, B_2)$ is stabilizable,
**A2** $(C_2, A)$ is detectable,
**A3** $D_{11} - D_{12}D_{12}^{\dagger}D_{11}D_{21}^{\dagger}D_{21} = 0$.

A1 and A2 are standard which guarantee that $G(s)$ is internally stabilizable. A3 is necessary for the $\mathcal{H}_2$ problem to be solvable(Zhou *et al.*, 1995), i.e. the direct feedthrough term in the closed loop transfer matrix must be zero in order for the $\mathcal{H}_2$ norm to be bounded. *Nonsingularity assumptions are not made* here.

In this section, we use the LMI characterization of $\mathcal{H}_2$ norm given in Lemma 1 to solve the $\gamma$-optimal $\mathcal{H}_2$ control problem. The solvability condition and a controller to $\gamma$-optimal $\mathcal{H}_2$ control problem are given in this section. The proof is summarized in Appendix A.

The derivation of Theorem 1 is strongly inspired by the Riccati inequality approach to $\mathcal{H}_\infty$ problem(Sampei *et al.*, 1990).

**Theorem 1** *Assume A1, A2, A3.*
*(1) () The $\gamma$-optimal $\mathcal{H}_2$ control problem is solvable iff there exist matrices $X = X'$, $M$, $Y = Y'$, $N$, $R$, $W = W'$ such that Eq. (10)~Eq. (14) below hold.*

$$D_{11} + D_{12}RD_{21} = 0 \tag{10}$$

$$\begin{bmatrix} AX + B_2M + (AX + B_2M)' & (C_1X + D_{12}M)' \\ C_1X + D_{12}M & -I \end{bmatrix} < 0 \tag{11}$$

$$\begin{bmatrix} (YA + NC_2)' + YA + NC_2 & (C_1 + D_{12}RC_2)' \\ C_1 + D_{12}RC_2 & -I \end{bmatrix} < 0 \tag{12}$$

$$\begin{bmatrix} W & (B_1 + B_2RD_{21})' & (YB_1 + ND_{21})' \\ B_1 + B_2RD_{21} & X & I \\ YB_1 + ND_{21} & I & Y \end{bmatrix} > 0 \tag{13}$$

$$\text{Trace}W < \gamma^2 \tag{14}$$

*(2) When Eq. (10)~Eq. (14) hold, a $\gamma$-optimal $\mathcal{H}_2$ controller $K(s)$ is given by*

$$K(s) = (A_K,\ B_K,\ C_K,\ D_K) \tag{15}$$

*in which*

$$\begin{aligned} D_K &= R \\ C_K &= -D_KC_2 + MX^{-1} \\ B_K &= B_2D_K - Z^{-1}(XN - B_2D_K) \\ A_K &= A + B_2D_KC_2 + B_2C_K - B_KC_2 + J \\ J &= Z^{-1}\{(AX + B_2M) + X(A + B_2D_KC_2)' \\ &\quad + X(C_1 + D_{12}D_KC_2)'(C_1X + D_{12}M)\}X^{-1} \\ Z &= XY - I. \end{aligned} \tag{16}$$

Note Eq. (10)~ Eq. (14) are coupled and not symmetric with respect to the pairs $(X, M)$, $(Y, N)$. This is the most significant difference from the decoupled and symmetric 2 Riccati-equations solution to nonsingular $\mathcal{H}_2$ problems.

Since Theorem 1 involves an equation Eq. (10), the condition is not easy to verify. So in the sequel we derive an equivalent condition in which Eq. Eq. (10) is removed.

**Corollary 1** *Assume A1, A2, A3 and define*

$$\hat{B}_2 = B_2(I - D_{12}^{\dagger}D_{12}),\ \hat{B}_1 = B_1 - B_2D_{12}^{\dagger}D_{11} \qquad \hat{C}_2 = (I - D_{21}D_{21}^{\dagger})C_2,\ \hat{C}_1 = C_1 - D_{11}D_{21}^{\dagger}C_2 \tag{17}$$

*(1) The $\gamma$-optimal $\mathcal{H}_2$ control problem is solvable iff there exist $X = X'$, $M$, $Y = Y'$, $N$, $R$, $W = W'$ such that Eq. (18)~Eq. (21) below hold.*

$$\begin{bmatrix} AX + B_2M + (AX + B_2M)' & (C_1X + D_{12}M)' \\ C_1X + D_{12}M & -I \end{bmatrix} < 0 \tag{18}$$

$$\begin{bmatrix} (YA + NC_2)' + YA + NC_2 & (\hat{C}_1 + D_{12}R\hat{C}_2)' \\ \hat{C}_1 + D_{12}R\hat{C}_2 & -I \end{bmatrix} < 0 \tag{19}$$

$$\begin{bmatrix} W & (\hat{B}_1 + \hat{B}_2RD_{21})' & (YB_1 + ND_{21})' \\ \hat{B}_1 + \hat{B}_2RD_{21} & X & I \\ YB_1 + ND_{21} & I & Y \end{bmatrix} > 0 \tag{20}$$

$$\mathrm{Trace} W < \gamma^2. \tag{21}$$

*(2) When Eq. (18)~Eq. (21) hold, a $\gamma$-optimal $\mathcal{H}_2$ controller $K(s)$ is given by Eq. (15) and Eq. (16) with $D_K = R$ replaced by*

$$D_K = -D_{12}^{\dagger}D_{11}D_{21}^{\dagger} + R - D_{12}^{\dagger}D_{12}RD_{21}D_{21}^{\dagger}. \tag{22}$$

(**Proof**) According to Lemma 2, all $R$ satisfying $D_{11} + D_{12}RD_{21} = O$ is parameterized by an arbitrary matrix $S$ as

$$R = -D_{12}^{\dagger}D_{11}D_{21}^{\dagger} + S - D_{12}^{\dagger}D_{12}SD_{21}D_{21}^{\dagger}. \tag{23}$$

Substitution of this equation into Theoerm 1 and rearrangment with respect to $S$ based on assumption A3 yields Corollary 1 immediately, in which $S$ has been rewritten as $R$. □

## 4. CONCLUSION

In this paper, a suboptimal singular $\mathcal{H}_2$ control problem has been solved based on LMI approach. The solvability conditions are shown to be given by 4 LMIs and the formula of a suboptimal $\mathcal{H}_2$ controller is provided.

Compared with the LMI solution to $\mathcal{H}_\infty$ control problems(Gahinet and Apkarian, 1994; Iwasaki and Skelton, 1994), both the derivation process and the LMI solution to singular $\mathcal{H}_2$ problems are not so elegant. This is partly due to the fact that the LMI characterization of $\mathcal{H}_2$ norm involves 3 LMIs in contrast to the 1 LMI involved in $\mathcal{H}_\infty$ norm characterization(Gahinet and Apkarian, 1994; Iwasaki and Skelton, 1994). There is still a need to search for more elegant solution.

It is believed that the LMI approach is promising and could yield insight in dealing with robust $\mathcal{H}_2$ control problems(Paganini, 1995; Stoorvogel, 1993).

## 5. REFERENCES


Bernstein, D. S. and W. H. Haddad (1989). LQG control with an $\mathcal{H}_\infty$ performance bound: A Riccati equation approach. *IEEE TAC* **34(3)**, 293–305.

Boullion, T. L. and P. L. Odell (1971). *Generalized Inverse Matrices.* Wiley-Interscience. New York.

Gahinet, P. and P. Apkarian (1994). A linear matrix inequality approach to $\mathcal{H}_\infty$ control. *Int. Journal of Robust and Nonlinear Control* **4**, 421–448.

Hall, S. and J. How (1993). Mixed $\mathcal{H}_2$ /$\mu$ performance bounds using dissapation theory. In: *Proc. of 1993 CDC.* pp. 1536–1541.

Iwasaki, T. and R.E. Skelton (1994). All controllers for the general $\mathcal{H}_\infty$ control problem: LMI existence conditions and state space formulas. *Automatica* **30(8)**, 1307–1317.

Khargonekar, P. and M. Rotea (1991). Mixed $\mathcal{H}_2$ /$\mathcal{H}_\infty$ control: A convex optimization approach. *IEEE TAC* **36(7)**, 824–837.

Paganini, F. (1995). Necessary and sufficient conditions for robust $\mathcal{H}_2$ performance. In: *Proc. of 1995 CDC.* pp. 1970–1975.

Sampei, M., T. Mita and M. Nakamichi (1990). An algebraic approach to $\mathcal{H}_\infty$ output feedback control problems. *Systems & Control Letters* **14(1)**, 13–24.

Stoorvogel, A. A. (1992). The singular $\mathcal{H}_2$ control problem. *Automatica* **28(3)**, 627–632.

Stoorvogel, A. A. (1993). The robust $\mathcal{H}_2$ control problem: A worst case design. *IEEE-TAC* **38(9)**, 1358–1370.

Zhou, K., J. Doyle and K. Glover (1995). *Robust and Optimal Control.* Prentice Hall. New Jersey.

Zhou, K., K. Glover, B. Bodenheimer and J. Doyle (1993). Mixed $\mathcal{H}_2$ and $\mathcal{H}_\infty$ performance objectives i: Robust performance analysis. *IEEE-TAC* **39(8)**, 1564–1574.

## Appendix A. PROOF OF THEOREM 1

### A.1 *Sufficiency*

Assume Eq. (10)$\sim$ Eq. (14), we prove that the controller given in Theorem 1 internally stabilizes $G$ and achieves $||H_{zw}||_2 < \gamma$.

For simplicity of presentation, we define the following notations.

$$\begin{bmatrix} A_R & B_{1R} \\ C_{1R} & 0 \end{bmatrix} = \begin{bmatrix} A + B_2RC_2 & B_1 + B_2RD_{21} \\ C_1 + D_{12}RC_2 & 0 \end{bmatrix} \quad (A.1)$$

$$\begin{bmatrix} \Sigma_1 & \Sigma_2 \\ \Sigma_3 & \Sigma_4 \end{bmatrix} = \begin{bmatrix} AX + B_2M & YA + NC_2 \\ C_1X + D_{12}M & YB_1 + ND_{21} \end{bmatrix} \quad (A.2)$$

From Eq. (13) we have

$$\begin{bmatrix} X & I \\ I & Y \end{bmatrix} > 0$$

which is equivalent to

$$X > 0, \quad X^{-1}Z > 0.$$

So the equations in Eq. (16) are well defined. With respect to the controller of Eq. (15), a realization of $H_{zw}$ is given by

$$H_{zw} = \left[\begin{array}{cc|c} A + B_2D_KC_2 & B_2C_K & B_1 + B_2D_KD_{21} \\ B_KC_2 & A_K & B_KD_{21} \\ \hline C_1 + D_{12}D_KC_2 & D_{12}C_K & D_{11} + D_{12}D_KD_{21} \end{array}\right]$$

Substituting Eq. (16) into this equation and making a similarity transformation with respect to $\begin{bmatrix} I & 0 \\ I & I \end{bmatrix}$, we obtain

$$H_{zw} = (A_c,\ B_c,\ C_c,\ 0)$$

where

$$A_c = \begin{bmatrix} A + B_2D_KC_2 + B_2C_K & B_2C_K \\ J & A + B_2D_KC_2 - B_KC_2 + J \end{bmatrix}$$

$$= \begin{bmatrix} A_{11} & A_{12} \\ A_{21} & A_{22} \end{bmatrix}$$

$$A_{11} = \Sigma_1X^{-1}$$

$$A_{12} = \Sigma_1X^{-1} - A_R$$

$$A_{21} = Z^{-1}(\Sigma_1 + XA_R' + XC_{1R}'\Sigma_3)X^{-1}$$

$$A_{22} = Z^{-1}(\Sigma_1 + X\Sigma_2X - A_RX + XA_R' + XC_{1R}'\Sigma_3)X^{-1}$$

$$B_c = \begin{bmatrix} B_1 + B_2D_KD_{21} \\ -B_1 - B_2D_KD_{21} + B_KD_{21} \end{bmatrix}$$

$$= \begin{bmatrix} B_{1R} \\ Z^{-1}(B_{1R} - X\Sigma_4) \end{bmatrix}$$

$$C_c = \begin{bmatrix} C_1 + D_{12}D_KC_2 + D_{12}C_K, & D_{12}C_K \end{bmatrix}$$

$$= \begin{bmatrix} \Sigma_3X^{-1}, & \Sigma_3X^{-1} - C_{1R} \end{bmatrix}.$$

Therefore, to show the sufficiency we need to prove $A_c$ is stable and

$$||H_{zw}||_2 = ||C_c(sI - A_c)^{-1}B_c||_2 < \gamma$$

based on the conditions of Eq. (10)$\sim$Eq. (14). That is, to show there exists $X_c = X_c' > 0$ such that

$$\Phi_A = \begin{bmatrix} X_cA_c + A_c'X_c & C_c' \\ C_c & -I \end{bmatrix} < 0 \quad (A.3)$$

$$\Phi_B = \begin{bmatrix} W & B_c'X_c \\ X_cB_c & X_c \end{bmatrix} > 0 \quad (A.4)$$

according to Lemma 1(3).

Let $\Phi_X$, $\Phi_Y$, $\Phi_{XY}$ be the left hand sides of Eq. (11), Eq. (12), Eq. (13). We show that

$$X_c = \text{blk diag}\,[X^{-1},\ X^{-1}Z] > 0 \quad (A.5)$$

is such a matrix. Firstly, it is a simple calculation to obtain

$$\Phi_A \sim \begin{bmatrix} \Phi_X & \\ & \Phi_Y \end{bmatrix} < 0$$

Therefore, we conclude $\Phi_A < 0$. Similarly,

$$\Phi_B \sim \Phi_{XY} > 0.$$

So the sufficiency has been proven.

### A.2 *Necessity*

Let $F$, $L$ be such that stabilize $A+B_2F$ and $A+LC_2$. By using the parameterization of stabilizing controllers(Zhou *et al.*, 1995), a realization of $H_{zw}$ is obtained as

$$H_{zw} = (A_c,\ B_c,\ C_c,\ D_c) \quad (A.6)$$

where

$$A_c = \begin{bmatrix} A + B_2F & -B_2C_Q & -B_2(F + D_QC_2) \\ 0 & A_Q & B_QC_2 \\ 0 & 0 & A + LC_2 \end{bmatrix}$$

$$B_c = \begin{bmatrix} B_1 - B_2D_QD_{21} \\ B_QD_{21} \\ B_1 + LD_{21} \end{bmatrix}$$

$$C_c = \begin{bmatrix} C_1 + D_{12}F, & -D_{12}C_Q, & -D_{12}(F + D_QC_2) \end{bmatrix}$$

$$D_c = D_{11} - D_{12}D_QD_{21}. \quad (A.7)$$

in which $(A_Q, B_Q, C_Q, D_Q)$ is the realization of a stable transfer matrix $Q(s)$. If the controller is a $\gamma$-optimal $\mathcal{H}_2$ one, $A_c$ must be stable and

$$||H_{zw}||_2 = ||C_c(sI - A_c)^{-1}B_c + D_c||_2 < \gamma.$$

Then by Lemma 1(1),(3)

$$D_c = 0 \tag{A.8}$$

must hold and there must exist $X_c = X_c'$, $W = W'$ such that

$$\begin{bmatrix} X_cA_c + A_c'X_c & C_c' \\ C_c & -I \end{bmatrix} < 0 \tag{A.9}$$

$$\begin{bmatrix} W & B_c' \\ B_c & X_c^{-1} \end{bmatrix} > 0 \tag{A.10}$$

$$\text{Trace}W < \gamma^2. \tag{A.11}$$

*Proof of equation Eq. (10)*: Define

$$R = -D_Q \tag{A.12}$$

and use Eq. (A.8), we obtain Eq. (10).

*Proof of equation Eq. (11)*: From Eq. (A.10) we have $X_c = X_c' > 0$. Multiplying blk diag $[X_c^{-1}, I]$ to the left and right sides of Eq. (A.9) gives

$$\begin{bmatrix} A_cX_c^{-1} + X_c^{-1}A_c' & X_c^{-1}C_c' \\ C_cX_c^{-1} & -I \end{bmatrix} < 0. \tag{A.13}$$

Partition $X_c^{-1}$ in accordance with $A_c$ as

$$\begin{bmatrix} X_{11} & X_{21}' & X_{31}' \\ X_{21} & X_{22} & X_{32}' \\ X_{31} & X_{32} & X_{33} \end{bmatrix} = X_c^{-1} > 0 \tag{A.14}$$

and define

$$X = X_{11} > 0 \tag{A.15}$$
$$M = FX_{11} - C_QX_{21} - (F + D_QC_2)X_{31}.$$

Then Eq. (11) is derived by multiplying $\begin{bmatrix} I & 0 & 0 & 0 \\ 0 & 0 & 0 & I \end{bmatrix}$ to the left side of Eq. (A.13) and its transpose to the right side.

*Proof of equation Eq. (12)*: Define $T = \begin{bmatrix} I & 0 & I \\ 0 & I & 0 \\ 0 & 0 & I \end{bmatrix}$ and

$$\tilde{A}_c = T^{-1}A_cT = \begin{bmatrix} A + B_2F & -B_2C_Q & -(L + B_2D_Q)C_2 \\ 0 & A_Q & B_QC_2 \\ 0 & 0 & A + LC_2 \end{bmatrix}$$

$$\tilde{C}_c = C_cT = \begin{bmatrix} C_1 + D_{12}F, & -D_{12}C_Q, & C_1 - D_{12}D_QC_2 \end{bmatrix}$$
$$\tilde{X}_c = T'X_cT.$$

Multiplying blk diag $[T, I]$ to the left side of Eq. (A.9) and its transpose to the right side yields

$$\begin{bmatrix} \tilde{X}_c\tilde{A}_c + \tilde{A}_c'\tilde{X}_c & \tilde{C}_c' \\ \tilde{C}_c & -I \end{bmatrix} < 0. \tag{A.16}$$

Partition $\tilde{X}_c$ in accordance with $\tilde{A}_c$ as

$$\begin{bmatrix} Y_{11} & Y_{21}' & Y_{31}' \\ Y_{21} & Y_{22} & Y_{32}' \\ Y_{31} & Y_{32} & Y_{33} \end{bmatrix} = \tilde{X}_c > 0 \tag{A.17}$$

and define

$$Y = Y_{33} \tag{A.18}$$
$$N = Y_{33}L + Y_{32}B_Q - Y_{31}(L + B_2D_Q).$$

Then Eq. (12) is obtained by multiplying $\begin{bmatrix} 0 & 0 & I & 0 \\ 0 & 0 & 0 & I \end{bmatrix}$ and its transpose to the left and right side of Eq. (A.16) respectively and by substitution of $D_Q = -R$.

*Proof of equation Eq. (13)*: Note $\begin{bmatrix} Y_{31} & Y_{32} & Y_{33} - Y_{31} \end{bmatrix} = \begin{bmatrix} 0 & 0 & I \end{bmatrix} \tilde{X}_cT^{-1}$ is row full rank and $X_c^{-1} = (T^{-1})'\tilde{X}_c^{-1}T^{-1} > 0$. The following equations can be derived from Eq. (A.14) and Eq. (A.17)

$$\begin{bmatrix} I & 0 & 0 \\ Y_{31} & Y_{32} & Y_{33} - Y_{31} \end{bmatrix} X_c^{-1} \begin{bmatrix} I & Y_{31}' \\ 0 & Y_{32}' \\ 0 & (Y_{33} - Y_{31})' \end{bmatrix}$$
$$= \begin{bmatrix} X_{11} & I \\ I & Y_{33} \end{bmatrix} > 0 \tag{A.19}$$
$$\begin{bmatrix} Y_{31} & Y_{32} & Y_{33} - Y_{31} \end{bmatrix} B_c = \Sigma_4.$$

Now multiplying $\begin{bmatrix} I & 0 & 0 & 0 \\ 0 & I & 0 & 0 \\ 0 & Y_{31} & Y_{32} & Y_{33} - Y_{31} \end{bmatrix}$ and its transpose to the left and right side of Eq. (A.10) respectively, and using Eq. (A.12), Eq. (A.14), Eq. (A.18), Eq. (A.19), it turns out that Eq. (13) is true.

# AN ORDER REDUCTION PROCEDURE TO COMPOSITE NASH SOLUTION OF SINGULARLY PERTURBED SYSTEMS

**Hua Xu* Hiroaki Mukaidani** Koichi Mizukami*****

* *Graduate School of Systems Management*
*The University of Tsukuba, Tokyo*
*3-29-1 Otsuka, Bunkyo-ku, Tokyo 112-0012, Japan*
** *Faculty of Information Sciences, Hiroshima City University*
*3-4-1, Ozuka-higashi, Asaminami-ku, Hiroshima, 731-3194, Japan*
*** *Faculty of Integrated Arts and Sciences, Hiroshima University*
*1-7-1, Kagamiyama, Higashi-Hiroshima, 739-0046, Japan*

Abstract: In this paper, a Nash game of singularly perturbed systems is studied in a viewpoint of descriptor system theory. A new method based on generalized coupled algebraic Riccati equations arising in descriptor systems is presented to find the composite Nash equilibrium solution of singularly perturbed systems. In this method, the full-order system is decomposed into a slow subsystem and a fast subsystem, and the slow subsystem is viewed as a special kind of descriptor systems. Through the solution of the slow Nash game, which is a Nash game for descriptor system, the composite Nash equilibrium solution is obtained. It is proven that the composite Nash equilibrium solution achieves a performance which is $O(\varepsilon)$ close to the full-order performance. Two numerical examples are given to show that the proposed method is valid for both standard and nonstandard singularly perturbed systems. 

Keywords: Singular perturbation method, descriptor systems, Nash games, differential game, reduced-order models

## 1. INTRODUCTION

The descriptor equation is a natural representation of dynamical systems which describe static and dynamic relations of physical variables explicitly. The descriptor systems has found many applications in large-scale systems, singularly perturbed systems, circuit theory, economics and other areas. As one of applications in singularly perturbed systems, the descriptor system is used in the study of various control problems and dynamic games for both standard and nonstandard singularly perburbed systems (Wang *et al.*, 1988, 1994; Xu *et al.*, 1997; Xu and Mizukami, 1997).

In this paper, a Nash game for singularly perturbed systems is studied in a viewpoint of descriptor system theory. A new method, based on generalized coupled algebraic Riccati equation arising in descriptor systems, is proposed to find the composite Nash equilibrium solution. In this method, the full-order system is decomposed into a slow subsystem and a fast subsystem and the slow subsystem is viewed as a special kind of descriptor systems. Through the solution of the slow Nash game, which is a Nash game for descriptor system, the composite Nash equilibrium solution is obtained. It is shown that the composite Nash equilibrium solution can be obtained simply by only revising the solution of the slow

Nash game. Therefore, the method is a reduced-order method and is independent of the small parameter $\varepsilon$. It is proven that the composite Nash equilibrium solution achieves a performance which is $O(\varepsilon)$ close to the full-order performance. Two numerical examples are given to show that the proposed method is valid for both standard and nonstandard singularly perturbed systems.

This paper is organized as follows. In next section, the closed-loop Nash equilibrium equation of Player 1 and 2 for the full-order Nash game is given. In Section 3, the full-order problem is decomposed into a slow subsystem Nash game and a fast subsystem Nash game. Their solutions are investigated. In Section 4, the composite Nash equilibrium solution is obtained, and its near-optimality property is studied. Section 5 discusses some conclusions.

## 2. FULL-ORDER NASH GAME

Consider a linear time-invariant singularly perturbed system

$$\dot{x} = A_{11}x + A_{12}z + B_{11}u_1 + B_{12}u_2, \quad (1a)$$

$$\varepsilon\dot{z} = A_{21}x + A_{22}z + B_{21}u_1 + B_{22}u_2, \quad (1b)$$

with $x(0) = x_0, z(0) = z_0$ and performance criteria

$$J_i = \frac{1}{2}\int_0^\infty \left( \begin{bmatrix} x \\ z \end{bmatrix}^T Q_i \begin{bmatrix} x \\ z \end{bmatrix} + u_i^T R_{ii} u_i \right.$$

$$\left. + u_j^T R_{ij} u_j \right) dt, \quad i,j = 1,2, \quad i \neq j \quad (2)$$

where $R_{ii} > 0, R_{ij} \geq 0$,

$$Q_i = \begin{bmatrix} Q_{i11} & Q_{i12} \\ Q_{i12}^T & Q_{i22} \end{bmatrix} = \begin{bmatrix} C_{i1}^T C_{i1} & C_{i1}^T C_{i2} \\ C_{i2}^T C_{i1} & C_{i2}^T C_{i2} \end{bmatrix}, \quad (3)$$

and $\varepsilon$ is a small positive parameter, $x(t) \in R^n$ and $z(t) \in R^m$ are states, $u_1(t) \in R^{r_1}$, $u_2(t) \in R^{r_2}$ are the controls of Player 1 and Player 2, and all matrices are of appropriate dimensions. The system (1) is called the nonstandard singularly perturbed system if the matrix $A_{22}$ is singular.

A Nash equilibrium solution is a pair $(u_1^*, u_2^*)$ such that

$$J_i(u_i^*, u_j^*) \leq J_i(u_i, u_j^*), \quad i = 1,2, \quad i \neq j. \quad (4)$$

for all admissible $u_i$.

Let us consider the closed-loop Nash equilibrium solution to the full-order problem. Before doing that, let us define the following matrices.

$$A_\varepsilon = \begin{bmatrix} A_{11} & A_{12} \\ A_{21}/\varepsilon & A_{22}/\varepsilon \end{bmatrix}; \; A = \begin{bmatrix} A_{11} & A_{12} \\ A_{21} & A_{22} \end{bmatrix}, \quad (5a)$$

$$B_{i\varepsilon} = \begin{bmatrix} B_{1i} \\ B_{2i}/\varepsilon \end{bmatrix}; \; B_i = \begin{bmatrix} B_{1i} \\ B_{2i} \end{bmatrix}, \quad (5b)$$

$$S_{i\varepsilon} = B_{i\varepsilon} R_{ii}^{-1} B_{i\varepsilon}^T; \; S_i = B_i R_{ii}^{-1} B_i^T, \quad (5c)$$

$$G_{j\varepsilon} = B_{j\varepsilon} R_{jj}^{-1} R_{ij} R_{jj}^{-1} B_{j\varepsilon}^T; \quad (5d)$$

$$G_j = B_j R_{jj}^{-1} R_{ij} R_{jj}^{-1} B_j^T, \quad (5e)$$

with $i,j = 1,2, i \neq j$. Furthermore, the $kl$th block of matrices $S_i, G_j$ are denoted henceforth by $S_{kli}, G_{klj}, k,l = 1,2$, and $S_{kli}^T = S_{lki}, G_{klj}^T = G_{lkj}, k \neq l$.

The closed-loop Nash equilibrium solution to the full-order problem is given by

$$u_i^* = -R_{ii}^{-1} B_{i\varepsilon} \hat{K}_i y, \quad i = 1,2, \quad i \neq j, \quad (6)$$

where $y := [x^T, z^T]^T$ and $\hat{K}_i$ is a unique stabilizing solution of the coupled algebraic Riccati equation

$$A_\varepsilon^T \hat{K}_i + \hat{K}_i A_\varepsilon + Q_i - \hat{K}_i S_{i\varepsilon} \hat{K}_i -$$

$$\hat{K}_i S_{j\varepsilon} \hat{K}_j - \hat{K}_j S_{j\varepsilon} \hat{K}_i + \hat{K}_j G_{j\varepsilon} \hat{K}_j = 0, \quad (7)$$

with $i,j = 1,2, i \neq j$. The above Nash equilibrium solution can also be expressed as, equivalently,

$$u_i^* = -R_{ii}^{-1} B_i K_i y, \quad (8)$$

where $K_i$ is a unique stabilizing solution of the generalized coupled algebraic Riccati equation

$$(i) \; A^T K_i + K_i^T A + Q_i - K_i^T S_i K_i -$$

$$K_i^T S_j K_j - K_j^T S_j K_i + K_j^T G_j K_j = 0, \quad (9a)$$

$$(ii) \; E_\varepsilon K_i = K_i^T E_\varepsilon, \quad (9b)$$

where

$$E_\varepsilon = \begin{bmatrix} I_n & 0 \\ 0 & \varepsilon I_m \end{bmatrix}. \quad (10)$$

We have the relation $\hat{K}_i = E_\varepsilon K_i$.

## 3. DECOMPOSITION OF SLOW AND FAST GAMES

Similar to the standard singularly perturbed systems, we decompose the full-order game into two subsystem games.

**Slow Nash game:** Find a Nash equilibrium solution $(u_{1s}^*, u_{2s}^*)$ of

$$J_{is} = \frac{1}{2}\int_0^\infty \left( \begin{bmatrix} x_s \\ z_s \end{bmatrix}^T Q_i \begin{bmatrix} x_s \\ z_s \end{bmatrix} + u_{is}^T R_{ii} u_{is} \right.$$

$$\left. + u_{js}^T R_{ij} u_{js} \right) dt, \quad i,j = 1,2, \quad i \neq j \quad (11)$$

for the slow subsystem

$$E\dot{y}_s = Ay_s + B_1 u_{1s} + B_2 u_{2s},$$
$$Ey_s(0) = Ey_0, \quad (12)$$

where $y_s := [x_s^T, z_s^T]^T$, $E = E_\varepsilon|_{\varepsilon=0}$, $A$, $B_i$ are defined in (5), and $Q_i$ in (3).

**Fast Nash game:** Find a Nash equilibrium solution $(u_{1f}^*, u_{2f}^*)$ of

$$J_{if} = \frac{1}{2}\int_0^\infty (z_f^T C_{i2}^T C_{i2} z_f + u_{if}^T R_{ii} u_{if}$$
$$+ u_{jf}^T R_{ij} u_{jf})dt, \quad i,j = 1,2, \; i \neq j \quad (13)$$

for the fast subsystem

$$\varepsilon \dot{z}_f = A_{22} z_f + B_{21} u_{1f} + B_{22} u_{2f},$$
$$z_f(0) = z_0 - z_s(0), \quad (14)$$

where $z_f = z - z_s$, $u_{1f} = u_1 - u_{1s}$ and $u_{2f} = u_2 - u_{2s}$.

The fast subsystem (14) is derived by assuming that the slow variables are constant during fast transients, that is, $\dot{z}_s = 0$ and $x_s =$ a constant.

Let us first consider the solution of the fast Nash game.

*Proposition 1.* The fast Nash game admits a Nash equilibrium solution

$$u_{if}^* = -R_{ii}^{-1} B_{2i}^T K_{if} z_f, \quad i = 1,2, \quad (15)$$

if there exists a unique stabilizing solution to the coupled algebraic Riccati equations

$$A_{22}^T K_{if} + K_{if} A_{22} + C_{i2}^T C_{i2}^T - K_{if} S_{22i} K_{if}$$
$$- K_{if} S_{22j} K_{jf} - K_{jf} S_{22j} K_{if}$$
$$+ K_{jf} G_{22j} K_{jf} = 0, \quad (16)$$

with $i,j = 1,2$, $i \neq j$.

Next let us consider the solution of the slow Nash game. Before doing that, let us first introduce the other generalized coupled algebraic Riccati equations.

$$(i)\; A^T P_{is} + P_{is}^T A + Q_i - P_{is}^T S_i P_{is} -$$
$$P_{is}^T S_j P_{js} - P_{js}^T S_j P_{is} + P_{js}^T G_j P_{js} = 0, \quad (17a)$$
$$(ii)\; EP_{is} = P_{is}^T E, \quad (17b)$$

with $i,j = 1,2$, $i \neq j$. The solution $P_{is}$ of (17) takes a lower-triangular block form

$$P_{is} = \begin{bmatrix} P_{11is} & 0 \\ P_{21is} & P_{22is} \end{bmatrix}, \quad P_{11is}^T = P_{11is}, \quad (18)$$

because of (17b). It is worthy to note that $P_{22is}$ may not be symmetric. The coupled algebraic Riccati equations (17) can be partitioned into

$$P_{11is} A_{11} + P_{21is}^T A_{21} + A_{11}^T P_{11is} + A_{21}^T P_{21is}$$
$$- P_{11is} S_{11i} P_{11is} - P_{21is}^T S_{12i}^T P_{11is}$$
$$- P_{11is} S_{12i} P_{21is} - P_{21is}^T S_{22i} P_{21is}$$
$$- P_{11js} S_{11j} P_{11is} - P_{21js}^T S_{12j}^T P_{11is}$$
$$- P_{11js} S_{12j} P_{21is} - P_{21js}^T S_{22j} P_{21is}$$
$$- P_{11is} S_{11j} P_{11js} - P_{21is}^T S_{12j}^T P_{11js}$$
$$- P_{11is} S_{12j} P_{21js} - P_{21is}^T S_{22j} P_{21js}$$
$$+ P_{11js} G_{11j} P_{11js} + P_{21js}^T G_{12j}^T P_{11js}$$
$$+ P_{11js} G_{12j} P_{21js} + P_{21js}^T G_{22j} P_{21js}$$
$$+ Q_{i11} = 0, \quad (19a)$$

$$P_{22is}^T A_{21} + A_{12}^T P_{11is} + A_{22}^T P_{21is}$$
$$- P_{22is}^T S_{12i}^T P_{11is} - P_{22is}^T S_{22i} P_{21is}$$
$$- P_{22js}^T S_{12j}^T P_{11is} - P_{22js}^T S_{22j} P_{21is}$$
$$- P_{22is}^T S_{12j}^T P_{11js} - P_{22is}^T S_{22j} P_{21js}$$
$$+ P_{22js}^T G_{12j}^T P_{11js} + P_{22js}^T G_{22j} P_{21js}$$
$$+ Q_{i12}^T = 0, \quad (19b)$$

$$P_{22is}^T A_{22} + A_{22}^T P_{22is} - P_{22is}^T S_{22i} P_{22is}$$
$$- P_{22js}^T S_{22j} P_{22is} - P_{22is}^T S_{22j} P_{22js}$$
$$+ P_{22js}^T G_{22j} P_{22js} + Q_{i22} = 0. \quad (19c)$$

Suppose that (19c) admit a real solution pair $(P_{221s}, P_{222s})$. Then, (19b) can be rewritten as

$$\hat{A}_{22}^T P_{21is} - W_{22j}^T P_{21js} + \hat{A}_{12}^T P_{11is} + W_{12j}^T P_{11js}$$
$$+ \hat{Q}_{i12}^T = 0, \quad i,j = 1,2, \; i \neq j, \quad (20)$$

where

$$\hat{A}_{12} = A_{12} - S_{121} P_{221s} - S_{122} P_{222s}, \quad (21a)$$
$$\hat{A}_{22} = A_{22} - S_{221} P_{221s} - S_{222} P_{222s}, \quad (21b)$$
$$W_{12i} = G_{12i} P_{22is} - S_{12i} P_{22js}, \quad (21c)$$
$$W_{22i} = S_{22i} P_{22js} - G_{22i} P_{22is}, \quad (21d)$$
$$\hat{Q}_{i12} = Q_{i12} + A_{21}^T P_{22is}. \quad (21e)$$

From (20),

$$\begin{bmatrix} \hat{A}_{22}^T & -W_{222}^T \\ -W_{221}^T & \hat{A}_{22}^T \end{bmatrix} \begin{bmatrix} P_{211s} \\ P_{212s} \end{bmatrix} =$$
$$- \begin{bmatrix} \hat{A}_{12}^T & -W_{122}^T \\ -W_{121}^T & \hat{A}_{12}^T \end{bmatrix} \begin{bmatrix} P_{111s} \\ P_{112s} \end{bmatrix}$$

$$-\begin{bmatrix} Q_{112}^T \\ Q_{212}^T \end{bmatrix}. \qquad (22)$$

Therefore, solving the equations (22) gives ($P_{211s}$, $P_{212s}$) which are expressed in terms of ($P_{111s}$, $P_{112s}$).

$$P_{21is} = -Q_{si}^T - \hat{W}_{ii}P_{11is} - \hat{W}_{ij}P_{11js},$$
$$i,j = 1,2,\ i \neq j, \qquad (23)$$

where, the inverse of the corresponding matrix is assumed to exist and the explicit expressions for $Q_{si}, \hat{W}_{ii}, \hat{W}_{ij}$ are omitted for brevity. Substituting (23) into (19a) arrive at nonstandard coupled algebaric Riccati equations of ($P_{111s}, P_{112s}$), that is,

$$\begin{aligned}
&P_{11is}A_{11} - (Q_{si}^T + \hat{W}_{ii}P_{11is} + \hat{W}_{ij}P_{11js})^T A_{21} \\
&+A_{11}^T P_{11is} - A_{21}^T(Q_{si}^T + \hat{W}_{ii}P_{11is} + \hat{W}_{ij}P_{11js}) \\
&-P_{11is}S_{11i}P_{11is} + (Q_{si}^T + \hat{W}_{ii}P_{11is} \\
&+\hat{W}_{ij}P_{11js})^T S_{12i}^T P_{11is} \\
&+P_{11is}S_{12i}(Q_{si}^T + \hat{W}_{ii}P_{11is} + \hat{W}_{ij}P_{11js}) \\
&-(Q_{si}^T + \hat{W}_{ii}P_{11is} + \hat{W}_{ij}P_{11js})^T S_{22i}(Q_{si}^T \\
&+\hat{W}_{ii}P_{11is} + \hat{W}_{ij}P_{11js}) \\
&-P_{11js}S_{11j}P_{11is} + (Q_{sj}^T + \hat{W}_{jj}P_{11js} \\
&+\hat{W}_{ji}P_{11is})^T S_{12j}^T P_{11is} \\
&+P_{11js}S_{12j}(Q_{si}^T + \hat{W}_{ii}P_{11is} + \hat{W}_{ij}P_{11js}) \\
&-(Q_{sj}^T + \hat{W}_{jj}P_{11js} + \hat{W}_{ji}P_{11is})^T S_{22j}(Q_{si}^T \\
&+\hat{W}_{ii}P_{11is} + \hat{W}_{ij}P_{11js}) \\
&-P_{11is}S_{11j}P_{11js} + (Q_{si}^T + \hat{W}_{ii}P_{11is} \\
&+\hat{W}_{ij}P_{11js})^T S_{12j}^T P_{11js} \\
&+P_{11is}S_{12j}(Q_{sj}^T + \hat{W}_{jj}P_{11js} + \hat{W}_{ji}P_{11is}) \\
&-(Q_{si}^T + \hat{W}_{ii}P_{11is} + \hat{W}_{ij}P_{11js})^T S_{22j}(Q_{sj}^T \\
&+\hat{W}_{jj}P_{11js} + \hat{W}_{ji}P_{11is}) \\
&+P_{11js}G_{11j}P_{11js} - (Q_{sj}^T + \hat{W}_{jj}P_{11js} \\
&+\hat{W}_{ji}P_{11is})^T G_{12j}^T P_{11js} \\
&-P_{11js}G_{12j}(Q_{sj}^T + \hat{W}_{jj}P_{11js} + \hat{W}_{ji}P_{11is}) \\
&+(Q_{sj}^T + \hat{W}_{jj}P_{11js} + \hat{W}_{ji}P_{11is})^T G_{22j}(Q_{sj}^T \\
&+\hat{W}_{jj}P_{11js} + \hat{W}_{ji}P_{11is}) \\
&+Q_{i11} = 0,\ \ i,j = 1,2,\ i \neq j. \qquad (24)
\end{aligned}$$

Clearly, the existence of a real solution to the coupled algebraic Riccati equations (19c) and a real solution to the nonstandard coupled algebraic Riccati equations (24) is sufficient for the existence of a real solution to the generalized coupled algebraic Riccati equations (17) suppose that the related inverse matrix exists.

*Proposition 2.* Suppose that the generalized coupled algebraic Riccati equation (17) admit an impulse-free and stabilizing solution $P_{is}$ and $EP_{is}$ is unique. Then, the slow Nash game admits a Nash equilibrium solution, given by

$$u_{is}^* = -R_{ii}^{-1}B_i^T P_{is}y_s, \quad i = 1,2. \qquad (25)$$

*Remark 3.* Similar to descriptor systems, an important feature of the slow Nash game is that the feedback Nash equilibrium solution is not unique. This fact is clear if we note that any real solution of the coupled algebraic Riccati equation (19c) is allowed in the feedback gain of (25). But $EP_{is} = P_{is}^T E$ is required to be unique.

## 4. NEAR-OPTIMALITY OF COMPOSITE NASH EQUILIBRIUM

In this section, let us construct a composite Nash equilibrium solution $u_{ic}^* = u_{is}^* + u_{if}^*$ and analyze its near-optimality. It has been known that the Nash equilibrium solution for the slow Nash game is not unique. However, the corresponding Nash equilibrium solution for the fast Nash game is unique. Comparing (16) and (19c), it is found that they have the same parameter matrices. This means that a unique stabilizing solution of (16) is also a solution of (19c). Taking the solution $P_{22is} = K_{if}$, we get a special Nash equilibrium solution to the slow Nash game, denoted by

$$u_{is}^{*+} = -R_{ii}^{-1}B_i^T P_{is}^+ y_s, \qquad (26)$$

where

$$P_{is}^+ = \begin{bmatrix} P_{11is}^+ & 0 \\ P_{21is}^+ & P_{22is}^+ \end{bmatrix}, \qquad (27)$$

$P_{22is}^+ = K_{if}$ and $P_{11is}^+$ is the corresponding solution of (24) when $P_{22is} = P_{22is}^+$. $P_{21is}^+$ is obtained from (22).

From the analyses above, it is obtained readily

$$\begin{aligned}
u_{ic}^* &= u_{is}^{*+} + u_{if}^* = -R_{ii}^{-1}[B_{1i}^T \ \ B_{2i}^T]\times \\
&\begin{bmatrix} P_{11is}^+ & 0 \\ P_{21is}^+ & P_{22is}^+ \end{bmatrix}\begin{bmatrix} x_s \\ z_s \end{bmatrix} - R_{ii}^{-1}B_{2i}^T K_{if}z_f = \\
&-R_{ii}^{-1}[B_{1i}^T \ \ B_{2i}^T]\begin{bmatrix} P_{11is}^+ & 0 \\ P_{21is}^+ & P_{22is}^+ \end{bmatrix}\begin{bmatrix} x \\ z \end{bmatrix}, \qquad (28)
\end{aligned}$$

where $x(t) \approx x_s(t)$ and $z(t) \approx z_s(t) + z_f(t)$.

*Remark 4.* Let us compare (25) with (28). it is found that $u_{ic}^*$ is obtained by computing and revising the Nash equilibrium solution $u_{is}^*$ to the slow Nash game.

Now, let us apply the composite Nash equilibrium solution $u_c^*$ to the full-order system (1) and compare it with the exact Nash equilibrium solution (8). In order to do that, let us first make the following remarks on the unique stabilizing solution $K_i$ of the generalized coupled algebraic Riccati equation (9).

Under certain conditions, for example, the existence of the solutions to the coupled algebraic Riccati equations (19c),(24) and the existence of the related inverse matrices, the generalized coupled algebraic Riccati equations (9) admit a unique stabilizing solution $K_i$ which possesses a power series expansion at $\varepsilon = 0$, that is,

$$K_i = \begin{bmatrix} K_{11i}^{(0)} & \varepsilon K_{21i}^{(0)T} \\ K_{21i}^{(0)} & K_{22i}^{(0)} \end{bmatrix}$$

$$+\sum_{k=1}^{\infty} \frac{\varepsilon^k}{k!} \begin{bmatrix} K_{11i}^{(k)} & \varepsilon K_{21i}^{(k)T} \\ K_{21i}^{(k)} & K_{22i}^{(k)} \end{bmatrix}, \quad i = 1,2. \quad (29)$$

Now, let us compare the composite Nash equilibrium solution $u_{ic}^*$ with the exact Nash equilibrium solution $u_i^*$ and show the $O(\varepsilon)$ approximation of $J_i^*$. Applying the composite Nash equilibrium solution $u_{ic}^*$ to the full-order system (1), we have

$$J_{ic}^* = \frac{1}{2} y^T(0) E_\varepsilon P_{ic} y(0), \quad (30)$$

where $P_{ic}$ is the solution of the generalized Lyapunov equations

$$(i)\ (A - S_1 P_{1s}^+ - S_2 P_{2s}^+)^T P_{ic}$$
$$+ P_{ic}^T (A - S_1 P_{1s}^+ - S_2 P_{2s}^+)$$
$$= -P_{is}^{+T} S_i P_{is}^+ - P_{js}^{+T} G_j P_{js}^+ - Q_i, \quad (31a)$$

$$(ii)\ E_\varepsilon P_{ic} = P_{ic}^T E_\varepsilon. \quad (31b)$$

Similar to $K_i$, the generalized Lyapunov equations (31) also admit a unique stabilizing solution $P_{ic}$ which possesses a power series expansion at $\varepsilon = 0$, that is,

$$P_{ic} = \begin{bmatrix} P_{11ic}^{(0)} & \varepsilon P_{21ic}^{(0)T} \\ P_{21ic}^{(0)} & P_{22ic}^{(0)} \end{bmatrix}$$

$$+\sum_{k=1}^{\infty} \frac{\varepsilon^k}{k!} \begin{bmatrix} P_{11ic}^{(k)} & \varepsilon P_{21ic}^{(k)T} \\ P_{21ic}^{(k)} & P_{22ic}^{(k)} \end{bmatrix}, \quad i = 1,2. \quad (32)$$

*Theorem 1.* The first term of the power series of $J_{ic}^*$ and $J_i^*$ at $\varepsilon = 0$ are the same, that is,

$$J_{ic}^* = J_i^* + O(\varepsilon), \quad i = 1,2, \quad (33)$$

and hence the composite Nash equilibrium solution is an $O(\varepsilon)$ near-optimal solution to the full-order Nash game (1),(2).

To the end of this section, the design procedure to find the composite Nash equilibrium solution will be summarized and two numerical examples will be solved.

**Step 1.** Find the unique stabilizing solution $K_{if}$ of the coupled algebraic Riccati equations (16).

**Step 2.** Based on $K_{if}$, find the impulse-free stabilizing solution $P_{is}^+$ of the generalized coupled algebraic Riccati equations (17).

**Step 3.** Substitute $y_s(t)$ of (26) by $y(t)$ to get the composite Nash equilibrium solution (28) directly.

Different from the optimal control problem and zero-sum differential game (Xu *et al.* 1997 and Xu and Mizukami, 1997), parallel computations for the solutions of (16),(17) are impossible.

*Example 5.* Consider a standard singularly perturbed system

$$\begin{bmatrix} \dot{x}(t) \\ \varepsilon \dot{z}(t) \end{bmatrix} = \begin{bmatrix} 1 & 2 \\ -1 & -2 \end{bmatrix} \begin{bmatrix} x(t) \\ z(t) \end{bmatrix} + \begin{bmatrix} 1 \\ 2 \end{bmatrix} u_1(t)$$

$$+ \begin{bmatrix} 1 \\ 2 \end{bmatrix} u_2(t), \quad x(0) = 1, \quad z(0) = 2. \quad (34)$$

The performance criteria are

$$J_1 = \frac{1}{2} \int_0^\infty \{ \begin{bmatrix} x \\ z \end{bmatrix}^T \begin{bmatrix} 2 & 1 \\ 1 & 2 \end{bmatrix} \begin{bmatrix} x \\ z \end{bmatrix}$$
$$+ u_1^2 + 2u_2^2 \} dt. \quad (35a)$$

$$J_2 = \frac{1}{2} \int_0^\infty \{ \begin{bmatrix} x \\ z \end{bmatrix}^T \begin{bmatrix} 2 & 1 \\ 1 & 2 \end{bmatrix} \begin{bmatrix} x \\ z \end{bmatrix}$$
$$+ 2u_1^2 + u_2^2 \} dt. \quad (35b)$$

Following the steps 1-3, the composite Nash equilibrium solution is obtained as follows.

$$u_{1c}^* = -[1 \ \ 2] \times$$
$$\begin{bmatrix} \sqrt{2}/2 & 0 \\ (\sqrt{2}+1)(\sqrt{3}-1)/4 & (\sqrt{3}-1)/2 \end{bmatrix} \begin{bmatrix} x \\ z \end{bmatrix}, \quad (36a)$$

$$u_{2c}^* = -[1 \ \ 2] \times$$
$$\begin{bmatrix} \sqrt{2}/2 & 0 \\ (\sqrt{2}+1)(\sqrt{3}-1)/4 & (\sqrt{3}-1)/2 \end{bmatrix} \begin{bmatrix} x \\ z \end{bmatrix}. \quad (36b)$$

The resulting performance values for several values of $\varepsilon$ when the full-order Nash equilibrium solution and the composite Nash equilibrium solution are applied respectively are given in Table 1, where $J_1 = J_2 = J$ since symmetric property.

Table 1 Comparing the numerical results

| $\varepsilon$ | 0.5 | 0.2 | 0.1 | 0.01 | 0.001 |
|---|---|---|---|---|---|
| $J^*(\varepsilon)$ | 1.30 | 0.73 | 0.54 | 0.37 | 0.36 |
| $J_c^*(\varepsilon)$ | 1.24 | 0.73 | 0.54 | 0.37 | 0.36 |

It is seen that $J^*(\varepsilon)$ and $J_c^*(\varepsilon)$ have the same limit as $\varepsilon \to 0$.

*Example 6.* Further consider a nonstandard singularly perturbed system

$$\begin{bmatrix} \dot{x}(t) \\ \varepsilon \dot{z}(t) \end{bmatrix} = \begin{bmatrix} 1 & 2 \\ -1 & 0 \end{bmatrix} \begin{bmatrix} x(t) \\ z(t) \end{bmatrix} + \begin{bmatrix} 1 \\ 2 \end{bmatrix} u_1(t)$$

$$+ \begin{bmatrix} 1 \\ 2 \end{bmatrix} u_2(t), \quad x(0) = 1, \quad z(0) = 2. \qquad (37)$$

The performance criteria are the same as (35). Following the steps 1-3, the composite Nash equilibrium solution is obtained as follows.

$$u_{1c}^* = -[1 \;\; 2]\times$$

$$\begin{bmatrix} (1+\sqrt{15})/4 & 0 \\ (3\sqrt{2}+\sqrt{30}-3-\sqrt{15})/8 & \sqrt{2}/2 \end{bmatrix} \begin{bmatrix} x \\ z \end{bmatrix}, \qquad (38a)$$

$$u_{2c}^* = -[1 \;\; 2]\times$$

$$\begin{bmatrix} (1+\sqrt{15})/4 & 0 \\ (3\sqrt{2}+\sqrt{30}-3-\sqrt{15})/8 & \sqrt{2}/2 \end{bmatrix} \begin{bmatrix} x \\ z \end{bmatrix}, \qquad (38b)$$

The resulting performance values for several values of $\varepsilon$ when the full-order Nash equilibrium solution and the composite Nash equilibrium solution are applied respectively are given in Table 2, where $J_1 = J_2 = J$ since symmetric property.

Table 2 Comparing the numerical results

| $\varepsilon$ | 0.5 | 0.1 | 0.01 | 0.001 | 0.0001 |
|---|---|---|---|---|---|
| $J^*(\varepsilon)$ | 1.77 | 0.85 | 0.63 | 0.61 | 0.61 |
| $J_c^*(\varepsilon)$ | 1.69 | 0.84 | 0.63 | 0.61 | 0.61 |

$J^*(\varepsilon)$ and $J_c^*(\varepsilon)$ have the same limit as $\varepsilon \to 0$.

## 5. CONCLUSIONS

In this paper, by using the generalized coupled algebraic Riccati equation arising in descriptor systems, the composite Nash game for singularly perturbed systems has been studied. The same problem has been treated by Gardner and Cruz (1978) in which the Hierarchical Reduction Scheme is proposed. An important observation in that paper is that the usual order reduction procedure for singularly perturbed optimal control systems does not lead to a well-posed problem when extended directly to the linear-quadratic nonzero-sum closed-loop Nash game. In this paper, the full-order system is decomposed into a slow subsystem and a fast subsystem, and the slow subsystem is viewed as a special kind of descriptor systems. It is shown that the proposed method really leads to a well-posed problem. Combining the results in this paper and those in authors's other papers (Xu *et al.* 1997 and Xu and Mizukami, 1997), it is claimed that the proposed order reduction procedure, that is, viewing the slow subsystem as a descriptor system, will lead to a well-posed solution no matter what a problem is, for example, an optimal control problem, a zero-sum differential game or a nonzero-sum differential game.

## REFERENCES


Başar, T. and G.J. Olsder (1995). *Dynamic Noncooperative Game Theory*, Academic Press, New York.

Dai, L. (1989). *Singular Control Systems*, Lecture Notes in Control and Information Sciences, Edited by M.Thoma and A.Wyner, Springer-Verlag, Berlin.

Gajic, Z. and X. Shen (1993). *Parallel Algorithms for Optimal Control of Large Scale Linear Systems*, Springer-Verlag, London.

Gardner, Jr., B. F. and J. B. Cruz, Jr. (1978). Well-posedness of singularly perturbed Nash games. *Journal of the Franklin Institute*, **306**, 355-374.

Khalil, H. K. and P. V. Kokotovic (1979). Feedback and well-posedness of singularly perturbed Nash games. *IEEE Transactions on Automatic Control*, **24**, 699-708.

Khalil, H. K. (1989). Feedback control of nonstandard singularly perturbed systems. *IEEE Transactions on Automatic Control*, **34**, 1052-1060.

Kokotovic, P. V., H. K. Khalil, and J. O'Reilly (1986) *Singular Perturbation Methods in Control: Analysis and Design*, Academic Press, New York.

Wang, Y.Y., S.J. Shi and Z.J.Zhang (1988). A descriptor-system approach to singular perturbation of linear regulators. *IEEE Transactions on Automatic Control*, **33**, 370-373.

Wang, Y. Y., P. M. Frank and N. E. Wu (1994). Near-optimal control of nonstandard singularly perturbed systems. *Automatica*, **30**, 277-292.

Xu, H. and K. Mizukami (1997). New method for composite optimal control of singularly perturbed systems. *International Journal of Systems Sciences*, **28**, 161-172.

Xu, H. and K. Mizukami (1997). Infinite-horizon differential games of singularly perturbed systems: a unified approach. *Automatica*, **33**, 273-276.

# GAME PROBLEM OF APPROACH UNDER FAILURE OF CONTROLLING DEVICES

**A.A. Chikrii** *

* *Institute of Cybernetics of NAS of Ukraine,*
*40 Prsp. Akad. Glushkov, 252650 MSP 252022 Kievv-22, Ukraine*
*E-mail: chik@d165.icyb.kiev.ua*

Abstract: The linear differential game of approaching given terminal set by a trajectory of the conflict controlled process is treated here. The process evolves in such a way that at some a priori unknown instant of time the control of the first player is shut off for a certain period of time, which is necessary for the elimination of the breakdown and which duration is known in advance. Then the process lasts up to the moment when its trajectory hits the terminal set. On the basis of technique, leaning upon the resolving function's construction, developed by the author, the sufficient conditions of two kinds for solvability of the approach problem are derived in the paper.
The results are illustrated by the model example of the game problem with "simple motions" dynamics. *Copyright © 1999 IFAC*

Keywords: differential game, set-values, control system, optimization problems, vehicle

## 1. INTRODUCTION

The differential game of pursuit and escape are given a wide coverage to in the monographs (Pontryagin, 1988), (Chikrii, 1997), (Nikolskii, 1984), (Krasovskii, 1970), (Isaacs, 1965), (Psheni-tchnyj and Ostapenko, 1992), (Grigorenko, 1990). The method of resolving functions, developed in (Chikrii, 1993), (Chikrii, 1996*a*), (Chikrii and Pilipenko, 1991), (Pilipenko and Chikrii, 1993), (Chikrii and Patlanzhoglu, 1997), (Patlanzhoglu and Chikrii, 1992), (Chikrii *et al.*, 1986), (Chikrii, 1994) (Baranovskaya *et al.*, 1997), (Chikrii, 1996*b*), (Chikrii, 1985), (Chikrii and Prokopovich, 1995), appears as one of the efficient ones. Here it is applied to solution of the problem of pursuit under failure of controlling devicies. Similar problem was studied with help of Pontryagin's direct methods in (Nikolskii, 1987), (Nikolskii and Pan, 1994), (Nikolskii, 1988).

## 2. PROBLEM STATEMENT

Let the conflict-controlled process be given:

$$\dot{z} = Az + \beta_\theta(t)u - v, z \in R^n, u \in U, v \in V \quad (1)$$

where $A$ is a square matrix of order $n$, $U, V \in K(R^n)$ are nonempty compacts of Euclidian space $E^n$, the numerical function $\beta_\theta(t)$ has the form

$$\beta_\theta(t) = \begin{cases} 0, & t \in [\theta, \theta + h] \\ 1, & t \notin [\theta, \theta + h] \end{cases}$$

and $\theta \geq 0$, $h = const > 0$.

The terminal set is given which has a cylindrical form

$$M^* = M_0 + M \quad (2)$$

where $M_0$ is a linear subspace in $R^n$ and $M \in \text{co}K(L)$ is a complex compact from the orthogonal complement to $M_0$ in $R^n$.

The goal of the first player ($u$) is to bring a trajectory of the system (1) to the set (2) in the shortest time. The goal of the second player ($v$) is to avoid the meeting of a trajectory with the terminal set or, if it is impossible, to maximally postpone the time instant of the meeting. In so doing the second player employs as controls arbitrary Lebesque measurable functions, taking its values in $V$, and the first one constructs his control on the basis of information on the initial state of the process and the prehistory of control of the second player that is

$$u(t) = u(z_0, v_t(\cdot))$$

where $v_t(\cdot) = \{v(s) : s \in [0,t]\}$. Here $u(t)$ is measurable function taking its values in $U$.

Taking the first player's side let us establish sufficient conditions for the game termination in some guaranteed time.

## 3. SCHEME OF THE METHOD AND THE PRINCIPAL RESULT

Denote by $\pi$ the orthoprojector acting from $R^n$ to $L$, where $L = M_0^{\perp}$, and by $e^{At}$ the fundamental matrix of the homogeneous system (1). Let us introduce the set-valued maps

$$W(t,v) = \pi e^{At}(U - v)$$

$$W(t) = \pi e^{At}U \overset{*}{-} \pi e^{At}V$$

Here $\overset{*}{-}$ is the operation of geometric subtraction of sets (Pontryagin, 1988), defined in the following way

$$X \overset{*}{-} Y = \{z : z + Y \subset X\} = \cap_{y \in Y}(X - y)$$

**Condition 1 (Pontryagin's condition).** $W(t) \neq \emptyset$ for all $t \geq 0$.

From the properties of the operation of geometric subtraction of sets there follows that the map $W(t)$ is upper semicontinuous in Hausdorff metric and therefore (Joffe and Tikhomirov, 1974) contains at least one Borel selection (Aubin and Aklund, 1984), (Kuratovskii, 1966). We fix one of such selections $\gamma(t)$, $\gamma(t) \in W(t)$, and introduce the function

$$\xi(t,z,\gamma(\cdot)) = \pi e^{At}z + \int_0^t \gamma(t-\tau)d\tau$$

Denote

$$t^* = t^*(\theta,t) = \tag{3}$$

$$= \min\Big\{t \geq s > \theta + h : \int_\theta^{\theta+h} \left[\pi e^{A(t-\tau)}V - \gamma(t-\tau)\right] d\tau \subset \subset \int_{\theta+h}^{s} [W(t-\tau) - \gamma(t-\tau)]d\tau\Big\}$$

In the sequel we shall examine only those $t$ and $\theta$ for which $t^*$ exists.

Let us introduce the resolving function by the formula

$$\alpha_\theta(t,\tau,z,v,\gamma(\cdot)) = $$

$$= \begin{cases} \sup\{\alpha \geq 0 : [W(t-\tau,v) - \gamma(t-\tau) \cap \\ \alpha[M - \xi(t,z,\gamma(\cdot))] \neq \emptyset\}, & \tau \notin [\theta,t^*] \\ 0, & \tau \in [\theta,t^*] \end{cases}$$

In other words, the resolving function vanishes on the section $[\theta, \theta+h]$ of failure of the controlling devices and on the section $(\theta+h, t^*]$ of elimination of the emergency consequences.

If $t \geq 0$ and $\xi(t,z,\gamma(\cdot)) \in M$ then $\alpha_\theta(t,\tau,z,v,\gamma(\cdot)) = \infty$ for all $\tau, v$, $0 \leq \tau \leq t$, $v \in V$. In the case when $\xi(\cdot,z,\gamma(\cdot)) \notin M$ the resolving function takes finite values enjoing all its properties studied in (Chikrii, 1997).

Set

$$T(z,\gamma(\cdot)) = \min\{t : \min_{t^*-h \geq \theta \geq 0} \int_0^t \inf_{v \in V} \alpha_\theta(t,\tau,z,v,\gamma(\cdot))d\tau \geq 1\}$$

**Theorem 1.** Let for the conflict controlled process (1), (2) Pontrygin's condition hold and let for some initial state $z_0$ of the process and some Borel selection $\gamma(\cdot))$ $T = T(z_0, \gamma(\cdot)) < +\infty$. Then a trajectory of the process (1) may be brought to the set $M^*$ at time $T$ despite the failure of the controlling devices during time $h$.

**Proof.** Let $v(t)$ be an arbitrary admissable control of the second player and $\theta$ be instant of the failure of the controlling devices of the first player, unknown to him in advance.

In this connection the control of the first player is to be deetermined on the intervals of time $[0,\theta]$, $[\theta+h, t^*]$, and $(t^*, T]$. Note, that on the interval $[\theta+h, t^*]$ the consequences of the controlling devices' failure are eliminated with account of the inclusion (3). On the other hand, according to the scheme of the method of the resolving functions the union of intervals $[0,\theta] \cup (t^*, t_*]$ appears as an active section, while the interval $(t_*, T]$ as a passive one, where $t_*$ is a instant of switching which will determined below.

Let us introduce the test function

$$h(t, v(\cdot)) = 1 - \int_0^t \alpha_\theta(T, \tau, z_0, v(\tau), \gamma(\cdot))d\tau$$

It is continuous in $t$, nonincreasing and $h(0, v(\cdot)) = 1$. From the definition of function $T(z_0, \gamma(\cdot))$ there follows that an instant $t_*$ exists such that $h(0, v(\cdot)) = 0$. Without loss of generality it may be assumed that $t^* = t^*(\theta, T) < t_* \leq T$.

Assume that $\xi(T, z_0, \gamma(\cdot)) \notin M$.

For $\tau \in [0, \theta] \cup (t^*, t_*]$ let us study the set-valued map

$$U_1(\tau, v) = \{u \in U : \pi e^{A(T-\tau)}(u - v) - \gamma(T - \tau) \in \alpha_\theta(T, \tau, z_0, v, \gamma(\cdot)) \times [M - \xi(T, z_0, \gamma(\cdot))]\} \quad (4)$$

It is jointly Borel (Joffe and Tikhomirov, 1974). The selection

$$u_1(\tau, v) = \text{lexmin} U_1(\tau, v)$$

is respectively a Borel function (Chikrii, 1997), (Joffe and Tikhomirov, 1974).

Set the control of the first player equal to

$$u(\tau) = u_1(\tau, v(\tau)), \ \tau \in [0, \theta] \cup (t^*, t_*] \quad (5)$$

In view of the known assertions on the superposition of functions $u(\tau)$ is Lebesque measurable function.

On the passive segment $\tau \in (t_*, T]$ let us set $\alpha_\theta(T, \tau, z_0, v, \gamma(\cdot)) \equiv 0$ in expression (4) and choose the control of the first player in accordance with the procedure suggested above

$$U_2(\tau, v) = \{u \in U : \pi e^{A(T-\tau)}(u - v) - \gamma(T - \tau) = 0\}$$
$$u_2(\tau, v) = \text{lexmin} U_2(\tau, v)$$
$$u(\tau) = u_2(\tau, v(\tau)), \ t \in (t_*, T] \quad (6)$$

If $\xi(T, z_0, \gamma(\cdot)) \in M$ let us choose the control of the first player for $\tau \in [0, \theta] \cup (t^*, T]$ using formula (6). It remains to choose the control on the segment $[\theta + h, t^*]$. In view of expression (3) the following inclusion holds

$$\int_\theta^{\theta+h} \left[\pi e^{A(T-\tau)} V - \gamma(T - \tau)\right] d\tau \subset \int_{\theta+h}^{t^*} [W(T - \tau) - \gamma(T - \tau)]d\tau$$

In view of the fact that the control of the second player on the interval $[\theta, \theta + h]$ is known to the first player at the instant $\theta + h$, denoting

$$\int_\theta^{\theta+h} \left[\pi e^{A(T-\tau)} v(\tau) - \gamma(T - \tau)\right] d\tau = \omega$$

one can deduce the inclusion

$$\omega \in \int_{\theta+h}^{t^*} [W(T - \tau) - \gamma(T - \tau)]d\tau$$

Hence, the map $W(t) - \gamma(t)$ has a Borel selection $\omega(t)$ such that

$$\omega = \int_{\theta+h}^{t^*} \omega(T - \tau)d\tau$$

Denote

$$U_3(\tau, v) = \{u \in U : \pi e^{A(T-\tau)}(u - v) - \omega(T - \tau) = 0\}$$
$$u_3(\tau, v) = \text{lexmin} U_3(\tau, v)$$

and set

$$u(\tau) = u_3(\tau, v(\tau)), \ \tau \in [\theta + h, t^*] \quad (7)$$

Let us now show that if the control of the first player is chosen in accordance with formulas (5)-(7) then a trajectory of the system (1) hits the set $M^*$ at the time $T$ for any admissible controls of the second player.

Let $\xi(T, z_0, \gamma(\cdot)) \notin M$.

Using Cauchy formula we derive the presentation

$$\pi z(T) = \pi e^{AT} z_0 + \int_{[0,\theta] \cup [t^*, t_*]} \pi e^{A(T-\tau)}(u(\tau) - v(\tau))d\tau + \int_\theta^{\theta+h} \pi e^{A(T-\tau)} v(\tau)d\tau + \int_{\theta+h}^{t^*} \pi e^{A(T-\tau)}(u(\tau) - v(\tau))d\tau + \int_{t_*}^{T} \pi e^{A(T-\tau)}(u(\tau) - v(\tau))d\tau \quad (8)$$

Adding to and subtracting from the right side of the equation (8) the term $\int_0^T \gamma(T - \tau)d\tau$ and taking into account he laws of the control choice, one can obtain

$$\pi z(T) = \xi(T, z_0, \gamma(\cdot)) \times \left[1 - \int_0^T \alpha_\theta(T, \tau, z_0, v(\tau), \gamma(\cdot))d\tau\right] + \int_0^T \alpha_\theta(T, \tau, z_0, v(\tau), \gamma(\cdot))M d\tau$$

Since $\alpha_\theta(T, \tau, z_0, v(\tau), \gamma(\cdot)) \equiv 0$ for $\tau \in [t_*, T]$ and therefore $\int_0^T \alpha_\theta(T, \tau, z_0, v(\tau), \gamma(\cdot))d\tau = 1$ then, by virtue of the fact that co$M = M$ (Rockafellar, 1970), $\pi z(T) \in M$.

Similar result may be deduced in the case $\xi(T, z_0, \gamma(\cdot)) \in M$ also. The theorem is proved.

Let us consider a somewhat different approach to solution of the original problem (1), (2).

Denote

$$M(t, \theta) = M \stackrel{*}{-} \int_\theta^{\theta+h} \left[\pi e^{A(t-\tau)}V - \gamma(t-\tau)\right] d\tau, t \geq \theta \geq 0$$

**Condition 2.** Sets $M(t, \theta) \neq \emptyset$ and $\theta$-independent for $t \geq 0$.

Let us introduce the resolving function by the following formula

$$\alpha_\theta^*(t, \tau, z, v, \gamma(\cdot)) = \begin{cases} \sup\{\alpha \geq 0 : [W(t-\tau, v) - \gamma(t-\tau)] \cap \\ \alpha[M(t, \theta) - \xi(t, z, \gamma(\cdot))] \not\in \emptyset\}, \\ \qquad \tau \not\in [\theta, \theta + h] \\ 0, \qquad \tau \in [\theta, \theta + h] \end{cases}$$

and the corresponding function of time

$$T^*(z, \gamma(\cdot)) = \min\{t \geq 0 : \min_{t-h \geq \theta \geq 0} \int_0^t \inf_{v \in V} \alpha_\theta^*(t, \tau, z, v, \gamma(\cdot))d\tau \geq 1\}$$

**Theorem 2.** Let for the conflict controlled process (1), (2) the conditions 1 and 2 hold and let for initial state $z_0$ and some Borel selection $\gamma(\cdot)$ $T^* = T^*(z_0, \gamma(\cdot)) < +\infty$.

Then the game (1), (2) can be terminated at time $T^*$ from the initial state $z_0$.

**Proof.** Let $v(t)$ be the control of the second player and $\theta$ be the instant of failure of the first player's controlling devices.

Let us consider the test function

$$h^*(t, v(\cdot)) = 1 - \int_0^t \alpha_\theta^*(T^*, \tau, z, v(\tau), \gamma(\cdot))d\tau$$

Obviously, it is continuous, nonincreasing, and $h^*(0, v(\cdot)) = 1$. Therefore an instant $t_*, \theta + h < t_* \leq T^*$, exists such that $h^*(t_*, v(\cdot)) = 0$.

Let $\xi(T^*, z_0, \gamma(\cdot)) \not\in M(T^*, \theta) = M(T^*)$. Set

$$U_1^*(\tau, v) = \{u \in U : \pi e^{A(T^*-\tau)}(u - v) - \gamma(T^* - \tau) \in \alpha_\theta^*(T^*, \tau, z_0, v, \gamma(\cdot)) \times [M(T^*) - \xi(T^*, z_0, \gamma(\cdot))]\}$$

$$u_1^*(\tau, v) = \text{lexmin} U_1^*(\tau, v)$$

$$U_2^*(\tau, v) = \{u \in U : \pi e^{A(T^*-\tau)}(u - v) - \gamma(T^* - \tau) = 0\}$$

$$u_2^*(\tau, v) = \text{lexmin} U_2^*(\tau, v)$$

$$u(\tau) = u_2^*(\tau, v(\tau)), \ \tau \in (t_*, T^*]$$

If $\xi(T^*, z_0, \gamma(\cdot)) \in M(T^*)$, then the control of the first player at $\tau \in [0, \theta] \cup [\theta + h, T^*]$ will be chosen in the form (9).

Then, analogously to the proof of Theorem 1, Cauchy formula yields the inclusion $\pi z(T^*) \in M$.

Let us illustrate the obtain results by a simple example.

## 4. EXAMPLE

Consider the conflict controlled process

$$\dot{z} = \beta_\theta(t)u - v, \ z \in R^n,$$

$$||u|| \leq a, \ ||v|| \leq 1, \ a > 1$$

with the terminal set

$$M^* = \{z : ||z|| \leq \varepsilon\}.$$

Firstly let us follow the first scheme. Obviously, $M_0 = \{0\}$, and $M = \{z : ||z|| \leq \varepsilon\}$. Therefore, $L = M_0 = R^n$, and the orthoprojector $\pi$ appears as the operator of identical transform and is given by the unit matrix. So far as $A = 0$ then $e^{At} = I$ is the unit matrix. Then

$$W(t, v) = aS - v, \ W(t) = (a-1)S$$

where $S$ is a sphere of unit radius centered at the origin of space $R^n$. As $a > 1$, Pontryagin's condition is satisfied and, what is more, $0 \in W(t)$, $t \geq 0$. Set $\gamma(t) \equiv 0$. Then

$$t^* = \min\{s : s \in [\theta + h, t], \\ hS \subset (s - \theta - h)(a-1)S\}$$

$$t^* = \theta + h + \frac{h}{a-1}$$

The resolving function is

$$\alpha(z,v) = \alpha_\theta(t,\tau,z,v,\gamma(\cdot)) = \frac{(z,v)+a\varepsilon}{||z||^2-\varepsilon^2} + \frac{\sqrt{[(z,v)+a\varepsilon]^2 + (||z||^2-\varepsilon^2)(a^2-||v||)^2}}{||z||^2-\varepsilon^2}$$

$$\min_{||v||\leq 1} \alpha(z,v) = \frac{a-1}{||z||-\varepsilon}$$

So far $\alpha_\theta(t,\tau,z,v,\gamma(\cdot))$ does not depend on $t$, $\tau$, and $\theta$, the duration of the process of pursuit (Chikrii, 1996*a*) equals to $\frac{||z_0||-\varepsilon}{a-1}$, which summed with the duration of elimination of the breakdown and its consequences, gives the total time of approach

$$T(z_0,0) = \frac{||z_0||+h-\varepsilon}{a-1} + h$$

By virtue of (4) the control of the first player on the active segment (Chikrii, 1996*a*) has the form

$$u(\tau) = v(\tau) - \alpha(z_0, v(\tau))(m(z_0, v(\tau)) - z_0)$$

where

$$m(z_0,v) = -\varepsilon\frac{v - \alpha(z_0,v)z_0}{||v-\alpha(z_0,v)z_0||}$$

is a current target point in the terminal set. On the passive segment $u(\tau) = v(\tau)$, provided the approach should occur at the time $T(z_0,0)$. Generally speaking, in the case of erronious actions of the evader the approach may occur earlier (Chikrii, 1997).

The second scheme produces the following result. Since $M(t,\theta) = \varepsilon S \overset{*}{-} hS = (\varepsilon - h)S$, then Condition 2 holds if $\varepsilon \geq h$. Set $\gamma^*(t) \equiv 0$, then

$$\alpha^*(\tau,v) = \alpha^*_\theta(t,\tau,z,v,\gamma^*(\cdot)) = \frac{(z,v)+a(\varepsilon-h)}{z_*} + \frac{\sqrt{[z,v)+a(\varepsilon-h)]^2 + z_*(a^2-||v||^2)}}{z_*}$$

where $z_* = ||z||^2 - (\varepsilon-h)^2$

$$\min_{||v||\leq 1} \alpha^*(\tau,v) = \frac{a-1}{||z||-\varepsilon+h}$$

Thus, the time of approach equals

$$T^*(z_0,0) = \frac{||z_0||+h-\varepsilon}{a-1} + h$$

and coincides with $T(z_0,0)$.

The control of the first player on the active segment (Chikrii, 1996*a*) is

$$u(\tau) = v(\tau) - \alpha^*(z_0, v(\tau))(m^*(z_0, v(\tau)) - z_0)$$

where

$$m^*(z_0,v) = -(\varepsilon - h)\frac{v-\alpha^*(z_0,v)z_0}{||v-\alpha(z_0,v)z_0||}$$

## 5. REFERENCES


Aubin, I.-P. and J. Aklund (1984). *Applied Nonlinear Analysis.* John Wiley. New York.

Baranovskaya, L.V., A.A. Chikrii and Al.A. Chikrii (1997). Inverse Minkowski functionals in nonstationary problem of group pursuit. *Izv. Rus. Akad. Sci., Theory and Systems of Control* (1), 165–176. (in Russian).

Chikrii, A.A. (1985). Differential games with many pursuers. In: *Mathematical Control Theory.* Vol. 14. pp. 81-107. Banach Center Publ., PWN, Warsaw.

Chikrii, A.A. (1993). Minkowski' functionals in the theory of pursuit. *Russian Acad. Sci. Dokl. Math.* **47**(2), 243–248.

Chikrii, A.A. (1994). The method of gauge function in pursuit theory. In: *VI Intern. Symp. on Dynamic Games and Appl.* pp. 318–325. Quebec, Canada.

Chikrii, A.A. (1996*a*). Multi-valued mappings and its selections in game problems of control. *Journal of Automation and Information Sciences* **28**(5 & 6), 36–50.

Chikrii, A.A. (1996*b*). Quasilinear controlled processes under conflict, dynamical systems, 2. *J. Math. Sci.* **80**(1), 1489–1518.

Chikrii, A.A. (1997). *Conflict-Controlled Processes.* Kluwer Academic Publishers. Boston-London-Dordrecht.

Chikrii, A.A. and O.M. Patlanzhoglu (1997). On one class of quasilinear differential games of pursuit. *Differentz. Uravn.* (6), 786–794. (in Russian).

Chikrii, A.A. and P.V. Prokopovich (1995). Linear avoidance in the case of interaction of controlled objects groups. In: *New Trends in Dynamic Games and Applications, Ann. Internat. Soc. Dynam. Games, 3.* pp. 259–269. Birkhauser. Boston, MA.

Chikrii, A.A. and Yu.V. Pilipenko (1991). Conflict-controlled processes with periodic failure of Pontryagin's condition. *Avtomatika* (4), 35–46. (in Russian).

Chikrii, A.A., M.V. Pittcyk and N.B. Shyshkina (1986). First direct method of Pontryagin and some efficient methods of pursuit. *Cybernetics* **22**(5), 627–636.

Grigorenko, N.L. (1990). *Mathematical Methods of Control of Several Dynamic Processes.* Izdat. Mosk. Gos. Univ. Moscow. (in Russian).

Isaacs, R. (1965). *Differential Games.* John Wiley. New York.

Joffe, A.D. and V.M. Tikhomirov (1974). *Theory of Extremal Problems.* Nauka. Moscow.

Krasovskii, N.N. (1970). *Game Problems on Motions Encounter.* Nauka. Moscow.

Kuratovskii, K. (1966). *Topology.* Vol. 1. Mir. Moscow.

Nikolskii, M.S. (1984). *First Direct Method of Pontryagin in Differential Games.* Izdat. Mosk. Gos. Univ.. Moscow.

Nikolskii, M.S. (1987). On some problem of control with failures. *Doklady AN USSR* **287**(6), 1317–1320. (in Russian).

Nikolskii, M.S. (1988). On one problem of control with failures in dynamics. *Trudy Matem. Inst. Akad. Nauk SSSR* **185**, 181–186. (in Russian).

Nikolskii, M.S. and Ch. Pan (1994). One differential game of pursuit with possible failure in dynamics. *Vestnik Mosk. Gos. Univ.* (2), 70–72. (in Russian).

Patlanzhoglu, O.M. and A.A. Chikrii (1992). Modified Pontryagin's checking example. *Avtomatika* (1), 63–76. (in Russian).

Pilipenko, Yu.V. and A.A. Chikrii (1993). Oscillatory conflict-controlled processes. *Prikl. Mat. Mekh* **57**(3), 407–417.

Pontryagin, L.S. (1988). *Selected Scientific Works.* Vol. 2. Nauka. Moscow.

Pshenitchnyj, B.N. and V.V. Ostapenko (1992). *Differential Games.* Naukova Dumka. Kiev, Ukraine.

Rockafellar, R.T. (1970). *Convex Analysis.* Princeton Univ. Press. N.Y.

# PERFORMANCE OPTIMIZATION OF GAS TURBINE ENGINES USING THE STUDGA

**Valceres Silva Wael Khatib Peter Fleming**

*cop95vvs@shef.ac.uk*
*Department of Automatic Control and Systems Engineering, University Of Sheffield, Mappin Street , Sheffield S1 3JD, UK*

Abstract: Control of gas turbine engines is concerned with producing adequate thrust while maintaining stable and safe operation. Additional control layers can allow improvements in engine performance. Performance optimization can be expressed in terms of: minimizing fuel consumption while maintaining nominal thrust output, maximizing thrust for the same fuel consumption and minimizing turbine blade temperature. A new evolutionary approach called the StudGA is used as the optimization framework to design for optimal performance in terms of the three categories above. This approach gives faster convergence and can potentially save on expensive simulations. Model-based experimental results are used to illustrate these approaches. *Copyright © 1999 IFAC*

Keywords: Efficient algorithms, Fuel control, Gas turbines, Genetic algorithms, Optimization.

## 1. INTRODUCTION

Gas turbine engines are highly non-linear plants with a complex range of operation described by a flight envelope (Fig. 1). The set points within a flight envelope are defined by altitude and Mach number each requiring a certain thrust demand level.

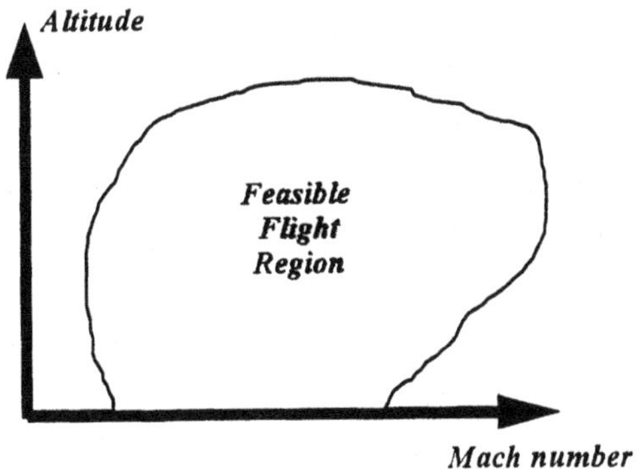


**Fig. 1.** A typical flight envelope.

Control systems for such plants reflect their inherent complexities. The main task for these systems is the production of adequate thrust while maintaining safe and stable operation. Performance requirements vary according to mission characteristics. Good passenger aircraft operation requires minimum running and maintenance costs. Military aircraft require maximum available thrust and speed of response for combat (Dash) missions. However, minimizing running cost by reducing fuel consumption, and thus increasing available range, is also vital for non-combat (Loiter) missions. Other potential benefits include reductions in engine loading. The outlet of the combustion chamber, for example, leads into the turbine stages which are subjected to high temperatures and pressures. Minimizing turbine blade temperature can reduce the effects of fatigue and creep. This leads to reductions in maintenance costs and enhances the life span of engine components. Dodd and Martin (1997) define (performance) optimization of a gas turbine engine to be one, or a combination of the following:

- Minimizing fuel consumption while maintaining thrust.
- Maximizing thrust.
- Maximizing engine life by reducing turbine blade temperature while maintaining thrust.

Reducing the required fuel consumption levels *per se* will lead to reductions in thrust. To maintain thrust levels, changes to other engine inputs controlling the inlet and outlet geometry parameters have to be employed.

Variable geometry devices are used for closed-loop multivariable engine control of the important variables to be controlled to ensure good performance. Nozzle efficiency has a direct effect upon thrust. The type and geometry of the inlet and inlet duct will determine the pressure loss and distortion of the air supplied to the engine, which will affect the installed thrust and fuel consumption. Performance optimization of this engine can use the effects of these variable geometry devices to establish compromise control configurations, to improve the working conditions.

Dodd and Martin (1997) approached this problem from a modeling point of view. They proposed a neural network approach to modeling the engine with these performance optimizing scenarios in mind. Small changes are applied successively to inlet and outlet settings to recover nominal thrust settings after a small reduction in fuel demand is introduced. The simulation results for the new settings are used by a back-propagation algorithm for training the neural network to find the optimum values.

Even a few percent decrease in fuel consumption translates to a significant saving of running costs over the life of a typical gas turbine.

## 2. THE STUD GENETIC ALGORITHM

The application of evolutionary computing (EC) techniques in general and genetic algorithms (GAs) in particular has seen wide ranging applications in control systems design (Fonseca and Fleming, 1995). GAs work on a population of solutions instead of single point search. Most GAs use binary representations for parameter encoding. The binary chromosome structure allows multiple type parameters to share the same string. No derivative information is required. Simple operators allow a typical GA to search for optimal solutions using stochastic transitions. These features allow GAs to work in difficult multi-modal spaces characterized by disjoint feasible domains. The main cost of using a GA is that of function evaluation. If the function evaluation is expensive, a non-linear gas turbine engine simulation for example, the potential benefits of using GAs can be somewhat reduced. Increasing the efficiency of the GA will mean reducing the number of trials (the number of generations multiplied by the number of individuals per population) needed to arrive at optimal or near-optimal solutions. Pre-mature convergence and entrapment in regions of local optima have to be avoided.

The need for making GAs more efficient for more expensive evaluations was the motivation for a new approach called the StudGA (Khatib and Fleming, 1998). This new approach uses no stochastic selection schemes for transition between successive generations. The best individual is chosen from each generation and made to share its genetic information with the other individuals which form the mating pool. Other simple operators such as the double point crossover are used in the usual way . Simple comparison operators are used to ensure only useful exchanges take place. A non-useful crossover resulting in little or no change is replaced by a random mutation. Application of this new technique to difficult test functions shows dramatic improvements in terms of efficiency over existing GA techniques.

The StudGA appears to offer the same benefits as most standard GAs. The flexible binary representation is maintained. The improvements in efficiency bring these methods to the level of standard hill-climbing methods in terms of convergence speeds. However, the pitfalls of these methods are avoided. StudGAs can deal with discontinuous functions and are amenable to parallelisation because they work on a population of solutions. No stochastic selection is used as in standard GAs. This feature appears to help improve the speed of convergence for the StudGA.

## 3. THE SPEY GAS TURBINE ENGINE

For the purposes of this work, a non-linear SIMULINK model for a SPEY gas turbine engine is used. This is a two-spool engine used mostly for powering military jets. There are three engine inputs:

- Fuel Flow
- Inlet guide vane angle
- Exhaust nozzle area

These inputs can be used to control various engine responses. Sensors provide various outputs for control purposes. The control loops are not unique. The main output to be controlled is thrust. No direct measurement of thrust is available. Other parameters are used to control thrust implicitly. Extensive work has been published on various control strategies for such an engine (Dadd, *et al.*, 1995; Samar, 1995; Silva and Fleming, 1998). The SIMULINK model allows the specification of different operating points in terms of altitude and Mach number. The model can also simulate dry and re-heated modes. The work described here is concerned with dry operation only. The same methods can be applied to re-heat modes.

## 4. PERFORMANCE OPTIMIZATION FRAMEWORK

The objective of this work is to establish a schedule of extra gains applied to the three engine inputs to decrease fuel consumption while maintaining adequate thrust levels, maximize thrust for the same fuel levels and minimize turbine blade temperature for the same thrust level. The definition of adequate thrust levels will depend on the particular mission requirements and resources in question. For the purposes of this work, the maximum possible fuel saving is investigated for a thrust level maintained within 1.0% of the nominal value for the set point in question. The modified engine configuration has to meet the following requirements:

- Minimum surge margin (stability requirement)
- Maximum turbine blade temperature (safety limit)

The engine simulation is configured to return null values in the event of any physical operational limits being breached. These limits are imposed on spool speeds and inlet and outlet geometry settings.

Designs for fuel reduction, thrust maximization and temperature minimization are presented for an engine at zero altitude and Mach number (zero-zero) conditions. This is done for a range of thrust demand from idle to maximum dry. The effect of varying Mach number is also investigated. The effect of Altitude variation is not included in this study. Future work will include this to map out a comprehensive design for the entire flight envelope.

The optimization performance of the StudGA is compared with that of a standard GA for the zero-zero point with maximum thrust demand for a minimum fuel scenario. The two GAs are implemented using the GA Toolbox for MATLAB (Chipperfield, *et al.*, 1995). Optimization is carried out using a single aggregating objective with a simple penalty function for the constraints. The thrust loss is penalized with an exponential penalty term. The 1.0% band for thrust requirement allows a more efficient equality constraint handling and reflects the numerical accuracy of the simulations. The optimization for fuel reduction can thus be described by:

**Minimize:** *fuel demand + constraint violation penalties*

- *Surge margin penalty:*

If below minimum requirement:

*penalty = 100*(limit-actual)/limit*

Else: *penalty = zero*

- *Temperature penalty:*

If more than limit:

*penalty = 100*(actual-limit)/limit*

Else: *penalty = zero*

- *Thrust penalty:*

If lower/higher than limit:

*penalty = exp[|actual-nominal|/nominal]*

Else: *penalty = zero*

Similar objective functions are used for the other two cases.

## 5. RESULTS

### *5.1 The StudGA vs. normal GA*

The StudGA and a standard GA were both applied to find minimum fuel consumption at zero-zero conditions maintaining nominal thrust levels. Mutation and crossover rates are the same for the two algorithms. 20 individuals were used for a total of 15 generations to search for solutions. The convergence graph (Fig. 2) shows that the StudGA quickly finds a fuel setting with 7% reduction achieving 99.2% of nominal thrust. The standard GA only manages to find a 5% saving. Another GA using 75 individuals was used in order to achieve better results. Although this latter GA appears to have the same convergence performance as the StudGA in terms of generations, the real cost is nearly 3.5 times more evaluations. This illustrates the potential efficiency gains possible using the StudGA. All the remaining analysis was conducted using the StudGA.

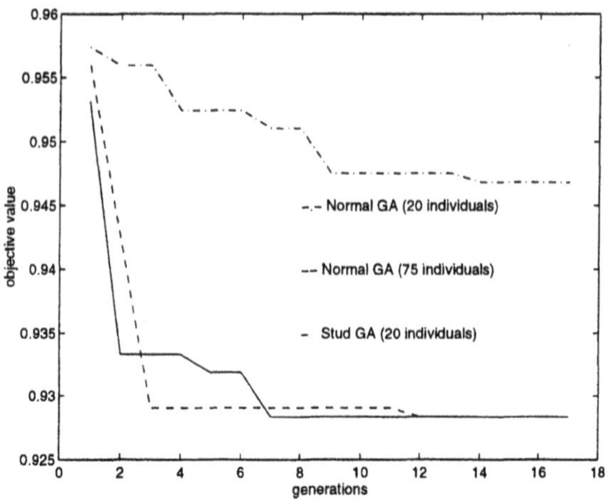


**Fig. 2.** Convergence performance for StudGA and normal GA.

*5.2 Fuel consumption minimization*

The StudGA is applied to find optimal fuel savings at zero-zero conditions for a range of thrust demand levels from idle to maximum. The results (Fig. 3) show that fuel savings are possible across the thrust range. However, as the thrust demand falls towards idle settings, no real gains are possible. This can be attributed to the fact that at those settings, the engine is running rather efficiently under low load conditions. For these demand settings, the main task is to keep the engine cycle in operation with very low net thrust.

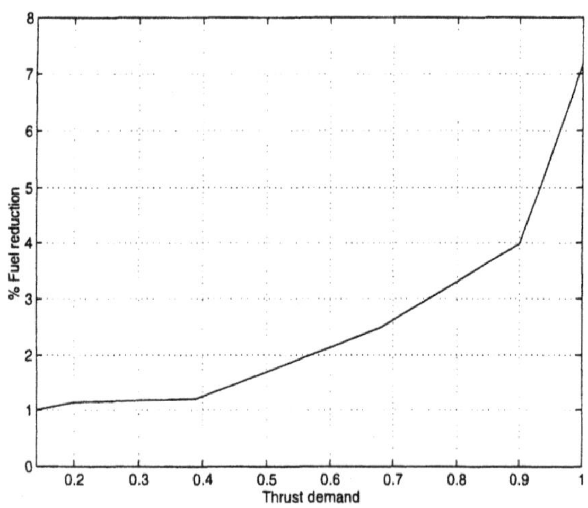


**Fig. 3.** Fuel savings vs. thrust demand

Fuel savings are also investigated for a range of Mach numbers from 0 to 1.0 at sea level conditions with full thrust demand. Again, reductions are possible but tend to fall slightly with increasing Mach number (Fig. 4). The result for the sonic point (Mach 1.0) has to be viewed with care as this is a highly non-linear region in real life. The accuracy of the simulation for points near unity Mach number is quite low. The performance at such a Mach number is studied usually for transonic performance as a barrier to overcome, and not as an operating point *per se*.

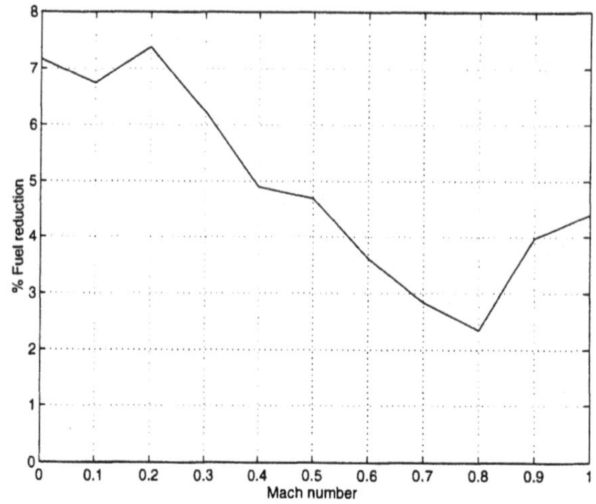


**Fig. 4.** Fuel saving vs. Mach number

*5.3 Thrust maximization*

The thrust maximization response appears to be similar to that in the fuel reduction case, in relation to variation in thrust demand. The percentage gains are, however, about half those for fuel (Fig. 5). This indicates that the engine is already designed to give near maximum possible thrust.

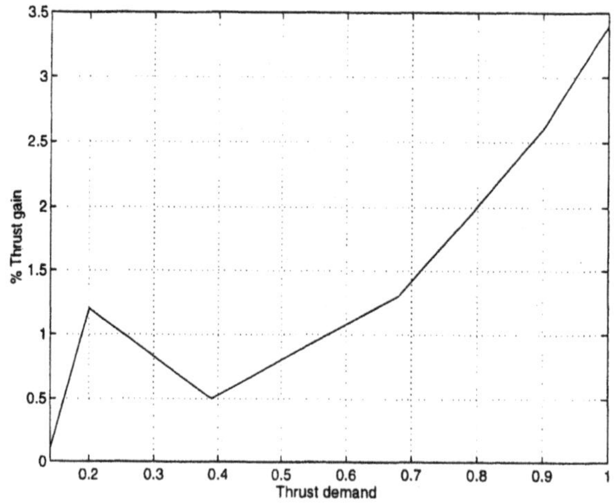


**Fig. 5.** Thrust gain vs. thrust demand

The thrust maximization is also tried for Mach number variations from 0 to 1.0 at sea level conditions (Fig. 6). The possible thrust increase varies slightly with increasing Mach number. A slight increase is observed for modest Mach numbers (~ 0.2) similar to the response for fuel savings (Fig. 4), but drops slightly for higher Mach numbers. The results for high Mach number (greater than 0.8) are not reliable for the reasons highlighted above.

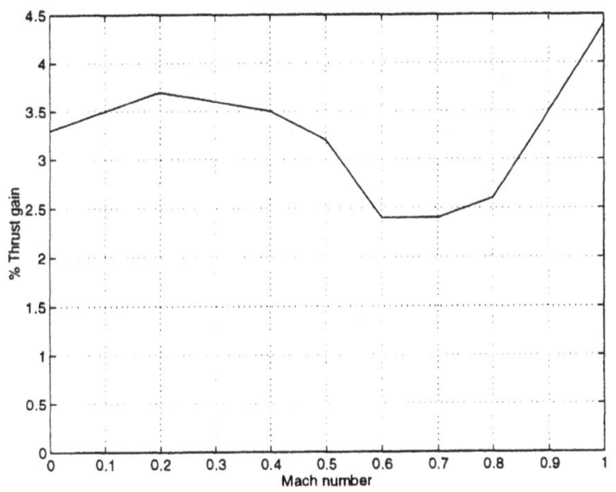


**Fig. 6.** Thrust gain vs. Mach number

*5.4 TBT minimization*

Reductions in TBT appear to be possible across the range of thrust demand (Fig. 7). These reductions tend to follow reductions in fuel which is injected into the combustion chamber to add energy to the high pressure gas coming in from the compressors. The response is thus similar to that in the fuel reduction case.

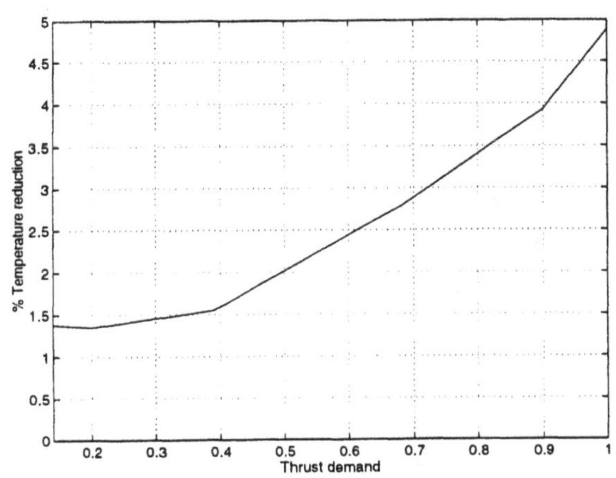


**Fig. 7.** TBT reduction vs. thrust demand

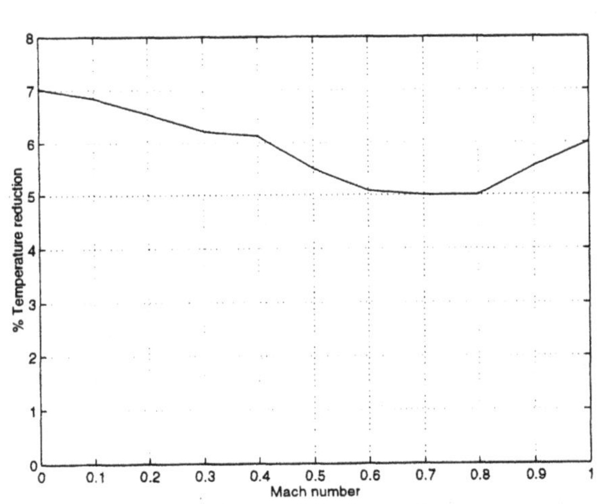


**Fig. 8.** TBT reduction vs. Mach number

The reduction in TBT was also possible for variations in Mach number, from 0 to 1.0 sea level condition, with full thrust demand. As in the fuel savings case, the reductions in TBT tend to fall slightly with increasing Mach number (Fig. 8).

# 6. CONCLUSIONS AND FUTURE WORK

It is possible to add control layers to a gas turbine engine in order to achieve improvements in performance. Modifications to inlet and outlet geometry parameters can be used to design for reductions in fuel while maintaining thrust levels at the nominal values. Thrust can be maximized for the same fuel settings. Turbine blade temperature can be reduced without affecting thrust levels. The performance gains are possible for various operating points. These gains in various performance metrics can be summarized by the following:

- reduce running costs through fuel savings;
- increase flight range;
- increase engine life through reducing high engine core temperatures;
- and increase engine power for vital aircraft maneuvers and critical missions.

The StudGA is a powerful optimization framework for these design scenarios. Improvements over traditional GAs appear to indicate a more efficient algorithm in terms of the number of function evaluations. Where a design case requires expensive and lengthy calculations, StudGAs can offer simple but effective optimization engines that lead to optimal or near-optimal solutions with a minimum number of evaluations. StudGAs are inherently simple to set up and implement for a variety of optimization scenarios. More work is planned to investigate other performance optimization criteria. The integration of such strategies within an overall control system needs to be studied. A multiobjective, multidisciplinary approach to optimal control of gas turbine engines is possible by designing to meet all of the performance, cost and safety requirements simultaneously. To coincide with these suggestions, an extension of the StudGA to handle multiobjective optimization cases can lead to a more robust design strategy. Initial results for implementing multiobjective StudGAs are encouraging.

A complete performance optimization design of this engine will involve establishing the required set of control gains. These gains will then be scheduled for each set point. The application of such schedule is envisaged for cases when the engine is in steady state conditions for a particular set point and no transitionary maneuvers are taking place.

## ACKNOWLEDGEMENTS

The first author gratefully acknowledges support of this research by CAPES, '*Coordenação de Aperfeiçoamento de Pessoal de Nível Superior*' and FUNREI, '*Fundação de Ensino Superior de São João del Rei*', Brazil. The second author wishes to acknowledge the support of EPSRC (CASE Award) and BAe PLC Sowerby Research Centre especially Dr Phil Greenway. The authors also wish to thank DERA Pyestock for the Spey engine model. We would also like to thank Dr Ian Pratt for his help and advice with the engine model.

## REFERENCES

Chipperfield, A., P.J. Fleming, H. Polheim and C. Fonseca (1995). *Genetic Algorithm Toolbox for use with MATLAB©*, Version 1.2, Department of Automatic Control and Systems Engineering, University of Sheffield.

Dadd, G.J., A.E. Sutton and A.W.M. Greig (1995). Multivariable control of military engines. *AGARD Conference Proceedings* N° **572**, Advanced aero-engine concepts and controls, pp.1-12.

Dodd, N. and J. Martin (1997). Using neural networks to optimise gas turbine aero engines. *Computing & Control Engineering Journal*, June, pp. 129-135.

Fonseca, C.M. and P.J.Fleming (1995). An overview of evolutionary algorithms in multiobjective optimisation. *Evolutionary Computation*, The Massachusetts Institute of Technology, **3(1)**, pp. 1-16.

Khatib, W. and P. Fleming (1998). The StudGA: A Mini Revolution?. *Proc. Fifth Int. Conf. on Parallel Problem Solving from Nature* (Eiben, A.E., Bäck, T., Schoenauer, M and Schwefel, H-P. Ed.), LNCS 1498, pp. 683-691. Springer, The Netherlands.

Samar, R. (1995). *Robust multi-mode control of high performance aero-engines*. PhD Thesis, University of Leicester. England.

Silva, V.V.R. and P. Fleming (1998). Multiobjective non-linear robust control design for gas turbine engines using evolutionary computing. *Proc. Mathematical Theory Networks Systems Conference* - MTNS'98, 'Universidade de Padova', Italy.

14th Triennial World Congress, Beijing, P.R. China

F-2d-09-2

# A GENETIC OPTIMIZATION METHOD FOR DYNAMIC PROCESSES

**B. de Andrés-Toro, J.M. Girón-Sierra, J.A. López-Orozco, J. Alvarez-Ruiz, P. Fernández Blanco**

*Departamento de Arquitectura de Computadores y Automática*
*Universidad Complutense de Madrid, 28040 Madrid, Spain*
*E-mail: deandres@eucmax.sim.ucm.es, Tf.: +34 91 394 43 84, Fax.: +34 91 394 46 87*

Abstract: The paper presents a dynamic optimization method based on Genetic Algorithms. The optimization problem considered is to find a temporal profile for the control variable, so the process, under such control, follows an optimal trajectory. The method encodes candidate solutions as chromosomes, using a piecewise linear approximation of the control profile. To show the general validity of the method, several examples are taken from optimization literature, and the method is applied, always having satisfactory results. The method has interesting operational and computational advantages. *Copyright © 1999 IFAC*

Keywords: Optimization, Genetic algorithms, Process control, Optimal trajectory.

## 1.- INTRODUCTION

There are important industrial and laboratory processes that can be described by state variables that follow a trajectory along time. This trajectory depends on control inputs, that drive the process to obtain some desired results. Usually there are constraints for the state variables and the control inputs. In general, we have here an optimization problem, that could be multiobjective: less time, more production. The solution is in the form of a trajectory of the control inputs, that should be applied to obtain the best behaviour of the process.

If no analytical solution is available for the dynamic optimization problem at hand, there are still some algorithmic methods that could be useful. Dynamic programming is a good candidate, but has some well-known problems when the grid to be explored has many nodes (Edgar and Himmelblau, 1988). There is a line of research (Dadebo and McAuley, 1995; Bojkov and Luus, 1994) that promotes iterative dynamic programming algorithms to cope with relatively complex problems. Also, there are alternatives in the form of search methods (Gupta,1995) that can get even better results. In general, these methods apply a discretization of the control variable(s) into a series of equal time intervals, and the optimal profile obtained consists of a series of constant horizontal steps.

Genetic Algorithms are showing interesting features to solve optimization problems, as a searching method that avoids to be trapped into local maxima. To be applied to a problem, a genetic encoding must be defined to allow for the search of solutions in terms of chromosomes.

This paper presents a method for the optimization of dynamic processes, based on genetic algorithms through the direct genetic encoding of piecewise linear aproximations of control profiles. The method is the result of some years of research related to beer fermentation (Andres-Toro et al., 1998). To show that the method has a general potential, some examples of other processes have been selected, and the method has been applied with successful results.

The first example is the beer fermentation, that constitutes a fairly complex problem. The second example has an analytical solution, as published in (Ramirez, 1994), so it is useful to show how good is the solution reached by our approximation. The rest of the examples are taken from the recent literature that deals with iterative dynamic programming or semiexhaustive search, and serve for comparison purposes (in every case, the method presented in this paper obtains better results).

## 2.- THE GENETIC-BASED METHOD

### *2.1.- Genetic Encoding:*

Genetic Algorithms (Goldberg, 1989) are search algorithms that mimic natural selection and genetic. It differ from other searching methods in several ways: it start from a population of solutions, not a single initial solution; it work with probabilistic rules, instead of deterministic rules; it search several solutions in parallel; etc. The algorithms proceeds along several generations. A performance index is employed to select parents for each generation, and to eliminate worst members of the population.

The first step to apply Genetic Algorithms is to represent the problem in terms of chromosomes. For the kind of problems considered here, the approach taken by our method is to use a linear piecewise approximation of the profile of the control variable, as described in figure 1. The total time of the process is divided into equal time intervals. The ordered series of values of the control variable at the breakpoints, from left to right, are taken in the same order as genes, forming a chromosome

Chromosome = [119, 142, 153, 118...]

(for a profile with real values of 11.9, 14.2, 15.3...)

To ease computations, the encoding is directly made in integer numbers (not a binary representation). These numbers are obtained from real numbers by a displacement of the decimal point and rounding.

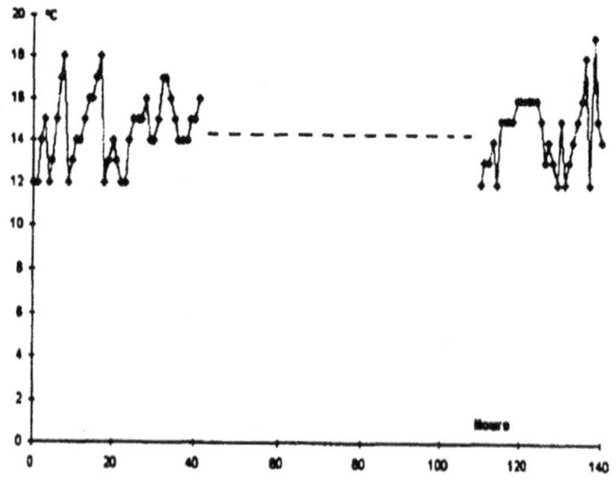

Fig. 1. a piecewise approximation of the profile of the control variable.

### *2.2.- Specifications of the Algorithm:*

The dynamic optimization problem to be solved in each case has two main components: a model of the process (differential equations that describe its temporal behaviour, as a response to the control inputs), and a performance index (an objective function) to measure how good is any trajectory of the process.

Each chromosome represents a candidate optimal solution for the control, to drive the dynamical process along a trajectory that exhibits best characteristics. The method we propose, uses the performance index as the fitting function of the genetic algorithm. The fitting function is employed to select best chromosomes. The evaluation of each chromosome is made by running a simulation of the dynamic process: the model is used to know the response to the control variable profile described by the chromosome, and to determine the value of the performance index.

After extensive studies on what are the more convenient specifications for the Genetic Algorithm, the following values were established: an initial population of 10 individuals, mutation probability of 0.007, and crossover probability of 0.7.

The steps of the algorithm are as follows:

1. Obtain 10 individuals as initial population.
2. Evaluate the individuals according to performance index J.
3. Select the four individuals as parents, using the roulette-wheel method.
4. Obtain new offspring employing crossover and mutation operators.
5. Evaluate the new individuals.
6. Substitute four worst individuals with the new offspring.
7. If last generation, end; else, return to step 3.

### *2.3.- Comments and Results*

The method has been implemented in Borland C++, to achieve rapid calculations. By using a modular approach to the software development, a general tool is obtained: easy to apply for different processes, changing only procedures with the model and the performance index. The method has been named Genetic Dynamic Optimization (GDO).

In all the cases we studied the results were encouraging from the very beginning. A discretization of the control variable profile as fine as desired can be applied, with a moderate computational cost. As shown in the following examples, using a Pentium II/233 the computation times are in the order of minutes, or even seconds, for the kind of discretizations employed in the literature.

## 3.- APPLICATION OF THE METHOD TO EXAMPLES:

Several examples have been selected, in order to show the viability of the GDO method. The first is related with our research with the optimization of beer fermentation, and includes a relatively complex model. The second has a published analytical solution, and is a good reference to know if the result obtained by GDO is far from the true optimum. The others belong to the line of research mentioned before, that propose alternative algorithms with results that are interesting for comparison purposes.

Table 1 summarizes the main aspects concerning the GDO algorithm, calculation time, and optimization results, for each of the examples considered in this study.

Table 1. Values of the variables and results obtained for the examples.

| | Ex. 1 | Ex. 2 | Ex 3 |
|---|---|---|---|
| Genes / Chromosome | 28 | 16 | 20 |
| Population Size | 10 | 10 | 10 |
| Number Generations | 1000 | 1000 | 5000 |
| Process Duration | 140h | 8m | 1m |
| Discretization (time) | 5h | 0.5m | 0.05m |
| dt (integration) | 0.1h | 0.01m | 0.01m |
| Computation Time | 10.1m | 1.38s | 0.349s |
| Optimum Value | 592.36 | 0.674 | 0.093 |

| | Ex. 4 | Ex. 5 | Ex 6 |
|---|---|---|---|
| Genes / Chromosome | 26 | 20 | 33 |
| Population Size | 10 | 10 | 10 |
| Number Generations | 3000 | 2000 | 1000 |
| Process Duration | 26h | 60m | 132h |
| Discretization (time) | 1h | 0.33m | 4h |
| dt (integration) | 0.01h | 0.01m | 0.005h |
| Computation Time | 14.3m | 3.98m | 2.76m |
| Optimum Value | 106.34 | 20911.1 | 93.019 |

*Example 1:* The case considered is the optimization of the conventional batch fermentation of beer, keeping industrial conditions: no stirring, industrial wort and yeast. The control variable we can manipulate is temperature along time. The objective is to determine the control input to be applied, a temperature profile, such that optimizes (maximize) the yield of ethanol. There are two constraints: the concentration of ethyl acetate and diacethyl can not surpass certain limits that degrades taste and smell.

Temperature is also limited under 18°C because of risk of spoilage by *Lactobacillus Plantarum*. The process model is explained in (Andres-Toro et al., 1997). This model distinguishes two different phases: lag phase, and fermentation phase, both with his own differential equations.

*Lag phase:*

$$x_{active} + x_{lag} = \text{constant} = 0.48 \cdot x_{initial} \quad (1)$$

$$dx_{active}/dt = \mu_{lag} \cdot (0.48 \cdot x_{initial} - x_{active}) \quad (2)$$

$$dx_{lag}/dt = \dot{x}_{active} = -\mu_{lag} \cdot x_{lag} \quad (3)$$

*Fermentation phase:*

$$dx_{active}/dt = \mu_x \cdot x_{active} - k_m \cdot x_{active} + \mu_L \cdot x_{lag} \quad (4)$$

$$\mu_x = \mu_{xo} s/(0.5\, s_{initial} + e) \quad (5)$$

$$dx_{bottom}/dt = \mu_D \cdot x_+ \quad (6)$$

$$ds_{cons}/dt = \mu_s \cdot x_{active} \quad (7)$$

$$\mu_s = \mu_{so} s/(k_s + s) \quad (8)$$

$$de/dt = \mu_a \cdot f \cdot x_{active} \quad (9)$$

$$f = 1 - e/(0.5 \cdot s_{initial}) \quad (10)$$

$$\mu_a = \mu_{ao} s/(k_a + s) \quad (11)$$

$$d(ea)/dt = Y_{eas} \cdot \frac{ds}{dt} = Y_{eas} \cdot \mu_s \cdot x_{active} \quad (12)$$

$$d(vdk)/dt = k_{dc} \cdot s\, x_{active} - k_{dm} \cdot (vdk) \cdot e \quad (13)$$

As temperature is the variable we can manipulate, the main effort of our experimental research was directed to establish how it affects the model parameters. The expressions obtained are the following:

$$\mu_{xo} = 1.095 10^{47} \cdot e^{\frac{-63720}{1.99536(T+273.15)}} \quad (14)$$

$$k_m = 3.373 10^{56} \cdot e^{\frac{-76450}{1.99536(T+273.15)}} \quad (15)$$

$$Y_{eas} = 1.1129 10^{39} \cdot e^{\frac{-53056}{1.99536(T+273.15)}} \quad (16)$$

$$\mu_{D0} = 4.889 10^{14} \cdot e^{\frac{-20020}{1.99536(T+273.15)}} \quad (17)$$

$$\mu_{S0} = 6.232 10^{-19} \cdot e^{\frac{23254}{1.99536(T+273.15)}} \quad (18)$$

$$\mu_{ao} = 26.3865\, e^{\frac{-2528.6}{1.99536(T+273.15)}} \quad (19)$$

$$\mu_{lag} = 2.2041 \cdot 10^{13} \cdot e^{\frac{-76450}{1.99536(T+273.15)}} \quad (20)$$

$$k_a = 1.1081 \cdot 10^{-52} \cdot e^{\frac{-76450}{1.99536(T+273.15)}} \quad (21)$$

Taking into account the main goals of the optimization, an objective function J, was formed adding five weighted terms:

$$J_1 = +10\, \text{ethanol}_{end} \quad (22)$$

$$J_2 = -5.73\, e^{(95\, \text{diacethyl} - 11.51)} \quad (23)$$

$$J_3 = -1.16\, e^{(460\, \text{acetate} - 66.77)} \quad (24)$$

$$J_4 = -\int_0^t 9.91 \cdot 10^{-17} \cdot e^{2.31 \cdot T} \cdot dt \quad (25)$$

$$J_5 = -\sum_{i=1}^{140} \text{abs}(T_{i+1} - T_i) \quad (26)$$

The last term, that is a simple calculation of the profile length, will smooth the temperature profile obtained by the GDO. Note also how the exponentials will forbid any limit violation in the final by-product concentrations. The time allowed for the fermentation is 140 hours.

The use of one objective function grouping all the aspects of the problem makes easy the application of Genetic Algorithms. It is immediate to employ it as the fitting function.

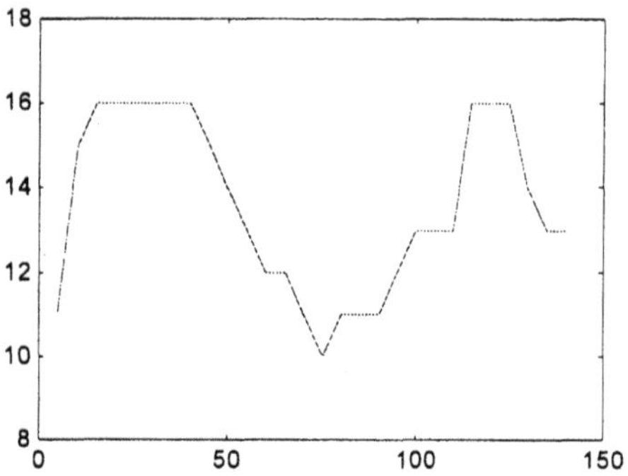


Fig.2. Optimal control policy for the example 1

The application of our method GDO renders a maximum value of 592.37. The optimal control profile obtained is represented in figure 2.

*Example 2*: The example is taken from the book (Ramirez, 1994), pp. 84-87. It consists of a batch reactor with two first order reactions, that was first studied by (Bilous and Amundson, 1956):

$$P \xrightarrow{k_1} Q \xrightarrow{k_2} R$$

The material balance equations for species $P(x_1)$, and $Q(x_2)$, are given by:

$$dx_1 / dt = f_1 \quad (27)$$

$$dx_2 / dt = f_2 \quad (28)$$

where,

$$f_1 = -k_1 x_1 \quad f_2 = k_1 x_1 - k_2 x_2 \quad (29)$$

$$k_1 = k_{10} e^{-E_1/RT} \quad k_2 = k_{20} e^{-E_2/RT} \quad (30)$$

In (Ramirez, 1994) there is a long description of the problem, with parameter values, analytical expressions, and a figure with the optimal solutions.

The time allowed for the process is $t_f$ = 8 min. The control variable is temperature. An optimal profile of the temperature u = T(t) must be found that maximizes the yield of species Q, $x_2$. at final time $t_f$.

Using our GDO method, a satisfactory solution was found, obtaining an optimum value of 0.6737 and spending only 1.38s of computation time. The optimum value obtained is similar to the value displayed in a figure of (Ramirez, 1994). The optimal control profile obtained with the GDO method is displayed in figure 3.

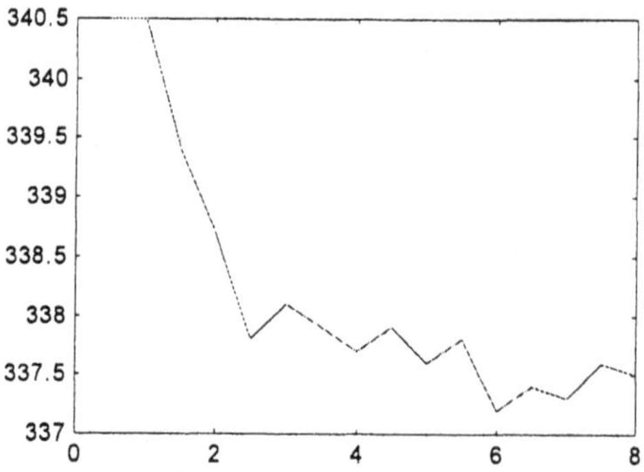


Fig.3.Optimal control policy for the example 2

*Example 3:* This example has been studied by (Yeo, 1980; Luus, 1990; Gupta, 1995). It is a process described by the following model:

$$dx_1 / dt = x_2 \quad (31)$$

$$dx_2 / dt = -x_3 u + 16t - 8 \quad (32)$$

$$dx_3 / dt = u \quad (33)$$

$$dx_4/dt = x_1^2 + x_2^2 + + 0.0005(x_2 + 16t - 8 - 0.1x_3u^2)^2 \quad (34)$$

$$x(0) = [0 \ \ -1 \ \ -\sqrt{5} \ \ 0]^T \quad \text{and} \quad t_f = 1$$

The control is bounded by $-4 \le u \le 10$

It is required to find a control trajectory over $t_f$ = 1 that minimizes $x_4(t_f)$.

Employing quasilinearization, (Yeo, 1980) obtained an optimum value of 0.14052. In (Luus, 1990), the same problem is attacked with IDP (iterative dynamic programming), taking 10 control steps, with 20 iterations, and computing during 11 s. on a Cray X-MP/24, to obtain an optimum value of 0.12012. And using semiexhaustive search, (Gupta, 1995) gets, with 24 iterations, very similar results: on a 486/33 computer, after 76 seconds of computation, the search reaches an optimum value of 0.12011.

In our case, employing GDO, in less than half second (0.349s) a minimum valour equal to $x_4$ = 0.092753 is obtained. The control profile (u) is represented in figure 4.

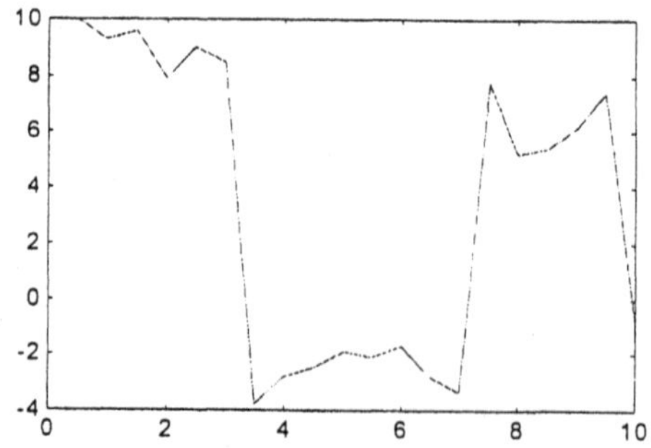


Fig.4. Optimal control for the example 3.

*Example 4:* The maximization of secreted protein in a fed-batch reactor was studied first by (Park and Ramírez , 1988), and later by (Luus, 1991a).The system is described by the following equations:

$$dx_1/dt = g_1(x_2 - x_1) - (u/x_5)x_1 \quad (35)$$

$$dx_2/dt = g_2x_3 - (u/x_5)x_2 \quad (36)$$

$$dx_3/dt = g_3x_3 - (u/x_5)x_3 \quad (37)$$

$$dx_4/dt = -7.3g_3x_3 + (u/x_5)(20 - x_4) \quad (38)$$

$$dx_5/dt = u \quad (39)$$

where, $x(0) = [0\ 0\ 1\ 5\ 1]$

$$0 \le u \le 10$$

$$g_1 = 4.75g_3/(0.12 + g_3) \quad (40)$$

$$g_2 = x_4e^{-5x_4}/(0.1 + x_4) \quad (41)$$

$$g_3 = 21.87x_4/((x_4 + 0.4)(x_4 + 62.5)) \quad (42)$$

The total amount of protein ($x_1$ ($t_f$) $x_5$ ($t_f$)) produced at $t_f$ = 24 h. must be maximized. For this problem the results obtained by dynamic programming were very close to those obtained by non linear programming, as described in (Luus and Rosen, 1991).

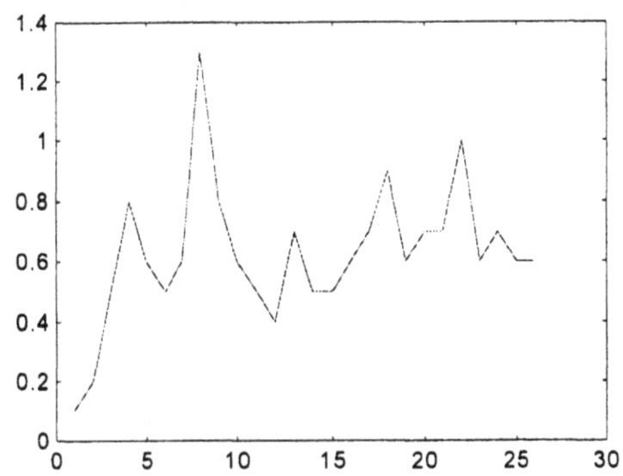


Fig.5. Optimal control policy for the example 4.

Using IDP, for a piecewise constant control trajectory over 20 time stages, and $t_f$ = 24, (Luus and Rosen, 1991) reached a maximum value of 106.27 (106.28 for non linear programming), spending less than 1min of computation time on a CRAY X-MP/24.

With the GDO method, with 14.32 min. of computation time, a maximum value of 106.34 is obtained. Figure 5 displays the optimal control profile.

*Example 5*: Chen and Hwang, 1990; Luus, 1991b; Gupta, 1995) consider this problem. It consists of a fed-batch reactor, in which ethanol is produced by *Sacharomyces cerevisiae* and the production rate of ethanol is inhibited by itself. Assuming that at in the beginning, the fermenter is filled with 10 litres of broth and that the maximum capacity of the fermenter is 200 litres, the control variable (u) in this case is the maximum feed rate, restricted up to 10 l/hr. The total amount of ethanol ($x_3(t_f)$ $x_4(t_f)$) produced at $t_f$=62h, must be maximized. The following differential equations describe the system:

$$dx_1/dt = g_1x_1 - ux_1/x_4 \quad (43)$$

$$dx_2/dt = -10.g_1x_1 + u.(150 - x_2)/x_4 \quad (44)$$

$$dx_3/dt = g_2x_1 - u.x_3/x_4 \quad (45)$$

$$dx_4/dt = u \quad (46)$$

where: $g_1 = 0.408.x_2/((1 + x_3/16)(0.22 + x_2))$ (47)

$$g_2 = x_2/((1 + x_3/71.5)(0.44 + x_2)) \quad (48)$$

$$x(0) = [1\ 150\ 0\ 10]$$

$$0 \le u \le 10$$

$$x_4(t) \le 200$$

(Chen and Hwang, 1990) found a maximum value of 20073. For a piecewise control trajectory over 20 time stages, using IDP (Luus, 1991b) reported 20814.8. And starting from the optimal trajectory obtained by Luus, but using semiexhaustive search, (Gupta, 1995), reached a maximum value of 20830. With the GDO method, we obtain a maximum value of 20911.08, in 3.98 min of computation time. The control profile obtained is shown in figure 6.

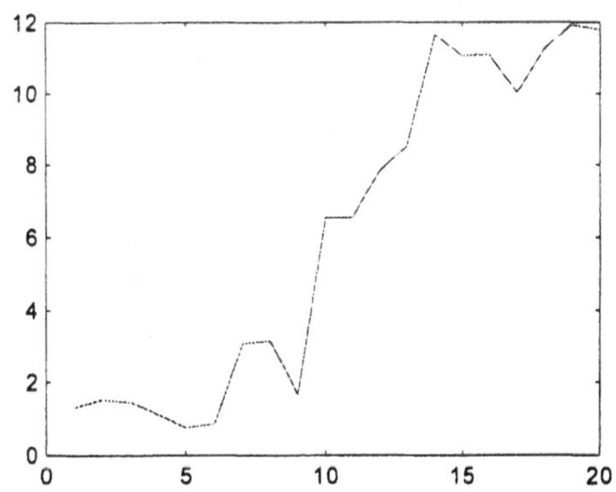


Fig.6. Optimal control policy for the example 5.

*Example 6:* This example consists in one fed-batch fermenter for biosynthesis of penicillin, with constraints on the state variables. Feed rate is the control variable, and we wish that the total amount of penicillin ($x_2(t_f)$ $x_4(t_f)$) produced at $t_f$=132h is maximized. The process is described by the following equations:

$$dx_1/dt = h_1x_1 - x_1u/(500x_4) \quad (49)$$

$$dx_2/dt = h_2x_1 - 0.01x_2 - x_2u/(500x_4) \quad (50)$$

$$dx_3/dt = -h_1x_1/0.47 - h_2x_1/1.2 - 0.029x_1x_3/(0.0001 + x_3) + (1 - x_3/500)u/x_4 \quad (51)$$

$$dx_4/dt = u/500 \quad (52)$$

where:

$$h_1 = 0.11x_3/(0.006x_1 + x_3) \quad (53)$$

$$h_2 = 0.0055x_3/(0.0001 + x_3(1 + 10x_3)) \quad (54)$$

$$x(0) = [1.5\ 0\ 0\ 7]^T$$

$$0 \le u \le 50$$

$$0 \le x_1 \le 40$$

$$0 \le x_3 \le 25$$

$$0 \le x_4 \le 10$$

For this problem, (Cuthrell and Biegler, 1989) found a maximum value of 87.69. Using IDP for a trajectory over 20 time stages, (Luus, 1993) reported a maximum value of 87.948. Beginning from a control trajectory obtained by Luus, and with semihexaustive search, (Gupta, 1995) found a maximum value of 88, spending for 15 iterations, 47 min on the 486/33 PC.

In this case, the GDO method can obtain a maximum value of 93.019. Figure 7 displays the optimal control policy.

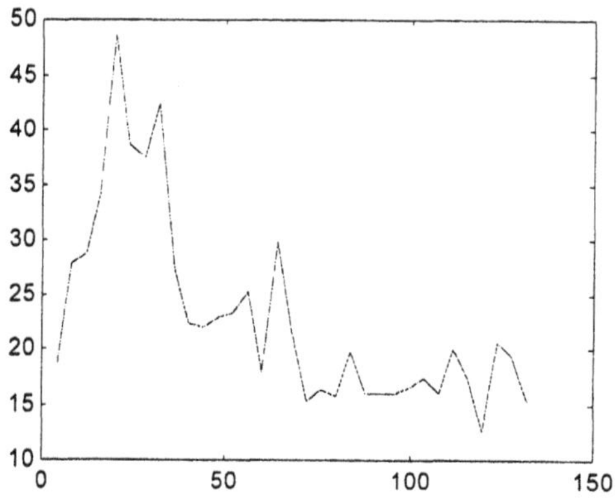


Fig.7. Optimal control policy for the example 6.

## 4.- CONCLUSIONS

The optimization method described in this paper provides an easy way to solve nonlinear optimal control problems. Based on Genetic Algorithms, the method is able to avoid local optimal solutions. Compared to other methods, it has advantages of simple, fast, application and moderate computational cost. By way of examples, the paper shows the potential of the method for the optimization of dynamic processes.

Because of less computation is needed, the method paves the way to study finer discretizations, in search of better approximations to the true optimal solutions. Fast versions of the method could be useful, in certain slow processes, for real-time optimization applications.

## REFERENCES

Andres-Toro, B., J.M. Giron-Sierra., C. Fernandez-Conde, J.M. Peinado, and F. Garcia-Ochoa (1997). A kinetic model for beer fermentation under industrial operational conditions, pp.203-208. *Proc. IMACS/IFAC 2th. Int. Symp.. Mathematical Modelling and Simulation in Agricultural and Bio-Industries*, Budapest, Hungary.

Andres-Toro, B., J.M. Giron-Sierra, C. Fernandez-Conde, J.M. Peinado, and F. Garcia-Ochoa (1998). A fast genetic optimization for batch fermentation processes, *Proc. 7th. Intl. Conf. On Computer Applications in Biotechnology CAB7*, IFAC, Osaka, Japan, pp. 61-66.

Bojkov, B., and R. Luus (1994). Application of iterative dynamic programming to time optimal control, *Trans. IChemE*, **72**, part A, pp. 72-80.

Chen, Ch., and Ch. Hwang (1990). Optimal control computation for differential-algebraic process systems with general constraints. *Chem. Eng. Comm.*, **97**, pp.9-26.

Cuthrell, J.E., and L.T. Biegler (1989). Simultaneous optimization and solution methods for batch reactor control profiles. *Comput. Chem. Eng.*, **13**, pp.49-62.

Dadebo, S.A., and K.B. McAuley (1995). A simultaneous iterative solution technique for time-optimal control using dynamic programming, *Ind. Eng. Chem.* Res., **34**, pp. 2077-2083.

Edgar, T.F., and D.M. Himmelblau (1988). Optimization of chemical processes. McGraw-Hill, New York.

Goldberg, D.E. (1989). Genetic Algorithms in Search, Optimization, and Machine Learning, Addison-Wesley.

Gupta, Y. P.(1995). Semiexhaustive search for solving nonlinear optimal control problems. *Ind. Eng. Chem. Res.*, **34**, pp.3878-3884.

Lim, H.C., Y.J. Tayeb, J.M. Modak, and P. Bonte (1986). Computational algorithms for optimal feed rates for a class of fed-batch fermentation: numerical results for penicillin and cell mass production. *Biotechnol. Bioeng.*, **28**, pp.1408-1420.

Luus, R.(1990).Optimal control by dynamic programming using systematic reduction in grid size. *Int. J.. Control,* **51**, pp.995-1013.

Luus, R., and O. Rosen (1991).Application of dynamic programming to final state constrained optimal control. *Ind. Eng. Chem. Res.*, **30**, pp.1525-1530.

Luus, R (1991a).Effect of the choice of final time in optimal control of non linear systems. *Can. J. Chem. Eng.*, **69**, pp.144-151.

Luus, R.(1991b). Application of iterative dynamic programming to state constrained optimal control problems. *Hung. J. Ind. Chem.*, **19**, pp.245-254.

Luus, R.(1993). Piecewise linear continuous optimal control by iterative dynamic programming. *Ind. Eng. Chem. Res.*, **32**, pp.859-865.

Park, S., and W.F. Ramirez (1988). Optimal production of secreted protein in fed-batch reactors. *AIChE J.*, **34**, pp. 1550-1558.

Ramirez , W. Fred (1994). Process control and identification. Academic Press, Inc, pp.84-86.

Yeo, B.P. (1980). Modified quasilinearization algorithm for the computation of optimal singular control. *Int. J. Control*, **32**, pp.723-730.

14th Triennial World Congress, Beijing, P.R. China





# REACHABILITY AND MINIMUM ENERGY CONTROL OF NONNEGATIVE 2D ROESSER TYPE MODELS

**TADEUSZ KACZOREK**

*Warsaw University of Technology, Faculty of Electrical Engineering*
*Institute of Control and Industrial Electronics*
*00-662 Warszawa, Koszykowa 75, Poland*
*e mail: kaczorek @ nov.isep.pw.edu.pl*

**Abstract.** A new class of nonnegative 2D Roesser type models is introduced. Necessary and sufficient conditions are established for the reachability and controllability of the nonnegative 2D Roesser type model for zero boundary conditions. It is shown that the nonnegative 2D Roesser type model having not nilpotent system matrix is unreachable for nonzero boundary conditions. The minimum energy control problem is formulated and solved for the nonnegative 2D Roesser type model with zero boundary conditions. The considerations are illustrated by means of a numerical example. 

Keywords: controllability, reachability, nonnegative 2D Roesser model, minimum energy control

## 1. INTRODUCTION

The most popular models of two-dimensional (2D) systems are the models introduced by Roesser (Roesse, 1975), Fornasini and Marchesini (Fornasini and Marchesini, 1976; 1978). The reachability and controllability of positive discrete-time linear systems have been considered in (Coxson, 1987; Fanti, et all, 1989; 1990). The reachability and controllability and the minimum energy control of 2D linear systems have been considered in many papers and books (Kaczorek, 1993, 1994a,b, 1995, Kaczorek and Klamka, 1986,1988; Klamka, 1988a,b, 1991a,b, 1993, 1994). The minimum energy control problem for the classical 2D Roesser model was formulated and solved by Klamka (Klamka 1991) and next the method was extended for 2D linear systems with variable coefficients (Kaczorek and Klamka , 1986) and other type of 2D models (Kaczorek and Klamka, 1988; Klamka 1988a,b, 1991a,b, 1993, 1994). Recently Valcher and Fornasini in (Valcher and Fornasini, 1995) have investigated some interesting properties of homogeneous 2D positive systems described by the second Fornasini-Marchesini type models. In this paper a nonnegative 2D Roesser type model is introduced. Necessary and sufficient conditions are established for the reachability of the nonnegative 2D Roesser type model for zero boundary conditions. It is shown that the nonnegative 2D Roesser type model having not nilpotent system matrix is unreachable for nonzero boundary conditions. The minimum energy control problem is solved for the nonnegative 2D Roesser type model with zero boundary conditions.

## 2. PRELIMINARIES

Let $R_+ := [0, +\infty)$ be the set of nonnegative numbers and let $Z_+ := \{0,1,2,...\}$ be the set of nonnegative integers. The set of nonnegative matrices of size n by m will be denoted by $R_+^{n\times m}$ and $R_+^n := R_+^{n\times 1}$.

Consider the 2D Roesser model

$$x_{ij}^{(1)} = Ax_{ij} + Bu_{ij}, \quad y_{ij} = Cx_{ij} + Du_{ij} \tag{1}$$

$$x_{ij}^{(1)} = \begin{bmatrix} x_{i+1,j}^h \\ x_{i,j+1}^v \end{bmatrix}, \quad x_{ij} = \begin{bmatrix} x_{ij}^h \\ x_{ij}^v \end{bmatrix}, \quad i,j \in Z_+$$

where $x_{ij}^h \in R^{n_1}$ is the horizontal state vector, $x_{ij}^v \in R^{n_2}$ is the vertical state vector, $u_{ij} \in R^m$ is the input vector, $y_{ij} \in R^p$ is the output vector,

$$A = \begin{bmatrix} A_1 & A_2 \\ A_3 & A_4 \end{bmatrix}, B = \begin{bmatrix} B_1 \\ B_2 \end{bmatrix}, C = [C_1 \ \ C_2]$$

$A_1 \in R^{n_1\times n_1}, A_4 \in R^{n_2\times n_2}, B_1 \in R^{n_1\times m}, B_2 \in R^{n_2\times m}$, $C_1 \in R^{p\times n_1}, C_2 \in R^{p\times n_2}, D \in R^{p\times m}$.

**Definition 1.** The 2D Roesser model (1) is called nonnegative 2D Roesser type model if for all boundary conditions

$$x_{0j}^h \in R_+^{n_1}, j \in Z_+, \quad x_{i0}^v \in R_+^{n_2}, i \in Z_+ \tag{2}$$

and for all $u_{ij} \in R_+^m$, $i,j \in Z_+$ we have

$x_{ij} \in R_+^n, n = n_1 + n_2$ and $y_{ij} \in R_+^p$ for $i,j \in Z_+$

**Proposition 1.** The 2D Roesser model (1) is nonnegative if and only if

$$A \in R_+^{n\times n}, B \in R_+^{n\times m}, C \in R_+^{p\times n}, D \in R_+^{p\times m} \tag{3}$$

**Proof.** From (1) it follows that $x_{ij} \in R_+^n$ *and* $y_{ij} \in R_+^p, i,j \in Z_+$ for all boundary conditions (2) and $u_{ij} \in R_+^m$, $i,j \in Z_+$ if and only if (3) holds. □

The transition matrix $T_{ij}$ for (1) is defined as follows (Roesser, 1975; Kaczorek, 1993; Klamka, 1991c)

$$T_{ij} = \begin{cases} I_n \ (\text{the identity matrix}) & \text{for } i = j = 0 \\ T_{10}T_{i-1,j} + T_{01}T_{i,j-1} & \text{for } i,j \geq 0 \ (i+j \neq 0) \\ T_{ij} = 0 \ (\text{the zero matrix}) & \text{for } i < 0 \text{ or/and } j < 0 \end{cases} \tag{4}$$

where

$$T_{10} := \begin{bmatrix} A_1 & A_2 \\ 0 & 0 \end{bmatrix}, T_{01} := \begin{bmatrix} 0 & 0 \\ A_3 & A_4 \end{bmatrix}$$

**Proposition 2.** The transition matrix $T_{ij}$ for the nonnegative 2D Roesser type model is a nonnegative matrix, i.e.

$$T_{ij} \in R_+^{n\times n} \text{ for all } i,j \in Z_+ \tag{5}$$

**Proof.** From (4) it follows that $T_{10} \in R_+^{n\times n}$, $T_{01} \in R_+^{n\times n}$, $T_{11} = T_{10}T_{01} + T_{01}T_{10} \in R_+^{n\times n}$ and recurrently $T_{ij} \in R_+^{n\times n}$ for all $i,j \in Z_+$. □

The solution to (1) with boundary conditions (2) is given by

$$x_{ij} = x_{bc}(i,j) + \sum_{(p,g)\in D_{ij}} M_{i-p,j-q} u_{pq} \tag{6}$$

where

$$x_{bc}(i,j) := \sum_{p=0}^{i} T_{i-p,j} \begin{bmatrix} 0 \\ x_{p0}^v \end{bmatrix} + \sum_{q=0}^{j} T_{i,j-q} \begin{bmatrix} x_{0q}^h \\ 0 \end{bmatrix},$$

$$M_{i-p,j-q} := T_{i-p-1,j-q} \begin{bmatrix} B_1 \\ 0 \end{bmatrix} + T_{i-p,j-q-1} \begin{bmatrix} 0 \\ B_2 \end{bmatrix} \tag{7}$$

Conditions will be established under which the nonnegative 2D Roesser type model is reachable for zero boundary conditions and the minimum energy control problem will be solved.

## 3. REACHABILITY

**Definition 2.** The nonnegative 2D Roesser type model (1) is called reachable for zero boundary conditions (ZBC) at the point $(h,k), h,k \in Z_+$, if for zero boundary conditions (2) and every $x_f \in R_+^n$ there exists a sequence of inputs $u_{ij} \in R_+^m$ for $(i,j) \in D_{hk}$, $D_{hk} :=$

$$\{(i,j) \in Z_+ \times Z_+, 0 \leq i \leq h, 0 \leq j \leq k \text{ and } i+j \neq h+k\} \tag{8}$$

such that $x_{hk} = x_f$.

**Definition 3.** The nonnegative 2D Roesser type model (1) is called reachable for any nonzero boundary conditions (NBC) at the point $(h,k), h,k \in Z_+$, if for any nonzero boundary conditions (2) and every $x_f \in R_+^n$ there exists a sequence of inputs $u_{ij} \in R_+^n$ for $(i,j) \in D_{hk}$ such that $x_{hk} = x_f$.

**Definition 4.** A set of all nonnegative linear combinations of columns of the matrix $A \in R^{n\times m}$ is called the positive image of $A$ (denoted by $\mathrm{Im}_+ A$)

$$\mathrm{Im}_+ A := \{y \in R_+^n : y = Ax \ \ for \ all \ \ x \in R_+^m\} \quad (9)$$

**Definition 5.** A matrix $R_n \in R^{n\times n}$ is called the generalised positive permutation matrix if and only if $R_n$ has only one positive entry in each row and column and the remaining entries are equal zero.

**Theorem 1.** The nonnegative 2D Roesser type model (1) is reachable for ZBC at the point $(h,k)$ if and only if

i) $\mathrm{Im}_+ R_{hk} = R_+^n$ (10)

or equivalently

ii) there exists a generalised positive permutation matrix $R_n$ consisting of $n$ columns of $R_{hk}$

where

$$R_{hk} := [M_{hk}, M_{h-1,k}, M_{h,k-1}, \ldots, M_{10}, M_{01}] \quad (11)$$

**Proof.** From (6) for $i = h, j = k, x_{hk} = x_f$ and $x_{bc}(h,k) = 0$

we have

$$x_f = R_{hk} u(h,k) \quad (12)$$

where

$$u(h,k) := \left[u_{00}^T, u_{10}^T, u_{01}^T, \ldots, u_{h-1,k}^T, u_{h,k-1}^T\right]^T$$

$T$ denotes the transposition.

Note that for a nonnegative 2D Roesser type model (1) $M_{pq} \in R_+^{n\times m}$ and $R(h,k) \in R_+^{n\times hkm}$. From (12) and (9) it follows that there exists $u_{ij} \in R_+^m$ for $(i,j) \in D_{hk}$ for any $x_f \in R_+^n$ if (10) holds.

Let $u_n \in R_+^m$ be a vector consisting of these entries of $u(h,k)$ which correspond to the columns selected in $R_n$. Assuming the remaining columns of $u(h,k)$ zero from (12) we obtain $x_f = R_n u_n$. It is well-known (Coxson, 1987) that $R_n^{-1} \in R_+^{n\times n}$ if and only if $R_n$ is a generalised positive permutation matrix. Hence $u_n = R_n^{-1} x_f \in R_+^n$ if and only if ii) is satisfied. The equivalence of conditions i) and ii) can be easily shown. □

**Remark 1.** It is well-known (Valcher and Fornasini, 1995) that a finite memory Roesser model has a nilpotent matrix $A$, i.e.

$$\det\begin{bmatrix} I_{n_1}z_1 - A_1, & -A_2 \\ -A_3 & , I_{n_2}z_2 - A_4 \end{bmatrix} = z_1^{n_1} z_2^{n_2} \quad (13)$$

Using (7) it is easy to show that if (13) holds then $x_{bc}(i,j) = 0$ for $i > n$, $j > n$.

**Theorem 2.** The nonnegative 2D Roesser type model (1) having not nilpotent matrix $A$ is unreachable at the point $(h,k)$ for NBC.

**Proof.** From (6) for $i = h, j = k$ and $x_{hk} = x_f$ and NBC we have

$$x_f - x_{bc}(h,k) = R(h,k)u(h,k) \quad (14)$$

By assumption $x_f \in R_+^n$ and $x_{bc}(h,k) \in R_+^n$ are any vectors and hence $x_f - x_{bc}(h,k) \notin R_+^n$.

If $A$ is not a nilpotent matrix then there does not exist a sequence $u_{ij} \in R_+^m$ for $(i,j) \in D_{hk}$ satisfying (14) for $x_f - x_{bc}(h,k)$ with at least one negative component. Therefore, the nonnegative 2D Roesser model is unreachable for NBC. □

**Theorem 3.** The nonnegative 2D Roesser type model (1) is reachable for ZBC at the point $(h,k)$ if $rank\, R_{hk} = n$ and there exists a $R_{hk}^r$ such that:

$$R_{hk}^r \in R_+^{(h+1)(k+1)m\times n} \quad (15)$$

where $R_{hk}^r$ is a right inverse of $R_{hk}$ $\left(R_{hk}R_{hk}^r = I_n\right)$.

**Proof.** By assumption $rank\, R_{hk} = n$ and there exists

$$R_{hk}^r = R_{hk}^T\left[R_{hk}R_{hk}^T\right]^{-1} \quad (16)$$

Hence from (12) we have $u(h,k) = R_{hk}^r x_f \in R_+^{(h+1)(k+1)m}$ for any $x_f \in R_+^n$ if (15) holds. □

**Example 1.** Consider the model (1) with

$$A = \begin{bmatrix} A_1 & A_2 \\ A_3 & A_4 \end{bmatrix} = \begin{bmatrix} 1 & 0 & 1 \\ 0 & 1 & 1 \\ 0 & 1 & 0 \end{bmatrix}, B = \begin{bmatrix} B_1 \\ B_2 \end{bmatrix} = \begin{bmatrix} 1 \\ 0 \\ 1 \end{bmatrix} \quad (17)$$

In this case $n_1 = 2, n_2 = 1$ and $m = 1$

To check the reachability at the point $(h,k) = (1,2)$ using (17), (7) and (11) we compute the matrix

$$R_{12} = [M_{12}, M_{02}, M_{11}, M_{10}, M_{01}] = \begin{bmatrix} 0 & 0 & 0 & 1 & 0 \\ 0 & 0 & 1 & 0 & 0 \\ 1 & 0 & 0 & 0 & 1 \end{bmatrix} \quad (18)$$

It is easy to verify that $\mathrm{Im}_+ R_{12} = 3$. Therefore, the condition (10) is satisfied and the model (1) with (17) is reachable at the point (1,2).

Note that the matrix

$$R_3 = [M_{11}, M_{10}, M_{01}] = \begin{bmatrix} 0 & 1 & 0 \\ 1 & 0 & 0 \\ 0 & 0 & 1 \end{bmatrix}$$

is a generalised positive permutation matrix and the condition ii) of Theorem 1 is also satisfied. Hence the model (1) with (17) is reachable at the point (1,2). The sequence $u(1,2)$ which transfers the model from ZBC to the any local state $x_f \in R_+^3$ at the point (1,2) is given by

$$u(1,2) = [u_{00}, u_{10}, u_{01}, u_{02}, u_{11}]^T = [0 \ 0 \ u_3^T]^T$$

where

$$u_3 = \begin{bmatrix} u_{01} \\ u_{02} \\ u_{11} \end{bmatrix} = R_3^{-1} x_f = \begin{bmatrix} 0 & 1 & 0 \\ 1 & 0 & 0 \\ 0 & 0 & 1 \end{bmatrix} x_f$$

Using (16) and (18) we obtain

$$R_{12}^r = R_{12}^T [R_{12} R_{12}^T]^{-1} = \begin{bmatrix} 0 & 0 & 0.5 \\ 0 & 0 & 0 \\ 0 & 1 & 0 \\ 1 & 0 & 0 \\ 0 & 0 & 0.5 \end{bmatrix} \in R_+^{6\times3}$$

Then the condition (15) is satisfied and by theorem 3 the model (1) with (17) is reachable at the point (1,2).

## 4. CONTROLLABILITY

**Definition 6.** The positive 2D Roesser type model (1) is called controllable (to zero) if and only if there exists a point $(h,k), h,k \in Z_+$, and a sequence of inputs $u_{ij} \in R_+^m$ for $(i,j) \in D_{hk}$ for any NBC such that $x_{hk} = 0$.

**Theorem 4.** The nonnegative 2D Roesser type model (1) is controllable (to zero) if and only if the system matrix $A$ of (1) is nilpotent matrix.

**Proof.** If $A$ is a nilpotent matrix then (13) holds and it can be easily shown that $x_{bc}(i,j) = 0$ for $i > n_1$ and $j > n_2$.
In this case the equation

$$x_{bc}(h,k) + R_{hk} u(h,k) = 0 \tag{19}$$

is satisfied for $u(h,k) = 0$, $h > n_1$ and $k > n_2$.

If $A$ is not a nilpotent matrix then there does not exist $u(h,k) \in R_+^{(h+1)(k+1)m}$ satisfying (19) for any $x_{bc}(h,k) \in R_+^n$. □

**Example 2.** Consider the model (1) with

$$A = \begin{bmatrix} A_1 & A_2 \\ A_3 & A_4 \end{bmatrix} = \left[\begin{array}{cc:c} 0 & 1 & 1 \\ 0 & 0 & 1 \\ \hdashline 0 & 0 & 0 \end{array}\right].$$

In this case we have $n_1 = 2, n_2 = 1$ and

$$\det \begin{bmatrix} I_{n_1} z_1 - A_1, & -A_2 \\ -A_3, & I_{n_2} z_2 - A_4 \end{bmatrix} = \begin{vmatrix} z_1, & -1, & -1 \\ 0, & z_1, & -1 \\ 0, & 0, & z_2 \end{vmatrix} = z_1^2 z_2$$

Using (4) we obtain

$$T_{10} = \begin{bmatrix} A_1 & A_2 \\ 0 & 0 \end{bmatrix} = \begin{bmatrix} 0 & 1 & 1 \\ 0 & 0 & 1 \\ 0 & 0 & 0 \end{bmatrix},$$

$$T_{01} = \begin{bmatrix} 0 & 0 \\ A_3 & A_4 \end{bmatrix} = \begin{bmatrix} 0 & 0 & 0 \\ 0 & 0 & 0 \\ 0 & 0 & 0 \end{bmatrix},$$

$$T_{20} = T_{10}^2 = \begin{bmatrix} 0 & 0 & 1 \\ 0 & 0 & 0 \\ 0 & 0 & 0 \end{bmatrix}$$

$$T_{ij} = \begin{bmatrix} 0 & 0 & 0 \\ 0 & 0 & 0 \\ 0 & 0 & 0 \end{bmatrix} \text{ for } \begin{array}{l} i > 2 \text{ and } j = 0 \\ i = 0 \text{ and } j > 1 \\ i > 0 \text{ and } j > 0 \end{array}$$

and

$$x_{bc}(i,j) = \sum_{p=0}^{i} T_{i-p,j} \begin{bmatrix} 0 \\ x_{p0}^v \end{bmatrix} + \sum_{q=0}^{j} T_{i,j-q} \begin{bmatrix} x_{0q}^h \\ 0 \end{bmatrix} = 0$$

for $i > 2, j > 1$ and any $x_{p0}^v$ and $x_{0q}^h$.
Therefore the system can be transfered to zero by $u(2,1) = 0$ for any boundary conditions $x_{p0}^v$ and $x_{0q}^h$.

## 5. MINIMUM ENERGY CONTROL

Consider the nonnegative 2-D Roesser type model (1) and the performance index

$$I(u) := \sum_{(p,q)\in D_{hk}} u_{pq}^T Q u_{pq} \tag{20}$$

where $Q$ is the $m \times m$ symmetric positive definite weighting matrix such that $Q^{-1} \in R_+^{m\times m}$.
The minimum energy control problem for the nonnegative 2D Roesser type model with zero boundary conditions (2) can be stated as follows. Given the matrices $A, B$ of (1), the weighting

matrix $Q$ and the point $(h,k)$, find a sequence $u_{ij} \in R_+^m$ for $(i,j) \in D_{hk}$ which transfers the model from zero boundary conditions to the desired local state $x_f = x_{hk}$ and minimizes the performance index (20).

To solve the problem we define the matrix

$$W_Q(h,k) := \sum_{(p,q)\in D_{hk}} M_{h-p,k-q} Q^{-1} M_{h-p,k-q}^T = \quad (21)$$
$$= R_{hk} Q_d R_{hk}^T$$

where $M_{h-p,k-q}$ and $R_{hk}$ are defined by (7) and (11), respectively.

$$Q_d := diag\left[Q^{-1},\ldots,Q^{-1}\right] \in R_+^{hkm\times hkm}$$

Using (21) it is easy to show that for the nonnegative 2D Roesser type model the matrix $W_Q(h,k) \in R_+^{n\times n}$ is nonsingular if and only if the matrix $R(h,k)$ has full row rank

Define the sequence of inputs

$$\hat{u}_{ij} := Q^{-1} M_{h-i,k-j}^T W_Q^{-1}(h,k) x_f, (i,j) \in D_{hk} \quad (22)$$

Note that $\hat{u}_{ij} \in R_+^m$ for any $x_f \in R_+^n$ if

$$W_Q^{-1}(h,k) \in R_+^{n\times n} \quad (23)$$

**Theorem 5.**

Let us assume that

i) the nonnegative 2D Roesser type model is reachable for ZBC at the point $(h,k)$,

ii) $Q^{-1} \in R_+^{m\times m}$ and (23) holds,

iii) $\bar{u}_{ij}, (i,j) \in D_{hk}$ is any sequence of inputs which transfers the model from zero boundary conditions to the desired local state $x_f = x_{hk}$.

Then the sequence of inputs (22) accomplishes the same task and

$$I(\hat{u}) \le I(\bar{u}) \quad (24)$$

Moreover, the minimum value of (20) is given by

$$I(\hat{u}) = x_f^T W_Q^{-1}(h,k) x_f \quad (25)$$

**Proof.** First we shall show that the sequence of inputs (22) provides $x_{hk} = x_f$.

From (22) it follows that

$$\hat{u}(h,k) := \left[\hat{u}_{00}^T, \hat{u}_{10}^T, \hat{u}_{01}^T, \ldots, \hat{u}_{h-1,k}^T, \hat{u}_{h,k-1}^T\right]^T = \quad (26)$$
$$= Q_d R_{hk}^T W_Q^{-1}(h,k) x_f$$

Substituting (26) into (12) and using (21) we obtain

$$x_{hk} = R_{hk}\hat{u}(h,k) = R_{hk} Q_d R_{hk}^T W_Q^{-1}(h,k) x_f = x_f$$

Since both $\bar{u}_{ij}$ and $\hat{u}_{ij}$ transfer the model from zero boundary conditions to $x_f$ then

$$R_{hk}\bar{u}(h,k) = R_{hk}\hat{u}(h,k)$$

and

$$R_{hk}\left[\bar{u}(h,k) - \hat{u}(h,k)\right] = 0 \quad (27)$$

From (27) and (26) we have

$$\left[\bar{u}(h,k) - \hat{u}(h,k)\right]^T R_{hk}^T W_Q^{-1}(h,k) x_f = \quad (28)$$
$$= \left[\bar{u}(h,k) - \hat{u}(h,k)\right]^T Q_d^{-1} \hat{u}(h,k) = 0$$

Using (28) it is easy to show that

$$\bar{u}^T(h,k) Q_d^{-1} \bar{u}(h,k) = \hat{u}^T(h,k) Q_d^{-1} \hat{u}(h,k) +$$
$$+ \left[\bar{u}(h,k) - \hat{u}(h,k)\right]^T Q_d^{-1} \left[\bar{u}(h,k) - \hat{u}(h,k)\right]$$

or

$$\sum_{(p,q)\in D_{hk}} \bar{u}_{pq}^T Q \bar{u}_{pq} = \sum_{(p,q)\in D_{hk}} \hat{u}_{pq}^T Q \hat{u}_{pq} + \quad (29)$$
$$+ \sum_{(p,q)\in D_{hk}} \left[\bar{u}_{pq} - \hat{u}_{pq}\right]^T Q \left[\bar{u}_{pq} - \hat{u}_{pq}\right]$$

The inequality (24) holds since the last term in (29) is always nonnegative.

To obtain the minimum value of (20) we substitute (26) into (20)

$$I(\hat{u}) = \sum_{(p,q)\in D_{hk}} \hat{u}_{pq}^T Q \hat{u}_{pq} = \hat{u}^T(h,k) Q_d^{-1} \hat{u}(h,k) =$$
$$\left[Q_d R^T(h,k) W_Q^{-1}(h,k) x_f\right]^T$$
$$Q_d^{-1}\left[Q_d R^T(h,k) W_Q^{-1}(h,k) x_f\right] =$$
$$= x_f^T W_Q^{-1}(h,k) R_{hk} Q_d R_{hk}^T W_Q^{-1}(h,k) x_f =$$
$$= x_f^T W_Q^{-1}(h,k) x_f$$

since by (21) $R_{hk} Q_d R_{hk}^T W_Q^{-1}(h,k) = I$. □

**Example 3.** Consider the nonnegative 2D Roesser type model (1) with

$$A = \begin{bmatrix} A_1 & A_2 \\ A_3 & A_4 \end{bmatrix} = \begin{bmatrix} 1 & 0 & 1 \\ 0 & 1 & 0 \\ 1 & 0 & 1 \end{bmatrix}, \quad B = \begin{bmatrix} B_1 \\ B_2 \end{bmatrix} = \begin{bmatrix} 0 \\ 1 \\ 1 \end{bmatrix}$$

for $h = k = 1$. Using (7) and (11) we obtain

$$R_{11} = \left[M_{11} M_{10} M_{01}\right] = \begin{bmatrix} 1 & 0 & 0 \\ 0 & 1 & 0 \\ 0 & 0 & 1 \end{bmatrix}, rankR_{11} = 3$$

By theorem 1. the model is reachable for ZBC at the point (1,1). Let the performance index has the form (20) with

$$h = k = 1 \text{ and } Q = \begin{bmatrix} 2 & 0 & 0 \\ 0 & 1 & 0 \\ 0 & 0 & 3 \end{bmatrix} \text{ and } x_f = \begin{bmatrix} \frac{1}{2} \\ 1 \\ \frac{1}{3} \end{bmatrix}.$$

In this case using (21) and (26) we obtain

$$W_Q(1,1) = R_{11}Q_d R_{11}^T = \begin{bmatrix} \frac{1}{2} & 0 & 0 \\ 0 & 1 & 0 \\ 0 & 0 & \frac{1}{3} \end{bmatrix}$$

$$\hat{u}(1,1) = \begin{bmatrix} u_{00} \\ u_{10} \\ u_{01} \end{bmatrix} = Q_d R^T(1,1) W_Q^{-1}(1,1) x_f = \begin{bmatrix} \frac{1}{2} \\ 1 \\ \frac{1}{3} \end{bmatrix}$$

and the minimum value of the performance index is equal

$$I(\hat{u}) = x_f^T W_Q^{-1}(1,1) x_f = \frac{11}{6}$$

## 6. CONCLUDING REMARKS

Necessary and sufficient conditions have been established for the reachability and controllability of the nonnegative 2D Roesser type model for zero boundary conditions. It has been shown that the nonnegative 2D Roesser type model having not nilpotent matrix $A$ is unreachable for nonzero boundary conditions (2). The minimum energy control problem has been formulated and solved for the model with zero boundary conditions. The results presented can be extended for the n-D (n>2) Roesser type model with constant and variable coefficients. An extension of the above considerations for singular 2D linear systems is also possible. It is well-known (Kaczorek 1993) that the first Fornasini-Marchesini model is a special case of the Roesser model. Therefore, the results obtained in this paper for the nonnegative 2D Roesser type model can be immediately extended for the nonnegative first 2D Fornasini-Marchesini type model.

## REFERENCES


Coxson P.G. (1987). Positive input reachability and controllability a positive systems, *Linear Algebra and its Applications*, vol. 94, pp.35-53.

Fanti M.P., Maione B. and Turchiano B. (1989). Controllability of linear single-input positive discrete-time systems, *Int. J. Control*, vol. 50, No 6, pp. 2523-2542.

Fanti M.P., Maione B. and Turchiano B. (1990). Controllability of multi-input positive discrete-time systems, *Int. J. Control*, vol. 51, No 6, pp. 1295-1308.

Fornasini E., Marchesini G. (1976). State space realization of two-dimensional filters, *IEEE Trans.Autom.Control*, **AC-21**, pp. 484-491.

Fornasini E., Marchesini G. (1978). Doubly indexed dynamical systems: State space models and structural properties, Math. Syst.Theory **12**.

Kaczorek T. (1993). Linear Control Systems, vol. 2, *Research Studies Press and J. Wiley*, New York.

Kaczorek T. (1994a). When the local controllability of the general model of 2-D linear systems implies its local reachability, *Systems and Control Letters*, vol. 23, pp. 445-452.

Kaczorek T. (1994b). When the local controllability of Roesser model implies its local reachability, *Bull. Pol. Acad. Techn. Sci.*, vol. 41, pp. 261-267.

Kaczorek T. (1995). U-Reachability and U-Controllability of 2-D Roesser Model, , *Bull. Pol. Acad. Techn. Sci.*, vol. 43, No 1, pp. 31-37.

Kaczorek T. and Klamka J. (1986). Minimum energy control of 2-D linear systems with variable coefficients, *Int. J. Control*, vol. 44, No. 3, pp. 645-650.

Kaczorek T. and Klamka J.(1988). Minimum energy control for general model of 2-D linear systems, *Int. J. Control*, vol. 47, No. 5, pp. 1555-1562.

Klamka J. (1988a). M-dimensional nonstationary linear discrete systems in Banach spaces, *Proc. 12 World IMACS Congress*, Paris, vol. 4, pp. 31-33.

Klamka J. (1988b). Constrained controllability of 2-D linear systems, *Proc. 12 World IMACS Congress*, Paris, vol. 2, pp. 166-169.

Klamka J. (1991a), Complete controllability of singular 2-D system, *Proc. 13 IMACS World Congress*, Dublin, pp. 1839-1840.

Klamka J. (1991b), Minimum energy control of singular 2-D linear systems with variable coefficients, *Proc. IMACS Symp.* Lille, vol. 2, pp. 155-159.

Klamka J. (1991c). Controllability of Dynamical Systems, *Kluwer Academic Publ.*, Dordrecht.

Klamka J. (1993). Minimum energy control problem for general linear 2-D systems in Hilbert spaces, *Proc. IEEE Symp.* Crete.

Klamka J. (1994). Constrained Controllability of Discrete 2-D Linear Systems, *Proc. IMACS Intern. Symp. Signal Processing, Robotics and Neural Networks*, April 25-27, Lille, pp. 166-169.

Roesser P.R. (1975). A discrete state-space model for linear image processing,, *IEEE Trans. Autom. Contr.*, vol. AC-20, No. 1, pp. 1-10.

Valcher M.E. and Fornasini E. (1995). State Models and Asymptotic Behaviour of 2D Positive Systems, *IMA Journal of Mathematical Control & Information*, No 12, pp. 17-36.

# APPLICATION OF THE VEHICLE OF A.N.KRYLOV'S PRODUCING VECTOR AND FACTORS IN THE ANALYSIS OF NON-LINEAR SYSTEMS.

**Podchukaev V.A., Orlov D.V.**

*Russia, 410054, Saratov city, Polytechnicheskaya, 77, Saratov State Technical University, faculty Engineering Cybernetics and Computer Sciences. E-mail: tki@sstu.saratov.su, fax: (+7)-(8452)-24-15-17.*

Abstract: The article is devoted to development of the mathematical vehicle of A.N.Krylov's factors (Podchukaev V.A., 1997) with reference to problems of designing necessary and sufficient conditions of a reducibility by A.M.Lyapunov and existence of limit cycles and them generalizations, which technical appendices in theory automatic controls - are well-known. The examples are resulted. 

Keywords: Non-linear systems, A.N.Krylov's factors, Reducibility, Limit cycle, Theory automatic controls.

## 1. INTRODUCTION.

Concerning thickening of automatic control systems and increase of requests to exactitude of operation parameters of these systems, there was a need of the point description of mathematical models of control systems. To satisfy these requests the most full models of the description of automatic controls systems noted by means of nonlinear systems of the differential equations can only.

However, analysis of such systems conjugates with the certain computing difficulties, this entail the large temporary costs. Also frequently, it is impossible or very difficulty to receive analytical expressions for solutions for such systems.

The given work is devoted to analytical methods of the analysis of nonlinear systems of the differential equations with use of A.N.Krylov's producing vector and factors. The application of this means has allowed finding necessary and sufficient conditions of a reducibility of nonlinear systems. (Reduction of nonlinear non-stationary systems to stationary systems.) In addition, it was possible to receive a criterion of existence of limit cycles and their generalizations at nonlinear systems.

The obtained criterions allow considerably simplifying the analysis of nonlinear systems of automatic control. In particular, by means of an offered analytical criterion of a reducibility deriving a more simple kind of a researched nonlinear model is possible by reduction it to possibility of undamped oscillations among solutions of a nonlinear system for a design stage. Knowing a possibility of existence of such oscillations, the designer can stimulate or to extinguish these oscillations depending on the purpose of management.

## 2. AUXILIARY INFORMATIONS.

Let us consider a system of ordinary differential equations of a general view

$$\dot{\mathbf{x}} = \mathbf{f}(\mathbf{x},t),\ \mathbf{x}\in R^n,\ t\in[0,\infty),\ \mathbf{x}(t_0) = \mathbf{x}_0 \tag{1}$$

where through $\mathbf{x}$ the vector of condition, $\mathbf{x}_0$ – vector of the entry conditions, t - time, $\mathbf{f}(\mathbf{x},t)$ - specific vector function of dimensionality n is designated, concerning which let's assume, that it is determined and is continuous together with the partial derivatives on x and t on some open set of space of variables $R^n\times[0,\infty)$, that guarantees existence of the unique solution $\mathbf{x}$: $\mathbf{x}(t,t_0,\mathbf{x}_0)$, passing through a point $(\mathbf{x}_0,t_0)$, not being a condition of balance.

Assume a fair identity of Bylov-Grobman

$$\mathbf{F}(\mathbf{x},t) \equiv \mathbf{P}(\mathbf{x},t)\mathbf{x}, \tag{2}$$

let us record an equation (1) in the vector-matrix form

$$\dot{\mathbf{x}} = [\mathbf{P(x,t)}]_{(1)}\mathbf{x}, \quad (3)$$

where through $[\mathbf{P(x,t)}]_{(1)}$ – the functional matrix of the size $n \times n$ is designated, which range are the solutions of an equation (1), and range of definition - any rule of allocation from vector $\mathbf{f(x,t)}$ of a matrix $\mathbf{P(x,t)}$, at which is executed (2).

It is necessary to mark, that hereinafter we consider an equation (3) as character representation of a non-linear system without a substitution in a functional matrix $[\mathbf{P(x,t)}]_{(1)}$ solutions of an equation (1), on which it is determined.

**Definition 1.** Let us name as producing A.N.Krylov's vector any vector $[\mathbf{B(x,t)}]_{(1)}$ (dimensionality n), which elements are determined, are continuous and are limited together with the partial derivatives on x and t down to (n-1)-th of the order on some open set of space of variables $R^n \times [0,\infty)$, satisfying to a condition,

$$\text{rank}\,[\mathbf{W(x,t)}]_{(1)} = n, \quad (4)$$

where $[\mathbf{W(x,t)}]_{(1)}$ – matrix of a controllability, defined expression

$$[\mathbf{W(x,t)}]_{(1)} = [\mathbf{B(x,t)}, D\mathbf{B(x,t)}, \ldots, D^{n-1}\mathbf{B(x,t)}]_{(1)} \quad (5)$$

in which through D the differential operator, induced by following differential expression is designated

$$D = [\mathbf{P(x,t)}]_{(1)} - d(.)/dt. \quad (6)$$

As shown in (Podchukaev V.A., 1997), non-linear non-stationary system (2) by means of nondegenerate non-linear transformation

$$x = [\mathbf{W(x,t)}]_{(1)}y, \; y \in R^n, \quad (7)$$

can be adduced to a system of a kind

$$\dot{y} = [\bar{\mathbf{P}}\mathbf{(x,t)}]_{(1)}y, \quad (8)$$

which in the matrix of a right member has the Luenberger's-Krylov's form

$$[\bar{\mathbf{P}}\mathbf{(x,t)}]_{(1)} = [e_2, e_3, \ldots, e_n, -\mathbf{H(x,t)}]_{(1)}, \quad (9)$$

where

$$\mathbf{e}_i = \text{colon}\,[0, \ldots, 0, \underset{1}{} \ldots \underset{i-1}{0}, \underset{i}{1}, \underset{i+1}{0}, \ldots, \underset{n}{0}],$$

$$-[\mathbf{H(x,t)}]_{(1)} = [\mathbf{W(x,t)}^{-1} D^n \mathbf{B(x,t)}]_{(1)}. \quad (10)$$

**Definition 2.** Let us name as A.N.Krylov's factors elements of vector

$$[\mathbf{H(x,t)}]_{(1)} = \text{colon}\,[h_n(x,t), \ldots, h_1(x,t)],$$

of a defined agrees (10).

As shown in (Podchukaev V.A., 1997), in a case of fixed linear systems the A.N.Krylov's factors coincide factors of a characteristic polynomial (we shall designate them through $\alpha_i$ (i= =1,...,n)), and in case of non-linear or non-stationary linear systems are connected to them by a ratio

$$[\alpha_i(x,t)]_{(1)} = \sum_{j=1}^{\infty} C_{n-j}^{i-j}\,[h_j(x,t)]_{(1)}^{(i-j)} \quad (11)$$

3. A.N.KRYLOV'S FACTORS IN THE ANALYSIS OF A REDUCIBILITY OF NON-LINEAR SYSTEMS

In a 1946 N.P.Erugin (afterwards academician of an Academy of sciences USSR) has protected a doctoral thesis on research of reducibility on A.M.Lyapunov of non-stationary linear systems. The outcomes of this activity have found reflection in the monography (Erugin N.P., 1946; Erugin N.P., 1979; Demidovich B.P., 1967), where the linear non-stationary systems were considered

$$\dot{x} = \mathbf{P(t)}x, \; x \in R^n, \quad (12)$$

which in elements of a matrix $\mathbf{P(t)}$ - continuous and limited on a semi-axis $[t_0,\infty)$ function of time. In the thesis, N.P.Erugin because of definitions of reducibility on A.M.Lyapunov has formulated necessary and sufficient conditions of a reducibility of systems in a general view (12) to a system with constants by factors

$$\dot{y} = \bar{P}y, \; y \in R^n, \quad (13)$$

in conditions of existence of A.M.Lyapunov's transformation matrix **Z** of dimensionality n×n such, that $\mathbf{Z}$, $\mathbf{Z}^{-1}$ and $\mathbf{dZ/dt}$ - are determined, are continuous and are limited on a semi-axis $[t_0,\infty)$. The finding of this transformation matrix was offered by use of the solution joined to a system (12) equations and matrix exponential curve, which scarching is connected to certain computing difficulties. The large attention in the book was given to systems of the second order with elements being periodic functions, and functions, defined

$$p_{ii}(t) = a_i + \sum_{k=2}^{\infty} p_{ii}^{(k)} / (t^k),$$

$$p_{ij}(t) = \sum_{k=1}^{\infty} p_{ij}^{(k)} / (t^k), \; (i,j = 1,2).$$

Hereinafter we shall name systems considered in (Erugin N.P., 1946), systems of a N.P.Erugin's type.

We shall demonstrate now application of producing vector and A.N.Krylov's factors for finding of conditions of a reducibility of non-linear systems alternate to conditions, offered by N.P.Erugin by (Erugin N.P., 1946; Erugin N.P., 1979). The main outcome can be formulated as the following theorem.

**Theorem 1.** Necessary and the sufficient A.M.Lyapunov's condition of a reducibility of a system (3) consists in existence of producing A.N.Krylov's vector $[\mathbf{B(x,t)}]_{(1)}$ (or vector making with a matrix $[\mathbf{P(x,t)}]_{(1)}$ this system completely controlled pair (when is satisfied condition (4)), for which the A.N.Krylov's factors satisfy to conditions:

$$[h_i(x,t)]_{(1)} = \text{const}, \ (i=1,...,n). \qquad (14)$$

**Proof of the necessity**. Let system (3) - is reduced, that is there are A.M.Lyapunov's transformations presenting it to a kind, at which the matrix of a system is a reduced matrix of numbers. Without loss of generality as such A.M.Lyapunov's transformations it is possible to take transformation (7), presenting an initial system (3) to Krylov's-Luenberger's form (8), which with the right of column is deviated by factors of a secular equation of a reducible system. As under the supposition the initial system - is presented, the factors of a secular equation cannot be anything by other, as with numbers, that is to satisfy to a condition (14). The necessity is proved.

**Proof of the sufficiency**. Let there is producing A.N.Krylov's vector, making with a matrix $[\mathbf{P(x,t)}]_{(1)}$ completely controlled pair, that is satisfying to a condition (4). Then the transformation (7) is results initial systems in Krylov's-Luenberger's form (8). It is uneasy to see, that fulfillment of ratios (14) it is enough for a reducibility of a system (3). Thus, the theorem is proved.

Above mentioned necessary and the sufficient conditions of a reducibility of non-linear systems are spreader both on hardly presented, and on weak reducible systems (Podchukaev V.A., 1997). Really, the system (3) is poorly presented to a system (13) (or is presented in a classical sense), as the transformation (7) is A.M.Lyapunov's transformation. If besides between vectors **x** and **y** has a place unequivocal conformity, which is, obtained from (7) equation,

$$[\mathbf{W(x,t)}^{-1}]_{(1)}\mathbf{x=y}$$

has among a set of the solutions one real solution $\mathbf{x(y)} \in R^n$, the system (3) hardly is presented to a system (13).

The validity of the theorem 1 was checked on N.P.Erugin's systems (Erugin N.P., 1946; Erugin N.P., 1979) second and third is ordinal, and then by means of it was possible to generalize systems of analytical evaluations MAPLE (Monzon B.M., 1998; Govorukhin V.N., 1998) these systems on n-dimensional case. Hereinafter as systems of a N.P.Erugin's type we shall name systems satisfying to a condition (Erugin N.P., 1946; Morozov V.M., et al., 1988):

$$g=\lim_{t\to\infty} \left(\int_0^t \sigma(P)dt\right)/t,$$

where g - some final number, σ(P) - sum of diagonal elements of a matrix $[\mathbf{P(x,t)}]_{(1)}$.

So, we shall consider n-dimensional a system of a kind (3) with a matrix **P(t)**, which elements have a structure

$$P_{ii}(t)=b_i+V_i(t), \quad P_{ij}(t)=q_i*(I_i(t)/I_j(t)),$$
$$(i\neq j), \quad q_i = \text{const},$$

where

$$V_k(t) = \sum_{i=1}^{\infty} r_i v_i(t), \ r_i=\text{const},$$
$$t$$

$$I_k = \exp\left(\int_0^t V_k(t)dt\right), \ b_k = \text{const}, \ (k=1, 2,..., n),$$

$v_i(t)$ - specific on a semi-axis [0, ∞) limited functions of time, i = 1,2..., n. In a course of activity as $v_i(t)$ the functions of completely various types, as that undertook: exponential, periodic, return trigonometrically functions, function of a kind

$$v_i(t)=1/t^s \quad \text{by } s=2, 3, ..., n,$$

and also various combinations of listed types of functions. The calculations have shown, that at producing vector of a kind

$$B(t) = \text{colon}\ [0, b_2*I_2, b_3*I_3, ..., b_n*I_n,]$$

the systems of a N.P.Erugin's type - are reduced.

**Example 1. Presented model of a gyrovertical with rotated vessels.** The considered device represents the gyro with a vertical axis of own rotation, which is made in a cover and is suspended (Morozov V.M., et al., 1988). On a cover of the platform is mounted which rotates round an axis geometrically conterminous to an axis of a curl, from a constant by angular rate ω. On platform are two communicated vessels, jointed capillary handset and in which is located mercury are installed. When the mercury is distributed equally in both vessels, the barycentre of all mobile system are coincides a rallying-point of the gyro. The platform rotates in the same leg, as curl of the gyro. The circumscribed device concerns to gyrovertical with radial correction presenting an axis of a curl of the gyro to a vertical in the shortest way.

At the description of motion of the gyro the coordinate system $O\xi\eta\zeta$ is used, the axes which of $O\xi$ and $O\eta$ are located in a horizontal plane and are directed accordingly on east and north, the axis $O\zeta$ is vertical. The position of an axis z of a curl of the gyro is determined by corners α, β.

If to neglect nutation motion of the gyro, it is possible to consider that it the vector of a moment of momentum is directed along an axis z of a curl and is a constant. Assuming, that the device is installed on the fixed basis and neglecting influence of rotation of the Earth, we shall have a matrix equation of motion of the gyro (Morozov V.M., et al., 1988) as (12) with a matrix **P(t)** of the following kind:

$$\mathbf{P(t)} = \begin{vmatrix} 0 & 0 & c\sin(wt) \\ 0 & 0 & -c\cos(wt) \\ -b\cos(wt) & -b\sin(wt) & -b \end{vmatrix}.$$

here **x** = colon(α, β, φ), φ - corner of declination of a mirror of fluid in vessels, b, c - positive constants parameters of a system.

It is uneasy to show that this system is presented. Really, with by means of producing A.N.Krylov's vector

$$\mathbf{B(t)}=\text{colon}[0, 0, \exp(b\,t)],$$

making evaluations under the formula (5), we shall receive a transformation matrix. Then the vector of A.N.Krylov's factors is determined by following expression:

$$\mathbf{H} = \text{colon}[bcw-2b^3-2bw^2, -5b^2-w^2, -4b],$$

that is the matrix **W(t)** results a system (3) in a kind (13). It is easy to notice, that the offered transformation hardly simplifies model of motion of the gyro, that, in turn, facilitates modeling of a system of automatic control, in which structure the circumscribed device can include.

## 4. A.N.KRYLOV'S FACTORS IN THE ANALYSIS OF LIMIT CYCLES

August 8, 1900 greatest germane mathematician David Hilbert, native of city Velau approach Konigsberg (foreign honorary corresponding member (1922) and foreign honorary real term (1934) Academies of Sciences USSR) on II International congress of mathematicians in Paris has read unique for a history of mathematics the report "Mathematical problems", enveloping problems of mathematics as a whole.

By addressing to areas of the analysis and geometry, Hilbert has formulated 23 problems, which and now, that already without small hundred years, have appeared by approaching objects to concentrate round itself creative efforts of the contributors of various scientific directions.

From a point of view of the theory of automatic control the special interest is represented by 16-th problem formulated by Hilbert or problem of topology algebraic curve and surfaces (Pogorelov A.V., 1969), which by most interesting from positions of the numerous technical appendices, are the closed curves or limit cycles.

We shall be set by the purpose to formulate a criterion of availability of a limit cycle among the solutions of a system of ordinary differential second-order equations (equation of a kind (2) at n = 2), using thus the vehicle of producing vector and A.N.Krylov's factors (Podchukaev V.A., 1997).

For maintenance of a wholeness of a stated material, we shall give the following explanations.

***Homeomorphism, homeomorphness map*** - one from main concepts of topology. Two topological spaces are named homeomorphness, if there is one-to-one a continuous map one from them on other, for which inverse mapping also continuously; thus map is named as a homeomorphism. In addition, the properties of figures, which do not vary at transition to homeomorphness figures, are named topological. Examples of absolute concepts: compactness (bicompactness) and connectivity. In our case, a **diffeomorphism** for an equation (2) is the nondegenerate non-linear transformation of vector of condition of this equation as (7).

For an adduced time scale an equation of small oscillations of a pendulum it is possible to record like

$$\ddot{z} = -z \qquad (15)$$

By introduction of labels $y_2 = z$, $y_1 = dz/dt$ we shall present an equation (12) in the form of identification canonical Krylov's-Luenberger's representation (Erugin N.P., 1946; Podchukaev V.A., 1997) (form of Cauchy)

$$\begin{aligned} \dot{y}_1 &= -y_2 \\ \dot{y}_2 &= y_1 \end{aligned} \qquad (16)$$

By dividing the second equation of a system (16) on the first equation, we shall receive the following equation of phase trajectory (Voronov A.A., et al., 1986)

$$Dy_2/dy_1 = -y_1/y_2, \qquad (17)$$

which presented to an equation with dividing variable

$$y_2dy_2 + y_1dy_1 = 0, \qquad (18)$$

and which is easily integrated in quadratures

$$y_1^2 + y_2^2 = R^2, \qquad (19)$$

where any constant R depends on the entry conditions

$$R = [y_1^2(0) + y_2^2(0)]^{1/2}. \tag{20}$$

It is uneasy to see that the equality (19) determines a circle on an affine (phase) plane (Penzov U.E., 1972). Thus, on the language of A.N.Krylov's factors at any entry conditions the circle is set by equalities

$$h_2 = 1, h_1 = 0. \tag{21}$$

Let us use these equalities for the formulation of a criterion of existence of a limit cycle. Obtained outcome we shall formulate as the following theorem.

**Theorem 2.** Necessary and the sufficient condition of existence of a limit cycle of a non-linear system (3) second order consists in existence of producing A.N.Krylov's vector (or vector satisfying to the condition (4), in which n = 2), at which the A.N.Krylov's factors are determined by equalities (21).

**Proof of the necessity.** Let among the solutions of a system (3) there is a limit cycle (closed trajectory of a phase plane or tame curve, homeomorphness circle). Therefore, there is a topological space, distinct from initial (what space formed by vector of condition $\mathbf{x} \in R^n$) is, in which the limit cycle of an initial system (3) is deformed in a circle. As that space, it is possible to take space, which main characteristics are the A.N.Krylov's factors, that is space formed by vector of the condition $\mathbf{y} \in R^n$. The diffeomorphism of both spaces in conditions (4) is set by equality (7). In deformed on the attitude to initial topological space $\mathbf{y} \in R^n$ information carriers about a limit cycle are the A.N.Krylov's factors, computed according to (10). But if the initial system had a limit cycle, for the deformed system, what is (8), these factors cannot be anything by other, as by defining by equalities (21), as other representation of a circle on a phase plane, defined A.N.Krylov's factors, - is not present.

**Proof of the sufficiency.** Let system is adduced (3) nondegenerate non-linear transformation (7) to a kind (8). Thus, in general case the necessary conditions of existence of a limit cycle can be recorded like

$$[h_1(x,t)](1) \equiv 0, [h_2(x,t)](1) = 1. \tag{22}$$

The identical equality to zero first from A.N.Krylov's factors eliminates a capability of character representation of this factor. For second of a factor is not so, as the following equality can be considered as in general case differential (in particular cases algebraic) an equation, which left-hand part for each equation (3) has unique character representation:

$$[h_2(x,t)](1) = 1 \tag{23}$$

Thus the sufficient condition of existence of a limit cycle (limit cycles) for an input equation (3) consists in existence of the solutions (radicals) of an equation (23).

The theorem is proved.

The validity of this theorem can be illustrated by the following examples. All calculations are executed in an shell of a mathematical packet of character evaluations MAPLE V (Monzon B.M., 1998; Govorukhin V.N., 1998).

**Example 2. Research of an Van-der-Poul's equation.** An Van-der-Poul's equation (Podchukaev V.A., 1997; Andreev U.N., 1976)

$$\ddot{z} - a(1 - z^2)\dot{z} + z = 0$$

by means of management of labels $x_1 = z$, $x_2 = dz/dt$ is possible to record in the form of Cauchy

$$\dot{x}_1 = x_2$$

$$\dot{x}_2 = -x_1 + a(1 - x_1^2)x_2 .$$

Let us show that the offered criterion allows giving an inference about availability of a limit cycle without a numerical integration of a Van-der-Poul's equation.

In this way producing A.N.Krylov's vector, ensuring reduction to the Krylov's-Luenberger's form with A.N.Krylov's factors, satisfying to a condition (21), is found as

$$\mathbf{B(x,t)} = \text{colon } [0, \exp\{0.5a \int_T^0 (1 - x_1^2)dt\}].$$

Thus, the equation (23) has a kind

$$-1 + 0.25a^2 - 0.5a^2x_1^2 + 0.25a^2x_1^4 + ax_1\,(dx_1/dt) = -1.$$

It is uneasy to be convinced, that the substitution

$$x_1 = \{2[2^{1/2}\,\text{th}(2^{1/2}\,t/a)] + 1\}^{1/2}$$

translates this equation in an identity, that is necessary and sufficient condition of existence at an Van-der-Poul's equation of a limit cycle.

**Example 2.** Let us consider a set of equations (look, for example, [1, s.134; 10, s.37])

$$\dot{x}_1 = x_1 - x_2 - ax_1$$

$$\dot{x}_2 = x_1 + x_2 - ax_2,$$

where $a = x_1^2 + x_2^2 - 1$.

It is known, that a limit cycle for this equation is the circle of a single radius $x_1^2 + x_2^2 = 1$. Let us show that the offered criterion also allows giving an inference about availability of a limit cycle, and, with smaller efforts, as known methods. In a considered example the producing A.N.Krylov's vector, ensuring reduction to the Krylov's-Luenberger's form with A.N.Krylov's factors, satisfying to a condition (21), is found as

$$\mathbf{B(x,t)} = \text{colon } [0, \exp(-\int_t^0 (x_1^2 + x_2^2 - 1))].$$

Thus the expression for the second A.N.Krylov's factor at once accept a kind (23):

$$[h_2(x,t)](1) = 1,$$

that is necessary and sufficient condition of existence of a limit cycle.

## 5. THE DESCRIPTION OF THE ALGORITHM FOR TWO-DIMENSIONAL SYSTEMS

In light of the analysis, explained offered algorithm, of limit cycles consists of the following sequence of operations:

1) Searching producing A.N.Krylov's vector and satisfying to a condition (4) (unformalized on the present moment operation);

2) Evaluation of A.N.Krylov's factors pursuant to second from expressions (10);

3) Check of fulfillment of equalities (21). Return to item 1, if the equalities (21) are not executed;

4) Find the Solution of an equation (23).

Thus, the algorithm of the analysis of limit cycles supposing natural generalization on many-dimensional case ($n > 2$) is offered. A weak place of algorithm is the problem of selection of producing A.N.Krylov's vector, which is difficulty formalizable and in many respects depends on intuition of the contributor.

## 6. GENERAL CASE N > 2.

Analog of a limit cycle in finite-dimensional space of dimensionality $n>2$ is the closed surface.

Let us construct a system, which is homeomorphness to an initial system (8), which solutions in affine space belong to a hyperspherical (closed hypersurface of the second order).

**Definition 3.** Let us name a hyperspherical as a generalized limit cycle.

Let us accept a vector of A.N.Krylov's factors (10) in follow kind

$$\mathbf{H}_i = \text{colon}[\underset{1}{0}, \ldots, \underset{k-1}{0}, \underset{k}{1}, \underset{k+1}{0}, \ldots \underset{n}{0}], \tag{24}$$

then a scalar entry of a system (8) in the form Krylov's-Luenberger's will be:

$$\begin{aligned}
\dot{y}_1 &= 0, \\
\dot{y}_2 &= y_1, \\
&\ldots\ldots\ldots\ldots \\
\dot{y}_{k-1} &= y_{k-2}, \\
\dot{y}_k &= -y_n + y_{k-1}, \\
\dot{y}_{k+1} &= y_k, \\
&\ldots\ldots\ldots\ldots \\
\dot{y}_{n-1} &= y_{n-2}, \\
\dot{y}_n &= y_{n-1}.
\end{aligned} \tag{25}$$

It is uneasy to see, that the solutions first (k-1) - equations of a system (25) are determined by a ratio

$$y_i(t) = \sum_{j=1}^{i} y_j(0) t^{i-j}/(i-j)!, \quad (i=1,\ldots,k-1) \tag{26}$$

however concerning the remaining equations of this system the following assertion can be formulated.

**Assertion 1.** In a phase space $R^{n-k+1}$ with coordinates ($y_k$, $y_{k+1}$, ..., $y_{n-1}$, $y_n$) the solutions of a system (25) belong to the closed hypersurface of the second order (hyperspherical), defined equality

$$\sum_{i=k}^{n} y_i^2 = R^2. \tag{27}$$

**Proof.** Multiplying both parts of the equations (from number k to n) systems (25) with dt, we shall receive

$$\begin{aligned}
dy_k &= (-y_n + y_{k-1})dt, \\
dy_{k+1} &= y_k dt, \\
&\ldots\ldots\ldots\ldots \\
dy_{n-1} &= y_{n-2}dt, \\
dy_n &= y_{n-1}dt.
\end{aligned}$$

Whence with allowance for (26) after an integration, erections of both parts of obtained expressions in quadrate and summation of both parts of obtained equalities we have (27), where is entered the designation

$$R^2 = \sum_{i=k}^{n-1} [\int_0^t y_i(\tau)d\tau + y_i(t_0)]^2 + [\int_0^t y_n(\tau)d\tau +$$

$$+ \sum_{j=1}^{k-1} y_j(t_0)t^{k-j}/(k-j)! + y_n(t_0)]^2.$$

With allowance, for stated the following theorem is fair.

**Theorem 3.** Necessary and sufficient condition of existence of a generalized limit cycle of a nonlinear system (1) order n > 2 consist in existence of a A.N.Krylov's producing vector, for want of which the factors of A.N.Krylov are certain by equalities

$$h_1(x,t)=0,\ldots, h_{k-1}(x,t)=0,$$
$$h_k(x,t)=1, h_{k+1}(x,t)=0,\ldots h_n(x,t)=0, (n \geq k \geq 2).$$

The proof of this theorem is similar to the proof of the theorem 2.

Let us demonstrate an obtained outcome on the following examples, for which the A.N.Krylov's producing vector, found as follows, will be general

$$B(x,t) = \text{colon}\{0,0,0, \exp[\int_0^t r(\tau)d\tau]\},$$

where $r = x_1^2 + x_2^2 + x_3^2 + x_4^2$.

**Example 3.**

$$x_1 = rx_1 + x_2 - x_3,$$

$$x_2 = rx_2 - x_3 + x_4,$$

$$x_3 = rx_3 + x_4,$$

$$x_4 = rx_4 - x_1.$$

$h_1(x,t)=0$, $h_2(x,t)=0$,
$h_3(x,t)=0$, $h_4(x,t)=1$ or
H = colon[1, 0, 0, 0].

**Example 4.**

$$x_1 = rx_1 - x_2 - x_3 - x_4,$$

$$x_2 = rx_2 - x_3,$$

$$x_3 = rx_3 + x_1,$$

$$x_4 = rx_4 - x_1.$$

H = colon[0, 1, 0, 0].

**Example 5.**

$$x_1 = rx_1 - x_2 - x_3 - x_4,$$

$$x_2 = rx_2 - x_1 - x_3,$$

$$x_3 = rx_3 + x_4,$$

$$x_4 = rx_4.$$

H = colon[0, 0, 1, 0].

## 7. ARGUING OF OUTCOMES

Summarizing explained it is possible to conclude that the vehicle of producing A.N.Krylov's vector and factors is the rather effective tool of the system's analysis of non-linear differential equations for a reducibility and existence of limit cycles and them generalizations on many-dimensional case (closed surfaces). Dignities of the given method consist of the following. The revealing of a property of reducibility at nonlinear systems allows, for example, it is essential to simplify a structure of a model of a considered control system. The detection of a limit cycle or its generalization at a nonlinear system on an analysis stage allows beforehand to accept to the designer appropriate measures. For example, becomes possible as required to stimulate or to extinguish undamped oscillations of a control system.

The essential difficulty of use of this vehicle is integrated to absence of the formalized procedure of searching of producing A.N.Krylov's vector, as any recommendations on it is not developed yet.

## REFERENCES


Podchukaev V.A. (1997). *The analytical theory automatic control*. 200. Publishing house of Saratov State technical university.

Pogorelov A.V. (1969). *Differential geometry*. 176. Science, Moscow.

Erugin N.P. (1946). *Reducible systems*. 94. Publishing house of an Academy of Sciences USSR, Moscow.

Erugin N.P. (1979). *The book for reading under the guideline of differential equations*. 743. Science and engineering, Minsk.

Demidovich B.P. (1967). *The lectures under the mathematical theory of stability*. 473. Science, Moscow.

Morozov V.M., Kalenova V.I. (1988). *An estimation and control in non-stationary linear systems*. 142. Publishing houses of Moscow State University. Moscow.

Andreev U.N. (1976). *Control finite-dimensional linear objects*. 424. Science, Moscow.

Voronov A.A., Kim D.P.,Lohin V.M. (1986). *Theory of automatic control: the manual on a specialty "Automation and telemechanics". The theory of non-linear and special systems of automatic control* (A.A.Voronov. (Ed.)). 504. Higher school, Moscow.

Penzov U.E. (1972). *Analytical geometry*. 365. Publishing house of Saratov State technical university, Saratov.

Monzon B.M. (1998). *Maple V. Power Edition*. 240. Publishing house "EAGLE OWL", Moscow.

Govoruhin V.N., Zibulin V.G. (1998). *Introduction in*

*Maple. A mathematical packet for all*. 208. The World, Moscow.

Bautin N.N., Leontovich E.A. (1956). ***Methods and receptions of qualitative research of dynamic systems on a plane***. 496. Science, Moscow.

# THE YAKUBOVICH-KALMAN-POPOV LEMMA AND STABILITY ANALYSIS OF DYNAMIC OUTPUT FEEDBACK SYSTEMS

**Rolf Johansson** * **Anders Robertsson** *

* *Department of Automatic Control, Lund Inst. Technology, Lund University, PO Box 118, SE-221 00 Lund Sweden; Fax (+46) 46 138118; E-mail Rolf.Johansson@control.lth.se*

Abstract: This paper presents theory for extension of the Yakubovich-Kalman-Popov (YKP) lemma for stability analysis relevant for observer-based feedback control systems. We show that minimality is not necessary for existence of Lur'e-Lyapunov functions. Implications for output feedback stabilization, positivity, factorization and passivity are given. *Copyright © 1999 IFAC*

Keywords: Stability, Nonlinear Systems, Output Feedback, Factorization

## INTRODUCTION

There are at least three contexts in which the Yakubovich-Kalman-Popov (YKP) lemma turns out to be instrumental for a solution: the absolute stability problem of nonlinear feedback control; *cf.* Lefschetz (1965), Meyer (1966), Zames (1966), Khalil (1996); Boyd et al. (1994); and adaptive control based on output error feedback; Parks (1966); and the covariance-matrix factorization of realization theory; Anderson and Moore (1979);. We quote the lemma deriving from Yakubovich (1962), Kalman (1963) and Popov (1961)

*YKP Lemma:* Let $G_0(s) = C(sI - A)^{-1}B + D$ be a $p \times p$ transfer function where $A$ is Hurwitzian, $(A, B)$ is controllable and $(A, C)$ is observable. Then, $G_0(s)$ is strictly positive real if and only if there exist a positive symmetric matrix $P$, matrices $L, R$ and a positive constant $\varepsilon$ such that

$$\begin{aligned} PA + A^T P &= -LL^T - \varepsilon P \\ PB - C^T &= -LR^T \\ D + D^T &= RR^T \end{aligned} \tag{1}$$

This important lemma supports results of stability theory of nonlinear feedback (circle criterion; Popov criterion) and adaptive system theory and for extensions of Lyapunov theory such as passivity theory; Khalil (1996), Ortega et al. (1995). Although the lemma is very powerful in its predictions, the strictly positive real (SPR) conditions imposed on the transfer function are, unfortunately, rather restrictive for application. As for the absolute stability problem of nonlinear feedback systems, the starting point is the Lur'e problem of Lur'e and Postnikov (1944) with a linear system and nonlinear feedback of cone-bounded nonlinear variation described by the function $\psi(\cdot)$

$$\dot{x} = Ax + Bu = Ax - B\psi(z) \tag{2}$$

$$z = Cx \tag{3}$$

$$u = -\psi(z), \quad \psi^T(z)(\psi(z) - Kz) \leq 0 \tag{4}$$

A Lyapunov function candidate is the Lur'e-Lyapunov function

$$V(x,z) = x^T P x + \alpha \int_0^z \psi^T(\zeta) K d\zeta, \tag{5}$$

$$P = P^T > 0, \quad \alpha > 0 \tag{6}$$

which satisfies requirements on 'positivity', 'radial growth', 'continuity' and 'differentiation'. As guaranteed by the YKP lemma, the stability condition $\dot{V} \leq 0$ holds under fairly restrictive SPR conditions. The circle theorem for SPR systems deals with the time-varying case under the assumption that $\psi(\cdot, t)$ belongs to the cone $[0, \infty]$ and that $\inf_\omega \mathrm{Re}\, G(j\omega) > 0$ for which the Lur'e system is $\mathcal{L}_2$−stable; Khalil (1996). As the Lyapunov functions generated by backstepping techniques apparently cannot be designed using the YKP lemma, there might exist other relevant classes of non-SPR systems for which extended regions of stability might exist; Krstić et al. (1995). In our results, we extend stability theory to include non-SPR transfer functions and dynamic output feedback or observer-based design violating the controllability condition of the YKP lemma. A constructive method to provide quadratic Lyapunov functions and Lur'e-Lyapunov function for nonlinear output feedback of non-SPR transfer functions is given.

## PROBLEM FORMULATION

We will consider dynamic output feedback described by the state-space system

$$\frac{dx_0}{dt} = A_0 x_0 + B_0 u \tag{7}$$

$$\frac{d\xi}{dt} = A_\xi \xi + B_\xi u + K_\xi y \tag{8}$$

$$z = (C_0 \ C_\xi)\begin{pmatrix} x_0 \\ \xi \end{pmatrix} + (D_0 \ D_\xi) u \tag{9}$$

$$u = -\psi(z) \tag{10}$$

where $\{A_0, B_0, C_0, D_0\}$ and the state $x_0$ represent the original system—*i.e.*, the control object—whereas $\{A_\xi, B_\xi, C_\xi, D_\xi\}$ and $\xi$ represent the observer dynamics added to the system description. As all observer states without restriction may be assumed available to measurement, the matrix $C_\xi$ may be chosen freely, say, as motivated by calculated output feedback parameters. This class of systems violates the controllability assumption of the YKP lemma. In shorter notation, we have

$$\frac{dx}{dt} = Ax + Bu, \quad x = \begin{pmatrix} x_0 \\ \xi \end{pmatrix} \in \mathbb{R}^n, \tag{11}$$

$$z = Cx + Du, \quad z \in \mathbb{R}^m, \quad u \in \mathbb{R}^m \tag{12}$$

$$u = -\psi(z) \tag{13}$$

and for the Lur'e-Lyapunov function

$$V(x) = \frac{1}{2} x^T P x + \alpha \int_0^z \psi^T(\zeta) K d\zeta \tag{14}$$

*A Lyapunov Equation*

Let $Q = Q^T \geq 0$ be given. The standard YKP matrix equation may then be formulated as

$$\begin{aligned} -Q &= \begin{pmatrix} PA + A^T P & PB - C^T \\ B^T P - C & -(D + D^T) \end{pmatrix} \\ &= \begin{pmatrix} P & 0 \\ 0 & I_m \end{pmatrix} \begin{pmatrix} A & B \\ -C & -D \end{pmatrix} \\ &+ \begin{pmatrix} A & B \\ -C & -D \end{pmatrix}^T \begin{pmatrix} P & 0 \\ 0 & I_m \end{pmatrix} \end{aligned} \tag{15}$$

Let

$$\mathcal{A} = \begin{pmatrix} A & B \\ -C & -D \end{pmatrix} \in \mathbb{R}^{(n+m)\times(n+m)}, \tag{16}$$

$$\mathcal{P}_o = \begin{pmatrix} P & 0 \\ 0 & I_m \end{pmatrix} \in \mathbb{R}^{(n+m)\times(n+m)} \tag{17}$$

Then, the YKP matrix equation (15) may be reformulated as the Lyapunov equation

$$\mathcal{P}_o \mathcal{A} + \mathcal{A}^T \mathcal{P}_o = -Q \tag{18}$$

Conversely, a matrix triple $\{\mathcal{A}, Q, \mathcal{P}\}$ resulting from such a Lyapunov equation may provide a Lyapunov function or a Lur'e-Lyapunov function relevant for stability analysis. This observation is pursued throughout the paper. We will show that the Lyapunov equation (18) may contribute to stability theory for non-SPR systems.

## RESULTS AND EXTENSIONS

*Lemma 1.* (YKP Lyapunov Equation). Given a linear system $\Sigma(A, B, C, D)$ with the system matrix

$$\mathcal{A} = \begin{pmatrix} A & B \\ -C & -D \end{pmatrix} \in \mathbb{R}^{(n+m)\times(n+m)} \tag{19}$$

and a matrix $Q \in \mathbb{R}^{(n+m)\times(n+m)}$, $Q = Q^T > 0$, there is solution $\mathcal{P} = \mathcal{P}^T > 0$ to the Lyapunov equation

$$-Q = \mathcal{P}\mathcal{A} + \mathcal{A}^T \mathcal{P} \tag{20}$$

$$\mathcal{P} = \begin{pmatrix} \mathcal{P}_{11} & \mathcal{P}_{12} \\ \mathcal{P}_{12}^T & \mathcal{P}_{22} \end{pmatrix} \in \mathbb{R}^{(n+m)\times(n+m)} \tag{21}$$

if and only if all eigenvalues $\{\lambda_k\}_{k=1}^{n+m}$ of $\mathcal{A}$ have negative real part. Moreover, every positive definite solution $\mathcal{P} = \mathcal{P}^T > 0$ to the Lyapunov equation (20) provides a solution $P \in \mathbb{R}^{n\times n}$ to the YKP matrix equation

$$-Q = \begin{pmatrix} PA + A^T P & PB - C^T \\ B^T P - C & -(D + D^T) \end{pmatrix} \tag{22}$$

as

$$P = \mathcal{P}_{11} - \mathcal{P}_{12}\mathcal{P}_{22}^{-1}\mathcal{P}_{12}^T, \quad P = P^T > 0 \quad (23)$$

under the feedback transformation

$$\mathcal{T} = \begin{pmatrix} I_n & 0 \\ -\mathcal{P}_{22}^{-1}\mathcal{P}_{12}^T & \mathcal{P}_{22}^{-1/2} \end{pmatrix} \quad (24)$$

The transfer function of the system matrix obtained from the feedback transformation $\mathcal{T}^{-1}\mathcal{A}\mathcal{T}$ is stable and positive real.

*Proof 1.* By the correspondence of Eq. (18) and Eq. (15), it is clear that the YKP matrix equation can be reformulated as Lyapunov equation. The converse is not obvious as the Lyapunov equation (18) resulting from Eq. (15) requires the solution $\mathcal{P}$ to be block diagonal. To that end, consider any matrix solution $\mathcal{P} = \mathcal{P}^T > 0$ with decomposition

$$\mathcal{P} = \begin{pmatrix} \mathcal{P}_{11} & \mathcal{P}_{12} \\ \mathcal{P}_{12}^T & \mathcal{P}_{22} \end{pmatrix} \quad (25)$$

If $\mathcal{P}$ is of full rank, then $\mathcal{P}_{22}$ is invertible and there always exists the similarity transformation matrix of Eq. (24) which renders

$$\mathcal{P}_o = \mathcal{T}^T\mathcal{P}\mathcal{T} = \begin{pmatrix} \mathcal{P}_{11} - \mathcal{P}_{12}\mathcal{P}_{22}^{-1}\mathcal{P}_{12}^T & 0 \\ 0 & I_m \end{pmatrix} > 0$$

Left multiplication of the Lyapunov equation by $\mathcal{T}^T$ and right multiplication by $\mathcal{T}$ preserves the Lyapunov equation structure but renders the solution $\mathcal{T}^T\mathcal{P}\mathcal{T}$ on the block-diagonal form of Eq. (17) with a resultant similarity transformation $\mathcal{T}^{-1}\mathcal{A}\mathcal{T}$.

*Remark 1.* The appropriate block-diagonal form can be obtained via feedback design—*e.g.*, as suggested by $\mathcal{T}$. In order to accomplish positive realness, the right multiplication implicitly suggests a state feedback control law to stabilize the system whereas the left multiplication suggests an observer structure. Also, by choice of $Q$ such that

$$Q = \mathcal{T}^{-T}Q_0\mathcal{T}^{-1} \quad (26)$$

the Lyapunov equation may have a solution on the block-diagonal form for $Q_0$ given.

*Example* Consider the state-space system

$$\dot{x} = x(t) - 2u(t) \quad (27)$$

$$y(t) = -2x(t) + 2u(t) \quad (28)$$

and $G_0(s) = 2(s+1)/(s-1)$. Although the system is unstable, the parameter matrix

$$\mathcal{A} = \begin{pmatrix} A & B \\ -C & -D \end{pmatrix} = \begin{pmatrix} 1 & -2 \\ 2 & -2 \end{pmatrix}, \quad (29)$$

has eigenvalues with negative real part. The feedback transformation matrix

$$\mathcal{T} = \begin{pmatrix} 1 & 0 \\ 1 & 1 \end{pmatrix} \quad (30)$$

gives the transformed system matrix. For

$$\mathcal{A}_\mathcal{T} = \mathcal{T}^{-1}\mathcal{A}\mathcal{T} = \begin{pmatrix} -1 & -2 \\ 1 & 0 \end{pmatrix}, \quad G_\mathcal{T}(s) = \frac{2}{s+1}$$

$$Q = \begin{pmatrix} 1 & 0 \\ 0 & 0 \end{pmatrix}, \mathcal{P}_o = \mathcal{P}_o^T = \begin{pmatrix} 0.5 & 0 \\ 0 & 1 \end{pmatrix} > 0, \quad (31)$$

there is a solution to the Lyapunov equation $\mathcal{P}_o\mathcal{A}_\mathcal{T} + \mathcal{A}_\mathcal{T}^T\mathcal{P}_o = -Q$. Note that the $\mathcal{A}_\mathcal{T}$ resulting from the feedback transformation of the unstable system provides a block-diagonal solution $\mathcal{P} = \mathcal{P}^T > 0$ and thus $P = P^T = 0.5 > 0$.

*Lemma 2.* (Positivity and Factorization). Let $\{A, B, C, D\}$ be a LTI state-space system and

$$\mathcal{A} = \begin{pmatrix} A & B \\ -C & -D \end{pmatrix} \quad (32)$$

If $(\mathcal{A}, Q^{1/2})$ is observable and if all eigenvalues of $\mathcal{A}$ are in the open left-half plane, then the Lyapunov equation

$$\mathcal{P}\mathcal{A} + \mathcal{A}^T\mathcal{P} = -Q^{T/2}Q^{1/2} = -Q \quad (33)$$

provide a unique positive definite solution $P = P^T > 0$ to the YKP matrix equation. Moreover, there exist rational functions $G(s) = C(sI - A)^{-1}B + D$ and

$$\Gamma(s) = (\Gamma_1(s)\ \Gamma_2(s)) \quad (34)$$

$$= (Q_1\ Q_1(sI - A)^{-1}B + Q_2) \quad (35)$$

with $Q_1, Q_2$ matrices satisfying

$$\Omega = \Gamma^T(-s)\Gamma(s) = \begin{pmatrix} -(PA + A^TP) & G_{12}(s) \\ G_{12}^T(-s) & G(s) + G^T(-s) \end{pmatrix} \quad (36)$$

$$0 \le \Gamma_1^T(-i\omega)\Gamma_1(i\omega) = G(i\omega) + G^T(-i\omega) \quad (37)$$

*Proof:* See Appendix.

## OBSERVERS AND NONLINEAR FEEDBACK

Consider the nonlinear feedback stabilization problem of Eqs. (7-10) and assume that $\{A_0, B_0, C_0, D_0\}$ be given and that the observer $\{A_\xi, B_\xi, C_\xi, D_\xi, K_\xi\}$ be chosen appropriately. Nonlinear feedback stabilization by means of observer design and SPR design of loop transfer function can be approached by:

*Algorithm: Lur'e-Lyapunov function design*

**1.** Find state-space matrices $A_0, B_0, C_0, D_0$;

**2.** Introduce augmented system dynamics and state-space system matrices $A_\xi$, $B_\xi$, $C_\xi$, $D_\xi$, $K_\xi$ and arrange the aggregate system matrix

$$\mathcal{A} = \begin{pmatrix} A & B \\ -C & -D \end{pmatrix} \quad (38)$$

If necessary, modify the augmented system dynamics $\{A_\xi, B_\xi, C_\xi, D_\xi, K_\xi\}$ so that the eigenvalues of $\mathcal{A}$ have negative real part.
**3.** Choose $Q = Q^{T/2}Q^{1/2} \geq 0$ such that $(\mathcal{A}, Q^{1/2})$ is observable and solve the Lyapunov equation

$$\mathcal{P}\mathcal{A} + \mathcal{A}^T\mathcal{P} = -Q, \quad Q, \mathcal{P} \in \mathbb{R}^{(n+m)\times(n+m)} \quad (39)$$

**4.** Find the positive definite solution $\mathcal{P}$ with block decomposition

$$\mathcal{P} = \begin{pmatrix} \mathcal{P}_{11} & \mathcal{P}_{12} \\ \mathcal{P}_{12}^T & \mathcal{P}_{22} \end{pmatrix}, \quad \mathcal{P}_{11} \in \mathbb{R}^{n\times n} \quad (40)$$

**5.** Find Schur complement matrix $P$ and the feedback transformation matrix

$$\mathcal{T} = \begin{pmatrix} I_n & 0 \\ -\mathcal{P}_{22}^{-1}\mathcal{P}_{12}^T & \mathcal{P}_{22}^{-1/2} \end{pmatrix} \quad (41)$$

$$P = \mathcal{P}_{11} - \mathcal{P}_{12}\mathcal{P}_{22}^{-1}\mathcal{P}_{12}^T, \quad (42)$$

**6.** Assign Lyapunov function candidate or Lur'e-Lyapunov function

$$V(x) = x^TPx + \alpha \int_0^z \psi^T(s)K ds \quad (43)$$

*Example (Rohrs et al. (1982))*

Consider the state-space system

$$\frac{dx}{dt} = \begin{pmatrix} -31 & -259 & -229 \\ 1 & 0 & 0 \\ 0 & 1 & 0 \end{pmatrix} x + \begin{pmatrix} 1 \\ 0 \\ 0 \end{pmatrix} u$$

$$y = \begin{pmatrix} 0 & 0 & 458 \end{pmatrix} x \quad (44)$$

with the transfer function

$$G_0(s) = \frac{2}{s+1} \cdot \frac{229}{s^2 + 30s + 229} \quad (45)$$

By the relative degree 3, the conditions of the YKP lemma are violated and the Popov criterion predicts a finite gain margin for output feedback. For

$$A = \begin{pmatrix} A_0 & 0_{3\times3} \\ K_\xi C_0 & A_\xi \end{pmatrix}, \quad K_\xi = \begin{pmatrix} -1 \\ 0 \\ 1 \end{pmatrix}, \quad Q = I_6$$

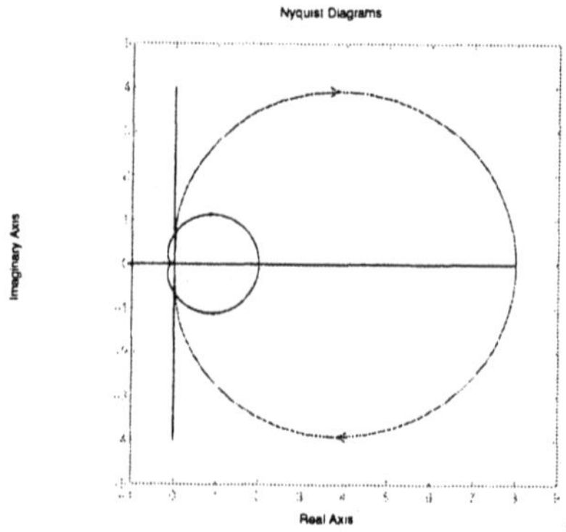


Fig. 1. Nyquist diagram of original system and transformed system.

we have a solution $P = P^T > 0$ to the Lyapunov equation $PA + A^TP = -Q$ found from $\mathcal{P}$ by the Schur-complement formula and with all eigenvalues of $A$ possessing negative real part. An observer-transformed positive real loop-transfer function can be obtained as

$$G(s) = \frac{(12s + 14)(s+2)^2(s^2 + 30s + 263)}{(s+1)(s+2)^3(s^2 + 30s + 229)}$$

with Nyquist diagram with the positive real property (Fig. 1); Popov diagram of the transformed system suggesting high gain-margin; stable system transients and Lyapunov function transients of system under feedback control (Fig. 2). Simulation of simple negative feedback connection of the loop-transfer function $40G_0(s)$ *via* a nonlinearity $\psi(\cdot)$ within the cone $[0, \infty)$ provides a limit cycle.(Fig. 2). As predicted by the theory presented, the observer-compensated system eliminates the limit cycle and provides asymptotically stable behavior with Lyapunov function $V(x)$—see Fig. 2.

## DISCUSSION

We have provided an interpretation of the YKP matrix equation as a Lyapunov equation and have shown that this equation might provide Lyapunov functions for stability analysis (asymptotic stability and absolute stability) also for a class of systems without PR properties. In particular, we have shown that Lyapunov functions may be provided also for observer-based feedback systems with possible pole-zeros cancellations introduced purposely. Such systems are not covered by regular formulations of the YKP lemma due the violation of minimality, controllability and observability. Actually, the minimality condition appears to be crucial only for the necessity part of the proof for the YKP lemma; *cf.* Khalil (1996). As pointed out by several

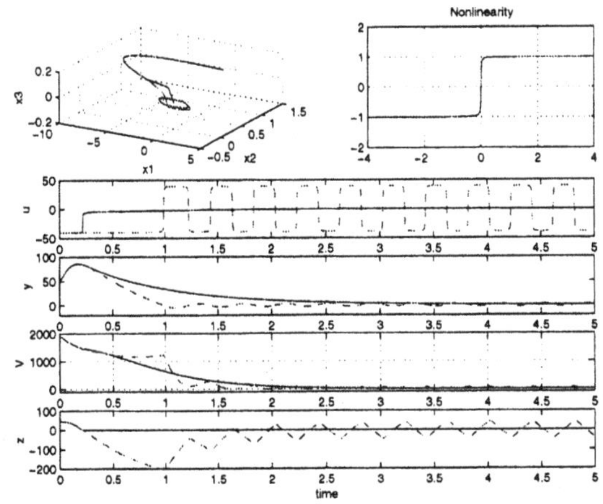


Fig. 2. State-space diagram (*upper left*) with state trajectory of stabilized system (*solid line*) and original limit-cycle (*dashed line*); Feedback nonlinearity $\psi(\cdot)$ (*upper right*); Input $u$ *vs.* time; output $y$ *vs.* time; Lyapunov function $V(x)$ *vs.* time and behavior of $V(x)$ for non-compensated system; feedback signal $z$—trajectories of original feedback system (*dashed*) and stabilized system (*solid*), respectively.

authors—*e.g.*, Meyer (1966), Narendra and Annaswamy (1989), Rantzer (1996)—there may exist $P = P^T \geq 0$ without any controllability assumption satisfied.

The nonstandard evaluation of the eigenvalues of $\mathcal{A}$ makes sense in that the eigenvalues are invariant under the similarity transformation of state-space transformation As for eigenvalue assignment for $\mathcal{A}$, it is relevant to evaluate other matrix operations on $\mathcal{A}$ such as left and right multiplication by

$$\mathcal{L} = \begin{pmatrix} I_n & K \\ 0 & I_m \end{pmatrix}, \text{ and } \mathcal{R} = \begin{pmatrix} I_n & 0 \\ -L & I_m \end{pmatrix} \quad (46)$$

with the resultant matrix $\mathcal{A}' = \mathcal{L}\mathcal{A}\mathcal{R}$. Such matrix multiplications relate to feedback transformations of observer design and state-feedback control stability, respectively as found by Kokotović and Sussman (1989). In fact, the similarity transformation provided by the algorithm implicitly suggest both a stabilizing state feedback control law (the right multiplication) and a linear combination of states and output to provide an SPR transfer function. By adding dynamics (integrators) up to the complexity of a full-order observer, it is possible to accomplish the SPR property needed for subsequent application of control algorithms of nonlinear control, adaptive control or realization theory; *cf.* Mazenc and Praly (1996), Coron and Praly (1991).

## CONCLUSIONS

We have shown that there are solutions to the Yakubovich-Kalman-Popov equation in the form of nonminimal positive real systems. Thus, the very restrictive conditions of positive realness relevant to observer-based feedback control systems can be significantly relaxed. Important applications of such systems can be found in nonlinear stability theory, observer design and design of feedback stabilization—*e.g.*, by means of the Popov criterion or circle criterion. A method for construction of Lur'e-Lyapunov functions for systems with observer-based feedback control is given.

## 1. REFERENCES


ANDERSON, B. D. O. and J. B. MOORE (1979): *Optimal Filtering.* Prentice Hall, Englewood Cliffs, NJ.

BOYD, S., L. EL GHAOUI, E. FERON, and V. BALAKRISHNAN (1994): *Linear Matrix Inequalities in System and Control Theory,* vol. 15. SIAM, Philadelphia, PA.

CORON, J. M. and L. PRALY (1991): "Adding an integrator for the stabilization problem." *Systems and Control Letters,* **17**, pp. 89–104.

KALMAN, R. E. (1963): "Lyapunov functions for the problem of Lur'e in automatic control." *Proc. Nat. Acad. Sci,* **49:2**.

KHALIL, H. K. (1996): *Nonlinear Systems.* Prentice Hall, Upper Saddle River, NJ.

KOKOTOVIĆ, P. V. and H. J. SUSSMAN (1989): "A positive real condition for global stabilization of nonlinear systems." *Systems and Control Letters,* **13**, pp. 125–133.

KRSTIĆ, M., I. KANELLAKOPOULOS, and P. KOKOTOVIĆ (1995): *Nonlinear and Adaptive Control Design.* John Wiley & Sons, New York.

LEFSCHETZ, S. (1965): *Stability of Nonlinear Control Systems.* Acad. Press, New York.

LOZANO LEAL, R. and S. R. JOSHI (1990): "Strictly positive real transfer functions revisited." *IEEE Transactions on Automatic Control,* **35:11**, pp. 1243–1245.

LUR'E, A. I. and V. N. POSTNIKOV (1944): "On the theory of stability of regulation system." *Prikladnaya Matematika i Mekhanika,* **VIII**, pp. 246–248.

LYAPUNOV, A. M. (1892): *The general problem of the stability of motion.* Kharkov Mathematical Society, Kharkov, Russia.

MAZENC, F. and L. PRALY (1996): "Adding integrations, saturated control and global asymptotic stabilization for feedforward

systems." *IEEE Trans. Autom. Control*, **41**.
MEYER, K. R. (1966): "On the existence of Lyapunov functions for the problem of Lur'e." *SIAM J. Control*, **3:3**.
NARENDRA, K. S. and A. M. ANNASWAMY (1989): *Stable Adaptive Systems.* Prentice Hall, Englewood Cliffs, NJ.
NARENDRA, K. S. and J. H. TAYLOR (1973): *Frequency Domain Criteria for Absolute Stability.* Academic Press, New York.
NIJMEIJER, H. and A. J. VAN DER SCHAFT (1990): *Nonlinear Dynamical Control Systems.* Springer-Verlag.
ORTEGA, R., A. LORIA, R. KELLY, and L. PRALY (1995): "On output feedback global stabilization of Euler-Lagrange systems." *Int. J. Robust and Nonlinear Cont.*, **5:4**, pp. 313–324.
PARKS, P. C. (1966): "Liapunov redesigns of model reference adaptive control systems." *IEEE Trans. Automatic Control*, **AC-11**, pp. 362–367.
POPOV, V. M. (1961): "Absolute stability of nonlinear systems of automatic control." *Avtomatika i Telemekhanika*, **22:8**.
RANTZER, A. (1996): "On the Kalman-Yakubovich-Popov lemma." *Systems and Control Letters*, **28:1**, pp. 7–10.
ROHRS, C. E., L. VALAVANI, M. ATHANS, and G. STEIN (1982): "Robustness of adaptive control algorithms in the presence of unmodeled dynamics." In *Proc. Conf. Decision and Control.* Orlando, FL.
YAKUBOVICH, V. A. (1962): "Solution of certain matrix inequalities occuring in the theory of automatic control." *Dokl. Akad. Nauk SSSR*, **143:6**, pp. 1304–1307.
ZAMES, G. (1966): "On the input-output stability of time-varying nonlinear feedback systems." *IEEE Trans. Automatic Control*, **AC-11:2 (Part I), 3 (Part II)**.

*Acknowledgement* This research was supported by the National Board for Industrial and Technical Development (NUTEK).

*Appendix—Proof of Lemma 2*

We make the following constructive proof of positive realness: Let

$$E = \begin{pmatrix} I_n & 0_{n\times m} \\ 0_{m\times n} & 0_m \end{pmatrix} \tag{47}$$

$$L(s) = \begin{pmatrix} I_n & 0 \\ C(sI_n - A)^{-1} & I_m \end{pmatrix} \tag{48}$$

$$R(s) = \begin{pmatrix} I_n & -(sI_n - A)^{-1}B \\ 0 & I_m \end{pmatrix} \tag{49}$$

First, the transmission zeros of the system $\{A, B, C, D\}$ are found from

$$sE - \mathcal{A} = \begin{pmatrix} (sI_n - A) & -B \\ C & D \end{pmatrix} \tag{50}$$

$$= L(s)\begin{pmatrix} sI_n - A & 0 \\ 0 & G(s) \end{pmatrix} R(s) \tag{51}$$

Let $Q^{1/2}$ denote a matrix factor so that

$$Q = Q^{T/2}Q^{1/2} = \begin{pmatrix} Q_1^T \\ Q_2^T \end{pmatrix} (Q_1 \; Q_2) \tag{52}$$

Assume that for $(\mathcal{A}, Q^{1/2})$ observable and eigenvalues of $\mathcal{A}$ with negative real part, the Lyapunov equations

$$\mathcal{P}\mathcal{A} + \mathcal{A}^T\mathcal{P} = -Q \tag{53}$$

$$PA + A^TP = -Q_1^TQ_1 \overset{\text{def}}{=} -Q_{11} \tag{54}$$

have provided the positive definite solutions $\mathcal{P}$ and $P$ and that the solution obtained has been brought to block-diagonal form of Eq. (17). Then, expand the Lyapunov equation

$$Q = \mathcal{P}(sE - \mathcal{A}) + (-sE - \mathcal{A}^T)\mathcal{P} \tag{55}$$

By Eq. (50), it follows that

$$\mathcal{P}(sE - \mathcal{A}) = \begin{pmatrix} P(sI_n - A) & 0 \\ C & G(s) \end{pmatrix} R(s)$$

Thus, for $\Gamma(s) = Q^{1/2}R^{-1}(s)$ and

$$\Gamma(s) = (Q_1 \; Q_1(sI - A)^{-1}B + Q_2)$$

$$G_{12}(s) = C^T + (-sI_n - A^T)P(sI_n - A)^{-1}B$$

$$= Q_{11}(sI_n - A)^{-1}B + C^T - PB \tag{56}$$

one finds that

$$\Omega = \Gamma^T(-s)\Gamma(s) = R^{-T}(-s)Q^TQR^{-1}(s)$$

$$= R^{-T}(\mathcal{P}(sE - \mathcal{A}) + (-sE - \mathcal{A}^T)\mathcal{P})R^{-1}$$

$$= \begin{pmatrix} -(PA + A^TP) & G_{12}(s) \\ G_{12}^T(-s) & G(s) + G^T(-s) \end{pmatrix} \tag{57}$$

with rank deficit only at the transmission zeros of $\{A, B, C, D\}$. By the matrix equations (15, 57), the simultaneous transfer-function positivity and the positive definite Lyapunov equation properties follow from the diagonal matrix equation blocks of $\Gamma^T(-s)\Gamma(s)$ that

$$-(PA + A^TP) = \Gamma_1^T(-s)\Gamma_1(s) = Q_1^TQ_1$$

$$G(s) + G(-s) = \Gamma_2^T(-s)\Gamma_2(s) \tag{58}$$

As for $s = i\omega$, it follows that the 'positive-real' condition be fulfilled.

  

# MINIMAL PARAMETRIZATION REGULATORS FOR MIMO "STRICT FEEDBACK FORM" SYSTEMS

**Giovanni L. Santosuosso**

*University of Cassino*
*Via Di Biasio 43, 03043 Cassino (FR), ITALY*
*Tel.+39-776-299689, Fax+39-776-310812,*
*email:santosuosso@ing.unicas.it*

Abstract: In this paper it will be considered a class of MIMO nonlinear systems in "strict feedback form" affected by a norm bounded matrix of time dependent disturbances. It will be assumed the matrix norm to be unknown, and it will be constructed a robust controller whose dynamics turn out to be minimal, yielding with this respect the best possible control strategy. It is also shown how this strategy can be applied to the elastic joint robot model simplified dynamic equations, and found a global state space coordinate transformation and a state feedback such that the system dynamics can be expressed in "strict feedback form". *Copyright ©1999 IFAC*

Keywords: MIMO Nonlinear systems, Robust control, Parametrization, Robot Control, Lyapunov function.

## 1. INTRODUCTION

Consider the MIMO nonlinear system in "strict feedback form"

$$\begin{aligned} \dot{\xi}_m &= \xi_{m+1} + \sum_{j=1}^{m} \varphi_{mj}(\xi_1, \ldots, \xi_m)\theta\xi_j, \\ & 1 \le m \le r-1 \\ \dot{\xi}_r &= v + \sum_{j=1}^{r} \varphi_{rj}(\xi_1, \ldots, \xi_r)\theta\xi_j, \\ y &= \xi_1 \end{aligned} \tag{1}$$

with state $\xi = [\xi_1^T \cdots \xi_r^T]^T$, $\xi_i \in (I\!R^n)$, $i = 1, \ldots, r$, $v \in I\!R^n$ is the control input and $y \in I\!R^n$ is the output and $\varphi_{mj}$, with $1 \le m \le j$ , and $1 \le j \le r$ are matrices with smooth entries. It is assumed this class of systems to be affected by time varying disturbances, represented in the $n \times n$ matrix $\theta$. whose entries are piecewise continuous functions of time, and such that for all $t > 0$, $\|\theta(t)\| < c_\theta$. If $c_\theta$ is unknown, then regulation can be achieved via an "integral " control law, i.e. a dynamic feedback

$$\begin{aligned} \dot{\lambda} &= \tau(\lambda, \xi), \ \ \tau(0,0) = 0 \\ v &= \alpha_r(\lambda, \xi), \ \ \alpha_r(0,0) = 0. \end{aligned} \tag{2}$$

with $\lambda \in I\!R^{n_\lambda}$, and $\tau(\cdot)$, $\alpha_r(\cdot)$ smooth functions on $I\!R^{n_\lambda} \times I\!R^{\mu} \times I\!R^r$ that estimates the unmodelled system dynamics in order to achieve robust regulation. This problem has been solved by Marino *et al.* (1993), for SISO systems, introducing a dynamic feedback such that $\dim(\lambda) = r$. An important quality of the control law is of course, being *not overparametrized*; in other words the property that the dimension $n_\lambda$ of $\lambda$ is "as small as possible" in particular not dependent on the dimension of the state. The problem of the regulation for a system with *constant* unknown parameters without overparametrization has been solved in Krstic *et al.* (1992). In this paper it will be established a correspondence between the regulation problem and the modelling of the process as a feedback interconnection of systems having the output sufficiently attenuated with respect to the input, solving the regulation problem with the *minimum* number of parameters. This can be formalized as follows.

*Problem 1.* **nonlinear global regulation with minimal parametrization.** Consider System (1). Assume that for all $t > 0$, $\|\theta(t)\| < c_\theta$, where $c_\theta$ is an unknown positive real number. Find a dynamic feedback (2) such that the dimension of the dynamics of the controller are minimal, and for every initial condition, are such that $\lim_{t\to\infty} \xi = 0$.

Problem above , in the case of SISO system, has been solved by Santosuosso (1995)-(1996). The contribution of this note is in the extension of the results in Santosuosso (1996) to MIMO systems. Note however that it is decided also to apply this new design strategy to the problem of the nonlinear set point robust regulation of an elastic joint robot with minimal parametrization. This approach helps filling the gap between the geometric theory of control and its applications to mechanical and electrical systems.

### 1.1 *Background material*

It will be introduced some preliminary definitions and results about nonlinear dissipative systems, which will be used to establish our main result. Consider a system $\Sigma_{\tilde{z}\tilde{w}}$

$$\begin{aligned} \dot{\tilde{x}} &= F(\tilde{x}) + G(\tilde{x})\tilde{w} \\ \dot{\tilde{z}} &= H(\tilde{x}) \end{aligned} \tag{3}$$

with state $\tilde{x} \in I\!R^{\tilde{n}}$, time dependent disturbance $\tilde{w} \in I\!R^{\tilde{m}}$ and output $\tilde{z} \in I\!R^{\tilde{p}}$. It will be assumed that $F(\cdot)$, $G(\cdot)$, $H(\cdot)$ are smooth functions globally on $I\!R^{\tilde{n}}$, and that there exists $\tilde{x} = \tilde{x}^e$, such that $F(\tilde{x}^e) = 0$, $H(\tilde{x}^e) = 0$. $\mathcal{B}(\Omega_{\tilde{w}})$ will denote the set of the functions $\tilde{w} : I\!R \to I\!R^{\tilde{m}}$ piecewise continuous and such that the trajectory of system (3) is well defined. It is interesting to report conditions implying dissipativeness, as in Willems (1972), (1972b); Hill *et al.* (1976), Van Der Shaft (1992), as follows.

*Proposition 2.* Consider system (3). Let $\gamma \in I\!R^+$, and $V : I\!R^{\tilde{n}} \to I\!R$, be a positive semidefinite smooth function. Set

$$\begin{aligned} D(V, \tilde{x}, \tilde{w}, \gamma) = {} & L_F V(\tilde{x}) + L_G V(\tilde{x})\tilde{w} \\ & -\gamma^2 \|\tilde{w}\|^2 + \|H(\tilde{x})\|^2 \end{aligned}$$

$$J(V, \tilde{x}, \gamma) = L_F V(\tilde{x}) + \frac{1}{4\gamma^2} \|L_G V(\tilde{x})\|^2 + \|H(\tilde{x})\|^2$$

Then

$$D(V, \tilde{x}, \gamma, \tilde{w}) \leq S(\tilde{x}) \Longleftrightarrow J(V, \tilde{x}, \gamma) \leq S(\tilde{x})$$

where $S(\tilde{x})$ is an arbitrary scalar mapping. Besides, (a) $\Leftrightarrow$ (b) $\Rightarrow$ (c), where

(a) The dissipation inequality $D(V, \tilde{x}, \gamma, \tilde{w}) \leq 0$ holds for all $\tilde{x} \in I\!R^{\tilde{n}}$ and all $\tilde{w} \in \mathcal{B}(\Omega_{\tilde{w}})$.

(b) The Hamilton Jacobi inequality $J(V, \tilde{x}, \gamma) \leq 0$ holds for all $\tilde{x} \in I\!R^{\tilde{n}}$.

(c) The system has $L_2$ gain less than or equal to $\gamma$, i.e. $\int_0^T \|\tilde{z}(t)\|^2 dt \leq \gamma^2 \int_0^T \|\tilde{w}(t)\|^2 dt$ for all $T \geq 0$ and all $\tilde{w} \in \mathcal{B}(\Omega_{\tilde{w}})$, $\tilde{z}(t) = H(\varphi(T, 0, \tilde{x}^e, \tilde{w}))$ output of system (3) with $\tilde{w} \in \mathcal{B}(\Omega_{\tilde{w}})$, and initial condition $\tilde{x}(0) = \tilde{x}^e$

Expression $s(\tilde{w}, \tilde{z}) = \gamma^2 \|\tilde{w}\|^2 - \|\tilde{z}(t)\|^2$ is called *dissipation rate,* associated with $\Sigma_{\tilde{z}\tilde{w}}$, and plays an important role when it is considered dissipativeness in a nonlinear system. It is finally reported a theorem, due to Hill *et al.* (1976), concerning the stability of the closed loop interconnection of two systems, both having " finite $L_2$ gain".

*Proposition 3.* (**Small Gain Theorem**). Let $\gamma, \mu \in I\!R^+$. Consider system (3) where $\Omega_{\tilde{x}} \subseteq I\!R^{\tilde{n}}$ is such that $H(\tilde{x}) = 0 \Rightarrow \tilde{x} \in \Omega_{\tilde{x}}$, and let (3) be dissipative with supply rate $s(\tilde{w}, \tilde{z}) = \gamma^2 \|\tilde{w}\|^2 - \|\tilde{z}\|$, with a smooth positive definite and globally proper storage function $V$, $V(\tilde{x}^e) = 0$. Consider the output feedback $\tilde{w} = K\tilde{z}$, where $K(\cdot)$ is a matrix whose entries are piecewise continuous functions of time, and $\|K(t)\| < \mu$ for all $t \geq 0$. If $\gamma\mu < 1$ then the trajectories of the closed loop system are globally bounded and converge asymptotically to $\Omega_{\tilde{x}}$. Besides, if the set $\Omega_{\tilde{x}}$ coincides with $\tilde{x} = \tilde{x}^e$, (i.e. the system is globally observable) then the closed loop system is globally asymptotically stable around $\tilde{x} = \tilde{x}^e$.

## 2. MAIN RESULT

### 2.1 *Systems with relative degree 1 for vanishing disturbances*

In order to prove the main result of this note, the control strategy will proceed inductively, i.e. starting to consider the case in which in system the system (1) $r = 1$, and this happens when the nominal system has relative degree 1 for vanishing disturbances.

$$\begin{aligned} \dot{\lambda} &= \tau \\ \dot{\xi}_1 &= v_1 + \varphi_{11}(\xi_1 + d_l)\theta\xi_1. \end{aligned} \tag{4}$$

Here appear also the scalar observer dynamics $\lambda$. Assume that it can be set $v_1 = v_1(\lambda, \xi_1)$, where the function $v_1(\lambda, \xi_1)$ together with $\tau = \tau(\lambda, \xi_1)$ has to be defined within the synthesis procedure. It will be introduced now two quantities to express system (4) in a more interesting form. To this purpose, set respectively $\tilde{g}_1 = (1 + \|\varphi_{11}\|) I_n$ and $\delta_1 = \frac{1}{c_\theta} \tilde{g}_1^{-1} \varphi_{11}\theta$.

Remark that the entries of the $n \times n$ matrix $\delta_1(\cdot)$ are piecewise continuous functions of time and such that $\|\delta_1\| < 1$ for all $t > 0$.

Define the $n$ dimensional row vector $\tilde{w}_1 = \delta_1\xi_1$. By setting $\tilde{x}_1 = (\lambda, \xi_1)$, system (4), can be written as the feedback interconnection of the two systems $\Sigma_1$ and $\Delta_1$, respectively

$$\Sigma_1 : \begin{cases} \dot{\tilde{x}}_1 = F_1(\tilde{x}_1) + P_1 v_1 + Q_1 \tau \\ \quad +c_\theta G_1(\tilde{x}_1)\tilde{w}_1, \\ \tilde{z}_1 = H_1(\tilde{x}_1) \end{cases}$$

$$\Delta_1 : \{\tilde{w}_1 = K_1 \tilde{z}_1.$$

where $F_1(\tilde{x}_1) = [0 \quad 0]^T$, $G_1(\tilde{x}_1) = [0 \quad \tilde{g}_1(\xi_1)]^T$, $H_1(\tilde{x}_1) = [\xi_1]$ are matrices with smooth entries, $P_1 = [0 \quad I_n]^T$, $Q_1 = [1, 0, \ldots, 0]^T$, and $K_1(\cdot)$ is a matrix whose entries are piecewise continuous functions of time, such that $\|K_1\| < 1$. Let $\gamma$ be positive scalar, $\gamma < 1$, and consider the expression

$$V_1(\lambda, \xi_1, c_\theta^2) = \frac{1}{2}(c_\theta^2 - \lambda)^2 + \frac{1}{2}\|\xi_1\|^2 \quad (5)$$

$$D_1(v_1, \tau, \tilde{w}_1) = \frac{dV_1}{dt} - \gamma^2\|\tilde{w}_1\|^2 + \|\tilde{z}_1\|^2 \quad (6)$$

and $\frac{dV_1}{dt}$ is computed along the trajectories of the closed loop system $\Sigma_1$. By expliciting the terms in the right hand side of (6), noting that $\frac{d\xi_1}{dt} = v_1 + c_\theta \tilde{g}_1 \tilde{w}_1$, and rearranging terms, it can be obtained

$$\begin{aligned} D_1(v_1, \tau, \tilde{w}_1) \le \xi_1^T \left[ v_1 + \frac{\lambda}{4\gamma^2}\tilde{g}_1\tilde{g}_1^T\xi_1 \right] \\ -(c_\theta^2 - \lambda)\left[ \frac{1}{4\gamma^2}\xi_1^T\tilde{g}_1\tilde{g}_1^T\xi_1 - \tau \right]. \end{aligned} \quad (7)$$

It is obvious, at this point, that the "natural" choice of $v_1 = \alpha_1$ where $\alpha_1 = -(\frac{\lambda}{4\gamma^2}\tilde{g}_1\tilde{g}_1^T\xi_1)$, yields

$$D_1(\alpha_1, \tau, \tilde{w}_1) \le (c_\theta^2 - \lambda)(\rho_1 \tilde{z}_1 - \tau) \quad (8)$$

with $\rho_1 = \left[\frac{\lambda}{4\gamma^2}\tilde{g}_1\tilde{g}_1^T\right]$.

If it is defined

$$\begin{aligned} J_1(v_1, \tau) = L_{F_1}V_1(\tilde{x}_1) + L_{Q_1\tau}V_1(\tilde{x}_1) \\ +L_{P_1 v}V_1(\tilde{x}_1) + \frac{c_\theta^2}{4\gamma^2}\|L_{G_1}V_1(\tilde{x}_1)\|^2 + \|\tilde{z}_1\|^2 \end{aligned} \quad (9)$$

then inequality (8) implies

$$J_1(\alpha_1, \tau) \le (c_\theta^2 - \lambda)(\rho_1\tilde{z}_1 - \tau). \quad (10)$$

If it is chosen $\tau = \rho_1\tilde{z}_1$ then the last expression is the Hamilton-Jacobi inequality associated with system $\Sigma_1$, and shows, choosing $\gamma < 1$, by proposition 3 that $\lim_{t\to\infty}(\xi_1(t)) = 0$.

### 2.2 *System with relative degree greater than 1 for vanishing disturbances*

If $r > 1$, then it has to be shown basically that if the problem it is considered is solvable when $r = i - 1$, $i$ being a positive integer greater than one, then when $r = i$ the synthesis of a control law which regulates the state of the given system can be constructed by modifying the solution when $r = i - 1$. In order to describe this property, consider system

$$S_{i-1} : \begin{cases} \dot{\lambda} = \tau \\ \dot{\xi}_m = \xi_{m+1} + \sum_{j=1}^{m} \varphi_{mj}(\xi_1, \ldots, \xi_m)\theta\xi_j, \\ \quad m = 1, \ldots, i-2 \\ \dot{\xi}_{i-1} = v_{i-1} \\ \quad + \sum_{j=1}^{i-1} \varphi_{(i-1)j}(\xi_1, \ldots, \xi_{i-1})\theta\xi_j, \end{cases}$$

Assume that it have been found $i-1$ smooth functions $\alpha_j = \alpha_j(\lambda, \xi_1, \ldots, \xi_j)$, $j = 1, \ldots, i-1$, $\alpha_j(\lambda, 0, \ldots, 0) = 0$, for all $j = 1, \ldots, i-1$, all $\lambda \in I\!R$, such that hold properties similar to the ones outlined in the previous section, when $i - 1 = 1$. To this purpose, set $\zeta_1 = \xi_1$, $\zeta_2 = \xi_2 - \alpha_1(\lambda, \xi_1)$, ..., $\zeta_{i-1} = \xi_{i-1} - \alpha_{i-2}(\lambda, \xi_1, \ldots, \xi_{i-2})$.

By introducing the $n(i-1)$ dimensional column vector $\tilde{z}_{i-1} = [\zeta_1^T \ \zeta_2^T \ \ldots \ \zeta_{i-1}^T]^T$, it is easy to infer that $\xi_m = \chi_{m(i-1)}\tilde{z}_{i-1}$ $m = 1, \ldots i-1$, where $\chi_{m(i-1)}(\lambda, \xi_1, \ldots, \xi_{i-1})$ are suitable $[n \times n(i-1)]$ matrices. Define also, for $m = 1, \ldots i-1$

$$\tilde{g}_m = \left(1 + \left\| \sum_{j=1}^{m} \varphi_{mj}(\xi_1, \ldots, \xi_m) \right\| \|\chi_{m(i-1)}\| \right) I_n$$

$$\delta_m = \frac{1}{c_\theta}\tilde{g}_m^{-1} \sum_{j=1}^{m} \varphi_{mj}(\xi_1, \ldots, \xi_m)\theta\chi_{m(i-1)}.$$

Setting $\tilde{x}_{i-1} = (\lambda, \xi_1, \ldots, \xi_{i-1})$, and $\tilde{w}_{i-1} = [\delta_1^T \ldots \delta_{i-1}^T]^T \tilde{z}_{i-1}$, system $S_{i-1}$ can be written as

$$\Sigma_{i-1} : \begin{cases} \dot{\tilde{x}}_{i-1} = F_{i-1}(\tilde{x}_{i-1}) + P_{i-1}v_{i-1} \\ \quad +Q_{i-1}\tau + c_\theta G_{i-1}(\tilde{x}_{i-1})\tilde{w}_{i-1} \\ \tilde{z}_{i-1} = H_{i-1}(\tilde{x}_{i-1}) \end{cases}$$

$$\Delta_{i-1} : \{\tilde{w}_{i-1} = K_{i-1}\tilde{z}_{i-1},$$

where $F_{i-1}(\tilde{x}_{i-1})$, $G_{i-1}(\tilde{x}_{i-1})$, and $H_{i-1}(\tilde{x}_{i-1})$are matrices with smooth entries, $P_{i-1} = [0, \ldots, 0, 1]^T$, $Q_{i-1} = [1, 0, \ldots, 0]^T$, and $K_{i-1}$, with $\|K_{i-1}\| < 1$, is a matrix whose entries are piecewise continuous functions. Let

$$V_{i-1}(\tilde{x}_{i-1}) = (c_\theta^2 - \lambda)^2 + \frac{1}{2}\|\tilde{z}_{i-1}(\tilde{x}_{i-1})\|^2 \quad (11)$$

be a positive definite and proper smooth function, and

$$\begin{aligned} J_{i-1}(v_{i-1}, \tau) = L_{F_{i-1}}V_{i-1} + L_{P_{i-1}v_{i-1}}V_{i-1} + \\ L_{Q_{i-1}\tau}V_{i-1} + \frac{c_\theta^2}{4\gamma^2}\|L_{G_{i-1}}V_{i-1}\|^2 + \|\tilde{z}_{i-1}\|^2. \end{aligned} \quad (12)$$

Assume that, as in (10) for $i-1=1$, $\sigma_1=0$,

**(h1)** $J_{i-1}(\alpha_{i-1},\tau) \le [(c_\theta^2-\lambda)+\sigma_{i-1}\tilde{z}_{i-1}]^T(\rho_{i-1}\tilde{z}_{i-1}-\tau)$,

where $\rho_{i-1}(\cdot)$, $\sigma_{i-1}(\cdot)$ are $\nu+i-1$ dimensional row vectors with smooth entries.

Consider now system $S_i$, which is obtained considering system $S_{i-1}$ setting $v_{i-1}=\xi_i$, together with the dynamics

$$\dot{\xi}_i = v_i + \left(\sum_{j=1}^{i}\varphi_{ij}(\xi_1,\ldots,\xi_{i-1})\theta\xi_j\right).$$

It is straightforward to verify that by letting $\zeta_i=\xi_i-\alpha_{i-1}(\lambda,\xi_1,\ldots,\xi_{i-1})$,and $\tilde{z}_i=\left[\tilde{z}_{i-1}^T\zeta_i^T\right]^T$, there exist suitable $n\times ni$ matrices $\chi_{mi}(\lambda,\xi_1,\ldots,\xi_i)$, $m=1,\ldots i-1$,such that $\xi_m=\chi_{mi}\tilde{z}_i$, $m=1,\ldots i-1$. If it is also defined

$$\tilde{g}_i = \left(1+\left\|\sum_{j=1}^{i}\varphi_{ij}(\xi_1,\ldots,\xi_i)\right\|\|\chi_{ii}\|\right)I_n$$

$$\delta_i = \frac{1}{c_\theta}\tilde{g}_i^{-1}\sum_{j=1}^{m}\varphi_{ij}(\xi_1,\ldots,\xi_i)\theta\chi_{ii}$$

$$\tilde{w}_i = \begin{bmatrix}\tilde{w}_{i-1}\\ \delta_i\tilde{z}_i\end{bmatrix} \tag{13}$$

then system $S_i$ can be written as

$$\Sigma_i : \begin{cases}\dot{\tilde{x}}_i = F_i(\tilde{x}_i)+P_iv_i \\ \qquad +Q_i\tau + c_\theta G_i(\tilde{x}_i)\tilde{w}_i, \\ \tilde{z}_i = H_i(\tilde{x}_i)\end{cases} \tag{14}$$

$$\Delta_i : \{\tilde{w}_i = K_i\tilde{z}_i,$$

where $\tilde{x}_i=(\lambda,\xi_1,\ldots,\xi_i)$, $P_i=[0,\ldots,0,1]^T$, $Q_i=[1,0,\ldots,0]^T$, $|K_i(\cdot)|\le 1$. Consider now the positive definite function

$$V_i = V_{i-1}+\frac{1}{2}\|\xi_i-\alpha_{i-1}\|^2 \tag{15}$$

and the expression

$$J_i(v_i,\tau) = L_{F_i}V_i + L_{P_iv_i}V_i + L_{Q_i\tau}V_i + \frac{c_\theta^2}{i\gamma^2}\|L_{G_i}V_i\|^2+\|\tilde{z}_i\|^2. \tag{16}$$

Set

$$\varrho_i(\lambda,\xi_1,\ldots,\xi_i) = \frac{L_{G_i}(\xi_i-\alpha_{i-1})}{4\gamma^2} \left\{\left[L_{G_i}(\xi_i-\alpha_{i-1})\right]^T(\xi_i-\alpha_{i-1}) \left[L_{G_{i-1}}V_{i-1}\ 0\right]^T\right\},$$

$$\rho_i = \left[\rho_{i-1}\ \varrho_i^T\right]$$

It is straightforward to verify that by choosing $v_i=\alpha_i$, where

$$\alpha_i = -\left[\left(L_{P_{i-1}}V_{i-1}\right)^T + L_{F_{i-1}}\left[(\xi_i-\alpha_{i-1})\right] + \frac{\partial\alpha_{i-1}}{\partial\lambda}\rho_i\tilde{z}_i + \lambda\varrho_i + (\xi_i-\alpha_{i-1}) - \varrho_i(\sigma_{i-1}\tilde{z}_{i-1})\right],$$

$$\sigma_i = \left[\sigma_{i-1}\ \frac{\partial\alpha_{i-1}}{\partial\lambda}\right]$$

expression (16) yields

**(t1)** $J_i(\alpha_i,\tau)\le[(c_\theta^2-\lambda)+\sigma_i\tilde{z}_i]^T(\rho_i\tilde{z}_i-\tau)$.

Previous arguments can be summarized in the following

*Lemma 4.* Consider system $S_{i-1}$, and assume that by a proper choice of $i-1$ smooth functions $\alpha_j=\alpha_j(\lambda,\xi_1,\ldots,\xi_j)$, $j=1,\ldots,i-1$, $\alpha_j(\lambda,0,\ldots,0)=0$, for all $j=1,\ldots,i-1$, all $\lambda\in I\!R$, system $S_{i-1}$ can be described as the feedback interconnection of the two systems $\sum_{i-1}$ and $\Delta_{i-1}$. Assume also that inequality **(h1)** holds.

Then it is possible to find a smooth function $\alpha_i=\alpha_i(\lambda,\xi_1,\ldots,\xi_i)$, $\alpha_i(0,\ldots,0,\lambda)=0$, for all $\lambda\in I\!R$, describing equivalently system $S_i$ as the feedback interconnection of two systems $\sum_i$ and $\Delta_i$, and such that inequality **(t1)** holds.

The arguments above basically show how to construct iteratively the robust controller expressions, and play a fundamental role in establishing the main result.

*Proposition 5.* Let $\gamma\in I\!R^+$, $\gamma<1$. Consider system (1), and suppose that $\theta$ is such that for all $t>0$, $\|\theta(t)\|<c_\theta$. If $c_\theta$ is unknown, then the global regulation with minimal parametrization problem is solvable for the system considered.

**PROOF.** The result has already been shown if $r=1$. If $r>1$ then by the arguments above the hypotheses of lemma 4 hold for $i=2$. Therefore, it can be applied lemma 4 $r-1$ times, the $r-1$-th time setting $v=\xi_{r+1}$, and obtain an iterative procedure for the synthesis of $r$ smooth functions $\alpha_j(\xi_1,\ldots,\xi_j,\lambda)$, $j=1,\ldots,r$ such that system (1) can be written as (14) with $i=r$, with $\|K_r\|<1$. It is straightforward to conclude that the control law $v=\alpha_r(\lambda,\xi_1,\ldots,\xi_r)$, together with the dynamics $\dot{\lambda}=\rho_r\tilde{z}_r$, is such that $J_r(\alpha_r,\rho_r\tilde{z}_r)\le 0$, and the application of the Small Gain Theorem, as stated in proposition 3, proves the result, noting in this case $H_1(\tilde{x}_r)=0\Rightarrow\xi=0$. □

## 3. ELASTIC JOINT ROBOT NONLINEAR GLOBAL SET POINT REGULATION WITH MINIMAL PARAMETRIZATION

It is presented now an example of application of the design strategy developed in previous sections, for the set point regulation of elastic joints robots. The dynamic model of robots with rigid links and elastic joints requires twice the number of coordinates to completely characterize the configuration of rigid bodies constituting the arm. It is referred to the flexible joint robot simplified dynamic equations (Spong 1987)

$$\begin{aligned} &H_l(q_l)\ddot{q}_l + C(q_l,\dot{q}_l)\dot{q}_l + g(q_l) \\ &+K(q_l - q_m) - \theta\dot{q}_l = 0 \\ &H_m\ddot{q}_m - K(q_l - q_m) = u \end{aligned} \tag{17}$$

where $q_l \in \mathbb{R}^n$ and $q_m \in \mathbb{R}^n$ denote respectively the link and motor angular positions, $H_l(q_l)$ and $H_m$ the inertia matrices, $C(q_l,\dot{q}_l)\dot{q}_l$ the centrifugal and Coriolis forces, $g(q_l)$ the gravity acting on the links, $K$ the joint stiffness matrix , $u(t)$ is the applied input torque and $\theta$ is a matrix whose entries are *unknown norm bounded* disturbances. It is assumed $\theta \in \mathcal{B}_\theta$ which will denote the set of matrices whose entries are piecewise continuous and such that for all $t > 0$, $\|\theta(t)\| < c_\theta$, where $c_\theta$ is an unknown positive real number. It is assumed the disturbances $\theta$ to affect the system only when the robot links are moving, which is reasonable for set point regulation. Note however that it is ignored the bound on this disturbance and it is *not required* in our design procedure the disturbance to vanish for $t \to \infty$. It will also be considered in design strategy of the invertibility property of the inertia matrices $H_l(q_l)$ , $H_m$ and of the joint stiffness matrix $K$, together with their products, i.e. terms like $H_l^{-1}(q_l)K$, or $H_m^{-1}K$ *et cetera.* The flexible joint robot model has attracted the attention of the academic community in the last decade, (see Battilotti, *et.al.*, 1995). It has been shown (Tomei, 1991) that a PID control is sufficient to ensure global stabilization when $\theta$ is set equal to zero. To the contrary this property is not proved when the disturbances cannot be neglected. Assume that it is required to regulate robustly in the robot model (17) the link and the motor variables $q_l, q_m$ to desired values $d_l$ and $d_m = d_l + K^{-1}g(d_l)$ achieved using control input $u$ applied at the motor side of the elastic joints. The solution to this problem can be found in the light of the arguments presented in previous section.

### 3.1 *Global feedback equivalence to a "strict feedback form" system*

It will be found now a global state space coordinate transformation $\xi = \Phi[q_{ld}, \dot{q}_l, q_{md}, \dot{q}_m]$, where $q_{ld} = q_l - d_l$, $q_{md} = q_m - d_m$, such that the robot model (17) can be expressed as in (1). This constructive procedure will work also as a pattern to abstract the general strategy to solve the problem consider for all the systems in " strict feedback form ". Set $\xi_1 = \Phi_1(q_{ld}) = q_{ld}$, $\xi_2 = \Phi_2(q_{ld},\dot{q}_l) = \dot{q}_l$, and compute $\dot{\xi}_2 = \ddot{q}_l$ from system (17), obtaining

$$\begin{aligned} \dot{\xi}_2 = &\varphi_3(q_{ld},\dot{q}_l) + H_l^{-1}(q_{ld}+d_l)Kq_{md} \\ &+H_l^{-1}(\xi_1+d_l)\theta\xi_2 \end{aligned}$$

where

$$\begin{aligned} \varphi_3(q_{ld},\dot{q}_l) = &-H_l^{-1}(q_{ld}+d_l)\left[C(q_{ld}+d_l,\dot{q}_l)\dot{q}_l\right. \\ &\left.+g(q_{ld}+d_l) + K(q_{ld} - K^{-1}g(d_l))\right]. \end{aligned}$$

Now define

$$\begin{aligned} \xi_3 &= \Phi_3(q_{ld},\dot{q}_l,q_{md}) \\ &= \varphi_3(q_{ld},\dot{q}_l) + H_l^{-1}(q_{ld}+d_l)Kq_{md} \end{aligned}$$

let's differentiate $\xi_3$ with respect to the time, obtaining

$$\begin{aligned} \dot{\xi}_3 = &\varphi_4(q_{ld},\dot{q}_l,q_{md}) + H_l^{-1}(q_{ld}+d_l)K\dot{q}_m \\ &+\frac{\partial\varphi_3}{\partial\dot{q}_l}H_l^{-1}(q_{ld}+d_l)\theta\xi_2 \end{aligned}$$

where

$$\begin{aligned} \varphi_4(q_{ld},\dot{q}_l,q_{md}) = &\frac{\partial\varphi_3}{\partial q_{ld}}\dot{q}_l + \frac{\partial\varphi_3}{\partial\dot{q}_l}\Phi_3(q_{ld},\dot{q}_l,q_{md}) \\ &+\sum_{i=1}^{n}\left[\frac{\partial H_l^{-1}(q_{ld}+d_l)}{\partial q_{ldi}}K\right]q_{md}. \end{aligned}$$

Let

$$\begin{aligned} \xi_4 &= \Phi_4(q_{ld},\dot{q}_l,q_{md},\dot{q}_m) \\ &= \varphi_4(q_{ld},\dot{q}_l,q_{md}) + H_l^{-1}(q_{ld}+d_l)K\dot{q}_m. \end{aligned}$$

Is straightforward to verify that

$$\begin{bmatrix} \xi_1 \\ \xi_2 \\ \xi_3 \\ \xi_4 \end{bmatrix} = \begin{bmatrix} \Phi_1(q_{ld}) \\ \Phi_2(q_{ld},\dot{q}_l) \\ \Phi_3(q_{ld},\dot{q}_l,q_{md}) \\ \Phi_4(q_{ld},\dot{q}_l,q_{md},\dot{q}_m) \end{bmatrix}$$

defines a global diffeomorphism between the variables $(\xi_1,\xi_2,\xi_3,\xi_4)$ and the variables $(q_{ld},\dot{q}_l,q_{md},\dot{q}_m)$ for all $d_l \in \mathbb{R}^n$.

In fact

$$\begin{aligned} q_{ld} &= \xi_1 \\ \dot{q}_l &= \xi_2 \\ q_{md} &= -\varphi_3(\xi_1,\xi_2) + K^{-1}H_l(\xi_1+d_l) \\ \dot{q}_m &= -\varphi_4(\xi_1,\xi_2,\xi_3) + K^{-1}H_l(\xi_1+d_l). \end{aligned}$$

The differentiation of $\xi_4$ with respect to the time yields

$$\dot{\xi}_4 = \varphi_5(q_{ld}, \dot{q}_l, q_{md}, \dot{q}_m) + H_l^{-1}(q_{ld}+d_l)K\ddot{q}_m + \frac{\partial \varphi_4}{\partial \dot{q}_l} H_l^{-1}(q_{ld}+d_l)\theta\xi_2$$

where

$$\varphi_5(q_{ld}, \dot{q}_l, q_{md}, \dot{q}_m) = \frac{\partial \varphi_4}{\partial q_{ld}}\dot{q}_l + \frac{\partial \varphi_4}{\partial \dot{q}_l}\Phi_3(q_{ld}, \dot{q}_l, q_{md}) \frac{\partial \varphi_4}{\partial q_{md}}\dot{q}_m + \sum_{i=1}^{n}\left[\frac{\partial H_l^{-1}(q_{ld}+d_l)}{\partial q_{ldi}}K\right]\dot{q}_m.$$

Let's substitute now $\ddot{q}_m$ from the second model identity in (17)

$$\ddot{q}_m = H_m^{-1}K(q_{ld} - q_{md} - K^{-1}g(d_l)) + H_m^{-1}u$$

in $\dot{\xi}_4$ obtaining

$$\begin{aligned}\dot{\xi}_4 = &\varphi_5(q_{ld}, \dot{q}_l, q_{md}, \dot{q}_m)\\ &+H_l^{-1}(q_{ld}+d_l)KH_m^{-1}K(q_{ld} - q_{md} - K^{-1}g(d_l))\\ &+\frac{\partial \varphi_4}{\partial \dot{q}_l}H_l^{-1}(q_{ld}+d_l)\theta\xi_2\\ &+H_l^{-1}(q_{ld}+d_l)KH_m^{-1}u.\end{aligned}$$

Let's introduce now the static feedback

$$\begin{aligned}u = &-K(q_{ld} - q_{md} - K^{-1}g(d_l)) \qquad (18)\\ &-H_mK^{-1}H_l(q_{ld}+d_l)\left[\varphi_5(q_{ld}, \dot{q}_l, q_{md}, \dot{q}_m) + v_4\right]\end{aligned}$$

yielding

$$\dot{\xi}_4 = v_4 + \frac{\partial \varphi_4}{\partial \dot{q}_l}H_l^{-1}(q_{ld}+d_l)\theta\xi_2.$$

Note that system (17) together with the feedback (18) can be expressed in the new coordinates $\xi$ as

$$\begin{aligned}\dot{\lambda} &= \tau\\ \dot{\xi}_1 &= \xi_2\\ \dot{\xi}_2 &= \xi_3 + H_l^{-1}(\xi_1 + d_l)\theta\xi_2\\ \dot{\xi}_3 &= \xi_4 + \frac{\partial \varphi_3}{\partial \dot{q}_l}H_l^{-1}(q_{ld}+d_l)\theta\xi_2 \qquad (19)\\ \dot{\xi}_4 &= v_4 + \frac{\partial \varphi_4}{\partial \dot{q}_l}H_l^{-1}(q_{ld}+d_l)\theta\xi_2.\end{aligned}$$

Observe that system above is a special instance of a "strict feedback form" system (1), and the procedure proposed in this note robustly regulates the plant with minimal parametrization.

## 4. CONCLUSION

It is generally assumed that the robust regulation of most nonlinear systems affected by disturbances with an unknown norm bound requires a dynamic regulator whose dimension is at least equal to the dimension of the state space. There are cases, anyway, when global regulation can be achieved more efficiently. It has been shown that for systems in strict feedback form (1) this can be achieved simply with by an integrator and a smooth nonlinear mapping. Note also there are classes of mechanical systems to which this design strategy can be applied, like in the problem of the set point global regulation of an elastic joint robot.

## REFERENCES


Battilotti, S. and Lanari, L. (1995)."Adaptive Disturbance Attenuation for Elastic Joint Robots" Proc. 34-th CDC Conf. pp. 2823-2828, New Orleans.

Hill, D. , and P. Moylan, (1976). " The Stability of Nonlinear Dissipative Systems ". *IEEE Trans. Aut. Control*, Vol. 21:pp. 377-382.

Isidori, A. (1995). *Nonlinear Control Systems.* Springer-Verlag, New York, 3-rd edition.

Krstic, M., I. Kanellakopoulos and P. V. Kokotovic (1992). "Adaptive Nonlinear Control without Overparametrization". *Systems and Control Letters* ,Vol.19: pp. 177–185.

Krstic, M., I. Kanellakopoulos and P. V. Kokotovic (1995). *Nonlinear and Adaptive Control Design.* John Wiley and Sons, New York.

Marino, R., and P. Tomei (1993). "Robust Stabilization of Feedback Linearizable Time-varying Uncertain Nonlinear Systems ". *Automatica* ,Vol. 29, N. 1, pp. 181-189.

Marino, R., and P. Tomei (1995). *Nonlinear Control Design - Geometric, Adaptive and Robust.* Prentice Hall, Hemel Hempstead.

Santosuosso, G. L. (1995). "Nonlinear Almost Disturbance Decoupling and Adaptive Stabilization without Overparametrization" *Proc. 34-th CDC Conf.* pp. 2970-2975, New Orleans.

Santosuosso, G. L. (1996). "Robust Regulation with Minimal Paramerization via a "Small Gain Approach" for a Class of Nonlinear Systems." *Proc. 13th World IFAC Congr.* Vol. E, pp. 443-448. San Francisco.

Spong, M. W. (1987) "Modeling and Control of Elastic Joint Robots", Trans. of the ASME, Vol. 109, pp. 310-319.

Van Der Shaft, A. (1992). $L_2$ Gain Analysis of Nonlinear Systems and Nonlinear State Feedback $H_\infty$ Control. *IEEE Trans. Aut. Control.*, Vol. 37 :pp. 770-784.

Willems, J. (1972). " Dissipative Dynamical Systems. - Part I : General Theory ". *Arch. Rational Mechanics and Analysis*, Vol. 45: pp. 321-351.

Tomei, P. (1991). "A simple PD Controller for Robots with Elastic Joints". IEEE Trans. Automat. Control ,Vol. 36: pp. 1208-1213.

F-2e-01-2
14th Triennial World Congress, Beijing, P.R. China

# ROBUST OUTPUT FEEDBACK STABILIZATION OF THE ANGULAR VELOCITY OF A RIGID BODY

**F. Mazenc and A. Astolfi**

*Department of Electrical and Electronic Engineering*
*and Centre for Process Systems Engineering,*
*Imperial College of Science, Technology and Medicine,*
*London SW7 2BY, UK*
*FAX:+44-171-5946606, E-mail:{f.mazenc,a.astolfi}@ic.ac.uk*

Abstract: The problem of semi-global stabilization of the angular velocity of an under-actuated rigid body in the presence of model errors is addressed and solved using a smooth, time-varying, dynamic, output feedback control law. The proposed scheme improves some of the existing results and provides guidelines for a general stabilization strategy applicable to systems which are not exponentially stabilizable. Simulations results are included. *Copyright ©1999 IFAC*

Keywords: Output feedback, Euler's equations, Semi-global stabilization.

## 1. INTRODUCTION

The issue of state feedback stabilization of the angular velocity of a rigid body with two control torques has been studied and solved using various approaches. The design method proposed in (Aeyels 1985*b*) relies on the center manifold theory, whereas the solution in (Outbib and Sallet 1992) makes use of control Lyapunov functions and La Salle invariance principle. Moreover, in (Aeyels 1985*a*) a control law which is robust against errors on the principal moments of inertia is presented. Robustness results are also discussed in (Morin 1996), where the exploitation of the homogeneity properties of the system and the use of homogeneous control laws lead to the design of a state feedback control law yielding robustness with respect to errors in the principal moments of inertia and in the location of the actuators. Finally, a somewhat different approach is pursued in (Astolfi and Rapaport 1998), where robustness against external disturbances is studied. We conclude that a lot of attention has been devoted to the stabilization problem for the angular velocity of a rigid body in the event of actuator failure. On the other hand, the *dual* problem, *i.e.* the stabilization in the event of sensor malfunction, has not been studied with the same interest.

To the best of our knowledge such a problem has been only recently studied in (Astolfi 1999), where a hybrid control law yielding exponential convergence has been derived using Lyapunov techniques and exploiting the fact that the system is linear in the unmeasured states. The result in (Astolfi 1999) presents two main limitations. First the measured states have to fulfill a strong structural assumption, namely the measured velocities are the ones about the principal axes with the smallest and the largest moments of inertia. Such restriction cannot be removed using the approach in (Astolfi 1999). Secondly, the feedback law requires the precise knowledge of the moments of inertia of the rigid body, and no robustness bound can be derived. In the present paper, we improve in some respects the aforementioned results. We design a time-varying, dynamic, output feedback control law for the rigid body with two controls and partial state information. The present construction, which uses completely different arguments from that in (Astolfi 1999), yields on one side weaker results, as only semi-global stability can be proved, but on the other hand, stronger result, as the proposed control law is smooth and is able to counteract model errors analogous to those considered in (Morin 1996). The leading idea of our design is as follows. We solve an output tracking problem for a bounded trajectory which goes to the origin when the time goes to infinity. It must be noticed that, even locally, such problem is non-trivial as the zero-dynamics of the system with the considered output are stable, but not asymptotically stable, hence the classical results of (Isidori 1995, Section 4.5), relying on partial output linearization, do not apply. Finally, we emphasize that the stabilization tool used in this work does not exploit the homogeneity property of the system to be controlled and does not make use of the fact that the system is linear in the unmeasured states. As a result, both asymptotic stabilization via partial state information and robustness properties are obtained. However, this does not hold globally.

*Remark 1.* It must be observed that the proposed construction has a general nature, *i.e.* it might be used to solve stabilization problems for classes of systems. These include nonholonomic chained forms, see (Murray and Sastry 1993) for the definition, and nonholonomic canonical systems with maximum growth, see (Murray and Sastry 1993) for the definition and (Mazenc and Astolfi n.d.) where the construction presented in this work is used, together with back-stepping and forwarding arguments, to asymptotically stabilize a local description of the *ball and plate* system (Brockett and Dai 1992).

**Preliminaries.**

(1) Throughout the paper, the symbol $c$ is used to denote generically a strictly positive number. For instance, we may write: $c + 3c^2 + 1 = c$.
(2) By $\dot{f}(t)$ we denote the first derivative of the function of time $f(t)$.

## 2. PROBLEM STATEMENT

Consider a rigid body in an inertial reference frame and let $\Omega_1$, $\Omega_2$ and $\Omega_3$ denote the angular velocity components along a body fixed reference frame having the origin at the center of gravity and consisting of the three principal axes. The Euler's equations for the rigid body, subject to two independent controls aligned with two principal axes, are (see (Aeyels and Szafranski 1988))

$$\begin{cases} \dot{\Omega}_3 = a_3\,\Omega_1\Omega_2 \\ \dot{\Omega}_2 = a_2\,\Omega_1\Omega_3 + b_2u_2\ ,\ b_2 \neq 0 \\ \dot{\Omega}_1 = a_1\,\Omega_2\Omega_3 + b_1u_1\ ,\ b_1 \neq 0 \end{cases} \tag{1}$$

where $(u_1, u_2)$ are the controls and the $a_i$'s and the $b_i$'s are constant parameters describing the inertial properties of the rigid body. If $a_3 = 0$, system (1) is not controllable, moreover, if $a_2 = a_1 = 0$, then $a_3 = 0$ as well. Hence, without lack of generality, we assume $a_3 \neq 0$ and $a_2 \neq 0$. If the rigid body possesses a symmetry axis then $a_1 = 0$ and $a_3$ and $a_2$ do not have the same sign (see (Outbib and Sallet 1992)), whereas in the non-symmetric case, all the $a_i$'s are nonzero and the $a_i$'s do not have the same sign. As a result we may assume, without loss of generality, that $a_3a_2 < 0$. Therefore, every rigid body with two controls can be described by equations of the form

$$\begin{cases} \dot{\omega}_3 = -\,\omega_1\omega_2 \\ \dot{\omega}_2 = \omega_1\omega_3\, +\, v_2 \\ \dot{\omega}_1 = s\,\omega_2\omega_3\, +\, v_1 \end{cases} \tag{2}$$

with $s \in \{-1,0,1\}$, $\omega_1 = \sqrt{-a_2a_3}\Omega_1$, $\omega_2 = \sqrt{|a_1a_3|}\Omega_2$, $\omega_3 = -|a_3|/a_3\sqrt{|a_1a_2|}\Omega_3$, $v_1 = \sqrt{-a_2a_3}b_1u_1$ and $v_2 = \sqrt{|a_1a_3|}b_2u_2$.

However, when there are errors on the principal moments of inertia and on the location of the actuators, as shown in (Morin 1996), the angular velocities of the rigid body are described by equations of the form

$$\begin{cases} \dot{\Omega}_3 = (a_3 + \mu_7)\Omega_1\Omega_2 + \mu_1u_1 + \mu_2u_2 \\ \dot{\Omega}_2 = (a_2 + \mu_8)\Omega_1\Omega_3 + b_2u_2 + \mu_3u_1 + \mu_4u_2 \\ \dot{\Omega}_1 = (a_1 + \mu_9)\Omega_2\Omega_3 + b_1u_1 + \mu_5u_1 + \mu_6u_2 \end{cases} \tag{3}$$

where the $\mu_i$'s are unknown parameters depending on the *size* of the model errors. Applying to (3) the change of coordinates and the change of feedback resulting in (2), we deduce that the equations of the rigid body with errors on the principal moments of inertia and on the location of the actuators can be described by the equations

$$\begin{cases} \dot{\omega}_3 = -\omega_1\omega_2 + \kappa_3(\omega_1,\omega_2,\omega_3) + \mu_3(v_1,v_2) \\ \dot{\omega}_2 = \omega_1\omega_3 + v_2 + \kappa_2(\omega_1,\omega_2,\omega_3) + \mu_2(v_1,v_2) \\ \dot{\omega}_1 = s\,\omega_2\omega_3 + v_1 + \kappa_1(\omega_1,\omega_2,\omega_3) + \mu_1(v_1,v_2) \end{cases} \tag{4}$$

where $s$ is unknown, but $s \in [-1,1]$, and the functions $\kappa_i(\omega_1,\omega_2,\omega_3)$ and $\mu_i(v_1,v_2)$, $i = 1,2,3$ are unknown but satisfy, for some $\varepsilon > 0$, the following bounds

$$\sum_{i=1}^{3} |\kappa_i(\cdot)| \leq \varepsilon\left(|\omega_1||\omega_2| + |\omega_1||\omega_3| + |\omega_2||\omega_3|\right) \tag{5}$$

and

$$\sum_{i=1}^{3} |\mu_i(\cdot)| \ \leq\ \varepsilon(|v_1| + |v_2|). \tag{6}$$

The objective of the present paper is the asymptotic stabilization of the origin of system (4) by means of an output feedback control law which makes use of the output $y_a = (\omega_3, \omega_1)^T$ or $y_b = (\omega_2, \omega_1)^T$. The case $y_c = (\omega_3, \omega_2)^T$ can be dealt with in an analogous manner and will not be discussed in detail.

## 3. MAIN RESULT

*Theorem 3.1.* Consider the system (4) with $y_a$ or $y_b$ as output. Assume that (5) and (6) are satisfied. Then there exists $\bar{\varepsilon} > 0$ such that if $\varepsilon \in [0, \bar{\varepsilon}]$ the system (4) is semi-globally stabilizable by dynamic time-varying output feedback.

*Remark 2.* It must be noted that an explicit estimate for $\bar{\varepsilon}$ can be determined from the arguments contained in the part (ii) of the proof. However, for simplicity we leave the details of the derivation.

*Proof.* The proof breaks up in two parts. First, we construct an output feedback stabilizing control law for the unperturbed system, *i.e.* system (4) with $\kappa_i(\cdot) = 0$ and $\mu_i(\cdot) = 0$ for $i = 1, 2, 3$, then we study the robustness properties of system (4) when this control is applied.

*(i) Design of a stabilizing control law for the nominal system.*

Consider the new time dependent coordinate

$$z = \omega_1 - \gamma(t), \tag{7}$$

where $\gamma(t)$ is a smooth function of time to be specified. As a result, system (4) with $\kappa_i(\cdot) = 0$ and $\mu_i(\cdot) = 0$ for $i = 1, 2, 3$ is described by the equations

$$\begin{cases} \dot{\omega}_3 = -\gamma(t)\omega_2 - z\omega_2 \\ \dot{\omega}_2 = \gamma(t)\omega_3 + z\omega_3 + v_2 \\ \dot{z} = -\dot{\gamma}(t) + s\,\omega_2\omega_3 + v_1. \end{cases} \tag{8}$$

Consider the reduced order observer

$$\begin{cases} \dot{\widehat{\omega}}_3 = -\gamma(t)\widehat{\omega}_2 - z\widehat{\omega}_2 + \theta_3 \\ \dot{\widehat{\omega}}_2 = \gamma(t)\widehat{\omega}_3 + z\widehat{\omega}_3 + v_2 + \theta_2, \end{cases} \tag{9}$$

where the functions $\theta_i$, $i = 2, 3$ are to be specified, and the error coordinates $\widetilde{\omega}_3 = \omega_3 - \widehat{\omega}_3$ and $\widetilde{\omega}_2 = \omega_2 - \widehat{\omega}_2$. Simple calculations yield

$$\begin{cases} \dot{\widetilde{\omega}}_3 = -\gamma(t)\widetilde{\omega}_2 - z\omega_2 + z\widehat{\omega}_2 - \theta_3 \\ \dot{\widetilde{\omega}}_2 = \gamma(t)\widetilde{\omega}_3 + z\omega_3 - z\widehat{\omega}_3 - \theta_2. \end{cases} \tag{10}$$

Let

$$\begin{aligned} v_1 &= -k(z)z + \dot{\gamma}(t) \\ v_2 &= -z\widehat{\omega}_3 - \gamma(t)\widehat{\omega}_2 - \theta_2, \end{aligned} \tag{11}$$

where $k(z)$ is a strictly positive function to be specified. As a result the closed loop system is described by the equations

$$\begin{cases} \dot{\widehat{\omega}}_3 = -\gamma(t)\widehat{\omega}_2 - z\widehat{\omega}_2 + \theta_3 \\ \dot{\widehat{\omega}}_2 = \gamma(t)\widehat{\omega}_3 - \gamma(t)\widehat{\omega}_2 \\ \dot{\widetilde{\omega}}_3 = -\gamma(t)\widetilde{\omega}_2 - z\widetilde{\omega}_2 - \theta_3 \\ \dot{\widetilde{\omega}}_2 = \gamma(t)\widetilde{\omega}_3 + z\widetilde{\omega}_3 - \theta_2 \\ \dot{z} = -k(z)z + s(\widehat{\omega}_2 + \widetilde{\omega}_2)(\widehat{\omega}_3 + \widetilde{\omega}_3). \end{cases} \tag{12}$$

We now consider separately the case with output $y_a$ and the case with output $y_b$.

*Case 1.* The output is $y_a = (\omega_3, \omega_1)^T$. Setting $\theta_3 = \gamma(t)\widetilde{\omega}_3$ and $\theta_2 = z\widetilde{\omega}_3$, the control laws become

$$\begin{aligned} v_1 &= -k(z)z + \dot{\gamma}(t) + \widehat{\omega}_2\widehat{\omega}_3 \\ v_2 &= -\gamma(t)\widehat{\omega}_2 - z\widehat{\omega}_3 + z\widetilde{\omega}_3 \end{aligned} \tag{13}$$

and system (12) simplifies as follows

$$\begin{cases} \dot{\widehat{\omega}}_3 = -\gamma(t)\widehat{\omega}_2 - z\widehat{\omega}_2 + \gamma(t)\widetilde{\omega}_3 \\ \dot{\widehat{\omega}}_2 = \gamma(t)\widehat{\omega}_3 - \gamma(t)\widehat{\omega}_2 \\ \dot{\widetilde{\omega}}_3 = -\gamma(t)\widetilde{\omega}_2 - \gamma(t)\widetilde{\omega}_3 - z\widetilde{\omega}_2 \\ \dot{\widetilde{\omega}}_2 = \gamma(t)\widetilde{\omega}_3 \\ \dot{z} = -k(z)z + s(\widehat{\omega}_2 + \widetilde{\omega}_2)(\widehat{\omega}_3 + \widetilde{\omega}_3). \end{cases}$$

*Case 2.* The output is $y_b = (\omega_2, \omega_1)^T$. Setting $\theta_3 = z\widetilde{\omega}_2$ and $\theta_2 = \gamma(t)\widetilde{\omega}_2$ the control laws become

$$\begin{aligned} v_1 &= -k(z)z + \dot{\gamma}(t) \\ v_2 &= -z\widehat{\omega}_3 - \gamma(t)\widehat{\omega}_2 - \gamma(t)\widetilde{\omega}_2 \end{aligned} \tag{14}$$

and system (12) simplifies as follows

$$\begin{cases} \dot{\widehat{\omega}}_3 = -\gamma(t)\widehat{\omega}_2 + z\widetilde{\omega}_2 \\ \dot{\widehat{\omega}}_2 = \gamma(t)\widehat{\omega}_3 - \gamma(t)\widehat{\omega}_2 \\ \dot{\widetilde{\omega}}_3 = -\gamma(t)\widetilde{\omega}_2 \\ \dot{\widetilde{\omega}}_2 = -\gamma(t)\widetilde{\omega}_2 + \gamma(t)\widetilde{\omega}_3 + z\widetilde{\omega}_3 \\ \dot{z} = -k(z)z + s(\widehat{\omega}_2 + \widetilde{\omega}_2)(\widehat{\omega}_3 + \widetilde{\omega}_3). \end{cases}$$

In the remainder of the proof we consider only the case where the output is $y_a$. An analogous demonstration can be carried out for the output $y_b$.

Let $\gamma(t)$ be a bounded, strictly positive and decreasing function. Consider the two positive definite and radially unbounded functions

$$P(l, m) = l^2 + m^2 + lm$$

and

$$Q(l, m) = P(-l, m)$$

and the function

$$U(\widehat{\omega}_3, \widehat{\omega}_2, \widetilde{\omega}_3, \widetilde{\omega}_2, z) = P(\widetilde{\omega}_3, \widetilde{\omega}_2) + z^2 + \lambda Q(\widehat{\omega}_3, \widehat{\omega}_2),$$

where $\lambda$ is a strictly positive parameter to be specified. It must be noted that this function is positive definite, radially unbounded, zero at zero and differentiable. Using the triangular inequality, one has

$$\dot{U} \le -\frac{3}{4}\gamma(t)P(\tilde{\omega}_3,\tilde{\omega}_2) - \frac{3}{4}\lambda\gamma(t)Q(\widehat{\omega}_3,\widehat{\omega}_2)$$
$$+\lambda c\gamma(t)P(\tilde{\omega}_3,\tilde{\omega}_2) - 2k(z)z^2$$
$$+c|z|\left(P(\tilde{\omega}_3,\tilde{\omega}_2) + Q(\widehat{\omega}_3,\widehat{\omega}_2)\right).$$

Choosing

$$\lambda \in \left]0, \frac{1}{1+4c\sup_{t\ge 0}\gamma(t)}\right],$$

using Joung's inequality and setting $k(z) = k_1(|z|+1)$ with $k_1 \ge 4$, yields

$$\dot{U} \le -\frac{1}{2}\gamma(t)P(\tilde{\omega}_3,\tilde{\omega}_2) - \frac{3}{4}\lambda\gamma(t)Q(\widehat{\omega}_3,\widehat{\omega}_2)$$
$$-k(z)z^2 + \frac{c}{k_1}[P(\tilde{\omega}_3,\tilde{\omega}_2) + Q(\widehat{\omega}_3,\widehat{\omega}_2)]^{\frac{3}{2}}.$$

As a result

$$\dot{U} \le \left[-\frac{1}{2}\gamma(t) + \frac{c}{k_1}\sqrt{U(t)}\right]U(t). \quad (15)$$

This inequality allows to establish the semi-global asymptotic stabilizability of the undisturbed system (4) if $\gamma(t)$ is properly selected. Many choices are possible and this flexibility is important from a qualitative and practical point of view to ensure a good convergence. In what follows we propose one such selection.

*Lemma 3.2.* Let $R \ge 5000$ and

$$E_R = \{(\widehat{\omega}_3,\widehat{\omega}_2,\tilde{\omega}_3,\tilde{\omega}_2,z) \mid\mid U(\cdot) \le R\}.$$

Consider an initial condition $(\widehat{\omega}_3,\widehat{\omega}_2,\tilde{\omega}_3,\tilde{\omega}_2,z)$ belonging to $E_R$. Let

$$\gamma(t) = \frac{4\gamma_0}{1+t},$$

where $\gamma_0 = \frac{1}{32}\sqrt{R}$ and $k_1 \ge 64(c+1)$, where $c$ is the positive number $c$ in equation (15).

Then for all $t \ge 0$ the inequality

$$\dot{U} \le -\frac{\gamma_0}{2(1+t)}U(t) \quad (16)$$

holds.

Inequality (16) yields

$$\lim_{t\to+\infty} U(t) = 0.$$

It follows that every trajectory converges to the origin. However, since the function $\gamma(t)$ depends on $R$, we do not yet have proved semi-global asymptotic stability. To complete the proof, consider $\mathcal{R} > 1$ and a trajectory with an initial condition $(\omega_1,\omega_2,\omega_3,\widehat{\omega}_2,\widehat{\omega}_3)$ such that

$$U(\widehat{\omega}_3,\widehat{\omega}_2,\omega_3,\omega_2,\omega_1) \le \mathcal{R}.$$

We now show that for a suitable choice of feedback the considered trajectory converges to the origin. To begin with observe that the following inequality

$$U(\widehat{\omega}_3,\widehat{\omega}_2,\tilde{\omega}_3,\tilde{\omega}_2,z) \le c\mathcal{R} + 2\gamma_0^2, \quad c \ge 1$$

is satisfied. Let $R = \frac{512}{511}c\mathcal{R}$. According to Lemma 3.2, if $\mathcal{R}$ is sufficiently large then for some feedback and $\gamma_0 = \frac{1}{32}\sqrt{R}$, the set $E_R$ is contained within the basin of attraction of the corresponding closed-loop system. Since $\gamma_0 = \frac{1}{32}\sqrt{R}$, we have $R = c\mathcal{R} + 2\gamma_0^2$. It then follows that the system (4) is semi-globally asymptotically stabilized by the family of control laws (13).

*(ii) Robustness analysis.*

In this part, we prove that the family of feedback (13) semi-globally asymptotically stabilizes (4) when disturbances satisfying the bounds (5) and (6) with $\varepsilon \in [0,\bar{\varepsilon}]$ are present.

*Step 1.* Consider the functions $\kappa_i(\omega_1,\omega_2,\omega_3)$, $i = 1,2,3$. Using inequality (15) and the bound (5), one has

$$\dot{U} \le \left[-\frac{1}{2}\gamma(t) + \frac{c}{k_1}\sqrt{U(t)}\right]U(t) + c\varepsilon\gamma(t)U(t)$$
$$+c\varepsilon\sqrt{U(t)}\left[|\omega_1||\omega_2| + |\omega_1||\omega_3| + |\omega_2||\omega_3|\right].$$

Therefore, if $\varepsilon \in \left[0, \frac{1}{16c}\right]$, we conclude

$$\begin{aligned}\dot{U} &\le \left[-\frac{3}{8}\gamma(t) + \left(\frac{c}{k_1} + c\varepsilon\right)\sqrt{U(t)}\right]U(t) \\ &\le \left[-\frac{3}{8}\gamma(t) + \frac{1}{8}\sqrt{U(t)}\right]U(t).\end{aligned} \quad (17)$$

*Step 2.* We now include also the functions $\mu_i(\cdot)$, $i = 1,2,3$. According to the bound (6), and using the inequality (17), we obtain

$$\dot{U} \le \left[-\frac{3}{8}\gamma(t) + \frac{1}{8}\sqrt{U(t)}\right]U(t) + k(z)c\varepsilon\sqrt{U(t)}|z|$$
$$+c\varepsilon\gamma(t)U(t) + c\varepsilon\sqrt{U(t)}|\dot{\gamma}(t)| + c\varepsilon U(t)^{\frac{3}{2}}.$$

As a result, if $\varepsilon \in \left[0, \frac{1}{16c}\right]$,

$$\dot{U} \le \left[-\frac{1}{4}\gamma(t) + \frac{3}{16}\sqrt{U(t)}\right]U(t)$$
$$+k(z)c\varepsilon\sqrt{U(t)}|z| + c\varepsilon\sqrt{U(t)}|\dot{\gamma}(t)|.$$

Moreover, since $k(z)$ does not depend on $R$, we obtain, if $k_1$ is sufficiently large

$$\dot{U} + \widehat{|z|^{\frac{3}{2}}} \le \left[-\frac{1}{4}\gamma(t) + \frac{1}{4}\sqrt{U(t)}\right]U(t)$$
$$-\frac{1}{2}k(z)|z|^{\frac{3}{2}} + c\varepsilon\sqrt{U(t)}|\dot{\gamma}(t)|.$$

It follows that the positive definite and radially unbounded function

$$\mathcal{U}(\widehat{\omega}_3,\widehat{\omega}_2,\tilde{\omega}_3,\tilde{\omega}_2,z) = U(\widehat{\omega}_3,\widehat{\omega}_2,\tilde{\omega}_3,\tilde{\omega}_2,z) + |z|^{\frac{3}{2}}$$

satisfies

$$\dot{\mathcal{U}} \leq \left[-\frac{1}{4}\gamma(t) + \frac{1}{4}\sqrt{\mathcal{U}(t)}\right]\mathcal{U}(t) + c\varepsilon\sqrt{\mathcal{U}(t)}|\dot{\gamma}(t)|.$$

Finally, the selection $\gamma(t) = \frac{4\gamma_0}{1+t}$, yields

$$\dot{\mathcal{U}} \leq \left[-\frac{\gamma_0}{1+t} + \frac{1}{2}\sqrt{\mathcal{U}(t)}\right]\mathcal{U}(t) + c\varepsilon\frac{\gamma_0^{\frac{3}{2}}}{(1+t)^3}$$

and as long as

$$\frac{\gamma_0}{1+t} \geq 4\sqrt{\mathcal{U}(t)}$$

$$\overbrace{\mathcal{U}(t) - \frac{\delta}{(1+t)^2}}^{\cdot} \leq -\frac{\gamma_0}{2(1+t)}\left[\mathcal{U}(t) - \frac{\delta}{(1+t)^2}\right] \quad (18)$$

where

$$\delta = -\frac{c\varepsilon\gamma_0^{\frac{3}{2}}}{\gamma_0 - 2}.$$

Observe now that for $\gamma_0$ sufficiently large $|\delta| \leq c\varepsilon\sqrt{\gamma_0}$. Moreover, if $\varepsilon$ is smaller than a constant number independent of $R$, and $\gamma_0$ is sufficiently large, then, using arguments similar to those invoked to prove Lemma 3.2, we conclude semi-global stabilizability of system (4) with model errors satisfying (5) and (6) directly from inequality (18). This concludes our proof. ◁

*Remark 3.* It is worth mentioning that there are many possible choices for $\gamma(t)$. For instance

$$\gamma(t) = \gamma_0\left[\rho(t) + \frac{1-\rho(t)}{1+t}\right]$$

with $\rho(t) = 1$ if $t \leq \chi(R)$ and $\rho(t) = e^{-t+\chi(R)}$ if $t \geq \chi(R)$ where $\chi(\cdot)$ is an increasing function suitably chosen is a possible alternative choice.

## 4. SIMULATIONS

In this section some simulations, showing the performance and the robustness properties of the proposed design, are presented. We consider the system (4) with $s = -1$ and output $y_a = (\omega_3, \omega_1)^{\mathsf{T}}$. The parameters $\eta_1$ through $\eta_9$ are randomly generated variables with zero mean and variance equal to 0.3. The control law has been built according to the result proved in Lemma 3.2 with $\gamma_0 = 1$.

Figure 1 displays the state variables $\omega_1, \omega_2, \omega_3, \hat{\omega}_2$ and $\hat{\omega}_3$, and the control actions $v_1$ and $v_2$, from the initial state[1] $(1, -1, 1, 0, 0)$ in the nominal case, *i.e.* $\eta_i = 0$ for $i = 1, \ldots, 9$; whereas Figures 2 and 3 show two simulation results for nonzero $\eta$'s. Observe the convergence of the trajectories of the closed loop system despite the presence of substantial model errors.

[1] The controller states are initialized at $(\hat{\omega}_2(0), \hat{\omega}_3(0)) = (0, 0)$.

## 5. CONCLUDING REMARKS

The problem of robust output feedback stabilization of the angular velocity of a rigid body has been addressed and solved via a dynamic, time-varying control law. The proposed result improves in several respects existing ones and provides a new tool for stabilization of systems with non-stabilizable linearization. Various problems are left open. First, the extension of the proposed semi-global stabilizer to a global stabilizer and secondly the much more challenging problem of output feedback stabilization with one control torque and one measured state.

## 6. REFERENCES


Aeyels, D. (1985*a*). Stabilization by smooth feedback of the angular velocity of a rigid body. *Systems and Control Letters* **5**, 59–63.

Aeyels, D. (1985*b*). Stabilization of a class of nonlinear systems by smooth feedback control. *Systems and Control Letters* **5**, 289–294.

Aeyels, D. and M. Szafranski (1988). Comments on the stabilizability of the angular velocity of a rigid body. *Systems and Control Letters* **10**, 35–39.

Astolfi, A. (1999). Output feedback control of the angular velocity of a rigid body. *Systems and Control Letters (to appear).*

Astolfi, A. and A. Rapaport (1998). Robust stabilization of the angular velocity of a rigid body. *Systems and Control Letters* pp. 257–264.

Brockett, R. W. and L. Dai (1992). *Nonholonomic Kinematics and the Role of Elliptic Functions in Constructive Controllability.* Nonholonomic Motion Planning. Kluwer Academic.

Isidori, A. (1995). *Nonlinear Control Systems. Third Edition.* Springer Verlag.

Mazenc, F. and A. Astolfi (n.d.). Global stabilization of a new class of nonholonomic systems. *In preparation.*

Morin, P. (1996). Robust stabilization of the angular velocity of a rigid body with two controls. *European Journal of Control* **2**, 51–56.

Murray, R. M. and S. S. Sastry (1993). Nonholonomic motion planning: steering using sinusoids. *IEEE Trans. Autom. Control* **38**(5), 700 - 716.

Outbib, R. and G. Sallet (1992). Stabilizability of the angular velocity of a rigid body revisited. *Systems and Control Letters* **18**, 93–98.

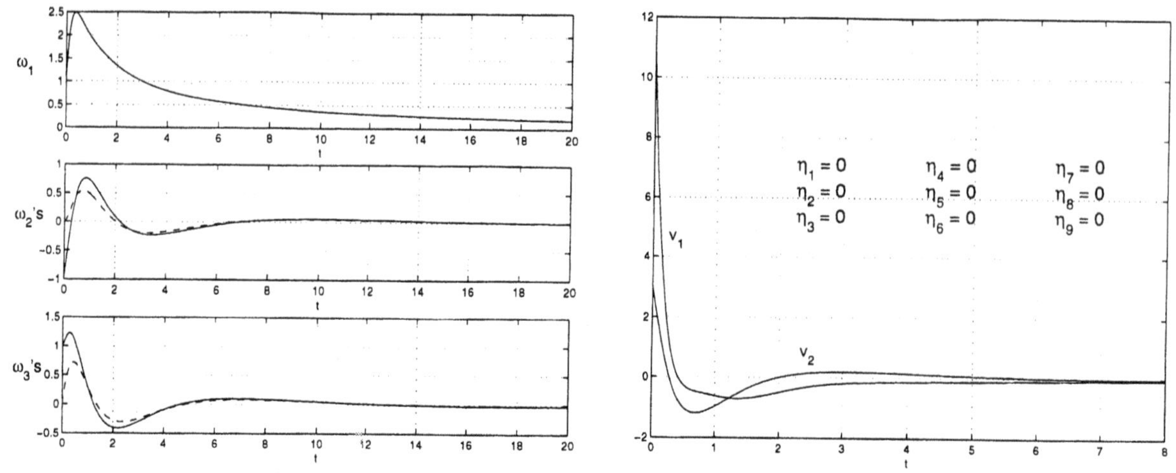


Fig. 1. Nominal simulations. State histories (left) and control actions (right) from the initial state $(\omega_1(0), \omega_2(0), \omega_3(0)) = (1, -1, 1)$. The states $\omega_2$ and $\omega_3$ are displayed together with their estimates (dashed lines) $\hat{\omega}_2$ and $\hat{\omega}_3$. The variables $\omega$'s are in rad/s.

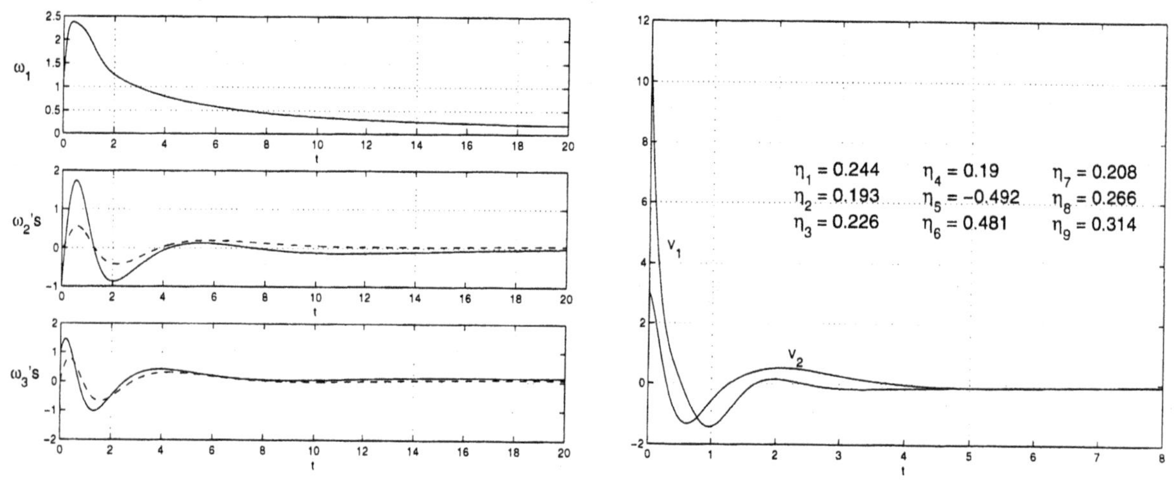


Fig. 2. Simulations with model errors. State histories (left) and control actions (right) from the initial state $(\omega_1(0), \omega_2(0), \omega_3(0)) = (1, -1, 1)$. The states $\omega_2$ and $\omega_3$ are displayed together with their estimates (dashed lines) $\hat{\omega}_2$ and $\hat{\omega}_3$. The variables $\omega$'s are in rad/s.

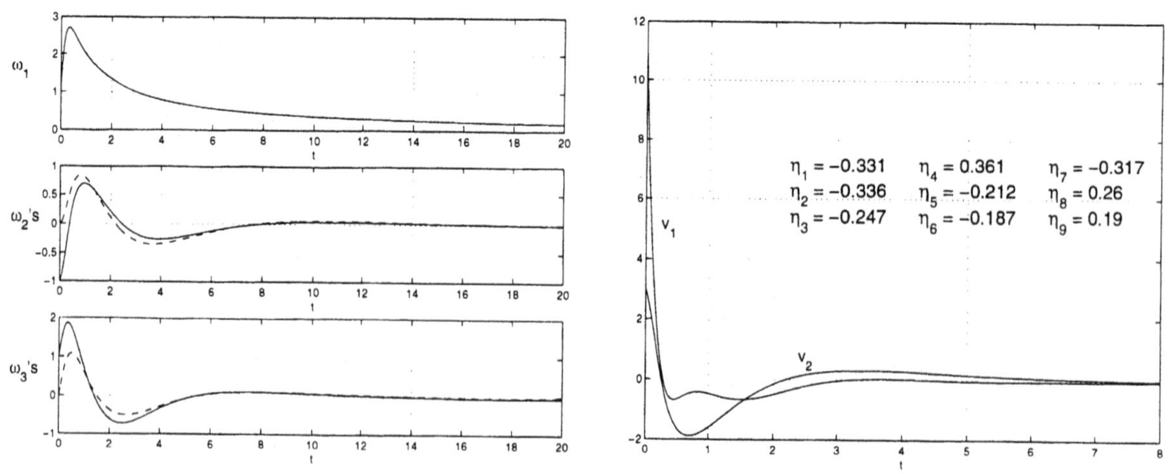


Fig. 3. Simulations with model errors. State histories (left) and control actions (right) from the initial state $(\omega_1(0), \omega_2(0), \omega_3(0)) = (1, -1, 1)$. The states $\omega_2$ and $\omega_3$ are displayed together with their estimates (dashed lines) $\hat{\omega}_2$ and $\hat{\omega}_3$. The variables $\omega$'s are in rad/s.

# STRUCTURE AND ROBUST STABILIZATION OF MULTIVARIABLE NONLINEAR INTERLACING SYSTEMS

**Xiaoping Liu** **Kemin Zhou** **Guoxiang Gu**

*Dept of Automatic Control*
*Northeastern University*
*Shenyang, 110006, PRC*
*liuxp@mail.neu.edu.cn*

*Dept of Electrical and Computer Eng.*
*Louisiana State University*
*Baton Rouge, LA 70803, USA*
*kemin@ee.lsu.edu ggu@ee.lsu.edu*

Abstract: An algorithm is proposed by which a new structure can be defined for nonlinear uncertain systems. Such a structure is called nested lower triangular form. It is observed that all nonlinear uncertain systems in the nested lower triangular form can be generated from a simplest autonomous subsystem via a sequence of augmentations through input channel or interconnection channel. It is shown that robust stabilizability is invariant under the augmentations, so all systems in the nested lower triangular form can be globally asymptotically stabilized for any addmissible uncertain parameters if their zero dynamics is globally exponentially stable. *Copyright ©1999 IFAC*

Keywords: Nonlinear systems, stabilization, robust control, state feedback, Lyapunov function.

## 1. INTRODUCTION

The problem of robust stabilization arises naturally when the nonlinear controlled plant includes uncertainties such as unknown time-varying parameters or disturbances. Over the last decade, various robust stabilization techniques have been developed for uncertain nonlinear systems in the literature. The method of Lyapunov redesign, developed in Leitmann (1978), Corless and Leitmann (1981), and Barmish *et al.* (1983) begins with a control Lyapunov function (CLF) for the nominal system and then attempts to use this CLF as a robust control Lyapunov function (RCLF) for the uncertain system. The success of this method is guaranteed under a restrictive matching condition. Backstepping, which is based on the nonlinear stabilization technique of "adding an integrator" introduced by Tsinias (1989) and Byrnes and Isidori (1989), and was first used in nonlinear adaptive control Kanellakopoulos *et al.* (1991), leads to the discovery of a structural strict feedback condition (much weaker than the matching condition) under which the systematic construction of an RCLF is always possible. These robust backstepping results first appeared in Freeman and Kokotovic (1992) and were obtained independently in Marino and Tomei (1993), Qu (1993), and Slotine and Hedrick (1993). For further information about backstepping, see books by Marino and Tomei (1995), Kristic *et al.* (1995), and Freeman and Kokotovic (1996) and references therein.

The extension of the backstepping design method to multivariable nonlinear systems has been studied in detail by Schwartz *et. al.* (1996), Lin (1997b), and Qu and Dawson (1994), yielding stabilizing and robustly stabilizing feedback control laws. Such an extension still falls into the category of systems with strict feedback form. Other related results about multivariable nonlinear systems with uncertainties are presented by Liu *et al* (1998, 1999), which discuss those systems which may not be in strict feedback form via combination of backstepping and augmantation.

It is easily seen from the previous results that structural conditions, such as matching condition and lower triangular conditions, play an important role in designing robustly stabilizing feedback laws for both linear and nonlinear systems. See also Wei (1990) for the antisymmetric stepwise configuration, and Sepulchre *et al.* (1997) for the interlaced condition.

In this paper, a new algorithm is developed for multivariable nonlinear uncertain systems. Based on this algorithm, a new concept "nested lower triangular form" is introduced. Two kinds of augmentations, *i.e.* augmentation through input channel and augmentation through interconnection channel, are defined. It is observed that all systems in the nested lower triangular form can be generated from a simplest autonomous subsystem via a sequence of augmentations through input channel or interconnection channel. It is proved that both augmentations maintain the robust stabilizability of the systems. Therefore, the globally exponential stability of the zero dynamics and the nested lower triangular structure of the system guarantee the existence of globally robustly stabilizing controller.

## 2. DEFINITION AND EXAMPLES

Consider nonlinear uncertain systems of the form:

$$\dot{\xi}^0 = \phi^0(\xi^0, \xi^1, ..., \xi^m, \theta) \qquad (1)$$

$$\begin{aligned} \dot{\xi}^1_1 &= \xi^1_2 + \phi^1_1(\xi^0, \xi^1, ..., \xi^m, \theta) \\ &\vdots \\ \dot{\xi}^1_{n_1-1} &= \xi^1_{n_1} + \phi^1_{n_1-1}(\xi^0, \xi^1, ..., \xi^m, \theta) \end{aligned} \qquad (2)$$

$$\begin{aligned} &\vdots \\ \dot{\xi}^m_1 &= \xi^m_2 + \phi^m_1(\xi^0, \xi^1, ..., \xi^m, \theta) \\ &\vdots \\ \dot{\xi}^m_{n_m-1} &= \xi^m_{n_m} + \phi^m_{n_m-1}(\xi^0, \xi^1, ..., \xi^m, \theta) \end{aligned} \qquad (3)$$

where $\xi^0 \in R^{n_o}$, $\xi^i = [\xi^i_1, ..., \xi^i_{n_i}]^T \in R^{n_i}$, and all the functions are smooth and vanishing at the origin for any $\theta \in \Omega$ with $\Omega$ a known compact set. Let $u_i = \xi^i_{n_i}$. Then (1)-(3) represents a system with $m$ inputs, which is composed of $m$ single-input subsystems and one autonomous subsystem with interconnections. In order to design a robustly stabilizing feedback control laws for the system, some assumptions are required on both the subsystems and the interconnections. Before introducing these assumptions, the following algorithm is presented.

**Algorithm 1**
**Step 1**: *Consider smooth functions* $\phi^i_j$ *and define* $M(1) = \{1, ..., m\}$, $N_i(1) = \{1, ..., n_i\}$, *and*

$$\begin{aligned} \phi^0_1(\xi^0, \xi^1, ..., \xi^m, \theta) &= \phi^0(\xi^0, \xi^1, ..., \xi^m, \theta) \\ \phi^{ij}_1(\xi^0, \xi^1, ..., \xi^m, \theta) &= \phi^i_j(\xi^0, \xi^1, ..., \xi^m, \theta) \end{aligned}$$

*Now the following notations are introduced:*

$$\begin{aligned} a^0_{ts}(1) &= \begin{cases} 0 & \text{if } \frac{\partial \phi^0_1}{\partial \xi^t_s} \equiv 0 \\ 1 & \text{otherwise} \end{cases} \\ a^{ij}_{ts}(1) &= \begin{cases} 0 & \text{if } \frac{\partial \phi^{ij}_1}{\partial \xi^t_s} \equiv 0 \\ 1 & \text{otherwise} \end{cases} \\ r^0_t(1) &= \max_s \{s \in N_t(1) | a^0_{ts}(1) \neq 0\} \\ r^{ij}_t(1) &= \max_s \{s \in N_t(1) | a^{ij}_{ts}(1) \neq 0\} \\ r^i_t(1) &= \max_j \{j \in N_i(1) | a^{ij}_t(1) \leq 1\} \end{aligned}$$

*Assume that there exists an integer* $t_1$ *so that*

1. $r^0_{t_1}(1) \leq 1$;
2. $r^i_{t_1}(1) \geq 1$ *for all* $i \in M(1)$;
3. $r^{ij}_{t_1}(1) \leq r^i_{t_1}(1) + 1$ *for all* $1 \leq j \leq r^i_{t_1}(1)$ *and* $i \in M(1)$;
4. $a^{ij}_{il}(1) = 0$ *for all* $r^i_{t_1}(1) + 1 \leq j \leq n_i - 1$, $j \leq l \leq n_i - 1$, *and* $i \in M(1)$;
5. *for* $r^i_{t_1}(1) + 1 \leq j \leq n_i - 1$ *and* $i \in M(1)$, *if* $a^{ij}_{ts}(1) = 1$ *for some* $t \neq i$, *then* $a^{ij}_{ts}(1) = 0$ *for all* $1 \leq p \leq s - 1$ *and* $j + 1 \leq q \leq n_i - 1$.

*Then, it is no difficult to see that* $\xi^{t_1}_2, ..., \xi^{t_1}_{n_{t_1}}$ *do not affect* $\phi^0_1$ *and* $\phi^{ij}_1$ *for all* $1 \leq j \leq r^i_{t_1}(1)$ *and* $i \in M(1)$. *As a result, it follows that*

$$\begin{aligned} \phi^0_1(X_1, \xi^{t_1}_1) &= \phi^0_1(X_1, 0) + \bar{\phi}^0_1(X_1, \xi^{t_1}_1)\xi^{t_1}_1 \\ \phi^{ij}_1(X_1, \xi^{t_1}_1) &= \phi^{ij}_1(X_1, 0) + \bar{\phi}^{ij}_1(X_1, \xi^{t_1}_1)\xi^{t_1}_1 \end{aligned}$$

*with* $X_1 = \mathrm{col}\{\xi^1, ..., \xi^{t_1-1}, \xi^{t_1+1}, ..., \xi^m\}$ *and*

$$\begin{aligned} \bar{\phi}^0_1(X_1, \xi^{t_1}_1) &= \int_0^1 \left[\frac{\partial \bar{\phi}^0_1(X_1, \beta)}{\partial \beta}\right]_{\beta = \gamma \xi^{t_1}_1} d\gamma \\ \bar{\phi}^{ij}_1(X_1, \xi^{t_1}_1) &= \int_0^1 \left[\frac{\partial \bar{\phi}^{ij}_1(X_1, \beta)}{\partial \beta}\right]_{\beta = \gamma \xi^{t_1}_1} d\gamma \end{aligned}$$

**Step k**: *Consider functions* $\phi^{ij}_{k-1}(X_{k-1}, 0)$ *and define* $M(k) = \{i \in M(k-1) | i \neq t_{k-1}\}$, $N_i(k) =$

$\{1,...,r^i_{t_{k-1}}(k-1)\}$, $\phi^0_k(X_{k-1}) = \phi^0_{k-1}(X_{k-1},0)$ *and* $\phi^{ij}_k(X_{k-1}) = \phi^{ij}_{k-1}(X_{k-1},0)$. *Now the following notations are introduced:*

$$
\begin{aligned}
a^0_{ts}(k) &= \begin{cases} 0 & \text{if } \frac{\partial\phi^0_k}{\partial\xi^t_s} \equiv 0 \\ 1 & \text{otherwise} \end{cases} \\
a^{ij}_{ts}(k) &= \begin{cases} 0 & \text{if } \frac{\partial\phi^{ij}_k}{\partial\xi^t_s} \equiv 0 \\ 1 & \text{otherwise} \end{cases} \\
r^0_t(k) &= \max_s\{s \in N_t(k) | a^0_{ts}(k) \neq 0\} \\
r^{ij}_t(k) &= \max_s\{s \in N_t(k) | a^{ij}_{ts}(k) \neq 0\} \\
r^i_t(k) &= \max_j\{j \in N_i(k) | a^{ij}_t(k) \leq 1\}
\end{aligned}
$$

*Assume that there exists an integer* $t_k$ *so that*

1. $r^0_{t_k}(k) \leq 1$;
2. $r^i_{t_k}(k) \geq 1$ *for all* $i \in M(k)$;
3. $r^{ij}_{t_k}(k) \leq r^i_{t_k}(k) + 1$ *for all* $1 \leq j \leq r^i_{t_k}(k)$ *and* $i \in M(k)$;
4. $a^{ij}_{il}(k) = 0$ *for all* $r^i_{t_k}(k) + 1 \leq j \leq r^i_{t_{k-1}}(k-1)$, $j \leq l \leq r^i_{t_{k-1}}(k-1)$, *and* $i \in M(k)$;
5. *for* $r^i_{t_k}(k) + 1 \leq j \leq r^i_{t_{k-1}}(k-1)$ *and* $i \in M(k)$, *if* $a^{ij}_{ts}(k) = 1$ *for some* $t \neq i$, *then* $a^{ij}_{ts}(k) = 0$ *for all* $1 \leq p \leq s-1$ *and* $j+1 \leq q \leq r^i_{t_{k-1}}(k-1)$.

*Then,*
*it is no difficult to see that* $\xi^{t_k}_2, ..., \xi^{t_k}_{r^i_{t_{k-1}}(k-1)}$ *do not affect* $\phi^0_k$ *and* $\phi^{ij}_k$ *for all* $1 \leq j \leq r^i_{t_k}(k)$ *and* $i \in M(k)$. *As a result, it follows that*

$$\phi^0_k(X_k, \xi^{t_k}_1) = \phi^0_k(X_k, 0) + \bar{\phi}^0_k(X_k, \xi^{t_k}_1)\xi^{t_k}_1$$
$$\phi^{ij}_k(X_k, \xi^{t_k}_1) = \phi^{ij}_k(X_k, 0) + \bar{\phi}^{ij}_k(X_k, \xi^{t_k}_1)\xi^{t_k}_1$$

*with* $X_k = \mathrm{col}\{\xi^i | i \neq t_j, j = 1, ..., k\}$ *and*

$$
\begin{aligned}
\bar{\phi}^0_k(X_k, \xi^{t_k}_1) &= \int_0^1 \left[\frac{\partial\bar{\phi}^0_k(X_k,\beta)}{\partial\beta}\right]_{\beta=\gamma\xi^{t_k}_1} d\gamma \\
\bar{\phi}^{ij}_k(X_k, \xi^{t_k}_1) &= \int_0^1 \left[\frac{\partial\bar{\phi}^{ij}_k(X_k,\beta)}{\partial\beta}\right]_{\beta=\gamma\xi^{t_k}_1} d\gamma
\end{aligned}
$$

Algorithm 1 is said to be feasible if there exist $t_1, ..., t_m$ so that Algorithm 1 terminates at step $m$. It is not difficult to check that Algorithm 1 is feasible for the following five examples.

**Example 1** *Consider a single-input system of the form*

$$
\begin{aligned}
\dot{\xi}^0 &= \phi^0(\xi^0) + \psi^0_1(\xi^0, \xi^1_1, \theta)\xi^1_1 \\
\dot{\xi}^1_1 &= \xi^1_2 + \phi^1_1(\xi^0, \xi^1_1, \theta) \\
&\vdots \\
\dot{\xi}^1_{n_1-1} &= \xi^1_{n_1} + \phi^1_{n_1-1}(\xi^0, \xi^1_1, ..., \xi^1_{n_1-1}, \theta)
\end{aligned}
$$

*with* $u_1 = \xi^1_{n_1}$.

**Example 2** *Consider a two-input system of the form*

$$
\begin{aligned}
\dot{\xi}^0 &= \phi^0(\xi^0) + \psi^0_1(\xi^0, \xi^1_1, \theta)\xi^1_1 \\
\dot{\xi}^1_1 &= \xi^1_2 + \phi^1_1(\xi^0, \xi^1_1, \theta) \\
&\vdots \\
\dot{\xi}^1_{n_1-1} &= \xi^1_{n_1} + \phi^1_{n_1-1}(\xi^0, \xi^1_1, ..., \xi^1_{n_1-1}, \theta) \\
\dot{\xi}^2_1 &= \xi^2_2 + \phi^2_1(\xi^0, \xi^1, \xi^2_1, \theta) \\
&\vdots \\
\dot{\xi}^2_{n_2-1} &= \xi^2_{n_2} + \phi^2_{n_2-1}(\xi^0, \xi^1, \xi^2_1, ..., \xi^2_{n_2-1}, \theta)
\end{aligned}
$$

*with* $u_i = \xi^i_{n_i}$ *for* $i = 1, 2$.

**Example 3** *Consider a two-input system of the form*

$$
\begin{aligned}
\dot{\xi}^0 &= \phi^0(\xi^0) + \psi^0_1(\xi^0, \xi^1_1, \theta)\xi^1_1 \\
&+ \psi^0_2(\xi^0, \xi^1, \xi^2_1, \theta)\xi^2_1 \\
\dot{\xi}^1_1 &= \xi^1_2 + \phi^1_1(\xi^0, \xi^1_1, \theta) + \psi^1_{12}(\xi^0, \xi^1, \xi^2_1, \theta)\xi^2_1 \\
&\vdots \\
\dot{\xi}^1_{n_1-1} &= \xi^1_{n_1} + \phi^1_{n_1-1}(\xi^0, \xi^1_1, ..., \xi^1_{n_1-1}, \theta) \\
&+ \psi^1_{n_1-1,2}(\xi^0, \xi^1, \xi^2_1, \theta)\xi^2_1 \\
\dot{\xi}^2_1 &= \xi^2_2 + \phi^2_1(\xi^0, \xi^1, \xi^2_1, \theta) \\
&\vdots \\
\dot{\xi}^2_{n_2-1} &= \xi^2_{n_2} + \phi^2_{n_2-1}(\xi^0, \xi^1, \xi^2_1, ..., \xi^2_{n_2-1}, \theta)
\end{aligned}
$$

*with* $u_i = \xi^i_{n_i}$ *for* $i = 1, 2$.

**Example 4** *Consider a two-input system of the form*

$$
\begin{aligned}
\dot{\xi}^0 &= \phi^0(\xi^0) + \psi^0_1(\xi^0, \xi^1_1, \theta)\xi^1_1 \\
&+ \psi^0_2(\xi^0, \xi^1_1, ..., \xi^1_{n_1}, \xi^2_1, \theta)\xi^2_1 \\
\dot{\xi}^1_1 &= \xi^1_2 + \phi^1_1(\xi^0, \xi^1_1, \theta) \\
&+ \psi^1_{12}(\xi^0, \xi^1_1, ..., \xi^1_{n_1}, \xi^2_1, \theta)\xi^2_1 \\
&\vdots \\
\dot{\xi}^1_{n_1-1} &= \xi^1_{n_1} + \phi^1_{n_1-1}(\xi^0, \xi^1_1, ..., \xi^1_{n_1-1}, \theta) \\
&+ \phi^1_{n_1-1,2}(\xi^0, \xi^1_1, ..., \xi^1_{n_1}, \xi^2_1, \theta)\xi^2_1 \\
\dot{\xi}^1_{n_1} &= \xi^1_{n_1+1} + \phi^1_{n_1}(\xi^0, \xi^1_1, ..., \xi^1_{n_1}, \xi^2_1, \theta) \\
&\vdots \\
\dot{\xi}^1_{n_1+n_2-1} &= \xi^1_{n_1+n_2} + \phi^1_{n_1+n_2-1}(\xi^0, \xi^1, ..., \\
&\quad \xi^1_{n_1+n_2-1}, \xi^2_1, ..., \xi^2_{n_2-1}, \theta) \\
\dot{\xi}^2_1 &= \xi^2_2 + \phi^2_1(\xi^0, \xi^1_1, ..., \xi^1_{n_1}, \xi^2_1, \theta) \\
&\vdots
\end{aligned}
$$

$$\dot{\xi}^2_{n_2-1} = \xi^2_{n_2} + \phi^2_{n_2-1}(\xi^0, \xi^1, ..., \xi^1_{n_1+n_2-1}, \xi^2_1, ..., \xi^2_{n_2-1}, \theta)$$

*with* $u_1 = \xi^1_{n_1+n_2}$ *and* $u_2 = \xi^2_{n_2}$.

**Example 5** *Consider a two-input system of the form*

$$\begin{aligned}
\dot{\xi}^0 &= \phi^0(\xi^0) + \psi^0_1(\xi^0, \xi^1_1, \theta)\xi^1_1 \\
&+ \phi^0_2(\xi^0, \xi^1_1, ..., \xi^1_{n_1}, \xi^2_1, \theta)\xi^2_1 \\
\dot{\xi}^1_1 &= \xi^1_2 + \phi^1_1(\xi^0, \xi^1_1, \theta) \\
&+ \psi^1_{12}(\xi^0, \xi^1_1, ..., \xi^1_{n_1}, \xi^2_1, \theta)\xi^2_1 \\
&\vdots \\
\dot{\xi}^1_{n_1-1} &= \xi^1_{n_1} + \phi^1_{n_1-1}(\xi^0, \xi^1_1, ..., \xi^1_{n_1-1}, \theta) \\
&+ \psi^1_{n_1-1,2}(\xi^0, \xi^1_1, ..., \xi^1_{n_1}, \xi^2_1, \theta)\xi^2_1 \\
\dot{\xi}^1_{n_1} &= \xi^1_{n_1+1} + \phi^1_{n_1}(\xi^0, \xi^1_1, ..., \xi^1_{n_1}, \\
&\qquad \xi^2_1, ..., \xi^2_{n_2-1}, \theta) \\
&\vdots \\
\dot{\xi}^1_{n_1+n_2-1} &= \xi^1_{n_1+n_2} + \phi^1_{n_1+n_2-1}(\xi^0, \xi^1_1, ..., \\
&\qquad \xi^1_{n_1+n_2-1}, \xi^2_1, ..., \xi^2_{n_2-1}, \theta) \\
\dot{\xi}^2_1 &= \xi^2_2 + \phi^2_1(\xi^0, \xi^1_1, ..., \xi^1_{n_1}, \xi^2_1, \theta) \\
&\vdots \\
\dot{\xi}^2_{n_2-1} &= \xi^2_{n_2}\phi^2_{n_2-1}(\xi^0, \xi^1_1, ..., \xi^1_{n_1}, \\
&\qquad \xi^2_1, ..., \xi^2_{n_2-1}, \theta)
\end{aligned}$$

*with* $u_1 = \xi^1_{n_1+n_2}$ *and* $u_2 = \xi^2_{n_2}$.

It follows from Example 1 and Example 2 that Algorithm 1 is feasible for systems in a lower triangular form. Examples 3-5 show that Algorithm 1 is feasible for systems in a special form, which is more general than the lower triangular form. It is easily seen that Examples 3-5 have a common feature that the subsytem $(\xi^1_1, ..., \xi^1_{n_1})$ in the lower triangular form is nested in the whole system which is also in the lower triangular form in the view point of $(\xi^1_{n_1+1}, ..., \xi^1_{n_1+n_2}, \xi^2_1, ..., \xi^2_{n_2})$. So, it is reasonable to give the following definition.

**Definition 1** *A system is said to be in a nested lower triangular form if Algorithm 1 is feasible.*

**Remark 1** *The robust stabilization problem for the system in Example 1 has been discussed in detail by Marino and Tomei (1995), Krstic et al. (1995), Freeman and Kokotovic (1996), Sepulchre et al. (1997), and Lin (1997a), for the system in Example 2, by Lin (1997b). The systems in Example 3 and Example 4 are investigated by Liu et al (1998) and Liu et al (1999). The system in Example 5 does not belong to any set of systems addressed by Liu et al (1998), Liu et al (1999) and in Lin (1997b), but do fall into the category of systems in the nested lower triangular form.*

## 3. MAIN RESULTS

In order to prove the main results, the following lemma is needed.

**Lemma 1** *Consider a system of the form*

$$\begin{aligned}
\dot{\eta} &= \phi^\eta_\eta(\eta) + \phi^\eta_\zeta(\eta, \zeta, \theta) \\
\dot{\zeta} &= \phi^\zeta_\zeta(\eta, \zeta, \theta)
\end{aligned} \qquad (4)$$

*Suppose that there exists a positive definite and proper smooth Lyapunov function* $V_\eta(\eta)$ *such that*

$$\begin{aligned}
L_{\phi^\eta_\eta(\eta)} V_\eta(\eta) &\leq -\beta^\eta_\eta \| \eta \|^2 \\
\| \frac{\partial V_\eta(\eta)}{\partial \eta} \| &\leq \gamma \| \eta \|
\end{aligned}$$

*for some* $\beta^\eta_\eta > 0$ *and* $\gamma > 0$, *and the function* $V_\zeta(\eta, \zeta) = V_\eta(\eta) + \frac{1}{2}\zeta^2$ *has time derivative*

$$\dot{V}_\zeta \leq -\beta^\zeta_\eta \| \eta \|^2 - \beta^\zeta_\zeta \zeta^2 \quad .$$

*for some* $\beta^\zeta_\eta > 0$ *and* $\beta^\zeta_\zeta > 0$. *Then,*

1. *there exists a smooth function* $\alpha(\eta, \zeta, \xi)$ *with* $\alpha(0,0,0) = 0$ *such that the function* $V_\xi(\eta, \zeta, \xi) = V_\eta(\eta) + \frac{1}{2}\zeta^2 + \frac{1}{2}\xi^2$ *has time derivative*

$$\dot{V}_\xi \leq -\beta^\xi_\eta \| \eta \|^2 - \beta^\xi_\zeta \zeta^2 - \beta^\xi_\xi \xi^2$$

   *for some* $\beta^\xi_\eta > 0$, $\beta^\xi_\zeta > 0$, *and* $\beta^\xi_\xi > 0$ *along the trajectories of the system*

$$\begin{aligned}
\dot{\eta} &= \phi^\eta_\eta(\eta) + \phi^\eta_\zeta(\eta, \zeta, \theta) + \phi^\eta_\xi(\eta, \zeta, \xi, \theta)\xi \\
\dot{\zeta} &= \phi^\zeta_\zeta(\eta, \zeta, \theta) + \phi^\zeta_\xi(\eta, \zeta, \xi, \theta)\xi \\
\dot{\xi} &= u + \phi^\xi_\xi(\eta, \zeta, \xi, \theta)
\end{aligned} \qquad (5)$$

   *with* $u = \alpha(\eta, \zeta, \xi)$;

2. *furthermore, there exists another smooth function* $\gamma(\eta, \zeta, \xi, u)$ *with* $\gamma(0,0,0,0) = 0$ *such that the function* $V_u(\eta, \zeta, \xi) = V_\eta(\eta) + \frac{1}{2}\zeta^2 + \frac{1}{2}\xi^2 + \frac{1}{2}[u - \alpha(\eta, \zeta, \xi)]^2$ *has time derivative*

$$\begin{aligned}
\dot{V}_\xi &\leq -\beta^u_\eta \| \eta \|^2 - \beta^u_\zeta \zeta^2 - \beta^u_\xi \xi^2 \\
&- \beta^u_u [u - \alpha(\eta, \zeta, \xi)]^2
\end{aligned}$$

*for some $\beta^u_\eta > 0$, $\beta^u_\zeta > 0$, $\beta^u_\xi > 0$ and $\beta^u_u > 0$ along the trajectories of the system*

$$\begin{aligned} \dot\eta &= \phi^\eta_\eta(\eta) + \phi^\eta_\zeta(\eta,\zeta,\theta) + \phi^\eta_\xi(\eta,\zeta,\xi,\theta)\xi \\ \dot\zeta &= \phi^\zeta_\zeta(\eta,\zeta,\theta) + \phi^\zeta_\xi(\eta,\zeta,\xi,\theta)\xi \\ \dot\xi &= u + \phi^\xi_\xi(\eta,\zeta,\xi,\theta) \\ \dot u &= v + \phi^u_u(\eta,\zeta,\xi,u,\theta) \end{aligned} \quad (6)$$

*with $v = \gamma(\eta,\zeta,\xi,u)$.*

For convenience, the operation of generating the system (5) from (4) is called augmentation through interconnection channel, and the one of generating (6) from (5) is called augmentation through input channel. Lemma 1 shows that both augmentations do maintain the robust stabilizability. It is worth noting that the dynamics $\zeta$ may not appear in system (4). In this case, $\zeta$ also disappears in both (5) and (6).

It is important that all systems in the nested lower triangular form can always be generated from the simplest subsystem

$$\dot\eta = \phi^\eta_\eta(\eta)$$

via a sequence of augmentations (through input channel or interconnection channel). This can be seen from the following example.

**Example 6** *It is not difficult to check that the following system is in the nested lower triangular form.*

$$\begin{aligned} \dot x^0 &= \phi^0_0(x^0) + \phi^0_1(x^0,x^1_1)x^1_1 \\ &+ \phi^0_2(x^0,x^1_1,x^1_2,x^2_1)x^2_1 \\ &+ \phi^0_3(x^0,x^1_1,x^1_2,x^2_1,x^2_2,x^3_1)x^3_1 \\ \dot x^1_1 &= x^1_2 + \phi^1_2(x^0,x^1_1,x^1_2,x^2_1)x^2_1 \\ &+ \phi^1_3(x^0,x^1_1,x^1_2,x^2_1,x^2_2,x^3_1)x^3_1 \\ \dot x^1_2 &= u_1 + \psi^1(x^0,x^1_1,x^1_2,x^2_1,x^2_2,x^3_1,x^3_2,u_2,u_3) \\ \dot x^2_1 &= x^2_2 + \phi^2_3(x^0,x^1_1,x^1_2,x^2_1,x^2_2,x^3_1)x^3_1 \\ \dot x^2_2 &= u_2 + \psi^2(x^0,x^1_1,x^1_2,x^2_1,x^2_2,x^3_1,x^3_2,u_3) \\ \dot x^3_1 &= x^3_2 + \phi^3_3(x^0,x^1_1,x^1_2,x^2_1,x^2_2,x^3_1,x^3_2) \\ \dot x^3_2 &= u_3 + \psi^3(x^0,x^1_1,x^1_2,x^2_1,x^2_2,x^3_1,x^3_2) \end{aligned}$$

*This system can be generated from $\dot x^0 = \phi^0_0(x^0)$ via augmentations through interconnection channel $\phi^0_1 x^1_1$, interconnection channels $\phi^0_2 x^2_1$ and $\phi^1_2 x^2_1$, interconnection channels $\phi^0_3 x^3_1$, $\phi^1_3 x^3_1$ and $\phi^2_3 x^3_1$, input channel $x^3_2$, input channel $x^2_2$, and input channel $x^1_2$.*

As a result, the following conclusion can be easily drawn.

**Theorem 1** *Suppose the system (1)-(3) is in the nested lower triangular form and its zero dynamics is exponentially stable. Then, it can be globally asymptotically stabilized for any $\theta \in \Omega$.*

*Acknowledgements*: This work was supported in part by grants from AFOSR (F49620-97-1-0064), ARO (DAAH04-96-1-0193), LEQSF (DOD/LEQSF (1996-99)-04), LEQSF (1995-98)-RD-A-14, Fok Ying Tung Education Foundation, and Trans-Century Training Program Foundation for the Talents by State Education Commission of China.

## RERERENCES

Barmish, B. R., M. J. Corless, and G. Leitmann (1983). A new class of stabilizing controllers for uncertain dynamical systems. *SIAM J. Control Optimiz.*, **21**, 246-255.

Byrnes, C. I. and A. Isidori (1989). New results and examples in nonlinear feedback stabilization. *Systems & Control Letters*, **12**, 437-442.

Corless, M. J. and G. Leitmann (1981). Continuous state feedback guaranteeing uniform ultimate boundedness for uncertain dynamic systems. *IEEE Trans. Automat. Control*, **26**, 1139-1144.

Freeman, R. A. and P. V. Kokotovic (1992). Backstepping design of robust controllers for a class of nonlinear systems. in *Proceedings of the IFAC Nonlinear Control Systems Design Symposium*, Bordeaux, France, pp.307-312.

Freeman, R. A. and P. V. Kokotovic (1996). *Robust Nonlinear Control Design*. Birkhauser, Boston.

Isidori, A. (1995). *Nonlinear Control Systems*, 3rd. ed. Springer-Verlag, Berlin, 1995.

Kanellakopoulos, I., P. V. Kokotovic, and A. S. Morse (1991). Systematic design of adaptive controllers for feedback linearizable systems. *IEEE Trans. Automat. Control*, **36**, 1241-1253.

Krstic, M., I. Kanellakopoulos, and P. V. Kokotovic (1995). *Nonlinear and Adaptive Control Design*. Wiley, New York.

Leitmann, G. (1978). Guaranteed ultimate boundedness for a class of uncertain linear dynamical systems. *IEEE Trans. Automat. Control*, **23**, 1109-1110.

Lin, W. (1997a). Global robust stabilization of minimum-phase nonlinear systems with uncertainties. *Automatica*, **33**, 453-462.

Lin, W. (1997b). Robust control of a class of MIMO minimum-phase nonlinear systems. in *Proc. of the American Control Conf.*, Albuquerque, New Mexico, pp.838-842.

Liu, X.P., K.M. Zhou and G.X. Gu (1998). Robust stabilization of MIMO uncertain nonlinear systems without strict feedback conditions. *Proc. of American Control Conf.*, 766-770. Philadelphia, PA, USA.

Liu, X. P., G. X. Gu and K. M. Zhou (1999). Robust stabilization of MIMO nonlinear systems by backstepping. *Automatica*, to appear.

Marino, R. and P. Tomei (1993). Robust stabilization of feedback linearizable time-varying uncertain nonlinear systems. *Automatica*, **29**, 181-189.

Marino, R. and P. Tomei (1995). *Nonlinear Control Design.* Prentice Hall Int. (UK).

Qu, Z. (1993). Robust control of nonlinear uncertain systems under generalized matching conditions. *Automatica*, **29**, 985-998.

Qu, Z. (1995). Robust control design for nonlinear uncertain systems without matching conditions. *IEEE Trans. Automat. Control*, **40**, 1453-1460.

Qu, Z. and D. Dawson (1993). Robust control of cascaded and individually feedback linearizable nonlinear systems. *Automatica*, **30**, 1057-1064.

Schwartz, B., A. Isidori, and T. J. Tarn (1996). Global normal forms for MIMO nonlinear systems with application to stabilization and disturbance attenuation. in *Proc. of the 35th Conf. on Decision and Control*, Kobe, Japan, pp.1041-1046.

Sepulchre, R., M. Jankovic, and P. Kokotovic, *Constructive Nonlinear Control*, Springer, 1997.

Slotine, J. J. E. and K. Hedrick (1993). Robust input-output feedback linearization. *Int. J. Control*, **57**, 1133-1139.

Tsinias, J. (1989). Sufficient Lyapunov like conditions for stabilization. *Mathematics of Control, Signals, and Systems*, **2**, 343-357.

Wei, K. (1990). Quadratic stabilizability of linear systems with structural independent time-varying uncertainties. *IEEE Trans. Automat. Control*, **35**, 268-277.

14th Triennial World Congress, Beijing, P.R. China

# Three-dimensional Nonlinear $H_\infty$ Guidance Law with Maneuvering Targets

Hsin-Yuan Chen* Ciann-Dong Yang†
Institute of Aeronautics and Astronautics,
National Cheng Kung University,
Tainan, Taiwan

## Abstract

This paper proposes a three-dimensional $H_\infty$ robust guidance law for homing missiles with nonlinear kinematics in the homing phase. Unlike conventional approaches where target's acceleration is often assumed to be known or needs to be estimated in real time, the proposed 3D robust guidance law can achieve performance robustness in the absence of target's acceleration information. The most difficult and challenging task involved in applying nonlinear $H_\infty$ control theory is the solution of the associated Hamilton-Jacobi partial differential inequality (HJPDI). In this paper we show that the HJPDI of the three-dimensional missile guidance problem can be solved analytically with simple manipulations. The numerical simulations show that the three-dimensional $H_\infty$ robust guidance law exhibits strong robustness properties against disturbances from target's maneuvers. *Copyright © 1999 IFAC*

## I. Introduction

Most analytic studies of the missile guidance problem in two-dimensional engagements are based on the assumption that the missile follows the proportional guidance command. Then, in three-dimensional engagements, Adler[1] applies the classical differential geometry surface theory to study the missile guidance problem and derives proportional navigation law in terms of geometry curvatures. Lin[2] applies linear quadratic Gaussian theory to design missile acceleration commands and then transforms these commands into normal acceleration and bank attitude commands for bank-to-turn missile maneuver. Cochran et al.[3] study ture proportional navigation guidance problems. The closed-form solutions is derived in the form of an elliptic function for a nonmaneuvering target. Kumar et al.[4] apply the optimal control technique to design three-phase feedback guidance laws. Lin and Lee[5] apply the linear exponential quadratic Gaussian method to derive the optimal missile pitch acceleration and roll rate commands with Poisson square wave target maneuvering model. In Ref. 6, Yang and Yang derive the analytic solutions for a missile following the generalized proportional navigation guidance command in terms of elliptic functions. In Ref. 7, the Frenet-Serret formula in classical differential geometry cure theory and the characteristics of a fictitious missile pointing velocity vector are used to design missile guidance curvature command. Qualitative analysis is conducted to study capture capability of the designed guidance command in three-dimensional engagements. In the literature listed for three-dimensional engagements, bank-to turn guidance commands are designed in Refs. 2, 4, and 5 and skid-to-turn guidance commands are designed in Refs. 3, 6. The present paper applies the nonlinear $H_\infty$ control theory to the real 3D guidance law design, which, as far as the authors can realize, has not been considered in the literature before. Under this approach, the exact nonlinear governing equations are considered, and the 3D guidance problem is formulated as a nonlinear disturbance attenuation $H_\infty$ control problem where the target's accelerations are regarded as unpredictable disturbances, and the purpose of the control design is to attenuate the influence of target's maneuvers on the performance of guidance law, such as, miss distance and energy consumption, etc.. The characterization of the nonlinear $H_\infty$ guidance law relies on the solution of HJPDI which is a second-order nonlinear partial differential inequality. At the first glance, this HJPDI is rather complicated and hardly to be solved. However, after detailed investigation, a nice analytical solution is found to be possible via simple manipulations.

## II. Nonlinear $H_\infty$ Control Theory

In this section, we introduce the standard results of the nonlinear $H_\infty$ control theory for later use. Consider a nonlinear state space system:

$$\dot{x} = f(x) + g(x)w, \quad w \in R^p, \quad f(0) = 0 \qquad (1a)$$

$$z = h(x), \quad z \in R^q, \quad h(0) = 0 \qquad (1b)$$

where $x$ is the state vector, and $w$ and $z$ are the exogenous disturbances to be rejected and the penalized

Ph.D. candidate,E-mail:p4884117@sparc2.cc.ncku.edu.tw
Professor, Tel : 886-6-2757575 ext. 63669 ; Fax : 886-6-2389940;
E-mail : cdyang@mail.ncku.edu.tw.

output signal, respectively. We assume that $f(x)$, $g(x)$, $h(x)$ are $C^\infty$ functions and $x = 0$ is the equilibrium point of the system, i.e., $f(0) = h(0) = 0$. If there exists a scalar $C^1$ function $U : R^n \longrightarrow R^+$ with $U(0) = 0$ such that

$$\frac{1}{2\gamma^2}\left(\frac{\partial U}{\partial x}\right) g(x)g(x)^T \left(\frac{\partial U}{\partial x}\right)^T + \left(\frac{\partial U}{\partial x}\right) f(x)$$

$$+\frac{1}{2}h^T(x)h(x) \leq 0 \tag{2}$$

, then the system is said to have $L_2$ gain $\leq \gamma$, i.e.

$$\int_0^\infty z^T z dt \leq \gamma^2 \int_0^\infty w^T w dt \tag{3}$$

where $\frac{\partial U}{\partial x} = \left[\frac{\partial U}{\partial x_1}\frac{\partial U}{\partial x_2}\cdots\frac{\partial U}{\partial x_n}\right]$. When control $u$ is applied to the system, we get the controlled system as:

$$\dot{x} = f(x) + g_1(x)w + g_2(x)u \tag{4a}$$

$$z = h_1(x) + k_{12}(x)u \tag{4b}$$

where $x \in R^n$, $w \in R^{p1}$, $u \in R^{p2}$, and $z \in R^q$. $f(x)$, $g_1(x)$, $g_2(x)$, $h_1(x)$, and $k_{12}(x)$ are $C^\infty$ functions, and we assume that $h_1^T k_{12} = 0$, and $k_{12}^T k_{12} = I$. These assumptions are introduced to simplify the following derivation, but as shown later these assumptions are satisfied automatically for the missile guidance problem. The nonlinear $H_\infty$ control problem is to find the control $u$ such that the $L_2$-gain of the closed-loop system is less than or equal to $\gamma$. By replacing $f(x)$, $g(x)$, and $h(x)$ in Eq.(2) with $f(x)+g_2(x)u$, $g_1(x)$, and $h_1(x)+k_{12}(x)u$, respectively, the condition that the $L_2$-gain of the closed-loop system is equal to or lower than $\gamma$ becomes

$$\frac{1}{2}\left(u^T + \frac{\partial U}{\partial x}g_2\right)\left(u^T + \frac{\partial U}{\partial x}g_2\right)^T$$

$$+\frac{1}{2}\left(\frac{\partial U}{\partial x}\right)\left(\frac{1}{\gamma^2}g_1 g_1^T - g_2 g_2^T\right)\left(\frac{\partial U}{\partial x}\right)^T$$

$$+\left(\frac{\partial U}{\partial x}\right) f + \frac{1}{2}h_1^T h_1 \leq 0 \tag{5}$$

The control $u$ minimizing the left-hand-side of the above inequality can be easily found as

$$u(x) = -g_2^T(x)\left(\frac{\partial U}{\partial x}\right)^T \tag{6}$$

Substituting this control into Eq.(5), we obtain the desired HJPDI as

$$\left(\frac{\partial U}{\partial x}\right) f + \frac{1}{2}\left(\frac{\partial U}{\partial x}\right)\left(\frac{1}{\gamma^2}g_1 g_1^T - g_2 g_2^T\right)\left(\frac{\partial U}{\partial x}\right)^T$$

$$+\frac{1}{2}h_1^T h_1 \leq 0 \tag{7}$$

Hence, solving the nonlinear $H_\infty$ control problem is equivalent to finding a positive function $U(x)$ satisfying HJPDI. The corresponding HJPDI for missile guidance problem will be derived in the next section. The $H_\infty$ guidance law given by Eq.(6) guarantees that the controlled system (4a) is internally stable, i.e., the system can return to the equilibrium point from an initial perturbation $x_0$ in the absence of external disturbance $w$. This can be verified by showing that if $U > 0$ is a solution of the HJPDI, then $U$ is a qualified Lyapunov function of the system (4a). To behave as a Lyapunov function, $U$ must satisfy the condition $\dot{U} \leq 0$. Substituting Eq.(6) into Eq.(4a) yields $\dot{x} = f(x) - g_2(x)g_2^T(x)(\partial U/\partial x)^T$ which can be used to evaluate $\dot{U}$ as:

$$\dot{U} = \left(\frac{\partial U}{\partial x}\right)\frac{dx}{dt} = \left(\frac{\partial U}{\partial x}\right) f(x) - \left(\frac{\partial U}{\partial x}\right) g_2(x)g_2^T(x)\left(\frac{\partial U}{\partial x}\right)$$

$$\leq -\frac{1}{2}\left(\frac{\partial U}{\partial x}\right)\left(\frac{1}{\gamma^2}g_1 g_1^T + g_2 g_2^T\right)\left(\frac{\partial U}{\partial x}\right)^T - \frac{1}{2}h_1 h_1^T \leq 0$$

The last inequality is from the HJPDI (7). The properties of $U > 0$ and $\dot{U} \leq 0$ show that $U$ is a qualified Lyapunov function, and hence the controlled system (4a) is stable in the sense of Lyapunov. In summary, the nonlinear $H_\infty$ control theory[8,9,10] guarantees: (A) state boundedness (Lyapunov stability); (B) performance boundedness: $\|z\|_2 \leq \gamma\|w\|_2$.

## III. Formulation of Robust Guidance Problems

The 3D pursuit geometry is described in the spherical coordinates (refer to Ref.19) where the relative position vector along the line of sight is expressed by $\vec{r}$. The differentiation of $\vec{r}$ gives the 3D relative velocity as

$$\dot{\vec{r}} = V_r\vec{e}_r + V_\theta\vec{e}_\theta + V_\phi\vec{e}_\phi$$

$$= \dot{r}\vec{e}_r + r\dot{\theta}\cos\phi\,\vec{e}_\theta + r\dot{\phi}\vec{e}_\phi \tag{8}$$

Differentiating both sides of the above equation yields the components of the relative acceleration as

$$\ddot{r} - r\dot{\phi}^2 - r\dot{\theta}^2\cos^2\phi = w_r - u_r \tag{9}$$

$$r\ddot{\theta}\cos\phi + 2\dot{r}\dot{\theta}\cos\phi - 2r\dot{\phi}\dot{\theta}\sin\phi = w_\theta - u_\theta \tag{10}$$

$$r\ddot{\phi} + 2\dot{r}\dot{\phi} + r\dot{\theta}^2\cos\phi\sin\phi = w_\phi - u_\phi \tag{11}$$

where $w = [w_r\ w_\theta\ w_\phi]$ and $u = [u_r\ u_\theta\ u_\phi]$ are the acceleration vectors of the target and the missile, respectively. To apply the nonlinear $H_\infty$ control theory, Eq.(9) to Eq.(11) need to be recast into the set of first-order differential equations:

$$\dot{x} = f(x) + g_1(x)w + g_2(x)u \tag{12}$$

where

$$x = \begin{bmatrix} r \\ \phi \\ V_r \\ V_\theta \\ V_\phi \end{bmatrix}, \quad f(x) = \begin{bmatrix} V_r \\ \frac{V_\phi}{r} \\ \frac{V_\theta^2+V_\phi^2}{r} \\ -\frac{V_rV_\theta}{r} + \frac{V_\theta V_\phi \tan\phi}{r} \\ -\frac{V_rV_\phi}{r} - \frac{V_\theta^2 \tan\phi}{r} \end{bmatrix}$$

$$g_1(x) = -g_2(x) = \begin{bmatrix} 0 & 0 & 0 \\ 0 & 0 & 0 \\ 1 & 0 & 0 \\ 0 & 1 & 0 \\ 0 & 0 & 1 \end{bmatrix} \tag{13}$$

The penalized output is

$$z = h_1(x) + k_{12}(x)u \tag{14}$$

where

$$h_1(x) = \begin{bmatrix} \rho h(r, V_\theta, V_\phi) \\ 0 \end{bmatrix}, \quad k_{12}(x) = \begin{bmatrix} 0 \\ I \end{bmatrix} \tag{15}$$

with the $h$ function defined as

$$h(r, V_\theta, V_\phi) = \frac{1}{r}(V_\theta^2 + V_\phi^2) = r(\dot\theta^2 \cos^2\phi + \dot\phi^2) \tag{16}$$

The problem of $H_\infty$ guidance law design now can be stated as: find the missile acceleration command $u = [u_r \; u_\theta u_\phi]^T$ such that the $L_2$-gain of the system described by Eq.(13) and Eq.(15) is lower than or equal to $\gamma$, i.e.,

$$\frac{\int_0^\infty z^T z dt}{\int_0^\infty w^T w dt} = \frac{\int_0^\infty (\rho^2 {h_1}^2 + u^T u)dt}{\int_0^\infty (w_r^2 + w_\theta^2)dt} \leq \gamma^2 \quad \forall w \in \mathbf{L_2} \tag{17}$$

## IV. Solutions of Hamilton-Jacobi PDI

In this section, the 3D $H_\infty$ guidance law is to design the missile acceleration $u$ such that the guidance system is internally stable and $\|z\|_2 \leq \gamma\|w\|_2$. The associated HJPDI is obtained by substituting $f(x)$, $g_1(x)$, $g_2(x)$, $h_1(x)$, $k_{12}(x)$ into Eq.(7) as

$$V_r\frac{\partial U}{\partial r} + \frac{V_\phi}{r}\frac{\partial U}{\partial \phi} + \frac{1}{r}(V_\theta^2 + V_\phi^2)\frac{\partial U}{\partial V_r}$$

$$+\frac{V_\theta}{r}(-V_r + V_\phi \tan\phi)\frac{\partial U}{\partial V_\theta} - \frac{1}{r}(V_rV_\phi + V_\theta^2 \tan\phi)\frac{\partial U}{\partial V_\phi}$$

$$+\frac{1}{2}\left(\frac{1}{\gamma^2} - 1\right)\left[\left(\frac{\partial U}{\partial V_r}\right)^2 + \left(\frac{\partial U}{\partial V_\theta}\right)^2 + \left(\frac{\partial U}{\partial V_\phi}\right)^2\right]$$

$$+\frac{\rho^2}{2r^2}(V_\theta^2 + V_\phi^2)^2 \leq 0 \tag{18}$$

This is a nonlinear second-order partial differential inequality in the unknown function $U(r, \phi, V_r, V_\theta, V_\phi)$. If a qualified $U$ can be found, the nonlinear $H_\infty$ guidance law is then given by Eq.(6) as

$$u(r, \phi, V_r, V_\theta, V_\phi) = -g_2^T\left(\frac{\partial U}{\partial x}\right)^T$$

$$= \begin{bmatrix} \partial U/\partial V_r \\ \partial U/\partial V_\theta \\ \partial U/\partial V_\phi \end{bmatrix} \tag{19}$$

,i.e., $u_r = \partial U/\partial V_r$, $u_\theta = \partial U/\partial V_\theta$, $u_\phi = \partial U/\partial V_\phi$. Hence, the main problem of $H_\infty$ guidance law design is to find a positive function $U$ satisfying the HJPDI in Eq.(18). It can be seen that the nature of the solution depends on the magnitude of the disturbance attenuation level $\gamma$: (i) $\gamma > 1$, the sixth term in the HJPDI is always negative, (ii) $\gamma < 1$, the sixth term in the HJPDI is always positive. These two kinds of solutions will be derived separately in the following.

### (A) $\gamma > 1$

In this case, the solution of the HJPDI may be non-unique, but a possible candidate can be found from the relations:

$$V_r\frac{\partial U}{\partial r} + \frac{V_\phi}{r}\frac{\partial U}{\partial \phi} + \frac{V_\theta}{r}(-V_r + V_\phi \tan\phi)\frac{\partial U}{\partial V_\theta}$$

$$+\frac{1}{2}\left(\frac{1}{\gamma^2} - 1\right)\left[\left(\frac{\partial U}{\partial V_r}\right)^2 + \left(\frac{\partial U}{\partial V_\theta}\right)^2 + \left(\frac{\partial U}{\partial V_\phi}\right)^2\right] \leq 0 \tag{20a}$$

$$\frac{1}{r}(V_\theta^2 + V_\phi^2)\frac{\partial U}{\partial V_r} + \frac{\rho^2}{2r^2}(V_\theta^2 + V_\phi^2)^2 \leq 0 \tag{20b}$$

It is noted that a function $U$ satisfying Eq.(20) is also a solution of the HJPDI in Eq.(18), but the inverse is not true. The solution of Eq.(20b) can be found readily as

$$U(r, \phi, V_r, V_\theta, V_\phi) = \frac{\lambda V_r(V_\theta^2 + V_\phi^2)}{r} \tag{21}$$

where

$$\lambda \leq -\rho^2/2 \tag{22}$$

. The simplest positive solution is obtained by taking $f(r, \phi, V_\theta, V_\phi)$ as a constant $c$. The magnitude of $c$ is determined to guarantee the positiveness of $U$. If the closing speed $V_r$ is negative, then the first term in $U$ is always positive, which implies that $c$ can be taken as 0. If we allow $V_r$ to be positive, a large enough $c$ must be chosen. To ensure that this $c$ does exist, we need to show that $\lambda V_r(V_\theta^2 + V_\phi^2)/r$ is bounded. $V_r$ is bounded since $V_r$ is one of the state $x = [r \; \phi \; V_r \; V_\theta \; V_\phi]$ which is stable in the sense of Lyapunov as proved in section II. $\frac{1}{r}(V_\theta^2 + V_\phi^2) = r(\dot\theta^2 \cos^2\phi + \dot\phi^2) = h$ is bounded for all bounded $w$ by noting Eq.(17). The boundedness of $V_r$ and $(V_\theta^2 + V_\phi^2)/r$ implies that there exists a finite $c$ such that the $U$ in Eq.(21) is positive. To check whether the $U$ given by Eq.(21) is feasible, we substitute this function into Eq.(20a), and the solution of the inequality is

$$\lambda \leq -\frac{3}{2\left(1-\frac{1}{\gamma^2}\right)} \tag{23a}$$

$$\lambda \leq \frac{1}{1-\frac{1}{\gamma^2}}\left[1-\sqrt{1+\rho^2(1-\frac{1}{\gamma^2})}\right] \tag{23b}$$

The final admissible range of $\lambda$ is determined by the intersection of the intervals in Eq.(22) and Eq.(23). The answer is Eq.(23a), and the corresponding weighting factor $\rho$ in the following range:

$$\rho \leq \sqrt{\frac{3}{1-\frac{1}{\gamma^2}}} \tag{24}$$

It can be seen that the 3D $H_\infty$ guidance law does not need the information of target's acceleration $w_r$ and $w_\theta$; nevertheless, the line-of-sight information provided by seeker is still required.

**(B) $\gamma < 1$**

A smaller value of the disturbance attenuation level $\gamma$ means that the interceptive performance is less sensitive to the target's maneuvering; however, more stringent conditions need to be imposed on the corresponding robust guidance law. Inspecting the HJPDI (18), we search for the possible solution candidate from the following auxiliary inequality:

$$\frac{1}{2}\left(\frac{1}{\gamma^2}-1\right)\left(\frac{\partial U}{\partial V_r}\right)^2+\frac{(V_\theta^2+V_\phi^2)}{r}\left(\frac{\partial U}{\partial V_r}\right) + \frac{\rho^2(V_\theta^2+V_\phi^2)^2}{2r^2} \leq 0 \tag{25}$$

By treating Eq.(25) as a second-order algebraic inequality in $\frac{\partial U}{\partial V_r}$, the solution set can be found as

$$\frac{\partial U}{\partial V_r} = \lambda\frac{(V_\theta^2+V_\phi^2)}{r} \tag{26}$$

where

$$\frac{-1-\sqrt{1-\rho^2(1/\gamma^2-1)}}{1/\gamma^2-1} \leq \lambda \leq \frac{-1+\sqrt{1-\rho^2(1/\gamma^2-1)}}{1/\gamma^2-1} \tag{27}$$

It can be seen that in this case the weighting factor $\rho$ can not be assigned arbitrarily, and must be within the interval

$$0 \leq \rho \leq \frac{1}{\sqrt{1/\gamma^2-1}} \tag{28}$$

The integration of Eq.(26) gives

$$U = \lambda\frac{V_r(V_\theta^2+V_\phi^2)}{r} \tag{29}$$

This $U$ is positive by noting that $\lambda$ is negative from Eq.(27) and the closing speed $V_r$ also must be negative to ensure successful interception. We substitute this $U$ into Eq.(18) to check whether this form of $U$ can satisfy the HJPDI. After some manipulations, we get

$$\frac{1}{2}\left(\frac{1}{\gamma^2}-1\right)\frac{\lambda^2(V_\theta^2+V_\phi^2)^2}{r^2}+\frac{\lambda(V_\theta^2+V_\phi^2)^2}{r^2}+\frac{\rho^2(V_\theta^2+V_\phi^2)^2}{2r^2} + \left(\frac{1}{\gamma^2}-1\right)\frac{2\lambda^2V_r^2(V_\theta^2+V_\phi^2)}{r^2}-\frac{3\lambda V_r^2(V_\theta^2+V_\phi^2)}{r^2} \leq 0 \tag{30}$$

Define the total velocity $V^2(t) = V_r^2(t)+V_\theta^2(t)+V_\phi^2(t)$, $V_\Theta^2(t) = V_\theta^2(t)+V_\phi^2(t)$, and then, expand the above equation into the following form:

$$-\frac{\lambda^2V_r^2V_\Theta^2}{2\gamma^2r^2}+\frac{\lambda^2V_\Theta^2V^2}{2\gamma^2\rho^2}+\frac{\lambda V_\Theta^4}{r^2}-\frac{\lambda^2V_\Theta^4}{2r^2}+\frac{\rho^2V_\Theta^4}{2r^2} -\frac{3\lambda V_r^2V_\Theta^2}{r^2}-\frac{2\lambda V_r^2V_\Theta^2}{r^2}+\frac{2\lambda^2V_r^2V_\Theta^2}{\gamma^2r^2} \leq 0 \tag{31}$$

Collecting the terms $V_r^2V_\Theta^2/r^2$, and substituting $V_\Theta^2(t) = V^2(t)-V_r^2(t)$ into the above equation.

$$\frac{V_r^2V_\Theta^2}{r^2}\left[-(2-\frac{3}{2\gamma^2})\lambda^2-3\lambda\right]+\frac{\lambda V_\Theta^2(V^2-V_r^2)}{r^2}+\frac{\lambda^2V_\Theta^2V^2}{2\gamma^2r^2} -\frac{\lambda^2V_\Theta^2(V^2-V_r^2)}{2r^2}+\frac{\rho^2V_\Theta^2(V^2-V_r^2)}{2r^2} \leq 0 \tag{32}$$

To combine the similar terms and the solution of the inequality needs to be satisfied the following:

$$\lambda \geq 0 \quad \text{or} \quad \lambda \leq \frac{-3}{2-\frac{3}{2\gamma^2}} \tag{33a}$$

$$\frac{-1-\sqrt{1-\rho^2(1/\gamma^2-1)}}{1/\gamma^2-1} \leq \lambda \leq \frac{-1+\sqrt{1-\rho^2(1/\gamma^2-1)}}{1/\gamma^2-1} \tag{33b}$$

$$1-\sqrt{1+\rho^2} \leq \lambda \leq 1+\sqrt{1+\rho^2} \tag{33c}$$

The final admissible range of $\lambda$ is determined by the intersection of the intervals in Eq.(33). The answer depends on the choice of the weighting factor $\rho$ for the interval as

$$\frac{2\sqrt{3(7\gamma^2-1)}\gamma}{4\gamma^2-3} \leq \rho \leq \sqrt{\frac{\gamma^2}{1-\gamma^2}} \tag{34}$$

The allowable interval of $\lambda$ is given by

$$1-\sqrt{1+\rho^2} \leq \lambda \leq \frac{-3}{2-\frac{3}{2\gamma^2}} \tag{35}$$

The corresponding range of $\gamma$ is

$$\frac{\sqrt{3}}{2} \leq \gamma \leq 1 \tag{36}$$

In summary, the solution of the HJPDI with $\gamma$ is given by Eq.(29) where the constant $\lambda$ locates within the interval of Eq.(35). The form of the associated 3D robust guidance law with $\gamma < 1$ is the same as it with $\gamma > 1$. The constant $\lambda$ can be regarded as the missile's navigation gain whose value depends on the disturbance level $\gamma$ to be achieved and the weighting factor $\rho$. The smaller $\gamma$ is, the more robust the missile will be with respect to the variations of the target's maneuvers.

## V. Numerical Results

In this section, we exploit numerical simulations to justify the feasibility of the above theoretical derivations. To make the analysis independent of the physical units, we normalize the governing equations to their dimensionless forms in terms of the following dimensionless variables. The relative distance $r$ is normalized to $\bar{r} = r/r_0$; the relative velocities $V_r$, $V_\theta$, and $V_\phi$ are normalized to $\bar{V}_r = V_r/V_0$, $\bar{V}_\theta = V_\theta/V_0$, and $\bar{V}_\phi = V_\phi/V_0$ respectively, where $V_0 = \sqrt{V_{r0}^2 + V_{\theta_0}^2 + V_{\phi_0}^2} = \sqrt{\dot{r}_0^2 + (r_0\dot{\theta}_0 \cos\phi_0)^2 + (r_0\dot{\phi}_0)^2}$; the acceleration components $u_r$, $u_\theta$, and $u_\phi$ are normalized to $\bar{u}_r = u_r/(V_0^2/r_0)$, $\bar{u}_\theta = u_\theta/(V_0^2/r_0)$, and $\bar{u}_\phi = u_\phi/(V_0^2/r_0)$ respectively; the time $t$ is normalized to $\tau = t/(r_0/V_0)$. In the following, the variables with the over-line symbol denote dimensionless variables. The performance and robustness of four missile guidance laws will be tested and compared. Their acceleration commands are listed below.

**(M1)** $H_\infty$ guidance law with $\gamma > 1$: $\bar{u}_r = \lambda(\bar{V}_\theta^2 + \bar{V}_\phi^2)/\bar{\rho}$, $\bar{u}_\theta = 2\lambda\bar{V}_r\bar{V}_\theta/\bar{\rho}$, and $\bar{u}_\phi = 2\lambda\bar{V}_r\bar{V}_\phi/\bar{\rho}$.

**(M2)** $H_\infty$ guidance law with $\gamma < 1$: $\bar{u}_r = \lambda(\bar{V}_\theta^2 + \bar{V}_\phi^2)/\bar{\rho}$, $\bar{u}_\theta = 2\lambda\bar{V}_r\bar{V}_\theta/\bar{\rho}$, and $\bar{u}_\phi = 2\lambda\bar{V}_r\bar{V}_\phi/\bar{\rho}$.

**(M3)** Realistic true proportional navigation[19] (RTPN): $\bar{u}_r = 0$, $\bar{u}_\theta = -\lambda(d\phi/d\tau)\bar{V}_r$, and $\bar{u}_\phi = -l(d\theta/d\tau)\cos\phi\bar{V}_r$.

**(M4)** Pure proportional navigation[20](PPN) guidance law: $\bar{u}_r = -\lambda\bar{V}'_M\sqrt{(d\phi/d\tau)^2 + (d\theta/d\tau\cos\phi)^2}$ $\bar{u}_\theta = \lambda\bar{V}'_M(d\theta/d\tau)\cos\phi\cos(\psi_0 - k\Psi)$, and $\bar{u}_\phi = \lambda\bar{V}'_M(d\phi/d\tau)\cos(\psi_0 - k\Psi)$ where $\bar{V}'_M$ is the normalized missile's velocity in pursuit plane; $k = \lambda - 1$ with $\lambda$ being the navigation gain; $\psi_0$ is the initial aspect angle of $\vec{V}'_M$ with respect to the inertial reference line $\vec{e_r}$.

- Maneuvering targets:

In this part, we employ seven maneuvering strategies for targets to test the robust capability of the 3D $H_\infty$ guidance laws. These six different targets which are extended from 2D plane to 3D space include:

**(T1)** Smart target: $\bar{w}_T = \frac{\lambda_T}{\bar{\rho}\bar{\Omega}}\vec{e_h}$. where $\bar{\Omega}$ is normalized angular velocity.

**(T2)** Modified smart target: $\bar{w}_T = \frac{\lambda_T}{(d\bar{\rho}/d\tau)\bar{\Omega}}\vec{e_h}$

**(T3)** Sinusoidal target: $\bar{w}_T = \lambda_T\sin(\bar{\Omega}\tau)\vec{e_r} + \lambda_T\sin(\bar{\Omega}\tau)\vec{e_h}$

**(T4)** Ramp target: $\bar{w}_T = \lambda_T\tau\vec{e_r} + \lambda_T\tau\vec{e_h}$

**(T5)** Step target: $\bar{w}_T = \lambda_T\vec{e_r} + \lambda_T\vec{e_h}$

**(T6)** RTPN target: $\bar{u}_r = 0$, $\bar{u}_\theta = -\lambda(d\phi/d\tau)\bar{V}_r$, and $\bar{u}_\phi = -l(d\theta/d\tau)\cos\phi\bar{V}_r$.

It should be emphasized that these target models are only for demonstration purpose and other more practical target models can be chosen to test the 3D $H_\infty$ guidance law. Whatever target model is used, the 3D $H_\infty$ guidance law can guarantee the disturbance attenuation level lower than $\gamma$ as long as target's acceleration is finite. Here, we choose the target's navigation gains $\lambda_T$, as the uncertainty disturbances to test the robustness properties of the $H_\infty$ guidance laws. From the definition of the system's $L_2$-gain (see Eq.(17)), we recognize that when $\gamma$ is small, it indicates that good interceptive performance can be preserved in the presence of large target maneuvers (i.e.large $\| w \|_2$). In the following, the simulations will show that using the 3D $H_\infty$ robust guidance law the system's $L_2$-gain can be kept below $\gamma$ under variations of the target's maneuvering strategy.In Fig.2 and Fig.3, the two $H_\infty$ guidance laws demonstrate robustness against the seven aforementioned maneuvering targets, being able to keep the $L_2$-gain smaller than $\gamma$ for different target's navigation gain and for different target's maneuvering strategies. A comparison between the results of Fig.2 and Fig.3 reveals that the robustness of the 3D $H_\infty$ guidance law with $\gamma < 1$ is better than that of the $H_\infty$ guidance law with $\gamma > 1$. On the other hand, if the missile's guidance law employs the conventional proportional navigations such as RTPN(see Fig.4), the disturbance attenuation level may diverge for some specific maneuvering targets. We can find from Fig.4 that the RTPN missile guidance law has poor ability to pursue the step target and the ramp target. In summary, among the four missile guidance laws, the 3D $H_\infty$ guidance laws, especially for $\gamma < 1$, exhibit the most robust performance with respect to the variation of the target's navigation gains.

## VII. Conclusions

The nonlinear $H_\infty$ control theory has been exploited in this paper to design guidance law for homing missiles. And in this paper, we extend 2D $H_\infty$ missile guidance laws into 3D $H_\infty$ missile guidance laws successfully. By regarding target maneuvers as disturbances inputs, we reformulate the missile guidance problem as a nonlinear disturbance attenuation control problem. After solving the associated Hamilton-Jacobi partial differential inequality, we obtain three kinds of 3D $H_\infty$ guidance laws which possess excellent performance robustness against maneuvering targets. As compared with the proportional navigation schemes, the performance of the proposed 3D $H_\infty$ robust guidance laws is shown to be more insensitive to variations of target's maneuvers.

## References


[1] Adler, F.P., "Missile Guidance by Three-Dimensional Proportional Navigation, " Journal of Applied Physics, Vol.27, No.5, 1956, pp.500-507. [2] Lin, C.F., "Classical Modern Control System Design for Terminal Guidance of Bank-to-Turn Intercept Missiles, "Proceedings of The AIAA Guidance and Control Conference, AIAA, New York, 1983, pp.283-301. [3] Cochran, J.E., Jr., No, T.S., and Thaxton, D.G., "Analytical Solution to a Guidance Problem, " Journal of Guidance, Control, and Dynamics, Vol.14, No.1, 1991, pp.117-122. [4] Kumar, R.R., Seywald, H., and Cliff, E.M., "Near-Optimal Three-Dimensional Air-to-Air Missile Guidance Against Maneuvering Target," Journal of Guidance, Control, and Dynamics, Vol.18, No.3, 1995, pp.457-464. [5] Lim, J., and Lee, S., "Bank-to Turn Optimal Guidance with Linear Exponential Quadratic Gaussian Performance Criterion, " Journal of Guidance, Control, and Dynamics, Vol.18, No.5, 1995, pp.951-958. [6] Yang, C.D., and Yang, C.C., "Analytical Solution of Generalized Three-Dimensional Proportional Navigation," Journal of Guidance, control, and Dynamics, Vol.19, no.3, 1996, pp.721-724. [7] Chiou, Y.C., and Kuo, C.Y., "Geometric Approach to Three-Dimensional Missile Guidance Problem," Journal of Guidance, Control, and Dynamics, Vol.21, No.2, March-April, 1998, pp.335-341.


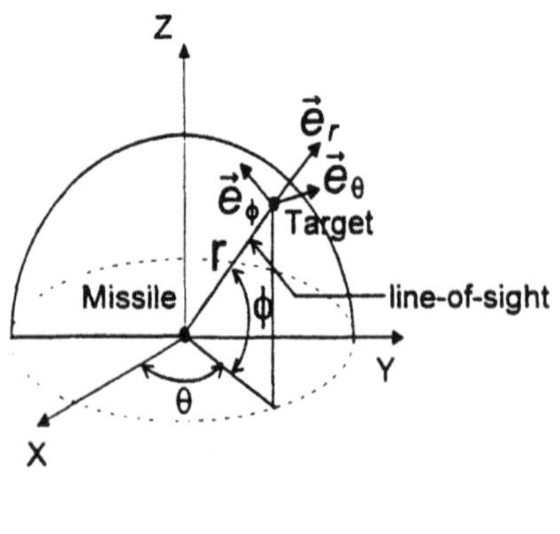


Fig. 1 Three-dimensional pursuit geometry

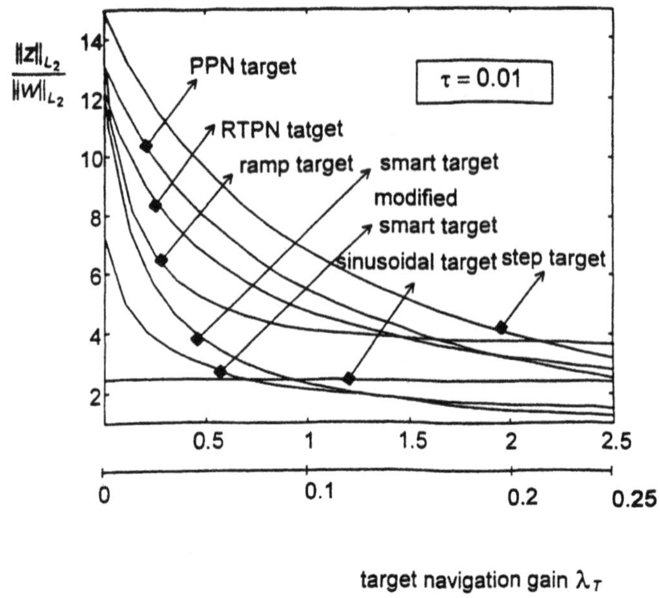


Fig. 2 Robustness of $L_2$-gain for $H_\infty$ guidance law with $\gamma > 1$

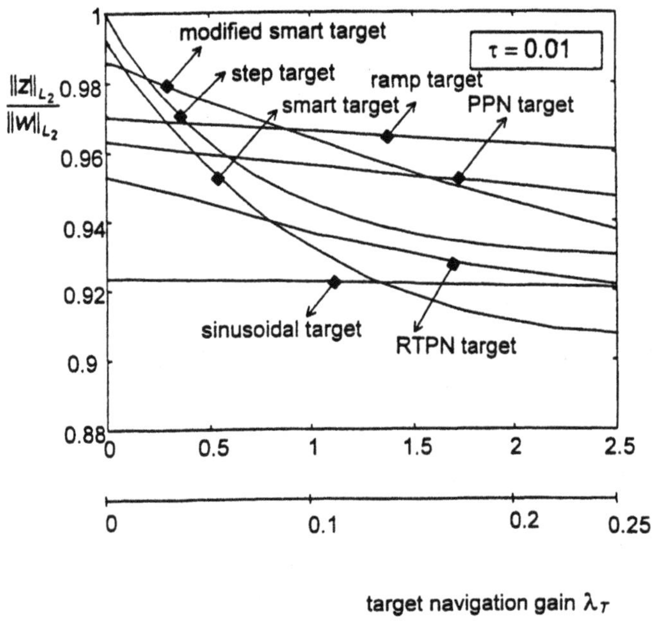


Fig. 3 Robustness of $L_2$-gain for $H_\infty$ guidance law with $\gamma < 1$

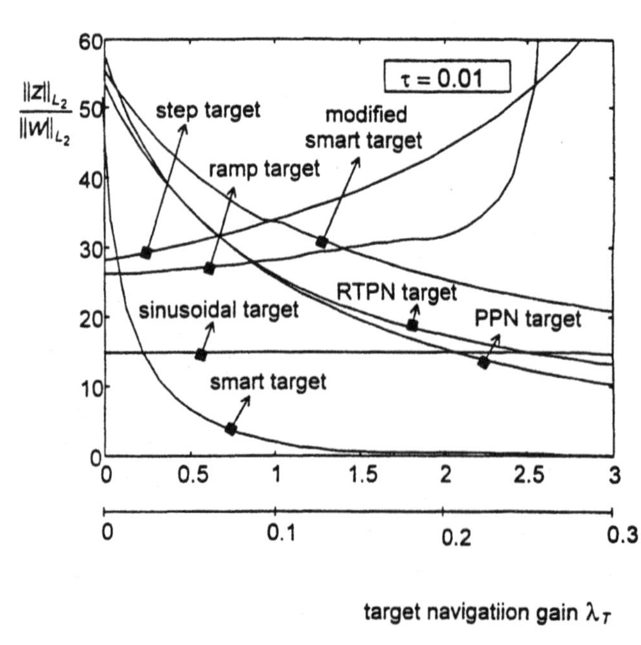


Fig. 4 Robustness of $L_2$-gain for RTPN guidance law

14th Triennial World Congress, Beijing, P.R. China



# NONLINEAR ROBUST CONTROLLER DESIGN BASED ON $H_\infty$ I/O LINEARIZATION AND $\mu$-SYNTHESIS

**Shr-Shiung Hu**

*202nd Arsenal Combined Service Forces P. O. Box 90582 Nankang, Taiwan JL89012@ms26.hinet.net*

***B. C. Chang****

*Department of Mechanical Engineering and Mechanics Drexel University, Philadelphia, PA 19104, U. S. A. bchang@coe.drexel.edu TEL: (215) 895-1790*

***H. H. Yeh***

*WL/FIGC-3, 2210 8th Street, Building 146 Wright-Patterson Air Force Base, Dayton, Ohio 45433-7531, U. S. A. yehh@wl.wpafb.af.mil*

***H. G. Kwatny***

*Department of Mechanical Engineering and Mechanics Drexel University, Philadelphia, PA 19104, U. S. A. Harry.kwatny@cbis.ece.drexel.edu*

Abstract: In this paper, an $H_\infty$ approximate I/O linearization and the $\mu$-synthesis will be employed to obtain a nonlinear robust controller for an aircraft longitudinal flight control problem. Simulations of the tracking responses of the closed-loop system will be also included. *Copyright © 1999 IFAC*

Keywords: nonlinear control, robust control, flight control

## 1. INTRODUCTION

A robust nonlinear control design approach using the $H_\infty$ approximate input/output (I/O) linearization (Hu *et al.*, 1998b,c; Zong *et al.*, 1995) in the inner-loop and $\mu$-synthesis (Balas *et al.*, 1993; Chang *et al.*, 1998; Zhou *et al.*, 1997) in the outer-loop is proposed and applied to a nonminimum-phase longitudinal flight control problem. Like the feedback I/O linearization (Chang *et al.*, 1998; Isidori 1995), the $H_\infty$ approximate I/O linearization modifies a nonlinear system to resemble that of a linear system. Unlike the feedback I/O linearization, the $H_\infty$ approximate I/O linearization does not require full state feedback or the plant being minimum phase. A nonlinear $H_\infty$ controller is to be found so that the compensated inner loop nonlinear system approximates a linear reference model. The $H_\infty$ approximate I/O linearization problem is formulated as an output feedback nonlinear $H_\infty$ control problem and hence the nonlinear $H_\infty$ controller formulas( Ball *et al.*, 1993; Isidori and Astolfi, 1992; van der Schaft, 1992, 1996) will be employed to obtain an inner-loop nonlinear $H_\infty$ controller. To construct a nonlinear $H_\infty$ controller one needs to solve the Hamilton-Jacobi inequalities (HJIs) or equations (HJEs). Up to date, there is no computational algorithm for the exact explicit solution of the HJE; however, an approximate solution can be obtained by using a successive computational approach (van der Schaft 1992; Hu *et al.*, 1998b,c) . The solution is in the form of a power series in which the first term is computed via the solution of the corresponding algebraic Riccati equation (ARE). A numerical difficulty arises in solving the ARE corresponding to the second HJE when $(-\overline{A}, B_1 D_\perp)$ is not stabilizable where $\overline{A}, B_1$, and $D_\perp$ are determined from the system matrices of the linearized model (Hu *et al.*, 1998c). A preliminary discussion on this computational issue for a simpler problem, with DGKF (Doyle *et al.*, 1989) kind of assumptions, was presented in (Hu *et al.*, 1998a). In (Hu *et al.* 1998c), We proposed modified nonlinear $H_\infty$ controller formulas to eliminate the numerical difficulty for a more general problem. We also showed that the $H_\infty$

* This research was supported in part by NASA Langley Research Center under Contract NCC-1-224 and in part by the Boeing Company under Contract NAS1-20220.

I/O linearization problem falls into the category of the "more general problem" in which $(-\bar{A}, B_1 D_\perp)$ is not stabilizable; hence, the modified $H_\infty$ controller formulas (Hu *et al.,* 1998c) are needed for the proposed nonlinear robust controller design.

In the inner-loop controller design, we first choose a linear reference, for which a convenient choice is the Taylor linearization of the nonlinear system at the equilibrium point of interest. Then, a nonlinear $H_\infty$ controller is designed to minimize the $H_\infty$ norm of the difference between the compensated nonlinear system and the linear reference model. In the outer-loop controller design, we start by representing the compensated inner-loop nonlinear system by the linear reference model with an uncertainty characterized by the $H_\infty$ norm. A linear $H_\infty$ optimization (Doyle *et al.*, 1989; Glover and Doyle, 1988) or $\mu$-synthesis problem formulation is then set up to construct a linear robust controller that addresses tracking, regulation, and robustness issues. The overall controller is the integration of the inner-loop nonlinear $H_\infty$ controller and the outer-loop linear robust controller. The proposed approach will be applied to an aircraft longitudinal flight control problem (Kwatny *et al.* 1991, 1998).

The rest of the paper is organized as follows. Section 2 includes the notations and a brief review of nonlinear $H_\infty$ control. The design of the inner-loop nonlinear controller and the outer-loop linear robust controller for the longitudinal flight control problem are demonstrated in Sections 3 and 4 respectively. Time responses of the tracking problem for the closed-loop system are also included in Section 4. Finally, Section 5 gives some concluding remarks.

## 2. NONLINEAR $H_\infty$ CONTROL

### 2.1 Notations

The notations used in this paper are fairly standard. $x^T$ is the transpose of vector $x$. $\|x\|^2 = x^T x$ is the squared Euclidean norm. $\mathbf{R}^n$ is the $n$-dimensional Euclidean space. $\xi$ is the state of the controller. Let $A \in \mathbf{R}^{n\times n}$, then $A > 0$ means that $A$ is positive definite and $A \geq 0$ means $A$ is positive semi-definite. ARE is the algebraic Riccati equation and ARI is the algebraic Riccati inequality. HJE and HJI are the Hamilton-Jacobi equation and inequality respectively.

### 2.2 The Nonlinear $H_\infty$ Control Problem

Consider the following nonlinear input-affine generalized plant:

$$G : \begin{cases} \dot{x} = f(x) + g_1(x)w + g_2(x)u \\ z = h_1(x) + D_{12}(x)u \\ y = h_2(x) + D_{21}(x)w \end{cases} \tag{1}$$

where $x \in \mathbf{R}^n$ is the state of the system, $z \in \mathbf{R}^{p1}$ is the controlled output, $w \in \mathbf{R}^{m1}$ is the exogenous input including all commands and disturbances, $u \in \mathbf{R}^{m2}$ represents the control input, and $y \in \mathbf{R}^{p2}$ is the measured output. The nonlinear system (1) is assumed to satisfy the following:

$$D_{12}^T(x)D_{12}(x) := E_1(x) > 0 \tag{2a}$$

$$D_{21}(x)D_{21}^T(x) := E_2(x) > 0 \tag{2b}$$

The problem is to find a controller

$$K : \begin{cases} \dot{\xi} = A_K(\xi) + B_K(\xi)y \\ u = C_K(\xi) \end{cases} \tag{3}$$

such that the closed-loop system is stable and $\gamma$-dissipative (Hu *et al.*, 1998b,c) or equivalently the $L_2$-gain (van der Schaft, 1992) of the system is less than or equal to $\gamma$, a positive prescribed number.

For the definition of the linearized model, the properties for the linearized model, the associated algebraic Riccati equations (AREs) or inequalities (ARIs), the computational issue regarding to the numerical difficulty encountered when we solve the second HJE, and the problem formulation for the $H_\infty$ approximate I/O linearization problem, the reader can refer to (Hu *et al.*, 1998b,c).

## 3. DESIGN OF AN INNER-LOOP NONLINEAR $H_\infty$ CONTROLLER

In (Hu *et al.*, 1998b,c), we formulate the nonlinear $H_\infty$ approximate I/O linearizaton problem as an output feedback nonlinear $H_\infty$ control problem. The objective is to find a nonlinear $H_\infty$ controller such that the compensated nonlinear system approximates the linear reference model in I/O behavior. The modified nonlinear $H_\infty$ controller formulas mentioned in (Hu *et al.*, 1998b,c) will be employed to design an inner-loop nonlinear $H_\infty$ controller to an aircraft longitudinal flight control problem.

The nonlinear model of the aircraft longitudinal dynamics (Kwatny *et al.*, 1991, 1998) is

$$\begin{bmatrix} \cos a & -V\sin\alpha & V\sin\alpha & 0 \\ \sin a & V\cos\alpha & -V\cos\alpha & 0 \\ 0 & 0 & 1 & 0 \\ 0 & 0 & 0 & 1 \end{bmatrix} \frac{d}{dt}\begin{bmatrix} V \\ \alpha \\ \theta \\ q \end{bmatrix} \quad (4)$$

$$= \begin{bmatrix} -\sin\theta + \Lambda_w \sin\alpha + \Lambda_t \sin\alpha_t + T - \Delta_f \cos\alpha \\ \cos\theta - \Lambda_w \cos\alpha - \Lambda_t \cos\alpha_t - \Delta_f \sin\alpha \\ q \\ \frac{V_*^2 l^*}{gr^2}\{\Sigma_w + \kappa\Lambda_w \cos\alpha - (1-\kappa)\Lambda_t \cos\alpha_t\} - \frac{cV_*}{mgr^2} q \end{bmatrix}$$

where $\theta$ - pitch angle, $q$ - pitch rate, $V$ - normalized speed, $\alpha$ - angle of attack, $T$ - normalized engine thrust, and $\delta_e$ - elevator deflection angle. For illustrative and computational purposes we complete the model by specifying the following functions and parameters.

$$\Lambda_w = f_w(\alpha)\bar{\rho}V^2,\ \Lambda_t = f_t(\alpha_t)\bar{\rho}V^2,$$

$$\Delta_f = \left(a + b[f_w(\alpha)]^2\right)\bar{\rho}V^2,\ \Sigma_w = \sigma_w(\alpha)\bar{\rho}V^2$$

$$f_w(\alpha) = \left[\alpha - 2.05(\alpha - .05)^3\right]/.05,\ cV_*/mgr^2 = 8$$

$$f_t(\alpha_t) = .1\left[(\alpha_t - .05) - 3(\alpha_t - .05)^3\right]/.05$$

$$\sigma_w(\alpha) = 0,\ \bar{\rho} = 1, a = b = .05,\ V_*^2 l^*/gr^2 = 300$$

Let the state of the system and the control input be

$$x = \begin{bmatrix} x_1 & x_2 & x_3 & x_4 \end{bmatrix}^T = \begin{bmatrix} \theta & q & V & \alpha \end{bmatrix}^T \quad (5a)$$

$$u = \begin{bmatrix} T & \delta_e \end{bmatrix}^T \quad (5b)$$

The outputs are $V$ and $\gamma_{fp}$ where $\gamma_{fp}$ represents the flight path angle, i.e.,

$$\gamma_{fp} = \alpha - \theta \quad (6)$$

We are particularly interested in the horizontal flight, i.e., $q = 0$ and $\gamma_{fp} = 0$. The nonlinear model of the aircraft longitudinal flight control problem can be rewritten as follows:

$$\dot{x} = F(x, u, \kappa) \quad (7a)$$

where $F(x,u,\kappa)$ is a nonlinear function of the state vector $x$, control input $u$, and the parameter $\kappa$ which represents the center of gravity of the aircraft. The function is too lengthy to be included in the paper; it can be found in (Hu, 1998b). The output equation is

$$\bar{y} = \begin{bmatrix} V & \gamma_{fp} \end{bmatrix}^T \quad (7b)$$

Next, we will design a nonlinear $H_\infty$ controller to approximately linearize the nonlinear model in (7) based on the proposed nonlinear $H_\infty$ approximate I/O linearization.

It is easy to verify that $\theta = 0.1 := \theta_0$, $q = 0 := q_0$, $V = 0.9671 := V_0$, $\alpha = 0.1 := \alpha_0$, $T = 0.1165 := T_0$, $\delta_e = 0.01 := \delta_{e0}$, and $\kappa = 0.05611 := \kappa_0$ is an equilibrium point corresponding to horizontal flight. The following shift rules will be applied to (7) to shift the equilibrium point to $\boldsymbol{x = 0}$

Shift rules =

$$\left\{ \begin{matrix} \theta \to \theta + \theta_0, q \to q + q_0, V \to V + V_0, \alpha \to \\ \alpha + \alpha_0, T \to T + T_0, \delta_e \to \delta_e + \delta_{e0}, \kappa \to \kappa_0 \end{matrix} \right\} \quad (8)$$

Then (7a) becomes

$$\dot{x} = F(x,u,\kappa)\Big|_{Shift\ rules} := F_1(x,u,\kappa) \quad (9)$$

Next, we approximate (9) to the following nonlinear input-affine system (Hu, 1998b),

$$\dot{x} = f_0(x) + g_0(x)u \quad (10a)$$

$$\bar{y} = \begin{bmatrix} V - V_0 \\ \gamma_{fp} \end{bmatrix} \quad (10b)$$

where

$$f_0(x) = F_1(x, 0, \kappa) \quad (10c)$$

$$g_0(x) = \frac{\partial F_1}{\partial u}(x, 0, \kappa) \quad (10d)$$

Note that there are right-half plane zeros in the linearized model of (10) which means the system is non-minimum phase. Hence, the conventional feedback input-output linearization can not be employed to this aircraft longitudinal flight control problem. However, one can follow the nonlinear $H_\infty$ approximate I/O linearization problem formulation (Hu, *et al.*, 1998b,c) to build a nonlinear generalized plant. The new state variables of the generalized plant are

$$\begin{bmatrix} x & \tilde{x} \end{bmatrix}^T = \begin{bmatrix} \theta & q & V & \alpha & \tilde{\theta} & \tilde{q} & \tilde{V} & \tilde{\alpha} \end{bmatrix}^T := \begin{bmatrix} x_1 & x_2 & x_3 & x_4 & x_5 & x_6 & x_7 & x_8 \end{bmatrix}^T \quad (11)$$

With $\gamma = 1.1$, one can compute the solutions of the AREs which are defined in (Hu et al., 1998b,c). Note that $Y_\infty$ is singular as expected and the rank of $Y_\infty$ is 4. Following the algorithm of finding approximate solution of the HJE in (Hu *et al.*, 1998b,c), one can obtain $V^{(3)}(x)$ by solving 120 linear equations and $V^{(3)}(x)$ is shown in Appendix B in (Hu, 1998b). Then the second-order approximate solution of the HJE can be obtained by:

$$X^{[2]}(x) = Xx + \frac{1}{2}\frac{\partial^T V^{(3)}(x)}{\partial x} \quad (12)$$

## 4. DESIGN OF AN OUTER-LOOP ROBUST CONTROLLER AND SIMULATIONS

In this section, we will formulate a $\mu$-synthesis control problem to achieve command tracking, disturbance reduction, and robustness improvement so that an outer-loop linear robust controller can be constructed to provide robust stability/performance against the inexact dynamics matching in the inner-loop approximate I/O linearization design. Fig. 1 shows the block diagram for the outer-loop controller design. The compensated inner-loop nonlinear system is represented by the linear reference model and an uncertainty block $\Delta(s)$ which is characterized by an $H_\infty$ bound. The linear $H_\infty$ control and $\mu$-synthesis technique can be employed to design an outer-loop linear robust controller to address the uncertainty. Finally, the overall controller will be the combination of the nonlinear $H_\infty$ controller and the linear robust controller as shown in Fig. 2.

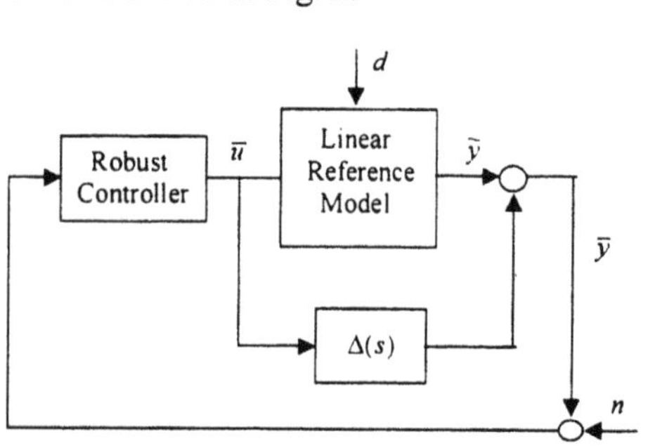


Fig. 1 Outer-loop robust controller design

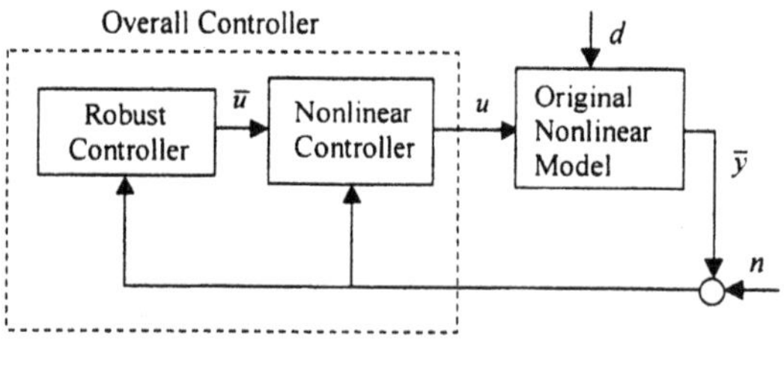


Fig. 2 Structure of the overall controller

In Fig. 3, $P_L$ stands for the linear reference model which is the linearized model of the nonlinear plant (7). The objective is to find a controller $K$ so that the closed-loop system is robustly stable and the outputs of the aircraft longitudinal flight controlled variables $V$ and $\gamma_{fp}$ follow $w_3$ and $w_4$, the reference tracking signals, respectively as closely as possible. $W_e$ is a weighting function for the tracking error, usually a low-pass filter; the combination of $W_\Delta$ and $\Delta_1$ represents the plant uncertainty, and usually $W_\Delta$ is a high-pass filter.

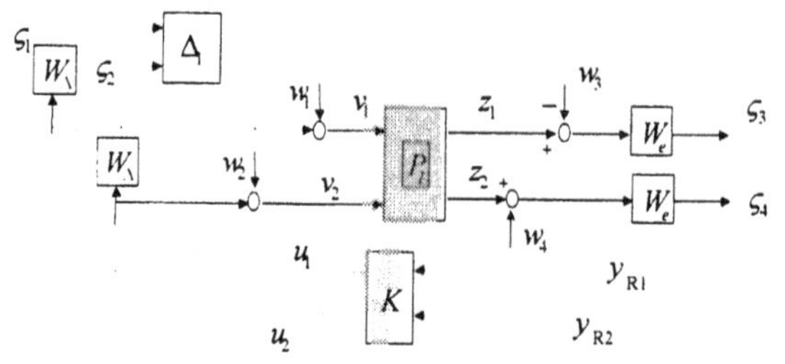


Fig. 3 Formulation of an outer-loop control problem.

$W_e$ and $W_\Delta$ are weighting functions chosen by the designers such that the design specifications can be met. We choose them as follows.

$$W_e = \frac{0.5}{s+0.03} \qquad W_\Delta = \frac{100(s+100)}{s+10000} \tag{13}$$

$W_e$ is a low-pass filter to emphasize the tracking accuracy at low frequencies. $W_\Delta$ is for plant uncertainties including the inexact dynamics matching in the inner-loop.

Combining the plant, $P_L$, and the weighting functions $W_e$ and $W_\Delta$, we have the generalized plant $G$ in Fig. 4. $\Delta_1$ is a $2\times2$ plant uncertainty block, and $\Delta_2$ a $2\times2$ fictitious performance block.

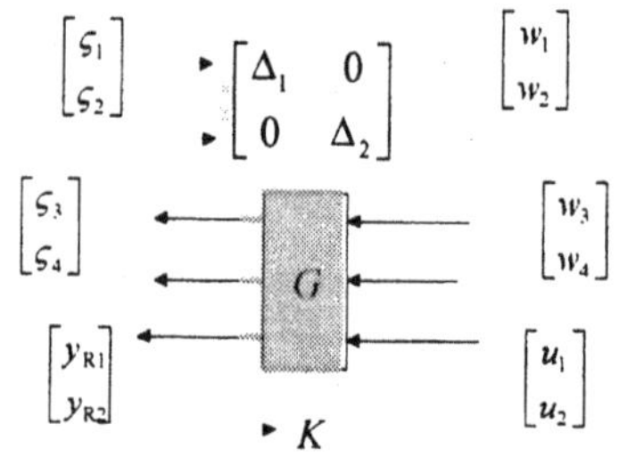


Fig. 4 Generalized plant for $\mu$-synthesis.

With the *D-K* iteration algorithm synthesis (Balas *et al.*, 1993; Chang *et al.*, 1998; Zhou *et al.*, 1997), we first obtained an $H_\infty$ controller $K_1(s)$ with optimal $H_\infty$ norm equal to 1.5463. Since $K_1(s)$ ignores the structure information of $\Delta$ and treats $\Delta$ as a full matrix, it gives a conservative solution to the problem. After four iterations, the process converges to a controller $K_4(s)$ with optimal $H_\infty$ norm reduced to 1.3701 which gives the $\bar{\sigma}$ plot and $\mu$ plot of $\mathfrak{F}_l[\hat{G}(s), K_4(s)]$ in Fig. 5, where $\mathfrak{F}_l[G(s), K(s)]$ is the lower linear fractional transformation.

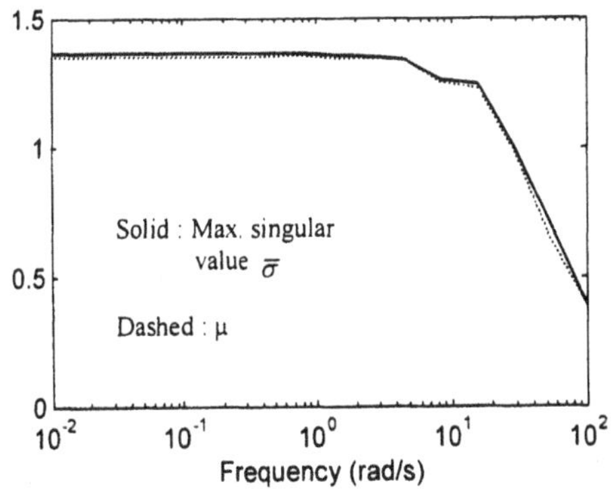


Fig. 5 $\bar{\sigma}$ and $\mu$ plots for the 4th D-K iteration $\mu$-synthesis control law

In the *D-K* iteration, we choose the order of the scaling function $D(s)$ to be 3 which implies that $\hat{G}(s)$, and hence $K_4(s)$, are of order 20. $K_4(s)$ can be reduced to a 10th order controller $K_r(s)$ by truncating its balanced realization.

Time response simulations for the closed-loop system with the nonlinear $H_\infty$ controller and reduced $\mu$-synthesis controller $K_r(s)$ will be given in the following.

*Time Response Simulations*

The simulation diagram is shown in Fig. 6 in which the original nonlinear model was given in (7). The inner-loop nonlinear controller was designed in Section 3. The outer-loop controller $K_r(s)$, of 10th order, was designed in this section by $\mu$-synthesis and model reduction.

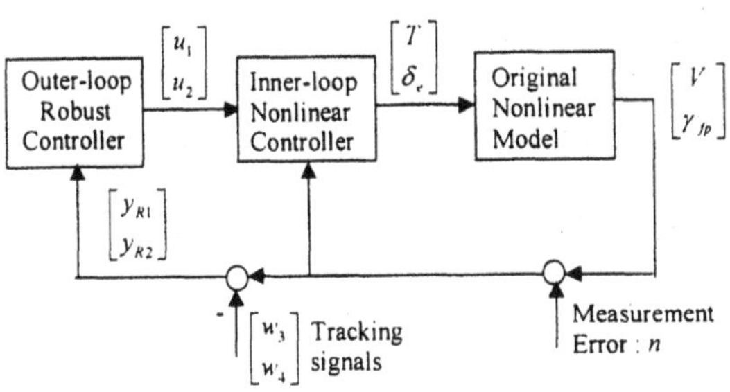


Fig. 6 Simulation Diagram.

First, we assume that there are no measurement errors. Let the tracking signals be

$$\begin{bmatrix} w_3 \\ w_4 \end{bmatrix} = \begin{bmatrix} A_V(1-e^{-t}) \\ A_\gamma(1-e^{-t}) \end{bmatrix} \tag{14}$$

where $A_V$ and $A_\gamma$ are the amplitude constants. The tracking signals can be regarded as a combination of low frequency signals or as the output of the low-pass filter $1/(s+1)$ driven by a step function. We start from the tracking signals with $A_V = -0.1$ and $A_\gamma = 0$. This means that the aircraft speed is reduced while maintaining the flight path angle zero (horizontal flight). Fig. 7 shows the tracking responses of the closed-loop system. We can see that the tracking error is very small for the $V$ response. Although the response for the flight path angle has oscillation at the beginning, it converges back to zero after 5 seconds.

Next, we assume that the tracking signals are $A_V = -0.1$ and $A_\gamma = 0.1$. Fig. 8 shows the tracking responses of the closed-loop system. We can see that the robust nonlinear controller works properly because the output time responses for the closed-loop system follow the tracking signals.

Now, we will apply the measurement errors to the system and check how they will affect the tracking responses. Let the measurement error be

$$n = \begin{bmatrix} 0.01\sin 100t \\ 0.01\sin 100t \end{bmatrix} \tag{15}$$

Fig. 9 shows that the measurement errors have only slight effect on the tracking response.

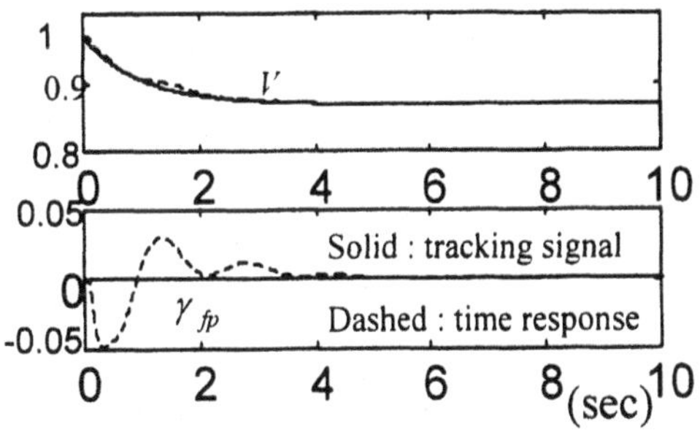


Fig. 7 Tracking responses for the closed-loop system with $A_V = -0.1$, $A_\gamma = 0$, and no measurement error.

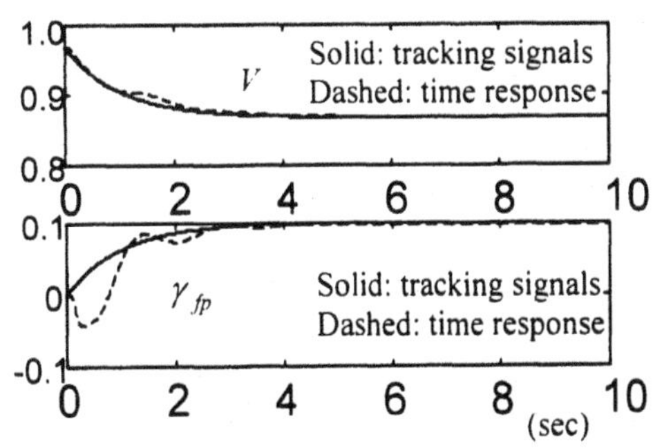


Fig. 8 Tracking responses for the closed-loop system with $A_V = -0.1$, $A_\gamma = 0.1$, and no measurement error.

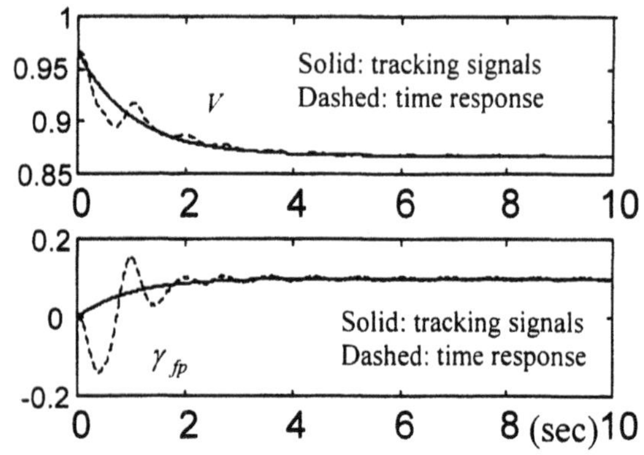


Fig. 9 Tracking responses for the closed-loop system with $A_V = -0.1$, $A_\gamma = 0.1$, and measurement error $n$.

## 5. CONCLUSIONS

In this paper, the $H_\infty$ approximate I/O linearization formulation was applied to design an inner-loop nonlinear controller for an aircraft longitudinal flight control problem. Then $\mu$-synthesis was employed to design an outer-loop robust controller and address tracking, regulation, and robustness issues. The time responses of the tracking signals for the closed-loop system revealed that the overall nonlinear robust controller is able to provide robust stability and robust performance for the plant uncertainties and state vector measurement errors.

## REFERENCES


Balas, G. J., Doyle, J. C., Glover, K., Packard, A., and Smith, R. (1993) $\mu$-Analysis and Synthesis Toolbox, The Mathworks, Inc., MA, MUSYN Inc., MN.

Ball, J. A., Helton, J. W., and Walker, M. L (1993). "Control for nonlinear systems with output feedback," *IEEE Trans. Automatic Control*, **Vol. 38**, pp. 545-959.

Chang, B. C., Kwatny, H. G., and Hu, S. S. (1998) " An Application of Robust Feedback Linearization to a Ball and Beam Control Problem," *1998 IEEE Conference on Control Applications*.

Doyle, J. C., Glover, K., Khargonekar, P. P., and Francis, B. A. (1989). "State space solutions to standard $H_2$ and $H_\infty$ control problems," *IEEE Trans. Automatic Control*, **Vol.34**, pp. 831-846.

Glover, K., and Doyle, J. C. (1988). "State-space formula for all stabilizing controllers that satisfy an $H_\infty$-norm bound and relations, *Systems & Control Letters* 11, pp. 167-172.

Hu, S. S., Yang, P. H., and Chang, B. C. (1998a). "A Computational Issue in Nonlinear $H_\infty$ Control ," *Proceedings of the 1998 American Control Conference*.

Hu, S. S. (1998b). " Computational Issues in the Design of Robust Nonlinear Controllers," Ph.D. Dissertation, Department of Mechanical Engineering and Mechanics, Drexel University, U. S. A.

Hu, S. S., Yang, P. H., and Chang, B. C. (1998c). "Modified Nonlinear $H_\infty$ Controller Formulas and the $H_\infty$ I/O Linearization Problem", *Proceedings of the 1998 IEEE Conference on Decision and Control*.

Isidori, A. and Astolfi, A. (1992). "Disturbance attenuation and $H_\infty$-Control Via measurement feedback in nonlinear systems," *IEEE Trans. Automatic Control*, **Vol. 37**., No. 9, 1992.

Isidori A. (1995). Nonlinear Control System, 3rd ed. Berlin: Spring-Verlag.

Kwatny, H. G., Bennett, W. H., and Berg, J. M., (1991). "Regulation of Relaxed Stability Aircraft," *IEEE Transactions on Automatic Control*, **Vol. AC-36**, pp. 1325-2323,.

Kwatny H. G., and Chang, B. C. (1998). " Constructing Linear Families from Parameter-Dependent Nonlinear Dynamics," *IEEE Trans. Automatic Control*, **Vol. 43**, No. 8, pp.1143-1147.

Van der Schaft, A. J. (1992). "$L_2$-gain analysis of nonlinear systems and nonlinear $H_\infty$ control," *IEEE Trans. Automatic Control*, **Vol. 37**, pp. 770-784.

Van der Schaft, A. J. (1996). $L_2$-Gain and Passivity Techniques in Nonlinear Control, Spriger, London.

Zhou, K., Doyle, J. C., and Glover, K (1997) Robust and Optimal Control, Prentice-Hall, Inc.

Zong, R., Chang, B. C., Yeh, H. H., and Banda, S. S. (1995). " An approximate I/O linearization via nonlinear $H_\infty$ model matching," *Proceedings of the 34th IEEE Conference on Decision and Control*.

# ROBUST STABILITY OF UNCERTAIN SYSTEMS WITH DIFFERENTIABLE NONLINEARITY

**Anatoly Gaiduk**

*Control System Department, Taganrog State University of Radioengineering, Nekrasovskii 44, 347928, Taganrog, Russia, Phone: 7 86344-53707 Fax: 86344-61657, Email: rector@trtu.rnd.su*

Abstract: The aim of this paper is to find the robust stability conditions of a closed-loop time invariant systems, which consist of the robust stable uncertain linear part and the differentiable nonlinearity as feedback. The nonlinearity satisfies usual sector limitations and some additional weak conditions. When the linear part is certain all that was said is close to known Kalman-Aizerman problem. The solution of the robust absolute stability problem is obtained by means of special Lyapunov function. 

Keywords: Uncertain, nonlinear systems, robust, absolute stability, Lyapunov function.

## 1. INTRODUCTION

Robust stability of the time invariant uncertain linear systems can be easy checked using the Kharitonov or the edge theorems (Bose and Shi, 1987; Ackermann, 1993). But real uncertain systems usually are nonlinear and to research its stability it is necessary to use more complicated methods (DeCarlo, et al., 1988). Nevertheless very often the system has only one nonlinearity as feedback. Similar systems with certain linear part are considered in absolute stability theory (Aizerman, 1949; Popov, 1962). There are many sufficient criteria of absolute stability (Barabanov, 1988), when nonlinearity $f(\sigma)$ is such that $0 \le \sigma f(\sigma) \le \mu_0 \sigma^2$, $\mu_0 < \infty$.

For this case Aizerman (1949) brought up hypothesis: a nonlinear system is stable, if the linear system to which nonlinear system is transferred when nonlinearity $f(\sigma)$ is replaced by function $\mu\sigma$ is stable for all $\mu \in [0, \mu_0]$. This hypothesis is convenient for practice, but it is not correct in general. Later R. Kalman brought up hypothesis that nonlinear system with differentiable nonlinearity is stable, if shown linear system is stable for all $\mu \in [0, \mu_1]$, where $|d f(\sigma) / d\sigma| \le \mu_1$. This suggestion is not correct in general case similarly.

Here the uncertain systems with differentiable nonlinearity as feedback are considered and it is expect to show that those systems are absolute stable, if uncertain linear systems with $f(\sigma) = \mu\sigma$ are robust stable for all $\mu \in [0, \mu_0]$ and some weak additional conditions are satisfied. The solution of this problem is obtained by means of Lyapunov direct method with applying some stage of evaluating of the Lyapunov function from (Kalman and Bertram, 1960).

## 2. PROBLEM STATEMENT

Let the time invariant uncertain system is described by equations

$$\dot{x} = A x - b f(\sigma), \qquad \sigma = c^T x, \tag{1}$$

where $x \in R^n$ is the state vector, $A, b, c$ are uncertain $n \times n$ -matrix and n-vectors, $f(\sigma)$ is the scalar differentiable function with properties

$$0 \le \sigma f(\sigma) \le \mu_0 \sigma^2,$$
$$|f'(\sigma)| \le \mu_1 < \infty, \quad f(0) = 0, \tag{2}$$

$$|f''(\sigma)| \le \mu_2 < \infty, \quad |\sigma| < \varepsilon_2, \tag{3}$$

where $f'(\sigma)$, $f''(\sigma)$ are the first and second derivatives, $\mu_i$, $i = 0, 1, 2$ are constants and $\varepsilon_2$ is as small number as possible.

Suppose the uncertain matrix $A$ and vectors $b, c$ belong to given multitude $\Omega$. Let

$$a(s) = \det(sI - A) =$$
$$= \alpha_n s^n + \alpha_{n-1} s^{n-1} + \cdots + \alpha_1 s + \alpha_0$$

$$Q = \left[ b \,\vdots\, Ab \,\vdots\, \ldots \,\vdots\, A^{n-1} b \right],$$

$$N = \begin{bmatrix} \alpha_1 & \alpha_2 & \cdots & \alpha_{n-1} & 1 \\ \alpha_2 & \alpha_3 & \cdots & 1 & 0 \\ \vdots & \vdots & \iddots & 0 & 0 \\ \alpha_{n-1} & 1 & 0 & 0 & 0 \\ 1 & 0 & 0 & 0 & 0 \end{bmatrix}, \tag{4}$$

$$\tilde{c}^T = c^T Q N = [\varsigma_1 \, \varsigma_2 \cdots \varsigma_n],$$

$$h(\mu, s) = \det\left(sI - A + \mu b c^T\right) =$$
$$= \eta_n s^n + \eta_{n-1} s^{n-1} + \cdots + \eta_1 s + \eta_0, \tag{5}$$

where $\alpha_n = \eta_n = 1$, $\eta_{i-1} = \alpha_{i-1} + \mu \varsigma_i$, $i = \overline{1, n}$. Suppose also that for all uncertain matrix and vectors there are expressions

$$\underline{\alpha}_{i-1} \le \alpha_{i-1} \le \bar{\alpha}_{i-1},$$

$$\underline{\varsigma}_i \le \varsigma_i \le \bar{\varsigma}_i, \quad i = \overline{1, n}, \tag{6}$$

$$a(s) \in \Gamma, \quad h(\mu, s) \in \Gamma, \quad \mu \in [0, \mu_0], \tag{7}$$

$$\det Q \ne 0, \tag{8}$$

where $\Gamma$ is a multitude of the Hurwitz polynomials. Note that in expression (5) $\underline{\eta}_{i-1} \le \eta_{i-1} \le \bar{\eta}_{i-1}$, $i = \overline{1, n}$ and the numbers $\underline{\eta}_{i-1}$, $\bar{\eta}_{i-1}$ can be easily computed from (5) – (7).

Expressions (6), (7) are the robust stability conditions of the linear uncertain system (1), if nonlinearity $f(\sigma)$ is replaced by function $\mu\sigma$ (Bose and Shi, 1987; Ackermann, 1993). The main purpose of this paper is to show that conditions (2) - (8) are sufficient to asymptotic stability at all of the nonlinear uncertain systems (1). As one can see, case in which the matrix $A$ and vectors $b$, $c$ are certain is the Kalman-Aizerman problem with the additional conditions (3), (8).

## 3. SOLUTION OF THE PROBLEM

Consider function

$$\mu(\sigma) = f(\sigma)/\sigma, \quad \sigma \ne 0,$$
$$\mu(0) = f'(0). \tag{9}$$

From expressions (2), (3) and (9) it is followed that

$$0 \le \mu(\sigma) \le \mu_0, \quad |\mu'(\sigma)| \le \mu_3, \tag{10}$$

where $\mu_3$ is a constant. In this case equations (1) can be written as

$$\dot{x} = H(x)\, x, \tag{11}$$

where

$$H(x) = A - \mu(\sigma)\, b c^T = [h_{ij}(x)],$$

since $\sigma = c^T x$.

Determine the polynomial

$$h(x, s) = \det(sI - H(x)) =$$
$$= \eta_n s^n + \eta_{n-1}(x) s^{n-1} + \cdots + \eta_1(x) s + \eta_0(x). \tag{12}$$

Evidently, $\eta_{i-1}(x) = \alpha_{i-1} + \mu(\sigma) \varsigma_i$ for all $i = \overline{1, n}$.

Let $r_i(x)$ $i=\overline{1,n+1}$ are the coefficients of the first column of the Routh array (Chen and Chu, 1966) computed for polynomial $h(x,s)$ (12). Because the conditions (2) - (7) are satisfied, there are constants $\varepsilon_\eta$, $\overline{\eta}$, $\varepsilon_r$, $\overline{r}$, and $\overline{h}$ such that

$$0<\varepsilon_\eta \le \eta_i(x) \le \overline{\eta} < \infty,$$
$$0<\varepsilon_r \le r_i(x) \le \overline{r} < \infty,$$
$$\left|h_{ij}(x)\right| \le \overline{h} < \infty, \qquad (13)$$

for all $A,b,c \in \Omega$ and $x \in R^n$.

Under condition (8) the matrix $P=QN$ is nonsingular for any $A,b \in \Omega$ and matrix $\overline{H}(x)=P^{-1}H(x)P$ is companion matrix (Kalman and Bertram, 1960) of the polynomial $h(x,s)$ (12) for all $A,b,c \in \Omega$, and $x \in R^n$. In this case by virtue conditions (6), (7) and (13) there is nonsingular matrix $T(x)$ such that $\tilde{H}(x)=T^{-1}(x)\,\overline{H}(x)\,T(x)$ is the Schwarz matrix (Kalman and Bertram, 1960; Chen and Chu, 1966), that is

$$\tilde{H}(x)=\begin{bmatrix} 0 & 1 & 0 & \cdots & 0 & 0 & 0 \\ \tilde{h}_{21} & 0 & 1 & \cdots & 0 & 0 & 0 \\ 0 & \tilde{h}_{32} & 0 & \cdots & 0 & 0 & 0 \\ \vdots & \vdots & \vdots & \ddots & \vdots & \vdots & \vdots \\ 0 & 0 & 0 & \cdots & 0 & 1 & 0 \\ 0 & 0 & 0 & \cdots & \tilde{h}_{n-1,n-2} & 0 & 1 \\ 0 & 0 & 0 & \cdots & 0 & \tilde{h}_{n,n-1} & \tilde{h}_{nn} \end{bmatrix},$$

for all $A,b,c \in \Omega$ and $x \in R^n$. Elements of this matrix are

$$\tilde{h}_{i+1,i}(x)=-\frac{r_{n+2-i}(x)}{r_{n-1}(x)}, \quad i=\overline{1,n-1},$$

$$\tilde{h}_{n,n}(x)=-\eta_{n-1}(x) \qquad (15)$$

limited and differentiable. Similarly the elements of the matrix $T(x)$ are the functions of the ratio $r_{i+2}(x)/r_i(x)$ limited and differentiable for all $A,b,c \in \Omega$ and $x \in R^n$.

Now it is possible to bring forward the main theorem of this paper as follows.

*Theorem 1.* If conditions (2), (3), (6) – (8) are satisfied for all $A,b,c \in \Omega$ and $\mu \in [0,\mu_0]$, then the zero state of the nonlinear system (1) is asymptotically stable at all.

*Proof:* Consider the Lyapunov function

$$V(x)=x^T M(x)x\exp(\varphi(x)). \qquad (16)$$

Here $x \in R^n, \|x\| < \infty$ and $\varphi(x)$ is a scalar limited unknown function chosen below. Matrix

$$M(x)=P_1^T(x)B(x)P_1(x),$$

where

$$P_1(x)=T^{-1}(x)P^{-1},$$

$$B(x)=\mathrm{diag}\{\beta_1(x)\cdots\beta_n(x)\}.$$

In last expression $\beta_i(x)=r_{n+2-i}(x)\,r_{n+1-i}(x)$, $i=\overline{1,n}$.

Under inequalities (13) the function $V(x)$ (16) is the positive definite (Kalman and Bertram, 1960; Barbashin, 1970) for all $A,b,c \in \Omega$ and $x \in R^n$, $\|x\| < \infty$.

For the system (11) the derivative $\dot{V}(x)$ is

(14)

$$\dot{V}(x)=\left\{x^T\left[H^T(x)M(x)+M(x)H(x)\right]x+\right.$$
$$\left.+x^T\dot{M}(x)x+\dot{\varphi}(x)x^T M(x)x\right\}\exp(\varphi(x))=$$
$$=x^T\left[H^T(x)M(x)+M(x)H(x)\right]x\exp(\varphi(x)) \qquad (17)$$

since $\varphi(x)$ is chosen specifically such that

$$\dot{\varphi}(x)=-x^T\dot{M}(x)x/x^T M(x)x,$$
$$\varphi(x(t))_{t=0}=\varphi_0. \qquad (18)$$

where $\varphi_0$ is some constant.

Under conditions (4), (8) and definitions of the matrices $P$, $T(x)$ and $M(x)$ it is follows from expressions (17), (14) and (15)

$$\dot{V}(x)=-2\eta_{n-1}^2(x)\left(P_{1n}(x)x\right)^2\exp(\varphi(x)). \qquad (19)$$

where $P_{1n}(x)$ is the n-th row of the matrix $P_1(x)$ (Kalman and Bertram, 1960; Chen and Chu, 1966).

In order to show that function $V(x) \to \infty$ when $\|x\| \to \infty$ and $V(x) \to 0$ when $\|x\| \to 0$ it is considered the derivative of $\ln V(x)$ with respect to $t$ by using (10) and (18):

$$d(\ln V(x))/dt = \left(\dot{x}^T M(x)x + + x^T M(x)\dot{x}\right)/x^T M(x)x. \quad (20)$$

Let $x/r = q$, $l(x) = M(x)q/q^T M(x)q$ $\|x\| = r$, $\|x_0\| = r_0$. Integrating equality (20) from zero to $t$ one obtain

$$\ln(V(x)/V(x_0)) = 2\ln(r/r_0) + 2\int_0^t l^T(x)\dot{q}\,dt. \quad (21)$$

Evidently

$$\left|\int_0^t l^T(x)\dot{q}\,dt\right| \le \sup\left\|l^T(x)\right\| \left|\int_0^t \dot{q}\,dt\right| \quad (22)$$

and $\left|\int_0^t \dot{q}\,dt\right| \le 2n$, since $\|q\| = 1$.

Let $\gamma_1(x)$, $\tau_1(x)$, $\nu_1$ are smallest and $\gamma_n(x)$, $\tau_n(x)$ $\nu_n$ are largest eigenfunctions of the symmetrical matrices $B(x)$, $\left(T^{-1}(x)\right)^T T^{-1}(x)$ and $\left(P^{-1}\right)^T P^{-1}$. It is known (Markus and Mink, 1964), that

$$l^T(x)l(x) \le \gamma_n^2(x)\tau_n^2(x)\nu_n^2/\gamma_1^2(x)\tau_1^2(x)\nu_1^2. \quad (23)$$

Since by virtue of conditions (13) all functions $r_i(x)$, $i = \overline{1, n+1}$ in (15) and (16) are strictly positive and bounded, hence there are numbers $q_2, q_3, q_4, q_5$ so that $q_2 \le \gamma_i(x) \le q_3$, $q_4 \le \tau_i(x) \le q_5$. Therefore from (20), (23) it is followed

$$\int_0^t l^T(x)\,\dot{q}\,dt \le q_6 < \infty, \quad (24)$$

for all $x \in R^n, \|x\| < \infty$, where $q_6 = 2n\nu_n q_3 q_5/q_2 q_4 \nu_1$. Therefore in accordance to (21), (24) Lyapunov function $V(x) \to \infty$ only if $r = \|x\| \to \infty$.

On another hand, from (21) (24) and (16) it is followed that $|\varphi(x)| < \infty$ for all $x \in R^n$ and $\|x\| \le q_7 < \infty$, because in opposite case module of the left part of (21) is infinite but right part is finite. Hence the Lyapunov function $V(x) \to 0$ if and only if $\|x\| \to 0$.

Since the derivative $\dot{V}(x)$ (19) is only semi-definite negative function (Kalman and Bertram, 1960; Barbashin, 1970) to complete proof, it is necessary to show that $\dot{V}(x) = 0$ only in the points, which are not solution of the system (11) (Barbashin, 1970).

From (19) it is followed that the derivative $\dot{V}(x) = 0$ if and only if

$$x = x^* = \sum_{j=1}^{n-1} \gamma_j P_{1j}^{-1}(x^*),$$

where $\gamma_j$ are not all zero arbitrary scalars, $P_{1j}^{-1}(x)$ is j-th column of the matrix $P_1^{-1}(x)$. But shown vectors $x^*$ can not be solution of the system $\dot{x} = H(x)x$. Really, if $x^*$ is solution of the system (11), then should be carry out equality

$$0 = H(x^*)x^* = \sum_{j=1}^{n-1} \gamma_j H(x^*)PT_j(x^*).$$

But this equality contradicts to non-singularity of the matrices $H(x)$, $P$ and $T(x)$ (Kalman and Bertram, 1960) for all $A, b, c \in \Omega$, and $x \in R^n$. Hence, under theorem 1 conditions the function $V(x)$ (16) satisfies to Krasovskii-Barbashin theorem (Barbashin, 1970) complete and the zero state of the system (11) or (1) is asymptotically stable at all.

As an specifically case, next statement follows from theorem 1.

*Theorem 2.* The nonlinear system (1) where $A, b, c$ and $f(\sigma)$ (2) are given certain matrix, vectors and differentiable nonlinearity is absolute stable, if

a) $|d(f(\sigma)/\sigma)/d\sigma| < \infty$ or there is $\varepsilon_2$ as small as possible such that

$$|f''(\sigma)| \le \mu_2 < \infty, \quad \text{for } |\sigma| < \varepsilon_2,$$

b) $\det\left[b \vdots Ab \vdots \ldots \vdots A^{n-1}b\right] \neq 0$,

c) polynomial $h(\mu, s) = \det\left(sI - A + \mu bc^T\right)$

is a Hurwitz polynomial for all $\mu \in [0, \mu_0]$.

Evidently conditions a) and b) are additional to Kalman-Aizerman hypothesis to be correct in case of differentiable nonlinearity.

## 4. EXAMPLE AND SIMULATION

In the example is considered system (1), where

$$A = \begin{bmatrix} 0 & 1 & 0 & 0 \\ 0 & 0 & 1 & 0 \\ 0 & 0 & 0 & 1 \\ -\alpha_0 & -\alpha_1 & -\alpha_2 & -\alpha_3 \end{bmatrix},$$

$$b = \begin{bmatrix} 0 \\ 0 \\ 0 \\ 1 \end{bmatrix}, \quad c = \begin{bmatrix} c_1 \\ 0 \\ 0 \\ c_4 \end{bmatrix},$$

$\alpha_0 \in [80, 100]$, $\alpha_1 \in [60, 70]$, $\alpha_2 \in [30, 37]$, $\alpha_3 \in [3, 6]$, $c_1 \in [3, 6]$, $c_4 \in [0.2, 2]$. Differentiable nonlinearity $f(\sigma)$ satisfies conditions (2), (3) or (10).

Let $f(\sigma) = \mu\sigma$. Then coefficients of the polynomial $h(\mu, s)$ (5) are $\eta_0 \in [80, 100 + 6\mu]$, $\eta_1 \in [60, 70]$, $\eta_2 \in [30, 37]$, $\eta_3 \in [3, 6 + 2\mu]$. Using Kharitonov theorem one can find that linear uncertain system is robust stable until one of four Kharitonov polynomials

$$h_3(\mu, s) = s^4 + (6 + 2\mu)s^3 + 30s^2 + 60s + 100 + 6\mu$$

is a Hurwitz polynomial. According to Hurwitz criterion that is if $0 \leq \mu < \bar{\mu}$, where $\bar{\mu}$ is a positive root of the equation

$$60((6 + 2\mu)30 - 60) - (100 + 6\mu)(6 + 2\mu)^2 = 0,$$

$(\bar{\mu} \approx 3.31778)$. Simulation by Matlab program of considered system with $\alpha_0 = 100 + 6\mu$, $\alpha_1 = 60$, $\alpha_2 = 30$, $\alpha_3 = 6 + 2\mu$, when $f(\sigma) = 0.5\mu\sigma(1 + \sin(q\sigma))$, or $f(\sigma) = \mu\sigma$ shows that the uncertain nonlinear system is asymptotically stable until $0 \leq \mu < \bar{\mu}$ according to statement of the theorem 1.

## 5. CONCLUSION

In this paper it is shown that if nonlinearity satisfies usual sector limitation and additional weak conditions are carried out the known robust stability criteria can be used to check the stability of the time invariant uncertain systems with differentiable nonlinearity as feedback. If linear part of the considered nonlinear system is certain, for that purpose the Routh or Hurwitz stability criteria can be used.

## REFERENCES


Ackermann, J.E. (1993). *Robust control systems with uncertain physical parameters.* Springer Verlag, London.

Aizerman, M.A. (1949). About one problem concerning with dynamic systems stability. *Uspehi Mathematical Sciences*, **4**, 186-188 (in Russian).

Barabanov, N.E. (1988). New absolute stability criterion of control system with one differentiable nonlinearity. Proc. *Acad. Sci. USSR*, Vol. 299, **3**, 570-572 (in Russian).

Barbashin, E.A. (1970). *Lyapunov functions.* Nauka, Moscow (in Russian).

Bose, N.K. and Y.Q. Shi (1987). A simple general proof of generalized stability criterion. *IEEE Transaction Circuits and Systems*, **10**, 57-63.

Chen, C.F. and H. Chu (1966). A matrix for evaluating Schwarz's form. *IEEE Transaction on Automatic Control*, **8**, 303-305.

DeCarlo, R.A., Zak S.H. and G.P. Matthews (1988). Variable structure control of nonlinear multivariable systems. *Proceeding IEEE*, **3**, 212-232.

Kalman, R.E. and J.B. Bertram (1960). Control system analysis and design via the second method of Lyapunov. *Transaction ASME*, D, **6**, 371-393.

Markus, M. and H. Mink (1964). *A survey matrix theory and matrix inequalities.* Allyn and Bacon, Boston

Popov, V.M. (1962). On absolute stability of nonlinear control systems. *Avtomatika i telemekhanika*, **1**, 3-18 (in Russian).

14th Triennial World Congress, Beijing, P.R. China



# Robust Nonlinear Forwarding with Smooth State Feedback Control

Weizhou Su and Minyue Fu

Department of Electrical and Computer Engineering
The University of Newcastle, NSW 2308, Australia
Email: eesu@ee.newcastle.edu.au; eemf@ee.newcastle.edu.au

**Abstract.** In this paper, a robust forwarding control technique is introduced for a class of uncertain nonlinear systems admitting the so-called *up-augmented structure.* This technique gives a recursive design procedure which yields a smooth controller. Further, it can be combined with the well-known back-stepping technique to solve the robust stabilization problem for a large class of uncertain nonlinear systems. In contrast to previous methods on nonlinear forwarding, our technique is able to handle systems with large parameter uncertainties.

**Keywords:** Nonlinear systems; Forwarding; Robust stabilization; State feedback.

## 1 Introduction

In this paper, we consider the robust stabilization problem for a class of uncertain nonlinear systems admitting the so-called *up-augmented structure.* This structure starts with a *base system* of the following form:

$$\dot{x}(t) = f(x(t), q) + b(q)u(t) \tag{1}$$

where $x(t) \in \mathbf{R}^n$ is the state, $u(t) \in \mathbf{R}$ is the control, $q \in \mathbf{R}^l$ is an uncertain parameter vector contained in a compact set $\Omega$, $b(q)$ is continuous in $q$, $f(x,q)$ is continuous in $q$ and smooth in $x$ with $f(0,q) = 0$.

An *up-augmented system* is given by

$$\begin{aligned} \dot{x}_0 &= f_0(x,q) \\ \dot{x} &= f(x,q) + b(q)[u + d(x, x_0, \eta, q)] \end{aligned} \tag{2}$$

where $x_0 \in \mathbf{R}$ is a new state variable, $t$ represents time, $f_0(x,q)$ is continuous in $q$ and smooth in $x$ with $f_0(0,q) = 0$, $d(x, x_0, \eta, q)$ is continuous in $q$ and smooth in $(x, x_0, \eta)$ with $d(0,0,0,q) = 0$. The parameter $\eta$ represents state variables other than $x_0$ and $x$ if (2) is a subsystem of a larger one, or void otherwise. We will denote $x^+ = [x_0, x^T]^T$ and $d(x^+, \eta, q) = d(x, x_0, \eta, q)$.

The stabilization problem for the up-augmented structure has been heavily studied recently for the case where the nonlinear system does not involve uncertain parameters; see e.g., [6], [1], [3] and [4]. However, these methods, called forwarding or feedforwarding, all involve some kind of cancellation of nonlinearities. Subsequently, they do not apply to systems with uncertain parameters, especially of large sizes. This is in great contrast to robust stabilization of linear systems. In fact, a seminal paper of Wei [7] in 1990 proposed a method capable to deal with the up-augmented structure involving large uncertain parameters.

In a recent paper by the authors [5], Wei's robust stabilization technique was generalized to the nonlinear case. Our solution to the robust stabilization problem involves a two-step controller. In the first step, a controller is designed to drive the base system to a local region. Subsequently, a second controller is used to maintain the state of the base system small while driving the augmented state to zero. This robust forwarding technique is conceptually different from the standard forwarding ones as no nonlinear cancellation is involved. When specialized to the linear case, this technique recovers the result of Wei [7]. Further, a recursive application of this technique leads to a solution to robust stabilization of systems with the so-called *upper-triangular structure*; see [5].

However, our robust forwarding technique in [5] has a weakness which serves as the motivation for this paper. That is, the two-step controller is nonsmooth. This implies that it is difficult to combine it with the well-known back-stepping technique for handling a larger class of uncertain nonlinear systems.

The back-stepping technique deals with the so-called *down-augmented structure*. For the same base system (1), a down-augmented system is of the form:

$$\begin{aligned} \dot{x} &= f(x,q) + b(q)x_{n+1} \\ \dot{x}_{n+1} &= \theta_{n+1}(q)[u + d(x, x_0, \eta, q)] \end{aligned} \qquad (3)$$

where $x_{n+1} \in \mathbf{R}$ is a new state variable, $\theta_{n+1}(q)$ is a continuous function bounded away from zero, and $d(\cdot)$ is the same as before. It is well-known now that this structure can be easily handled using the standard back-stepping technique; see, e.g., [2]. A striking feature of the back-stepping technique is that it is capable to treat uncertain parameters of large sizes. A key assumption in the back-stepping technique is that the base system has a smooth stabilizer, say $u_n(x)$. This is because a state transformation

$$z = x; \quad z_{n+1} = x_{n+1} - u_n(x) \qquad (4)$$

is used and the differentiability of the new state is required. It is worth to note that the back-stepping technique yields a smooth stabilizer, provided that the down-augmented system is smooth.

The need for a smooth stabilizer for the up-augmented structure is now clear: Such a stabilizer would allow the base system to be an up-augmented structure. That is, robust stabilization can be achieved for a much larger class of uncertain nonlinear systems generated using recursive up-augmentations and down-augmentations. This structure, when specialized to the linear case, corresponds to the so-called *anti-symmetric stepwise configuration*, coined by Wei [7].

With the above introduction, the objective of this paper is simply stated: We aim to propose a robust forwarding technique that would yield a smooth stabilizing controller for the up-augmented system in (2).

We point out that all the results allow the system to be time-varying. But for notational convenience, the time-dependence is suppressed.

## 2 Robust Forwarding with Non-smooth Control

In this section we revisit the robust forwarding technique in [5]. This technique does not yield a smooth controller, but will serves as the basis for a smooth controller later. As mentioned in the previous section, the key mechanism involved is a two-step control law. In the first step, a nonlinear controller is applied to the base system so that its state $x$ converges to a "small" bounded set $\Omega$ while $x_0$ is not regulated. In the second step, a nonlinear controller is designed to maintain $x$ within $\Omega$ while driving the augmented state $x^+$ to zero. Overall, this two-step control law achieves robust global asymptotic stabilization (RGAS) and robust local exponential stabilization (RLES).

ASSUMPTIONS

**Assumption 2.1** (*Local Exponential Stabilizability*): There exists a smooth controller $u_n(x)$ for the base system (1) such that, with

$$u(t) = u_n(x(t)), \qquad (5)$$

the state of the system (1) is RLES.

**Assumption 2.2** (*Local Smoothness Properties*): For the same local region $\Omega$ and local controller $u_0(x)$ as above, any $x \in \Omega$ and $q \in Q$, there holds

$$\begin{aligned} &f_0(x,q) = a(x,q)x; \\ &f(x,q) + b(q)u_0(x) = A(x,q)x \\ &b(q) = \theta_n(q)\begin{pmatrix} 0 \\ 1 \end{pmatrix} = \theta_n(q)b \end{aligned} \qquad (6)$$

where

$$\begin{aligned} A(x,q) &= \begin{bmatrix} 0 & A^-(x,q) \\ d_{n1}(x,q) & \star \end{bmatrix} \\ a(x,q) &= [\theta_0(x,q) \quad \star] \end{aligned} \qquad (7)$$

with $*$ representing an arbitrary term, $1 \geq \theta_0(x,q) \geq \underline{\theta}_0 > 0$; $1 \geq \theta_n(x,q) > \underline{\theta}_n > 0$, where $\underline{\theta}_0$ and $\underline{\theta}_n$ are constants. □

**Remark 2.1** With the local smoothness assumption, the RLES property in Assumption 2.1 can be modified to the following: there exists $\mu > 0, \varepsilon > 0$ and matrix $P = P^T > 0$ such that

$$PA(x,q) + A^T(x,q)P \leq -\varepsilon I, \ \forall x \in \Omega, q \in Q. \quad (8)$$

where

$$\Omega = \{x : \ x^T P x < \mu\}. \qquad (9)$$

**Assumption 2.3** (*Global Properties*): Consider the following system derived from (2):

$$\dot{x} = f(x,q) + b(q)[u + d(x^+, \eta, q)] \qquad (10)$$

Given any smooth function $\eta(t)$ and $0 < \rho < 1$, there exists a smooth controller $u_d(x^+, \eta)$ such that, with

$$u(t) = u_d(x^+(t), \eta(t)), \tag{11}$$

the state of the system (2) will be driven into $\rho\Omega$ in a finite time $T$, where $\Omega$ is given in (9), and $\rho\Omega = \{\rho x : x \in \Omega\}$.

**Remark 2.2** Assumption 2.3 appears to be strong. However, we note that it is automatically satisfied for first order systems because a "high-gain" $u(t)$ can be designed to "overcome" both $f(x,q)$ and $d(x^+, \eta, q)$, forcing the state to converge to $\rho\Omega$. In Section 4, we will show that this property can be preserved under robust forwarding design. □

LYAPUNOV FUNCTION AND CONTROLLER DESIGN

Now we pay attention to controller design for (2). First, we utilize Assumption 2.3 and apply (11) to drive $x(t)$ into $\rho\Omega$ in a finite time $T$. In this step, $x_0(t)$ is not regulated. Once $x(t) \in \rho\Omega$, we switch to a local mode where a different controller $u^+(x^+, \eta)$ will be applied. This controller will maintain $x(t)$ in $\Omega$ while driving $x^+(t)$ to zero. The design of $u^+(x^+, \eta)$ relies on a local Lyapunov function for (2)

$$\begin{aligned} V^+(x^+) &= (x_0 - (\gamma\ 0)Px)^2 \\ &\quad + \int_0^{V(x)} s(w)dw > 0,\ \forall x \in \Omega \end{aligned} \tag{12}$$

where $\gamma < 0$ is a constant to be specified and $s(\cdot)$ is a locally smooth function satisfying: $s(w) > 0$ $\forall w \in [0, \mu)$;

$$\int_0^v s(w)dw < \infty,\ \forall v \in [0, \mu); \tag{13}$$

and

$$\lim_{v \to \mu} \int_0^v s(w)dw \to \infty. \tag{14}$$

**Remark 2.3** Note that (12) includes a quite large set of the Lyapunov functions. A particular choice of $s(\cdot)$ is given by [5]:

$$s(w) = \frac{\mu}{\mu - w}.$$

For linear systems we can take $\mu = \infty$ and $s(w)$ a constant, which yields a quadratic Lyapunov function. It can be verified that this is the same Lyapunov function used in Wei [7]. In generally, these Lyapunov functions are non-quadratic. However, as $x \to 0$, $V^+(x^+)$ becomes quadratic in $x^+$ because $s(0) > 0$.

We also note that the function $\int_0^{V(x)} s(w)dw$ resembles a "potential barrier" and the Lyapunov function (12) is valid only for $x \in \Omega$, i.e.,

$$V^+(x^+) \to \infty \quad as \quad x^T P x \to \mu \tag{15}$$

This implies that future $x \in \Omega$ as long as that $V^+(x^+)$ remains bounded.

For notational simplicity, we will denote $s(V(x))$ by $s(x)$. Defining

$$P^+ = \begin{bmatrix} 1 & -(\gamma\ 0)P \\ -P\begin{pmatrix} \gamma \\ 0 \end{pmatrix} & s(x)P + P\begin{pmatrix} \gamma \\ 0 \end{pmatrix}(\gamma\ 0)P \end{bmatrix} \tag{16}$$

which is positive definite for all $x \in \Omega$. The inverse of $P^+$ is given by

$$S^+ = \begin{bmatrix} s(x) + (\gamma\ 0)P\begin{pmatrix} \gamma \\ 0 \end{pmatrix} & (\gamma\ 0) \\ \begin{pmatrix} \gamma \\ 0 \end{pmatrix} & P^{-1} \end{bmatrix}. \tag{17}$$

Also define a (nonlinear) state transformation

$$z^+ = (z_0,\ z^T)^T = P^+ x^+. \tag{18}$$

To simplify the analysis, we also assume in this section that $\eta$ is void, i.e., $d(x^+, \eta, q) = d_1(x^+, q)$. Since $d_1(x^+, q)$ is smooth in $x^+$ and $d_1(0, q) = 0$, we can rewrite

$$d_1(x^+, q) = D^+(x^+, q)x^+ = D^+(x^+, q)S^+ z^+ \tag{19}$$

for some $D^+(x^+, q)$ smooth in $x^+$ and continuous in $q$.

Differentiating $V^+(x^+(t))$ along the trajectory of (2), we have

$$\begin{aligned} \dot{V}^+ &= 2[x_0 - (\gamma\ 0)Px][\dot{x}_0 - (\gamma\ 0)P\dot{x}] \\ &\quad + 2s(V(x))x^T P\dot{x} \\ &= 2(x^+)^T P^+ \dot{x}^+ \end{aligned} \tag{20}$$

**Theorem 2.1** For the up-augmented system (2) satisfying Assumptions 2.1-2.3 and $d(x^+, \eta, q) = d_1(x^+, q)$, there exist $\gamma < 0$ and $\alpha(x^+) > 0$ such that the nonlinear controller

$$u(t) = u^+(x^+) = u_n(x) - \alpha(x^+)z^T b \tag{21}$$

will render

$$\dot{V}^+(x^+) \le -\varepsilon^+(x)V^+(x^+), \quad \forall x \in \Omega, x_0 \in \mathbf{R} \tag{22}$$

for some continuous $\varepsilon^+(x) > 0$, $x \in \Omega$.

Moreover, the following choice of $\gamma$, $\alpha^+(x^+)$ and $\varepsilon^+(x)$ will suffice :

$$0 < \bar{\varepsilon} < \bar{\varepsilon}_{max} = \min_{q\in Q; x\in\Omega} \lambda_{min}\left[-P^{-1}\left(A^T(x,q)P + PA(x,q)\right)P^{-1}\right] \quad (23)$$

$$\gamma < \gamma_{max} = \min_{q\in Q;\ x\in\Omega} \frac{1}{2\theta_0(q)}\left[a(x,q)\left(A^T(x,q)P + PA(x,q) + \bar{\varepsilon}P^2\right)^{-1} a^T(x,q) - \bar{\varepsilon}\right] \quad (24)$$

$$\varepsilon^+(x) = \frac{\bar{\varepsilon}}{2}\lambda_{min}\left(s^{-1}(x)P^+(x)\right) > 0 \quad (25)$$

$$\alpha(x^+) = s^{-1}(x)\frac{\delta^2(x^+)}{\bar{\varepsilon}} \quad (26)$$

where $\delta(x^+)$ is any smooth function satisfying

$$\delta(x^+) \geq \max_{q\in Q}\left\| s(x)D(x^+,q)S^+(s) + \gamma\frac{d_{n1}(x,q)}{\theta_n(q)}\right\| \quad (27)$$

*Proof :* Using Assumptions 2.1-2.2, we obtain

$$\dot{V}^+ = (x^+)^T\left[P^+A^+(x,q) + A^+(x,q)^T P^+\right]x^+ + 2(x^+)^T P^+ b^+(q)[u + d_1(x^+,t,q)] \quad (28)$$

where $A^+(x,q)$ is defined as

$$A^+(x,q) = \begin{bmatrix} 0 & a(x,q) \\ 0 & A(x,q)\end{bmatrix};\ b^+(q) = \begin{bmatrix} 0 \\ b(q)\end{bmatrix}.$$

It is easy to verify that

$$A^+(x,q)S^+ + S^+A^+(x,q)^T = s^{-1}(x)\begin{bmatrix} 2a(x,q)\begin{pmatrix}\gamma\\0\end{pmatrix} & a(x,q)P^{-1} + (\gamma\ \ 0)A^T(x,q) \\ P^{-1}a^T(x,q) + A(x,q)\begin{pmatrix}\gamma\\0\end{pmatrix} & P^{-1}A^T(x,q) + A(x,q)P^{-1}\end{bmatrix}. \quad (29)$$

It follows from (28) and (29) that

$$\dot{V}^+ = s^{-1}(x)\left(z^+\right)^T\begin{bmatrix} 2\theta_0(x,q)\gamma & a(x,q)P^{-1} \\ P^{-1}a^T(x,q) & P^{-1}A^T(x,q) + A(x,q)P^{-1}\end{bmatrix} z^+ + 2s^{-1}(x)z^T A(x,q)\begin{pmatrix}\gamma\\0\end{pmatrix} z_0 + 2\left(z^+\right)^T b^+(q)[u + d_1(x^+,q)]. \quad (30)$$

The choice of $\bar{\varepsilon}$ and $\gamma$ in (23) and (24) assures that, for $\forall q \in Q,\ \ x \in \Omega$,

$$\begin{bmatrix} 2\theta_0(x,q)\gamma & a(x,q)P^{-1} \\ P^{-1}a^T(x,q) & P^{-1}A^T(x,q) + A(x,q)P^{-1}\end{bmatrix} \leq -\bar{\varepsilon}I.$$

Also, with Assumption 2.2 and $(z^+)^T b^+ = z^T b$, we have

$$z^T A(x,q)\begin{pmatrix}\gamma\\0\end{pmatrix} = z^T\gamma d_{n1}(x,q)b = \gamma\frac{d_{n1}(x,q)}{\theta_n(q)}(z^+)^T b^+(q).$$

Therefore, from the above discussion and (30), we have

$$\dot{V}^+ \leq -s^{-1}(x)\bar{\varepsilon}\left(z^+\right)^T z^+ + 2s^{-1}(x)(z^+)^T b^+(q) \cdot \left[s(x)u + \gamma\frac{d_{n1}(x,q)}{\theta_n(q)}z_0 + s(x)d_1\right]. \quad (31)$$

Using (19) and (27), we have

$$\left\|\gamma\frac{d_{n1}(q)}{\theta_n(q)}z_0 + s(x)d_1(x^+,q)\right\| \leq \delta(x^+)\|z^+\|. \quad (32)$$

Then substituting (32), (26) and the controller (21) into (31) results in

$$\begin{aligned}\dot{V}^+ &\leq -s^{-1}(x)\frac{\bar{\varepsilon}}{2}(z^+)^T z^+ \\ &\leq -\frac{\bar{\varepsilon}}{2}\left(P^{+1/2}x^+\right)^T\left(s^{-1}(x)P^+\right)\left(P^{+1/2}x^+\right) \\ &\leq -\varepsilon^+(x^+)V(x^+) \end{aligned} \quad (33)$$

In particular, $\varepsilon^+(0) > 0$. Hence, we have RGAS and RLES for the closed-loop system (2). $\nabla\nabla\nabla$

# 3 Robust Forwarding with Smooth Control

In this section, we show how to modify the non-smooth controller in the previous section so that it becomes a smooth one.

Let $\xi(\cdot) : \mathbf{R} \to \mathbf{R}$ be any monotonic smooth function satisfying the following properties:

$$\begin{aligned}\xi(v) &= 0,\ v \leq 0 \\ \xi(v) &= 1,\ v \geq 1\end{aligned} \quad (34)$$

Denote

$$F(v) = \int_0^v s(w)dw \quad (35)$$

Note that $F(v)$ is monotonic.

Then, define the new smooth stabilizing controller:

$$u(t) = \begin{cases} u_d(x^+,t) & t < T; \\ (1-\xi(\tau))u_d + \xi(\tau)u^+ & t \geq T\end{cases} \quad (36)$$

where $T \geq 0$ is the first instant when $x \in \rho\Omega$ (see Assumption 2.3), and

$$\tau = \kappa_1 \left(F(V(x(t))) - F(V(x(T)))\right) + \kappa_2(t-T) \tag{37}$$

with any $\kappa_i > 0$. The smoothness of this controller is due to the definition of $\xi(\cdot)$ and $\tau = 0$ at switching time $t = T$. The key feature of this new controller is that it becomes $u^+$ when either $t$ or $V(x)$ becomes sufficiently large. Combining with the properties of $u^+$ in Theorem 2.1, this will assure that $x$ remains $\Omega$ for $t \geq T$ and that $x^+ \to 0$ as $t \to \infty$. This result is summarized in the following theorem:

**Theorem 3.1** For the up-augmented system (2) satisfying Assumptions 2.1-2.3 and $d(x^+, \eta, q) = d_1(x^+, q)$, the smooth controller in (36) will render the closed-loop system RGAS and RLES.

*Proof*: Follows from Theorem 2.1 and the argument above. ∇∇∇

# 4 Recursive Application of Robust Forwarding

The purpose of this section is to show that the robust nonlinear forwarding technique studied in the previous sections can be applied recursively. A simple motivation for this is the need to deal with the so-called *upper-triangular structure*:

$$\begin{array}{rcl} \dot{x}_1 & = & f_1(x_2, \cdots, x_n, q) \\ & \vdots & \\ \dot{x}_{n-1} & = & f_{n-1}(x_n, q) \\ \dot{x}_n & = & f_n(x_1, \cdots, x_n, q) + u \end{array} \tag{38}$$

where $x_1, \cdots, x_n$ are state variables, $q$ is an uncertain parameter vector as before and $f_i$, $i = 1, \cdots, n$ are smooth in $x$ and continuous in $q$.

But the significance of the recursive applicability goes beyond the upper-triangular structure. As we will point out in the next section, a much richer class of nonlinear systems can be robustly stabilized by combining robust nonlinear forwarding and backstepping.

Technically speaking, we need to show that the robust nonlinear forwarding technique used in the previous sections can preserve Assumptions 2.1-2.3 for the up-augmented structure. The results below apply to both smooth and non-smooth controllers.

First, we return to the general case where $\eta$ is not void. Since $d(x^+, \eta, q) = 0$ when $(x^+, \eta) = 0$, the function $d(x^+, \eta, q)$ can be decomposed into two smooth terms as follows:

$$d(x^+, \eta, q) = d_1(x^+, q) + d^+(x^+, \eta, q)\eta \tag{39}$$

with $d_1(0, q) = 0$.

Consider the following control law

$$u(x^+, \eta) = u_1(x^+) + u_2(x^+, \eta) \tag{40}$$

where $u_1(x^+)$ is as in (36). Then, the system

$$\begin{array}{rcl} \dot{x}_0 & = & f_0(x, q) \\ \dot{x} & = & f(x, q) + b(q)[u_1(x^+) + d_1(x^+, q)] \end{array} \tag{41}$$

is RGAS and RLES.

We first justify Assumption 2.2. Note that $f_0(x, q)$, $f(x, q)$ and $b(q)[u_1(x^+) + d(x^+, q)]$ are locally smooth. It follows that

$$\begin{array}{rl} & \left[ \begin{array}{c} f_0(x, q) \\ f(x, q) + b(q)[u_1(x^+) + d_1(x^+, q)] \end{array} \right] \\ = & A_c^+(x^+, q)x^+ \end{array} \tag{42}$$

where

$$A_c^+(x^+, q) = \left[ \begin{array}{cc} 0 & a^+(x, q) \\ \left( \begin{array}{c} 0 \\ \star \end{array} \right) & A_c(x^+, q) \end{array} \right].$$

Hence , the system (41) satisfies Assumption 2.2.

For Assumptions 2.1 and 2.3, we have the following two theorems, respectively:

**Theorem 4.1** For the system (41), take $s_0 > s(0) > 0$ and let

$$P_0^+ = \left[ \begin{array}{cc} 1 & -(\gamma \ \ 0)P \\ -P\left( \begin{array}{c} \gamma \\ 0 \end{array} \right) & s_0 P + P\left( \begin{array}{c} \gamma \\ 0 \end{array} \right)(\gamma \ \ 0)P \end{array} \right] \tag{43}$$

Then, there exists $\mu^+ > 0$ such that

$$P_0^+ A_c^+(x^+, q) + \left(A_c^+(x^+, q)\right)^T P_0^+ \leq -\varepsilon_0^+ I \tag{44}$$

for all $x^+ \in \Omega^+ = \{x^+ : (x^+)^T P_0^+ x^+ < \mu^+\}$.

*Proof:* This follows from the RLES property of the system. The details are omitted.

**Theorem 4.2** Suppose the up-augmented system (2) satisfies Assumptions 2.1-2.3. For any given $\beta > 0$, the controller (40) with

$$u_2(x^+, \eta) = -\frac{1}{\beta s(x)}(1+\eta^2)z^T b\delta_2^2(x^+, \eta) \tag{45}$$

will locally render

$$\dot{V}^+(x^+) \le -\varepsilon^+ V^+(x^+) + \beta \tag{46}$$

where $\delta_2(x^+, \eta)$ is a smooth function satisfying

$$\delta_2(x^+, \eta) \ge \max_{q \in Q} |d_2(x^+, \eta, q)|. \tag{47}$$

and $\varepsilon^+ = \varepsilon_0^+/2$ with $\varepsilon_0^+$ given in Theorem 4.1.

*Proof*: Modified from Theorem 2.1. Details are omitted.

# 5 Concluding Remarks

In this paper, we have proposed a new design technique, robust nonlinear forwarding, for robust stabilization of nonlinear systems with an up-augmented structure. This technique allows us to deal with nonlinear systems with large parameter uncertainties, can provide either smooth or non-smooth (but simpler) stabilizing controllers.

The importance of smooth controllers is that the robust nonlinear forwarding technique can be combined with the back-stepping technique to produce a rich class of uncertain nonlinear systems which can be robustly stabilized. Roughly speaking, the structure of such a system is mainly characterized by its locally linearized version and admits the so-called *anti-symmetric stepwise configuration* (ASSC) which has been studied by Wei in [7] for uncertain linear systems.

To explain the ASSC, we consider the following system

$$\dot{x} = f(x, q) + b(q)u \tag{48}$$

where $f(x,q)$ is smooth in $x$ and continuous in $q$, $b(q)$ is continuous in $q$, and $q \in Q$ is an uncertain parameter vector as before. Define

$$M(x,q) = \begin{bmatrix} \frac{\partial f(x,q)}{\partial x} & b(q) \end{bmatrix} \tag{49}$$

and adopt the following convention:

- $*$ = any scalar function of $x$ and $q$ with a known bound over $\Omega \times Q$;
- $\theta$ = any scalar function of $x$ and $q$ with $1 \ge |\theta| \ge \underline{\theta} > 0$ over $\Omega \times Q$.

Then, examples of ASSC are given as follows:

$$\begin{bmatrix} 0 & \theta & * & 0 \\ 0 & 0 & \theta & 0 \\ * & * & * & \theta \end{bmatrix}$$

$$\begin{bmatrix} 0 & \theta & * & 0 & 0 \\ 0 & 0 & \theta & 0 & 0 \\ * & * & * & \theta & 0 \\ * & * & * & * & \theta \end{bmatrix}$$

$$\begin{bmatrix} 0 & \theta & * & * & 0 \\ 0 & 0 & \theta & 0 & 0 \\ 0 & 0 & * & \theta & 0 \\ * & * & * & * & \theta \end{bmatrix}$$

These examples are all generated via a sequence of up- and down-augmentations. For example, the first example is generated via an up-augmentation from the lower-right $2 \times 3$ structure. A precise definition of the ASSC can be found in Wei [7] with the exception that the matrix $M(x,q)$ in [7] is independent of $x$.

# References


[1] M. Jankovic, R. Sepulchre and P. V. Kokotovic, "Constructive Lyapunov stabilization of nonlinear cascade systems". *IEEE Trans. Aut. Contr*, **AC-41**, (1996) 1723-1735.

[2] M. Krstic,P. V. Kokotovic and I. Kanellakopoulos, "Nonlinear And Adaptive Control Design," Wiley, 1995.

[3] F. Mazenc and L. Praly, "Adding integrations, saturated controls, and stabilization for feedforwarding systems". *IEEE Trans. Aut. Contr*, **AC-41**, (1996) 1559-1578.

[4] R. Sepulchre, M. Jankovic and P. V. Kokotovic, "Interlaced system and recursive designs for global stabilization".(1996)

[5] W. Su and M. Fu, "Robust stabilization of a class of nonlinear systems upper-triangular structure," *Proceedings of IEEE Conference on Decision and Control*, Tampa, Florida, December, 1998.

[6] A.Teel, "Feedback stabilization: Nonlinear solutions to inherently nonlinear problems," Memorandum no. UCB/ERL M92/65, June 12, 1992.

[7] K. Wei, "Quadratic stabilizability of linear systems with structural independent time-varying uncertainties". *IEEE Trans. Aut. Contr*, **AC-35**, (1990) 268-277.

14th Triennial World Congress, Beijing, P.R. China



F-2e-02-2

# LIMIT CYCLES IN UNCERTAIN SYSTEMS WITH MULTIPLE NONLINEARITIES

**Serkan T. Impram* and Neil Munro**

*Control Systems Centre, UMIST, PO Box 88, Manchester, UK*
** impram@csc.umist.ac.uk*

Abstract: The describing function method can be used to analyze control systems containing separable nonlinear elements. In this paper, the method is employed in conjunction with some of the robustness tools for uncertain linear systems in order to predict the existence of limit cycles in a feedback loop with multiple nonlinearities connected in series. Examples are provided to illustrate the procedures which are, of course, subject to the usual errors and restrictions of the describing function method. *Copyright © 1999 IFAC*

Keywords: multiple nonlinearities, describing functions, parameter uncertainty.

## 1. INTRODUCTION

Despite many years of intensive research that has been devoted to the analysis of nonlinear systems, the lack of a unified approach maintains the issue as a challenging problem with numerous, still unanswered questions. At present, the literature is rich enough in publications citing a host of techniques for the study of such systems. However, no one method of analysis has general applicability and the most successful approach to a specific problem will depend upon several factors including the system order and structure, the number of nonlinear effects and so on.

Stability is of paramount importance in any control system. Among the various techniques developed to investigate this particular aspect of nonlinear systems, much emphasis has been placed on describing function methods. It has been suggested that the approach is not a stability method but a procedure for detecting nonlinear oscillations known as limit cycles. However, a limit cycle is more often than not considered as a characteristic of instability and knowledge of its existence is critical. In this regard, together with the Nyquist stability condition, describing functions may not only give approximate solutions, but also provide good insight for further simulation studies (Atherton and Dorrah, 1980).

The describing function method has been widely employed in the analysis and design of control systems involving a single nonlinearity (see (Atherton, 1975; Gelb and Vander Velde, 1968) and references therein). However, a great deal of its simplicity is lost when nonlinear effects are present at more than one point in a system, and over the last three decades, a considerable amount of work has been done to extend its applicability to such cases. Initial attempts have either been restricted to single-loop feedback systems with two nonlinearities (Davison, 1968; Negoescu and Sebastian, 1971), or have resorted to some sort of a trial and error procedure (Gran and Rimer, 1965). Due to the importance of computer graphics for practicing engineers, several researchers have tried to advance these techniques to cover systems with multiple nonlinearities. Two of the papers which fall into this category are (Garg and Aleman, 1972; Grayson, 1965). A major drawback of both of these approaches is that they become quite unwieldy when the number of nonlinear elements goes beyond three. In this respect, the first systematic procedure that can deal with $n$ nonlinearities in a feedback loop appears to be the one given in (Bhargava and Rao, 1970). Surprisingly, this very simple and straightforward approach seems not to be known to most of the researchers in this field. Systems which involve series or parallel combinations of multiple single-valued and/or multi-valued nonlinear elements are treated in (Davison and Constantinescu, 1971). Although, this can be considered as a generalization of the principles given in (Bhargava and Rao, 1970), from quite a similar point of view, it is the most complete reference available, to date.

Almost all of the research into describing functions, including the aforementioned works, has been carried out assuming systems with known structures. Unfortunately, in real life, there is often a mismatch between the actual system and its derived model since a precise mathematical description of a process is rarely available.

Within the realm of uncertain systems, describing functions have received much less attention. There are only a few papers available in the literature and they all seem to be the fruit of very recent efforts in this direction. In (Ferreres and Fromion, 1998; Tierno, 1997), the describing function method is employed within the context of $\mu$–synthesis. While the latter deals with a single nonlinearity, the former reformulates the problem using linear fractional transformations and allows for multiple nonlinear elements. References (Fadali and Chachavalvoong, 1995; Impram and Munro, 1998), on the other hand, recast the problem in the generalized interval polynomial framework. The overall picture turns out to be that of determining the stability of a polynomial, which can be performed using the concept of value sets (Barmish, 1994). The main advantage of the approach is that parametric uncertainties in both the linear and nonlinear parts of the system can be handled with equal ease. The limitation of this approach is due to the overly conservative results that are inevitable as the number of uncertain elements increases.

The aim of this paper is to combine the available describing function methods with some of the now well established robustness tools for linear systems, in order to predict the existence of sustained oscillations in uncertain control systems with multiple nonlinearities. The proposed procedures are, basically, extensions of the ideas of (Bhargava and Rao, 1970; Davison and Constantinescu, 1971) to systems with parameter variations in the linear parts. It is shown, through examples, that bounds on the possible limit cycle frequencies and amplitudes can be obtained with minimal effort and that the corresponding values for the nominal system lie within these extremes.

## 2. SYSTEMS WITH MULTIPLE NONLINEARITIES

The feedback configuration that is considered throughout this paper is depicted in Fig. 1 where the $\mathbf{q}_i$ are vectors of uncertain parameters. For the time being, suppose that the linear plants are not subject to parameter variations, so that the $\mathbf{q}_i$ are fixed at their nominal values $\mathbf{q}_i^o$. The strictly proper transfer functions $G_i(s, \mathbf{q}_i^o)$ for $i = 1, 2, \ldots, n$ are

$$G_i(s, \mathbf{q}_i^o) = B_i(s, \mathbf{q}_i^o) D_i^{-1}(s, \mathbf{q}_i^o) \quad (1)$$

where $B_i(s, \mathbf{q}_i^o)$ and $D_i(s, \mathbf{q}_i^o)$ are polynomials without common factors. As usual, the $G_i(s, \mathbf{q}_i^o)$ are assumed to possess low-pass characteristics, so that the describing function technique is applicable. The nonlinearities $f_i(u_i)$ are restricted to be static (output depends only on magnitude of input) and symmetric elements. Replacing these by their respective describing functions $N_i(A_i)$, the input/output relations for $i = 1, 2, \ldots, n$ become

$$\begin{aligned} u_i(t) &= A_i \sin(\omega t + \psi_{i-1}) \\ y_i(t) &= A_i |N_i(A_i)| \sin(\omega t + \psi_{i-1} + \phi_i) \end{aligned} \quad (2)$$

where $\phi_i = \angle N_i(A_i)$ and $\psi_0 = 0$.

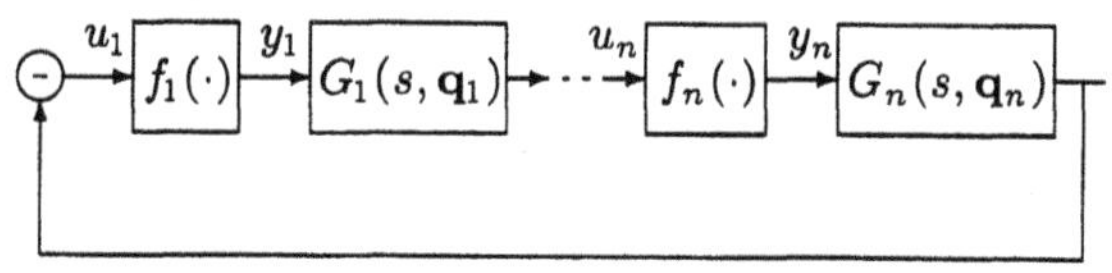


Fig. 1. Feedback system with $n$ nonlinearities.

Applying the conventional Nyquist criterion, the necessary condition for the existence of limit cycles is obtained as

$$\prod_{i=1}^{n} G_i(j\omega, \mathbf{q}_i^o) N_i(A_i) = -1 \quad (3)$$

This can be decomposed into the gain and phase conditions

$$\prod_{i=1}^{n} |G_i(j\omega, \mathbf{q}_i^o)| |N_i(A_i)| = 1 \quad (4a)$$

$$\sum_{i=1}^{n} \angle G_i(j\omega, \mathbf{q}_i^o) + \phi_i = \mp(2k+1)180^\circ \quad (4b)$$

where $k = 0, 1, 2, \ldots$. By observing the manner of interconnections of elements in the system and balancing the signal magnitudes at the nonlinearity inputs using (2), (4a) can be further split into

$$\begin{aligned} A_2 &= A_1 |G_1(j\omega, \mathbf{q}_1^o)| |N_1(A_1)| \\ &\vdots \\ A_n &= A_{n-1} |G_{n-1}(j\omega, \mathbf{q}_{n-1}^o)| |N_{n-1}(A_{n-1})| \end{aligned} \quad (5)$$

If there exist $A_i$ and $\omega$ values which satisfy (4b) and (5) simultaneously, then it can be concluded that the system will probably sustain oscillations. Note that, $A_1 = A_n|G_n(j\omega, \mathbf{q}_n^o)||N_n(A_n)|$ has not been included in (5) as it will automatically be satisfied when (4b) and (5) hold.

## 3. LINEAR SYSTEMS WITH UNCERTAIN PARAMETERS

Parametric uncertainty in a plant transfer function may arise from a variety of factors. It may be because of the inherent imprecisions in the modelling process or it may manifest itself as actual parameter variations that occur during the operation of the system. One way to represent it is in terms of the transfer function coefficients.

Let the numerators and denominators of $G_i(s, \mathbf{q}_i)$ be respectively given by

$$B_i(s, \mathbf{q}_i) = \sum_{k=1}^{b} \beta_{i,k}(\mathbf{q}_i)s^k$$
$$D_i(s, \mathbf{q}_i) = \sum_{k=1}^{d} \delta_{i,k}(\mathbf{q}_i)s^k \tag{6}$$

where $\mathbf{q}_i \in \mathbb{R}^{l_i}$ with components $\mathbf{q}_{i,j}$ defining axis parallel uncertainty boxes

$$\mathbf{Q}_i = \{\mathbf{q}_i \in \mathbb{R}^{l_i} : \mathbf{q}_{i,j}^- \leq \mathbf{q}_{i,j} \leq \mathbf{q}_{i,j}^+, j = 1, \ldots, l_i\}$$

As in (1), $B_i(s, \mathbf{q}_i)$ and $D_i(s, \mathbf{q}_i)$ are coprime and $b < d$ for all $\mathbf{q}_i \in \mathbf{Q}_i$.

Depending upon the form in which the $\mathbf{q}_i$ appear in the coefficient functions $\beta_{i,k}(\mathbf{q}_i)$ and $\delta_{i,k}(\mathbf{q}_i)$, $B_i(s, \mathbf{q}_i)$ and $D_i(s, \mathbf{q}_i)$ will belong to one of the following polynomial families (Barmish, 1994; Bhattacharyya *et al.*, 1995): interval, affine linear, multilinear, or polynomic. Here, the analysis is restricted to linear systems with numerator and denominator polynomials which possess either an interval or an affine linear uncertainty structure in their coefficients. The reason for ruling out the other two possibilities is due to the fact that, in general, there is no systematic way of eliminating conservative results (neither in the time, nor in the frequency domain) in the face of such perturbation structures.

The frequency domain properties of linear systems with coefficients depending in an interval or affine linear fashion on the uncertain parameters, are well documented. For a detailed exposition of the concepts summarized in the rest of the section, the reader is referred to (Bhattacharyya *et al.*, 1995).

Denote the minimum and maximum magnitude points of $G_i(j\omega, \mathbf{q}_i)$ at each $\omega$ by

$$g_i^-(\omega) = \min |G_i(j\omega, \mathbf{q}_i)|$$
$$g_i^+(\omega) = \max |G_i(j\omega, \mathbf{q}_i)| \tag{7}$$

The extreme values of the phase can be expressed similarly as

$$p_i^-(\omega) = \min \angle G_i(j\omega, \mathbf{q}_i)$$
$$p_i^+(\omega) = \max \angle G_i(j\omega, \mathbf{q}_i) \tag{8}$$

If $g_i^o(\omega)$ and $p_i^o(\omega)$ represent the corresponding quantities for the nominal system (obtained by setting $\mathbf{q}_i = \mathbf{q}_i^o$), then the following inequalities are valid:

$$g_i^-(\omega) \leq g_i^o(\omega) \leq g_i^+(\omega)$$
$$p_i^-(\omega) \leq p_i^o(\omega) \leq p_i^+(\omega) \tag{9}$$

The relationships in (7) and (8) define the gain and phase envelopes of the uncertain plant $G_i(s, \mathbf{q}_i)$. Consequently, for nonlinear systems involving such linear parts, it is not pertinent to talk about specific values $A_{ic}$ and $\omega_c$ which satisfy (4b) and (5) simultaneously, but rather, the intervals in which these solutions lie.

## 4. UNCERTAIN SYSTEMS WITH MULTIPLE NONLINEARITIES

This section presents graphical methods, which are slight modifications of each other, for determining the existence of limit cycles and computing bounds on the $A_{ic}$ and $\omega_c$. More specifically, three different system configurations are considered. First, a system with $n$ memoryless nonlinearities is treated. Next, one of the nonlinear elements is allowed to be multi-valued. Finally, the most general case where all nonlinearities possess memory, is studied.

*Case 1:* When all the nonlinearities have single-valued input/output characteristics, the phase condition of (4b) becomes

$$\sum_{i=1}^{n} \angle G_i(j\omega, \mathbf{q}_i) = \mp(2k+1)180^\circ \tag{10}$$

and the method of approach is as follows:

1. Using (10), obtain all possible oscillation frequency ranges $[\omega_c^-, \omega_c^+]$, and determine the corresponding intervals $[g_i^-(\omega_c), g_i^+(\omega_c)]$ in which the plant magnitudes vary.
2. For each of these $\omega_c$ ranges repeat steps 3–6.
3. Starting from several different $A_i$ values go through the recursion

$$[A_2^-, A_2^+] = M_1[g_1^-, g_1^+]$$
$$\vdots \tag{11}$$
$$[A_n^-, A_n^+] = [M_{n-1}^-, M_{n-1}^+] \otimes [g_{n-1}^-, g_{n-1}^+]$$

and store the result of

$$[K^-, K^+] = [M_n^-, M_n^+] \otimes [g_n^-, g_n^+] \quad (12)$$

where $M_i = A_i N_i(A_i)$ and $\otimes$ denotes multiplication within the context of interval mathematics (Moore, 1966).

4. Plot $K^-$ and $K^+$ against $A_1$.
5. Note the intersections with the $K = A_1$ line. These points give the possible intervals for $A_{1c}$.
6. Use these intervals in (11) starting with

$$[A_2^-, A_2^+] = [M_1^-, M_1^+] \otimes [g_1^-, g_1^+] \quad (13)$$

and obtain the bounds on $A_{2c}, \dots, A_{nc}$.

*Remark:* $M_i^-$ and $M_i^+$ are simply the minimum and maximum values of $A_i N_i(A_i)$ for a given range $[A_i^-, A_i^+]$.

*Case 2:* For a system which contains $n-1$ memoryless nonlinearities and a multi-valued nonlinear element, the preceding procedure needs to be modified such that the phase contribution of this element is taken into account.

Without loss of generality, let $f_m(u_m)$ be the nonlinearity with memory. Then, the phase condition around the loop becomes

$$\sum_{i=1}^{n} \angle G_i(j\omega, \mathbf{q}_i) = \mp(2k+1)180^\circ - \phi_m \quad (14)$$

where $\phi_m$ is defined as in (2), and the analysis is carried out accordingly:

1. Choose an arbitrary $A_m$.
2. Compute $\phi_m$ and determine all possible frequency ranges which satisfy (14).
3. Obtain the corresponding intervals for the plant magnitudes.
4. For each of the frequency ranges, go through the recurrence relationship

$$[A_{m+1}^-, A_{m+1}^+] = M_m[g_m^-, g_m^+]$$

$$\vdots$$

$$[A_n^-, A_n^+] = [M_{n-1}^-, M_{n-1}^+] \otimes [g_{n-1}^-, g_{n-1}^+]$$
$$[A_1^-, A_1^+] = [M_n^-, M_n^+] \otimes [g_n^-, g_n^+] \quad (15)$$

$$\vdots$$

$$[A_{m-1}^-, A_{m-1}^+] = [M_{m-2}^-, M_{m-2}^+] \otimes [g_{m-2}^-, g_{m-2}^+]$$

where

$$M_i = \begin{cases} A_i|N_i(A_i)| & , \quad i = m \\ A_i N_i(A_i) & , \quad i \neq m \end{cases}$$

and keep the result of

$$[K^-, K^+] = [M_{m-1}^-, M_{m-1}^+] \otimes [g_{m-1}^-, g_{m-1}^+] \quad (16)$$

5. Repeat steps 1–4 for different $A_m$ values.
6. Plot $K^-$ and $K^+$ versus $A_m$.
7. Note the intersections with the $K = A_m$ line. These points yield the possible ranges for $A_{mc}$.
8. Using these intervals go over steps 2–4. Step 2 gives the possible oscillation frequency ranges $[\omega_c^-, \omega_c^+]$. Starting (15) with

$$[A_{m+1}^-, A_{m+1}^+] = [M_m^-, M_m^+] \otimes [g_m^-, g_m^+]$$

and omitting (16), yields intervals for the remaining amplitudes.

*Remark:* In step 8 above, $\omega_c^-$ and $\omega_c^+$ are determined according to the extreme values of $\phi_m$ for the range $[A_{mc}^-, A_{mc}^+]$.

*Case 3:* The method of analysis outlined in the previous case, may be extended to systems with two or more nonlinear elements with memory. The steps are as follows:

1. Assume an arbitrary $A_1$.
2. For an $\omega$ range of interest repeat steps 3–5.
3. Compute the intervals $[g_i^-(\omega), g_i^+(\omega)]$.
4. Using the recursion given in (11) where now $M_i = A_i|N_i(A_i)|$, obtain the bounds on $A_2, \dots, A_n$.
5. For these bounds, calculate and store

$$[\Theta^-, \Theta^+] = \sum_{i=1}^{n} [p_i^-, p_i^+] \oplus [\phi_i^-, \phi_i^+]$$

where $p_i^-$ and $p_i^+$ are as in (8), $\phi_1^- = \phi_1^+$ and $\oplus$ is the interval addition operator.

6. Plot $\Theta^-$ and $\Theta^+$ against $\omega$.
7. Determine the intersections with the lines $\mp(2k+1)180^\circ$ and get all possible $\omega$ ranges.
8. For each of these frequency ranges repeat steps 3–4 and keep the result of

$$[K^-, K^+] = [M_n^-, M_n^+] \otimes [g_n^-, g_n^+] \quad (17)$$

9. Repeat steps 1–8 for different $A_1$ values.
10. Plot $K^-$ and $K^+$ versus $A_1$ and note the intersections with the $K = A_1$ line. These give the possible bounds on $A_{1c}$.
11. Using these bounds, go over steps 2–8, starting step 4 with (13). The possible ranges for the limit cycle frequency $\omega_c$ are obtained in step 7, whereas step 8 yields the intervals for $A_{2c}, \dots, A_{nc}$. Note that, there is no need to evaluate (17).

The procedures described in this section can also be applied to the nominal systems, with $\mathbf{q}_i = \mathbf{q}_i^o$, when all the intervals collapse down to specific values. For example, $[g_i^-, g_i^+]$ is replaced by $g_i$ and $[A_i^-, A_i^+]$ by $A_i$. Moreover, the inequalities in (9) and the extreme point nature of the methods guarantee that the frequency and amplitude ranges obtained contain the corresponding quantities of the nominal system.

## 5. STABILITY CONSIDERATIONS

In all three cases, the stability of any resulting limit cycle is determined from the slopes of the K curves at the intersections with $K = A_i$ where $A_i$ is the amplitude which acts as the independent variable. When the slope lies in the interval $(-1, 1)$, the limit cycle is stable. Whenever it is in the intervals $(-\infty, -1]$ or $[1, \infty)$, the oscillations grow without bound. On the other hand, if there is no intersection between the curves and the straight line, the stability of the overall system is assessed as follows: If the K curves lie completely above the $K = A_i$ line, then the system is unstable. If they are completely below, then the system is stable (Bhargava and Rao, 1970).

## 6. NUMERICAL EXAMPLES

*Example 1:* Consider the feedback system of Fig. 1 with $n = 3$. The input/output characteristics of the nonlinearities are shown in Fig. 2. The linear transfer functions have coefficients which vary in independent intervals:

$$G_1(s, \mathbf{q}_1) = \frac{a_1}{s^2 + b_1 s + c_1}$$

$$a_1 \in [19.8, 20.1], b_1 \in [6.9, 7.1], c_1 \in [11.7, 12.3]$$

$$G_2(s, \mathbf{q}_2) = \frac{a_2}{s^2 + b_2 s + c_2}$$

$$a_2 \in [28.5, 31.3], b_2 \in [5.1, 6.2], c_2 \in [4.8, 5.2]$$

$$G_3(s, \mathbf{q}_3) = \frac{a_3}{s + b_3}$$

$$a_3 \in [9.9, 10.1], b_3 \in [7.5, 8.4]$$

The nominal system is obtained by setting $\mathbf{q}_1^o = [a_1^o\ b_1^o\ c_1^o]^T = [20\ 7\ 12]^T$, $\mathbf{q}_2^o = [a_2^o\ b_2^o\ c_2^o]^T = [30\ 6\ 5]^T$ and $\mathbf{q}_3^o = [a_3^o\ b_3^o]^T = [10\ 8]^T$.

Since the nonlinearities are memoryless, the first procedure outlined in Section 4 is appropriate. Fig. 3 shows the results after step 4 (the dashed curve is for $\mathbf{q}_i = \mathbf{q}_i^o$), whereas Table 1 gives a comparison between the experimental and predicted limit cycles. It is clear that the $A_{ic}$ and $\omega_c$ for $\mathbf{q}_i = \mathbf{q}_i^o$ fall into the intervals obtained when the

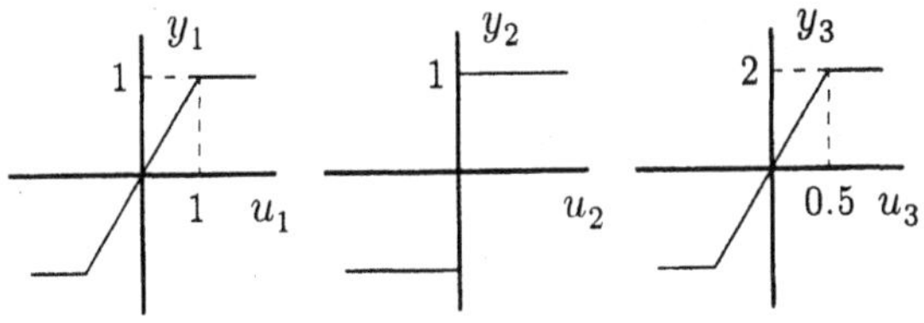


Fig. 2. Nonlinear input/output characteristics.

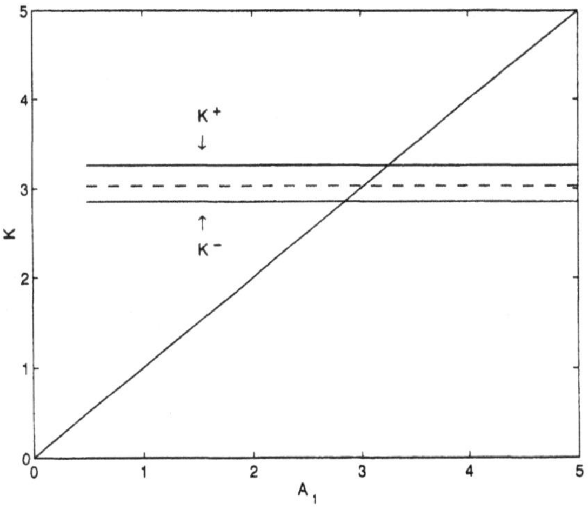


Fig. 3. $K$ curves of Example 1.

linear parts are subject to parameter uncertainties. Note, however, that some of the experimental values are outside of these ranges. This discrepancy is, of course, a natural consequence of the approximations involved in the describing function method and does not disprove the validity of the approach.

Table 1. Limit cycle results for Example 1.

| | Experimental | Nominal | Uncertain |
|---|---|---|---|
| $\omega_c$ | 2.62 | 2.41 | [2.34, 2.48] |
| $A_{1c}$ | 2.51 | 3.03 | [2.86, 3.26] |
| $A_{2c}$ | 1.39 | 1.38 | [1.32, 1.46] |
| $A_{3c}$ | 2.64 | 2.64 | [2.35, 3.34] |

*Example 2:* When the second nonlinearity of the previous example is replaced by a hysteretic relay (with an output level of 1 and a total hysteresis gap of 2), $A_2$ becomes the independent variable and the second procedure given in Section 4 needs to be implemented.

The possible range for $A_{2c}$ is obtained from Fig. 4 by projecting the intersection points of $K^-$ and

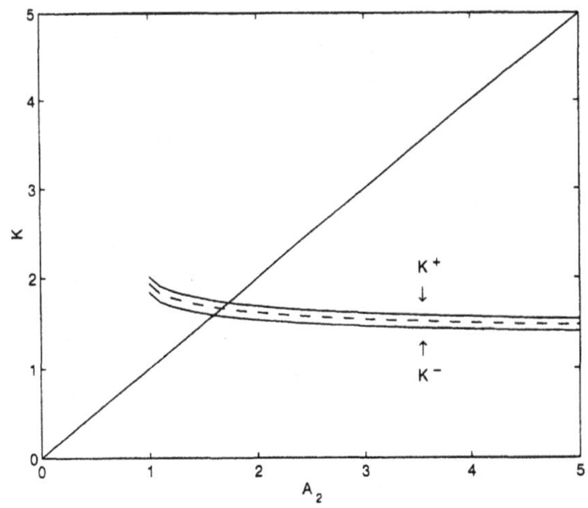


Fig. 4. $K$ curves of Example 2.

Table 2. Limit cycle results for Example 2.

| | Experimental | Nominal | Uncertain |
|---|---|---|---|
| $\omega_c$ | 1.84 | 1.72 | [1.61, 1.81] |
| $A_{1c}$ | 2.49 | 3.11 | [2.92, 3.35] |
| $A_{2c}$ | 1.59 | 1.69 | [1.59, 1.71] |
| $A_{3c}$ | 3.75 | 3.63 | [3.21, 4.69] |

$K^+$ with the $K = A_2$ line onto the horizontal axis. The results, summarized in Table 2, justify the approach.

*Example 3:* Assume now that the configuration of Example 2 is altered such that the first and third nonlinearities are also hysteretic relays. The first one has the same output level and total hysteresis gap of 2, whereas the respective parameters of the third one are 1.5 and 1. Application of the relevant procedure generates the numbers given in Table 3 below.

*Remark:* In all three of the examples, the limit cycles are predicted to be stable.

Table 3. Limit cycle results for Example 3.

| | Experimental | Nominal | Uncertain |
|---|---|---|---|
| $\omega_c$ | 1.88 | 1.51 | [1.42, 1.64] |
| $A_{1c}$ | 1.88 | 2.35 | [2.21, 2.52] |
| $A_{2c}$ | 3.18 | 3.55 | [3.34, 3.72] |
| $A_{3c}$ | 3.67 | 4.04 | [3.46, 5.14] |

## 7. CONCLUSION

This paper has described three systematic procedures for predicting the existence of limit cycles and obtaining bounds on their frequency and amplitude values, in uncertain control systems with multiple nonlinearities. They are all extreme point based approaches and depend, as do all describing function techniques, on how accurate is the assumption that the inputs to the nonlinearities are approximately sinusoidal. If the linear parts are sufficiently low-pass, the difference between the actual and calculated limit cycles will be negligible, despite the approximate nature of the describing functions. However, if this is not true, the results are very likely to lead to an incorrect assessment of the system behaviour.

*Acknowledgement:* This work was supported by the EPSRC under Grant GR/L63303.

## REFERENCES

Atherton, D. P. (1975). *Nonlinear Control Engineering.* Van Nostrand Reinhold. London.

Atherton, D. P. and H. T. Dorrah (1980). A survey on nonlinear oscillations. *Int. J. Control* **31**, 1041–1105.

Barmish, B. R. (1994). *New Tools for Robustness of Linear Systems.* Macmillan. New York.

Bhargava, V. and D. N. Rao (1970). Application of describing function technique to memoryless nonlinearities in tandem. *Int. J. Control* **12**, 1–8.

Bhattacharyya, S. P., H. Chapellat and L. H. Keel (1995). *Robust Control: The Parametric Approach.* Prentice-Hall. Upper Saddle River, New Jersey.

Davison, E. J. (1968). Application of the describing function technique in a single-loop feedback system with two nonlinearities. *IEEE Trans. Automat. Control* **AC-13**, 168–170.

Davison, E. J. and D. Constantinescu (1971). A describing function technique for multiple nonlinearities in a single-loop feedback system. *IEEE Trans. Automat. Control* **AC-16**, 56–60.

Fadali, M. S. and N. Chachavalvoong (1995). Describing function analysis of uncertain nonlinear systems using the Kharitonov approach. In: *Proc. American Control Conference.* Seattle, USA. pp. 2908–2912.

Ferreres, G. and V. Fromion (1998). Nonlinear analysis in the presence of parametric uncertainties. *Int. J. Control* **69**, 695–716.

Garg, D. P. and R. Aleman (1972). A limit cycle analysis of control systems with multiple nonlinearities. *Int. J. Control* **69**, 695–716.

Gelb, A. and W. E. Vander Velde (1968). *Multiple Input Describing Functions and Nonlinear System Design.* McGraw-Hill. New York.

Gran, R. and M. Rimer (1965). Stability analysis of systems with multiple nonlinearities. *IEEE Trans. Automat. Control* **AC-10**, 94–97.

Grayson, L. P. (1965). Stability and oscillations in systems containing one or more single-valued symmetric nonlinearities. *J. Franklin Inst.* **280**, 52–64.

Impram, S. T. and N. Munro (1998). Describing function analysis of nonlinear systems with parametric uncertainties. In: *Proc. UKACC Int. Conference CONTROL 98.* Swansea, UK. pp. 112–116.

Moore, R. E. (1966). *Interval Analysis.* Prentice-Hall. Englewood Cliffs, New Jersey.

Negoescu, N. and L. Sebastian (1971). Grapho-analytical method for the determination of self-sustained oscillations for a class of automatic nonlinear systems. *IEEE Trans. Automat. Control* **AC-16**, 272–274.

Tierno, J. E. (1997). Describing function analysis in the presence of uncertainty. *J. Guidance, Control, Dynamics* **20**, 956–961.

# ROBUST FEEDBACK DESIGN OF A CLASS OF NONLINEAR CASCADED SYSTEMS WITH STRUCTURAL UNCERTAINTY

**Tielong shen* Katsutoshi Tamura* Peter N. Nikiforuk****

* *Department of Mechanical Engineering, Sophia University, Kioicho 7-1, Chiyoda-ku, Tokyo, 102-8554, Japan, tetu-sin@hoffman.cc.sophia.ac.jp*
** *Department of Mechanical Engineering, University of Saskatchenwan, Saskatoon, Saskatchewan, Canada*

Abstract: This paper is concerned with the robust stabilization and robust $L_2$ disturbance attenuation of a class of cascaded nonlinear systems with structural uncertainties. The design tools are construction of a Lyapunov function for robust stabilization and a storage function for robust $L_2$ gain performance. It is shown that if for each subsystem there exist Lyapunov functions satisfying Hamilton-Jacobi inequalities related with nominal subsystems, then a robust stabilizing controller can be constructed such that the sum of the functions becomes a Lyapunov function for the cascaded system with uncertainties. This concept is extended to the construction of the storage function and the controller for the robust $L_2$ disturbance attenuation problem by introducing an appropriate weight into the sum of the subsystem's Lyapunov functions.

Keywords: Robust stabilization, robust $L_2$ synthesis, Lyapunov-based design, cascaded systems.

## INTRODUCTION

This paper addresses the robust stabilization and robust disturbance attenuation of a class of cascaded nonlinear systems via state feedback. The system under consideration consists of two nonlinear subsystems with a cascade connection and the uncertainties are represented by gain bounded nonlinear functions. The system is of the form

$$\Sigma : \begin{cases} \dot{x} = f_1(x) + c(x,\xi,y) + \Delta f_1(x,\xi,y) \\ \dot{\xi} = f_2(\xi) + G(\xi)u + \Delta f_2(\xi) \\ y = h(\xi) \end{cases} \tag{1}$$

When $\Delta f_1 = 0$ and $\Delta f_2 = 0$ the system is said to be nominal and is denoted by $\Sigma_0$.

The global stabilization of known cascaded nonlinear systems in which $\Delta f_1 = 0, \Delta f_2 = 0$ has been well established by many researchers, see (Kokotovic, *et al.* 1989; Saberi, *et al.*, 1990; Ortega, 1990; Krstic, *et al.*, 1995). For the nominal system $\Sigma_0$, it has been shown by Saberi, *et al.*(1990) that if the $\xi$-subsystem is linear and stabilizable, and the unforced $x$-subsystem $\dot{x} = f_1(x) + c(x,0,0)$ is asymptotically stable, then the cascaded system $\Sigma_0$ may be stabilized by state feedback control under the condition that the connection $c$ is spanned by $y$. This result is established based on the fact that if we can find a Lyapunov function $V(x)$ for the unforced $x$-subsystem and a storage function $W(\xi)$ for the linear subsystem, then the sum

$$U(x,\xi) = V(x) + W(\xi) \tag{2}$$

will be a Lyapunov candidate for the cascaded system. As shown in (Ortega, 1990) and (Krstic it et al., 1995), this result can be extended to the case in which the $\xi$-subsystem is nonlinear but strictly passive with a positive definite storage function $W(\xi)$. In this case, the sum of $V(x)$ and $W(\xi)$ plays the role of a Lyapunov function for the cascaded system. However, in general, the sum given in (2) can not be a Lyapunov function for the known cascaded system $\Sigma_0$, even if the subsystems are stable with Lyapunov functions $V(x)$ and $W(\xi)$ respectively. In this case, an additional cross term $\psi(x, \xi)$ is required in the sum. An approach has been presented by Mazenc, *et al.* (1997) for constructing the $\psi(x, \xi)$ such that the sum plays the role of a Lyapunov function.

In the recent past Lyapunov theory has been primarily used to sythesis the robust feedback control of uncertain nonlinear systems. In the Lyapunov-based robust synthesis framework, the key is to seek such a Lyapunov function based on the nominal system with a bounding function for uncertainties that ensures stability for all uncertainties under consideration. Some robust design issues have been investigated in this line of research. For example, robust stabilization problems were studied by Lin (1997) and Su and Xie (1996) for parametric uncertainties and by Qu, *et al.* (1994) and Lin and Shen (1997) for structural uncertainties. The robust $L_2$ disturbance attenuation problem was addressed by Shen and Tamura (1995, 1997), Shen, *et al.* (1998). In this case, the Lyapunov function played the role of the storage function satisfying a dissipation inequality, which is necessary to guarantee the $L_2$ disturbance attenuation performance. In (Shen and Tamura,1995), the desired Lyapunov function was found by solving the Hamilton-Jacobi inequality, and in (Shen *et al.*, 1998) a constructive way was shown without solving the Hamilton-Jacobi inequality.

In this paper, we focus our attention on a cascaded nonlinear system with structural uncertainties. First, we study the *robust stabilization problem* of the system. The structure of the system is similar to the one in (Kokotovic, *et al.*), 1989; Saberi, *et al.*, 1990; Kristic, *et al.*, 1995) except that uncertainties are considered in this paper. We will show that if there exists an appropriate Lyapunov function for the nominal $x$-subsystem and a storage function for the $\xi$-subsystems, then we can derive the robust stabilizing controller such that the sum of the functions will be the Lyapunov function for all uncertainties. It is also shown that by employing the robust passivation techniques proposed by Lin and Shen (1997), this approach can be used to design a robust stabilizing controller for the case in which $\xi$-subsystem is of triangular structure. Another contribution of this paper is to extend this approach to solve the *robust disturbance attenuation problem* in the sense of the $L_2$ gain. The $L_2$ gain synthesis problem is typically solved using the dissipativity conception (van der Schft, 1996; Shen and Tamura, 1995, 1997). The key point in this extension is how to construct the storage function without solving the Hamilton-Jacobi inequality, which ensures the $\gamma$-dissipativity of the overall system. We will show that if both of the nonminal subsystems are asymptotically stable, then the storage function can be constructed based on the Lyapunov functions with appropriate weightings.

## LYAPUNOV FUNCTIONS FOR CASCADED SYSTEM

Consider first known systems described by

$$\Sigma_0 : \begin{cases} \dot{x} = f_1(x) + C(x, \xi, y)y \\ \dot{\xi} = f_2(\xi) + G(\xi)u \\ y = h(\xi) \end{cases} \tag{3}$$

where $u$ is the control input and $y$ is the output of the $\xi$-subsystem, $f_1, f_2, h, C$ and $G$ are nonlinear functions with appropriate dimensions and assumed to satisfy $f_1(0) = 0, f_2(0) = 0$.

Our goal is to find an appropriate Lyapunov function with which we can construct a stabilizing feedback control law $u = \alpha(x, \xi)$ such that the closed loop system of (3) is asymptotically stable. To achieve this goal, we make the following assumptions for each subsystems.

**H1** There exists a positive definite function $V(x)$ ($V(0) = 0$) such that $L_{f_1}V < 0$, $\forall x \neq 0$.
**H2** The $\xi$-subsystem with a $C^1$ storage function $W(\xi)$ is passive, i.e. there exists a $W(\xi) \in C^1$ such that $\dot{W} \leq u^T y$ holds for all $u$.
**H3** $\{f_2, h\}$ is zero state detectable.

Under these assumptions, we construct Lyapunov function candidates of the form

$$\tilde{U}(x, \xi) = \Psi_o(s), \quad s = \Psi_i(V) + W(\xi) \tag{4}$$

where $\Psi_o$ and $\Psi_i$ are any given $\kappa_\infty$ functions.

**Proposition 1.** Suppose that $H1 \sim H3$ holds. For any given $\kappa_\infty$ functions $\Psi_o$ and $\Psi_i$, there exists a state feedback $u = \alpha(x, \xi)$ such that the cascaded system (3) with the control law is globally asymptotically stable and the positive definite function $\tilde{U}$ defined by (4) is a Lyapunov function. Furthermore, a desired control law is given by

$$\alpha = -\frac{d\Psi_i}{dV} L_C^T V - y \tag{5}$$

**Proof.** Calculating the derivative of $\tilde{U}$ along any trajectory of the system, gives

$$\dot{\tilde{U}} = \frac{d\Psi_o}{ds}\{\frac{d\Psi_i}{dr}(L_{f_1}V + L_C V y) + \dot{W}\} \quad (6)$$

Since $\Psi_o, \Psi_i$ are $\kappa_\infty$ functions, we have $\frac{d\Psi_o}{ds} > 0$ and $\frac{d\Psi_i}{dV} > 0$. Hence, with the assumptions $H1 \sim H3$ and noting the control law (5),

$$\begin{aligned}\dot{\tilde{U}} &\leq \frac{d\Psi_o}{ds}\{\frac{d\Psi_i}{dr}(L_{f_1}V + L_C V y) + u^T y\} \\ &< -\frac{d\Psi_o}{ds} y^T y, \ \forall x \neq 0 \end{aligned} \quad (7)$$

Therefore, the stability of equilibrium $(x, \xi) = (0, 0)$ follows from the detectability condition. □

**Remark 1.** In Proposition 1, by setting $\Psi_o(s) = s$ and $\Psi_i(r) = r$ one obtains the result given by Saberi, *et al.* (1990) and Ortega (1990). In this paper, the *free weighting functions* $\Psi_o$ and $\Psi_i$ will be used to meet additional performance specifications.

**Remark 2.** It should be noted that the assumptions have been relaxed by Jankovic, *et al.* (1996) to require the $x$-subsystem to be Lyapunov stable and the interconnection term $C(x, \xi, y)y$ to be globally Lipschitz on $x$ for any fixed $\xi$. This relaxation was achieved by adding a cross term $\phi(x, \xi)$ to $\tilde{U}$. However, we will continue using the Lyapunov function without the cross term and require the subsystems to be asymptotically stable. Our goal is robust synthesis which now be described.

## ROBUST STABOLIZATION

Consider the robust stabilization of the system

$$\begin{cases} \dot{x} = f_1(x) + C(x, \xi, y)y + \Delta f_1(x, \xi, y) \\ \dot{\xi} = f_2(\xi) + G(\xi)u + \Delta f_2(\xi) \\ y = h(\xi) \end{cases} \quad (8)$$

where $\Delta f_1$ and $\Delta f_2$ are described by

$$\Delta f_1(x, \xi, y) = E_1(x, y)\delta_1(x, \xi, y)$$
$$\Delta f_2(\xi) = E_2(\xi)\delta_2(\xi)$$

$E_1$ and $E_2$ are known mappings, and unknown mappings $\delta_1$ and $\delta_2$ are bounded by

$$||\delta_1(x, \xi, y)||^2 \leq \rho_1(x, y), \ ||\delta_2(\xi)||^2 \leq \rho_2(\xi)$$

with the known functions $\rho_1$ and $\rho_2$.

The smooth functions $||L_{E_1}V||^2$ and $||\rho_1||^2$ can be decomposed as follows

$$||L_{E_1}V||^2 = ||L_{E_*}V||^2 + M_e(x, y)y \quad (9)$$
$$||\rho_1(x, y)||^2 = ||\rho_*(x)||^2 + M_\rho(x, y)y \quad (10)$$

where $E_*(x) = E_1(x, 0)$, $\rho_*(x) = \rho_1(x, 0)$ and $M_e$ and $M_\rho$ are known functions.

**Theorem 1.** Suppose that system (8) satisfies the following assumptions.

**A1** There exist a $V(x) > 0$ ($V(0) = 0$) and a scale $\epsilon_1 > 0$ which satisfy the HJI inequality

$$L_{f_1}V + \frac{1}{2}||L_{E_*}V||^2 + \frac{1}{2}||\rho_*(x)||^2 \leq -\epsilon_1 V(x)$$

**A2** There exists a positive definite function $W(\xi)$ which satisfies the HJI inequality

$$L_{f_2}W + \frac{1}{2}||L_{E_2}W||^2 + \frac{1}{2}||\rho_2(\xi)||^2 \leq 0$$

and $L_{G_2}W = h^T(\xi)$ holds for all $x$.

**A3** $\{f_2 + E_2\delta_2, h\}$ is detectable for all $\delta_2$.

Then, the system is globally asymptotically stabilizable by state feedback $u = -\alpha(x, \xi, y)$ where

$$\alpha(x, \xi, y) = L_C^T V + \frac{1}{2}M_e^T(x, y) + \frac{1}{2}M_\rho^T(x, y) + y$$

**Proof.** Consider the Lyapunov function (4) with $\Psi_o(s) = s$ and $\Psi_i(r) = r$, i.e.

$$U(x, \xi) = V(x) + W(\xi) \quad (11)$$

For the closed loop system, applying the conditions $A1$ and $A2$ yields

$$\begin{aligned}\dot{U} &= L_{f_1}V + L_C V y + L_{E_1}\delta_1 + L_{f_2}W \\ &\quad + L_{G_2}Wu + L_{E_2}W\delta_2 \\ &\leq -\epsilon_1 V(x) - \frac{1}{2}||L_{e_*}V||^2 - \frac{1}{2}||\rho_*(x)||^2 \\ &\quad -\frac{1}{2}||L_{E_2}W||^2 - \frac{1}{2}||\rho_2(\xi)||^2 \\ &\quad + L_C V y + L_{E_1}\delta_1 + L_{G_2}Wu + L_{E_2}W\delta_2 \end{aligned} \quad (12)$$

The substitution of (9) and (10) and completing the squares gives

$$\begin{aligned}\dot{U} &\leq -\epsilon_1 V(x) - \frac{1}{2}||L_{E_1}^T V - \delta_1||^2 - \frac{1}{2}||L_{E_2}^T W - \delta_2||^2 \\ &\quad \frac{1}{2}\{||\rho_1||^2 - ||\delta_1||^2\} \\ &\quad -\frac{1}{2}\{||\rho_2||^2 - ||\delta_2||^2\} + y^T u \\ &\quad + L_C V y + \frac{1}{2}M_e(x, y)y + \frac{1}{2}M_\rho(x, y)y \\ &\leq -\epsilon_1 V(x) - y^T y \end{aligned} \quad (13)$$

□

**Remark 3.** According to the definition of *robust passivity* and the *robust KYP lemma* given by Lin and Shen (1997), assumption $A2$ is a sufficient condition for the $\xi$-subsystem being robust passive with the storage function $W(\xi)$.

This result shows that if the subsystems are robust stable and robust passive, then the cascaded system can be robustly stabilized by state feedback. This is a natural extension of the result for nominal systems mentioned in section 2. However, in this case of uncertainty, in order to obtain the robustness we have to add gain functions $M_e$ and $M_\rho$, which relate to the interconnection term to the control law. Of course, if the $\xi$-subsystem is not robust passive, we can first achieve the robust passivity with preliminary feedback using the techniques in (Lin and Shen, 1997).

Under the assumption of relative degree one for the $\xi$-subsystem, there exists the change of coordinates with a preliminary feedback that transforms the subsystem into normal form, so that the cascaded system is represented in the form

$$\begin{cases} \dot{x} = f_1(x)+C(x,\eta,y)y+E_1(x,y)\delta_1(x,\eta,y) \\ \dot{\eta}_1 = f_{21}(\eta_1,\eta_2)+E_{21}(\eta_1,\eta_2)\delta_2(\eta) \\ \dot{\eta}_2 = v+E_{22}(\eta_1,\eta_2)\delta_2(\eta) \\ y = \eta_2 \end{cases} \quad (14)$$

where $\eta^T = [\eta_1^T \ \eta_2^T]$.

In the following, we will consider the robust stabilization of this system. In order to design the preliminary feedback, we make the following assumption.

**A2′** The $\xi$-subsystem is robust minimum phase (Lin and Shen, 1997), i.e. there exists $W_0(\eta_1) > 0$ such that

$$\frac{\partial W_0}{\partial \eta_1}\{f_{21}(\eta_1,0)+E_{21}(\eta_1,0)\delta_2(\eta_1,0)\} < 0$$

holds for all $\delta_2$.

Note that under condition $A2'$, there exists a positive scaling function $\lambda(\eta_1) > 0$ such that

$$L_{f_2^*}W_0+\frac{\lambda(\eta_1)}{2}||L_{E_2^*}W_0||^2+\frac{1}{2\lambda(\eta_1)}||\rho_2^*(\xi)||^2 < 0$$

holds (Lin and Shen, 1997; Shen, *et al.*, 1997) and the smooth functions $f_2$, $||L_{E_{21}}W_0||^2$ and $||\rho_2(\eta)||^2$ can be decomposed as

$$f_2(\eta_1,\eta_2) = f_2^*(\eta_1)+\tilde{f}(\eta_1,\eta_2)\eta_2 \quad (15)$$

$$||L_{E_{21}}W_0||^2 = ||L_{E_{21}^*}W_0||^2+N_e(\eta_1,\eta_2)\eta_2 \quad (16)$$

$$||\rho_2(\eta)||^2 = ||\rho_2^*(\eta_1)||^2+N_\rho(\eta_1,\eta_2)\eta_2 \quad (17)$$

where $f_2^*(\eta_1) = f_{21}(\eta_1,0)$, $E_{21}^*(x) = E_{21}(\eta_1,0)$ and $\rho_2^*(\eta_1) = \rho_2(\eta_1,0)$.

**Theorem 2.** Suppose that the system assumptions $A1$ and $A2'$ and that the $\eta$-subsystem is zero state detectable for all uncertainties $\delta_2$. Then the cascaded system (14) is globally asymptotically stabilizable by state feedback

$$v = -\alpha(x,\eta,y)-\beta(\eta_1,\eta_2) \quad (18)$$

where $\alpha$ is given in Theorem 1 and

$$\beta = L_{\tilde{f}_2}^T+\frac{\lambda(\eta_1)}{2}N_e^T(\eta_1,\eta_2)+\frac{1}{2\lambda(\eta_1)}N_\rho^T(\eta_1,\eta_2) + \frac{\lambda_1(\eta_1)}{2}\{2E_{22}L_{E_{21}}^TW_0+E_{22}E_{22}^T\eta_2\}$$

**Proof.** As shown in (Lin and Shen, 1997), under the assumption $A2'$ one can find a preliminary feedback such that the $\eta$-subsystem is robust passive with the storage function $W(\eta_1,\eta_2) = W_0(\eta_1)+\frac{1}{2}\eta_2^T\eta_2$. We can then construct a Lyapunov function candidate for the overall system

$$U(x,\eta_1,\eta_2) = V(x)+W(\eta_1,\eta_2) \quad (19)$$

and derive the desired feedback directly using this function.

Using the same techniques as in (13) with assumption $A1$, it is easy to show that

$$\begin{aligned} \dot{U} &= L_{f_1}V+L_CVy+L_{E_1}\delta_1+\dot{W} \\ &\le -\epsilon_1V(x)-\frac{1}{2}||L_{E_1}^TV-\delta_1||^2-\frac{1}{2}\{||\rho_1||^2-||\delta_1||^2\} \\ &\quad +L_CVy+M_e(x,y)y+M_\rho(x,y)y+\dot{W} \end{aligned} \quad (20)$$

Noting that $y = \eta_2$ and using assumption $A2'$ and (15)~(17) yields

$$\begin{aligned} \dot{U} &\le -\epsilon_1V(x)+L_CVy+M_e(x,y)y+M_\rho(x,y)y \\ &\quad +L_{f_2^*}W_0+L_{\tilde{f}_2}W_0\eta_2+L_{E_{21}}W_0\delta_2+\eta_2^T\{u+E_{22}\delta_2\} \\ &\le -\epsilon_1V(x)+L_CVy+\frac{1}{2}M_e(x,y)y+\frac{1}{2}M_\rho(x,y)y \\ &\quad +\eta_2^Tu+L_{\tilde{f}_2}\eta_2-\frac{\lambda}{2}||L_{E_{21}}^TW_0+E_{22}^T\eta_2-\frac{1}{\lambda}\delta_2||^2 \\ &\quad -\frac{1}{2\lambda}\{||\rho_2||^2-||\delta_2||^2\}+\frac{\lambda(\eta_1)}{2}N_e\eta_2+\frac{1}{2\lambda}N_\rho\eta_2 \\ &\quad +\frac{\lambda_1}{2}\{2L_{E_{21}}W_0E_{22}^T+\eta_2E_{22}E_{22}^T\}\eta_2 \\ &\le -\epsilon_1V(x)-y^Ty \end{aligned} \quad (21)$$

for any $\delta_1$ and $\delta_2$. □

**Remark 4.** It should be noted that this result is an application of Theorem 1. Indeed, using the robust KYP lemma (Lin and Shen, 1997), it is easy to show that applying the preliminary feedback $v = -\beta(\eta_1,\eta_2)$ makes the $\eta$-subsystem with the storage function $W$ robust passive.

## ROBUST $L_2$ DISTURBANCE ATTENUATION

Consider the nonlinear system

$$\Sigma: \begin{cases} \dot{x} = f_1(x)+C(x,\xi,y)y+\Delta f_1(x,\xi,y) \\ \dot{\xi} = f_2(\xi)+G(\xi)u+D(\xi)w+\Delta f_2(\xi) \\ y = h(\xi) \\ z = p(x,\xi) \end{cases} \quad (22)$$

where $w$ is the disturbance, and $z$ is the penalty signal given by

$$z = p(x,\xi) = \begin{bmatrix} p_1(x) \\ p_2(\xi) \end{bmatrix} \qquad (23)$$

Our goal is to find a state feedback such that the closed loop system is globally asymptotically stable and the $L_2$ gain of the closed loop system from $w$ to $z$ is less than a given level $\gamma$ for all uncertainties. This goal can be achieved by rendering the closed loop system with a positive storage function $U(x,\xi)$ dissipative with respect to the supply rate $\frac{\epsilon}{2}\{\gamma^2\|w\|^2 - \|z\|^2\}$, i.e.

$$\dot{U} \leq \frac{\epsilon}{2}\{\gamma^2\|w\|^2 - \|z\|^2\}, \ \forall w \qquad (24)$$

holds for all uncertainties, where $\epsilon > 0$.

Suppose that $V(x) > 0$ is proper and radially unbounded. Then, for any given scale $k > 0$ it is possible to select a $\mathcal{K}_\infty$ function $\kappa(V)$ satisfying

$$\kappa(V) \geq \frac{k}{2}\|p_1(x)\|^2 \qquad (25)$$

Using $\kappa$, we can construct a function as follows

$$\Psi_i(V) = V \sup_{0\leq\theta\leq 1} \frac{d\kappa(\theta)}{d\theta} + \int_V^{2V} \kappa(\theta)d\theta \qquad (26)$$

It is then easy to check that $\Psi_i$ satisfies

$$V\frac{d\Psi_i}{dV} \geq \kappa(V) \qquad (27)$$

With this weighting function and $\Psi_o(s) = s$, one can construct a desired feedback control law such that the Lyapunov function candidate (4) satisfies the dissipation inequality. To this end, we make the following assumption.

**B1** For the $x$-subsystem, there exist a positive function $V(x)$ and a scalar $\epsilon_1 > 0$ such that

$$L_{f_1+E_1\delta_1}V \leq -\epsilon_1 V$$

**B2** For the $\xi$-subsystem, there exist a feedback law $u = -\alpha_l(\xi)$, a positive function $W(\xi)$ and a given scalar $\epsilon_2 > 0$ such that

$$\frac{\partial W}{\partial \xi}\{f_\alpha+\Delta f_2+D_1 w\}\leq\frac{\epsilon_2}{2}\{\gamma^2\|w\|^2-\|p_2\|^2\}$$

holds for all $\delta_2$, and $L_G W = h^T(\xi)$, where $f_\alpha = f_2 - G\alpha_l$. .

Then, there exists a $\lambda_1(x) > 0$ such that

$$\frac{d\Psi_i}{dV}L_{f_*}V + \frac{\lambda_1(x)}{2}\|\frac{d\Psi_i}{dV}L_{E_*}V\|^2 + \frac{1}{2\lambda_1(x)}\|\rho_*(x)\|^2 \leq -\epsilon_1\frac{d\Psi}{dV}$$

**Theorem 3.** Given $\gamma > 0$. Suppose that the system satisfies assumptions $B1$, $B2$ and $A3$. Let $k = \frac{\epsilon_2}{\epsilon_1}$ and the state feedback controller be given by $u = -\alpha_l(\xi) - \beta(x,\xi,y)$ where

$$\beta = \frac{d\Psi_i}{dV}L_C^T V + \frac{1}{2}M_e^T + \frac{1}{2}M_\rho^T + y \qquad (28)$$

Then, the closed loop system is globally asymptotically stable and for any given $T > 0$,

$$\int_0^T \|z\|^2 dt \leq \gamma^2 \int_0^T \|w\|^2 dt + Q(x_0,\xi_0), \ \forall w \qquad (29)$$

holds for all $\delta_1$ and $\delta_2$, where $Q$ is a positive definite function and $x_0 = x(0)$, $\xi_0 = \xi(0)$.

**Proof.** Consider the positive function

$$U(x,\xi) = \Psi_i(V(x)) + W(\xi) \qquad (30)$$

Closing the loop with the state feedback and calculating

$$H(x_0,\xi_0,w) = \dot{U} + \frac{\epsilon_2}{2}\{\|z\|^2 - \gamma^2\|w\|^2\} \qquad (31)$$

along the trajectories for any $w$, yields

$$\begin{aligned} H = & \frac{d\Psi_i}{dV}\{L_{f_1}V + L_C V y + L_{E_1}\delta_1\} + L_{f_\alpha}W \\ & -L_G W\beta + L_D W w + L_{E_2}W\delta_2 + \frac{\epsilon_2}{2}\{\|z\|^2 - \gamma^2\|w\|^2\} \\ \leq & -\epsilon_1\kappa(V) + \frac{\epsilon_2}{2}\|p_1(x)\|^2 - y^T\beta \\ & + y^T\{+\frac{d\Psi_i}{dV}L_C^T V + \frac{1}{2}M_e^T(x,y) + \frac{1}{2}M_\rho^T(x,y)\} \\ \leq & -y^T y \end{aligned} \qquad (32)$$

Hence, for any given $T > 0$, integrating the above inequality shows that

$$\int_0^T \|z\|^2 dt \leq \int_0^T \|w\|^2 dt + \frac{1}{\epsilon_2}U(x_0,\xi_0)$$

holds for any uncertainty $\delta_1$ or $\delta_2$.

Consider the case of $w = 0$. From the zero-state detectability condition, $y = 0$ implies that $\xi \to 0$, and that $x \to 0$ by $B1$. Hence, the robust stability follows by the Lyapunov function $U$ with the inequality (32) for the case $w = 0$. □

This result shows that if the $\xi$-subsystem's $L_2$ gain is less than the given level $\gamma$ for all $\delta_1$, then one can find a feedback control law by adjusting the weighting function such that the Lyapunov function takes on the role of a storage function which ensures the robust $L_2$ gain of the cascaded system. The problem is how to find the preliminary feedback $\alpha_l(\xi)$. Under the conditions that the $\xi$-subsystem is of relative degree one and $D(\xi)$ is matched by $G(\xi)$, one can find the desired $\alpha_l$ by using the recursive design method for dissipation inequality (Shen, *et al.*, 1998).

Assume that the $\xi$-subsystem is of relative degree one for all $\delta_1$ and in addition that $D(\xi)$ satisfies the following condition.

**B3** There exists a $D_0(\xi)$ such that $D(x) = G(\xi)D_0(\xi)$.

Then, as shown in the previous section, one can seek a preliminary feedback and the change of coordinates which transforms the system into

$$\begin{cases} \dot{x} = f_1(x) + C(x,\eta,y)y + E_1(x,y)\delta_1(x,\eta,y) \\ \dot{\eta}_1 = f_{21}(\eta_1,\eta_2) + E_{21}(\eta_1,\eta_2)\delta_2(\eta) \\ \dot{\eta}_2 = v + D_0(\xi)w + E_{22}(\eta_1,\eta_2)\delta_2(\eta) \\ y = \eta_2 \\ z = \begin{bmatrix} p_1(z) \\ p_2(\eta) \end{bmatrix} \end{cases} \tag{33}$$

We will now construct a storage function $W(\eta_1,\eta_2)$ which satisfies assumption $(B2)$. To do this, it is necessary to introduce the weighting function

$$\Psi_w(W_0) = W_0 \sup_{0\le\theta\le 1} \frac{d\kappa_w(\theta)}{d\theta} + \int_{W_0}^{2W_2} \kappa_w(\theta)d\theta \tag{34}$$

with any given $\mathcal{K}_\infty$ function $\kappa_w$ satisfying

$$\kappa_w(W_0) \geq \frac{k}{2}\|p_2^*(\eta_1)\|^2 \tag{35}$$

where $\|\rho_*(\eta_1)\|^2 = \|\rho_2(\eta_1,0)\|^2 + \tilde{\rho}_2(\eta_1,\eta_2)\eta_2$.

**Theorem 4.** Given $\gamma > 0$. Suppose that the system (33) satisfies $B1$, $B2'$, $B3$ and the $\eta$-subsystem is zero-state detectable for all $\delta_2$. Let the smooth function $\alpha_l(\eta_1,\eta_2)$ be given by

$$\begin{aligned} \alpha_l(\eta_1,\eta_2) = & \frac{\lambda_1}{2}N_e^T + \frac{1}{2\lambda_1}N_\rho^T \\ & + \frac{\lambda_1}{2}E_{22}\{2\frac{d\psi_w}{dW_0}L_{E_{21}}^T W_0 + E_{22}E_{22}^T\eta_2\} \\ & + \frac{2}{\epsilon_2\gamma^2}D_0D_0^T\eta_2 + \frac{\epsilon_2}{2}\tilde{p}^T(\eta_1,\eta_2) \end{aligned} \tag{36}$$

Then the closed loop system of (??) with the feedback $v = \alpha_l(\eta_1,\eta_2) + \beta(\eta,x,y)$ is asymptotically stable and has a robust $L_2$ gain less than $\gamma$, where the function $\beta$ is given in Theorem 3.

## CONCLUSION

This paper described the robust stabilization and robust $L_2$ disturbance attenuation via state feedback for a class of cascaded nonlinear systems. It was shown that if there exist Lyapunov functions for both unforced subsystems, then a Lyapunov function can be constructed with which a desired state feedback law can be designed for the robust stabilization problem, and for the robust $L_2$ disturbance attenuation a storage function can be constructed with the Lyapunov functions by introducing nonlinear weighting.

## REFERENCES


Kokotovic,P.V. and H.J.Sussann(1989). A positive real condition for global stabilization of nonlinear systems. *Syst. and Contr. Lett.*, **13**,pp.125-133.

Saberi,A., P. V. Kokotovic and H.J.Sussann (1990). Global stabilization of partially linear composite systems. *SIAM J. Contr. Optimization*, **28**,pp.1491-1503.

Ortega, R.(1990). On global stabilization of cascaded nonlinear systems, *Proc. 29th IEEE CDC*, pp.3388-3389.

itemKrstic, M., I. Kabellakopoulos and P. V. Kokotovic(1995). Nonlinear and adaptive control design, New York, Wiley.

Jankovic,M., R. Sepulchre and P. V. Kokotovic(1996). Constructive Lyapunov stabilization of nonlinear cascade systems, *IEEE Trans. on Auto. Contr.*,**41**,pp.1723-1735.

Mazenc,F., R. Sepulchre and M. Jankovic(1997). Lyapunov function for stable cascades and applications to global stabilization, *Proc. 36th IEEE CDC*,pp.2843-2846.

Qu, Q. and D. M. Dawson (1994). Robust control of cascaded and individually feedback linearizablle nonlinear systems, *Automatica*, **30**pp.1057-1064.

Lin,W.(1997). Global Robust Stabilization of Minimum-Phase Nonlinear Systems with Uncertainty, *Automatica*,**33**,pp.453-462.

Lin, W. and T. Shen(1997). Robust passivity and global stabilization of nonlinear systems with gain bounded uncertainty", *Proc. 36th IEEE CDC*, San Diego.

Shen, T. and K. Tamura (1995). Robust $H_\infty$ control of uncertain nonlinear system via state feedback, *IEEE Trans.* **AC-40**,pp.766-768.

Shen, T. and K. Tamura(1997). Robust dissipativity of nonlinear systems with state driven uncertainty, *Proc. 36th IEEE CDC*, San Diego.

Shen,T., L. Xie and K. Tamura(1998) Robust almost disturbance decoupling and $H_\infty$ control for nonlinear systems with gain bounded uncertainty, *IEEE CDC98*, Tampa.

van der Schaft,A. J.(1996). L2-gain and passivity techniques in nonlinear control, Lecture notes in control and information sciences, *Spinger*, NY.

Su,W. and L. Xie(1996). Robust control of nonlinear feedback passive systems, *Sys. and Contr. Lett.*, **28**,pp.85-93.

# UNIVERSAL DYNAMIC CONTROL LAW AND RATE FEEDBACKS

**S. H. Lam** [1]

*Department of Mechanical and Aerospace Engineering*
*Princeton University, Princeton, NJ 08544, USA*
*www.princeton.edu/~lam*
*lam@princeton.edu*

Abstract: A common approach to control problems is to look for a static control law through a detailed study of the known mathematical model of the system, assuming that the needed measurements of the state vector are available for closed-loop feedbacks. Recently, this author showed that a special *universal dynamic control law* can control a class of systems *without* the need of detailed knowledge of their mathematical models—provided good quality measured *rate* data of certain components of the state vector are available. The present paper shows that when additional knowledge beyond the minimum is available, the amount of needed rate data can be reduced. Numerical simulations are presented to elucidate the effects of measurement noises. *Copyright © 1999 IFAC*

Keywords: Control, Nonlinear, Robust, Adaptive, Noise.

## 1. INTRODUCTION

An often repeated conventional wisdom is that a control engineer must have an accurate and detailed mathematical model of the system to be controlled in order to be able to design an appropriate controller. In most cases, static control laws are used and full state feedbacks (with the help of estimators) are needed. Recently, Lam (1997a, 1997b, 1998) showed that for certain problems a *universal dynamic control law* (UDCL) is available which needs a very small amount of general information about the system—provided the feedbacks to the controller include certain *rate* data. The present paper examines the trade-offs between the need for detailed knowledge of the system model and the need for time derivatives of the outputs variables via an example. Simulation results are presented to elucidate the impact of finite measurement errors.

[1] Edwin Wilsey '03 Professor of Engineering.

## 2. THE EXAMPLE PROBLEM

Consider the ODE system for $\mathbf{x}(t)$:

$$\frac{dx_1}{dt} = f_1(x_1, x_2, t), \tag{1a}$$

$$\frac{dx_2}{dt} = f_2(x_1, x_2, x_3, t), \tag{1b}$$

$$\frac{dx_3}{dt} = f_3(x_1, x_2, x_3, t; u), \tag{1c}$$

where $u$ is a scalar control "force." The functions $f_1$, $f_2$ and $f_3$ are assumed to be $O(1)$ smooth functions of their arguments, and that $\mathbf{x}$, $u$ and $t$ have been normalized so that they are all $O(1)$ entities. The goal is to find $u$—assuming whatever needed sensor measurements are available—such that the system state vector $\mathbf{x}$ is asymptotically stable about the origin. The desired "settling time" of the controlled system is $O(\tau)$ seconds.

This paper confines attention to the case when $f_1$, $f_2$ and $f_3$ have the following general properties:

$$\frac{\partial f_1}{\partial x_3} = 0, \quad (2a)$$
$$\frac{\partial f_1}{\partial x_2} \neq 0, \quad \text{(sign is known)}, \quad (2b)$$
$$\frac{\partial f_2}{\partial x_3} \neq 0, \quad \text{(sign is known)}, \quad (2c)$$
$$\frac{\partial f_3}{\partial u} \neq 0, \quad \text{(sign is known)}. \quad (2d)$$

The questions of interest are:

(1) what sensor measurements are needed as feedbacks when the controller's knowledge of the system is minimum?
(2) what benefits can be reaped if additional informations are available?

## 2.1 *Measurement Errors*

Since the controller is not fully informed about the system, it must totally trust the reliability and accuracy of the sensor measurements. In practice, measured data always contain noise. To distinguish measured data from the actual variables, all sensor measurements shall be marked by an asterisk superscript. Hence:

$$x_n^*(t) = x_n(t) + \delta_{n,0}, \quad (3a)$$
$$\dot{x}_n^*(t) = \dot{x}_n(t) + \delta_{n,1}, \quad (3b)$$
$$\ddot{x}_n^*(t) = \ddot{x}_n(t) + \delta_{n,2}, \quad (3c)$$
$$\ldots$$

where $\delta_{n,k}$, $\delta_{n,k}$, $\delta_{n,k}$, etc. represent the respective measurement noises—the second subscript represents the order of derivative involved. The UDCL, understandably, expects all measurement noises to be not only small but also to have zero mean values.

# 3. THE LEAST-INFORMED CONTROLLER

Let $x_1$ be the main output variable of interest. An alternative state vector $\mathbf{y}$ is introduced to replace $\mathbf{x}$, and its components $y_1$, $y_2$ and $y_3$ are defined by:

$$y_1 \equiv x_1, \quad (4a)$$
$$y_2 \equiv \frac{dy_1}{dt} + \frac{\Lambda_{11}}{\tau} y_1, \quad (4b)$$
$$y_3 \equiv \frac{dy_2}{dt} + \frac{\Lambda_{22}}{\tau} y_2, \quad (4c)$$

where $\Lambda_{11}$ and $\Lambda_{22}$ are $O(1)$ user-chosen positive numbers. The values of $\mathbf{y}(t)$ and its time derivatives computed from measured data shall also be marked by asterisk superscripts.

For this problem, Lam's *universal dynamic control law* (UDCL) is:

$$\frac{du}{dt} = -\frac{1}{B_{123}\Delta t}\left(\dot{y}_3^* + \frac{\Lambda_{33}}{\tau}(y_3^* - \varpi)\right), \quad (5)$$

where $\Delta t$ is a "sufficiently small" positive time constant to be chosen, $\Lambda_{33}$ is another user-chosen $O(1)$ positive number, and $\varpi$ is yet another user-chosen function of $t$ (called the *residual control* which will be exploited later), and

$$B_{123} \equiv \frac{\partial f_1}{\partial x_2}\frac{\partial f_2}{\partial x_3}\frac{\partial f_3}{\partial u} \quad (6)$$

which is assumed to be $O(1)$ and non-zero. Note that for the class of problems described by (1a,b,c), $B_{123}$ is the only information about the system needed by the UDCL. In fact, since $\Delta t$ is a free positive parameter to be chosen, only the sign of $B_{123}$ is needed. In order to use (5), the task of the control engineer is reduced to choosing a sufficiently small $\Delta t$, pick $\Lambda_{11}$, $\Lambda_{22}$ $\Lambda_{33}$ and $\tau$ to obtain the desired transient properties, and pick $\varpi(t)$ for whatever purposes one may have in mind—provided reliable and accurate $y_3^*(t)$ *and* $\dot{y}_3^*(t)$ are available to be used for the evaluation of the right hand side of (5). Note that measurements of up to the third order time derivatives of $x_1(t)$ are needed to compute $\dot{y}_3^*(t)$.

In actual applications, Nature integrates (1a,b,c) using the real world as an analog computer, while the controller collaborates by numerical integration of (5) using a digital computer (Euler's method is preferred). Lam (1997a, 1997b, 1998) showed that (5) is stable in the small $\Delta t$ limit for any reasonable initial condition $u(0)$ when $f_3$ depends on $u$ linearly. The present paper's generalization that (5) remains stable even when the $u$ dependence of $f_3$ is nonlinear is straightforward and can easily be verified by the readers. Whenever the computed $u(t)$ is finite and smooth, the associated bracketed term on the right hand side of (5) must be $O(\Delta t)$ when $\Delta t$ is sufficiently small, *i.e.*:

$$\dot{y}_3^* + \frac{\Lambda_{33}}{\tau}(y_3^* - \varpi) = O(\Delta t). \quad (7)$$

Equation (7) is normally derived by applying the so-called *quasi-steady* approximation to (5), a technique well known in the theory of stiff ODEs (Lam, 1997a). Let $u_\infty(\mathbf{x},t)$ denote the exact solution of (5) in the $\Delta t \to 0$ limit when all measurements are error-free (*i.e.* $\mathbf{y}^* = \mathbf{y}$). It can easily be shown that $u_\infty(\mathbf{x},t)$ is identical to the exact solution to this problem obtained by the method of *feedback linearization* (Isidori,1995; Marino and Tomei, 1995). Unlike feedback linearization, which needs detailed knowledge of $f_1$, $f_2$ and $f_3$ in order to evaluate $u_\infty(\mathbf{x};t)$, the UDCL as given by (5) needs only the sign of $B_{123}$—provided the properties listed in (2a,b,c,d) are satisfied. In other words, any smooth solution generated by (5)—with error-free measured data—is automatically an approximation to $u_\infty(\mathbf{x};t)$:

$$u(t) = u_\infty(\mathbf{x}(t),t) + O(\Delta t). \tag{8}$$

The disadvantage is that measurements of up to the third order time derivatives of $x_1(t)$ are needed. It is of interest to note that the order of the highest time derivative needed by UDCL is called the *relative degree* in the nonlinear control literature. To show the effects of finite measurement errors, the quasi-steady approximation (7) is expressed in terms of the actual $x_1$ to yield:

$$\dddot{x}_1 + \frac{\Lambda_{11}+\Lambda_{22}+\Lambda_{33}}{\tau}\ddot{x}_1 + \frac{\Lambda_{11}\Lambda_{22}+\Lambda_{22}\Lambda_{33}+\Lambda_{33}\Lambda_{11}}{\tau^2}\dot{x}_1 + \frac{\Lambda_{11}\Lambda_{22}\Lambda_{33}}{\tau^3}x_1 - \frac{\Lambda_{33}\varpi}{\tau} + \delta = O(\Delta t), \tag{9}$$

where $\delta$ represents the net consequence of all the (small) measurement errors incurred in approximating $y_3(t)$ and $\dot{y}_3(t)$ by $y_3^*(t)$ and $\dot{y}_3^*(t)$, respectively. Equation (9) represents the dynamics of $x_1(t)$ of the controlled system when $t >> \Delta t$. Note that as far as (9) is concerned, the impacts of $\delta$ and $\varpi$ are indistinguishable. This observation shall be exploited in the simulations presented later.

Since $\varpi$ is a forcing term in (9), it can be used by the controller to "steer" $x_1(t)$ to approximately track some user-specified desired trajectory for $t >> \tau$ —provided that the desired trajectory is sufficiently smooth.

## 4. EXPLOITING ADDITIONAL INFORMATION

When certain additional informations about $f_1$ and $f_2$ are known, significant benefits can be reaped.

### 4.1 *If $f_1(x_1,x_2,t)$ is precisely known*

If $f_1(x_1,x_2,t)$ is precisely known, (4a,b) yield:

$$y_1^* = x_1^*, \qquad y_2^* = f_1(x_1^*,x_2^*,t) + \frac{\Lambda_{11}x_1^*}{\tau}. \tag{10}$$

Equation (4c) yields:

$$y_3^* = \frac{D_1 f_1}{Dt} + \frac{\Lambda_{11}f_1}{\tau} + \frac{\partial f_1}{\partial x_2}\dot{x}_2^* + \frac{\Lambda_{22}y_2^*}{\tau}. \tag{11}$$

where

$$\frac{D_1}{Dt} \equiv \frac{\partial}{\partial t} + f_1(x_1^*,x_2^*,t)\frac{\partial}{\partial x_1^*}, \tag{12}$$

The recommended UDCL remains (5) without modification. By the use of (10) and (11), the bracketed term on the right hand side of (5) becomes:

$$\frac{D_1}{Dt}\left(\frac{D_1 f_1}{Dt} + \frac{\Lambda_{11}f_1}{\tau} + \frac{\Lambda_{22}y_2^*}{\tau}\right) + \frac{D_1}{Dt}\left(\frac{\partial f_1}{\partial x_2^*}\right)\dot{x}_2^* + \frac{\partial f_1}{\partial x_2^*}\ddot{x}_2^* + \frac{\Lambda_{33}}{\tau}y_3^* - \frac{\Lambda_{33}\varpi}{\tau}, \tag{13}$$

which the controller is capable of evaluation when $x_1^*(t)$, $x_2^*(t)$, $\dot{x}_2^*(t)$ and $\ddot{x}_2^*(t)$ are provided. In other words, the precise knowledge of $f_1(x_1,x_2,t)$ can remove the need of the third order time derivative, $\dddot{x}_1^*(t)$.

#### 4.1.1. *If $f_2(x_1,x_2,x_3,t)$ is also precisely known*

If in addition $f_2(x_1,x_2,x_3,t)$ is also precisely known, it is easily shown that $\ddot{x}_2^*$ can be evaluated in terms of $x_1^*(t)$, $x_2^*(t)$, $x_3^*(t)$ and $\dot{x}_3^*(t)$.

For example, if $f_1 = \alpha x_2$ and $f_2 = \beta x_3$ (where $\alpha$ and $\beta$ are constants), then we have $\dot{x}_1^*(t) = \alpha x_2^*(t)$, $\ddot{x}_1^*(t) = \alpha\beta x_3^*(t)$ and $\dddot{x}_1^*(t) = \alpha\beta\dot{x}_3^*(t)$.

### 4.2 *If $\frac{\partial f_1}{\partial x_1}$ is sufficiently negative*

If $\partial f_1/\partial x_1$ is known to be sufficiently negative, then $x_1$ is expected to be inherently stable without the assistance of $u$. Advantage can be taken of this observation to ignore $x_1$—provided the requirement that $x_1$ should asymptotically decay toward zero is relaxed.

For this case, $x_2$ replaces $x_1$ as the main output variable, and accurate measurements of $x_2^*(t)$ and its first and second order time derivative ($\dot{x}_2^*(t)$ and $\ddot{x}_2^*(t)$) are assumed available. The vector $\mathbf{y}$ is now defined by :

$$y_1 \equiv x_1, \tag{14a}$$

$$y_2 \equiv x_2, \tag{14b}$$

$$y_3 \equiv \frac{dy_2}{dt} + \frac{\Lambda_{22}}{\tau} y_2. \tag{14c}$$

Equation (5) remains the recommended dynamic control law except that $B_{123}$ is now replaced by $B_{23} = (\partial f_2/\partial x_3)(\partial f_3/\partial u)$. The bracketed term on the right hand side of (5) is:

$$\ddot{x}_2^* + (\frac{\Lambda_{22}+\Lambda_{33}}{\tau})\dot{x}_2^* + \frac{\Lambda_{22}\Lambda_{33}}{\tau^2}x_2^* - \frac{\Lambda_{33}\varpi}{\tau} \tag{15}$$

In other words, the additional knowledge removes the need for measurement of the third order time derivative of $x_2(t)$ in the UCDL.

For this case, $\varpi$ is allowed to depend on $x_1^*$ and can serve as a residual control force to exert some influence on $x_1(t)$. For example, if it turns out that the magnitude of $\partial f_1/\partial x_1$ is not up to the task of guaranteeing asymptotic stability for $x_1$, the option of using the residual control $\varpi(x_1^*, t)$ can be explored (Lam, 1998).

### 4.3 *If both $\frac{\partial f_1}{\partial x_1}$ and $\frac{\partial f_2}{\partial x_2}$ are sufficiently negative*

If both $\partial f_1/\partial x_1$ and $\partial f_2/\partial x_2$ are known to be sufficiently negative, then $x_1$ and $x_2$ are expected to be inherently stable without the assistance of $u$. We can similarly take advantage of this observation.

For this case, $x_3$ is the sole variable of interest. The vector $\mathbf{y}$ is now identical to $\mathbf{x}$. The recommended dynamic control law remains to be (5) except that $B_{123}$ is replaced by $B_3 = \partial f_3/\partial u$. The actual ODE to be numerically integrated by the controller is now simply:

$$\frac{du}{dt} = -\frac{1}{B_3\Delta t}\left(\dot{x}_3^* + \frac{\Lambda_{33}}{\tau}(x_3^* - \varpi)\right). \tag{16}$$

In other words, the additional knowledge removes the need for the second and third order time derivatives.

For this case, $\varpi$ is allowed to depend on $x_1^*$ and $x_2^*$ in addition to $t$. Similar comments at the end of the previous section are also applicable here.

## 5. THE ISSUE OF MEASUREMENT NOISE

Theoretically, the impact of noise (as represented by $\delta$) is most felt by $x_3$ and $u$, less by $x_2$, and least by $x_1$. So long as all the needed measurements have good signal-to-noise ratios (*i.e.* $\delta << 1$), the UDCL will do just fine since their impacts are expected to be $O(\delta)$ (see (9)).

It is well known that numerical differentiation amplifies noise. Hence in general its use degrades the performance of UDCL. Consequently, the UDCL strongly advocates direct measurements of rate data. If for practical reasons direct measurements are not possible, then the option of using numerical differentiation could be considered. Some averaging and/or filtering of the data before and/or after numerical differentiations may be needed to modulate the increase of $\delta_{i,k}$ with increasing $k$. Note that (5) is essentially a low-pass filter for $u$. Hence, "high frequency" components of noise in $\delta$ can be somewhat muted by increasing the value of $\Delta t$—and in so doing sacrificing some "accuracy" of the resulting performance.

## 6. NUMERICAL SIMULATIONS

The following example is used to demonstrate the UDCL for the least-informed case:

$$f_1 = \alpha x_2 + g_1(x_1, t), \tag{17a}$$

$$f_2 = \beta x_3 + g_2(x_1, x_2, t), \tag{17b}$$

$$f_3 = g_3(x_1, x_2, x_3, t) + \gamma u. \tag{17c}$$

where $\alpha$, $\beta$ and $\gamma$ are $O(1)$ constants or functions of $t$ which are known not to change signs. Hence, $B_{123} = \alpha\beta\gamma \neq 0$. To simulate the controlled system, both (1a,b,c) and (5) are numerically integrated together on a desktop computer. It is assumed that $x_1^*$ and $\dot{x}_1^*$ are directly measured, so that $y_2^*$ can be computed via (4b) without the use of numerical differentiation. However, the higher time derivatives are computed by numerical differentiations. For example, $y_3^*(t)$ is computed via (4c) using the backward difference approximation for $\dot{y}_2^*(t)$:

$$y_3^*(t) = \frac{y_2^*(t) - y_2^*(t-\delta t)}{\delta t} + \frac{\Lambda_{22}}{\tau}y_2^*(t) \tag{18}$$

where $\delta t$ is the numerical integration time-step used in the simulation. The needed $\dot{y}_3^*(t)$ on the right hand side of (5) is computed similarly—numerical differentiations is used twice in this evaluation.

According to (9), the measurement noise term $\delta$ can be emulated by using $\varpi$ (instead of $\delta$). Hence, in the simulations, no measurement noises are included in $x_1^*$ and $\dot{x}_1^*$; the net effect of all measurement noises (and numerical differentiation errors) is emulated by choosing $\varpi(t) \neq 0$ (with $\delta = 0$ always).

Theoretically, the UDCL can handle any arbitrary differentiable $g_1$, $g_2$ and $g_3$, *deterministic or stochastic*. For concreteness, the following specific deterministic functions were chosen:

$$g_1 = x_1 - x_1^2 + x_1^3, \tag{19a}$$

$$g_2 = x_1 + 2x_2 + x_1 x_2, \tag{19b}$$

$$g_3 = x_1 - x_2 + 3x_3 + x_1 x_2 - 1. \tag{19c}$$

The initial conditions were $x_1(0) = -0.05$, $x_2(0) = -0.5$, $x_3(0) = -0.55$ and $u(0) = -2$. The integration stepsize was $\delta t = 0.01$, and $\Lambda_{11}$, $\Lambda_{22}$, and $\Lambda_{33}$ were chosen to be 1.0, 1.0 and 3.0, respectively. The simulation results are presented in Figs. 1, 2, 3, 4 and 5. Fig. 1, which uses $\alpha = \beta = \gamma = 1$, $\Delta t = 2\delta t$ and $\tau = 0.5$, shows the ideal case when the measurements are noise-free ($\varpi = 0$)—even though numerical differentiations were used twice in evaluating the right hand side of the UDCL. Fig. 2 shows the effects of noise as emulated by $\varpi = 0.1\sin(100t)$—all other parameters are held fixed. Since $\Lambda_{33}/\tau = 6$, the amplitude of the emulated $\delta$ term is actually 0.6. The impacts of this noise on $x_1(t)$, $x_2(t)$ and even $x_3(t)$ are seen to be very minor indeed, but $u(t)$ now oscillates with a significant amplitude. Fig. 3 shows an identical run (as Fig. 2) except the magnitude of $\Delta t$ is increased by a factor of three. It is seen that the amplitude of the undesirable oscillations is substantially reduced. Fig. 4 shows an identical run (as Fig. 3) except the value of $\tau$ is doubled. Fig.5 shows an identical run (as Fig. 4) except $\gamma = 1 + u^2$—in other words, $f_3(, x_1, x_2, x_3, t; u)$ now depends on $u$ nonlinearly. The UDCL handles this complication in stride. When different $g_1(x_1, t)$, $g_2(x_1, x_2, t)$ and $g_3(x_1, x_2, x_3, t)$ (and different $\alpha$, $\beta$, $\gamma$ and initial conditions) were used than those shown in (19a,b,c), the simulated behaviors of the controlled system remained the same—only $u(t)$ was affected.

The UDCL can tolerate $f_1$ and $f_2$ to have a slight dependence on $x_3$ and $u$, respectively, and/or $f_1$ to have a *very* slight dependence on $u$—provided a lower bound for $\Delta t$ is respected and the somewhat degraded resulting accuracy is considered acceptable. See (Lam, 1998) for a more detailed discourse on "nearly singular" problems.

The theory and the simulations performed above assumed that the actuator is capable of delivering any $u(t)$ computed by the UDCL. When the initial value for $u(0)$ is arbitrarily chosen, the UDCL

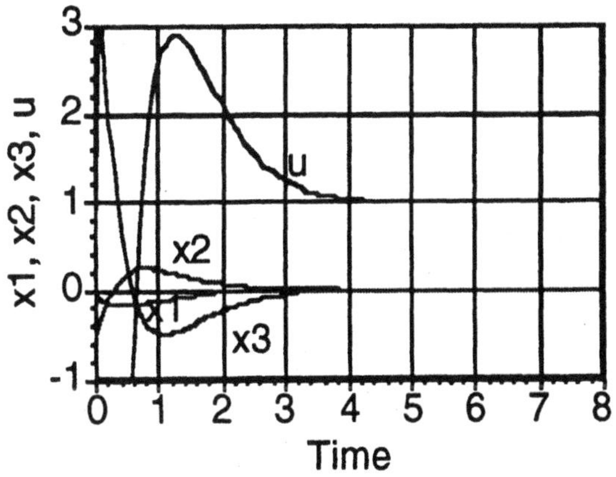


Fig. 1. $x_1$, $x_2$ $x_3$ and $u$ vs. time with $\varpi = 0$
$\alpha = \beta = \gamma = 1$, $\Lambda_{11} = \Lambda_{22} = 1, \Lambda_{33} = 3$.
$\delta t = 0.01$, $\Delta t = 2\delta t$, $\tau = 1.0$.

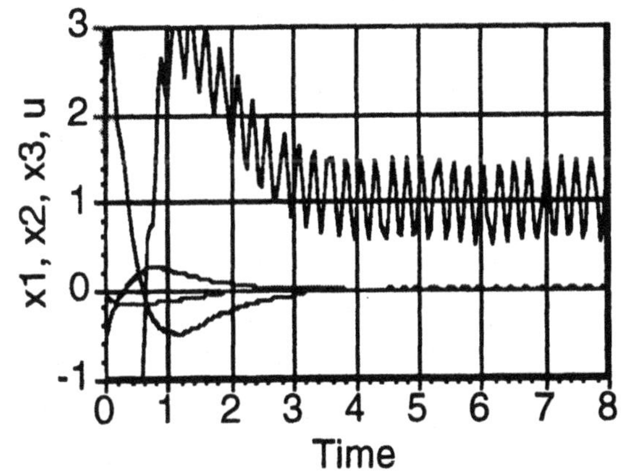


Fig. 2. $x_1$, $x_2$ $x_3$ and $u$ vs. time with $\varpi = 0.1\sin(100t)$. All other parameters are identical to Fig. 1.

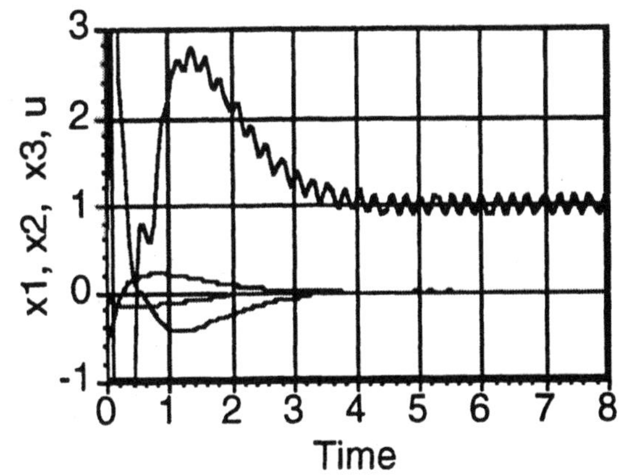


Fig. 3. $x_1$, $x_2$ and $x_3$ and $u$ vs. time with $\varpi = 0.1\sin(100t)$. All other parameters are identical to Fig. 2 except $\Delta t = 6\delta t$.

needs some time to approach $u_\infty$, and during this brief period the system could behave somewhat erratically. Experience shows that it is often a good idea to initially use a somewhat larger value of $\Delta t$ and later reduce it to the desired sufficiently small value after the confusion is over. This was not done in the simulations presented here.

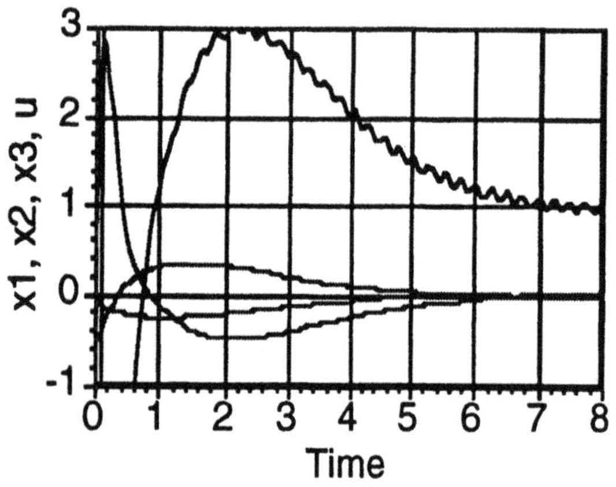


Fig. 4. $x_1$, $x_2$ and $x_3$ and $u$ vs. time with $\varpi = 0.1\sin(100t)$. All other parameters are identical to Fig. 3 except $\tau = 1.0$.

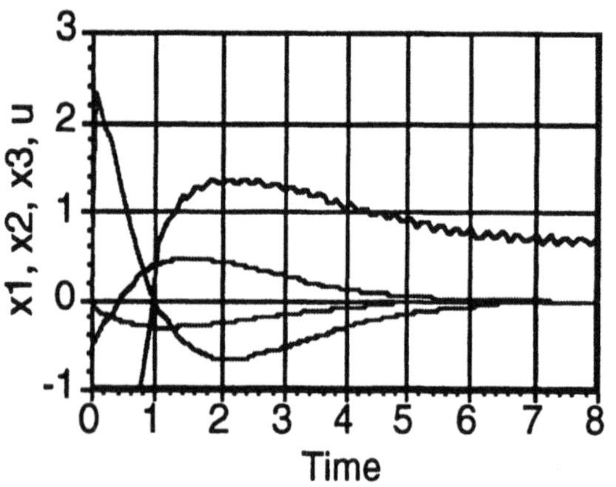


Fig. 5. $x_1$, $x_2$ and $x_3$ and $u$ vs. time with $\varpi = 0.1\sin(100t)$. All other parameters are identical to Fig. 4 except $\gamma = 1 + u^2$.

## 7. CONCLUDING REMARKS

The main feature of the UDCL is that very little knowledge of the system to be controlled is needed. Consequently, real disturbances to the system or changes in the real system, whether deterministic or stochastic, are readily accommodated by the UDCL. The perceptive readers may have noticed that linearity of the system model is totally irrelevant from the point of view of the theory of UDCL. The basic ideas of UDCL can be applied to a general nonlinear system. In other words, the generic methodology can remain applicable even if conditions (2a,b,c,d) were not satisfied.

The information about the system most needed by UDCL is the so-called *pulse-response matrix*, represented by $B_{123}$ and $B_{23}$, etc. in this paper. In the design of the system, which should include the design of the actuators and sensors in addition to the system itself, the designers should try to make the pulse-response matrix a constant matrix if at all possible. In the design of the sensors, direct measurements of rate data should be given high priority—making sure the needed highest time derivatives are measured reliably and accurately. From the vantage point of UDCL, the fabled successes of "inertial guidance" can be attributed to the measurement of accelerations in the control of spacecrafts and aircrafts. Once the pulse-response matrix is known and all needed rate data are guaranteed to have good signal-to-noise ratios, the UDCL does not need a mathematical model of the actual system at all.

Conceptually, both the UDCL and the feedback linearization approach use $u$ to cancel out all the relevant actual terms and replace them with user-chosen terms. The UDCL approach uses the measured rate data to compute the relevant actual terms via the quasi-steady approximation (Lam, 1997b), while the conventional feedback linearization approach uses the detailed knowledge of the system. The price for the robustness of UDCL is the need for measured rate data with good signal-to-noise ratios, and some tolerance for finite tracking errors. Since detailed knowledge of the system is often uncertain, out-of-date, or perhaps even faulty, the option of using UDCL should be seriously considered whenever the needed sensor technology for good rate data measurements is available and high control accuracy is not a prime requirement.

## 8. REFERENCES

Isidori, A. (1995). *Nonlinear Control Systems*, Prentice Hall.

Lam, S. H. (1997a). "Quasi-steady approximation and adaptive nonlinear control," *Proceedings of the 15th IMACS World Congress on Scientific Computation, Modeling and Applied Mathematics, Vol. 3, Computational Chemistry and Biology, Edited by A. Sydow*, pp. 127-132. Wissenschaft & Technik Verlag, Berlin.

Lam, S. H. (1997b) "A robust universal controller," DSC-Vol. 61, *Proceedings of the ASME Dynamic Systems and Control Division, Editor: Giorgio Rizzoni*, Book No. H01105-1997, pp. 571-577. ASME International, New York.

Lam, S. H. (1998). "A new approach to robust nonlinear controls," to appear in *Journal of Systems Analysis Modelling Simulation*, Gordon and Breach.

Marino, R. and Tomei, P. (1995). *Nonlinear Control Design, Geometric, Adaptive and Robust*, Prentice Hall.

14th Triennial World Congress, Beijing, P.R. China

# LOCAL STABILIZATION OF LINEAR SYSTEMS WITH POSITION AND RATE BOUNDED ACTUATORS

**Sophie** TARBOURIECH, **Germain** GARCIA*, **Didier** HENRION

*L.A.A.S.-C.N.R.S., 7 Avenue du Colonel Roche, 31077 Toulouse cedex 4, France.*
*E-mail : tarbour@laas.fr - Fax : +33 (0) 561 33 69 69*

Abstract: The robust stabilization of uncertain systems subject to position and rate limited actuators is addressed. A local approach is chosen because no open-loop stability assumption is a priori considered. A saturating stabilizing control law and a guaranteed domain of stability are derived from two design frameworks. The first one is based on the ARE-approach, while a second formulation via LMIs is described. The main strength of our methods is that the position and rate system inputs are allowed to saturate. *Copyright ©1999 IFAC*

Keywords: Position and Rate Limited Actuators, Uncertain Systems, Robust Stabilization, Riccati Equations, LMI-formulation.

## 1. INTRODUCTION

During the last years, due to its practical interest the problem of stabilizing linear systems with saturating actuators has widely been studied : see (Bernstein and Michel, 1995) for a chronological bibliography. Significant results have emerged addressing the global (Sontag and Sussmann, 1990) or semi-global (Lin and Saberi, 1993) stabilization. They inherently require some stability assumptions for the open-loop system. Relaxing these assumptions, some results addressing the problem of local stabilization were obtained (Gomes da Silva Jr and Tarbouriech, 1998).

Another ubiquitous feature, which must be taken into account in the design of a control law, concerns the presence of uncertainty. Many results are available in the uncertain system context, see (Zhou *et al.*, 1996) for a good panel.

Several new results recently appeared in the study of systems subject to both uncertainty and actuator saturations, see (Tarbouriech and Garcia, 1997) for an overview. Nevertheless, really a very few works have been published so far to combine both features in spite of their practical relevance.

Furthermore, another situation which has to be taken into account concerns the actuator rate limitations. Some solutions for avoiding the saturation have been recently proposed (Shewchun and Feron, 1997), (Saberi and Stoorvogel, 1997).

Motivated by the previous observations, the aim of this note is to study the problem of stabilization of linear systems that are simultaneously uncertain and subject to both input position and rate constraints. We focus more precisely on a robust local stabilization approach because we relax open-loop stability assumptions. In other words, we want to compute both a stabilizing controller and the associated region of stability in which the stability of the closed-loop saturated system is ensured. The control design is proposed by combining a polytopic representation of saturation nonlinearities and standard quadratic sta-

* This author is also with INSA, Complexe Scientifique de Rangueil, 31077 Toulouse, France.

bilization results. The computation of both controller and stability region are first based on a Riccati equation approach and next on an LMI formulation.

**Notation.** For any vector $x \in \Re^n$, $x_{(i)}$ denotes the $i$th component of $x$. The elements of a matrix $A \in \Re^{m\times n}$ are denoted by $A_{(i,l)}$, $i = 1,\dots,m$, $l = 1,\dots,n$. $A_{(i)}$ denotes the $i$th row of matrix $A$. For two symmetric matrices, $A$ and $B$, $A > B$ means that $A - B$ is positive definite. $A'$ denotes the transpose of $A$. $I_m$ denotes the m-order identity matrix. $co\{\cdot\}$ denotes a convex hull. For any vector $\nu \in \Re^m$, one defines $\overline{sat}(\nu_{(i)}) = sign(\nu_{(i)})min(u_{0(i)}, |\nu_{(i)}|)$ and $sat(\nu_{(i)}) = sign(\nu_{(i)})min(u_{1(i)}, |\nu_{(i)}|)$, $i = 1,\dots,m$.

## 2. PROBLEM STATEMENT

Consider a linear continuous-time uncertain system :

$$\dot{x}(t) = (A + \Delta A)x(t) + (B + \Delta B)u(t) \quad (1)$$

where $x \in \Re^n$ and $u \in \Re^m$ are respectively the state and the input vectors. Matrices $A$ and $B$ are real constant matrices of appropriate dimensions. For system (1) we suppose that the following assumptions hold :

**A1.** The input vector is subject to position and rate constraints. That is, one has $u \in \mathcal{U}_0 \subset \Re^m$ and $\dot{u} \in \mathcal{U}_1 \subset \Re^m$ with

$$\mathcal{U}_0 = \{u \in \Re^m ; -u_{0(i)} \le u_{(i)} \le u_{0(i)}, u_{0(i)} > 0, i = 1,\dots,m\} \quad (2)$$

$$\mathcal{U}_1 = \{u \in \Re^m ; -u_{1(i)} \le \dot{u}_{(i)} \le u_{1(i)}, u_{1(i)} > 0, i = 1,\dots,m\} \quad (3)$$

**A2.** Pair $(A, B)$ is stabilizable.

**A3.** Matrices $\Delta A$ and $\Delta B$ are defined as follows

$$[\Delta A \ \Delta B] = D_1 F(t) [E_1 \ E_2] \quad (4)$$

where $D_1$, $E_1$ and $E_2$ are constant matrices of appropriate dimensions defining the structure of uncertainty. Matrix $F(t)$ takes values in $\mathcal{F} \subset \Re^{l\times r}$ and is the parameter uncertainty :

$$\mathcal{F} = \{F(t) \in \Re^{l\times r}; F(t)'F(t) \le I_r, \forall t\}. \quad (5)$$

This kind of uncertainty is the so-called *norm-bounded uncertainty* and was considered frequently in the robust control literature : see (Dorato, 1987), (Garcia *et al.*, 1994) and references therein.

The objective of this paper consists in determining a control law which stabilizes system (1) under hypotheses A1, A2 and A3. Because no particular assumption was made regarding stability of open-loop system matrix $A + \Delta A$ and due to possible actuator saturations from constraints (2) and (3), the domain of attraction of the origin constitutes the maximal set in which the closed-loop system can be initialized. Actually, the problem of determining this domain, that is, the set of all initial conditions that can be driven to the origin, is a very challenging issue which is not exactly solved in the general case.

In order to solve our control objective, we consider as in (Shewchun and Feron, 1997) the following augmented system[1] :

$$\dot{z} = (\mathcal{A} + \Delta\mathcal{A})z + \mathcal{B}v \quad (6)$$

with

$$\mathcal{A} = \begin{bmatrix} A & B \\ 0 & 0 \end{bmatrix}, \Delta\mathcal{A} = \begin{bmatrix} \Delta A & \Delta B \\ 0 & 0 \end{bmatrix}, \quad \mathcal{B} = \begin{bmatrix} 0 \\ I_m \end{bmatrix}, z = \begin{bmatrix} x \\ u \end{bmatrix}, v = \dot{u} \quad (7)$$

For system (6)-(7), the uncertainty term $\Delta\mathcal{A}$ is defined as $\Delta\mathcal{A} = \mathcal{D}_1 F(t)\mathcal{E}$ with $\mathcal{D}_1 = \begin{bmatrix} D_1 \\ 0 \end{bmatrix}$ and $\mathcal{E} = [E_1 \ E_2]$.

The main advantage of this formulation is that the control rate now appears as the system input, the original control being a state of the system. Thus, the constrained control rate and position problem can be formulated as a control position and state constraint problem for system (6).

In order to solve our objective from constraints (2) and (3) we propose to study the following problem.

*Problem 1.* Find both a matrix $\mathcal{K}$ and a set of admissible initial conditions $\mathcal{X}_0 \subset \Re^{n+m}$ such that for any $z(0) \in \mathcal{X}_0$ the resulting trajectories of the closed-loop system :

$$\dot{z} = (\mathcal{A} + \Delta\mathcal{A}) \begin{bmatrix} [I_n \ 0]z \\ \overline{sat}([0 \ I_m]z) \end{bmatrix} + \mathcal{B}sat(\mathcal{K} \begin{bmatrix} [I_n \ 0]z \\ \overline{sat}([0 \ I_m]z) \end{bmatrix}) \quad (8)$$

are steered to the origin.

In system (8), the saturation terms $\overline{sat}(.)$ and $sat(.)$ mean that $u$ and $\dot{u}$ are allowed to saturate. In summary, our objective consists in solving Problem 1 with a set $\mathcal{X}_0$ as large as possible.

## 3. POLYTOPIC REPRESENTATION

In this section, we express for nonlinear system (8) a locally equivalent polytopic representation. Indeed, system (8) equivalently writes :

[1] By simplicity in the sequel of the paper we omit the dependence on time $t$ in vector and matrix functions.

$$\dot{z} = (\mathcal{A}+\Delta\mathcal{A})\begin{bmatrix} I_n & 0 \\ 0 & \Gamma(\alpha(z)) \end{bmatrix} z + \mathcal{B}\Gamma(\beta(z))\mathcal{K}\begin{bmatrix} I_n & 0 \\ 0 & \Gamma(\alpha(z)) \end{bmatrix} z = (\mathcal{A}(\alpha(z)) + \Delta\mathcal{A}(\alpha(z)))z + \mathcal{B}(\beta(z))\mathcal{K}(\alpha(z))z \quad (9)$$

where $\Gamma(\alpha(z))$ and $\Gamma(\beta(z))$ are diagonal matrices whose the diagonal elements are respectively defined for $i = 1, ..., m$ as :

$$\alpha_{(i)}(z) = sign(\frac{u_{0(i)}}{\begin{bmatrix} 0 & I_m \end{bmatrix}_{(i)} z})min(1, \frac{u_{0(i)}}{|\begin{bmatrix} 0 & I_m \end{bmatrix}_{(i)} z|}) \quad (10)$$

and

$$\beta_{(i)}(z) = sign(\frac{u_{1(i)}}{\mathcal{K}_{(i)}\nu})min(1, \frac{u_{1(i)}}{|\mathcal{K}_{(i)}\nu|}) \quad (11)$$

where $\nu$ denotes the vector $\begin{bmatrix} [I_n \; 0] z \\ sat([0 \; I_m] z) \end{bmatrix}$. Note that for any $z \in \Re^{n+m}$ one gets : $0 < \alpha_{(i)}(z) \leq 1$ and $0 < \beta_{(i)}(z) \leq 1$, $\forall i = 1, ..., m$.

Recalling the formulation of our control objective, the domain of evolution of the state $z$ has to be limited. Therefore if we consider any compact set $\mathcal{S}_0 \subset \Re^{n+m}$ then for $z \in \mathcal{S}_0$ components of vector $\alpha(z)$ and $\beta(z)$ admit a lower bound denoted respectively :

$$\alpha_{\ell(i)} = \min\{\alpha_{(i)}(z); z \in \mathcal{S}_0\}, \; i = 1, ..., m \quad (12)$$

$$\beta_{\ell(i)} = \min\{\beta_{(i)}(z); z \in \mathcal{S}_0\}, \; i = 1, ..., m. \quad (13)$$

Therefore for any $z \in \mathcal{S}_0$ one gets, $\forall i = 1, ..., m$,

$$0 < \alpha_{\ell(i)} \leq \alpha_{(i)}(z) \leq 1 \quad (14)$$

$$0 < \beta_{\ell(i)} \leq \beta_{(i)}(z) \leq 1 \quad (15)$$

Define thus two vectors $\alpha_\ell$ and $\beta_\ell$. Therefore, we can define the following vertex matrices :

$$\mathbb{A}_j = \mathcal{A}\begin{bmatrix} I_n & 0 \\ 0 & \Gamma_j(\alpha_\ell) \end{bmatrix} ; \; \Delta\mathbb{A}_j = \Delta\mathcal{A}\begin{bmatrix} I_n & 0 \\ 0 & \Gamma_j(\alpha_\ell) \end{bmatrix}$$
$$\mathbb{B}_k = \mathcal{B}\Gamma_k(\beta_\ell) \; ; \; \mathbb{K}_j = \mathcal{K}\begin{bmatrix} I_n & 0 \\ 0 & \Gamma_j(\alpha_\ell) \end{bmatrix} \quad (16)$$

for $j = 1, ..., 2^m$ and $k = 1, ..., 2^m$. The $\Gamma_j(\alpha_\ell)$ (resp. $\Gamma_j(\beta_\ell)$) are the $2^m$ vertex diagonal matrices whose the elements can take the value 1 or $\alpha_{\ell(i)}$, $i = 1, ..., m$ (resp. 1 or $\beta_{\ell(i)}$, $i = 1, ..., m$). Hence, if $z \in \mathcal{S}_0$ then $\dot{z}$ can be determined from the following polytopic model :

$$\dot{z} = \sum_{j=1}^{2^m} \lambda_j(\mathbb{A}_j + \Delta\mathbb{A}_j)z + \sum_{k=1}^{2^m} \mu_k \mathbb{B}_k \sum_{j=1}^{2^m} \lambda_j \mathbb{K}_j z \quad (17)$$

with $\sum_{j=1}^{2^m} \lambda_j = 1$, $\lambda_j \geq 0$ and $\sum_{k=1}^{2^m} \mu_k = 1$, $\mu_k \geq 0$. Note that for $z \in \mathcal{S}_0$ one gets $(\mathbb{A}(\alpha(z)) + \Delta\mathbb{A}(\alpha(z))) \in co\{\mathbb{A}_1 + \Delta\mathbb{A}_1, ..., \mathbb{A}_{2^m} + \Delta\mathbb{A}_{2^m}\}$ and $\mathbb{B}(\beta(z)) \in co\{\mathbb{B}_1, ..., \mathbb{B}_{2^m}\}$. Furthermore note that vectors $\alpha_\ell$ and $\beta_\ell$ allow to define two polyhedral sets :

$$S(u_0, \alpha_\ell) = \{z \in \Re^{n+m}; -\frac{u_{0(i)}}{\alpha_{\ell(i)}} \leq \begin{bmatrix} 0 & I_m \end{bmatrix}_{(i)} z \leq \frac{u_{0(i)}}{\alpha_{\ell(i)}}, i = 1, ..., m\} \quad (18)$$

and

$$S(u_1, \beta_\ell) = \{z \in \Re^{n+m}; -\frac{u_{1(i)}}{\beta_{\ell(i)}} \leq \mathbb{K}_{j(i)} z \leq \frac{u_{1(i)}}{\beta_{\ell(i)}}, \; i = 1, ..., m, j = 1, ..., 2^m\} \quad (19)$$

Actually, the set $S(u_0, \alpha_\ell) \cap S(u_1, \beta_\ell)$ contains $\mathcal{S}_0$ and corresponds to the maximal set in which model (17) represents system (8) or (9).

## 4. ARE-BASED APPROACH

In order to solve our objective (Problem 1), we consider as a convenient choice for $\mathcal{X}_0$, ellipsoids defined from a symmetric positive definite matrix $P$ as :

$$\mathfrak{E}(P, \rho) = \{z \in \Re^{n+m}; z'Pz \leq \rho, \; \rho > 0\} \quad (20)$$

Given $R = R' > 0$ and $Q = Q' > 0$. Consider $X_\epsilon = X'_\epsilon > 0$ solution to the Riccati equation :

$$\mathcal{A}'X_\epsilon + X_\epsilon\mathcal{A} + X_\epsilon(\mathcal{D}_1\mathcal{D}'_1 - \mathcal{B}R_\epsilon^{-1}\mathcal{B}')X_\epsilon + \mathcal{E}'\mathcal{E} + Q_\epsilon = 0 \quad (21)$$

with $\epsilon > 0$, $R_\epsilon = \epsilon R$ and $Q_\epsilon = \epsilon Q$. Then, a first result solving our objective is stated in the following proposition.

*Proposition 2.* Given $X_\epsilon = X'_\epsilon > 0$ solution to (21). Then the state feedback gain :

$$\mathcal{K} = -R_\epsilon^{-1}\mathcal{B}'X_\epsilon \quad (22)$$

and the set $\mathfrak{E}(X_\epsilon, \rho)$, defined as in (20), solve Problem 1 if

$$\mathbb{A}'_j X_\epsilon + X_\epsilon\mathbb{A}_j + X_\epsilon\mathcal{D}_1\mathcal{D}'_1 X_\epsilon + \mathbb{E}'_j\mathbb{E}_j + \mathbb{K}'_j\mathbb{B}'_k X_\epsilon + X_\epsilon\mathbb{B}_k\mathbb{K}_j < 0 \quad \forall j = 1, ..., 2^m, \; \forall k = 1, ..., 2^m \quad (23)$$

with $\mathbb{A}_j$ and $\mathbb{B}_k$ defined in (16) and $\mathbb{E}_j = \mathcal{E}\begin{bmatrix} I_n & 0 \\ 0 & \Gamma_j(\alpha_\ell) \end{bmatrix}$, $\forall j = 1, ..., 2^m$.

**Proof.** By considering the controller proposed in (22) and the ellipsoid $\mathfrak{E}(X_\epsilon, \rho)$ defined for a given positive scalar $\rho$, we can define the vectors $\alpha_\ell$ and $\beta_\ell$ as follows for $i = 1, ..., m$ and $j = 1, ..., 2^m$:

$$\alpha_{\ell(i)} = \min(1, \frac{u_{0(i)}}{\sqrt{\rho}\sqrt{\left[\,0\ I_m\,\right]_{(i)} X_\epsilon^{-1}(\left[\,0\ I_m\,\right]_{(i)})'}}) \tag{24}$$

$$\beta_{\ell(i)} = \min_j(1, \frac{u_{1(i)}}{\sqrt{\rho}\sqrt{R_{\epsilon(i)}^{-1}\mathcal{B}'X_\epsilon Z_j X_\epsilon \mathcal{B}(R_{\epsilon(i)}^{-1})'}}) \tag{25}$$

with $Z_j = \begin{bmatrix} I_n & 0 \\ 0 & \Gamma_j(\alpha_\ell) \end{bmatrix} X_\epsilon^{-1} \begin{bmatrix} I_n & 0 \\ 0 & \Gamma_j(\alpha_\ell) \end{bmatrix}$. In order to prove that the set $\mathfrak{E}(X_\epsilon, \rho)$ is a domain of local stability for the saturated system (8) it suffices to prove the decreasing of the time-derivative of $V(z) = z'X_\epsilon z$ along the trajectories of the polytopic system (17) resulting from (24) and (25). Thus, one gets $\dot{V}(z) = \dot{z}'X_\epsilon z + z'X_\epsilon \dot{z} = \sum_{j=1}^{2^m} \lambda_j z'[(\mathbb{A}_j + \Delta\mathbb{A}_j)'X_\epsilon + X_\epsilon(\mathbb{A}_j + \Delta\mathbb{A}_j)]z + \sum_{k=1}^{2^m} \mu_k z'[\sum_{j=1}^{2^m} \lambda_j \mathbb{K}_j' \mathbb{B}_k' X_\epsilon + X_\epsilon \mathbb{B}_k \sum_{j=1}^{2^m} \lambda_j \mathbb{K}_j]z$.

If $z'[(\mathbb{A}_j + \Delta\mathbb{A}_j)'X_\epsilon + X_\epsilon(\mathbb{A}_j + \Delta\mathbb{A}_j)]z + z'[\mathbb{K}_j'\mathbb{B}_k'X_\epsilon + X_\epsilon\mathbb{B}_k\mathbb{K}_j]z < 0$, $\forall j = 1, ..., 2^m$, $\forall k = 1, ..., 2^m$, then one gets $\dot{V}(z) < 0$. From Petersen (Petersen, 1987) for any $F(t) \in \mathcal{F}$ one gets

$z'[(\mathbb{A}_j + \Delta\mathbb{A}_j)'X_\epsilon + X_\epsilon(\mathbb{A}_j + \Delta\mathbb{A}_j)]z + z'[\mathbb{K}_j'\mathbb{B}_k'X_\epsilon + X_\epsilon\mathbb{B}_k\mathbb{K}_j]z \leq z'[\mathbb{A}_j'X_\epsilon + X_\epsilon\mathbb{A}_j + X_\epsilon\mathcal{D}_1\mathcal{D}_1'X_\epsilon + \mathbb{E}_j'\mathbb{E}_j + \mathbb{K}_j'\mathbb{B}_k'X_\epsilon + X_\epsilon\mathbb{B}_k\mathbb{K}_j]z$

with $\mathbb{A}_j$ and $\mathbb{B}_k$ defined in (16) and $\mathbb{E}_j = \mathbb{E}\begin{bmatrix} I_n & 0 \\ 0 & \Gamma_j(\alpha_\ell) \end{bmatrix}$, $\forall j = 1, ..., 2^m$. Therefore if condition (23) is met for $j = 1, ..., 2^m$ and $k = 1, ..., 2^m$ one gets $\dot{V}(z) < 0$ in $\mathfrak{E}(X_\epsilon, \rho)$. Hence in $\mathfrak{E}(X_\epsilon, \rho)$ the polytopic model (17) represents system (8) and therefore the local stability of the saturated system (8) is ensured for any $z(0) \in \mathfrak{E}(X_\epsilon, \rho)$. □

Recall that our final objective summarizes that we have to find $P$ and $\rho$ such that the closed-loop system (8) is locally stable in the largest Lyapunov level set $\mathfrak{E}(P, \rho)$. Note furthermore that the set of admissible initial conditions in which we are interested is the set $\mathcal{D}_0 = \{x(0) \in \Re^n; \begin{bmatrix} x(0) \\ 0 \end{bmatrix} \in \mathfrak{E}(X_\epsilon, \rho)\} = \{x(0) \in \Re^n; x(0)'X_{\epsilon 1}x(0) \leq \rho\}$ where $X_{\epsilon 1}$ is the left block of $n \times n$ of $X_\epsilon$, which can be partitioned as : $X_\epsilon = \begin{bmatrix} X_{\epsilon 1} & X_{\epsilon 2} \\ X_{\epsilon 2}' & X_{\epsilon 3} \end{bmatrix}$. In order to address Problem 1 we propose now some sketches of an algorithm.

**ARE-algorithm**

(1) Given arbitrary positive definite matrices $R$, $Q$, solve ARE (21) for $X$ and $\epsilon$.
(2) Derive controller from relations (22).
(3) Compute the scaling parameters $\phi_{(i)} = \frac{u_{0(i)}}{\sqrt{\left[\,0\ I_m\,\right]_{(i)} X_\epsilon^{-1}(\left[\,0\ I_m\,\right]_{(i)})'}}$, $i = 1, ..., m$ and consider $\zeta_0 = \min(\phi_{(i)})$. Thus, define $\alpha_{\ell(i)}$ as functions of a parameter $\zeta \geq \zeta_0$ as $\alpha_{\ell(i)} = \min(1, \frac{\phi_{(i)}}{\zeta})$. Build matrices $\Gamma_j(\alpha_\ell)$, $j = 1, ..., 2^m$, as described in section 3.
(4) Compute the scaling parameters $\psi_{(i,j)} = \frac{u_{1(i)}}{\sqrt{R_{\epsilon(i)}^{-1}\mathcal{B}'X_\epsilon Z_j X_\epsilon \mathcal{B}(R_{\epsilon(i)}^{-1})'}}$, $i = 1, ..., m$, $j = 1, ..., 2^m$, with $Z_j = \begin{bmatrix} I_n & 0 \\ 0 & \Gamma_j(\alpha_\ell) \end{bmatrix} X_\epsilon^{-1} \begin{bmatrix} I_n & 0 \\ 0 & \Gamma_j(\alpha_\ell) \end{bmatrix}$. Thus, define $\beta_{\ell(i)}$ as functions of parameter $\zeta$ as $\beta_{\ell(i)} = \min_j(1, \frac{\psi_{(i,j)}}{\zeta})$, $j = 1, ..., 2^m$. Build matrices $\Gamma_k(\beta_\ell)$, $k = 1, ..., 2^m$, as described in section 3.
(5) Find the maximal value $\zeta_{max}$ such that condition (23) is satisfied. A bisection method can be used for example. Hence, the ellipsoid in which the local stability of system (8) is guaranteed is defined from (20) by $P = X_\epsilon$ obtained in step 1 and $\rho = \zeta_{max}^2$.

*Remark 3.* The existence of matrix $X_\epsilon$ solution to ARE (21) is independent of the choice of $R$ and $Q$. It only relies on the existence of a small enough $\epsilon$ (Petersen, 1987). However, the *a priori* choice of these matrices influences the size of $\mathfrak{E}(P, \rho)$. Nevertheless, the way to take into account this fact is not easy to characterize. Some global nonlinear optimization problems can be set but their solution is not easily attainable (Tarbouriech and Garcia, 1997).

*Example 4.* In order to illustrate this first ARE-based result we consider the single integrator system studied in (Shewchun and Feron, 1997). System (1) is described by the following data :

$$A = 0;\ B = 1;\ D_1 = 0$$
$$E_1 = 0;\ E_2 = 0;\ u_0 = 1;\ u_1 = 1$$

which gives matrices $\mathcal{A}$ and $\mathcal{B}$ of system (6) : $\mathcal{A} = \begin{bmatrix} 0 & 1 \\ 0 & 0 \end{bmatrix}$ and $\mathcal{B} = \begin{bmatrix} 0 \\ 1 \end{bmatrix}$. From (21) the stabilizing state feedback is $\mathcal{K} = -R_\epsilon^{-1}\mathcal{B}'X_\epsilon$ and by computing $X_\epsilon$ and $\mathcal{K}$ from $R = 50000$ and $Q = \begin{bmatrix} 1 & 0 \\ 0 & R \end{bmatrix}$ (Shewchun and Feron, 1997) one obtains $P = X_\epsilon = 10^4 \begin{bmatrix} 0.0225 & 0.0224 \\ 0.0224 & 5.0223 \end{bmatrix}$ and $K = \begin{bmatrix} -0.0045 & -1.0045 \end{bmatrix}$. By testing the validity of condition (23) one gets $\rho = 7.8940\ 10^5$, $\alpha_\ell = 0.2517$ and $\beta_\ell = 0.2511$. In Figure 1, we compare our stability domain (solid line) in the space of $z$ with the one obtained in (Shewchun and Feron, 1997) (dotted line).

In the above example even if we obtain a set of admissible condition with an interesting size, it is clear, due to the implicit dependence of the choice of $R$ and $Q$, for the size $\mathfrak{E}(P, \rho)$ that our result may conduct to conservative results.

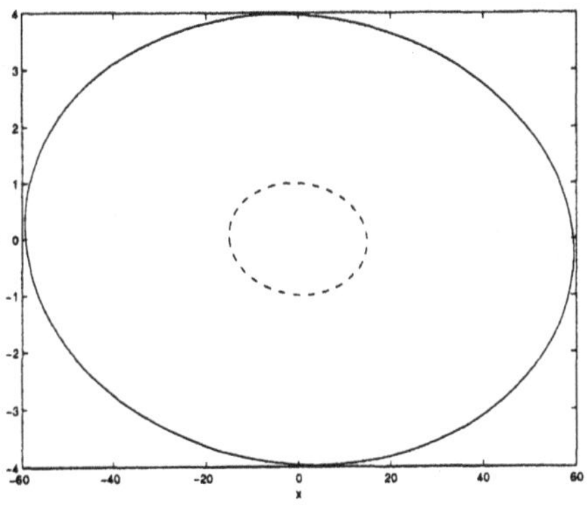

Fig. 1. Stability domains

## 5. LMI-BASED APPROACH

Set

$$\mathcal{Q}(j,k) = \begin{bmatrix} I_n & 0 \\ 0 & \Gamma_j(\alpha_\ell) \end{bmatrix} \mathcal{A}'P + P\mathcal{A} \begin{bmatrix} I_n & 0 \\ 0 & \Gamma_j(\alpha_\ell) \end{bmatrix} + \begin{bmatrix} I_n & 0 \\ 0 & \Gamma_j(\alpha_\ell) \end{bmatrix} \mathcal{K}'\Gamma_k(\beta_\ell)\mathcal{B}'P + P\mathcal{B}\Gamma_k(\beta_\ell)\mathcal{K} \begin{bmatrix} I_n & 0 \\ 0 & \Gamma_j(\alpha_\ell) \end{bmatrix} \tag{26}$$

*Proposition 5.* Consider matrix $\mathcal{Q}(j,k)$ defined in (26). If there exist a symmetric positive definite matrix $P$, a matrix $\mathcal{K}$, two vectors $\alpha_\ell$, $\beta_\ell$, a positive scalar $\gamma$ such that

$$\begin{bmatrix} \mathcal{Q}(j,k) & \begin{bmatrix} I_n & 0 \\ 0 & \Gamma_j(\alpha_\ell) \end{bmatrix} \mathcal{E}' & P\mathcal{D}_1 \\ \mathcal{E} \begin{bmatrix} I_n & 0 \\ 0 & \Gamma_j(\alpha_\ell) \end{bmatrix} & -I_l & 0 \\ \mathcal{D}_1'P & 0 & -I_r \end{bmatrix} < 0, \quad \forall j = 1,...,2^m, \forall k = 1,...,2^m \tag{27}$$

$$\begin{bmatrix} P & \alpha_{\ell(i)} \left( \begin{bmatrix} 0 & I_m \end{bmatrix}_{(i)} \right)' \\ \alpha_{\ell(i)} \begin{bmatrix} 0 & I_m \end{bmatrix}_{(i)} & \gamma u_{0(i)}^2 \end{bmatrix} \geq 0, \quad \forall i = 1,...,m \tag{28}$$

$$\begin{bmatrix} P & \beta_{\ell(i)} \left( \mathcal{K}_{(i)} \begin{bmatrix} I_n & 0 \\ 0 & \Gamma_j(\alpha_\ell) \end{bmatrix} \right)' \\ \beta_{\ell(i)} \mathcal{K}_{(i)} \begin{bmatrix} I_n & 0 \\ 0 & \Gamma_j(\alpha_\ell) \end{bmatrix} & \gamma u_{1(i)}^2 \end{bmatrix} \geq 0, \ \forall i = 1,...,m, j = 1,...,2^m \tag{29}$$

$$0 < \alpha_{\ell(i)} \leq 1, \ i = 1,...,m \tag{30}$$

$$0 < \beta_{\ell(i)} \leq 1, \ i = 1,...,m \tag{31}$$

then the feedback gain $\mathcal{K}$ and the set $\mathfrak{E}(P,\gamma^{-1})$ solve Problem 1.

**Proof.** If relation (27) is satisfied one can prove that the state feedback gain $\mathcal{K}$ thus obtained is such that the time-derivative of the Lyapunov function $V(z) = z'Pz$ is strictly negative along the trajectories of the closed-loop system. Furthermore the polytopic representation (17) of the saturated system is only valid in the set $S(u_0,\alpha_\ell) \cap S(u_1,\beta_\ell)$, defined in (18) and (19). Hence relations (28) and (29) correspond to the satisfaction of the inclusion $\mathfrak{E}(P,\gamma^{-1}) \subset S(u_0,\alpha_\ell) \cap S(u_1,\beta_\ell)$. □

The major difficulty in the application of Proposition 5 resides in the fact that the variables to be found are simultaneously $\gamma$, $\alpha_\ell$, $\beta_\ell$, $P$ and $\mathcal{K}$. Thus relations (27) and (29) are multilinear in the variables. Only relation (28) is an LMI condition in $P$, $\alpha_\ell$ and $\gamma$. Relations (30) and (31) are LMIs in $\alpha_\ell$ and $\beta_\ell$. However, some efficient methods for solving simultaneously relations (27), (28), (29), (30) and (31) can be proposed. Indeed, if we fix four variables among five then relations (27), (28) and (29) become LMI conditions in the remaining variable. In this sense, some relaxation techniques can be used (see (Henrion *et al.*, 1997)). Based on the above comments we propose the following possible relaxations :

- Relax 1 : given $P$, $\mathcal{K}$ and $\alpha_\ell$. Solve (27), (28) and (29) for $\gamma$, and $\beta_\ell$.
- Relax 2 : given $P$, $\mathcal{K}$ and $\beta_\ell$. Solve (27), (28) and (29) for $\alpha_\ell$ and $\gamma$.
- Relax 3 : given $P$, $\alpha_\ell$ and $\beta_\ell$, solve (27), (28) and (29) for $\mathcal{K}$ and $\gamma$.
- Relax 4 : given $\mathcal{K}$, $\alpha_\ell$ and $\beta_\ell$, solve (27), (28) and (29) for $P$ and $\gamma$.

Recall that our objective consists in obtaining a set $\mathfrak{E}(P,\gamma^{-1})$ as large as possible, Thus we can seek to optimize the size of $\mathfrak{E}(P,\gamma^{-1})$ via the maximization of its volume, or equivalently minimization of $log(det(P^{-1}))$, $\gamma$ or $trace(P)$. Moreover, in view of inclusion $\mathfrak{E}(P,\gamma^{-1}) \subset S(u_0,\alpha_\ell) \cap S(u_1,\beta_\ell)$, vectors $\alpha_\ell$ and $\beta_\ell$ have to be minimized in order to enlarge $S(u_0,\alpha_\ell) \cap S(u_1,\beta_\ell)$, and thus allow a maximal saturation tolerance in $\mathfrak{E}(P,\gamma^{-1})$. Therefore we can consider the following criteria :

- Criterion 1 : $\min \gamma$
- Criterion 2 : $\min trace(P)$
- Criterion 3 : $\min log(det(P^{-1}))$
- Criterion 4 : $\min \sum_{i=1}^{m} \alpha_{\ell(i)}$
- Criterion 5 : $\min \sum_{i=1}^{m} \beta_{\ell(i)}$.

Valid associations of criteria and relaxation schemes give rise to the following LMI optimization problems :

- LMIP 1: optimize criterion 1 subject to relax 1.
- LMIP 2: optimize criterion 1 subject to relax 2.
- LMIP 3: optimize criterion 1 subject to relax 3.
- LMIP 4: optimize criterion 2 subject to relax 4.
- LMIP 5: optimize criterion 3 subject to relax 4.
- LMIP 6: optimize criterion 4 subject to relax 2.

- LMIPi 7: optimize criterion 5 subject to relax 1.

Then, we can propose the following type of algorithm.

**LMI-algorithm**

(1) Solve relations (27), (28) and (29) for $P$, $K$ and $\gamma$ by choosing $\alpha_{\ell(i)} = \beta_{\ell(i)} = 1$, $\forall i = 1, ..., m$. Relations (27), (28) and (29) are expressed as equivalent LMI conditions in $W$, $Y$ and $\gamma$ by using the classical transformation $P = W^{-1}$ and $\mathcal{K} = YP$.
(2) Solve LMIP 1, 2, 3, 4, 5, 6 or 7 for its global optimum.
(3) Go to step 2 until no significant change occurs in any variable $P$, $\mathcal{K}$, $\gamma$, $\alpha_\ell$ and $\beta_\ell$.

*Remark 6.* Note that the initial values of $P$ and $K$ used to begin LMI-algorithm is very important. Unfortunately, there exists not an easy way to optimize this step. We can begin the algorithm by considering as in (Shewchun and Feron, 1997) the values $P$ and $K$ derived from the ARE-algorithm. This important fact could be investigated by meaning maybe some global optimization problems subject to multilinear constraints.

*Example 7.* Consider an uncertain model of the dynamics of a helicopter in a vertical plane, as described in (Schmitendorf, 1988). System (1) is defined by the following data :

$$A = \begin{bmatrix} -0.0366 & 0.0271 & 0.0188 & -0.4555 \\ 0.0482 & -1.01 & 0.0024 & -4.0208 \\ 0.1002 & 0.2855 & -0.707 & 1.3229 \\ 0 & 0 & 1 & 0 \end{bmatrix}$$

$$B = \begin{bmatrix} 0.4422 & 0.1761 \\ 3.0447 & -7.5922 \\ -5.52 & 4.99 \\ 0 & 0 \end{bmatrix}$$

$$D_1 = \begin{bmatrix} 0 & 0 & 0 \\ 0 & 0 & 2.0673 \\ 0.2192 & 1.2031 & 0 \\ 0 & 0 & 0 \end{bmatrix}$$

$$E_1 = \begin{bmatrix} 0 & 1 & 0 & 0 \\ 0 & 0 & 0 & 1 \\ 0 & 0 & 0 & 0 \end{bmatrix} ; E_2 = \begin{bmatrix} 0 & 0 \\ 0 & 0 \\ 1 & 0 \end{bmatrix}$$

Input and rate limits are $u_0 = \begin{bmatrix} 5 \\ 2 \end{bmatrix}$ and $u_1 = \begin{bmatrix} 2 \\ 5 \end{bmatrix}$. After 4 repetitions of the LMI optimization problem sequence 6, 7, 3, 4, 1, 2, 3, we eventually get: $\alpha_\ell = \begin{bmatrix} 0.8937 \\ 0.8808 \end{bmatrix}$, $\beta_\ell = \begin{bmatrix} 0.8926 \\ 1 \end{bmatrix}$ and $\gamma = 0.4200$ with

$$P = \begin{bmatrix} 0.0864 & 0.0184 & 0.0046 & -0.0506 & -0.0465 & -0.0427 \\ 0.0184 & 0.5865 & 0.2803 & -0.2248 & -0.2130 & -0.6300 \\ 0.0046 & 0.2803 & 0.4796 & 0.4334 & -0.6198 & -0.0076 \\ -0.0506 & -0.2248 & 0.4334 & 1.5928 & -0.6006 & 1.0523 \\ -0.0465 & -0.2130 & -0.6198 & -0.6006 & 2.0484 & -0.3001 \\ -0.0427 & -0.6300 & -0.0076 & 1.0523 & -0.3001 & 1.8475 \end{bmatrix}$$

$$\mathcal{K} = \begin{bmatrix} -0.0036 & 0.5088 & 0.8208 & 0.4524 & -1.9080 & 0.0018 \\ 0.0067 & 1.4675 & -0.0731 & -2.4675 & 0.9427 & -4.2418 \end{bmatrix}$$

## 6. CONCLUSION

In spite of the potentialities of these approaches and the investigations done up to now, it is not possible to conclude on the superiority of the one over the other. Investigation on this particular point will be done in a near future.

## 7. REFERENCES


Bernstein, D.S. and A.N. Michel (1995). A chronological bibliography on saturating actuators. *Int. J. of Robust and Nonlinear Control* **5**, 375–380.

Dorato, P. (1987). *Robust Control.* IEEE Press Book.

Garcia, G., J. Bernussou and D. Arzelier (1994). Robust stabilization of discrete-time linear systems with norm-bounded time varying uncertainty. *Systems and Control Letters* **22**, 327–339.

Gomes da Silva Jr, J.M. and S. Tarbouriech (1998). Local stabilization of discrete-time linear systems with saturating controls: an lmi-based approach. *ACC, Philadelphia (USA),* pp. 92–96.

Henrion, D., S. Tarbouriech and G. Garcia (1997). Output feedback robust stabilization of uncertain linear systems with saturating controls. *36th IEEE-CDC, San Diego (USA),* pp. 193–198.

Lin, Z. and A. Saberi (1993). Semi-global exponential stabilization of linear systems subject to input saturation vi linear feedback. *Systems & Control Letters* **21**, 225–239.

Petersen, I. R. (1987). A stabilization algorithm for a class of uncertain linear systems. *Systems and Control Letters* **8**, 351–357.

Saberi, A. and A.A. Stoorvogel (1997). Stabilization and regulation of linear systems with saturated and rate-limited actuators. *ACC, Albuquerque (USA),* pp. 3920–3921.

Schmitendorf, W.E. (1988). Designing stabilizing controllers for uncertain systems using the riccati equation approach. *IEEE Trans. Autom. Control* **33, no.4**, 376–379.

Shewchun, J.M. and E. Feron (1997). High performance bounded control. *ACC, Albuquerque (USA),* **5**, 3250–3254.

Sontag, E.D. and H.J. Sussmann (1990). Nonlinear output feedback design for linear systems with saturating control. *29th IEEE-CDC, Honolulu (USA),* pp. 3414–3416.

Tarbouriech, S. and G. Garcia (1997). *Control of Uncertain Systems with Bounded Inputs.* Editors of the book in Springer-Verlag (LNCIS), vol.227. London.

Zhou, K., J.C. Doyle and K. Glover (1996). *Robust and optimal control.* Prentice Hall.

14th Triennial World Congress, Beijing, P.R. China

F-2e-02-6

# $H_\infty$ PERFORMANCE OF LINEAR SYSTEMS SUBJECT TO INPUT SATURATION

**Xuequan Xiong, Yongyan Cao and Youxian Sun**

*Institute of Industrial Process Control*
*Zhejiang University, Hangzhou, P. R. China*

*e-mail: xqxiong@iipc.zju.edu.cn*

Abstract: In this paper, the $H_\infty$ performance problems of linear systems subject to two types of input saturation are considered respectively. A sufficient condition, which makes the closed-loop systems to be quadratic stable and satisfy the $H_\infty$ performance prior given, is derived. According to this sufficient condition, a method is presented to design a static feedback controller for the plant in the present of input saturation. All the results are expressed by matrix inequalities. *Copyright © 1999 IFAC*

Keywords: linear systems, saturation control, $H_\infty$ control, quadratic stability, state feedback.

## 1. INTRODUCTION

In practical engineering, dynamic systems are all subject to input constraints. The analysis and synthesis of controller for dynamic systems subject to input saturation have been received increasing attention recently (for example, see Bernstein *et al.* (1995a, b), and references therein). Mainly there are two approaches to deal with control subject to input saturation. One approach is to take into account control constraints in the process of analysis and design. Considerable improvement has been made in this area. For null-controllable systems, a nested feedback design technique was proposed by Sussman *et al.* (1994) to design nonlinear globally asymptotically stabilizing controller; a low-high gain method was present by Saberi *et al.* (1996) to design linear semi-globally stabilizing controllers. Gilbert *et al.* (1991) proposed the method of maximal output admissible sets to analyze the attraction region for linear systems in the present of control and state constraints. Tyan *et al.* (1995) got a LQG-type Riccati equation to be a sufficient condition for local asymptotically stability, and minimized the LQG-type cost to design linear or anti-windup controller. These methods tend to focus more on the stability and less on performance. Another approach is to design a linear controller first, ignoring the saturation, and then design an anti-windup compensator to weaken the influence of saturation. This method was widely used in the early control engineering. There are many techniques to design the anti-windup compensator, such as GCT(generalized conditioning technique) (see Walgama *et al.* (1992)), IMC(internal model control) (see Campo *et al.* (1990)), ..., and a unified framework for the study of anti-windup design was presented by Kothare *et al.* (1994). Through this method, the well-established linear control theory can be well applied in practical use. However, in most case the approach is only applicable to the open-loop stable plant.

In this paper, the $H_\infty$ performance problems of linear systems subject to input saturation are considered. The input constraints have two types, one is energy bounded, and another is amplitude bounded. A sufficient condition to satisfy stability and an $H_\infty$ performance prior given for the closed-loop systems subject to input constraint is derived, the closed-loop systems are in a unified framework similar to Tyan *et al.* (1995). According to the sufficient condition, a method to design static feedback controller is presented.

*1.1 NOTATION*

$P>0$ $P$ is positive-definite symmetry matrix.

$\|u\|$ Euclidian norm of $u$, that is

$$\|u\| = \sqrt{u^T u}\,.$$

$\|w\|_2$ $L_2$ norm of $w$, that is

$$\|w\|_2 = \sqrt{\int_0^{+\infty} w^T w dt}\,.$$

## 2. SYSTEMS AND PRELIMINARIES

Consider the plant with input saturation

$$\Sigma 1:\quad \begin{aligned} \dot{x}_p &= A_p x_p + B_p \sigma(u) + E_p w \\ y_p &= C_{p1} x_p \\ z_p &= C_{p2} x_p + D_p u \end{aligned} \qquad (1)$$

where $x_p \in \Re^{n_p}$, is the state and $x_p(0)=0$, $u \in \Re^m$ is the control, $y_p \in \Re^{q_1}$ is the measured output, $z_p \in \Re^{q_2}$ is the controlled output, $w \in \Re^r$ is the disturb input, and $\sigma(\cdot)$ is one type of the two saturation functions as following:

i) $\sigma(u)=\beta(u)u \qquad \beta(u)=\begin{cases} 1 & \|u\|\le 1 \\ 1/\|u\| & \|u\|>1 \end{cases}$

ii) $\sigma(u)=\beta(u)u$

$$\beta(u) = diag(\beta_1(u),\cdots,\beta_i(u),\cdots,\beta_m(u))$$

$$\beta_i(u_i) = \begin{cases} 1 & |u_i|<1 \\ 1/|u_i| & |u_i|<1 \end{cases} \quad i = 1, 2, \cdots, m$$

These two saturation functions are common in practice. The first one is constraint by energy, the second one is constraint by amplitude. When $u$ is a scalar, i) and ii) are the same.

Assume the controller with anti-windup compensator is:

$$\begin{aligned} \dot{x}_c &= A_c x_c + B_c y_p + E_c(\sigma(u)-u) \\ u &= C_c x_c \end{aligned} \qquad (2)$$

where $x_c \in \Re^{n_c}$ and $x_c(0)=0$. Then the closed-loop system is:

$$\Sigma:\quad \begin{aligned} \dot{x} &= Ax + B(\sigma(u)-u) + Ew \\ u &= Cx \\ z &= C_\infty x \end{aligned} \qquad (3)$$

where

$$x = \begin{pmatrix} x_p \\ x_c \end{pmatrix} \in \Re^{n_p+n_c} \text{ and } x(0)=0,$$

$$z = z_p,\quad A = \begin{pmatrix} A_p & B_p C_c \\ B_c C_{p_1} & A_c \end{pmatrix},$$

$$B = \begin{pmatrix} B_p \\ E_c \end{pmatrix},\qquad C = \begin{pmatrix} 0 & C_c \end{pmatrix},$$

$$E = E_p,\quad C_\infty = \begin{pmatrix} C_{P_2} & D_p C_c \end{pmatrix}$$

The system $\Sigma$ can be looked as the unified framework of the close-loop system subject to input saturation.

## 3. $H_\infty$ PERFORMANCE ANALYSIS OF CLOSED-LOOP SYSTEMS

In this section, the $L_2$ gain performance (*i.e.* $H_\infty$ performance, when saturation is absent) of the system $\Sigma$ is considered. The following two lemmas are well known and are important in the following of this paper.

Lemma 1: for any $\lambda > 0$

$$2x^T XYy \le \lambda x^T XX^T x + \lambda^{-1} y^T Y^T Yy \qquad (4)$$

Lemma 2: If $V(0)=0$ and for any $0<t<T$, $\dot{V}(t) < -aV(t) + b$, where $a>0$, $b>0$, then for any $0<t<T$, $V(t)<b/a$.

Theorem 1: For the system $\Sigma$, $\sigma(u)$ is in the form of i), and $\|w(t)\| \le 1$. If there exist $\lambda>0$, $\alpha>0$, $0\le l \le 1$, and the following inequalities have the solution of $P>0$, $Q>0$:

$$A^T P + PA + P(BB^T + \lambda^{-1}EE^T)P + l^2 C^T C + \alpha P < 0 \qquad (5)$$

$$(\alpha/\lambda)P - (1-l)^2 C^T C > 0 \qquad (6)$$

$$A^T Q + QA + Q(BB^T + \gamma^{-2} EE^T)Q + l^2 C^T C + C_\infty^T C_\infty < 0 \qquad (7)$$

then the system $\Sigma$ is quadratic stable and $\|z\|_2 < \gamma \|w\|_2$.

Proof: Select Lyapunov function as

$$V(t) = x^T Px. \qquad (8)$$

From (6), it is easy to see that if $V \le \lambda/\alpha$, then

$$x^T C^T Cx < 1/(1-l)^2. \qquad (9)$$

$$\begin{aligned} \dot{V} = dV/dt &= x^T(A^T P + PA)x \\ &\quad + 2x^T PB(\sigma(u)-u) + 2x^T PEw \\ &\le x^T(A^T P + PA + PBB^T P + \lambda^{-1}PEE^T P)x \\ &\quad + (\sigma(u)-u)^T(\sigma(u)-u) + \lambda w^T w \\ &< x^T(-l^2 C^T C - \alpha P)x \\ &\quad + (1-\beta(u))^2 x^T C^T Cx + \lambda w^T w \\ &= -\alpha V + ((1-\beta(u))^2 - l^2)x^T C^T Cx \\ &\quad + \lambda w^T w. \end{aligned} \qquad (10)$$

Assume there exist $T_1 > 0$, $1-\beta(u(T_1)) > l$, then

$$x(T_1)^T C^T Cx(T_1) > 1/(1-l)^2 . \quad (11)$$

Because $x(0)=0$, there must exist $0<T<T_1$, for any $0<t<T$, $x^T C^T Cx < 1/(1-l)^2$ and $x(T)^T C^T Cx(T) = 1/(1-l)^2$. Then for any $0<t<T$,

$$\dot{V} < -\alpha V + \lambda w^T w, \quad (12)$$

$$V < \lambda / \alpha, \quad (13)$$

and then

$$V(T) \le \lambda / \alpha, \quad (14)$$

$$x(T)^T C^T Cx(T) < 1/(1-l)^2, \quad (15)$$

which is a contradiction with the assumption. From the above discussion, the conclusion can be made that for any $t>0$, $1-\beta(u)<l$ and $V(t)<\lambda/\alpha$, which means the system is quadratic stable.

Select $V_1 = x^T Qx$,

$$\begin{aligned} & dV_1/dt + z^T z - \gamma^2 w^T w \\ &= x^T(A^T Q + QA)x + 2x^T QB(\sigma(u)-u) \\ &\quad + 2x^T QEw + x^T C_\infty^T C_\infty x - \gamma^2 w^T w \\ &\le x^T(A^T Q + QA + QBB^T Q + \gamma^{-2} QEE^T Q \\ &\quad + C_\infty^T C_\infty + (1-\beta(u))^2 C^T C)x \\ &< x^T(A^T Q + QA + QBB^T Q + \gamma^{-2} QEE^T Q \\ &\quad + C_\infty^T C_\infty + l^2 C^T C)x \\ &< 0 \end{aligned} \quad (16)$$

then $\|z\|_2 < \gamma \|w\|_2$. □

From the proof, we can see that (5) and (6) guarantee the stability, and (7) guarantees the $H_\infty$ performance of the system.

Remark 1: In theorem 1, if $l$=0, (5) and (6) hold, then (7) changed to the inequality as following can guarantee quadratic stability and $\|z\|_2 < \gamma \|w\|_2$.

$$A^T Q + QA + \gamma^{-2} QEE^T Q + C_\infty^T C_\infty < 0 \quad (17)$$

In this case, for any $t$>0, $\|u\| < 1$.

Remark 2: In theorem 1, if $l$>1, then only (5) and (7) can guarantee quadratic stability and $\|z\|_2 < \gamma \|w\|_2$.

In the following, the $H_\infty$ performance of the system Σ which saturation part is in the form of ii) is considered. $C$ can be written as

$$C^T = \begin{pmatrix} C_1^T & C_2^T & \cdots & C_m^T \end{pmatrix}^T \quad (18)$$

where $C_i$ is a row vector.

Theorem 2: For system Σ, assume the $\sigma(u)$ is in the form of ii), and $w^T w \le 1$. If there exist $\lambda$>0, $\alpha$>0, 0≤$l$≤ 1, and the following inequalities have the solution of $P$>0, $Q$>0 :

$$\begin{aligned} & A^T P + PA + P(BB^T + \lambda^{-1} EE^T)P \\ &\quad + l^2 C^T C + \alpha P < 0 \end{aligned} \quad (19)$$

$$(\alpha/\lambda)P - (1-l)^2 C_i^T C_i > 0 \quad \forall i = 1 \ \cdots \ m \quad (20)$$

$$\begin{aligned} & A^T Q + QA + Q(BB^T + \gamma^{-2} EE^T)Q \\ &\quad + l^2 C^T C + C_\infty^T C_\infty < 0 \end{aligned} \quad (21)$$

then the system Σ is quadratic stable and $\|z\|_2 < \gamma \|w\|_2$

Proof: Select Lyapunov function as

$$V(t) = x^T Px . \quad (22)$$

From (20), it is easy to see that if $V \le \lambda/\alpha$, then for any $1 \le i \le m$,

$$|u_i| = |C_i x| < 1/(1-l) . \quad (23)$$

$$\begin{aligned} \dot{V} &= dV/dt = x^T(A^T P + PA)x \\ &\quad + 2x^T PB(\sigma(u)-u) + 2x^T PEw \\ &\le x^T(A^T P + PA + PBB^T P + \lambda^{-1} PEE^T P)x \\ &\quad + (\sigma(u)-u)^T(\sigma(u)-u) + \lambda w^T w \\ &< x^T(-l^2 C^T C - \alpha P)x + \\ &\quad + \sum_{i=1}^{m} (1-\beta_i(u_i))^2 x^T C_i^T C_i x + \lambda w^T w \\ &\le -\alpha V + \max_{i=1,\cdots,m}((1-\beta_i(u_i)^2 - l^2)) x^T C^T Cx \\ &\quad + \lambda w^T w \end{aligned} \quad (24)$$

Assume there exists $T_1 > 0$ satisfying $\max_{i=1,\cdots m}(1-\beta(u(T_1))) > l$, then

$$\max_{i=1,\cdots,m} |C_i x(T_1)| > 1/(1-l) . \quad (25)$$

Because $x(0)=0$, there must exist $0<T<T_1$, 1≤$j$≤$m$, for any $0<t<T$, $\max_{i=1,\cdots,m} |C_i x| < 1/(1-l)$ and $|C_j x(T)| = 1/(1-l)$. Then for any $0<t<T$,

$$\dot{V} < -\alpha V + \lambda w^T w, \quad (26)$$

$$V < \lambda/\alpha, \quad (27)$$

and then

$$V(T) \le \lambda/\alpha, \quad (28)$$

$$|C_j x(T)| < 1/(1-l)^2, \quad (29)$$

which is a contradiction with the assumption. From the above discussion, the conclusion can be made that for any $t>0$, $\max_{i=1,\cdots m}(1-\beta(u(T_1))) < l$ and $V(t) < \lambda/\alpha$, which means the system is quadratic stable.

Select $V_1 = x^T Qx$,

$$\begin{aligned} & dV_1/dt + z^T z - \gamma^2 w^T w \\ &= x^T(A^T Q + QA)x + 2x^T QB(\sigma(u)-u) \end{aligned}$$

$$+2x^TQEw+x^TC_\infty^TC_\infty x-\gamma^2w^Tw$$
$$\le x^T(A^TQ+QA+QBB^TQ+\gamma^{-2}QEE^TQ$$
$$+C_\infty^TC+\max_{i=1,\cdots,m}(1-\beta_i(u_i))^2C^TC)x$$
$$<x^T(A^TQ+QA+QBB^TQ+\gamma^{-2}QEE^TQ$$
$$+C_\infty^TC_\infty+l^2C^TC)x$$
$$<0 \tag{30}$$

then $\|z\|_2<\gamma\|w\|_2$. □

Remark 3: As the same of theorem 1, we can get the similar result as the Remark 1 and Remark 2.

## 4. THE SYNTHESIS OF STATIC STATE FEEDBACK CONTROLL

In this section, the synthesis of static state feedback controller for the open-loop system $\Sigma 1$ is considered. For simplicity, in the following the subscript of '$p$' in $\Sigma 1$ will be omitted. According to the theorem 1, the follow result can be got:

Theorem 3: For the system $\Sigma 1$, $D$=0, $\sigma(u)$ is in the form of i), and $w^Tw\le 1$. If there exist $\lambda$>0 , $\alpha$>0, 0≤$v$≤ 1, and any group of the following two group matrix inequalities have the solution of $Q$>0 :

1) $$\begin{pmatrix} QA^T+AQ-vBB^T+\alpha Q & E \\ E^T & -\lambda I\end{pmatrix}<0 \tag{31}$$

$$\frac{\alpha}{\lambda}Q-vBB^T>0 \tag{32}$$

$$\begin{pmatrix} QA^T+AQ-vBB^T & E & QC_\infty^T \\ E^T & -\gamma^2 & 0 \\ C_\infty Q & 0 & -I\end{pmatrix}<0 \tag{33}$$

2) $$\begin{pmatrix} QA^T+AQ-BB^T+\alpha Q & E \\ E^T & -\lambda I\end{pmatrix}<0 \tag{34}$$

$$\frac{\alpha}{\lambda}Q-BB^T>0 \tag{35}$$

$$\begin{pmatrix} QA^T+AQ-2BB^T & E & QC_\infty^T \\ E^T & -\gamma^2 & 0 \\ C_\infty Q & 0 & -I\end{pmatrix}<0 \tag{36}$$

then the static feedback controller: $u=-B^TQ^{-1}x$ can guarantee the close-loop system to be quadratic stable and $\|z\|_2<\gamma\|w\|_2$.

Proof: First, let's proof 1). Substitute $u=Kx$ to system $\Sigma 1$, then using theorem 1's result, one sufficient condition to be quadratic stable and $\|z\|_2<\gamma\|w\|_2$ for the closed-loop system is: there exist $\lambda$>0, $\alpha$>0, 0≤$l$≤ 1, the following inequalities have the solution of $P$>0 :

$$(A+BK)^TP+P(A+BK)+P(BB^T$$
$$+\lambda^{-1}EE^T)P+l^2K^TK+\alpha P<0 \tag{37}$$
$$(\alpha/\lambda)P-(1-l)^2K^TK>0 \tag{38}$$
$$(A+BK)^TP+P(A+BK)+P(BB^T$$
$$+\gamma^{-2}EE^T)P+l^2K^TK+C_\infty^TC_\infty<0 \tag{39}$$

Suppose $K=-B^TP$, then (37)-(39) is equal to

$$A^TP+PA-(1-l^2)PBB^TP$$
$$+\lambda^{-1}PEE^TP+\alpha P<0 \tag{40}$$
$$(\alpha/\lambda)P-(1-l)^2PBB^TP>0 \tag{41}$$
$$A^TP+PA-(1-l^2)PBB^TP$$
$$+\gamma^{-2}PEE^TP+C_\infty^TC_\infty<0 \tag{42}$$

Define $v=1-l^2$, $Q=P^{-1}$. Since 0≤$l$≤ 1, 0<$v$<1 and $(1-l)^2<1-l^2=v$, it is easy to see that (31)-(33) can guarantee (40)-(42) to hold, then 1) is sufficient to guarantee quadratic stable and $\|z\|_2<\gamma\|w\|_2$.

Second, as the same method above, if $l$=0, from Remark 1, it is easy to see that 2) also is sufficient to guarantee quadratic stability and $\|z\|_2<\gamma\|w\|_2$. □

Remark 4: If $\sigma(u)$ is in the form of ii), using the same method, the similar result as theorem 3 can be reached: only needed to change (13) and (16) as:

$$\frac{\alpha}{\lambda}Q-vB_iB_i^T>0,\ \forall i=1\ \cdots\ m \tag{43}$$

where $B=(B_1\ \ B_2\ \ \cdots\ \ B_m)$.

Because we let $P$=$Q$ in applying Theorem 1 and define $v=1-l^2$ in the process of the above proof, the result is conservative.

## 5. EXAMPLE

Let us consider the linear system with saturation actuator (Nguyen *et al.* (1997))

$$\dot{x}(t)=\begin{pmatrix}0 & 1\\ -10 & -10\end{pmatrix}x(t)+\begin{pmatrix}0\\6\end{pmatrix}\sigma(u(t))$$
$$+\begin{pmatrix}0.5\\1\end{pmatrix}w(t)$$
$$z(t)=\begin{pmatrix}5 & 0\end{pmatrix}x(t)$$

with $x(0)=0$, $\|w(t)\|<1$.

If the controller has been known as $u=Kx$, $K=(\text{-2.0008}\ \ \text{-0.1985})$. Let's consider of the $H_\infty$ performance of the closed-loop system. In the absence of saturation, the $L_2$ gain form $w$ to $z$ is 1.5. In the present of saturation, when $\alpha$=1.2, $\lambda$=0.1 and

$l$=0.2 $\gamma$=2.4, (5)-(7) have the solution $P$>0, $Q$>0, then the $L_2$ gain form $w$ to $z$ is less than 2.4.

Considering the synthesis problem for $\gamma$<2.5, according to Theorem 3, via MATLAB LMI toolbox, we can get the static state feedback can be $u = Kx$, $K = (-0.5209 \;\; -0.1332)$.

## 6. CONCLUSION

In this note, the $H_\infty$ performance problem of linear system subject to input saturation was considered. First a sufficient condition for the closed-loop systems to be quadratic stable and $L_2$ gain to be smaller than one value prior given is derived. The sufficient condition was expressed by three matrix inequalities. According to the result of closed-loop systems, a sufficient condition of existing a static state feedback controller for the open-loop systems was derived. these sufficient conditions were all expressed by matrix inequalities which can be solved by MATAB LMI toolbox.

## ACKNOWLEDGEMENT

This work was supported by the National Science Foundation of China, No.69635010.

## REFERENCES

Bernstein, D. S. and Michel, A. N.(1995a). A chronological Bibliography on Saturating Actuators, *Int. Jour. of Robust and Nonlinear Control*, **5**, pp.375-380.

Bernstein, D. S. and Michel, A. N., Editors(1995b). Special Issue on Saturating Actuators, *Int. Jour. of Robust and Nonlinear Control*, **5**.

Campo, P. J. and Morari, M.(1990). Robust control of process subject to saturation nonlinearities, *Comput. Chem. Engng.*, **14**, pp.48-51.

Gilbert, E. G. and Tan, K. T.(1991). Linear System with State and Control Constraints: The Theory and Application of Maximal Output Admissible Sets, *IEEE Trans. AC*, **36**, pp1009-1020.

Kothare, M. V., Campo, P. J., Morari, M and Nett, C. N. (1994). A Unified Framework for Study of Anti-windup Designs, *Automatic*, **30**, pp.1869-1883

Nguyen, T. and Jabbari, F.(1997). $H_\infty$ Design for System with Input Saturation: an LMI approach, *Proceeding of the American Control Conference*.

Sussmann, H. J., E. D. Sontag and Y. Yang,(1994). A General Result on the Stabilization of Linear System Using Bound Controls, *IEEE Trans. AC*, **39**, pp.2411-2425.

Saberi, A. Lin, Z. and Teel, A. R.(1996). Control of Linear Systems subject to input saturation, *IEEE Trans. AC*, **41**, pp.368-378.

Tyan, F. and Bernstein, D. S.(1995). Anti-windup Compensator Synthesis for System with Saturation Actuators, *Int. Jour. of Robust and Nonlinear Control*, **5**, pp.521-537.

Walgama, K. S., Rönnbäck, S. and Sternby, J.(1992). Generalization of conditioning technique for anti-windup compensators, *IEE Proc. D.*, **139**, pp.109-118.

# EVOLUTION STRATEGIES FOR SYNTHESIS OF MIXED $H_2 / H_\infty$ FIXED-STRUCTURE CONTROLLERS

**Renato A. Krohling *, Leandro dos S. Coelho ** and Antonio Augusto R. Coelho ****

** Lehrstuhl fuer Systemtheorie der Elektrotechnik*
*Universitaet des Saarlandes, Geb. 13, Postfach 151150*
*D - 66041 Saarbruecken, Germany - E-mail: renato@lse.uni-sb.de*

*** Federal University of Santa Catarina*
*Department of Automation and Systems, C.P. 476 - 88040.900*
*Florianópolis, SC, Brazil - E-mail: {lscoelho, aarc}@lcmi.ufsc.br*

Abstract: A new method for mixed $H_2 / H_\infty$ controllers synthesis is presented in this paper. The problem is formulated as the minimization of a nominal $H_2$ performance index subject to a $H_\infty$ robust stability constraint. The novel method, based on two derandomized Evolution Strategies (*ESs*) with individual step-sizes and correlated mutations using appropriate operators and a penalty function, is proposed for the solution of the constrained optimization problem. Simulation results are to illustrate the new approach, and to show the validity of the proposed synthesis method. 

Keywords: robust control, $H_\infty$ control, optimization problems, genetic algorithms, evolution strategies.

## 1. INTRODUCTION

The mixed $H_2 / H_\infty$ control problem consists of the design of a controller which internally stabilizes the plant and satisfies the robust stability constraint on the $H_\infty$ norm. In the last years the mixed $H_2 / H_\infty$ problem has received attention from the theoretical viewpoint of design methods (Doyle, *et al.*, 1994; Snaizer, 1994). In the recent literature, some researches interesting have been presented to develop practical design method for this problem (Chen, *et al.*, 1995; Lo Bianco and Piazzi, 1997).

Recently Kröhling (1997) has presented a novel formulation for the mixed $H_2 / H_\infty$ optimal control problem. In this paper, the method in Kröhling (1997) is extended to consider the design of robust controllers for uncertainty plants. Firstly, the synthesis of mixed $H_2 / H_\infty$ optimal fixed-structure controller is formulated as the minimization of a nominal $H_2$ performance index subject to a $H_\infty$ robust stability constraint. Next, for the solution of the highly non-linear constrained minimization problem for which conventional optimization techniques cannot be employed to obtain a close solution, a new method based on *ESs* is proposed. Two derandomized *ESs* using appropriate operators and a penalty function are employed: one for minimizing the *ISE* performance index, and another for maximizing the robust stability constraint.

This paper is organized as follows: section 2 describes the problem; section 3 is devoted to a brief discussion of *ESs*; in section 4 the design method using *ESs* is presented; in section 5 a detailed design example is given and section 6 presents the conclusion.

## 2. PROBLEM DESCRIPTION

In the context of single-input single-output linear time-invariant systems (Merz and Jaschek, 1996), consider the feedback control system shown in Fig. 1. The fixed-structure controller is described by a rational transfer function $C(s,K)$ where $K := [k_1, k_2, \ldots, k_m]^T$ designates the vector of the controller parameters. It is assumed that the plant to be controlled is described by the nominal transfer function $G(s)$ and undergoes a perturbation described by $\Delta G(s)$, which is assumed to be stable and limited by

$$|\Delta G(jw)| < |W(jw)|, \quad \forall w \in [0, \infty) \tag{1}$$

where the function $W(s)$ is stable and known. In this paper is considered that the plant is described by the multiplicative uncertainty model $G$ given by

$$G := \{G(s,\Delta) = G(s)\,(1+\Delta G(s))\} \tag{2}$$

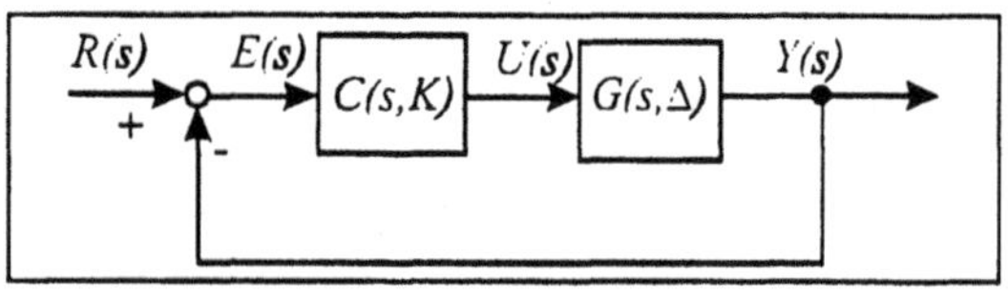


Fig. 1. Unity-feedback control system.

It is required that $\Delta G(s)$ does not cancel unstable poles of $G(s)$ in forming $G(s,\Delta)$. The robust stability result in Doyle, *et al.* (1992) reveals that if the nominal feedback system ($\Delta G(s) = 0$) is internally stable for a controller $C(s,K)$, then $C(s,K)$ provide robust stability if and only if

$$\left\| \frac{C(s,K)G(s)\Delta G(s)}{1+C(s,K)G(s)} \right\|_\infty < 1 \tag{3}$$

From the definition of the $H_\infty$ norm, the robust stability constraint (3) can be written as:

$$\max_{w \in [0,\infty)} \sqrt{\alpha(w,K)} < 1, \text{ where} \tag{4}$$

$$\alpha(w,K) = \frac{\text{num}(w,K)}{\text{den}(w,K)} = \frac{\left(C(jw,K)G(jw)W(jw)\right)\left(C(-jw,K)G(-jw)W(-jw)\right)}{\left(1+C(jw,K)G(jw)\right)\left(1+C(-jw,K)G(-jw)\right)}$$

The robust stability constraint (4) can be rewritten as:

$$\sqrt{\max_{w \in [0,\infty)} \alpha(w,K)} < 1 \tag{5}$$

The function $\alpha(w,K)$ can be expressed in the form:

$$\alpha(w,K) = \frac{\text{num}(w,K)}{\text{den}(w,K)} = \frac{\sum_{i=0}^{p} a_i(K) w^{2i}}{\sum_{i=0}^{q} b_i(K) w^{2i}} \tag{6}$$

Both polynomials $\text{num}(w,K)$ and $\text{den}(w,K)$ have both only even power of $w$, and the coefficients $a_i(.)$ and $b_i(.)$ are functions of $K$.

Assumption 1: In the equation (6) the degree of $\text{num}(w,K) < \text{den}(w,K)$, i.e., $p < q$, so that the function $\alpha(w,K)$ have finite value for $w \geq 0$ and tends toward zero for $w \to \infty$. This implicates that the product of $C(s,K) . G(s)$ and $W(s)$ must be strictly proper and proper rational functions, respectively.

The mixed $H_2 / H_\infty$ optimal control problem consists of minimizing the $H_2$ nominal performance index which internally stabilizes a plant $G(s)$ and satisfies a robust stability constraint (5). Let the nominal $H_2$ performance be the *ISE* given by

$$J = \int_0^\infty e^2(t)\,dt \tag{7}$$

The error signal $E(s)$ for a control system as the one shown in Fig. 1 is given by

$$E(s) = \frac{R(s)}{1+C(s,K)G(s)} \tag{8}$$

where $R(s)$ is the unit step reference signal. Using the Parseval theorem an alternative representation in the frequency domain for $J$ is given by

$$J = -\frac{1}{2\pi j} \int_{-j\infty}^{+j\infty} E(s)\,E(-s)\,ds \tag{9}$$

Let $E(s) = \frac{C(s)}{D(s)}$ where $D(s) = \sum_{i=0}^{n} d_i s^i$, $C(s) = \sum_{i=0}^{n-1} c_i s^i$ and $n$ is the degree of the polynomial $D(s)$.

Assumption 2: The degree of the polynomial $C(s)$ must be smaller or equal to $n-1$ for the integral $J$ in the equation (9) to be finite. The equation (9) can be written as

$$J_n - \frac{1}{2\pi j} \int_{-j\infty}^{+j\infty} \frac{\left(\sum_{i=0}^{n-1} c_i s^i\right)}{\left(\sum_{i=0}^{n} d_i s^n\right)} \frac{\left(\sum_{i=0}^{n-1} c_i (-s)^i\right)}{\left(\sum_{i=0}^{n} d_i (-s)^i\right)} ds \tag{10}$$

Equation (10) can be solved using the residue theorem. Expressions for $J_n$ as functions of the coefficients of the error transfer function $E(s)$, are found in Newton, *et al.* (1957).

Remark: The evaluation of the performance index *ISE* hinges on the stability of the closed loop system. Consequently, the value of $J_n$ is always positive as it should be for stable systems. In Jury (1974) is indicated that stability and evaluation of $J_n$ can be performed simultaneously. If the stability test fails, then the computation of $J_n$ can be terminated.

Since $E(s)$ contains the parameters of the controller, the value of $J_n$ can be minimized adjusting the vector of parameters. Then the mixed $H_2 / H_\infty$ control problem can be stated as follows:

$$\min_{K \text{ stabilizing}} J_n(K) \quad \text{subject to} \quad \sqrt{\max_{w\in[0,\infty)} \alpha(w,K)} < 1 \qquad (11)$$

Hence, the problem of the synthesis of the controller is now one of how to solve the constrained minimization problem above. In the following the employment of *ESs* for the solution of the optimization problem (11) is investigated.

## 3. EVOLUTION STRATEGIES

Similar to *GAs*, *ESs* are algorithms which imitate the principles of natural evolution as a method to solve parameter optimization problems. However, *ES* adopts biological features of the poligeny and pleiotropy, it uses the mutation operator as main operator, it works directly with real numbers, and it allows self-adaptation of strategy parameters through standard deviation and covariance (Beyer, 1995).

Comparable to other optimization techniques, the performance of *ESs* depends on a suitable choice of internal strategy control parameters. Apart from a fixed setting, *ESs* facilitate an adjustment of such parameters within a self-adaptation process, while in *GAs* the control parameters are adjust by trial and error method.

The self-adaptation of strategy parameters provides one of the main features of the success of *ESs*, because *ESs* use evolutionary principles to search in the space of object variables and strategy parameters simultaneously. In the $(\mu+\lambda)$-*ES* case, the parental generation is taken into account during selection, while in the $(\mu,\lambda)$-*ES* case only offspring undergoes selection, and the parent die off. In this paper, two derandomized $(1+\lambda)$-*ESs* with individual step-sizes and correlated mutations are used (Coelho and Coelho, 1998; Ostermeier, *et al.*, 1993).

The implemented $(1+\lambda)$-*ESs* try to derandomize the adaptation process by exploiting information gathered in preceding iterations. Derandomized adaptation usually takes place without mutation of the strategy parameters itself. It uses selected points (more specific: selected mutation steps) in object parameter space for strategy parameter adjustment.
The derandomized $(1+\lambda)$-*ESs* have start point chosen randomly with uniform probability distribution, and they use two operators: selection and mutation. The recombination operator is not utilized because $\mu$=1 is adopted.

*Selection:* The selection (adaptation) operator is completely deterministic; it just chooses the best fitted $\mu$ individuals ($1 \le \mu < \lambda$) out of the set of $\lambda$ offspring individuals. The individual ($\mu$=1) with the highest fitness advances to the next generation, i.e.,

$$K_\mu(t+1) = K_{\lambda sel}(t) \qquad (12)$$

where $K_\mu(t)$ is the parameter vector of generation $t$, *sel* marks the quantities of the *sel*ected offspring of generation $t$.

*Mutation:* The mutation operator is the main operator and introduces random modification into the population. The implemented algorithm realizes mutation ellipsoids in the direction of the selected offspring. The creation of $\lambda$ offspring is given by:

$$K_{\lambda_i}(t) = K_\mu(t) + \delta_{scal}(t)\cdot(\xi_i\delta_{iso}(t)Z_i + \psi_i\delta_{ani}(t)A_0(t)) \qquad (13)$$

where $i$=[1;$\lambda$]; $\delta_{scal}(t)$ are *scal*ing factors of generation $t$, $\xi_i$ is the step-size changing factor (1,0.1)-normally distributed, $\delta_{iso}(t)$ is the step-size of *iso*tropic mutations of generation $t$; $Z_i$ is the random vector with components (0,1)-normally distributed, $\psi_i$ is (0,1)-normally distributed, $\delta_{ani}(t)$ is the step-size of anisotropic (correlated) mutations of generation $t$, $A_0(t)$ is the orientation of long axis of mutation ellipse in generation $t$ (Ostermeier, *et al.*, 1993). The domain of the variable $k_j$ (object variable vector) is given by $k_j = [\underline{k}_j \ \overline{k}_j]$, where $\underline{k}_j$ and $\overline{k}_j$ denote the lower and upper bound of the variable $k_j$, respectively.

## 4. OPTIMIZATION USING EVOLUTION STRATEGIES AND THE PROPOSED METHOD

Two derandomized $(1+\lambda)$-*ESs* are employed for the solution of the constrained optimization problem, i.e., *ES_1* to minimize the nominal $H_2$ performance index $J_n(K)$, and *ES_2* to maximize the robust stability constraint $\alpha(w,K)$ as depicted in Fig. 2. Initially, *ES_1* is started with the controller parameters within the search domain as specified by the designer. These parameters are then send to *ES_2* which is initialized with the variable frequency $w$.

*ES_2* maximizes the restriction function (robust stability) during a fixed number of generations for each individual of *ES_1*. Next, if the maximum value of the robust stability constraint is larger than one, a penalizing value will be associated to the corresponding individual of *ES_1*.

Individuals of *ES_1* who satisfy the robust stability constraint, will not be penalized. In the evaluation of the fitness function of *ES_1* individuals with higher fitness value are automatically selected and those penalized will not survive the evolutionary process. Before a *ES* is utilized for the purpose of a design it is necessary to take the following points into account: i) an evolutionary representation of solutions to the problem at hand, ii) a way to create the initial population and, iii) an evaluation function that determines the fitness of each individual.

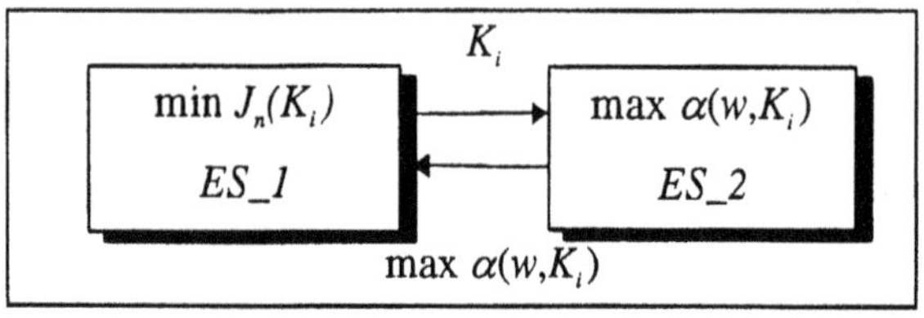


Fig. 2. Optimization with two $(1+\lambda)$-*ESs*.

*4.1 Representation*

In the real-coded representation each vector of individuals is codified as a vector of object variables (representations of solutions) and a set of strategy parameters denoted by equation (13). For the design problem at hand, the object variables (parameters of the controller) have been codified in floating point and concatenated in a individual for *ES_1*. For *ES_2* an individual consists of only a frequency. The *ESs* have been initialized randomly.

*4.2 Fitness function*

In this work a penalty function is employed. The technique of penalty functions transforms the constrained optimization problem into an unconstrained optimization problem, penalizing infeasible individuals. Since *ES* maximize the performance, it is necessary to convert the minimization problem into a maximization problem, i.e., $\min(J_n) = -\max(-J_n)$. Let the integral performance index be $J_n(K)$ then the value of the fitness of each individual of *ES_1*, $K_i (i = 1, \ldots, N)$, is determined by the evaluation function, denoted by $Eval\_1\,(K_i)$ given by:

$$Eval\_1\,(K_i) = -J_n(K_i) - P(K_i) \qquad (14)$$

and $N$ denotes the population size of *ES_1*. The value of the penalty term is discussed in the following. Let the robust stability constraint be $\alpha(w, K)$ then the value of the fitness of each chromosome of *ES_2*, $w_j\ (j = 1, \ldots, M)$, is determined by the evaluation function, denoted by $Eval\_2\,(w_j)$ given by:

$$Eval\_2\,(w_j) = \alpha(w_j, K_i) \qquad (15)$$

and $M$ denotes the population size of *ES_2*. The penalty value for the individual $K_i$ is calculated based on squared-distance, i.e.,

$$P(K_i) = \begin{cases} 0, & \text{if } \sqrt{\max \alpha(w, K_i)} < 1 \\ C_1 \left(\sqrt{\max \alpha(w, K_i)}\right)^2, & \text{if } \sqrt{\max \alpha(w, K_i)} > 1 \\ C_2, & \text{if } K_i \text{ is unstable} \end{cases} \qquad (16)$$

If the individual $K_i$ does not satisfy the stability test applied to the characteristic equation of the system, then $K_i$ is an unstable individual and it is penalized with a very large positive constant $C_2$.and automatically $K_i$ do not survive the evolutionary process. In case the individual $K_i$ satisfies the stability test, but not the robust stability constraint, then it is an infeasible individual, and it is penalized with $C_1 \left(\sqrt{\max \alpha(w, K_i)}\right)^2$, where $C_1$ is a positive constant to be adjusted. Otherwise the individual $K_i$ is feasible and it is not penalized.

*4.3 Proposed method*

In the following the proposed method is described in the form of an algorithm.

*Step 1:* Given a plant with transfer function *G(s)* and a fixed-structure controller with transfer function $C(s,K)$, obtain the transfer function of the error *E(s)*. Given a weighting function $W(s)$, determine the robust stability constraint $\alpha(w, K)$

*Step 2:* Specify the lower and upper bounds of the controller parameters.

*Step 3:* Set up *ES_1* and *ES_2* parameters: number of parents, number of offspring, population size, and maximum number of generations.

*Step 4:* Initialize the populations of *ES_1*, $K_i (i = 1, \ldots, N)$, and *ES_2*, $w_j\ (j = 1, \ldots, M)$, and set the generation number of *ES_1* to $g$=1, (the index $g$ denotes the number of generations for *ES_1*).

*Step 5:* For each individual $K_i$ of the *ES_1* population, calculate the maximum value of $\alpha(w, K_i)$ using *ES_2*. If no individuals of the *ES_1* population satisfy the constraint $\sqrt{\max \alpha(w; K_i)} < 1$ then a feasible solution is assumed to be non-existent and the algorithm stops. In this case, a new controller structure has to be assumed.

*Step 6:* Calculate the fitness value for each individual $K_i$ of *ES_1* by using equations (14) and (16).

*Step 7:* Select individuals using deterministic selection and apply evolutionary operator of mutation to the individuals of *ES_1* population.

*Step 8:* For each individual $K_i$ of the *ES_1*, calculate $(\max\ \alpha(w, K_i))$ using *ES_2*.

*Step 9:* If maximum number of generation $g$ of *ES_1* is achieved, stop, otherwise set $g$=$g$+1, and go to step 6.

The calculation of the maximum value of the constraint $(\max\ \alpha(w, K_i))$ is done by *ES_2* and consists of the following sub-steps:

*Sub-step a:* Initialize the gene of each individual $w_j\ (j = 1,....,M)$ in the population and set the generation number to $h$=1, (the index $h$ indicates the number of generations for *ES_2*).

*Sub-step b:* Evaluate the fitness of each individual by using equation (15), select individuals using deterministic selection and apply evolutionary operator of mutation to the individuals.

*Sub-step c:* If maximum number of generation $h$ is achieved, stop and return the fitness of the best individual $(\max\ \alpha(w, K_i))$ to *ES_1*, otherwise set $h$=$h$+1, and go to sub-step b.

## 5. DESIGN EXAMPLE

For the purpose of comparison the recent example given by Lo Bianco and Piazzi (1997) is considered. The plant is described by the transfer function

$$G(s) = \frac{1.8}{s^2(s+2)} \tag{17}$$

The plant uncertainty is expressed by

$$\Delta G(s) = W(s) = \frac{0.1}{s^2 + 0.1s + 10} \tag{18}$$

The fixed-structure controller examined is given by the following transfer function

$$C(s) = k_1 \frac{s^2 + 2k_4k_5s + k_5^2}{(s+k_2)(s+k_3)} \tag{19}$$

For the control system as shown in Fig. 1, the error transfer function, *E(s)*, the performance index, $J_5$, and the robust stability constraint are given in the appendix. The parameters of the controller are searched into the following bounds:

$k_1$ = *[1, 1000]*; $k_2$ = *[1, 100]*; $k_3$ = *[1, 100]*;
$k_4$ = *[0, 1]*; $k_5$ = *[0.1, 100]*.

<u>*ES parameters setting:*</u>

- *ES_1:* $\mu_1$=1, $\lambda_1$=19, $N$=20, $g$=70;
- *ES_2:* $\mu_2$=1, $\lambda_2$=14, $M$=15, $h$=30,
- penalty constant: $C_1$=100 and $C_2$=1000000.

The proposed method using (1+$\lambda$)-*ES*, as described in the previous section, has been applied to the design of the mixed $H_2 / H_\infty$ fixed-structure controller. The best controller parameters for this example obtained by the minimization of the performance index, $J_5$, subject to the robust stability constraint $\sqrt{\max \alpha(w, K_i)} < 1$ using the proposed method is given by: $K^* = [1000; 14.515; 14.523; 1; 0.541]$

The convergence of the performance index (best individual) by using *ES_1* is shown in Fig. 3. The minimal value of the performance index after 70 generations is $J_5(K^*)$=0.1507. The results obtained by the proposed method correspond to those of Lo Bianco and Piazzi (1997).

For the solution of the constrained optimization problem the use of *ES* has shown particularly suitable for the exploration of a search space with continuous variables on large domains without numerical precision problems. The proposed method has been used with different number of parents and offspring and the same results are obtained showing a very robust characteristic.

In order to test controller design based on the proposed method, a unit step signal is applied to the system in Fig. 1. The closed-loop response for both cases: a) nominal plant and b) with plant uncertainty $\Delta G(s)$ is shown in Fig. 4. The results of the Fig. 4 show that the controller obtained by the proposed method can control the perturbed plant $G(s, \Delta)$ successfully.

## 6. CONCLUSION

This paper presented an intelligent method for synthesis of mixed $H_2 / H_\infty$ fixed-structure controllers. The problem was formulated as an optimization problem with constraint of type $H_\infty$ norm. A method based on two derandomized (1+$\lambda$)-*ESs* with appropriate operators and a penalty function was proposed for the solution of the problem. An *ES* was used to minimize the nominal $H_2$ performance index and another calculates the maximum value of the $H_\infty$ robust stability constraint.

The validity of the novel method was demonstrated by a detailed design example and the results compared to those obtained by Lo Bianco and Piazzi (1997). The results obtained so far demonstrated clearly the potential of intelligent methods based on evolutionary computation for synthesis of mixed $H_2 / H_\infty$ fixed-structure controller. The method presented can be applied to the design of controllers for plants with structured uncertainties, i.e., interval plants. Future work will be dedicated in this direction.

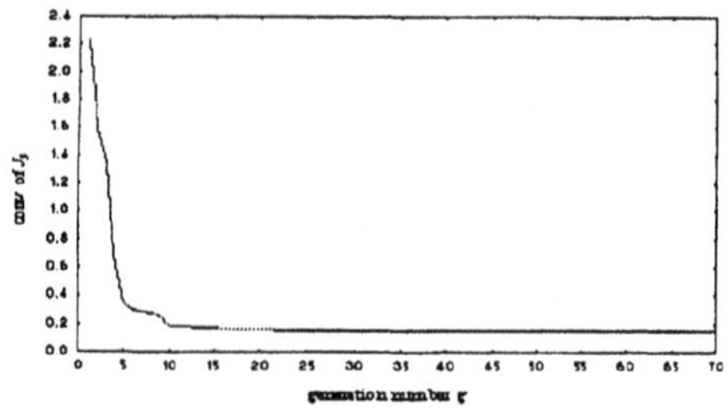


Fig. 3. Convergence of the *ISE* (*ES_1*).

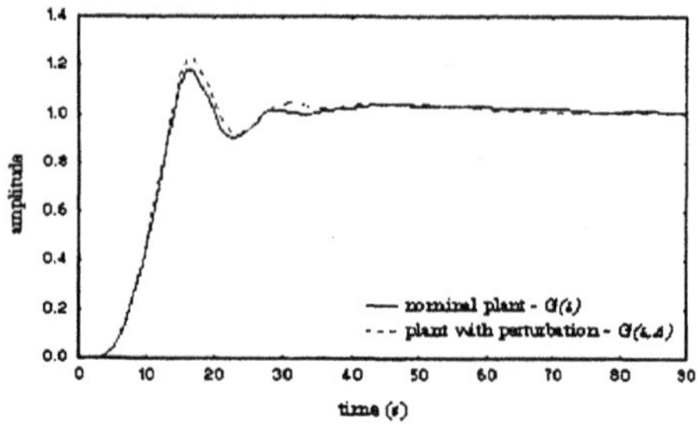


Fig. 4. Closed-loop step response of Fig. 1.

## ACKNOWLEDGMENTS

The authors wish to express their gratitude to Prof. Dr.-Ing. H. Jaschek.

## REFERENCES

Beyer, H.-G. (1995). Toward a Theory of Evolution Strategies: Self-Adaptation. *Evolutionary Computation*, **3**, 311-348.

Coelho, L.S. and A.A.R. Coelho (1998). Genetic Algorithms and Evolution Strategies Applied in Identification and Control: Case Study, In: *Soft Computing in Eng. Design and Manufacturing* (Chawdhry, P.K., R. Roy and R.K. Pant (Ed.)), 430-438. Springer-Verlag, London.

Chen, B.-S., Y.-M. Cheng and C.-H. Lee (1995). A Genetic Approach to Mixed $H_2 / H_\infty$ Optimal PID Control. *IEEE Control Systems*, **15**, 51-60.

Doyle, J.C., K. Zhou, K. Glover and B. Boddenheimer (1994). Mixed $H_2$ and $H_\infty$ Performance Objectives II: Optimal Control. *IEEE Trans. on Automatic Control*, **39**, 1575-1587.

Doyle, J.C., B.A. Francis and A.R. Tannenbaum (1992). *Feedback Control Theory.* Macmillan Publishing, New York, USA.

Jury, E.I. (1974). *Inners and the Stability of Dynamic Systems*. John Wiley, USA.

Krohling, R.A. (1997). Synthesis of PID Controllers for Disturbance Rejection: a Real-Coded Genetic Algorithms Approach. *European Conf. on Intelligent Tech.*, Aachen, Germany, 862-868.

Lo Bianco, C.G. and A. Piazzi (1997). Mixed $H_2 / H_\infty$ Fixed-Structure Control via Semi-Infinite Optimization. *7th IFAC Int. Symp. on CACSD*, Gent, Belgium, 329-334.

Merz, L. and H. Jaschek (1996). *Grundkurs der Regelungstechnik.* R. Oldenburg Verlag, 13 edition, München, Germany.

Newton, G.C., L.A. Gould and J.F. Kaiser (1957). *Analytic Design of Linear Feedback Controls.* John Wiley & Sons. New York, USA.

Ostemeier, A., A. Gawelczyk and N. Hansen (1993). A Derandomized Approach to Self-Adaptation of Evolution Strategies. *Evolutionary Computation*, **2**, 369-380.

Snaizer, M. (1994). An Exact Solution to General SISO Mixed $H_2 / H_\infty$ Problems via Convex Optimization. *IEEE Trans. on Automatic Control*, **39**, 2511-2517.

## APPENDIX

The error signal is expressed as:

$$E(s) = \frac{c_4 s^4 + c_3 s^3 + c_2 s^2 + c_1 s + c_0}{d_5 s^5 + d_4 s^4 + d_3 s^3 + d_2 s^2 + d_1 s + d_0}, \text{ where}$$

$$c_0 = 0, c_1 = 2k_2k_3, c_2 = 2(k_2 + k_3) + k_2k_3,$$
$$c_3 = 2 + k_2 + k_3, c_4 = 1$$

$$d_0 = 1.8k_1k_5^2, d_1 = 3.6k_1k_4k_5, d_2 = 2k_2k_3 + 1.8k_1,$$
$$d_3 = 2(k_2 + k_3) + k_2k_3, d_4 = 2 + k_2 + k_3, \; d_5 = 1$$

The performance index, $J_5$, for this case is given by (Newton, *et al.*, 1957):

$$J_5 = \frac{A}{2m_5}, \text{ where}$$

$$A = c_4^2 m_0 + (c_3^2 - 2c_2c_4)m_1 + (c_2^2 - 2c_1c_3 + 2c_0c_4)m_2 +$$
$$(c_1^2 - 2c_0c_2)m_3 + c_0^2 m_4,$$
$$m_1 = -d_0d_3 + d_1d_2, \; m_2 = -d_0d_5 + d_1d_4,$$
$$m_0 = (d_3m_1 - d_1m_2)/d_5, \; m_3 = (d_2m_2 - d_4m_1)/d_0,$$
$$m_4 = (d_2m_3 - d_4m_2)/d_0, \; m_5 = d_0(d_1m_4 - d_3m_3 + d_5m_2).$$

The robust stability constraint, $\alpha(w, K)$, is determined by using the symbolic manipulator Mathematica and is given by:

$$\alpha(w,K) = \frac{num\,(w,K)}{den\,(w,K)}, \text{ where}$$

$$num(w,K) = 0.0324k_1^2k_5^4 + (-0.0648k_1^2k_5^2 +$$
$$0.1296k_1^2k_4^2k_5^2)w^2 + (0.0324k_1^2)w^4$$

$$den(w,K) = \left(100 + (-20 + 4k_1^2k_4^2k_5^2)w^2 + w^4\right) *$$
$$(3.24k_1^2k_5^4 + (-6.48k_1^2k_5^2 - 7.2k_1k_2k_3k_5^2 +$$
$$12.96k_1^2k_4^2k_5^2)w^2 + (3.24k_1^2 + 7.2k_1k_2k_3 +$$
$$4k_2^2k_3^2 - 14.4k_1k_2k_4k_5 - 14.4k_1k_3k_4k_5 -$$
$$7.2k_1k_2k_3k_4k_5 + 7.2k_1k_5^2 + 3.6k_1k_2k_5^2 + 3.6k_1k_3k_5^2)\,w^4 +$$
$$(-7.2k_1 - 3.6k_1k_2 + 4k_2^2 -$$
$$3.6k_1k_3 + 4k_3^2 + k_2^2k_3^2 + 7.2k_1k_4k_5)w^6 + (4k_2^2k_3^2)w^8 + w^{10})$$

# A NECESSARY AND SUFFICIENT CONDITION FOR ROBUST $H_\infty$ DECENTRALIZED CONTROL VIA STATIC OUTPUT FEEDBACK

**Chen Bing*** **Wei Yongde**** **Zhang Siying***

*P.O.Box 135, Depart of Automation of Northeastern University
Shenyang, Liaoning 110006, P.R.China
** Liaoning University, Shenyang, Liaoning, 110000, P.R.China

**Abstract:** This note focuses on linear large scale systems which are subject to both time-varying norm-bounded parameter uncertainty and exogenous disturbance, andaddresses the following robust $H_\infty$ control problem: presenting a necessary and sufficient condition for the existence of a stabilizing decentralized static output feedback controller that makes the closed-loop systems to be quadratically stable and achieves a prescribed level of disturbance attenuation for all admissible parameter uncertainties.

**Keywords:** output feedback, robust $H_\infty$ control, decentralized quadratic stabilization.

## 1. INTRODUCTION

In recent years, it seems that much attention have been paid to designing $H_\infty$ controller which would be robust against time-varying real parameter uncertain in the controlled system. Since this Problem involves the notions of both robust stabilization and $H_\infty$ control , we refer to it as robust $H_\infty$ control ( Xie,L.1992 ). Such a problem in the linear state feedback setting has been completely solved by using a Riccati equation approach (Xie,L 1990;1992). In the case of output feedback, sufficient conditions for designing robust $H_\infty$ controller with a fixed order have beenderived (Madjwale,1989), while Veillette (1989)has developed $H_\infty$ controller for systems with uncertainty in the state matrix only. Then, for the case of the linear dynamic output feedback, uncertain system with parameter uncertainties appearing in all state, input and output

This work was supported in part by the National_Natural Science Foundation of China(69774005),in part by the University Doctoral Foundation of chinese State Education Commission(97014508),in part by the National Key Project of China and in part by the Natural Science Foundation of Liaoning Education Commission (9709211121) matrix were considered, and a necessaryand sufficient condition was proposed(Xie,1992). To date, little attention has been paid to the problem of robust $H_\infty$ control via static output feedback, and to the problem of robust $H_\infty$ decentralized control for uncertain large-scale systems. In this note, we consider uncertain linear composite systems with parameter uncertainties appearing the all state, input and interconnections matrices. The attention is focused on the robust $H_\infty$ decentralized control via static output feedback . Our main result presents a necessary and sufficient condition of the existtence of robust $H_\infty$ decentralized static output feedback conytoller.

## 2. PROPLEM STATEMENT AND DEFINITIONS

Consider the large-scale systems composed of N interconnected subsystems

$$\dot{x}_i = (A_i + \Delta A_i)x_i + (B_i + \Delta B_i)u_i + \sum_{j=1,j\neq i}^{N} H_{ij}x_j + B_{i1}\omega_i \qquad (2.1\text{ a})$$

$$z_i = C_{i1}x_i + D_{i1}u_i \qquad (2.1\text{b})$$

$$y_i = C_{i2}x_i \qquad (2.1\text{c})$$

where, $x_i \in R^n$ is the state, $u_i \in R^m$ is the control input, $\omega_i \in R^q$ is the disturbance input, $y_i \in R^m$ is the measured output, $z_i \in R^p$ is the controlled output, respectively. $A_i, B_i, B_{io}, C_{i1}, C_{i2}, M_i$ and $H_j$ are real constant matrices of appropriate dimensions that described the ith subsystem. $\Delta A_i, \Delta B_i$ and $L_{ij}(t)$ are real-valued matrix functions representing time-varying parameter uncertainties. The parameter uncertainties considered here are norm-bounded and of the form

$$\left[\Delta A_i \; \Delta B_i\right] = D_i F_i(.)\left[E_{i1} \; E_{i2}\right] \qquad (2.2\text{a})$$

$$H_{ij} = M_i L_{ij}(.)H_j \qquad (2,2\text{b})$$

where, $D_i, E_{i1}, E_{i2}, M_i$ and $H_j$ are all known constant matrices with appropriate dimensions and $F_i(t)$ $L_{ij}(t)$are unknown matrix functions satisfying the following inequalities

$$F_i^T(t)F_i(t) \le I \qquad (2.\ 3\text{a})$$

$$G_i^T(t)G_i(t) \le I \qquad (2.\ 3\text{b})$$

where, $G_i(t) = \left[L_{i1}(t) \ldots L_{ii-1}(t) \; L_{ii+1}(t) \ldots L_{iN}(t)\right]$ with the element of $F_i(t)$ and $L_{ij}(t)$ being Lebesgue measurable. Then, for the sake of convenience, system (2.1) can be rewritten as the following form

$$x = (A + \Delta A + A_o)x + (B + \Delta B)u + B_1\omega \qquad (2.1\text{a}')$$

$$z = C_1 x + D_1 u \qquad (2.1\text{b}')$$

$$y = C_2 x \qquad (2.1\text{c}')$$

where,

$$A_o = \begin{bmatrix} 0 & A_{12} & \cdots & A_{1N-1} & A_{1N} \\ & \ddots & & & \\ & & & \ddots & \\ A_{N1} & A_{N2} & \cdots & A_{NN-1} & 0 \end{bmatrix}$$

$$A_{ij} = M_i F_{ij}(t) H_i \quad (\text{i,j}=1,2,\ldots,\text{N};\; j \neq i)$$

his note addresses the problem of designing the static output feedback controller for the system (2.1) such that the closed-loop system is quadratically stable and achieves a prescribed level of disturbance attenuation in the $H_\infty$-norm sense for all admissible uncertainties which satisfying (2.3). To motivate the technique used in this note, we first recall the notions of quadratic stability and disturbance attenuation. Let us consider a system with a parameter uncertainty of the following form

$$\Sigma: \begin{cases} x = (A + \Delta A)x + B\omega \\ z = Cx \end{cases}$$

**Definition 2.1** : The system $\Sigma$ is said to be quadratically stable if there exist a positive definite symmetric matrix $P$ such that for all admissible uncertainty $\Delta A$

$$(A + \Delta A)^T P + P(A + \Delta A) < 0$$

Similarly, a uncertain system is said to be quadratically stablizable via state output feedback if there exists a matrix $K$ such that with $u = Ky$ the resulting closed-loop system is quadratically stable. In order to guarantee an performance of system for all admissible parameter uncertainties, we adopt the technique used in quadratic stability, i.e., we incorporate $\Delta A(.)$ in system $\Sigma$. This leads to the notion of quadratic stability with disturbance attenuation.

**Definition 2.2** : give a scalar $r > 0$, the system $\Sigma$ is said to be quadratic stable with disturbance attenuation $r$ if there exists a symmetric definite matrix $P$ such that for all admissible uncertainty $\Delta A$

$$(A+\Delta A)^T P + P(A+\Delta A) + r^{-2}PBB^TP + C^TC < 0$$

Similarly, give a scalar $r>0$ , a uncertain system is said to be quadratically stabilizable with disturbance attenuation $r$ via static output feedback if there exists a matrix $K$ such that $u=Ky$ with the resulting closed-loop system is quadratically stable with disturbance attenuation $r$.

**Remark 2.1:** The notion of quadratic stability with disturbance attenuation is a natural extension of quadratic stability to incorporate $H_\infty$ performance and according to the result (Xie, 1992), suppose the system $\Sigma$ is quadratically stable with disturbance attenuation $\dot{r}>0$, then, this system is quadratically stable. Moreover, with zero-initial condition for $x(t)$, $\|z\|_2 < r\|\omega\|_2$ for all admissible $\Delta A(.)$ and all non zero $\omega \in L_2[0,\infty)$. where, $\|.\|$ denotes the usual $L_2[0,\infty)$-norm.

## 3. MAIN RESULT

The approach adopted in this note to solve the robust $H_\infty$ decentralized control problem involves solving a set of parameter-certain algebraic Riccati equations associated with disturbance attenuation $r>0$. given the system (2.1) and any desired $H_\infty$-norm bound constraint $r>0$(or disturbance attenuation), we define the following algebraic Riccati equation corresponding to the problem of quadratic stabilization with disturbance attenuation $r$,for i=1,2,...,N

$$\begin{aligned}&A_i^TP_i + P_iA_i + r^{-2}P_iB_{i1}B_{i1}^TP_i \\ &\quad + \varepsilon_iP_iD_iD_i^TP_i + \varepsilon_i^{-1}E_{i1}^TE_{i1} + C_i^TC_i \\ &\quad + \eta_iP_iM_iM_i^TP_i + (\sum_{j\neq i}^{N}\eta_i^{-1})H_i^{-1}H_i < 0\end{aligned} \tag{3.1}$$

where, $\varepsilon_i>0, \eta_i>0$ are the design parameters. We first establish a key theorem which presents a necessary and sufficient condition for the free system of system (2.1) to be quadratically stable with disturbance attenuation $r$.

**Theorem 3.1:** Let $r>0$ be given. Then, the free system of system (2.1) is to be quadratically stable with disturbance attenuation $r$ if and only if there exist positive constants $\varepsilon_i>0, \eta_i>0$ (i=1,2,...,N)such that inequalities (3.1) have a positive definite symmetric solution $P_i$ (i=1,2,...,N)

**Proof:** Let $P=diag[P_1 \dots P_N]$, then

$$\begin{aligned}&A_o^TP+PA_o = A_o^T\{diag[P_10\dots0]+\dots+diag[0\dots0P_N]\} \\ &\quad +\{diag[P_10\dots0]+\dots+diag[0\dots0P_N]\}A_o \\ &\le \eta_1 diag[P_1M_1M_1^TP_10\dots0]+\frac{1}{\eta_1}diag[0H_2^TH_2\dots H_N^TH_N] \\ &+\dots\dots \\ &+\eta_N diag[0\dots0P_NM_NM_N^TP_N]+\frac{1}{\eta_N}diag[H_1^TH_1\dots H_{N-1}^TH_{N-1}0 \\ &= diag[\eta_1I\dots\eta_NI]PMM^TP+diag[\sum_{j\neq1}^{N}\frac{1}{\eta_j}I\dots\sum_{j\neq N}^{N}\frac{1}{\eta_j}I]H^TH\end{aligned} \tag{3.2}$$

It follows from (3.2) that

$$\begin{aligned}&\left[A^T+\Delta A+A_o\right]^TP+P\left[A^T+\Delta A+A_o\right]+r^{-2}PB_1B_1^TP+C_1^TC_1 \\ &= diag[A_1^TP_1+P_1A_1\dots A_N^TP_N+P_NA_N]+A_o^TP+PA_o \\ &+r^{-2}diag[P_1B_{i1}B_{i1}^TP_1\dots P_NB_{iN}B_{iN}^TP_N] \\ &+diag[C_{11}^TC_{11}\dots C_{N1}^TC_{N1}]+diag[(D_1F_1E_{11})^TP_1 \\ &+P_1(D_1F_1E_{11})\dots(D_NF_NE_{N1})^TP_N+P_N(D_NF_NE_{N1})] \\ &\le diag[A_i^TP_i+P_iA_i+r^{-2}P_iB_{i1}B_{i1}^TP_i+\varepsilon_iP_iD_iD_i^TP_i \\ &\quad +\varepsilon^{-1}E_{i1}^TE_{i1} \\ &\quad +C_{i1}^TC_{i1}+\eta_iP_iM_iM_i^TP_i+\sum_{j\neq i}^{N}\eta_i^{-1}H_i^TH_i]\end{aligned} \tag{3.3}$$

by (3.1) and (3.3), we have the free system of (2.1) is quadratically stable with disturbance attenuation.To show the necessity, we suppose that there exists a matrix $P>0$ such that

$$\begin{aligned}&[A+\Delta A+A_o]^TP+P[A+\Delta A+A_o] \\ &\quad +r^{-2}PB_1B_1^TP+C_1^TC_1<0\end{aligned} \tag{3.4}$$

for all admissible uncertainties holds. Let

$$Z=[A+\Delta A]^TP+P[A+\Delta A]+r^{-2}PB_1B_1^TP+C_1^TC_1 \tag{3.5}$$

then, it follows from (3.4) and (3.5) that

$$Z < -A_o^TP-PA_o \tag{3.6}$$

Let

$$\Pi_i = \begin{bmatrix} 0 & \cdots & & \cdots & & \cdots & 0 \\ \vdots & & & & & & \vdots \\ 0 & \cdots & & \cdots & & \cdots & 0 \\ L_{i+1,1} & \cdots & L_{i+1,i} & 0 & L_{i+1,i+2} & \cdots & L_{i+1,N} \\ \vdots & & & & & & \vdots \\ L_{N,1} & \cdots & L_{N,i} & \cdots & \cdots & L_{N,N-1} & 0 \end{bmatrix}$$

$$P = \begin{bmatrix} P_{11} & P_{12} & \cdots & P_{1N} \\ P_{21} & P_{22} & \cdots & P_{2N} \\ \vdots & \vdots & \vdots & \vdots \\ P_{N1} & P_{N2} & \cdots & P_{NN} \end{bmatrix}$$

$$\overline{G}_i = [L_{i1}\dots L_{ii-1}0L_{ii+1}\dots L_{iN}] \quad P^i = [P_{1i}^T\dots P_{Ni}^T]^T$$

$$\bar{x}_i = [x_1^T \ldots x_{i-1}^T, x_{i+1}^T \ldots x_N^T]^T$$

thus, for any $x(\neq 0) \in R^{n \times N}$, we have

$$x^T Z x < -2x^T P A_o = -2x^T P M \Pi_1 H x - 2x^T P M \begin{bmatrix} \bar{G}_1 \\ 0 \end{bmatrix} H x \quad (3.7)$$

or

$$x^T (Z + 2xPM\Pi_1 H) x < -2x^T PM \begin{bmatrix} \bar{G}_1 \\ 0 \end{bmatrix} Hx$$

$$= -2x^T P^1 M_1 G_1 diag[H_2 \ldots H_N] \bar{x}_1 \quad (3.8)$$

for all $\Delta A_i$ and $G_i$ holds. Thus, by inequality (3.8), we have

$$x^T (Z + 2PM\Pi_1 H) x < -2\max\{x^T P^1 M_1 G_1 diag[H_2 \ldots H_N]\bar{x}_1;\ \text{for all } G_1;\ G_1^T G_1 \leq I\} \leq 0 \quad (3.9)$$

Thus, it follows that

$$[x^T (Z + 2PM\Pi_1 H) x]^2 > 4\max\{x^T P^1 M_1 G_1 diag[H_2 \ldots H_N]\bar{x}_1;\ \text{for all } G_1;\ G_1^T G_1 \leq I\}^2$$

$$= 4x^T P^1 M_1 M_1^T [P_{11} \ldots P_{N1}] x x^T diag[0 H_2^T H_2 \ldots H_N^T H_N] x$$

Now, by using lemma 4(Petersen,1986), there exists a constant $\eta_1 > 0$ such that

$$Z + 2PM\Pi_1 H + \eta_1 P^1 M_1 M_1^T [P_{11} \ldots P_{1N}] + \frac{1}{\eta_1} diag[0 H_2^T H_2 \ldots H_N^T H_N] < 0 \quad (3.10)$$

Now, let

$$Z_1 = \eta_1 P^1 M_1 M_1^T [P_{11} \ldots P_{1N}] + \frac{1}{\eta_1} diag[0 H_2^T H_2 \ldots H_N^T H_N] < 0$$

Then

$$Z_1 < -2PM\Pi_1 H$$

so for $x(\neq 0)$, we have

$$x^T Z_1 x < -2x^T PM\Pi_1 Hx = -2x^T PM\Pi_2 Hx - 2x^T PM \begin{bmatrix} 0 \\ \bar{G}_2 \\ 0 \end{bmatrix} Hx$$

using the same way used to prove (3.10), we can find a constant $\eta_2 > 0$ such that

$$Z + \eta_1 P^1 M_1 M_1^T [P_{11} \ldots P_{1N}] + \frac{1}{\eta_1} diag[0 H_2^T H_2 \ldots H_N^T H_N]$$
$$+ 2PM\Pi_2 H + \eta_2 P^2 M_2 M_2^T [P_{21} \ldots P_{2N}]$$
$$+ \frac{1}{\eta_2} diag[H_1^T H_1 0 H_3^T H_3 \ldots H_N^T H_N] < 0$$

Using this way for N times, we can find constants $\eta_1 > 0, \ldots, \eta_N > 0$ such that

$$Z + \sum_{i=1}^{N} \eta_i P^i M_i M_i^T [P_{i1} \ldots P_{iN}] + diag[\sum_{j \neq 1}^{N} \frac{1}{\eta_j} H_1^T H_1 \ldots \sum_{j \neq N}^{N} \frac{1}{\eta_j} H_N^T H_N] < 0 \quad (3.11)$$

Now, take the (i,i) block of (3.11), it follows that

$$(A_i + \Delta A_i)^T P_{ii} + P_{ii}(A_i + \Delta A_i) + r^{-2} P_{ii} B_{i1} B_{i1}^T P_{ii} + C_{i1}^T C_{i1} + \eta_i P_{ii} M_i M_i^T P_{ii} + \sum_{j \neq i}^{N} \frac{1}{\eta_j} H_i^T H_i < 0 \quad (3.12)$$

Again let

$$Z = A_i^T P_{ii} + P_{ii} A_i + r^{-2} P_{ii} B_{i1} B_{i1}^T P_{ii} + C_{i1}^T C_{i1} + \eta_i P_{ii} M_i M_i^T P_{ii} + \sum_{j \neq i}^{N} \frac{1}{\eta_j} H_i^T H_i$$

then, we have

$$Z < -\Delta A_i^T P_{ii} - P_{ii} \Delta A_i$$

Again using the method used in the proof of (3.10), it follows that there exists a constants $\varepsilon_i > 0$ (i=1,2,...,N)such that

$$A_i^T P_{ii} + P_{ii} A_i + r^{-2} P_{ii} B_{i1} B_{i1}^T P_{ii} + \varepsilon_i P_{ii} D_i D_i^T P_{ii} + \varepsilon_i^{-1} E_{i1}^T E_{i1} + C_{i1}^T C_{i1} + \eta_i P_{ii} M_i M_i^T P_{ii} + \sum_{j \neq i}^{N} \eta_j^{-1} H_i^T H_i < 0$$

take $P_{ii} = P_i$,then (3.1) holds.

**Theorem 3.2:** Let $r > 0$ be given. Then system (2.1) is quadratically stabilizable with disturbance attenuation $r > 0$ via a decentralized output feedback law $u_i = K_i y_i$ (i=1,2,...N) iff there exists matrix $Q_i$ and constants $\varepsilon_i > 0$, $\eta_i > 0$ and $\delta_i > 0$ such that the following Riccati inequality has a solution $P_i = P_i^T > 0$ (i=1,2,...,N)

$$A_i^T P_i + P_i A_i - \delta_i^2 P_i B_i B_i^T P_i + \varepsilon_i P_i D_i D_i^T P_i + r^{-2} P_i B_{i1} B_{i1}^T P_i + \eta_i P_i M_i M_i^T P_i + \sum_{j \neq i}^{N} \eta_j^{-1} H_i^T H_i + Q_i^T Q_i < 0 \quad (3.14)$$

where

$$Q_i = \begin{bmatrix} C_{i1} + D_{i1} F_i C_{i2} \\ \frac{1}{\sqrt{\varepsilon_i}}(E_{i1} + E_{i2} F_i C_{i2}) \\ \delta_i B_i^T P_i + \frac{1}{\delta_i} F_i C_{i2} \end{bmatrix} \quad (3.15)$$

**Proof :** we suppose that decentralized output feedback law $u_i = K_i y_i$ such that the resulting closed-loop system is quadratically stable with disturbance attenuation $r$. It follows from theorem 3.1 that there exist $P_i$ (i=1,2,...,N) such that

$$
\begin{aligned}
0 > {} & [A_i + B_i K_i C_{i2}]^T P_i + P_i[A_i + B_i K_i C_{i2}] \\
& + r^{-2} P_i B_{i1} B_{i1}^T P_i + \varepsilon_i P_i D_i D_i^T P_i \\
& + \varepsilon_i^{-1}[E_{i1} + E_{i2} K_i C_{i2}]^T [E_{i1} + E_{i2} K_i C_{i2}] \\
& + \eta_i P_i M_i M_i^T P_i + \sum_{j \neq i}^{N} \eta_i^{-1} H_i^T H_i \\
& + [C_{i1} + D_{i1} K_i C_{i2}]^T [C_{i1} + D_{i1} K_i C_{i2}] \qquad (3.16)
\end{aligned}
$$

By applying lemma 3 in [4] to (3.16), there exists a constant $\delta_i > 0$ such that

$$
\begin{aligned}
0 > {} & [A_i + B_i K_i C_{i2}]^T P_i + P_i[A_i + B_i K_i C_{i2}] \\
& + r^{-2} P_i B_{i1} B_{i1}^T P_i + \varepsilon_i P_i D_i D_i^T P_i + \\
& + \varepsilon_i^{-1}[E_{i1} + E_{i2} K_i C_{i2}]^T [E_{i1} + E_{i2} K_i C_{i2}] \\
& + \eta_i P_i M_i M_i^T P_i + \sum_{j \neq i}^{N} \eta_i^{-1} H_i^T H_i \\
& + [C_{i1} + D_{i1} K_i C_{i2}]^T [C_{i1} + D_{i1} K_i C_{i2}] \\
& + \delta_i^{-1} C_{i2}^T K_i^T K_i C_{i2} \\
= {} & A_i^T P_i + P_i A_i - \delta_i^2 P_i B_i B_i^T P_i + \varepsilon_i P_i D_i D_i^T P_i \\
& + r^{-2} P_i B_{i1} B_{i1}^T P_i \\
& + \eta_i P_i M_i M_i^T P_i + \sum_{j \neq i}^{N} \eta_i^{-1} H_i^T H_i + Q_i^T Q_i < 0
\end{aligned}
$$

To show the sufficiency, we propose that (3.14) holds. Thus

$$
\begin{aligned}
& [A_i + B_i K_i C_{i2}]^T P_i + P_i[A_i + B_i K_i C_{i2}] \\
& + r^{-2} P_i B_{i1} B_{i1}^T P_i + \varepsilon_i P_i D_i D_i^T P_i + \\
& + \varepsilon_i^{-1}[E_{i1} + E_{i2} K_i C_{i2}]^T [E_{i1} + E_{i2} K_i C_{i2}] \\
& + \eta_i P_i M_i M_i^T P_i + \sum_{j \neq i}^{N} \eta_i^{-1} H_i^T H_i + \\
& + [C_{i1} + D_{i1} K_i C_{i2}]^T [C_{i1} + D_{i1} K_i C_{i2}] \\
\le {} & A_i^T P_i + P_i A_i - \delta_i^2 P_i B_i B_i^T P_i + \varepsilon_i P_i D_i D_i^T P_i + r^{-2} P_i B_{i1} B_{i1}^T P_i \\
& + \eta_i P_i M_i M_i^T P_i + \sum_{j \neq i}^{N} \eta_i^{-1} H_i^T H_i + Q_i^T Q_i < 0 \qquad (3.17)
\end{aligned}
$$

By theorem 3.1, it follows that the closed-loop system is quadratically stable with disturbance attenuation $r$.

## 4. CONCLUSION

This note has considered the problem of quadratical stabilization with disturbance attenuation via decentralized output feedback for a parameter uncertain large-scale systems. A necessary and sufficient condition for quadratically stabilizable with disturbance attenuation via decentralized output feedback had been obtained. And based on this condition a algorithm to calculating one particular output stabilizing gain is proposed. The convergence remains to be proved, however.

## REFERENCE


Derese, I. and E.Noldus. (1980). Design of linear feedback law for bilinear systems. Int.J.Contr., Vol.31, pp. 219-237

Iwasaki, T.and R.E.Skelton, (1994). All controller for the general $H_\infty$ control problem: LMI Existence Condition and state space formulas. Automatica, Vol.30, pp. 1307-1317

Madjwale,A.N. W.M. Haddad and D.S. Bernstein, (1989). Robust control design for systems with structured parameter uncertainty, Syst. Contr. Lett., Vol.12, pp. 393-407

Petersen, I.R. and C.V.Hollot, (1986). A Riccati equation approach to the stabilization of uncertain linear systems, Automatica,Vol.22, pp. 397-411

Petersen,I.R. (1987). A stabilization algorithm for a class of uncertain linear systems, Syst. Contr. Lett., Vol.8, pp. 351-357

Xie, L.Xie, M.Fu and C.E.deSouza, (1992 ). $H_\infty$ control and quadratic stabilization of systems with parameter uncertainty via output feed-back, IEEE Trans. Automata. Contr., Vol.37, pp. 1253-1256

Xie, L.Xie and C.E.deSouza, (1990). Robust $H_\infty$ control for linear time-invariant systems with norm-bounded uncertainty in the input matrix, Syst.Contr. Lett., Vol.14, pp. 398-396

Xie,L. M.Fu and C.E.deSouza, (1992). Robust $H_\infty$ control for linear systems with norm-bounded time-varying uncertainty, IEEETrans.Automat. Contr., Vol.37, pp. 1188-1191

Veillette,R.J. J.V.Medanic, and W.R.Perkins,(1989). Robust stabilization and disturbance rejection for systems with structure uncertainty , in Proc. 28th IEEE Conf. Decision Contr., Tampa, FL, Dec.

# $H_\infty$ ROBUST PERFORMANCE DESIGN TO MISSILE AUTOPILOT

**Zheng-Xin Weng , Song-Jiao Shi**

*Department of Automation, Shanghai Jiao Tong University, Shanghai, 200030, PRC*
*Fax: 86-21-62813155, Email: zxweng@mail1.sjtu.edu.cn*

Abstract: This paper presents a novel $H_\infty$ robust performance design for a missile autopilot that includes both unmodelled dynamics and real parametric uncertainties. It is shown that this problem can be solved by $H_\infty$ mixed sensitivity method. The main idea is as follows: Firstly, the original problem is transformed into a robust performance design problem for a pure parameter uncertain system. Then, the latter problem is reduced to a standard $H_\infty$ control problem that can be solved by existing methods. The simulation results show that the robust performance of the missile autopilot designed by the presented method is satisfactory. *Copyright © 1999 IFAC*

Keywords: Missiles, H-infinity control, Robust performance, Robust stability, Uncertainty, Mixed sensitivity problem

## 1. INTRODUCTION

Future tactical missile will be required to operate over an expanded flight envelope to meet the challenge of highly maneuverable tactical aircraft. A particular challenge is that of the missile endgame. This involves the final few seconds before delivery of ordnance. During this phase, a missile autopilot can expect large and rapidly time-varying acceleration commands from the guidance law. In turn, the missile is operating at a high and rapidly changing angle of attack. In such a scenario, the aerodynamic coefficients of the missile will be changed greatly. In the process of control, the high-frequency unmodelled flexible-mode dynamics and the varying of aerodynamic coefficients must be considered. Several papers (Shamma and Cloutier, 1993) have discussed this problem.

In this decade, the $H_\infty$ control problem has become one of the dominant streams in the realm of control systems since the pioneering work of Zames (1981). The original formulation of $H_\infty$ optimal control theory was in an input-output setting. Most solution techniques available at that time involved analytic functions or operator-theoretic methods. Hence, the frequency domain methods were strongly promoted. The actual cornerstone for the state-space approach was marked by the celebrated two Riccati equation method which was presented by Doyle, *et al.* (1989) and was commonly referred to as the "DGKF" method. Since this method is both computationally robust and theoretically transparent, it has surpassed all other approaches and provided a basis for studying $H_\infty$ control theory and its applications.

Recently, interest is being focused on the $H_\infty$ robust performance design problem for plants with both unmodelled dynamics and real parametric uncertainties. It is a well-known fact that $H_\infty$ optimal control theory offers a direct and efficient methodology for designing highly practical robust multivariable control systems. However, when there is parameter uncertainty in plant models, no robust behavior on $H_\infty$ performance along with stability can be guaranteed by the standard $H_\infty$ control method. It is now known that the $\mu$-method presented by Doyle (1982) may be used to treat the problem considered here. However, the computations involved in $\mu$-analysis and $\mu$-synthesis are complicated. From the point of view of engineering application, this paper deeply investigates the $H_\infty$ robust performance design to a missile autopilot that includes both unmodelled dynamics and real parametric uncertainties, and first shows that the problem can be solved by $H_\infty$ mixed sensitivity method. The train of thoughts of this paper is as follows: Firstly, the original problem is transformed into a robust performance design problem for a pure parameter uncertain system. Then, the latter problem is reduced to a standard $H_\infty$ control problem which can be solved by existing methods, such as, "DGKF" method.

## 2. MISSILE DYNAMICS

The missile dynamics considered here are taken from the paper of Shamma and Cloutier (1993). These dynamics are representative of a missile traveling at Mach 3 at an altitude of 20000 ft. However, they do not correspond to any particular missile airframe. The nonlinear state equations are as follows:

$$\dot{\alpha} = f\frac{g\cos(\alpha/f)}{WV}Z + q,\ \dot{q} = fm/I_{yy}, \quad (1)$$

where $f = 180/\pi$ is the radians-to-degrees conversion, $g = 9.8m/s^2$ is the acceleration of gravity, $I_{yy} = 247.64kg\cdot m^2$ is the pitch moment of inertia, $m = C_m QSd(N\cdot m)$ is the pitch moment, $d = 0.2286m$ is the reference diameter, $Q = 293445.8677N/m^2$ is the dynamic pressure, $q$ (*rad/s*) is the pitch rate, $S = 0.040877m^2$ is the reference area, $V = 947.71464m/s$ is the speed, $W = 2000.376N$ is the weight, $Z = C_z QS(N)$ is the normal force, and $\alpha$(deg) is the angle of attack.

The normal force and pitch moment aerodynamic coefficients are approximated by

$$C_z = \Phi_z(\alpha) + b_z\delta, C_m = \Phi_m(\alpha) + b_m\delta, \quad (2)$$

where $\delta$(deg) is the fin deflection, and

$$b_m = -0.206, b_z = -0.034,$$

$$\Phi_m(\alpha) = 0.000215\alpha^3 - 0.0195\alpha|\alpha| + 0.051\alpha,$$

$$\Phi_z(\alpha) = 0.000103\alpha^3 - 0.00945\alpha|\alpha| - 0.170\alpha.$$

These approximations are accurate for $\alpha$ in the range of $\pm 20\deg$.

The missile tailfin actuator is modeled as the following second-order system:

$$\delta(s) = \frac{\omega_a^2}{s^2 + 1.4\omega_a s + \omega_a^2}\delta_c(s), \quad (3)$$

where $\delta_c$(deg) is the commanded fin deflection, $\omega_a = 150rad/s$ is the actuator bandwidth.

The autopilot will be required to control the normal acceleration (expressed in *g*) $\eta_z = Z/W$ via commanded fin deflection $\delta_c$. The general performance objective is to track acceleration step commands with a steady state accuracy of less than 0.5% and a time constant of 0.2s. The controller must provide robust performance over a wide range of angles of attack and avoid high-frequency instabilities caused by unmodelled flexible-body modes.

## 3. $H_\infty$ ROBUST PERFORMANCE SYNTHESIS

### *3.1 Robust Performance Problem*

The system considered in this paper is depicted as Fig.1. In this diagram, ***d*** is an external disturbance vector; ***u*** is the control input; ***y*** is the output; ***K*** is the controller to be designed; and $\bar{G}$(s) is the real plant that can be expressed as:

$$\bar{G}(s) = [I + \Delta G(s)]G(s,\Sigma), \quad (4)$$

where ***G***(s, Σ) represents the plant with parametric uncertainty, and has the form of

$$G(s,\Sigma) = \left[\begin{array}{c:c} A_g + \Delta A_g & B_g \\ \hdashline C_g + \Delta C_g & 0 \end{array}\right], \quad (5)$$

with

$$\begin{bmatrix}\Delta A_g \\ \Delta C_g\end{bmatrix} = \begin{bmatrix}E_a \\ E_c\end{bmatrix}\Sigma F_a, \Sigma \in \Omega \overset{\Delta}{=} \{\Sigma \mid \Sigma^T\Sigma \le I\}.$$

Here, $E_a, E_c$ and $F_a$ are known constant matrices. ***G***(s,0) is referred to as the nominal plant, and $(A_g, B_g, C_g)$ is the state-space realization of the nominal plant ***G***(s,0). **ΔG**(s) represents the

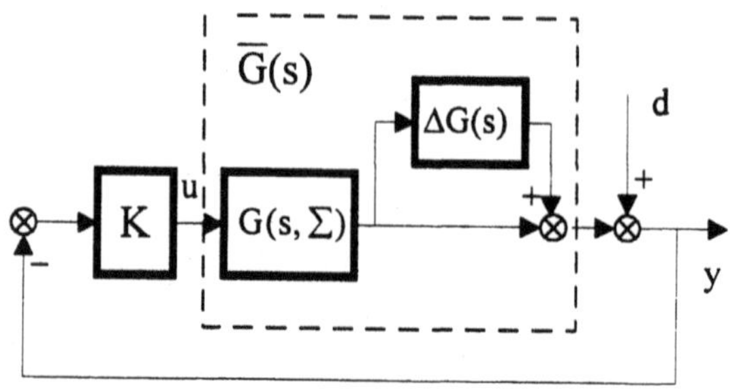


Fig. 1. Feedback system.

unstructured uncertainty (i.e., output multiplicative uncertainty) which characterizes the unmodelled dynamics, and satisfies

$$\bar{\sigma}[\Delta G(j\omega)] < |W_2(j\omega)|, \forall s = j\omega, \omega \geq 0 . \qquad (6)$$

Here $\bar{\sigma}[\bullet]$ represents the maximum singular values of a matrix, and $W_2(j\omega)$ is the uncertainty weighting function.

**Definition 1** *If* $\Delta G(s)$ *satisfies (6) and* $\Sigma \in \Omega$, *then* $\Delta G(s)$ *and* $\Sigma$ *will be said to be admissible.*

The robust performance problem considered in this paper can be expressed as the following problem:

**Problem 1** *Given plant (4), find a controller* $K(s)$ *such that the system in Fig.1 is stable for all admissible parameter perturbations* $\Sigma$ *and unstructured uncertainties* $\Delta G(s)$, *and* $y$ *remains in a desired set of response for exogenous disturbance* $d$ *and all admissible uncertainties* $\Delta G(s)$ *and* $\Sigma$.

Assume that $\bar{G}(s)$ has the same number of unstable modes as $G(s, \Sigma)$, and that there exists a controller $K$ that stabilizes $G(s, \Sigma)$ for all admissible parameter perturbations $\Sigma$.

Let $\Sigma_1 \in \Omega$ be given, consider the nominal performance design and robust stability problem for the plant $G(s, \Sigma_1)$ with unstructured uncertainty $\Delta G(s)$. It is well known that the closed loop system in Fig.1 is stable for all admissible $\Delta G(s)$ and the given $\Sigma_1 \in \Omega$ if and only if

$$\left\| W_2 G(s, \Sigma_1) K (I + G(s, \Sigma_1) K)^{-1} \right\|_\infty \leq 1 . \qquad (7)$$

Besides the requirements for robust stability, the performance specification is necessary for the system design. Usually, it is required to minimize the effect of external disturbance $d$ on the output $y$. In the $H_\infty$ control theory, the performance specifications are usually denoted by the $H_\infty$ norm of the weighted sensitivity function, i.e.

$$\left\| W_1 (I + G(s, \Sigma_1) K)^{-1} \right\|_\infty \leq 1 . \qquad (8)$$

The requirements (7) ~ (8) can be readily handled within the framework of $H_\infty$ control theory since they are approximately equivalent to (Safonov and Chiang, 1988)

$$\left\| M(s, \Sigma_1) \right\|_\infty \leq 1 , \qquad (9)$$

where

$$M(s, \Sigma_1) = \begin{bmatrix} W_1 S \\ W_2 T \end{bmatrix} . \qquad (10)$$

Here $S = (I + G(s, \Sigma_1) K)^{-1}$ is the so-called sensitivity function, and $T = G(s, \Sigma_1) K (I + G(s, \Sigma_1) K)^{-1}$ is the complementary sensitivity function, $W_1(s)$ is the performance weighting function that reflects the performance requirement. $W_2(s)$ is the uncertainty weighting function.

In the $H_\infty$ control theory, $M(s, \Sigma_1)$ in (10) corresponds to the closed-loop transfer function of the system in Fig.2. In the diagram, $P(s, \Sigma_1)$ is called generalized plant.

The object is to find a stabilizing controller $K$. such that the inequality (9) holds for a given $\Sigma_1 \in \Omega$. This optimization problem is called $H_\infty$ mixed sensitivity problem, and it can be reduced to the standard one by setting the parameter uncertainty $\Sigma_1 \in \Omega$ to zero.

Up to here, this paper only discusses the nominal performance design and robust stability problem for the plant $G(s, \Sigma_1)$ with unstructured uncertainty $\Delta G(s)$, and reduces it to the $H_\infty$ mixed sensitivity problem. It is a common sense that $H_\infty$ method cannot handle the robust performance problem directly (Doyle, 1982). But, in the following, it will be shown that the $H_\infty$ mixed sensitivity method can

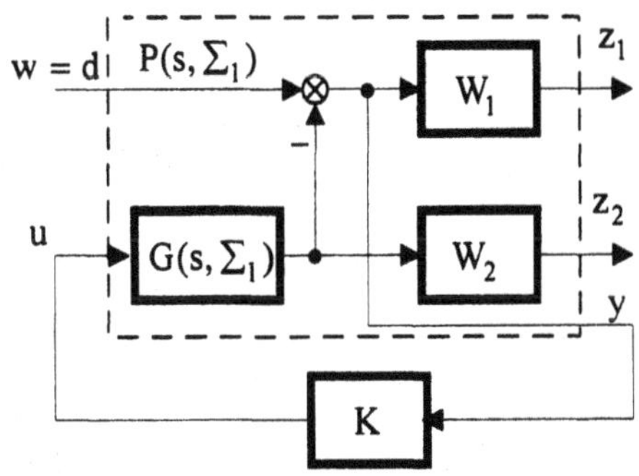


Fig. 2. Standard structure for S/T mixed sensitivity problem

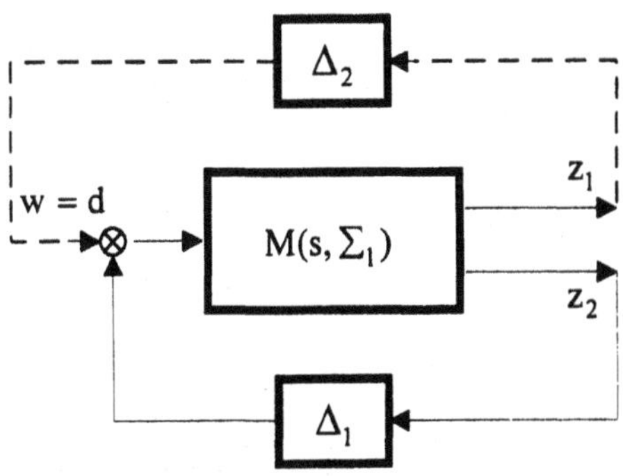


Fig. 3. Diagram for robust performance problem

be used to tackle the robust performance design problem for some systems with unstructured uncertainty.

It is known that the optimization problem (9) can be handled within the framework of robust performance problem (see Fig.3). In the diagram, $\Delta_1$ denotes the uncertainty perturbation block, $\Delta_2$ is the fictitious performance block, and satisfy $\bar{\sigma}[\Delta_1]<1$, $\bar{\sigma}[\Delta_2]<1$

$$M(s,\Sigma_1)=\begin{bmatrix} M_1(s,\Sigma_1) \\ M_2(s,\Sigma_1) \end{bmatrix}, \tag{11}$$

and

$$F_l(M(s,\Sigma_1),\Delta_1)=M_1(s,\Sigma_1)(I-\Delta_1 M_2(s,\Sigma_1))^{-1} \tag{12}$$

is the closed-loop transfer function from $w$ to $z_1$. Then, have the following Lemma about the robust performance.

**Lemma 1** *$F_l(M(s,\Sigma_1),\Delta_1)$ is stable, and $\|F_l(M(s,\Sigma_1),\Delta_1)\|_\infty \le 1$, if and only if*

$$\|M(s,\Sigma_1)\|_\infty \le 1. \tag{13}$$

*Proof* From the robust stability theorem of Doyle (1988).

$$\begin{gathered} \|F_l(M(s,\Sigma_1),\Delta_1)\|_\infty \le 1 \Leftrightarrow \\ \det[I-F_l(M(s,\Sigma_1),\Delta_1)\Delta_2]\neq 0, \\ \forall s=j\omega, \forall\Delta_2, \bar{\sigma}(\Delta_2)<1. \end{gathered}$$

Similarly,

$$\begin{gathered} F_l(M(s,\Sigma_1),\Delta_1) \text{ is stable } \forall\Delta_1,\bar{\sigma}(\Delta_1)<1 \Leftrightarrow \\ \det[I-M_2(s,\Sigma_1)\Delta_1]\neq 0, \forall s=j\omega, \forall\Delta_1,\bar{\sigma}(\Delta_1)<1, \end{gathered}$$

$$\begin{aligned} &\|M(s,\Sigma_1)\|_\infty \le 1 \Leftrightarrow \det[I-M(s,\Sigma_1)\Delta]\neq 0 \\ &\Leftrightarrow \det[I-\Delta M(s,\Sigma_1)]\neq 0 \\ &\Leftrightarrow \det(I-[\Delta_2 \quad \Delta_1]\begin{bmatrix} M_1(s,\Sigma_1) \\ M_2(s,\Sigma_1) \end{bmatrix})\neq 0 \\ &\Leftrightarrow \det[I-\Delta_2 M_1(s,\Sigma_1)-\Delta_1 M_2(s,\Sigma_1)]\neq 0 \\ &\Leftrightarrow \det[I-\Delta_1 M_2(s,\Sigma_1)]\cdot \\ &\quad \det[I-\Delta_2 M_1(s,\Sigma_1)(I-\Delta_1 M_2(s,\Sigma_1))^{-1}]\neq 0 \\ &\Leftrightarrow \det[I-M_2(s,\Sigma_1)\Delta_1]\cdot \\ &\quad \det[I-F_l(M(s,\Sigma_1),\Delta_1)\Delta_2]\neq 0 \\ &\quad \forall s=j\omega, \forall\Delta=[\Delta_2 \Delta_1], \bar{\sigma}(\Delta)<1. \end{aligned}$$

According to Safonov and Chiang (1988),

$$\frac{1}{\sqrt{2}}\bar{\sigma}([A\,B])\le \max\{\bar{\sigma}(A),\bar{\sigma}(B)\}\le \bar{\sigma}([A\,B]),$$

so $\bar{\sigma}(\Delta)=\bar{\sigma}([\Delta_2\,\Delta_1])<1$ can be satisfied approximately, and the result follows immediately. □

Lemma 1 shows that the condition of (9) is a necessary and sufficient condition for the robust performance of the plant $G(s,\Sigma_1)$ with unstructured uncertainty $\Delta G(s)$. Since the parameter perturbation $\Sigma_1\in\Omega$ is given arbitrarily, Lemma 1 should hold for all $\Sigma\in\Omega$, i.e.

**Lemma 2** *$F_l(M(s,\Sigma),\Delta_1)$ is stable, and $\|F_l(M(s,\Sigma),\Delta_1)\|_\infty \le 1$ for all $\Sigma\in\Omega$, if and only if*

$$\|M(s,\Sigma)\|_\infty \le 1, \quad \forall\Sigma\in\Omega. \tag{14}$$

According to the Lemma 2, the robust performance design problem for systems with both unmodelled dynamics and real parametric uncertainties(i.e., Problem 1) can be reduced to the following robust performance design problem for a pure parameter uncertain system.

**Problem 2** *Find a controller $K$, such that the generalized plant $P(s,\Sigma)$ can be stabilized for all $\Sigma\in\Omega$, and the closed-loop transfer function $M(s,\Sigma)$ in (10) satisfies*

$$\|M(s,\Sigma)\|_\infty \le 1, \quad \forall\Sigma\in\Omega. \tag{15}$$

### *3.2 Conditions For Robust Performance*

To find the solution of Problem 2, it is necessary to know the state-space realization of the generalized plant $P(s,\Sigma)$. Here, the state-space form of $P(s,\Sigma)$ is given as follow:

$$P(s,\Sigma)=\left[\begin{array}{c|cc} A+\Delta A & B_1+\Delta B_1 & B_2+\Delta B_2 \\ \hline C_1+\Delta C_1 & D_{11}+\Delta D_{11} & D_{12}+\Delta D_{12} \\ C_2+\Delta C_2 & D_{21}+\Delta D_{21} & D_{22}+\Delta D_{22} \end{array}\right], \tag{16}$$

with

$$\begin{bmatrix} \Delta A & \Delta B_1 & \Delta B_2 \\ \Delta C_1 & \Delta D_{11} & \Delta D_{12} \\ \Delta C_2 & \Delta D_{21} & \Delta D_{22} \end{bmatrix}=\begin{bmatrix} E_1 \\ E_2 \\ E_3 \end{bmatrix}\Sigma[F_1 \quad F_2 \quad F_3], \Sigma\in\Omega. \tag{17}$$

The generalized plant can also be characterized by the following state equations:

$$\dot{x} = Ax + B_1 w + B_2 u + E_1 \Sigma (F_1 x + F_2 w + F_3 u)$$
$$z = C_1 x + D_{11} w + D_{12} u + E_2 \Sigma (F_1 x + F_2 w + F_3 u) \quad (18)$$
$$y = C_2 x + D_{21} w + D_{22} u + E_3 \Sigma (F_1 x + F_2 w + F_3 u)$$

Let

$$z_p \stackrel{\Delta}{=} F_1 x + F_2 w + F_3 u, \quad w_p \stackrel{\Delta}{=} \Sigma z_p$$

then

$$\dot{x} = Ax + \begin{bmatrix} E_1 & B_1 \end{bmatrix} \begin{bmatrix} w_p \\ w \end{bmatrix} + B_2 u,$$
$$\begin{bmatrix} z_p \\ z \end{bmatrix} = \begin{bmatrix} F_1 \\ C_1 \end{bmatrix} x + \begin{bmatrix} 0 & F_2 \\ E_2 & D_{11} \end{bmatrix} \begin{bmatrix} w_p \\ w \end{bmatrix} + \begin{bmatrix} F_3 \\ D_{12} \end{bmatrix} u, \quad (19)$$
$$y = C_2 x + \begin{bmatrix} E_3 & D_{21} \end{bmatrix} \begin{bmatrix} w_p \\ w \end{bmatrix} + D_{22} u.$$

The above system (19) is referred to as the augmented plant, and denoted by $\boldsymbol{P}_{aug}(s)$. Here, $w_p$ and $z_p$ are considered as the fictitious input and output, respectively. The robust performance problem for system (19) can be described as Fig.4*a*: Find a controller $\boldsymbol{K}$, such that the closed-loop system is robustly stable, and the closed-loop transfer function $\boldsymbol{T}_{zw}(s,\Sigma)$ from $\boldsymbol{w}$ to $z$ satisfies $\|\boldsymbol{T}_{zw}(s,\Sigma)\|_\infty \le 1$, i.e. $\|\boldsymbol{M}(s,\boldsymbol{\Sigma})\|_\infty \le 1, \forall \boldsymbol{\Sigma} \in \Omega$. To analyze the robust stability and the robust performance of the system in Fig.4*a*, regard the controller in Fig.4*a* as a part of the plant, and absorb it into the plant. So the system in Fig.4*a* can be reduced to the one in Fig.4*b*, where $\boldsymbol{T}$ is the closed-loop transfer function from $(w_p{}^T \ \ w^T)^T$ to $(z_p{}^T \ \ z^T)^T$. The following theorem provides a sufficient condition for robust performance.

**Theorem 1** *$F_u(T,\Sigma)$ is stable, and $\|F_u(T,\Sigma)\|_\infty < 1$, $\forall \Sigma \in \Omega$, if*

$$\|T\|_\infty < 1, \quad (20)$$

*where $F_u(T,\Sigma)$ is the closed-loop transfer function $\boldsymbol{T}_{zw}(s,\Sigma)$ in Fig.4b from $\boldsymbol{w}$ to $z$.*

On the basis of Lemma 2 and Theorem 1, and the fact of $F_u(T,\Sigma) = \boldsymbol{T}_{zw}(s,\Sigma) = \boldsymbol{M}(s,\Sigma)$, the following theorem can be derived.

**Theorem 2** *$F_l(M(s,\Sigma),\Delta_1)$ is stable, and $\|F_l(M(s,\Sigma),\Delta_1)\|_\infty \le 1, \forall \Sigma \in \Omega$, if*

$$\|T\|_\infty < 1. \quad (21)$$

On the basis of Theorem 2 , the robust performance design problem for systems with both unmodelled dynamics and real parametric uncertainties(i.e., Problem 1) may be reduced to the following problem:

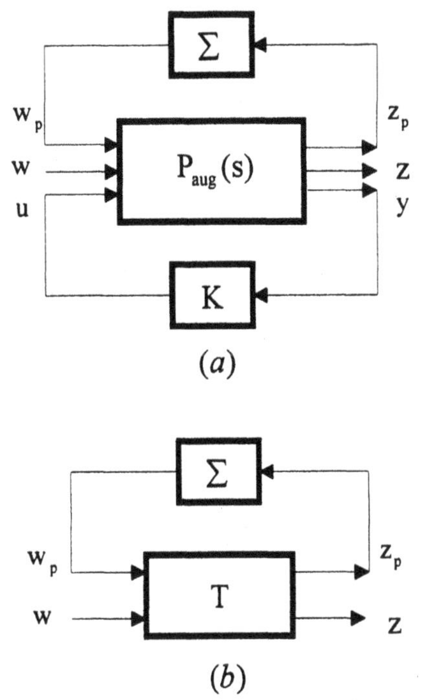


Fig. 4. Diagram for robust performance of the augmented plant

**Problem 3** *Find a controller $\boldsymbol{K}$, such that the system (19) can be stabilized and the closed-loop transfer function $\boldsymbol{T}$ from $\begin{pmatrix} w_p \\ w \end{pmatrix}$ to $\begin{pmatrix} z_p \\ z \end{pmatrix}$ satisfies the condition (21).*

For simplicity, define

$$z \leftarrow \begin{bmatrix} z_p \\ z \end{bmatrix}, w \leftarrow \begin{bmatrix} w_p \\ w \end{bmatrix}, C_1 \leftarrow \begin{bmatrix} F_1 \\ C_1 \end{bmatrix},$$
$$D_{11} \leftarrow \begin{bmatrix} 0 & F_2 \\ E_2 & D_{11} \end{bmatrix}, D_{12} \leftarrow \begin{bmatrix} F_3 \\ D_{12} \end{bmatrix},$$
$$B_1 \leftarrow \begin{bmatrix} E_1 & B_1 \end{bmatrix}, D_{21} \leftarrow \begin{bmatrix} E_3 & D_{21} \end{bmatrix}.$$

Then , system (19) can be rewritten as

$$\dot{x} = Ax + B_1 w + B_2 u,$$
$$z = C_1 x + D_{11} w + D_{12} u, \quad (22)$$
$$y = C_2 x + D_{21} w + D_{22} u,$$

and the Problem 3 can be expressed as: For the system (22), find a stabilizing controller $\boldsymbol{K}$, such that the closed-loop transfer function $\boldsymbol{T}_{zw}$ from $\boldsymbol{w}$ to $z$ satisfies $\|T_{zw}\|_\infty < 1$. This is just the so-called standard $H_\infty$ control problem, it can be solved by the standard $H_\infty$ methods, for example, the "DGKF" method.

## 4. MISSILE AUTOPILOT DESIGN

The nonlinear state equations in (1) are linearized about trim operating points to form linear state-space equations of the form:

$$\dot{x} = Ax + Bu, \quad y = Cx + Du, \qquad (23)$$

where $x = [\alpha \ q]^T, u = \delta, y = [q \ \eta_z]^T$. As the operating angle of attack ranging over 0 to 20 degrees, a number of linearized models can be obtained. In the design process, the following average model will be used, and the difference between the linearized models and the average model will be considered as parameter uncertainty.

$$A = \begin{bmatrix} -0.9 & 1 \\ -134.0 & 0 \end{bmatrix}, \quad B = \begin{bmatrix} -0.117 \\ -130.8 \end{bmatrix},$$

$$C = \begin{bmatrix} 0 & 1 \\ -1.78 & 0 \end{bmatrix}, \quad D = \begin{bmatrix} 0 \\ -0.203 \end{bmatrix}. \qquad (24)$$

Use the *S/T* mixed sensitivity structure in Fig. 2. to solve this problem. According to the performance specifications and the limitation of the high frequency gain, choose the sensitivity weighting function $W_1(s)$ as:

$$W_1(s) = \begin{bmatrix} \dfrac{14.9451s + 200\rho_1}{42.7003s + 1} & 0 \\ 0 & \dfrac{14.9451s + 200\rho_2}{42.7003s + 1} \end{bmatrix}, \qquad (25)$$

where $\rho_1$ and $\rho_2$ are tunable parameters. The complementary sensitivity weighting function $W_2(s)$ is determined by following two factors. One is the high-frequency unmodelled flexible modes beyond 200*rad/s*, the other one is the omitted dynamics of the missile tailfin actuator. So the weighting function $W_2(s)$ is chosen to be

$$W_2(s) = \frac{s}{100} I_{2\times 2}. \qquad (26)$$

Since the average model in (24) is not strictly proper, a precompensator 1/(*s*+0.001) is introduced. Based on the presented method, this design problem can be reduced to the standard $H_\infty$ control problem which can be solved by the existing methods. In the design process, the values of $\rho_1$ and $\rho_2$ are increased until no solution exists for any larger $\rho_1$ and $\rho_2$. The final values of $\rho_1$ and $\rho_2$ are 1 and 10, respectively, and the resulting controller is given as:

$$\dot{x}_c = A_{cp}x_c + B_{cp}y, \quad u = C_{cp}x_c. \qquad (27)$$

Fig. 5 shows a nonlinear simulation response of the missile autopilot system to a 25*g* step acceleration command, exhibiting good robust performance. The results also show the effectiveness of the $H_\infty$ robust performance design method presented in this paper.

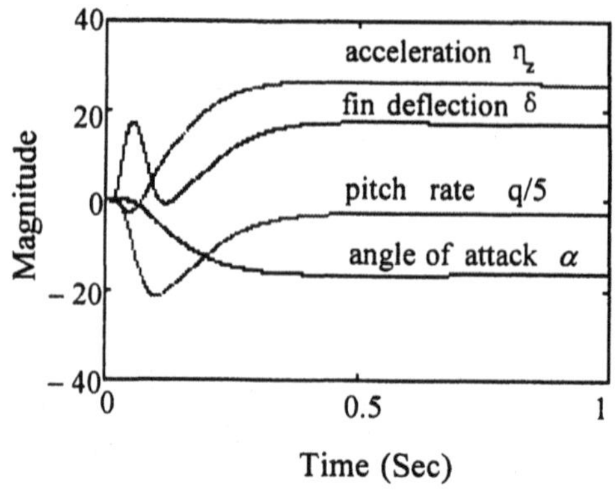


Fig. 5. Nonlinear simulation response

## 5. CONCLUSIONS

This paper thoroughly studied the $H_\infty$ robust performance design problem for systems with both unmodelled dynamics and real parametric uncertainties, and reduced this complicated design problem to a standard $H_\infty$ control problem. The technique presented in this paper has been applied to the missile autopliot design problem successfully, and the results show that the technique presented here is effective.

## REFERENCES


Doyle, J.C. (1982). Analysis of feedback systems with structured uncertainties. *IEE Proceeding~D*, **129**, 242-250.

Doyle, J.C. (1988). A review of $\mu$ for case studies in robust control. In: *Proceedings of IFAC 10th Triennial World Congress* (R. Isermann, Ed.). Vol. 8, pp. 365-372. Pergamon Press, Oxford, England.

Doyle, J.C., K. Glover, P.P. Khargonekar and B.A. Francis (1989). State-space solutions to standard $H_2$ and $H_\infty$ control problems. *IEEE Transactions on Automatic Control*, **34**, 831-847.

Safonov, M.G. and R.Y. Chiang (1988). CACSD using the state-space $L^\infty$ theory — a design example. *IEEE Transactions on Automatic Control*, **33**, 477-479.

Shamma, J.S. and J.R. Cloutier (1993). Gain-scheduled missile autopilot design using linear parameter varing transformations. *Journal of Guidance, Control, and Dynamics*, **16**, 256-263.

Zames, G. (1981). Feedback and optimal sensitivity: model reference transformations, multiplicative seminorms, and approximate inverses. *IEEE Transactions on Automatic Control*, **26**, 301-320.

14th Triennial World Congress, Beijing, P.R. China

F-2e-03-4

# ROBUST $H_\infty$ FILTERING FOR UNCERTAIN PARAMETER SYSTEM

**Ming Zhang Hongwei Xie Dinghan Shi**

*Department of Automatic Control, National University of Defense Technolog, Changsha , P. R. China 410073* E-mail: zhangming@at.nudt.edu.cn

Abstract: In this paper, the necessary and sufficient condition is derived for the existence of a solution to $H_\infty$ filter for the nominal system, but the sufficient condition is derived for the existence of a solution to robust $H_\infty$ filter for uncertain parameter systems. These conditions are in terms of Algebraic Riccati Inequality (**ARI**) so that a set of $H_\infty$ filters and robust $H_\infty$ filters can be designed. *Copyright © 1999 IFAC*

Key word: Filtering, Uncertain linear systems, Robust stability, $H_\infty$ optimization, Algebraic Riccati equation,

## 1. INTRODUCTION

For linear system in terms of state space with certain parameter, Kalman Filtering method can be used to estimate the system states via output measures if the noise is Gauss white and the power spectral density is known; $H_\infty$ filter can also be designed to estimate the system states if the energy of noise is bounded. $H_\infty$ filtering and smoothing problem for linear system in terms of state-space has been studied in Nagpal and Khargonekar (1991). For $H_\infty$ filtering problem of linear system with certain parameter, Nagpal and Khargonekar (1991) gave the necessary and sufficient condition for the existence of $H_\infty$ filter and proposed the design method based on Algebraic Riccati Equation (**ARE**). But by this method, only one $H_\infty$ filter can be obtained. It is difficult to optimize the multiple system performance indexes simultaneously.

For uncertain parameter system, how can robust filter be designed? For linear system with norm-bounded time-varying uncertainty, Algebraic Riccati Inequality (**ARI**) is frequently used to study quadratic stability in recent results (Khargonekar et al. 1990; Packard and Doyle, 1990; Xie and Souza, 1992; Xie et. al.,1992). Barmish (1985) suggested the necessary and sufficient condition for quadratic stabilizability of uncertain systems. But quadratic stability of the uncertain systems is only the sufficient condition for robust stability of the uncertain systems. In this paper only the sufficient condition for the existence of robust $H_\infty$ filter for uncertain parameter system will be studied.

Using the technique of **ARI**, this paper obtains not only the necessary and sufficient condition for the existence of $H_\infty$ filter for the nominal system of uncertain parameter systems but the sufficient condition for the existence of robust $H_\infty$ filter for linear systems with norm-bounded time-varying uncertainty. More over, a set of $H_\infty$ filters and robust $H_\infty$ filters can be designed based on **ARI** so that it is possible to optimize the multiple system performance indexes simultaneously.

## 2. PROBLEM STATEMENT

Consider the following state space of linear uncertain parameter system:

$$\begin{aligned} \dot{x}(t) &= (A + \Delta A)x(t) + (B + \Delta B)w(t) \\ y(t) &= (C + \Delta C)x(t) + v(t) \\ z(t) &= Lx(t) \end{aligned} \tag{1}$$

where $x(t)$ is the state vector of the system, $y(t)$ is the measurement, $w(t)$ denotes the system noise, $v(t)$ denotes the measurement noise and $z(t)$ is the estimated state.

For the analysis of system (1), the following conditions are assumed:

$$\Delta A = H_1 FE, \Delta C = H_2 FE, \Delta B \Delta B^T \le M_B \tag{2}$$

$$F^T F \le I, M_B = M_B^T \ge 0 \tag{3}$$

where $H_1, H_2, E, M_B$ are constant matrices, $F$ is perturbed matrix.

For the system (1) , a linear filter will be given as follows:

$$\begin{aligned} \dot{\hat{x}}(t) &= (A - KC)\hat{x}(t) + Ky(t) \\ \hat{z}(t) &= L\hat{x}(t) \end{aligned} \tag{4}$$

so that $H_\infty$ performance

$$\|T(s)\|_\infty \le \gamma \tag{5}$$

is satisfied with the transfer function

$$T(s): \binom{w}{v} \to e = z - \hat{z} \tag{6}$$

where $K$ is a suitable gain to be found and $\gamma > 0$ is a given value.

In the case of the (partly) uncertain system given by (1) with (2) (3),

$$T(s) = C_C\left[sI - (A_C + H_C F E_C)\right]^{-1} B_C \tag{7}$$

where

$$A_C = \begin{bmatrix} A & 0 \\ KC & A-KC \end{bmatrix}, B_C = \begin{bmatrix} B+\Delta B & 0 \\ 0 & K \end{bmatrix}, H_C = \begin{bmatrix} H_1 \\ KH_2 \end{bmatrix} \tag{8}$$

$$C_C = \begin{bmatrix} L & -L \end{bmatrix}, E_C = \begin{bmatrix} E & 0 \end{bmatrix} \left( or\, B_C = \begin{bmatrix} B & 0 \\ 0 & K \end{bmatrix} \right)$$

In the case of the nominal system given by (1) with $\Delta A \equiv 0, \Delta B \equiv 0, \Delta C \equiv 0$ ,

$$T(s) = L\left[sI - (A - KC)\right]^{-1}\begin{bmatrix} B & -K \end{bmatrix} \tag{9}$$

In order to define $H_\infty$ filter and robust $H_\infty$ filter, the following definitions are given.

*Definition 1* (Xie et al., 1992): Given a scalar $\gamma > 0$ , the system given by (1) with $\Delta B \equiv 0$ and $\Delta C \equiv 0$ is said to be quadratically stable with disturbance attenuation $\gamma$ if there exists a symmetric positive-definite matrix $P$ such that for all admissible uncertainty $\Delta A(\cdot)$:

$$[A + \Delta A(\cdot)]^T P + P[A + \Delta A(\cdot)] + \gamma^{-2} PBB^T P + C^T C < 0$$

*Definition 2* (Petersen, 1987): Given a scalar $\gamma>0$, the system given by (1) with $\Delta A \equiv 0, \Delta B \equiv 0, \Delta C \equiv 0$ is stable with disturbance attenuation $\gamma$ if the following conditions are satisfied:
1) A is a stable matrix; 2) $\|C(sI - A)^{-1} B\|_\infty < \gamma$ .

*Definition 3*: Given a scalar $\gamma>0$, the system given by (1) (or given by (1) with $\Delta B \equiv 0$ and $\Delta C \equiv 0$ ) is robust stable with disturbance attenuation $\gamma$ if for all admissible uncertainty $\Delta A(\cdot)(or \Delta B(\cdot), \Delta C(\cdot))$:

1) $A+\Delta A(\cdot)$ is robust stable;

2) $\|R(s)\|_\infty < \gamma$ with transfer function

$$R(s)=C[sI-(A+\Delta A(\cdot))]^{-1}B$$

$$\left(or R(s)=(C+\Delta C(\cdot)[sI-(A+\Delta A(\cdot))]^{-1}(B+\Delta B(\cdot))\right)$$

*Definition 4*: Given a scalar $\gamma>0$, the filter (4) is said to be $H_\infty$ filter at level $\gamma$ for nominal system given by (1) with $\Delta A\equiv 0, \Delta B\equiv 0, \Delta C\equiv 0$ if the system given by (9) is stable with disturbance attenuation $\gamma$.

*Definition 5*:Given a scalar $\gamma>0$, the filter (4) is said to be robust $H_\infty$ filter at level $\gamma$ for (partly) uncertain system given by (1) with (2) and (3) (or $\Delta B\equiv 0$) if the system given by (7) with (8) is robust stable with disturbance attenuation $\gamma$.

In this paper, '$A\le B$' or '$B\ge A$' means '$A-B$' is semi-positive definite; and '$A<B$' or '$B>A$' means '$A-B$' is positive definite.

## 3. MAIN RESULTS

*Lemma 1*: Given a scalar $\gamma>0$, system (1) with $\Delta A\equiv 0, \Delta B\equiv 0, \Delta C\equiv 0$ is stable with disturbance attenuation $\gamma$ if and only if there exists symmetric positive-definite $P>0$ such that:

$$A^T P+PA+\gamma^{-2}PBB^T P+C^T C<0 \tag{10}$$

Proof: See in Khargonekar et al. (1990).

Remark 1: Denote $Q=\gamma P^{-1}$, then (10) becomes:

$$AQ+QA^T+\gamma^{-2}QC^T CQ+BB^T<0 \tag{11}$$

*Lemma 2*: Given a scalar $\gamma>0$, the system given by (1) with (2), $\Delta B\equiv 0$ and $\Delta C\equiv 0$ is quadratically stable with disturbance attenuation $\gamma$ if and only if there exists a symmetric positive-definite $P>0$ and a scalar $\delta>0$ such that

$$A^T P+PA+\gamma^{-2}PBB^T P+\delta^2 PHH^T P+\delta^2 E^T E+C^T C<0 \tag{12}$$

Proof: see in Xie et al. (1992).

Remark 2: (12) can be rewritten as:

$$A^T P+PA+\gamma^{-2}P[B \quad \gamma\delta^{-1}H][B \quad \gamma\delta^{-1}H]^T P+\begin{bmatrix} C \\ \delta E\end{bmatrix}^T\begin{bmatrix} C \\ \delta E\end{bmatrix}<0 \tag{13}$$

Thus if there exists a scalar $\delta>0$ such that $R_\delta(s)$ is stable with disturbance attenuation $\gamma$, then system given by (1) with (2), $\Delta B\equiv 0$ and $\Delta C\equiv 0$ is quadratically stable with disturbance attenuation $\gamma$.

where

$$R_\delta(s)=\begin{pmatrix} C \\ \delta E\end{pmatrix}(sI-A)^{-1}\begin{pmatrix} B & \gamma\delta^{-1}H\end{pmatrix} \tag{14}$$

*Lemma 3*: Let $W$, $R$, $A$ be square matrixes with the same dimension, and $W$, $R$ be symmetric positive-definite matrixes, there exists one and only one symmetric positive-definite matrix $P$ such that:

$$A^T P+PA-PWP+R=0 \tag{15}$$

Proof: See in Willems (1971).

Remark 3: Denote $Q=P^{-1}$, (15) becomes

$$AQ+QA^T-W+QRQ=0 \tag{16}$$

*Theorem 1*: Given a scalar $\gamma>0$, there exists $K$ such that (4) is $H_\infty$ filter at level $\gamma$ for nominal system given by (1) with $\Delta A\equiv 0, \Delta B\equiv 0, \Delta C\equiv 0$ if and only if there exists a symmetric positive-definite $P$ such that

$$AP+PA^T+P(C^T C-\gamma^{-2}L^T L)P+BB^T<.0 \tag{17}$$

Moreover when $K=PC^T$, (4) is $H_\infty$ filter at level $\gamma$ for nominal system.

*Proof*: ($\Leftarrow$) Let $K=PC^T$, (17) can be rewritten as

$$\begin{aligned}&(A-KC)P+P(A-KC)^T+\gamma^{-2}PL^T LP\\&+[B \quad -K][B \quad -K]^T<0\end{aligned} \tag{18}$$

Thus from Lemma.1 and Definition 4, the sufficient condition is established.

($\Rightarrow$) If there exists $K$ such that (4) is $H_\infty$ filter at level $\gamma$ for nominal system, then there exists a symmetric positive-definite matrix $P$ such that (18) holds and:

$$KCP+PC^T K-KK^T-PC^T CP=-(K \quad PC)(K \quad PC)^T\le 0$$

Thus $P$ will make (17) true.

For robust $H_\infty$ filtering problem for linear system with (partly) uncertain parameter, the following results can be obtained from Lemma 2.

*Theorem 2*: Given a scale $\gamma > 0$, if $A$ is stable and there exist $\delta > 0, \gamma_1 > 0$ $(\sqrt{2}\gamma_1 < \gamma)$, and a symmetric positive-definite matrix $P>0$ such that:

$$\left\| E(sI - A)^{-1}\begin{bmatrix}\delta B & 0 & \gamma H_1\end{bmatrix}\right\|_\infty < \gamma_1 \quad (19)$$

and

$$A_\delta P + PA_\delta^T + PW_\delta P + R_\delta < 0 \quad (20)$$

where

$$\begin{aligned}&A_\delta = A - \gamma^2\delta^{-2}H_1H_2^TC\\ &W_\delta = \gamma_2^{-2}L^TL - C^TC + \gamma^2\delta^{-2}C^TH_2H_2^TC(\gamma_2^2 = 0.5\gamma^2 - \gamma_1^2)\\ &R_\delta = BB^T + \gamma^2\delta^{-2}H_1H_1^T\end{aligned} \quad (21)$$

Then when $K = PC^T$, (4) is $H_\infty$ filter at level $\gamma$ for linear system with partly uncertain parameter (system given by (1) with (2) and $\Delta B \equiv 0$).

*Proof*: From Lemma2.2 the following results can be obtained: if there exists a scalar $\delta > 0$ and matrix $K$ such that $T_\delta(s)$ is robust stable with disturbance attenuation $\gamma$ where

$$T_\delta(s) = \begin{bmatrix} L & -L \\ \delta E & 0\end{bmatrix}\left(sI - \begin{bmatrix} A & 0 \\ KC & A-KC\end{bmatrix}\right)^{-1}\begin{bmatrix} B & 0 & \gamma\delta^{-1}H_1 \\ 0 & K & \gamma\delta^{-1}KH_2\end{bmatrix} \quad (22)$$

then filter (4) is $H_\infty$ filter at level $\gamma$ for linear system with partly uncertain parameters

So it is only to proof the following results:

1. $\begin{bmatrix} A & 0 \\ KC & A-KC\end{bmatrix}$ is stable;
2. $\left\|T_\delta(s)\right\|_\infty < \gamma$

By linear transform: $T_L = \begin{bmatrix} I & 0 \\ I & -I\end{bmatrix}$,

$$T_\delta(s) = \begin{bmatrix} L(sI - A + KC)^{-1}\begin{bmatrix} B & -K & \gamma\delta^{-1}(H_1 - KH_2)\end{bmatrix} \\ \delta E(sI - A)^{-1}\begin{bmatrix} B & 0 & \gamma\delta^{-1}H_1\end{bmatrix}\end{bmatrix}. \quad (23)$$

From $K = PC^T$ and (20) with (21), the following **ARI** can be obtained:

$$\begin{aligned}&(A-KC)P + P(A-KC)^T + \gamma_2^{-2}PL^TLP\\ &+\begin{bmatrix} B & -K & \gamma\delta^{-1}(H_1 - KH_2)\end{bmatrix}\begin{bmatrix} B & -K & \gamma\delta^{-1}(H_1 - KH_2)\end{bmatrix}^T < 0\end{aligned} \quad (24)$$

From Lemma 1,

$$\left\| L(sI - A + KC)^{-1}\begin{bmatrix} B & -K & \gamma\delta^{-1}(H_1 - KH_2)\end{bmatrix}\right\|_\infty < \gamma_2 \quad (25)$$

can be obtained and (A-KC) is stable.

From (19) and (25),the following formation holds:

$$\begin{aligned}&\left\|T_\delta(s)\right\|_\infty^2 =\\ &\left\|\begin{bmatrix} L(sI - A + KC)^{-1}\begin{bmatrix} B & -K & \gamma\delta^{-1}(H_1 - KH_2)\end{bmatrix} \\ \delta E(sI - A)^{-1}\begin{bmatrix} B & 0 & \gamma\delta^{-1}H_1\end{bmatrix}\end{bmatrix}\right\|_\infty^2\\ &\le 2(\gamma_1^2 + \gamma_2^2) = \gamma^2\end{aligned}$$

As $A$ and $(A\text{-}KC)$ are stable, $\begin{bmatrix} A & 0 \\ KC & A-KC\end{bmatrix}$ is stable.

Thus this ends the proof.

*Theorem 3*: Let $\gamma > 0$ be given, if $A$ is stable and there exist $\delta > 0, \gamma_1 > 0$ $(\sqrt{2}\gamma_1 < \gamma)$, and a symmetric positive-definite matrix $P>0$ such that:

$$\left\| E(sI - A)^{-1}\begin{bmatrix}\sqrt{2}\delta B & \sqrt{2}\delta M_B^{\frac{1}{2}} & \gamma H_1\end{bmatrix}\right\|_\infty < \gamma_1 \quad (26)$$

and

$$A_\delta P + PA_\delta^T + PW_\delta P + R_\delta < 0 \quad (27)$$

where

$$\begin{aligned}&A_\delta = A - \gamma^2\delta^{-2}H_1H_2^TC\\ &W_\delta = \gamma_2^{-2}L^TL - C^TC + \gamma^2\delta^{-2}C^TH_2H_2^TC(\gamma_2^2 = 0.5\gamma^2 - \gamma_1^2)\\ &R_\delta = 2BB^T + 2M_B + \gamma^2\delta^{-2}H_1H_1^T\end{aligned} \quad (28)$$

Then when $K = PC^T$, (4) is $H_\infty$ filter at level $\gamma$ for linear system with uncertain parameter (system given by (1) with (2)and (3)).

*Proof*: The result follows along the same lines as Theorem 2.

## 4 APPLICATION

In order to use the main results, the following theorem will be established.

*Theorem 4*: If there exists a symmetric positive-definite $P_0 > 0$ such that

$$A_0P_0 + P_0A_0^T + P_0(W_1 - W_2)P_0 + R < 0 \quad (29)$$

with

$$W_1 \geq 0, W_2 \geq 0, R \geq 0 \quad (30)$$

Then there exists a symmetric positive-definite $P_Q > 0$ such that $P_0 + P_Q$ satisfies:

$$A_0 P + P A_0^T + P(W_1 - W_2)P + R < 0 \quad (31)$$

*Proof:* It is known from conditions that there exists a symmetric positive-definite $Q_0^{'} > 0$ such that:

$$A_0 P_0 + P_0 A_0^T + P_0(W_1 - W_2)P_0 + R + Q_0^{'} < 0 \quad (32)$$

Denote $Q_0 = P_0^{-1} Q_0^{'} P_0^{-1}$, (32) can be rewritten as

$$A_0 P_0 + P_0 A_0^T + P_0(W_1 - W_2 + Q_0)P_0 + R < 0 \quad (33)$$

For any $Q$ with $0 < Q < Q_0$, it follows that there exists $P_Q > 0$ such that:

$$A_Q P_Q + P_Q A_Q + P_Q R_Q P_Q - W_Q = 0 \quad (34)$$

with

$$\begin{aligned} &A_Q = A_0 + P_0(W_1 - W_2 + Q); \\ &W_Q = P_0(Q_0 - Q)P_0 > 0; R_Q = Q + W_1 > 0; \end{aligned} \quad (35)$$

Thus $P = P_0 + P_Q$ make the following inequality true:

$$\begin{aligned} &A_0 P + P A_0^T + P(W_1 - W_2)P + R \\ &= -PQP - P_Q W_2 P_Q < 0 \end{aligned} \quad (36)$$

This ends the proof.

It is known from the proof procedure of Theorem 4 that if there exists one solution satisfying (29), then there exists a set of solutions satisfying (29). So based on Theorem 1, 2 and 3, a set of $H_\infty$ filters for nominal system and robust $H_\infty$ filters for (partly) uncertain systems can be designed.

*Algorithm 1*: To design $H_\infty$ filter to optimize multiple system performance indexes:

Denote

$$A_0 = A, W_1 = \gamma^{-2} L^T L, W_2 = C^T C, R = BB^T$$

1.Find a matrix $P_0 > 0$ of **ARI** (17);

2.Compute

$$Q_0 = -P_0^{-1}\left[A_0 P_0 + P_0 A_0^T + P_0(W_1 - W_2)P_0 + R\right]P_0^{-1}$$

3.Select $Q$ with $0 < Q < Q_0$. Find the solution $P_Q$ from **ARE** (34) with (35).

4.Compute $P = P_0 + P_Q$. Let $K = PC^T$ and obtain filter (4). Examine if other performances are satisfied. If yes, success; else return to Step 3.

*Remark 4*: If a solution $P_0 > 0$ can't be found from **ARI** (17), then we can't design $H_\infty$ filter at level $\gamma$ by Theorem 1.

*Algorithm 2*: To design robust filter to optimize multiple system performance indexes

Denote

$$\begin{aligned} &A_0 = A - \gamma^2 \delta^{-2} H_1 H_2^T C; W_2 = C^T C \\ &W_1 = \gamma_2^{-2} L^T L + \gamma^2 \delta^{-2} C^T H_2 H_2^T C; \\ &R = BB^T + \gamma^2 \delta^{-2} H_1 H_1^T \end{aligned}$$

(or $R = 2BB^T + 2M_B + \gamma^2 \delta^{-2} H_1 H_1^T$)

1:Find $\delta > 0$ such that (19) or (26) holds and $\sqrt{2}\gamma_1 < \gamma$;

2:Find a solution $P_0 > 0$ from **ARI** (20) with $\gamma_2 = \sqrt{\frac{1}{2}\gamma^2 - \gamma_1^2}$; (if failure, return to Step 1)

3:Compute

$$Q_0 = -P_0^{-1}\left[A_0 P_0 + P_0 A_0^T + P_0(W_1 - W_2)P_0 + R\right]P_0^{-1}$$

4 Select $Q$ with $0 < Q < Q_0$. Find the solution $P_Q$ from **ARE** (34) with (35)

5:Compute $P = P_0 + P_Q$. Let $K = PC^T$ and obtain filter (4). Examine if other performances are satisfied. If yes, success; else return to Step 3

*Remark 6*: If Step1 is failure, the Algorithm 2 is failure.

To design filter according to Algorithm 1 or 2, **ARI** (17), (20) or (27) need to be solved. .But **ARE** with a parameter $\lambda$ can be used to replaced **ARI**. For example, instead of solving **ARI** (17) the following **ARE** with parameter $\lambda$ can be solved

$$AP + PA^T + P(C^T C - \gamma^{-2} L^T L)P + BB^T + \lambda I = 0$$

## 5.CONCLUSION

Based on basic connection between **ARE** and $H_\infty$ norm inequality, the paper has studied $H_\infty$ filtering problem for nominal system and robust $H_\infty$ filtering problem for uncertain parameter systems. By using the technique of **ARI** neatly, a bridge between the certain parameter system and uncertain parameter systems can be built. Moreover a set of $H_\infty$ filters for nominal system and robust $H_\infty$ filters for uncertain system can be designed.

Using the technique of **ARI** neatly, it is possible to optimize multiple system performance indexes simultaneously. Moreover, to design (robust) $H_\infty$ filter to optimize multiple system performance indexes according to Algorithm 1(or 2), a series of **ARE** need to be solved.

## REFERENCES


.Barmish B.R(1985) "Necessary and Sufficient conditions for quadratic stabilizability of an uncertain system" *J. Optimiz. Theory Appl.* **46,** 399-408

Khargonekar P.P, I.R.Petersen and K.Zhou(1990) "Robust stabilization of uncertain linear system: Quadratic stabilizability and $H_\infty$ control theory" *IEEE. Trans. Automat. Control,* **35**, 356-361

Nagpal K M. and Khargonekar P.P. (1991) "Filtering and Smoothing in an $H_\infty$ Setting " *IEEE. Trans. Automat. Control.*, **36**,152-166

Packard A. and Doyle J. (1990) "Quadratic stability with real and complex perturbations" *IEEE. Trans. Automat. Control,* **35,**198-201

Petersen I.R (1987) "Disturbance attenuation and $H_\infty$ optimization: A design method based on the algebraic Riccati Equation", *IEEE. Trans. Automat. Control,* **32**, 427-429

Willems J. C. (1971) "Least squares stationary optimal control and the algebraic Riccati equation" *IEEE. Trans. Automat. Control,* **16** 621-634

Xie L and Souza C E D (1992)"Robust $H_\infty$ control for Linear System with Norm-bounded Time-varying Uncertainty" *IEEE. Trans. Automat.Control.***37**, 1188-1191

Xie L., Minyue Fu and Souza C E D (1992,) "$H_\infty$ control and Quadratic stabilization of system with Parameter Uncertainty Via Output Feedback" *IEEE. Trans. Automat. Control,* **37**, 1253-1256

# THE ROBUST $H_\infty$ PERFORMANCE PROBLEM FOR SYSTEMS WITH GENERAL NONLINEAR UNCERTAINTIES *

**Lei Guo, Chun-Bo Feng and Kanjian Zhang**

*Research Institute of Automation, Southeast University, Nanjing, 210096, P. R. CHINA ,*
*Email: leiguo@seu.edu.cn*

Abstract: In this paper robust performance analysis and $H_\infty$ controller design for a class of systems with time-varying and nonlinear uncertainties are considered. The LMI approach is extended to solve the $H_\infty$ control problem for the general uncertain systems, which have nonlinear perturbations not only in the state equation, but also in the controlled output equation and the measurement output equation. Meanwhile, these uncertainties not only can be involved in state, but also in control input, measurement output, exogenous input and derivative of state. A new general bounded real lemma is presented and both the robust $H_\infty$ state feedback and output feedback controller design are discussed for this general uncertain systems. Moreover, LMI-based feasible algorithms based on LMI for controller design are also obtained. *Copyright © 1999 IFAC*

Keywords: $H_\infty$ control, robust control, LMI, nonlinear control.

## 1 Introduction

$H_\infty$ control theory has made significant advancement in both linear system and nonlinear system context. The linear $H_\infty$ control problem has been solved via 2-Riccati-Equation approach and Linear-Matrix-Inequality (LMI) approach (Doyle, *et al.,* 1989; Iwasaki and Skelton, 1994; Gahinet, 1996). During the past several years, the Hamilton-Jaccobi-Inequality based results have been obtained for the nonlinear $H_\infty$ problem (See Isidori and Kang, 1995 and references therein). However, these HJIs are difficult to solve in general. Some special nonlinear systems were presented to simplify the HJI-based results. For example, in recent papers (Zhan *et al.*, 1996; Guo *et al.*, 1997) the $H_\infty$ control problem was discussed for Lur'e systems with single and multiple nonlinear loops respectively.

On the other hand, the $H_\infty$ robust performance synthesis problem for uncertain systems has been an important research subject in control engineering. For linear plants with norm-bounded parameter perturbations, many results based on algebraic Riccati equation (ARE) or inequality (ARI) were presented.(see e.g. Xie *et al*, 1992; Xie, 1996). Furthermore, in Xie (1996) based on LMI approach the corresponding results was extended to the uncertain systems with perturbations of a linear fractional form in all matrices of the state and output equations. Another type of structured uncertainty described by the polytope of vertex matrices was considered by Zhou et al (1994). In Wang (1992), Shen and Tamura (1994), the $H_\infty$ control problem for systems with nonlinear uncertainty satisfying Lipschitz condition was discussed. The robust $H_\infty$ problem was reduced to the nominal one of a extended linear system with scaling parameters considered, and simple ARE-based solutions were provided. However, most previous uncertain models did not include any uncertainty in the term of derivative of state, which often appears in practical plants such as robot, except those in Shen and Tamura (1994). Even in Shen and Tamura (1994), the uncertainties did not appear in all system matrices and only state feedback controllers were designed. Hence, it is important to consider the robust $H_\infty$ control problem for systems with uncertainties in all terms of dynamical equations including that of the derivative of the state. Our objective in this paper is to extend the LMI approach proposed in Iwasaki and Skelton (1994), Gahinet (1996) to solve the $H_\infty$ control problem for the general uncertain systems, which have nonlinear perturbations not only in the state equation, but also in the controlled output equation and the measurement output equation.

## 2 Problem Statement

Consider the nonlinear system described by

$$\begin{aligned}\dot{x} & +E_0 f_0(\dot{x},t) = Ax + E_1 f_1(x,t) + B_1 w \\ & +L_1 h_1(x,w,t) + B_2 u + M_1 g_1(x,t)u,\end{aligned} \quad (1)$$

$$\begin{aligned}z = & \ C_1 x + E_2 f_2(x,t) + D_{11} w + L_2 h_2(x,w,t) \\ & +D_{12} u + M_2 g_2(x,t)u,\end{aligned} \quad (2)$$

$$y = C_2 x + E_3 f_3(x,t) + D_{21} w + L_3 h_3(x,w,t), \quad (3)$$

where $x$ denotes the state vector defined in a neighborhood of the origin in $R^n$, $w \in R^{n_w}$ is the exogenous input, $u \in R^{n_m}$ control input, $z \in R^{n_z}$ controlled output, $y \in R^{n_y}$ measurement output, $A, B_1, B_2, C_1, C_2, D_{12}$, $E_0 \in R^n \times R^{n_0}, E_1 \in R^n \times R^{n_1}, E_2 \in R^{n_z} \times R^{n_2}, E_3 \in R^{n_y} \times R^{n_3}, L_1 \in R^n \times R^{n_4}$, $L_2 \in R^{n_z} \times R^{n_5}$, $L_3 \in R^{n_y} \times R^{n_6}$, $M_1 \in R^n \times R^{n_7}, M_2 \in R^{n_z} \times R^{n_8}$, are known matrices with appropriate dimensions. The mappings $f_0(\dot{x},t)$ : $R^n \times R \rightarrow R^{n_0}$, $f_i(x,t)$ :$R^n \times R \rightarrow R^{n_i}, i = 1,2,3$, $h_i(x,t)$ :$R^n \times R \rightarrow R^{n_{i+3}}, i = 1,2,3$, $g_i(x,t)$ : $R^n \times R \rightarrow R^{n_{i+6}} \times R^{n_u}, i = 1,2$, are unknown nonlinear smooth functions.

Similar to Shen and Tamura (1994) and Xie *et al* (1996), the following assumptions are made for the unknown functions:

Assumption 1(A1): The uncertain functions $f_i(x,t)(i = 0,1,2,3), h_j(x,w,t)(j = 1,2,3)$, $g_k(x,t)(k = 1,2)$ satisfy

$$\begin{gathered}\|f_0(\dot{x},t)\| \leq \|W_0 \dot{x}\|, \|f_i(x,t)\| \leq \|W_i x\|, \\ \|h_j(x,w,t)\| \leq \|W_{hj} w\|, \|g_k(x,t)u\| \leq \|W_{gk} u\|, \\ \forall \dot{x} \in R^n, \forall x \in R^n, w \in R^{n_w}. \forall t \in R, \\ i = 1,2,3. j = 1,2,3, k = 1,2\end{gathered} \quad (4)$$

where $W_i(i = 0,1,2,3), W_{hj}(j = 1,2,3), W_{gk}(k = 1,2)$ are given weighting matrices.

Assumption 2(A2): The uncertain functions $f_i(x,t)(i = 0,1,2,3)$, $h_j(x,w,t)(j = 1,2,3)$, $g_k(x,t)(k = 1,2)$ satisfy

$$\begin{gathered}f_i(0,t) = 0, h_j(0,0,t) = 0, g_k(0,t) = 0, \\ i = 0,1,2,3; j = 1,2,3; k = 1,2,\end{gathered}$$

and the origin $x = 0$ is the equilibrium point of the free system.

The model not only can describe some practical systems (e.g., some robot systems), but can be considered as an approximation of general nonlinear systems. In fact, most previous uncertain models can be coped with as special cases of this model. Firstly, the so-called norm bounded uncertainty discussed in Xie *et al* (1992) and Xie (1996), which is the most popular model in the recent papers, can be considered as a special case of ours. In fact, their models are required to satisfy a matching property (see Xie (1996) for details). Secondly, the models considered in Wang *et al.* (1992), Xie *et al.* (1996), Wang and Zhan (1996) are also special cases of ours naturally. On the other hand, the uncertain plants discussed can be considered as a class of nonlinear plants, where the nonlinear terms satisfy norm-bounded conditions Xie *et al.* (1996). It is obvious that the Lur'e systems discussed in Zhan *et al.* (1996), Guo *et al.* (1997) can also be dealt with as a special case of ours.

Consider the following free linear system with nonlinear perturbations

$$\begin{aligned}\dot{x} + E_0 f_0(\dot{x},t) = & \ Ax + E_1 f_1(x,t) + B_1 w \\ & +L_1 h_1(x,w,t),\end{aligned} \quad (5)$$

$$z = C_1 x + E_2 f_2(x,t) + D_{11} w + L_2 h_2(x,w,t), \quad (6)$$

If the system is globally asymptotically stable about the origin under zero initial conditions with $w = 0$, and $\|z\|_T < \gamma^2 \|w\|_T, \forall T > 0$, the system is said to be ***robust stable and*** $L_2$***-gain finite.*** In the following, $\gamma$ is supposed to be 1 without loss of generality. In the following, the independent variables in $f, g, h$ are omitted for brevity in case of no confusion.

The Robust $H_\infty$ Performance Problem is to find controllers for the plant (1–3) such that the closed-loop system is robust stable and $L_2$-gain finite. If state (output) feedback controllers are considered, then the above problem will be denoted by P1 (P2).

## 3 Basic Lemmas and State Feedback Controller Design

The following lemma (Lemma 1) gives a new

sufficient criterion to judge the robust $H_\infty$ performance for the free system described by (5) and (6). Lemma 2 illustrates that the robust $H_\infty$ performance problem of (5) and (6) can be cast into that of an corresponding auxiliary certain system . It is a direct extension of Theorem 2 in Shen and Tamura (1994).

**Lemma 1:** If there exist $\lambda_i > 0, i = 0,1,2, \lambda_{hj} > 0, j = 1,2$ and $P > 0$ satisfying

$$\begin{bmatrix} A^TP+PA & PB_H & C_H^T \\ B_H^TP & -I & D_H^T \\ C_H & D_H & -I \end{bmatrix} < 0, \qquad (7)$$

then the uncertain system described by (5) and (6) is stable and $L_2$-gain finite, where

$$\begin{bmatrix} B_H \\ C_H^T \\ D_H \end{bmatrix} := \left[\begin{array}{ccc} B_1 & B_2 & 0 \\ \hline C_1^T & C_2^T & C_3^T \\ \hline D_{11} & 0 & D_{H04} \\ D_{H11} & D_{H12} & 0 \\ D_{H21} & O & O \end{array}\right], \qquad (8)$$

and

$$B_2 := \begin{bmatrix} \lambda_0 E_0 & \lambda_1 E_1 & \lambda_{h1} L_1 \end{bmatrix},$$
$$C_2^T := \begin{bmatrix} \frac{1}{\lambda_0}(W_0A)^T & \frac{1}{\lambda_1}W_1^T & 0 \end{bmatrix},$$
$$C_3^T := \begin{bmatrix} \frac{1}{\lambda_2}W_2^T & 0 \end{bmatrix},$$
$$D_{H04} := \begin{bmatrix} \lambda_2 E_2 & \lambda_{h2} L_2 \end{bmatrix},$$
$$D_{H12} := \begin{bmatrix} W_0E_0 & \frac{\lambda_1}{\lambda_0}W_0E_1 & \frac{\lambda_{h1}}{\lambda_0}W_0L_1 \end{bmatrix},$$
$$D_{H21}^T := \begin{bmatrix} 0 & \frac{1}{\lambda_{h1}}W_{h1}^T & 0 & \frac{1}{\lambda_{h2}}W_{h2}^T \end{bmatrix}$$

The proof is omitted for brevity.

If the uncertain system described by (5) and (6) satifies (7), it is said to have *strong $H_\infty$ performance* (Zhou *et al.*, 1994; Shen and Tamura, 1994). This notion is obviously equivalent to 'quadratic stabilization with disturbance attenuation' introduced by Xie *et al.* (1992).

**Lemma 2:** If there exist $\lambda_i > 0, i = 0,1,2, \lambda_{hj} > 0, j = 1,2$ and $P > 0$ such that the following auxiliary linear system

$$\begin{cases} \dot{x} = Ax + B_Hw \\ z = C_Hx + D_Hw \end{cases} \qquad (9)$$

is stable and satisfies $\|z\|_2 \le \|w\|_2$, then the uncertain system described by (5) and (6) has strong $H_\infty$ performance.

The result can be proved via bounded real lemma and Lemma 1 directly.

On the other hand, it is noted that (7) is equivalent to

$$\begin{bmatrix} QA_H^T + A_HQ & B_H & QC_H^T \\ B_H^T & -I & D_H^T \\ C_HQ & D_H & -I \end{bmatrix} < 0, \qquad (10)$$

with transformation $Q := P^{-1}$. In fact, (10) is also equivalent to a Riccati inequality. It is easy to verify that this criterion, as a new generalized bounded real lemma, extends some previous results.

Substitute the state feedback controller $u = Kx$ to the plant (1) and (3). The closed-loop system is

$$\dot{x} + E_0f_0 = (A+B_2K)x + \bar{E}_1\bar{f}_1 + B_1w + L_1h_1, \quad (11)$$

$$z = (C_1 + D_{12}K)x + \bar{E}_2\bar{f}_2 + D_{11}w + L_2h_2, \quad (12)$$

where $(i = 1,2)$

$$\bar{E}_i := \begin{bmatrix} E_i & \lambda_{gi}M_i \end{bmatrix}, \bar{f}_i := \begin{bmatrix} f_i \\ \frac{1}{\lambda_{gi}}g_iKx \end{bmatrix}.$$

Denote

$$\bar{W}_i := \begin{bmatrix} W_i \\ \frac{1}{\lambda_g}W_{gi}K \end{bmatrix}, i = 1,2, \qquad (13)$$

then

$$\begin{aligned} \|\bar{f}_i\|^2 &= \|f_i\|^2 + \tfrac{1}{\lambda_{gi}{}^2}\|g_iKx\|^2 \\ &\le \|W_ix\|^2 + \tfrac{1}{\lambda_{gi}{}^2}\|W_{gi}Kx\|^2 \,. \\ &= \|\bar{W}_ix\|^2 \end{aligned}$$

Substituting the coefficient matrices of (11-13) to (10) yields

$$\begin{bmatrix} M(Q,R) & \bar{B} & \bar{C}^T(R) \\ \bar{B}^T & -I & \bar{D}^T \\ \bar{C}(R) & \bar{D} & -I \end{bmatrix} < 0, \qquad (14)$$

where

$$M(Q,R) := AQ + B_2R + QA^T + R^TB_2,$$
$$R := KQ, \bar{C}(R) := \hat{C} + \hat{D}R,$$

and

$$\begin{bmatrix} \bar{B} \\ \hat{C}^T \\ \hat{D}^T \end{bmatrix} := \begin{bmatrix} N_{B1} & N_{B2} \\ N_{C1} & N_{C2} \\ N_{D1} & N_{D2} \end{bmatrix},$$
$$N_{B1} := \begin{bmatrix} B_1 & \lambda_0E_0 & \lambda_1E_1 & \lambda_1\lambda_{g1}M_1 \end{bmatrix},$$
$$N_{B2} := \begin{bmatrix} \lambda_{h1}L_1 & 0 & 0 & 0 \end{bmatrix},$$
$$N_{C1} := \begin{bmatrix} C_1^T & \frac{1}{\lambda_0}(W_0A)^T & \frac{1}{\lambda_1}W_1^T & 0 \end{bmatrix},$$
$$N_{C2} := \begin{bmatrix} 0 & \frac{1}{\lambda_2}W_2^T & 0 & 0 \end{bmatrix},$$
$$N_{D1} := \begin{bmatrix} D_{12}^T & 0 & 0 & \frac{1}{\lambda_1\lambda_{g1}}(W_{g1})^T \end{bmatrix},$$
$$N_{D2} := \begin{bmatrix} 0 & 0 & \frac{1}{\lambda_1\lambda_{g2}}W_{g2}^T & 0 \end{bmatrix},$$

$$\bar{D} := \begin{bmatrix} N_{D3} & N_{D4} \end{bmatrix},$$
$$N_{D3} := \begin{bmatrix} D_{11} & 0 & 0 & 0 \\ \frac{1}{\lambda_0}W_0B_1 & W_0E_0 & \bar{D}_{of23} & \frac{\lambda_{h1}}{\lambda_0}W_0L_1 \\ \bar{D}_{sf31} & O & O & O \end{bmatrix},$$
$$N_{D4} := \begin{bmatrix} \lambda_2E_2 & \lambda_2\lambda_{g2}M_2 & \lambda_{h2}L_2 \\ 0 & 0 & 0 \\ O & O & O \end{bmatrix},$$

$$\bar{D}_{sf31}^T := \begin{bmatrix} 0 & 0 & \frac{1}{\lambda_{h1}}W_{h1}^T & 0 & 0 & \frac{1}{\lambda_{h2}}W_{h2}^T \end{bmatrix}^T,$$
$$\bar{D}_{of23} := \begin{bmatrix} \frac{\lambda_1}{\lambda_0}W_0E_1 & \frac{\lambda_1\lambda_{g1}}{\lambda_0}W_0M_1 \end{bmatrix}.$$

(14) is an LMI with respect to $Q > 0$ and $R$. The theorems below result from the above derivation.

**Theorem 1:** For the uncertain plants (1) and (3), there exists a state feedback controller such that the closed-loop system has strong $H_\infty$ performance if and only if there exist $Q > 0$ and $R$ satisfying (14). In this case, the desired controller is $K = RQ^{-1}$.

**Theorem 2:** For the uncertain plants (1) and (3), there exists a state feedback controller $u = Kx$ such that the closed-loop system has strong $H_\infty$ performance if and only if $u = Kx$ is an $H_\infty$ controller for the following auxiliary plant with given parameters $\lambda_i, \lambda_{gj}.\lambda_{kj}, i = 0, 1, 2; j = 1, 2$,

$$\begin{cases} \dot{x} & = Ax + \bar{B}w + B_2u \\ z & = \hat{C}x + \bar{D}w + \hat{D}u \end{cases}. \quad (15)$$

Theorem 1 provides a procedure to design robust $H_\infty$ controllers, while Theorem 2 reduces the robust problems to that of an auxiliary plant without uncertainty. These results are direct extensions of those in Shen and Tamura (1994).

## 4 Output Feedback Controller: General Case

Suppose $\Sigma_c$ is a output feedback controller with order of $n_c$, which can be described as

$$\begin{cases} \dot{x}_c & = A_cx_c + B_cy \\ u & = C_cx_c \end{cases}. \quad (16)$$

In the following we apply Lemma 1 and 2 to the closed-loop system combined with (5),(6) and (16).

Define

$$\mathsf{A} := A, \mathsf{B}_2 := B_2, \mathsf{C}_2 := C_2,$$
$$\begin{bmatrix} \mathsf{B}_1 \\ \mathsf{D}_{21} \\ \mathsf{C}_1{}^T \\ \mathsf{D}_{12}{}^T \\ \mathsf{D}_{11}{}^T \end{bmatrix} = \begin{bmatrix} P_{B1} & P_{B2} & P_{B3} \\ P_{D1} & P_{D2} & P_{D3} \\ P_{C1} & P_{C2} & P_{C3} \\ P_{D4} & P_{D5} & P_{D6} \\ P_{D7} & P_{D8} & P_{D9} \end{bmatrix}, \quad (17)$$

where

$$P_{B1} := \begin{bmatrix} B_1 & \lambda_0E_0 & 0 \end{bmatrix},$$
$$P_{B2} := \begin{bmatrix} \lambda_1E_1 & \lambda_1M_1 & 0 \end{bmatrix},$$
$$P_{B3} := \begin{bmatrix} \lambda_{h1}L_1 & 0 & 0 & 0 & 0 \end{bmatrix},$$
$$P_{D1} := \begin{bmatrix} D_{21} & 0 & 0 \end{bmatrix},$$
$$P_{D2} := \begin{bmatrix} 0 & 0 & \lambda_1E_3 \end{bmatrix},$$
$$P_{D3} := \begin{bmatrix} 0 & \lambda_{h1}L_3 & 0 & 0 & 0 \end{bmatrix},$$
$$P_{C1} := \begin{bmatrix} C_1^T & \frac{1}{\lambda_0}A^TW_0^T & 0 \end{bmatrix},$$
$$P_{C2} := \begin{bmatrix} \frac{1}{\lambda_1}W_1^T & 0 & \frac{1}{\lambda_1}W_3^T \end{bmatrix},$$
$$P_{C3} := \begin{bmatrix} 0 & 0 & \frac{1}{\lambda_2}W_2^T & 0 & 0 \end{bmatrix},$$
$$P_{D4} := \begin{bmatrix} D_{12}^T & 0 & 0 \end{bmatrix},$$
$$P_{D5} := \begin{bmatrix} 0 & \frac{1}{\lambda_1}W_{g1}^T & 0 \end{bmatrix},$$
$$P_{D6} := \begin{bmatrix} 0 & 0 & 0 & \frac{1}{\lambda_2}W_{g2}^T & 0 \end{bmatrix},$$
$$P_{D7} := \begin{bmatrix} D_{11} & 0 & 0 \\ \frac{1}{\lambda_0}W_0B_1 & W_0E_0 & 0 \\ 0 & 0 & 0 \\ D_{of}^T & 0 & 0 \end{bmatrix},$$
$$P_{D8} := \begin{bmatrix} 0 & 0 & 0 \\ \frac{\lambda_1}{\lambda_0}W_0E_1 & \frac{\lambda_1}{\lambda_0}W_0M_1 & 0 \\ 0 & 0 & \frac{\lambda_1}{\lambda_0}W_0B_cE_3 \\ 0 & 0 & 0 \end{bmatrix},$$
$$P_{D9} := \begin{bmatrix} P_{D91} & P_{D92} \end{bmatrix},$$

and

$$P_{D91} := \begin{bmatrix} 0 & 0 \\ \frac{\lambda_1}{\lambda_0}W_0L_1 & 0 \\ 0 & \frac{\lambda_{h1}}{\lambda_0}W_0B_cL_3 \\ 0 & 0 \end{bmatrix},$$
$$P_{D92} := \begin{bmatrix} \lambda_2E_2 & \lambda_2M_2 & \lambda_{h2}L_2 \\ 0 & 0 & 0 \\ 0 & 0 & 0 \\ 0 & 0 & 0 \end{bmatrix},$$
$$D_{of}^T = \begin{bmatrix} 0 \mid \frac{1}{\lambda_{h1}}W_{h1}^T & \frac{1}{\lambda_{h1}}W_{h3}^T & 0 & 0 & \frac{1}{\lambda_{h2}}W_{h2} \end{bmatrix}.$$

It can be verified that the matrices in the closed-loop systems corresponding to $A_H, B_H, C_H, D_H$ of (8) are

$$\bar{A}_H = \begin{bmatrix} \mathsf{A} & \mathsf{B}_2C_c \\ B_c\mathsf{C}_2 & A_c \end{bmatrix}, \bar{B}_H := \begin{bmatrix} \mathsf{B}_1 \\ B_c\mathsf{D}_{21} \end{bmatrix},$$
$$\bar{C}_H := \begin{bmatrix} \mathsf{C}_1 & \mathsf{D}_{12}C_c \end{bmatrix}, \bar{D}_H = \mathsf{D}_{11}. \quad (18)$$

In fact, if the non-strictly proper controllers are concerned, the methods in Iwasaki and Skelton (1994) are not available here. Fortunately, we can separate the Lyapunov parameter $P$ and the controller parameter $A_c, B_c, C_c$ and get the strictly proper controllers by means of the method in Gahinet (1996) and Guo *et al.* (1998). Since there are controller parameter $B_c$ included in $\mathsf{D}_{11}$, the following result involves LMIs with an algebraic constraints.

**Theorem 3:** For the general uncertain plants described by (1) and (2), there exists an output feedback controller such that the closed-loop systems have strong $H_\infty$ performance if and only if the following two LMIs

$$\begin{bmatrix} R & I \\ I & S \end{bmatrix} > 0 \quad (19)$$

$$\begin{bmatrix} L_{11} & L_{12} \\ L_{12}^T & L_{22} \end{bmatrix} < 0 \quad (20)$$

are solvable about $S > 0, R > 0, F, G, H$, and $B_c$, and there exist matrices $M$ with full-row rank and $N$ with full-column rank satisfying

$$MN^T = I - RS, NB_c = F. \quad (21)$$

In this case, $C_c, A_c$ can be given via the following algebraic equation

$$\begin{aligned} C_c = & \; GM^+ \\ A_c = & \; N^+[H - NB_cC_2R + SB_2C_cM^T \\ & +SAR]M^+. \end{aligned} \quad (22)$$

In (20), the shorthand notations are

$$\begin{aligned} L_{11} &:= \begin{bmatrix} L_{111} & H^T \\ H & L_{112} \end{bmatrix}, \\ L_{111} &:= AR + RA^T + B_2G + G^TB_2^T, \\ L_{112} &:= SA + A^TS + FC_2 + C_2^TF^T \end{aligned} \quad (23)$$

$$\begin{aligned} L_{12} &:= \begin{bmatrix} B_1 & RC_1^T + G^TD_{12}^T \\ B_1^TS + D_{21}^TF^T & C_1^T \end{bmatrix}, \\ L_{22} &:= \begin{bmatrix} -I & D_{11}^T \\ D_{11} & -I \end{bmatrix}. \end{aligned} \quad (24)$$

Proof: The proof can be obtained in connection with the main results in Gahinet (1996) and Lemma 2. We only give a brief illustration in order to save space.

Suppose that $A, B_i, C_i, D_{1i}, D_{21}, i = 1, 2$ are known matrices. From (18) we can consider the closed-loop system as the following plant combined with the same SPC

$$\begin{cases} \dot{\bar{x}} &= A\bar{x} + B_1w + B_2u \\ z &= C_1\bar{x} + D_{11}w + D_{12}u \\ y &= C_2\bar{x} + D_{21}w \end{cases} . \quad (25)$$

Thus, the same procedure as that in Gahinet (1996) can be used to disjoin $P$ and the controller parameter. However, $B_c$ appearing in $D_{11}$ does not make the separation completely. That is, $F, R, S$ and $B_c$ need satisfy (21) besides (19)and (20). Thus, we complete the proof.

This theorem reduces the output feedback problem to the solvability of LMIs (20) and (19) with an algebraic constraint (21). In the following, we will simplify the algorithm for the general case. It is first noted that if $N$ with full-column rank satisfying

$$NB_c = F, \quad (26)$$

then there must exists a desired $M$ satisfying $MN^T = I - RS$ with $N$.

Without loss of generality, we assume

$$n_y \leq n_c \leq n. \quad (27)$$

In this case, for the key algebraic equation $NB_c = F$ about $N$, we can show that there must exist an $N$ with full-column rank satisfying (26) under the above assumption.

If $n_y < n_c \leq n$, then there exist infinite solutions of (26) since the number of unknown qualities ($n_c$) in each row of $N$ is larger than that of equations ($n_y$) involved. Then we can always seek the full-column rank ones from these solutions. For example, supposing $n = n_c > n_y$, when we get an singular $N$ from (26), we can choose a tiny parameter $a$ such that $N + aI$ is nonsingular and $F + aB_c$ also satisfies (20) and (19) together with other variables.

If $n_y = n_c \leq n$, we can always obtain an nonsingular matrix $B_c$ satisfying (20) and (19) joint with other variables. For instance, we can choose $B_c + aI$ instead of the former $B_c$. Hence, $N = FB_c^{-1}$ holds and $N$ has full-column rank.

In conclusion, Theorem 3 can be simplified as follows under the general assumption (27).

**Theorem 4**: Suppose the assumption (27) holds. For the general uncertain plants described by (1) and (2), there exists an output feedback controller such that the closed-loop systems have strong $H_\infty$ performance if and only if the LMIs (19) and (20) have a group of solutions on $S > 0, R > 0, F, G, H$, and $B_c$. In this case, there must exist an $N$ with full-column rank satisfying (26) and $A_c, C_c$ can be given via (22), where

$$M := (I - RS)(N^T)^+.$$

Proof can be obtained based on the above argument.

Suppose $n_c = n > n_y$. The design steps for so-called full order controllers can be summarized briefly as follows:

(1) Solve $S, R, F, G, H$ and $B_c$ via LMIs (19) and (20);

(2) Solve algebraic equation (26) to get $N$. If $N$ is nonsingular, turn to step (3). If not, modify $N$ to $N + aI$ such that $N + aI$ is nonsingular and $F + aB_c$ still satisfies (19) and (20) together with other variables, where $a$ is a tiny constant.

(3) Compute nonsingular matrices $M$ such that $M = (I - RS)(N^T)^{-1}$;

(3) Solve $A_c, C_c$ via linear algebraic equations (22) with the superscript + changed to −1.

## 5 Conclusion

This paper considers robust performance analysis and $H_\infty$ controller design for a class of systems with time-varying and nonlinear uncertainties. These uncertainties not only appear in state, but in control input, measurement output, exogenous input and derivative of state. A new sufficient condition based on LMI is firstly provided to analysis the robust $H_\infty$ performance problem of the free systems. Solvable conditions are presented for both the state feedback and output feedback controller design problems. The controller design of the robust $H_\infty$ control problem are given in terms of the solvability of a class of LMIs.

## References


Doyle, J.C., K. Glover, P. P. Khargonekar, B. A. Francis (1989). State space solutions to standard $H^\infty$ control problems, *IEEE Trans. Auto. Contr.*, 39, 350-360.

Iwasaki, T. and R. E. Skelton (1994). All controllers for the general $H^\infty$ control problem: LMI existence conditions and state space formulas. *Automatica*, Vol.30:1307-1317.

Gahinet, P. (1996). Explicit controller formulas for LMI-based $H^\infty$ synthesis, *Automatica*, Vol.32:1007-1014.

Isidori, A. and W. Kang (1995). $H^\infty$ control via measurement feedback for general nonlinear systems. *IEEE Trans. Automat.Contr.*, vol.40,pp466-472.

Zhan, W., L.Y. Wang, Y. Liu and J. Sun. (1996). $H^\infty$ control for systems with sector bound nonlinearities. In:*Proc.of IFAC'96,1996*.

Guo, L., X Xin and C B Feng (1997). Robust Stabilization and H-infinity Control Problems for Lur'e Systems. In: *Proc. of 2nd ASCC'97*, Vol.1:733-736, Seoul.

Xie, L., M. Fu, and C. de Souza (1992). $H^\infty$ control and quadratic stabilization for systems with parameter uncertainty via output feedback. *IEEE Trans. Automat. Contr.*, 37, 1253-1256.

Gu, K. (1994). $H_\infty$ control of systems under norm bounded uncertainties in all system matrices. *IEEE Trans. Automat. Contr.*, 39, 1320-1322.

Xie, L.(1996). Output feedback $H_\infty$ control of systems with parameter unceratinty. *Int. J. Control*, 63, 741-750.

Zhou, K., P.P. Khargonekar, J. Stoustrup, et al.(1995). Robust Performance of Systems with Structured Uncertainties in state space. *Automatica*, Vol.31, 249–258.

Wang, Y.Y., L. Xie and C.E. Souza (1992). Robust control of a class of uncertain nonlinear systems. *Syst. Contr. Lett.*, Vol.19, 139-144.

Shen, T. and K.Tamura (1994). Robust $H^\infty$ control of linear systems with nonlinear perturbation. In:*Proc.of ACC*, pp.201-204.

Xie, L., S. Xie and C.E.de Souza (1996). $H_\infty$ Control of A Class of Uncertain Nonlinear Systems: An Observer-based Controller Design. In:*Proc. of IFAC'96*.

Wang, L.Y. and W. Zhan (1996). Robust disturbance attenuation with stability for linear systems with norm-bounded nonlinear uncertainties. *IEEE Trans. Automat. Contr.*, Vol.41(6):886-888.

Guo, L., C. B. Feng and X. Xin (1998). Existence Conditions of Discrete-Time Strictly Proper $H_\infty$ Controllers. In:*Proc. of ACC'98*, Philadelphia, 1985-1986.

14th Triennial World Congress, Beijing, P.R. China

F-2e-03-6

# ROBUST RELIABLE $H_\infty$ CONTROLLERS DESIGN VIA LMIS

**Dong Hu**

*Lab.305, Dept. of Automatic Control, Nanjing University of Aero. & Astro, CHINA*
*Email: cwhac@nuaa.edu.cn*

**Wenhua Chen**

*Center for Control and System, University of Glasgow G12 8QQ, UK*
*Email: wenchen@mech.gla.ac.uk*

Abstract: The normal $H_\infty$ control design deals with both plant modeling uncertainties and exogenous signal uncertainties by constructing a controller which stabilizes uncertain linear systems while satisfying an $H_\infty$-norm bound constraint on disturbance attenuation for all admissible uncertainties. However, the control design may result in unsatisfactory performances or even instabilities in the event of sensor failures in practical plants. This paper presents a novel technique to design robust reliable $H_\infty$ controllers for a class of time varying uncertainty system with sensor failures by solving three linear matrix inequalities (LMIs). *Copyright © 1999 IFAC*

Keywords: Reliability, sensor failures, $H_\infty$ control, uncertainty, robust stability, linear matrix inequalities (LMIs).

## 1. INTRODUCTION

Recently, the robust $H_\infty$ control theory has been developed rapidly. Since a solution to $H_\infty$ controllers by solving two algebraic Riccati equations was presented (AREs) in 1988 (Doyle *et al.*,1989), the robust $H_\infty$ control theory has been applied in many practical designs. However, some inherent constraint conditions of DGKF's solution tend to limit the scope and performance of application of the state-space design technique of $H_\infty$ controllers. The convex optimization theory has been introduced to $H_\infty$ synthesis problems via solving three linear matrix inequalities (LMIs) so as to simply the whole design procedure (Gahinet and Apkarian.,1994a). Many control problems can be recast into the versions of LMIs (Boyd *et al.*, 1994). LMIs play an important role in $H_\infty$ controllers design of many control systems including linear vary parameter systems (Apkarian and Gahinet, 1995). Thus, versions of numerical solutions instead of analytical versions of controllers for $H_\infty$ problems can be acquired via solving LMIs using convex optimization algorithm (Gahinet *et al.*, 1994b).

Both DGKF's and LMI technique of computing $H_\infty$ controllers deal with both plant modeling and exogenous signal uncertainties by synthesizing a controller which stabilizes uncertain linear systems while satisfying an $H_\infty$-norm bound constraint on disturbance attenuation for all admissible uncertainties. The robust LQG controllers for a class of systems under parameter and noise uncertainties has been designed (Chen, 1994). The relatively few people attempts to develop methodologies for the design of robust reliable $H_\infty$ controllers with measuring device failures. Plants with practical control environments where there are both system uncertainties and control component failures become lower performance or even instability. In particular, it is necessary to design autopilots of underwater vehicle accounting for sensor failures. A technique for the design of reliable $H_\infty$ control systems using the observer-based output feedback was introduced (Veillette *et al.,* 1992). In this paper, the new approach to computing reliable $H_\infty$ controllers via solving three linear matrix inequalities (LMIs) is presented which could be regarded as the extension of the results of Veillette *et al* (1992).

The paper is organized as follows. Section 1 and Section 2 review the robust reliable design for uncertain system. Section 3 prensents a class of time-varying systems. The design technique of based-observer robust reliable $H_\infty$ controller is presented in Section 4 The last section is conclusion.

## 2. RELIABILITY OF SYSTEMS WITH SENSOR FAILURES

After some sensors of the system with time-varying uncertainties fail in Fig.1. and Fig.2., it is impossible to acquire satisfactory performance using general MIMO $H_\infty$ control design technique. Sometimes, sensor failures may result in unexpected instabilities of the designed closed-loop system. In flight control systems such as missile, underwater vehicle, sensors in plants often fail because of practical operation environments. In order to acquire necessary reliabilities of autopilots, the sensor failures must be taken into account in the design of autopilots of underwater vehicle.

A representative class of time-varying uncertain systems with failures of part sensors is given in Fig. 1. and Fig. 2. Let $\Omega \subseteq \{1,2,\cdots,dim(y)\}$ correspond to a selected subset of sensors susceptible to failures. Let $C = C_\Omega + C_{\bar{\Omega}}$, where $C_\Omega$ denotes the measurement matrix associated with $\Omega$; $C_{\bar{\Omega}}$ denotes the measurement matrix associated with the complentmentary subset of measurements. Let $\omega \subseteq \Omega$ correspond to a particular subset of the susceptible sensors that actually experience a failure. So, $C = C_\omega + C_{\bar{\omega}}$, where $C_\omega$ and $C_{\bar{\omega}}$ have meanings analogous to those $C_\Omega$ and $C_{\bar{\Omega}}$ such that $C_\omega^T C_\omega \le C_\Omega^T C_\Omega$. In addition, the decomposed

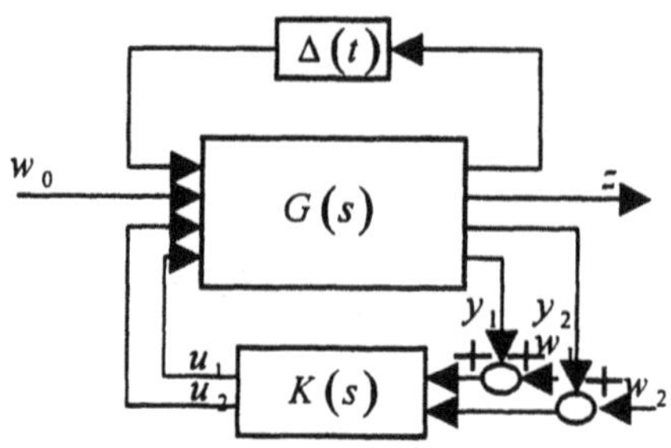


Fig. 1. Normal case of uncertain systems

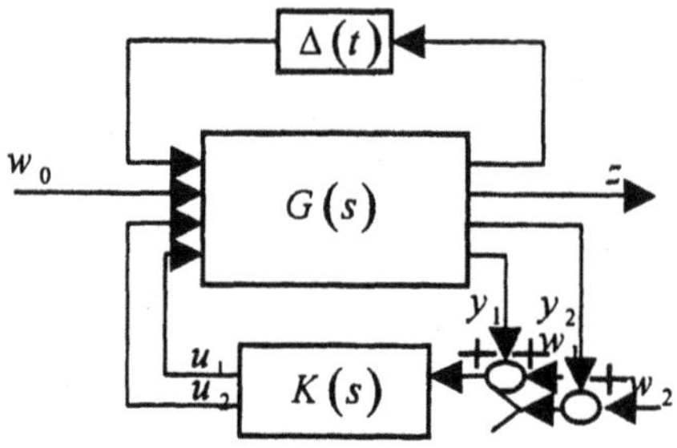


Fig. 2 . Sensor failures case of uncertain systems

versions of the observer matrix are $L = L_\omega + L_{\bar{\omega}}$, then $LC = L_\omega C_\omega + L_{\bar{\omega}} C_{\bar{\omega}}$. More details see (Veillette *et al.*, 1992).

## 3. A CLASS OF TIME-VARYING UNCERTAIN SYSTEMS

Consider a class of uncertain linear systems described by the following state-space models:

$$\begin{aligned} \dot{x}(t) &= [A + \Delta A(t)]x(t) + Bu(t) + Gw_1(t) \\ y(t) &= Cx(t) + w_2(t) \\ z(t) &= \begin{bmatrix} z_1(t) \\ z_2(t) \end{bmatrix} = \begin{bmatrix} Hx(t) \\ u(t) \end{bmatrix} \\ \Delta A(t) &= DF(t)E \end{aligned} \tag{1}$$

where $x(t) \in \mathbf{R}^n$ are the state variables, $u(t) \in \mathbf{R}^m$ is the control input, $y(t) \in \mathbf{R}^p$ is the measured output, $w_1(t) \in \mathbf{R}^r$ and $w_2(t) \in \mathbf{R}^p$ are the disturbance inputs, $z(t) \in \mathbf{R}^q$ is the controlled output. ***A***, ***B***, ***C***, ***D***, ***H*** are known real constant matches with appropriate dimensions. $\Delta A(t)$ represents time-varying parameter uncertainty. Other is $D \in \mathbf{R}^{n \times i}$, $E \in \mathbf{R}^{j \times n}$, $F(t) \in \mathbf{R}^{i \times j}$ $F(t)^T F(t) \le I$, see (Xie and Souza, 1992).

Theorem 1. (Veillette *et al.*, 1992): Let the constant $\gamma > 0$ is given. Uncertain system (1) is quadratically stabilizable with $\|T(s)\|_\infty \le \gamma$ via linear output feedback $u = K(s)y$ and real symmetric positive definite matrix $X \in \mathbf{R}^{n\times n}$ such that the matrix inequality.

$$A_{cl}^T X + XA_{cl} + \frac{1}{\gamma^2} XB_{cl}B_{cl}^T X + C_{cl}C_{cl}^T < 0 \qquad (2)$$

where $[A_{cl}, B_{cl}, C_{cl}]$ is a state-space realization of the closed-loop system.

Lemma 1. (Gahinet and Apkarian, 1994a): The block matrix $\begin{bmatrix} P & M \\ M^T & Q \end{bmatrix}$ is negative definite if and only if

$$\begin{cases} Q < 0 \\ P - MQM^T < 0 \end{cases}.$$

## 4. ROBUST RELIABLE $H_\infty$ CONTROLLERS

Consider the system (1) with sensor failures when all states are not the inputs of the controller. $(A, B)$ is a controllable pair; $(A, C_{\bar{\Omega}})$ is an observable pair, where $C_{\bar{\Omega}}$ is the decomposition matrix of $C$ for $\bar{\Omega}$. When $\gamma > 0$ is given, design a fixed output feedback reliable controller $K(s)$ to stabilize the system and have $\|T_{wz}(s)\|_\infty \le \gamma$ for all admissible uncertainties and part sensor failures.

The output-feedback controller for the system (1) is based on a state observer of the form:

$$\dot{x}_K = Ax_K + Bu + D\hat{f} + G\hat{w}_1 + L(y - Cx_K) \qquad (3)$$

where $\hat{f} = K_u x_K$, $\hat{w}_1 = K_d x_K$.
$L$ is the observer gain, $K_u$ is the uncertain estimation gain, $K_d$ is the disturbance estimation gain. $\hat{f}$ and $\hat{w}_1$ account for the uncertainty and disturbance input, respectively. The form of the controller becomes:

$$\begin{aligned} \dot{x}_K &= (A + BK + DK_u + GK_d - LC)x_K + Ly \\ u &= Kx_K \end{aligned} \qquad (4)$$

where $K$ is the control feedback gain. Let sensors corresponding to any set $\omega \subseteq \Omega$ fail. The measured output, that is the sensor output, becomes: $y = C_{\bar{\omega}}x + w_{2\bar{\omega}} + y_\omega^F$ where $w_{2\bar{\omega}}$ is the decomposition matrix of $w_2$ for $\bar{\omega}$, $w_2 = w_{2\bar{\omega}} + w_{2\omega}$. Then, a closed-loop system of 2n is as follow:

$$\begin{aligned} \dot{x}_{cl} &= A_{cl}x_{cl} + G_{cl}w_\omega^F \\ z &= H_{cl}x_{cl} \end{aligned} \qquad (5)$$

where $x_{cl} = \begin{bmatrix} x \\ x_K \end{bmatrix}$, $w_\omega^F = \begin{bmatrix} w_1 \\ w_{2\bar{\omega}} \\ y_\omega^F \end{bmatrix}$.

$$A_{cl} = \begin{bmatrix} A + DF(t)E & BK \\ LC_{\bar{\omega}} & A + BK + DK_u + GK_d - LC \end{bmatrix}$$

$$G_{cl} = \begin{bmatrix} G & 0 & 0 \\ 0 & (L^T)_{\bar{\omega}}^T & (L^T)_\omega^T \end{bmatrix}, \quad H_{cl} = \begin{bmatrix} H & 0 \\ 0 & K \end{bmatrix}.$$

$(L^T)_{\bar{\omega}}^T$ and $(L^T)_\omega^T$ are the decomposition matrices of $L^T$ for $\bar{\omega}$ and $\omega$.

The observer error is $e = x_K - x$, then,

$$\begin{aligned} \dot{x}_e &= A_e x_e + G_e w_\omega^F \\ z &= H_e x_e \end{aligned} \qquad (6)$$

where

$$A_e = \begin{bmatrix} A + DF(t)E + BK & BK \\ DK_u + GK_d - DF(t)E - (L^T)_\omega^T C_\omega & A + DK_u + GK_d - LC \end{bmatrix}$$

$$G_e = \begin{bmatrix} G & 0 & 0 \\ -G & (L^T)_{\bar{\omega}}^T & (L^T)_\omega^T \end{bmatrix}, H_e = \begin{bmatrix} H & 0 \\ K & K \end{bmatrix}.$$

Our problem may become into selecting $K$, $L$, $K_u$ and $K_d$ such that system (6) is quadratically stable with $\|T_{wz}(s)\|_\infty \le \gamma$.

Theorem 2: Given $\gamma > 0$, system (1) with sensor failures for $\omega \subseteq \Omega$. The observer-based controller (4) quadratically stabilizes the system (1) with $\|T_{wz}(s)\|_\infty \le \gamma$ in the sense of $\|z\|_2 < \gamma\|w_\omega^F\|_2$, then if and only if there exists a positive definite matrix pair $[R, S]$ such that three linear matrix inequities (LMIs):

$$\begin{bmatrix} AR + RA^T - BB^T & RH^T & \gamma RC_\Omega^T & \gamma RE^T & G & D \\ HR & -I & 0 & 0 & 0 & 0 \\ \gamma C_\Omega R & 0 & -I & 0 & 0 & 0 \\ \gamma ER & 0 & 0 & -I & 0 & 0 \\ G^T & 0 & 0 & 0 & -\gamma^2 I & 0 \\ D^T & 0 & 0 & 0 & 0 & -\gamma^2 I \end{bmatrix} < 0 \qquad (7)$$

$$\begin{bmatrix} A^T S + SA - \gamma^2 C_{\bar{\Omega}}^T C_{\bar{\Omega}} & SG & SD & \gamma E^T & \gamma C_{\Omega}^T & H^T \\ G^T S & -\gamma^2 I & 0 & 0 & 0 & 0 \\ SD^T & 0 & -\gamma^2 I & 0 & 0 & 0 \\ \gamma E & 0 & 0 & -I & 0 & 0 \\ \gamma C_{\Omega} & 0 & 0 & 0 & -I & 0 \\ H & 0 & 0 & 0 & 0 & -I \end{bmatrix} < 0 \tag{8}$$

$$\begin{bmatrix} R & I \\ I & S \end{bmatrix} > 0 \tag{9}$$

then, the parameters of controller are $K = -B^T X$, the uncertain estimation gain $K_u = \frac{1}{\gamma^2} D^T X$, the disturbance estimation gain $K_d = \frac{1}{\gamma^2} G^T X$, and $L = \gamma^2 (S - X)^{-1} C^T = \gamma^2 (RS - I)^{-1} RC^T$, $R = X^{-1}$.

Proof. Corresponding to sensor failures any $\omega \subseteq \Omega$, we can find a $2n \times 2n$ matrix $X_e > 0$ such that a quadratical matrix inequality from theorem 1.

$$A_e^T X_e + X_e A_e + \frac{1}{\gamma^2} X_e G_e G_e X_e + H_e H_e < 0 \tag{10}$$

By substituting the parameters of controller into (10) according to Section 2:

$$A_e = \begin{bmatrix} A + DF(t)E - BB^T X & -BB^T X \\ \frac{1}{\gamma^2}(DD^T + GG^T)X - DF(t)E - (L^T)_{\omega}^T C_{\omega} & A + \frac{1}{\gamma^2}(DD^T + GG^T)X - LC \end{bmatrix}$$

$$G_e = \begin{bmatrix} G & 0 & 0 \\ -G & (L^T)_{\bar{\omega}}^T & (L^T)_{\omega}^T \end{bmatrix},$$

$$H_e = \begin{bmatrix} H & 0 \\ -B^T X & -B^T X \end{bmatrix}.$$

$\because X_e > 0$ is a $2n \times 2n$ positive definite real symmetric matrix, so define $X_e = \begin{bmatrix} X & 0 \\ 0 & Y \end{bmatrix}$ where $X$, $Y$ is $n \times n$ positive definite matrix respectively.

So, (10) can be transformed,

$$\begin{bmatrix} A^T X + XA - XBB^T X + \frac{1}{\gamma^2} XGG^T X + E^T F(t)^T D^T X + XDF(t)E + H^T H & \frac{1}{\gamma^2} XDD^T Y - C_{\omega}(L^T)_{\omega} Y - E^T F(t)^T D^T Y \\ \frac{1}{\gamma^2} YDD^T X - Y(L^T)_{\omega}^T C_{\omega} - YDF(t)E & A^T Y + YA + \frac{1}{\gamma^2} X(GG^T + DD^T)Y + \frac{1}{\gamma^2} Y(GG^T + DD^T)X + \frac{1}{\gamma^2} YGG^T Y - C^T L^T Y - YLC + \frac{1}{\gamma^2} YLL^T Y + XBB^T X \end{bmatrix} < 0 \tag{11}$$

(11) can be modified to

$$\begin{bmatrix} A^T X + XA - XBB^T X + \frac{1}{\gamma^2} X(GG^T + DD^T)X + \gamma^2 EE^T + \gamma^2 C_{\Omega}^T C_{\Omega} + H^T H & 0 \\ 0 & A^T Y + YA + XBB^T X + \frac{1}{\gamma^2} Y(GG^T + DD^T)Y - \gamma^2 C_{\bar{\Omega}}^T C_{\bar{\Omega}} + \gamma^2 E^T E + \gamma^2 C_{\Omega}^T C_{\Omega} + H^T H \end{bmatrix}$$
$$-\gamma^2 \begin{bmatrix} E^T \\ 0 \end{bmatrix} [I - F^T(t)F(t)] [E \quad 0] - \gamma^2 \begin{bmatrix} C_{\omega}^T \\ C_{\omega}^T \end{bmatrix} [C_{\omega} \quad C_{\omega}]$$
$$-\gamma^2 \begin{bmatrix} C_{\Omega-\omega}^T & 0 \\ 0 & C_{\Omega-\omega}^T \end{bmatrix} \begin{bmatrix} C_{\Omega-\omega} & 0 \\ 0 & C_{\Omega-\omega} \end{bmatrix}$$
$$-\begin{bmatrix} \gamma E^T F^T(t) - \frac{1}{\gamma} XD \\ \frac{1}{\gamma} YD \end{bmatrix} \begin{bmatrix} \gamma F(t)E - \frac{1}{\gamma} XD & \frac{1}{\gamma} D^T Y \end{bmatrix} < 0 \tag{12}$$

where $C_{\Omega-\omega}$ represents the decomposition matrix of $C$ for $\Omega - \omega$.

Because of $F^T(t)F(t) \le I$ the matrix inequality (12) is satisfied when (10) is satisfied, if following two matrix inequalities are satisfied:

$$A^T X + XA - XBB^T X + \frac{1}{\gamma^2} X(GG^T + DD^T)X + \gamma^2 EE^T + \gamma^2 C_{\Omega}^T C_{\Omega} + H^T H < 0 \tag{13}$$

$$A^T Y + YA + XBB^T X + \frac{1}{\gamma^2} Y(GG^T + DD^T)Y - \gamma^2 C_{\bar{\Omega}}^T C_{\bar{\Omega}} + \gamma^2 E^T E + \gamma^2 C_{\Omega}^T C_{\Omega} + H^T H < 0 \tag{14}$$

For (13), let $R = X^{-1}$, the following result can be obtained:

$$AR + RA^T + R\left(\gamma^2 EE^T + \gamma^2 C_\Omega^T C_\Omega + H^T H\right)R - BB^T + \frac{1}{\gamma^2}\left(GG^T + DD^T\right) < 0 \tag{15}$$

According to Lemma 1 Schur complement, (7) can be obtained.

Add the two sides of (13) and (14), let $S = X + Y$ Gives,

$$A^T S + SA - \gamma^2 C_{\bar{\Omega}}^T C_{\bar{\Omega}} + \frac{1}{\gamma^2} S\left(GG^T + DD^T\right)S + \gamma^2 E^T E + \gamma^2 C_\Omega^T C_\Omega + H^T H < 0 \tag{16}$$

Similarly, according to Lemma 1, (8) can be obtained.

Because $Y > 0$, i.e. $S - R^{-1} > 0$, $RS > I$ from (Packard *et al.*, 1991), then (9) can be obtained.

From the proof as above, Theorem 2 can be obtained

For a given $\gamma$, linear matrix inequalities (7), (8), (9) define a convex set for a positive matrix pair (***R***, ***S***). It is easy to solve these three Limas using strong convex optimization algorithms (Gahinet *et al.*, 1994b). Fortunately, our results is similar to Gahinet' s results (Gahinet and Apkarian, 1994a).

Reduce case: The problem becomes a non-convex optimization problem because of the rank constraint $Rank(I - RS) < k$. At present, there are no valid algorithms to solve non-convex optimization problems.

## 5. CONCLUSIONS

The design technique of a kind of observer-based output feedback $H_\infty$ controller for a class of time-varying uncertain system with sensor failures is given in this paper, which is simplified to solve three linear matrix inequalities (LMIs) using convex optimization algorithms. When computing controllers with multi-objective cost, our technique based on LMIs which is considered as the extension of the results of Veillette *et al.* (1992) is better and easier than the conventional approach based-on algebraic Riccati equations (AREs). Using our Theorem 2, the designed controller tolerates sensor failures within a prespecified set of susceptible sensors, and deduces the effect of exogenous disturbance inputs and unexpected sensor failures outputs under a predefined level. This work is the previous theoretical study of the design of autopilots of underwater vehicle.

## REFERENCES


Apkarian, P. and P. Gahinet (1995). A convex charaterization of gain-scheduled $H_\infty$ controllers. *IEEE Trans. Aut. Cont.*, **AC-40**, 853-864.

Boyd, S., P. Ghaoui, L. El. Feron and E. Balakrishnan (1994). *Linear matrix inequalities in system and control theory*. SIAM Studies in Applied Mathematics, Philadelpphia.

Chen, W-H. (1994). Design of minimax robust LQG controllers under parameter and noise uncertainties. *Int. J. Robust and Nonlinear Control*, **4**, 713-722.

Doyle, J.C., K. Glover, P.P. Khargonekar and B.A. Francis (1989). State-space solutions to standard $H_2$ and $H_\infty$ control problems. *IEEE Trans. Aut. Control*. **AC-34**, 831-847.

Gahinet, P. and P. Apkarian (1994a). A Linear matrix inequality approach to $H_\infty$ control. *Int. J. Robust and Nonlinear Control*, **4**, 421-448.

Gahinet, P., A. Nemirovskii, A. Laub and M. Chilali (1994b). *LMI control toolbox*. The Math Works Inc, Natick, MA.

Packard, A., K. Zhou, P. Pandey and G. Becker (1991). A collection of robust control problem leading to LMIs. *Proc. IEEE Conf. Decision and Control*, Brighton, U.K., 1245-1450.

Veillette, R.J., J.V. Medanic and W.R. Perkins (1992). Design of reliable control systems. *IEEE Trans. Aut. Control.* **AC-37**, 290-303.

Xie, L. and C. E. Souza (1992). Robust $H_\infty$ control for linear systems with norm_bounded time-varying uncertainty. *IEEE Trans. Aut.* Control. **AC-37**, 1188-1191.

14th Triennial World Congress, Beijing, P.R. China

# ROBUST $\mathcal{H}_2$ AND GENERALIZED $\mathcal{H}_2$ PERFORMANCE ANALYSIS

**Mario Sznaier*, Takeshi Amishima* and Jorge Tierno****

*Department of Electrical Engineering, The Pennsylvania State University, University Park, Pennsylvania 16802, USA. E-mail: {msznaier,takeshi}@strider.ee.psu.edu

**Honeywell Technology Center, MN65-2810, 3660 Technology Dr., Minneapolis, Minnesota 55418, USA. E-mail: jtierno@htc.honeywell.com

**Abstract.** The problem of assessing robust $\mathcal{H}_2$ performance (as a first step towards robust $\mathcal{H}_2$ synthesis) has been open for the last two decades. In this paper we give sufficient conditions for robust $\mathcal{H}_2$ and generalized $\mathcal{H}_2$ performance under LTI, not necessarily causal, model uncertainty. These conditions have complexity roughly comparable to that of $\mathcal{H}_\infty$ conditions for the same problem. Moreover, we show that in the case of MISO and SIMO systems these conditions are indeed necessary and sufficient, when the uncertainty structure has one scalar or two full blocks.


**Key Words.** Robust Performance, $\mathcal{H}_2$, Generalized $\mathcal{H}_2$

## 1. INTRODUCTION

$\mathcal{H}_2$ control is appealing since there is a well established connection between the performance index being optimized and performance requirements encountered in practical situations. Moreover, the resulting controllers are easily found by solving two Riccati equations, and in the state-feedback case exhibit good robustness properties (Anderson and Moore, 1990). However, these margins vanish in the output feedback case (Doyle, 1978).

Following this paper, several attempts were made to incorporate robustness into the $\mathcal{H}_2$ framework (Stein and Athans, 1987; Zhang and Freudenberg, 1990). More recently these efforts led to the mixed $\mathcal{H}_2/\mathcal{H}_\infty$ problem (Bernstein and Haddad, 1989; Doyle *et al.*, 1994; Zhou *et al.*, 1994; Kaminer *et al.*, 1993; Sznaier, 1994; Scherer, 1995; Chen and Wen, 1995). While these results represent significant progress towards obtaining robust $\mathcal{H}_2$ controllers, they suffer from the fact that only nominal performance is guaranteed.

Robust $\mathcal{H}_2$ performance was analyzed in (Stoorvogel, 1993) where bounds on the worst-case performance were obtained, leading to tractable synthesis problems. However, they assume non-causal, non-linear time varying model uncertainty. Recently, both state-space (Feron, 1997) and frequency domain (Paganini, 1995a; Paganini, 1995b) convex upper bounds on the worst case $\mathcal{H}_2$ norm have been proposed. The state-space based bound (obtained using an impulse response based interpretation of the $\mathcal{H}_2$ norm and dynamic (non-causal) stability multipliers) is appealing since it takes into account, to some extent, causality. However, in order to obtain tractable problems, these multipliers must be restricted to the span of some basis, selected a-priori. Moreover, the complexity of this basis is limited by the fact that the computational complexity of the resulting LMI problem grows roughly as the 10-th power of the state dimension (Paganini and Feron, 1997). On the other hand, frequency-domain methods cannot impose causality, but lead to simple LMI based conditions for robust performance. Unfortunately, as shown in (Sznaier and Tierno, 1998) these conditions are only sufficient, and potentially very conservative, in the case of multivariable noise.

In this paper we consider the problems of assessing the worst case $\mathcal{H}_2$ and *generalized* $\mathcal{H}_2$ performance. The main results of the paper

[0] This work was supported in part by NSF under grant ECS-9625920.

provide sufficient conditions for robust performance in both cases. Further, these conditions are necessary and sufficient for MISO and SIMO systems when the uncertainty structure has a single scalar or two full blocks. As a corollary of this derivation, we show that the condition in (Paganini, 1995a) is indeed necessary and sufficient for robust $\mathcal{H}_2$ performance in the case of SISO systems for these uncertainty structures (a fact conjectured but not proved in (Paganini, 1995a)) and that the worst possible gap for MISO systems having $n_w$ inputs is $\sqrt{n_w}$. For simplicity, in the sequel all the derivations are carried out in the continuous-time case. However, the formulae apply as well to discrete-time systems with minimal modifications.

## 2. PRELIMINARIES

### 2.1. *Notation and Definitions*

By $\mathcal{H}_2$ we denote the space of complex valued matrix functions $G(s)$ with analytic continuation in $Re(s) > 0$ and square integrable on the $j\omega$ axis, equipped with the usual $\mathcal{H}_2$ norm:

$$||G||_2^2 \doteq \frac{1}{2\pi} \int_{-\infty}^{\infty} \text{Trace}\,[G(j\omega)G(j\omega)^*]\,d\omega,$$

In the same space, the *generalized* $\mathcal{H}_2$ norm is defined as

$$||G||^2_{\text{gen}\mathcal{H}_2} \doteq \frac{1}{2\pi}\bar{\sigma}\left\{\int_{-\infty}^{\infty} G(j\omega)G(j\omega)^* d\omega\right\}$$

It can be shown that $||.||_{\text{gen}\mathcal{H}_2}$ is the induced norm from $\mathcal{L}^2$ to $\mathcal{L}^\infty$ (Wilson, 1989). Obviously, in the case of MISO systems it coincides with the regular $\mathcal{H}_2$ norm, while for systems having $n_z$ outputs we have that

$$||G||^2_{\text{gen}\mathcal{H}_2} \leq ||G||_2^2 \leq n_z ||G||^2_{\text{gen}\mathcal{H}_2}$$

Given two matrices $M$ and $\Delta$ of compatible dimensions we denote by $M \star \Delta$ the upper LFT $\mathcal{F}_u(M, \Delta)$.

Let $\mathcal{L}(\mathcal{L}^2)$ denote the set of linear bounded operators in $\mathcal{L}^2$. In the sequel we will consider the following set of operators in $\mathcal{L}(\mathcal{L}^2)$:

$$\begin{aligned}&\mathcal{B}\mathbf{\Delta} = \{\Delta \in \mathcal{L}(\mathcal{L}^2):\\ &\Delta = \text{diag}\,[\delta_1 I_{r_1}, \ldots, \delta_S I r_S, \Delta_{S+1}, \ldots, \Delta_{S+F}]:\\ &||\Delta||_{\mathcal{L}^2 \to \mathcal{L}^2} \leq 1\}\end{aligned}$$

The subsets of $\mathcal{B}\mathbf{\Delta}$ formed by linear time invariant and causal linear time invariant operators will be denoted by $\mathcal{B}\mathbf{\Delta}^{\text{LTI}}$ and $\mathcal{B}\mathbf{\Delta}^{\text{LTI}}_{\text{causal}}$ respectively. For ease of notation we also introduce a set of *constant* complex matrices having a structure similar to that of the operators in $\mathcal{B}\mathbf{\Delta}$:

$$\begin{aligned}&\mathcal{B}\mathbf{\Delta}_m = \{\Delta \in C^{n \times n}:\\ &\Delta = \text{diag}\,[\delta_1 I_{r_1}, \ldots, \delta_S I r_S, \Delta_{S+1}, \ldots, \Delta_{S+F}]:\\ &||\Delta||_2 \leq 1\}\end{aligned}$$

Finally, we will also make use of the following set of scaling matrices which commute with the elements in $\mathcal{B}\mathbf{\Delta}$:

$$\begin{aligned}&\mathbf{X} = \{X:\\ &X = \text{diag}\,[X_1, \ldots, X_S, x_{S+1}I_{m_1}, \ldots, x_{S+F}I_{m_F}],\\ &X = X^* > 0\}\end{aligned}$$

*Definition 1* (Robust $\mathcal{H}_2$ and generalized $\mathcal{H}_2$ performance) The uncertain system $(M, \Delta)$ has robust $\mathcal{H}_2$ performance if it is robustly stable and

$$\sup_{\Delta \in \mathcal{B}\mathbf{\Delta}^{\text{LTI}}} ||M \star \Delta||_2 \leq 1 \tag{1}$$

Similarly, it has robust generalized $\mathcal{H}_2$ performance if it is stable and

$$\sup_{\Delta \in \mathcal{B}\mathbf{\Delta}^{\text{LTI}}} ||M \star \Delta||_{\text{gen}\mathcal{H}_2} \leq 1 \tag{2}$$

Similar definitions apply when $\Delta$ is restricted to the subset $\mathcal{B}\mathbf{\Delta}^{\text{LTI}}_{\text{causal}}$.

### 2.2. *A Simple Condition for Robust Generalized $\mathcal{H}_2$ Performance*

*Lemma 1* The interconnection $M \star \Delta$ achieves robust generalized $\mathcal{H}_2$ performance against $\Delta \in \mathcal{B}\mathbf{\Delta}^{\text{LTI}}$ (not necessarily causal) if there exists $X(\omega) \in \mathbf{X}$, and a real $y(\omega) > 0$, such that

$$M(j\omega)^* \begin{bmatrix} X(\omega) & 0 \\ 0 & I_{n_z \times n_z} \end{bmatrix} M(j\omega) - \begin{bmatrix} X(\omega) & 0 \\ 0 & y(\omega) I_{n_w \times n_w} \end{bmatrix} < 0;\ \forall\omega \tag{3}$$

$$\int_{-\infty}^{\infty} y(\omega)\frac{d\omega}{2\pi} \leq 1 \tag{4}$$

Proof: Using standard singular value inequalities we have that:

$$\sup_{\Delta \in \mathcal{B}\mathbf{\Delta}^{\text{LTI}}} ||M \star \Delta||^2_{\text{gen}\mathcal{H}_2} \leq \int_{-\infty}^{\infty} \sup_{\Delta(j\omega) \in \mathcal{B}\mathbf{\Delta}_m} \left\{\bar{\sigma}\left[(M \star \Delta)(j\omega)\right]^2\right\} \frac{d\omega}{2\pi} \tag{5}$$

From the Main Loop Theorem (Packard and Doyle, 1993) it follows that

$$\sup_{\Delta \in \mathcal{B}\mathbf{\Delta}_m} \left\{ \bar{\sigma}\left[ (M \star \Delta)(j\omega) \right]^2 \right\} < y(\omega) \qquad (6)$$

if there exists $X(\omega) \in \mathbf{X}$ such that

$$M(j\omega)^* \begin{bmatrix} X(\omega) & 0 \\ 0 & I \end{bmatrix} M(j\omega) - \begin{bmatrix} X(\omega) & 0 \\ 0 & y(\omega) I \end{bmatrix} < 0 \qquad (7)$$

The proof follows now by integrating (6) over frequency.

## 3. ROBUST $\mathcal{H}_2$ ANALYSIS

### 3.1. *The MISO and SIMO Cases*

In this section we exploit Lemma 1 to obtain conditions for robust $\mathcal{H}_2$ performance in the simpler MISO and SIMO cases.

*Lemma 2* Conditions (3)–(4) are sufficient for robust $\mathcal{H}_2$ performance against LTI (not necessarily causal) perturbations $\Delta \in \mathcal{B}\mathbf{\Delta}^{\mathrm{LTI}}$ for MISO or SIMO systems. Moreover, if the uncertainty structure $\mathbf{\Delta}$ satisfies $2S + F \leq 2$ then they are also necessary.

Proof: Since for MISO systems $\bar{\sigma}(MM^*) = \mathrm{Trace}(MM^*)$ and since the Trace and $\int$ operators commute, equation (5) can be made an equality by optimizing the integrand frequency by frequency. Moreover, in this case $\|M \star \Delta\|_2 = \|M \star \Delta\|_{\mathrm{gen}\mathcal{H}_2}$. The proof follows now from the fact that for the uncertainty structures considered here, $\mu$ coincides with its upper bound. A similar proof holds for the SIMO case.

Next we compare our results with a similar condition proposed in (Paganini, 1995a).

*Condition 1* If there exists $X(\omega) \in \mathbf{X}$, and $Y(\omega) = Y^*(\omega) \in \mathbf{C}^{m \times m}$, such that

$$M(j\omega)^* \begin{bmatrix} X(\omega) & 0 \\ 0 & I \end{bmatrix} M(j\omega) - \begin{bmatrix} X(\omega) & 0 \\ 0 & Y(\omega) \end{bmatrix} < 0, \ \forall \omega \qquad (8)$$

$$\int_{-\infty}^{\infty} \mathrm{Trace}(Y(\omega)) \frac{d\omega}{2\pi} \leq 1 \qquad (9)$$

then the inteconnection $M \star \Delta$ achieves robust $\mathcal{H}_2$ performance against $\Delta \in \mathcal{B}\mathbf{\Delta}^{\mathrm{LTI}}$.

Unfortunately, as shown with the simple counterexample in (Sznaier, 1998), this condition is only sufficient for systems having $n_w > 1$ inputs, where the gap can be at least as large as $\sqrt{n_w}$. As we show next, this is indeed the worst case gap. To this effect note that (8) implies:

$$M \star \Delta(\omega)^* M \star \Delta(\omega) < Y(\omega) \qquad (10)$$

Thus for MISO systems

$$\begin{aligned} \sup_{\Delta \in \mathcal{B}\mathbf{\Delta}^{\mathrm{LTI}}} \|M \star \Delta\|_2^2 &= \int_{-\infty}^{\infty} \bar{\sigma}[Y(\omega)] \tfrac{d\omega}{2\pi} \\ &\leq \int_{-\infty}^{\infty} \mathrm{Trace}\,[Y(\omega)] \tfrac{d\omega}{2\pi} \leq n_w \int_{-\infty}^{\infty} \bar{\sigma}[Y(\omega)] \tfrac{d\omega}{2\pi} \\ &\leq n_w \sup_{\Delta \in \mathcal{B}\mathbf{\Delta}^{\mathrm{LTI}}} \|M \star \Delta\|_2^2 \end{aligned} \qquad (11)$$

*Remark 1* From equation (11) it follows that Condition 1 is indeed necessary and sufficient in the SISO case.

### 3.2. *The MIMO Case*

In this section we obtain a sufficient condition for robust $\mathcal{H}_2$ performance of a MIMO system by decomposing it into a collection of MISO systems and exploiting the results of section 3.1

*Theorem 1* Consider a MIMO system having $n_w$ inputs and $n_z$ outputs. Let $M_{ij}^{(i)}$ denote the $i^{th}$ row of the operator $M_{ij}$ and define:

$$\tilde{M} = \left[ \begin{array}{cccc|cccc} M_{11} & 0 & \dots & 0 & M_{12} & 0 & \dots & 0 \\ 0 & M_{11} & \dots & 0 & 0 & M_{12} & \dots & 0 \\ \vdots & & \ddots & \vdots & \vdots & & \ddots & \vdots \\ 0 & 0 & \dots & M_{11} & 0 & 0 & \dots & M_{12} \\ M_{21}^{(1)} & 0 & \dots & 0 & M_{22}^{(1)} & 0 & \dots & 0 \\ 0 & M_{21}^{(2)} & \dots & 0 & 0 & M_{22}^{(2)} & \dots & 0 \\ \vdots & & \ddots & \vdots & \vdots & & \ddots & \vdots \\ 0 & \dots & \dots & M_{21}^{(n_z)} & 0 & 0 & \dots & M_{22}^{(n_z)} \end{array} \right] \qquad (12)$$

Then $M \star \Delta$ achieves robust $\mathcal{H}_2$ performance if there exists $\tilde{X}$ and $Y(\omega)$ such that

$$\tilde{M}(\omega)^* \begin{bmatrix} \tilde{X}(\omega) & 0 \\ 0 & I \end{bmatrix} \tilde{M}(\omega) - \begin{bmatrix} \tilde{X}(\omega) & 0 \\ 0 & Y(\omega) \end{bmatrix} < 0 \tag{13}$$

$$\int_{-\infty}^{\infty} \sum_{i=1}^{n_z} y_i(\omega) \frac{d\omega}{2\pi} \leq 1 \tag{14}$$

where:

$$\begin{array}{l} \tilde{X}(\omega) = \tilde{X}^*(\omega) = [X_{ij}(\omega)], \; X_{ij} \in \mathbf{X}, \tilde{X} > 0 \\ Y(\omega) = \mathrm{diag}\{y_1 I, y_2 I \ldots, y_{n_z} I\} \\ y_j(\omega) = y_j(\omega)^* > 0 \end{array} \tag{15}$$

Proof:

$$\sup_{\Delta \in \mathcal{B}\mathbf{\Delta}^{LTI}} \|M \star \Delta\|_2^2 = \int_{-\infty}^{\infty} \sup_{\Delta(\omega) \in \mathcal{B}\mathbf{\Delta}_m} \left\{ \sum_{i=1}^{n_z} \bar{\sigma}^2 \left[ M^{(i)}(\Delta, \omega) \right] \right\} \frac{d\omega}{2\pi} \tag{16}$$

where

$$M^{(i)}(\Delta, \omega) = \begin{bmatrix} M_{11} & M_{12}^{(i)} \\ M_{21}^{(i)} & M_{22}^{(i)} \end{bmatrix} \star \Delta \tag{17}$$

Next, note that

$$\begin{array}{l} \bar{\sigma}^2 \left[ M^{(i)}(\Delta, \omega) \right] < y_i, \; \forall \Delta \in \mathcal{B}\mathbf{\Delta}^{LTI} \\ \Longleftrightarrow \\ \det(I - \tilde{M}\tilde{\Delta}) \neq 0 \; \forall \Delta \in \mathcal{B}\mathbf{\Delta}^{LTI} \end{array} \tag{18}$$

where

$$\begin{array}{l} \tilde{\Delta} = \left[ \begin{array}{c|c} \mathrm{diag}(\Delta) & 0 \\ \hline 0 & \mathrm{diag}(y_i^{-1} \Delta_i) \end{array} \right] \\ \Delta \in \mathcal{B}\mathbf{\Delta}^{LTI}, \; \Delta_i \in \mathcal{C}^{n_w \times 1}, \|\Delta_i\|_\infty \leq 1 \end{array} \tag{19}$$

Thus, assessing robust $\mathcal{H}_2$ performance leads to a $\mu$ problem with $n_z$ repeated full blocks. It can be easily shown that a sufficient condition for (18) to hold is the existence of matrix functions $\tilde{X}(\omega)$ and $Y(\omega)$ such that

$$\tilde{M}^* \begin{bmatrix} \tilde{X} & 0 \\ 0 & I \end{bmatrix} \tilde{M} - \begin{bmatrix} \tilde{X} & 0 \\ 0 & Y \end{bmatrix} < 0 \tag{20}$$

Thus

$$\sup_{\Delta \in \mathcal{B}\mathbf{\Delta}^{LTI}} \mathrm{Trace}\left[ (M \star \Delta)(M \star \Delta)^* \right] \leq \sum_{i=1}^{n_z} y_i(\omega) \tag{21}$$

and the proof follows now by integrating frequency by frequency.

*Remark 2* Clearly, a similar sufficient condition can be obtained by decomposing the system into a collection of SIMO (rather than MISO) subsystems.

## 4. EXAMPLES AND COMPARISONS

*Example 1* Consider the discrete–time system used in (Sznaier, 1998) to illustrate the gap between $W^\eta$ and $\mathcal{H}_2$ performance:

$$\begin{bmatrix} \begin{pmatrix} q_1 \\ q_2 \end{pmatrix} \\ z \end{bmatrix} = \begin{bmatrix} 0 & 0 & 1 & 0 \\ 0 & 0 & 0 & 1 \\ 1 & 0 & 0 & 0 \end{bmatrix} \begin{bmatrix} p_1 \\ p_2 \\ \begin{pmatrix} f_1 \\ f_2 \end{pmatrix} \end{bmatrix} \tag{22}$$

$$z = (M \star \Delta) \begin{bmatrix} f_1 \\ f_2 \end{bmatrix} = [\Delta_{11} \quad \Delta_{12}] \begin{bmatrix} f_1 \\ f_2 \end{bmatrix}$$

Clearly in this case we have that

$$\sup_{\Delta \in \mathcal{B}\mathbf{\Delta}^{LTI}} \|M \star \Delta\|_2 = 1 \tag{23}$$

and the worst case norm can be achieved with causal LTI perturbations. Since $\Delta$ is a full block, $X = xI_{2\times 2}$, and equation (3) becomes

$$\begin{bmatrix} 1 & 0 & 0 & 0 \\ 0 & 0 & 0 & 0 \\ 0 & 0 & x & 0 \\ 0 & 0 & 0 & x \end{bmatrix} - \begin{bmatrix} x & 0 & 0 & 0 \\ 0 & x & 0 & 0 \\ 0 & 0 & y & 0 \\ 0 & 0 & 0 & y \end{bmatrix} < 0. \tag{24}$$

Equation (24) implies the following inequalities

$$1 - x < 0, \qquad x - y < 0 \tag{25}$$

from which it follows that $y > 1$. Since $y$ does not depend on $\omega$,

$$\int_0^{2\pi} y(\omega) \frac{d\omega}{2\pi} > 1 \tag{26}$$

and the inequality can be achieved up to arbitrarily small $\epsilon$. Thus the system achieves robust $\mathcal{H}_2$ performance.

An alternative, state–space based, upper bound on the worst case $\mathcal{H}_2$ norm has been recently proposed in (Feron, 1997). Proceeding as in (Paganini, 1995b) it can be shown that in the case of LTI uncertainty this approach always gives a tighter upper bound than Lemma 2 *provided* that constant multipliers (i.e. X) are used in both cases. Moreover, this approach takes into account, to some extent, causality. However, the approach that we propose here allows for using dynamic multipliers (of arbitrarily high order) with a very modest computational complexity increase, since the problem is solved frequency by frequency and no explicit state space realization of the multipliers is required. Rather, a matrix $X(\omega)$ of dimensions determined by the dimensions of the plant is computed at each frequency point. On the other hand using dynamic multipliers in (Feron, 1997) entails a growth in computational time of roughly the 10-th power the order of the multipliers (Paganini, 1997). These issues are illustrated in the following example.

*Example 2* Consider the following 2 inputs and 2 outputs continuous–time system:

$$\begin{bmatrix} q \\ z_1 \\ z_2 \end{bmatrix} = M \begin{bmatrix} p \\ w_1 \\ w_2 \end{bmatrix} \tag{27}$$

where

$$\begin{aligned} M &\doteq \begin{bmatrix} M_{11} & M_{12}^{11} & 0 \\ M_{21}^{11} & 0 & 0 \\ 0 & 0 & M_{22}^{22} \end{bmatrix} \\ &= \begin{bmatrix} -0.845\frac{s+1.5}{s+2} & \frac{s-1.5}{(s+2)^2} & 0 \\ \frac{s-1.5}{(s+2)^2} & 0 & 0 \\ 0 & 0 & \frac{1}{s+1} \end{bmatrix} \end{aligned} \tag{28}$$

In this case Theorem 1 leads to:

$$\begin{aligned} X(\omega) &= \begin{bmatrix} x_1 & x_2 \\ x_2^* & x_3 \end{bmatrix} \\ Y(\omega) &= \begin{bmatrix} y_1 & 0 & 0 & 0 \\ 0 & y_1 & 0 & 0 \\ 0 & 0 & y_2 & 0 \\ 0 & 0 & 0 & y_2 \end{bmatrix} \end{aligned} \tag{29}$$

and the corresponding LMI can be solved analytically, yielding:

$$\frac{M_{21}^{11*}(jw)M_{21}^{11}(jw)M_{12}^{11*}(jw)M_{12}^{11}(jw)}{1 - M_{11}^{*}(jw)M_{11}(jw)} < y_1$$

$$M_{22}^{22*}(jw)M_{22}^{22}(jw) < y_2$$

Integrating over $\omega$ we get $||T_{zw}||^2_{\mathcal{H}_2} = 0.5304$. Table 1 shows the results of applying the method proposed in (Feron, 1997) to this problem. Here $N$ and $n_{dec}$ denote the order of the dynamic multipliers and the number of decision variables in the associated LMI problem [1]. From this table it follows that the causality constraints embedded in this formulation outweight the ability to use arbitrarily high order multipliers when $N \geq 2$. However, note that the rather modest cost improvement is achieved at the price of a substantial increase in computational complexity.

Table 1 $||T_{zw}||^2_{\mathcal{H}_2}$ vs. the order $N$ of the dynamic multipliers.

| $N$ | Bound | $n_{dec}$ |
|---|---|---|
| 0 | 0.5383 | 4 |
| 2 | 0.5300 | 61 |
| 4 | 0.5300 | 135 |

## 5. MULTIOBJECTIVE PROBLEMS

In this section we briefly consider the extension of the theory to accommodate mixed $\mathcal{H}_2/\mathcal{H}_\infty$ performance specifications. While this mixed problem first originated as an attempt to incorporate robustness into the LQG framework, it is also interesting in its own, since in many situations it is desired to accommodate inputs containing both bounded energy and bounded spectral density signals (see for instance (Rotstein *et al.*, 1996)). In its most general formulation, the input $w$ to the system is partitioned as $w = [w_1^T \;\; w_2^T]^T$ where $w_1 \in \mathcal{L}^2$ and $w_2 \in \mathcal{S}$ (the space of bounded spectral density signals), and the output is partitioned into $z = [z_1^T \;\; z_2^T]^T$. The goal is to design a controller to minimize $||T_{z_2w_2}||_2$ subject to the constraint $||T_{z_1w_1}||_\infty \leq \gamma$, where $T_{z_iw_i}$ denotes the closed loop transfer function from the input $w_i$ to the output $z_i$. In the sequel we will assume that such a controller has been designed (using for instance the methods in (Sznaier, 1994)) and used to close the loop around the nominal plant. Our goal is to assess the worst case value of performance over $\Delta \in \mathcal{B}\mathbf{\Delta}^{\mathrm{LTI}}$. Note that the worst case value of *each* transfer matrix can be found separately using $\mu$–analysis (in the $\mathcal{H}_\infty$) case or the techniques proposed in this paper (for the $\mathcal{H}_2$ component). However, in some cases it is desirable to have a measure of worst case joint performance (or constraint violation). This can be accomplished by defining $J(\Delta) = \alpha_1||T_{z_1w_1}(\Delta)||^2_\infty + \alpha_2||T_{z_2w_2}(\Delta)||^2_2$, where $\alpha_i$ are fixed weights, and finding the worst possible case of $J(\Delta)$. As we show in the sequel,

[1] recall that the computational time grows as $n^5_{dec}$ (Paganini, 1997)

this can be accommodated in our formalism with minimal modifications. For simplicity, we consider the case where $z_2$ is a scalar, but the results can be easily modified to accommodate MIMO systems proceeding as in section 3.2.

*Lemma 3* Consider a system $M$ with input $w = [w_1^T \;\; w_2^T]^T$, $w_1 \in \mathcal{L}^2$, $w_2 \in \mathcal{S}$ and output $z = [z_1^T \;\; z_2^T]^T$, where $z_2$ is a scalar. Let $J(\Delta) = \alpha_1 \|M_{z_1 w_1}(\Delta)\|_\infty^2 + \alpha_2 \|M_{z_2 w_2}(\Delta)\|_2^2$, where $M_{z_i, w_i}(\Delta)$ denotes the transfer matrix from $w_i$ to $z_i$ obtained when interconnecting $M$ with $\Delta \in \mathcal{B}\mathbf{\Delta}^{\mathrm{LTI}}$. If there exists $X(\omega) \in \mathbf{X}$, a real transfer function $y(\omega) > 0$, and a positive number $\gamma$, such that

$$M(j\omega)^* \begin{bmatrix} X(\omega) & 0 & 0 \\ 0 & I & 0 \\ 0 & 0 & 1 \end{bmatrix} M(j\omega) - \begin{bmatrix} X(\omega) & 0 & 0 \\ 0 & \gamma^2 I & 0 \\ 0 & 0 & y(\omega) I \end{bmatrix} < 0, \; \forall \, \omega \tag{30}$$

and

$$\alpha_1 \gamma^2 + \alpha_2 \int_{-\infty}^{\infty} y(\omega) \frac{d\omega}{2\pi} \leq 1 \tag{31}$$

then $\sup_{\Delta \in \mathcal{B}\mathbf{\Delta}^{\mathrm{LTI}}} J(\Delta) < 1$

Proof: The proof follows by using an argument similar to that in the proof of Lemma 2 and Theorem 1, by noting that the additional row(s) in (30) imply that $\|M_{z_1 w_1}(\Delta)\|_\infty < \gamma$.

## 6. REFERENCES


Anderson, B. D. O. and J. B. Moore (1990). *Optimal Control: Linear Quadratic Methods.* Prentice Hall, New Jersey.

Bernstein, D. and W. Haddad (1989). LQG control with an $\mathcal{H}_\infty$ performance bound: A Riccati equation approach. *IEEE Trans. Autom. Contr.*, **34**(2), 293–305.

Chen, X. and J. T. Wen (1995). A linear matrix inequality approach to the general mixed $\mathcal{H}_2/\mathcal{H}_\infty$ control problem. *Proc. 1995 ACC*, **2**, 1443–1447.

Doyle, J.(1978). Guaranteed Margins for LQG Regulators. *IEEE Trans. Autom. Contr.*, **23**(4), 756–757.

Doyle, J., K. Zhou, K. Glover and B. Bodenheimer (1994). Mixed $\mathcal{H}_2$ and $\mathcal{H}_\infty$ performance objectives II: Optimal control. *IEEE Trans. Autom. Contr.*, **39**(8), 1575–1587.

Feron, E.(1997). Analysis of Robust $\mathcal{H}_2$ Performance Using Multiplier Theory. *SIAM J. Contr. and Opt.*, **35**(1), 160-177.

Kaminer, I., P. Khargonekar and M. Rotea (1993). Mixed $\mathcal{H}_2/\mathcal{H}_\infty$ control for discrete-time systems via convex optimization. *Automatica*, **29**(1), 57–70.

Packard, A. and J. Doyle (1993). The complex structured singular value. *Automatica*, **29**(1), 71–109 .

Paganini, F.(1995a). *Sets and Constraints in the Analysis of Uncertain Systems.* Ph.D. Dissertation, Caltech, Pasadena, California.

Paganini, F.(1995b). Necessary and Sufficient Conditions for Robust $\mathcal{H}_2$ Performance. *Proc. 1995 CDC*, 1970–1975.

Paganini, F. and E. Feron (1997). Analysis of Robust $\mathcal{H}_2$ Performance: Comparisons and Examples. *Proc. 36th CDC*, **2**, 1000-1006.

Rostein, H., M. Sznaier and M. Idan (1996). $\mathcal{H}_2/\mathcal{H}_\infty$ filtering: Theory and Aerospace Application. *Int. J. Robust and Non-Linear Contr., special issue of $\mathcal{H}_\infty$ and robust estimation*, **6**(4),347-366.

Scherer, C. W. (1995). Multiobjective $\mathcal{H}_2/\mathcal{H}_\infty$ control. *IEEE Trans. Autom. Contr.*, **40**(6), 1054–1062.

Stein, G. and M. Athans (1987). The LQG/LTR procedure for multivariable feedback control design. *IEEE Trans. Autom. Contr.*, **32**(2), 105–114.

Stoorvogel, A. A. (1993). The Robust $\mathcal{H}_2$ Problem: A Worst-Case Design. *IEEE Trans. Autom. Contr.*, **38**(9), 1358–1370.

Sznaier, M. (1994). An exact solution to general SISO mixed $\mathcal{H}_2/\mathcal{H}_\infty$ problems via convex optimization. *IEEE Trans. Autom. Contr.*, **39**(12), 2511–2517.

Sznaier, M. and J. Tierno (1998). Is set modeling of white noise a good tool for robust $\mathcal{H}_2$ analysis?. *Automatica*, to appear.

Wilson, D. A. (1989). Convolution and Hankel operator norms for linear systems. *IEEE Trans. Autom. Contr.*, **34**(1), 94–97.

Zhang, Z. and J. S. Freudenberg (1990). LTR for nonminimum phase plants. *IEEE Trans. Autom. Contr.*, **35**(5),547–553.

Zhou, K., K. Glover, B. Bodenheimer and J. Doyle (1994). Mixed $\mathcal{H}_2$ and $\mathcal{H}_\infty$ performance objectives I: Robust performance analysis. *IEEE Trans. Autom. Contr.*, **39**(8), 1564–1574.

14th Triennial World Congress, Beijing, P.R. China
F-2e-04-2

# A DYNAMIC GAMES APPROACH TO DISTURBANCE ATTENUATION CONTROL OF DISCRETE-TIME DESCRIPTOR SYSTEMS

**Hua Xu**

*Graduate School of Systems Management*
*The University of Tsukuba, Tokyo*
*3-29-1 Otsuka, Bunkyo-ku, Tokyo 112-0012, Japan*
*Email:xuhua@gssm.otsuka.tsukuba.ac.jp*

Abstract: In this paper, the $H_\infty$-optimal control problem for discrete-time descriptor systems under perfect state measurements is studied. By using a dynamic games approach, some sufficient conditions on a disturbance attenuation level $\gamma$ are derived such that controllers exist to guarantee such a performance level. Upper and lower bounds of the optimum disturbance attenuation level are obtained which are characterized in terms of a parameterized discrete-time Riccati equation and parameterized continuous-time algebraic Riccati equations. A design procedure is provided to find the disturbance attenation controllers for a prespecified achievable disturbance attenuation level $\gamma$. A numerical example is included to illustrate the results obtained in the paper. *Copyright©1999 IFAC*

Keywords: Discrete-time system, descriptor systems, H-infinity control, differential game, disturbance rejection

## 1. INTRODUCTION

As one of the most natural settings, the dynamic games approach has been used in the study of $H_\infty$-optimal control problems for state space systems (Başar and Berhard, 1995 and the references therein). In this paper, based on some recent results obtained by the authors on the linear quadratic dynamic game of discrete-time descriptor systems, the $H_\infty$ optimal control problem for discrete-time descriptor systems under the perfect state measurements is studied. Some sufficient conditions on a disturbance attenuation level $\gamma$ are derived such that the disturba nce attenuation controllers exist. These conditions are described in terms of the existence conditons of linear feedback saddle-point solutions of linear quadratic dynamic games of discrete-time descriptor systems, that is, the solvability of some parameterized reduced-order Riccati equations, the invertibility of a parameterized Hamiltonian matrix and the positive definiteness of parameterized symmetric matrices. It is shown that, if a prespecified disturbance attenuation level is achieva ble, there exist uncountably many central $H_\infty$-controllers to guarantee such a performance level. An example is included to describe the results obtained in the paper.

## 2. PROBLEM FORMULATION

Consider the linear discrete-time descriptor system

$$Ex_{k+1} = Ax_k + Bu_k + Dw_k, \quad x_0 = 0, \quad (1)$$

for $k \in \mathsf{K} := \{0, 1, 2, ..., N-1\}$, where $x_k$ is the n-dimensional state vector, $u_k$ is the m-dimensional control vector. $w_k$ is the disturbance, a $l^2$ sequence, belonging to $\mathsf{W_k}$, where $\mathsf{W_k}$ is a l-dimensional Euclidean space. The matrix $E$ is a square matrix of rank $r \leq n$. The pencil $(sE - A)$

is assumed to be regular (i.e., $|(sE - A)| \not\equiv 0$). It is also assumed that the system triplet $(E, A, B)$ is causally controllable.

Associated with the desciptor system (1) is a performance index

$$L(u, w) = |Ex_N|^2_{Q_N} + \sum_{k=0}^{N-1} \{|x_k|^2_Q + |u_k|^2\}, \quad (2)$$

$$Q_N \geq 0; \quad Q \geq 0.$$

where

$$u := (u_0, u_1, ..., u_{N-1}),$$

$$w := (w_0, w_1, ..., w_{N-1}),$$

and the system triplet $(E, A, Q)$ is assumed to be causally reconstructible. Let us also introduce the notation $J(\mu, w)$ to denote $L(u, w)$ when $u$ is substituted by a permissible controller $\mu$. In this paper, the case is considered in which the state can be measured completely, that is, the closed-loop memoryless information on $x_k, k \in \mathbf{K}$(in terms of the dynamic game theory) is available. Hence, the attention is restricted to linear feedback controllers of the form $u_k = \mu_k(x_k) = K_k x_k$ at each stage. A permissible controller $\mu := (\mu_0, \mu_1, ..., \mu_{N-1})$ is restricted in the linear feedback controller class such that the closed-loop system obtained is causal when $w \equiv 0$. Let us denote the class of all these controllers by $\mathsf{M}$.

Following Başar and Bernhard (1995), the $H_\infty$-optimal control problem is the minimization of the quantity

$$\sup_{\{w_k \in \mathsf{W}_k; k \in \mathsf{K}\}} \{J(\mu, w)\}^{1/2}/\|w\|, \quad (3)$$

over all permissible controllers $\mu \in \mathsf{M}$, where $\|w\|$ denotes an appropriate $l^2$ norm of $w$. In the case that a minimum does not exist, the problem becomes the derivation of a controller $\mu$ which will ensure a performance in a given neighborhood of the infimum of (3), denoted by

$$\inf_{\mu \in \mathsf{M}} \sup_{\{w_k \in \mathsf{W}_k; k \in \mathsf{K}\}} \{J(\mu, w)\}^{1/2}/\|w\| := \gamma^*. \quad (4)$$

The nonnegative quantity $\gamma^*$ is the optimum (minimax) disturbance attenuation level. A more general problem of interest is the characterization of all controllers $\mu^\gamma$, such that

$$\sup_{\{w_k \in \mathsf{W}_k; k \in \mathsf{K}\}} \{J(\mu^\gamma, w)\}^{1/2}/\|w\| \leq \gamma, \quad (5)$$

for a given $\gamma > \gamma^*$.

The above $H_\infty$-optimal control problem is closely related to a so-called soft-constrained linear quadratic zero-sum dynamic game under the closed-loop memoryless information structure, which has the cost functional

$$L_\gamma(u, w) = L(u, w) - \gamma^2 \sum_{k=0}^{N-1} |w_k|^2, \quad (6)$$

where $\gamma > 0$ is a fixed parameter.

Toward the end of this section, let us introduce some transformations for later uses. There exist nonsingular matrices $M$ and $H$ such that

$$MEH = \begin{bmatrix} I_r & 0 \\ 0 & 0 \end{bmatrix}, \quad r = rank\ E. \quad (7)$$

Based on the transformation (7), let us define

$$MAH = \begin{bmatrix} A_{11} & A_{12} \\ A_{21} & A_{22} \end{bmatrix}, \quad (8a)$$

$$MB = \begin{bmatrix} B_1 \\ B_2 \end{bmatrix}, \quad MD = \begin{bmatrix} D_1 \\ D_2 \end{bmatrix}, \quad (8b)$$

$$H^T Q H = \begin{bmatrix} C_1^T \\ C_2^T \end{bmatrix} [C_1 \ C_2], \quad (8c)$$

$$M^{-T} Q_N M^{-1} = \begin{bmatrix} Q_{11N} & Q_{12N} \\ Q_{12N}^T & Q_{22N} \end{bmatrix}. \quad (8d)$$

Let us further define some other matrices.

$$T_1^\gamma = \begin{bmatrix} A_{11} & -S_{11}^\gamma \\ Q_{11} & A_{11}^T \end{bmatrix}, \quad T_2^\gamma = \begin{bmatrix} A_{12} & -S_{12}^\gamma \\ Q_{12} & A_{21}^T \end{bmatrix}, \quad (9a)$$

$$T_3^\gamma = \begin{bmatrix} A_{21} & -(S_{12}^\gamma)^T \\ -Q_{12}^T & -A_{12}^T \end{bmatrix}, \quad (9b)$$

$$T_4^\gamma = \begin{bmatrix} A_{22} & -S_{22}^\gamma \\ -Q_{22} & -A_{22}^T \end{bmatrix}, \quad (9c)$$

where

$$S_{11}^\gamma = B_1 B_1^T - \gamma^{-2} D_1 D_1^T, \quad (10a)$$

$$S_{12}^\gamma = B_1 B_2^T - \gamma^{-2} D_1 D_2^T, \quad (10b)$$

$$S_{22}^\gamma = B_2 B_2^T - \gamma^{-2} D_2 D_2^T. \quad (10c)$$

## 3. THE SOFT-CONSTRAINED LINEAR QUADRATIC ZERO-SUM DYNAMIC GAME

In this section, the soft-constrained linear quadratic zero-sum dynamic game described by (1) and (6) is studied under the closed-loop memoryless information structure, where the assumption that the system triplet $(E, A, B)$ is causally controllable is relaxed to the assumption that the system quaternary $(E, A, B, D)$ is causally controllable. Also, the initial state $x_0$ is taken to be known, but not necessarily zero. Moreover, let us assume

*Assumption 1.* The matrix $T_4^\gamma$ is invertible.

*Assumption 2.* The following two dual algebraic Riccati equations admit real symmetric solutions $X^\gamma$ and $Y^\gamma$ respectively ,

$$X^\gamma A_{22} + A_{22}^T X^\gamma - X^\gamma S_{22}^\gamma X^\gamma + Q_{22} = 0, \quad (11a)$$

$$A_{22} Y^\gamma + Y^\gamma A_{22}^T - Y^\gamma Q_{22} Y^\gamma + S_{22}^\gamma = 0. \quad (11b)$$

If $(T_4^\gamma)^{-1}$ exists, it can be calculated that

$$\begin{bmatrix} A_r^\gamma & -S_r^\gamma \\ Q_r^\gamma & (A_r^\gamma)^T \end{bmatrix} = T_1^\gamma - T_2^\gamma (T_4^\gamma)^{-1} T_3^\gamma, \quad (12)$$

directly by some numerical methods. However, for the purpose of theoretical analysis, the explicit expressions of $A_r^\gamma, S_r^\gamma, Q_r^\gamma$ will be required.

*Lemma 1.* Suppose that Assumptions 1,2 are satisfied. Then, there exist matrices $B_r^\gamma \in R^{r\times m}$, $D_r^\gamma \in R^{r\times l}$ such that $S_r^\gamma = B_r^\gamma (B_r^\gamma)^T - \gamma^{-2} D_r^\gamma (D_r^\gamma)^T$. Moreover, $Q_r^\gamma$ is a positive semidefinite matrix.

*Theorem 1.* For the soft-constrained linear quadratic zero-sum dynamic game of the discrete-time descriptor systems described by (1) and (6), let Assumptions 1 and 2 be satisfied. Then,

(i) the dynamic game admits a linear feedback saddle-point solution if, and only if,

$$\gamma^2 I_r + (B_r^\gamma)^T P_{k+1}^\gamma B_r^\gamma > 0, \ (k \in \mathsf{K}), \quad (13a)$$

$$\gamma^2 I_r - (D_r^\gamma)^T P_{k+1}^\gamma D_r^\gamma > 0, \ (k \in \mathsf{K}), \quad (13b)$$

where $P_{k+1}^\gamma$ satisfies the discrete-time Riccati equation

$$P_k^\gamma = Q_r^\gamma + (A_r^\gamma)^T P_{k+1}^\gamma (I_r + S_r^\gamma P_{k+1}^\gamma)^{-1} A_r^\gamma,$$
$$P_N^\gamma = Q_{11N}. \quad (14)$$

(ii) There exist uncountably many linear feedback saddle-point solutions, with the family of the equilibriu m strategies given by

$$\mu_k^\gamma(x_k) = -B^T M^T \times$$
$$\begin{bmatrix} L_k^{\gamma 1} & 0 \\ L_k^{\gamma 2} - F_k^1 Z_k^{\gamma 2} & F_k^1 \end{bmatrix} H^{-1} x_k, \ (k \in \mathsf{K}), \quad (15a)$$

$$\nu_k^\gamma(x_k) = \gamma^{-2} D^T M^T \times$$
$$\begin{bmatrix} L_k^{\gamma 1} & 0 \\ L_k^{\gamma 2} - F_k^2 Z_k^{\gamma 2} & F_k^2 \end{bmatrix} H^{-1} x_k, \ (k \in \mathsf{K}), \quad (15b)$$

where $F_k^1$, $F_k^2$ are arbitrary two $(n-r)\times(n-r)$ matrices making $A_{22} - B_2 B_2^T F_k^1 + \gamma^{-2} D_2 D_2^T F_k^2$ invertible, and

$$\begin{bmatrix} Z_k^{\gamma 2} \\ L_k^{\gamma 2} \end{bmatrix} = -(T_4^\gamma)^{-1} T_3^\gamma \begin{bmatrix} I_r \\ L_k^{\gamma 1} \end{bmatrix}, \quad (16)$$

$$L_k^{\gamma 1} = P_{k+1}^\gamma (I_r + S_r^\gamma P_{k+1}^\gamma)^{-1} A_r^\gamma. \quad (17)$$

(iii) All the linear feedback saddle-point solutions generate a unique state trajectory

$$x_k^* = H \begin{bmatrix} z_k^{1*} \\ z_k^{2*} \end{bmatrix}, \ (k \in \mathsf{K}), \quad (18)$$

where $z_k^{1*}$ satisfies

$$z_{k+1}^{1*} = (I_r + S_r^\gamma P_{k+1}^\gamma)^{-1} A_r^\gamma z_k^{1*}, \quad (19)$$

with $z_0^1$ is given and

$$z_k^{2*} = Z_k^{\gamma 2} z_k^{1*}. \quad (20)$$

The saddle-point value is

$$L_\gamma(u^\gamma, w^\gamma) = z_0^{1T} P_0^\gamma z_0^1. \quad (21)$$

(iv) If the matrix $\gamma^2 I_r - (D_r^\gamma)^T P_{k+1}^\gamma D_r^\gamma$ has a negative eigenvalue for some $k \in \mathsf{K}$, then the game does not admit a saddle-point solution under the closed-loop memoryless information structure, and its upper value becomes unbounded.

**Proof (sketch)** First, under Assumptions 1,2, a soft-constrained linear quadratic zero-sum game for a reduced-order state space system is formulated, that is, finding the linear feedback saddle-point solution of the cost function

$$J_r^\gamma = z_N^{1T} Q_{11N} z_N^1$$
$$+ \sum_{k=0}^{N-1} \{ z_k^{1T} Q_r^\gamma z_k^1 + u_k^T u_k - \gamma^2 w_k^T w_k \}, \quad (22)$$

subject to the system equation

$$z_{k+1}^1 = A_r^\gamma z_k^1 + B_r^\gamma u_k + D_r^\gamma w_k, \quad (23)$$

where $[z_k^{1T} \ z_k^{2T}]^T = H^{-1} x_k$. The soft-constrained linear quadratic zero-sum game of the reduced-order state space system admits a unique linear feedback saddle-point solution if, and only if the conditions (13a,b) are satisfied (Başar and Olsder, 1982).

Then, some relations are obtained from the above problem, that is,

(i) two discrete-time Riccati equations

$$P_k^\gamma = (A_r^\gamma - B_r^\gamma (B_r^\gamma)^T L_k^{\gamma 1})^T P_{k+1}^\gamma (I_r$$
$$- D_r^\gamma (D_r^\gamma)^T P_{k+1}^\gamma)^{-1} (A_r^\gamma - B_r^\gamma (B_r^\gamma)^T L_k^{\gamma 1})$$
$$+ Q_r^\gamma + (L_k^{\gamma 1})^T B_r^\gamma (B_r^\gamma)^T L_k^{\gamma 1},$$
$$P_N^\gamma = Q_{11N}, \quad (24a)$$

and

$$P_k^\gamma = (A_r^\gamma + D_r^\gamma (D_r^\gamma)^T L_k^{\gamma 1})^T P_{k+1}^\gamma (I_r$$

$$+B_r^\gamma(B_r^\gamma)^T P_{k+1}^\gamma)^{-1}(A_r^\gamma + D_r^\gamma(D_r^\gamma)^T L_k^{\gamma 1})$$

$$+Q_r^\gamma + (L_k^{\gamma 1})^T D_r^\gamma (D_r^\gamma)^T L_k^{\gamma 1},$$

$$P_N^\gamma = Q_{11N}, \quad (24b)$$

are equivalent to the discrete-time Riccati equation (14), in other words , they have the same solution $P_k^\gamma$;

(ii)

$$(I_r + B_r^\gamma(B_r^\gamma)^T P_{k+1}^\gamma)^{-1}(A_r^\gamma + D_r^\gamma(D_r^\gamma)^T L_k^{\gamma 1})$$

$$= (I_r + S_r^\gamma P_{k+1}^\gamma)^{-1} A_r^\gamma, \quad (25a)$$

$$(I_r - D_r^\gamma(D_r^\gamma)^T P_{k+1}^\gamma)^{-1}(A_r^\gamma - B_r^\gamma(B_r^\gamma)^T L_k^{\gamma 1})$$

$$= (I_r + S_r^\gamma P_{k+1}^\gamma)^{-1} A_r^\gamma. \quad (25b)$$

Finally, using the relations above, it is proven that (15a),(15b) constitute the linear feedback saddle-point solutions to the soft-constrained zero-sum game after relatively lengthy derivations.

## 4. DESIGN OF THE DISTURBANCE ATTENUATION CONTROLLER

Let us now study the original disturbance attenuation problem formulated in Section 2 under the following assumption.

*Assumption 3.* The following two dual algebraic Riccati equations admit real symmetric solutions $X$ and $Y$ respectively,

$$XA_{22} + A_{22}^T X - XB_2B_2^T X + Q_{22} = 0, \quad (26a)$$

$$A_{22}Y + YA_{22}^T - YQ_{22}Y + B_2B_2^T = 0. \quad (26b)$$

Note that the objective of this section is to find a controller $\mu^* \in \mathsf{M}$ such that

$$\sup_{\{w_k \in \mathsf{W}_k; k \in \mathsf{K}\}} \{J(\mu^*, w)\}^{1/2}/\|w\| = \gamma^*, \quad (27)$$

where $\gamma^*$ is defined by (4), or a controller $\mu^\gamma \in \mathsf{M}$ such that

$$\sup_{\{w_k \in \mathsf{W}_k; k \in \mathsf{K}\}} \{J(\mu^\gamma, w)\}^{1/2}/\|w\| \le \gamma, \quad (28)$$

for any $\gamma > \gamma^*$. In view of (21) and under the conditions of Theorem 1, we have

$$J_\gamma(\mu^\gamma, w) \le J_\gamma(\mu^\gamma, \nu^\gamma) = z_0^{1T} P_0^\gamma z_0^1 \quad (29)$$

with the inequality holding for all $z_0^1$, and all $l^2$ sequences $w$. Using the relationship (6) between $J_\gamma$ and $J$, the inequality (29) can be written equivalently as

$$J(\mu^\gamma, w) \le z_0^{1T} P_0^\gamma z_0^1 + \gamma^2 \sum_{k=0}^{N-1} |w_k|^2, \quad (30)$$

and taking $z_0^1 = 0$, we arrive at

$$\sup_{\{w_k \in \mathsf{W}_k; k \in \mathsf{K}\}} \{J(\mu^\gamma, w)\}^{1/2}/\|w\| \le \gamma. \quad (31)$$

Hence, given any $\gamma$ satisfying Assumptions 1,2 and the condition (13) of Theorem 1, the controllers $\mu_k^\gamma(x_k), k \in \mathsf{K}$ given by (15a), which is further required to ensure that the obtained closed-loop system are causal by selecting $F_k^1$ when $w_k \equiv 0, k \in \mathsf{K}$, guarantee the prespecified disturbance attenuation level $\gamma$.

Let us introduce a set of performance levels.

$$\Gamma_0 := \{\gamma > 0 :$$

$$\text{Hamiltonian matrix } T_4^\gamma \text{ is invertible}\}. \quad (32)$$

$\Gamma_0$ is a nonempty set because the system quaternary $(E, A, B, Q)$ is causally controllable and reconstructible. Let $\gamma_0$ be the maximum of $\Gamma_0$, that is,

$$\gamma_0 := \max\{\gamma : \gamma \in \Gamma_0\}. \quad (33)$$

Taking into account $\gamma_0$, let us introduce two other sets.

$$\Gamma_{xy} := \{\gamma \ge \gamma_0 :$$

$$\text{conditions (11) are satisfied}\}, \quad (34a)$$

$$\Gamma := \{\gamma \ge \gamma_0 :$$

$$\text{conditions (13) are satisfied}\}. \quad (34b)$$

$\Gamma_{xy}$ and $\Gamma$ are nonempty open sets. The former is because Assumption 3 is satisfied, and the latter because $\Gamma_0$ and $\Gamma_{xy}$ are nonempty sets. Moreover, there is no any direct relation between $\Gamma_{xy}$ and $\Gamma$ because the computation of (12) is in fact independent of $X^\gamma$ and $Y^\gamma$. Let $\gamma_{xy}$ and $\hat{\gamma}$ be the infimums of $\Gamma_{xy}$ and $\Gamma$ respectively, that is,

$$\gamma_{xy} := \inf\{\gamma : \gamma \in \Gamma_{xy}\}, \quad (35a)$$

$$\hat{\gamma} := \inf\{\gamma : \gamma \in \Gamma\}. \quad (35b)$$

*Theorem 2.* Consider the disturbance attenuation control problem formulated in Section 2 under Assumption 3. The following results hold.

(i) There exist uncountably many central $H_\infty$-controllers to achieve the prespecified performance level $\gamma \ge \max\{\hat{\gamma}, \gamma_{xy}\}$, with the family of the controllers given by

$$\mu_k^\gamma(x_k) = -B^T M^T \times$$

$$\begin{bmatrix} L_k^{\gamma 1} & 0 \\ L_k^{\gamma 2} - F_k Z_k^{\gamma 2} & F_k \end{bmatrix} H^{-1} x_k, \ (k \in \mathsf{K}), \quad (36)$$

where $F_k$ is an arbitrary $(n-r) \times (n-r)$ matrix making $A_{22} - B_2 B_2^T F_k$ invertible, and $L_k^{\gamma 1}$, $Z_k^{\gamma 2}$ and $L_k^{\gamma 2}$ are defined by (16),(17) respectively.

(ii) If $\hat{\gamma} \geq \gamma_{xy}$, then $\gamma^* = \hat{\gamma}$.

(iii) Otherwise, $\hat{\gamma} \leq \gamma^* \leq \gamma_{xy}$.

**Proof** (i) From the definitions of $\hat{\gamma}$ and $\gamma_{xy}$, it is known that Assumptions 1,2 and Condition (i) of Theorem 1 are satisfied by any $\gamma \geq \max\{\hat{\gamma}, \gamma_{xy}\}$. Hence,

$$\mu_k^{\gamma}(x_k) = -B^T M^T \times$$

$$\begin{bmatrix} L_k^{\gamma 1} & 0 \\ L_k^{\gamma 2} - F_k Z_k^{\gamma 2} & F_k \end{bmatrix} H^{-1} x_k, \quad (37a)$$

$$\nu_k^{\gamma}(x_k) = D^T M^T \begin{bmatrix} L_k^{\gamma 1} & 0 \\ L_k^{\gamma 2} & 0 \end{bmatrix} H^{-1} x_k, \quad (37b)$$

constitute the linear feedback saddle-point solutions from Theorem 1. As the result, any controller given by (36) with invertible $A_{22} - B_2 B_2^T F_k$ forms a central $H_\infty$ controller.

(ii) For a disturbance attenuation level $\gamma > \hat{\gamma}$, there exists a central $H_\infty$ controller to achieve this performance level as in (i). On the other hand, for a $\gamma < \hat{\gamma}$, the condition (13b) will not be satisfied, which means that, by Theorem 1(iii), there does not exist any closed-loop controller that will delive r a bound in (31) that is smaller than $\hat{\gamma}$, and hence $\gamma^* = \hat{\gamma}$.

(iii) This is obvious from (i) and (ii).

*Remark 1.* $\gamma^* \neq \gamma_{xy}$ in (iii) of Theorem 2 because, although one of the algebraic Riccati equations (11) does not admit a real symmetric solution when $\gamma < \gamma_{xy}$, two dual quadratic equations

$$(\hat{X}^{\gamma})^T A_{22} + A_{22}^T \hat{X}^{\gamma} -$$

$$(\hat{X}^{\gamma})^T S_{22}^{\gamma} \hat{X}^{\gamma} + Q_{22} = 0, \quad (38a)$$

$$A_{22} \hat{Y}^{\gamma} + (\hat{Y}^{\gamma})^T A_{22}^T -$$

$$(\hat{Y}^{\gamma})^T Q_{22} \hat{Y}^{\gamma} + S_{22}^{\gamma} = 0. \quad (38b)$$

may admit a real solution even if $\gamma < \gamma_{xy}$. In this case, the results of Lemma 1 still hold true except that Assumption 2 is substituted by the existence of real solutions to two dual quadratic equations (38) respectively, that is, $\hat{X}^{\gamma}$ and $\hat{Y}^{\gamma}$. This fact implies that the related soft-constrained dynamic game may still admit a linear feedback saddle-point solution for $\gamma < \gamma_{xy}$, and hence the possible existence of the $H_\infty$-controller.

*Remark 2.* It is our conjecture that the condition (13) is more stringent than the condition (11), and hence $\hat{\gamma} \geq \gamma_{xy}$. If this conjecture is correct, then the optimum disturbance attenuation level $\gamma^*$ can be determined completely by the condition (13b) which is similar to the corresponding condition in state space systems (Başar and Bernhard, 1995).

For any $\gamma > \gamma^*$, the disturbance attenuation controllers can be constructed by following the steps given below.

**Step 1.** Calculate $A_r^{\gamma}, S_r^{\gamma}, Q_r^{\gamma}$ by using the formula (12).

**Step 2.** Solve the discrete-time Riccati equation (14).

**Step 3.** Calculate $L_k^{\gamma 1}$, $L_k^{\gamma 2}$ and $Z_k^{\gamma 2}$ by using (16), (17).

**Step 4.** Construct the controllers according to (36).

## An illustrative example

Consider the following descriptor system and the parameterized performance index

$$\begin{bmatrix} 1 & 0 \\ 0 & 0 \end{bmatrix} \begin{bmatrix} z_{k+1}^1 \\ z_{k+1}^2 \end{bmatrix} = \begin{bmatrix} 1 & 2 \\ 1 & 0 \end{bmatrix} \begin{bmatrix} z_k^1 \\ z_k^2 \end{bmatrix}$$

$$+ \begin{bmatrix} 2 \\ 1 \end{bmatrix} u_k + \begin{bmatrix} 0 \\ 1 \end{bmatrix} w_k, \quad (39)$$

$$L_{\gamma}(u, w) = \sum_{k=0}^{N-1} \{2(z_k^1)^2 + 2(z_k^2)^2$$

$$+ u_k^2 - \gamma^2 w_k^2\}. \quad (40)$$

In the following, the small letters are used to denote the corresponding scalars. The Hamiltonian matrix $T_4^{\gamma}$ of this example is

$$T_4^{\gamma} = \begin{bmatrix} 0 & -1 + \gamma^{-2} \\ -2 & 0 \end{bmatrix}, \quad (41)$$

which is invertible for every $\gamma > 1$. Two dual algebraic Riccati equations (11a),(11b) are

$$2\gamma^2 + (1 - \gamma^2)(x^{\gamma})^2 = 0, \quad (42a)$$

$$-2\gamma^2 (y^{\gamma})^2 - (1 - \gamma^2) = 0, \quad (42b)$$

whose solutions are $x_\gamma = \sqrt{2}\gamma / \sqrt{\gamma^2 - 1}$ and $y_\gamma = \sqrt{2}\sqrt{\gamma^2 - 1}/2\gamma$ respectively. Obviously,

$x^{\gamma}, y^{\gamma}$ are real if $\gamma > 1$. Hence, $\gamma_0 = \gamma_{xy} = 1$. From (12), it is obtained that

$$a_r^{\gamma} = \frac{1 + \gamma^2}{1 - \gamma^2}, \ q_r^{\gamma} = \frac{2 - 3\gamma^2}{1 - \gamma^2},$$

$$s_r^{\gamma} = \frac{6 - 2\gamma^2}{1 - \gamma^2}, \quad (43)$$

where $s_r^\gamma$ can be decomposed to

$$s_r^\gamma = \frac{(\sqrt{2\gamma^2(\gamma^2-1)}-2)^2}{(\gamma^2-1)^2} - \frac{(\sqrt{2\gamma^2(\gamma^2-1)}-2\gamma^2)^2}{\gamma^2(\gamma^2-1)^2} \quad (44)$$

The corresponding discrete-time Riccati equation becomes

$$p_k^\gamma = \frac{2-3\gamma^2}{1-\gamma^2} + \frac{(1+\gamma^2)^2}{(1-\gamma^2)^2}[1 + \frac{6-2\gamma^2}{1-\gamma^2} p_{k+1}^\gamma]^{-1} p_{k+1}^\gamma, \quad p_N^\gamma = 0. \quad (45)$$

Now, let us take $N = 3$ and $s = \gamma^2$. The condition (13b) becomes

$$s > 1, \quad (46a)$$

$$s + \frac{(\sqrt{2s(s-1)}-2s)^2(2-3s)}{(s-1)^3} > 0, \quad (46b)$$

$$s + \frac{(\sqrt{2s(s-1)}-2s)^2(2-3s)}{(s-1)^3} \times \frac{(8s^2-22s+14)}{(7s^2-24s+13)} > 0, \quad (46c)$$

which can be further simplified to

$$s > 1, \quad (47a)$$

$$(s-1)^3 + (2-3s)(6s-2 - 4\sqrt{2s(s-1)}) > 0, \; s > 1, \quad (47b)$$

$$(s-1)^3(7s^2-24s+13) + (2-3s)\times (8s^2-22s+14)(6s-2-4\sqrt{2s(s-1)}) > 0, \; s > 3. \quad (47c)$$

By numerical computation, the results are obtained that

$$s > 1, \quad s > 3.80849, \quad s > 4.68831, \quad (48)$$

which implies that the optimum disturbance attenuation level is $\gamma^* \approx \sqrt{4.68831} \approx 2.16525$. Therefore, for a prespecified disturbance attenuation level $\gamma \geq 2.16525$, uncountably many feedback minimax controllers can be constructed by using (15a).

## 5. CONCLUSIONS

In this paper, the finite-horizon $H_\infty$-optimal control problem for discrete-time descriptor systems has been studied. By using the dynamic game theoretic approach, the upper and lower bounds of the optimum disturbance attenuation level have been provided in terms of the solutions of the parameterized discrete-time Riccati equation and the parameterized continues-time algebraic Riccati equations. It is shown that, if a prespecified disturbance attenuation level is achievable, then there exist uncountably many central $H_\infty$-controllers to guarantee such a performance level. In this paper, only the finite-horizon problem is studied. The same results are also available for the infinite-horizon problem suppose that some stabilizable and observable conditions are satisfied.

## REFERENCES


Başar, T. and G.J. Olsder (1982). *Dynamic Noncooperative Game Theory*, Academic Press, New York.

Başar, T. and P. Bernhard (1995). *$H^\infty$-Optimal Control and Related Minimax Design Problems: A Dynamic Game Approach*, Birkhäuser, Boston.

Bender, D.J. and A.J. Laub (1987a). The linear-quadratic optimal regulator for descriptor systems. *IEEE Transactions on Automatic Control*, **32**, 672-688.

Bender, D.J. and A.J. Laub (1987b). The linear-quadratic optimal regulator for descriptor systems: discrete-time case. *Automatica*, **23**, 71-85.

Dai, L. (1989). Singular Control Systems. *Lecture Notes in Control and Information Sciences*, Edited by M.Thoma and A.Wyner, Springer-Verlag, Berlin.

Başar (1991). A dynamic games Approach to controller design: disturbance rejection in discrete-time. *IEEE Transactions on Automatic Control*, **36**, 936-952.

Xu, H. and K. Mizukami (1994). Linear-quadratic zero-sum differential games for generalized state space systems, *IEEE Transactions on Automatic Control*, **39**, 143-147.

Xu, H. and K. Mizukami (1994). On the Isaacs equation of differential games for descriptor systems. *Journal of Optimization Theory and Applications*, **83**, 405-419.

Xu, H. and K. Mizukami (1996). Upper and lower bounds of $H_\infty$-optimal performance for a class of continuous-time descriptor systems. *Proceeding of the 35th IEEE Conference on Decision and Control*, **1**, 1021-1026.

14th Triennial World Congress, Beijing, P.R. China



# SOME PROPERTIES OF $l^1$ OPTIMAL CONTROLLER*

**Huajing Fang**

*Department of Automatic Control Engineering,*
*Huazhong University of Science and Technology*
*Wuhan 430074, P.R. China*
*E_mail: hjfang@blue.hust.edu.cn, Tel: +86 27 87541109*

Abstract: Two properties of $l^1$ optimal controller are pointed out and discussed in this paper. First, it is pointed out that some $l^1$ optimal controller may be noncausal (anticipatory), and can not be implemented with a real physical plant. A method for modifying a noncausal $l^1$ controller to a causal one is presented. Secondly, two simple examples show that the optimal solution of $l^1$ norm minimization problem does not always exist for systems with zero and (or) pole on the unit circle. Some preliminary results for determining the existence are presented. *Copyright © 1999 IFAC*

Key words: Robust control; optimization; discrete-time systems; disturbance rejection; bounded disturbances;control theory; linear systems; minimization; suboptimal systems

## 1. INTRODUCTION

The $l^1$ optimal control problem was first presented by Vidyasagar(1986), and solved by Dahleh and Pearson (1987,1988), Diaz-bobillo and Dahleh (1993). In this paper some properties of $l^1$ optimal controller will be further discussed.

Here, the symbol $A$ denotes the set of z-transfer functions of all stable discrete-time systems $\hat{f} = \sum_{i=0}^{\infty} f_i z^i$, where $f = \{f_i\}_{i=0}^{i=\infty} \in l^1$ and z is a delay operator. The set $A$ can be viewed as a space of bounded operators on space $l^\infty$, and the induced norm on $A$ is given by $\|\hat{f}\|_A = \|f\|_1$.

The $l^1$ sensitivity minimization problem of a discrete time, linear, time-invariant feedback control system can be described as an optimization problem

$$\mu_0 = \inf_{\hat{c} \in S(\hat{p})} \left\| (1+\hat{p}\hat{c})^{-1}\hat{w} \right\|_A \tag{1}$$

Where $S(\hat{p})$ denotes the set of all controllers which stabilize the plant $\hat{p}$,

$$S(\hat{p}) = \left\{ \hat{c}(z) = \frac{\hat{x}+\hat{q}\hat{d}_p}{\hat{y}-\hat{q}\hat{n}_p} \,|\, \hat{q} \in A,\ \hat{y}-\hat{q}\hat{n}_p \neq 0 \right\} \tag{2}$$

$\hat{p} = \hat{n}_p / \hat{d}_p$, $\hat{n}_p, \hat{d}_p \in A$, and coprime; $\hat{x}, \hat{y} \in A$
$\hat{x}\hat{n}_p + \hat{y}\,\hat{d}_p = 1$ (Dahleh and Pearson, 1987, 1988).

With the controller corresponding the optimal solution of problem(1), the closed loop system can optimally reject persistent bounded disturbances. However the example presented in this paper shows that sometimes the optimal feedback controller has some poles at the origin of z-plane, namely

$$\hat{c}^* = \frac{u(z)}{y(z)} = \frac{a_0 + a_1 z + \ldots + a_n z^n}{z^l (b_0 + b_1 z + \ldots\ b_m z^m)}$$

* This work was supported by the National Natural Science Foundation of China under Grant 69274003 and 69774014

The difference equation corresponding to this z transfer function is

$$b_0 u(k-l) + b_1 u(k-l-1) + \dots b_m u(k-l-m)$$
$$= a_0 y(k) + a_1 y(k-1) + \dots a_n y(k-n)$$

then

$$u(k) = \{[a_0 y(k+l) + \dots a_n y(k+l-n)]$$
$$-[b_1 u(k-1) + \dots b_m u(k-m)]\}/b_0$$

The output signal $u(k)$ of this controller is related to the input signal $y(i)$ of time $i = k+1, k+2, \dots, k+l$. This fact means that this controller is noncausal, and can not be implemented with a real physical plant.

Substitute (2) in to (1), the $l^1$ optimization problem can be phrased as

$$\mu_0 = \inf_{\hat{q} \in A} \left\| \hat{h} - \hat{q}\hat{v} \right\|_A \tag{3}$$

where $\hat{h} = \hat{d}_p \hat{y} \hat{w}$ $\hat{v} = \hat{d}_p \hat{n}_p \hat{w}$. The zeros and poles of $\hat{p}$ is the zeros of $\hat{v}$. The optimal solution $\hat{q}^*$ for the $l^1$ optimal problem (3) always exists when $\hat{v}$ has no zeros on the unit circle (Dahleh and Pearson, 1987, 1988), but the situation will become complicated when $\hat{v}$ has zeros on the unit circle. Examples are presented in this paper , the first one has solution and another one has no solution.

The remainder of this paper is organized as follows. Section 2 discusses the possibility to modify a noncausal $l^1$ optimal controller to a causal one such that the $l^1$ norm performance function is not influenced too much, then presents a simple modifying method and an example that illustrate the results. Section 3 presents some results about the existence problem of $l^1$ optimal solution, and some examples which show the existence problem for system with zero and (or) pole on the unit circle. Section 4 contains some concluding remarks.

## 2. THE MODIFICATION OF THE NONCAUSAL $l^1$ OPTIMAL CONTROLLER

### *2.1 Modification Possibility*

From above discussion, It is known that there is possibility that some $l^1$ optimal controller for the sensitivity minimization problem is noncausal. In order to avoid such situation, the feasible region of $l^1$ optimal sensitivity problem should be restricted on the set of causal controllers. Let $S_r(\hat{p})$ denote the set of all causal controllers which stabilize the plant $\hat{p}$. There are two problems that should be discussed first. Is the set $S_r(\hat{p})$ empty for any causal plant $\hat{p}$? How is the $l^1$ norm performance index influenced when the feasible region of $l^1$ optimal problems is restricted on the set $S_r(\hat{p})$ i.e. what difference are there between the infimum $\mu_r$ of optimization problem

$$\mu_r = \inf_{\hat{c} \in S_r(\hat{p})} \left\| 1 + \hat{p}\hat{c})^{-1} \hat{w} \right\|_A \tag{4}$$

and the $\mu_0$ of the problem (1)?

Theorem 1. The set $S_r(\hat{p})$ is not empty for any causal plant $\hat{p}$. and $S_r(\hat{p}) = S(\hat{p})$ if $\hat{p}$ has at least one zero at the origin of z-plane.

Proof: A controller $\hat{c} \in S(\hat{p})$ is causal if $\hat{c}$ has no poles at the origin of z-plane, with equation (2), this is the same to say

$$\hat{y}(0) - \hat{q}(0)\hat{n}_p(0) \neq 0 \tag{5}$$

It is immediate from $\hat{x}\hat{n}_p + \hat{y}\hat{d}_p = 1$ and causality of $\hat{p}$ that $\hat{x}(0)\hat{n}_p(0) + \hat{y}(0)\hat{d}_p(0) = 1$ and $\hat{d}_p(0) \neq 0$, then $\hat{y}(0) = 1/\hat{d}_p(0) \neq 0$ if $\hat{p}$ has at least one zero at the origin of z-plane (that is $\hat{n}_p(0) = 0$) .This implies the (5) hold for any $\hat{q} \in A$, namely $S_r(\hat{p}) = S(\hat{p})$. When $\hat{p}$ has no zero at the origin of z-plane (that is $\hat{n}_p(0) \neq 0$), the (5) holds if choosing $\hat{q} \in A$ such that $\hat{q}(0) = q_0 \neq \hat{y}(0)/\hat{n}_p(0) = \beta$, this is an easy job and implies the set $S_r(\hat{p})$ is not empty.

Let $A_r \subset A$, and

$$A_r = \left\{ \hat{q} \in A \mid \hat{q}(0) = q_0 \neq \beta, \ when \ \hat{n}_p(0) \neq 0 \right\}$$

then

$$S_r(\hat{p}) = \left\{ \hat{c}(z) = \frac{\hat{x} + \hat{q}\hat{d}_p}{\hat{y} - \hat{q}\hat{n}_p} \mid \hat{q} \in A_r, \hat{y} - \hat{q}\hat{n}_p \neq 0 \right\}$$

Lemma 1. The set $A_r$ is dense in $A$. (The proof is omitted)

Theorem 2 . For any causal $\hat{p}$, $\mu_0 = \mu_r$

Proof: By employing (2),

$$\mu_0 = \inf_{\hat{c} \in S(\hat{p})} \left\| (1 + \hat{p}\hat{c})^{-1} \hat{w} \right\|_A = \inf_{\hat{q} \in A} \left\| \hat{h} - \hat{q}\hat{v} \right\|_A$$

Using the definition of infimum, it is known that for any $\varepsilon > 0$ there exists a $\hat{q}_1 \in A$ such that $\left\| \hat{h} - \hat{q}_1\hat{v} \right\|_A \leq \mu_0 + \varepsilon/2$ . Assume $\hat{q}_1 \notin A_r$. It is follows from Lemma 1 that $A_r$ is dense in $A$ and there exists a $\hat{q}_2 \in A_r$ such that $\left\| \hat{q}_1 - \hat{q}_2 \right\|_A \leq \varepsilon / 2 \left\| \hat{v} \right\|_A$ . Then

$$\left\| \hat{h} - \hat{q}_2\hat{v} \right\|_A = \left\| \hat{h} - \hat{q}_1\hat{v} + (\hat{q}_1 - \hat{q}_2)\hat{v} \right\|_A$$
$$\leq \left\| \hat{h} - \hat{q}_1\hat{v} \right\|_A + \left\| \hat{q}_1 - \hat{q}_2 \right\|_A \left\| \hat{v} \right\|_A$$
$$\leq \mu_0 + \frac{\varepsilon}{2} + \frac{\varepsilon}{2\left\| \hat{v} \right\|_A} * \left\| \hat{v} \right\|_A = \mu_0 + \varepsilon .$$

This means $\mu_r = \mu_0$.

*2.2 Modification Method*

A method will be presented here, which modifies a noncausal $l^1$ optimal controller to a causal one such that the performance index of closed system with this modified controller can arbitrarily approach the infimum $\mu_0$.

Let $\hat{p}$ have $l$ zeros $\alpha_i$, $i = 1, 2, \ldots l$, and $m$ poles $\alpha_{l+j}$, $j = 1, 2, \ldots m$ ($l + m = n$) inside the closed unit disk. For simplicity and without loss of generality, it is assumed that $\alpha_i$'s and $\alpha_{l+j}$'s are all real and distinct. It is well know, see (Dahleh and Pearson 1988; Fang,1990), that the optimal weighted sensitivity function $\hat{b}^* = (1 + \hat{p}\hat{c}^*)^{-1}\hat{w}$ for the $l^1$ sensitivity minimization problem (1) is a polynomial and have

$$\begin{aligned} \hat{b}^*(\alpha_i) &= \hat{w}(\alpha_i) \qquad & i = 1, 2, \ldots l \\ \hat{b}^*(\alpha_{l+j}) &= 0 \qquad & j = 1, 2, \ldots m \end{aligned} \tag{6}$$

From the expression of $\hat{c}^* = (\hat{w} - \hat{b}^*)/\hat{b}^*\hat{p}$, it can be seen that the zeros $\alpha_i$, $i = 1, 2, \ldots l$, of $\hat{p}$ are canceled by the zeros of $\hat{w} - \hat{b}^*$, and the zeros $\alpha_{l+j}$, $j = 1, 2, \ldots m$, of $\hat{b}^*$ are canceled by the poles of $\hat{p}$, then the unstable zeros of $\hat{b}^*$ with $\alpha_{l+j}$, $j = 1, 2, \ldots m$, excluded are the unstable poles of $\hat{c}^*$. Using this fact and noticing that $\hat{p}$ is causal and has no poles at the origin of z-plane, it can be concluded that the $\hat{c}^*$ is causal if and only if $\hat{b}^*(0) = b_0^* \neq 0$, then a noncausal $\hat{c}^*$ can be modified to a causal one by modifying the $\hat{b}^*$ into $\hat{b}'$ such that $b'(0) = b_0' \neq 0$ under the restriction relation (6).

Suppose the degree of polynomial $\hat{b}^*$ is $N - 1$, and define $M = \max\{2n, N\}$. The equation (6) can be expanded into linear algebra equations

$$\mathbf{EB} = \mathbf{H} \tag{7}$$

where

$$\mathbf{B} = \begin{bmatrix} b_0 \\ b_1 \\ \vdots \\ b_{N-1} \end{bmatrix}, \quad \mathbf{E} = \begin{bmatrix} 1 & \alpha_1 & \alpha_1^2 & \cdots & \alpha_1^{M-1} \\ 1 & \alpha_2 & \alpha_2^2 & \cdots & \alpha_2^{M-1} \\ \vdots & \vdots & \vdots & \vdots & \vdots \\ 1 & \alpha_n & \alpha_n^2 & \cdots & \alpha_n^{M-1} \end{bmatrix}$$

and $\mathbf{H} = \left(\hat{w}(\alpha_1), \ldots \hat{w}(\alpha_l), 0 \ldots 0\right)^T$.

Blocking matrix $\mathbf{E}$ into $\mathbf{E} = [\mathbf{E}_1 \quad \mathbf{E}_2]$, where $\mathbf{E}_1 = [\mathbf{E}_{1,0}, \mathbf{E}_{1,1}, \ldots \mathbf{E}_{1,n-1}]$ is the first $n$ column vectors of $\mathbf{E}$ and $\mathbf{E}_2 = [\mathbf{E}_{2,n}, \mathbf{E}_{2,n+1}, \ldots \mathbf{E}_{2,M-1}]$ is the last $M - n$ column vectors. Let $\mathbf{x} = (x_n, x_{n+1}, \ldots x_{M-1})^T$ be any $M - n$ degree real vector, and let $\mathbf{I}$ denote a $M - n$ degree unit matrix, then all solution of equation (7) are

$$\mathbf{B} = \begin{bmatrix} \mathbf{E}_1^{-1}\mathbf{H} \\ 0 \end{bmatrix} - \begin{bmatrix} \mathbf{E}_1^{-1}\mathbf{E}_2 \\ -\mathbf{I} \end{bmatrix}\mathbf{x}, \quad \mathbf{x} \in R^{M-n}$$

Assume $\mathbf{B}^* = (b_0^*, b_1^*, \ldots b_{M-1}^*)^T$ corresponding to $\mathbf{x}^* = (b_n^*, b_{n+1}^*, \ldots, b_{M-1}^*)^T$ is the parameters vector of the $l^1$ optimal sensitivity function $\hat{b}^*$. It is follows the above discussion that if $b_0^* = 0$, the corresponding controller $\hat{c}^*$ is noncausal, so the only thing needed to do is to modify the $\mathbf{x}^*$ to $\mathbf{x}'$, such that the corresponding $\hat{b}'$ satisfy (7) and

$$\|\mathbf{B}'\|_1 \leq \|\mathbf{B}^*\|_1 + \varepsilon, \quad b_0' \neq 0$$

then the controller $\hat{c}'$ is causal and

$$\left\|(1 + \hat{p}\hat{c}')^{-1}\hat{w}\right\|_A = \|\mathbf{B}'\|_1 \leq \mu_0 + \varepsilon$$

It is not difficult to show that the first row of matrix $\mathbf{E}_1^{-1}\mathbf{E}_2$ is not a zero vector, so this modification is always possible.

Define $\Delta = \left\| \begin{pmatrix} \mathbf{E}_1^{-1}\mathbf{E}_2 \\ -\mathbf{I} \end{pmatrix} \right\|_1$, then

Theorem 3. For any given positive real $\varepsilon > 0$, choose $\mathbf{x}'$ such that $\|\mathbf{x}^* - \mathbf{x}'\|_1 \leq \varepsilon/\Delta$, then $\|\mathbf{B}'\|_1 \leq \mu_0 + \varepsilon$

Proof.

$$\begin{aligned} \|\mathbf{B}'\|_1 &= \left\| \begin{bmatrix} \mathbf{E}_1^{-1}\mathbf{H} \\ \mathbf{0} \end{bmatrix} - \begin{bmatrix} \mathbf{E}_1^{-1}\mathbf{E}_2 \\ -\mathbf{I} \end{bmatrix}\mathbf{x}' \right\|_1 \\ &\leq \left\| \begin{bmatrix} \mathbf{E}_1^{-1}\mathbf{H} \\ 0 \end{bmatrix} - \begin{bmatrix} \mathbf{E}_1^{-1}\mathbf{E}_2 \\ -\mathbf{I} \end{bmatrix}\mathbf{x}^* \right\|_1 + \left\| \begin{bmatrix} \mathbf{E}_1^{-1}\mathbf{E}_2 \\ -\mathbf{I} \end{bmatrix}(\mathbf{x}^* - \mathbf{x}') \right\|_1 \\ &\leq \mu_0 + \Delta\|\mathbf{x}^* - \mathbf{x}'\|_1 \leq \mu_0 + \varepsilon \end{aligned}$$

The selection of $\mathbf{x}'$ can be further simplified in real computation. Using $\mathbf{Q} = (q_n, q_{n+1}, \ldots q_{M-1})$ to denote the first row of matrix $\mathbf{E}_1^{-1}\mathbf{E}_2$, assuming that the subscript of the $q_i$ which has the largest absolute value among all the $q_i, i = n, n+1 \ldots M-1$ is $k$, and defining $\delta = \varepsilon \Big/ \left\| \begin{bmatrix} \mathbf{E}_1^{-1}\mathbf{E}_{2,k} \\ 1 \end{bmatrix} \right\|_1$, then $\mathbf{x}'$ can be selected, such that $\mathbf{x}_k' = \mathbf{x}_k^* + \delta$ and $\mathbf{x}_i' = \mathbf{x}_i^*$ for all $i \neq k$. The proof is similar to the one for theorem 3 and is omitted here.

*2.3 Example*

The plant $\hat{p}=(1-3z)(1-2z)$ and the weighting function $\hat{w}=\dfrac{2+z}{12-17z+6z^2}$. Solve the $l^1$ sensitivity minimization problem (1).

Solving this problem with the theory presented by Dahleh and Pearson (1987, 1988), Mendlovitz(1989) and Fang(1990), the solution is $\hat{b}^*=z$, $\mu_0=1$ and $\hat{c}^*=\dfrac{2-z}{z(12-17z+6z^2)}$. This controller has a pole at the origin of z-plane and is noncausal. With the modification method presented above,

$$\mathbf{B}=\begin{bmatrix}\mathbf{E}_1^{-1}\mathbf{H}\\ \mathbf{0}\end{bmatrix}-\begin{bmatrix}\mathbf{E}_1^{-1}\mathbf{E}_2\\ -\mathbf{I}\end{bmatrix}\mathbf{x}$$

$$=\begin{bmatrix}0\\1\\0\\0\end{bmatrix}-\begin{bmatrix}-1/6\\5/6\\-1\\0\end{bmatrix}\mathbf{x}_2-\begin{bmatrix}-5/36\\19/36\\0\\-1\end{bmatrix}\mathbf{x}_3 \tag{8}$$

It is from $\mathbf{B}^*=\begin{pmatrix}0 & 1 & 0 & 0\end{pmatrix}$ that $x_2^*=x_3^*=0$. Choose $\varepsilon=1/10$ and define $\delta=\varepsilon\Big/\left\|\begin{matrix}\mathbf{E}_1^{-1}\mathbf{E}_{2,2}\\ 1\end{matrix}\right\|_1=\dfrac{1}{20}$, then let $x_2'=x_2^*+\delta$, $x_3'=x_3^*=0$. Substituting $x_2', x_3'$ into (8), it follows that

$$\mathbf{B}'=\begin{pmatrix}\frac{1}{120} & \frac{23}{24} & \frac{1}{20} & 0\end{pmatrix}, \hat{b}'=\frac{1}{120}+\frac{23}{24}z+\frac{1}{20}z^2,$$

$$\mu'=1\frac{1}{60},\quad \hat{c}'=\frac{228-103z-6z^2}{(1+115z+6z^2)(12-17z+6z^2)}$$

which is causal, and

$$\mu'-\mu_0=1/60<\varepsilon,\ (\mu'-\mu_0)/\mu_0=1.667\,\%$$

## 3. SOME RESULTS ABOUT THE SOLUTION EXISTENCE OF $l^1$ OPTIMAL PROBLEMS

The solution existence of the $l^1$ optimal problem (3) has been proved by Dahlch and Pearson(1987,1988) when the $\hat{v}$ has no zeros on the unit circle. These existence problems will be discussed again, two simple examples given below show that the $l^1$ optimal solution does not always exist when the $\hat{v}$ has zeros on the unit circle, and some results for determine the existence are presented.

Consider the general $l^1$ optimal problem

$$\mu_0=\inf_{\hat{q}\in A}\left\|\hat{h}-\hat{q}\hat{v}\right\|_A \tag{9}$$

Assume thc $\hat{v}$ has $n$ zeros $\alpha_i$, $i=1,2,\dots n$ in the closed unit disk and all these zeros are distinct for simplicity. Define

$$S=\left\{k\in l^1 \mid \hat{k}(\alpha_i)=0,\ i=1,2,\dots,n\right\},$$

$$g_j=\sum_{i=1}^{n}\left\{\lambda_i \operatorname{Re}\left[\alpha_i^j\right]+\lambda_{n+i}\operatorname{Im}\left[\alpha_i^j\right]\right\},\quad j=0,1,\dots.$$

where $\Lambda=(\lambda_1,\lambda_2,\dots,\lambda_{2n})$ is an arbitrary real vector, and $g=\{g_i\}_{i=0}^{\infty}\in l^\infty$ due to $|\alpha_i|\le 1$ for all $i$. Let $\hat{k}=\hat{q}\hat{v}$, then $\hat{q}\in A$ is equivalent to $k\in S$

With the optimization theory(Luenberger, 1969), and the results of Dahlch and Preason (1987, 1988)

$$\mu_0=\inf_{k\in S}\left\|\hat{h}-\hat{k}\right\|_A$$

$$=\max\sum_{i=1}^{n}\left\{\lambda_i\operatorname{Re}\left[\hat{h}(\alpha_i)\right]+\lambda_{n+i}\operatorname{Im}\left[\hat{h}(\alpha_i)\right]\right\}, \tag{10}$$

$$s.t.\left|g_j\right|\le 1,\quad j=0,1,\dots$$

Let Opt.1 be the left optimal problem in (10). The right optimal problem in (10) is the dual problem of Opt.1, which always has solution. When $\hat{v}$ has no zero on the unit circle the Opt.1 has solution that can reach the infimum $\mu_0$ (Dahlch and Preason, 1987, 1988), but the situation becomes complicated when $\hat{v}$ has some zeros on the unit circle. Some examples show the optimal solution does not always exist. We will discuss this problem preliminary and present some results.

### *3.1 Existence of Optimal Solution*

Theorem 4. 1) The optimal solution $\hat{k}^*$of Opt.1 exists such that $\left\|\hat{h}-\hat{k}^*\right\|_A=\mu_0$ if and only if the following algebraic equations have at least one solution

$$\sum_{i=0}^{\infty}(x_i+y_i)=\mu_0,$$

$$\sum_{i=0}^{\infty}x_i\alpha_j^i-\sum_{i=0}^{\infty}y_i\alpha_j^i=\hat{h}(\alpha_j),\quad j=1,2,\dots,n, \tag{11}$$

$$x_i\ge 0,\ y_i\ge 0,\ i=0,1,\dots.$$

2) If $\mathbf{X}^*=(x_0^*,x_1^*,\dots),\mathbf{Y}^*=(y_0^*,y_1^*,\dots)$ is a solution of equations (11), then the optimal solution of Opt.1 is $\hat{k}^*=\hat{h}-\sum_{i=0}^{\infty}(x_i^*-y_i^*)z^i$

Proof. First, it is shown that if $x_i^*, y_i^*, (i=0,1,2,\dots)$ is a solution of equations (11), the relations

$$x_i^*y_i^*=0 \tag{12}$$

hold for all $i=1,2,\dots$. Prove it by contradiction. If there is a $t$, such that $x_t^*y_t^*\neq 0$, assuming $x_t^*\ge y_t^*$, without loss of generality, and defining

$\mathbf{\bar{X}} = (\bar{x}_0, \bar{x}_1, \ldots)$, $\mathbf{\bar{Y}} = (\bar{y}_0, \bar{y}_1, \ldots)$, where $\bar{x}_t = x_t^* - y_t^*$ $\bar{y}_t = 0$ and $\bar{x}_i = x_i^*$ $\bar{y}_i = y_i^*$ for all $i \neq t$, then $\mathbf{\bar{X}}, \mathbf{\bar{Y}}$ satisfy equations (11). Defining $\mathbf{\bar{B}} = \mathbf{\bar{X}} - \mathbf{\bar{Y}}$ $\bar{b}_i = \bar{x}_i - \bar{y}_i$ $\hat{b}_1 = \sum_{i=0}^{\infty} \bar{b}_i z^i$, $\hat{k}_1 = \hat{h} - \hat{b}_1$, it can be gotten that $\hat{k}_1(\theta_i) = 0$ $i = 1, 2, \ldots, n$, $k_1 \in S$, and

$$\left\|\hat{h} - \hat{k}_1\right\|_A = \left\|\hat{b}_1\right\|_A = \sum_{i=0}^{\infty} |\bar{x}_i + \bar{y}_i|$$

$$= \sum_{i=0}^{\infty} (\bar{x}_i + \bar{y}_i) = \sum_{i=0}^{\infty} (x_i^* + y_i^*) - 2y_t^* < \mu_0$$

This is a contradiction with the assumption that $\mu_0$ is the infimum of the Opt. 1, then the (12) holds

Sufficiency. Define $\mathbf{B}^* = \mathbf{X}^* - \mathbf{Y}^*$, $\hat{b} = \sum b_i^* z^i$,

$\hat{k}^* = \hat{h} - \hat{b}^*$, then $k^* \in S$, and

$$\left\|\hat{h} - \hat{k}^*\right\|_A = \left\|\hat{b}^*\right\|_A = \sum_{i=0}^{\infty} |x_i^* - y_i^*| = \sum_{i=0}^{\infty} (x_i^* + y_i^*) = \mu_0$$

This $\hat{k}^*$ is an optimal solution of Opt. 1.

Necessity. Assume Opt. 1 has a optimal solution $\hat{k}^*$, let $\hat{b}^* = \hat{h} - \hat{k}^* = \sum_{i=0}^{\infty} b_i^* z^i$ and $\mathbf{X}' = (x_0', x_1', \ldots)$, $\mathbf{Y}' = (y_0', y_1', \ldots)$ where $x_i' = b_i^*, y_i' = 0$ for $b_i^* \geq 0$, $x_i' = 0, y_i' = -b_i^*$ for $b_i^* < 0$, then $X', Y'$ satisfy the equations (11), and have

$$\sum_{i=0}^{\infty} (x_i' + y_i') = \sum_{i=0}^{\infty} |b_i^*| = \left\|\hat{h} - \hat{k}^*\right\|_A = \mu_0$$

This implies that the equations (11) has at least one solution $\mathbf{X}', \mathbf{Y}'$.

With this theorem, one can solve first the dual problem of Opt. 1 to get the infimun $\mu_0$, then determine the existence of Opt. 1 by analyzing the existence of the solution of equations (11) (see following examples ).

Let Opt. 2 be a new optimal problem which is the dual problem of Opt. 1 with the first $N(N \geq n)$ constrictions.

Theorem 5. If there exists an integer N such that the infimum of Opt. 2 $\mu(N) = \mu_0$, then Opt. 1 has at least one optimal solution $\hat{k}^*$ and $\hat{b}^* = \hat{h} - \hat{k}^*$ is a polynomial with a degree smaller than $N$.

Proof: Choose a $n$-1 degree polynomial $\hat{h}_p$ such that

$$\hat{h}_p(\alpha_i) = \hat{h}(\alpha_i), \quad i = 1, 2, \ldots, n. \tag{13}$$

Define another optimal problem Opt. 3

$$\mu_p = \inf_{k_p \in S_p} \left\|\hat{h}_p - \hat{k}_p\right\|_A$$

where

$$S_p = \left\{ k_p \in R^n \middle| \hat{k}_p = \sum_{i=0}^{n-1} k_i z^i, \hat{k}_p(\alpha_i) = 0; i = 1, 2, \ldots, n \right\}$$

By the equation (13), it is easy to see that the dual problem of Opt. 3 is identical with the Opt. 2, then $\mu_p = \mu(N) = \mu_0$. The Opt. 3 is a minimum norm problem in a $N$ dimension real vector space, which always has optimal solution $k_p^*$ (Luenberger, 1969). Let $\hat{b}^* = \hat{h}_p - \hat{k}_p^*$, then $\hat{b}^*$ is a polynomial with degree less than $N$ and $\left\|\hat{b}^*\right\|_A = \mu_0$. Let $\hat{k}^* = \hat{h} - \hat{b}^*$, then $k^* \in S$ and is an optimal solution of Opt. 1.

Corollary 1. Suppose there exists a positive integer $N$, such that $|g_i| \leq 1$ for all $i$ if $|g_i| \leq 1$ for $i = 0, 1, \ldots, N-1$, then Opt. 1 has at least one optimal solution $\hat{k}^*$ and $\hat{b}^* = \hat{h} - \hat{k}^*$ a polynomial with degree less than $N$.

Proof. By the assumption of this theorem, the all restrictions of the dual problem of Opt. 1 will be satisfied if the first $N$ restrictions are satisfied, so the infimum of Opt. 1 is equal to the supremum of dual problem with first $N$ restrictions. Then the conclusion follows immediately from Theorem 5.

### *3.2 Examples*

Examples 1. Let the $\hat{v}$ has two zeros in closed unit disk $\alpha_1 = 1, \alpha_2 = \theta, 0 < \theta < 1$ and $\hat{h}(\alpha_1) = 1$, $\hat{h}(\alpha_2) = 1$. Find the solution of Opt. 1.

Define $f_1(\lambda_1, \lambda_2) = \lambda_1 \hat{h}(\alpha_1) + \lambda_2 \hat{h}(\alpha_2) = \lambda_1 + \lambda_2$, then the dual problem of Opt. 1 is given by

$$\begin{aligned} &\mu_0 = \max f_1(\lambda_1 + \lambda_2) \\ &s.t. \left|\lambda_1 + \lambda_2 \theta^i\right| \leq 1, i = 0, 1, 2, \ldots \end{aligned} \tag{14}$$

Figure 1 shows that the straight line $\lambda_1 + \lambda_2 \theta^i = 1$ and $\lambda_1 + \lambda_2 \theta^i = -1$ approaches the vertical line $\lambda_1 = 1$ and $\lambda_1 = -1$ respectively as $i$ approaches infinity, so the feasible region decreases as the number of resections increase, finally the shadow part is the feasible region of the dual problem (14). Also, it can be seen in this figure that $\mu(N)|_{N=1} = \mu_0 = 1$ due to the factor that the isoline of function $f_1(\lambda_1, \lambda_2)$ is parallel to the straight line $\lambda_1 + \lambda_2 = 1$. It is follows from theorem 5 that Opt. 1 has at least one optimal solution $\hat{k}^*$, and it is easy to sea that $\hat{b}^* = 1$ satisfies $\|\hat{b}^*\|_A = \mu_0 = 1$ and $\hat{k}^* = \hat{h} - \hat{b}^*$ $k^* \in S$ is the optimal solution.

Example 2. The $\hat{v}$ has the same closed unit disk zeros as example 1 ,but $\hat{h}(\alpha_1) = 1$, $\hat{h}(\alpha_2) = 0$. Solve the problem Opt.1. again

Define $f_2(\lambda_1, \lambda_2) = \lambda_1 \hat{h}(\alpha_1) + \lambda_2 \hat{h}(\alpha_2) = \lambda_1$ , then the dual problem of Opt.1 is given by

$$\mu_0 = \max f_2(\lambda_1 + \lambda_2)$$
$$s.t. \left|\lambda_1 + \lambda_2 \theta^i\right| \le 1, i = 0,1,2,\ldots, \tag{15}$$

Figure 1 shows that the optimal $\mu(N)$ of the above optimization problem with first $N$ restrictions is the $\lambda_1$ which is at the intersection point of the straight line $\lambda_1 + \lambda_2 \theta^{N-1} = 1$ and $\lambda_1 + \lambda_2 = -1$. Solve the equations

$$\begin{cases} \lambda_1 + \lambda_2 \theta^{N-1} = 1 \\ \lambda_1 + \lambda_2 = -1 \end{cases}$$

get $\mu(N) = \lambda_1 = \dfrac{1+\theta^{N-1}}{1-\theta^{N-1}}$, which is large then $\mu_0 = 1$ for any $N$, so there exists no $N$ for Theorem 5. Substitute $\mu_0 = 1, \alpha_1 = 1, \alpha_2 = \theta$, $\hat{h}(\alpha_1) = 1$, $\hat{h}(\alpha_2) = 0$ into equations (10) and solve them, then get

$$\begin{bmatrix} x_0 \\ y_0 \\ x_1 \end{bmatrix} = \frac{1}{2(1-\theta)}\{ \begin{bmatrix} -2\theta \\ 0 \\ 2 \end{bmatrix} + \sum_{i=2}^{\infty} \begin{bmatrix} 2\theta(1-\theta^{i-1}) \\ 0 \\ -2(1-\theta^i) \end{bmatrix} x_i + \sum_{i=1}^{\infty} \begin{bmatrix} -2(1-\theta^i) \\ -2(1-\theta) \\ 2(1-\theta^i) \end{bmatrix} y_i \}$$

For $y_0 \ge 0, x_0 \ge 0, x_1 \ge 0$ ,there should be $y_i = 0, \forall i$,

$$\sum_{i=2}^{\infty} x_i - \sum_{i=2}^{\infty} \theta^{i-1} x_i \ge 1$$

$$\sum_{i=2}^{\infty} x_i - \theta \sum_{i=2}^{\infty} \theta^{i-1} x_i \le 1$$

but these inequalities have no solution since $x_i \ge 0, 0 < \theta < 1$ and $\sum_{i=2}^{\infty} x_i > \sum_{i=2}^{\infty} \theta^{i-1} x_i > \theta \sum_{i=2}^{\infty} \theta^{i-1} x_i$. this implies the equations (11) have no solution, then it follows from Theorem 4 that this example has no optimal solution.

## 4. CONCLUSIONS

It is pointed out in this paper that not all $l^1$ optimal controllers are causal. A method for modifying a noncausal optimal controller to a causal one is presented, which insures that the modified controller is causal and the optimal performances are not lost obviously. Although these results are presented based on the $l^1$ sensitivity minimization problem, the idea can be used for other $l^1$ optimal problem and the results are similar.

The existence of $l^1$ optimal solution for plant with zero and (or) pole on the unit circle has been studied. Examples show that the $l^1$ optimal solution does not always exist. Some preliminary results for determining the existence are presented.

## REFERENCES


Dahleh, M. A. and J.B.Pearson (1987) $l^1$- optimal feedback controllers for MIMO discrete-time systems. *IEEE Trans. on Automatic Control*, **AC-31**, 314-322.

Dahleh, M. A. and J. B. Pearson (1988), Optimal rejection of persistent disturbances, robust stability,and mixed sensitivity minimization.*IEEE Trans. on Automatic Control, AC-33*, 722-731.

Diaz-bobillo, I.J. and M.A.Dahleh(1993), Minimization of the maximum peak-to-peak gain: The general multiblock problem. *IEEE Trans. on Automatic Control*. **.38**, 1459-1482.

Fang H. J.(1990). Designing $l^1$ Optimal Feedback Controller by Simplex Method(in Chinese), *Control and Decision*, **5** 44-47

Luenberger, D.C.(1969). *Optimization by Vector space Methods*. John Wiley & Sons, New York.

Mendlovitz, M.A.(1989). A simple solution to the $l^1$ optimization problem. *Systems & Control Letters*, **12**, 461-463.

Vidyasagar, M. (1986). Optimal rejection of persistent bounded disturbances. *IEEE Trans. on Automatic Control.*, **AC-31**, 527-534.


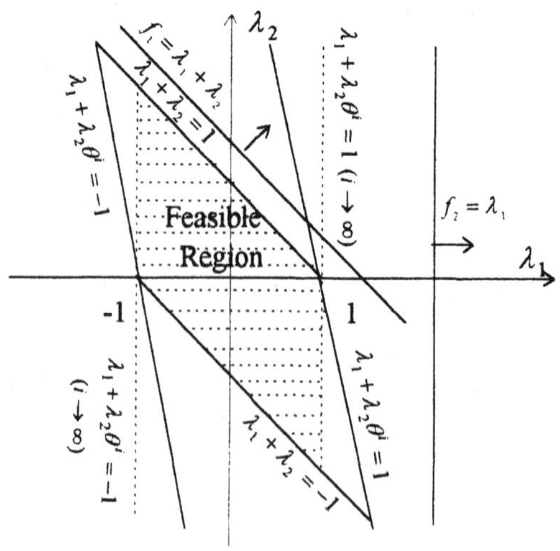


Fig. 1. The dual problems of example 1,2

14th Triennial World Congress, Beijing, P.R. China



F-2e-04-4

# AN IQC-BASED STABILITY CRITERION FOR SYSTEMS WITH SLOWLY VARYING PARAMETERS

**Anders Helmersson** *,[1]

* *Department of Electrical Engineering*
*Linköping University, SE-581 83 Linköping, Sweden,*
*email:* andersh@isy.liu.se

Abstract:
An integral quadratic constraints (IQC) is introduced for stability analysis of linear systems with slowly varying parameters. The parameters are assumed to be bounded and with bounded derivatives. Other types of uncertainties can be included in the problem. The new criterion yields less conservative bounds than previously proposed criteria. *Copyright © 1999 IFAC*
**Keywords:** Stability analysis, robustness, linear, time varying systems, integral quadratic constraints.

## 1. INTRODUCTION

The analysis of parameter-dependent system plays an important part in many control applications, for instance gain scheduling problems, which has attracted a great deal of attention during recent years.

In this paper we will consider the stability of linear parameter varying (LPV) systems. We can also include performance bounds in the analysis. The parameters and their derivatives are assumed to bounded. The plant to be analyzed depends as a linear fractional transformation (LFT) of the parameters or functions thereof.

The analysis will be based on integral quadratic constraints (IQC) Yakubovich (1967); Megretski and Rantzer (1997). An IQC-based stability criterion for slowly-varying parameters (uncertainties) has been proposed by Jönsson and Rantzer Jönsson and Rantzer (1994); Jönsson (1996). The IQC proposed and elaborated in this paper exploits more of the structure of the problem than the previous criterion. Since higher flexibility in the IQC multiplier can be allowed the new criterion gives less conservative bounds. This is illustrated by two examples.

The paper is organized as follows. In section 2 the integral quadratic constraints (IQC) is briefly introduced. Linear parameter varying systems are discussed in section 3. In section 4, we apply IQCs to the systems with slowly varying parameters. This includes a review of the swapping lemma. The method is applied to two examples in section 5. The paper is concluded in section 6.

### 1.1 *Notation*

For matrices $A^*$ denotes the complex conjugate transpose.

$\mathcal{RH}_\infty$ denotes denotes stable real-rational transfer functions; $\mathcal{L}_2^n[0,\infty)$ denotes the Hilbert space of measurable functions $f : [0,\infty) \to \mathbb{R}^n$ satisfying

$$||f||^2 = \int_0^\infty |f(t)|^2 dt < \infty$$

This is a subspace of $\mathcal{L}_{2e}^n[0,\infty)$, whose members only need to be square integrable on finite intervals.

[1] This work was supported by the Swedish National Board for Industrial and Technical Development (NUTEK), which is gratefully acknowledged

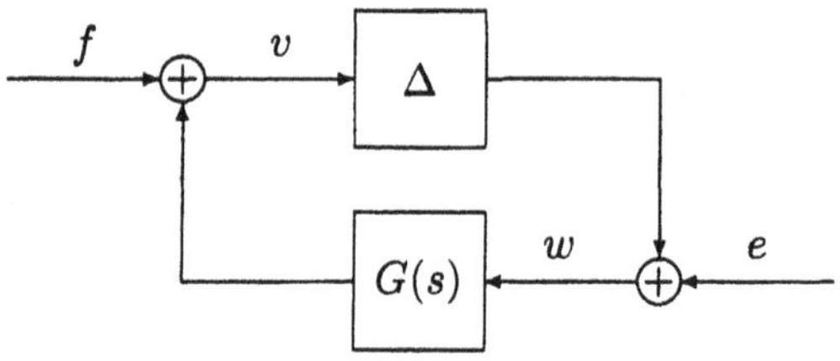


Fig. 1. Basic feedback configuration $(G(s), \Delta)$.

With a (dynamic) *operator* we mean a function $F$ from one $\mathcal{L}_{2e}[0,\infty)$ space to another. An operator is said to be *causal* if $(Ff)(t)$ does not depend on future values, $f(\tau), \tau > t$ and *parametric* if it only depends on the current value $f(t)$. An operator is *bounded* if bounded input energy $||f||$ also gives bounded output energy $||Ff||$.

For appropriately dimensioned matrices $K$ and $M = \left[\begin{smallmatrix} M_{11} & M_{12} \\ M_{21} & M_{22} \end{smallmatrix}\right]$ and assuming the inverse exists, the upper linear fractional transformation (LFT) is defined as

$$K \star M = M_{22} + M_{21}K(I - M_{11}K)^{-1}M_{12}.$$

## 2. INTEGRAL QUADRATIC CONSTRAINTS

The integral quadratic constraints (IQCs) has been proposed for robustness analysis Yakubovich (1967); Megretski and Rantzer (1997). The IQC states a stability criterion for the interconnection of a stable system $G \in \mathcal{RH}_\infty$ and a bounded causal operator $\Delta$, see figure 1.

$$\begin{cases} v = G(s)w + f \\ w = \Delta v + e. \end{cases} \tag{1}$$

We say that the interconnection of $G(s)$ and $\Delta$ is *well-posed* if the map $(v, w) \rightarrow (e, f)$ defined by (1) has a causal inverse on $\mathcal{L}_{2e}$. The interconnection is *stable* if, in addition, the inverse is bounded, that is if there exists a constant $C$ such that

$$\int_0^T (|v(t)|^2 + |w(t)|^2)dt \leq C\int_0^T (|f(t)|^2 + |e(t)|^2)dt$$

for any $T \geq 0$ and for any solution of (1).

Depending on the particular application, various versions of IQCs are available. Two signals $w \in \mathcal{L}_2[0,\infty)$ and $v \in \mathcal{L}_2[0,\infty)$ are said to satisfy the IQC defined by $\Pi$, if

$$\int_{\infty}^{\infty} \begin{bmatrix} \hat{v}(j\omega) \\ \hat{w}(j\omega) \end{bmatrix}^* \Pi(j\omega) \begin{bmatrix} \hat{v}(j\omega) \\ \hat{w}(j\omega) \end{bmatrix} d\omega \geq 0 \tag{2}$$

where absolute integrability is assumed. Here $\hat{v}(j\omega)$ and $\hat{v}(j\omega)$ represent the harmonic spectrum of the signals $v$ and $w$ at the frequency $\omega$. In principle $\Pi : j\mathbb{R} \rightarrow \mathbb{C}$ can be any measurable Hermitian-valued function. In most applications, however, it is sufficient to use rational functions that are bounded on the imaginary axis.

A time-domain form of (2) is

$$\int_0^\infty \sigma(x_\pi(t), v(t), w(t))dt \geq 0 \tag{3}$$

where $\sigma$ is a quadratic form, and $x_\pi$ is defined by

$$\dot{x}_\pi(t) = A_\pi x_\pi(t) + B_v v(t) + B_w w(t), \; x_\pi(0) = 0$$

where $A_\pi$ is a Hurwitz matrix.

The main theorem from Megretski and Rantzer (1997) goes as follows

*Theorem 1.* (Megretski and Rantzer (1997)). Let $G \in \mathcal{RH}_\infty$ and let $\Delta$ be a bounded causal operator. Assume that:

i) for every $\tau \in [0,1]$, the interconnection of $G(s)$ and $\tau\Delta$ is well-posed;
ii) for every $\tau \in [0,1]$, the IQC defined by $\Pi$ is satisfied by $\tau\Delta$;
iii) there exists $\epsilon > 0$ such that

$$\begin{bmatrix} G(j\omega) \\ I \end{bmatrix}^* \Pi(j\omega) \begin{bmatrix} G(j\omega) \\ I \end{bmatrix} \leq -\epsilon I, \; \forall\omega \in \mathbb{R}. \tag{4}$$

Then the feedback interconnection of $G(s)$ and $\Delta$ is stable.

Note that if the upper left corner, $\Pi_{11}$, of $\Pi$ is positive semidefinite then $\Delta = 0$ satisfies (2). If further the lower right corner, $\Pi_{22}$, is negative semidefinite then any convex combination of $\Delta$'s satisfying (2) also satisfies the IQC. Thus, $\Pi_{11} \geq 0$ and $\Pi_{22} \leq 0$ imply that $\tau\Delta$ satisfies (2) for $\tau \in [0,1]$ if and only if $\Delta$ does so. This simplifies assumption ii).

The search for multipliers, $\Pi$, can be carried out as a convex optimization problem by parametrizing

$$\Pi(j\omega) = \sum_i x_i \Pi_i(j\omega)$$

where $x_i$ are positive real parameters and $\Pi_i$ is a set of basis multipliers. By applying the Kalman-Yakubovich-Popov lemma Willems (1971); Yakubovich (1974); Rantzer (1995), the search for $x_i$, can be implemented using linear matrix inequalities (LMIs).

## 3. LINEAR PARAMETER VARYING SYSTEMS

The problem addressed in this paper is the analysis of linear parameter varying (LPV) systems with a state-space realization

$$\dot{x} = A(\theta)x + B(\theta)w \tag{5a}$$
$$z = C(\theta)x + D(\theta)w \tag{5b}$$

where $w$ is the disturbance signal and $z$ the error signal. The time-varying parameters, $\theta = (\theta_1, \ldots, \theta_l)$, as well as their rates of variation, $\dot{\theta}$, are assumed to be bounded.

### 3.1 *LFT parametrized systems*

In (5) the structure of the parametrization is not explicitly stated. In order to facilitate the analysis we will make a further assumption on the structure by assuming

$$\begin{bmatrix} A(\theta) & B(\theta) \\ C(\theta) & D(\theta) \end{bmatrix} = \Delta(\theta) \star \begin{bmatrix} \tilde{A} & \tilde{B} \\ \tilde{C} & \tilde{D} \end{bmatrix} \qquad (6)$$

where $\Delta$ is a function of the parameter vector $\theta$ and $\tilde{A}$, $\tilde{B}$, $\tilde{C}$ and $\tilde{D}$ are constant matrices. The upper linear fractional transformation is denoted by $\star$:

$$\Delta \star \begin{bmatrix} \tilde{A} & \tilde{B} \\ \tilde{C} & \tilde{D} \end{bmatrix} = \tilde{D} + \tilde{C}(I - \Delta\tilde{A})^{-1}\Delta\tilde{B}.$$

The parameter $\theta$ is assumed to bounded in terms of bounds on $\Delta$ and possibly also its time derivative, $\dot{\Delta}$:

$$\begin{bmatrix} I \\ \Delta \end{bmatrix}^* \Pi \begin{bmatrix} I \\ \Delta \end{bmatrix} \geq 0 \qquad (7)$$

or

$$\begin{bmatrix} I & 0 \\ 0 & I \\ \hline \Delta & 0 \\ 0 & \dot{\Delta} \end{bmatrix}^* \Pi \begin{bmatrix} I & 0 \\ 0 & I \\ \hline \Delta & 0 \\ 0 & \dot{\Delta} \end{bmatrix} \geq 0. \qquad (8)$$

where

$$\Pi = \begin{bmatrix} \Pi_{11} & \Pi_{12} \\ \Pi_{12}^* & \Pi_{22} \end{bmatrix}.$$

We will further assume that $\Pi_{11} \geq 0$ and $\Pi_{22} \leq 0$. This assures that the set of $\Delta$ that satisfies (8) is convex and that $\tau\Delta$ is included for all $\tau \in [0,1]$ and $\Delta$ is included.

## 4. IQC ANALYSIS

We will here propose an IQC for verifying stability of linear system interconnected with slowly varying parameters. We define slowly varying parameters, $\Delta$, such that their rate of change $\dot{\Delta}$ is norm-bounded by $d$, that is $\|\dot{\Delta}\| \leq d$.

The main tool for this analysis is the swapping lemma that tells how a linear operator on state-space form commutes with a time-varying parameters.

### 4.1 *Swapping Lemma*

The following swapping lemma is essentially from Jönsson (1996).

*Lemma 1.* (Swapping Lemma). Suppose that $\hat{\Delta}$ has the derivative $\dot{\hat{\Delta}}$ and $u \in \mathcal{L}_2$. Further let $T \in \mathcal{RH}_\infty$:

$$T(s) = \hat{D} + \hat{C}(sI - \hat{A})^{-1}\hat{B}$$
$$U(s) = \hat{C}(sI - \hat{A})^{-1}$$
$$V(s) = (sI - \hat{A})^{-1}\hat{B}$$

such that

$$\begin{bmatrix} \hat{A} & \hat{B} \\ \hat{C} & \hat{D} \end{bmatrix}\begin{bmatrix} \hat{\Delta} & 0 \\ 0 & \Delta \end{bmatrix} = \begin{bmatrix} \hat{\Delta} & 0 \\ 0 & \bar{\Delta} \end{bmatrix}\begin{bmatrix} \hat{A} & \hat{B} \\ \hat{C} & \hat{D} \end{bmatrix}. \qquad (9)$$

Then

$$\bar{\Delta}T(s)u = \left(T(s)\Delta + U(s)\dot{\hat{\Delta}}V(s)\right)u. \qquad (10)$$

**Proof:** *Let $s = \frac{d}{dt}$. We note that*

$$(sI - \hat{A})\hat{\Delta}V(s)u = \dot{\hat{\Delta}}V(s)u + \hat{\Delta}\hat{B}u$$
$$= \dot{\hat{\Delta}}V(s)u + \hat{B}\Delta u. \qquad (11)$$

*Let $U(s)$ operate from the left on (11). After addition of $\bar{\Delta}\hat{D}u = \hat{D}\Delta u$, we get*

$$\left(\bar{\Delta}\hat{D} + \hat{C}\hat{\Delta}V(s)\right)u$$
$$= \left(U(s)\dot{\hat{\Delta}}V(s) + \hat{D}\Delta + U(s)\hat{B}\Delta\right)u,$$

*which is equivalent to (10).* □

Note that the block-diagonal structure of $T$, $V$ and $U$ is implicitly defined by the commuting equation (9). The variable $\Delta$ is the original uncertainty block, which can include time-varying parameters as well as other operators. The variable $\hat{\Delta}$ only includes the time-varying parameters, possibly repeated, while $\bar{\Delta}$ has the same structure as $\Delta$, but the time-varying parameters may be repeated.

In the paper the structure of $\mathbf{\Delta}$ is implicit, and instead the structure is defined by commuting properties (9).

For instance, if we assume that $\Delta = \text{diag}\,[\delta_1 I_{n_1}, \Delta_2]$ where $\delta_1$ is a slowly varying parameter and $\Delta_2$ contains remaining (dynamic) operators, a natural choice of $T$ would be to use

$$\begin{bmatrix} \hat{A} & \hat{B} \\ \hat{C} & \hat{D} \end{bmatrix} = \left[\begin{array}{c|cc} \hat{A} & \hat{B}_1 & 0 \\ \hline \hat{C}_1 & \hat{D}_1 & 0 \\ 0 & 0 & I \end{array}\right]. \qquad (12)$$

Then $\hat{\Delta} = \delta_1 I_{k_1}$, $\bar{\Delta} = \text{diag}\,[\delta_1 I_{k_1}, \Delta] = \text{diag}\,[\delta_1 I_{n_1+k_1}, \Delta_2]$, where $k_1$ is the dimension of $\hat{A}$. In general $\hat{\Delta}$ contains the slowly varying parameters in $\Delta$, but possibly (differently) repeated.

### 4.2 *IQC Formulation*

We will study an IQC stability criterion for the interconnected system $(G(s), \Delta)$, where $\Delta$ and $\dot{\Delta}$ are norm-bounded. We use the same notation as in

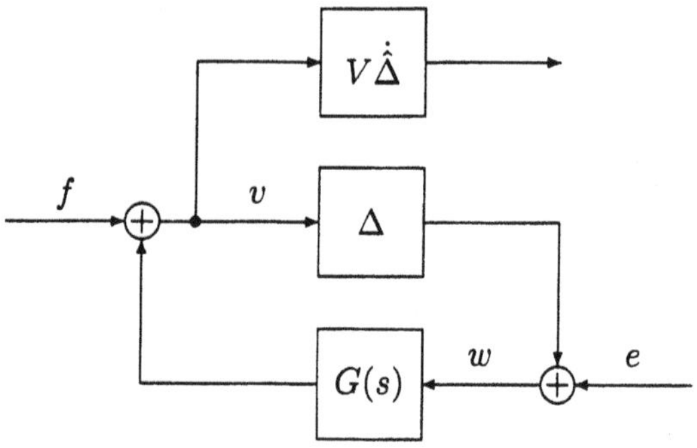


Fig. 2. Augmented feedback configuration.

lemma 1. Instead of applying the IQC directly to $G$, we use it on the augmented system $\begin{bmatrix} G(s) & 0 \end{bmatrix}$ and the corresponding augmented operator $\tilde{\Delta} = \begin{bmatrix} \Delta \\ V(s)\dot{\Delta} \end{bmatrix}$, see figure 2.

*Theorem 2.* Assume that $\Delta$ is a bounded, slowly varying parameter uncertainty such that $\dot{\Delta}$ is bounded and such that

$$\begin{bmatrix} I & 0 \\ 0 & I \\ \hline \bar{\Delta}(t) & 0 \\ 0 & \dot{\Delta}(t) \end{bmatrix}^* P \begin{bmatrix} I & 0 \\ 0 & I \\ \hline \bar{\Delta}(t) & 0 \\ 0 & \dot{\Delta}(t) \end{bmatrix} \geq 0. \quad (13)$$

holds for a given static symmetric matrix $P = P^*$ and for any combination of $\bar{\Delta}(t)$ and $\dot{\Delta}(t)$. Consider the IQC for the augmented system defined by the following multiplier:

$$\Pi(j\omega) = \left[\;\right]^* P \begin{bmatrix} T(j\omega) & 0 & 0 \\ V(j\omega) & 0 & 0 \\ \hline 0 & T(j\omega) & U(j\omega) \\ 0 & 0 & I \end{bmatrix}. \quad (14)$$

The interconnection of $G(s)$ and $\Delta$ is stable if

$$\begin{bmatrix} G(j\omega) & 0 \\ I & 0 \\ 0 & I \end{bmatrix}^* \Pi(j\omega) \begin{bmatrix} G(j\omega) & 0 \\ I & 0 \\ 0 & I \end{bmatrix} < 0 \quad (15)$$

holds for all $\omega \in \mathbb{R} \cup \{\infty\}$.

**Proof:** *Consider the augmented system* $\begin{bmatrix} G(s) & 0 \end{bmatrix}$ *interconnected with the corresponding augmented operator* $\tilde{\Delta} = \begin{bmatrix} \Delta \\ V(s)\dot{\Delta} \end{bmatrix}$, *see figure 2. We observe that the augmented operator is bounded since* $T \in \mathcal{RH}_\infty$ *and, consequently,* $\dot{\Delta}V(s)$ *is bounded if* $\dot{\Delta}$ *is such. Regarding stability, the augmented interconnection is identical to the original interconnection* $(G(s), \Delta)$. *Using the swapping lemma,* $T(s)\Delta + U(s)\dot{\Delta}V(s) = \bar{\Delta}T(s)$, *we obtain*

$$z = \begin{bmatrix} T(s) & 0 & 0 \\ V(s) & 0 & 0 \\ 0 & T(s) & U(s) \\ 0 & 0 & I \end{bmatrix} \begin{bmatrix} I \\ \Delta \\ \dot{\Delta}V(s) \end{bmatrix} v$$

$$= \begin{bmatrix} T(s) \\ V(s) \\ \bar{\Delta}T(s) \\ \dot{\Delta}V(s) \end{bmatrix} v = \begin{bmatrix} I & 0 \\ 0 & I \\ \bar{\Delta} & 0 \\ 0 & \dot{\Delta} \end{bmatrix} \begin{bmatrix} T(s) \\ V(s) \end{bmatrix} v.$$

*In state-space form this becomes*

$$\dot{\hat{x}}(t) = \hat{A}\hat{x}(t) + \hat{B}v(t), \qquad \hat{x}(0) = 0 \quad (16a)$$

$$z(t) = \begin{bmatrix} I & 0 \\ 0 & I \\ \hline \bar{\Delta}(t) & 0 \\ 0 & \dot{\Delta}(t) \end{bmatrix} \begin{bmatrix} \hat{C} & \hat{D} \\ I & 0 \end{bmatrix} \begin{bmatrix} \hat{x}(t) \\ v(t) \end{bmatrix}. \quad (16b)$$

*In this case the time-domain form of the IQC-constraint (2) as defined in (3) becomes*

$$\int_0^\infty z(t)^* P z(t) dt = \quad (17a)$$

$$\int_0^\infty \left[\;\right]^* \left[\;\right]^* P \begin{bmatrix} I & 0 \\ 0 & I \\ \hline \bar{\Delta}(t) & 0 \\ 0 & \dot{\Delta}(t) \end{bmatrix} \begin{bmatrix} z_T(t) \\ z_V(t) \end{bmatrix} dt \geq 0 \quad (17b)$$

*where* $z_T = Tv$ *and* $z_V = Vv$ *are defined by (16a) and*

$$\begin{bmatrix} z_T(t) \\ z_V(t) \end{bmatrix} = \begin{bmatrix} \hat{C} & \hat{D} \\ I & 0 \end{bmatrix} \begin{bmatrix} \hat{x}(t) \\ v(t) \end{bmatrix}.$$

*The inequality in (17b) follows by (13), which concludes the proof.* □

### 4.3 *Comparison with Previous Criterion*

Jönsson and Rantzer proposed an IQC for the stability analysis of linear system with slowly varying parameters Jönsson and Rantzer (1994); Jönsson (1996):

$$\begin{bmatrix} M \\ I \end{bmatrix}^* \begin{bmatrix} X & S - S^* \\ S^* - S & -R^*(I - U_R U_R^*)R \end{bmatrix} \begin{bmatrix} M \\ I \end{bmatrix} < 0, \quad (18)$$

where

$$\begin{aligned} X &= R^*R + V_R^*V_R + d^2 V_S^* V_S + U_S U_S^* \\ R &= D_R + C_R(sI - A_R)^{-1}B_R \\ S &= D_S + C_S(sI - A_S)^{-1}B_S \\ U_R &= C_R(sI - A_R)^{-1} \\ U_S &= C_S(sI - A_S)^{-1} \\ V_R &= (sI - A_R)^{-1}B_R \\ V_S &= (sI - A_S)^{-1}B_S \end{aligned}$$

such that $R$ and $S$ are both stable. We will now show that (18) is more conservative than theorem 2. Let

$$P = \left[\begin{array}{ccccc|ccccc} I & 0 & 0 & 0 & 0 & 0 & 0 & 0 & 0 & 0 \\ 0 & 0 & 0 & 0 & 0 & 0 & 0 & -I & 0 & 0 \\ 0 & 0 & 0 & 0 & 0 & 0 & I & 0 & 0 & 0 \\ 0 & 0 & 0 & I & 0 & 0 & 0 & 0 & 0 & 0 \\ 0 & 0 & 0 & 0 & d^2I & 0 & 0 & 0 & 0 & 0 \\ \hline 0 & 0 & 0 & 0 & 0 & -I & 0 & 0 & 0 & 0 \\ 0 & 0 & I & 0 & 0 & 0 & 0 & 0 & 0 & 0 \\ 0 & -I & 0 & 0 & 0 & 0 & 0 & 0 & 0 & 0 \\ 0 & 0 & 0 & 0 & 0 & 0 & 0 & 0 & -I & 0 \\ 0 & 0 & 0 & 0 & 0 & 0 & 0 & 0 & 0 & -I \end{array}\right]$$

and

$$T = \begin{bmatrix} R \\ S \\ I \end{bmatrix}, \; U = \begin{bmatrix} U_R & 0 \\ 0 & U_S \\ 0 & 0 \end{bmatrix}, \; V = \begin{bmatrix} V_R \\ V_S \end{bmatrix}.$$

This together with (14) in theorem 2 and using the Schur complements yields

$$\begin{bmatrix} M \\ I \end{bmatrix}^* \begin{bmatrix} X & S - S^* \\ S^* - S & -R^*(I + U_R U_R^*)^{-1} R \end{bmatrix} \begin{bmatrix} M \\ I \end{bmatrix} < 0. \tag{19}$$

Note that (19) is very close to the IQC (18) proposed by Jönsson and Rantzer Jönsson and Rantzer (1994); Jönsson (1996) for slowly varying parameters, which is slightly more conservative, since

$$I - U_R U_R^* \leq (I + U_R U_R^*)^{-1}.$$

for all $U_R$.

## 5. EXAMPLES

We will show the stability criteria on two examples. In both examples we improve the bounds compared to previous IQC-based criteria.

### 5.1 *Example I*

This example is from Jönsson and Rantzer (1994). A simple SISO system is considered:

$$\begin{aligned} \dot{x}(t) &= (A + C\delta(t)B)\,x(t) \\ &= \begin{bmatrix} -0.21 & -1 + 0.8\delta(t) \\ 1 & 0 \end{bmatrix} x(t) \end{aligned}$$

where it is assumed that $\delta(t) \in [-\gamma, \gamma]$. In the case the parameter is constant, the system is asymptotically stable if and only if $\delta < 1.25$, that is, when $\gamma < 1.25$.

If $\dot{\delta}$ is unbounded, then stability is assured if $|\delta| < 0.261$. This bound is obtained as the inverse of the $\mathcal{H}_\infty$-norm of

$$G(s) = D + C(sI - A)^{-1}B$$

where

$$\begin{bmatrix} A & B \\ C & D \end{bmatrix} = \left[\begin{array}{cc|c} -0.21 & -1 & 0.8 \\ 1 & 0 & 0 \\ \hline 0 & 1 & 0 \end{array}\right].$$

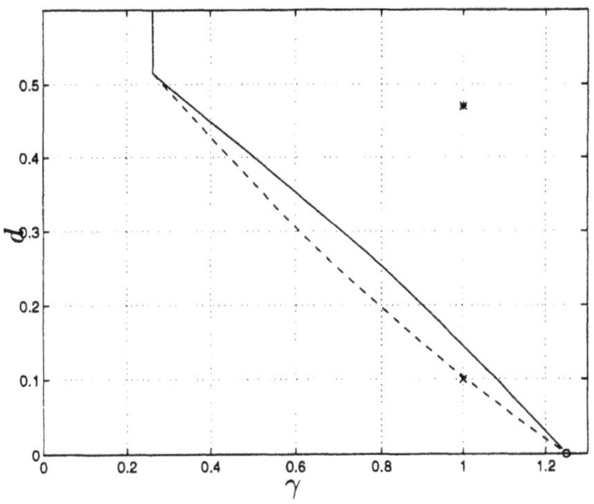


Fig. 3. Example I: Stability bounds $d$ as a function of $\gamma$. The system is stable if $\delta(t) \in [-\gamma, \gamma]$ and $|\dot{\delta}(t)| \leq d$. The lower, dashed line uses $T$ as defined in (20) while the upper, solid line uses (21). The cross ($\times$) marks the bound given in Jönsson and Rantzer (1994). The circle ($\circ$) at $\gamma = 1.25$, denotes the bound obtained with constant $\delta$. Arbitrarily fast variations in $\delta$ can be allowed if $|\delta| < 0.261$, marked by the vertical line. The criteria show stability in the area below and to the left of the lines. The asterisk ($*$) at $\gamma = 1$ and $d = 0.46$, shows a case for which instability has been verified.

We now apply theorem 2 using

$$T(s) = \begin{bmatrix} \dfrac{1}{s+1} \\ 1 \end{bmatrix}. \tag{20}$$

This multiplier has the same pole as that in Jönsson and Rantzer (1994). There the analysis yields a stability bound of $|\dot{\delta}| < d = 0.1$ for $\gamma = 1$. Using (20) yields $d = 0.1036$, which is in close agreement with Jönsson and Rantzer (1994).

If we instead choose

$$T(s) = \begin{bmatrix} \dfrac{1}{s+10} \\ 1 \end{bmatrix}, \tag{21}$$

we obtain $d = 0.1456$ for $\gamma = 1$, assuming $P_{11} \geq 0$ and $P_{22} \leq 0$. Figure 3 shows the stability bounds, $d$, as a function of $\gamma$. Instability has been verified for $\gamma = 0.41$ without any bounds on $\dot{\delta}$ and $d = 0.46$ for $\gamma = 1$; these bounds are (probably) not the true bounds though.

### 5.2 *Example II*

This example has been analyzed in Jönsson and Rantzer (1996) using IQCs. We consider the ship steering dynamics as in Example 9.6 in Åström and Wittenmark (1989). The dynamics can be approximated by the Nomoto model

$$\begin{aligned} \dot{x}(t) &= v(t)(-ax(t) + bv(t)\nu(t)) \\ \dot{\psi}(t) &= x(t) \end{aligned}$$

where $\psi$ denotes the heading of the ship, $\nu$ denotes the rudder angle and $v$ is the speed of the ship. It is assumed that $v(t) \geq 0$. We will, as in Åström and Wittenmark (1989), study the stability of the ship dynamics for an unstable tanker with $a = -0.3$ and $b = 0.8$, which is controlled by a PD regulator

$$v = -K\psi \text{ where } K(s) = k(1 + T_d s)$$

with $k = 2.5$ and $T_d = 0.86$.

The closed loop system can be described by

$$x(t) = (v(t)I_2)G_0(s)x(t)$$

with

$$G_0(s) = \begin{bmatrix} -a/s & b \\ -K(s)/s^2 & 0 \end{bmatrix}.$$

This system is not strictly stable and instead we replace it for the purpose of analysis by

$$\begin{aligned} G(s) &= (I + G_0(s))(I - G_0(s))^{-1} \\ &= D + C(sI - A)^{-1}B \end{aligned}$$

where

$$\begin{bmatrix} A & B \\ C & D \end{bmatrix} = \left[\begin{array}{cc|cc} -(a + bkT_d) & -bk & -2 & -2b \\ 1 & 0 & 0 & 0 \\ \hline a + bkT_d & bk & 1 & 2b \\ kT_d & k & 0 & 1 \end{array}\right]$$

and

$$\delta(t) = \frac{v(t) - v_{\text{nom}}}{v(t) + v_{\text{nom}}} \in [-1, 1],$$

where $v_{\text{nom}}$ is the nominal speed of the ship. For the transformed system, $d$ is solved as a function of $\gamma$ such that the system is verified to be stable when

$$|\delta| \leq \gamma \qquad \text{and} \qquad |\dot{\delta}| \leq d.$$

We analyze the system using theorem 2 with

$$T = \begin{bmatrix} I_2 \\ (s+1)^{-1} I_2 \end{bmatrix}. \tag{22}$$

Assuming that $\delta$ is constant, stability can be assured for $|\delta| \leq 0.7028$ or, equivalently, $0.1745 \leq v \leq 5.7296$. This is close to the true stability bound, $v/v_{\text{nom}} \geq -a/(bkT_d) = 0.1744$.

Figure 4 shows the area in the $(v, |\dot{v}|)$-plane where stability is proven using the different criteria.

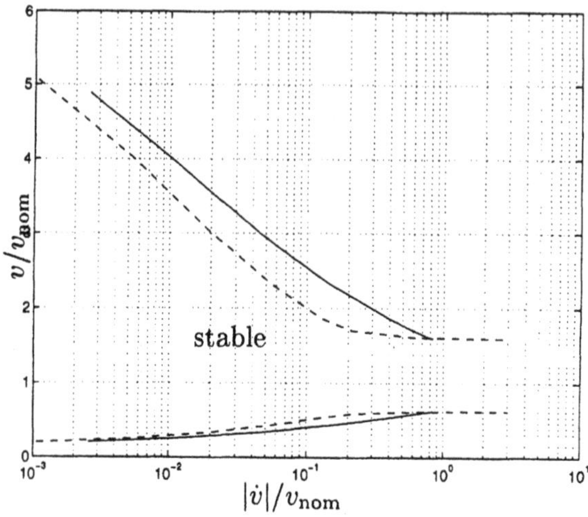


Fig. 4. Example II: Stability bounds for $v$ and $\dot{v}$ using different IQC criteria. The solid lines show the proven stability bounds using $T$ as defined in (22). The dashed lines mark the bounds given in Jönsson and Rantzer (1996).

## 6. CONCLUSIONS

A new criterion for verifying stability of linear systems with slowly varying parameters is derived.

It is based on an integral quadratic constraint (IQC). The method is illustrated on two examples. The new criteria outperforms previous methods documented in the literature. An advantage with the new formulation is its simpler structure, which allows an easier search for the appropriate multipliers.

## References


Åström, K. J. and B. Wittenmark (1989). *Adaptive Control.* Addison Wesley Series in Electrical and Computer Engineering: Control Engineering. Addison Wesley.

Jönsson, U. (1996). Robustness Analysis of Uncertain and Nonlinear Systems. PhD thesis. Lund Institute of Technology. Lund, Sweden.

Jönsson, U. and A. Rantzer (1994). System with uncertain parameters — time-variations with bounded derivatives. In: *Proceedings of the 33rd Conference on Decision and Control.* Vol. 3. Lake Buena Vista, Florida. pp. 3074–3079.

Jönsson, U. and A. Rantzer (1996). Systems with uncertain parameters – time variations with bounded derivatives. *International Journal of Robust and Nonlinear Control* **6**(9-10), 969–982.

Megretski, A. and A. Rantzer (1997). System analysis via integral quadratic constraints. *IEEE Transactions on Automatic Control* **42**(6), 819–830.

Rantzer, A. (1995). A note on the Kalman-Yacubovich-Popov lemma. In: *Proceedings of the 3rd European Control Conference.* Vol. 3, part 1. Rome, Italy. pp. 1792–1795.

Willems, J. C. (1971). *The Analysis of Feedback Systems.* MIT Press. Cambridge, MA.

Yakubovich, V. A. (1967). Frequency conditions for the absolute stability of control systems with several nonlinear or linear nonstationary units. *Atomat. Telemech.* pp. 5–30.

Yakubovich, V. A. (1974). A frequency theorem for the case in which the state and control spaces are Hilbert spaces with an application to some problems of optimal controls — Part I-II. *Sibirskii Mat. Zh.* **15**(3), 639–668. English translation in *Siberian Math. J.*

# PERFORMANCE LIMIT: A $\mu$-SYNTHESIS APPROACH

**Yufen He[#] Da-Wei Gu[*] Guangxiong Wang[+]**
**Ian Postlethwaite[*] Zhang Xiong[#]**

*[#]Group 601, Department of Computer Science and Engineering, Beijing University of Aeronautics and Astronautics, Beijing 100083, China, PR. Email: heyf@buaa.edu.cn*
*[+]Department of Control Engineering, Harbin Institute of Technology, Harbin 150001, China, PR*
*[*]Control Systems Research, Department of Engineering, University of Leicester, Leicester LE1 7RH, UK*

Abstract: An important aspect of linear controller design is to clarify whether the design reaches the performance limit. Based on an industrial project—a high precision three-axis gyro test table (servo system), a paradigm is presented to demonstrate that the performance limit of the disturbance attenuation design of the servo system can be reached in a systematic way by using the $\mu$-synthesis. Moreover, this paper reveals that the performance limit is determined by the bandwidth. This finding will help the designer to judge whether any given performance specification is achievable.


Keywords: Performance limits, Servo systems, Disturbance attenuation, $\mu$-Synthesis, Bandwidth.

## 1. INTRODUCTION

The existing linear controller design methods are often successful in finding an appropriate controller so that the closed-loop system can match the prescribed performance specifications and robust stability specifications. In practice if a designer fails to obtain a suitable controller, then the designer may either relax the design goals or might suspect whether a suitable controller can be found by another design approach. Despite this, clarifying whether the design achieves the performance limit (Boyd and Barratt, 1991) i.e. the best performance, is a significant research area in the control systems design. In general, reaching the performance limit depends on an experienced control engineer and often costs a lot of computation to verify it. Based on a servo system design—a high precision three-axis gyro test table, this paper gives a systematic scheme to reach the performance limit of the disturbance attenuation design of the servo system.

The main concern in the servo system design is the input disturbance attenuation, which is definitely different from the general sensitivity scheme. Usually the general sensitivity only reflects the output disturbance rejection, and it has nothing to do with the servo system design requirements. In the gyro test table, the input torque disturbance is likely to interrupt the process of the gyro drift's test, of which its normal action cycle is 24 hours or even longer.

In order to rigorously formulate the performance specification and robust stability specification, the $\mu$-synthesis approach (Stein and Doyle, 1991) is employed for the reason that this design framework is flexible to handle many engineering situations. This paper reveals that the performance limit of the gyro test table is determined by the bandwidth and independent of the performance weighting functions selected in the $\mu$-synthesis.

This paper is organized as follows: Section 2 outlines the requirements of the gyro test table design. Section 3 formulates the problem into the $\mu$-synthesis framework. The main results of the paper are presented in Section 4. The conclusions are given in Section 5.

## 2. THREE-AXIS GYRO TEST TABLE

Gyro test table is a servo system used to assess the accuracy of gyros. There are three working modes of the three-axis gyro test table (TAGTT) which is being studied.

- *Position mode*: to position a test package at any direction through gimbals' rotation, i.e. to provide an accurate pointing reference (reference direction) for test use.
- *Rate mode*: to rotate a test package at a precision rate. This is an obvious requirement of a servo system. In the rate mode, the reference signal is merely a velocity signal $r = v_0 t$, where $v_0$ is generally in the order of 0.08°/s.
- *Gyro servo mode*: In this mode, each of the mutually perpendicular three-gimbal can be slaved to a gyro. It is used to measure the drift of a gyro. The body of the TAGTT provides a fixed position. The input axis of the gyro is normally set in parallel with the axis of the Earth. When the base of the turntable rotates along with the rotation of the Earth relative to the inertial space, the turntable will make the gyro rotate at the direction of its input axis. This generates the gyro torque at the direction of the gyro's output axis. Subsequently, the rotor of the gyro rotates around the output axis due to the torque. The sensor then detects the rotation and outputs the error signal which drives the servo system to turn the turntable in the opposite direction of the base rotation, in order to keep constant direction of the gyro's input axis relative to the inertial space. Therefore the measured angle between the turntable and the base is the angle relative to the inertial space if the gyro has no drift. In fact, the gyro does drift, then the measured angle includes both the drift of the gyro and the base relative to the inertial space. Because the angular velocity is the angle divided by the measurement time and the angular velocity of the base is that of the Earth, the drift velocity of the gyro can thus be deducted.

A distinct aspect of the TAGTT design is the high level of input disturbance torque which severely interferes with the stability of the TAGTT and consequently leads to breakdown of the test (Wang, 1977). The servoed slipring assemblies are not available to the TAGTT because of the space and structure limitations of the TAGTT. Therefore the sliprings are on the same axis with the gimbal. This means that the large friction torque of the sliprings applies to the main servo loop, which is in the order of 0.981 $\mathrm{N \cdot m}$ or even higher. In addition to the large static friction torque, there are torque disturbances caused by the other two gimbals in the rate mode.

In summary, the main concern of the TAGTT (servo system) design is the input disturbance attenuation, not the output disturbance rejection in the general sensitivity problems.

## 3. THE $\mu$-SYNTHESIS DESIGN PROCEDURE

In this section, the servo design problem of the TAGTT is formulated by using the $\mu$-synthesis approach. Since the theme of this paper is on the performance limit of the disturbance attenuation design of the servo system, it is therefore not stressed in particular on how to design the TAGTT to meet the specified performance specification. This section forms the basis for finding the performance limit.

### *3.1 Problem Description*

As being exposed in Section 2: the reference input in Fig. 1 is merely a velocity signal, the tracking error is mainly caused by the torque disturbance $T_d$ which is at the input of the perturbed plant $\tilde{G}(s)$. The main design requirement is to minimize the deflection $\theta$ of each of the three-axis subjected to various torque disturbance $T_d$, i.e.

$$\min_{K \text{ stabilizing}} \bar{\sigma}\left(T_{\theta T_d}\right) \tag{1}$$

where $T_{\theta T_d}$ denotes the transfer function from $T_d$ to $\theta$, $\bar{\sigma}\left(T_{\theta T_d}\right)$ denotes the maximum singular value of $T_{\theta T_d}$.

Because the whole structure of the TAGTT is symmetric, the coupling between each axis is not significant. The design of the TAGTT is carried out at each of the axes. The coupling effect will not be the main factor in considering the achievement of the performance limit.

### *3.2 Formulating the $\mu$-Synthesis General Framework*

For tight control of the servo system, not only the disturbance attenuation (performance specification) but also the uncertainty should be matched precisely. The commonly used general sensitivity minimization (Kwakernaak, 1993), however, generally treats the output disturbance rejection. The $\mu$-synthesis approach provides a rigorous framework for analyzing and solving the servo control problem.

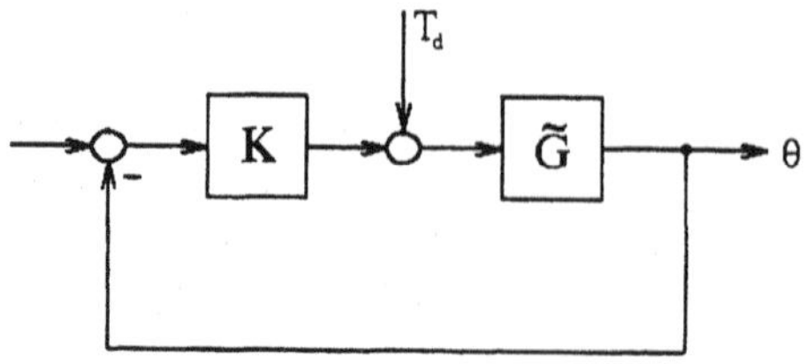


Fig. 1 Block diagram of servo loop

Necessary and sufficient conditions for robust stability (RS) and robust performance (RP) can be formulated in terms of the structured singular value denoted as $\mu$.

*General Framework.* The uncertainty is represented as multiplicative output uncertainty lumped into one full-block. This is also the practical rule of uncertainty description in the $H_\infty/\mu$ design (Skogestad and Postlethwaite, 1996). From Fig. 2, it gives

$$\tilde{G} = \left[I + \Delta W_{\text{del}}\right]G, \; \|\Delta\|_\infty < 1$$

where the uncertainty weighting function $W_{\text{del}}(s)$ describes the magnitude of the perturbation $\Delta$ block which parametrizes all of the assumed model uncertainty in the problem. $G(s)$ is the nominal plant.

Rearranging the system in Fig. 2 so as to match Fig. 3 gives the augmented plant $P(s)$ in Eq.(2).

$$P(s) = \begin{bmatrix} 0 & W_{\text{del}}G & -W_{\text{del}}G \\ W_{\text{p}} & W_{\text{p}}G & -W_{\text{p}}G \\ I & G & -G \end{bmatrix} = \begin{bmatrix} P_{11} & P_{12} \\ P_{21} & P_{22} \end{bmatrix} \qquad (2)$$

There are three pairs of input/output in the augmented system *P*. The first pair consists of the outputs of the perturbation $\Delta$: $v \in \mathbf{R}^{m_1}$ and the inputs of the perturbation: $z \in \mathbf{R}^{q_1}$. The second pair consists of the exogenous inputs: the torque disturbance $T_{\text{d}} \in \mathbf{R}^{m_2}$ and the errors: $e \in \mathbf{R}^{q_2}$ regulated by the performance weighting function $W_{\text{p}}(s)$. The third pair consists of the control signal: $u \in \mathbf{R}^{m_3}$ and the measurements: $\theta \in \mathbf{R}^{q_3}$.

*Robust Performance Formulation.* Assume the compensator $K(s)$ in Fig. 3 is manipulated into the feedback loop diagram in Fig. 4. The system $M \in \mathbf{C}^{(m_1+m_2)\times(q_1+q_2)}$ is the 2 × 2 block-structured transfer function matrix, which is derived by the lower linear fractional transformation. The notation $F_\ell(P,K)$ indicates that the lower loop of *P* is closed with *K*.

$$\begin{aligned} M &= F_\ell(P,K) \\ &= P_{11} + P_{12}K(I - P_{22}K)^{-1}P_{21} \\ &= \begin{bmatrix} -W_{\text{del}}GK(I+GK)^{-1} & W_{\text{del}}(I+GK)^{-1}G \\ W_{\text{p}}(I+GK)^{-1} & W_{\text{p}}(I+GK)^{-1}G \end{bmatrix} \\ &= \begin{bmatrix} M_{11} & M_{12} \\ M_{21} & M_{22} \end{bmatrix} \end{aligned}$$

Robust Performance in Fig. 4 is equivalent to

$$\|F_{\text{u}}(M,\Delta)\|_\infty \le 1$$

where $F_{\text{u}}(P,K)$ is the upper linear fractional transformation

$$F_{\text{u}}(P,K) = M_{22} + M_{21}\Delta(I - M_{11}\Delta)^{-1}M_{12}$$

The following theorem addresses the RP problem in Fig. 4.

**Theorem 1** (Stein and Doyle, 1991)

- Nominal performance (NP) is satisfied if and only if $\|M_{22}\|_\infty \le 1$;
- Robust stability (RS) is satisfied if and only if $\|M_{11}\|_\infty \le 1$;
- Robust performance (RP) is satisfied if and only if $\sup_\omega \mu(M) \le 1$.

Therefore the robust performance design of the gyro test table is that

$$\begin{aligned} \text{RS} &\Leftrightarrow \left\| -W_{\text{del}}GK(I+GK)^{-1} \right\|_\infty \le 1 \\ \text{NP} &\Leftrightarrow \left\| W_{\text{p}}(I+GK)^{-1}G \right\|_\infty \le 1 \qquad (3) \\ \text{RP} &\Leftrightarrow \left\| \mu(M) \right\|_\infty \le 1 \end{aligned}$$

## 4. PERFORMANCE LIMIT

Before discussing the performance limit of the disturbance attenuation design of the gyro test table (servo system), the uncertainty weighting function $W_{\text{del}}(s)$ and the performance weighting function

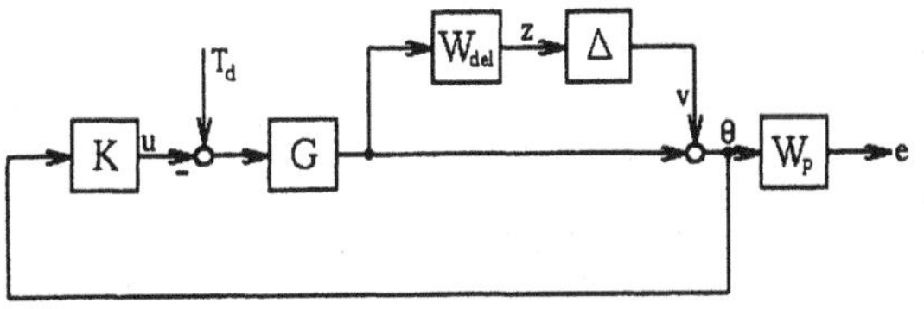


Fig. 2 Disturbance attenuation interconnection structure

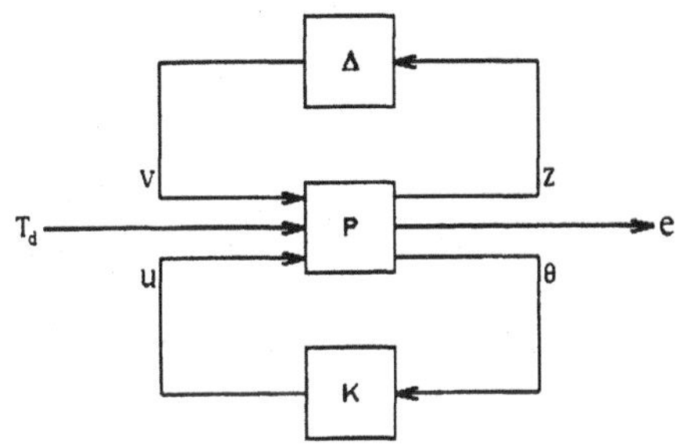


Fig. 3 General control framework

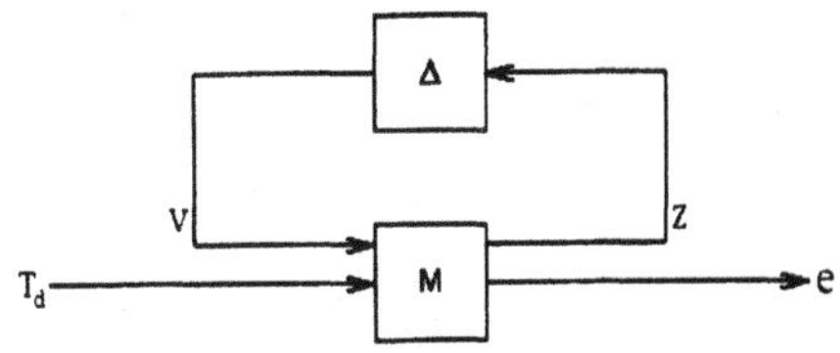


Fig. 4 Closed-loop *M* system

$W_p(s)$ should be selected first. This is a normal step in the $H_\infty/\mu$ design. Although the performance limit is based on the gyro test table under design, choosing the weighting functions is not confined to the physical feasibility of the system. However, the discovery of the performance limit in the gyro test table design will assist the designer to judge whether any desired performance specification is feasible to achieve and how to reach it. Moreover, this result can be generalized onto other similar systems.

*4.1 Uncertainty Weighting Functions*

The rotational dynamics of one gimbal of the TAGTT can be modelled as

$$G(s) = \frac{2.0626\times10^5}{Js^2}\ \hat{\sec}/(\mathrm{N}\cdot\mathrm{m}),\ J = 2.62\ \mathrm{kg}\cdot\mathrm{m}^2$$

where $J$ denotes the moment of inertia. There are two plant zeroes at infinity, which makes $P_{12}$ lose rank on the $j\omega$-axis (Glover and Doyle, 1988). This can be settled by selecting a nonproper uncertainty weight $W_{del}$. In general, it is reasonable to allow model errors of 10% (static perturbations) at low frequencies. The magnitude of the output multiplicative perturbations continues to increase steadily beyond the frequency $BWa$ where it crosses 0dB-axis. $BWa$ imposes an upper bound on the bandwidth of the servo system. For $BWa$=87.3 rad/s, $W_{del}$ can be selected as

$$W_{del}(s) = (0.0114s+0.1)(0.001s+1) \tag{4}$$

Different $W_{del}$ under $BWa$=100 rad/s, 115 rad/s, 130 rad/s, 145 rad/s, 175 rad/s are chosen so as to investigate how the uncertainty affects the performance of the disturbance attenuation.

*4.2 Performance Weighting Functions*

An important characteristic of the $H_\infty/\mu$ design is that the closed-loop performance is given by the performance weighting function selected in the design. However an optimal controller in the result of the $H_\infty/\mu$ design is usually only with respect to a performance weight specified. If the performance weight were not chosen properly, the best performance of the system would be restrained realization. In order to cope with the static friction, the control law must include an integral at least, i.e. the $W_p$ must have an integrator in it at least. Although the disturbance attenuation curve would be lower over the low frequencies with two integrators in $W_p$, it results in a type-4 servo system. That will make the implementation more difficult due to that a more complicated conditional stability problem has to be considered, which is, however, beyond the scope of this paper.

Four different forms of $W_p$ are chosen to seek the relationship between $W_p$ and the performance of disturbance attenuation.

$$W_{p1} = g\frac{s+\rho}{s}\ \ \mathrm{N}\cdot\mathrm{m}/\hat{\sec}$$

$$W_{p2} = 0.084\frac{s+\rho}{s}\cdot\frac{s+20}{s+1000}\ \ \mathrm{N}\cdot\mathrm{m}/\hat{\sec}$$

$$W_{p3} = \frac{\rho}{s}\ \ \mathrm{N}\cdot\mathrm{m}/\hat{\sec}$$

$$W_{p4} = \rho\ \ \mathrm{N}\cdot\mathrm{m}/\hat{\sec}$$

where $\rho$ should be selected as large as possible during the $\mu$-synthesis so as to increase the frequency, of which the disturbance torque could be attenuated. $g$ in $W_{p1}$ is also a parameter to be designed. However, provided one of $g$ and $\rho$ is determined, the other one can be determined by using the $\mu$-synthesis.

*4.3 Performance Limit*

The performance limit in the gyro test table (servo system) design has a more distinct and practical significance than that in the general sensitivity consideration. Eq.(1) shows that the achievable minimum disturbance attenuation, i.e. the performance limit includes not only the peak value of disturbance attenuation but also its singular value Bode plot over the low and middle frequency range. In general, it is not easy to reach the performance limit. The trial and error method is beset with certain disadvantage that a designer is often left with the haunting thought that if he tried this or that form of compensation he might have been able to meet the specifications. The $H_\infty$ synthesis has the facility to detect an inconsistent set of specifications, however, it is difficult to clarify whether the design has reached the performance limit. This section will show that the $\mu$-synthesis approach is an effective method to reach the performance limit in a systematic way.

Four different forms of $W_p$ chosen in Section 4.2 are employed in this subsection to carry out the disturbance attenuation design using $\mu$-synthesis under different $W_{del}$ with respect to six frequencies $BWa$. $\mu$-Synthesis is an optimal design scheme in the sense that it can make the performance weight reach its maximum value, i.e. $\rho$ in the $W_p$ chosen in Section 4.2. The disturbance attenuation curves thus obtained under $W_{p1}$, $W_{p2}$, $W_{p3}$ and $W_{p4}$ are shown in Fig. 5, Fig. 7, Fig. 6 and Fig. 8 respectively. Fig. 5—Fig. 8 manifest themselves that the disturbance attenuation curve drops along with the increment of $BWa$, i.e. it decreases along with the increment of the bandwidth of the system. For convenience of further analysis, the disturbance attenuation related data are given in Table 1—Table 4.

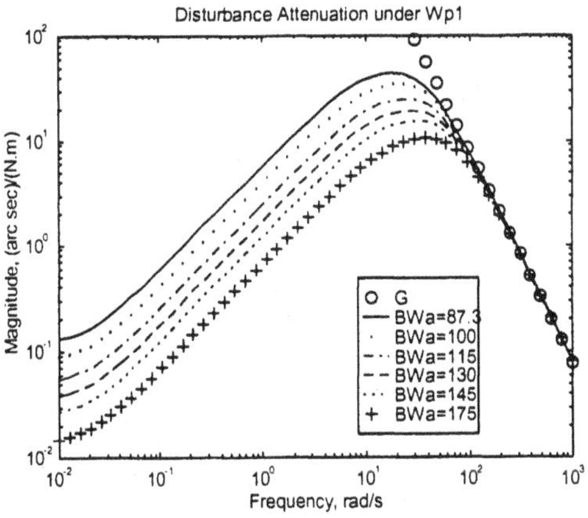


Fig. 5 Disturbance attenuation under $W_{p1}$

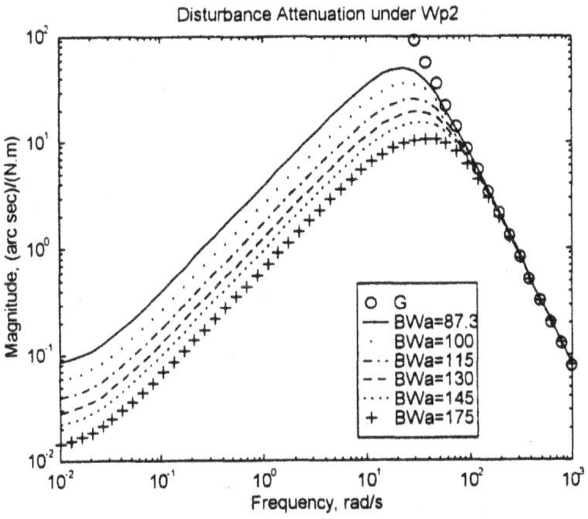


Fig. 6 Disturbance attenuation under $W_{p2}$

Table 1 Data of disturbance attenuation under $W_{p1}$

| $BWa$, rad/s | $\|T_{\theta T_d}\|_\infty$, sêc/(N·m) | $\omega_{cg}$, rad/s | $\|\mu(M)\|_\infty$ |
|---|---|---|---|
| 87.3 | 43.76 | 75.8 | 1.01 |
| 100 | 34.33 | 85.34 | 0.9984 |
| 115 | 24.87 | 101.3 | 0.9974 |
| 130 | 18.97 | 115.6 | 0.9955 |
| 145 | 15.34 | 128.8 | 0.9966 |
| 175 | 10.65 | 152.4 | 0.9971 |

Table 2 Data of disturbance attenuation under $W_{p2}$

| $BWa$, rad/s | $\|T_{\theta T_d}\|_\infty$, sêc/(N·m) | $\omega_{cg}$, rad/s | $\|\mu(M)\|_\infty$ |
|---|---|---|---|
| 87.3 | 50.13 | 70.94 | 0.9611 |
| 100 | 35.66 | 84.02 | 0.9939 |
| 115 | 25.6 | 99.85 | 0.9925 |
| 130 | 19.3 | 115.1 | 0.9921 |
| 145 | 15.37 | 128.1 | 0.9971 |
| 175 | 10.76 | 153.4 | 0.9918 |

$\|\mu(M)\|_\infty \le 1$ in Table 1—Table 4 reflects the optimal design results of the $\mu$-synthesis. $\omega_{cg}$ is the gain crossover frequency of the open loop, which is in the same order of the closed-loop bandwidth (Franklin *et al.*, 1994). $\|T_{\theta T_d}\|_\infty$ equals the maximum of $|T_{\theta T_d}|$ for the SISO system, i.e. the peak value of disturbance attenuation curve. $T_{\theta T_d} = (I+GK)^{-1}G$ is the transfer function from $T_d$ to $\theta$.

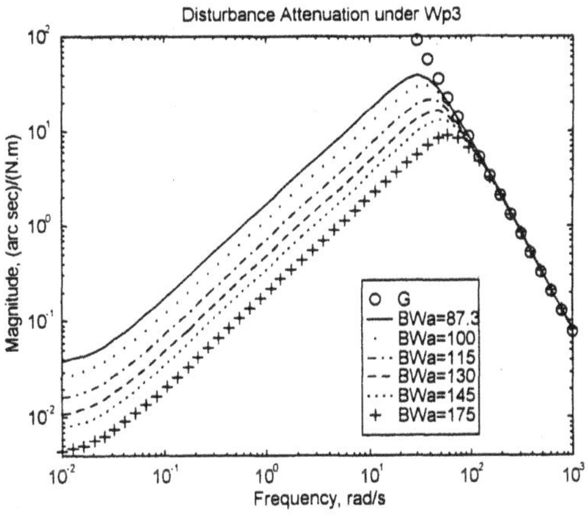


Fig. 7 Disturbance attenuation under $W_{p3}$

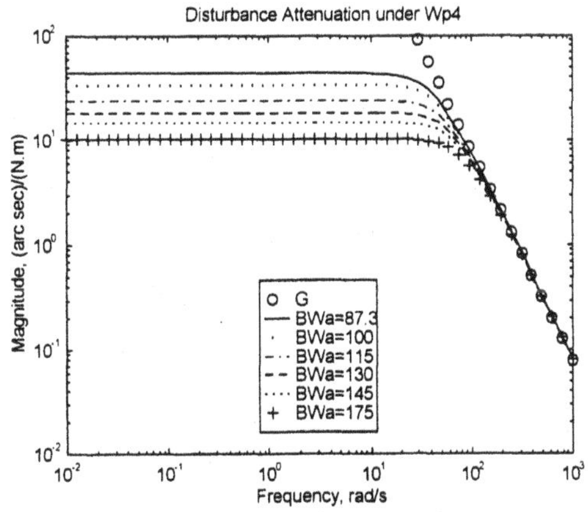


Fig. 8 Disturbance attenuation under $W_{p4}$

Table 3 Data of disturbance attenuation under $W_{p3}$

| $BWa$, rad/s | $\|T_{\theta T_d}\|_\infty$, sêc/(N·m) | $\omega_{cg}$, rad/s | $\|\mu(M)\|_\infty$ |
|---|---|---|---|
| 87.3 | 39.07 | 80.6 | 0.9961 |
| 100 | 30.03 | 94.57 | 0.9921 |
| 115 | 21.57 | 107.3 | 0.9948 |
| 130 | 16.43 | 124.9 | 0.9948 |
| 145 | 13.28 | 138.3 | 0.9939 |
| 175 | 9.23 | 165.9 | 0.9976 |

Table 4 Data of disturbance attenuation under $W_{p4}$

| $BWa$, rad/s | $\|T_{\theta T_d}\|_\infty$, sêc/(N·m) | $\omega_{cg}$, rad/s | $\|\mu(M)\|_\infty$ |
|---|---|---|---|
| 87.3 | 44.02 | 74.56 | 0.9934 |
| 100 | 33.8 | 85.02 | 0.9948 |
| 115 | 24.22 | 100.9 | 0.9981 |
| 130 | 18.47 | 115.8 | 0.9966 |
| 145 | 14.94 | 128.6 | 0.9969 |
| 175 | 10.37 | 154 | 0.9966 |

Firstly, the data of disturbance attenuation under $BWa$=87.3 rad/s in the above four tables will be analyzed. Table 1 shows that $\|T_{\theta T_d}\|_\infty$=43.76 $\hat{sec}/(N \cdot m)$. To expect a better disturbance attenuation, an additional zero is added into $W_{p1}$ resulting the performance weight in the form of $W_{p2}$. However, $\|T_{\theta T_d}\|_\infty$=50.13 $\hat{sec}/(N \cdot m)$ does not vary much in comparison with $W_{p1}$'s. Furthermore, eliminating zeros from $W_{p1}$ and $W_{p2}$ leads to $W_{p3}$, though $\|T_{\theta T_d}\|_\infty$=39.07 $\hat{sec}/(N \cdot m)$ still remains similar. $W_{p4}$ is a constant performance weight ($\|T_{\theta T_d}\|_\infty$=44.02 $\hat{sec}/(N \cdot m)$), which is employed just to show further $\|T_{\theta T_d}\|_\infty$ is independent of performance weight, where in a sense, the meaning of *systematic way* can be easily accepted. With a simple form of the performance weight like $W_{p1}$, the performance limit can be easily achieved using the $\mu$-synthesis procedure.

Detailed checks of the peak value of disturbance attenuation under $BWa$=100 rad/s, 115 rad/s, 130 rad/s, 145 rad/s, 175 rad/s reveal that *the performance limit is determined by the bandwidth.* The multiplicative perturbation is the most essential perturbation form in this practical design. In the gyro test table design, all structured and unstructured uncertainties are lumped into a stable multiplicative perturbation. For example, various dynamics is deliberately ignored in order to obtain a simpler nominal model for the design. The bandwidth of the system is a measure of speed of response. It is a design specification, which is limited by the robust stability requirement. The larger bandwidth yields the better disturbance attenuation. Table 1—Table 4 show that the peak value of disturbance attenuation is determined by the bandwidth. Under the same maximum allowable bandwidth $BWa$, the performance limit remains to be (approximately) invariable no matter what kinds of performance weights are chosen in the $\mu$-synthesis. Obviously, all the performance is robust, though only the nominal disturbance attenuation curves are shown in the figures.

## 5. CONCLUSIONS

The $\mu$-synthesis approach has been presented to provide a systematic strategy in reaching the performance limit in robust control system design. Based on an industrial project—a high precision three-axis gyro test table, this paper shows that in the servo system design which is distinct from the general sensitivity problem, the performance limit, i.e. the minimum achievable disturbance attenuation is determined by the bandwidth and its peak value is independent of the actual performance weighting functions selected. Although the result is derived from the design of one axis case, it can be applied onto the design of any other axis of the TAGTT.

## REFERENCE


Boyd, S.P. and C.H. Barratt (1991). *Linear Controller Design: Limits of Performance.* Prentice Hall, Singapore.

Franklin, G.F., J.D. Powell and A. Emami-Naeini (1994). *Feedback Control of Dynamic Systems, 3rd ed.*. Addison-Wesley, Reading, Massachusetts.

Glover, K. and J.C. Doyle (1988). State-space formulae for all stabilizing controllers that satisfy an $H_\infty$-norm bound and relations to risk sensitivity. *Syst. Contr. Lett.*, **11**, pp. 167-172.

Kwakernaak, H. (1993). Robust control and $H_\infty$-optimization—tutorial paper. *Automatica*, **29**, pp. 255-273.

Skogestad, S. and I. Postlethwaite (1996). *Multivariable Feedback Control: Analysis and Design.* John Wiley & Sons, Chichester.

Stein, G. and J.C. Doyle (1991). Beyond singular values and loop shapes. *AIAA J. Guidance, Contr., and Dynamics*, **14**, pp. 5-16.

Wang, G.-X. (1977). On the design of the servo loop of a high precision gyro test table. *J. Harbin Institute of Technology*, **9**, no. 4, pp. 1-7. (In Chinese)

14th Triennial World Congress, Beijing, P.R. China



# PARAMETER DEPENDENT LYAPUNOV FUNCTIONS BASED ON LINEAR FRACTIONAL TRANSFORMATION

**Anders Helmersson** *,[1]

* *Department of Electrical Engineering*
*Linköping University, SE-581 83 Linköping, Sweden,*
*email:* andersh@isy.liu.se

Abstract:
A parameter dependent Lyapunov function is introduced for stability analysis of linear systems with slowly varying parameters. The Lyapunov function is a quadratic form using a linear fractional transformation (LFT) of the parameters.
The parameters are assumed to be bounded and with bounded derivatives. Other types of uncertainties can be included in the problem. The new criterion yields less conservative bounds than previously proposed criteria. *Copyright © 1999 IFAC*

Keywords: Stability analysis, robustness, linear, time varying systems, parametrization, Lyapunov functions

## 1. INTRODUCTION

The analysis of parameter-dependent system plays an important part in many control applications, for instance gain scheduling problems, which has attracted a great deal of attention during recent years.

In this paper we will consider the stability of linear parameter varying (LPV) systems. We can also include performance bounds in the analysis. The parameters and their derivatives are assumed to bounded. The plant to be analyzed depends as a linear fractional transformation (LFT) of the parameters or functions thereof.

The analysis is based on quadratic Lyapunov functions that are parametrized with respect to the parameters in the plant to be analyzed. A new parametrization is used, where the Lyapunov functions is a quadratic form of an LFT-parametrized basis function. This yields a quadratic criterion, which can be solved using linear matrix inequalities, in very much the same way as when using IQCs.

In this paper we focus on the concept of using LFT-parametrized Lyapunov functions. This includes techniques for verifying that quadratic constraints are satisfied for the complete set of allowable parameters. The key lemma presented in this paper is related to the celebrated Kalman-Yakubovich-Popov lemma.

The paper is organized as follows. Linear parameter varying systems are discussed in section 2. In section 3, the parametrized Lyapunov functions are introduced and applied to systems with slowly varying parameters. The method is applied to two examples in section 4. The paper is concluded in section 5.

### 1.1 *Notation*

For matrices $A^*$ denotes the complex conjugate transpose.

[1] This work was supported by the Swedish National Board for Industrial and Technical Development (NUTEK), which is gratefully acknowledged

For appropriately dimensioned matrices $K$ and $M = \begin{bmatrix} M_{11} & M_{12} \\ M_{21} & M_{22} \end{bmatrix}$ and assuming the inverse exists, the upper linear fractional transformation (LFT) is defined as

$$K \star M = M_{22} + M_{21}K(I - M_{11}K)^{-1}M_{12}.$$

## 2. LINEAR PARAMETER VARYING SYSTEMS

The problem addressed in this paper is the analysis of linear parameter varying (LPV) systems with a state-space realization

$$\dot{x} = A(\theta)x + B(\theta)w \quad (1a)$$
$$z = C(\theta)x + D(\theta)w \quad (1b)$$

where $w$ is the disturbance signal and $z$ the error signal. The time-varying parameters, $\theta = (\theta_1, \ldots, \theta_l)$, as well as their rates of variation, $\dot{\theta}$ are assumed to be bounded.

### 2.1 *LFT parametrized systems*

In (1) the structure of the parametrization is not explicitly stated. In order to facilitate the analysis we will make a further assumption on the structure by assuming

$$\begin{bmatrix} A(\theta) & B(\theta) \\ C(\theta) & D(\theta) \end{bmatrix} = \Delta(\theta) \star \begin{bmatrix} \tilde{A} & \tilde{B} \\ \tilde{C} & \tilde{D} \end{bmatrix} \quad (2)$$

where $\Delta$ is a function of the parameter vector $\theta$ and $\tilde{A}$, $\tilde{B}$, $\tilde{C}$ and $\tilde{D}$ are constant matrices. The upper linear fractional transformation is denoted by $\star$:

$$\Delta \star \begin{bmatrix} \tilde{A} & \tilde{B} \\ \tilde{C} & \tilde{D} \end{bmatrix} = \tilde{D} + \tilde{C}(I - \Delta\tilde{A})^{-1}\Delta\tilde{B}.$$

The parameter $\theta$ is assumed to bounded in terms of bounds on $\Delta$ and possibly also its time derivative, $\dot{\Delta}$:

$$\begin{bmatrix} I \\ \Delta \end{bmatrix}^* \Pi \begin{bmatrix} I \\ \Delta \end{bmatrix} \geq 0 \quad (3)$$

or

$$\left[\begin{array}{cc} I & 0 \\ 0 & I \\ \hline \Delta & 0 \\ 0 & \dot{\Delta} \end{array}\right]^* \Pi \left[\begin{array}{cc} I & 0 \\ 0 & I \\ \hline \Delta & 0 \\ 0 & \dot{\Delta} \end{array}\right] \geq 0. \quad (4)$$

where

$$\Pi = \begin{bmatrix} \Pi_{11} & \Pi_{12} \\ \Pi_{12}^* & \Pi_{22} \end{bmatrix}.$$

We will further assume that $\Pi_{11} \geq 0$ and $\Pi_{22} \leq 0$. This assures that the set of $\Delta$ that satisfies (4) is convex and that $\tau\Delta$ is included for all $\tau \in [0, 1]$ and $\Delta$ is included.

Specifically, if $\bar{\sigma}(\Delta(t)) \leq \gamma$ and $\bar{\sigma}(\dot{\Delta}) \leq d$, where $\bar{\sigma}$ denotes the maximum singular value, we may use

$$\Pi = \begin{bmatrix} \gamma^2 P_1 & 0 & G_1 & 0 \\ 0 & d^2 P_2 & 0 & G_2 \\ G_1^* & 0 & -P_1 & 0 \\ 0 & G_2^* & 0 & -P_2 \end{bmatrix}$$

where $P_1 = P_1^*, P_2 = P_2^* \geq 0$ and $G_1 = -G_1^*$ and $G_2 = -G_2^*$ are matrices of compatible dimension and structure. If $\Delta$ is a diagonally repeated scalar parameter, then $P_1$, $P_2$, $G_1$ and $G_2$ are full matrices; if $\Delta$ has a block-diagonal structure $P_1$, $P_2$, $G_1$ and $G_2$ will then have a corresponding block-diagonal structure as well, such that $P_1$ and $G_1$ commute with $\Delta$ and $P_2$ and $G_2$ commute with $\dot{\Delta}$.

We can also define $\Pi$ as a set of LMIs:

$$\left[\begin{array}{cc} I & 0 \\ 0 & I \\ \hline \Delta_i & 0 \\ 0 & \dot{\Delta}_i \end{array}\right]^* \Pi \left[\begin{array}{cc} I & 0 \\ 0 & I \\ \hline \Delta_i & 0 \\ 0 & \dot{\Delta}_i \end{array}\right] \geq 0, \qquad i = 1, .., q$$

In this case the set of allowable $(\Delta, \dot{\Delta})$ is all convex combinations of $(\Delta_i, \dot{\Delta}_i)$. Here we must include the constraints $\Pi_{11} \geq 0$ and $\Pi_{22} \leq 0$ in order to assure convexity.

## 3. PARAMETER DEPENDENT LYAPUNOV FUNCTION

In this section we will study a parameter-dependent Lyapunov function, $P$, for verifying stability of systems with slowly-varying parameters.

### 3.1 *Performance Bounds*

The linear system under consideration is described by

$$\dot{x} = Ax + Bw$$
$$z = Cx + Dw$$

where $A$, $B$ $C$ and $D$ may be parameter-dependent. Based on the input $w$ and output $z$ a performance criterion, $J$, can be expressed as

$$J = \int_0^\infty \begin{bmatrix} z(t) \\ w(t) \end{bmatrix}^* \Pi \begin{bmatrix} z(t) \\ w(t) \end{bmatrix} dt$$

This criterion includes the $\mathcal{L}_2$-induced gain of a system, which is obtained using

$$\Pi = \begin{bmatrix} \gamma^{-1} I & 0 \\ 0 & -\gamma I \end{bmatrix}. \quad (5)$$

If $J \leq 0$ then the $\mathcal{L}_2$-induced gain is less than $\gamma > 0$, that is, $||z|| \leq \gamma||w||$.

Introduce the quadratic Lyapunov function

$$V(x) = x^* P x$$

where $P = P^* > 0$ is a symmetric, positive definite matrix, which may be parameter-dependent. To assure internal asymptotic stability it is assumed that $P$ is positive definite ($P > 0$). Introduce the Hamiltonian

$$\begin{aligned} H &= \dot{V} + \begin{bmatrix} z \\ w \end{bmatrix}^* \Pi \begin{bmatrix} z \\ w \end{bmatrix} \\ &= x^* \dot{P} x + x^* P \dot{x} + \dot{x}^* P x \\ &\quad + \begin{bmatrix} Cx + Dw \\ w \end{bmatrix}^* \Pi \begin{bmatrix} Cx + Dw \\ w \end{bmatrix} \\ &= \begin{bmatrix} x \\ w \end{bmatrix}^* \left( \begin{bmatrix} \dot{P} + PA + A^*P & PB \\ B^*P & 0 \end{bmatrix} \right. \\ &\quad \left. + \begin{bmatrix} C & D \\ 0 & I \end{bmatrix}^* \Pi \begin{bmatrix} C & D \\ 0 & I \end{bmatrix} \right) \begin{bmatrix} x \\ w \end{bmatrix}. \end{aligned}$$

If we assume that

$$\begin{bmatrix} \dot{P} + PA + A^*P & PB \\ B^*P & 0 \end{bmatrix} + \begin{bmatrix} C & D \\ 0 & I \end{bmatrix}^* \Pi \begin{bmatrix} C & D \\ 0 & I \end{bmatrix} < 0 \qquad (6)$$

and integrating $H$ from $t = 0$ to infinity and assuming $x(0) = 0$ yield

$$\begin{aligned} 0 > \int_0^\infty H dt &= \lim_{t\to\infty} V(t) - V(0) \\ &\quad + \int_0^\infty \begin{bmatrix} z(t) \\ w(t) \end{bmatrix}^* \Pi \begin{bmatrix} z(t) \\ w(t) \end{bmatrix} dt \\ &\geq \int_0^\infty \begin{bmatrix} z(t) \\ w(t) \end{bmatrix}^* \Pi \begin{bmatrix} z(t) \\ w(t) \end{bmatrix} dt \end{aligned}$$

for any non-zero $w \in \mathcal{L}_2[0, \infty)$.

When (5) applies, we can rewrite (6) using the Schur complement as (5) as

$$\begin{bmatrix} \dot{P} + PA + A^*P & PB & C^* \\ B^*P & -\gamma I & D^* \\ C & D & -\gamma I \end{bmatrix} < 0, \qquad (7)$$

which is known as the bounded real lemma.

The inequality (6) can also be rewritten as

$$\begin{bmatrix} A & B \\ C & D \\ I & 0 \\ 0 & I \end{bmatrix}^* \begin{bmatrix} 0 & 0 & P & 0 \\ 0 & \Pi_{11} & 0 & \Pi_{12} \\ P & 0 & \dot{P} & 0 \\ 0 & \Pi_{12}^* & 0 & \Pi_{22} \end{bmatrix} \begin{bmatrix} A & B \\ C & D \\ I & 0 \\ 0 & I \end{bmatrix} < 0. \qquad (8)$$

### 3.2 *LFT Parametrization*

If we consider time-varying or parameter-dependent systems this inequality must hold for any possible value of the parameter and time. In order to facilitate the verification we will assume a certain form of the parametrization of the Lyapunov function.

One parametrization that is common in the literature is affine, see for instance Apkarian *et al.* (1994); Gahinet *et al.* (1994); Feron *et al.* (1995); Gahinet *et al.* (1995); Apkarian *et al.* (1995)

$$P(\theta) = P_0 + \sum_i \theta_i P_i$$

We will here instead employ a parametrization based on the linear fractional transformation (LFT) Helmersson (1995):

$$P = T(\hat{\Delta})^* Q T(\hat{\Delta}) \qquad (9)$$

where $\hat{\Delta}$ is a function of the parameters, $\theta$, in the system, $Q = Q^*$ is a symmetric and static matrix and

$$\begin{aligned} T(\hat{\Delta}) &= \hat{\Delta} \star \begin{bmatrix} \hat{A} & \hat{B} \\ \hat{C} & \hat{D} \end{bmatrix} = \hat{D} + \hat{C}\hat{\Delta}(I - \hat{A}\hat{\Delta})^{-1}\hat{B} \\ &= \hat{D} + \hat{C}(I - \hat{\Delta}\hat{A})^{-1}\hat{\Delta}\hat{B}. \end{aligned}$$

Note that (9) also includes affine parametrizations. To see this let for instance

$$T = \begin{bmatrix} \delta_1 I & 0 \\ 0 & \delta_2 I \end{bmatrix} \star \left[\begin{array}{cc|c} 0 & 0 & I \\ 0 & 0 & I \\ \hline I & 0 & 0 \\ 0 & I & 0 \\ 0 & 0 & I \end{array}\right] = \begin{bmatrix} \delta_1 I \\ \delta_2 I \\ I \end{bmatrix}$$

$$Q = \begin{bmatrix} 0 & 0 & \frac{1}{2}P_1 \\ 0 & 0 & \frac{1}{2}P_2 \\ \frac{1}{2}P_1 & \frac{1}{2}P_2 & P_0 \end{bmatrix}.$$

Then

$$P = P_0 + \delta_1 P_1 + \delta_2 P_2.$$

In order to find $\dot{P} = \dot{T}^* Q T + T^* Q \dot{T}$ we need to know $\dot{T}$. We can express the time-derivative of $T$ as an LFT:

$$\begin{aligned} \dot{T} &= \hat{C}\dot{\hat{\Delta}}(I - \hat{A}\hat{\Delta})^{-1}\hat{B} \\ &\quad + \hat{C}\hat{\Delta}(I - \hat{A}\hat{\Delta})^{-1}\hat{A}\dot{\hat{\Delta}}(I - \hat{A}\hat{\Delta})^{-1}\hat{B} \\ &= \hat{C}(I + \hat{\Delta}(I - \hat{A}\hat{\Delta})^{-1}\hat{A})\dot{\hat{\Delta}}(I - \hat{A}\hat{\Delta})^{-1}\hat{B} \\ &= \hat{C}(I - \hat{\Delta}\hat{A})^{-1}\dot{\hat{\Delta}}(I - \hat{A}\hat{\Delta})^{-1}\hat{B} \end{aligned}$$

The inequality (8) includes

$$\begin{bmatrix} 0 & P \\ P & \dot{P} \end{bmatrix} = \begin{bmatrix} T & \dot{T} \\ 0 & T \end{bmatrix}^* \begin{bmatrix} 0 & Q \\ Q & 0 \end{bmatrix} \begin{bmatrix} T & \dot{T} \\ 0 & T \end{bmatrix}$$

where

$$\begin{bmatrix} T & \dot{T} \\ 0 & T \end{bmatrix} = \begin{bmatrix} \hat{\Delta} & & \\ & \hat{\Delta} & \\ & & \dot{\hat{\Delta}} \end{bmatrix} \star \left[\begin{array}{ccc|cc} \hat{A} & 0 & \hat{A} & \hat{B} & 0 \\ 0 & \hat{A} & 0 & 0 & \hat{B} \\ 0 & \hat{A} & 0 & 0 & \hat{B} \\ \hline \hat{C} & 0 & \hat{C} & \hat{D} & 0 \\ 0 & \hat{C} & 0 & 0 & \hat{D} \end{array}\right]$$

This has the same structure as (9). Thus

$$\begin{bmatrix} A & B \\ C & D \\ I & 0 \\ 0 & I \end{bmatrix}^* \begin{bmatrix} 0 & 0 & P & 0 \\ 0 & \Pi_{11} & 0 & \Pi_{12} \\ P & 0 & \dot{P} & 0 \\ 0 & \Pi_{12}^* & 0 & \Pi_{22} \end{bmatrix} \begin{bmatrix} A & B \\ C & D \\ I & 0 \\ 0 & I \end{bmatrix} < 0$$

is equivalent to

$$\begin{bmatrix} TA+\dot{T} & TB \\ C & D \\ T & 0 \\ 0 & I \end{bmatrix}^* \begin{bmatrix} 0 & 0 & Q & 0 \\ 0 & \Pi_{11} & 0 & \Pi_{12} \\ Q & 0 & 0 & 0 \\ 0 & \Pi_{12}^* & 0 & \Pi_{22} \end{bmatrix} \times \begin{bmatrix} TA+\dot{T} & TB \\ C & D \\ T & 0 \\ 0 & I \end{bmatrix} < 0 \qquad (10)$$

where $A, B, C$ and $D$ are LFT parametrized with respect to $\Delta$, $T$ with respect to $\hat{\Delta}$ and $\dot{T}$ with respect to $\hat{\Delta}$ and $\dot{\hat{\Delta}}$; both $Q$ and $\Pi$ are constant matrices. In order to assure that $P = T^*QT$ is a valid Lyapunov function it must be bounded from above and below such that

$$\beta I \geq P \geq \alpha I \qquad (11)$$

where $\alpha$ and $\beta$ are positive numbers. The upper bound can be verified if $T$ is bounded for all allowable $\hat{\Delta}$. To show this it suffices to check that

$$I - \hat{A}\hat{\Delta}$$

is non-singular for all allowable $\hat{\Delta}$. We may use the following lemma:

*Lemma 1.* If

$$\begin{bmatrix} A \\ I \end{bmatrix}^* \Pi \begin{bmatrix} A \\ I \end{bmatrix} < 0 \qquad (12)$$

then $I - A\Delta$ is non-singular for all $\Delta$ such that

$$\begin{bmatrix} I \\ \Delta \end{bmatrix}^* \Pi \begin{bmatrix} I \\ \Delta \end{bmatrix} \geq 0 \qquad (13)$$

holds.

**Proof:** *Let be a solution to $x = A\Delta x$. Post-multiplying (13) by $x$ and pre-multiplying by its transpose yields*

$$0 \leq x^* \begin{bmatrix} I \\ \Delta \end{bmatrix}^* \Pi \begin{bmatrix} I \\ \Delta \end{bmatrix} x = \begin{bmatrix} A\Delta x \\ \Delta x \end{bmatrix}^* \Pi \begin{bmatrix} A\Delta x \\ \Delta x \end{bmatrix}$$

$$= (\Delta x)^* \begin{bmatrix} A \\ I \end{bmatrix}^* \Pi \begin{bmatrix} A \\ I \end{bmatrix} \Delta x$$

*from which we conclude that $\Delta x = 0$ and consequently there exist no $x \neq 0$ such that $x = A\Delta x$.* □

An alternative proof is given in Corollary 1 of Iwasaki and Hara (1998).

The inequalities, (10) and (11), can be compressed into one:

$$\tilde{G}(\tilde{\Delta})^* M \tilde{G}(\tilde{\Delta}) < 0 \qquad (14)$$

where $\tilde{G}$ is an LFT parametrization of $\tilde{\Delta}$, which includes $\Delta$, $\hat{\Delta}$, $\hat{\Delta}$ and $\dot{\hat{\Delta}}$. Note that $\hat{\Delta}$ occurs twice in $\tilde{\Delta}$.

To show that the inequality above is satisfied for any $\tilde{\Delta}$ belonging to a specified set we will use the following new lemma, which is related Kalman-Yakubovich-Popov lemma.

*Lemma 2.* Given $A \in \mathbb{R}^{n\times k}, B \in \mathbb{R}^{n\times m}, M = M^T \in R^{(k+m)\times(k+m)}$, and $\Pi \in \mathbb{R}^{(n+k)\times(n+k)}$ such that

$$M + \begin{bmatrix} A & B \\ I & 0 \end{bmatrix}^* \Pi \begin{bmatrix} A & B \\ I & 0 \end{bmatrix} < 0. \qquad (15)$$

Then

$$\begin{bmatrix} \Delta(I - A\Delta)^{-1}B \\ I \end{bmatrix}^* M \begin{bmatrix} \Delta(I - A\Delta)^{-1}B \\ I \end{bmatrix} < 0 \qquad (16)$$

holds for any $\Delta \in \mathbb{R}^{k\times n}$ such that

$$\begin{bmatrix} I \\ \Delta \end{bmatrix}^* \Pi \begin{bmatrix} I \\ \Delta \end{bmatrix} \geq 0 \qquad (17)$$

**Proof:** *Post-multiply* (15) *by* $\begin{bmatrix} \Delta(I-A\Delta)^{-1}B \\ I \end{bmatrix}$ *and pre-multiply by its transpose. Using* (17), *this yields*

$$0 > \begin{bmatrix} \Delta(I - A\Delta)^{-1}B \\ I \end{bmatrix}^* M \begin{bmatrix} \Delta(I - A\Delta)^{-1}B \\ I \end{bmatrix}$$

$$+ B^*(I - A\Delta)^{-*} \begin{bmatrix} I \\ \Delta \end{bmatrix}^* \Pi \begin{bmatrix} I \\ \Delta \end{bmatrix} (I - A\Delta)^{-1}B$$

$$\geq \begin{bmatrix} \Delta(I - A\Delta)^{-1}B \\ I \end{bmatrix}^* M \begin{bmatrix} \Delta(I - A\Delta)^{-1}B \\ I \end{bmatrix},$$

*which concludes the proof.* □

In order to verify (14), we let $\tilde{G}(\tilde{\Delta}) = \tilde{D} + \tilde{C}\tilde{\Delta}(I - \tilde{A}\tilde{\Delta})^{-1}\tilde{B}$. If

$$[\,\tilde{C}\ \tilde{D}\,]^* M [\,\tilde{C}\ \tilde{D}\,] + \begin{bmatrix} \tilde{A} & \tilde{B} \\ I & 0 \end{bmatrix}^* \Pi \begin{bmatrix} \tilde{A} & \tilde{B} \\ I & 0 \end{bmatrix} < 0$$

or, equivalently,

$$\begin{bmatrix} \tilde{A} & \tilde{B} \\ I & 0 \\ \tilde{C} & \tilde{D} \end{bmatrix}^* \begin{bmatrix} \Pi_{11} & \Pi_{12} & 0 \\ \Pi_{12}^* & \Pi_{22} & 0 \\ 0 & 0 & M \end{bmatrix} \begin{bmatrix} \tilde{A} & \tilde{B} \\ I & 0 \\ \tilde{C} & \tilde{D} \end{bmatrix} < 0$$

then (14) holds for any $\tilde{\Delta}$ satisfying

$$\begin{bmatrix} I \\ \tilde{\Delta} \end{bmatrix}^* \Pi \begin{bmatrix} I \\ \tilde{\Delta} \end{bmatrix} \geq 0,$$

which is essentially the same as (4) but with different repetitions of the blocks in $\Delta$.

Note that for a given $\tilde{G}$, that is to say for given $A$, $B$, $C$, $D$, $T$ and $\dot{T}$, we can search for $M$ and $\Pi$ using LMIs.

## 4. EXAMPLES

We will show the stability criteria on two examples. In both examples we improve the bounds compared to previous IQC-based criteria Jönsson (1996); Jönsson and Rantzer (1994, 1996)

### 4.1 *Example I*

This example is from Jönsson and Rantzer (1994). A simple SISO system is considered:

$$\dot{x}(t) = (A + C\delta(t)B)\,x(t) = \begin{bmatrix} -0.21 & -1 + 0.8\delta(t) \\ 1 & 0 \end{bmatrix} x(t)$$

where it is assumed that $\delta(t) \in [-\gamma, \gamma]$. In the case the parameter is constant, the system is asymptotically stable if and only if $\delta < 1.25$, that is, when $\gamma < 1.25$.

If $\dot{\delta}$ is unbounded, then stability is assured if $|\delta| < 0.261$. This bound is obtained as the inverse of the $\mathcal{H}_\infty$-norm of

$$G(s) = D + C(sI - A)^{-1}B$$

where

$$\left[\begin{array}{c|c} A & B \\ \hline C & D \end{array}\right] = \left[\begin{array}{cc|c} -0.21 & -1 & 0.8 \\ 1 & 0 & 0 \\ \hline 0 & 1 & 0 \end{array}\right].$$

We now apply the parameter-dependent Lyapunov criterion with

$$T(\delta) = \begin{bmatrix} \delta & 0 \\ 0 & \delta \end{bmatrix} \star \left[\begin{array}{cc|cc} -0.5 & 0 & 1 & 0 \\ 0 & 0.5 & 0 & 1 \\ \hline 1 & 0 & 0 & 0 \\ 0 & 1 & 0 & 0 \\ 0 & 0 & 1 & 0 \\ 0 & 0 & 0 & 1 \end{array}\right] = \begin{bmatrix} \frac{2\delta}{2+\delta} & 0 \\ 0 & \frac{2\delta}{2-\delta} \\ 1 & 0 \\ 0 & 1 \end{bmatrix}. \tag{18}$$

The lower bound could probably be improved by using a basis multiplier, $T$, of higher order. However, this leads to an increased computational load, as well as that the condition number of the required $P$ usually increases. This may cause numerical problems, which could result in worse bounds even if this is not theoretically true.

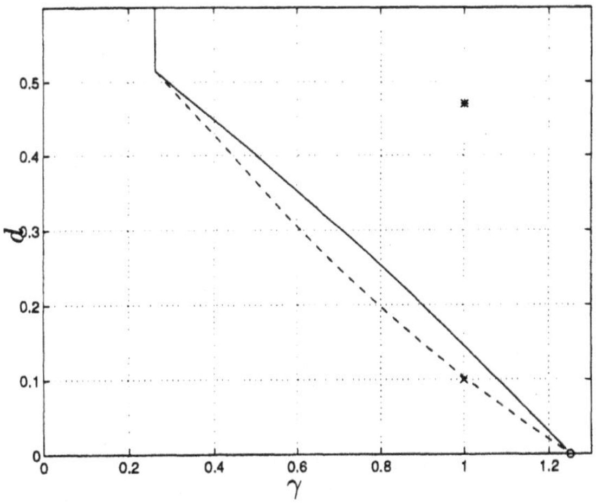


Fig. 1. Example I: Stability bounds $d$ as a function of $\gamma$. The system is stable if $\delta(t) \in [-\gamma, \gamma]$ and $|\dot{\delta}(t)| \leq d$. The lower, dashed line uses an IQC-derived criterion while the upper, solid line uses (18). The cross ($\times$) marks the bound given in Jönsson and Rantzer (1994). The circle ($\circ$) at $\gamma = 1.25$, denotes the bound obtained with constant $\delta$. Arbitrarily fast variations in $\delta$ can be allowed if $|\delta| < 0.261$, marked by the vertical line. The criteria show stability in the area below and to the left of the lines. The asterisk ($*$) at $\gamma = 1$ and $d = 0.46$, shows a case for which instability has been verified.

### 4.2 *Example II*

This example is has been analyzed in Jönsson and Rantzer (1996) using IQCs. We consider the ship steering dynamics as in Example 9.6 in Åström and Wittenmark (1989). The dynamics can be approximated by the Nomoto model

$$\dot{x}(t) = v(t)(-ax(t) + bv(t)\nu(t))$$
$$\dot{\psi}(t) = x(t)$$

where $\psi$ denotes the heading of the ship, $\nu$ denotes the rudder angle and $v$ is the speed of the ship. It is assumed that $v(t) \geq 0$. We will, as in Åström and Wittenmark (1989), study the stability of the ship dynamics for an unstable tanker with $a = -0.3$ and $b = 0.8$, which is controlled by a PD regulator

$$\nu = -K\psi \qquad \text{where} \qquad K(s) = k(1 + T_d s)$$

with $k = 2.5$ and $T_d = 0.86$.

The closed loop system can be described by

$$x(t) = (v(t)I_2)G_0(s)x(t)$$

with

$$G_0(s) = \begin{bmatrix} -a/s & b \\ -K(s)/s^2 & 0 \end{bmatrix}.$$

This system is not strictly stable and instead we replace it for the purpose of analysis by

$$G(s) = (I + G_0(s))(I - G_0(s))^{-1} = D + C(sI - A)^{-1}B$$

where

$$\left[\begin{array}{c|c} A & B \\ \hline C & D \end{array}\right] = \left[\begin{array}{cc|cc} -(a + bkT_d) & -bk & -2 & -2b \\ 1 & 0 & 0 & 0 \\ \hline a + bkT_d & bk & 1 & 2b \\ kT_d & k & 0 & 1 \end{array}\right]$$

and

$$\delta(t) = \frac{v(t) - v_{\text{nom}}}{v(t) + v_{\text{nom}}} \in [-1, 1],$$

where $v_{\text{nom}}$ is the nominal speed of the ship. For the transformed system, $d$ is solved as a function

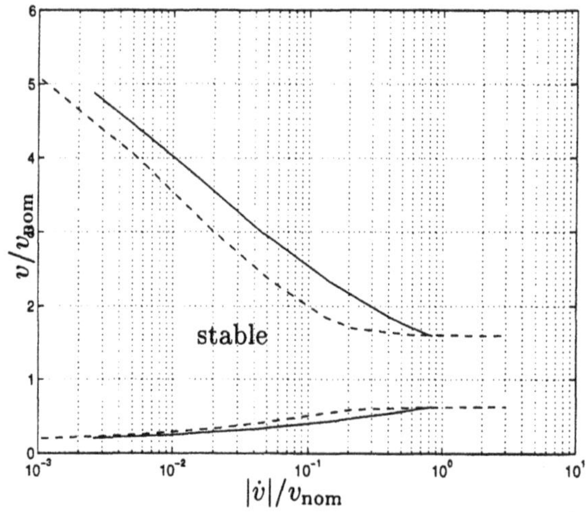


Fig. 2. Example II: Stability bounds for $v$ and $\dot{v}$ using different IQC criteria. The solid lines show the proven stability bounds using $T$ as defined in (18). The dashed lines mark the bounds given in Jönsson and Rantzer (1996).

of $\gamma$ such that the system is verified to be stable when

$$|\delta| \leq \gamma \qquad \text{and} \qquad |\dot{\delta}| \leq d.$$

Then

$$v = \frac{1+\delta}{1-\delta} v_{\text{nom}} \qquad \text{and} \qquad \dot{v} = \frac{2v_{\text{nom}}}{(1-\delta)^2}\dot{\delta},$$

and the system is assured to be stable when

$$\frac{1-\gamma}{1+\gamma} \leq \frac{v(t)}{v_{\text{nom}}} \leq \frac{1+\gamma}{1-\gamma}$$
$$\frac{|\dot{v}|}{v_{\text{nom}}} \leq \frac{2d}{(1+\gamma)^2}.$$

We analyze the system using an LFT-parametrized Lyapunov function generated by (18) yields the same bound within an accuracy of three digits.

Figure 2 shows the area in the $(v, |\dot{v}|)$-plane where stability is proven using the different criteria.

## 5. CONCLUSIONS

A new criterion for verifying stability of linear systems with slowly varying parameters is derived. It is based on a parameter-dependent quadratic Lyapunov function. A parametrization based on linear fractional transformations (LFTs) is used. This specific parametrization gives a flexible choice of Lyapunov functions and yet allows us to verify stability criteria using linear matrix inequalities.

The method is illustrated on two examples. The new criteria outperforms previous methods documented in the literature.

## References

Apkarian, P., E. Feron and P. Gahinet (1995). Parameter-dependent Lyapunov functions for robust control of systems with real parametric uncertainty. In: *Proceedings of the 3rd European Control Conference.* Vol. 3, part 2. Rome, Italy. pp. 2275–2280.

Apkarian, P., P. Gahinet and G. Becker (1994). Self-scheduled $H_\infty$ control of linear parameter-varying systems. In: *Proceedings of the American Control Conference.* Vol. 1. Baltimore, Maryland. pp. 856–860.

Åström, K. J. and B. Wittenmark (1989). *Adaptive Control.* Addison Wesley Series in Electrical and Computer Engineering: Control Engineering. Addison Wesley.

Feron, E., P. Apkarian and P. Gahinet (1995). $\mathcal{S}$-procedure for the analysis of control systems with parametric uncertainties via parameter-dependent Lyapunov functions. In: *Proceedings of the 3rd European Control Conference.* Vol. 1. Rome, Italy. pp. 141–146.

Gahinet, P., P. Apkarian and M. Chilali (1994). Affine parameter-dependent Lyapunov functions for real parametric uncertainty. In: *Proceedings of the 33rd Conference on Decision and Control.* Vol. 3. Lake Buena Vista, Florida. pp. 2026–2031.

Gahinet, P., P. Apkarian, M. Chilali and E. Feron (1995). Affine parameter-dependent Lyapunov functions and real parametric uncertainty. In: *Proceedings of the 3rd European Control Conference.* Vol. 3, part 2. Rome, Italy. pp. 2262–2267.

Helmersson, A. (1995). An LFT parametrization of Lyapunov functions. In: *2nd Russian Swedish Control Conference.* St. Petersburg, Russia. pp. 23–27.

Iwasaki, T. and S. Hara (1998). Well-posedness of feedback systems: Insights into exact robustness analysis and approximate computations. *IEEE Transactions on Automatic Control* **43**, 619–630.

Jönsson, U. (1996). Robustness Analysis of Uncertain and Nonlinear Systems. PhD thesis. Lund Institute of Technology. Lund, Sweden.

Jönsson, U. and A. Rantzer (1994). System with uncertain parameters — time-variations with bounded derivatives. In: *Proceedings of the 33rd Conference on Decision and Control.* Vol. 3. Lake Buena Vista, Florida. pp. 3074–3079.

Jönsson, U. and A. Rantzer (1996). Systems with uncertain parameters – time variations with bounded derivatives. *International Journal of Robust and Nonlinear Control* **6**(9-10), 969–982.

# AUTHOR INDEX

# KEYWORD INDEX

www.ingramcontent.com/pod-product-compliance
Ingram Content Group UK Ltd.
Pitfield, Milton Keynes, MK11 3LW, UK
UKHW052301280926
15471UKWH00006B/1043

9 780080 432175